TRENDS OF APPLIED MICROBIOLOGY FOR SUSTAINABLE ECONOMY

Developments in Applied Microbiology and Biotechnology

TRENDS OF APPLIED MICROBIOLOGY FOR SUSTAINABLE ECONOMY

Edited by

RAVINDRA SONI
Department of Agricultural Microbiology, College of Agriculture,
Indira Gandhi Krishi Vishwavidyalaya, Raipur, Chhattisgarh, India

DEEP CHANDRA SUYAL
Department of Microbiology, Akal College of Basic Sciences, Eternal University,
Sirmour, Himachal Pradesh, India

AJAR NATH YADAV
Department of Biotechnology, Dr. Khem Singh Gill Akal College of Agriculture,
Eternal University, Baru Sahib, Sirmour, Himachal Pradesh, India

REETA GOEL
Department of Biotechnology, Institute of Applied Sciences and Humanities,
GLA University Mathura, Mathura, Uttar Pradesh, India

Academic Press is an imprint of Elsevier
125 London Wall, London EC2Y 5AS, United Kingdom
525 B Street, Suite 1650, San Diego, CA 92101, United States
50 Hampshire Street, 5th Floor, Cambridge, MA 02139, United States
The Boulevard, Langford Lane, Kidlington, Oxford OX5 1GB, United Kingdom

Copyright © 2022 Elsevier Inc. All rights reserved.

No part of this publication may be reproduced or transmitted in any form or by any means, electronic or mechanical, including photocopying, recording, or any information storage and retrieval system, without permission in writing from the publisher. Details on how to seek permission, further information about the Publisher's permissions policies and our arrangements with organizations such as the Copyright Clearance Center and the Copyright Licensing Agency, can be found at our website: www.elsevier.com/permissions.

This book and the individual contributions contained in it are protected under copyright by the Publisher (other than as may be noted herein).

Notices

Knowledge and best practice in this field are constantly changing. As new research and experience broaden our understanding, changes in research methods, professional practices, or medical treatment may become necessary.

Practitioners and researchers must always rely on their own experience and knowledge in evaluating and using any information, methods, compounds, or experiments described herein. In using such information or methods they should be mindful of their own safety and the safety of others, including parties for whom they have a professional responsibility.

To the fullest extent of the law, neither the Publisher nor the authors, contributors, or editors, assume any liability for any injury and/or damage to persons or property as a matter of products liability, negligence or otherwise, or from any use or operation of any methods, products, instructions, or ideas contained in the material herein.

ISBN 978-0-323-91595-3

For information on all Academic Press publications
visit our website at https://www.elsevier.com/books-and-journals

Publisher: Stacy Masucci
Acquisitions Editor: Linda Versteeg-Buschman
Editorial Project Manager: Franchezca A. Cabural
Production Project Manager: Omer Mukthar
Cover Designer: Matthew Limbert

Typeset by STRAIVE, India

Contents

Contributors xix
Preface xxvii

Chapter 1 Trends of agricultural microbiology for sustainable crops production and economy: An introduction 1
Tanvir Kaur, Divjot Kour, and Ajar Nath Yadav

1.1 Introduction 1
1.2 Role of beneficial microbiomes in plant growth promotion 2
1.3 Microbiomes for mitigation of biotic and abiotic stresses 15
1.4 Beneficial microbiomes for sustainable crop production and protection ... 23
1.5 Beneficial microbiomes in organic agriculture 27
1.6 Implication of beneficial microbes sustainable economic 27
1.7 Conclusion 28
References 28

Chapter 2 Phytobiome research: Recent trends and developments 45
V.T. Anju, Madhu Dyavaiah, and Busi Siddhardha

2.1 Introduction 45
2.2 Plant-associated microbiomes 46
2.3 Plant-microbial interactions 48
2.4 Functional roles of phytobiome for sustainable agriculture 50
2.5 Plant microbiome in the field of translation and commercialization 58
2.6 Conclusions 58
References 59

Chapter 3 An overview of microbial diversity under diverse ecological niches in northeast India 65
Krishna Giri, Bhanushree Doley, Gaurav Mishra,
Deep Chandra Suyal, Rupjyoti C. Baruah, and R.S.C. Jayaraj

- 3.1 Introduction 65
- 3.2 Microbial diversity in different ecological niches 66
- 3.3 Northeast microbial database (NEMiD) 102
- 3.4 Conclusion and future prospects 102
- References 103

Chapter 4 Microbial consortium and crop improvement: Advantages and limitations 109
Dibyajit Lahiri, Moupriya Nag, Sougata Ghosh, Ankita Dey, and Rina Rani Ray

- 4.1 Introduction 109
- 4.2 Microbial consortia 110
- 4.3 Plant microbiota existing above the ground 111
- 4.4 Managed microbial consortia 111
- 4.5 Plant growth-promoting microorganisms 113
- 4.6 Microbial interactions within consortium 114
- 4.7 Stimulation of plant growth under stressed condition ... 116
- 4.8 Application of microbial consortium in agriculture 116
- 4.9 Conclusion and future scope 119
- References 119

Chapter 5 Revisiting soil-plant-microbes interactions: Key factors for soil health and productivity 125
Subhadeep Mondal, Suman Kumar Halder, and
Keshab Chandra Mondal

- 5.1 Introduction 125
- 5.2 Important types of plant-microbes interactions 127

5.3 Major pathways of improved soil health and productivity attributed by different plant-microbe interactions 130
5.4 Biofertilizers 142
5.5 Outlook and conclusion 144
References 145

Chapter 6 Biological control of forest pathogens: Success stories and challenges 155
Ratnaboli Bose, Aditi Saini, Nitika Bansal, M.S. Bhandari, Amit Pandey, Pooja Joshi, and Shailesh Pandey

6.1 Introduction 155
6.2 The importance of forest pathogens 156
6.3 Biological control: A brief exposition 160
6.4 Success stories of biological control with special reference to *Heterobasidion annosum* 166
6.5 Challenges 169
6.6 Future directions 175
6.7 Conclusion 176
References 177

Chapter 7 Cold-tolerant and cold-loving microorganisms and their applications 185
Gayan Abeysinghe, H.K.S. De Zoysa, T.C. Bamunuarachchige, and Mohamed Cassim Mohamed Zakeel

7.1 Introduction 185
7.2 Diversity of cold-tolerant mutants in cold ecosystems 186
7.3 Mechanisms of cold tolerance in microorganisms 186
7.4 Aspects of cold-tolerant enzymes 191
7.5 Future prospects 200
References 200

Chapter 8 Plant growth-promoting diazotrophs: Current research and advancements 207
Chanda Vikrant Berde, P. Veera Bramhachari, and Vikrant Balkrishna Berde

- 8.1 Introduction 207
- 8.2 Nitrogen fixation in diazotrophs 208
- 8.3 Terrestrial nitrogen-fixing diazotrophs 209
- 8.4 Nitrogen fixation in the ocean 210
- 8.5 Genomic and transcriptomics of diazotrophs 212
- 8.6 Beneficial mechanisms other than N-fixation provided by diazotrophs 214
- 8.7 Application in global agriculture 215
- 8.8 Future challenges in agriculture: Application of diazotrophs to nonlegumes 217
- 8.9 Conclusions and future perspectives 218
- References 219

Chapter 9 Role of mycorrhizae in plant-parasitic nematodes management 225
H.K. Patel, Y.K. Jhala, B.L. Raghunandan, and J.P. Solanki

- 9.1 Introduction 225
- 9.2 Role of arbuscular mycorrhiza in the suppression of plant-parasitic nematodes 230
- 9.3 Interaction of arbuscular mycorrhiza with soil-borne nematodes associated with plants 233
- 9.4 Mycorrhizal mode of action to manage plant-parasitic nematode 234
- 9.5 Mycorrhizal approaches for the management of plant pathogens 240
- 9.6 Production and commercialization of AMF 243
- 9.7 Conclusion 246
- References 246

Chapter 10 Plant growth-promoting and biocontrol potency of rhizospheric bacteria associated with halophytes 253
Kalpna D. Rakholiya, Mital J. Kaneria, Paragi R. Jadhav, and Satya P. Singh

10.1 Introduction 253
10.2 Plant-microbe interactions and mitigation of abiotic stress .. 253
10.3 Concluding remarks and future prospects 263
Acknowledgments... 263
References 263

Chapter 11 Nanotechnology for plant growth promotion and stress management..................................... 269
Pooja Sharma, Ashutosh Shukla, Mamta Yadav, Anuj Kumar Tiwari, Ravindra Soni, Sudhir Kumar Srivastava, and Surendra Pratap Singh

11.1 Introduction 269
11.2 Synthesis, absorption, and translocation of nanoparticles in plants 271
11.3 Nanoparticles in plant growth and stress tolerance......... 273
11.4 Green nanoparticle 276
11.5 Genetic engineering in nanoparticle 276
11.6 Conclusion 278
References 278

Chapter 12 Role of microbial biotechnology for strain improvement for agricultural sustainability 285
Akhila Pole, Anisha Srivastava, Mohamed Cassim Mohamed Zakeel, Vijay Kumar Sharma, Deep Chandra Suyal, Anup Kumar Singh, and Ravindra Soni

12.1 Introduction 285
12.2 Microbial inoculants in agriculture for sustainability........ 287

12.3	Microbial biotechnology for sustainable agriculture	301
12.4	Advantages and limitation	306
12.5	Concluding remarks	307
References		307

Chapter 13 Harnessing the potential of genetically improved bioinoculants for sustainable agriculture: Recent advances and perspectives ... **319**

Vinay Kumar, Anisha Srivastava, Lata Jain, Sorabh Chaudhary, Pankaj Kaushal, and Ravindra Soni

13.1	Introduction	319
13.2	Importance of manipulation of microbial bioinoculants	320
13.3	Genetic traits need to be manipulated for enhancing efficacy of microbes	322
13.4	Genetic engineering techniques used for genetic manipulation of microbes	323
13.5	Genetic manipulations/improvement of bioinoculants for different traits with examples	323
13.6	Conclusion	334
References		334

Chapter 14 Omics technologies for agricultural microbiology research .. **343**

Jagmohan Singh, Dinesh K. Saini, Ruchika Kashyap, Sandeep Kumar, Yuvraj Chopra, Karansher S. Sandhu, Mankanwal Goraya, and Rashmi Aggarwal

14.1	Introduction	343
14.2	Genomics	346
14.3	Transcriptomics	353
14.4	Proteomics	357
14.5	Metabolomics	366
14.6	Integrated omics	376
14.7	Conclusion	381
References		381

Chapter 15 Plant growth-promoting microorganism-mediated abiotic stress resilience in crop plants 395
Sonth Bandeppa, Priyanka Chandra, Savitha Santosh, Saritha M, Seema Sangwan, and Samadhan Yuvraj Bagul

15.1 Introduction 395
15.2 Diverse abiotic stress affecting plants.................. 397
15.3 Microbe-mediated abiotic stress alleviation 400
15.4 Mechanisms of microbe-mediated abiotic stress tolerance.................. 401
15.5 Conclusion 409
References 410

Chapter 16 Phosphate biofertilizers: Recent trends and new perspectives........................ 421
Mohammad Saghir Khan, Asfa Rizvi, Bilal Ahmed, and Jintae Lee

16.1 Introduction 421
16.2 Importance of P and rationale for using phosphate biofertilizers in agrosystems................. 423
16.3 Current status of phosphate biofertilizers 424
16.4 Development of phosphate biofertilizers: An overview................. 427
16.5 Overview of P solubilization mechanisms................. 432
16.6 Phosphate biofertilizers: Phyto-beneficial and eco-physiological perspective 434
16.7 Trends of phosphate biofertilizers use: A key for sustainable agriculture 437
16.8 Molecular engineering of phosphate biofertilizers 445
16.9 Challenges and future prospects of phosphate biofertilizers 447
References 447

Chapter 17 Plant-microbe interactions: Beneficial role of microbes for plant growth and soil health 463
Raghu Shivappa, Mathew Seikholen Baite, Prabhukarthikeyan S. Rathinam, Keerthana Umapathy, Prajna Pati, Anisha Srivastava, and Ravindra Soni

17.1 Introduction 463
17.2 Rhizobacteria 464
17.3 Ecological considerations for plant beneficial function of microbes in the field 465
17.4 Secondary metabolites in plant-microbe interaction 465
17.5 Microbial-plant defense genes involved in interaction 467
17.6 RNAi and CRISPR/Cas9 technology to explore plant-microbe interaction 470
References 473

Chapter 18 Potash biofertilizers: Current development, formulation, and applications 481
Shiv Shanker Gautam, Manjul Gondwal, Ravindra Soni, and Bhanu Pratap Singh Gautam

18.1 Introduction 481
18.2 Potash biofertilizers and potassium solubilizing microorganisms (KSMs) 483
18.3 Current development of potash biofertilizers 487
18.4 Formulation of potash biofertilizers 490
18.5 Field applications and crop improvement 493
18.6 Conclusions and future perspectives 494
References 494

Chapter 19 Phosphate-solubilizing microbial inoculants for sustainable agriculture 501
Sonth Bandeppa, Kiran Kumar, P.C. Latha, P.G.S. Manjusha, Amol Phule, and C. Chandrakala

19.1 Introduction 501

19.2 Status and availability of soil phosphorus 503
19.3 Importance of phosphate-solubilizing microorganism
      in agriculture 503
19.4 Diversity of phosphate-solubilizing microorganisms........ 505
19.5 Mechanisms of P solubilization by PSMs 505
19.6 Types of phosphate biofertilizers...................... 508
19.7 Production, quality standards, evaluation, and marketing
      of phosphate biofertilizers 509
19.8 Plant growth-promoting activities of PSMs 513
19.9 Influence of PSMs on plant growth and yield............ 513
19.10 Genetics of phosphate solubilization by PSMs........... 513
19.11 Impact of application of phosphate biofertilizers on
      native soil microorganisms 516
19.12 Constraints in using phosphate biofertilizers 517
19.13 Future prospects 518
19.14 Conclusion 518
References 519

Chapter 20 Trichoderma: Improving growth and tolerance to biotic and abiotic stresses in plants 525

Bahman Fazeli-Nasab, Laleh Shahraki-Mojahed, Ramin Piri, and Ali Sobhanizadeh

20.1 Introduction 525
20.2 Trichoderma... 526
20.3 Role of *Trichoderma* in stimulating plant growth........... 529
20.4 Role of *Trichoderma* in increasing germination indices...... 532
20.5 Significance of seed germination and seedling
      establishment in seed production 534
20.6 Factors affecting germination......................... 535
20.7 Seedling establishment............................... 535
20.8 Effect of environmental stresses on seedling
      establishment....................................... 535
20.9 Significance of regarding stresses of heavy metals......... 536

20.10 Application of seed biological treatments (biopriming) in tolerance induction to environmental stresses. 537
20.11 Biopriming steps based on fungi microstructure 539
20.12 Effect of *Trichoderma* species on increasing growth and antioxidant activity 542
20.13 Role of *Trichoderma* in dealing with biological stresses (biotic). 544
20.14 Induction of plant resistance to nonbiological stresses (abiotic). 545
20.15 Biological mechanisms of *Trichoderma*. 547
20.16 Colonization as interactions between *Trichoderma* and plant. 549
20.17 Conclusion 552
References 553

Chapter 21 Bacterial biofertilizers for bioremediation: A priority for future research 565
Asfa Rizvi, Bilal Ahmed, Shahid Umar, and Mohammad Saghir Khan
21.1 Introduction 565
21.2 Heavy metal contamination of agronomic soils: An overview. 567
21.3 Bioremediation: Concepts and prospects 567
21.4 Biofertilizer technology in bioremediation 569
21.5 Biofertilizers—A general perspective. 584
21.6 Conclusion and future perspectives 594
Acknowledgment. 594
References 595

Chapter 22 Biopesticides: A key player in agro-environmental sustainability 613
H. R Archana, K Darshan, M Amrutha Lakshmi, Thungri Ghoshal, Bishnu Maya Bashayal, and Rashmi Aggarwal
22.1 Introduction 613
22.2 What is sustainable environment and agriculture? 614

22.3	Why must chemical pesticides not be used?	616
22.4	What are biopesticides?	618
22.5	Sources and types of biopesticides	619
22.6	Mode of action of biopesticides on different pests/pathogens	626
22.7	Role of biopesticides in sustainable environment and agricultural production	635
22.8	Market of biopesticides	640
22.9	Future prospects and conclusion	642
References		643

Chapter 23 Plant-pathogen interaction: Mechanisms and evolution ... 655
U.M. Aruna Kumara, P.L.V.N. Cooray, N. Ambanpola, and N. Thiruchchelvan

23.1	Introduction	655
23.2	Plants	656
23.3	Plant pathogens	657
23.4	Plant-pathogen interaction	659
23.5	Mechanisms involved in plant-pathogen interactions	661
23.6	Evolution of plant-pathogen interactions	675
23.7	Conclusions	678
References		679

Chapter 24 Global biofertilizer market: Emerging trends and opportunities ... 689
Sanjay Kumar Joshi and Ajay Kumar Gauraha

24.1	Introduction: The market and opportunities	689
24.2	The market dynamics and growth drivers	690
24.3	Growth-share matrix	691
24.4	Biofertilizer market Segmentation	692
24.5	The challenges	692
24.6	Market ecosystem	693

24.7 Biofertilizers and Indian agriculture 694
24.8 Market potential and constraints 695
24.9 Impact of COVID-19 on biofertilizer market and the future 696
References 697

Chapter 25 Organic agriculture for agro-environmental sustainability 699
Neelam Thakur, Simranjeet Kaur, Tanvir Kaur, Preety Tomar, Rubee Devi, Seema Thakur, Nidhi Tyagi, Rajesh Thakur, Devinder Kumar Mehta, and Ajar Nath Yadav

25.1 Introduction 699
25.2 Standards of organic agriculture 702
25.3 Status of organic agriculture 702
25.4 Pros and cons of organic agriculture 716
25.5 Some of the products used in organic farming 718
25.6 Impact of organic agriculture 722
25.7 Regulatory policies in organic agriculture 727
25.8 Market potential of organic produce 730
25.9 Conclusion and future prospects 731
References 731

Chapter 26 Contributing effects of vermicompost on soil health and farmers' socioeconomic sustainability 737
Pallabi Mishra and Debiprasad Dash

26.1 Introduction 737
26.2 Literature review 740
26.3 Effect of vermicompost on paddy-grown soil health 740
26.4 Role of vermicompost usage on farmers' socioeconomic sustainability 743
26.5 Methodology 747

26.6 Brief profile of the villages taken for the study............ 747
26.7 Results and discussion............................ 747
26.8 Conclusion..................................... 754
References.. 754

Index... 759

Contributors

Gayan Abeysinghe Department of Biological Sciences, Faculty of Applied Sciences, Rajarata University of Sri Lanka, Mihintale, Sri Lanka

Rashmi Aggarwal Division of Plant Pathology, ICAR-Indian Agricultural Research Institute, New Delhi, India

Bilal Ahmed School of Chemical Engineering, Yeungnam University, Gyeongsan, Republic of Korea

N. Ambanpola Board of Study in Plant Protection, Postgraduate Institute of Agriculture, Faculty of Agriculture, University of Peradeniya, Preadeniya, Sri Lanka

M Amrutha Lakshmi ICAR-Indian Institute of Oilpalm Research, Pedavagi, Andhra Pradesh, India

V.T. Anju Department of Biochemistry and Molecular Biology, School of Life Sciences, Pondicherry University, Pondicherry, India

H. R Archana Division of Seed Science and Technology, ICAR-Indian Agricultural Research Institute, New Delhi, India

U.M. Aruna Kumara Department of Agricultural Technology, Faculty of Technology, University of Colombo, Homagama, Sri Lanka

Samadhan Yuvraj Bagul ICAR-Directorate of Medicinal and Aromatic Plant Research, Boriavi, Gujarat, India

Mathew Seikholen Baite Crop Protection Division, ICAR-National Rice Research Institute (ICAR-NRRI), Cuttack, Odisha, India

T.C. Bamunuarachchige Department of Bioprocess Technology, Faculty of Technology, Rajarata University of Sri Lanka, Mihintale, Sri Lanka

Sonth Bandeppa ICAR-Indian Institute of Rice Research, Hyderabad, Telangana, India

Nitika Bansal Forest Pathology Discipline, Forest Protection Division, Forest Research Institute, Dehradun, India

Rupjyoti C. Baruah Rain Forest Research Institute, Jorhat, Assam, India

Bishnu Maya Bashayl Division of Plant Pathology, ICAR-Indian Agricultural Research Institute, New Delhi, India

Chanda Vikrant Berde Marine Microbiology, School of Earth, Ocean and Atmospheric Sciences, Goa University, Taleigao Plateau, Goa, India

Vikrant Balkrishna Berde Department of Zoology, Arts Commerce and Science College, Lanja, Maharashtra, India

M.S. Bhandari Genetics and Tree Improvement Division, Forest Research Institute, Dehradun, India

Ratnaboli Bose Forest Pathology Discipline, Forest Protection Division, Forest Research Institute, Dehradun, India

P. Veera Bramhachari Department of Biotechnology, Krishna University, Machilipatnam, Andhra Pradesh, India

Priyanka Chandra ICAR-Central Soil Salinity Research Institute, Karnal, Haryana, India

C. Chandrakala ICAR-Indian Institute of Rice Research, Hyderabad, Telangana, India

Sorabh Chaudhary Department of Crop Protection, Central Potato Research Institute, Meerut, India

Yuvraj Chopra College of Agriculture, Punjab Agricultural University, Ludhiana, India

P.L.V.N. Cooray Board of Study in Plant Protection, Postgraduate Institute of Agriculture, Faculty of Agriculture, University of Peradeniya, Preadeniya, Sri Lanka

K Darshan Forest Protection Division, ICFRE-Tropical Forest Research Institute (TFRI), Jabalpur, Madhya Pradesh, India

Debiprasad Dash KVK Bhadrak, OUAT, Bhubaneswar, India

Rubee Devi Department of Biotechnology, Dr. Khem Singh Gill Akal College of Agriculture, Eternal University, Baru Sahib, Sirmour, Himachal Pradesh, India

Ankita Dey Department of Biotechnology, Maulana Abul Kalam Azad University of Technology, Haringhata, West Bengal, India

Bhanushree Doley Rain Forest Research Institute, Jorhat, Assam, India

Madhu Dyavaiah Department of Biochemistry and Molecular Biology, School of Life Sciences, Pondicherry University, Pondicherry, India

Bahman Fazeli-Nasab Research Department of Agronomy and Plant Breeding, Agricultural Research Institute, University of Zabol, Zabol, Iran

Ajay Kumar Gauraha Agri-Business and Rural Management, Indira Gandhi Agricultural University, Raipur, India

Bhanu Pratap Singh Gautam Department of Chemistry, Laxman Singh Mahar Govt. P.G. College, Pithoragarh, Uttarakhand, India

Shiv Shanker Gautam Serve India Inter College, Roshanpur, Gadarpur, Udham Singh Nagar, Uttarakhand, India

Sougata Ghosh Department of Microbiology, School of Science, RK. University, Rajkot, Gujarat, India

Thungri Ghoshal Division of Plant Pathology, ICAR-Indian Agricultural Research Institute, New Delhi, India

Krishna Giri Rain Forest Research Institute, Jorhat, Assam, India

Manjul Gondwal Department of Chemistry, Laxman Singh Mahar Govt. P.G. College, Pithoragarh, Uttarakhand, India

Mankanwal Goraya Department of Plant Pathology, North Dakota State University, Fargo, ND, United States

Suman Kumar Halder Department of Microbiology, Vidyasagar University, Midnapore, West Bengal, India

Paragi R. Jadhav Department of Biosciences (UGC-CAS), Saurashtra University, Rajkot, Gujarat, India

Lata Jain ICAR—National Institute of Biotic Stress Management, Raipur, India

R.S.C. Jayaraj Rain Forest Research Institute, Jorhat, Assam, India

Y.K. Jhala Department of Agricultural Microbiology, Anand Agricultural University, Anand, Gujarat, India

Pooja Joshi Forest Pathology Discipline, Forest Protection Division, Forest Research Institute, Dehradun, India

Sanjay Kumar Joshi Agri-Business and Rural Management, Indira Gandhi Agricultural University, Raipur, India

Mital J. Kaneria Department of Biosciences (UGC-CAS), Saurashtra University, Rajkot, Gujarat, India

Ruchika Kashyap Department of Agronomy, Horticulture, and Plant Sciences, South Dakota State University, Brookings, SD, United States

Simranjeet Kaur Department of Zoology, Akal College of Basic Sciences, Eternal University, Baru Sahib, Himachal Pradesh, India

Tanvir Kaur Department of Biotechnology, Dr. Khem Singh Gill Akal College of Agriculture, Eternal University, Baru Sahib, Sirmour, Himachal Pradesh, India

Pankaj Kaushal ICAR—National Institute of Biotic Stress Management, Raipur, India

Mohammad Saghir Khan Department of Agricultural Microbiology, Faculty of Agricultural Sciences, Aligarh Muslim University, Aligarh, Uttar Pradesh, India

Divjot Kour Department of Biotechnology, Dr. Khem Singh Gill Akal College of Agriculture, Eternal University, Baru Sahib, Sirmour, Himachal Pradesh, India

Kiran Kumar Crop Production Unit, ICAR-Directorate of Groundnut Research, Junagadh, India

Sandeep Kumar ICAR - Indian Institute of Natural Resins and Gums, Ranchi, Jharkhand, India

Vinay Kumar ICAR—National Institute of Biotic Stress Management, Raipur, India

Dibyajit Lahiri Department of Biotechnology, University of Engineering & Management, Kolkata, West Bengal, India

P.C. Latha ICAR-Indian Institute of Rice Research, Hyderabad, Telangana, India

Jintae Lee School of Chemical Engineering, Yeungnam University, Gyeongsan, Republic of Korea

P.G.S. Manjusha ICAR-Indian Institute of Rice Research, Hyderabad, Telangana, India

Devinder Kumar Mehta Dr. Y.S. Parmar University of Horticulture and Forestry, Nauni-Solan, India

Gaurav Mishra Rain Forest Research Institute, Jorhat, Assam, India

Pallabi Mishra Department of Business Administration, Utkal University, Bhubaneswar, India

Keshab Chandra Mondal Department of Microbiology, Vidyasagar University, Midnapore, West Bengal, India

Subhadeep Mondal Centre for Life Sciences, Vidyasagar University, Midnapore, West Bengal, India

Moupriya Nag Department of Biotechnology, University of Engineering & Management, Kolkata, West Bengal, India

Amit Pandey Forest Pathology Discipline, Forest Protection Division, Forest Research Institute, Dehradun, India

Shailesh Pandey Forest Pathology Discipline, Forest Protection Division, Forest Research Institute, Dehradun, India

H.K. Patel Department of Agricultural Microbiology, Anand Agricultural University, Anand, Gujarat, India

Prajna Pati Department of Agricultural Entomology, Institute of Agricultural Sciences (IAS), Siksha 'O' Anusandhan Deemed to be University, Bhubaneswar, Odisha, India

Amol Phule ICAR-Indian Institute of Rice Research, Hyderabad, Telangana, India

Ramin Piri Department of Agronomy and Plant Breeding, Faculty of Agriculture, University of Tehran, Tehran, Iran

Akhila Pole Department of Agricultural Microbiology, College of Agriculture, Indira Gandhi Krishi Vishwavidyalaya, Raipur, Chhattisgarh, India

B.L. Raghunandan AICRP-Biological Control of Crop Pests, Anand Agricultural University, Anand, Gujarat, India

Kalpna D. Rakholiya Department of Biosciences (UGC-CAS), Saurashtra University, Rajkot, Gujarat, India

Prabhukarthikeyan S. Rathinam Crop Protection Division, ICAR-National Rice Research Institute (ICAR-NRRI), Cuttack, Odisha, India

Rina Rani Ray Department of Biotechnology, Maulana Abul Kalam Azad University of Technology, Haringhata, West Bengal, India

Asfa Rizvi Department of Botany, School of Chemical and Life Sciences, Jamia Hamdard, New Delhi, India

Aditi Saini Forest Pathology Discipline, Forest Protection Division, Forest Research Institute, Dehradun, India

Dinesh K. Saini Department of Genetics and Plant Breeding, Chaudhary Charan Singh University, Meerut, Uttar Pradesh, India

Karansher S. Sandhu Department of Crop and Soil Sciences, Washington State University, Pullman, WA, United States

Seema Sangwan Division of Microbiology, Indian Agricultural Research Institute, New Delhi, Delhi, India

Savitha Santosh ICAR-Central Institute for Cotton Research, Nagpur, Maharashtra, India

M. Saritha ICAR-Central Arid Zone Research Institute, Jodhpur, Rajasthan, India

Laleh Shahraki-Mojahed Department of Biochemistry, School of Medicine, Zabol University of Medical Sciences, Zabol, Iran

Pooja Sharma Department of Environmental Microbiology, School for Environmental Sciences, Babasaheb Bhimrao Ambedkar Central University, Lucknow, Uttar Pradesh, India

Vijay Kumar Sharma Department of Post-harvest Science, Agricultural Research Organization-Volcani Centre, Rishon LeZion, Israel

Raghu Shivappa Crop Protection Division, ICAR-National Rice Research Institute (ICAR-NRRI), Cuttack, Odisha, India

Ashutosh Shukla Plant Molecular Biology Laboratory, Department of Botany, Dayanand Anglo-Vedic (PG) College, Chhatrapati Shahu Ji Maharaj University, Kanpur, Uttar Pradesh, India

Busi Siddhardha Department of Microbiology, School of Life Sciences, Pondicherry University, Pondicherry, India

Anup Kumar Singh Department of Agricultural Microbiology, College of Agriculture, Indira Gandhi Krishi Vishwavidyalaya, Raipur, Chhattisgarh, India

Jagmohan Singh Division of Plant Pathology, ICAR-Indian Agricultural Research Institute, New Delhi; Guru Angad Dev Veterinary and Animal Sciences University, Ludhiana, India

Satya P. Singh Department of Biosciences (UGC-CAS), Saurashtra University, Rajkot, Gujarat, India

Surendra Pratap Singh Plant Molecular Biology Laboratory, Department of Botany, Dayanand Anglo-Vedic (PG) College, Chhatrapati Shahu Ji Maharaj University, Kanpur, Uttar Pradesh, India

Ali Sobhanizadeh Department of Horticultural Science, Faculty of Agriculture, University of Zabol, Iran

J.P. Solanki Department of Agricultural Microbiology, Anand Agricultural University, Anand, Gujarat, India

Ravindra Soni Department of Agricultural Microbiology, College of Agriculture, Indira Gandhi Krishi Vishwavidyalaya, Raipur, Chhattisgarh, India

Anisha Srivastava Department of Agricultural Microbiology, College of Agriculture, Indira Gandhi Krishi Vishwavidyalaya, Raipur, Chhattisgarh, India

Sudhir Kumar Srivastava Chemical Research Laboratory, Department of Chemistry, Dayanand Anglo-Vedic (PG) College, Chhatrapati Shahu Ji Maharaj University, Kanpur, Uttar Pradesh, India

Deep Chandra Suyal Department of Microbiology, Akal College of Basic Sciences, Eternal University, Sirmour, Himachal Pradesh, India

Neelam Thakur Department of Zoology, Akal College of Basic Sciences, Eternal University, Baru Sahib, Himachal Pradesh, India

Rajesh Thakur Krishi Vigyan Kendra Kandaghat, Solan, India

Seema Thakur Krishi Vigyan Kendra Kandaghat, Solan, India

N. Thiruchchelvan Department of Agricultural Biology, Faculty of Agriculture, University of Jaffna, Kilinochchi, Sri Lanka

Anuj Kumar Tiwari Department of Botany, Bhavan's Mehta Mahavidyalaya, Bharwari, Uttar Pradesh, India

Preety Tomar Department of Zoology, Akal College of Basic Sciences, Eternal University, Baru Sahib, Himachal Pradesh, India

Nidhi Tyagi Krishi Vigyan Kendra Kandaghat, Solan, India

Keerthana Umapathy Crop Protection Division, ICAR-National Rice Research Institute (ICAR-NRRI), Cuttack, Odisha, India

Shahid Umar Department of Botany, School of Chemical and Life Sciences, Jamia Hamdard, New Delhi, India

Ajar Nath Yadav Department of Biotechnology, Dr. Khem Singh Gill Akal College of Agriculture, Eternal University, Baru Sahib, Sirmour, Himachal Pradesh, India

Mamta Yadav Plant Molecular Biology Laboratory, Department of Botany, Dayanand Anglo-Vedic (PG) College, Chhatrapati Shahu Ji Maharaj University, Kanpur, Uttar Pradesh, India

Mohamed Cassim Mohamed Zakeel Department of Plant Sciences, Faculty of Agriculture, Rajarata University of Sri Lanka, Puliyankulama, Anuradhapura, Sri Lanka; Centre for Horticultural Science, Queensland Alliance for Agriculture and Food Innovation, The University of Queensland, Ecosciences Precinct, Dutton Park, QLD, Australia

H.K.S. De Zoysa Department of Bioprocess Technology, Faculty of Technology, Rajarata University of Sri Lanka, Mihintale, Sri Lanka

Preface

This book, besides discussing challenges and opportunities in the field of microbiology, also reveals the advancement and world scenario for microbial interactions in various environments. It further enlightens applied research in the field of agriculture for sustainability and management. Furthermore, important microbial genes and enzymes, gene modeling and engineering, genetically engineered bioinoculants, next-generation technologies, omics and nano-based technologies, etc., are other significant domains. Moreover, the book also covers recent trends in industrial and economic microbiology. Nevertheless, it holds the promise to discuss present agricultural practices employed across the world that affect microbial diversity, which should be discussed when considering sustainable developmental practices.

This book consists of 26 chapters based on applications of microbial methods and their importance in agriculture and biotechnology. We trust that readers of this book will find it precise and up to date as far as applied microbiology, especially agriculturally important microbes, are concerned. The editors want to thank all contributors to this book who took time to contribute their valuable thoughts and experiences as chapters for this book.

Trends of agricultural microbiology for sustainable crops production and economy: An introduction

Tanvir Kaur, Divjot Kour, and Ajar Nath Yadav

Department of Biotechnology, Dr. Khem Singh Gill Akal College of Agriculture, Eternal University, Baru Sahib, Sirmour, Himachal Pradesh, India

1.1 Introduction

A new trend in agriculture is the use of microbes for sustainable crop production. It is an alternative method for greener agriculture as conventional agricultural uses huge amounts of chemically synthesized products such as fertilizers and pesticides. The use of pesticides and fertilizers has been shown to impact the environment. The use of pesticides and fertilizers has adversely affected the quality of water as well as soil. Their use has also been proven to affect other organisms such as plants, small organisms, and microbial species. The use of microbes in agriculture has solved all related issues of the environment along with the farmer's expectation of crop productivity (Kumar et al., 2021b).

Microbes undergo different types of mechanisms that help in increasing the growth, development, and productivity of the crop plant (Yadav, 2021a). The biological fixation of nitrogen; the solubilization of nutrients such as phosphorus, potassium, and zinc; the chelation of iron through the production of iron-chelating agents such as siderophores; and the production of phytohormones are some of the major mechanisms that help in directly promoting plant growth (Suman et al., 2016; Verma et al., 2017a; Yadav et al., 2018). These mechanisms are used to provide the entire basic growth nutrients that helps plants to develop such as nitrogen, phosphorus, potassium, and zinc. On the other hand, microbes also regulate agriculture productivity indirectly by protecting them from biotic and abiotic factors. Crop plants are host to many pathogens and pests that adversely affect productivity (Kaur et al., 2020b). Pests and pathogens, that is, biotic factors, affecting plant growth are serious destroyers of productivity. To control them, microbes are used to produce toxic compounds that inhibit pathogen growth (Singh et al., 2020c).

Ammonia, hydrogen cyanide, hydrolytic enzymes such as chitinase and β-1,3, and glucanase are some toxic compounds that inhibit pathogen growth and development. On the other hand, abiotic stress is also a major constraint that affects plant productivity. Drought, salinity, and extreme temperature are some abiotic stresses that affect crop growth. Such types of stresses in plant could be alleviated by the production of phytohormones and 1-aminocyclopropane-1-carboxylate deaminase (Verma et al., 2017b, c). All the above-mentioned microbial mechanisms play significant roles in plant growth and productivity and microbes can be applied as biofertilizer and biopesticides. Microbes as biofertilizers and biopesticides could also be used to conduct organic agriculture. Usually, organic agriculture is practiced without chemicals and only natural products are applied. Therefore, the use of these wonder organisms can also increase crop yield while leading to sustainable agriculture (Kour et al., 2019c).

Microbial products as biofertilizers and biopesticides are commonly available in the markets, which have been drastically increasing with the increase of environmental health awareness. Presently, in 24 countries, 170 organizations are involved in the production of commercial biofertilizers. This is an introductory chapter for the book *Trends of Agricultural Microbiology for Sustainable Crops Production and Economy*, which deals with the complete basic functions of microbes in agriculture to enhance crop productivity, including mechanisms and their applications as well as the effect on the worldwide economy.

1.2 Role of beneficial microbiomes in plant growth promotion

Beneficial microbes are ubiquitous in nature and have been sorted out as soil and plant microbiomes. These microbes have the capability to promote plants using diverse plant growth-promoting mechanisms while protecting the plant from phytopathogenic microbes and diverse abiotic stresses. They have been recognized as potential crop enhancers that increase crop productivity in a sustainable way by undergoing various mechanisms including nitrogen fixation; phosphorus, potassium, and zinc solubilization; and the production of siderophores, ammonia, HCN, and phytohormones (Kaur et al., 2021; Yadav, 2021b). The use of microbes as bioinoculants is a sustainable tool for agroenvironmental sustainability (Fig. 1.1).

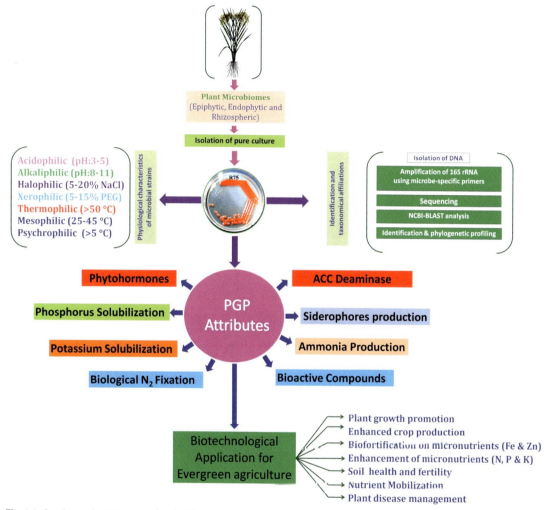

Fig. 1.1 A schematic representation for the characterization of microbes and their agricultural applications for sustainable development. Adapted with permission from Kour, D., Kaur, T., Devi, R., Rana, K.L., Yadav, N., Rastegari, A.A., et al., 2020a. Biotechnological applications of beneficial microbiomes for evergreen agriculture and human health. In: Rastegari, A.A., Yadav, A.N., Yadav, N. (Eds.) Trends of Microbial Biotechnology for Sustainable Agriculture and Biomedicine Systems: Perspectives for Human Health. Elsevier, Amsterdam, pp. 255–279. https://doi.org/10.1016/B978-0-12-820528-0.00019-3.

1.2.1 Nitrogen fixation

The most important constituent of living organisms (plants, animals, and microbes) is nitrogen. It is found in biological molecules such as amino acids and nucleic acids (RNA and DNA). On Earth, nitrogen is the most abundant element and is present as di-nitrogen

(N_2) gas in the Earth's atmosphere. However, the element is not available for the majority of organisms, especially plants and microbes, because of nitrogen molecule inertness (high strength of triple bound and stable electron configuration) (Valentine et al., 2018). As a result, modern agriculture is largely dependent upon chemically synthesized nitrogen fertilizer to meet nitrogen requirements and increase crop yields (Yadav, 2021c).

On the other hand, microbes, especially bacteria containing the *nif* gene (coding gene of nitrogenase, a protein complex), fix nitrogen via the ATP-dependent process of biological nitrogen fixation. Bacteria reduce the inert nitrogen gas molecules and convert it into simpler compounds such as ammonia and nitric oxide in the presence of protein complex, nitrogenase to fulfill its own nitrogen requirement (Dos Santos et al., 2012). In nature, three different types of bacteria are known to fix nitrogen: associative, free-living, and symbiotic (Mahmud et al., 2020) (Table 1.1). Free-living nitrogen-fixing bacteria exist in soil. Bacterial species such as *Novosphingobium* sp., *Pseudomonas* sp., *Serratia* sp. (Islam et al., 2013), *Azotobacter vinelandi* (Bellenger et al., 2011), *Azotobacter chroococcum* (Basak and Biswas, 2010), *Paenibacillus riograndensis* (Beneduzi et al., 2010), and *Paenibacillus jilunlii* (Jin et al., 2011) are some examples of free-living nitrogen-fixing bacteria.

Symbiotic bacteria develop symbiotic relationships with leguminous plants by forming nodules with the help of a *nod* gene. Genera such as *Rhizobium, Bradyrhizobium, Mesorhizobium,* and *Sinorhizobium*, which are collectively known as rhizobia, are classified as symbiotic nitrogen-fixing bacteria (Mahmud et al., 2020). The third type of nitrogen-fixing bacteria, associative bacteria, is endophytic and fixes nitrogen. *Pantoea agglomerans* (Rana et al., 2021), *Pseudomonas stutzeri* (Pham et al., 2017a), and *Gluconacetobacter diazotrophicus* (Meneses et al., 2011) are some endophytic nitrogen-fixing bacteria.

In agriculture, these microbes could be used as potent nitrogen biofertilizers for the enhancement of sustainable plant growth yield while replacing chemical-based fertilizers (Farrar et al., 2014; Vandana et al., 2017). Much evidence is available showing that that these microbes enhance plant growth and promotion along with crop productivity. In the study from Basak and Biswas (2010), it was reported that the coinoculation of potassium (K) solubilizing and nitrogen (N) fixing bacteria, *Bacillus mucilaginosus* and *A. chroococcum* respectively, in Sudan grass (*Sorghum vulgare*) enhanced plant growth. It was reported that coinoculation also enhanced the biomass of the plant acquisition of nutrients and the K and N content in the soil. In a similar study, the coinoculation of phosphorus-solubilizing and N-fixing bacteria, identified as *Pseudomonas chlororaphis* and *Arthrobacter pascens* respectively, in walnut seedlings increased P and N uptake and biomass

Table 1.1 Beneficial plant growth-promoting microbiomes and their roles.

Types of biofertilizer	Subgroups	Examples	References
Nitrogen	Free living	Azohydromonas caseinilytica	Dahal et al. (2021)
		Azotobacter chroococcum	Farah Ahmad et al. (2006)
		Azotobacter vinelandi	Bellenger et al. (2011)
		Bacillus subtilis	Satapute et al. (2012)
		Cellvibrio diazotrophicus	Suarez et al. (2014)
		Niveispirillum cyanobacteriorum	Cai et al. (2015)
		Paenibacillus azotifixans	Islam et al. (2012)
		Paenibacillus graminis	Navarro-Noya et al. (2012)
		Paenibacillus jilunlii	Jin et al. (2011)
		Paenibacillus riograndensis	Beneduzi et al. (2010)
		Paraburkholderia azotifigens	Choi and Im (2018)
	Symbiotic	Bradyrhizobium brasilense	Da Costa et al. (2017)
		Bradyrhizobium campsiandrae	Michel et al. (2021)
		Bradyrhizobium forestalis	da Costa et al. (2018)
		Bradyrhizobium oligotrophicum	Hara et al. (2019)
		Bradyrhizobium subterraneum	Grönemeyer et al. (2015)
		Bradyrhizobium vignae	Grönemeyer et al. (2016)
		Ensifer meliloti	Bessadok et al. (2021)
		Mesorhizobium atlanticum	Helene et al. (2019)
		Paraburkholderia atlantica	Paulitsch et al. (2020)
		Paraburkholderia guartelaensis	Paulitsch et al. (2019)
		Rhizobium azibense	Mnasri et al. (2014)
		Rhizobium leguminosarum	Nohwar et al. (2019)
		Sinorhizobium meliloti	Pawlicki-Jullian et al. (2010)
	Associative	Agrobacterium tumefaciens	My et al. (2015)
		Burkholderia caballeronis	Martínez-Aguilar et al. (2013)
		Burkholderia phymatum	Lowman et al. (2016)
		Burkholderia vietnamiensis	Tang et al. (2010)
		Enterobacter cloacae	Swamy et al. (2016)
			Meneses et al. (2011)

Continued

Table 1.1 Beneficial plant growth-promoting microbiomes and their roles—cont'd

Types of biofertilizer	Subgroups	Examples	References
Phosphorus	P-solubilizing	Klebsiella variicola	Wei et al. (2014)
		Nitrospirillum amazonense	Schwab et al. (2018)
		Pantoea agglomerans	Rana et al. (2021)
		Pseudomonas stutzeri	Pham et al. (2017a)
		Acinetobacter baumannii	Collavino et al. (2010)
		Variovorax boronicumulans	Collavino et al. (2010)
		Exiguobacterium mexicanum	Collavino et al. (2010)
		Stenotrophomonas maltophilia	Collavino et al. (2010)
		Bacillus magaterium	Xiang et al. (2011)
		Pseudomonas aeruginosa	Srinivasan et al. (2012)
		Halococcus hamelinensis	Yadav et al. (2015)
		Aspergillus niger	Li et al. (2016)
		Penicillium oxalicum	Li et al. (2016)
		Bacillus cereus	Wang et al. (2017)
		Talaromyces aurantiacus	Zhang et al. (2018)
		Aspergillus neoniger	Zhang et al. (2018)
		Pseudomonas libanensis	Kour et al. (2019a)
		Streptomyces laurentii	Kour et al. (2020b)
		Acinetobacter calcoaceticus	Kour et al. (2020c)
		Tsukamurella tyrosinosolvens	Zhang et al. (2021)
	P-mobilizing	Acinetobacter pittii	Liu et al. (2014)
Potassium	K-solubilizing	Bacillus megaterium	Keshavarz Zarjani et al. (2013)
		Bacillus pseudomycoides	Pramanik et al. (2019)
		Burkholderia cenocepacia	Raji and Thangavelu (2021)
		Burkholderia cepacia	Zhang and Kong (2014)
		Enterobacter cloacae	Liu et al. (2014)
		Frateuria aurantia	Subhashini (2015)
		Microbacterium foliorum	Zhang and Kong (2014)
		Paenibacillus glucanolyticus	Liu et al. (2012)
		Pantoea ananatis	Bakhshandeh et al. (2017)
		Rahnella aquatilis	Bakhshandeh et al. (2017)
Zinc	Zn-solubilizing	Burkholderia cepacia	Gontia-Mishra et al. (2017)
		Bacillus aryabhattai	Mumtaz et al. (2017)
		Bacillus megaterium	Dinesh et al. (2018)
		Acinetobacter baumannii	Upadhyay et al. (2021)

Table 1.1 Beneficial plant growth-promoting microbiomes and their roles—cont'd

Types of biofertilizer	Subgroups	Examples	References
Iron	Siderophore-producing	*Paracoccus sphaerophysae*	Deng et al. (2011)
		Bacillus amyloliquefaciens	Gaonkar and Bhosle (2013)
		Enterococcus faecalis	Ong et al. (2016)
		Aneurinibacillus aneurinilyticus	Kumar et al. (2018)
		Pseudomonas weihenstephanensis	Sinha et al. (2019)
		Sinorhizobium meliloti	Sepehri and Khatabi (2021)
Phytohormones	Phytohormone production	*Acetebacter diazotrophicus*	Patil et al. (2011)
		Bacillus subtilis	Liu et al. (2013)
		Planomicrobium chinense	Das and Tiwary (2014)
		Pseudomonas fluorescence	Hussein and Joo (2015)
		Lysinibacillus endophyticus	Yu et al. (2016)
		Porostereum spadiceum	Hamayun et al. (2017b)
		Pseudomonas koreensis	Kang et al. (2021)

(Yu et al., 2012). In a report, the free-living nitrogen-fixing bacterium *Novosphingobium* sp. inoculated in red pepper roots significantly enhanced chlorophyll and the uptake of different macro- and micronutrients from the soil compared to noninoculated plants (Islam et al., 2013).

Another study reported four different species of dizaotrophic bacteria belonging to the genera *Bacillus, Klebsiella, Microbacterium,* and *Paenibacillus* that were reported to enhance the plant growth of rice seeds (Ji et al., 2014). An endophytic diazotroph of the lodgepole pine, *Paenibacillus polymyxa*, was reported as enhancing the plant growth of the gymnosperm tress species (Tang et al., 2017). In another study, the nitrogen-fixing and plant growth-promoting rhizobacterium *P. stutzeri* was reported to increase the plant growth of rice seedlings compared to chemically fertilized rice seedlings (Pham et al., 2017b). In a study, three different species belonging rice genera *Sphingomonas, Psychrobacillus,* and *Enterobacter*, namely *Sphingomonas trueperi, Psychrobacillus psychrodurans* and *Enterobacter oryzae*, were reported as significant fixers of biological nitrogen. The inoculation of these strains in maize and wheat plants was reported to promote early growth, root length, root structure, and nutrient uptake in maize as well as the dry weight of roots and the number of roots tips in wheat plants (Xu et al., 2018b). In a similar report, *Pseudomonas stutzeri* inoculated in maize was reported to improve plant growth by fixing nitrogen (Ke et al., 2019).

1.2.2 Phosphorus solubilization

Phosphorus is the second most required nutrient for structural and metabolic functions. It builds up about 0.2% of the dry weight of plants. This nutrient is a major component of nucleic acids and phospholipids and it is also involved in energy transfer mechanisms. In plants, it also helps in the formation and elongation of roots, along with tolerance to cold and diseases (Kour et al., 2021b). Phosphorus is the second mineral nutrient after nitrogen that is commonly a limiting factor for growth plant, as only 0.1% of the phosphorus is available for plant use (Zhu et al., 2011a). Traditionally, phosphorus deficiency is addressed through synthetically synthesized phosphorus fertilizer (diammonium phosphate and monoammonium phosphate). However, phosphorus-based fertilizers are not used by the plant completely, which causes environmental problems such as fertility loss and the contamination of soil and groundwater as well as eutrophication (Kang et al., 2011; Kumar et al., 2021; Kour et al., 2019b).

Various phosphorus-solubilizing microbes (PSB) are used to solubilize both P present in nature, that is, organic (Po), and the inorganic (Pi) form of phosphorus through different mechanisms (Kour et al., 2020d). The inorganic form of P is solubilized by the principle mechanism, that is, the production of mineral-dissolving compounds such as carbon dioxide, protons, organic acids, hydroxyl ions, and siderophores (Sharma et al., 2011). The solubilization of P via organic acid production is the well-known mechanism in which the P is released by lowering the soil pH (Usha and Padmavathi, 2012). The solubilization of the organic form of P, which is also referred to as phosphorus mineralization, is done through the enzymatic action of the enzyme phytase (Singh et al., 2020b). This enzyme mineralizes inositol phosphate (soil phytate), which is the most abundant form of organic P. These two mechanism of P solubilization and mineralization avail the phosphorus in soil which could be utilized by the plants for their development (Alori et al., 2017).

Therefore, the use of plant growth-promoting microbes is an efficient strategy for P supplementation to plants for their development. Several microbes such as actinomycetes, archaea, algae, bacteria, and fungi mineralize, solubilize, and mineralize the insoluble form (apatite, fluoroapatite, francolites, strengite, variscite, and wavellite) to the soluble form of P ($H_2PO_4^-$ and HPO_4^{2-}). In a study, various bacterial isolates were sorted from the root-associated soil and the bulk soil of orchards. From all the bacterial isolates, 21 were reported as efficient solubilizers of phosphorus. Among all the isolates, *Enterobacter aerogenes*, *Burkholderia* spp., and *Acinetobacter baumannii* were reported as efficient solubilizers of tricalcium phosphate and recognized to promote the growth of *Phaseolus vulgaris* (Collavino et al., 2010). In another study, a halo-tolerant bacterium was reported for solubilizing phosphorus via organic acid production, namely *Bacillus magaterium* (Xiang et al., 2011).

In a similar report, the bacterium *Pseudomonas aeruginosa* and the fungus *Aspergillus* sp. were reported for solubilizing the inorganic form of phosphorus under saline conditions (Srinivasan et al., 2012). A thermo-tolerant bacterium, *Brevibacillus* sp., was also reported for solubilizing P (rock phosphate). This bacterium was reported for solubilizing P via acid production such as citric, gluconic, formic, and malic acid (Yadav et al., 2013). In another report, three potential phosphate-solubilizing bacteria, *Acinetobacter pittii*, *Escherichia coli*, and *Enterobacter cloacae*, isolated from the rhizosphere and the bulk soil of a field cultivated with betel nut were reported as an efficient mobilizer P nutrient (Liu et al., 2014). Archaea, unique microbes that inhabit places with extreme temperature, pH, and salinity, have also been reported for solubilizing phosphorus. In a report, haloarchaea isolated from the water, sediments, and rhizospheric soil of plants growing in the Rann of Kutch, *Natrinema* sp. and *Halococcus hamelinensis*, were reported as potential solubilizers of P. These archeal strains were reported for solubilizing phosphorus via lowering the soil pH through organic acid production (Yadav et al., 2015).

The fungal species *Aspergillus niger* and *Penicillium oxalicum* were also reported for solubilizing P. In this study, it was reported that both fungal strains were solubilizing phosphorus through the production of organic acids such as citric, formic, and oxalic acids (Li et al., 2016). A Gram-positive bacterium named *Bacillus cereus*, sorted out from Chinese cabbage root-associated soil, was reported as solubilizing P via the production of acetic, ascorbic, citric, lactic, and tartaric acids (Wang et al., 2017). In a report, phosphorus-solubilizing fungi identified as *Aspergillus neoniger* and *Talaromyces aurantiacus* isolated from the moso bamboo were reported as solubilizing P in acidic stress conditions (Zhang et al., 2018). In another report, drought stress-tolerant microbes were also reported for solubilizing P nutrients *Pseudomonas libanensis* (Kour et al., 2019a), *Streptomyces laurentii*, *Penicillium* sp. (Kour et al., 2020b), and *Acinetobacter calcoaceticus* (Kour et al., 2020c). *Tsukamurella tyrosinosolvens*, which was isolated from roots associated with the soil of tea, was also reported in a study for solubilizing phosphorus macronutrients (Zhang et al., 2021).

1.2.3 Potassium solubilization

Potassium (K) is another essential plant growth-promoting macronutrient. It the third most required nutrient of plants and in higher plants, it is the most abundantly absorbed cation. K plays an important role in plant growth, development, and metabolism. It is also used to activate several plant enzymes, increase photosynthesis, reduce respiration, maintain cell turgor, and help in the uptake and transportation of nitrogen and sugars. Moreover, potassium also plays a key role in crop quality as it helps increase disease resistance, grain filling, and kernel weight (Ahmad et al., 2016; Rajawat et al., 2020; Teotia et al., 2016). If plants don't get the necessary amount of potassium, the plant may be poorly developed and its roots and seeds will be not developed, which may lead to lower crop productivity. In earlier times, the potassium in soil was rich. But now, the amount of K has been depleted due to the overexploitation of the land (Sindhu et al., 2014).

To fulfill the plant requirement of K, chemical fertilizer such as potash is being used, but it has various environmental issues. Agriculturally important microbes are also reported for solubilizing K in soil in insoluble form such as feldspars, muscovite, orthoclase, biotite, illite, vermiculite, micas, and smectite (Sparks and Huang, 1985). Various potassium-solubilizing microbes are known to solubilize the K in soil by different mechanisms. Acid production is a mechanism through which K can be solubilized. In acid production, microbes produce various types of organic acids such as citric, oxalic, and tartaric acids. The production of acidic compounds is the predominant mechanism of K solubilization. Another known mechanism of potassium solubilization through microbes is the production of exopolysaccharides. This mechanisms helps in releasing K from silicates as produced exopolysaccharides absorb SiO_2 and lead to a reaction toward SiO_2 and K^+ solubilization (Sindhu et al., 2014).

Numerous microbes have been reported to solubilize K in soil through listed mechanisms. In a report, *Paenibacillus glucanolyticus* isolated from the black pepper rhizosphere was reported as a potential solubilizer of K in soil (Liu et al., 2012). Isolated bacteria from Iranian soil, identified as *Bacillus megaterium* and *Arthrobacter* sp., were also reported for K solubilization (Keshavarz Zarjani et al., 2013). In a similar report, KSB strains were isolated from the tobacco plant rhizosphere, namely *Agrobacterium tumefaciens, Burkholderia cepacia, E. aerogenes, Klebsiella variicola, Microbacterium foliorum*, and *P. agglomerans*. Among these strains, *K. variicola* was reported as a potential KSB and it also helped in the plant growth promotion of the tobacco plant (Zhang and Kong, 2014).

In a similar report, *Frateuria aurantia* was reported as a potential solubilizer of K; it also promotes tobacco plant growth after inoculation compared to a control (Subhashini, 2015). *Pseudomonas azotoformans* was also reported as an efficient solubilizer of K. This strain was also reported as degrading pollution and rejuvenating land for agriculture benefits (Saha et al., 2016a). In another report, three bacterial strains identified as *Enterobacter* sp., *Pantoea ananatis*, and *Rahnella aquatilis*, were reported for solubilizing both K and P via organic acid production. All these strains individually enhanced the plant growth, biomass, height, diameter, and leaf area of rice seedlings (Bakhshandeh et al., 2017). *A. tumefaciens* was recognized as a pronounced K solubilizer. The inoculation of this strain on maize also enhanced nutrient assimilation and plant growth (Meena et al., 2018). *Bacillus pseudomycoides* (Pramanik et al., 2019), *A. pittii*, *Ochrobactrum ciceri* (Ashfaq et al., 2020), and *Burkholderia cenocepacia* (Raji and Thangavelu, 2021) were also reported as potential solubilizers of K (Raji and Thangavelu, 2021).

1.2.4 Zinc solubilization

Zinc (Zn) solubilization is another mechanism through which agriculturally important microbes promote plant growth. Zinc is an essential nutrient required in minute concentrations for humans, animals, and plants. In plants, zinc plays a key role in photosynthesis; the metabolism of carbohydrates, auxins, and proteins; the formation of sucrose and starch; and reproduction (Saravanan et al., 2011). Plants absorb this nutrient in zinc ion (Zn^{2+}) form. In a soil solution, zinc ions are present in an insufficient amount as zinc is in the insoluble form. The insufficient supply of this nutrient could result in plant abnormalities that affect crop productivity (Hafeez et al., 2013; Sahu et al., 2018).

Plant growth-promoting microbes used to solubilize and mineralize these nutrients also help promote plant growth and development. The zinc-solubilizing microbes (ZSB) could convert the insoluble form of Zn to a soluble form through single and multiple mechanisms. Like other nutrient solubilizations, ZSB could solubilize insoluble zinc by producing organic acids such as 2-ketogluconic acid (Fasim et al., 2002) and 5-ketogluconic acid (Saravanan et al., 2007). The organic acid production helps in lowering pH and releases zincs ions. The other mechanism that solubilizes zinc is mineral chelation through zinc-chelating compounds released by microbes (Obrador et al., 2003). Chelation of zinc is the dominant process for zinc availability then organic acids production mechanism (Hussain et al., 2018).

Earth's crust has abundant zinc solubilizers. Many studies have reported various microbial species for Zn solubilization. In a report, *Bacillus aryabhattai* was reported as an efficient solubilizer of zinc. The strains also exhibited other plant growth-promoting attributes such as IAA, ammonia, and siderophore-producing traits; they reportedly enhanced the yield of wheat and soybeans (Ramesh et al., 2014). In another investigation, the rhizobacteria of wheat and sugarcane were identified as potential zinc solubilizers, that is, *E. cloacae*, *Pseudomonas fragi*, *Pantoea dispersa*, and *P. agglomerans*. These strains also significantly increased zinc in wheat after inoculation compared to control (Kamran et al., 2017).

Other zinc-solubilizing bacteria were identified as *Bacillus* sp., *B. substilis*, and *B. aryabhattai*. These strains also contributed to increasing the maize root and shoot length as well as the biomass (Mumtaz et al., 2017). *B. cepacia*, *Klebsiella pneumoniae*, and *P. aeruginosa* were reported as potential zinc solubilizers (Gontia-Mishra et al., 2017). In another report, different strains of *A. tumefaciens* and *Rhizobium* sp. were also reported as solubilizers of zinc in a study by Khanghahi et al. (2018). Microbes such as *B. megaterium* (Dinesh et al., 2018; Bhatt and Maheshwari, 2020), *Pseudomonas* sp. (Zaheer et al., 2019), *Bacillus* sp., (Mumtaz et al., 2017; Ahmad et al., 2021), *A. baumannii*, and *B. cepacia* (Upadhyay et al., 2021) were also reported as efficient solubilizers of zinc; they were isolated from different soils including bulk and rhizospheric.

1.2.5 Siderophore production

Siderophore production is the plant growth-promoting mechanism of microbes that helps in the acquisition of iron nutrients from the soil. These are the small organic compounds produced under the iron-limiting conditions by plant growth-promoting microbes. Siderophores, a metal-chelating agent, have a high affinity toward ferric ions. Microbes use to produce these compounds for their own survival in iron depleted environment as iron plays a key role in electron transport chain, oxidative phosphorylation and tri-carboxylic acid cycles. Iron (Fe) is also an important nutrients for plants as it plays an important role in photosynthesis, oxidation–reduction reaction, biosynthesis of vitamins, antibiotics, toxins, cytochromes, nucleic acid pigments, and aromatic compounds (Fardeau et al., 2011; Messenger and Barclay, 1983). In soil, iron is a limiting factor that cannot be uptaken by plants, so microbes are used for iron acquisition through siderophore production.

Microbes produce three different categories of siderophores: catecholates, carboxylates, and hydroxamates (Saha et al., 2016b). These types of siderophores vary on the oxygen ligand for Fe(III) coordination. Hydroxamate siderophores are the most common type of siderophore in nature. This type of siderophore is usually produced by bacteria, and even fungi. This type of siderophore consists of $C(=O)N-(OH)$ R, where R could be an amino acid or its derivative. Hydroxamates have the capability to form an octahedral complex with Fe^{3+}. The hydroxamates form a strong bond with ferric ions, which prevents hydrolysis and enzymatic degradation in the natural environment (Winkelmann, 2007). Another type of siderophore, catecholate, is mostly produced by bacteria. In this type of siderophores form a hexadentate octahedral complex by chelating iron with two oxygen atoms. Carboxylate siderophores are produced mostly by bacteria such as *Staphylococcus* and *Rhizobium* and fungi such as mucorales. Carboxylate siderophores form bonds with iron through carboxyl and hydroxyl groups (Dave and Dube, 2000).

Siderophore-producing capability has been reported by various microbial species such as *Paracoccus sphaerophysae*, an endophyte of the root nodule of *Sphaerophysa salsalu*. Additionally, this strain also shows antifungal activity (Deng et al., 2011). In another report, a species of the genera *Streptomyces* was reported as a producer of siderophores and auxin. The strain inoculated on wheat plants also enhanced the shoot length, dry weight, and nutrient concentration of nitrogen, phosphorus, iron, and manganese compared to the control (Sadeghi et al., 2012). Another investigation reported *Bacillus amyloliquefaciens* as a potent strain for producing siderophores which exquisite iron from soil. *B. amyloliquefaciens* has also been found to remediate soil contaminated with metals such as arsenic, lead, aluminum, and cadmium (Gaonkar and Bhosle, 2013). Siderophore production also has another benefits for plants in addition to iron availability, that is, pest biocontrol. The iron deficiency in soil also helps in killing pathogens and pests of plants. It was reported that a siderophore-producing bacteria, *Pseudomonas* sp., has been an antagonist against the pathogenic fungi *Rhizoctonia solani* (Solanki et al., 2014).

In a similar report, a pronounced producer of siderophores, the bacteria *B. cepacia*, was reported as an efficient biocontrol agent of pathogens such as *Staphylococcus aureus* and *Enterococcus faecalis* (Ong et al., 2016). In a report, two bacteria produced the hydroxamate type of siderophore, which helps in the acquisition of iron from soil (Ghavami et al., 2017). In another report, the siderophore-producing bacteria *Aneurinibacillus aneurinilyticus*, *Aeromonas* sp., and *Pseudomonas* sp. were reported to enhance plant growth when inoculated as a consortium by inhibiting the phytopathogen *Fusarium solani* (Kumar et al., 2018). In another investigation, microbial species such as *Enterococcus casseliflavus*, *Pseudomonas weihenstephanensis*, and *Psychrobacter piscatorii* reportedly produced siderophores (Sinha et al., 2019). Microbial species such as *Bacillus halotolerans*, *Bacillus subtilis*, *Bacillus safensis* (Sarwar et al., 2020), and *Sinorhizobium meliloti* were also reported as efficient producers of siderophores (Sepehri and Khatabi, 2021).

1.2.6 Phytohormone production

Plant growth and development are dependent on mineral nutrients, whether macro- or micronutrients. Along with nutrients, they also require phytohormones for their proliferation and development. Plants require phytohormones such as auxin, abscisic acid, cytokinin, ethylene, and gibberellins, which play various roles in plant systems (Tiwari et al., 2020). Auxin is an important class of phytohormones that plays a significant role in the regulation of various processes related to plant growth. Generally, auxins are synthesized by the plants at the root, shoot, and expanding leaves. Auxins are transported across the plant from the site of biosynthesis via auxin transporters. They are used to regulate other hormones such as strigolactones, which help maintain the organ founder cell population and trigger organ primordia initiation. They also play an important cellular role in the induction of CycD3 and CDKs, which govern various cell cycle checkpoints. Another important plant hormone, cytokinin, also plays a role in plant cell division activation and the regulation of the cellular level. Cytokinins also have an important role in the phosphoregulation of CDK at the G2/M checkpoint. These phytohormones are also the main driver of zeatin and chlorophyll production, leaf expansion, nutrition signaling, and root growth.

Gibberellins are another important phytohormone produced by microbes. This phytohormone helps in cellular elongation and division in plants. It also helps in internodium elongation (Davies, 2004). Similarly, abscisic acid also helps in the physiological functioning of plants under stress conditions (Rai et al., 2011). Ethylene is another important phytohormone of plants that helps in plant sharpening. Further, it regulates floral transition and root architecture (Vandenbussche and Van Der Straeten, 2018). All these phytohormones are usually synthesized by the plant itself, but in some harsh conditions, plants are not able to produce phytohormones. Microbes are also helpful in plant growth promotion by producing several types of plant growth hormones. They are reported for producing all five types of phytohormones required for plant development.

In a study, *Acetebacter diazotrophicus* from sugarcane was reported for the production of indole 3-acetic acid (IAA) (Patil et al., 2011). In another investigation, an epiphytic pink-pigmented methylotrophic bacterium, *Methylobacterium* sp., was isolated from different crops such as sugarcane, pigeonpea, potato, radish, and mustard. This strain was reported for producing cytokinin and its inoculation in wheat crops improves the seed germination, growth, and productivity of plants (Meena et al., 2012). In another study, a plant growt-promoting rhizobacteria was isolated from soil. It was identified as

Promicromononspora sp. and was reported for secreting gibberellins. The inoculation of this strain on tomatoes enhanced crop production and salicylic acid (Kang et al., 2012). *B. subtilis* was reported for producing cytokinin and its inoculation in *Platycladus orientalis* (oriental thuja) seedlings under drought conditions increased organic acid production and alleviated drought stress (Liu et al., 2013).

The indole acetic acid type of phytohormone-producing bacterium called *Planomicrobium chinense* was isolated from a site contaminated with diesel oil; it was reported to enhance the growth of *Vigna radiata* (Das and Tiwary, 2014). In a similar report, the rhizobacteria of *Panax ginseng, Pseudomonas fluorescence*, and *A. chroococcum*, were also reported for producing IAA; they were also reported as potent species for the development of biofertilizer (Hussein and Joo, 2015). In another investigation, *Lysinibacillus endophyticus*, an endophytic bacterium isolated from corn root, was also reported for the production of indole-3-acetic (Yu et al., 2016). An endophytic fungus, *Porostereum spadiceum*, was reported for producing gibberellins; it also enhance soybean growth under salt stress (Hamayun et al., 2017b). In a report, two different species—*B. subtilis* and *Azospirillum brasilens*-were reported for the production of the phytohormone abscisic acid. These strains also help in alleviating heavy metal stress and their uptake through plant roots (Xu et al., 2018c).

In a study, *Streptomyces fradiae* was determined to be an efficient producer of IAA. The strain inoculation on tomato plant enhanced the root and shoot weight and length. Additionally, this strain increased the plant growth parameters almost two-fold compared to the control (Myo et al., 2019). An indole acetic acid-producing bacterium named *Providencia* sp. was sorted out from the rhizospheric region of the tomato. This bacterium also helps in promoting plant growth after inoculation (Rushabh et al., 2020). In another report, the rhizobacterium *Pseudomonas koreensis*, was reported for the production of gibberellins while also enhancing the growth of lettuce and Chinese cabbage (Kang et al., 2021).

1.3 Microbiomes for mitigation of biotic and abiotic stresses

Agricultural productivity is regulated by two major constituents: abiotic and biotic factors. The major abiotic factors affecting crop productivity include stress alkalinity, drought, flooding, salinity, and high and low temperatures (Kumar et al., 2019a, b). The biotic factors include pathogenic attacks, which have been known to account for about a 30% reduction in annual agricultural productivity (Fisher et al., 2012). There are a number of biotechnological tools that have been expansively applied to improve crops under stress conditions. Among these, the exploitation of plant growth-promoting microbes is of chief importance for their imminent role (Bhardwaj et al., 2014; Gangwar et al., 2017; Yang et al., 2009). Environmentally friendly approaches instigate a broad range of microbes for the promotion of plant growth, the improved uptake of nutrients, and tolerance to biotic and abiotic stresses. Inoculating crops with stress-tolerant, plant growth-promoting microbes is an effectual as well as novel way to alleviate stress conditions. This section deals with diverse stresses to which the plants are exposed and how stress-tolerant, plant growth-promoting microbes mitigate the adverse affects of abiotic stress.

1.3.1 Abiotic stress management

1.3.1.1 Salinity stress

Soil salinity is a major problem as far as soil health is concerned and is in fact increasing day by day, especially in arid and semiarid areas (Al-Karaki, 2006; Giri et al., 2003). According to FAO 2008 data, it was estimated that more than 800 million hectares of land throughout the world are affected by salinity (Munns and Tester, 2008). The major reasons for salinity include natural causes as well as the accumulation of salts for longer periods of time where evaporation greatly exceeds precipitation and salts dissolved in groundwater reach and accumulate at the soil surface through capillary movement (Estrada et al., 2013; Kohler et al., 2007). This stress is one of the major constraints limiting crop productivit. The development of salt-tolerant crop varieties could be one alternative, but the strategy is not economical for sustainable agriculture. However, inoculating crops with PGP microbes would be a better choice, as it minimizes production costs and environmental hazards (Arora et al., 2012). Upadhyay et al. (2012) showed the plant growth promotion and antioxidant activity of wheat with an inoculation of *Arthrobacter* sp. and *B. subtilis* under saline conditions. Ahmad et al. (2015) investigated *Trichoderma harzianum*'s role in mitigating the effect of salinity stress in Indian mustard. The treated seedlings showed enhanced shoot and root length, plant dry weight, proline content, oil content, and enzymatic activities as well as reduced malondialdehyde content. Metwali et al. (2015) reported the alleviation of salinity stress in the faba bean with an inoculation of *B. subtilis, Pseudomonas fluorescens*, and *Pseudomonas putida*.

Zhang et al. (2016) evaluated the role of *Trichoderma longibrachiatum* on wheat growth under salinity stress. The strain increased the relative water content in the roots and leaves, the chlorophyll content, the root activity, the leaf proline content, and antioxidant enzyme activity while decreasing the leaf malondialdehyde content under saline conditions. Hamayun et al. (2017a) evaluated the potential of a novel gibberellin-producing basidiomycetous endophytic fungus, *P. spadiceum*, to alleviate salinity stress and promote the health benefits of soybeans. Kumar et al. (2017) evaluated the role of salt-tolerant *Trichoderma* sp. on the growth of maize under salinity stress. The study clearly revealed the increase in shoot and root length, leaf area, total biomass, stem and leaf fresh weight, total chlorophyll, and proline and phenol content as well as a lower accumulation of malondialdehyde content. Wang et al. (2018) examined the effects of colonization with two AMF, *Funneliformis mosseae* and *Diversispora versiformis*, alone and in combination on the growth and nutrient uptake of *Chrysanthemum morifolium* in a greenhouse experiment under salinity stress. The study revealed that the root length, dry weight of the shoot and root, and root N concentration were higher in the mycorrhizal plants under conditions of moderate salinity, especially with the colonization of *D. versiformis*.

Xu et al. (2018a), studied the influence of *Glomus tortuosum* on the morphology, photosynthetic pigments, chlorophyll fluorescence, photosynthetic capacity, and rubisco activity of maize under saline stress in a pot experiment. The study validated that maize plants appeared to have a high dependency on AMF, which improved physiological mechanisms by raising the chlorophyll content, efficiency of light energy utilization, gas exchange, and rubisco activity under salinity stress. Kouadria et al. (2018), examined the roles of *Alternaria chlamydospora*, *Chaetomium coarctatum*, *Embellisi aphragmospora*, *Fusarium graminearum*, *Fusariume quseti*, and *Phoma betae* for their capability to improve the germination of durum wheat under salinity stress. Fungal strains considerably enhanced the germination and growth of durum wheat, with the highest germination percentage being shown by *A. chlamydospora*. Asaf et al. (2018) evaluated the potential of *Aspergillus flavus* to promote the growth of soybeans under salinity stress. The strain appreciably increased the growth of plants and mitigated salinity stress by downregulating abscissic acid and jasmonic acid while elevating antioxidant activities.

A study from Cordero et al. (2018) suggested PGPR as a viable, economical, and ecofriendly alternative to chemical fertilization in salinity soils. Vimal et al. (2019) reported the potential of *Curtobacterium albidum* in the alleviation of salinity stress in paddies. Nawaz et al. (2020) showed improved yield and growth in wheat with single as well as coinoculation of *Bacillus pumilus*, *Exiguobacterium aurantiacum*, and *P. fluorescence* under saline conditions. Sapre et al. (2021) reported the amelioration of salinity stress in peas with treatment of *Acinetobacter bereziniae*, *Alcaligenes faecalis*, and *Enterobacter ludwigii*. The utilization of PGP microbes in the alleviation of salinity stress is gaining worldwide interest and is a feasible option to maintain crop productivity under salinity conditions.

1.3.1.2 Drought stress

Drought stress is another major abiotic stress that causes a considerable decline in the yield and growth of most plants. Drought occurs slowly and silently and does not cause any prior short-term impact. This makes it harder for timely detection and preparation (Kerry et al., 2018; Sena et al., 2017). This situation is rapidly worsening by the widespread use of nondegrading chemical fertilizers, overgrazing, the overuse of land, and unsustainable exploitation of natural resources (Ruggiero et al., 2017). The deficiency of water greatly influences crop yield but improved yield under water-fed conditions is very important for food security (Ergen and Budak, 2009; Khan et al., 2018). Drought stress critically affects biochemical mechanisms and photosynthetic pigments, ultimately reducing crop growth and productivity (Moghadam et al., 2011). When plants are exposed to water-deficit conditions, reactive oxygen species (ROS) such as superoxides, hydrogen peroxide, and hydroxyl radicals accumulate (Chen et al., 2011). ROS are cytotoxic for cells and have an effect on proteins, nucleic acids, and lipids of the cell; they also stop the natural metabolism when present in high concentrations (Muthukumar et al., 2001). The application of PGP microbes is a strategy to mitigate drought stress and improve plant growth.

Sohrabi et al. (2012) investigated the effects of *Glomus* sp. on the physiological characteristics of the chickpea under nonstress as well as drought conditions. The inoculation of chickpea by arbuscular mycorrhizal (AM) considerably increased guaiacol peroxidases and polyphenol oxidase activities compared with the noninoculated chickpea. In general, the most guaiacol peroxidase and polyphenol oxidase activities were observed in plants inoculated with *Glomus etunicatum* and *Glomus versiform* species, and the most ascorbate peroxidase activity was observed in plants inoculated with *Glomus intraradices*. Shukla et al. (2012) studied the effect of drought-tolerant isolates of the endophytic fungus *T. harzianum* on rice. Rice seedlings colonized with *Trichoderma* were slower to wilt in response to drought. Drought conditions varying from 3 to 9 days of water holding increased the concentration of stress-induced metabolites in rice leaves. Zhu et al. (2012) studied the influence of AM fungus on growth, gas exchange, chlorophyll concentration, chlorophyll fluorescence, and water status of maize in pot culture under well-watered and drought stress conditions. AM symbioses remarkably increased the net photosynthetic rate and transpiration rate. Mycorrhizal plants had higher chlorophyll content, stomatal conductance, maximal fluorescence, maximum quantum efficiency of PSII photochemistry and potential photochemical efficiency, higher relative water content, and water use efficiency under drought stress when compared with nonmycorrhizal plants. Habibzadeh et al. (2013) evaluated the effects of the inoculation of *Glomus mosseae* and *G. intraradices* and water-deficit stress on mung plants. The inoculation improved the yield, leaf P and N, protein percentage, harvest index of protein, protein yield, seed yield, and ecosystem water use. The study suggested that both *G. mosseae* and *G. intraradices* appreciably improved the yield and reduced the water-deficit stress in mung bean plants.

Yaghoubian et al. (2014) investigated the effects of *G. mosseae* and *Piriformospora indica* on lipid peroxidation, antioxidant enzyme activity, and growth of *Triticum aestivum* under drought stress in a greenhouse. The results indicated lower levels of hydrogen peroxide and lipid peroxidation rate and increased activities of antioxidant enzymes and leaf chlorophyll content. Gusain et al. (2014) investigated the effect of *T. harzianum* and *Fusarium pallidoroseum* on biomass production, catalase, superoxide dismutase, and the peroxidase activities of rice under conditions of drought stress. The plants inoculated with plant growth-promoting fungi showed higher catalase, superoxide dismutase, and peroxidase activities as well as increased dry weight of the shoot and root. The study concluded that higher biomass as well as enhanced enzymatic activities may be the mechanism involved in the alleviation of drought stress and maintaining plant homeostasis under stress conditions. Gusain et al. (2015) reported drought stress amelioration in different cultivars of rice with the application of different PGPRs.

Guler et al. (2016) studied the effect of *Trichoderma atroviride* in maize seedlings under drought stress. The root colonization of fungus increased the fresh and dry weight of maize roots, prevented lipid peroxidation, and induced antioxidant enzyme activity while less hydrogen peroxide content was observed in response to drought stress in inoculated plants. Kanwal et al. (2017) concluded that PGPR in combination with compost and mineral fertilizer significantly reduced the effect of drought stress on wheat by positively influencing the physiological and biochemical parameters of wheat plants.

Ganjeali et al. (2018) conducted an experiment to examine the impacts of *G. mosseae* on improving the drought tolerance of the common bean. The inoculated plants showed increased dry weight of the root and shoot, phosphorus content, CO_2 assimilation, relative water content, transpiration rate, superoxide dismutase, polyphenol oxidase and peroxidase activities, proline content, and leaf soluble proteins along with lower stomatal resistance as compared to the non-AM plants. Kour et al. (2019a) reported the alleviation of drought stress in wheat. Kour et al. (2020b) reported the mitigation of drought stress in the great millet with the inoculation of drought-adaptive, P-solubilizing *S. laurentii*. A study from Sood et al. (2020) reported the alleviation of drought stress in maize treated with *B. subtilis*. Kour et al. (2020c) showed the amelioration of drought stress in the fox tail millet treated with *A. calcoaceticus* and *Penicillium* sp. Thus, drought-adaptive microbes hold great potential for more sustainable agricultural activities and better food choices.

1.3.1.3 High temperature stress

Temperature is another major abiotic stress determining the growth and productivity of plants across the globe (Allakhverdiev et al., 2008; Archna et al., 2015; Zhu et al., 2011b). High-temperature stress causes a series of morphological, physiological, and biochemical changes in plants. Each species of plants has its own most favorable range of temperature for growth and reproduction. Therefore, extreme variations can exert a thermodynamic influence on nucleic acids and proteins as well as substructures of plant cells (Niu and Xiang, 2018; Ruelland and Zachowski, 2010; Verma et al., 2019). Plants have evolved diverse adaptation mechanisms for survival under such adverse conditions, among which the most common is the stress-induced gene expression activation of the adjustment of the plant metabolism (Iba, 2002). Thus, breeding efforts are also being focused on gene transformation to efficiently improve plant adaptability to extreme temperatures (Grover et al., 2000; Sharkey, 2000), but breeding techniques are not cost-effective and are time-consuming. Alternatively, there are reports of PGP microbes that could be used for inoculating crops under extremes of temperature so that the adverse effects of high temperature could be mitigated.

Zhu et al. (2011b) studied the effects of *G. etunicatum* symbiosis on gas exchange, chlorophyll fluorescence, and pigment concentration of maize plants under high-temperature stress. AM symbiosis improved the stomatal conductance and transpiration rate, chlorophyll content, carotenoids, water use efficiency, holding capacity of water, and relative water content. Wu (2011) evaluated the effects of *G. mosseae* on the growth, root morphology, superoxide dismutase and catalase activities, and soluble protein content of *Poncirus trifoliata* seedlings at low (15°C), optimum (25°C), and high (35°C) temperatures. Mycorrhizal colonization significantly increased superoxide dismutase (SOD) and catalase activities as well as soluble protein content at high temperature. The study suggested that the mycorrhizal alleviation of temperature stress in *P. trifoliata* seedlings was at high temperature while the alleviation was noticeably weakened at low temperature. The study from Sarkar et al. (2018) revealed that wheat inoculation with *Ochrobactrum pseudogrignonense* and *B. safensis* protected the plants from temperature stress.

1.3.1.4 Low temperature

Low-temperature stress is the most severe abiotic stress limiting crop growth and productivity worldwide (Bunn et al., 2009; Chen et al.,

2013; Zhang et al., 2009). It could cause many injuries to plants including reduced cellular milieu osmotic potential, immediate mechanical constraints, macromolecule activity changes (Wu and Zou, 2010), increased accumulation of hydrogen peroxide (Zhang et al., 2009), and noteworthy alternation of the plasma membrane (Janicka-Russak et al., 2012). The use of PGP microbes has opened up new possibilities for the mitigation of chilling stress in plants (Verma et al., 2016, 2015; Yadav et al., 2019). Chen et al. (2013) revealed the potential of *F. mosseae* in the alleviation of low-temperature stress for cucumber seedlings. The inoculated seedlings had noteworthy higher fresh and dry weight; improved content of flavonoids, lignin, phenols, and phenolic compounds; and DPPH activity. Additionally, large increments have been observed in caffeic acid peroxidase, chlorogenic acid peroxidase, cinnamyl alcohol dehydrogenase, glucose-6-phosphate dehydrogenase, guaiacol peroxidase, phenylalanine ammonia-lyase, polyphenol oxidase, and shikimate dehydrogenase. Liu et al. (2017) studied the effect of *G. mosseae* on growth, antioxidant, osmoregulation, and nutrition in *Vaccinium ashei* and *Vaccinium corymbosum* plants exposed to low temperature. The study showed that inoculation with AM fungi enhanced ascorbate peroxidase, guaiacol peroxidase, glutathione, and superoxide dismutase activities in leaves while decreasing the concentrations of hydrogen peroxide, malondialdehyde, and superoxide anion radicals. Further, the inoculated plants had higher concentrations of soluble sugars and proline as well as phosphorus and potassium content.

In the study by Ghorbanpour et al. (2018), the effects of *T. harzianum* as a biocontrol agent were demonstrated on the tolerance of *Solanum lycopersicum* L. exposed to chilling stress. The study revealed that the strain mitigated the adverse effects of chilling stress. The study by Singh et al. (2020d) suggested that the coinoculation of *Achromobacter xylosoxidans* and *T. harzianum* could be a green initiative that could improve growth and mitigate cold stress in *Ocimum sanctum*. Thus, stress conditions greatly halt agricultural production. During stress conditions, plant growth is greatly affected. There could be nutritional and hormonal imbalances, increased susceptibility to disease, and much more. Plant growth could be increased by using microbial inoculants with PGP traits, which is in fact a novel as well as environmentally safe strategy.

1.3.2 Biotic stress management

Plant diseases act as biotic stress factors, causing farmers to suffer economic loss. Biotic stress also causes food spoilage during storage through toxin production. To combat these diseases, farmers rely on chemical fungicides and pesticides, which in turn are not safe for human health and the environment (Latha et al., 2019). Biotic stresses deprive their host of its nutrients and can lead to the death of plants (Gull et al., 2019). To get rid of such problems, the use of biocontrol agents for protecting plants from diseases has gained greater attention. The biocontrol agents utilizing PGP microbes appear to be an outstanding approach to maintain plant productivity and growth under biotic stress (Hashem et al., 2019). PGP microbes use a wide range of mechanisms to act as alleviators of biotic stress such as the production of antibiotics, hydrolytic enzymes, siderophores, HCN, and ammonia (Yadav et al., 2020b).

Antibiotics are low molecular weight organic compounds that play an active role in plant disease biocontrol and often act in concert with competition and parasitism (Jha, 2018). The important antibiotics known to play a major role in antibiosis include pyoluteorin, phenazines, and volatile HCN (Haas and Défago, 2005). Other identified antibiotics include 2,3-butanediol, 6-pentyl-α-pyrone, 2-hexyl-5- propyl resorcinol, and D-gluconic acid produced by endophytes (Dandurishvili et al., 2011). *Pseudomonas* strains are known to produce different types of antibiotics, including phenezine-1-carboxylate, pyoluteorin, and pyrolnitrin (Kour et al., 2019d; Kraus and Loper, 1995). Other antibiotics such as circulin, colistin, and polymyxin effective against Gram-positive and Gram-negative bacteria as well as phytopathogens have been reported from *Bacillus* sp. (Mahmood et al., 2019). The antibiotics fengycin, iturin, and surfactin have been reported from *B. subtilis*. Bacteriocins with a lower killing spectrum produced by bacteria have also played roles in the defense systems of plants against bacterial infection. Cell wall -egrading enzymes produced by bacteria can limit the activities of other microbes (Shoda, 2000). Chitinase and laminarinase producing *P. stutzeri* have been found to digest and lyse the mycelia of *F. solani*, preventing root rot (Lim et al., 1991).

β-1,3 glucanase producer *Pseudomonas cepacia* was shown to decrease the incidence of disease caused by *Pythium ultimum*, *R. solani*, and *Sclerotium rolfsii* by damaging the mycelia (Fridlender et al., 1993; Govindasamy et al., 2008). Siderophores are low molecular weight iron-chelating compounds with affinity for ferric ions; they are produced by microbes. There are different categories of siderophores with aerobactin, ferribactin, ferrioxamine, francobactin, and Schizokinen falling under the hydroxamate type whereas agrobactin, enterochelin, and parabactin are catecholates (Sayyed et al., 2013). Siderophores play a major role in biocontrol by binding most of the available Fe^+ and preventing the growth of any pathogens. Some strains of *Pseudomonas* have been shown to suppress the activities of plant pathogens by producing siderophores or pseudobactin (Shoda, 2000). A study by Xia et al. (2020) reported the potential of *Bacillus xiamenensis* in suppressing the red rot of sugarcane and enhancing plant growth. Shahzad et al. (2021) reported the role of *B. aryabhattai* in protecting tomato plants from *Fusarium* wilt and plant growth promotion. PGP microbes as biocontrol agents are an efficient strategy against plant pathogens and in reducing the use of chemical agents.

1.4 Beneficial microbiomes for sustainable crop production and protection

Microbes, little wonders inhabiting diverse environmental conditions, are known to play various roles such as mineralization, biogeochemical cycles (phosphorus and nitrogen), and decomposition. These roles tend to maintain environmental cycles. Crop production is one such role that benefits the agriculturist to produce high-quality crops in a sustainable way. A beneficial microbial population or PGP microbes can improve crop production either by fulfilling nutrient requirements or by protecting the crop from pests and pathogens by inhibiting their growth (Yadav et al., 2018). PGP microbes can be used as biofertilizers and biopesticides for sustainable crop production, which can also help in achieving organic farming under normal and stress environmental conditions (Fig. 1.2).

1.4.1 Biofertilizers

Fertilizer is a basic need of the agricultural system. It is widely used in huge amounts all over the globe. It is mainly used for increasing crop productivity as necessary soil nutrients have been depleted drastically due to the overutilization of land resources and human activities. Currently, fertilizers used in agricultural fields are mostly chemically synthesized, which have various harmful aspects. The accumulation of harmful chemicals, pollution, and the loss of soil fertility as well as the biodiversity of animals, insects, and microbes are some negative effects of chemical-based fertilizers. Fertilizers will becomes an enemy of earth in sooner time. After years of studies, plant growth microbes have been recognized as a suitable product for agriculture (Singh et al., 2020a). Beneficial microbes in diverse habitats could be used as biofertilizers to fulfill required plant nutrients. They use different mechanisms through which they convert the insoluble form of nutrients to soluble forms of macro- and micronutrients by fixing, solubilizing, or chelating. The fixing mechanism is mainly used for availing nitrogen nutrients, which is known as biological nitrogen fixation. In this, atmospheric dinitrogen is fixed by microbes by secreting nitrogenase enzyme activity (Subrahmanyam et al., 2020).

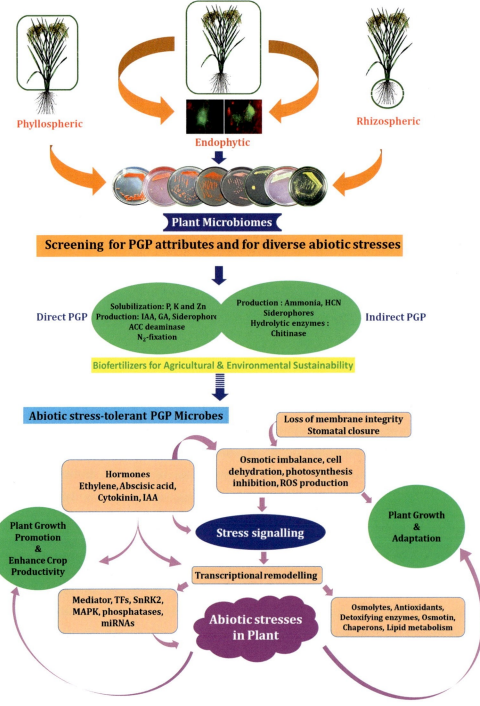

Fig. 1.2 Beneficial microbiomes with plant growth-promoting attributes as biofertilizers for PGP and soil fertility under natural and stressed conditions. Adapted with permission from Kour, D., Rana, K.L., Yadav, A.N., Yadav, N., Kumar, M., Kumar, V., et al., 2020d. Microbial biofertilizers: bioresources and eco-friendly technologies for agricultural and environmental sustainability. Biocatal. Agric. Biotechnol. 23, 101487. https://doi.org/10.1016/j.bcab.2019.101487.

Another microbe mechanism is solubilization, which can avail nutrients such as phosphorus, potassium, and zinc mainly by lowering the soil pH (Kaur et al., 2020a; Kour et al., 2020a). The lowering of pH is mainly achieved by the secretion of organic acids such as citric, gluconic, lactic, acetic, and oxalic acids (Kour et al., 2021b). Chelation is another microbial mechanism used to get iron from its ore. This mechanism usually fulfills the iron requirements of a plant. All these microbial mechanisms help in fertilizing crops and increasing the yield in a sustainable way (Rai et al., 2020). Much evidence has been published in reputed peer-reviewed journals. In a study, the rhizobacterium *P. fluorescens* was inoculated in wheat under a greenhouse. The inoculation resulted in enhanced grain yield and biomass (Smyth et al., 2011). In another investigation, the endophytic bacterium *Bacillus* sp. was reported to increase the leaf number, shoot length, and dry biomass of strawberries after inoculation (de Melo Pereira et al., 2012).

An actinomycetes identified as *Streptomyces* sp. was also recognized as a potent strain to be used in agriculture. Its use in inoculation has increased rice growth because it has the capability of secreting the most important class of phytohormones, that is, auxin (Gopalakrishnan et al., 2014). In a similar way, endophytes associated with sweet sorghum identified as *Bacillus* sp. and *Pantoea* sp. were reported as effective inoculum when tested on sorghum (Mareque et al., 2015). In another investigation, the IAA-producing bacterium *Enterobacter* sp. was reported as increasing wheat plant growth under salinity stress (Sorty et al., 2016). Rhizospheric microbes of Seabuckthorn and *E. ludwigii* were also reported as a potential microbial inoculum in stress regions, as it tested positive for solubilizing phosphorus (Dolkar et al., 2018). In another report, *Acinetobacter guillouiae*, an endophytic nitrogen fixing, K, Zn, P-solubilizing, IAA, ammonia producing bacteria was reported for enhancing the plant growth of wheat plants (Rana et al., 2020). The phosphorus-solubilizing bacterium *Pseudomonas* sp. was also reported as a PGPM when isolated from the high altitudes of Himalayan soil (Adhikari et al., 2021).

1.4.2 Biopesticides

Beneficial microbes also have been reported as efficient biocontrol agents. In agriculture, others inputs used in fields are pesticides, as various crops—mainly horticulture crops—are attacked by different pests and pathogens. These pests and pathogens tend to decrease crop productivity and sometimes complete erase the plant products. In order to prevent this, farmers use different types of biocontrol agents that are synthesized chemically such as fungicides, pesticides, and nematocides. These chemical products can adversely affect the health of the environment (Kour et al., 2021a). Pesticides have also been marked as dangerous pollutants of the Earth's crust because they do not easily degrade and can remain on the Earth for a very long time. A main concern of environmentalists is also the bioremediation of such pesticides in soil for longer periods, as they also contribute to pollution and land degradation (Thakur et al., 2020).

The use of microbes as biocontrol agents is another remarkable facet of these little wonders that helps crop productivity in an ecofriendly way (Kumar et al., 2021a). Microbes exhibit different mechanisms for alleviating crop stress caused by pests and pathogens such as the production of ammonia, hydrogen cyanide (HCN), hydrolytic enzymes, and siderophores. Ammonia, HCN, and hydrolytic enzymes are used to kill microbes directly whereas siderophore production limits iron nutrients in the soil, killing pathogens because of nutrient deficiency (Yadav et al., 2020a). Different microbial species have been reported that have the potential to be biopesticides. In a study, *Burkholderia pyrrocinia* was reported for controlling the pathogens of poplar canker, that is, *Cytospora chrysosperma*, *Fusicoccum aesculi*, and *Phomopsis macrospora*, as they have the capability of producing different hydrolytic enzymes such as β-1, 3-glucanases, chitinases, and proteases (Ren et al., 2011). Another report found that *Talaromyces flavus* is a potential antagonistic against the causal agent of wilt disease (Naraghi et al., 2012). In another report, actinomycetes species identified as *Streptomyces cyaneofuscatus*, *S. kanamyceticu*, *S. flavotricini* and *S. rochei* were found to inhibit the growth of the cotton pathogen causing wilt, that is, *Verticillium dahlia* (Xue et al., 2013).

B. subtilis was also reported as a biocontrol agent that inhibits the growth of *Fusarium* sp. when tested in apple seedlings (Ju et al., 2014). In another report, *P. aeruginosa* was reported as a cogent strain that promotes plant growth and inhibits the growth of *A. flavus* and *Fusarium oxysporum* (Goswami et al., 2015). The fusarium wilt causing pathogen *F. oxysporum* also has been reported to be inhibited by the biocontrol agent identified as *P. polymyxa* (Du et al., 2017). In a similar report, *Streptomyces* sp. was also reported for inhibiting the causal agent of *Fusarium* wilt and the bacterial wilt of the tomato (Zheng et al., 2019). Another pathogenic species of genera *Fusarium*, that is, *F. graminearum*, a causal agent of corn stalk rot, was effectively inhibited by *Bacillus velezensis* (Wang et al., 2020). One of the most renowned biocontrol agent genera *Trichoderma*, that is, *T. asperellum*, was also reported to suppress the soilborne fungal phytopathogen *F. oxysporum* (Win et al., 2021).

1.5 Beneficial microbiomes in organic agriculture

Organic farming is a sustainable technique for the agro-ecosystem. This technique is used to enhance the Earth's resources, genetic diversity, and crop production. Developing countries are bestowed with lot of potential to produce all varieties of organic products due to their various agro-climatic regions (Willer and Lernoud, 2019). Asia is endowed with various types of naturally available organic forms of nutrients in different parts of the continent, which will help in the organic cultivation of crops substantially. In India, only 40% of the cultivable area is affected by fertilizers where irrigation facilities are available and the remaining 60% of arable land is mainly rain-fed, so a negligible amount of fertilizers is used (Reganold and Wachter, 2016). Farmers in these areas often apply organic manure as a source of nutrients but microbes could also be utilized in organic farming to produce and protect plants and maintain soil health and fertility (Bhattacharjee and Dey, 2014).

1.6 Implication of beneficial microbes sustainable economic

The production of biofertilizers is always demand-driven; demand creation among farmers is the most important step toward biofertilizer promotion (Mazid and Khan, 2015). The commercial biofertilizer history began in 1895 with the launch of nitragin, a rhizobia culture. About 170 organizations in 24 countries are involved in the commercial production of biofertilizers. In India, the first commercial production started in 1956 at the Indian Agricultural Research Institute, New Delhi, and the Agricultural College and Research Institute, Coimbatore. The Ministry of Agriculture under the 9th plan put effort in promoting and popularizing the input by setting up the National Project on Development and Use of Biofertilizers (Ghosh, 2004). The commonly used biofertilizers in India include *Azospirillum*, *Azotobacter*, blue green algae, *Rhizobium*, and phosphate solubilizing and mobilizing biofertilizers. The current global market for agricultural products raised through organic farming is valued at around $30 billion with a growth rate of around 8% (Mishra and Dash, 2014). The government of India and the different state governments have been promoting biofertilizers through extensions, grants, and subsidies on sales with varying degrees of emphasis. Farmers are also learning about the technology. The enterprise of firms operating through their marketing, research, and development efforts will lead to the widespread use of resources as soon as the prospect of making a profit is felt.

1.7 Conclusion

Microbes are a miracle found by world researchers. These invisible miracles are known to work for the Earth from the time they originate. Now, their role in maintaining the environment is largely recognized. Every day, new role are found that have a wide variety of applications in various fields. The role of microbes in agriculture has been known for quite a while. They have been recognized as potential crop enhancers that increase crop productivity in a sustainable way by undergoing various mechanisms. Various plant growth promoting mechanisms that enhance crop productivity include nitrogen fixation; phosphorus, potassium, and zinc solubilization; and the production of siderophores, ammonia, HCN, and phytohormones. Microbes are known as a sustainable technique that heals the depleted environment. Along with increasing crop productivity, microbial-based products have also established huge markets worldwide and affect the economy in better ways. In conclusion, it can be said that the use of microbial-based inputs is the best alternative to increase crop production without affecting environmental health.

References

Adhikari, P., Jain, R., Sharma, A., Pandey, A., 2021. Plant growth promotion at low temperature by phosphate-solubilizing *Pseudomonas* Spp. Isolated from high-altitude himalayan soil. Microb. Ecol. https://doi.org/10.1007/s00248-021-01702-1.

Ahmad, P., Hashem, A., Abd-Allah, E.F., Alqarawi, A., John, R., Egamberdieva, D., et al., 2015. Role of *Trichoderma harzianum* in mitigating NaCl stress in Indian mustard (*Brassica juncea* L) through antioxidative defense system. Front. Plant Sci. 6, 1–15.

Ahmad, M., Nadeem, S.M., Naveed, M., Zahir, Z.A., 2016. Potassium-solubilizing bacteria and their application in agriculture. In: Meena, V.S., Maurya, B.R., Verma, J.P., Meena, R.S. (Eds.), Potassium Solubilizing Microorganisms for Sustainable Agriculture. Springer India, New Delhi, pp. 293–313, https://doi.org/10.1007/978-81-322-2776-2_21.

Ahmad, I., Ahmad, M., Hussain, A., Jamil, M., 2021. Integrated use of phosphate-solubilizing *Bacillus subtilis* strain IA6 and zinc-solubilizing *Bacillus* sp. strain IA16: a promising approach for improving cotton growth. Folia Microbiol. 66, 115–125. https://doi.org/10.1007/s12223-020-00831-3.

Al-Karaki, G.N., 2006. Nursery inoculation of tomato with arbuscular mycorrhizal fungi and subsequent performance under irrigation with saline water. Sci. Hortic. 109, 1–7.

Allakhverdiev, S.I., Kreslavski, V.D., Klimov, V.V., Los, D.A., Carpentier, R., Mohanty, P., 2008. Heat stress: an overview of molecular responses in photosynthesis. Photosynth. Res. 98, 541–550.

Alori, E.T., Glick, B.R., Babalola, O.O., 2017. Microbial phosphorus solubilization and its potential for use in sustainable agriculture. Front. Microbiol. 8, 1–8.

Archna, S., Priyank, V., Nath, Y.A., Kumar, S.A., 2015. Bioprospecting for extracellular hydrolytic enzymes from culturable thermotolerant bacteria isolated from Manikaran thermal springs. Res. J. Biotechnol. 10, 33–42.

Arora, N.K., Tewari, S., Singh, S., Lal, N., Maheshwari, D.K., 2012. PGPR for protection of plant health under saline conditions. In: Maheshwari, D. (Ed.), Bacteria in Agrobiology: Stress Management. Springer, Berlin, Heidelberg, pp. 239–258.

Asaf, S., Hamayun, M., Khan, A.L., Waqas, M., Khan, M.A., Jan, R., et al., 2018. Salt tolerance of *Glycine max*. L induced by endophytic fungus *Aspergillus flavus* CSH1, via regulating its endogenous hormones and antioxidative system. Plant Physiol. Biochem. 128, 13–23.

Ashfaq, M., Hassan, H.M., Ghazali, A.H.A., Ahmad, M., 2020. Halotolerant potassium solubilizing plant growth promoting rhizobacteria may improve potassium availability under saline conditions. Environ. Monit. Assess. 192, 697. https://doi.org/10.1007/s10661-020-08655-x.

Bakhshandeh, E., Pirdashti, H., Lendeh, K.S., 2017. Phosphate and potassium-solubilizing bacteria effect on the growth of rice. Ecol. Eng. 103, 164–169. https://doi.org/10.1016/j.ecoleng.2017.03.008.

Basak, B.B., Biswas, D.R., 2010. Co-inoculation of potassium solubilizing and nitrogen fixing bacteria on solubilization of waste mica and their effect on growth promotion and nutrient acquisition by a forage crop. Biol. Fertil. Soils 46, 641–648.

Bellenger, J.P., Wichard, T., Xu, Y., Kraepiel, A., 2011. Essential metals for nitrogen fixation in a free-living N_2-fixing bacterium: chelation, homeostasis and high use efficiency. Environ. Microbiol. 13, 1395–1411.

Beneduzi, A., Costa, P.B., Parma, M., Melo, I.S., Bodanese-Zanettini, M.H., Passaglia, L.M., 2010. *Paenibacillus riograndensis* sp. nov., a nitrogen-fixing species isolated from the rhizosphere of *Triticum aestivum*. Int. J. Syst. Evol. Microbiol. 60, 128–133.

Bessadok, K., Navarro-Torre, S., Fterich, A., Caviedes, M.A., Pajuelo, E., Rodríguez-Llorente, I.D., et al., 2021. Diversity of rhizobia isolated from Tunisian arid soils capable of forming nitrogen-fixing symbiosis with *Anthyllis henoniana*. J. Arid Environ. 188. https://doi.org/10.1016/j.jaridenv.2021.104467, 104467.

Bhardwaj, D., Ansari, M.W., Sahoo, R.K., Tuteja, N., 2014. Biofertilizers function as key player in sustainable agriculture by improving soil fertility, plant tolerance and crop productivity. Microb. Cell Fact. 13, 1–10.

Bhatt, K., Maheshwari, D.K., 2020. Zinc solubilizing bacteria (*Bacillus megaterium*) with multifarious plant growth promoting activities alleviates growth in *Capsicum annuum* L. 3 Biotech 10, 1–10. https://doi.org/10.1007/s13205-019-2033-9.

Bhattacharjee, R., Dey, U., 2014. Biofertilizer, a way towards organic agriculture: a review. Afr. J. Microbiol. Res. 8, 2332–2343.

Bunn, R., Lekberg, Y., Zabinski, C., 2009. Arbuscular mycorrhizal fungi ameliorate temperature stress in thermophilic plants. Ecology 90, 1378–1388.

Cai, H., Wang, Y., Xu, H., Yan, Z., Jia, D., Maszenan, A.M., et al., 2015. *Niveispirillum cyanobacteriorum* sp. nov., a nitrogen-fixing bacterium isolated from cyanobacterial aggregates in a eutrophic lake. Int. J. Syst. Evol. Microbiol. 65, 2537–2541.

Chen, A.C., Arany, P.R., Huang, Y.-Y., Tomkinson, E.M., Sharma, S.K., Kharkwal, G.B., et al., 2011. Low-level laser therapy activates NF-kB *via* generation of reactive oxygen species in mouse embryonic fibroblasts. PLoS ONE 6, e22453.

Chen, S., Jin, W., Liu, A., Zhang, S., Liu, D., Wang, F., et al., 2013. Arbuscular mycorrhizal fungi (AMF) increase growth and secondary metabolism in cucumber subjected to low temperature stress. Sci. Hortic. 160, 222–229.

Choi, G.-M., Im, W.-T., 2018. *Paraburkholderia azotifigens* sp. nov., a nitrogen-fixing bacterium isolated from paddy soil. Int. J. Syst. Evol. Microbiol. 68, 310–316.

Collavino, M.M., Sansberro, P.A., Mroginski, L.A., Aguilar, O.M., 2010. Comparison of in vitro solubilization activity of diverse phosphate-solubilizing bacteria native to acid soil and their ability to promote *Phaseolus vulgaris* growth. Biol. Fertil. Soils 46, 727–738.

Cordero, I., Balaguer, L., Rincón, A., Pueyo, J.J., 2018. Inoculation of tomato plants with selected PGPR represents a feasible alternative to chemical fertilization under salt stress. J. Plant Nutr. Soil Sci. 181, 694–703.

Da Costa, E.M., Guimarães, A.A., Vicentin, R.P., de Almeida Ribeiro, P.R., Leão, A.C.R., Balsanelli, E., et al., 2017. *Bradyrhizobium brasilense* sp. nov., a symbiotic nitrogen-fixing bacterium isolated from Brazilian tropical soils. Arch. Microbiol. 199, 1211–1221.

da Costa, E.M., Guimarães, A.A., de Carvalho, T.S., Rodrigues, T.L., de Almeida Ribeiro, P.R., Lebbe, L., et al., 2018. *Bradyrhizobium forestalis* sp. nov., an efficient nitrogen-fixing bacterium isolated from nodules of forest legume species in the Amazon. Arch. Microbiol. 200, 743–752.

Dahal, R.H., Chaudhary, D.K., Kim, D.-U., Kim, J., 2021. *Azohydromonas caseinilytica* sp. nov., a nitrogen-fixing bacterium isolated from forest soil by using optimized culture method. Front. Microbiol. 12, 1–10.

Dandurishvili, N., Toklikishvili, N., Ovadis, M., Eliashvili, P., Giorgobiani, N., Keshelava, R., et al., 2011. Broad-range antagonistic rhizobacteria *Pseudomonas fluorescens* and *Serratia plymuthica* suppress *Agrobacterium* crown gall tumours on tomato plants. J. Appl. Microbiol. 110, 341–352.

Das, R., Tiwary, B.N., 2014. Production of indole acetic acid by a novel bacterial strain of *Planomicrobium chinense* isolated from diesel oil contaminated site and its impact on the growth of *Vigna radiata*. Eur. J. Soil Biol. 62, 92–100. https://doi.org/10.1016/j.ejsobi.2014.02.012.

Dave, B., Dube, H., 2000. Chemical characterization of fungal siderophores. Indian J. Exp. Biol. 38, 56–62.

Davies, P.J., 2004. Plant Hormones: Biosynthesis, Signal Transduction, Action!, third ed. vol. 3 Kluwer, Dordrecht, p. 750.

de Melo Pereira, G.V., Magalhães, K.T., Lorenzetii, E.R., Souza, T.P., Schwan, R.F., 2012. A multiphasic approach for the identification of endophytic bacterial in strawberry fruit and their potential for plant growth promotion. Microb. Ecol. 63, 405–417. https://doi.org/10.1007/s00248-011-9919-3.

Deng, Z.-S., Zhao, L.-F., Xu, L., Kong, Z.-Y., Zhao, P., Qin, W., et al., 2011. *Paracoccus sphaerophysae* sp. nov., a siderophore-producing, endophytic bacterium isolated from root nodules of *Sphaerophysa salsula*. Int. J. Syst. Evol. Microbiol. 61, 665–669.

Dinesh, R., Srinivasan, V., Hamza, S., Sarathambal, C., Anke Gowda, S.J., Ganeshamurthy, A.N., et al., 2018. Isolation and characterization of potential Zn solubilizing bacteria from soil and its effects on soil Zn release rates, soil available Zn and plant Zn content. Geoderma 321, 173–186. https://doi.org/10.1016/j.geoderma.2018.02.013.

Dolkar, D., Dolkar, P., Angmo, S., Chaurasia, O.P., Stobdan, T., 2018. Stress tolerance and plant growth promotion potential of *Enterobacter ludwigii* PS1 isolated from Seabuckthorn rhizosphere. Biocatal. Agric. Biotechnol. 14, 438–443. https://doi.org/10.1016/j.bcab.2018.04.012.

Dos Santos, P.C., Fang, Z., Mason, S.W., Setubal, J.C., Dixon, R., 2012. Distribution of nitrogen fixation and nitrogenase-likes sequences amongst microbial genomes. BMC Genomics 13, 1–18. https://doi.org/10.1186/1471-2164-13-162.

Du, N., Shi, L., Yuan, Y., Sun, J., Shu, S., Guo, S., 2017. Isolation of a potential biocontrol agent *Paenibacillus polymyxa* NSY50 from vinegar waste compost and its induction of host defense responses against *Fusarium* wilt of cucumber. Microbiol. Res. 202, 1–10. https://doi.org/10.1016/j.micres.2017.04.013.

Ergen, N.Z., Budak, H., 2009. Sequencing over 13 000 expressed sequence tags from six subtractive cDNA libraries of wild and modern wheats following slow drought stress. Plant Cell Environ. 32, 220–236.

Estrada, B., Aroca, R., Maathuis, F.J., Barea, J.M., Ruiz-Lozano, J.M., 2013. Arbuscular mycorrhizal fungi native from a Mediterranean saline area enhance maize tolerance to salinity through improved ion homeostasis. Plant Cell Environ. 36, 1771–1782.

Farah Ahmad, I.A., Aqil, F., Wani, A.A., Sousche, Y.S., 2006. Plant growth promoting potential of free-living diazotrophs and other rhizobacteria isolated from northern Indian soil. Biotechnol. J. 1, 1112–1123.

Fardeau, S., Mullie, C., Dassonville-Klimpt, A., Audic, N., Sonnet, P., 2011. Bacterial iron uptake: A promising solution against multidrug resistant bacteria. In: Science against Microbial Pathogens: Communicating Current Research and Technological Advances. Formatex Research Center, Badajoz, pp. 695–705.

Farrar, K., Bryant, D., Cope-Selby, N., 2014. Understanding and engineering beneficial plant-microbe interactions: plant growth promotion in energy crops. Plant Biotechnol. J. 12, 1193-1206.

Fasim, F., Ahmed, N., Parsons, R., Gadd, G.M., 2002. Solubilization of zinc salts by a bacterium isolated from the air environment of a tannery. FEMS Microbiol. Lett. 213, 1-6.

Fisher, M.C., Henk, D.A., Briggs, C.J., Brownstein, J.S., Madoff, L.C., McCraw, S.L., et al., 2012. Emerging fungal threats to animal, plant and ecosystem health. Nature 484, 186-194.

Fridlender, M., Inbar, J., Chet, I., 1993. Biological control of soilborne plant pathogens by a β-1, 3 glucanase-producing *Pseudomonas cepacia*. Soil Biol. Biochem. 25, 1211-1221.

Gangwar, M., Saini, P., Nikhanj, P., Kaur, S., 2017. Plant growth-promoting microbes (PGPM) as potential microbial bio-agents for eco-friendly agriculture. In: Adhya, T., Mishra, B., Annapurna, K., Verma, D., Kumar, U. (Eds.), Advances in Soil Microbiology: Recent Trends and Future Prospects. Springer, pp. 37-55.

Ganjeali, A., Ashiani, E., Zare, M., Tabasi, E., 2018. Influences of the arbuscular mycorrhizal fungus *Glomus mosseae* on morphophysiological traits and biochemical compounds of common bean (*Phaseolus vulgaris*) under drought stress. South Afr. J. Plant Soil 35, 121-127.

Gaonkar, T., Bhosle, S., 2013. Effect of metals on a siderophore producing bacterial isolate and its implications on microbial assisted bioremediation of metal contaminated soils. Chemosphere 93, 1835-1843. https://doi.org/10.1016/j.chemosphere.2013.06.036.

Ghavami, N., Alikhani, H.A., Pourbabaei, A.A., Besharati, H., 2017. Effects of two new siderophore-producing rhizobacteria on growth and iron content of maize and canola plants. J. Plant Nutr. 40, 736-746.

Ghorbanpour, A., Salimi, A., Ghanbary, M.A.T., Pirdashti, H., Dehestani, A., 2018. The effect of *Trichoderma harzianum* in mitigating low temperature stress in tomato (*Solanum lycopersicum* L.) plants. Sci. Hortic. 230, 134-141.

Ghosh, N., 2004. Promoting biofertilisers in Indian agriculture. Econ. Pol. Wkly 5, 5617-5625.

Giri, B., Kapoor, R., Mukerji, K., 2003. Influence of arbuscular mycorrhizal fungi and salinity on growth, biomass, and mineral nutrition of *Acacia auriculiformis*. Biol. Fertil. Soils 38, 170-175.

Gontia-Mishra, I., Sapre, S., Tiwari, S., 2017. Zinc solubilizing bacteria from the rhizosphere of rice as prospective modulator of zinc biofortification in rice. Rhizosphere 3, 185-190. https://doi.org/10.1016/j.rhisph.2017.04.013.

Gopalakrishnan, S., Vadlamudi, S., Bandikinda, P., Sathya, A., Vijayabharathi, R., Rupela, O., et al., 2014. Evaluation of *Streptomyces* strains isolated from herbal vermicompost for their plant growth-promotion traits in rice. Microbiol. Res. 169, 40-48. https://doi.org/10.1016/j.micres.2013.09.008.

Goswami, D., Patel, K., Parmar, S., Vaghela, H., Muley, N., Dhandhukia, P., et al., 2015. Elucidating multifaceted urease producing marine *Pseudomonas aeruginosa* BG as a cogent PGPR and bio-control agent. Plant Growth Regul. 75, 253-263. https://doi.org/10.1007/s10725-014-9949-1.

Govindasamy, V., Senthilkumar, M., Upendra-Kumar, A.K., 2008. PGPR-biotechnology for management of abiotic and biotic stresses in crop plants. In: Maheshwari, D.K., Dubey, R.C. (Eds.), Potential Microorganisms for Sustainable Agriculture. IK International Publishing, New Delhi, pp. 26-48.

Grönemeyer, J.L., Chimwamurombe, P., Reinhold-Hurek, B., 2015. *Bradyrhizobium subterraneum* sp. nov., a symbiotic nitrogen-fixing bacterium from root nodules of groundnuts. Int. J. Syst. Evol. Microbiol. 65, 3241-3247.

Grönemeyer, J.L., Hurek, T., Bünger, W., Reinhold-Hurek, B., 2016. *Bradyrhizobium vignae* sp. nov., a nitrogen-fixing symbiont isolated from effective nodules of *Vigna* and *Arachis*. Int. J. Syst. Evol. Microbiol. 66, 62-69.

Grover, A., Agarwal, M., Katiyar-Agarwal, S., Sahi, C., Agarwal, S., 2000. Production of high temperature tolerant transgenic plants through manipulation of membrane lipids. Curr. Sci. 79, 557–559.

Guler, N.S., Pehlivan, N., Karaoglu, S.A., Guzel, S., Bozdeveci, A., 2016. *Trichoderma atroviride* ID20G inoculation ameliorates drought stress-induced damages by improving antioxidant defence in maize seedlings. Acta Physiol. Plant. 38, 1–9.

Gull, A., Lone, A.A., Wani, N.U.I., 2019. Biotic and abiotic stresses in plants. In: Abiotic and Biotic Stress in Plants. Intech Open.

Gusain, Y.S., Singh, U., Sharma, A., 2014. Enhance activity of stress related enzymes in rice (*Oryza sativa* L.) induced by plant growth promoting fungi under drought stress. Afr. J. Agric. Res. 9, 1430–1434.

Gusain, Y.S., Singh, U., Sharma, A., 2015. Bacterial mediated amelioration of drought stress in drought tolerant and susceptible cultivars of rice (*Oryza sativa* L.). Afr. J. Biotechnol. 14, 764–773.

Haas, D., Défago, G., 2005. Biological control of soil-borne pathogens by fluorescent pseudomonads. Nat. Rev. Microbiol. 3, 307–319.

Habibzadeh, Y., Pirzad, A., Zardashti, M.R., Jalilian, J., Eini, O., 2013. Effects of arbuscular mycorrhizal fungi on seed and protein yield under water-deficit stress in mung bean. Agron. J. 105, 79–84.

Hafeez, B., Khanif, Y., Saleem, M., 2013. Role of zinc in plant nutrition—a review. J. Exp. Agric. Int. 3, 374–391.

Hamayun, M., Hussain, A., Khan, S.A., Kim, H.-Y., Khan, A.L., Waqas, M., et al., 2017a. Gibberellins producing endophytic fungus *Porostereum spadiceum* AGH786 rescues growth of salt affected soybean. Front. Microbiol. 8, 1–13.

Hamayun, M., Hussain, A., Khan, S.A., Kim, H.-Y., Khan, A.L., Waqas, M., et al., 2017b. Gibberellins producing endophytic fungus *Porostereum spadiceum* AGH786 rescues growth of salt affected soybean. Front. Microbiol. 8, 1–13. https://doi.org/10.3389/fmicb.2017.00686.

Hara, S., Morikawa, T., Wasai, S., Kasahara, Y., Koshiba, T., Yamazaki, K., et al., 2019. Identification of nitrogen-fixing *Bradyrhizobium* associated with roots of field-grown sorghum by metagenome and proteome analyses. Front. Microbiol. 10, 1–15. https://doi.org/10.3389/fmicb.2019.00407.

Hashem, A., Tabassum, B., Abd_Allah, E.F., 2019. *Bacillus subtilis*: a plant-growth promoting rhizobacterium that also impacts biotic stress. Saudi J. Biol. Sci. 26, 1291–1297.

Helene, L.C.F., Dall'Agnol, R.F., Delamuta, J.R.M., Hungria, M., 2019. *Mesorhizobium atlanticum* sp. nov., a new nitrogen-fixing species from soils of the Brazilian Atlantic Forest biome. Int. J. Syst. Evol. Microbiol. 69, 1800–1806.

Hussain, A., Zahir, Z.A., Asghar, H.N., Ahmad, M., Jamil, M., Naveed, M., et al., 2018. Zinc solubilizing bacteria for zinc biofortification in cereals: a step toward sustainable nutritional security. In: Meena, V. (Ed.), Role of Rhizospheric Microbes in Soil. Springer, Singapore, pp. 203–227.

Hussein, K.A., Joo, J.H., 2015. Isolation and characterization of rhizomicrobial isolates for phosphate solubilization and indole acetic acid production. J. Korean Soc. Appl. Biol. Chem. 58, 847–855. https://doi.org/10.1007/s13765-015-0114-y.

Iba, K., 2002. Acclimative response to temperature stress in higher plants: approaches of gene engineering for temperature tolerance. Annu. Rev. Plant Biol. 53, 225–245.

Islam, M.R., Sultana, T., Cho, J.-C., Joe, M.M., Sa, T., 2012. Diversity of free-living nitrogen-fixing bacteria associated with Korean paddy fields. Ann. Microbiol. 62, 1643–1650.

Islam, M.R., Sultana, T., Joe, M.M., Yim, W., Cho, J.C., Sa, T., 2013. Nitrogen-fixing bacteria with multiple plant growth-promoting activities enhance growth of tomato and red pepper. J. Basic Microbiol. 53, 1004–1015.

Janicka-Russak, M., Kabała, K., Wdowikowska, A., Kłobus, G., 2012. Response of plasma membrane H+-ATPase to low temperature in cucumber roots. J. Plant Res. 125, 291–300.

Jha, Y., 2018. Induction of anatomical, enzymatic, and molecular events in maize by PGPR under biotic stress. In: Meena, V. (Ed.), Role of Rhizospheric Microbes in Soil. Springer, Singapore, pp. 125–141.

Ji, S.H., Gururani, M.A., Chun, S.-C., 2014. Isolation and characterization of plant growth promoting endophytic diazotrophic bacteria from Korean rice cultivars. Microbiol. Res. 169, 83–98. https://doi.org/10.1016/j.micres.2013.06.003.

Jin, H.-J., Zhou, Y.-G., Liu, H.-C., Chen, S.-F., 2011. *Paenibacillus jilunlii* sp. nov., a nitrogen-fixing species isolated from the rhizosphere of *Begonia semperflorens*. Int. J. Syst. Evol. Microbiol. 61, 1350–1355.

Ju, R., Zhao, Y., Li, J., Jiang, H., Liu, P., Yang, T., et al., 2014. Identification and evaluation of a potential biocontrol agent, *Bacillus subtilis*, against *Fusarium* sp. in apple seedlings. Ann. Microbiol. 64, 377–383. https://doi.org/10.1007/s13213-013-0672-3.

Kamran, S., Shahid, I., Baig, D.N., Rizwan, M., Malik, K.A., Mehnaz, S., 2017. Contribution of zinc solubilizing bacteria in growth promotion and zinc content of wheat. Front. Microbiol. 8, 1–14. https://doi.org/10.3389/fmicb.2017.02593.

Kang, J., Amoozegar, A., Hesterberg, D., Osmond, D.L., 2011. Phosphorus leaching in a sandy soil as affected by organic and inorganic fertilizer sources. Geoderma 161, 194–201.

Kang, S.-M., Khan, A.L., Hamayun, M., Hussain, J., Joo, G.-J., You, Y.-H., et al., 2012. Gibberellin-producing *Promicromonospora* sp. SE188 improves *Solanum lycopersicum* plant growth and influences endogenous plant hormones. J. Microbiol. 50, 902–909. https://doi.org/10.1007/s12275-012-2273-4.

Kang, S.-M., Adhikari, A., Lee, K.-E., Park, Y.-G., Shahzad, R., Lee, I.-J., 2021. Gibberellin producing rhizobacteria *Pseudomonas koreensis* mu2 enhance growth of lettuce (*Lactuca sativa*) and Chinese cabbage (*Brassica rapa*, chinensis). J. Microbiol. Biotechnol. Food Sci. 2021, 166–170.

Kanwal, S., Ilyas, N., Batool, N., Arshad, M., 2017. Amelioration of drought stress in wheat by combined application of PGPR, compost, and mineral fertilizer. J. Plant Nutr. 40, 1250–1260.

Kaur, T., Rana, K.L., Kour, D., Sheikh, I., Yadav, N., Kumar, V., et al., 2020a. Microbe-mediated biofortification for micronutrients: present status and future challenges. In: Rastegari, A.A., Yadav, A.N., Yadav, N. (Eds.), Trends of Microbial Biotechnology for Sustainable Agriculture and Biomedicine Systems: Perspectives for Human Health. Elsevier, Amsterdam, pp. 1–17, https://doi.org/10.1016/B970-0-12-820528-0.00002-8.

Kaur, T., Yadav, A.N., Sharma, S., Singh, N., 2020b. Diversity of fungal isolates associated with early blight disease of tomato from mid Himalayan region of India. Arch. Phytopathol. Plant Prot. 53, 612–624. https://doi.org/10.1080/03235408.2020.1785098.

Kaur, T., Devi, R., Kour, D., Yadav, A., Yadav, A.N., Dikilitas, M., et al., 2021. Plant growth promoting soil microbiomes and their potential implications for agricultural and environmental sustainability. Biologia. https://doi.org/10.1007/s11756-021-00806-w.

Ke, X., Feng, S., Wang, J., Lu, W., Zhang, W., Chen, M., et al., 2019. Effect of inoculation with nitrogen-fixing bacterium *Pseudomonas stutzeri* A1501 on maize plant growth and the microbiome indigenous to the rhizosphere. Syst. Appl. Microbiol. 42, 248–260.

Kerry, R.G., Patra, S., Gouda, S., Patra, J.K., Das, G., 2018. Microbes and their role in drought tolerance of agricultural food crops. In: Patra, J., Das, G., Shin, H.S. (Eds.), Microbial Biotechnology. Springer, Singapore, pp. 253–273.

Keshavarz Zarjani, J., Aliasgharzad, N., Oustan, S., Emadi, M., Ahmadi, A., 2013. Isolation and characterization of potassium solubilizing bacteria in some Iranian soils. Arch. Agron. Soil Sci. 59, 1713–1723.

Khan, N., Bano, A., Shahid, M.A., Nasim, W., Babar, M.A., 2018. Interaction between PGPR and PGR for water conservation and plant growth attributes under drought condition. Biologia 73, 1083–1098.

Khanghahi, M.Y., Ricciuti, P., Allegretta, I., Terzano, R., Crecchio, C., 2018. Solubilization of insoluble zinc compounds by zinc solubilizing bacteria (ZSB) and optimization of their growth conditions. Environ. Sci. Pollut. Res. 25, 25862–25868. https://doi.org/10.1007/s11356-018-2638-2.

Kohler, J., Caravaca, F., Carrasco, L., Roldan, A., 2007. Interactions between a plant growth-promoting rhizobacterium, an AM fungus and a phosphate-solubilising fungus in the rhizosphere of *Lactuca sativa*. Appl. Soil Ecol. 35, 480–487.

Kouadria, R., Bouzouina, M., Azzouz, R., Lotmani, B., 2018. Salinity stress resistance of durum wheat (*Triticum durum*) enhanced by fungi. Int. J. Biosci. 12, 70–77.

Kour, D., Rana, K.L., Sheikh, I., Kumar, V., Yadav, A.N., Dhaliwal, H.S., et al., 2019a. Alleviation of drought stress and plant growth promotion by *Pseudomonas libanensis* EU-LWNA-33, a drought-adaptive phosphorus-solubilizing bacterium. Proc. Natl Acad. Sci. India Sec. B Biol. Sci. 90, 785–795.

Kour, D., Rana, K.L., Yadav, N., Yadav, A.N., 2019b. Bioprospecting of phosphorus solubilizing bacteria from Renuka lake ecosystems, lesser Himalayas. J. Appl. Biol. Biotechnol. 7, 1–6. https://doi.org/10.7324/JABB.2019.70501.

Kour, D., Rana, K.L., Yadav, N., Yadav, A.N., Kumar, A., Meena, V.S., et al., 2019c. Rhizospheric microbiomes: biodiversity, mechanisms of plant growth promotion, and biotechnological applications for sustainable agriculture. In: Kumar, A., Meena, V.S. (Eds.), Plant Growth Promoting Rhizobacteria for Agricultural Sustainability: From Theory to Practices. Springer, Singapore, pp. 19–65, https://doi.org/10.1007/978-981-13-7553-8_2.

Kour, D., Rana, K.L., Yadav, N., Yadav, A.N., Kumar, A., Meena, V.S., et al., 2019d. Rhizospheric microbiomes: biodiversity, mechanisms of plant growth promotion, and biotechnological applications for sustainable agriculture. In: Kumar, A., Meena, V.S. (Eds.), Plant Growth Promoting Rhizobacteria for Agricultural Sustainability: From Theory to Practices. Springer Singapore, Singapore, pp. 19–65, https://doi.org/10.1007/978-981-13-7553-8_2.

Kour, D., Kaur, T., Devi, R., Rana, K.L., Yadav, N., Rastegari, A.A., et al., 2020a. Biotechnological applications of beneficial microbiomes for evergreen agriculture and human health. In: Rastegari, A.A., Yadav, A.N., Yadav, N. (Eds.), Trends of Microbial Biotechnology for Sustainable Agriculture and Biomedicine Systems: Perspectives for Human Health. Elsevier, Amsterdam, pp. 255–279, https://doi.org/10.1016/B978-0-12-820528-0.00019-3.

Kour, D., Rana, K.L., Kaur, T., Sheikh, I., Yadav, A.N., Kumar, V., et al., 2020b. Microbe-mediated alleviation of drought stress and acquisition of phosphorus in great millet (*Sorghum bicolour* L.) by drought-adaptive and phosphorus-solubilizing microbes. Biocatal. Agric. Biotechnol. 23, 101501. https://doi.org/10.1016/j.bcab.2020.101501.

Kour, D., Rana, K.L., Yadav, A.N., Sheikh, I., Kumar, V., Dhaliwal, H.S., et al., 2020c. Amelioration of drought stress in foxtail millet (*Setaria italica* L.) by P-solubilizing drought-tolerant microbes with multifarious plant growth promoting attributes. Environ. Sustain. 3, 23–34. https://doi.org/10.1007/s42398-020-00094-1.

Kour, D., Rana, K.L., Yadav, A.N., Yadav, N., Kumar, M., Kumar, V., et al., 2020d. Microbial biofertilizers: bioresources and eco-friendly technologies for agricultural and environmental sustainability. Biocatal. Agric. Biotechnol. 23, 101487. https://doi.org/10.1016/j.bcab.2019.101487.

Kour, D., Kaur, T., Devi, R., Yadav, A., Singh, M., Joshi, D., et al., 2021a. Beneficial microbiomes for bioremediation of diverse contaminated environments for environmental sustainability: present status and future challenges. Environ. Sci. Pollut. Res. 28, 24917–24939. https://doi.org/10.1007/s11356-021-13252-7.

Kour, D., Rana, K.L., Kaur, T., Yadav, N., Yadav, A.N., Kumar, M., et al., 2021b. Biodiversity, current developments and potential biotechnological applications of phosphorus-solubilizing and -mobilizing microbes: a review. Pedosphere 31, 43–75. https://doi.org/10.1016/S1002-0160(20)60057-1.

Kraus, J., Loper, J.E., 1995. Characterization of a genomic region required for production of the antibiotic pyoluteorin by the biological control agent *Pseudomonas fluorescens* Pf-5. Appl. Environ. Microbiol. 61, 849–854.

Kumar, K., Manigundan, K., Amaresan, N., 2017. Influence of salt tolerant *Trichoderma* spp. on growth of maize (*Zea mays*) under different salinity conditions. J. Basic Microbiol. 57, 141–150.

Kumar, P., Thakur, S., Dhingra, G.K., Singh, A., Pal, M.K., Harshvardhan, K., et al., 2018. Inoculation of siderophore producing rhizobacteria and their consortium for growth enhancement of wheat plant. Biocatal. Agric. Biotechnol. 15, 264–269. https://doi.org/10.1016/j.bcab.2018.06.019.

Kumar, M., Kour, D., Yadav, A.N., Saxena, R., Rai, P.K., Jyoti, A., et al., 2019a. Biodiversity of methylotrophic microbial communities and their potential role in mitigation of abiotic stresses in plants. Biologia 74, 287–308. https://doi.org/10.2478/s11756-019-00190-6.

Kumar, V., Joshi, S., Pant, N.C., Sangwan, P., Yadav, A.N., Saxena, A., et al., 2019b. Molecular approaches for combating multiple abiotic stresses in crops of arid and semi-arid region. In: Singh, S., Upadhyay, S., Pandey, A., Kumar, S. (Eds.), Molecular Approaches in Plant Biology and Environmental Challenges. Energy, Environment, and Sustainability. Springer, Singapore, pp. 149–170, https://doi.org/10.1007/978-981-15-0690-1_8.

Kumar, M., Yadav, A.N., Saxena, R., Rai, P.K., Paul, D., Tomar, R.S., 2021. Novel methanotrophic and methanogenic bacterial communities from diverse ecosystems and their impact on environment. Biocatal. Agric. Biotechnol. 33. https://doi.org/10.1016/j.bcab.2021.102005.

Kumar, M., Yadav, A.N., Saxena, R., Paul, D., Tomar, R.S., 2021a. Biodiversity of pesticides degrading microbial communities and their environmental impact. Biocatal. Agric. Biotechnol. 31. https://doi.org/10.1016/j.bcab.2020.101883, 101883.

Kumar, M., Yadav, A.N., Saxena, R., Paul, D., Tomar, R.S., 2021b. Biodiversity of pesticides degrading microbial communities and their environmental impact. Biocatal. Agric. Biotechnol. 31. https://doi.org/10.1016/j.bcab.2020.101883, 101883.

Latha, P., Karthikeyan, M., Rajeswari, E., 2019. Endophytic bacteria: prospects and applications for the plant disease management. In: Ansari, R., Mahmood, I. (Eds.), Plant Health Under Biotic Stress. Springer, Singapore, pp. 1–50.

Li, Z., Bai, T., Dai, L., Wang, F., Tao, J., Meng, S., et al., 2016. A study of organic acid production in contrasts between two phosphate solubilizing fungi: *Penicillium oxalicum* and *Aspergillus niger*. Sci. Rep. 6, 1–8.

Lim, H.-S., Kim, Y.-S., Kim, S.-D., 1991. *Pseudomonas stutzeri* YPL-1 genetic transformation and antifungal mechanism against *Fusarium solani*, an agent of plant root rot. Appl. Environ. Microbiol. 57, 510–516.

Liu, D., Lian, B., Dong, H., 2012. Isolation of *Paenibacillus* sp. and assessment of its potential for enhancing mineral weathering. Geomicrobiol J. 29, 413–421.

Liu, F., Xing, S., Ma, H., Du, Z., Ma, B., 2013. Cytokinin-producing, plant growth-promoting rhizobacteria that confer resistance to drought stress in *Platycladus orientalis* container seedlings. Appl. Microbiol. Biotechnol. 97, 9155–9164. https://doi.org/10.1007/s00253-013-5193-2.

Liu, F.-P., Liu, H.-Q., Zhou, H.-L., Dong, Z.-G., Bai, X.-H., Bai, P., et al., 2014. Isolation and characterization of phosphate-solubilizing bacteria from betel nut (*Areca catechu*) and their effects on plant growth and phosphorus mobilization in tropical soils. Biol. Fertil. Soils 50, 927–937. https://doi.org/10.1007/s00374-014-0913-z.

Liu, X.M., Xu, Q.L., Li, Q.Q., Zhang, H., Xiao, J.X., 2017. Physiological responses of the two blueberry cultivars to inoculation with an arbuscular mycorrhizal fungus under low-temperature stress. J. Plant Nutr. 40, 2562–2570.

Lowman, S., Kim-Dura, S., Mei, C., Nowak, J., 2016. Strategies for enhancement of switchgrass (*Panicum virgatum* L.) performance under limited nitrogen supply based on utilization of N-fixing bacterial endophytes. Plant and Soil 405, 47–63. https://doi.org/10.1007/s11104-015-2640-0.

Mahmood, I., Rizvi, R., Sumbul, A., Ansari, R.A., 2019. Potential role of plant growth promoting Rhizobacteria in alleviation of biotic stress. In: Ansari, R., Mahmood, I. (Eds.), Plant Health Under Biotic Stress. Springer, Singapore, pp. 177–188.

Mahmud, K., Makaju, S., Ibrahim, R., Missaoui, A., 2020. Current progress in nitrogen fixing plants and microbiome research. Plan. Theory 9, 1–17.

Mareque, C., Taulé, C., Beracochea, M., Battistoni, F., 2015. Isolation, characterization and plant growth promotion effects of putative bacterial endophytes associated with sweet sorghum (*Sorghum bicolor* (L) Moench). Ann. Microbiol. 65, 1057–1067. https://doi.org/10.1007/s13213-014-0951-7.

Martínez-Aguilar, L., Salazar-Salazar, C., Méndez, R.D., Caballero-Mellado, J., Hirsch, A.M., Vásquez-Murrieta, M.S., et al., 2013. *Burkholderia caballeronis* sp. nov., a nitrogen fixing species isolated from tomato (*Lycopersicon esculentum*) with the ability to effectively nodulate *Phaseolus vulgaris*. Antonie Van Leeuwenhoek 104, 1063–1071.

Mazid, M., Khan, T.A., 2015. Future of bio-fertilizers in Indian agriculture: an overview. Int. J. Agric. Food Res. 3, 10–23.

Meena, K.K., Kumar, M., Kalyuzhnaya, M.G., Yandigeri, M.S., Singh, D.P., Saxena, A.K., et al., 2012. Epiphytic pink-pigmented methylotrophic bacteria enhance germination and seedling growth of wheat (*Triticum aestivum*) by producing phytohormone. Antonie Van Leeuwenhoek 101, 777–786. https://doi.org/10.1007/s10482-011-9692-9.

Meena, V.S., Zaid, A., Maurya, B.R., Meena, S.K., Bahadur, I., Saha, M., et al., 2018. Evaluation of potassium solubilizing rhizobacteria (KSR): enhancing K-bioavailability and optimizing K-fertilization of maize plants under Indo-Gangetic Plains of India. Environ. Sci. Pollut. Res. 25, 36412–36424. https://doi.org/10.1007/s11356-018-3571-0.

Meneses, C.H., Rouws, L.F., Simões-Araújo, J.L., Vidal, M.S., Baldani, J.I., 2011. Exopolysaccharide production is required for biofilm formation and plant colonization by the nitrogen-fixing endophyte *Gluconacetobacter diazotrophicus*. Mol. Plant Microbe Interact. 24, 1448–1458.

Messenger, A.J., Barclay, R., 1983. Bacteria, iron and pathogenicity. Biochem. Educ. 11, 54–63.

Metwali, E.M., Abdelmoneim, T.S., Bakheit, M.A., Kadasa, N.M., 2015. Alleviation of salinity stress in faba bean (*Vicia faba* L.) plants by inoculation with plant growth promoting rhizobacteria (PGPR). Plant Omics 8, 449–460.

Michel, D.C., da Costa, E.M., Guimaraes, A.A., de Carvalho, T.S., de Castro Caputo, P.S., Willems, A., et al., 2021. *Bradyrhizobium campsiandrae* sp. nov., a nitrogen-fixing bacterial strain isolated from a native leguminous tree from the Amazon adapted to flooded conditions. Arch. Microbiol. 203, 233–240.

Mishra, P., Dash, D., 2014. Rejuvenation of biofertilizer for sustainable agriculture and economic development. Consilience 2014, 41–61.

Mnasri, B., Liu, T.Y., Saidi, S., Chen, W.F., Chen, W.X., Zhang, X.X., et al., 2014. *Rhizobium azibense* sp. nov., a nitrogen fixing bacterium isolated from root-nodules of *Phaseolus vulgaris*. Int. J. Syst. Evol. Microbiol. 64, 1501–1506.

Moghadam, H.R.T., Zahedi, H., Ghooshchi, F., 2011. Oil quality of canola cultivars in response to water stress and super absorbent polymer application. Pesqui. Agropecu. Trop. 41, 579–586.

Mumtaz, M.Z., Ahmad, M., Jamil, M., Hussain, T., 2017. Zinc solubilizing *Bacillus* spp. potential candidates for biofortification in maize. Microbiol. Res. 202, 51–60. https://doi.org/10.1016/j.micres.2017.06.001.

Munns, R., Tester, M., 2008. Mechanisms of salinity tolerance. Annu. Rev. Plant Biol. 59, 651–681.

Muthukumar, T., Udaiyan, K., Rajeshkannan, V., 2001. Response of neem (*Azadirachta indica* A. Juss) to indigenous arbuscular mycorrhizal fungi, phosphate-solubilizing and asymbiotic nitrogen-fixing bacteria under tropical nursery conditions. Biol. Fertil. Soils 34, 417–426.

My, P.T., Manucharova, N., Stepanov, A., Pozdnyakov, L., Selitskaya, O., Emtsev, V., 2015. *Agrobacterium tumefaciens* as associative nitrogen-fixing bacteria. Moscow Univ. Soil Sci. Bull. 70, 133–138.

Myo, E.M., Ge, B., Ma, J., Cui, H., Liu, B., Shi, L., et al., 2019. Indole-3-acetic acid production by *Streptomyces fradiae* NKZ-259 and its formulation to enhance plant growth. BMC Microbiol. 19, 155. https://doi.org/10.1186/s12866-019-1528-1.

Naraghi, L., Heydari, A., Rezaee, S., Razavi, M., 2012. Biocontrol agent *Talaromyces flavus* stimulates the growth of cotton and potato. J. Plant Growth Regul. 31, 471–477. https://doi.org/10.1007/s00344-011-9256-2.

Navarro-Noya, Y.E., Hernández-Mendoza, E., Morales-Jiménez, J., Jan-Roblero, J., Martínez-Romero, E., Hernández-Rodríguez, C., 2012. Isolation and characterization of nitrogen fixing heterotrophic bacteria from the rhizosphere of pioneer plants growing on mine tailings. Appl. Soil Ecol. 62, 52–60.

Nawaz, A., Shahbaz, M., Asadullah, A.I., Marghoob, M.U., Imtiaz, M., Mubeen, F., 2020. Potential of salt tolerant PGPR in growth and yield augmentation of wheat (*Triticum aestivum* L.) under saline conditions. Front. Microbiol. 11, 1–12.

Niu, Y., Xiang, Y., 2018. An overview of biomembrane functions in plant responses to high-temperature stress. Front. Plant Sci. 9, 1–18.

Nohwar, N., Khandare, R., Desai, N., 2019. Isolation and characterization of salinity tolerant nitrogen fixing bacteria from *Sesbania sesban* (L) root nodules. Biocatal. Agric. Biotechnol. 21, 101325.

Obrador, A., Novillo, J., Alvarez, J., 2003. Mobility and availability to plants of two zinc sources applied to a calcareous soil. Soil Sci. Soc. Am. J. 67, 564–572.

Ong, K.S., Aw, Y.K., Lee, L.H., Yule, C.M., Cheow, Y.L., Lee, S.M., 2016. *Burkholderia paludis* sp. nov., an antibiotic-siderophore producing novel *Burkholderia cepacia* complex species, isolated from Malaysian tropical peat swamp soil. Front. Microbiol. 7, 1–14. https://doi.org/10.3389/fmicb.2016.02046.

Patil, N.B., Gajbhiye, M., Ahiwale, S.S., Gunjal, A.B., Kapadnis, B.P., 2011. Optimization of indole 3-acetic acid (IAA) production by *Acetobacter diazotrophicus* L1 isolated from sugarcane. Int. J. Environ. Sci. 2, 295.

Paulitsch, F., Dall'Agnol, R.F., Delamuta, J.R.M., Ribeiro, R.A., da Silva Batista, J.S., Hungria, M., 2019. *Paraburkholderia guartelaensis* sp. nov., a nitrogen-fixing species isolated from nodules of *Mimosa gymnas* in an ecotone considered as a hotspot of biodiversity in Brazil. Arch. Microbiol. 201, 1435–1446. https://doi.org/10.1007/s00203-019-01714-z.

Paulitsch, F., Dall'Agnol, R.F., Delamuta, J.R.M., Ribeiro, R.A., da Silva Batista, J.S., Hungria, M., 2020. *Paraburkholderia atlantica* sp. nov. and *Paraburkholderia franconis* sp. nov., two new nitrogen-fixing nodulating species isolated from Atlantic forest soils in Brazil. Arch. Microbiol. 202, 1369–1380.

Pawlicki-Jullian, N., Courtois, B., Pillon, M., Lesur, D., Le Flèche-Mateos, A., Laberche, J.-C., et al., 2010. Exopolysaccharide production by nitrogen-fixing bacteria within nodules of Medicago plants exposed to chronic radiation in the Chernobyl exclusion zone. Res. Microbiol. 161, 101–108.

Pham, V.T., Rediers, H., Ghequire, M.G., Nguyen, H.H., De Mot, R., Vanderleyden, J., et al., 2017a. The plant growth-promoting effect of the nitrogen-fixing endophyte *Pseudomonas stutzeri* A15. Arch. Microbiol. 199, 513–517.

Pham, V.T.K., Rediers, H., Ghequire, M.G.K., Nguyen, H.H., De Mot, R., Vanderleyden, J., et al., 2017b. The plant growth-promoting effect of the nitrogen-fixing endophyte *Pseudomonas stutzeri* A15. Arch. Microbiol. 199, 513–517. https://doi.org/10.1007/s00203-016-1332-3.

Pramanik, P., Goswami, A., Ghosh, S., Kalita, C., 2019. An indigenous strain of potassium-solubilizing bacteria *Bacillus pseudomycoides* enhanced potassium uptake in tea plants by increasing potassium availability in the mica waste-treated soil of North-east India. J. Appl. Microbiol. 126, 215–222.

Rai, M.K., Shekhawat, N., Gupta, A.K., Phulwaria, M., Ram, K., Jaiswal, U., 2011. The role of abscisic acid in plant tissue culture: a review of recent progress. Plant Cell Tiss. Org. Cult. 106, 179–190.

Rai, P.K., Singh, M., Anand, K., Saurabhj, S., Kaur, T., Kour, D., et al., 2020. Role and potential applications of plant growth promotion rhizobacteria for sustainable agriculture. In: Rastegari, A.A., Yadav, A.N., Yadav, N. (Eds.), Trends of Microbial Biotechnology for Sustainable Agriculture and Biomedicine Systems: Diversity and Functional Perspectives. Elsevier, Amsterdam, pp. 49–60, https://doi.org/10.1016/B978-0-12-820526-6.00004-X.

Rajawat, M.V.S., Singh, R., Singh, D., Yadav, A.N., Singh, S., Kumar, M., et al., 2020. Spatial distribution and identification of bacteria in stressed environments capable to weather potassium aluminosilicate mineral. Braz. J. Microbiol. 51, 751–764. https://doi.org/10.1007/s42770-019-00210-2.

Raji, M., Thangavelu, M., 2021. Isolation and screening of potassium solubilizing bacteria from saxicolous habitat and their impact on tomato growth in different soil types. Arch. Microbiol. https://doi.org/10.1007/s00203-021-02284-9.

Ramesh, A., Sharma, S.K., Sharma, M.P., Yadav, N., Joshi, O.P., 2014. Inoculation of zinc solubilizing *Bacillus aryabhattai* strains for improved growth, mobilization and biofortification of zinc in soybean and wheat cultivated in Vertisols of Central India. Appl. Soil Ecol. 73, 87–96. https://doi.org/10.1016/j.apsoil.2013.08.009.

Rana, K.L., Kour, D., Kaur, T., Sheikh, I., Yadav, A.N., Kumar, V., et al., 2020. Endophytic microbes from diverse wheat genotypes and their potential biotechnological applications in plant growth promotion and nutrient uptake. Proc. Natl. Acad. Sci. India Sec. B. Biol. Sci. 90, 969–979. https://doi.org/10.1007/s40011-020-01168-0.

Rana, K.L., Kour, D., Kaur, T., Devi, R., Yadav, A., Yadav, A.N., 2021. Bioprospecting of endophytic bacteria from the Indian Himalayas and their role in plant growth promotion of maize (*Zea mays* L.). J. Appl. Biol. Biotechnol. 9, 41–50.

Reganold, J.P., Wachter, J.M., 2016. Organic agriculture in the twenty-first century. Nat. Plants 2, 1–8.

Ren, J.H., Ye, J.R., Liu, H., Xu, X.L., Wu, X.Q., 2011. Isolation and characterization of a new *Burkholderia pyrrocinia* strain JK-SH007 as a potential biocontrol agent. World J. Microbiol. Biotechnol. 27, 2203–2215. https://doi.org/10.1007/s11274-011-0686-6.

Ruelland, E., Zachowski, A., 2010. How plants sense temperature. Environ. Exp. Bot. 69, 225–232. https://doi.org/10.1016/j.envexpbot.2010.05.011.

Ruggiero, A., Punzo, P., Landi, S., Costa, A., Van Oosten, M., Grillo, S., 2017. Improving plant water use efficiency through molecular genetics. Horticulturae 3, 1–22.

Rushabh, S., Kajal, C., Prittesh, P., Amaresan, N., Krishnamurthy, R., 2020. Isolation, characterization, and optimization of indole acetic acid-producing *Providencia* species (7MM11) and their effect on tomato (*Lycopersicon esculentum*) seedlings. Biocatal. Agric. Biotechnol. 28. https://doi.org/10.1016/j.bcab.2020.101732, 101732.

Sadeghi, A., Karimi, E., Dahaji, P.A., Javid, M.G., Dalvand, Y., Askari, H., 2012. Plant growth promoting activity of an auxin and siderophore producing isolate of Streptomyces under saline soil conditions. World J. Microbiol. Biotechnol. 28, 1503–1509. https://doi.org/10.1007/s11274-011-0952-7.

Saha, M., Maurya, B.R., Meena, V.S., Bahadur, I., Kumar, A., 2016a. Identification and characterization of potassium solubilizing bacteria (KSB) from Indo-Gangetic Plains of India. Biocatal. Agric. Biotechnol. 7, 202–209. https://doi.org/10.1016/j.bcab.2016.06.007.

Saha, M., Sarkar, S., Sarkar, B., Sharma, B.K., Bhattacharjee, S., Tribedi, P., 2016b. Microbial siderophores and their potential applications: a review. Environ. Sci. Pollut. Res. 23, 3984–3999.

Sahu, A., Bhattacharjya, S., Mandal, A., Thakur, J.K., Atoliya, N., Sahu, N., et al., 2018. Microbes: a sustainable approach for enhancing nutrient availability in agricultural soils. In: Meena, V.S. (Ed.), Role of Rhizospheric Microbes in Soil: Volume 2: Nutrient Management and Crop Improvement. Springer Singapore, Singapore, pp. 47–75, https://doi.org/10.1007/978-981-13-0044-8_2.

Sapre, S., Gontia-Mishra, I., Tiwari, S., 2021. Plant growth-promoting Rhizobacteria ameliorates salinity stress in pea (*Pisum sativum*). J. Plant Growth Regul. https://doi.org/10.1007/s00344-021-10329-y.

Saravanan, V., Madhaiyan, M., Thangaraju, M., 2007. Solubilization of zinc compounds by the diazotrophic, plant growth promoting bacterium *Gluconacetobacter diazotrophicus*. Chemosphere 66, 1794–1798.

Saravanan, V.S., Kumar, M.R., Sa, T.M., 2011. Microbial zinc solubilization and their role on plants. In: Maheshwari, D.K. (Ed.), Bacteria in Agrobiology: Plant Nutrient Management. Springer, Berlin Heidelberg, pp. 47–63, https://doi.org/10.1007/978-3-642-21061-7_3.

Sarkar, J., Chakraborty, B., Chakraborty, U., 2018. Plant growth promoting rhizobacteria protect wheat plants against temperature stress through antioxidant signalling and reducing chloroplast and membrane injury. J. Plant Growth Regul. 37, 1396–1412.

Sarwar, S., Khaliq, A., Yousra, M., Sultan, T., Ahmad, N., Khan, M.Z., 2020. Screening of Siderophore-producing PGPRs isolated from groundnut (*Arachis hypogaea* L.) rhizosphere and their influence on iron release in soil. Commun. Soil Sci. Plant Anal. 51, 1680–1692.

Satapute, P., Olekar, H., Shetti, A., Kulkarni, A., Hiremath, G., Patagundi, B., et al., 2012. Isolation and characterization of nitrogen fixing *Bacillus subtilis* strain as-4 from agricultural soil. Int. J. Rec. Sci. Res. 3, 762–765.

Sayyed, R., Chincholkar, S., Reddy, M., Gangurde, N., Patel, P., 2013. Siderophore producing PGPR for crop nutrition and phytopathogen suppression. In: Maheshwari, D. (Ed.), Bacteria in Agrobiology: Disease Management. Springer, Berlin, Heidelberg, pp. 449–471.

Schwab, S., Terra, L.A., Baldani, J.I., 2018. Genomic characterization of *Nitrospirillum amazonense* strain CBAmC, a nitrogen-fixing bacterium isolated from surface-sterilized sugarcane stems. Mol. Genet. Genomics 293, 997–1016.

Sena, A., Ebi, K.L., Freitas, C., Corvalan, C., Barcellos, C., 2017. Indicators to measure risk of disaster associated with drought: implications for the health sector. PLoS ONE 12, e0181394.

Sepehri, M., Khatabi, B., 2021. Combination of Siderophore-producing bacteria and *Piriformospora indica* provides an efficient approach to improve cadmium tolerance in alfalfa. Microb. Ecol. 81, 717–730. https://doi.org/10.1007/s00248-020-01629-z.

Shahzad, R., Tayade, R., Shahid, M., Hussain, A., Ali, M.W., Yun, B.-W., 2021. Evaluation potential of PGPR to protect tomato against *Fusarium* wilt and promote plant growth. PeerJ 9, e11194.

Sharkey, T.D., 2000. Some like it hot. Science 287, 435–437.

Sharma, S., Kumar, V., Tripathi, R.B., 2011. Isolation of phosphate solubilizing microorganism (PSMs) from soil. J. Microbiol. Biotechnol. Res. 1, 90–95.

Shoda, M., 2000. Bacterial control of plant diseases. J. Biosci. Bioeng. 89, 515–521.

Shukla, N., Awasthi, R., Rawat, L., Kumar, J., 2012. Biochemical and physiological responses of rice (*Oryza sativa* L.) as influenced by *Trichoderma harzianum* under drought stress. Plant Physiol. Biochem. 54, 78–88.

Sindhu, S.S., Parmar, P., Phour, M., 2014. Nutrient cycling: potassium solubilization by microorganisms and improvement of crop growth. In: Parmar, N., Singh, A. (Eds.), Geomicrobiology and Biogeochemistry. Springer, Berlin Heidelberg, Berlin, Heidelberg, pp. 175–198, https://doi.org/10.1007/978-3-642-41837-2_10.

Singh, A., Kumar, R., Yadav, A.N., Mishra, S., Sachan, S., Sachan, S.G., 2020a. Tiny microbes, big yields: microorganisms for enhancing food crop production sustainable development. In: Rastegari, A.A., Yadav, A.N., Yadav, N. (Eds.), Trends of Microbial Biotechnology for Sustainable Agriculture and Biomedicine Systems: Diversity and Functional Perspectives. Elsevier, Amsterdam, pp. 1–15, https://doi.org/10.1016/B978-0-12-820526-6.00001-4.

Singh, B., Boukhris, I., Pragya Kumar, V., Yadav, A.N., Farhat-Khemakhem, A., et al., 2020b. Contribution of microbial phytases to the improvement of plant growth and nutrition: a review. Pedosphere 30, 295–313. https://doi.org/10.1016/S1002-0160(20)60010 8.

Singh, C., Tiwari, S., Singh, J.S., Yadav, A.N., 2020c. Microbes in Agriculture and Environmental Development. CRC Press, Boca Raton.

Singh, S., Tripathi, A., Chanotiya, C.S., Barnawal, D., Singh, P., Patel, V.K., et al., 2020d. Cold stress alleviation using individual and combined inoculation of ACC deaminase producing microbes in *Ocimum sanctum*. Environ. Sustain. 3, 289–301.

Sinha, A.K., Parli Venkateswaran, B., Tripathy, S.C., Sarkar, A., Prabhakaran, S., 2019. Effects of growth conditions on siderophore producing bacteria and siderophore production from Indian Ocean sector of Southern Ocean. J. Basic Microbiol. 59, 412–424.

Smyth, E., McCarthy, J., Nevin, R., Khan, M., Dow, J., O'gara, F., et al., 2011. In vitro analyses are not reliable predictors of the plant growth promotion capability of bacteria; a *Pseudomonas fluorescens* strain that promotes the growth and yield of wheat. J. Appl. Microbiol. 111, 683–692.

Sohrabi, Y., Heidari, G., Weisany, W., Golezani, K.G., Mohammadi, K., 2012. Changes of antioxidative enzymes, lipid peroxidation and chlorophyll content in chickpea types colonized by different *glomus* species under drought stress. Symbiosis 56, 5–18.

Solanki, M.K., Singh, R.K., Srivastava, S., Kumar, S., Kashyap, P.L., Srivastava, A.K., et al., 2014. Isolation and characterization of siderophore producing antagonistic rhizobacteria against *Rhizoctonia solani*. J. Basic Microbiol. 54, 585–597.

Sood, G., Kaushal, R., Sharma, M., 2020. Alleviation of drought stress in maize (*Zea mays* L.) by using endogenous endophyte *Bacillus subtilis* in north West Himalayas. Acta Agric Scand. Sec. B. Soil Plant Sci. 70, 361–370.

Sorty, A.M., Meena, K.K., Choudhary, K., Bitla, U.M., Minhas, P.S., Krishnani, K.K., 2016. Effect of plant growth promoting bacteria associated with halophytic weed (*Psoralea corylifolia* L) on germination and seedling growth of wheat under saline conditions. Appl. Biochem. Biotechnol. 180, 872–882. https://doi.org/10.1007/s12010-016-2139-z.

Sparks, D., Huang, P., 1985. Physical chemistry of soil potassium. In: Munson, R.D. (Ed.), Potassium in Agriculture. American Society of Agronomy, Crop Science Society of America, and Soil Science Society of America, Madison, WI, pp. 201–276.

Srinivasan, R., Yandigeri, M.S., Kashyap, S., Alagawadi, A.R., 2012. Effect of salt on survival and P-solubilization potential of phosphate solubilizing microorganisms from salt affected soils. Saudi J. Biol. Sci. 19, 427–434. https://doi.org/10.1016/j.sjbs.2012.05.004.

Suarez, C., Ratering, S., Kramer, I., Schnell, S., 2014. *Cellvibrio diazotrophicus* sp. nov., a nitrogen-fixing bacteria isolated from the rhizosphere of salt meadow plants and emended description of the genus *Cellvibrio*. Int. J. Syst. Evol. Microbiol. 64, 481–486.

Subhashini, D., 2015. Growth promotion and increased potassium uptake of tobacco by potassium-mobilizing bacterium *Frateuria aurantia* grown at different potassium levels in vertisols. Commun. Soil Sci. Plant Anal. 46, 210–220.

Subrahmanyam, G., Kumar, A., Sandilya, S.P., Chutia, M., Yadav, A.N., 2020. Diversity, plant growth promoting attributes, and agricultural applications of rhizospheric microbes. In: Yadav, A.N., Singh, J., Rastegari, A.A., Yadav, N. (Eds.), Plant Microbiomes for Sustainable Agriculture. Springer, Cham, pp. 1–52, https://doi.org/10.1007/978-3-030-38453-1_1.

Suman, A., Yadav, A.N., Verma, P., 2016. Endophytic microbes in crops: diversity and beneficial impact for sustainable agriculture. In: Singh, D.P., Singh, H.B., Prabha, R. (Eds.), Microbial Inoculants in Sustainable Agricultural Productivity. Research Perspectives, vol. 1. Springer India, New Delhi, pp. 117–143, https://doi.org/10.1007/978-81-322-2647-5_7.

Swamy, C.T., Gayathri, D., Devaraja, T.N., Bandekar, M., D'Souza, S.E., Meena, R.M., et al., 2016. Plant growth promoting potential and phylogenetic characteristics of a lichenized nitrogen fixing bacterium, *Enterobacter cloacae*. J. Basic Microbiol. 56, 1369–1379.

Tang, S.-Y., Hara, S., Melling, L., Goh, K.-J., Hashidoko, Y., 2010. *Burkholderia vietnamiensis* isolated from root tissues of Nipa palm (*Nypa fruticans*) in Sarawak, Malaysia, proved to be its major endophytic nitrogen-fixing bacterium. Biosci. Biotechnol. Biochem. 74, 1008052106.

Tang, Q., Puri, A., Padda, K.P., Chanway, C.P., 2017. Biological nitrogen fixation and plant growth promotion of lodgepole pine by an endophytic diazotroph *Paenibacillus polymyxa* and its GFP-tagged derivative. Botany 95, 611-619.

Teotia, P., Kumar, V., Kumar, M., Shrivastava, N., Varma, A., 2016. Rhizosphere microbes: Potassium solubilization and crop productivity-present and future aspects. In: Meena, V.S., Maurya, B.R., Verma, J.P., Meena, R.S. (Eds.), Potassium Solubilizing Microorganisms for Sustainable Agriculture. Springer, pp. 315-325.

Thakur, N., Kaur, S., Tomar, P., Thakur, S., Yadav, A.N., 2020. Microbial biopesticides: current status and advancement for sustainable agriculture and environment. In: Rastegari, A.A., Yadav, A.N., Yadav, N. (Eds.), Trends of Microbial Biotechnology for Sustainable Agriculture and Biomedicine Systems: Diversity and Functional Perspectives. Elsevier, Amsterdam, pp. 243-282, https://doi.org/10.1016/B978-0-12-820526-6.00016-6.

Tiwari, P., Bajpai, M., Singh, L.K., Mishra, S., Yadav, A.N., 2020. Phytohormones producing fungal communities: metabolic engineering for abiotic stress tolerance in crops. In: Yadav, A.N., Mishra, S., Kour, D., Yadav, N., Kumar, A. (Eds.), Agriculturally Important Fungi for Sustainable Agriculture, Volume 1: Perspective for Diversity and Crop Productivity. Springer, Cham, pp. 1-25, https://doi.org/10.1007/978-3-030-45971-0_8.

Upadhyay, S.K., Singh, J.S., Saxena, A.K., Singh, D.P., 2012. Impact of PGPR inoculation on growth and antioxidant status of wheat under saline conditions. Plant Biol. 14, 605-611.

Upadhyay, H., Gangola, S., Sharma, A., Singh, A., Maithani, D., Joshi, S., 2021. Contribution of zinc solubilizing bacterial isolates on enhanced zinc uptake and growth promotion of maize (*Zea mays* L.). Folia Microbiol. https://doi.org/10.1007/s12223-021-00863-3.

Usha, S., Padmavathi, T., 2012. Phosphate solubilizers from the rhizosphere of *Piper nigrum* L. in Karnataka. India. Chil. J. Agric. Res. 72, 397-403.

Valentine, A.J., Benedito, V.A., Kang, Y., 2018. Legume nitrogen fixation and soil abiotic stress: From physiology to genomics and beyond. In: Foyer, C.H., Zhang, H. (Eds.), Nitrogen Metabolism in Plants in the Post-Genomic Era. vol. 42. Blackwell Publishing, Wiley-Blackwell, Oxford, UK, pp. 207-248. Ann plant rev.

Vandana, U.K., Chopra, A., Bhattacharjee, S., Mazumder, P.B., 2017. Microbial biofertilizer: a potential tool for sustainable agriculture. In: Panpatte, D.G., Jhala, Y.K., Vyas, R.V., Shelat, H.N. (Eds.), Microorganisms for Green Revolution: Volume 1: Microbes for Sustainable Crop Production. Springer Singapore, Singapore, pp. 25-52, https://doi.org/10.1007/978-981-10-6241-4_2.

Vandenbussche, F., Van Der Straeten, D., 2018. The role of ethylene in plant growth and development. In: McManus, M.T. (Ed.), Annual Plant Reviews: The Plant Hormone Ethylene. Wiley-Blackwell, Oxford.

Verma, P., Yadav, A.N., Khannam, K.S., Panjiar, N., Kumar, S., Saxena, A.K., et al., 2015. Assessment of genetic diversity and plant growth promoting attributes of psychrotolerant bacteria allied with wheat (*Triticum aestivum*) from the northern hills zone of India. Ann. Microbiol. 65, 1885-1899. https://doi.org/10.1007/s13213-014-1027-4.

Verma, P., Yadav, A.N., Khannam, K.S., Kumar, S., Saxena, A.K., Suman, A., 2016. Molecular diversity and multifarious plant growth promoting attributes of *Bacilli* associated with wheat (*Triticum aestivum* L.) rhizosphere from six diverse agro-ecological zones of India. J. Basic Microbiol. 56, 44-58. https://doi.org/10.1002/jobm.201500459.

Verma, P., Yadav, A.N., Khannam, K.S., Saxena, A.K., Suman, A., 2017a. Potassium-solubilizing microbes: diversity, distribution, and role in plant growth promotion. In: Panpatte, D.G., Jhala, Y.K., Vyas, R.V., Shelat, H.N. (Eds.), Microorganisms for Green Revolution: Volume 1: Microbes for Sustainable Crop Production. Springer, Singapore, pp. 125–149, https://doi.org/10.1007/978-981-10-6241-4_7.

Verma, P., Yadav, A.N., Kumar, V., Singh, D.P., Saxena, A.K., 2017b. Beneficial plant-microbes interactions: biodiversity of microbes from diverse extreme environments and its impact for crop improvement. In: Singh, D.P., Singh, H.B., Prabha, R. (Eds.), Plant-Microbe Interactions in Agro-Ecological Perspectives. Microbial Interactions and Agro-Ecological Impacts, vol. 2. Springer, Singapore, pp. 543–580, https://doi.org/10.1007/978-981-10-6593-4_22.

Verma, P., Yadav, A.N., Kumar, V., Singh, D.P., Saxena, A.K., 2017c. Beneficial plant-microbes interactions: biodiversity of microbes from diverse extreme environments and its impact for crop improvement. In: Singh, D.P., Singh, H.B., Prabha, R. (Eds.), Plant-Microbe Interactions in Agro-Ecological Perspectives. vol. 2, pp. 543–580, https://doi.org/10.1007/978-981-10-6593-4_22.

Verma, P., Yadav, A.N., Khannam, K.S., Mishra, S., Kumar, S., Saxena, A.K., et al., 2019. Appraisal of diversity and functional attributes of thermotolerant wheat associated bacteria from the peninsular zone of India. Saudi J. Biol. Sci. 26, 1882–1895. https://doi.org/10.1016/j.sjbs.2016.01.042.

Vimal, S.R., Patel, V.K., Singh, J.S., 2019. Plant growth promoting *Curtobacterium albidum* strain SRV4: an agriculturally important microbe to alleviate salinity stress in paddy plants. Ecol. Indic. 105, 553–562.

Wang, Z., Xu, G., Ma, P., Lin, Y., Yang, X., Cao, C., 2017. Isolation and characterization of a phosphorus-solubilizing bacterium from rhizosphere soils and its colonization of Chinese cabbage (*Brassica campestris* ssp. chinensis). Front. Microbiol. 8, 1270.

Wang, Y., Wang, M., Li, Y., Wu, A., Huang, J., 2018. Effects of arbuscular mycorrhizal fungi on growth and nitrogen uptake of *Chrysanthemum morifolium* under salt stress. PLoS ONE 13, e0196408.

Wang, S., Sun, L., Zhang, W., Chi, F., Hao, X., Bian, J., et al., 2020. *Bacillus velezensis* BM21, a potential and efficient biocontrol agent in control of corn stalk rot caused by *Fusarium graminearum*. Egypt. J. Biol. Pest Co. 30, 1–10. https://doi.org/10.1186/s41938-020-0209-6.

Wei, C.-Y., Lin, L., Luo, L.-J., Xing, Y.-X., Hu, C.-J., Yang, L.-T., et al., 2014. Endophytic nitrogen-fixing *Klebsiella variicola* strain DX120E promotes sugarcane growth. Biol. Fertil. Soils 50, 657–666.

Willer, H., Lernoud, J., 2019. The world of organic agriculture. In: Statistics and Emerging Trends 2019. Research Institute of Organic Agriculture FiBL and IFOAM Organics International.

Win, T.T., Bo, B., Malec, P., Khan, S., Fu, P., 2021. Newly isolated strain of *Trichoderma asperellum* from disease suppressive soil is a potential bio-control agent to suppress *Fusarium* soil borne fungal phytopathogens. J. Plant Pathol. 103, 549–561. https://doi.org/10.1007/s42161-021-00780-x.

Winkelmann, G., 2007. Ecology of siderophores with special reference to the fungi. Biometals 20, 379. https://doi.org/10.1007/s10534-006-9076-1.

Wu, Q., 2011. Mycorrhizal efficacy of trifoliate orange seedlings on alleviating temperature stress. Plant Soil Environ. 57, 459–464.

Wu, Q.-S., Zou, Y.-N., 2010. Beneficial roles of arbuscular mycorrhizas in citrus seedlings at temperature stress. Sci. Hortic. 125, 289–293.

Xia, Y., Farooq, M.A., Javed, M.T., Kamran, M.A., Mukhtar, T., Ali, J., et al., 2020. Multi-stress tolerant PGPR *Bacillus xiamenensis* PM14 activating sugarcane (*Saccharum officinarum* L.) red rot disease resistance. Plant Physiol. Biochem. 151, 640–649.

Xiang, W.-L., Liang, H.-Z., Liu, S., Luo, F., Tang, J., Li, M.-Y., et al., 2011. Isolation and performance evaluation of halotolerant phosphate solubilizing bacteria from the rhizospheric soils of historic Dagong Brine Well in China. World J. Microbiol. Biotechnol. 27, 2629–2637. https://doi.org/10.1007/s11274-011-0736-0.

Xu, H., Lu, Y., Tong, S., 2018a. Effects of arbuscular mycorrhizal fungi on photosynthesis and chlorophyll fluorescence of maize seedlings under salt stress. Emirates J. Food Agric., 199–204.

Xu, J., Kloepper, J.W., Huang, P., McInroy, J.A., Hu, C.H., 2018b. Isolation and characterization of N_2-fixing bacteria from giant reed and switchgrass for plant growth promotion and nutrient uptake. J. Basic Microbiol. 58, 459–471.

Xu, Q., Pan, W., Zhang, R., Lu, Q., Xue, W., Wu, C., et al., 2018c. Inoculation with *Bacillus subtilis* and *Azospirillum brasilense* produces abscisic acid that reduces Irt1-mediated cadmium uptake of roots. J. Agric. Food Chem. 66, 5229–5236.

Xue, L., Xue, Q., Chen, Q., Lin, C., Shen, G., Zhao, J., 2013. Isolation and evaluation of rhizosphere actinomycetes with potential application for biocontrol of *Verticillium* wilt of cotton. Crop Prot. 43, 231–240. https://doi.org/10.1016/j.cropro.2012.10.002.

Yadav, A.N., 2021a. Beneficial plant-microbe interactions for agricultural sustainability. J Appl Biol Biotechnol 9, 1–4. https://doi.org/10.7324/JABB.2021.91ed.

Yadav, A.N., 2021b. Biodiversity and bioprospecting of extremophilic microbiomes for agro-environmental sustainability. J. Appl. Biol. Biotechnol. 9, 1–6. https://doi.org/10.7324/JABB.2021.9301.

Yadav, A.N., 2021c. Microbial biotechnology for bio-prospecting of microbial bioactive compounds and secondary metabolites. J Appl Biol Biotechnol 9, 1–6. https://doi.org/10.7324/JABB.2021.92ed.

Yadav, H., Gothwal, R.K., Nigam, V.K., Sinha-Roy, S., Ghosh, P., 2013. Optimization of culture conditions for phosphate solubilization by a thermo-tolerant phosphate-solubilizing bacterium *Brevibacillus* sp. BISR-HY65 isolated from phosphate mines. Biocatal. Agric. Biotechnol. 2, 217–225. https://doi.org/10.1016/j.bcab.2013.04.005.

Yadav, A.N., Sharma, D., Gulati, S., Singh, S., Dey, R., Pal, K.K., et al., 2015. Haloarchaea endowed with phosphorus solubilization attribute implicated in phosphorus cycle. Sci. Rep. 5, 1–10.

Yadav, A.N., Kumar, V., Dhaliwal, H.S., Prasad, R., Saxena, A.K., 2018. Microbiome in crops: diversity, distribution, and potential role in crop improvement. In: Prasad, R., Gill, S.S., Tuteja, N. (Eds.), Crop Improvement through Microbial Biotechnology. Elsevier, Amsterdam, pp. 305–332. https://doi.org/10.1016/B978-0-444-63987-5.00015-3.

Yadav, A.N., Yadav, N., Sachan, S.G., Saxena, A.K., 2019. Biodiversity of psychrotrophic microbes and their biotechnological applications. J. Appl. Biol. Biotechnol. 7, 99–108. https://doi.org/10.7324/JABB.2019.70415.

Yadav, A.N., Mishra, S., Kour, D., Yadav, N., Kumar, A., 2020a. Agriculturally Important Fungi for Sustainable Agriculture, Volume 2: Functional Annotation for Crop Protection. Springer International Publishing, Cham.

Yadav, A.N., Singh, J., Rastegari, A.A., Yadav, N., 2020b. Plant Microbiomes for Sustainable Agriculture. Springer International Publishing, Cham.

Yaghoubian, Y., Goltapeh, E.M., Pirdashti, H., Esfandiari, E., Feiziasl, V., Dolatabadi, H.K., et al., 2014. Effect of *Glomus mosseae* and *Piriformospora indica* on growth and antioxidant defense responses of wheat plants under drought stress. Agric. Res. 3, 239–245.

Yang, J., Kloepper, J.W., Ryu, C.-M., 2009. Rhizosphere bacteria help plants tolerate abiotic stress. Trends Plant Sci. 14, 1–4.

Yu, X., Liu, X., Zhu, T.-H., Liu, G.-H., Mao, C., 2012. Co-inoculation with phosphate-solubilzing and nitrogen-fixing bacteria on solubilization of rock phosphate and their effect on growth promotion and nutrient uptake by walnut. Eur. J. Soil Biol. 50, 112–117. https://doi.org/10.1016/j.ejsobi.2012.01.004.

Yu, J., Guan, X., Liu, C., Xiang, W., Yu, Z., Liu, X., et al., 2016. *Lysinibacillus endophyticus* sp. nov., an indole-3-acetic acid producing endophytic bacterium isolated from corn root (*Zea mays* cv. Xinken-5). Antonie Van Leeuwenhoek 109, 1337–1344. https://doi.org/10.1007/s10482-016-0732-3.

Zaheer, A., Malik, A., Sher, A., Mansoor Qaisrani, M., Mehmood, A., Ullah Khan, S., et al., 2019. Isolation, characterization, and effect of phosphate-zinc-solubilizing bacterial strains on chickpea (*Cicer arietinum* L.) growth. Saudi J. Biol. Sci. 26, 1061–1067. https://doi.org/10.1016/j.sjbs.2019.04.004.

Zhang, C., Kong, F., 2014. Isolation and identification of potassium-solubilizing bacteria from tobacco rhizospheric soil and their effect on tobacco plants. Appl. Soil Ecol. 82, 18–25. https://doi.org/10.1016/j.apsoil.2014.05.002.

Zhang, W., Jiang, B., Li, W., Song, H., Yu, Y., Chen, J., 2009. Polyamines enhance chilling tolerance of cucumber (*Cucumis sativus* L.) through modulating antioxidative system. Sci. Hortic. 122, 200–208.

Zhang, S., Gan, Y., Xu, B., 2016. Application of plant-growth-promoting fungi *Trichoderma longibrachiatum* T6 enhances tolerance of wheat to salt stress through improvement of antioxidative defense system and gene expression. Front. Plant Sci. 7, 1405.

Zhang, Y., Chen, F.-S., Wu, X.-Q., Luan, F.-G., Zhang, L.-P., Fang, X.-M., et al., 2018. Isolation and characterization of two phosphate-solubilizing fungi from rhizosphere soil of moso bamboo and their functional capacities when exposed to different phosphorus sources and pH environments. PLoS ONE 13, e0199625.

Zhang, H., Han, L., Jiang, B., Long, C., 2021. Identification of a phosphorus-solubilizing *Tsukamurella tyrosinosolvens* strain and its effect on the bacterial diversity of the rhizosphere soil of peanuts growth-promoting. World J. Microbiol. Biotechnol. 37, 1–14. https://doi.org/10.1007/s11274-021-03078-3.

Zheng, X., Wang, J., Chen, Z., Zhang, H., Wang, Z., Zhu, Y., et al., 2019. A *Streptomyces* sp. strain: isolation, identification, and potential as a biocontrol agent against soil-borne diseases of tomato plants. Biol. Control 136. https://doi.org/10.1016/j.biocontrol.2019.104004, 104004.

Zhu, F., Qu, L., Hong, X., Sun, X., 2011a. Isolation and characterization of a phosphate-solubilizing halophilic bacterium *Kushneria* sp. YCWA18 from Daqiao Saltern on the coast of Yellow Sea of China. Evid. Based Complement. Alternat. Med. 2011, 615032. https://doi.org/10.1155/2011/615032.

Zhu, X.-C., Song, F.-B., Liu, S.-Q., Liu, T.-D., 2011b. Effects of arbuscular mycorrhizal fungus on photosynthesis and water status of maize under high temperature stress. Plant and Soil 346, 189–199.

Zhu, X., Song, F., Liu, S., Liu, T., Zhou, X., 2012. Arbuscular mycorrhizae improves photosynthesis and water status of *Zea mays* L. under drought stress. Plant Soil Environ. 58, 186–191.

Phytobiome research: Recent trends and developments

V.T. Anju[a], Madhu Dyavaiah[a], and Busi Siddhardha[b]
[a]Department of Biochemistry and Molecular Biology, School of Life Sciences, Pondicherry University, Pondicherry, India, [b]Department of Microbiology, School of Life Sciences, Pondicherry University, Pondicherry, India

2.1 Introduction

The environment containing macro- and microorganisms surrounding the plants and those living inside the plants is collectively called phytobiome. These organisms are from different taxa such as bacteria, fungi, viruses, oomycetes, and archaea. These organisms exhibit diverse biological interactions among themselves and others such as predation, competition, mutualism, symbiosis, pathogenesis, etc. (Leach et al., 2017). Phytobiome interactions eventually end up with the growth and development of their hosts. Plant-associated microbes are involved in several applications such as increased nutrient uptake by plants through biogeochemical cycles, enhanced crop production, reduced plant diseases, less chemical inputs, causing sustainable agriculture practices (Parakhia, 2018). The abundance and types of several species residing in different parts of plants and surrounding areas are different, which regulates the different functional roles. For example, the kind of microbiome abundance in the phyllosphere and rhizosphere region of a plant is different. The phyllosphere region, which is the aerial plant surface, harbors a more distinct microbiome. The rhizosphere region corresponds to the area near the plant roots, and it is the most environmentally stable zone. Several metataxonomic studies are conducted to gain knowledge on the formation of the microbiome in plants. These studies observed more bacterial abundance and diversity in the external root surface (rhizoplane region) (Remus-Emsermann et al., 2012; Hacquard et al., 2015). The phytobiome plays a crucial part in meeting the need of the increasing population. This microbiome can be manipulated to enhance the soil quality and for sustainable agriculture. The area of phytobiome engineering and research focuses mainly on microbiome engineering in an environment-friendly approach and to study the interaction of plant-microbiome communities to understand functional roles (Kumari et al., 2020).

2.2 Plant-associated microbiomes

Plants serve as a host for a rich and diverse source of microorganisms. Plants harbor some microbial communities inside the tissues, whereas some organisms are found in the surrounding environment. The plant-associated wide array of microorganisms, comprising bacteria, fungi, viruses, archaea, nematodes, and protists, is collectively termed plant microbiome or phytobiome (Wang et al., 2015). Many microbial species are beneficial to the plants, whereas other members of phytobiome develop parasitic or detrimental associations. Several nematode genera cause plant infections. The control and management of nematode infections are one of the biggest threat in farms, which drops the crop productivity. Among the nematode category, root-knot and cyst nematodes are biotrophic plant pathogens. They induce changes in the morphology and physiology of plants (Williamson and Gleason, 2003; Kaloshian and Teixeira, 2019). *Heterodera* sp. and *Globodera* sp. are the two most common cyst nematodes, whereas the common root-knot pathogens are *Meloidogyne hapla*, *M. javanica*, *M. incognita* and *M. arenaria* (Shukla et al., 2016).

Plant-associated microbiome takes part in both beneficial and detrimental associations. Specifically, organisms isolated from the leaf, root endosphere, rhizosphere, or phyllosphere region of the host plant enabled an augmented tolerance to abiotic stresses, indirect pathogen shield, and improved nutrient acquisition. Plant growth-promoting microorganisms (PGPMs) are the microorganisms found associated with plants for their growth. PGPMs include plant growth-promoting rhizobacteria (PGPR), nonrhizobacterial growth promotors, and plant growth-promoting fungi (Schlaeppi and Bulgarelli, 2015). The dominant bacterial phyla that influence plant metabolism by living in or surrounding plants are *Proteobacteria*, *Actinobacteria*, and *Bacteroidetes*. The major fungal phyla that colonize below and/or above plant tissues are *Ascomycota* and *Basidiomycota* (Singh et al., 2020b) (Fig. 2.1).

There is another category of microorganism that thrive inside plants to complete a part or entire life cycle. They are called endophytes. Obligate endophytes are those live inside the plants to complete their life cycle. There are both bacterial and fungal endophytes. Fungi from the family of *Ascomycota*, genera *Balansia*, *Epichloe*, and *Neotyphodium*, and some mycorrhizal fungi belong to the obligate endophytes. A few microbial genera from *Trichoderma* and *Hypocrea* and rhizosphere-competent bacterial colonizers (*Pseudomonas* and *Azospirillum*) are opportunistic endophytes. These opportunistic

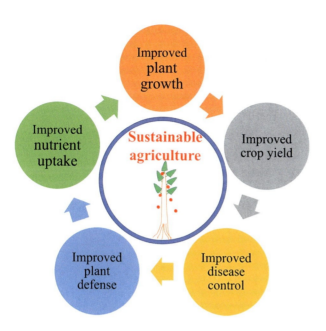

Fig. 2.1 The significant functions of phytobiome for improved productivity and sustainable agriculture.

endophytes are also known as epiphytes (who live outside the plants) but enter the plant endosphere to complete their life cycle. And most of the endophytes come under the group of facultative endophytes. The interaction between facultative endophytes and plants is unknown; however, there is an assumption that they must be in mutualistic interactions (Hardoim et al., 2015). There are plant endophytes from the archaea, such as phyla *Thaumarchaeota, Crenarchaeota*, and *Euryarchaeota* (Müller et al., 2015).

Several protists are regarded as the chief component of the soil microbiome, which improves the growth of plants. Among them, Stramenopiles-Alveolata-Rhizaria (SAR) is a group of protists from the lineage of *Oomycota* (Stramenopiles) and *Cercozoa* (Rhizaria). These organisms are one of the significant groups associated with the plants (Ruggiero et al., 2015). Protists increased the delivery of nutrients to the plant roots. Also, they maintain the balance of soil microbiome by parasitizing bacteria, fungi, and other eukaryotic organisms (Geisen et al., 2018).

The distribution of endophytes is majorly based on their ability to colonize plant parts and allocate plant resources. Root endophytes inhabit the root cracks and epidermis region of lateral root emergence below the root hair. The colonizers can emerge as intra- and intercellular populations. These root colonizers can migrate to other plant parts through the vascular tissues. Studies conducted observed that these endophytes can move continuously throughout the root microbiome. The composition of endophytes can be influenced by the kind of plant tissues or plants. For instance, the abundance of *Pseudomonas* sp. is more in the stems of potatoes than in the roots. As the crown region of carrots has more concentrations of photosynthate, it is speculated that there are a higher number of endophytes in that region than in the metaxylem tissues of carrots (Gaiero et al., 2013).

Soil virome and viruses of phytobiome were less investigated by researchers. The narrow research in plant virome is due to the inability to screen the viruses, and most of the viruses are considered pathogenic. As few viral genera are helpful for plants, they can be best utilized for sustainable agriculture in the future. Likewise, some viruses are beneficial to plants, some are viruses of plant pathogenic fungi, and some affect plant-associated bacteria. Though viruses are rich in phytobiome, they are possibly the most neglected group among them (Roossinck, 2019).

2.3 Plant-microbial interactions

The communications among plants and their microbiome are affected by several factors. These communications are highly diverse that shape the assembly and functioning of communities. Improved plant-microbe relations enhance the benefits of phytobiome to plants. The plant-microbe interactions mediated by signaling, mutualism, antagonism, cooperation, or commensalism provide significant advantages (Agler et al., 2016). The plant-microbe interaction can be pathogenic or nonpathogenic based on the nature of microorganisms. Parasitism and amensalism come under pathogenic interactions, whereas commensalism, mutualism, neutralism, and cooperation are nonpathogenic interactions (Bulgarelli et al., 2013). There are biotrophic, necrotrophic, and hemibiotrophic pathogens of plant hosts. Biotrophic pathogens feed on living cells, whereas necrotrophic pathogens derive nutrients from dead tissues or cells. Hemibiotrophs are the intermediate group that exhibits both forms of nutrient retrieval from living and decaying cells (Laluk and Mengiste, 2010).

The production of communication signals by microorganisms enable the plants to identify the organism as beneficial or pathogen. The plants are adapted to employ unique and effective defense strategies to recognize and fight the self and foreign molecules near them. Generally, two classes of immune receptors are employed by the plants as pattern recognition receptors. First, pathogen- or microbe-associated molecular recognition patterns recognize the signals produced by many of the microorganisms. The second type of receptors detects the effector molecule (such as flagellin) produced by the pathogen (Jones and Dangl, 2006). These receptors are related to a vast number of microbial-associated molecular patterns that prevent the colonization of soil endophytes to the interior regions of the root. The local and systemic immune responses are activated based on the kind of microbe and production of plant signaling hormones like ethylene, jasmonic acid, and salicylic acid due to the plant-microbe interactions. Thus, it is imperative to comprehend and study the type and mechanism of plant-microbe interactions to know about the effect of interactions (Kumar et al., 2017).

Plants may release some chemical signals into the surroundings when interacting with the microorganisms. These chemical signaling may positively or negatively impact other plants, members of its microbiome, or other phytobiome. Plant-plant and plant-microbe communication can be facilitated by the root exudates containing allelochemicals released by the plants during interactions (Sasse et al., 2018). Some signals produced by bacteria in the phytobiome are quorum sensing signals such as acylated homoserine lactones (HSL), quinolones, diffusible signal factors, and autoinducing peptides. These bacterial QS molecules are essential for developing and establishing biofilm colonization and releasing cell wall-deteriorating enzymes (Schaefer et al., 2013). These bacterial signals are generally involved in interspecies competition or cooperation. Some bacterial species of phytobiome cooperatively produce polybacterial disease in the plant host. Knot disease of the olive tree is contributed by the stable quorum sensing mechanism of three bacterial species. The bacterium *Pseudomonas savastanoi* pv. *savastanoi* (Psv) is the causative agent of knot disease. This bacterium can interact with its neighboring nonpathogenic bacteria *Pantoea agglomerans* and *Erwinia toletana* (epiphytes and endophytes of the olive tree). All the bacterial consortium together produce acyl homo serine-based quorum sensing signals and causes the infection in olive trees (Hosni et al., 2011).

Quorum sensing mechanism present in several rhizobacteria stimulates the secondary metabolism to secrete several antimicrobial metabolites and lytic enzymes. These compounds hold antagonistic properties against pathogenic bacteria, fungi, protozoa, and nematodes. These antimicrobial metabolites produced by them protect the host plant and plant parts from other microbial attacks. There are rhizospheric bacteria capable of QS, which antagonizes the root-infecting fungi and stops the infection by stimulating host plant defense mechanisms (Dubuis et al., 2007).

Induced systemic resistance was stimulated in bean and tomato plants by the QS signals (C4 and C6 HSL) of *Serratia plymuthica* HRO-C48 to protect from the fungal leaf pathogen (*Botrytis cinnera*). The same kind of HSL molecules perform different functions in *Arabidopsis thaliana*. They are involved in the improved production of auxin/cytokine. Longer HSL (C10) changed the expression of cell division and differentiation genes that caused developmental changes in the roots of *Arabidopsis thaliana* (Hartmann et al., 2014).

Similarly, bacteria, fungi, and plants produce several extracellular products, which mediate the signaling and plant development and impact stress responses. Some plants can secrete proteins of broad functions to prevent the development of diseases. At the same time, pathogens acquired mechanisms to breach the cell wall and defense mechanisms to colonize and infect plants. Pathogens produce several inhibitors of the cell wall, whereas plants secrete inhibitors of those enzymes. Eventually, the successful plant-pathogen interactions develop into plant infections (Vincent et al., 2020). Likewise, plant immunity is triggered by the cell wall component, pectin-degraded product (oligogalacturonides), which performs as a damage-associated molecular pattern signal. This causes the inhibition and control of plant pathogens, *Pectobacterium carotovorum*, *Botrytis cinerea*, and *Pseudomonas syringae*. Thus, the secretion of pectin degradation products can stimulate the immune mechanisms of the plant to fight phytopathogens (Benedetti et al., 2015).

2.4 Functional roles of phytobiome for sustainable agriculture

2.4.1 Role in nutrient mobilization

Nitrogen is a significant and vital element for plant growth. Nitrogen fixation by specific microorganisms containing nitrogenase enzyme is crucial for the mineralization of soil nitrogen for the use by plants. Atmospheric nitrogen is not readily uptaken by plants. Biological nitrogen fixation helps in the uptake of nitrogen. Nitrogen present in atmosphere is converted to ammonia by diazotrophs for the uptake of plants. These diazotrophs include free-living bacteria, archaea, and symbiotic bacteria. The major symbiotic bacteria involved in nitrogen fixation are from the family Rhizobiaceae, such as *Rhizobium*, *Bradyrhizobium*, *Azorhizobium*, *Mesorhizobium*, and *Sinorhizobium* (Moreira-Coello et al., 2019; Shamseldin et al., 2017). The mutualistic behavior among legumes and *Rhizobia* causes the formation of nodules in the roots of plants. These nodules help in the atmospheric nitrogen fixation (Masson-Boivin and Sachs, 2018) (Fig. 2.2).

The requirement of phosphorus for plant nutrition is very important after nitrogen. The microbial genera directing the mineralization and mobilization of insoluble phosphorus to soluble phosphorus are *Bacillus, Pseudomonas*, and *Rhizobium, Penicillium* and *Aspergillus*, actinomycetes, and arbuscular mycorrhizal fungi (Kalayu, 2019). The molecular studies demonstrated the importance of mycorrhizal fungi for the effective delivery of several vital nutrients.

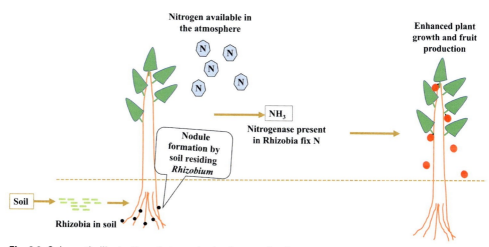

Fig. 2.2 Schematic illustration of atmospheric nitrogen fixation and its role in plant growth.

This fungus improved phosphate uptake from soil by several plants except for *Arabidopsis thaliana* and other *Brassicaceae* (Bonfante, 2010). Phosphorus is available in limited amounts and primarily insoluble, which is unavailable to most of the plants. Mineralization and phosphorus mobilization by soil microorganisms such as phosphorus-solubilizing microorganisms help in the phosphorus assimilation by plants (Wei et al., 2018).

Potassium is another vital nutrient demanded by plants after nitrogen and phosphorus. This element is available in large quantities in soil but primarily associated with other minerals, making them unavailable to plants. Thus, microorganisms present in soil and plants provide enhanced nutrient mobilization and delivery to the plants. A broad range of organisms such as fungi, saprophytic bacteria, and actinomycetes possess the potential to solubilize the insoluble form of potassium to readily accessible and soluble form. The occurrence of potassium-solubilizing bacteria in soil improved the nutrient delivery to plants. These bacteria can solubilize the inorganic and insoluble potassium available in the soil for utilization (Etesami et al., 2017). More diversity and population of these bacteria were observed in the rhizosphere region of plants than in nonrhizosphere soil. Several organisms are found to be effective potassium solubilizers of silicate rocks, like *B. circulanscan, B. mucilaginosus, B. edaphicus, Burkholderia, A. ferrooxidans, Arthrobacter* sp., *Enterobacter hormaechei, Paenibacillus mucilaginosus, P. frequentans, Cladosporium, Aminobacter, Sphingomonas, Burkholderia*, and *Paenibacillus glucanolyticus* (Meena et al., 2016).

The extended use of traditional fertilizers in farm lands improved agricultural yield. But the traditional chemical fertilizers caused a significant harm to the ecosystem. Thus, the use of biofertilizers containing various beneficial microorganisms was recommended for agricultural practices. Shen and co-workers studied bacteria capable of nitrogen fixation, solubilization, and mobilization of potassium and phosphorus from purple soil, which improved the growth of kiwifruit plants. They employed phenotypic and 16S rRNA sequencing methods to identify this PGPR from purple soil in China. The study revealed three *Bacillus* species (*B. amyloliquefaciens*, *B. pumilus*, and *B. circulans*) from the soil. The mixed inoculum of three bacterial isolates promoted the availability of nutrients and plant growth (Shen et al., 2016). Additionally, *Rhizobia* and *Pseudomonas* sp. capable of nitrogen fixation can participate in the solubilization of phosphorus and potassium (Bashir, 2017).

Another nutrient available in soil-bound organic molecule is sulfur. Thus, the plants are unable to uptake sulfur in its available format in the soil for their growth and development. The requirement of sulfur by plants differs depending on the host and its developmental stage. For instance, the requirement of sulfur by sunflower is much more than that of soybean and wheat, whereas rice, maize, and field beans demand less sulfur in their early growth phase (Senthilkumar et al., 2021). Sulfur-oxidizing bacteria containing cysteine desulfhydrase and 3-mercaptopyruvate sulfurtransferase enzymes help to metabolize organic sulfur through cysteine catabolism and release hydrogen sulfide for the uptake of plants (Xia et al., 2017). A study emphasized the requirement of sulfur-oxidizing bacteria for the development of canola plants on Saskatchewan soils. This soil is usually deficient in sulfur and is required for the vegetative growth of canola plants. Thus, the study confirmed the occurrence of sulfur-oxidizing bacteria in the rhizosphere and rhizoplane region of canola plants. Together with the enhanced nutrient uptake and plant growth, these bacteria showed considerable antagonistic activities toward the fungal pathogens of canola plants such as *Rhizoctonia solani* and *Leptosphaeria maculans* (Grayston and Germida, 1991).

As similar to the terrestrial ecosystem, plant-microbe interactions can be seen in the aquatic ecosystem also. In the perspective of marine or freshwater plants, some symbiotic microbial interactions help in the fixation of nitrogen and supplements phosphorus for growth. A typical example is observed between *Azolla filiculoides* and bacteria (*Anabaena azollae* and *Arthrobacter* sp.), which participate in nitrogen fixation in aquatic environments. Very limited data are available on different roles of microbes for the nutrient uptake by the aquatic plants. However, the studies on the vast potential of plant-microbe communications in aquatic ecosystems can be best utilized from the perspective of plant development and environment (Srivastava et al., 2017).

2.4.2 Effect on plant growth and health

Plant-microbial associations impact the plant growth and health through increased nutrient acquisition. Currently, several biofertilizers are available for improved plant growth based on the studies conducted by researchers on plant growth-promoting microorganisms. Also, recent developments in plant biotechnology yielded plant varieties with better nutritional value than before (Berg, 2009). The advancement occurred in the identification and imaging techniques added advantage to recognize and use novel soil microorganisms involved in plant growth and health. The antagonistic properties of plant-microbial associations help in the suppression of phytopathogens and promote plant growth during environmental stresses. The enhanced plant growth is contributed by the secretion of plant hormones required for growth and development. Phytohormones can perform as growth simulators and defense regulators. Though the plants can synthesize the hormones, the amount required for growth is provided through plant-microbe interactions (Berg, 2009). For example, ethylene is a plant hormone required by several plant species but at lower concentrations. Thus, the hormone balance among plants is maintained by plant-microbe interactions. At high concentrations, ethylene is a senescence phytohormone. The remarkable production of 1-aminocyclopropane-1-carboxylic acid deaminase by plant-associated bacteria maintains the concentrations of ethylene in plant roots by degrading the substrate of ethylene synthesis. Consequently, bacteria having this specific enzyme improves the root growth by decreasing the ethylene level (Glick, 2005). Some bacteria can maintain the balance of phytohormones. Ethylene is a phytohormone whose maintenance is important for plants, and some bacteria can perform this function. Ethylene improves the growth of *Arabidopsis thaliana* at low levels. But this hormone usually inhibits the development of plants at high concentrations and acts as senescence hormone (Pierik et al., 2007).

Another phytohormone secreted by the species of *Azospirillum*, *Pseudomonas*, and *Bacillus* is auxin. This is an important phytohormone, especially indole 3-acetic acid, which aids the root initiation process, cell division, and expansion (Nihorimbere et al., 2011). The hormone signals produced from the interactions of PGPR with plants can mediate the plant growth, induce resistance to infections, and maintain soil health for plants. They produce auxins, cytokinins, gibberellins, abscisic acid, salicylic acid, and jasmonic acid. The role of certain volatile organic compounds of these bacteria in improving the growth of the plant is discussed previously. In vitro co-cultivation of *Bacillus subtilis* GB03 and *B. amyloliquefaciens* IN937a along with *Arabidopsis thaliana* enhanced the health of plants. The study found that two compounds, 3-hydroxy-2-butanone and 2,3-butanediol, were present in both the bacteria (Fahad et al., 2015) (Mhlongo et al., 2018).

The less-explored category of phytobiome so far for their pathogenic properties in hosts are protists. Among them, most of the oomycete sp. from *Pythium* develop pathogenic interactions on plants. The latest profiling of oomycetes communities provided knowledge on the nonpathogenic benefits of oomycete-associated plant interactions, such as plant growth (Hassani et al., 2018). In a study, three PGPRs were recognized from the rhizosphere region of Indian mustard plants. The bacteria, namely, *Bacillus paramycoides* KVS27, *Bacillus thuringiensis* KVS25, and *Pseudomonas species* KVS20, verified their ability to enhance the growth. There was a cumulative increase in plant growth through the synergistic interactions among them. This is due to the adequate fixation of nitrogen, solubilization of phosphorus, and synthesis and release of phytohormones like auxins (Vishwakarma et al., 2018). In a study, qRT-PCR analysis of samples obtained from wheat rhizosphere showed *Bacillus* and *Pseudomonas* sp. These bacteria were able to produce acyl-homoserine lactones and take part in quorum sensing in the phytobiome of wheat for better growth rate and disease resistance (Moshynets et al., 2019).

2.4.3 Role in crop productivity

Phytobiome plays a significant part in productivity rate and sustainable agriculture. The emergence of high-throughput methods such as metagenomics facilitated the study of plant-microbiome interactions and their contribution in crop productivity. Smart agricultural practices combined with the phytobiome favored more crop yield. The mutualistic bacteria and fungi components of phytobiome promoted crop growth and fruit production (Kim and Lee, 2020). Field studies showed that enhanced canola yields were observed with the microbial genera from *Amycolatopsis* sp., *Serratia proteamaculans*, *Pedobacter* sp., *Arthrobacter* sp., *Stenotrophomonas* sp., *Fusarium merismoides*, and *Fusicolla* sp. (Lay et al., 2018).

The phytobiome engineering results in more productivity. New strategies are employed to enhance the survival and activities of phytobiome communities for better functions. The emergence of advanced gene-editing tools and biochemical or molecular techniques enabled microbiome engineering even in situ for better and effective fruit yield. Also, plant biotechnology has initiated with novel methods to improve crop performance through nutrient mobilization, resistance to environmental stress conditions, and sustainable and highly productive agriculture (Qiu et al., 2019). Sustainable agriculture is the need of the hour that should not compromise on nutrient-rich and quality foods. Thus, sustainable development goals have introduced several alternatives to channelize crop production effectively. One such initiative was the replacement of chemical fertilizers with biofertilizers rich in beneficial microbial inoculants.

The development of molecular techniques and molecular markers for culturable and nonculturable phytobiome explored a diverse number of soil microorganisms and their associations with hosts. Polymerase chain reaction and other DNA-based methods exposed the phylogenetic diversity and functional roles of several microbial populations. The incorporation of gene editing methods for the development of genetically engineered microbes improved the functional traits of microbes. This promoted food security and sustainability (Etesami et al., 2014). Recently, the widespread application of the gene-editing tool, CRISPR/Cas, enabled the activation of silent gene clusters and upgraded the metabolic products for more yield. In the era of synthetic biology and systems biology, CRISPR/Cas is gaining more attention in agricultural sectors for potential applications (Sekar and Prabavathy, 2014).

Through genome editing methods, the genes of microbial species can be modified for superior microorganisms. In a study, a bacterium symbiont of the bee gut was modified to induce RNA interference in the bee. This protected honeybees from parasites and viral infections. This technique was adopted in plant biotechnology to genetically modify the phytobiome, especially the bacterial endophytes, to activate the defense mechanisms and to fight pests and plant pathogens (Leonard et al., 2020).

2.4.4 Enhanced plant defense mechanisms

The role of phytobiome in disease control and pest management of plants is crucial. The plants or microorganisms can secrete molecules with the capacity for antagonism and to stimulate defense mechanisms. The demand for plant growth-promoting organisms to control plant diseases and pest management has increased for successful and effective agricultural production. Together, these microorganisms were manipulated through high-throughput techniques for enhanced nutrient availability and production of phytohormones. The siderophores produced by these bacteria control the infections by plant pathogens (Kumari et al., 2020).

The significant molecules produced by the bacteria through different interactions like phytohormones (jasmonic acid, ethylene, and salicylic acid) and volatile organic compounds can elicit defense mechanisms to control pests and pathogens. These hormones act and produce reactions according to the type of pathogenic organisms. For instance, jasmonic acid and ethylene pathways work to protect from necrotrophic pathogens, herbivores, and chewing insects, whereas salicylic acid is for insects that feed on the phloem and biotrophic pathogens (Pieterse et al., 2012; Glazebrook, 2005).

Some multipartite interactions are beneficial/detrimental for plants but yet to explore the exact mechanisms. For instance, some bacteria release during the feeding period of beetle larvae and found to be harmful to the host plants. These bacteria were able to manipulate the jasmonic acid plant defenses that enhanced larvae feeding on plants and facilitated herbivory (Shikano et al., 2017). Thus, some microbial interactions also cause harmful or unfavorable effects on hosts. In the same way, endophytes of plants exhibit plant defense mechanisms other than promoting plant growth. They are efficient in the stimulation of plant defense. Plant endophytes can protect from pathogen attack and abiotic and biotic stresses (Singh et al., 2020a). It has been studied that a few strains of endophytic actinomycetes like *Streptomyces diastaticus*, *S. fradiae*, *S. ossamyceticus*, *S. collinus*, and *S. olivochromogenes* of several medicinal plants were able to protect from *Sclerotium rolfsii*. The interactions of these endophytes with chickpea alleviated oxidative stress and activated systemic resistance for disease resistance (Singh and Gaur, 2016, 2017).

Endophytes improved the tolerance to drought and salt in wild grasses and reduced *Stagonospora* infection in wheat along with enhanced germination (Hubbard et al., 2012). In millets, mycotoxin production and fungal pathogen attack were reduced by some bacterial endophytes (Mousa et al., 2016). Gdanetz and Trail explored wheat microbiome analysis for the identification and analysis of biocontrol agents against the wheat pathogen *Fusarium graminearum* (Gdanetz and Trail, 2017). The microbiome associated with protection against fungal pathogens include *Alternaria* sp., *Cochliobolus* sp., *Bipolaris* sp., *Fusarium* sp., *Talaromyces* sp., *Colletotrichum* sp., *Parastagnospora* sp., *Trichoderma* sp., and *Penicillium* sp. (Gdanetz and Trail, 2017). Some fungi from the subcategory of mycorrhiza and endophytes can produce metabolites against several insects. They are able to elicit induced systemic defense responses against herbivores, parasitoids, and predators. Interestingly, a few entomopathogens may also act as endophytes of several plants. These endophytes exhibited antagonistic behavior toward herbivorous arthropods through the secretion of fungal metabolic products and induced systemic plant resistance (Bell et al., 2019). The significance of protists and bacteriophages in antagonism against the pathogen is also studied. The occurrence of *Ralstonia solanacearum* infection of tomato plants was reduced significantly by bacteriophages. In another study, the role of rarely studied protists in disease control was discussed. They found that protists reduced the abundance and prevalence of tomato pathogens by direct feed on the tomato pathogens or exert plant protection in an indirect way by predation (Xiong et al., 2020; Wang et al., 2019). Table 2.1 includes few examples of beneficial plant-microbe interactions.

Table 2.1 Major plant-microbe interactions and their functional roles.

Microorganisms	Class of phytobiome	Host	Activities	References
Methanococcoides burtonii	Archaea	Tobacco plants	Increased photosynthesis rate and plant growth due to the synthesis of archaeal Rubisco	Wilson et al. (2016)
Malassezia globose and *Neorhizobium galegae*	Fungi and bacteria	-	Protected from root-knot nematodes and promoted plant growth	Elhady et al. (2017)
Rhizosphere microbiome	Broad range of microorganisms	Soybean and tomato plants	Reduced the abundance of root lesion nematode, *Pratylenchus penetrans*	Elhady et al. (2018)
Arbuscular Mycorrhiza	Fungi	Tomato plants	Enhanced plant resistance to *Fusarium oxysporum* f. sp. *lycopersici*	Chialva et al. (2018)
Atractiella rhizophila	Fungi	*Populus*	Enhanced plant growth	Bonito et al. (2017)
Pseudomonas putida	Bacteria	*Nicotiana benthamiana*	Inhibited the pathogen *Xanthomonas campestris*	Bernal et al. (2017)
Pseudomonas piscium	Bacteria	Wheat plants	Inhibited the pathogen *Fusarium graminearum*	Chen et al. (2018)
Rhizobia sp	Bacteria	Legumes	Facilitates nitrogen fixation for plant growth	Glick (2012)
Azospirillum brasilense	Rhizobacteria	Strawberry	Siderophore production by rhizobacteria, decreased the symptoms of anthracnose caused by the fungal pathogen *Colletotrichum acutatum*	Tortora et al. (2011)
Bacillus sp., *Ochrobactrum* sp., and *Alcaligenes* sp.	Bacteria	Rice plants	Stimulated salt tolerance by ACC deaminase enzyme in rice growing in salinity stress	Bal et al. (2013)
Burkholderia sp.	Bacteria	Sunflower plants	Enhanced phosphate solubilization	Ambrosini et al. (2012)
Cedecea sp., *Burkholderia* sp., *Enterobacter* sp., *Pantoea* sp., *Cronobacter* sp., *Herbaspirillum* sp., and *Pseudomonas* sp.	Bacteria	Rice plants	Enhanced phosphate solubilization and nitrogen fixation	de Souza et al. (2013)
B. megaterium	Bacteria	*Arabidopsis*	Enhanced plant biomass production, changed root architecture, high photosynthesis and drought tolerance.	Zhou et al. (2016)

2.5 Plant microbiome in the field of translation and commercialization

The use of natural rhizobial inoculants mixed with seeds for improved crop yield was initiated in the 19th century. Later, in 1896 the first commercialized biofertilizer made of *Rhizobium* is Nitragin and was developed by Nobbe and Hiltner. These inoculants were initially formulated as liquid fertilizers. Later on, the type of formulations was modified for application in leguminous plants. These marketed formulations were lyophilized (freeze-dried) inoculants, gel-based inoculants such as polyacrylamide, alginate, or xanthan. There are a vast number of rhizobial inoculants that got patented and marketed. They are Gold CoatTM (vermiculite-based), Cell-Tech® (liquid seed-applied soybean LIFT (inoculant), liquid in-furrow inoculant), and Nitragin® Gold (air-dried clay powder for alfalfa) (Bellabarba et al., 2019).

The ability of *Aspergillus flavus* to effectively decrease aflatoxin in groundnut, cotton, and maize plants was studied previously. Later, a concoction of four nontoxigenic strains of *A. flavus* was proven to reduce the contamination of toxin in maize plants during trials performed in Senegal. The biocontrol agent was named Aflasafe SN01 (Bandyopadhyay et al., 2016; Dorner, 2004; Senghor et al., 2020). There are several *Bacillus*-based formulations approved in markets. These formulations are easy to produce due to their ability to form spores capable of enduring adverse conditions and enhanced shelf life (Cawoy et al., 2011). *Bacillus*-based biocontrol agents are in great demand due to their ability to replace chemical fertilizers. Kodiak is a combination of *Bacillus subtilis* GB03 with some fungicides used in the treatments of cotton seeds and plants. Kodiak has gained more attention in the cotton market as it exhibited enhanced disease control properties together with the fungicides (Jacobsen et al., 2004).

Microbial inoculants containing single or mixed genera of *Rhizobia* sp. and other few genera have been widely used in the agricultural field in several countries. The biofertilizer agents like *Bradyrhizobium* sp., *Pseudomonas* sp., *Azospirillum* sp., *Bacillus* sp., and *Azotobacter* sp. have been applied for the crops like soybean, common beans, cowpea, faba beans, chickpea, guar, rice, maize, wheat, sugar cane, tomato, and lettuce (Santos et al., 2019).

2.6 Conclusions

The phytobiome comprises bacteria, archaea, fungi, viruses, and protists associated with plants and their environment. There are phytobiome communities involved in beneficial interactions, whereas some are engaged in detrimental associations on plants.

The majority of plant-microbe communications that improve the health and growth of hosts are endophytic fungi, endophytic bacteria, and plant growth-promoting microorganisms. These organisms enhance the mobilization of nutrients; are involved in disease control and pest management; and elicit plant defense mechanisms for sustainable agriculture. Numerous bacterial and fungal strains were identified and studied for their functional roles in plants. The least explored members are protists and viruses. It is recommended for further studies on unexplored protist and viral communities associated with plants. High-throughput techniques and next-generation sequencing methods are employed to unravel the unculturable soil and plant microbiome. Unraveling the hidden world of phytobiome is the key to exploit the potential of crop yield and agricultural productivity.

References

Agler, M.T., Ruhe, J., Kroll, S., Morhenn, C., Kim, S.T., Weigel, D., et al., 2016. Microbial hub taxa link host and abiotic factors to plant microbiome variation. PLoS Biol. 14. https://doi.org/10.1371/journal.pbio.1002352, e1002352.

Ambrosini, A., Beneduzi, A., Stefanski, T., Pinheiro, F.G., Vargas, L.K., Passaglia, L.M.P., 2012. Screening of plant growth promoting Rhizobacteria isolated from sunflower (*Helianthus annuus* L.). Plant Soil 356, 245–264. https://doi.org/10.1007/s11104-011-1079-1.

Bal, H.B., Nayak, L., Das, S., Adhya, T.K., 2013. Isolation of ACC deaminase producing PGPR from rice rhizosphere and evaluating their plant growth promoting activity under salt stress. Plant Soil 366, 93–105. https://doi.org/10.1007/s11104-012-1402-5.

Bandyopadhyay, R., Ortega-Beltran, A., Akande, A., Mutegi, C., Atehnkeng, J., Kaptoge, L., et al., 2016. Biological control of aflatoxins in Africa: current status and potential challenges in the face of climate change. World Mycotoxin J. 9, 771–789. https://doi.org/10.3920/WMJ2016.2130.

Bashir, Z., 2017. Potassium solubilizing microorganisms: mechanism and diversity. Int. J. Pure Appl. Biosci. 5, 653–660. https://doi.org/10.18782/2320-7051.5446.

Bell, T.H., Hockett, K.L., Alcalá-Briseño, R.I., Barbercheck, M., Beattie, G.A., Bruns, M.A., et al., 2019. Manipulating wild and tamed phytobiomes: challenges and opportunities. Phytobiomes J. 3, 2471–2906. https://doi.org/10.1094/PBIOMES-01-19-0006-W.

Bellabarba, A., Fagorzi, C., DiCenzo, G.C., Pini, F., Viti, C., Checcucci, A., 2019. Deciphering the symbiotic plant microbiome: translating the most recent discoveries on rhizobia for the improvement of agricultural practices in metal-contaminated and high saline lands. Agronomy 9, 529. https://doi.org/10.3390/agronomy9090529.

Benedetti, M., Pontiggia, D., Raggi, S., Cheng, Z., Scaloni, F., Ferrari, S., et al., 2015. Plant immunity triggered by engineered in vivo release of oligogalacturonides, damage-associated molecular patterns. Proc. Natl. Acad. Sci. 112, 5533–5538. https://doi.org/10.1073/pnas.1504154112.

Berg, G., 2009. Plant-microbe interactions promoting plant growth and health: perspectives for controlled use of microorganisms in agriculture. Appl. Microbiol. Biotechnol. 84, 11–18. https://doi.org/10.1007/s00253-009-2092-7.

Bernal, P., Allsopp, L.P., Filloux, A., Llamas, M.A., 2017. The *Pseudomonas putida* T6SS is a plant warden against phytopathogens. ISME J. 11, 972–987. https://doi.org/10.1038/ismej.2016.169.

Bonfante, P., 2010. Plant-fungal interactions in mycorrhizas. In: Encyclopedia of Life Sciences. John Wiley & Sons, Ltd, Chichester, UK, https://doi.org/10.1002/9780470015902.a0022339.

Bonito, G., Hameed, K., Toome-Heller, M., Healy, R., Reid, C., Liao, H.L., et al., 2017. Atractiella rhizophila, sp. Nov., an endorrhizal fungus isolated from the Populus root microbiome. Mycologia 109, 18–26. https://doi.org/10.1080/00275514.2016.1271689.

Bulgarelli, D., Schlaeppi, K., Spaepen, S., van Themaat, E.V.L., Schulze-Lefert, P., 2013. Structure and functions of the bacterial microbiota of plants. Annu Rev Plant Biol 64, 807–838. https://doi.org/10.1146/annurev-arplant-050312-120106.

Cawoy, H., Bettiol, W., Fickers, P., Onge, M., 2011. Bacillus-based biological control of plant diseases. In: Pesticides in the Modern World - Pesticides Use and Management., https://doi.org/10.5772/17184.

Chen, Y., Wang, J., Yang, N., Wen, Z., Sun, X., Chai, Y., et al., 2018. Wheat microbiome bacteria can reduce virulence of a plant pathogenic fungus by altering histone acetylation. Nat. Commun. 9, 3429. https://doi.org/10.1038/s41467-018-05683-7.

Chialva, M., Salvioli di Fossalunga, A., Daghino, S., Ghignone, S., Bagnaresi, P., Chiapello, M., et al., 2018. Native soils with their microbiotas elicit a state of alert in tomato plants. New Phytol. 220, 1296–1308. https://doi.org/10.1111/nph.15014.

de Souza, R., Beneduzi, A., Ambrosini, A., da Costa, P.B., Meyer, J., Vargas, L.K., et al., 2013. The effect of plant growth-promoting rhizobacteria on the growth of rice (Oryza sativa L.) cropped in southern Brazilian fields. Plant Soil 366, 585–603. https://doi.org/10.1007/s11104-012-1430-1.

Dorner, J.W., 2004. Biological control of aflatoxin contamination of crops. J. Toxicol., Toxin Rev. 23, 425–450. https://doi.org/10.1081/TXR-200027877.

Dubuis, C., Keel, C., Haas, D., 2007. Dialogues of root-colonizing biocontrol pseudomonads. Eur. J. Plant Pathol. 119, 311–328. https://doi.org/10.1007/s10658-007-9157-1.

Elhady, A., Giné, A., Topalovic, O., Jacquiod, S., Sørensen, S.J., Sorribas, F.J., et al., 2017. Microbiomes associated with infective stages of root-knot and lesion nematodes in soil. PLoS One 12. https://doi.org/10.1371/journal.pone.0177145, e0177145.

Elhady, A., Adss, S., Hallmann, J., Heuer, H., 2018. Rhizosphere microbiomes modulated by pre-crops assisted plants in defense against plant-parasitic nematodes. Front. Microbiol. 9, 1133. https://doi.org/10.3389/fmicb.2018.01133.

Etesami, H., Mirseyed Hosseini, H., Alikhani, H.A., 2014. Bacterial biosynthesis of 1-aminocyclopropane-1-caboxylate (ACC) deaminase, a useful trait to elongation and endophytic colonization of the roots of rice under constant flooded conditions. Physiol. Mol. Biol. Plants 20, 425–434. https://doi.org/10.1007/s12298-014-0251-5.

Etesami, H., Emami, S., Alikhani, H.A., 2017. Potassium solubilizing bacteria (KSB): mechanisms, promotion of plant growth, and future prospects - a review. J. Soil Sci. Plant Nutr. 17, 897–911. https://doi.org/10.4067/S0718-95162017000400005.

Fahad, S., Hussain, S., Bano, A., Saud, S., Hassan, S., Shan, D., et al., 2015. Potential role of phytohormones and plant growth-promoting rhizobacteria in abiotic stresses: consequences for changing environment. Environ. Sci. Pollut. Res. 22, 4907–4921. https://doi.org/10.1007/s11356-014-3754-2.

Gaiero, J.R., McCall, C.A., Thompson, K.A., Day, N.J., Best, A.S., Dunfield, K.E., 2013. Inside the root microbiome: bacterial root endophytes and plant growth promotion. Am. J. Bot. 100, 1738–1750. https://doi.org/10.3732/ajb.1200572.

Gdanetz, K., Trail, F., 2017. The wheat microbiome under four management strategies, and potential for endophytes in disease protection. Phytobiomes J. 1, 158–168. https://doi.org/10.1094/PBIOMES-05-17-0023-R.

Geisen, S., Mitchell, E.A.D., Adl, S., Bonkowski, M., Dunthorn, M., Ekelund, F., et al., 2018. Soil protists: a fertile frontier in soil biology research. FEMS Microbiol. Rev. 42, 293–323. https://doi.org/10.1093/femsre/fuy006.

Glazebrook, J., 2005. Contrasting mechanisms of defense against biotrophic and necrotrophic pathogens. Annu. Rev. Phytopathol. 43, 205–227. https://doi.org/10.1146/annurev.phyto.43.040204.135923.

Glick, B.R., 2005. Modulation of plant ethylene levels by the bacterial enzyme ACC deaminase. FEMS Microbiol. Lett. 251, 1–7. https://doi.org/10.1016/j.femsle.2005.07.030.

Glick, B.R., 2012. Plant growth-promoting Bacteria: mechanisms and applications. Scientifica (Cairo) 2012, 963401. https://doi.org/10.6064/2012/963401.

Grayston, S.J., Germida, J.J., 1991. Sulfur-oxidizing bacteria as plant growth promoting rhizobacteria for canola. Can. J. Microbiol. 37, 521–529. https://doi.org/10.1139/m91-088.

Hacquard, S., Garrido-Oter, R., González, A., Spaepen, S., Ackermann, G., Lebeis, S., et al., 2015. Microbiota and host nutrition across plant and animal kingdoms. Cell Host Microbe 17, 603–616. https://doi.org/10.1016/j.chom.2015.04.009.

Hardoim, P.R., van Overbeek, L.S., Berg, G., Pirttilä, A.M., Compant, S., Campisano, A., et al., 2015. The hidden world within plants: ecological and evolutionary considerations for defining functioning of microbial endophytes. Microbiol. Mol. Biol. Rev. 79, 293–320. https://doi.org/10.1128/mmbr.00050-14.

Hartmann, A., Rothballer, M., Hense, B.A., Schröder, P., 2014. Bacterial quorum sensing compounds are important modulators of microbe-plant interactions. Front. Plant Sci. 5, 131. https://doi.org/10.3389/fpls.2014.00131.

Hassani, M.A., Durán, P., Hacquard, S., 2018. Microbial interactions within the plant holobiont. Microbiome 6, 58. https://doi.org/10.1186/s40168-018-0445-0.

Hosni, T., Moretti, C., Devescovi, G., Suarez-Moreno, Z.R., Fatmi, M.B., Guarnaccia, C., et al., 2011. Sharing of quorum-sensing signals and role of interspecies communities in a bacterial plant disease. ISME J. 5, 1857–1870. https://doi.org/10.1038/ismej.2011.65.

Hubbard, M., Germida, J., Vujanovic, V., 2012. Fungal endophytes improve wheat seed germination under heat and drought stress. Botany 90, 137–149. https://doi.org/10.1139/b11-091.

Jacobsen, B.J., Zidack, N.K., Larson, B.J., 2004. The role of *Bacillus*- based biological control agents in integrated Pest management systems: plant diseases. Phytopathology 94, 1272–1275. https://doi.org/10.1094/PHYTO.2004.94.11.1272.

Jones, J.D.G., Dangl, J.L., 2006. The plant immune system. Nature 444, 323–329. https://doi.org/10.1038/nature05286.

Kalayu, G., 2019. Phosphate solubilizing microorganisms: promising approach as biofertilizers. Int. J. Agron. 4917256. https://doi.org/10.1155/2019/4917256.

Kaloshian, I., Teixeira, M., 2019. Advances in plant-nematode interactions with emphasis on the notorious nematode genus *Meloidogyne*. Phytopathology 109, 1988–1996. https://doi.org/10.1094/PHYTO-05-19-0163-IA.

Kim, H., Lee, Y.H., 2020. The rice microbiome: a model platform for crop holobiome. Phytobiomes J. 4, 5–18. https://doi.org/10.1094/PBIOMES-07-19-0035-RVW.

Kumar, J., Singh, D., Ghosh, P., Kumar, A., 2017. Endophytic and epiphytic modes of microbial interactions and benefits. In: Plant-Microbe Interactions in Agro-Ecological Perspective, pp. 227–253, https://doi.org/10.1007/978-981-10-5813-4_12.

Kumari, B., Mani, M., Solanki, A.C., Solanki, M.K., Hora, A., Mallick, M.A., 2020. Phytobiome engineering and its impact on next-generation agriculture. In: Phytobiomes: Current Insights and Future Vistas. Springer Singapore, Singapore, pp. 381–403, https://doi.org/10.1007/978-981-15-3151-4_15.

Laluk, K., Mengiste, T., 2010. Necrotroph attacks on plants: wanton destruction or covert extortion? In: The Arabidopsis Book. 8., e0136.

Lay, C.Y., Bell, T.H., Hamel, C., Harker, K.N., Mohr, R., Greer, C.W., et al., 2018. Canola root-associated microbiomes in the Canadian prairies. Front. Microbiol. 9, 1188. https://doi.org/10.3389/fmicb.2018.01188.

Leach, J.E., Triplett, L.R., Argueso, C.T., Trivedi, P., 2017. Communication in the Phytobiome. Cell 169, 587–596. https://doi.org/10.1016/j.cell.2017.04.025.

Leonard, S.P., Powell, J.E., Perutka, J., Geng, P., Heckmann, L.C., Horak, R.D., et al., 2020. Engineered symbionts activate honey bee immunity and limit pathogens. Science 367, 573–576. https://doi.org/10.1126/science.aax9039.

Masson-Boivin, C., Sachs, J.L., 2018. Symbiotic nitrogen fixation by rhizobia—the roots of a success story. Curr. Opin. Plant Biol. 44, 7–15. https://doi.org/10.1016/j.pbi.2017.12.001.

Meena, V.S., Maurya, B.R., Verma, J.P., Meena, R.S., 2016. Potassium Solubilizing Microorganisms for Sustainable Agriculture. Springer India, New Delhi, p. 331, https://doi.org/10.1007/978-81-322-2776-2.

Mhlongo, M.I., Piater, L.A., Madala, N.E., Labuschagne, N., Dubery, I.A., 2018. The chemistry of plant–microbe interactions in the rhizosphere and the potential for metabolomics to reveal signaling related to defense priming and induced systemic resistance. Front. Plant Sci. 9, 112. https://doi.org/10.3389/fpls.2018.00112.

Moreira-Coello, V., Mouriño-Carballido, B., Marañón, E., Fernández-Carrera, A., Bode, A., Sintes, E., et al., 2019. Temporal variability of diazotroph community composition in the upwelling region off NW Iberia. Sci. Rep. 9, 3737. https://doi.org/10.1038/s41598-019-39586-4.

Moshynets, O.V., Babenko, L.M., Rogalsky, S.P., Iungin, O.S., Foster, J., Kosakivska, I.V., et al., 2019. Priming winter wheat seeds with the bacterial quorum sensing signal N-hexanoyl-L-homoserine lactone (C6-HSL) shows potential to improve plant growth and seed yield. PLoS One 14. https://doi.org/10.1371/journal.pone.0209460, e0209460.

Mousa, W.K., Shearer, C., Limay-Rios, V., Ettinger, C.L., Eisen, J.A., Raizada, M.N., 2016. Root-hair endophyte stacking in finger millet creates a physicochemical barrier to trap the fungal pathogen *fusarium graminearum*. Nat. Microbiol. 1, 16167. https://doi.org/10.1038/nmicrobiol.2016.167.

Müller, H., Berg, C., Landa, B.B., Auerbach, A., Moissl-Eichinger, C., Berg, G., 2015. Plant genotype-specific archaeal and bacterial endophytes but similar *Bacillus* antagonists colonize Mediterranean olive trees. Front. Microbiol. 6, 138. https://doi.org/10.3389/fmicb.2015.00138.

Nihorimbere, V., Ongena, M., Smargiassi, M., Thonart, P., 2011. Beneficial effect of the rhizosphere microbial community for plant growth and health. Biotechnol. Agron. Soc. Environ. 15, 327–337.

Parakhia, M.V., 2018. Manipulation of phytobiome: a new concept to control the plant disease and improve the productivity. J. Bacteriol. Mycol. 6, 322–324. https://doi.org/10.15406/jbmoa.2018.06.00227.

Pierik, R., Sasidharan, R., Voesenek, L.A.C.J., 2007. Growth control by ethylene: adjusting phenotypes to the environment. J. Plant Growth Regul. 26, 188–200. https://doi.org/10.1007/s00344-006-0124-4.

Pieterse, C.M.J., Van der Does, D., Zamioudis, C., Leon-Reyes, A., Van Wees, S.C.M., 2012. Hormonal modulation of plant immunity. Annu. Rev. Cell Dev. Biol. 28, 489–521. https://doi.org/10.1146/annurev-cellbio-092910-154055.

Qiu, Z., Egidi, E., Liu, H., Kaur, S., Singh, B.K., 2019. New frontiers in agriculture productivity: optimised microbial inoculants and in situ microbiome engineering. Biotechnol. Adv. 37, 107371. https://doi.org/10.1016/j.biotechadv.2019.03.010.

Remus-Emsermann, M.N.P., Tecon, R., Kowalchuk, G.A., Leveau, J.H.J., 2012. Variation in local carrying capacity and the individual fate of bacterial colonizers in the phyllosphere. ISME J. 6, 756–765. https://doi.org/10.1038/ismej.2011.209.

Roossinck, M.J., 2019. Viruses in the phytobiome. Curr. Opin. Virol. 5, 93–111. https://doi.org/10.1016/j.coviro.2019.06.008.

Ruggiero, M.A., Gordon, D.P., Orrell, T.M., Bailly, N., Bourgoin, T., Brusca, R.C., et al., 2015. A higher level classification of all living organisms. PLoS One 10. https://doi.org/10.1371/journal.pone.0119248, e0130114.

Santos, M.S., Nogueira, M.A., Hungria, M., 2019. Microbial inoculants: reviewing the past, discussing the present and previewing an outstanding future for the use of beneficial bacteria in agriculture. AMB Express 9, 205. https://doi.org/10.1186/s13568-019-0932-0.

Sasse, J., Martinoia, E., Northen, T., 2018. Feed your friends: do plant exudates shape the root microbiome? Trends Plant Sci. 23, 25–41. https://doi.org/10.1016/j.tplants.2017.09.003.

Schaefer, A.L., Lappala, C.R., Morlen, R.P., Pelletier, D.A., Lu, T.Y.S., Lankford, P.K., et al., 2013. LuxR- and luxI-type quorum-sensing circuits are prevalent in members of the *Populus deltoides* microbiome. Appl. Environ. Microbiol. 79, 5745–5752. https://doi.org/10.1128/AEM.01417-13.

Schlaeppi, K., Bulgarelli, D., 2015. The plant microbiome at work. Mol. Plant-Microbe Interact. 28, 212–217. https://doi.org/10.1094/MPMI-10-14-0334-FI.

Sekar, J., Prabavathy, V.R., 2014. Novel Phl-producing genotypes of finger millet rhizosphere associated pseudomonads and assessment of their functional and genetic diversity. FEMS Microbiol. Ecol. 89, 32–46. https://doi.org/10.1111/1574-6941.12354.

Senghor, L.A., Ortega-Beltran, A., Atehnkeng, J., Callicott, K.A., Cotty, P.J., Bandyopadhyay, R., 2020. The Atoxigenic biocontrol product Aflasafe SN01 is a valuable tool to mitigate aflatoxin contamination of both maize and groundnut cultivated in Senegal. Plant Dis. 104, 510–520. https://doi.org/10.1094/PDIS-03-19-0575-RE.

Senthilkumar, M., Amaresan, N., Sankaranarayanan, A., 2021. Isolation of Sulfur-Oxidizing Bacteria. pp. 69–70, https://doi.org/10.1007/978-1-0716-1080-0_14.

Shamseldin, A., Abdelkhalek, A., Sadowsky, M.J., 2017. Recent changes to the classification of symbiotic, nitrogen-fixing, legume-associating bacteria: a review. Symbiosis 71, 91–109. https://doi.org/10.1007/s13199-016-0462-3.

Shen, H., He, X., Liu, Y., Chen, Y., Tang, J., Guo, T., 2016. A complex inoculant of N2-fixing, P- and K-solubilizing bacteria from a purple soil improves the growth of kiwifruit (*Actinidia chinensis*) plantlets. Front. Microbiol. 7, 841. https://doi.org/10.3389/fmicb.2016.00841.

Shikano, I., Rosa, C., Tan, C.W., Felton, G.W., 2017. Tritrophic interactions: microbe-mediated plant effects on insect herbivores. Annu. Rev. Phytopathol. 55, 313–331. https://doi.org/10.1146/annurev-phyto-080516-035319.

Shukla, N., Kaur, P., Kumar, A., 2016. Molecular aspects of plant-nematode interactions. Indian J Plant Physiol 21, 477–488. https://doi.org/10.1007/s40502-016-0263-y.

Singh, S.P., Gaur, R., 2016. Evaluation of antagonistic and plant growth promoting activities of chitinolytic endophytic actinomycetes associated with medicinal plants against *Sclerotium rolfsii* in chickpea. J. Appl. Microbiol. 121, 506–518. https://doi.org/10.1111/jam.13176.

Singh, S.P., Gaur, R., 2017. Endophytic *Streptomyces* spp. underscore induction of defense regulatory genes and confers resistance against *Sclerotium rolfsii* in chickpea. Biol. Control 7, 1079–1090. https://doi.org/10.1016/j.biocontrol.2016.10.011.

Singh, S.P., Bhattacharya, A., Gupta, R., Mishra, A., Zaidi, F.A., Srivastava, S., 2020a. Endophytic Phytobiomes as defense elicitors. In: Solanki, M., Kashyap, P., Kumari, B. (Eds.), Phytobiomes: Current Insights and Future Vistas. Springer, Singapore, https://doi.org/10.1007/978-981-15-3151-4_12.

Singh, M.P., Singh, P., Singh, R.K., Solanki, M.K., Bazzer, S.K., 2020b. Plant microbiomes: Understanding the aboveground benefits. In: Solanki, M., Kashyap, P., Kumari, B. (Eds.), Phytobiomes: Current Insights and Future Vistas. Springer, Singapore, https://doi.org/10.1007/978-981-15-3151-4_12.

Srivastava, J.K., Chandra, H., Kalra, S.J.S., Mishra, P., Khan, H., Yadav, P., 2017. Plant-microbe interaction in aquatic system and their role in the management of water quality: a review. Appl Water Sci. https://doi.org/10.1007/s13201-016-0415-2.

Tortora, M.L., Díaz-Ricci, J.C., Pedraza, R.O., 2011. *Azospirillum brasilense* siderophores with antifungal activity against *Colletotrichum acutatum*. Arch. Microbiol. 193, 275–286. https://doi.org/10.1007/s00203-010-0672-7.

Vincent, D., Rafiqi, M., Job, D., 2020. The multiple facets of plant–fungal interactions revealed through plant and fungal Secretomics. Front. Plant Sci. 10, 1626. https://doi.org/10.3389/fpls.2019.01626.

Vishwakarma, K., Kumar, V., Tripathi, D.K., Sharma, S., 2018. Characterization of rhizobacterial isolates from Brassica juncea for multitrait plant growth promotion and their viability studies on carriers. Environ. Sustain. 1, 253–265. https://doi.org/10.1007/s42398-018-0026-y.

Wang, N., Jin, T., Trivedi, P., Setubal, J., Tang, J., Machado, M., Triplett, E., Coletta-Filho, H., Cubero, J., Deng, X., Wang, X., Zhou, C., Ancona, V., Lu, Z., Dutt, M., Borneman, J., Rolshausen, P., Rope, Y., 2015. Announcement of the international citrus microbiome (phytobiome) consortium. J. Citrus Pathol. 2.

Wang, X., Wei, Z., Yang, K., Wang, J., Jousset, A., Xu, Y., et al., 2019. Phage combination therapies for bacterial wilt disease in tomato. Nat. Biotechnol. 37, 1513–1520. https://doi.org/10.1038/s41587-019-0328-3.

Wei, Y., Zhao, Y., Shi, M., Cao, Z., Lu, Q., Yang, T., et al., 2018. Effect of organic acids production and bacterial community on the possible mechanism of phosphorus solubilization during composting with enriched phosphate-solubilizing bacteria inoculation. Bioresour. Technol. 247, 190–199. https://doi.org/10.1016/j.biortech.2017.09.092.

Williamson, V.M., Gleason, C.A., 2003. Plant–nematode interactions. Curr. Opin. Plant Biol. 6, 327–333. https://doi.org/10.1016/S1369-5266(03)00059-1.

Wilson, R.H., Alonso, H., Whitney, S.M., 2016. Evolving *Methanococcoides burtonii* archaeal rubisco for improved photosynthesis and plant growth. Sci. Rep. 6, 22284. https://doi.org/10.1038/srep22284.

Xia, Y., Lü, C., Hou, N., Xin, Y., Liu, J., Liu, H., et al., 2017. Sulfide production and oxidation by heterotrophic bacteria under aerobic conditions. ISME J 11, 2754–2766. https://doi.org/10.1038/ismej.2017.125.

Xiong, W., Song, Y., Yang, K., Gu, Y., Wei, Z., Kowalchuk, G.A., et al., 2020. Rhizosphere protists are key determinants of plant health. Microbiome 8, 27. https://doi.org/10.1186/s40168-020-00799-9.

Zhou, C., Ma, Z., Zhu, L., Xiao, X., Xie, Y., Zhu, J., et al., 2016. Rhizobacterial strain *Bacillus megaterium* BOFC15 induces cellular polyamine changes that improve plant growth and drought resistance. Int. J. Mol. Sci. 17, 976. https://doi.org/10.3390/ijms17060976.

An overview of microbial diversity under diverse ecological niches in northeast India

Krishna Giri[a], Bhanushree Doley[a], Gaurav Mishra[a], Deep Chandra Suyal[b], Rupjyoti C. Baruah[a], and R.S.C. Jayaraj[a]

[a]*Rain Forest Research Institute, Jorhat, Assam, India,* [b]*Department of Microbiology Akal College of Basic Sciences, Eternal University, Sirmour, Himachal Pradesh, India*

3.1 Introduction

The northeast (NE) India is situated at 25.5736°N latitude to 93°2473°E longitude and constitutes eight states, viz., Arunachal Pradesh, Assam, Manipur, Meghalaya, Mizoram, Nagaland, Sikkim, and Tripura. The region is represented by rich biodiversity of the Himalayas and the Indo-Burma region covering almost 8% of total geographical area of the country. Because of a wide range of physiographic divisions and altitudinal variation, NE India is the home for various endemic flora and fauna and also known for diverse cultures and ethnic groups (Giri, 2019). The diverse physiographic features provide suitable habitats to microorganisms.

This variation, in turn, supports a wide variety of habitats involving the colonization of microorganisms. Microorganisms inhabiting varying environments have a wide-scale bioprospecting potential. Particularly, enzymes produced by these microbes have potential contributions in agricultural, industrial, and health sectors. Exploitation of microbial diversity (bacteria, fungi, and actinomycetes), which are mostly inhabitants of the soil environment, has an important role in industries, agriculture, and economic development (Hassan, 2015). Microbes play a vital role in multiple ecosystem services such as decomposition of organic matter, biogeochemical cycling, biostimulation, plant growth promotion, biocontrol of pathogens, and ecosystem productivity in a sustainable manner. At the same time, they can be used as an alternative to chemical fertilizers in the form of biofertilizer formulations (Gortari et al., 2020). Microbial processes of soil organic matter mineralization improve soil physical properties and mitigate heavy metal toxicity (Hassan et al., 2017). Thus, microbial diversity is known as the key driver of ecosystem functioning through numerous essential processes in the biosphere. Ecological niches such as hot water spring to the wettest place, vegetation ranging from tropical to temperate, flood plains of the mighty Brahmaputra River to high altitudinal areas have been explored to document microbial diversity of Indo-Burma biodiversity hotspot. However, a few studies have been undertaken on microbial diversity from NE India, and the information is relatively lesser than that of inland desert or coastal locations (Vishniac, 1993). Unfortunately, climate change, erratic precipitation patterns, unsustainable land-use practices, and land degradation have threatened the microbial diversity associated with the pristine ecological niches. Attempts have been made to explore soil microbial biodiversity using both conventional and next-generation microbiome sequencing methods. Culture-based approach for soil bacteria identification requires special growth conditions such as culture media and incubation periods (Kirk et al., 2004; Vartoukian et al., 2010). Development of high-throughput metagenome sequencing methods has enabled the microbiologists to explore uncultivable soil microbial communities (Ellis et al., 2003; Fierer et al., 2012), which could possibly further strengthen the understanding of unstudied and unexplored microbes. In view of the paramount significance of soil microbes in ecological processes, the studies carried out across the NER region targeting diverse ecological niches have been compiled in this chapter; few of these studies have been able to describe novel genera, species, and strains of ecological and economic importance. The literature synthesis on microbial diversity in NER will not only present a detailed account of microbial resources but also identify the areas of further research to explore them for sustainable development of the region. The most studied environments and specific ecological niches therein are depicted in Fig. 3.1.

3.2 Microbial diversity in different ecological niches

3.2.1 Bacterial diversity in hot water springs

Hot water springs and hydrothermal vents are the ecological niche for extremophilic microbes that can thrive at very high temperatures and other extremities. These habitats have been the area of interest for microbiologists since the discoveries of microorganisms because the organisms thriving in such hostile environments have unique adaptive mechanism and ecological importance. Moreover, these organisms

Fig. 3.1 Ecological niches and associated microbial diversity in northeast India.

possess industrially important enzymes. For example, the heat-stable enzyme *Taq polymerase* was isolated from the extremophile *Thermus aquaticus*, inhabitant of hot springs. A hot water spring of Assam was explored to decipher bacterial diversity using polyphasic approach, viz., biochemical, physiological, chemotaxonomic, and phylogenetic analyses. The study reported that bacterial isolate *Aquimonas voraii* gen. Nov., sp. nov., belongs to a novel genus *Aquimonas*, which was quite different from other members of genera γ-proteobacteria. 16S rRNA sequence similarities of *A. voraii* 20 (T) were found to be less than 91.8% with other members of the class γ-proteobacteria. Hence, the bacterium was classified under a new genera and species, represented by type strain GPTSA 20(T) = MTCC 6713 (T) = JCM 12896(T) (Saha et al., 2005a). Similarly, bacterial isolate GPTSA 11^T was characterized as *Paenibacillus assamensis* sp. nov., having maximum 95.85% sequence similarity with *Paenibacillus apiarius* and less than 95% similarity with *Paenibacillus alvei*, *Paenibacillus cineris*, *Paenibacillus favisporus*, *Paenibacillus chibensis*, and *Paenibacillus azoreducens* belonging to phylum Firmicutes. The type strain of this bacterial isolate was designated as GPTSA 11^T (= MTCC 6934^T = JCM 13186^T) (Saha et al., 2005b). Identification of GPTSA-6^T, a Gram-negative, facultative anaerobic bacteria as *Aeromonas sharmana* sp. nov., from hot water environment was also a novel species discovery representing genus *Aeromonas* and phylum proteobacteria. 16S rRNA gene sequences of this strain revealed 99.23% similarity with an uncultured bacterial clone, A-8, and 95.13% with a cultured bacterium *Aeromonas sobria* ATCC 43979T. This bacterial isolate was named after Dr. Manju Sharma, a renowned microbiologist of India, and the type strain of this novel species was designated as GPTSA-6T (= MTCC 7090T = DSM 17445T) (Saha and Chakrabarti, 2006a). The study on hot water spring microbiology also reported *Flavobacterium indicum* sp. nov., a Gram-negative, strictly aerobic rod-shaped, oligotrophic bacterium. The bacterial strain GPTSA100-9^T was close descent of Cytophaga, Fusobacterium, and Bacteroides (CFB) group bacterium, A0653 (AF236016), having 93.4% similarity with *Flexibacter aurantiacus* subsp. *excathedrus* followed by 93.2%–92.0% with *Flavobacterium saliperosum*, *Flavobacterium soli*, *Flavobacterium aquatile*, and *Flavobacterium columnare*. Based on the genetic characterization, the bacterial strain GPTSA100-9^T was reported to be a novel species under genus Flavobacterium, type strain GPTSA100-9T (= MTCC 6936^T = DSM 17447^T) (Saha and Chakrabarti, 2006b).

Similarly, Saikia et al. (2011) isolated a bacterium BPM3 from hot water spring of Nambor Wildlife Sanctuary, Assam. The isolate was characterized using morphological, biochemical, and molecular methods and identified to be *Brevibacillus laterosporus* belonging to the phylum Firmicutes. This bacterial strain was evaluated for biocontrol of *Fusarium oxysporum* f. sp. *ciceri*, *F. semitectum*, *Magnaporthe grisea*, *Rhizoctonia oryzae*, and *Staphylococcus aureus* and found effective at 30°C, 8.5 pH. The results of greenhouse experiment revealed that *B. laterosporus* was able to suppress 30%–67% rice blast disease and protected the weight loss by 35%–56.5%. Moreover, this bacterium also showed strong antifungal and antibacterial activities, indicating the presence of C–H, carbonyl group, dimethyl group, –CH (2), and methyl group. Najar et al. (2018) reported *Geobacillus yumthangensis* sp. nov, a thermophilic bacterium for the first time from Yumthang hot spring of Sikkim. The thermostable characteristics of *Geobacillus* members make them attractive to the biotechnology industry as sources of thermostable enzymes (Champdore et al., 2007). An another study on Yumthang hot spring in north Sikkim by Panda et al. (2017) using Illumina sequencing reported the presence of Proteobacteria (54.33%), *Actinobacteria* (32.19%), *Firmicutes* (6.03%), *Bacteroidetes* (2.87%), and unclassified bacteria (2.91%) from the total reads. The study further opined that many bacterial and archaeal sequences are still unclassified, indicating the possibilities of novel microorganism possessing industrially desirable traits.

3.2.2 Metagenome analysis of Phumdi (a floating island) in Loktak Lake, Manipur

Metagenome analysis of Phumdi (a floating island) in Loktak Lake, Manipur, was performed to characterize soil bacterial diversity inhabiting the fresh water lake. The Phumdi was dominated by the phylum Proteobacteria (51%), Acidobacteria (10%), Actinobacteria (9%), and Bacteroidetes (7%). Upon comparing the data with four other aquatic habitats, *Candidatus solibacter, Bradyrhizobium, Candidatus koribacter, Pedosphaera, Methylobacterium, Anaeromyxobacter, Sorangium, Opitutus,* and *Acidobacterium* genera were reported as selectively dominant at this habitat (Puranik et al., 2016).

3.2.3 Microbial diversity in forest ecosystem

A forest ecosystem is a functional unit consisting of soil, trees, insects, animals, birds, and man as its interacting units. The forest is a large and complex ecosystem, which harbors rich diversity of flora and fauna as compared to the other ecosystems. The forest ecosystem provides several ecological niches to soil biota, viz., bulk soil, plant rhizosphere, leaf litter, and dead debris for their colonization and subsequent proliferation. Northeast India harbors tropical, subtropical, and temperate forest types due to vast physiographic and pedoclimatic variations in the region. These variations make the region a natural refuge to various living organisms. The forest ecosystem may provide tremendous opportunity for researchers to explore beneficial microbes that may pave the way to discoveries of novel compounds for mankind. Due to anthropogenic effect, soil microbes of this region are affected, resulting in alterations of soil nutrient status. Studies on soil fungi isolated from forests of different stages of regeneration from different altitudes showed variations in population number and enzyme activities. Dibru-Saikhowa Biosphere Reserve in Assam is considered as home for many endangered species of flora and fauna. A study carried out by Das et al. (2013) reported 26 fungal genera dominated by *Aspergillus, Penicillium, Trichoderma, Curvularia,* and *Rhizopus* from different sites of the Biosphere Reserve. The study also observed that the topsoil layer (0–10 cm) had greater microbial populations than the subsoil layers (10–20) and (20–30 cm) except during rainy season. The higher bacterial and fungal population was attributed to more nutrient availability. Seasonal variation was found to influence microbial population with the maximum count during spring and minimum in winter. Fungal population and their species composition have been reported to be influenced by the soil depth, seasonal variation, organic fertilizers, and vermicompost (Bhattacharyya and Jha, 2011; Swer et al., 2011). Similarly, a study on fungal population and enzyme assay carried out in northeast India reported higher dehydrogenase, urease, and phosphatase activities, indicating sound soil health and microbial population. The study explored *Aspergillus, Chromocleista, Penicillium, Trichoderma, Talaromyces, Nectria, Fusarium, Hypocrea, Pleurostomophora, Cladosporium, Mortierella, Thysanophora,* and *Chamaeleomyces* from phylum Ascomycota and

Zygomycota corresponding to seven orders (Eurotiales, Hypocreales, Calosphaeriales, Capnodiales, Pleosporales, Mucorales, and Mortierellales) and Incertae sedis in which Eurotiales as well as Hypocreales were designated as the most varied and abundant group of fungi along the entire altitudinal stretch in the eastern Himalayas (Devi et al., 2012).

The recent studies have found that the potent human pathogens such as *S. aureus* and *Pseudomonas aeruginosa* are becoming resistant to nearly all antibiotics forcing researchers to put more efforts in discovery of new antimicrobial compounds from microorganisms inhabiting different ecological niches. *Streptomycetes*, a group of Gram-positive soil actinobacteria, are well known for antibiotic production in the microbial world. This group of actinobacteria provided more than half of the naturally occurring antibiotics discovered to date. Actinobacteria from poorly explored habitats proved to be a potential producer of novel biological compounds. In view of this, Sharma and Thakur (2020) isolated 107 actinobacteria from different sites of Pobitora Wildlife Sanctuary and Kaziranga National Park of Assam, India, with a promising antimicrobial activity. The observations revealed that the isolated actinobacteria belong to the genus *Streptomyces*, *Nocardia*, and *Kribbella*, of which 51 isolates exhibited an antagonistic action against *S. aureus* MTCC 96, 49 isolates against *S. aureus* MRSA ATCC 43300, 59 isolates against *Escherichia coli* MTCC 40, and 60 isolates against *Candida albicans* MTCC 227. Among the 77 antagonistic actinobacteria, 24 isolates produced amylase, cellulase, protease, lipase, and esterase enzymes.

Saikia and Joshi (2012) reported the reduced population of fungal community in degraded soil of Cherrapunji than sacred groves. *Penicillium perpurogenum*, *Aspergillus* sp., *Fusarium* sp., and *Trichoderma* sp. were reported from both the sites. The study found that the management practices in the region influenced soil microbial population in different environmental conditions. The effect of deforestation, land degradation, organic carbon, pH, and temperature on bacterial and fungal colony-forming units (CFU) was studied in the conserved sacred forest, highly degraded, and highest rainfall receiving soil system of Cherrapunji, Meghalaya. The studied environments revealed a variation in soil physicochemical properties and bacterial population. The CFU count per gram soil in sacred forest was found to be higher than in the degraded land. A total of 39 bacterial isolates (26 from sacred forest and 13 from degraded land) were identified following Bergey's Manual of Systematic Bacteriology. Among the identified bacteria, *Bacillus* was the most dominating genus in sacred grove. The other bacterial genera were *Staphylococcus*, *Serratia*, *Pseudomonas*, and *Chromobacterium* (Joshi et al., 2009). A study from Eastern Himalayas was carried out to explore soil actinobacteria using the culture-dependent method, and morphological, biochemical, and molecular approaches of identification reported six actinomycetes *Streptomyces aureofaciens* (GL2), *S. chattanoogensis* (GP4), *S. niveoruber* (MA1), *S. cacaoi* subsp. *asoensis* (MB2), *S. galbus* (NG4) and *S. griseoruber* (NG5) (Bhattacharjee et al., 2012). Similarly, soil samples from altitudinal ranges of 34–1000, 1001–2000, 2001–3000, and 3001–3990 m a.s.l. in eastern Himalayan environment were studied for bacterial diversity analysis. Bacteria were initially isolated using the serial dilution plating method, and the representative isolates were identified through 16S rDNA gene sequencing. Firmicutes, Proteobacteria, and Bacteroidetes were the common phyla with *Bacillus* and *Pseudomonas* being the most abundant genera (Lyngwi et al., 2012).

About 99% microorganisms inhabiting different environments are not cultivable on artificial growth medium; thus, culture-independent metagenome sequencing for genes of interest has been developed to overcome this limitation, which has changed the scenario of environmental microbiology (Suenaga, 2012). This discipline has its roots in the analysis of 16S rRNA genes extracted from the environmental samples (Olsen et al., 1986). Instead of cataloguing only rRNAs, next-generation sequencing technique is capable of handling genomic DNA retrieved directly from environmental samples and referred to as metagenomics (Handelsman et al., 1998). Metagenome sequencing and denaturing gradient gel electrophoresis (DGGE) profiling of *Rhododendron arboreum* rhizospheric soil samples from Tawang, Arunachal Pradesh, characterized Actinobacteria (42.6%) as the most dominant phyla followed by Acidobacteria (24.02%), α- and β-proteobacteria (16%), and Chloroflexi (1.48%) in the core microbiome. Moreover, unique phylotypes of *Bradyrhizobium* and uncultured Rhizobiales were also found in significant proportions (Table 3.1). DGGE fingerprinting of rhizosphere soil revealed distinct profiles and higher band intensity as compared to the bulk soil 16S DNA fragments depicting higher diversity and abundance in rhizosphere of *R. arboreum* (Debnath et al., 2016).

A similar study using Illumina-based analysis of V4 region of bacterial 16S rRNA genes from Murlen National Park, Mizoram, reported 29 bacterial phyla. Illumina MiSeq platform and QIIME data analysis package were employed for forest microbiome analysis (De Mandal et al., 2015) and revealed Acidobacteria (39.45%) as the dominant phylum. The other major phyla reported were Proteobacteria (26.95%), Planctomycetes (7.81%), Actinobacteria (7.18%), Bacteroidetes (6.65%), Chloroflexi (4.11%), and Nitrospirae (3.33%). About 0.51% reads were not classified at the phylum level, and 84.44% reads could not be identified at the genus level, which revealed the uniqueness and unidentified bacterial community structure having novel bacterial populations with some unique properties (De Mandal et al., 2015).

Table 3.1 Summary of microbial diversity in the North-eastern region of India.

S. no.	Geographical location (latitude/ longitude)	Altitude/climate type	Source (rhizospheric/nonrhizospheric/forest/etc.) include plant name in case of rhizospheric samples	Type of study (culture dependent/independent)	Methodology used	Identified microbes	References
1	Upper and Central Brahmaputra Valley Zone of Assam	Tropical climate	Rhizosphere of rice grown in acidic soils of Assam	Culture dependent	Rhizobacteria were isolated using selective culture media. Identification of the selected isolates up to species level was carried out by Microbial Type Culture Collection (MTCC) at IM-Tech, Chandigarh	*Bacillus circulans, Streptomyces anthocysnicus, Bacillus sp., Pseudomonas aeroginosa, Bacillus pantothenticus, Pseudomonas pieketti, Bacillus megaterium, A. brasilense, A. amazonense, P. fluorescence*	Thakuria et al. (2004)
2	26°23′N and 93°52′E Assam	Elevation-111 m a.s.l. Temperature: 38°C. Tropical climate	Hot water spring sample	Culture dependent	Dilution-plating method using tryptic soy broth agar (TSBA composition TSB- 3% and agar 1.5%). The strain was identified using morphological and 16S rRNA gene sequencing	Phylum: γ- proteobacteria. Gram-variable, strictly aerobic, sporulating motile rods *Aquimonasvoraii* gen. Nov., sp. nov.	Saha et al., (2005a)

3	26°23′N and 93°52′E Assam	Elevation-111 m a.s l. Temperature: 38°C. Tropical climate	Hot water spring sample	Culture dependent	Dilution-plating method using tryptic soy broth agar (TSBA composition TSB-3% and agar 1.5%). The strain was identified using morphological and 16S rRNA gene sequencing	Phylum: Firmicutes aerobic, mesophilic gram variable, endospore-forming bacterium, *Paenibacillusassamensis* sp. nov. type strain GPTSA 11^T	Saha et al. (2005b)
4	26°23′N and 93°52′E Assam	Elevation-111 m a.s l. Temperature: 38°C. Tropical climate	Hot water spring sample	Culture dependent	Dilution-plating method using tryptic soy broth agar (TSBA composition TSB-3% and agar 1.5%). The strain was identified using phenotypic, chemotaxonomic, and phylogenetic analysis	Phylum: Proteobacteria Gram-negative, facultative anaerobic bacterium *Aeromonas sharmana* sp. nov. Type strain GPTSA-6^T	Saha and Chakrabarti (2006a)
5	26°23′N and 93°52′E Assam	Elevation-111 m a.s l. Temperature: 38°C. Tropical climate	Hot water spring sample	Culture dependent	Serial dilution plating on nutritionally poor medium, TSBA100 tryptic soy broth (Difco). The strain was identified using polyphasic taxonomic approach	Phylum: Bacteroidetes Gram-negative, strictly aerobic rods *Flavobacteriumindicum* sp. nov. Type strain GPTSA100-9^T	Saha and Chakrabarti (2006b)

Continued

Table 3.1 Summary of microbial diversity in the North-eastern region of India—cont'd

S. no.	Geographical location (latitude/ longitude)	Altitude/climate type	Source (rhizo-spheric/nonrhi-zospheric/forest/ cold desert/etc.) include plant name in case of rhizospheric samples	Type of study (culture dependent/ independent)	Methodology used	Identified microbes	References
6	25°15′N and 91°43′E Meghalaya	Elevation-1400 m a.s.l. Mawsynram Average annual rainfall 10,000 mm. Average monthly temperature ranges from around 11°C in January to just above 20°C in August	Soil samples of sacred forests and degraded areas of Cherrapunjee, Meghalaya, India	Culture dependent	Serial-dilution plating method using 0.85% normal physiological saline as diluents. Bacterial population was counted as colony-forming units per gram of dry soil (CFU/g)	Phylum: Firmicutes, *Bacillus* sp. and *Staphylococcus* sp. Phylum γ – proteo-bacteria: *Serratia* and *Pseudomonas sp.* Phylum β-protebacteria: *Chromobacterium* sp.	Joshi et al. (2009)
7	26°40′N and 92°58′E. Nameri forest soil, Assam	The average annual rainfall varies from 670 to 1100 mm, and the mean minimum and maximum soil temperatures range from 17 to 36°C. Humid tropical climate	Soil was sampled from three different spots at six depths: OA (1–9 cm), B (10–15 cm), C (16–30 cm), D (31–50 cm), E (51–100 cm), and F (101–200 cm)	Culture dependent	Fungal isolates were characterized based on cultural and morphological char-acteristics of spore and hyphae mounted in lactophenol and identified by taxo-nomic monographs (Barnett and Hunter, 1972; Domesch et al., 1980; Gilman, 1957; Subramanian, 1971)	Twenty-one fungal species belonging to 14 genera were recovered. *Aspergillus flavus* (8.4%) was dominant followed by *Penicillium chrysogenum* (8.0%) and lowest by *Rhizopus oryzae, R. nodosus,* and *Trichophyton* sp. (2.8% each). Phycomycetes (80.1%) were dominant in the study site followed by Zygomycetes (14.1%) and Ascomycetes (3.7%)	Bhattacharyya and Jha (2011)

8	27°6′N and 93°49′E longitude Papum Pare District of Arunachal Pradesh	Altitude: 392 m above mean sea level. Soil temperature: 22°C, average annual rainfall: 2782 mm and moisture content: 35%	Soil samples were collected after the completion of fire operation under jhum cultivation	Culture dependent	Phenotypic characteristic was observed on Tryptone yeast extract agar plates followed by 48-h incubation at 25°C, and microbes were identified using 16S rDNA	16S rRNA analysis revealed their maximum similarity with *Bacillus clausii*, *B. licheniformis*, *B. megaterium*, *B. subtilis*, *B. thuringiensis*, *P. aeruginosa* and *P. stutzeri*	Pandey et al. (2011)
9	27°6′N and 93°49′E Banderdewa forest reserve in the Papum Pare District of Arunachal Pradesh	Elevation 350 m a.s.l. Average annual maximum temperature 26°C and average total annual rainfall 2609 mm. Vegetation type: evergreen to semi evergreen mixed natural forests	Surface and subsurface soil samples from degraded forest (DF) and moderately degraded forest (MDF) due to shifting cultivation practice as well as undegraded natural forest (UDF)	Culture independent	Denaturation gradient gel electrophoresis (DGGE) profiling of the polymerase chain reaction (PCR) amplified 16S rDNA fingerprints	The cluster analysis of the DGGE bands of 16S rDNA fragments revealed a significant variation in bacterial community DNA content in degraded and moderately degraded forest sites than in undegraded sites. Shifting cultivation was found to be adversely affecting the bacterial diversity in the soil	Singh et al. (2011)

Continued

Table 3.1 Summary of microbial diversity in the North-eastern region of India—cont'd

S. no.	Geographical location (latitude/ longitude)	Altitude/climate type	Source (rhizospheric/nonrhizospheric/forest/ cold desert/etc.) include plant name in case of rhizospheric samples	Type of study (culture dependent/ independent)	Methodology used	Identified microbes	References
10	Jorhat (26°43′03.8″N-94°11′40.2″E), Tinsukia (27°20′34.5″N 110 95°42′33.2″E) and Lakhimpur (26°57′38.5″N 93°51′53.5″E) districts of Assam	Tropical climate	Bacterial endophytes were isolated from leaf, stem, and root section of both cultivated and wild rice plants	Culture dependent	The isolates were characterized using biochemical and molecular methods	Major phyla: Firmicutes (57.1%), Actinobacteria (20.0%), and Proteobacteria (22.8%) Genera: *Bacillus, Brevibacillus, Lysenibacillus, Microbacterium, Microbacteriaceae, Staphylococcus, Pantoea, Burkholderia, Acinetobacter, Pseudomonas,* and *Ralstonia* were isolated from cultivated rice However, *Bacillus, Stenotrophomonas, Microbacterium, Cellulosimicrobium, Proteus, Staphylococcus, Erwinia, Ochrobactrum,* and *Enterobacter* were isolated from wild rice	Kamala and Indira (2011)

11	26°23′N and 93°52′E Nambor Wild Life Sanctuary, Assam	Elevation-111 m a.s.l. Tropical climate	Mud samples of natural hot water spring	Culture dependent	Serial dilution of mud sample suspended in 100 mL physiological water (NaCl 9 g/L) followed by incubation at 28°C with shaking at 200 rpm for 30 min. The bacterium was characterized by morphological, biochemical, and molecular methods	Phylum: Firmicutes *Brevibacillus laterosporus*, strain BPM3	Saikia et al. (2011)
12	26°55′N to 28°40′N and 92°40′E to 94°21′E northeast	Eastern Himalayan range	Soil samples (5–15 cm depth) collected from altitudinal gradient of northeast India	Culture dependent	Serial dilution followed by the spread plate method using starch casein agar, Actinomycetes isolation agar, ISP-2 and ISP-3 medium. Morphological, biochemical, and 16S rRNA gene amplification and sequencing	Phylum actinobacteria: *Streptomyces aureofaciens* GL2, *S. chattanoogensis* GP4, *S. niveoruber* MA1, *S. cacaoi* subsp. *Asoensis* MB2, *S. galbus* NG4 and *S. griseoruber* NG5	Bhattacharjee et al. (2012)

Continued

Table 3.1 Summary of microbial diversity in the North-eastern region of India—cont'd

S. no.	Geographical location (latitude/ longitude)	Altitude/climate type	Source (rhizospheric/nonrhizospheric/forest/ cold desert/etc.) include plant name in case of rhizospheric samples	Type of study (culture dependent/ independent)	Methodology used	Identified microbes	References
13	22°3′0N and 89°97′E.northeast	Altitude from 24 m above sea level to 2000 m above sea level	Soil samples were collected from various vegetational and climatic zones spread over different altitudes of NE India under Eastern Himalayan range	Culture dependent	Molecular characterization of the isolates was done by PCR amplification of 18S rDNA using universal primers	A total of 107 isolates were characterized belonging to the phyla Ascomycota and Zygomycota, corresponding to seven orders (Eurotiales, Hypocreales, Calosphaeriales, Capnodiales, Pleosporales, Mucorales, and Mortierellales) and Incertae sedis. Species of *Penicillium* and *Aspergillus* were found to have the highest diversity index followed by Talaromyces and Fusarium	Devi et al. (2012)

14	25°17'N–25°18'N and 91°41'E–91°42'E Cherrapunjee, Meghalaya	Average annual rainfall and temperature 12,000 mm and 12–15°C, respectively	Biofilms substrata in water bodies of Cherrapunjee, Meghalaya	Culture dependent	Serial dilution plating method using nutrient agar (NA), meat agar (MA), and brain heart infusion agar (BHI) at 37°C. Genomic DNA isolation and 16S rDNA gene amplification and sequencing	Phylum Firmicutes: *Bacillus subtilis*. (JN695728) Phylum γ-proteobacteria: *Serratia marcescen* (JN566135), *Pseudomonas aeruginosa* (JN653472), *Proteus vulgaris* (JN566137) Phylum β-proteobacteria: *Chromobacterium piscinae* (JN566136) and *Achromobacter ruhlandii* (JN653473)	Banerjee et al. (2012)
15	Cherrapunji, Meghalaya	Average elevation of 1484 meters (4869 ft)	Soil samples were collected from degraded areas of Cherrapunji and adjacent sacred groves	Culture dependent	Identification of the fungal isolates was carried out on the basis of their macro- and microscopic characteristics	Zygomycota (7 species) and Ascomycota (56 species) Most prominent genera were *Penicillium* (17 species), *Aspergillus* (7 species), *Fusarium* (4 species), and *Trichoderma* (3 species)	Saikia and Joshi (2012)
16	Dibru-Saikhowa Biosphere Reserve forest 27°35'–27°50'N and 95°10'–95°40'E, Assam	Humid tropical climate	Soil samples collected from five different sampling sites of Dibru-Saikhowa Biosphere Reserve (DSBR) Forest of Assam	Culture dependent	Serial dilution plating method at 10^3 dilution level	Identified fungi: *Penicillium, Aspergillus, Trichoderma viride, Rhizopus,* and *Curvularia*	Das et al. (2013)

Continued

Table 3.1 Summary of microbial diversity in the North-eastern region of India—cont'd

S. no.	Geographical location (latitude/longitude)	Altitude/climate type	Source (rhizospheric/nonrhizospheric/forest/cold desert/etc.) include plant name in case of rhizospheric samples	Type of study (culture dependent/independent)	Methodology used	Identified microbes	References
17	Latitude 27°06′50.2″N, 93°36′40.8″E, Papum Pare District, Arunachal Pradesh	Altitude 392 m above mean sea level	Soil samples were collected after the completion of fire operations under shifting cultivation	Culture dependent	The soil was serially diluted, and appropriate dilutions were plated on Tryptone yeast extract agar (TYA) and actinomycetes isolation agar (AIA). Identified using 16S rDNA sequence analysis	Phylogenetic analysis revealed the maximum similarity of nine isolates to genus *Streptomyces*, and one each to *Kitasatospora* sp. and *Nocardia*	Malviya et al. (2013)
18	22°11′–28°23′N and 89°86′–97°42′E Tropical eastern Himalayan ranges of northeast	Altitudinal ranges: 34–1000 m, 1001–2000 m, 2001–3000 m, and 3001–3990 m	Soil samples collected from four altitudinal gradients in tropical eastern Himalaya	Culture dependent	Serial dilution and pour plate methods were used for isolation, and representative isolates were identified using 16S rDNA sequencing analysis	Firmicutes was the most common group followed by Proteobacteria and Bacteroidetes. Species belonging to the genera *Bacillus* and *Pseudomonas* were the most abundant	Lyngwi et al. (2013)

19	27°30′N, 91°51′E Tawang, Arunachal Pradesh	Temperature varied from 5°C during daytime to −2°C during the night. Altitude of 4175 m above the sea level. Average humidity: 35%	The soil and water samples were collected from Tawang district	Culture dependent	Phylogenetic analysis based on 16S rDNA sequences revealed the taxonomic affiliation of the 33 strains as species of Streptomyces	A total of 210 Streptomyces were isolated. Based on antifungal activity, a total of 33 strains, putatively Streptomyces spp., were selected and isolated	Debnath et al. (2013)
20	27°23′N–27.38′N and 95°38′E to 95.63°E. Sasoni and Balipara oil well, in oil drilling sites of Digboi Crude oil Refinery Assam	Average elevation of 165 meters above m s.l	Soil samples were collected from the crude oil-contaminated site in the depths ranging from 5 to 25 cm	Culture dependent	Enrichment culture method using nitrogen-deficient mineral media (NDM was followed for the isolation of bacteria. The dominant bacterial isolates were identified using 16S rDNA sequencing	Phylum γ- proteobacteria: *Acinetobacter junii*, Phylum β- proteobacteria: *Achromobacter* sp. and *Alcaligenes faecalis*	Mazumdar and Deka (2013)
21	24°8′N to 25°8′N, 92°15′E to 93°15′E The Barak valley region of Assam	Tropical climate	Soil samples were collected from industrial effluents, sewages, garages, and nearby petrol pump	Culture dependent	Samples were streaked on Pseudomonas Isolation Agar (PIA), Klebsellia Isolation Agar (KIA), and Starch Agar and incubated at 37°C for 24 h for recovery of potent isolates. Identified using morphological and biochemical methods	*Pseudomonas* sp., *Bacillus* sp. and *Klebsiella* sp.	Nath et al. (2013)

Continued

Table 3.1 Summary of microbial diversity in the North-eastern region of India—cont'd

S. no.	Geographical location (latitude/ longitude)	Altitude/climate type	Source (rhizospheric/nonrhizospheric/forest/etc.) include plant name in case of rhizospheric samples	Type of study (culture dependent/ independent)	Methodology used	Identified microbes	References
22	26°40′N and 92°58′E Brahmautra valley, Assam	Brahmaputra valley, Assam. The average annual rainfall: 670–1100 mm Mean minimum and maximum soil temperatures in between 17 and 36°C. Humid tropical climate	Soil samples from five geochemically and hydrologically different surface and subsurface soil habitats. i.e., agricultural land, tea garden, disturbed and undisturbed forests, and active flood plain of Brahmaputra valley, Assam, were collected for the study	Culture independent	The diversity of soil bacterial community was determined through sequence analysis of 16S-23S intergenic spacer regions (ISR) and polymerase chain reaction (PCR) using universal primers for bacterial domain	Major phylogenetic groups of bacteria: α-, β-, and γ-Proteobacteria, Acidobacterium, and Comamonadaceae	Bhattacharyya et al. (2014)
23	21°49′N and 88°36′E, Murlen National Park Champhai, Mizoram	Altitude: 400–1900 m	Soil samples collected from Murlen forest, Mizoram	Culture independent	Paired-end Illumina Mi-Seq sequencing of 16S rRNA gene amplification was carried out to study the bacterial community in the soil of Murlen National Park	Dominant phyla Acidobacteria, Proteobacteria, Planctomycetes, Actinobacteria, Bacteroidetes, Chloroflexi, and Nitrospirae	De Mandal et al. (2015)

24	26°11′N and 91°14′E, (Monapur) and 26°14′N and 91°18′E, (Senga) of Barpeta district, Assam, India	North Eastern parts of high-altitude hills	Soil samples were collected from Eastern Himalayan foot hills, Assam	Culture dependent and culture independent	Serial dilution of soil samples followed by colony morphology and pigmentation patterns, and distinct bacterial isolates were selected. The isolated strains were categorized into psychrophilic (4–20 °C) and psychrotolerant (4–37 °C). Amplified ribosomal DNA restriction analysis was used for identification	Isolated microbes belonged to divisions Actinobacteria, Firmicutes, and Proteobacteria	Venkatachalam et al. (2015)
25	Sonapur tea estate (26°06′56.40″N 91°58′33.18″E), Khetri tea estate (26°06′53.81″N 92°05′27.74″E), Tocklai tea growing area (26°45′18.40″N 94°13′16.92″E), Difaloo tea estate (26°36′29.41″N 93°35′03.96″E), Teok Tata tea estate (26°36′29.41″N 94°25′42.59″E), and Hathikuli tea estate (26°34′55.94″N 93°24′43.15″E), Assam, India	Tropical climate	Tea rhizosphere soil samples were collected from six different tea estates of Assam, India	Culture dependent	Serial plate dilution followed by 16S rDNA sequencing for identification	The isolates were identified as: *P. aeruginosa* strain KH45, *Enterobacter lignolyticus* strain TG1, *Bacillus pseudomycoides* strain SN29, and *Burkholderia* sp. Strain TT6	Dutta et al. (2015)

Continued

Table 3.1 Summary of microbial diversity in the North-eastern region of India—cont'd

S. no.	Geographical location (latitude/longitude)	Altitude/climate type	Source (rhizospheric/nonrhizospheric/forest/cold desert/etc.) include plant name in case of rhizospheric samples	Type of study (culture dependent/independent)	Methodology used	Identified microbes	References
26	25°–61′N and 91°–89′E Arunachal Pradesh, Assam, Manipur, Meghalaya, Mizoram, Nagaland, Tripura, Sikkim, and parts of North Bengal	—	Repository of culturable soil microbes isolated from different habitats covering entire northeast India	Culture dependent	Identification of culturable microbes was carried out using polyphasic approach	The database of culturable microbes is available online at http://mblabnehu.info/nemid/. This database is the first kind of online repository of microorganisms from Indo-Burma megabiodiversity hotspot (northeast India)	Joshi et al. (2015a, b)
27	Central Munga Eri Research and Training Institute (CMERTI), Ladhoigarh, Jorhat, Assam	Warm and temperate climate	Rhizosphere soil samples of Som plant *Machilus bombycina* cultivated in CMERTI, Ladhoigarh, Jorhat	Culture dependent	Isolation of microorganisms was done using five culture media, viz. nutrient agar, potato dextrose agar, King agar B, yeast malt agar, and actinomyces agar	Firmicutes: *Bacillus cereus*RB1 (MTCC 8297), Proteobactera: *Pseudomonas rhodesiae* RB4 (MTCC 8299), *P. rhodesiae*RB5 (MTCC 8300), *Chromobacterium violaceum* RB8 (MTCC 8071). Actinobacteria: *Streptomyces luteireticuli*RB3 (MTCC 8298)	Kalita et al. (2015)

28	Khuangcherapuk Cave (23°41'30"N, 92°37'5"E), Ailawng village, Mizoram	–	Ten individual composite sediment samples were collected from different places of the cave floor	Culture independent	V3 hypervariable region of 16S rDNA was sequenced using paired-end Illumina Mi-Seq technology, and the sequence was analyzed using QIIME data analysis package	The most dominant prokaryotic phylum was Actinobacteria, Firmicutes, Proteobacteria, Bacteroidetes, and *Chloroflexi*. At the family level, Mycobacteriaceae was dominant followed by Bacillaceae, Sphingomonadaceae, Alteromonadaceae, Salinisphaeraceae, Xanthomonadaceae, Flavobacteriaceae, and Moraxellaceae. The leading genera were *Mycobacterium*, *Rhodococcus*, *Alteromonas*, *Holomonas*, and *Salinisphaera*	De Mandal et al. (2015)
29	(25°57'22.70"N, 91°55'43.10"E), Pnahkyndeng Cave, Ri-Bhoi district, Meghalaya	–	Bat Guano sample, Pnahkyndeng Cave	Culture independent	V4 hypervariable region of 16S rDNA was sequenced using paired-end Illumina Mi-Seq technology, and the sequence was analyzed using QIIME data analysis package	18 different phyla dominated by Chloroflexi, Crenarchaeota, Actinobacteria, Bacteroidetes, Proteobacteria, and Planctomycetes	De Mandal et al. (2015)

Continued

Table 3.1 Summary of microbial diversity in the North-eastern region of India—cont'd

S. no.	Geographical location (latitude/ longitude)	Altitude/climate type	Source (rhizospheric/nonrhizospheric/forest/ cold desert/etc.) include plant name in case of rhizospheric samples	Type of study (culture dependent/ independent)	Methodology used	Identified microbes	References
30	Papum Pare district, Arunachal Pradesh, India	High-altitude hills. Temperate climate	Soil samples of post-fire operation were analyzed	Culture dependent	Colony and cell morphology method	*Cladosporium* sp., *Coniella* sp., *Paecilomyces* sp., *Penicillium* sp., *Rhizoctonia* sp., and *Trichoderma* sp.	Jain et al. (2016)
31	Three districts of Meghalaya, viz., Ri-bhoi, East Khasi Hills, and West Garo Hills	Subtropical climate	Soil samples were collected from healthy field-grown rice and maize crops	Culture dependent	Fluorescent pseudomonads were confirmed based on morphological and biochemical tests	*Pseudomonas fluorescens*	Kipgen et al. (2016)
32	Cachar, Hailakandi, Karimganj, NC Hills and Tezpur, Assam	Tropical climate	Root nodules of *Crotolaria pallida*	Culture dependent	PCR-RFLP analysis of 16S rDNA was carried out to study genetic diversity of bacterial isolates	5 isolates, viz, RCP2, KACP2, SICP2, NCP1, and NCP2, showed similarity with the reference strain *Rhizobium leguminoserum* MTCC-99, 3 isolates, viz., GCP1, SCP1 and TECP1, showed similarity with the reference strain *Bradyrhizobium japonicum* MTCC-120; and 2 isolates MKCP1 and DCP1, showed close similarity with the reference strain *Mesorhizobium thiogangeticum* MTCC-7001	Singha et al. (2016)

	Location	Site characteristics	Sample	Method	Methodology	Major findings	References
33	*Jhum* sites of northeast India	North-eastern parts of high-altitude hills	Soils collected from differently aged fallows of the *jhum* sites of northeast India	Culture dependent	Identification of bacteria was carried out using 16S rRNA gene sequencing and potent fungal isolates by ITS sequencing method	Bacterial isolates: *Curtobacterium oceanosedimentum*, *Bacillus methylotrophicus*, and *B. cereus* Fungal isolates: *Penicillium virgatum*, *Metarhizium pinghaense*, and *Penicillium stratisporum*	Banerjee et al. (2017)
34	23°19'08"N and 92°45'00"E Mizoram	Elevation: 520 m above m.s.l in an intermontane valley along Mat River	Rhizospheric soil sample were collected from wet land paddy field of Thenzawl, Mizoram, India	Culture dependent	Phosphate-solubilizing bacteria were isolated from rhizospheric soil sample using serial dilution and spread plate methods. Molecular characterization of potential bacterial isolates using 16S rRNA gene amplification	Major genus *Bacillus* (52.94%), *Burkholderia* (29.41%), *Alcaligenes* (11.76%), and *Staphylococcus* (5.88%)	Khiangte (2017)
35	Manipur	Subtropical to temperate climate	Leaves and stem of *Achyranthes aspera* L. were collected from different locations	Culture dependent	16S rRNA gene sequencing	*Serratia marcescens* AL2-16	Devi and Pandey (2017)
36	27°47'N latitude and 88°42'E longitude The Yumthang hot spring, Sikkim	Elevation 3597 m Average temperature 39–41°C	Microbial mat samples of the spring were collected for the community analysis	Culture independent	16S rRNA gene amplification using metagenome sequencing method was employed	Major phyla Proteobacteria Actinobacteria Firmicutes Bacteroidetes	Panda et al. (2017)

Continued

Table 3.1 Summary of microbial diversity in the North-eastern region of India—cont'd

S. no.	Geographical location (latitude/longitude)	Altitude/climate type	Source (rhizospheric/nonrhizospheric/forest/cold desert/etc.) include plant name in case of rhizospheric samples	Type of study (culture dependent/independent)	Methodology used	Identified microbes	References
37	Cachar, Assam	Tropical climate	Soil samples from petroleum industrial effluent site with high richness of PAH contamination	Culture dependent	16S rDNA gene sequencing was employed. The qualitative estimation of biofilm formation of the isolate was done by the test tube method and congo red agar method	*Acinetobacter* sp. PDB$_4$	Kotoky et al. (2017a)
38	Silchar, Assam	Tropical climate	Rhizospheric soil of contaminated site near automobile workshop	Culture dependent	The complete genomic DNA was sequenced using a next-generation sequencing system	*Klebsiella pneumoniae* Strain AWD5 Draft genome sequence	Rajkumari et al. (2017)
39	23°83′N–25°68′N and 93°03′E–94°78′E. Manipur	–	Rhizospheric soil of *Allium hookeri* Thwaites	Culture dependent	Bacterial strains identified through 16S rDNA analysis	*Arthrobacter luteolus* S4C7, *K. pneumoniae* S4C9, *K. pneumoniae* S4C10, *Enterobacter asburiae* S5C7, *K. pneumoniae* S6C1, and *K. quasipneumoniae* S6C2	Kshetri et al. (2017)

40	Silchar Assam	Tropical climate	Petroleum contaminated soil	Culture dependent	Whole-genome shotgun sequencing was performed using paired-end sequencing libraries and a TruSeq Nano DNA library prep kit (Illumina)	*S. marcescens* S2I7 Draft genome sequence	Kotoky et al. (2017b)
41	Silchar Assam	Tropical climate	Rhizospheric soil of contaminated sites	Culture dependent	Whole-genome shotgun sequencing was performed using paired-end sequencing libraries and a TruSeq Nano DNA library prep kit (Illumina)	*B. subtilis* SR1 Draft genome sequence	Kotoky et al. (2017c)
42	Northeast India	–	Crude oil-contaminated soil	Culture dependent	Whole-genome shotgun sequencing was performed using paired-end sequencing libraries and a TruSeq Nano DNA library prep kit (Illumina)	High-molecular-weight polyaromatic hydrocarbon-degrading *Pseudomonas fragi* strain *DBC* draft genome sequence	Singha et al. (2017)

Continued

Table 3.1 Summary of microbial diversity in the North-eastern region of India—cont'd

S. no.	Geographical location (latitude/ longitude)	Altitude/climate type	Source (rhizospheric/nonrhizospheric/forest/ cold desert/etc.) include plant name in case of rhizospheric samples	Type of study (culture dependent/ independent)	Methodology used	Identified microbes	References
43	The two forest ecosystems, viz., Nameri National Park (NNP) (27°0′36″N, 92°47′24″E) and Panidehing Wildlife Sanctuary (PWS) (27°7′19″N, 94°35′47″E), are located in the Sonitpur and Sivasagar districts, respectively, of Assam	Tropical climate	Soil samples were collected from seven different sites, each covering several diverse habitats, in both the forest ecosystems	Culture dependent	Isolation of microbes was carried out by the serial dilution technique and identified by 16S rRNA gene sequencing	172 morphologically distinct presumptive actinobacterial isolates were obtained from two sites. Antimicrobial isolates dominantly belonged to *Streptomyces*, followed by *Nocardia* and *Streptosporangium*	Das et al. (2018)
44	Cachar, Assam	Different regions of Cachar, Assam, India. Tropical climate	Rhizospheric soil of *Oryza sativa*	Culture dependent	Identification using phenotypic, physiological features, and 16S rDNA gene sequence analysis	*Stenotrophomonas maltophilia* RSD_6	Nevita et al. (2018)

45	Jorhat (26°43′03.8″N 94°11′40.2″E), Tinsukia (27°20′34.5″N 110 95°42′33.2″E) and Lakhimpur (26°57′38.5″N 93°51′53.5″E) districts of Assam	Tropical climate	Bacterial endophytes were isolated from leaf, stem, and root sections of 208 both cultivated and wild rice plant	Culture dependent	The bacterial isolates were characterized both morphologically and biochemically through various tests (gram staining, starch hydrolysis, casein hydrolysis, catalase reaction, 125 citrate and malate utilization, nitrate reduction, H2S production, and gelatin liquefaction) 126 according to the Bergey's Manual of Determinative Bacteriology Isolates were further characterized at the molecular level using 16S rRNA 216 gene sequence data	The bacteria isolated in this study belonged to 3 major phyla, viz., Firmicutes (57.1%), Actinobacteria (20.0%), and Proteobacteria (22.8%). Bacteria isolated from different parts of cultivated rice, the isolates belonged to 11 different genera, viz., *Bacillus*, *Brevibacillus*, *Lysenibacillus* *Microbacterium*, *Microbacteriaceae*, *Staphylococcus*, *Pantoea*, *Burkholderia*, *Acinetobacter*, *Pseudomonas*, and *Ralstonia*. However, *Bacillus*, *Stenotrophomonas*, *Microbacterium*, *Cellulosimicrobium*, *Proteus*, *Staphylococcus*, *Erwinia*, *Ochrobactrum*, and *Enterobacter* were the genera isolated from plant parts of wild rice	Borah et al. (2018)
46	27°47′34.50″N 88°42′30.96″E Sikkim	Temperature between 42 and 45°C pH 7.5 to 8	Water samples were collected from the source of the Yumthang hot spring, North Sikkim	Culture dependent	16S The 16S rRNA gene sequencing method using universal primers (27F and 1492R)	*Geobacillus yumthangensis* AYN2T	Najar et al. (2018)

Continued

Table 3.1 Summary of microbial diversity in the North-eastern region of India—cont'd

S. no.	Geographical location (latitude/ longitude)	Altitude/climate type	Source (rhizospheric/nonrhizospheric/forest/ cold desert/etc.) include plant name in case of rhizospheric samples	Type of study (culture dependent/ independent)	Methodology used	Identified microbes	References
47	Manipur	—	Fermented bamboo shots	Culture dependent	Whole-genome shotgun sequencing was performed using paired-end sequencing libraries and a TruSeq Nano DNA library prep kit (Illumina)	*B. subtilis Strain* FB6-3 draft genome sequence	Khunjan and Pandey (2018)
48	25°38′12.2″N, 94°10′29.6″E, Nagaland	Elevation 1668 m a.s.l. Subtropical climate	Alder-based jhum land soil samples (0–15 cm depth)	Culture dependent	Serial plate dilution method. Identification using 16S rDNA spacer sequence analysis	Phylum γ-Proteobacteria, Family: Enterobacteriaceae *Kosakonia sacchari* strain KhAn	Giri (2019)
49	Mizoram	Jhum fields Subtropical	Dry forest litter falls were collected from Jhum fields of Mizoram	Culture dependent	The isolates obtained after screening for halo formation in carboxymethylcellulose agar media were grown in the liquid broth of CMC media in which estimation of total soluble sugar (carbohydrate) and cellulase activity was performed. The estimation of total soluble sugar was carried out by anthrone method	Unidentified cellulose-degrading microorganisms	Sangma and Thakuria (2019)

50	Darchawi, Kumarghat (N24°08.277'E 092°03.313') East Betchara, (N 24°06.917' E 092°01.364') and Chinibagan, Kailashahar, (N24°18.544' E092°04.316'), Unokoti district, Tripura	Subtropical climate	Green leaves and roots were collected from healthy plants during fruiting season (FS) and nonfruiting season (NFS)	Culture dependent	The microscopic identification of the isolates was carried out by lacto phenol staining technique. Identification of selected fungal strain for bioactive potential was authenticated by Molecular identification (ITS sequence of rDNA) by Agharkar Research Institute (NFCCI, Pune, India)	Total twenty-four endophytic fungal strains and four nonsporulating forms were isolated. Highest relative frequency was recorded in *Pseudopestalotiopsis theae* from leaf, *Trichoderma asperellum* from root, and *Neopestalotiopsis piceana* from both leaf and root endophytes	Bhattacharya et al. (2019)
51	26°24.10.1"– 26°24.59.9"N, and 94°22.40.9"– 94°22.53.9"E Mokokchung district of Nagaland	Wet, warm and humid tropical climate with annual rainfall from 1800 to 2600mm and altitude ranges from 592 to 733 m a.s.l.	Root-associated soils of *Thysanolaena maxima*	Culture dependent	Serial plate dilution method	63 unidentified isolates associated with *T. maxima* were evaluated for plant growth-promoting traits	Deka et al. (2019)
52	Pobitora Wildlife Sanctuary (26°12' to 26°16'N and 91°58' to 92°05'E) and Kaziranga National Park (26°30' to 26°45'N; 93°08' to 93°36'E) of Assam, India	The pH range of the soil samples was 4.5–6.0	Soil samples were collected from Pobitora Wildlife Sanctuary and Kaziranga National Park of Assam, India	Culture dependent	Analysis of genetic diversity of the actinobacterial isolates was carried out by 16S rDNA-ARDRA	*Streptomyces* was the predominant genus, followed by *Nocardia* and *Kribbella*	Sharma and Thakur (2020)

Continued

Table 3.1 Summary of microbial diversity in the North-eastern region of India—cont'd

S. no.	Geographical location (latitude/longitude)	Altitude/climate type	Source (rhizospheric/nonrhizospheric/forest/cold desert/etc.) include plant name in case of rhizospheric samples	Type of study (culture dependent/independent)	Methodology used	Identified microbes	References
53	Cultivated jhum land N 26°20′54.4 and E 94°29′58.5. Abandoned jhum land was N 26°21′05.3 and E 94°28′16.5. Nagaland	Cultivated jhum land altitude was 1023 m, while the abandoned jhum land was 1044 m. Subtropical climate	Soil samples were collected from cultivated and abandoned jhum lands of Mokokchung district, Nagaland	Culture dependent	Serial dilution method was used to isolate soil fungi on RBA (Rose Bengal Agar) and PDA (Potato Dextrose Agar) plates and identified by studying their macro- and micromorphological characteristics	*Absidia* sp., *Alternaria* sp. *Aspergillus* sp., *Cladosporium* sp., *Fusarium* sp., *Geotrichum*, *Mortierella* sp., *Mucor* sp., *Penicillium* sp., and *Trichoderma* sp.	Kichu et al. (2020)
54	Natural habitat of northeast India	—	Roots of wild rice variety *Zizania latifolia*	Culture dependent	The endophyte was identified by 16S rDNA gene sequencing	*Serratia nematodiphila* (22WE)	Das et al. (2020)
55	Different districts of Meghalaya	—	Isolated from the roots of tomato	Culture dependent	Isolated and characterized by molecular method	Endophytic *Bacillus* spp.	Devi et al. (2020)

Water samples from extremely high rainfall areas of Cherrapunji, Meghalaya, were explored to characterize biofilm-forming bacteria using the culture-dependent method, and six isolates, viz., *Bacillus subtilis.* (JN695728), *Serratia marcescen* (JN566135), *Chromobacterium piscinae* (JN566136), *Proteus vulgaris* (JN566137), *P. aeruginosa* (JN653472), and *Achromobacter ruhlandii* (JN653473), were identified through 16S rDNA gene sequencing. However, 12 isolates (belonging to the genera *Bacillus, Proteus, Serratia, Pseudomonas, Chromobacterium,* and *Achromobacter*) were tentatively identified following Bergey's Manual of Determinative Bacteriology. Bacterial isolates *Proteus* and *Pseudomonas* were the predominant genera in biofilm samples. The pattern of biofilm-forming bacterial diversity in Cherrapunji, Meghalaya, was quite comparable with the findings of Olapade and Leff (2005) in other conditions, indicating nonsignificant influence of heavy rainfall in bacterial biofilm composition (Banerjee et al., 2012).

3.2.4 Microbial diversity in *jhum* agroecosystem

Shifting cultivation traditionally known as jhum is the most primitive form of agriculture by the tribes of hilly regions in northeast India. The jhum cycle of 15–25 years was a sustainable form of crop cultivation in the region, while drastic shortening of traditional jhum cycle (4–5 years) has several adverse effects in the environment (Bhagawati et al., 2015). The increasing population and accelerated land degradation have forced to reduce the fallow period of jhum and resulted in detrimental effects on the soil nutrients and resident microbiota (Binarani and Yadava, 2010). Soil microbial biomass, a key indicator of soil health, is highly influenced by soil temperature, moisture content, and organic carbon content. Burning operations in jhum fields affect the physicochemical and biological properties of soil, thus resulting in a decrease in microbial population (Miah et al., 2010). Soil microbiota are the drivers of litter decomposition and nutrient cycling in an ecosystem; hence, decreased active microbial population due to massive fire in jhum fields hampers the soil ecosystem and its fertility. A study on post-fire soil bacterial population by Pandey et al. (2011) reported the recovery of spore-forming *Bacillus* sp. and nonspore-forming *Pseudomonas* sp. in the soil. Similarly, post-fire soil analysis of microbial population carried out in Papum Pare district in Arunachal Pradesh reported *Streptomyces, Kitasatospora,* and *Nocardia* as dominant genera. The reported bacteria exhibited chitinase, glucanase, amylase, lipase, and protease activities. DGGE profiling and polymerase chain reaction (PCR) analysis of degraded and moderately degraded jhum land soil as well as undegraded forest from Banderdewa, Arunachal Pradesh, showed higher bacterial diversity in undegraded sites than in disturbed sites. The lower bacterial diversity in degraded soils was attributed to the negative effect of shifting cultivation land-use practice (Singh et al., 2011). Studies on post-fire survival of fungal species in jhum farming system reported *Aspergillus, Cladosporium, Paecilomyces, Penicillium,* and *Trichoderma* as dominant genera in northeast Himalaya. The identified fungi were reported to grow in a wide range of pH and temperatures. Moreover, the fungal isolates demonstrated ligninolytic activity, phosphate solubilization, lytic enzymes, volatile and diffusible antimicrobials, and ammonia production (Jain et al., 2016).

The jhum land soil is not only a repository of unique type of microbes capable of thriving in extreme conditions but also potential bioinoculants for fertility and productivity enhancement of jhum farming system. Banerjee et al. (2017) isolated 85 microbes from jhum sites of northeast India and 10 each of the bacterial and fungal isolates were screened for plant growth-promoting traits. Paddy plants were inoculated with selected microbial isolates exhibiting indole, siderophores, and ammonia production, phosphate solubilization, and catalase positive. The organisms resulted in better plant growth as compared to untreated ones.

Sangma and Thakuria (2019) isolated cellulose-degrading microorganisms (CDMs) from leaf litter mass of four jhum cycles of 2, 5, 10, and 20 years length in Mizoram and screened for their carboxymethylcellulose (CMC) degradation ability. Bacterial colony-forming units (cfu) counts on agar plates were recorded based on halo formation, and the potential decomposers were recorded in the order of 20 > 10 > 2 > 5 years fallow period. The bacterial cfu count (g^{-1}) litter mass was recorded in the range of 5.2×10^6 and 9.9×10^7, with the lowest in 5 years of fallow length and the highest in 2 years of fallow length. The cellulase activity ranged from 3.06 to 227.8 $\mu g\,mL^{-1}\,h^{-1}$. The study on jhum soil microbial activities again revealed substantial beneficial population of microbes. *Thysanolaena maxima* (TM) rhizosphere and bulk soil collected from 5 and 20 years of fallow phases in Mokokchung, Nagaland, was analyzed for soil enzyme profiling bacterial population status. The rhizobacterial and root endophytic bacterial count $(0.74 \pm 0.056 \times 10^7\,cfu\,g^{-1}$ soil and $0.083 \pm 0.004 \times 104\,cfu\,g^{-1}$ roots, respectively) was significantly higher in 20 years of fallow length than in 5 years of fallow length. Moreover, 63 bacterial isolates were screened for pectinase and cellulase, IAA, 1-aminocyclopropane-1-carboxylate deaminase (ACCD) production, N_2-fixation, phosphate (iP) solubilization, and organic P mineralization. The study suggested that cultivation of *T. maxima* in fallow jhum lands is helpful in rejuvenation of biological activities in terms of higher enzyme activities and active rhizobacterial population in frequently burnt soils under shorter *Jhum* cycles (Deka et al., 2019).

Himalayan alder (*Alnus nepalensis*) tree rhizosphere soil collected from Khonoma village in Kohima, Nagaland, was characterized for soil nutrients and microbial diversity using culture-dependent and qPCR gene quantification methods. Among the isolated bacteria, a novel diazotroph *Kosakonia sacchari* was also identified and reported for the first time from India. The bacterial strain KhAn Genbank accession number MG881883 holds promising biofertilization potential through free living nitrogen fixation in the soil (Giri, 2019). Similarly, the soil under alder-based jhum farming system was found enriched with major soil nutrients, and real-time PCR (qPCR) performed to quantify 16S rDNA gene showed 9.59×10^{12} copy numbers of bacterial genes present per gram of soil. Higher gene copy numbers in the soil indicate very high bacterial abundance in the soil, making it much more productive than conventional jhum farming practice (Giri et al., 2018). Similarly, a most recent study on cultivated and abandoned jhum fields of Mokokchung, Nagaland, reported nine fungal species, viz., *Absidia cylindrospora, Aspergillus flavus, Aspergillus niger, Cladosporium cladosporioides, Fusarium* sp., *Geotrichum candidum, Mortierella* sp., *Mucor circinelloides*, and *Trichoderma harzianum*, from the cultivated fields. However, only five fungi, viz., *Absidia glauca, Alternaria* sp., *Aspergillus* sp., *Mucor racemosus*, and *Penicillium* sp., were identified from abandoned jhum soil (Kichu et al., 2020).

3.2.5 Microbial diversity in agroecosystem

The ethnic diversity in northeast India is represented by more than 200 tribes, subtribes, and distinct nontribal communities. The sociocultural and linguistic diversity has a very strong influence on the farming practices across the region. The diverse farming systems evolved through the indigenous knowledge and wisdom of the communities have resulted in diverse land-use patterns in NER. These communities depend on agrarian economy: paddy is the principal food crop followed by maize, pulses, a variety of vegetables, and fruits. Excessive use of chemical fertilizers and pesticides has deleterious effects in the environment and human health. Therefore, at present, organic farming and exploitation of microbial resources for biofertilization potential have received huge attention of agricultural scientists, farmers, and environmental conservationists of the world. Studies on plant growth-promoting rhizobacteria and biofertilizer application in agricultural fields have shown promising results with potential as a substitute to chemical fertilizers for sustainable farming practices. Moreover, organic farming has an ability to improve soil quality and health of deteriorating agroecosystems (Droogers and Bouma, 1996; Mader et al., 2002; Girvan et al., 2004). Applications of farm yard manure, compost, vermicompost etc. have shown higher active soil microbial population compared to the chemical fertilizer (Swer et al., 2011; Pansombat et al., 1997; Tokuda and Hayatsu, 2001). Similarly, Thakuria et al. (2004) undertook a study using three selective culture media to characterize rhizobacterial communities in 15 rice-grown areas of Assam. The phosphate solubilizers and fluorescent bacteria were identified as eight different species. However, the three isolates of azospirilla group were identified to be *Azospirillum brasilense*, and the fourth one as *Azospirillum amazonense*. Among the characterized rhizobacteria, *A. amazonense* A10 (MTCC 4716), *Bacillus pantothenticus* P4 (MTCC 4695), and *Pseudomonas pieketti* Psd6 (MTCC 4715) were found to increase the rice yield by 55.5%, 12.2%, and 76.9%, respectively, over control and also found superior than *A. brasilense* CDJA and *Pseudomonas fluorescence* MTCC 103 in yield enhancement. Kamala and Indira (2011) from Imphal, Manipur, studied the *Trichoderma* spp. inhabiting agricultural fields and reported 32% isolates as potential biocontrol agents for control of damping-off disease in beans.

Another study in this line was carried out in Thenzawl, Mizoram, to isolate potential phosphate-solubilizing bacteria inhabiting paddy rhizosphere, where among the 37 isolates, 17 were characterized as elite phosphate solubilizers belonging to genera *Bacillus*, *Burkholderia*, *Alcaligenes*, and *Staphylococcus*. Soil of Brahmaputra Valley, Assam, has distinct physicochemical characteristics, which form unique ecological niche for the microbial communities (Bhattacharyya and Jha, 2011). An investigation on bacterial community analysis in the surface and subsurface soil using metagenomic sequence analysis of 16S-23S intergenic spacer regions (ISR) reported five major phylogenetic groups, viz., α-, β-, and γ-Proteobacteria, Acidobacterium, and Comamonadaceae (Bhattacharyya et al., 2014). Tea rhizosphere is inhabited by the diverse group of soil beneficial microbes, which perform essential ecosystem processes in the tea gardens. In a study carried out by Dutta et al. (2015), 217 bacteria were isolated from tea rhizosphere of Assam, and four of them revealed promising PGPR traits, viz., siderophore production, IAA production, phosphate solubilization and ammonia production, and subsequent positive effects on growth of tea plants. Among the four rhizobacteria, *Enterobacter lignolyticus* strain TG1 was found to have a strong plant growth-promoting potential.

The 22 bacterial isolates belonging to root nodules of *Crotolaria pallida* growing in five different locations of Assam were characterized through the comparison of morphological and physiological traits with reference strains *Rhizobium leguminoserum* MTCC-99, *Bradyrhizobium japonicum* MTCC-120, and *Mesorhizobium thiogangeticum* MTCC-7001. The isolates were very close to the reference strains in terms of morphological characteristics but varied in physiological features. The results of PCR-RFLP (restriction fragment length polymorphism) analysis of 10 isolates grouped them into three different 16S rDNA types, where 2 of them were close to the *B. japonicum* MTCC-120, 6 were related to *R. leguminoserum* MTCC-99, and 2 were related to *M. thiogangeticum* MTCC-7001 (Singha et al., 2016).

Kipgen et al. (2016) isolated 30 fluorescent pseudomonads from rice rhizosphere of Ri-bhoi, East Khasi Hills, and West Garo Hills districts in Meghalaya, which were characterized using morphological and biochemical methods. Of the 30 fluorescent pseudomonads, 21 were identified as *Pseudomonas fluorescens* and designated as I, II, III, and V biovars. These biovars were screened for their biocontrol potential on the basis of HCN production and tested for in vitro antagonism with *Ralstonia solanacearum*. Five phosphate-solubilizing bacteria (PSB), viz., *Arthrobacter luteolus* S4C7, *Klebsiella pneumoniae* S4C9, *K. pneumoniae* S4C10, *Enterobacter asburiae* S5C7, *K. pneumoniae* S6C1, and *K. quasipneumoniae* S6C2), were reported to inhabit the rhizospheric soil of *Allium hookeri* Thwaites in Manipur. Among these PSB strains, *K. pneumoniae* S4C10 was found to be the most efficient strain followed by *K. quasipneumoniae* S6C2. The phosphate solubilization capabilities of these strains were further optimized using varying carbon sources and found that the bacteria preferred fructose as the best carbon source as compared to the glucose (Kshetri et al., 2017). Endophytic bacteria *Microbacteriaceae bacterium*, *Microbacterium testaceum*, and *B. subtilis* isolated from rice plants grown in different sites of Assam are reported to be potential biofertilizers for productivity improvement of rice agroecosystem (Borah et al., 2018).

A plant-probiotic bacterium *Stenotrophomonas maltophilia* RSD6 having lignocellulolytic activity was isolated from rice plant rhizosphere and identified on the basis of its phenotypic and physiological features, and 16S rDNA gene sequencing. The bacterium was found to possess biostimulation attributes, viz., siderophore, indole acetic acid production, and inorganic phosphate solubilization. Eventually being lignocellulosic by the production of lignases, xylanase, and cellulase enzymes, *S. maltophilia* RSD6 was used for composting of rice straw residues. As a result of rice residue composting by this bacterium, nitrogen, phosphorus, and potassium contents in mature compost were increased. Moreover, the bacterium showed endoglucanase, exoglucanase, cellobiose, and xylanase activities and proved to be potential strain of interest for crop residue management (Nevita et al., 2018). Due to abiotic and biotic stress tolerance imparting nature of endophytic bacteria, a strain 22WE was isolated from the wild rice variety *Zizania latifolia* for bio-inoculating rice crop grown in acid soils. 16S rDNA gene sequencing identified it as *Serratia nematodiphila*, and rice plants were inoculated with this endophyte. Inoculation of *S. nematodiphila* resulted in more root surface, root volume, and shoot biomass under aluminum (Al)-toxic and phosphorus (P)-deficient conditions. The inoculated endophyte conferred plant protection from reactive oxygen against Al-induced stress (Das et al., 2020). Similarly, Rocky tomato seeds were treated using inoculum of endophytic *Bacillus* isolate for enhanced seedling vigor and germination in Meghalaya, India. Among the 12 endophytic *Bacillus* isolates, *Bacillus velezensis* ERBS51 gave maximum 95% seed germination and vigor index (1073.50 and 1472.5) after 7 and 14 days of treatment, respectively (Devi et al., 2020). An investigation on bacterial endophytes isolated from wild and cultivated rice plants was carried out to evaluate the plant growth promotion of rice grown in Eastern Himalayas. A total of 20 isolates—10 each from wild and cultivated rice—were tested and found that the isolate 34WE (*P. fluorescens*) showed better performance in plant physiological parameters and yield attributes than other isolates (Abdelbaset Hassan et al., 2020). Bhattacharya et al. (2019) reported 24 fungal endophytes and 4 nonsporulating forms from tea leaves and roots. *Pseudopestalotiopsis theae* was isolated from leaf samples and *Trichoderma asperellum* from roots, while *Neopestalotiopsis piceana* was found to inhabit both leaves and roots. These studies clearly indicate that the bacterial isolates from diverse ecological niches, including endophytes, have enormous potential to boost farm productivity in northeast India. Consolidated efforts in this area can be a boon for marginalized farming communities and sustainability of agriculture sector.

3.2.6 Contaminated habitats

Soil and water pollution due to crude oil contamination and abandoned drill sites in different parts of Assam is a serious environmental issue. Microorganisms have an unique ability to utilize high molecular weight petroleum components such as polycyclic aromatic hydrocarbons (PAHs) as an energy source. Subsequently, these pollutants are degraded and mineralized into environmentally innocuous chemicals through complex microbial enzyme systems (Kotoky and Pandey, 2020). Studies on bioremediation of oil-contaminated habitats have been extensively carried out to find out potential microbial strains. The bacterial strains *Acinetobacter junii* (AM14), *Achromobacter* sp. (AM02), and *Alcaligenes faecalis* (AM07) isolated from crude oil-contaminated soil of Digboi crude oil refinery, Assam, were reported to be free-living heterotrophic diazotrophs, which can potentially accelerate the bioremediation of hydrocarbon-contaminated soil through nitrogen fixation because crude oil has been found a good source of energy for nitrogen fixation (Mazumdar and Deka, 2013). Heavy metal contamination in the environment is another challenging issue due to its toxic effect in flora and fauna and human beings. Twenty copper-, zinc-, and lead-resistant bacteria were isolated from industrial effluents, sewages, garages, and petrol pumps of Barak valley region of Assam, India. Among the 20 isolates, *Pseudomonas* sp., *Klebsellia* sp., and *Bacillus* sp. were the predominant bacteria with the minimum inhibitory concentration as 60, 180, and 1800 μg/mL for copper, lead, and zinc, respectively. Also, rice shoot length was reported to increase significantly in contaminated soils inoculated with resistant bacteria (Nath et al., 2013). Similarly, *Acinetobacter* sp. PDB_4 has been reported as an excellent bacterium for biodegradation of anthracene, pyrene, and benzo (α) pyrene (BaP) PAHs with high emulsification index and biofilm formation under stress conditions (Kotoky et al., 2017a).

The Department of Microbiology, Assam University, Silchar, has carried out a series of investigations on bacterial isolates recovered from oil-contaminated environment and plant rhizosphere of contaminated soil. The researchers have employed next-generation sequencing tools very effectively and sequenced whole genome of six potential bacterial strains during the year 2017–18. Among these recent studies, the draft genome sequences of *Alcaligenes fecalis* BDB_4 isolated from crude oil-contaminated soil revealed the presence of important genes for PAH degradation, chemotaxis, membrane transport, and biofilm formation, giving insight into the complete PAH mineralization potential of this bacterium (Singha et al., 2017). The whole-genome shotgun sequencing of heavy metal-resistant, polyaromatic hydrocarbon-degrading bacterium *Serratia marcescens* S2I7 isolated from petroleum-contaminated sites in Assam consists of one circular chromosome (5,241,555 bp; GC content 60.1%) with 4533 coding sequences. The draft genome includes specific genetic elements for heavy metal resistance and PAH degradation (Kotoky et al., 2017b). In the series of whole-genome sequencing, *K. pneumoniae* strain AWD5 was isolated from rhizospheric soil of a contaminated site in Silchar, Assam. The genetic analysis revealed a 4,807,409 bp genome containing 25 rRNA genes, 81 tRNAs, 4636 coding sequences (CDS), PAH and benzoate-degrading genes (Rajkumari et al., 2017). The draft genome of metal-resistant and PAH-degrading *B. subtilis* SR1 consisted of a circular chromosome with 4,093,698 bp. The bacterium has plant growth-promoting attributes along with metal resistance and PAH degradation (Kotoky et al., 2017c). Another draft genome of *Pseudomonas fragi* strain DBC isolated from crude oil-contaminated soil comprised 5,072,304 bp with 54.09% GC content and PAH-degrading genes. Moreover, the organism is reported to have genetic elements for physiological functions such as chemotaxis, detoxification, and quorum sensing (Singha et al., 2017).

In addition to the whole-genome sequencing of bacterial strains capable of PAH degradation and resistant to heavy metal toxicity, a draft genome of *B. subtilis* strain FB6-3 has also been sequenced.

The bacterium was isolated from fermented bamboo shoot samples (Soibum) from Manipur. The genome was further analyzed for evolution of genetic code and de novo assembly showed the chromosome size of 4,192,717 bp and 3885 coding sequences (Khunjan and Pandey, 2018). The studies on genome sequencing using high-throughput sequencing methods revealed that the microbes inhabiting diverse ecological niches have huge potential in environmental cleanup, nutrient cycling, plant growth promotion, metal detoxification, and eco-restoration of degraded habitats. The attempts made by the researchers from Assam University, Silchar, are exemplary in the utilization of modern molecular biology tools to exploit the infinitely small organism for infinitely large role. A summary of microbial diversity is presented in Table 3.1.

3.3 Northeast microbial database (NEMiD)

North-Eastern Hill University, Shillong, Meghalaya, has developed Northeast Microbial Database (NEMiD), which is accessible at web portal http://mblabnehu.info/nemid without any login requirements. This is the first digital database on culturable microbes (bacteria, fungi, and actinomycetes) isolated from various ecosystems of northeast India (Joshi et al., 2015a, b). The database was created after cultural, biochemical, and molecular characterizations of microbial isolates leading to its taxonomic position up to the species, which provides different types of information from more than 140 study sites, profiling more than 229 microbial isolates with 229 16S rRNA partial sequences (Bhattacharjee and Joshi, 2014).

3.4 Conclusion and future prospects

Himalayas are the most pristine natural ecosystems in the world, which offer tremendous opportunities for discovery of hitherto unknown microbes from diverse environments and their bioprospecting for novel enzymes and chemicals of economic and industrial importance. Soil is one of the diverse ecosystems on the earth which harbor rich diversity of bacteria, fungi, archaea, viruses, and protists. The Eastern Himalayan region has diverse habitats comprising altitudinal gradients, climate, soil, vegetation types, and dynamic land-use patterns offering a wide scope for exploration and characterization of unexplored microbial wealth. The overview on exploration of microbial diversity under diverse ecological niches in NER revealed that the most of the studies have been undertaken following culture-dependent methods and thus able to report a very limited microbial isolates. Though, a few attempts have been made to document microbial diversity in different ecological niches using next-generation gene sequencing methods in this region focusing on bacterial diversity. The present endeavor to assemble the studies on diversity and documentation of microbes in NER will attract the attention of microbiologists to employ high-throughput shotgun sequencing techniques to decipher microbial repository existing under diverse environmental conditions. Studies on soil microbiome associated with different ecological niche in NER will help to understand the potential of microorganisms in bioremediation, eco-restoration, nutrient cycling, ecosystem productivity enhancement, sustainable agriculture, and synthesis of novel antibacterial compounds for human health protection from infectious and harmful diseases. It is recommended that much more studies should be undertaken following modern molecular techniques to explore microbial wealth of unique and pristine ecological niches of the north-eastern region.

References

Banerjee, S., Rai, S., Sarma, B., Joshi, S.R., 2012. Bacterial biofilm in water bodies of Cherrapunjee: the rainiest place on planet earth. Adv. Microbiol. 2 (4), 465–475.

Abdelbaset Hassan, M., Majumder, D., Thakuria, D., Rangappa, K., 2020. Evaluation of physiological parameters of rice bacterial endophytes in Eastern Himalaya region of India. Sudan J. Agric. Sci. 6 (1), 1–14.

Banerjee, A., Bareh, D.A., Joshi, S.R., 2017. Native microorganisms as potent bioinoculants for plant growth promotion in shifting agriculture (*Jhum*) systems. J. Soil Sci. Plant Nutr. 17 (1), 127–140.

Barnett, H.L., Hunter, B.B., 1972. Illustrated Genera of Imperfect Fungi, second ed. Burgess Publication, Minneapolis, p. 241.

Bhagawati, K., Bhagawati, G., Das, R., Bhagawati, R., Ngachan, S.V., 2015. The structure of Jhum (Traditional Shifting Cultivation System): prospect or threat to climate. Int. Lett. Nat. Sci. 46, 16–30.

Bhattacharjee, K., Joshi, S.R., 2014. NEMiD: a web-based curated microbial diversity database with Geo-based plotting. PLoS ONE 2014 (9), e94000.

Bhattacharjee, K., Banerjee, S., Joshi, S.R., 2012. Diversity of *Streptomyces* spp. in Eastern Himalayan region-computational RNomics approach to phylogeny. Bioinformation 8 (12), 548.

Bhattacharya, S., Debnath, S., Das, P., Saha, A.K., 2019. Diversity of fungal endophyte of L. var. kew from Unokoti district, *Ananus comosus* Tripura with bioactive potential of *Neopestalotiopsis piceana*. Asian J. Pharm. Pharmacol. 5 (2), 353–360.

Bhattacharyya, P.N., Jha, D.K., 2011. Seasonal and depth-wise variation in microfungal population numbers in Nameri forest soil, Assam, Northeast India. Mycosphere 2 (4), 297–305.

Bhattacharyya, P.N., Tanti, B., Barman, P., Jha, D.K., 2014. Culture-independent metagenomic approach to characterize the surface and subsurface soil bacterial community in the Brahmaputra Valley, Assam, North-East India, an Indo-Burma mega-biodiversity hotspot. World J. Microbiol. Biotechnol. 30, 519–528.

Binarani, R.K., Yadava, P.S., 2010. Effect of shifting cultivation on soil microbial biomass C, N and P under the shifting cultivations systems of Kangchup Hills, Manipur, North-East India. J. Exp. Sci. 1 (10).

Borah, M., Das, S., Baruah, H., Boro, R.C., Barooah, M., 2018. Diversity of culturable endophytic bacteria from wild and cultivated rice showed potential plant growth promoting activities. bioRxiv, 310797.

Champdore, M.D., Staiano, M., Rossi, M., D'Auria, S., 2007. Proteins from extremophiles as stable tools for advanced biotechnological applications of high social interest. J. R. Soc. Interface 4 (13), 183–191.

Das, K., Nath, R., Azad, P., 2013. Soil microbial diversity of Dibru-Saikhowa biosphere reserve forest of Assam, India. Global J. Sci. Front. Res. C Biol. Sci. 13 (3), 7–13.

Das, R., Romi, W., Das, R., Sharma, H.K., Thakur, D., 2018. Antimicrobial potentiality of actinobacteria isolated from two microbiologically unexplored forest ecosystems of Northeast India. BMC Microbiol. 18 (1), 1–16.

Das, J., Sultana, S., Rangappa, K., Kalita, M., Thakuria, D., 2020. Endophyte bacteria alter physiological traits and promote growth of rice (*Oryza sativa* L.) in aluminium toxic and phosphorus deficient acid inceptisols. J. Pure Appl. Microbiol. 14 (1), 627–639. https://doi.org/10.22207/JPAM.14.1.65.

De Mandal, S., Zothansanga, Lalremsanga, H.T., Kumar, N.S., 2015. Bacterial diversity of Murlen National Park located in Indo-Burman Biodiversity hotspot region: a metagenomic approach. Genom Data 5, 25–26.

Debnath, R., Saikia, R., Sarma, R.K., Yadav, A., Bora, T.C., Handique, P.J., 2013. Psychrotolerant antifungal *Streptomyces* isolated from Tawang, India and the shift in chitinase gene family. Extremophiles 17 (6), 1045–1059.

Debnath, R., Yadav, A., Gupta, V.K., Singh, B.P., Handique, P.J., Saikia, R., 2016. Rhizospheric bacterial Community of endemic *Rhododendron arboreum* Sm. Ssp. delavayi along Eastern Himalayan Slope in Tawang. Front. Plant Sci. 7, 1345. https://doi.org/10.3389/fpls.2016.01345.

Deka, J., Thakuria, D., Khyllep, A., Ahmed, G., 2019. Isolation and functional characterization of beneficial bacteria associated with roots of *Thysanolaena maxima* and rhizospheric soil enzymatic activities in jhum agriculture. Curr. Agric. Res. J. 7 (2), 189.

Devi, K.A., Pandey, P., 2017. Role of Bacterial Endophytes in Growth Promotion of *Achyranthes aspera* L (Ph.D. thesis). Assam University, Silchar.

Devi, L.S., Khaund, P., Nongkhlaw, F.M., Joshi, S.R., 2012. Diversity of culturable soil micro-fungi along altitudinal gradients of Eastern Himalayas. Mycobiology 40 (3), 151–158.

Devi, N.O., Devi, R.T., Rajesh, T., Thakuria, D., Ningthoujam, K., Debbarma, M., Hajong, M., 2020. Evaluation of endophytic *Bacillus* spp. isolates from tomato roots on seed germination and seedling vigour of tomato. J. Pharmacogn. Phytochem. 9 (2), 2402–2406.

Domesch, K.H., Gams, W., Anderson, T.H., 1980. Compendium of Soil Fungi. Academic Press, London.

Droogers, P., Bouma, J., 1996. Biodynamic vs. conventional farming effects on soil structure expressed by simulated potential productivity. Soil Sci. Soc. Am. J. 60 (5), 1552–1558.

Dutta, J., Handique, P.J., Thakur, D., 2015. Assessment of culturable tea rhizobacteria isolated from tea estates of Assam, India for growth promotion in commercial tea cultivars. Front. Microbiol. 6, 1252.

Ellis, R.J., Morgan, P., Weightman, A.J., Fry, J.C., 2003. Cultivation-dependent and -independent approaches for determining bacterial diversity in heavy-metal-contaminated soil. Appl. Environ. Microbiol. 69 (6), 3223.

Fierer, N., Leff, J.W., Adams, B.J., Nielsen, U.N., Bates, S.T., Lauber, C.L., Owens, S., Gilbert, J.A., Wall, D.H., Caporaso, J.G., 2012. Cross-biome metagenomic analyses of soil microbial communities and their functional attributes. Proc. Natl. Acad. Sci. 109 (52), 21390–21395.

Gilman, J.C., 1957. A Manual of Soil Fungi. Iowa State College Press, Ames, IA, p. 450.

Giri, K., 2019. The first report of indigenous free-living diazotroph *Kosakonia sacchari* isolated from Himalayan alder-based shifting cultivation system in Nagaland, India. J. Soil Sci. Plant Nutr. 19 (3), 574–579.

Giri, K., Mishra, G., Jayaraj, R.S.C., Kumar, R., 2018. Agrobio-cultural diversity of alder-based shifting cultivation practiced by Angami tribes in Khonoma village, Kohima, Nagaland. Curr. Sci. 115 (4), 598–599.

Girvan, M.S., Bullimore, J., Ball, A.S., Pretty, J.N., Osborn, A.M., 2004. Responses of active bacterial and fungal communities in soils under winter wheat to different fertilizer and pesticide regimens. Appl. Environ. Microbiol. 70 (5), 2692.

Gortari, F., Nowosad, M.I.P., Laczeski, M.E., Onetto, A., Cortese, I.J., Castrillo, M.L., Bich, G.A., Alvarenga, A.E., Lopez, A.C., Villalba, L., Zapata, P.D., 2020. Biofertilizers and biocontrollers as an alternative to the use of chemical fertilizers and fungicides in the propagation of yerba mate by mini-cuttings. Rev. Arvore 43.

Handelsman, J., Rondon, M.R., Brady, S.F., Clardy, J., Goodman, R.M., 1998. Molecular biological access to the chemistry of unknown soil microbes: a new frontier for natural products. Chem. Biol. 5, 245–249.

Hassan, Q.P., 2015. ENVIS Newsletter on Himalayan Ecology. vol. 12 (4) G.B. Pant Institute of Himalayan Environment and Development (GBPIHED), Kosi-Katarmal, Almora, Uttarakhand, pp. 1–12.

Hassan, N., Nakasuji, S., Elsharkawy, M.M., Naznin, H.A., Kubota, M., Ketta, H., Shimizu, M., 2017. Biocontrol potential of an endophytic *Streptomyces* sp. strain MBCN152-1 against *Alternaria brassicicola* on cabbage plug seedlings. Microbes Environ., ME17014.

Jain, R., Chaudhary, D., Dhakar, K., Pandey, A., 2016. A consortium of beneficial fungi survive the fire operations under shifting cultivation in northeast Himalaya, India. Natl. Acad. Sci. Lett. 39 (5), 343–346.

Joshi, S.R., Saikia, P., Koijam, K., 2009. Characterization of microbial indicators to assess the health of degraded soil in Cherrapunjee, India highest rainfall area of the world. Int. J. Biotechnol. Biochem. 5 (4), 379–391.

Joshi, S.R., Banerjee, S., Bhattacharjee, K., Lyngwi, N.A., Koijam, K., Khaund, P., Devi, L.S., Nongkhlaw, F.M.W., 2015a. Northeast microbial database: a web-based databank of culturable soil microbes from Northeast India. Curr. Sci. 108 (09), 1702–1706. http://mblabnehu.info/nemid/.

Joshi, S.R., Bhattacharjee, K., Banerjee, A., Bareh, D.A., 2015b. Microbial diversity distribution in the lower belt of eastern Himalaya. In: Biodiversity in Trop Ecosystems. Today & Tomorrow's Printers and Publishers, pp. 261–288.

Kalita, M., Bharadwaz, M., Dey, T., Gogoi, K., Dowarah, P., Unni, B.G., Ozah, D., Saikia, I., 2015. Developing novel bacterial based bioformulation having PGPR properties for enhanced production of agricultural crops. Indian J. Exp. Biol. 53 (1), 56–60.

Kamala, T., Indira, S., 2011. Evaluation of indigenous *Trichoderma* isolates from Manipur as biocontrol agent against *Pythium aphanidermatum* on common beans. 3 Biotech 1 (4), 217–225.

Khiangte, L., 2017. 16S rRNA Gene Profiling of Phosphorus Solubilizing Bacteria From Paddy Fields of Thenzawl (Doctoral dissertation). Mizoram University.

Khunjan, O., Pandey, P., 2018. Draft genome sequence of *Bacillus subtilis* strain FB6-3, isolated from fermented bamboo shoot. Microbiol. Resour. Announc. 7 (19), e01319-18.

Kichu, A., Ajungla, T., Nyenthang, G., Yeptho, L., 2020. Colonial and morphological characteristics of soil fungi from jhum land. Ind. J. Agric. Res. 54 (1).

Kipgen, T.L., Majumdar, D., Tyagi, W., Thakuria, D., Nath, B.C., 2016. Characterization of fluorescent pseudomonads from maize and rice rhizosphere of Meghalaya and screened for their antagonistic ability against *Ralstonia solanacearum*. Bioscan 11 (2), 699–704.

Kirk, J.L., Beaudette, L.A., Hart, M., Moutoglis, P., Klironomos, J.N., Lee, H., Trevors, J.T., 2004. Methods of studying soil microbial diversity. J. Microbiol. Methods 58 (2), 169–188.

Kotoky, R., Pandey, P., 2020. Rhizosphere assisted biodegradation of benzo (α) pyrene by cadmium resistant plantprobiotic *Serratia marcescens* S2I7, and its genomic traits. Sci. Rep. https://doi.org/10.1038/s41598-020-62285-4.

Kotoky, R., Das, S., Singha, L.P., Pandey, P., Singha, K.M., 2017a. Biodegradation of Benzo (a) pyrene by biofilm forming and plant growth promoting *Acinetobacter* sp. strain PDB4. Environ. Technol. Innov. 8, 256–268.

Kotoky, R., Singha, L.P., Pandey, P., 2017b. Draft genome sequence of heavy metal-resistant soil bacterium *Serratia marcescens* S2I7, which has the ability to degrade polyaromatic hydrocarbons. Genome Announc. 5 (48), e01338-17.

Kotoky, R., Singha, L.P., Pandey, P., 2017c. Draft genome sequence of polyaromatic hydrocarbon-degrading bacterium *Bacillus subtilis* SR1, which has plant growth-promoting attributes. Genome Announc. 5 (49), e01339-17.

Kshetri, L., Pandey, P., Sharma, G.D., 2017. Solubilization of inorganic rock phosphate by rhizobacteria of *Allium hookeri* Thwaites and influence of carbon and nitrogen sources amendments. J. Pure Appl. Microbiol. 11, 1899–1908.

Lyngwi, N.A., Koijam, K., Sharma, D., Joshi, S.R., 2012. Cultivable bacterial diversity along the altitudinal zonation and vegetation range of tropical Eastern Himalaya. Rev. Biol. Trop. 61 (1), 467–490.

Lyngwi, N.A., Koijam, K., Sharma, D., Joshi, S.R., 2013. Cultivable bacterial diversity along the altitudinal zonation and vegetation range of tropical Eastern Himalaya. Rev. Biol. Trop. 61 (1), 467–490.

Mader, P., Fliessbach, A., Dubois, D., Gunst, L., Fried, P., Niggli, U., 2002. Soil fertility and biodiversity in organic farming. Science 296 (5573), 1694–1697.

Malviya, M.K., Pandey, A., Sharma, A., Tiwari, S.C., 2013. Characterization and identification of actinomycetes isolated from fired plots under shifting cultivation in Northeast Himalaya, India. Ann. Microbiol. 63 (2), 561–569.

Mazumdar, A., Deka, M., 2013. Isolation of free living nitrogen fixing bacteria from crude oil contaminated soil. Int. J. Biol. Technol. 3 (4), 69–76.

Miah, S., Dey, S., Haque, S.S., 2010. Shifting cultivation effects on soil fungi and bacterial population in Chittagong Hill Tracts, Bangladesh. J. For. Res. 21 (3), 311–318.

Najar, I.N., Sherpa, M.T., Das, S., Verma, K., Dubey, V.K., Thakur, N., 2018. *Geobacillus yumthangensis* sp. nov., a thermophilic bacterium isolated from a north-east Indian hot spring. Int. J. Syst. Evol. Microbiol. 68 (11), 3430–3434.

Nath, S., Deb, B., Sharma, I., Pandey, P., 2013. Isolation and characterization of heavy metal resistant bacteria and its effect on shoot growth of *Oryza sativa* inoculated in industrial soil. Ann. Plant Sci. 2, 188–193.

Nevita, T., Sharma, G.D., Pandey, P., 2018. Composting of rice-residues using lignocellulolytic plant-probiotic *Stenotrophomonas maltophilia*, and its evaluation for growth enhancement of *Oryza sativa* L. Environ. Sustain. 1 (2), 185–196.

Olapade, O.A., Leff, L.G., 2005. Seasonal response of stream biofilm communities to dissolved organic matter and nutrient enrichments. Appl. Environ. Microbiol. 71 (5), 2278.

Olsen, G.J., Lane, D.J., Giovannoni, S.J., Pace, N.R., Stahl, D.A., 1986. Microbial ecology and evolution: a ribosomal RNA approach. Annu. Rev. Microbiol. 40, 337–365.

Panda, A.K., Bisht, S.S., Kaushal, B.R., De Mandal, S., Kumar, N.S., Basistha, B.C., 2017. Bacterial diversity analysis of Yumthang hot spring, North Sikkim, India by Illumina sequencing. Big Data Anal. 2 (1), 1–7.

Pandey, A., Chaudhry, S., Sharma, A., Choudhary, V.S., Malviya, M.K., Chamoli, S., Rinu, K., Trivedi, P., Palni, L.M.S., 2011. Recovery of *Bacillus* and *Pseudomonas* spp. from the 'Fired Plots' under shifting cultivation in Northeast India. Curr. Microbiol. 62 (1), 273–280.

Pansombat, K., Kanazawa, S., Horiguchi, T., 1997. Microbial ecology in tea soils: I. Soil properties and microbial populations. Soil Sci. Plant Nutr. 43 (2), 317–327.

Puranik, S., Pal, R.R., More, R.P., Purohit, H.J., 2016. Metagenomic approach to characterize soil microbial diversity of Phumdi at Loktak Lake. Water Sci. Technol. 74 (9), 2075–2086.

Rajkumari, J., Singha, L.P., Pandey, P., 2017. Draft genome sequence of *Klebsiella pneumoniae* AWD5. Genome Announc. 5 (5), e01531-16.

Saha, P., Chakrabarti, T., 2006a. *Aeromonas sharmana* sp. nov., isolated from a warm spring. Int. J. Syst. Evol. Microbiol. 56, 1905–1909.

Saha, P., Chakrabarti, T., 2006b. *Flavobacterium indicum* sp. nov., isolated from warm spring water in Assam, India. Int. J. Syst. Evol. Microbiol. 56, 2617–2621.

Saha, P., Krishnamurthi, S., Mayilraj, S., Prasad, G.S., Bora, T.C., Chakrabarti, T., 2005a. *Aquimonas voraii* gen. nov., sp. nov., a novel gamma proteobacterium isolated from a warm spring of Assam, India. Int. J. Syst. Evol. Microbiol. 55 (4), 1491–1495.

Saha, P., Mondal, A.K., Mayilraj, S., Krishnamurthi, S., Bhattacharya, A., Chakrabarti, T., 2005b. *Paenibacillus assamensis* sp. nov., a novel bacterium isolated from a warm spring in Assam, India. Int. J. Syst. Evol. Microbiol. 55, 2577–2581.

Saikia, P., Joshi, S.R., 2012. Changes in microfungal community in Cherrapunji—the wettest patch on Earth as influenced by heavy rain and soil degradation. Adv. Microbiol. 2 (4), 456–464.

Saikia, R., Gogoi, D.K., Mazumder, S., Yadav, A., Sarma, R.K., Bora, T.C., Gogoi, B.K., 2011. *Brevibacillus laterosporus* strain BPM3, a potential biocontrol agent isolated from a natural hot water spring of Assam, India. Microbiol. Res. 166 (3), 216–225.

Sangma, C.B., Thakuria, D., 2019. Isolation and screening of cellulose degrading microorganisms from forest floor litters of Jhum fallows. Proc. Natl. Acad. Sci. India Sect. B: Biol. Sci. 89 (3), 999–1006.

Sharma, P., Thakur, D., 2020. Antimicrobial biosynthetic potential and diversity of culturable soil actinobacteria from forest ecosystems of Northeast India. Sci. Rep. 10 (1), 1–18.

Singh, S.S., Schloter, M., Tiwari, S.C., Dkhar, M.S., 2011. Diversity of community soil DNA and bacteria in degraded and undegraded tropical forest soils of North-Eastern India as measured by ERIC-PCR fingerprints and 16S rDNA-DGGE profiles. J. Biodivers. Environ. Sci. 5 (15), 183–194.

Singha, B., Mazumder, P.B., Pandey, P., 2016. Characterization of plant growth promoting rhizobia from root nodule of *Crotolaria pallida* grown in Assam. Indian J. Biotechnol. 15 (2), 210–216.

Singha, L.P., Kotoky, R., Pandey, P., 2017. Draft genome sequence of *Pseudomonas fragi* strain DBC, which has the ability to degrade high molecular weight polyaromatic hydrocarbons. Genome Announc. 5 (49), e01347-17.

Subramanian, C.V., 1971. Hypomycetes an Account of Indian Species Except Cercospora. Indian Council of Agricultural Research Publication, New Delhi.

Suenaga, H., 2012. Targeted metagenomics: a high-resolution metagenomics approach for specific gene clusters in complex microbial communities. Environ. Microbiol. 14 (1), 13–22.

Swer, H., Dkhar, M.S., Kayang, H., 2011. Fungal population and diversity in organically amended agricultural soils of Meghalaya. Ind. J. Org. Syst. 6 (2), 3–12.

Thakuria, D., Talukdar, N.C., Goswami, C., Hazarika, S., Boro, R.C., Khan, M.R., 2004. Characterization and screening of bacteria from rhizosphere of rice grown in acidic soils of Assam. Curr. Sci., 978–985.

Tokuda, S.I., Hayatsu, M., 2001. Nitrous oxide emission potential of 21 acidic tea field soils in Japan. Soil Sci. Plant Nutr. 47 (3), 637–642.

Vartoukian, S.R., Palmer, R.M., Wade, W.G., 2010. Strategies for culture of 'unculturable' bacteria. FEMS Microbiol. Lett. 309 (1), 1–7. https://doi.org/10.1111/j.1574-6968.2010.02000.x.

Venkatachalam, S., Gowdaman, V., Prabagaran, S.R., 2015. Culturable and culture-independent bacterial diversity and the prevalence of cold-adapted enzymes from the Himalayan mountain ranges of India and Nepal. Microb. Ecol. 69 (3), 472–491.

Vishniac, H.S., 1993. The microbiology of Antarctic soils. In: Antarctic Microbiology. Wiley-Liss, pp. 297–341.

Microbial consortium and crop improvement: Advantages and limitations

Dibyajit Lahiri[a], Moupriya Nag[a], Sougata Ghosh[b], Ankita Dey[c], and Rina Rani Ray[c]

[a]Department of Biotechnology, University of Engineering & Management, Kolkata, West Bengal, India, [b]Department of Microbiology, School of Science, RK. University, Rajkot, Gujarat, India, [c]Department of Biotechnology, Maulana Abul Kalam Azad University of Technology, Haringhata, West Bengal, India

4.1 Introduction

The present need of the society is to produce ample quantity of food for more than 7 billion people, and the size is projected to reach approximately 9.5 billion by the year 2050 (Green et al., 2005). Further reports conferred that in the year 2020, approximately 900 million people were malnourished, which mainly occurred due to the inappropriate use of various agrochemicals like pesticides and fertilizers for the enhancement of agricultural yield (Gomiero et al., 2011). It has been observed that the soil associated with agriculture is gradually losing their physical and chemical properties and biological health and thus having a direct impact on the quality of vegetation (Tilman et al., 2002). Microbial communities that are present on earth surface play a pragmatic role in the maintenance of biogeochemical cycles. Rapid increase in the size of human population resulted in the enhancement of consumption of vegetables and fruits, and the demand of food materials would increase many-fold in years to come (Martin, 2018). The mechanism of sustainable agriculture and enhancement of agricultural productivity to ensure such demand to reach the plates of the consumers comprises various agrarian techniques that sometime have a detrimental effect upon the environment. At recent times, large quantities of insecticides, chemical fertilizers, and herbicides have been used to enhance the yield of crop and protect them from various types of pathogens. But the rampant use of these harsh chemicals has not only brought about undesired degradation of the environmental resources (Agostini et al., 2010), but also resulted in the destabilization of various microbiota associated with soil due to the deposition of heavy metals, nitrates, and toxins (Babin et al., 2019).

Such disastrous environmental issues have resulted in the development of newer approaches to facilitate the production of food without the use of much of genetically modified crops and agrochemicals (Mitchell et al., 2016). The alternate mechanism of enhancing the production of food is by the use of various types of plant-growth-promoting microorganisms (Cedeño et al., 2021) and various types of genetic modifications that can be brought about in plants (Gong et al., 2020). The applications of two or more compatible microbial cells belonging to different species help in the enhancement of beneficial additives, and their synergism will help to bring about a number of desired positive effects (Louca et al., 2018). The mechanism of innovative management of crop is an important technique in meeting the supply and demand of food, thereby bridging the gap between the yield of crop and enhancement in the size of the consumer population. This leads to the adoption of various eco-friendly strategies like the utilization of microbial consortia serving manifold functions like breakdown of nutrients for the easy uptake by plants, removal of various types of pathogenic microbes, reduction in the loss of crops, and production of molecules for crop plant protection. It has been observed that consortia made up of varied species are able to provide better additives to the soil that are beneficial for the yield of crops (Louca et al., 2018). The new technique of "plant-microbiome engineering" results in the addition of various types of effective bioinoculants that induce numerous types of biological networks within diverse soil types. This also helps in the recovery of various types of useful microbes, associated with the enhancement of the fertility of the soil. Thus, the microbial consortia play an important role in promoting the growth of the plants (Woo and Pepe, 2018).

4.2 Microbial consortia

The natural environment comprises microbial consortia that likely belongs to mammalian gut microbiota (Clavel et al., 2017), microbiota existing within the soil (Van Der Heijden et al., 2008), within food, biological wastes (Bayer et al., 2007), and aquatic ecosystem (Paerl and Pinckney, 1996). The natural environment is comprised of various types of microbes containing balanced communities, which help in maintenance of benefit between the members of the consortium and the host. The studies on the role of the microbial consortia have been a keen point of interest for the researches in the field of medicine and biotechnology. The microbial communities that exist naturally in the environment have been used for the purpose of waste water treatment, bioremediation, microbial fuel cells, and methanogenic digestion. Biofilm and lichens are considered to be natural consortia of microbial cells. The growth of the microbial consortia is also being dependent upon the various biotic and abiotic factors of natural environment. This concept can be utilized for the purpose of targeting various types of artificial assemblage that can be used in the processes of bioproduction, bioremediation, and agriculture (Kouzuma and Watanabe, 2014).

The time of existence of the microbial colony is dependent on the interaction between its members. Such interactions can be categorized into symbiotic comprising commensalism and mutualism, and antagonistic comprising parasitic and predatory (Santos and Reis, 2014). It has been further observed that the existence of the microbial species tends to remain in clusters to protect themselves from adverse conditions of the environment, incursions, and habitat. Actually, a microbial consortium is constituted by two or more microbial groups coexisting symbiotically that may be endo-symbiotically or ecto-symbiotically, or by both (Madigan et al., 2019).

4.3 Plant microbiota existing above the ground

Vegetative foliar tissues and floral parts that are the aerial components of the plants act as an important region to allow the growth of various types of epiphytic and endophytic microbial cells (Compant et al., 2020). The different parts of the plants like stem, leaves, and fruits comprise a systemic distribution of endophytes that is facilitated by the xylem (Compant et al., 2010). Various parts of the plants comprise distinct species of endophytes that are categorized on the basis of their source allocation. These mechanisms are associated with the distribution of microbial species at various genus and species levels within the phyllospheric and endospheric regions. Studies showed that the phyllospheric region possesses a grapevine-like structure comprising various types of microbial species like *Pseudomonas*, *Frigoribacterium*, *Sphingomonas*, *Bacillus*, *Pantoea*, *Acinetobacter*, *Methylobacterium*, and *Pantoea* (Zarraonaindia et al., 2015) (Fig. 4.1).

4.4 Managed microbial consortia

The group of managed consortia is made up of various microbes in assemblage which are either isolated from the natural environment or are manipulated and evolved selectively to meet up with the desired trait (Padmaperuma et al., 2018). The major objective to prepare a consortium is for the purpose of improving crop productivity by the enhancement in biomanufacturing. At natural environment, the

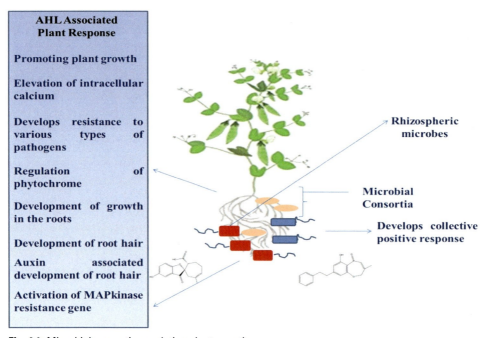

Fig. 4.1 Microbial consortia regulating plant growth.

consortia have taken years to evolve but this can be achieved within a small period of time through planned alternations in the laboratory. This mechanism of manipulation is dependent on factors comprising backgrounds of the microbial community, competitions, ratio of various component partners, and underyielding and overyielding effects (Schmidtke et al., 2010). Although the efficacy of the natural consortium is dependent on the environmental conditions, the artificial consortium behaves the same inside and outside the laboratory and is therefore independent of environmental parameters (Preininger et al., 2018). Hence, greater studies may facilitate the use of these consortia in agricultural fields for the enhancement in the yields (Enebe and Babalola, 2019).

4.4.1 Synthetic consortia

The mechanism of programmed communications existing between the members of the consortia plays a pivotal role in governing the engineering of microbial consortia. Various types of diffusible molecules can be used for the communication purpose. The mechanism of quorum sensing (QS) is the most common type of communication process that is dependent on the density of the microbial population (Smith et al., 2013). Naturally, the cells within the population produce small diffusible signal molecules in the process of QS.

Higher concentrations of the QS molecules result in the activation of the targeted genes that support the enhancement in the growth of the population. The application of synthetic biology helps in the activation of the gene circuit under the regulation of QS-activated promoter that takes place in the presence of high density of bacteria (Khalil and Collins, 2010). The microbial species existing within the synthetic consortia are unable to produce metabolites that are made by the replacement of the genes, responsible for the synthesis of metabolites in the presence of antibiotic-resistant marker.

4.5 Plant growth-promoting microorganisms

Plant growth-promoting microorganisms (PGPM) that include fungi, bacteria, algae, and actinomycetes have an important role in the promotion of plant growth by the involvement of direct or indirect mechanisms like ethylene reduction, mineral nutrition, and suppression of disease (Abhilash et al., 2016). These microorganisms play a significant function in sustainable agriculture by improving the fertility of the soil, production of various types of resistant crops, promotion of biodiversity through interaction with numerous beneficial microbes, and thereby enhancing the crop productivity. The efficacies of PGPMs depend on the interactions that take place between the single microbial cell and plants, thereby enhancing different parameters of growth and health of the plants comprising weight and length of the plants, nutritional content, and chlorophyll content that are associated with the tissues and fruits (Khan et al., 2019; Mahmud et al., 2021). A number of microbial species inhabiting various niches interact with each other in several ways. The plant has substantial influence in the growth of these microbial species at the rhizospheric region due to the exudation of various types of compounds like sugars, organic acids, and amino acids resulting in species colonization. A large number of interactions would take place following the inoculation of single organism in the field that results in the facilitation of growth of the plants (Babalola, 2010). The mechanism of seed germination also involves a constant interaction with the microbial species that are present above and below the ground. The dynamics of interaction varies with microbial structures that are associated with various phases of plant growth and development (Aulakh et al., 2001). The variation of the interaction dynamics is due to the presence of several types of volatile compounds, produced in the form of exudates of roots or from the aerial parts of the plants, thereby allowing the growth of rhizospheric microbiome (Mercado-Blanco and Bakker, 2007). Such interactions between the microbiome of soil and plants are also maintained by different abiotic factors also like availability of water, pH, temperature, and nutrient availability (Santoyo et al., 2017). Studies have shown that microbial consortia is more effective in comparison with individual strains in metabolizing various types of complex compounds and degrading various types if plant polymeric substances in two or more steps (Brenner et al., 2008). The premier commercial bioinoculant comprising living cells like rhizobial bacteria inducing the formation of nodules upon the roots help in the process of nitrogen fixation. Other groups of nitrogen-fixing organism like *Sinorhizobium meliloti* regulate the development of root nodules within *Medicago truncatula* (Mahmud et al., 2020). Studies have also shown that the promotion of biological nitrogen fixation by Clostridiales predominates the anaerobic consortia of microbes by humin at various pH that serve as extracellular mediator of electrons and can be used as an alternative of chemical fixation of nitrogen that leads to an effective growth of the plants (Dey et al., 2021).

At present scenario, a large number of synthetic fertilizers are applied to crops in a large scale to meet up the enhanced demand for food (Goswami and Deka, 2020). The use of synthetic fertilizers proved to be detrimental for environmental health, and thus, the use of plant growth-promoting bacteria (PGPB) played an important role in the enhancement of the fertility of the soil. They help in the regulation of the plant growth, development, and the productivity of crops. These growth-promoting effects are due to the presence of various types of macro- and micronutrients that are present within the soil (Solis et al., 2020). Some microbial consortia also possess the ability of fixing nitrogen, thereby transforming various types of commonly unavailable nutrients and making them available for the production of the phytohormones. Some of them can chelate iron that helps in the maintenance of the quality and health of the soil (Gosal and Kaur, 2017). The involvement of higher number of species with an eventually higher rate of interactions among the members constitutes a complex consortium, while simple consortium may be made up of only two types of microbial species (Bashan et al., 2020). The success of the microbial consortia is dependent on the type of strains that are associated with it (Bashan et al., 2020) and relative rate of the production of various types of metabolites.

4.6 Microbial interactions within consortium

Large diversity of bacterial species forms a part of plant microbiota that possess the ability to promote the growth and development in plants during both stress and optimal environmental conditions (van Deynze et al., 2018). Interactions that take place between the members of the consortia is the key factor that regulates the development of microbial consortium (Singh et al., 2019). The interactions taking place between the bacterial species can be classified into three effects: stimulatory, inhibitory, and neutral (Singh et al., 2019). Positive interactions result in the creation of network that helps in supporting individual members by the process of cross-feeding that involves the utilization of metabolites produced by one bacteria by the other. The mechanisms like protocooperation, mutualism, and commensalism are the instances of positive responses.

The negative interactions involve the suppression of the bacterial members living within the consortium, thereby destroying the architecture and functions of the community (Loccoz et al., 2015), which may be exemplified by the phenomenon of ammensalism, a type of unidirectional interaction, where one of the members gets affected by the toxic substance produced by another member (Roell et al., 2019). The mechanisms of parasitism and predation also denote the negative regulation of the growth of one species by the other species. Competition occurs among the members of consortium with the same niche requirement, resulting in the dominance of the faster growing member (Loccoz et al., 2015).

The mechanism of neutral interactions involves members of the consortium that do not affect or influence another member. The mechanism of neutralism occurs when two species consume different substances and do not produce compounds that have an inhibitory effect on other members of the consortium (Singh et al., 2019).

In agricultural practices, the microbial consortia showing a positive mutualistic interaction among the members are desirable for the stable performance and yield of the agricultural crops (Liu et al., 2012). The communication that takes place within the members of the consortium is highly dependent on various types of molecular signals, out of which the mechanism of quorum sensing plays an important role (Mukherjee and Bassler, 2019). The signal molecule acyl homoserine lactone (AHL) is the commonly understood chemical substance that aids in the mechanism of cell-to-cell communication (Song et al., 2011). The AHLs that are produced by bacterial cells like *Serratia liquefaciens* and *S. phymuthica* play an important role in the development of roots and the biomass of the plants. Other types of bacterial species like *Sinorhizobium fredii* and *Pantoea ananatis* result in the stimulation of the development of biofilm within the roots of *Oryza sativa* and *Phaseolus vulgaris* (Montaño et al., 2013). Studies have also shown that AHL helps in the induction of Ca^{2+} levels and development of primary growth of the roots (Montaño et al., 2013).

Apart from AHLs, other type of signaling molecules that are associated with the bacterial consortia is volatile organic compounds (VOCs) that are used for the purpose of interactions between plant and bacteria as well as bacteria and bacteria (Bukhat et al., 2020). The VOCs include compounds like terpenoids, alkenes, alkanes, sulfur-containing compounds, and alcohols (Insam and Seewald, 2010). The VOCs that are produced by the bacterial species have a positive effect on the growth of nearby or distant bacterial species that are present within the consortium. The antibacterial property of VOCs helps in controlling plant pathogens. It also helps in the stimulation of genes that are responsible for hormonal signaling, thereby controlling the growth of the plants (Bohm et al., 2017).

Fungal species that aid the promotion of plant growth comprise various types of mycorrhizal fungi that includes *Rhizophagus*, *Laccaria*, and *Funneliformis* that develop an obligate relationship with the roots of the plants (Pringle et al., 2009; Woo and Pepe, 2018). These consortia are involved in the process of exchange of carbon, thereby enhancing the capacity of plants to absorb water along with the nutrients, thereby bringing about a negative effect on abiotic and biotic stresses.

4.7 Stimulation of plant growth under stressed condition

Various physical conditions like climate change, excess agrochemicals, and modifications in land result in the development of various stress conditions for the interactions between the microbes in agro-ecosystem (Vimal et al., 2017). Different types of stress conditions result in the impairment in the morphology of plants, regulation of genes, and soil physicochemical and microbial properties resulting in the loss of yield (Goswami and Deka, 2020). Abiotic stresses like drought, extremes of temperature, salinity, and environmental contaminants and biotic stresses including interactions with living organisms like fungi, bacteria, viruses, insects, weeds and arachnids (Kumar and Verma, 2018) on agricultural plants are to be managed by plant-associated microbial consortia. Beneficial groups of bacteria help in the sustenance of various types of intrinsic resistances of plants; thereby, the persistence of microbial consortia protects the plants from various types of negative effects (Vimal et al., 2017).

4.8 Application of microbial consortium in agriculture

4.8.1 Nutrient mobilization and management of soil by the rhizospheric microbes

A successful farming is primarily dependent on the presence of high-yielding crop. The improvisation in the quality of soil is achieved by the mechanism of crop rotation, use of fertilizers and pesticides but the last two are found to have a detrimental effect on the environment. On the other hand, the biofertilizers made up of living microbes lead to the enhancement of crop yield in an ecofriendly way (Trivedi et al., 2017). Inoculation with concoction of microbes helps in supplying nutrients to the plants by the release of various compounds from the complex substances. Combination of microbial cells like *Glomus mosseae, Azospirillum, Pseudomonas striata, Tolypothrix tenuis, Nostoc muscorum, Anabaena variabilis,* and *Aulosira fertilissima* showed an enhancement in the yield of the plants (Chinnusamy et al., 2006). It was further observed that consortia of *Azospirillum brasilense, Punica granatum,* and *G. mosseae* when added to bananas and custard apples showed a marked increase in its productivity (Aseri et al., 2008). The soil microbiota forms an important part of the ecosystem as they provide essential elements like N, P, and C to the plant, which can be used as integral components needed for the growth of the plants by the utilization of organic substances that are associated with the soil. Phosphorous also acts as an essential nutrient required for the maintenance of the plants and enhancing the yield within the plants. Thus, various types of bacteria and fungi help in providing soluble phosphorus for the plant uptake (Baas et al., 2016). These microbial species help in providing soluble phosphorus by the mechanism of transformation of insoluble phosphate to soluble phosphorus by the secretion of organic acids for lowering the pH and thereby release various forms of phosphates (Alori et al., 2017). Bacterial consortia comprising four taxa *Enterobacter cloacae, Pseudomonas putida, Citrobacter freundii,* and *Comamonas testosterone* showed an effective role in solubilizing the phosphate, thus enhancing the productivity of the crops twofold especially within tomatoes (Baas et al., 2016).

4.8.2 Biostimulants

Biostimulants are substances different from the plant nutrients that are used for the purpose of enhancing the growth and development of the plants. Microbial consortia act as biostimulants for various types of food crops (Caradonia et al., 2018). Studies have shown that these biostimulants can enhance the productivity and efficiency by 5%–25% (Caradonia et al., 2018). There are five major categories of biostimulants: humic acids, microbial inoculants, fulvic acids, amino acids, and hydrolysates (Calvo et al., 2014). A major role played by biostimulants is that they possess the ability of altering the microbial flora (Le Mire et al., 2016) (Table 4.1).

Table 4.1 Microbial consortia acting as biostimulants.

Type of Consortium	Stress	Crop	Effect of the consortium	Reference
Pseudomonas sp., *Ochrobactrum pseudogrignonense* and *Bacillus subtilis*	Drought	Pea and black gram	Promotes plant growth, elevates various types of ROS oxygen-scavenging enzymes, and various types of cellular osmolytes.	Saikia et al. (2018)
Curtobacterium sp., *Rhizobium tropici*, *Acinetobacter calcoaceticus*, *Enterobacter asburiae*, *Rhodotorula graminis*, *Burkholderia vietnamiensis*, *Rahnella* sp., *Burkholderia* sp., *Pseudomonas* sp. and *Sphingomonas yanoikuyae*	Drought	Poplar	Reduction in the reactive oxygen species and promotes the growth of the plants.	Khan et al. (2016)
P. agglomerans and *B. megaterium*	Drought and aluminum	Mung bean	Promotion of plant growth, reduces the uptake of Al by plants, enhances the proline content, and also increases the superoxide dismutase activity	Silambarasan et al. (2019)
Alcaligenes sp., *Pseudomonas* sp., *Serratia proteamaculans* and *Bacillus* sp.	Petroleum hydrocarbons and petrol	Saltgrass	Promotes the plant growth	Xia et al. (2020)
Bacillus paralicheniformis, *Brevibacillus agri* and *Brevibacillus fluminis*	Salinity	Potato, tomato, chili, and brinjal	Promotion of plant growth	Goswami et al. (2019)
Pseudomonas sp.	*Phytophthora infestans*	Potato	Reduction in the release of zoospores, reduction in the growth of myecelium, and activation of systemic resistance.	de Vrieze et al. (2018)
Stenotrophomonas sp., *Xanthomonas* sp., and *Microbacterium* sp.	*Hyaloperonospora arabidopsidis*	*Arabidopsis thaliana*	Provides resistance to the plants, and promotes growth in the plants	Berendsen et al. (2018)
Bacillus subtilis and *Pseudomonas putida*	*Macrophomina phaseolina*	Mung bean	Provides resistance to the plants, and promotes growth in the plants	Sharma et al. (2018)
Mesorhizobium sp. and *Pseudomonas aeruginosa*	*Sclerotium rolfsii*	Chick pea	Promotes the growth of the plants	Singh et al. (2014)

4.9 Conclusion and future scope

The consortia of microbes find their applicability in the enhancement of agricultural yield. The mechanism of inoculating managed consortia with the plants also helps in the enhancement of the agricultural yield. The interactions between the microbes and plants take place by a network of complex signal molecules that include various types of metabolites, nonvolatile and volatile organic compounds that help in the maintenance of the gene expression (León et al., 2015). Thus, the consortia of microbes play a vital role in the increase in soil fertility by improved nitrogen fixation, solubilization of phosphorous, and the enhancement of the yield of the plants by providing plentiful supply of various nutrients and making them stress resistant.

References

Abhilash, P.C., Dubey, R.K., Tripathi, V., Gupta, V.K., Singh, H.B., 2016. Plant growth-promoting microorganisms for environmental sustainability. Trends Biotechnol. 34, 847–850.

Agostini, F., Tei, F., Silgram, M., Farneselli, M., Benincasa, P., Aller, M.F., 2010. Decreasing nitrate leaching in vegetable crops with better N management. In: Genetic Engineering, Biofertilisation, Soil Quality and Organic Farming. Sustainable Agriculture Reviews. vol. 4. Springer, Dordrecht, pp. 147–200.

Alori, E.T., Glick, B.R., Babalola, O.O., 2017. Microbial phosphorus solubilization and its potential for use in sustainable agriculture. Front. Microbiol. 8, 971. https://doi.org/10.3389/fmicb.2017.00971.

Aseri, G.K., Jain, N., Panwar, J., Rao, A.V., Meghwal, P.R., 2008. Biofertilizers improve plant growth, fruit yield, nutrition, metabolism and rhizo- sphere enzyme activities of pomegranate (*Punica granatum* L.) in Indian Thar Desert. Sci. Hortic. 117, 130–135.

Aulakh, M.S., Wassmann, R., Bueno, C., Kreuzwieser, J., Rennenberg, H., 2001. Characterization of root exudates at different growth stages of ten rice (*Oryza sativa* L.) cultivars. Plant Biol. 3, 139–148.

Baas, P., Bell, C., Mancini, L.M., Lee, M.N., Conant, R.T., Wallenstein, M.D., 2016. Phosphorus mobilizing consortium Mammoth P (TM) enhances plant growth. PeerJ 4, e2121. https://doi.org/10.7717/peerj.2121.

Babalola, O.O., 2010. Beneficial bacteria of agricultural importance. Biotechnol. Lett. 32, 1559–1570.

Babin, D., Deubel, A., Jacquiod, S., Sorensen, S.J., Geistlinger, J., Grosch, R., Smalla, K., 2019. Impact of long-term agricultural management practices on soil prokaryotic communities. Soil Biol. Biochem. 129, 17–28.

Bashan, Y., Prabhu, S.R., de Bashan, L.E., Kloepper, J.W., 2020. Disclosure of exact protocols of fermentation, identity of microorganisms within consortia, formation of advanced consortia with microbe-based products. Biol. Fertil. Soils 56, 443–445.

Bayer, E.A., Lamed, R., Himmel, M.E., 2007. The potential of cellulases and cellulosomes for cellulosic waste management. Curr. Opin. Biotechnol. 18, 237–245.

Berendsen, R.L., Vismans, G., Yu, K., Song, Y., de Jonge, R., Burgman, W.P., Burmølle, M., Herschend, J., Bakker, P.A.H.M., Pieterse, C.M.J., 2018. Disease-induced assemblage of a plant-beneficial bacterial consortium. ISME J. 12, 1496–1507.

Bohm, K.S., Sánchez, L.M., Garbeva, P., 2017. Microbial volatiles: small molecules with an important role in intra- and inter-kingdom interactions. Front. Microbiol. 8, 2484.

Brenner, K., You, L., Arnold, F.H., 2008. Engineering microbial consortia: a new frontier in synthetic biology. Trends Biotechnol. 26, 483–489.

Bukhat, S., Imran, A., Javaid, S., Shahid, M., Majeed, A., Naqqash, T., 2020. Communication of plants with microbial world: exploring the regulatory networks for PGPR mediated defense signaling. Microbiol. Res. 238, 126486.

Calvo, P., Nelson, L., Kloepper, J.W., 2014. Agricultural uses of plant biostimulants. Plant Soil 383, 3–41.

Caradonia, F., Battaglia, V., Righi, L., Pascali, G., La, A., 2018. Plant biostimulant regulatory framework: prospects in Europe and current situation at international level. J. Plant Growth Regul. 38 (2), 438–448.

Cedeño, L.R.M., Mosqueda, M.D.C.O., Lara, P.D.L., Cota, F.I.P., Villalobos, S.D.L.S., Santoyo, G., 2021. Plant growth-promoting bacterial endophytes as biocontrol agents of pre- and post-harvest diseases: fundamentals, methods of application and future perspectives. Microbiol. Res. 242, 126612.

Chinnusamy, M., Kaushik, B.D., Prasanna, R., 2006. Growth, nutritional, and yield parameters of wetland rice as influenced by microbial consortia under controlled conditions. J. Plant Nutr. 29, 857–871.

Clavel, T., Lagkouvardos, I., Stecher, B., 2017. From complex gut communities to minimal microbiomes via cultivation. Curr. Opin. Microbiol. 38, 148–155.

Compant, S., Clément, C., Sessitsch, A., 2010. Plant growth-promoting bacteria in the rhizo- and endosphere of plants: their role, colonization, mechanisms involved and prospects for utilization. Soil Biol. Biochem. 42, 669–678.

Compant, S., Cambon, M.C., Vacher, C., Mitter, B., Samad, A., Sessitsch, A., 2020. The plant endosphere world – bacterial life within plants. Environ. Microbiol. 23 (4), 1812–1829.

de Vrieze, M., Germanier, F., Vuille, N., Weisskopf, L., 2018. Combining different potato-associated Pseudomonas strains for improved biocontrol of Phytophthora infestans. Front. Microbiol. 9, 2573.

Dey, S., Awata, T., Mitsushita, J., 2021. Promotion of biological nitrogen fixation activity of an anaerobic consortium using humin as an extracellular electron mediator. Sci. Rep. 11, 6567.

Enebe, M.C., Babalola, O.O., 2019. The impact of microbes in the orchestration of plants' resistance to biotic stress: a disease management approach. Appl. Microbiol. Biotechnol. 103, 9–25.

Gomiero, T., Pimentel, D., Paoletti, M.G., 2011. Is there a need for a more sustainable agriculture? Crit. Rev. Plant Sci. 30, 6–23.

Gong, Z., Xiong, L., Shi, H., Yang, S., Estrella, L.R.H., Xu, G., Chao, D.Y., Li, J., Wang, P.Y., Qin, F., 2020. Plant abiotic stress response and nutrient use efficiency. Sci. China Life Sci. 63, 635–674.

Gosal, S.K., Kaur, J., 2017. Microbial inoculants: A novel approach for better plant microbiome interactions. In: Probiotics in Agroecosystem. Springer, Singapore, pp. 269–289.

Goswami, M., Deka, S., 2020. Plant growth-promoting rhizobacteria—alleviators of abiotic stresses in soil: a review. Pedosphere 30, 40–61.

Goswami, S.K., Kashyap, P.L., Awasthi, S., 2019. Deciphering rhizosphere microbiome for the development of novel bacterial consortium and its evaluation for salt stress management in solanaceous crops in India. Indian Phytopathol. 72, 479–488.

Green, R.E., Cornell, S.J., Scharlemann, J.P.W., Balmford, A., 2005. Farming and the fate of wild nature. Science 307, 550–555.

Insam, H., Seewald, M.S.A., 2010. Volatile organic compounds (VOCs) in soils. Biol. Fertil. Soils 46, 199–213.

Khalil, A.S., Collins, J.J., 2010. Synthetic biology: applications come of age. Nat. Rev. Genet. 11, 367–379.

Khan, Z., Rho, H., Firrincieli, A., Hung, S.H., Luna, V., Masciarelli, O., Kim, S.-H., Doty, S.L., 2016. Growth enhancement and drought tolerance of hybrid poplar upon inoculation with endophyte consortia. Curr. Plant Biol. 6, 38–47.

Khan, A., Singh, J., Upadhayay, V.K., Singh, A.V., Shah, S., 2019. Microbial biofortification: a green technology through plant growth promoting microorganisms. In: Sustainable Green Technologies for Environmental Management. Springer, Singapore, pp. 255–269.

Kouzuma, A., Watanabe, K., 2014. Microbial ecology pushes frontiers in biotechnology. Microbes Environ. 29, 1–3.

Kumar, A., Verma, J.P., 2018. Does plant—microbe interaction confer stress tolerance in plants: a review? Microbiol. Res. 207, 41–52.

Le Mire, G., Nguyen, M.L., Fassotte, B., Du Jardin, P., Verheggen, F., Delaplace, P., Jijakli, M.H., 2016. Review: implementing plant biostimulants and biocontrol strategies in the agroecological management of cultivated ecosystems. Biotechnol. Agron. Soc. Environ. 20, 299–313.

León, R.H., Solís, D.R., Pérez, M.C., Mosqueda, M.C.D.C., Rodríguez, L.I.M., la Cruz, H.R.D., Cantero, E.V., Santoyo, G., 2015. Characterization of the antifungal and plant growth-promoting effects of diffusible and volatile organic compounds produced by Pseudomonas fluorescens strains. Biol. Control 81, 83–92.

Liu, F., Bian, Z., Jia, Z., Zhao, Q., Song, S., 2012. The GCR1 and GPA1 participate in promotion of Arabidopsis primary root elongation induced by N-acyl-homoserine lactones, the bacterial quorum-sensing signals. Mol. Plant-Microbe Interact. 25, 677–683.

Loccoz, Y.M., Mavingui, P., Combes, C., Normand, P., Steinberg, C., 2015. Microorganisms and biotic interactions. In: Environmental Microbiology: Fundamentals and Applications. Springer, Dordrecht, The Netherlands, pp. 395–444.

Louca, S., Polz, M.F., Mazel, F., Albright, M.B.N., Huber, J.A., O'Connor, M.I., Ackermann, M., Hahn, A.S., Srivastava, D.S., Crowe, S.A., et al., 2018. Function and functional redundancy in microbial systems. Nat. Ecol. Evol. 2, 936–943.

Madigan, M., Bender, K., Buckley, D., Sattley, W., Stahl, D., 2019. Brock Biology of Microorganisms, Fifteenth, Global ed. Pearson, New York, NY, p. 173.

Mahmud, K., Makaju, S., Ibrahim, R., Missaoui, A., 2020. Current Progress in nitrogen fixing plants and microbiome research. Plan. Theory 9, 97.

Mahmud, K., Franklin, D., Ney, L., Cabrera, M., Habteselassie, M., Hancock, D., Newcomer, Q., Subedi, A., Dahal, S., 2021. Improving inorganic nitrogen in soil and nutrient density of edamame bean in three consecutive summers by utilizing a locally sourced bio-inocula. Org. Agric. 1–11.

Martin, C., 2018. A role for plant science in underpinning the objective of global nutritional security? Ann. Bot. 122, 541–553.

Mercado-Blanco, J., Bakker, P.A.H.M., 2007. Interactions between plants and beneficial Pseudomonas spp.: exploiting bacterial traits for crop protection. Antonie Leeuwenhoek 92, 367–389.

Mitchell, C., Brennan, R.M., Graham, J., Karley, A.J., 2016. Plant defense against herbivorous pests: exploiting resistance and tolerance traits for sustainable crop protection. Front. Plant Sci. 7, 1132.

Montaño, F.P., Guerrero, I.J., Matamoros, R.C.S., Baena, F.J.L., Ollero, F.J., Rodríguez-Carvajal, M.A., Bellogín, R.A., Espuny, M.R., 2013. Rice and bean AHL-mimic quorum-sensing signals specifically interfere with the capacity to form biofilms by plant-associated bacteria. Res. Microbiol. 164, 749–760.

Mukherjee, S., Bassler, B.L., 2019. Bacterial quorum sensing in complex and dynamically changing environments. Nat. Rev. Microbiol. 17, 371–382.

Padmaperuma, G., Kapoore, R.V., Gilmour, D.J., Vaidyanathan, S., 2018. Microbial consortia: a critical look at microalgae co-cultures for enhanced biomanufacturing. Crit. Rev. Biotechnol. 38, 690–703.

Paerl, H.W., Pinckney, J.L., 1996. A mini-review of microbial consortia: their roles in aquatic production and biogeochemical cycling. Microb. Ecol. 31, 225–247.

Preininger, C., Sauer, U., Bejarano, A., Berninger, T., 2018. Concepts and applications of foliar spray for microbial inoculants. Appl. Microbiol. Bio- Technol. 102, 7265–7282.

Pringle, A., Bever, J.D., Gardes, M., Parrent, J.L., Rillig, M.C., Klironomos, J.N., 2009. Mycorrhizal symbioses and plant invasions. Annu. Rev. Ecol. Syst. 40, 699–715. https://doi.org/10.1146/annurev.ecolsys.39.110707.173454.

Roell, G.W., Zha, J., Carr, R.R., Koffas, M.A., Fong, S.S., Tang, Y.J., 2019. Engineering microbial consortia by division of labor. Microb. Cell Factories 18, 35.

Saikia, J., Sarma, R.K., Dhandia, R., Yadav, A., Bharali, R., Gupta, V.K., Saikia, R., 2018. Alleviation of drought stress in pulse crops with ACC deaminase producing rhizobacteria isolated from acidic soil of Northeast India. Sci. Rep. 8, 1–16.

Santos, C.A., Reis, A., 2014. Microalgal symbiosis in biotechnology. Appl. Microbiol. Biotechnol. 98, 5839–5846.

Santoyo, G., Pacheco, C.H., Salmerón, J.H., León, R.H., 2017. The role of abiotic factors modulating the plant-microbe-soil interactions: toward sustainable agriculture. A review. Span. J. Agric. Res. 15, e03R01.

Schmidtke, A., Gaedke, U., Weithoff, G., 2010. A mechanistic basis for underyielding in phytoplankton communities. Ecology 91, 212–221.

Sharma, C.K., Vishnoi, V.K., Dubey, R.C., Maheshwari, D.K., 2018. A twin rhizospheric bacterial consortium induces systemic resistance to a phytopathogen Macrophomina phaseolina in mung bean. Rhizosphere 5, 71–75.

Silambarasan, S., Logeswari, P., Cornejo, P., Kannan, V.R., 2019. Role of plant growth-promoting rhizobacterial consortium in improving the Vigna radiata growth and alleviation of aluminum and drought stresses. Environ. Sci. Pollut. Res. 26, 27647–27659.

Singh, A., Jain, A., Sarma, B.K., Upadhyay, R.S., Singh, H.B., 2014. Rhizosphere competent microbial consortium mediates rapid changes in phenolic profiles in chickpea during Sclerotium rolfsii infection. Microbiol. Res. 169, 353–360.

Singh, R., Ryu, J., Kim, S.W., 2019. Microbial consortia including methanotrophs: some benefits of living together. J. Microbiol. 57, 939–952.

Smith, R.P., Tanouchi, Y., You, L., 2013. Synthetic microbial consortia and their applications. Synth. Biol., 243–258.

Solis, D.R., Guzmán, M.Á.V., Sohlenkamp, C., Santoyo, G., 2020. Antifungal and plant growth-promoting Bacillus under saline stress modify their membrane composition. J. Soil Sci. Plant Nutr. 20, 1549–1559.

Song, S., Jia, Z., Xu, J., Zhang, Z., Bian, Z., 2011. N-butyryl-homoserine lactone, a bacterial quorum-sensing signaling molecule, induces intracellular calcium elevation in Arabidopsis root cells. Biochem. Biophys. Res. Commun. 414, 355–360.

Tilman, D., Cassman, K.G., Matson, P.A., Naylor, R., Polasky, S., 2002. Agricultural sustainability and intensive production practices. Nature 418, 671–677.

Trivedi, P., Schenk, P.M., Wallenstein, M.D., Singh, B.K., 2017. Tiny microbes, big yields: enhancing food crop production with biological solutions. Microb. Biotechnol. 10, 999–1003.

Van Der Heijden, M.G.A., Bardgett, R.D., Van Straalen, N.M., 2008. The unseen majority: soil microbes as drivers of plant diversity and productivity in terrestrial ecosystems. Ecol. Lett. 11, 296–310.

van Deynze, A., Zamora, P., Delaux, P.-M., Heitmann, C., Jayaraman, D., Rajasekar, S., Graham, D., Maeda, J., Gibson, D., Schwartz, K.D., 2018. Nitrogen fixation in a landrace of maize is supported by a mucilage-associated diazotrophic microbiota. PLoS Biol. 16, e2006352.

Vimal, S.R., Singh, J.S., Arora, N.K., Singh, S., 2017. Soil-plant-microbe interactions in stressed agriculture management: a review. Pedosphere 27, 177–192.

Woo, S.L., Pepe, O., 2018. Microbial consortia: promising probiotics as plant biostimulants for sustainable agriculture. Front. Plant Sci. 9, 1801. https://doi.org/10.3389/fpls.2018.01801.

Xia, M., Chakraborty, R., Terry, N., Singh, R.P., Fu, D., 2020. Promotion of saltgrass growth in a saline petroleum hydrocarbons contaminated soil using a plant growth promoting bacterial consortium. Int. Biodeterior. Biodegrad. 146, 104808.

Zarraonaindia, C., Owens, I.M., Weisenhorn, S.M., West, P., Hampton-Marcell, K., Lax, J., 2015. The soil microbiome influences grapevine-associated microbiota. mBio 6, e02527-14.

Revisiting soil-plant-microbes interactions: Key factors for soil health and productivity

Subhadeep Mondal[a], Suman Kumar Halder[b], and Keshab Chandra Mondal[b]

[a]Centre for Life Sciences, Vidyasagar University, Midnapore, West Bengal, India, [b]Department of Microbiology, Vidyasagar University, Midnapore, West Bengal, India

5.1 Introduction

A gradual rise in the global human population has led to increased demand for food, and it generates tremendous pressure on agriculture to produce more and more crops for consumption. The production of food by the agriculture of a region depends mainly on the soil health, climatic conditions, and process of cultivation. The term "soil," which comes from the Latin word *solum* (floor or ground), is the prime base of our agricultural assets, universal economy, survival, and sustainability (Mishra and Arora, 2019). Over billions of years, the nature of soil is continuously renewable. Due to nonhomogeneous climatic conditions and surface nature of the earth, the soil is asymmetrically spread among the various topographical regions of the earth (Lal, 2015). The erosion of soil is a natural continuous phenomenon that occurs by natural catastrophic consequences or anthropogenic activities worldwide that collectively affects the soil quality. According to UNCCD (Tóth et al., 2018), nearly 2.18×10^7 metric tons of productive soil is disappears annually from the global agricultural system. Soil productivity is an indispensable parameter for successful agricultural yield rather than soil fertility (Bhargava et al., 2017). The nutrient profile of the soil and its physical ability regulate the fertility of the soil. The soil condition of a particular region regulates people's health by providing crops nourishments, which along with the surrounding environment determines the quality of breathable air and drinkable water. Thus, a strong relationship exists between soil quality and the health of people in its surrounding environment (Bhaduri et al., 2015).

The plant-microbe relationship serves a crucial role in ensuring sustainable agriculture and refurbishing the ecosystem (Badri and Vivanco, 2009). Usually, soil microbes play an indispensable role in maintaining soil health and plant growth. Soil is the largest operative terrestrial ecosystem, and its activity is maintained by a wide array of interactions among its biotic and abiotic components (Barea et al., 2005). Based on these interactions, various associations are observed between microbes and microbes or microbes and plants. Plant-microbe interactions are broadly classified as positive, negative, or neutral. Positive interactions maintain the vigor of both the plant and soil by recycling minerals, transporting organic nutrients and water, inducing disease resistance, prompting tolerance to stress conditions, and biodegrading a wide range of pollutants. Negative interactions comprise host-pathogen relationships leading to many plant diseases and antagonistic effects on plant growth. In addition, some microbes exist in the rhizosphere to collect their nutrition from the root exudates (REs), and they do not affect plants positively or negatively; therefore, they carry out neutral interactions (Akram et al., 2017). Hiltner coined the term "rhizosphere" in 1904 to designate the constricted zone of extreme microbial activity near legume roots, influenced by the REs.

The microbes that reside in various forms in the rhizosphere are influenced by the REs that enrich the soil with nutrients and microbial biomass. This leads to a change in the rhizosphere environment resulting from communications at the biochemical, physiological, and molecular levels between the microbes themselves or between the microbes and plants and animals (Mondal et al., 2020a). Plant REs are usually enriched with amino acids, polysaccharides, organic acids, root border cells, dead root cap cells, etc. Additionally, plant roots secrete several phyto-siderophores that isolate minerals from the soil, thereby improving plant nutrition. Plant REs also contain secondary metabolites that influence plant-microbe interactions (Weir et al., 2010).

Therefore, there is an intricate relationship between soil with the inhabiting plants and microbes. Though there is no clear image of the overall impact of these interactions, the interactions is beneficial for plants, microbes, and soil health. Till date there is limited evidence regarding the involvement of these interactions in mineral sequestration, infection control, enhancement of nutrient and water availability, improvement of soil aggregation (to stabilize top-soil), ecological detoxification of soil, stimulates stress resistance, etc. How plant-microbial interactions impact soil health, which in turn improves plant growth, leading to increased productivity as a collective outcome, has still not received significant attention. This chapter sums up the different kinds of interactions between the plants and microbes operating in soil and their impact to improve soil health and the overall physiology of the plant.

5.2 Important types of plant-microbes interactions

5.2.1 Mutualisms

Mutualisms are distinguished from other beneficial relations as they are generally species-specific, indispensable for the existence of one or both partners, and always both the partners co-evolutionary adapted (Brown and Ogle, 1997; Lipson and Kelley, 2014; Harman and Uphoff, 2019).

5.2.1.1 Rhizobia-legume mutualism

Rhizobia are the collections of nitrogen-fixing bacteria, which participate in nodule formation with the leguminous plant (Leguminosae/Fabaceae families). These bacteria include the members of α-proteobacteria (*Rhizobium, Azorhizobium, Bradyrhizobium, Mesorhizobium, Sinorhizobium* spp.), β-proteobacteria (*Burkholderia* sp.), and γ-proteobacteria (*Azotobacter, Pseudomonas* spp.) (Shiraishi et al., 2010). In addition to formation of nodules on roots, a few variants of Rhizobium are capable of forming the nodule on stem like mutualism between *Azorhizobium* sp. with *Sesbania rostrata*. Rhizobium usually forms nodules on roots but sometimes on stems, such as mutualism within the tropical tree *Sesbania rostrata*, and its associate *Azorhizobium* sp. The nod and nif genes are typically required for nodule formation, and nitrogenase enzyme-mediated nitrogen fixation is frequently found on a Sym plasmid. The nod genes are evolutionary transferred through the horizontal gene transfer mechanism among the different nodulating bacteria of the proteobacterial classes (Masson-Boivin et al., 2009). REs, mainly flavonoid compounds of the host plant, play an essential role in nodule formation (Peters et al., 1986).

Inside the root hair, the nod factors stimulate a series of complex physiological changes that lead to the root hair curling and subsequent formation of the cellulosic infection thread through which bacteria reach root cells and infect nearby cells. Uninterrupted cell proliferation causes the development of the root nodule. Rhizobia especially form a symbiotic association with their aquatic host plant by crack entry, where rhizobia conduct host entry through cracks at the lateral root. The bacteria in the nodule are morphologically converted into bacteroides, which assimilate atmospheric ammonium into nitrogen utilizing the nitrogenase enzyme. Before transport into the host plant, the assimilated nitrogen is converted into amino acids such as glutamine and asparagine, and the host plant transports organic acids and oxygen-bound leghemoglobin to its symbiotic partner for food and cellular respiration, respectively. The leghemoglobin maintains the oxygen-deficient environment in the nodule to sustain nitrogenase activity (Gage, 2004).

5.2.1.2 Actinorhizal associations

Compared to rhizobia-legume mutualism, the actinorhizal nitrogen-fixing relationship has not been carefully studied, but globally it accounts for around 25% of total biotic nitrogen fixation (Dilworth et al., 2008). The symbiotic association between *Frankia* (Actinobacteria) with different families (like Fagales, Cucurbitales, Rosales, etc.) of dicotyledonous plants is known as actinorhizal association. Compared to rhizobia, *Frankia* sp. assimilates the NH^{4+} produced in nitrogen fixation and exports amino acids such as arginine as nitrogen to host cells (Berry et al., 2011). To ensure and establish symbiosis, both *Rhizobium* sp. and *Frankia* sp. infect root through root hair curling before nodule formation, which is induced by different phenolics (viz. benzoic and cinnamic acids) and flavonoids (viz. flavanone and isoflavanone) (Ishimaru et al., 2011).

5.2.1.3 Plant-cyanobacterial mutualisms

The filamentous photosynthetic cyanobacteria fix nitrogen in the specialized cells heterocysts, and they form a mutualistic association with the water-fern *Azolla* (Pteridophyta), cycads (Gymnosperms), and the flowering plant *Gunnera* (Angiosperm). Cycads and *Nostoc* sp. form one of the earliest nitrogen-fixing mutualism, in which infection of cycads by *Nostoc* sp. introduces coralloid roots in which the latter live in a mucilaginous extracellular space. Another type of cyanobacteria, *Nostoc azollae*, develops in cavities below the leaves of the aquatic fern *Azolla*, representing the simplest type of nitrogen fixation among the plant-microbial associations.

5.2.2 Mycorrhizae

A mycorrhizal relationship develops between a wide range of plant roots and fungi. Three significant classes of mycorrhizae are recognized based on the colonization of fungal mycelia in or around plant root cells. Endomycorrhizae cause penetration inside the plant cell cytoplasm, while ectomycorrhizae form a thick mantle around lateral roots and develop between root cells without penetrating the cell. On the contrary, the third category, ectendomycorrhizae, penetrates inside the host cells and also creates a mantle (van der Heijden et al.,

2015). By increasing the periphery of the root surface, ectendomycorrhizae provide improved nutrient uptake and relief from a pool of biotic and environmental stresses (Harman and Uphoff, 2019).

5.2.2.1 Arbuscular mycorrhizae

Arbuscular mycorrhizae (AM), under the phylum Glomeromycota, are the most important mycorrhizae, developing an endomycorrhizal relationship with around 80% of angiosperms and about two-thirds of all plant species. Moreover, AM fungi extended their relationships with the gymnosperms, bryophytes, and ferns. AM fungi proliferate as highly branched tree-like structures known as arbuscules within the plant cortical cells, which act as hotspots of nutrient exchange in between the host plant and AM fungi. Within the host plant root, AM fungi also produce vesicles (vesicular-arbuscular mycorrhizae), which represent the fungi's dormant state, and gain the ability to infect new plant roots. To persist in soil, some AM fungi form spores. AM fungi are strictly dependent on their hosts (obligate symbiosis) and unable to live autonomously as saprotrophs in soil, but are present in a dormant state until they meet a suitable host root. AM fungi supply phosphorus to their hosts, reduce nitrogen loss by inhibiting the process of denitrification and leaching (Cavagnaro et al., 2012), and alleviate CO_2 emissions from agricultural soils (Solaiman, 2014). Besserer et al. (2006) reported that flavonoids are the main compound for altering nonsymbiotic AM fungi into symbiotic (Besserer et al., 2006); however, CO_2 accelerates the process (Poulin et al., 1993). Another compound, strigolactone, was reported to stimulate branching in *Gigaspora margarita*, cell proliferation in *Gigaspora rosea*, and spore germination in *Glomus intraradices* (Akiyama et al., 2005).

5.2.2.2 Ectomycorrhizae, ectendomycorrhizae, and arbutoid mycorrhizae

These three types of mycorrhiza have similar morphological features; thus, they are placed in a group together. EM association is present in most tree species, and is therefore dominant in forested ecosystems. These types of association are also found in nonwoody plants like *Kobresia myosuroides*. Different fungal strains capable of forming EM association belong to Basidiomycetes and Ascomycetes. In contrast with AM fungi, EM fungi are able to exist independently in the soil as saprotrophs, degrading complex organic matter; therefore, they transfer nitrogen, phosphorous, and other nutrients to the host plant. Apart from the formation of mantles of hyphae around stunted lateral roots, the fungal hyphae extend inwards, penetrating between the epidermis and cortex layer of root to form a network termed a Hartig net. Similarly, ectendomycorrhizae form a mantle as well as penetrating plant epidermal and cortical cells and forming a Hartig net. These are the members of Ascomycetes and form associations with the species of Pinus and Larix. Arbutoid mycorrhizae are associated with the Ericaceae subfamily Arbutoideae, and are likely to proliferate similarly to ectendomycorrhizae; however, they only infect epidermal cells.

5.2.2.3 Ericoid mycorrhizae

Ericoid relationships are found among the members of the plant family Ericaceae. They are considered to be the predominant mycorrhizal type in wetlands. The fungal associates belong to the members of the groups Ascomycetes and Deuteromycetes. Due to the complex organic nature of the bottom of the wetland soil, ericoid fungi adapted to extract nitrogen and make it available to the host plant. Ericoid mycorrhizae form intracellular coils, i.e., they are functionally analogous to arbuscules, which is useful in nutrient exchange.

5.2.2.4 Orchid and monotropoid mycorrhizae

On the basis of pattern of nutrient acquisition by the plant species through their fungal partner, these two mycorrhizal types are placed in one group. In these associations, the nonphotosynthetic plants (e.g., the family Orchidaceae) rely on their fungal partner (e.g., Basidiomycete, Ascomycetes) to obtain organic carbon. Orchids depend on the fungal partner for seed germination because orchid seeds are too small to have sufficient storage reserves. Certain orchids are nonphotosynthetic and mycoheterotrophic, and depend on their mycorrhizal association throughout their lifetime, while the mixotrophic members obtain organic carbon from both photosynthesis and mycorrhizal fungi. Nonphotosynthetic, parasitic plants form a monotropoid mycorrhizal association, and the fungal partner provides food for them but cannot obtain any kind of benefit from the plant partner (Brown and Ogle, 1997; Lipson and Kelley, 2014).

5.3 Major pathways of improved soil health and productivity attributed by different plant-microbe interactions

5.3.1 Mineral acquisition

Different agricultural lands are scarce in iron, phosphorus, and nitrogen, leading to plant growth retardation. Nitrogen is the principal nutrient required for plant vegetative and reproductive growth, and is usually present in soil in either organic or inorganic forms, transformed by soil-borne microorganisms into the utilizable form (as ammonium or nitrate) to plants. In nodules of leguminous plants, *Rhizobium* sp. fix atmospheric nitrogen in a form that is readily utilized by the host plant, and in turn, the host plant supplies photosynthates and some nitrogen fixation-related genes (Hunter, 2016). Aldonic acid and phenolics secreted by the root of leguminous plants stimulate nod genes to facilitate nitrogen fixation. Rhizobia-mediated biological nitrogen fixation is the most attractive source for economic and environmental alternatives to chemical fertilizers. Several other nitrogen-fixing endophytic and free-living rhizobacteria belonging to *Achromobacter, Azospirillum, Azotobacter, Bacillus, Bradyrhizobium, Burkholderia,* and *Pseudomonas* genera have been explored for their positive influences on crop production (Igiehon and Babalola, 2018). Phosphorus is a critical growth-limiting nutrient, and its largest reservoir is insoluble rock phosphates. In most soil, a lower amount of phosphorous is available to plants because a large part of the soil's organic or inorganic phosphates becomes immobilized as they form metal complexes with iron, aluminum, and calcium (Gyaneshwar et al., 2002). While the amount of plant-acceptable nitrogen is present in the millimolar range, plant-acceptable phosphorus is typically within the micromolar range. In terms of enhancing crop growth and yield when phosphate fertilizers are added to soil, the plant does not properly absorb these because it is quickly immobilized by binding to the soil particles or becomes moderately soluble; therefore, only a small fraction is available to the plant (Gyaneshwar et al., 2002). A pool of phosphate-solubilizing bacteria (belonging to *Alcaligenes, Aerobactor, Bacillus,* and *Pseudomonas* genera) and fungi (belonging to *Aspergillus, Cephalosporium, Chaetomium, Fusarium,* and *Penicillium* genera) are present in the rhizosphere, enhancing inorganic phosphate solubilization by liberating protons, hydroxyl ions, CO_2, and organic acid anions such as citrate, malate, and oxalate as well as mineralizing organic phosphate by secreting different phosphatases (Sharma et al., 2013). Rhizospheric microbes also assist the uptake of micronutrients like iron (Fe) and zinc (Zn). Microbes (bacteria, fungi) are liberating organic acid anions or siderophores (enterobactin, pyoverdine, ferrioxamines, ferrichromes) that after chelating with ferric ion (Fe^{3+}) are transported to the cell surface where subsequent conversion into soluble ferrous ion (Fe^{2+}) takes place (Marschner et al., 2010; Mendes et al., 2013). Rhizospheric microorganisms (e.g., *Curtobacterium, Pseudomonas, Stenotrophomonas, Streptomyces*) are found to mobilize Zn by acidification of surrounding soil by the action of secreted gluconic acid (Costerousse et al., 2018).

Aseri et al. (2008) performed experiments to determine the effectiveness of plant growth-promoting rhizobacteria (*Azotobacter chroococcum* and *Azospirillum brasilence*) and AM fungi (*Glomus mosseae* and *Glomus fasciculatum*) on the biomass yield, growth, and nutrient uptake of pomegranate (*Punica granatum* L.) individually or in combination. It has been noticed that combinatorial inoculation of PGPR and AM fungi results in higher biomass yield and improved uptake of nitrogen, phosphate, potassium, calcium, and magnesium. An increment in the uptake of nitrogen and phosphorous was found to improve symbiotic nitrogen fixation and phosphatase activity. Similarly, Khan (2005) found that multiple PGPR inoculation of crop plants through *Pseudomonas* and *Acinetobacter* strains leads to enhanced uptake of phosphorus, iron, potassium, calcium, zinc, and magnesium.

5.3.2 Biocontrol agent

Being primary producers, plants play an imperative role in the energy flow of the living environment of earth. However, soil-borne pathogenic microbes and harsh environmental conditions cause diseases in plants, leading to a decline in growth and productivity. To avoid such consequences, chemical pesticides or stimulants are usually employed for the long term. However, the toxic effects of these chemicals on the environment, humans, and other animals restrict their acceptance; as a remedy to this, biocontrol agents of natural origin are now being exploited to overcome these detrimental effects.

The rhizosphere microbes form an antagonistic relationship with the pathogens to restricts their propagation either by the action synthesized antimicrobial compounds (like butyrolactones, oligomycin A, phenazine-1-carboxylic acid, pyoluterin, etc.) and extracellular hydrolytic enzymes (like chitinase, glucanase, etc.), or by contending them for the acquisition of available nutrients and habitat (Mohanram and Kumar, 2019).

For instance, 2,4-diacetylphloroglucinol is biosynthesized by *Pseudomonas fluorescens*, which acts as a weapon to combat soil-borne pathogens like *Meloidogyne incognita* and *Fusarium oxysporum* (Meyer et al., 2016). Production of more than one antimicrobial compound is very common among the potential biocontrol strains; for example, *Agrobacterium radiobacter* was used to biosynthesize narrow spectrum agrocin 84 and broad spectrum polyketide antibiotics, which act against different plant pathogens (Raaijmakers et al., 2010). In addition, an array of rhizobacteria liberate extracellular chitinase and β-1,3 glucanase, which trigger the degradation of cell wall of fungal mycelia and spore and consequently control their propagation (Halder et al., 2013). Different members of the *Trichoderma* genus imparted their biocontrol efficiency by different means like mycoparasitism, by the action of secreted secondary metabolites (antibiotics) and hydrolytic enzymes (peroxidases and phenol oxidases), and also by competition for nutrients. Martínez-Medina et al. (2017)

conducted a split-root experiment in tomatoes with *Trichoderma harzianum* T-78 and found that it protects tomato roots by eliciting an induced systemic response in the local and systemic root tissues against *Meloidogyne incognita*. In some plants, species of the genus *Pseudomonas* and *Bacillus* could serve as elicitors in stimulating systemic resistance against pathogens.

Furthermore, sequestering the available iron by microbial siderophores restricts the growth phytopathogens. For instance, wilt caused by *Fusarium oxysporum* is restricted by the action of siderophores of *Bacillus subtilis* (Mohanram and Kumar, 2019). Fungal siderophores synthesized by *Aspergillus niger*, *Trichoderma harzianum*, and *Penicillium citrinum* were found to be efficient biocontrol pathogens and improved the growth of chickpeas (*Cicer arietinum*) (Janardan et al., 2011). *Pseudomonas aeruginosa* secreted siderophores which confer resistance against different fungus and viruses such as *Botrytis cinerea*, *Colletotrichum lindemuthianum*, and mosaic virus (Mohanram and Kumar, 2019). *Serratia marcescens* 90–166 secreted catechol-type siderophore that provides resistance to cucumber against bacterial, fungal, and viral pathogens (Press et al., 2001). The nonpathogenic strain of *Fusarium oxysporum* 47 was competent in triggering systemic acquired resistance by expression of genes that encode extracellular pathogenesis-related proteins and thereby control tomato wilt (Aimé et al., 2013). Ton et al. (2002) reported that the rhizobacterial relationship with the host plant stimulates either the salicylic acid-mediated signal transduction pathway or the jasmonic acid and ethylene-based signaling pathway for defense against pathogens. Mycorrhizae colonizing the rice root stimulate the enhanced expression of pathogenesis-related genes in their leaves and showed improved resistance against *Magnaporthe oryzae* (Campos-Soriano et al., 2012). The mycorrhizal colonization also induced resistance against herbivorous insects in the host plants; for example, *Glomus mosseae* colonization mediated a plant defense response against *Helicoverpa armigera* (Song et al., 2013). It was observed that mycorrhizal colonization resulted in the enhanced expression of the AOX, LOXD, PI-I, and PI-II genes in the leaves that elicits defence against caterpillar. Therefore, in the rhizosphere, varieties of microbial interactions of varying degrees of efficiency deliver greater biocontrol against plant pathogens and modify a plant's immune system.

5.3.3 Extenuating abiotic stresses

Climatic factors such as drought, extreme temperature, and excessive rainfall prompt abiotic stresses, while the edaphic factors are mainly soil correlated. Plants adapt some tactics such as stress tolerance and stress avoidance to deal with abiotic stresses. Plants cannot manage abiotic stress efficiently alone; therefore, they are genetically modified by either up- or downregulating the gene expression or utilizing the benefits of rhizosphere microbes to handle unfavorable environmental stresses (Brahmaprakash et al., 2017). The intrinsic genetic and metabolic abilities of rhizosphere microbes help to relieve the abiotic stresses in plants. A number of species of the genera *Achromobacter, Azotobacter, Bacillus, Burkholderia, Enterobacter, Pseudomonas, Trichoderma*, etc. have been thoroughly investigated for plant growth promotion through alleviating various types of abiotic stresses (Mohanram and Kumar, 2019). Naveed et al. (2014) conducted a field experiment in water deficit conditions where wheat inoculated with *Burkholderia phytofirmans* PsJN exhibited enhanced photosynthesis, high chlorophyll content, and better grain yield. Similarly, *Trichoderma harzianum* inoculated Indian mustard (*Brassica juncea*) showed increased uptake of essential nutrients, improved build-up of antioxidants and osmolytes, and reduced sodium ion uptake in saline conditions (Ahmad et al., 2015). In another experiment on evaluation of salt tolerance, arabidopsis inoculated with *Bacillus subtilis* GB03 resulted in the reduced accumulation of sodium ion throughout the plant body, leading to a salt-tolerant host plant (Zhang et al., 2008). Sen and Chandrasekhar (2015) found that improved root colonizing and exopolysaccharides secreting abilities of *Pseudomonas* were responsible for the better tolerance of the host plant against high salt concentration. Srivastava et al. (2008) reported that a thermotolerant strain *Pseudomonas putida* NBR10987 isolated from the drought-stressed rhizosphere of chickpea was able to secrete an exopolysaccharide having high water holding capacity that protected the host plant from water stress. Theocharis et al. (2012) reported that the rhizosphere microbes triggered tolerance to low temperatures (nonfreezing), leading to higher and faster gathering of stress-linked proteins and metabolites.

A huge number of soil-borne microbes assists plant stress response by curtailing ethylene production by degrading its metabolic precursor 1-aminocyclopropane-1-carboxylate (ACC) into α-ketobutyrate and ammonia (Stearns et al., 2012). Thus, a reduced level of ethylene enhanced the plants' ability to combat different abiotic and biotic stresses. The activity of ACC deaminase was helpful in enhancing plant responses toward drought stress, water stress, salinity stress, and enhanced plant growth (Arshad et al., 2008; Mayak et al., 2004a; Yang et al., 2009). For instance, the soil-borne *Achromobacter piechaudii* ARV8, having ACC deaminase activity, was competent to improve tomato and pepper seedlings (Mayak et al., 2004b). *Bradyrhizobium elkanii* synthesizes another ethylene inhibitor, rhizobitoxine, which reduces the negative impacts on nodulation during stress-stimulated ethylene production and triggers foliar chlorosis in soybeans (Vijayan et al., 2013).

Due to geological distribution and anthropogenic activities, vast portions of the earth are contaminated with nonbiodegradable and tenacious heavy metals. Different bacteria (like *Pseudomonas, Microbacterium, Verrucomicrobia, Actinobacteria*) and fungi (like *Aspergillus, Penicillium, Trichoderma, Lewia,* and the mycorrhizal fungi) havee the ability to bioleach heavy metals from soils, sludge and sediments through different metabolic activities such as adsorption, dissolution, complexation, oxidation, reduction and complexation, and thus directly or indirectly mitigates metal toxicity (Mulligan and Galvez-Cloutier, 2003; Pathak et al., 2009; Mohanram and Kumar, 2019). The heavy metal resistant *Bacillus* sp. SC2b could lower a significant fraction of heavy metals such as cadmium, lead, and zinc through biosorption and bacterial inoculation in toxic land areas, leading to protection of host plant growth by reducing toxicity (Ma et al., 2015). Chatterjee et al. (2009) found that the chromium-resistant bacterium *Cellulosimicrobium cellulans* could reduce the toxic and mobile Cr^{6+} in the soil to nontoxic and immobile Cr^{3+}. Thus, the bacterium inoculation in green chili cultivated chromium-contaminated land resulted in reduced Cr^{6+} uptake in the root and shoot of green chili of up to 56% and 37%, compared to the control. In addition, root exudates play an indispensable role in phytoremediation, and organic acids, especially citric acid and oxalic acid present in the root exudate of *Echinochloa crusgalli*, significantly improved mobilization and accumulation of heavy metals like cadmium, copper, and lead; thus, organic acids served as natural chelating agents to increase phytoextraction (Kim et al., 2010). The rhizosphere of hyperaccumulating plant *Amaranthus hypochondriacus* and *Amaranthus mangostanus* exists in the heavy metal-contaminated soil inoculated with the endophytic bacterium *Rahnella* sp. JN27 and triggered cadmium solubilization through the secretion of siderophores, facilitating cadmium uptake by the host plant (Yuan et al., 2014). AM fungi can synthesize an insoluble metal absorbing glycoprotein (glomalin) that causes metal sequestration and immobilizes metal; thus, it serves as a metal biostabilizer in soil (Vodnik et al., 2008).

5.3.4 Improved physiochemical nature of soil

Soil is the base of the terrestrial ecosystem. A better quality of soil is characterized by well-aggregated soil structure, easy to tillage, suitable water infiltration rates, aeration, root penetrability, and sufficient level of organic matter. The microbial role in forming and stabilizing rhizosphere soil aggregates has been well established (Miller and Jastrow, 2000). During the plant's normal growth, a wide array of inorganic and organic chemicals are secreted from its roots as root exudates profoundly affect the biochemical and physical nature of the surrounding soil (Walker et al., 2003). Details on the composition of root exudates are provided in Table 5.1. The numerous compounds of root exudate aid several functions such as maintaining the soil-root contact, root tip lubrication, protecting root surface from desiccation, chemoattractant of microbes, stabilizing soil-microbes aggregation, chelation or absorption of metal ions, growth promoteion, energy source, supply of nutrients, and antibiotics or inhibitors (Brahmaprakash et al., 2017; Godheja et al., 2017; Hawes et al., 2000; Koo et al., 2005; Ma et al., 2016;

Table 5.1 Common components of root exudates.

Compositions	Specific substance
Amino acids	α-Alanine, β-alanine, arginine, asparagine, aspartic acid, cystine/cysteine, glutamine, glycine, histidine, isoleucine, leucine, lysine, methionine, phenylalanine, proline, serine/homoserine
Enzymes	Amylase, dehydrogenase, dehalogenase, invertase, acid/alkaline phosphatase, peroxidase, phenolase, polygalacturonase, protease
Fatty acids	Linoleic, linolenic, oleic, palmitic, stearic
Flavonoids	Chalcone, flavones, flavonols, flavanones, flavonones, isoflavones
Growth factors and vitamins	Auxin, biotin, choline, ethanol, inositol, niacin, para-amino benzoic acid, pantothenate, pyridoxine, thiamine, strigolactones, n-methyl nicotinic acid
Nucleic acids	Adenine, cytidine, guanine, uridine
Organic acids	Acetic acid, aconitic acid, aldonic acid, butyric acid, chorismic acid, citric acid, erythronic acid, glutaric acid, lactic acid, maleic acid, malic acid, malonic acid, oxalic acid, piscidic acid, propionic acid, pyruvic acid, succinic acid, tartaric acid, tetronic acid, valeric acid
Phenolics	Caffeic acid, catechol, cinnamic acid, coumarin, eriodictyol, ferulic acid, genistein, isoliquiritigenin, liquiritigenin, luteolin, salycilic acid, sinapic acid, syringic acid, vanillic acid, syringic, transcinnamic acid
Carbohydrates	Arabinose, deoxyribose, fructose, fucose, galactose, glucose, maltose, mannitol, mucilage of various compositions, oligosaccharide, raffinose, rhamnose, ribose, sucrose, xylose
Sterols	Campesterol, cholesterol, sitosterol, stigmasterol
Others	Alcohols, alkyl sulfides, camalexin, dihydroquinone, glucosides, glucosinolates, glycinebetaine, inorganic ions and gaseous molecules (e.g., CO_2, H_2, H^+, OH^-, HCO_3), isothiocyanates, scopoletin, sorgoleone, volatile compounds (e.g., acetone, acetaldehyde, ethanol, formaldehyde, methanol, propionaldehyde)

Based on Koo, B.J., Adriano, D.C., Bolan, N.S., Barton, C.D., 2005. Root exudates and microorganisms. In: Hillel, D. (Ed.), Encyclopedia of Soils in the Environment, pp. 421–428; Vranova, V., Rejsek, K., Formanek, P., 2013. Aliphatic, cyclic, and aromatic organic acids, vitamins, and carbohydrates in soil: a review. Sci. World J. 2013; Ma, Y., Oliveira, R.S., Freitas, H., Zhang, C., 2016. Biochemical and molecular mechanisms of plant-microbe-metal interactions: relevance for phytoremediation. Front. Plant Sci. 7, 918; Godheja, J., Shekhar, S.K., Modi, D.R., 2017. Bacterial rhizoremediation of petroleum hydrocarbons (PHC). In: Plant-microbe interactions in agro-ecological perspectives. Springer, Singapore, pp. 495–519; Brahmaprakash, G.P., Sahu, P.K., Lavanya, G., Nair, S.S., Gangaraddi, V.K., Gupta, A., 2017. Microbial functions of the rhizosphere. In: Plant-Microbe Interactions in Agro-Ecological Perspectives. Springer, Singapore, pp. 177–210. Vives-Peris, V., de Ollas, C., Gómez-Cadenas, A., Pérez-Clemente, R.M., 2020. Root exudates: from plant to rhizosphere and beyond. Plant Cell Rep. 39(1), 3–17.

Vives-Peris et al., 2020; Vranova et al., 2013). The rhizosheath soil enriched with the root exudates is considerably wetter than the surrounding soil; therefore, such soil has improved water holding capacity (Young, 1995). Except for the macrofauna and plant root, the microbial biomass is the living entity of soil organic matter (Jenkinson, 1981). The rhizosphere microbiome plays an essential role in the decomposition of soil that improves soil fertility and eventually enhances plant productivity. Lignocellulolytic fungi such as *Phanerochaete chrysosporium*, *Pleurotus ostreatus*, *Polyporus ostriformis*, and *Trichoderma harzianum*, and bacteria such as *Cellulomonas* sp., *Chryseobacterium gleum*, *Cytophaga* sp., *Pseudomonas* sp., *Sporocytophaga* sp., and *Streptomyces* sp., are well-known to break down plant biomass, thereby recycling the nutrients and making these available for themselves and plants (Mohanram and Kumar, 2019).

Humus is mainly present in the soil in four forms as humin, humic substances, other nonhumic substances, and polysaccharides, and is responsible for maintaining the physicochemical nature of soil by enhancing the texture and arrangement of the soil, providing the buffering ability, improving water holding capacity, and boosting soil productivity. Soil structure determines the nature and distribution of pore space and water accessibility, thereby managing the heterogeneous nature of microbial habitats (Mukherjee, 2017). Soil particles are associated together by bacterial products and by the hyphae of saprophytic and AM fungi, into steady microaggregates (2–20 μm in diameter) that in turn are further associated by microbial products (organic acids, polysaccharides) into larger microaggregates (20–250 μm in diameter) and finally to macroaggregates (>250 μm in diameter) with bacterial polysaccharides and hyphae of AM fungi. The AM fungi, in cooperation with other microbes, produce water soil aggregates in different ecological situations (Requena et al., 2001). Regarding this, the contribution of glomalin, a glycoprotein synthesized by the external hyphae of AM fungi, has been noticed (Wright and Upadhyaya, 1998). The plant-microbial interactions support plant growth in nutrient-deficit soil; therefore, microbial symbionts inoculated in the rhizosphere soil of an indigenous species of plant have proven to be a successful strategy to prevent desertification.

5.3.5 Phytostimulation

Phytohormones synthesized by the rhizosphere bacteria and fungi induce plant growth, thereby increasing productivity. Different abiotic stress conditions change the phytohormonal level in plants, particularly auxins, abscisic acid, cytokinin, ethylene, gibberellic acid, jasmonic acid, and salicylic acid; therefore, they hamper plants' normal metabolic activities (Egamberdieva et al., 2017). Microbes synthesize such phytohormones as secondary metabolites, which are nonessential for their metabolic activities (Shi et al., 2017). Kurosawa (1926), for the first time, identified the phytohormone gibberellins that were metabolically synthesized in the fungus *Gibberella fujikuroi* that cause disease in rice plants. Gibberellic acid is normally involved in different plant physiological processes such as seed germination, stem elongation, fruit formation, and sex determination, and also participates in metal uptake and ion segregation in a plant, resulting in improved growth and sustaining plant metabolism under normal and various stress conditions (Bömke and Tudzynski, 2009; Iqbal and Ashraf, 2013). Miceli et al. (2020) reported that the external foliar application of gibberellic acid (10^{-5}) on the tomato and sweet pepper seedling cultivated in a hypersaline land showed a salt tolerance of 25 and 50 mM NaCl in tomato and sweet pepper, respectively. The control plants are protected from the negative effects of salinity such as biomass, leaf diameter, leaf number, plant height, relative water content, shoot/root ratio, and stomatal conductance. Different microorganisms were found to synthesize gibberellic acid; the most essential fungal genera include *Aspergillus fumigatus*, *Phoma glomerata*, *Trichoderma asperellum*, and bacterial genera such as *Azospirillum lipoferum*, *Acetobacter diazotrophicus*, *Bacillus aryabhattai*, *Herbaspirillum seropedicae*, and *Bacillus* sp. (Creus et al., 2004; Khan et al., 2011; Lei and Zhang, 2015; Meleigy and Khalaf, 2009).

Although auxin typically endorses cell division, elongation, and differentiation in the plant, its prime role is to promote root growth and development that enable the plant to adapt to normal and various stress environments (Asgher et al., 2015; Kudoyarova et al., 2019). Several stress conditions such as salinity, heavy metal toxicity, and deposition of organic pollutants reduce the auxin level in plants. Auxin at a low concentration (10^{-10} M) mitigates the toxic effect of lead on sunflower growth by enhancing root diameter, surface area, and volume (Fässler et al., 2010). Salt stress endurance in plants promoted by auxin is made possible by crosstalk between auxin and salicylic acid. For example, Iqbal and Ashraf (2007) reported that wheat seeds inoculated with auxin resulted in the significant alleviation of salt stress-mediated conditions. Plants with a well-developed root system can easily take up phosphate available by the rhizospheric microbes in the soil. Kudoyarova et al. (2017) experimented by inoculating wheat seed with *Pseudomonas extremaustralis* (IB-K13-1A) having the potentiality of both auxin producing and phosphate solubilization, and showed increased wheat production compared to seed inoculated with either phosphate solubilizer (*Advenella kashmirensis* IB-K1) or auxin producer (*Bacillus subtilis* IB-21). Xun et al. (2015) reported that *Acinetobacter* sp. inoculated oat (*Avena sativa*) plants exhibited improved breakdown of total petroleum hydrocarbons in contaminated soil up to 45%, and this effect was contributed by the auxin synthesized by the microbial inoculum (Benson et al., 2017).

During drought conditions, plant root elongation into the deeper layer of moister soil is important, and ethylene normally inhibits the plant root elongation. Microbes are synthesizing ACC deaminase, degrading the precursor of ethylene synthesis ACC, resulting in the reduced level of ethylene concentration in plants (Belimov et al., 2015). Belimov et al. (2009) showed that *Pisum sativum* inoculated with a strain of *Variovorax paradoxus* led to an increased level of ACC deaminase in xylem sap, and thus reduced the level of ethylene prompt root elongation instead of high auxin concentrations. It has been observed that ACC deaminase synthesizing plant growth-promoting microbes maintained plants' growth in a variety of stress conditions such as drought (Niu et al., 2018), high salt (Singh and Jha, 2016), the existence of organic toxicants and total petroleum hydrocarbons (Hong et al., 2011), and the presence of heavy metals (Tiwari and Lata, 2018).

Cytokinins usually maintain cellular proliferation and differentiation, and inhibit premature leaf senescence (Schmülling, 2002). Application of a high cytokine-producing strain of *Bacillus subtilis* IB 22 into the rhizospheres of lettuce and wheat resulted in increased leaf area (Arkhipova et al., 2019) due to shoot cell division and elongation promoted by cytokinin (Werner et al., 2003), and alternatively inhibits root growth (Werner et al., 2010). Wilkinson et al. (2012) reported that wheat seed inoculated with a cytokinin synthesizing strain of *Bacillus subtilis* IB 22 cultivated under moderate drought conditions resulted in earlier canopy closure and increased production by 40%.

Abscisic acid enhanced plant adaptation in various stress responses. Cohen et al. (2015) reported that Arabidopsis plants inoculated with ABA producing *Aspergillus brasilense* showed several antioxidative physiological consequences such as reduced lipid damage enumerated by malondialdehyde levels, early stomatal closure, improved leaf water relations, and detoxified free radicals.

Salicylic acid improved plant stress responses by managing activities of the antioxidative enzymes (Silva et al., 2017). Khan et al. (2014) showed that the application of external SA in salt-stressed *Vigna radiata* led to a reduction in ethylene level. Plants treated with SA exhibited improved growth through biomass accumulation, increased antioxidative enzyme activity, cell proliferation, photosynthetic rate, and membrane stability, decreased electrolyte outflow, prevented lipid peroxidation, and sustained transpiration rate (Azooz et al., 2011; Silva et al., 2017). Tang et al. (2017) reported that SA-treated plants exhibited better tolerance to drought conditions by retaining a lower level of free radicals. The beneficial rhizobacteria stimulate the induction of defensive systemic resistance response in plants by synthesizing the phytohormones jasmonic acid and ethylene (Zamioudis and Pieterse, 2012). Thus, microbes-mediated phytohormone biosynthesis is an effective mechanism to change plants' physiology, resulting in diverse outcomes from plant growth to protection against biotic or abiotic agents (Spaepen, 2015).

5.3.6 Rhizoremediation

Soil is essential for conserving environmental quality, determining the quality of food produced, land utilization, and maintaining human health (Gomiero, 2016). However, the tremendous level of increment in urbanization, industrialization, and exhaustive agricultural practices leads to soil quality deterioration and reduced productivity globally (Juan et al., 2008). Many land sites around the world are contaminated with several xenobiotic compounds like polyaromatic hydrocarbons (PAH), polychlorinated biphenyls (PCB), heavy metals, chemical pesticides, and other pollutants (Mishra and Arora, 2019). According to the Food and Agriculture Organization of the United Nations FAO (2017), around one-third soil of the earth is contaminated because of unsustainable practices.

Rhizoremediation is a subset of phytoremediation, where plant stimulates rhizosphere microbes to remediate environmental pollutants. The process of rhizoremediation is dependent on:

 (i) the hydrogeology and structure of soil
 (ii) the nature of the pollutants
(iii) the nutritional status and microbial composition of the polluted site (Blackburn and Hafker, 1993)

Plant root exudates play an essential role in the efficacious colonization of microbes surrounding the rhizosphere resulting in the successful biodegradation of soil pollutants. Yi and Crowley (2007) reported that the linoleic acid that exists in the root exudates secreted by plant roots serves as a surfactant to increase the bioavailability of pyrene, a PAH compound, by forming a layer on soil particles, leading to its improved attachment to the bacterial surface. The rhizoremediation strategy has been verified to be superior over bioremediation and phytoremediation techniques due to its ability to utilize a consortium of rhizospheric microbes, which are much more concentrated than the bulk soil (Marschner et al., 2001). The sloppy nature of rhizospheric soil triggered a strain (UT26) of *Sphingomonas* sp. to degrade lindane, a highly persistent pesticide (Segura et al., 2009). Exposure of PAH in the environment occurs by natural phenomena such as volcanic eruptions or forest fires and anthropogenic activities such as automobile emission, combustion of wood, effluent from petroleum industries, and coal tar production (Ravindra et al., 2008). According to Wilson and Jones (1993), various PAH compounds are carcinogenic, mutagenic, and teratogenic; therefore, consuming these compounds leads to serious consequences in living beings. According to Peng et al. (2008), the rhizospheric microbes cause biodegradation of PAH compounds through the process of mineralization, where the microbial dioxygenase enzymes carry out the cleavage of the aromatic ring structure of PAH into diols and organic acids. The following bacterial genera are involved in PAH degradation: *Alcaligenes, Bacillus, Burkholderia, Mycobacterium, Pseudomonas, Rhodococcus,* and *Sphingomonas* (Bisht et al., 2014). Cébron et al. (2011) reported that *Arthobacter* and *Pseudomonas* spp. in the ryegrass rhizosphere use the rhizospheric energy sources to degrade the phenanthrene class of PAHs. Pesticides (insecticides, fungicides, and herbicides) are used to control pests and polluted soil, air, and water after their exposure from agricultural practices and manufacturing sites. The rhizospheric bacterial genus *Acinetobacter, Alcaligenes, Bacillus, Burkholderia, Pseudomonas, Rhizobium, Serratia, Streptococcus,* and *Stenotrophomonas* spp. and the fungal genera *Phanerochaete* and *Ganoderma* spp. along with mycorrhizae are involved in degradation of pesticides (Velázquez-Fernández et al., 2012).

An experiment carried out by Yadav and Krishna (2015) showed that a PGPR strain RB1 was resistant to methyl parathion (an organophosphate pesticide) at 500 ppm and utilized it as a carbon and nitrogen source, and was reported to improve the growth of mungbean (*Vigna radiata*). Dubey and Fulekar (2012) demonstrated that *Stenotrophomonas maltophilia* MHF ENV 22 alone in the rhizosphere of *Pennisetum pedicellatum* could degrade cypermethrin of up to 58% in 192 h, and at a concentration of 100 mg kg^{-1}. Rajkumar et al. (2012) reported that the cyanobacteria such as *Anacystis nidulans, Microcystis aeruginosa,* and *Synechococcus elongatus* can degrade organochlorine and organophosphates classes of insecticides and in turn deliver hormones, enzymes, and vitamins to the host plants for their growth. Polychlorinated biphenyl (PCB) compounds are carcinogenic, recalcitrant, and have the property of bioaccumulation in the environment (Godheja et al., 2016). The rhizospheric bacterium *Sinorhizobium meliloti* degrades 2,4,4′-TCB, a PCB compound, up to 77% within 6 days (Tu et al., 2011). Zhang et al. (2015) reported that the cyanobacterium *Anabaena* PD-1 showed tolerance against PCB (Aroclor 1254) and degraded up to 84% within 25 days. This particular strain can also degrade 12 dioxin-like PCBs within the range of 37%–68% after 25 days. Azo dyes are very heavily utilized in the textile and leather industries, and their exposure to environmental causes serious problems to living beings due to their persistent and mutagenic nature.

Sinha et al. (2019) demonstrated that alleviation and refurbishment of azo dyes contaminated soil executed by cultivating *Alternanthera philoxeroides* inoculated with a PGPR strain *Klebsiella* sp. VITAJ23 and found that the degradation of 79% reactive green dye and concomitantly improve plant growth. Shafqat et al. (2017) conducted a similar type of study and observed that the three rhizospheric bacteria—*Achromobacter xylosoxidans*, *Burkholderia ginsengisoli*, and *Pseudomonas alcaligenes*—significantly eliminate azo dyes and improve plant growth. Mineral oil contamination of soil, water from the oil refineries, storage sites, and accidental oil spills impacted the water and terrestrial ecosystem significantly (Lacalle et al., 2018). Huang et al. (2005) reported that the utilization of ryegrass and PGPR led to the achievement of 60%–95% degradation of pollutants like total petroleum hydrocarbons (TPH) compared to the bioremediation and phytoremediation approaches. Andria et al. (2009) demonstrated that the endophytic bacteria *Pseudomonas* ITRI53 and *Rhodococcus* ITRH43 better colonized in the rhizospheric soil of *Lolium multiflorum* (Italian ryegrass) when the diesel contamination was raised from 1% to 2% and additionally reported that a hydrocarbon-degrading gene (alkB) expression responsible for synthesizing alkane monoxygenase was enhanced in high diesel-contaminated soil. A similar study conducted by Arslan et al. (2014) showed that upon inoculation of suitable endophytes and nutrients, supplementation at the rhizospheric soil of Italian grass led to 85% degradation of total petroleum hydrocarbons. Likewise, Hou et al. (2015) exhibited a 50% reduction of contamination polluted by aliphatic hydrocarbon-contaminated soil upon inoculation of endophytes in rhizospheric soil of *Festuca arundinacea*. Therefore, plant-microbial interaction in the rhizosphere degrades potential environmental pollutants and restores soil health, and enhances plant growth simultaneously.

5.4 Biofertilizers

Biofertilizers are the live microbes, when administrated in sufficient quantity in a carrier or liquid-based preparations, that promote plant growth and nutrition (Motsara et al., 1995). Different microbes that are typically utilized as biofertilizers are mentioned in Table 5.2. The application of biofertilizer can lessen the requirements for nitrogen, phosphorous, and potassium during crop cultivation from the sources of chemical fertilizer. They are principally recognized as low-cost, long-lasting, and eco-friendly substitutes to chemical fertilizers (Sahoo et al., 2014). When they are utilized as inoculants in an agricultural field, they multiply, are involved in nutrient cycling, and increase crop growth and productivity (Motsara et al., 1995). Among the different types of traditional chemical fertilizers applied for cultivation, 60%–90% of them are usually lost, and plants absorb only the residual 10%–40% (Bhardwaj et al., 2014). Thus, biofertilizers play a significant role in integrated nutrient management systems (INMS) for sustainable agricultural productivity and maintaining a healthy environment (Adesemoye and Kloepper, 2009).

Table 5.2 List of microbes utilized as biofertilizer and their efficiency.

List of microbe(s)	Beneficial attributes	Benefited plant(s)	Reference(s)
Athrobacter sp.	Induces phosphate solubilization, improved growth and yield	*Brassica oleracea*	Altuntaş (2018)
Acinetobacter sp.	Production of ACC deaminase, indole acetic acid, phosphate solubilization, and influences growth	*Triticum aestivum*	Patel and Archana (2017)
Acidothiobacillus ferooxidans	Phosphate solubilization, better growth, yield, and oil composition	*Cucurbita pepo*	Ansari et al. (2017)
Azotobacter chroococcum	Prompt nitrogen fixation, phosphate solubilization, and improved salt tolerance, production of indole acetic acid, improved growth and production	*Lycopersicon esculentum*, *Curcuma longa*	Kumar et al. (2014) and Van Oosten et al. (2018)
Azospirillum brasilense	Improved nutrient uptake, better yield, and triggered phosphate solubilization	*Triticum aestivum*	Pathak et al. (2016) and Boleta et al. (2020)
Acaulospora lacunosa	Enhances nutrient uptake and productivity	*Fragaria ananassa*	Chiomento et al. (2019)
Aspergillus niger	Enhanced availability of micronutrients and dry weight	*Lolium multiflorum*	Klaic et al. (2021)
Bradyrhizobium sp.	Production of siderophore, indole acetic acid, prompt nitrogen fixation, phosphate solubilization, enhanced growth parameters and yield	*Vigna radiata*	Alkurtany et al. (2018)
Bacillus spp.	Production of phytohormone, such as auxin, prompt phosphate solubilization and increased production	*Fragaria ananassa*	Rahman et al. (2018)
Burkholderia spp.	Solubilization of phosphate. Enhanced growth and yield	*Trigonella foenum-graecum*	Kumar et al. (2017)
Enterobacter cloacae	Influences nitrogen fixation, phosphate solubilization, siderophore production, ameliorated growth and yield	*Solanum tuberosum*	Verma et al. (2018)
Erwinia sp.	Triggers phosphate solubilization and better yield	*Triticum aestivum*	Sagar et al. (2018)
Glomus versiforme, *G. mosseae*, *G. etunicatum*, *G. deserticola*, *G. mosseae*, and *G. geosporus*	Soil quality improvement through increasing the availability of phosphorous and other nutrients, improved growth and yield of tomato under water stress conditions, enhanced chlorophyll content and nutrient uptake in maize, increases the total dry matter, chlorophyll content in strawberry, and improves snapdragon resistance to water stress	*Lycopersicon esculentum*, *Zea mays*, *Fragaria ananassa*, *Antirrhinum majus*	Xu et al. (2019), El Maaloum et al. (2020), Boyer et al. (2015), and Tognon et al. (2016)

Continued

Table 5.2 List of microbes utilized as biofertilizer and their efficiency—cont'd

List of microbe(s)	Beneficial attributes	Benefited plant(s)	Reference(s)
Herbaspirillum sp.	Synthesis of indole acetic acid, prompt nitrogen fixation, increased mineral uptake and yield	*Zea mays*	Curá et al. (2017) and Ávila et al. (2020)
Paenibacillus glucanolyticus	Synthesis of indole acetic acid and improved nutrient uptake	*Piper nigrum*	Sangeeth et al. (2012)
Phyllobacterium sp.	Production of siderophore	*Sorghum bicolor*	Shinde and Borkar (2018)
Pseudomonas sp.	Production of ACC deaminase, IAA, ammonium, induces phosphate solubilization, ameliorated growth and yield	*Lycopersicon esculentum*	Hernández-Montiel et al. (2017)
Rhizophagus irregularis	Improved tolerance to salt stress protects plants against pathogens (*Sclerotinia sclerotiorum*), improves nutrient uptake in plants, and increases production	*Triticum aestivum*, *Zea mays*, *Lycopersicon esculentum*	Krishnamoorthy et al. (2016) and Heydarian et al. (2018)
Rhizobium meliloti, *R. leguminosarum*	Production of siderophore, triggers nitrogen fixation, phosphate solubilization, and increases growth and yield of peanut; improved nitrogen utilization, growth, and production of soybean under drought stress	*Arachis hypogaea*, *Glycine max*	Mondal et al. (2020a,b) and Igiehon et al. (2019)
Streptomyces sp.	Production of siderophore and IAA, improved growth and metabolic activity	*Lycopersicon esculentum*	Dias et al. (2017)
Trichoderma sp.	Improved germination rate, dry weight, chlorophyll content, enhanced nitrogen and phosphorous content in soil	*Brassica campestris*	Ji et al. (2020)

5.5 Outlook and conclusion

There is still a significant lack of knowledge at present about the plant-microbes-soil interactions. For instance, we have gathered much information about the rhizobia-legume mutualism, but our understanding is restricted in the case of nonrhizobial nitrogen fixation; furthermore, there is a lack of studies in the field plant-mycorrhizal association except regarding arbuscular mycorrhiza. We are still far behind in terms of the knowledge of the actual diversity of microbes present in the rhizosphere, how crosstalk among the microbes-microbes and plant-microbes shape the nature of rhizosphere soil, how such interactions influence the biogeochemical cycle of the earth, how these interactions control the genetical and molecular nature of plant and microbes, etc.

There is a positive relationship between plants and microbes. Both benefit from each other; therefore, choosing an appropriate plant-microbial combination to improve soil health and productivity might be an emerging strategy toward sustainable agricultural practices. To understand the efficiency of plant-microbial interactions, a comprehensive study about the transient members of the microbial community and their functions, which are active at various stages of plant growth and development, is required. With the advancement in molecular biology and bioinformatics, better knowledge of the plant-microbe interactions could be achieved that will play a vital role in prescribing future biofertilizers. These future biofertilizers should provide the desired level of plant growth and productivity and improved biotic and abiotic tolerance. Future research should be focused on maintaining healthy and fertile soil by utilizing the benefits of plant-microbial interactions to attain economic and sustainable crop production.

References

Adesemoye, A.O., Kloepper, J.W., 2009. Plant-microbes interactions in enhanced fertilizer-use efficiency. Appl. Microbiol. Biotechnol. 85 (1), 1-12.

Ahmad, P., Hashem, A., Abd-Allah, E.F., Alqarawi, A.A., John, R., Egamberdieva, D., Gucel, S., 2015. Role of *Trichoderma harzianum* in mitigating NaCl stress in Indian mustard (*Brassica juncea* L) through antioxidative defense system. Front. Plant Sci. 6, 868.

Aimé, S., Alabouvette, C., Steinberg, C., Olivain, C., 2013. The endophytic strain *Fusarium oxysporum* Fo47: a good candidate for priming the defense responses in tomato roots. Mol. Plant-Microbe Interact. 26 (8), 918-926.

Akiyama, K., Matsuzaki, K.I., Hayashi, H., 2005. Plant sesquiterpenes induce hyphal branching in arbuscular mycorrhizal fungi. Nature 435 (7043), 824-827.

Akram, M.S., Shahid, M., Tahir, M., Mehmood, F., Ijaz, M., 2017. Plant-microbe interactions: current perspectives of mechanisms behind symbiotic and pathogenic associations. In: Plant Microbe Interactions in Agro-Ecological Perspectives. Springer Nature, Singapore, pp. 97-126.

Alkurtany, A.E.S., Ali, S.A.M., Mahdi, W.M., 2018. The efficiency of prepared biofertilizer from local isolate of *Bradyrhizobium* sp on growth and yield of mungbean plant. Iraqi J. Agric. Sci. 49 (5), 722-730.

Altuntaş, Ö., 2018. A comparative study on the effects of different conventional, organic and bio-fertilizers on broccoli yield and quality. Appl. Ecol. Environ. Res. 16 (2), 1595-1608.

Andria, V., Reichenauer, T.G., Sessitsch, A., 2009. Expression of alkane monooxygenase (alkB) genes by plant-associated bacteria in the rhizosphere and endosphere of Italian ryegrass (*Lolium multiflorum* L.) grown in diesel contaminated soil. Environ. Pollut. 157 (12), 3347-3350.

Ansari, M.H., Hashemabadi, D., Kaviani, B., 2017. Effect of cattle manure and sulfur on yield and oil composition of pumpkin (*Cucurbita pepo* var. *Styriaca*) inoculated with *Thiobacillus thiooxidans* in calcareous soil. Commun. Soil Sci. Plant Anal. 48 (18), 2103-2118.

Arkhipova, T., Galimsyanova, N., Kuzmina, L., Vysotskaya, L., Sidorova, L., Gabbasova, I., Melentiev, A., Kudoyarova, G., 2019. Effect of seed bacterization with plant growth-promoting bacteria on wheat productivity and phosphorus mobility in the rhizosphere. Plant Soil Environ. 65 (6), 313–319.

Arshad, M., Shaharoona, B., Mahmood, T., 2008. Inoculation with Pseudomonas spp. containing ACC-deaminase partially eliminates the effects of drought stress on growth, yield, and ripening of pea (*Pisum sativum* L.). Pedosphere 18 (5), 611–620.

Arslan, M., Afzal, M., Amin, I., Iqbal, S., Khan, Q.M., 2014. Nutrients can enhance the abundance and expression of alkane hydroxylase CYP153 gene in the rhizosphere of ryegrass planted in hydrocarbon-polluted soil. PLoS One 9 (10), e111208.

Aseri, G.K., Jain, N., Panwar, J., Rao, A.V., Meghwal, P.R., 2008. Biofertilizers improve plant growth, fruit yield, nutrition, metabolism and rhizosphere enzyme activities of pomegranate (*Punica granatum* L.) in Indian Thar Desert. Sci. Hortic. 117 (2), 130–135.

Asgher, M., Khan, M.I.R., Anjum, N.A., Khan, N.A., 2015. Minimising toxicity of cadmium in plants—role of plant growth regulators. Protoplasma 252 (2), 399–413.

Ávila, J.S., Ferreira, J.S., Santos, J.S., Rocha, P.A.D., Baldani, V.L., 2020. Green manure, seed inoculation with *Herbaspirillum seropedicae* and nitrogen fertilization on maize yield. Rev. Bras. de Eng. Agricola e Ambient. 24 (9), 590–595.

Azooz, M.M., Youssef, A.M., Ahmad, P., 2011. Evaluation of salicylic acid (SA) application on growth, osmotic solutes and antioxidant enzyme activities on broad bean seedlings grown under diluted seawater. Int. J. Plant Physiol. Biochem. 3 (14), 253–264.

Badri, D.V., Vivanco, J.M., 2009. Regulation and function of root exudates. Plant Cell Environ. 32 (6), 666–681.

Barea, J.M., Pozo, M.J., Azcon, R., Azcon-Aguilar, C., 2005. Microbial co-operation in the rhizosphere. J. Exp. Bot. 56 (417), 1761–1778.

Belimov, A.A., Dodd, I.C., Hontzeas, N., Theobald, J.C., Safronova, V.I., Davies, W.J., 2009. Rhizosphere bacteria containing 1-aminocyclopropane-1-carboxylate deaminase increase yield of plants grown in drying soil via both local and systemic hormone signalling. New Phytol. 181 (2), 413–423.

Belimov, A.A., Dodd, I.C., Safronova, V.I., Shaposhnikov, A.I., Azarova, T.S., Makarova, N.M., Davies, W.J., Tikhonovich, I.A., 2015. Rhizobacteria that produce auxins and contain 1-amino-cyclopropane-1-carboxylic acid deaminase decrease amino acid concentrations in the rhizosphere and improve growth and yield of well-watered and water-limited potato (*Solanum tuberosum*). Ann. Appl. Biol. 167 (1), 11–25.

Benson, A., Ram, G., John, A., Melvin Joe, M., 2017. Inoculation of 1-aminocyclopropane-1-carboxylate deaminase–producing bacteria along with biosurfactant application enhances the phytoremediation efficiency of *Medicago sativa* in hydrocarbon-contaminated soils. Biorem. J. 21 (1), 20–29.

Berry, A.M., Mendoza-Herrera, A., Guo, Y.Y., Hayashi, J., Persson, T., Barabote, R., Demchenko, K., Zhang, S., Pawlowski, K., 2011. New perspectives on nodule nitrogen assimilation in actinorhizal symbioses. Funct. Plant Biol. 38 (9), 645–652.

Besserer, A., Puech-Pagès, V., Kiefer, P., Gomez-Roldan, V., Jauneau, A., Roy, S., Portais, J.C., Roux, C., Bécard, G., Séjalon-Delmas, N., 2006. Strigolactones stimulate arbuscular mycorrhizal fungi by activating mitochondria. PLoS Biol. 4 (7), e226.

Bhaduri, D., Pal, S., Purakayastha, T.J., Chakraborty, K., Yadav, R.S., Akhtar, M.S., 2015. Soil quality and plant-microbe interactions in the rhizosphere. In: Sustainable Agriculture Reviews. Springer, Cham, pp. 307–335.

Bhardwaj, D., Ansari, M.W., Sahoo, R.K., Tuteja, N., 2014. Biofertilizers function as key player in sustainable agriculture by improving soil fertility, plant tolerance and crop productivity. Microb. Cell Factories 13 (1), 1–10.

Bhargava, P., Singh, A.K., Goel, R., 2017. Microbes: bioresource in agriculture and environmental sustainability. In: Plant-Microbe Interactions in Agro-Ecological Perspectives. Springer, Singapore, pp. 361–376.

Bisht, S., Pandey, P., Kaur, G., Aggarwal, H., Sood, A., Sharma, S., Kumar, V., Bisht, N.S., 2014. Utilization of endophytic strain *Bacillus sp*. SBER3 for biodegradation of polyaromatic hydrocarbons (PAH) in soil model system. Eur. J. Soil Biol. 60, 67–76.

Blackburn, J.W., Hafker, W.R., 1993. The impact of biochemistry, bioavailability and bioactivity on the selection of bioremediation techniques. Trends Biotechnol. 11 (8), 328–333.

Boleta, E.H.M., Shintate Galindo, F., Jalal, A., Santini, J.M.K., Rodrigues, W.L., Lima, B.H.D., Arf, O., Silva, M.R.D., Buzetti, S., Teixeira Filho, M.C.M., 2020. Inoculation with growth-promoting bacteria *Azospirillum brasilense* and its effects on productivity and nutritional accumulation of wheat cultivars. Front. Sustain. Food Syst. 4, 265.

Bömke, C., Tudzynski, B., 2009. Diversity, regulation, and evolution of the gibberellin biosynthetic pathway in fungi compared to plants and bacteria. Phytochemistry 70 (15–16), 1876–1893.

Boyer, L.R., Brain, P., Xu, X.M., Jeffries, P., 2015. Inoculation of drought-stressed strawberry with a mixed inoculum of two arbuscular mycorrhizal fungi: effects on population dynamics of fungal species in roots and consequential plant tolerance to water deficiency. Mycorrhiza 25 (3), 215–227.

Brahmaprakash, G.P., Sahu, P.K., Lavanya, G., Nair, S.S., Gangaraddi, V.K., Gupta, A., 2017. Microbial functions of the rhizosphere. In: Plant-Microbe Interactions in Agro-Ecological Perspectives. Springer, Singapore, pp. 177–210.

Brown, J.F., Ogle, H.J., 1997. Plant Pathogens and Plant Diseases. Published by Rockvale Publications for the Division of Botany, University of New England.

Campos-Soriano, L., García-Martínez, J., Segundo, B.S., 2012. The arbuscular mycorrhizal symbiosis promotes the systemic induction of regulatory defence-related genes in rice leaves and confers resistance to pathogen infection. Mol. Plant Pathol. 13 (6), 579–592.

Cavagnaro, T.R., Barrios-Masias, F.H., Jackson, L.E., 2012. Arbuscular mycorrhizas and their role in plant growth, nitrogen interception and soil gas efflux in an organic production system. Plant Soil 353 (1), 181–194.

Cébron, A., Louvel, B., Faure, P., France-Lanord, C., Chen, Y., Murrell, J.C., Leyval, C., 2011. Root exudates modify bacterial diversity of phenanthrene degraders in PAH-polluted soil but not phenanthrene degradation rates. Environ. Microbiol. 13 (3), 722–736.

Chatterjee, S., Sau, G.B., Mukherjee, S.K., 2009. Plant growth promotion by a hexavalent chromium reducing bacterial strain, *Cellulosimicrobium cellulans* KUCr3. World J. Microbiol. Biotechnol. 25 (10), 1829–1836.

Chiomento, J.L.T., Stürmer, S.L., Carrenho, R., da Costa, R.C., Scheffer-Basso, S.M., Antunes, L.E.C., Nienow, A.A., Calvete, E.O., 2019. Composition of arbuscular mycorrhizal fungi communities signals generalist species in soils cultivated with strawberry. Sci. Hortic. 253, 286–294.

Cohen, A.C., Bottini, R., Pontin, M., Berli, F.J., Moreno, D., Boccanlandro, H., Travaglia, C.N., Piccoli, P.N., 2015. *Azospirillum brasilense* ameliorates the response of *Arabidopsis thaliana* to drought mainly via enhancement of ABA levels. Physiol. Plant. 153 (1), 79–90.

Costerousse, B., Schönholzer-Mauclaire, L., Frossard, E., Thonar, C., 2018. Identification of heterotrophic zinc mobilization processes among bacterial strains isolated from wheat rhizosphere (*Triticum aestivum* L.). Appl. Environ. Microbiol. 84 (1), 1–16.

Creus, C.M., Sueldo, R.J., Barassi, C.A., 2004. Water relations and yield in Azospirillum-inoculated wheat exposed to drought in the field. Can. J. Bot. 82 (2), 273–281.

Curá, J.A., Franz, D.R., Filosofía, J.E., Balestrasse, K.B., Burgueño, L.E., 2017. Inoculation with *Azospirillum sp.* and *Herbaspirillum sp.* bacteria increases the tolerance of maize to drought stress. Microorganisms 5 (3), 41.

Dias, M.P., Bastos, M.S., Xavier, V.B., Cassel, E., Astarita, L.V., Santarém, E.R., 2017. Plant growth and resistance promoted by Streptomyces spp. in tomato. Plant Physiol. Biochem. 118, 479–493.

Dilworth, M.J., James, E.K., Sprent, J.I., Newton, W.E. (Eds.), 2008. Nitrogen-Fixing Leguminous Symbioses. vol. 7. Springer Science & Business Media.

Dubey, K.K., Fulekar, M.H., 2012. Chlorpyrifos bioremediation in Pennisetum rhizosphere by a novel potential degrader *Stenotrophomonas maltophilia* MHF ENV20. World J. Microbiol. Biotechnol. 28 (4), 1715–1725.

Egamberdieva, D., Wirth, S.J., Alqarawi, A.A., Abd_Allah, E.F., Hashem, A., 2017. Phytohormones and beneficial microbes: essential components for plants to balance stress and fitness. Front. Microbiol. 8, 2104.

El Maaloum, S., Elabed, A., Alaoui-Talibi, Z.E., Meddich, A., Filali-Maltouf, A., Douira, A., Ibnsouda-Koraichi, S., Amir, S., El Modafar, C., 2020. Effect of arbuscular mycorrhizal fungi and phosphate-solubilizing bacteria consortia associated with phospho-compost on phosphorus solubilization and growth of tomato seedlings (*Solanum lycopersicum* L.). Commun. Soil Sci. Plant Anal. 51 (5), 622–634.

FAO, 2017. Available at: http://www.fao.org/faostat/en/#home.

Fässler, E., Evangelou, M.W., Robinson, B.H., Schulin, R., 2010. Effects of indole-3-acetic acid (IAA) on sunflower growth and heavy metal uptake in combination with ethylene diamine disuccinic acid (EDDS). Chemosphere 80 (8), 901–907.

Gage, D.J., 2004. Infection and invasion of roots by symbiotic, nitrogen-fixing rhizobia during nodulation of temperate legumes. Microbiol. Mol. Biol. Rev. 68 (2), 280–300.

Godheja, J., Shekhar, S.K., Siddiqui, S.A., Modi, D.R., 2016. Xenobiotic compounds present in soil and water: a review on remediation strategies. J. Environ. Anal. Toxicol. 6 (392). 2161-0525.

Godheja, J., Shekhar, S.K., Modi, D.R., 2017. Bacterial rhizoremediation of petroleum hydrocarbons (PHC). In: Plant-microbe Interactions in Agro-ecological Perspectives. Springer, Singapore, pp. 495–519.

Gomiero, T., 2016. Soil degradation, land scarcity and food security: reviewing a complex challenge. Sustainability 8 (3), 281.

Gyaneshwar, P., Kumar, G.N., Parekh, L.J., Poole, P.S., 2002. Role of soil microorganisms in improving P nutrition of plants. Plant Soil 245 (1), 83–93.

Halder, S.K., Maity, C., Jana, A., Das, A., Paul, T., Das Mohapatra, P.K., Pati, B.R., Mondal, K.C., 2013. Proficient biodegradation of shrimp shell waste by *Aeromonas hydrophila* SBK1 for the concomitant production of antifungal chitinase and antioxidant chitosaccharides. Int. Biodeterior. Biodegradation 79, 88–97.

Harman, G.E., Uphoff, N., 2019. Symbiotic root-endophytic soil microbes improve crop productivity and provide environmental benefits. Scientifica 2019, 1–25.

Hawes, M.C., Gunawardena, U., Miyasaka, S., Zhao, X., 2000. The role of root border cells in plant defense. Trends Plant Sci. 5 (3), 128–133.

Hernández-Montiel, L.G., Chiquito Contreras, C.J., Murillo Amador, B., Vidal Hernández, L., Quiñones Aguilar, E.E., Chiquito Contreras, R.G., 2017. Efficiency of two inoculation methods of *Pseudomonas putida* on growth and yield of tomato plants. J. Soil Sci. Plant Nutr. 17 (4), 1003–1012.

Heydarian, A., Tohidi Moghadam, H.R., Donath, T.W., Sohrabi, M., 2018. Study of effect of arbuscular mycorrhiza (Glomus intraradices) fungus on wheat under nickel stress. Agron. Res. 16, 1660–1667.

Hong, S.H., Ryu, H., Kim, J., Cho, K.S., 2011. Rhizoremediation of diesel-contaminated soil using the plant growth-promoting rhizobacterium *Gordonia sp.* S2RP-17. Biodegradation 22 (3), 593–601.

Hou, J., Liu, W., Wang, B., Wang, Q., Luo, Y., Franks, A.E., 2015. PGPR enhanced phytoremediation of petroleum contaminated soil and rhizosphere microbial community response. Chemosphere 138, 592–598.

Huang, X.D., El-Alawi, Y., Gurska, J., Glick, B.R., Greenberg, B.M., 2005. A multi-process phytoremediation system for decontamination of persistent total petroleum hydrocarbons (TPHs) from soils. Microchem. J. 81 (1), 139–147.

Hunter, P., 2016. Plant microbiomes and sustainable agriculture: deciphering the plant microbiome and its role in nutrient supply and plant immunity has great potential to reduce the use of fertilizers and biocides in agriculture. EMBO Rep. 17 (12), 1696–1699.

Igiehon, N.O., Babalola, O.O., 2018. Rhizosphere microbiome modulators: contributions of nitrogen fixing bacteria towards sustainable agriculture. Int. J. Environ. Res. Public Health 15 (4), 574.

Igiehon, N.O., Babalola, O.O., Aremu, B.R., 2019. Genomic insights into plant growth promoting rhizobia capable of enhancing soybean germination under drought stress. BMC Microbiol. 19 (1), 1–22.

Iqbal, M., Ashraf, M., 2007. Seed treatment with auxins modulates growth and ion partitioning in salt-stressed wheat plants. J. Integr. Plant Biol. 49 (7), 1003–1015.

Iqbal, M., Ashraf, M., 2013. Gibberellic acid mediated induction of salt tolerance in wheat plants: growth, ionic partitioning, photosynthesis, yield and hormonal homeostasis. Environ. Exp. Bot. 86, 76–85.

Ishimaru, Y., Kakei, Y., Shimo, H., Bashir, K., Sato, Y., Sato, Y., Uozumi, N., Nakanishi, H., Nishizawa, N.K., 2011. A rice phenolic efflux transporter is essential for solubilizing precipitated apoplasmic iron in the plant stele. J. Biol. Chem. 286 (28), 24649–24655.

Janardan, Y., Verma, J.P., Tiwari, K.N., 2011. Plant growth promoting activities of fungi and their effect on chickpea plant growth. Asian J. Biol. Sci. 4 (3), 291–299.

Jenkinson, D.S., 1981. Microbial biomass in soil: measurement and turnover. Soil Biochem. 5, 415–471.

Ji, S., Liu, Z., Liu, B., Wang, Y., Wang, J., 2020. The effect of Trichoderma biofertilizer on the quality of flowering Chinese cabbage and the soil environment. Sci. Hortic. 262, 109069.

Juan, L.I., Zhao, B.Q., Li, X.Y., Jiang, R.B., Bing, S.H., 2008. Effects of long-term combined application of organic and mineral fertilizers on microbial biomass, soil enzyme activities and soil fertility. Agric. Sci. China 7 (3), 336–343.

Khan, A.G., 2005. Role of soil microbes in the rhizospheres of plants growing on trace metal contaminated soils in phytoremediation. J. Trace Elem. Med. Biol. 18 (4), 355–364.

Khan, A.L., Hamayun, M., Kim, Y.H., Kang, S.M., Lee, J.H., Lee, I.J., 2011. Gibberellins producing endophytic *Aspergillus fumigatus* LH02 influenced endogenous phytohormonal levels, isoflavonoids production and plant growth in salinity stress. Process Biochem. 46 (2), 440–447.

Khan, M.I.R., Asgher, M., Khan, N.A., 2014. Alleviation of salt-induced photosynthesis and growth inhibition by salicylic acid involves glycinebetaine and ethylene in mungbean (*Vigna radiata* L.). Plant Physiol. Biochem. 80, 67–74.

Kim, S., Lim, H., Lee, I., 2010. Enhanced heavy metal phytoextraction by *Echinochloa crusgalli* using root exudates. J. Biosci. Bioeng. 109 (1), 47–50.

Klaic, R., Guimarães, G.G., Giroto, A.S., Bernardi, A.C., Zangirolami, T.C., Ribeiro, C., Farinas, C.S., 2021. Synergy of *Aspergillus niger* and components in biofertilizer composites increases the availability of nutrients to plants. Curr. Microbiol. 78 (4), 1529–1542.

Koo, B.J., Adriano, D.C., Bolan, N.S., Barton, C.D., 2005. Root exudates and microorganisms. In: Hillel, D. (Ed.), Encyclopedia of Soils in the Environment. Elsevier, Oxford, pp. 421–428.

Krishnamoorthy, R., Kim, K., Subramanian, P., Senthilkumar, M., Anandham, R., Sa, T., 2016. Arbuscular mycorrhizal fungi and associated bacteria isolated from salt-affected soil enhances the tolerance of maize to salinity in coastal reclamation soil. Agric. Ecosyst. Environ. 231, 233–239.

Kudoyarova, G.R., Vysotskaya, L.B., Arkhipova, T.N., Kuzmina, L.Y., Galimsyanova, N.F., Sidorova, L.V., Gabbasova, I.M., Melentiev, A.I., Veselov, S.Y., 2017. Effect of auxin producing and phosphate solubilizing bacteria on mobility of soil phosphorus, growth rate, and P acquisition by wheat plants. Acta Physiol. Plant. 39 (11), 1–8.

Kudoyarova, G., Arkhipova, T., Korshunova, T., Bakaeva, M., Loginov, O., Dodd, I.C., 2019. Phytohormone mediation of interactions between plants and non-symbiotic growth promoting bacteria under edaphic stresses. Front. Plant Sci. 10, 1368.

Kumar, A., Singh, R., Giri, D.D., Singh, P.K., Pandey, K.D., 2014. Effect of *Azotobacter chroococcum* CL13 inoculation on growth and curcumin content of turmeric (*Curcuma longa* L.). Int. J. Curr. Microbiol. App. Sci. 3 (9), 275–283.

Kumar, H., Dubey, R.C., Maheshwari, D.K., 2017. Seed-coating fenugreek with Burkholderia rhizobacteria enhances yield in field trials and can combat Fusarium wilt. Rhizosphere 3, 92–99.

Kurosawa, E., 1926. Experimental studies on the nature of the substance secreted by the "bakanae" fungus. Nat. Hist. Soc. Formosa 16, 213–227.

Lacalle, R.G., Gómez-Sagasti, M.T., Artetxe, U., Garbisu, C., Becerril, J.M., 2018. *Brassica napus* has a key role in the recovery of the health of soils contaminated with metals and diesel by rhizoremediation. Sci. Total Environ. 618, 347–356.

Lal, R., 2015. Restoring soil quality to mitigate soil degradation. Sustainability 7 (5), 5875–5895.

Lei, Z.H.A.O., Zhang, Y.Q., 2015. Effects of phosphate solubilization and phytohormone production of *Trichoderma asperellum* Q1 on promoting cucumber growth under salt stress. J. Integr. Agric. 14 (8), 1588–1597.

Lipson, D.A., Kelley, S.T., 2014. Plant-microbe interactions. In: Ecology and the Environment. Springer, New York, pp. 177–204.

Ma, Y., Oliveira, R.S., Wu, L., Luo, Y., Rajkumar, M., Rocha, I., Freitas, H., 2015. Inoculation with metal-mobilizing plant-growth-promoting rhizobacterium Bacillus sp. SC2b and its role in rhizoremediation. J. Toxicol. Environ. Health A 78 (13–14), 931–944.

Ma, Y., Oliveira, R.S., Freitas, H., Zhang, C., 2016. Biochemical and molecular mechanisms of plant-microbe-metal interactions: relevance for phytoremediation. Front. Plant Sci. 7, 918.

Marschner, P., Yang, C.H., Lieberei, R., Crowley, D.E., 2001. Soil and plant specific effects on bacterial community composition in the rhizosphere. Soil Biol. Biochem. 33 (11), 1437–1445.

Marschner, P., Crowley, D.B., Rengel, Z., 2010. Interactions between rhizosphere microorganisms and plants governing iron and phosphorus availability. In: 19th World Congress of Soil Science, Soil Solutions for a Changing World, Brisbane, Australia.

Martínez-Medina, A., Fernandez, I., Lok, G.B., Pozo, M.J., Pieterse, C.M., Van Wees, S.C., 2017. Shifting from priming of salicylic acid- to jasmonic acid-regulated defences by Trichoderma protects tomato against the root knot nematode *Meloidogyne incognita*. New Phytol. 213 (3), 1363–1377.

Masson-Boivin, C., Giraud, E., Perret, X., Batut, J., 2009. Establishing nitrogen-fixing symbiosis with legumes: how many rhizobium recipes? Trends Microbiol. 17 (10), 458–466.

Mayak, S., Tirosh, T., Glick, B.R., 2004a. Plant growth-promoting bacteria confer resistance in tomato plants to salt stress. Plant Physiol. Biochem. 42 (6), 565–572.

Mayak, S., Tirosh, T., Glick, B.R., 2004b. Plant growth-promoting bacteria that confer resistance to water stress in tomatoes and peppers. Plant Sci. 166 (2), 525–530.

Meleigy, S.A., Khalaf, M.A., 2009. Biosynthesis of gibberellic acid from milk permeate in repeated batch operation by a mutant *Fusarium moniliforme* cells immobilized on loofa sponge. Bioresour. Technol. 100 (1), 374–379.

Mendes, R., Garbeva, P., Raaijmakers, J.M., 2013. The rhizosphere microbiome: significance of plant beneficial, plant pathogenic, and human pathogenic microorganisms. FEMS Microbiol. Rev. 37 (5), 634–663.

Meyer, S.L., Everts, K.L., Gardener, B.M., Masler, E.P., Abdelnabby, H.M., Skantar, A.M., 2016. Assessment of DAPG-producing *Pseudomonas fluorescens* for management of *Meloidogyne incognita* and *Fusarium oxysporum* on watermelon. J. Nematol. 48 (1), 43.

Miceli, A., Vetrano, F., Moncada, A., 2020. Effects of foliar application of gibberellic acid on the salt tolerance of tomato and sweet pepper transplants. Horticulturae 6 (4), 93.

Miller, R.M., Jastrow, J.D., 2000. Mycorrhizal fungi influence soil structure. In: Arbuscular Mycorrhizas: Physiology and Function. Springer, Dordrecht, pp. 3–18.

Mishra, I., Arora, N.K., 2019. Rhizoremediation: a sustainable approach to improve the quality and productivity of polluted soils. In: Phyto and Rhizo Remediation. Springer, Singapore, pp. 33–66.

Mohanram, S., Kumar, P., 2019. Rhizosphere microbiome: revisiting the synergy of plant-microbe interactions. Ann. Microbiol. 69 (4), 307–320.

Mondal, S., Halder, S.K., Yadav, A.N., Mondal, K.C., 2020a. Microbial consortium with multifunctional plant growth-promoting attributes: future perspective in agriculture. In: Yadav, A., Rastegari, A., Yadav, N., Kour, D. (Eds.), Advances in Plant Microbiome and Sustainable Agriculture. Microorganisms for Sustainability. vol 20. Springer, Singapore.

Mondal, M., Skalicky, M., Garai, S., Hossain, A., Sarkar, S., Banerjee, H., Kundu, R., Brestic, M., Barutcular, C., Erman, M., El Sabagh, A., 2020b. Supplementing nitrogen in combination with rhizobium inoculation and soil mulch in peanut (*Arachis hypogaea* L.) production system: part II. Effect on phenology, growth, yield attributes, pod quality, profitability and nitrogen use efficiency. Agronomy 10 (10), 1513.

Motsara, M.R., Bhattacharyya, P., Srivastava, B., 1995. Biofertiliser Technology, Marketing and Usage: A Sourcebook-Cum-Glossary. Fertiliser Development and Consultation Organization.

Mukherjee, D., 2017. Microorganisms: role for crop production and its interface with soil agroecosystem. In: Plant-Microbe Interactions in Agro-Ecological Perspectives. Springer, Singapore, pp. 333–359.

Mulligan, C.N., Galvez-Cloutier, R., 2003. Bioremediation of metal contamination. Environ. Monit. Assess. 84 (1), 45–60.

Naveed, M., Mitter, B., Reichenauer, T.G., Wieczorek, K., Sessitsch, A., 2014. Increased drought stress resilience of maize through endophytic colonization by *Burkholderia phytofirmans* PsJN and *Enterobacter* sp. FD17. Environ. Exp. Bot. 97, 30–39.

Niu, X., Song, L., Xiao, Y., Ge, W., 2018. Drought-tolerant plant growth-promoting rhizobacteria associated with foxtail millet in a semi-arid agroecosystem and their potential in alleviating drought stress. Front. Microbiol. 8, 2580.

Patel, J.K., Archana, G., 2017. Diverse culturable diazotrophic endophytic bacteria from Poaceae plants show cross-colonization and plant growth promotion in wheat. Plant Soil 417 (1), 99–116.

Pathak, A., Dastidar, M.G., Sreekrishnan, T.R., 2009. Bioleaching of heavy metals from sewage sludge: a review. J. Environ. Manag. 90 (8), 2343–2353.

Pathak, A., Chakrabarti, S.K., Das, R., Mandal, M.K., 2016. Response of different wheat varieties towards *Azospirillu*m and phosphate solubilizing bacteria (PSB) seed inoculation. J. Appl. Nat. Sci. 8 (1), 213–217.

Peng, R.H., Xiong, A.S., Xue, Y., Fu, X.Y., Gao, F., Zhao, W., Tian, Y.S., Yao, Q.H., 2008. Microbial biodegradation of polyaromatic hydrocarbons. FEMS Microbiol. Rev. 32 (6), 927–955.

Peters, N.K., Frost, J.W., Long, S.R., 1986. A plant flavone, luteolin, induces expression of *Rhizobium meliloti* nodulation genes. Science 233 (4767), 977–980.

Poulin, M.J., Bel-Rhlid, R., Piché, Y., Chênevert, R., 1993. Flavonoids released by carrot (*Daucus carota*) seedlings stimulate hyphal development of vesicular-arbuscular mycorrhizal fungi in the presence of optimal CO_2 enrichment. J. Chem. Ecol. 19 (10), 2317–2327.

Press, C.M., Loper, J.E., Kloepper, J.W., 2001. Role of iron in rhizobacteria-mediated induced systemic resistance of cucumber. Phytopathology 91 (6), 593–598.

Raaijmakers, J.M., De Bruijn, I., Nybroe, O., Ongena, M., 2010. Natural functions of lipopeptides from *Bacillus* and *Pseudomonas*: more than surfactants and antibiotics. FEMS Microbiol. Rev. 34 (6), 1037–1062.

Rahman, M., Sabir, A.A., Mukta, J.A., Khan, M.M.A., Mohi-Ud-Din, M., Miah, M.G., Rahman, M., Islam, M.T., 2018. Plant probiotic bacteria Bacillus and Paraburkholderia improve growth, yield and content of antioxidants in strawberry fruit. Sci. Rep. 8 (1), 1–11.

Rajkumar, M., Sandhya, S., Prasad, M.N.V., Freitas, H., 2012. Perspectives of plant-associated microbes in heavy metal phytoremediation. Biotechnol. Adv. 30 (6), 1562–1574.

Ravindra, K., Sokhi, R., Van Grieken, R., 2008. Atmospheric polycyclic aromatic hydrocarbons: source attribution, emission factors and regulation. Atmos. Environ. 42 (13), 2895–2921.

Requena, N., Perez-Solis, E., Azcón-Aguilar, C., Jeffries, P., Barea, J.M., 2001. Management of indigenous plant-microbe symbioses aids restoration of desertified ecosystems. Appl. Environ. Microbiol. 67 (2), 495–498.

Sagar, A., Thomas, G., Rai, S., Mishra, R.K., Ramteke, P.W., 2018. Enhancement of growth and yield parameters of wheat variety AAI-W6 by an organic farm isolate of plant growth promoting *Erwinia* species (KP226572). Int. J. Agric. Environ. Biotechnol. 11 (1), 159–171.

Sahoo, R.K., Ansari, M.W., Pradhan, M., Dangar, T.K., Mohanty, S., Tuteja, N., 2014. Phenotypic and molecular characterization of native *Azospirillum* strains from rice fields to improve crop productivity. Protoplasma 251 (4), 943–953.

Sangeeth, K.P., Bhai, R.S., Srinivasan, V., 2012. *Paenibacillus glucanolyticus*, a promising potassium solubilizing bacterium isolated from black pepper (*Piper nigrum* L.) rhizosphere. J. Spices Aromat. Crops 21 (2), 118–124.

Schmülling, T., 2002. New insights into the functions of cytokinins in plant development. J. Plant Growth Regul. 21 (1), 40–49.

Segura, A., Rodríguez-Conde, S., Ramos, C., Ramos, J.L., 2009. Bacterial responses and interactions with plants during rhizoremediation. Microb. Biotechnol. 2 (4), 452–464.

Sen, S., Chandrasekhar, C.N., 2015. Effect of PGPR on enzymatic activities of rice (*Oryza sativa* L.) under salt stress. Asian J. Plant Sci. Res. 5, 44–48.

Shafqat, M., Khalid, A., Mahmood, T., Siddique, M.T., Han, J.I., Habteselassie, M.Y., 2017. Evaluation of bacteria isolated from textile wastewater and rhizosphere to simultaneously degrade azo dyes and promote plant growth. J. Chem. Technol. Biotechnol. 92 (10), 2760–2768.

Sharma, S.B., Sayyed, R.Z., Trivedi, M.H., Gobi, T.A., 2013. Phosphate solubilizing microbes: sustainable approach for managing phosphorus deficiency in agricultural soils. Springerplus 2 (1), 1–14.

Shi, T.Q., Peng, H., Zeng, S.Y., Ji, R.Y., Shi, K., Huang, H., Ji, X.J., 2017. Microbial production of plant hormones: opportunities and challenges. Bioengineered 8 (2), 124–128.

Shinde, K.S., Borkar, S., 2018. Seed bacterialization induced proline content in *Sorghum bicolor* crop under severe drought condition. Int. J. Chem. Stud. 6 (2), 1191–1194.

Shiraishi, A., Matsushita, N., Hougetsu, T., 2010. Nodulation in black locust by the Gammaproteobacteria *Pseudomonas sp.* and the Betaproteobacteria *Burkholderia sp*. Syst. Appl. Microbiol. 33 (5), 269–274.

Silva, A.C.D., Suassuna, J.F., Melo, A.S.D., Costa, R.R., Andrade, W.L.D., Silva, D.C.D., 2017. Salicylic acid as attenuator of drought stress on germination and initial development of sesame. Rev. Bras. de Eng. Agricola e Ambient. 21 (3), 156–162.

Singh, R.P., Jha, P.N., 2016. Alleviation of salinity-induced damage on wheat plant by an ACC deaminase-producing halophilic bacterium *Serratia sp*. SL-12 isolated from a salt lake. Symbiosis 69 (2), 101–111.

Sinha, A., Lulu, S., Vino, S., Osborne, W.J., 2019. Reactive green dye remediation by *Alternanthera philoxeroides* in association with plant growth promoting *Klebsiella sp.* VITAJ23: a pot culture study. Microbiol. Res. 220, 42–52.

Solaiman, Z.M., 2014. Contribution of arbuscular mycorrhizal fungi to soil carbon sequestration. In: Mycorrhizal Fungi: Use in Sustainable Agriculture and Land Restoration. Springer, Berlin, Heidelberg, pp. 287–296.

Song, Y.Y., Ye, M., Li, C.Y., Wang, R.L., Wei, X.C., Luo, S.M., Zeng, R.S., 2013. Priming of anti-herbivore defense in tomato by arbuscular mycorrhizal fungus and involvement of the jasmonate pathway. J. Chem. Ecol. 39 (7), 1036–1044.

Spaepen, S., 2015. Plant hormones produced by microbes. In: Principles of Plant-Microbe Interactions. Springer, Cham, pp. 247–256.

Srivastava, S., Yadav, A., Seem, K., Mishra, S., Chaudhary, V., Nautiyal, C.S., 2008. Effect of high temperature on *Pseudomonas putida* NBRI0987 biofilm formation and expression of stress sigma factor RpoS. Curr. Microbiol. 56 (5), 453–457.

Stearns, J.C., Woody, O.Z., McConkey, B.J., Glick, B.R., 2012. Effects of bacterial ACC deaminase on *Brassica napus* gene expression. Mol. Plant-Microbe Interact. 25 (5), 668–676.

Tang, Y., Sun, X., Wen, T., Liu, M., Yang, M., Chen, X., 2017. Implications of terminal oxidase function in regulation of salicylic acid on soybean seedling photosynthetic performance under water stress. Plant Physiol. Biochem. 112, 19–28.

Theocharis, A., Bordiec, S., Fernandez, O., Paquis, S., Dhondt-Cordelier, S., Baillieul, F., Barka, E.A., 2012. *Burkholderia phytofirmans* PsJN primes *Vitis vinifera* L. and confers a better tolerance to low nonfreezing temperatures. Mol. Plant-Microbe Interact. 25 (2), 241–249.

Tiwari, S., Lata, C., 2018. Heavy metal stress, signaling, and tolerance due to plant-associated microbes: an overview. Front. Plant Sci. 9, 452.

Tognon, G.B., Sanmartín, C., Alcolea, V., Cuquel, F.L., Goicoechea, N., 2016. Mycorrhizal inoculation and/or selenium application affect post-harvest performance of snapdragon flowers. Plant Growth Regul. 78 (3), 389–400.

Ton, J., Van Pelt, J.A., Van Loon, L.C., Pieterse, C.M., 2002. Differential effectiveness of salicylate-dependent and jasmonate/ethylene-dependent induced resistance in Arabidopsis. Mol. Plant-Microbe Interact. 15 (1), 27–34.

Tóth, G., Hermann, T., da Silva, M.R., Montanarella, L., 2018. Monitoring soil for sustainable development and land degradation neutrality. Environ. Monit. Assess. 190 (2), 1–4.

Tu, C., Teng, Y., Luo, Y., Li, X., Sun, X., Li, Z., Liu, W., Christie, P., 2011. Potential for biodegradation of polychlorinated biphenyls (PCBs) by *Sinorhizobium meliloti*. J. Hazard. Mater. 186 (2–3), 1438–1444.

van der Heijden, M.G., Martin, F.M., Selosse, M.A., Sanders, I.R., 2015. Mycorrhizal ecology and evolution: the past, the present, and the future. New Phytol. 205 (4), 1406–1423.

Van Oosten, M.J., Di Stasio, E., Cirillo, V., Silletti, S., Ventorino, V., Pepe, O., Raimondi, G., Maggio, A., 2018. Root inoculation with *Azotobacter chroococcum* 76A enhances tomato plants adaptation to salt stress under low N conditions. BMC Plant Biol. 18 (1), 1–12.

Velázquez-Fernández, J.B., Martínez-Rizo, A.B., Ramírez-Sandoval, M., Domínguez-Ojeda, D., 2012. Biodegradation and bioremediation of organic pesticides. In: Pesticides—Recent Trends in Pesticide Residue Assay. IntechOpen, London, pp. 253–272.

Verma, P., Agrawal, N., Shahi, S.K., 2018. Effect of rhizobacterial strain *Enterobacter cloacae* strain pglo9 on potato plant growth and yield. Plant Arch. 18 (2), 2528–2532.

Vijayan, R., Palaniappan, P., Tongmin, S.A., Elavarasi, P., Manoharan, N., 2013. Rhizobitoxine enhances nodulation by inhibiting ethylene synthesis of *Bradyrhizobium elkanii* from Lespedeza species: validation by homology modeling and molecular docking study. World J. Pharm. Sci. 2, 4079–4094.

Vives-Peris, V., de Ollas, C., Gómez-Cadenas, A., Pérez-Clemente, R.M., 2020. Root exudates: from plant to rhizosphere and beyond. Plant Cell Rep. 39 (1), 3–17.

Vodnik, D., Grčman, H., Maček, I., Van Elteren, J.T., Kovačevič, M., 2008. The contribution of glomalin-related soil protein to Pb and Zn sequestration in polluted soil. Sci. Total Environ. 392 (1), 130–136.

Vranova, V., Rejsek, K., Formanek, P., 2013. Aliphatic, cyclic, and aromatic organic acids, vitamins, and carbohydrates in soil: a review. Sci World J. https://doi.org/10.1155/2013/524239.

Walker, T.S., Bais, H.P., Grotewold, E., Vivanco, J.M., 2003. Root exudation and rhizosphere biology. Plant Physiol. 132 (1), 44–51.

Weir, T.L., Perry, L.G., Gilroy, S., Vivanco, J.M., 2010. The role of root exudates in rhizosphere interactions with plants and other organisms. Annu. Rev. Plant Biol. 57, 233–266.

Werner, T., Motyka, V., Laucou, V., Smets, R., Van Onckelen, H., Schmülling, T., 2003. Cytokinin-deficient transgenic Arabidopsis plants show multiple developmental alterations indicating opposite functions of cytokinins in the regulation of shoot and root meristem activity. Plant Cell 15 (11), 2532–2550.

Werner, T., Nehnevajova, E., Köllmer, I., Novák, O., Strnad, M., Krämer, U., Schmülling, T., 2010. Root-specific reduction of cytokinin causes enhanced root growth, drought tolerance, and leaf mineral enrichment in Arabidopsis and tobacco. Plant Cell 22 (12), 3905–3920.

Wilkinson, S., Kudoyarova, G.R., Veselov, D.S., Arkhipova, T.N., Davies, W.J., 2012. Plant hormone interactions: innovative targets for crop breeding and management. J. Exp. Bot. 63 (9), 3499–3509.

Wilson, S.C., Jones, K.C., 1993. Bioremediation of soil contaminated with polynuclear aromatic hydrocarbons (PAHs): a review. Environ. Pollut. 81 (3), 229–249.

Wright, S.F., Upadhyaya, A., 1998. A survey of soils for aggregate stability and glomalin, a glycoprotein produced by hyphae of arbuscular mycorrhizal fungi. Plant Soil 198 (1), 97–107.

Xu, H., Shao, H., Lu, Y., 2019. Arbuscular mycorrhiza fungi and related soil microbial activity drive carbon mineralization in the maize rhizosphere. Ecotoxicol. Environ. Saf. 182, 109476.

Xun, F., Xie, B., Liu, S., Guo, C., 2015. Effect of plant growth-promoting bacteria (PGPR) and arbuscular mycorrhizal fungi (AMF) inoculation on oats in saline-alkali soil contaminated by petroleum to enhance phytoremediation. Environ. Sci. Pollut. Res. 22 (1), 598–608.

Yadav, P., Krishna, S.S., 2015. Plant growth promoting rhizobacteria: an effective tool to remediate residual organophosphate pesticide methyl parathion, widely used in Indian agriculture. J. Environ. Res. Develop. 9 (4), 1138–1149.

Yang, J., Kloepper, J.W., Ryu, C.M., 2009. Rhizosphere bacteria help plants tolerate abiotic stress. Trends Plant Sci. 14 (1), 1–4.

Yi, H., Crowley, D.E., 2007. Biostimulation of PAH degradation with plants containing high concentrations of linoleic acid. Environ. Sci. Technol. 41 (12), 4382–4388.

Young, I.M., 1995. Variation in moisture contents between bulk soil and the rhizosheath of wheat (*Triticum aestivum* L. cv. Wembley). New Phytol. 130 (1), 135–139.

Yuan, M., He, H., Xiao, L., Zhong, T., Liu, H., Li, S., Deng, P., Ye, Z., Jing, Y., 2014. Enhancement of Cd phytoextraction by two *Amaranthus* species with endophytic *Rahnella sp.* JN27. Chemosphere 103, 99–104.

Zamioudis, C., Pieterse, C.M., 2012. Modulation of host immunity by beneficial microbes. Mol. Plant-Microbe Interact. 25 (2), 139–150.

Zhang, H., Jiang, X., Lu, L., Xiao, W., 2015. Biodegradation of polychlorinated biphenyls (PCBs) by the novel identified cyanobacterium *Anabaena* PD-1. PLoS One 10 (7), e0131450.

Zhang, H., Kim, M.S., Sun, Y., Dowd, S.E., Shi, H., Paré, P.W., 2008. Soil bacteria confer plant salt tolerance by tissue-specific regulation of the sodium transporter HKT1. Mol. Plant-Microbe Interact. 21 (6), 737–744.

Biological control of forest pathogens: Success stories and challenges

Ratnaboli Bose[a], Aditi Saini[a], Nitika Bansal[a], M.S. Bhandari[b], Amit Pandey[a], Pooja Joshi[a], and Shailesh Pandey[a]

[a]Forest Pathology Discipline, Forest Protection Division, Forest Research Institute, Dehradun, India, [b]Genetics and Tree Improvement Division, Forest Research Institute, Dehradun, India

6.1 Introduction

Forests cover 31% of the world's surface, storing 296 gigatons of carbon. Approximately 2.4 billion people use forest-based energy for cooking (FAO and UNEP, 2020). The United Nations Strategic Plan for Forests (2020) goal calls for the reversal of global forest cover loss through protection, restoration, afforestation, and reforestation under the sustainable forest management program (The Global Forest Goals Report, 2021). Further, efforts to prevent forest degradation and dedicated contribution into addressing climate change are mentioned in forest goals as well. The target is to increase forest area by 3% worldwide by 2030. However, the world is not on track to meet this target. The depredation of forest from diseases also contributes to this loss. Rising demand for wood, wood derivatives, and non-wood forest products has positively influenced plantation forestry. The area of planted forests has increased by 123 million hectares since 1990 and now covers 294 million hectares. Roughly 45% of planted forests, which is 3% of all forests, are plantations. Primarily, they are composed of one or two tree species, native or exotic, even aged, regularly spaced for production forestry (FAO, 2020). However, globally, diseases caused by various biotic factors have contributed significantly to the ill health of natural as well as plantation forests (Boyd et al., 2013; Wingfield et al., 2015). Sincere efforts have been made worldwide to manage forest diseases to ensure the regular supply of forest produce, feed, and fiber (Mills et al., 2011; Franklin et al., 2018). Therefore, forest disease management is such a massive concern at present.

Forest sanitation, cultural control, and trenching are important silvicultural practices that find application in disease management, especially in plantation forestry (Fujimori, 2001). A limited range of chemical pesticides find heavy reliance in the prevention, mitigation, and control of forest diseases at the nursery level (Okorski et al., 2015). Application of chemicals ensure the production of quality planting material until transplanted to the field conditions (for plantation forests), but such control is not conceivable for natural forests. Importantly, chemical inputs have significantly contributed to effective disease control in forest-based business enterprises (FAO, 2001; Sturrock et al., 2011). Unfortunately, excessive chemical usage has been reported as a major cause of environmental pollution, and therefore, lobbying against chemical pesticides has led to major public attitudinal shifts with regard to their usage in disease management (Özkara et al., 2016). In view of these facts, there are accelerated efforts by the forest scientists to develop and promote eco-friendly disease management strategies.

Biological control, mostly augmentative biocontrol (ABC), is one of the best pest management options widely used in agricultural sector (Joshi et al., 2021), over 30 million hectares worldwide. North America is the biggest market for microbe-based biocontrol, while Latin America and Asia are emerging microbial markets (van Lenteren et al., 2018). Likewise, the application of biocontrol agents to manage diseases of forestry crops also gained importance (Cazorla and Mercado-Blanco, 2016; De Silva et al., 2019). Unlike chemical control, there is a complex variety of biological control available in the forestry sector, but the choice of biocontrol agents, their further development, and uptake into nursery practices would require an in-depth understanding of the complex interaction between plants, people, and environment (Barea, 2015). This article discusses the nature and practice of biological control as applied to forestry crop disease suppression. Also, it looks into the following areas:
- Key mechanisms of biocontrol
- Describes the current status of research and application of biological control against forestry crop diseases
- Prescribes future directions for furtherance of research in effective biocontrol product development.

6.2 The importance of forest pathogens

Plant pathology was helmed by German botanist Anton De Bary, who propounded the germ theory of disease and introduced the terms "parasitism" and "symbiosis" (Kutschera and Hossfeld, 2011). In his pathbreaking work, he defined the life cycle of the phytopathogenic oomycete *Phytophthora infestans*, which caused late blight in potatoes

(*Solanum tuberosum*) and was responsible for severe famines during the 1840s.

A few years later, in 1878, forest pathology was pioneered by Robert Hartig. He ushered in the concept of tree decay, as disease, caused by fungi that were both pathogenic and parasitic. This concept was earlier rejected by pathologists, who thought heartwood tissues in trees were dead (Merril and Shigo, 1979). Forest ecosystems are the hub of evolutionary processes, many of which are driven by the ecology that connects abiotic and biotic stresses. Forest pathology, a component of forest protection, deals with research of both biotic and abiotic stressors adversely affecting the health and productivity of a forest ecosystem. It encompasses both forestry and plant pathology. The concept of tree decline whereby biotic and abiotic factors act as predisposing, inciting, or contributing causes was given by Manion (1991). Biotic stresses for plants include pathogens (fungi, bacteria, viruses, nematodes, phytoplasma, viroids) and competing weeds, while abiotic stresses are drought, salinity, and extreme temperatures. Plantation and esthetic forestry is hampered by abiotic and biotic stresses.

Hepting (1963) asserted that climate factors influence known diseases or create new disease problems in trees. According to him, climate concerns influencing the propagation of fungal pathogens, largely responsible for forest diseases, were precipitation, temperature, humidity, fog and dew, wind, and radiation. Climate change may create different situations for forest disease and their management, and hence their productivity. It is thus pertinent to assess the impact of climate change on vital tree diseases to reduce yield losses (Chakraborty et al., 2000). Climate change is a major driver of introduced pathogens establishment and affects the distribution and severity of diseases (Shaw and Osborne, 2011; Sturrock et al., 2011, Trogisch et al., 2021). Major forest disease outbreaks through the introduction of an exotic pathogen leading to Chestnut blight were milestones for forest pathology. Chestnut blight is incited by *Cryphonectria parasitica* (formerly *Endothia parasitica*) and attacks American chestnut trees (*Castanea dentata*) throughout the United States and Canada. The fungal pathogen reached United States from infected imports of Japanese chestnut trees in the late 19th century. Over three and a half billion chestnut trees died by 1940. Chestnut continues to survive till date, by means of root sprouts, since root and root collar tissue may be unaffected by the fungus. These sprouts are short-lived and become blight infested, and the tree never reaches sexual maturity (Anagnostakis, 1987). In Europe, *Cr. parasitica* has caused widespread mortality of European chestnut (*Castanea sativa*) (Rigling and Prospero, 2018). White pine blister rust caused by *Cronartium ribicola* in North America and Europe, oak-*Quercus robur*

(European oak) and *Q. ilex* (holm oak) decline by *Phytophthora* in Europe, and Swiss needle cast of Douglas Fir caused by *Phaeocrytopus gauemanii* (Anagnostakis, 1987; Delatour and Guillamin, 1985; Solla et al., 2021) are important epidemics. Exotic plantations, on the other hand, of major tree crops like *Eucalyptus* spp. in China, India, Brazil, South America, and Indonesia are threatened by Chrysoporthe canker disease, Calonectria leaf blight (CLB), and Teratosphaeria leaf disease (Crous et al., 2019; Bose et al., 2020).

Prior knowledge of risk factors, hazard assessment, and diligent silvicultural measures are important forest disease management tools (Tainter and Baker, 1996). Risk mapping using epidemiological models and climate change scenarios has been attempted for commercial forestry crops like *Eucalyptus*. For CLB of eucalypts, a simple model of CqLB risk (i.e., mean annual precipitation ≥ 1400 mm and mean minimum temperature of the coldest month $\geq 16°C$) aided in evaluating extremely vulnerable regions in mainland South-East Asia, India, Africa, Australia, and Latin America. Apart from known regions from where CLB has been reported, viz., southern and central Vietnam, Kerala (India), Madagascar, and northern Australia, the model warned of potentially vulnerable areas (e.g., Central America and the Congo) (Booth et al., 2000).

Forest losses through disease and pest infestations can be minimized by way of forest site hazard evaluation (Park and Chung, 2006). It is useful for efficient identification and prediction of future hazards in forested areas. Forest sites may be classified as being low-, intermediate- to high-risk regions relative to probability of disease occurrence of susceptible species. The parameters used include climatic, edaphic, biotic, site factors, silvicultural practices, disease incidence, and severity (Tainter and Baker, 1996). For instance, in a Korea-based study, hazard ratings of pine trees and pine stands to pine wilt disease (PWD) incited by the pine wood nematode, one of the most destructive global threats to pines, were assessed considering the environmental factors both at pine community and at the individual tree level. Trees with greater diameter at breast height (DBH) were at higher risk rate than those with smaller DBH, indicating that large trees are more vulnerable to vector beetle attack given their tall height and large crown volume. Further, they found that reduced tree vigor and susceptibility to PWD were correlated. Geographical factors showed high correlation with PWD occurrence. Disease occurrence at high altitudes was less, but was more frequent on steep and south-facing slopes. Two computational models, self-organizing map (characterizing relations among variables) and random forest models (predicted ecological variables such as hazard rating of trees to nematode disturbance), supported these patterns reliably (Park et al., 2013).

Forest ecosystems especially under anthropogenic pressures are challenged by the threat of invasive and alien pathogens, which necessitates evolutionary biology of tree-pathogen relationships and forest ecology to be studied in harmony (Ennos, 2015; Hessenauer et al., 2021). With an increase in global trade and travel in forest products, the threat to biosecurity of forestry species is on the rise (Cleary et al., 2019). Anecdotal records showed that the fungus *Dothistroma septosporum* entered New Zealand through the footwear of a visiting forester, while fungal spores may have been wind-disseminated fungal spores across from Australia. This pathogen costs the New Zealand forestry industry tens of millions of dollars in lost productivity annually, and has curbed options for planting other pine species (Dyck, 2010). Likewise, *Fusarium circinatum* causing pine pitch canker is an invasive fungal pathogen in Europe distributed in *Pinus radiata* and *P. pinaster* forests in northern Spain and Portugal, and a threat to all pine-growing regions worldwide (Drenkhan et al., 2020; Elvira-Recuenco et al., 2021; Wingfield et al., 2008). Many countries in the Southern Hemisphere and South-East Asia cultivate *Eucalyptus*, *Acacia*, and *Pinus* monocultures for solid wood, pulp and paper, and other extractives (Attiwill, 1994; Inagaki et al., 2010; Paquette and Messier, 2010). However, as disease emergences continue to threaten productivity in these new forest ecosystems, viz., leaf blight of *Eucalyptus* by *Teratosphaeria destructans* in Malaysia or damping off, root rots and wilting of *Acacia crassicarpa* in Indonesia (Havenga et al., 2021; Oliveira et al., 2021), forest managers are faced with financial uncertainties.

Diagnostics of fungal and bacterial diseases of forestry species have been revolutionized through high-throughput sequencing (HTS). Population genetic studies using metabarcoding may be carried out for disease surveillance of soilborne, seedborne, and airborne pathogens, for identifying new pathogens and detection of outbreak disease epicenters (Piombo et al., 2021). *Phytophthora*, an aggressive agricultural and forest pathogen, requires early detection and identification of infection routes and spread is of high importance to minimize the threat they pose to natural ecosystems. For instance, environmental DNA (eDNA) was extracted from soil and water from forests and plantations in the northern Spain. *Phytophthora*-specific primers were adapted from the internal transcribed spacer (ITS) for use in HTS that revealed 13 terrestrial- and 35 aquatic-dwelling species of *Phytophthora*, respectively (Català et al., 2015), thus providing valuable information regarding possible routes of *Phytophthora* infection in that region. Tree species composition in natural forest stands may be influenced by fluctuating pathogen populations (Holah et al., 1993). Tree host-pathogen relationships necessitate proper understanding of the biodiversity of pathogens. In a China-based study, seven evergreen broadleaved keystone tree species, namely,

Castanopsis fabri (Fagaceae), *C. fissa* (Fagaceae), *Cyclobalanopsis fleuryi* (Fagaceae), *Engelhardia fenzelii* (Juglandaceae), *Artocarpus styracifolius* (Moraceae), *Canarium album*, and *Cryptocarya concinna* (Lauraceae), was used for molecular analysis of root-associated fungi. The HTS for the ITS region using Illumina MiSeq platform was performed followed by sequencing data analysis using QIIME data analysis package. Functional characterization revealed the most abundant genera of phytopathogens to be *Pestalotiopsis*, *Mycosphaerella*, and *Calonectria*. Further, the study found that host species showed the greatest impact on the fungal community composition, while neighboring plants were important for the plant pathogen communities and environmental factors (canopy openness, slope, aspect, soil pH, soil organic carbon, nitrogen, phosphorus, and potassium contents) showed lesser impacts upon fungal and phytopathogenic fungal community composition. Phylogenetically close host trees were more likely to harbor similar fungal or phytopathogenic fungal communities (Cheng and Yu, 2020). Therefore, trees are the important determinants of the microbial community composition in their vicinity.

Certain tree-associated microbial species may behave as mutualistic organisms, or even as symbionts. They may pose useful attributes as biological control agents (BCAs), and would reduce the population of another organism (pest/pathogen) (van Lenteren et al., 2018). Balancing natural microbial diversity and forest pathogens presents numerous obstacles and benefits. Deciphering plant-microbe interaction is essential for informed forest product utilization. In view of the global threat to forests from various pathogens, the role of environmentally friendly avirulent hyperparasites as biocontrol agents for the management of tree diseases, seems like an environmentally smart solution.

6.3 Biological control: A brief exposition

Biological control is a technique in which living organisms are used for the management of disease, weed, and insects. It has been used for centuries but their successful and extensive use started in the late 18th century. There are several definitions of biological control, an important one being "To inhibit the growth of pest/pathogen by the use of living beneficial organisms and making it less pathogenic than earlier it would be" (Eilenberg et al., 2001). According to a recent definition "Biological control is the exploitation of living agents (including viruses) to combat pestilential organisms (pests and pathogens), directly or indirectly, for human good. Biological control must always involve the following three separate players: 1) a biocontrol agent, 2) a pest, and 3) a human stakeholder benefitting from the pest control service provided by the biocontrol agent" (Stenberg et al., 2021).

The global importance of the concept of biocontrol is emphasized by the existence of the International Organization for Biological Control (IOBC) established in 1955. The IOBC promotes environmentally safe methods of pest and disease control. It is a voluntary organization of biological- control workers and regularly publishes research allied to biocontrol. To understand why there is need for biological control is important from an ecological, environmental, and sustainable point of view.

6.3.1 Why adopt biological control?

Three billion kilograms pesticides worth $40 billion is consumed globally per annum (Sharma et al., 2020). Risk to non-target organisms, including humans, beneficial microorganisms, beneficial insects, birds, earthworms, and aquatic life forms, is immense on account of nonbiodegradability and toxicity. As the use of chemicals increases, it causes drastic effects on the environment as well as on human health. Chemical interventions are herbicidal, as fertilizer, and microbicidal. Not only is even spray coverage a problem with contact fungicides, regular and sustained use of them may induce fungicide resistance in pathogens. For instance, copper-based fungicides, mancozeb and captan, have longer persistence times, as they can arrest fungal pathogen biochemical pathways at multiple points (Vallières et al., 2018). Other fungicides affect only single sites, and when these sites undergo mutation, the pathogens may become resistant to that fungicide. A European study states that most devastating and difficult to control highly adaptable, broad host range fungal pathogens persisting in soils, endangering *Humulus* saplings belong to the genera: *Phytophthora* (Pythiaceae, Peronosporales), *Pythium* (Pythiaceae, Peronosporales), *Fusarium* (Nectriaceae, Hypocreales), and *Cylindrocarpon* (Nectriaceae, Hypocreales), *Erysiphe* viz.—*Erysiphe alphitoides*, *E. quercicola*, *E. hypophylla* (Erysiphaceae, Erysiphales), *Peronospora* (Peronosporales, Peronosporaceae). Further, the study highlights *Lophodermium pinastri* (Rhytismatales, Rhytismataceae) causing pine needle cast and *Meria laricis* causing needle blight of larch as other important pathogens. Since these pathogens persist in soil as infectious and perennating structures, mass mortality of seedlings in forest nurseries is inevitable (Okorski et al., 2015). Fungicidal diversity is severely limited to tackle such forest disease problems, since fungicides are mainly carboxylic acid amides, dithiocarbamates, carbamates, phosphonics, ketamines, DMI (triazole), pyrimidines, inorganic salts, aromatic derivatives, pyridine carboxamides, strobilurins, and benzimidazoles (Głowacka et al., 2013).

So, to offset the deleterious impacts of chemical managment of disease and overcome difficulties, the accelerated use of biological control is imperative. Highly pathogen specific biocontrol is preferable to chemicals (Marfetán et al., 2020), which aim to be a broad spectrum. In forestry, plant growth promotion and disease reduction using biocontrol microbes is an area of interest (Himaman et al., 2016). Biocontrol measures as a support to forestry crop development are essential. The seedling stage, juvenile stages, and nursery propagation stages are those that are aided most by biocontrol. A range of biocontrol-based measures have been developed and are practicable for forestry crops. For instance, antagonism of 298 rhizobacteria isolated from rhizosphere and rhizoplane of tomato and *Eucalyptus* plants was determined against bacterial wilt of *Eucalyptus* caused by *Ralstonia solanacearum*. Various methods for antagonist delivery and pathogen inoculation were evaluated: (1) seeds biopriming by soaking seeds for 12 h in a BCA suspension after which germinated seedlings were immersed in the pathogen inoculum suspension, (2) seedlings germinated from bioprimed seeds were transplanted to sick soil inoculated with *R. solanacearum*, and (3) seedling roots were submerged in a suspension of *R. solanacearum*. Nine isolates, viz., (UFV-11, 32, 40, 56, 62, 101, 170, 229, and 270), were selected as potential antagonists to *R. solanacearum* as they showed a suppression of bacterial wilt in at least one of the methods assessed. These nine antagonists in vitro and in vivo (inoculated *Eucalyptus*) were pitted against two isolates of *R. solanacearum*. Isolates UFV-56 (*Bacillus thuringiensis*), UFV-62 (*Bacillus cereus*), and a commercial formulation of several rhizobacteria (Rizolyptus) effectively suppressed bacterial wilt in *Eucalyptus*, protecting the plants during the early stages of development (Santiago et al., 2015). Another Thailand-based study revealed that an actinomycete *Streptomyces ramulosus* strain EUSKR2S82 possessed strong antagonistic abilities against *Cryptosporiopsis eucalypti*, *Cylindrocladium* sp. (now *Calonectria* sp.), and *Teratosphaeria destructans*. Plant growth promotion activity, root colonization, and enhanced root length of *Eucalyptus* roots were seen as well (Himaman et al., 2016). Biocontrol is an absolute term, while management is what is aimed at (Hajek and Eilenberg, 2018). With passing time and advancement in research, various types of methods have been developed for the controlling of pests. A different type of biological control is accessible for work, but it is very important to understand the mechanism of multiplex interactions between the plants and the BCAs for more development.

6.3.2 Types of interactions contributing to biological control

Plants and pathogens interact with different types of organisms during their lifecycle. In several ways, such interactions can affect the health of the plant. These interactions help decipher the mechanisms of biological control. Protocooperation, neutralism, mutualism, commensalism, amensalism, competition, predation, and parasitism are some of the types of interactions between plant and pest. These interactions are visible micro- as well as macroscopic levels. Certain host-pest interactions beyond the threshold of tolerance for the host lead to its disease development. And for managing those interactions, biological control is required at multiple stages (Odum, 1953).

Biological control is a beneficial interaction, when seen from the perspective of plant host. Since forest diseases are predominantly fungal in origin (Tainter and Baker, 1996), the perceived efficacy and thrust area of research should be devoted to fungal biocontrol. Agricultural research has responded to fungal biocontrol since the 1800s. Eminent Dutch research groups are promoting "conscious agriculture" whereby biocontrol is an important pillar (van Lenteren et al., 2018).

Fungal biocontrol offers a wide array of organisms, widely found in nature—of nature, by nature, and for nature. There is an increasing interest in the exploitation of fungi for the control of invertebrate pests and diseases, as evident from the number of commercial products available or under development. Fungal biological control is an exciting and rapidly developing research area with implications in plant productivity, animal and human health and food production. The attractiveness of fungi as biocontrol agents, in particular, is due to their general ubiquitousness, high degree of host specificity, destruction of the pathogen, persistence, dispersal efficiency, an ease of culture, and maintenance in the laboratory. For instance, the most widely researched versatile biocontrol fungal genus *Trichoderma* (Ascomycota, Hypocreales, Hypocreaceae) is ubiquitous and frequently encountered in soil as plant symbionts, saprotrophs, and mycoparasites. It exhibits antagonism against diverse plant pathogens with a few *Trichoderma* species commercialized as biopesticides or biofertilizers (Alfiky and Weisskopf, 2021).

The research, development, and final commercialization of fungal BCAs continue to confront a number of obstacles, ranging from elucidating important basic biological knowledge to socioeconomic factors. Considerable advances have been made in separate areas but it is important to integrate and communicate these new findings (Butt et al., 2001). Whether tree disease biocontrol strategies can be singularly used or in conjunction with other approaches needs exploration. The practical and economic feasibility of use under field conditions must also be examined carefully.

6.3.3 Mechanisms of biocontrol

6.3.3.1 Direct mechanisms: Disease biocontrol is based primarily on three mechanisms

Parasitism

Fungi parasitizing fungi is termed as mycoparasitism. Living BCAs parasitize the plant pathogen, thus antagonizing and finally killing it. Viruses parasitize bacteria where, for instance, phage ϕsp1 (Myoviridae) was effective against wilt-causing pathogenic isolates *R. solanacearum* affecting *Solanum lycopersicum* (tomato) seedlings and *Solanum tuberosum* (potato). It reduced 81.39% and 87.75% reduction in pathogen load, in potato tuber and tomato seedlings, respectively, with 100% reversal of disease in pot assay of tomato and potato, when compared to pathogen-inoculated controls. Further, it reduced biofilm formation of *R. solancearum* by 73.68% (Umrao et al., 2021). *R. solanacearum* is an important *Eucalyptus* wilt pathogen, and maybe this method would be beneficial for disease management. A single-stranded circular DNA mycophagous virus Botrytis cinerea genomovirus 1 (BcGV1) was characterized in plant pathogen *Botrytis cinerea* (Hao et al., 2021) reported to cause gray mold, leaf blight, blossom blight, and stem rot in a total of 596 genera of vascular plants, dicotyledonous plants, and monocotyledonous plants (Elad et al., 2016). Wattle rust epidemic on *Acacia mearnsii* caused by rust fungus *Uromycladium acaciae* in southern Africa was observed in 2013. Three years later, conidiomata of a fungus *Sphaerellopsis macroconidia* appeared to be parasitizing telial spores of *U. acaciae* on *A. mearnsii* leaves from plantations in Mpumalanga, South Africa. *Sphaerellopsis* had less impact during epiphytotic intervals of the year but could reduce the overwintering of *U. acaciae* (Fraser et al., 2021) and may be employed as a natural biocontrol agent in field conditions.

Antibiosis

Endophytic *Pseudomonas protegens* isolated from Iranian oak showed antimicrobial activities like siderophore, hydrogen cyanide and protease production. These isolates inhibited the growth of bacterial plant pathogen *Pseudomonas syringae* pv. *syringae* in vitro, thereby posing biocontrol potential. Strain *Ps. protegens* Pp95 had the greatest inhibition on the pathogen growth, with the mean inhibition zone diameter of > 10 mm (Tashi-Oshnoei et al., 2017). Siderophores are low molecular weight molecules that chelate iron (Fe^{3+}) from the surrounding environment, an important component of cytochrome, heme, and nonheme proteins, and cofactor for various enzymes (Gupta et al., 2019). Phytopathogenic fungal or bacterial growth is inhibited under iron-challenged conditions (Tiwari et al., 2019). Proteases are extracellular enzymes that interfere with pathogenicity factors.

Competition

Biocontrol agents compete for space and nutrient with pathogen. *Serratia marcescens*, a potential BCA and rhizocompetent plant growth-promoting rhizobacterium, was examined for hydrolytic enzymatic activities, viz., lipolytic, amylase, cellulase, protease, mannanase, and chitinase showed 5.52, 4.17, 1.50, 0.69, 0.35, and 0.27, respectively (Putri et al., 2021). *Serratia marcescens*, *Pseudomonas azotoformans*, and *Trichoderma virens* were also found effetive against Fusarium wilt of Shisham (Banerjee et al., 2020). Forestry crops can benefit if such microbes are soil amended, as soil nutrient solubilization and protection from soil-borne plant pathogens can be ensured.

Direct biocontrol screening is measured by means of dual culture widely practiced. Bacterial, yeast, fungal, and oomycete BCA is co-cultivated on artificial agar-based nutrient media with pathogen. Antagonism of the BCA is measured in terms of measuring inhibition of the target pathogen (Sales et al., 2016). Ocular and microscopic examination of full-grown dual culture plates is done using parameters, viz., mutually intermingling hyphal growth of pathogen and antagonist, overgrowth by antagonist, inhibition at line of contact, aversion, and mycoparasitism (Skidmore and Dickinson, 1976). Volatiles secreted by the antagonists inhibit pathogen growth, and efficacy of volatiles is evaluated by inverted plate assays (Dennis and Webster, 1971). For instance, in a Thailand-based study, *Trichoderma asperelloides* produced antifungal volatile organic compounds (VOCs) inhibiting the mycelial growth of fungal pathogens *Corynespora cassiicola*, *Fusarium incarnatum*, *Neopestalotiopsis clavispora*, *N. cubana*, and *Sclerotium rolfsii*, with a percentage inhibition range of 38.88%–68.33%. Solid-phase microextraction (SPME) trapping of VOCs followed by gas chromatography-mass spectrometry (GC/MS) revealed 17 compounds. The major molecules were fluoro (trinitro) methane (18.192% peak area) and 2-phenylethanol (9.803% peak area) (Ruangwong et al., 2021). In vivo assays for direct antagonism on forestry crops have been done, for instance, colonizing infection and colonization efficiency of biocontrol fungus *Phlebiopsis gigantea* on precommercially cut stumps of *Picea abies* and *Pinus sylvestris* against Heterobasidion rot (Gaitnieks, 2020). These methods of *in planta* evaluation may be useful for the evaluation of BCAs.

6.3.3.2 Indirect mechanisms of biocontrol

Plants possess constitutive defense, yet induced resistance may be launched by plants in response to herbivores, insects, and microbes. These responses are called priming, immunization, or "vaccination" (Navarro et al., 2017) and forms of indirect mechanisms of biological control if the priming organism is a biocontrol agent, hence harmless to the host. The increase in resistance protects the host against subsequent attacks by more harmful pests. Induced resistance in plants is mainly of two kinds: systemic acquired resistance (SAR) and induced systemic resistance (ISR). Systemic acquired resistance is induced locally by pathogen attack, while beneficial microbes locally induce ISR, affording the plant systemic protection against a wide range of pathogens. For example, the fungal biocontrol agent *Trichoderma* has been shown to activate defense-related genes through VOC emission (Malmierca et al., 2012) and defenses may be local or systemic (Contreras-Cornejo et al., 2011). Dutch elm disease (DED), originally incited by *Ophiostoma ulmi*, was accidentally introduced into Europe and North America through world trade, during the 1920s, becoming a major alien disease-threatening native elm forests. Presently, the DED pandemic is caused by *Ophiostoma novo-ulmi* (Sordariomycetes) vector infected into disease-free *Ulmus* trees through bark beetles *Scolytus* sp. and *Hylurgopinus* sp. (Martínez-Arias et al., 2019). Vascular embolism as a result of *O. novo-ulmi* spread within the tree xylem leads to mortality (Martín et al., 2019).

Recently, a Spanish study revealed that endophytic BCA, Dothideomycetes isolate YCB36a, in vitro strongly inhibited *O. novo-ulmi* growth, released antipathogenic VOCs, chitinases and siderophores, nutritional overlap with pathogen. In vivo, preinoculated trees showed 40% less leaf wilt in Elm. Importantly, *Ulmus minor* inoculated with this BCA led to increased foliar stomatal conductance, flavonoid and total phenolic content rise in wood—indicative of BCA-mediated ISR activation (Martínez-Arias et al., 2021). Correct knowledge of biocontrol mechanisms involved in reducing crop loss due to pathogen attack is imperative for optimum results. Whether the BCA actually is reducing crop loss due to disease or if physiological improvement in plant health independent of pathogen reduction measures, needs to be examined by studying the mechanisms of biocontrol critically. For instance, nutrient solubilization from soil, a form of plant-growth promotion activity, leads to an improvement in plant health. However, this is not mediated by pest control, and hence, it cannot be regarded as resulting from biological control. A variety of mechanisms that do not target pathogen control have been excluded from the purview of being BCA mechanisms (Stenberg et al., 2021). A pictorial representation of mechanisms involved in biocontrol mechanisms is elucidated in Fig. 6.1.

6.4 Success stories of biological control with special reference to *Heterobasidion annosum*

Heterobasidion (Basidiomycota, Russulales), a notorious forest pathogen, includes many economically important species like

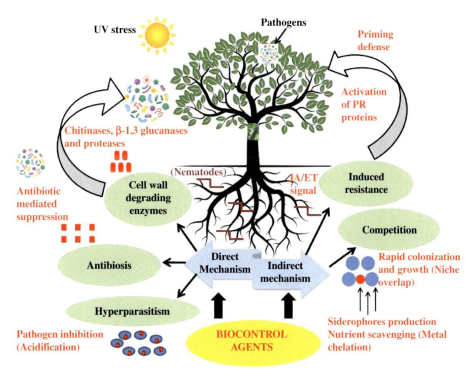

Fig. 6.1 Mechanisms involved in biocontrol.

H. annosum, H. insulare, H. rutilantiforme, H. pahangense, and *H. araucariae*. More than 1700 publications have been published on *H. annosum* in the last 40 years, indicating its importance as the most widely researched forest fungi. Further, *Heterobasidion* is the first tree pathogen whose genome sequences are completely known (Garbelotto and Gonthier, 2013; Olson et al., 2012). Moreover, to regulate and maintain the ecosystem services of forests, it is very important to develop reliable control strategies against *Heterobasidion*.

The coniferous forests of Northern Hemisphere that include North America, Europe, Japan, China, and Russia have a wide variety of *H. annosum* (Korhonen and Stenlid, 1998; Korhonen et al., 1998; Dai et al., 2003, 2006; Dai and Korhonen, 1999; Otrosina and Garbelotto, 2010; Worrall et al., 2010). For a very long time, *H. annosum* was considered a single species, until researchers disclosed the origin of intersterile groups (ISGs) (Korhonen, 1978; Capretti et al., 1990; Chase and Ullrich, 1988; Korhonen et al., 1998). Till now, three intersterile groups that have been identified in Eurasia (*H. annosum*) as well as two intersterile group have been found in North America. The various species are designated by different host priorities, e.g., some species of pines like Scots pine are attacked by *H. annosum*, but it is also related to other conifers and some broad-leaved species of trees. *H. irregulare* mostly infect species of junipers, incense cedar, and pines, while *H. parviporum* attacks spruce of Norway.

Among various species in the Mediterranean, *Pinus pinea* is naturally introduced in coastal regions (CAB International, 2000). As discussed above, *H. annosum* is a forest pathogen that attacks several species of *P. pinea* in forests of Europe (Capretti, 1998). Several reports suggest that *H. annosum* causes serious diseases like group-dying, mortality, failures, windthrows, and mostly infects gigantic *P. pinea* (120–160 years) (Amorini et al., 2000). Hence, a control plan has been developed against *H. annosum* so as to prevent coniferous forests from widespread attack in the future. In many countries, for several species of conifers, a fungus *Phlebiopsis gigantea* (Agaricomycetes, Polyporales) has been reported as a very potential biological control agent for preventing the growth of pathogenic *Heterobasidion* sp. (Korhonen et al., 1994; Capretti, 1998; Holdenrieder and Greig, 1998; Soutrenon et al., 1998; Roy et al., 2003). An experiment was performed in Castel Fusano (Italy) pinewood forest where an indigenous isolate of *P. gigantea* was used for controlling disease caused by *Heterobasidion*. From the same forest, a branch of fallen pine was collected to check the wood decay capability of *H. annosum* on pine wood. The experiment was performed in restricted conditions in which Rotstop formulation of *P. gigantea* was compared with the selected targeted strain. Then, the efficiency of domestic strain was tested in forest to know if this fungus showed remarkable properties as a biocontrol agent for *Heterobasidion* infection on *P. pinea*.

For carrying out experimental trials, in vitro logs of stone pine were taken to observe the efficacy of native isolate of *P. gigantea* against *Heterobasidion* as well as Rotstop on *P. pinea*. In this, logs were divided into two parts in which one half of the log remained covered and treatment was given to the other half of the log but within 2 h of cutting of stem—Rotstop and oidial suspension of selected strains from Castel Fusano (ISPaVe-PF52) was prepared. For these two treatments, estimated three (3) different concentrations were taken for testing (10,000, 20,000, and 50,000 viable oidia/mL). Within 1 hour (1 h), conidial suspension of *H. annosum* was sprayed on the treated as well as on untreated surfaces by using the strain ISPaVe-PF-101 taken from similar pine. Then, the logs were potted onto wet sand kept at room temperature of 18–25°C under natural light. After 4 weeks, the availability of fungal pathogen *H. annosum* as well as BCA *P. gigantea* was quantified. The disperse presence of *H. annosum* was observed in all the exposed subset of samples. In all the treated disk halves, it was observed that all the three concentrations of Rotstop and ISPaVe-PF-52 were helpful in 100% surface colonization, thus preventing infection of *H. annosum*. This data represents the efficiency of *P. gigantea* as a biocontrol agent on *P. pinea* tree even after a year of treatment (Annesi et al., 2005). A New Zealand-based research group listed out commercial bioinoculants based on *T. harzianum*, *T. polysporum*, *T. viride*, *T. atroviride*, *Bacillus subtilis*, *B. amyloliquefaciens*, *B. pumilus*, *Pseudomonas syringae*, *Ps. fluorescens*, and *Ps. chloroaphis* with potential to be used in forest nursery disease management (Reglinksi and Dick, 2005). Table 6.1 enlists some recent success stories highlighting biocontrol of forest pathogens.

6.5 Challenges

Drawbacks do not discourage the use of microbial biocontrol from becoming effective. These considerations simply assist consumers in selecting appropriate microbial products and taking the requisite steps to achieve optimal outcomes.

6.5.1 Spectrum of specificity

Because a single microbial antagonist is harmful to only a certain species of pathogens, each application can only control a subset of the pests in a field. Pathogens that escape biocontrol will live and may begin to cause damage if they are present throughout the treated environment. To tackle this, studies must also be fine-tuned to screen antagonism and rhizocompetence against diverse population of the same fungal pathogen. For instance, under greenhouse conditions, a strain XY21 of *Serratia* sp. showed antagonism ranging from 19% to 70% of tomato wilt caused by seven genetically different *R. solanacearum* strains. Root colonization of XY21 and its impact on tomato rhizospheric bacteria was monitored by denaturing gradient gel electrophoresis of 16S rRNA gene fragments PCR-amplified from total community DNA (Xue et al., 2013).

6.5.2 Risk of introduction into new habitats

The BCAs, while rhizocompetent and/or hyperparasitic and avirulent, are potential weed organisms that may outcompete native microbial diversity, which may have beneficial roles in maintenance of plant health in an introduced ecosystem (Collinge et al., 2019).

6.5.3 Sensitivity of biocontrol agents to stresses

On the other hand, biological commodities are affected by various biotic and abiotic stresses. The efficacy of some forms of microbial antagonist is reduced by heat, drying out, or exposure to ultraviolet

Table 6.1 List of selected diseases and biological control agents applied in forestry species.

Disease/symptom	Species of pathogen/parasite	Host forestry species	Biocontrol agent	Country/geographical region	Mechanism of interaction of BCA and pathogen	References
Populus tremuloides	Ophiostoma crassivaginatum	Populus tremuloides	Phialemonium curvatum	Canada	Antifungal secretion from the BCA	Hiratsuka and Chakravarty (1999)
Tsuga heterophylla Heterobasidion Rot Disease (HRD)	Arceuthobium tsugense (parasitic plant) Heterobasidion irregulare	Tsuga heterophylla Picea abies	Colletotrichum gloeosporioides Phlebiopsis gigantea	Canada Finland	In vitro wood growth and chip colonization Stump colonization by the BCA, prevention of pathogen from attacking	Askew et al. (2011) Terhonen et al. (2013)
HRD of Pinus resinosa	H. irregulare	Pinus resinosa	P. gigantea	Ontario, Canada	Stump colonization by the BCA prevention of pathogen from attacking	Dumas and Laflamme (2013)
Dothistroma Needle Blight	Dothistroma septosporum	Pinus contorta	Antibiotic-producing Bacillus Aneurinibacillus migulanus	Ireland	Antimicrobial cyclic peptide secretion, viz., gramicidin S (GS), by BCA	Alenezi et al. (2016)
Pitch Canker	Fusarium circinatum	Pinus radiata	Ps. fluorescens, Bacillus simplex, Erwinia billingiae	Spain	Pseudomonas may have produced several antifungal metabolites, including cyclic lipopeptides, phloroglucinols, phenazines, pyoluteorin, and pyrrolnitrin; other rhizospheric bacterial isolates screened may have similarly secreted antifungals, inhibiting the pathogen in vitro and reduced lesioning in young Pinus radiata trees	Iturritxa et al. (2017)
Armillaria Root diseases	A. cepistipes, A. ostoyae, A. mellea, A. gallica	Quercus sp.	T. virens, T. atrobrunneum	Hungary, Austria	Mycoparasitism, siderophore activity, extracellular enzymes β-glucosidase, cellobiohydrolase, β-xylosidase, and phosphatase secretion. Indole acetic acid secretion as plant growth promotion activity as well	Chen et al. (2019)

Disease	Host	Antagonist	Location	Mechanism	Reference
Brown Rot Disease	Ficus microcarpa	Phellinus noxious		Mycoparasitism	Chou et al. (2019)
Shoot dieback	Populus alba	Venturia tremulae (Pleosporales, Venturiaceae)	T. asperellum	Spain	Mycoparasitism, faster substrate utilization, weak antibiosis (in vitro)	Martínez-Arias et al. (2019)
Blight	Chestnut	Cryphonectria parasitica	Endophytes of order Pleosporales, Dothideales and Eurotiales	Europe	Induction of hypovirulence of C. parasitica in stem cankers of chestnut (in vivo)	Suzuki et al. (2021)
Wilt	Pinus spp.	Bursaphelenchus xylophilus (nematode)	Hypovirus, Group CHV1	South Korea	Nematophagy since it is an endoparasitic fungus	Yin et al. (2020)
Dieback	Fraxinus excelsior	Hymenoscyphus fraxineus	Esteya vermicola	Germany	Phomopsidin and 10-hydroxyphomopsidin, antifungal metabolite secreted by endophyte, H. rubiginosum	Halecker et al. (2020)
Powdery Mildew	Quercus robur	Erysiphe alphitoides	Hypoxylon rubiginosum	Poland	Gas chromatography-mass spectroscopy revealed 49 volatile compounds in the headspace of pure cultures of T. asperellum. Sesquiterpenes like daucene, dauca-4(11),8-diene, and isodaucene were maximally emitted, which may have involvement in induced systemic resistance in oak seedling	Oszako et al. (2021)
Numerous pathogens confrontation	F. camptocerus, F. oxysporum, Alternaria alternata, F. solani, Colletotrichum gloeosporioides, Ganoderma applanatum, Botrytis cinerea, Cytospora chrysosperma	—	T. asperellum isolated from Juglans mandshurica rhizosphere	China	Mycoparasitism shown against eight pathogens, plant growth promotion seen on Solanum lycopersicon seedlings as well, by enhanced nitrogen uptake, continuous upregulation of jasmonate(s) (JAR1 and MYC2) and salicylic acid (PR1 and NPR1) genes at 8 and 48 h	Yu et al. (2021)

Continued

Table 6.1 List of selected diseases and biological control agents applied in forestry species—cont'd

Disease/symptom	Species of pathogen/parasite	Host forestry species	Biocontrol agent	Country/geographical region	Mechanism of interaction of BCA and pathogen	References
Shoot blight disease	*Sphaeropsissa pinea*	*Pinus massoniana*	*B. pumilus*	China	The BCA was able to inhibit fungal mycelial growth and spore germination of *S. sapinea*, disturb cell membrane integrity of the pathogen cell (electrolyte leakage, nucleic acid extrusion) as well as the normal synthesis of cell contents. Also BCA successfully colonized pine shoots to retard *S. sapinea* infection and eventual Sphaeropsis shoot blight disease.	Dai et al. (2020)
HRD of conifers	*Heterobasidion* sp. (3 native to Europe) + 1 *H. irregulare* exotic strain	Conifers	*Pseudomonas protegens*	Europe	Antibiosis (in vitro) by dual culture and cell-free filtrate of *P. protegens*-amended media in which inoculated pathogen was inhibited	Pellicciaro et al. (2021)
Sapstain	*Ophiostoma piceae*, *Leptographium procerum*, and *Sphaeropsissa pinea*	*Pinus radiata*	*T. harzianum* amended with chitosan	New Zealand	Fungistasis from chitosan and *T. harzianum* combination and mycoparasitism	Singh and Chittenden (2021)

radiation. As a result, appropriate timing and application protocols are critical for biocontrol agents (Usta, 2013).

6.5.4 Laborious laboratory screening and field screening of BCAs

The process of bringing bacteria-based BCAs to market starts in the field, with the detection of the target crop and pathogen, as well as an understanding of disease epidemiology and existing disease-control techniques (Köhl et al., 2011). The 16S rRNA gene sequencing and multilocus sequence analysis (MLSA) are the best available tools for these processes (Glaeser and Kämpfer, 2015; Joseph and Forsythe, 2012). The potency of bacterial isolates determined antagonism studies accompanied by greenhouse and controlled field trials. The BCAs must be tested against a variety of pathogens in different geographical locations, under different climatic conditions, and on different crops to assess their potential for broad-spectrum activity. The most commonly known indirect multiple pathways for biological regulation include the synthesis of antibiotics, viz., phenazines and 2,4-diacetylphloroglucinol by *Pseudomonas* sp., lipopeptides such as iturin and fengycin by *Bacillus* sp., competition for nutrients, lytic enzyme production (chitinase and glucanase), and induced systemic resistance (ISR) in the host plant (Ghyselinck et al., 2013; Glick, 2012; Köhl et al., 2011). Therefore, the screened BCAs must be preserved in reputed culture collections (Velivelli et al., 2014) for their further use by different end-users in due course.

6.5.5 Paucity of funding and poor policy support

Another constant threat is that public education and funding for biological control technologies are vital for preserving the economic, legislative, and technological viability of this important method for invasive species management (Bigler et al., 2006). Farmers obtain pest management information from multiple of outlets. The costs of ads and salesmanship by chemical sector executives are easily recouped from pesticide industry earnings. Implementation of biological control agents and procedure has no comparable sales force or budget, unless one counts the severely understaffed and underfunded university extension programs. Competition with chemical pesticide companies for the interest and time commitment of farmers and other pest managers has been a persistent threat for biological control for decades (Messing and Brodeur, 2018). There are inherent challenges and external entity, prospects for cost sharing, capability development and maintenance, and more rigorous research programs are vital to the long-term success of biological control. If governments and companies want to effectively protect their forest resources, they will need to increase their strategic commitment in mechanisms explicitly planned to foster such emphasis (Garnas et al., 2012). Additionally, for economic threshold, other management techniques besides should be considered and implemented in combination with BCAs to effectively minimize disease production and crop yield loss in various crop systems.

6.5.6 Complex anatomy

Trees are woody, large, bulky and long duration. Unlike annuals, crop rotation by seasonal shifts to prevent overwintering of pathogens makes it more difficult to develop and implement effective biological control approaches. Persistent dormant inoculum especially for wide host range pathogens in the form of chlamydospores/sclerotia (*Calonectria* spp.; *Fusarium* spp.) or fruiting bodies in case of basidiomycetous fungi (*Ganoderma* spp.) builds up over time. Multiple infection events, over a season or consecutive ones, are favored by complex root architecture of mature trees, creating more infection courts. The rhizocompetence of BCAs would be compromised under such conditions.

6.5.7 Potential to become human pathogens

A problem arises when screened isolates show DNA sequence similarity (16S rRNA or ITS) to genera which contain human pathogens. For instance, a China based study found that a rhizocompetent bacterial antagonistic strain against *R. solanacearum* wilt on tomato, called XY21, showed affiliation with *Serratia marcescens*, which contains strains known to be human pathogens. Certain *S. marcescens* strains can form biofilms which may be deleterious in the case of immunosuppressed patients whose lung tissue may become colonized by these bacteria (Xue et al., 2013).

6.5.8 Nagoya protocol

The BCAs may be of native or exotic origin. With rapidly increasing accidental invasive species, deliberate introduction of nonnative BCAs is often required for tackling invasive pests. Recent regulations have stalled the screening for novel, nonnative biocontrol agents. Risk analyses for commercial biocontrol, especially nonnative species, came under the first phase of regulation, in the 1980s. Environmental risk assessments for BCAs led to lesser utilization of new nonnative natural enemies due to higher application costs. These regulations aimed at minimizing risks of releasing nonnative BCAs, to ensure and enhance the reliability of biological disease management. In the second phase of regulation, in the 1993 Rio Summit, at the Convention on Biological Diversity (CBD), one of the outcomes was the Nagoya Protocol. This agreement provides the basis for an effective implementation of fair and equitable benefit sharing arising out of utilization of genetic resources such as biocontrol agents. Since the ownership of biological control agents is at stake, signatories must comply with legally binding guidelines before accessing genetic resources and sharing benefits. This aspect of CBD creates bottlenecks in the usage of and research on nonnative BCAs against invasive pathogens in several countries (Van Lenteren, 2021).

6.5.9 Fluctuating trends

The global biological control market, comprising microbials, as a component has a predicted compound annual growth rate (CAGR) of 14.7% between the period (2021–26). The COVID-19 pandemic has severely impacted the biological control market. Periodic lockdowns worldwide and travel restrictions have lowered access and availability of BCAs in the market. Production is also compromised due to labor, raw material paucity, and poor transportation. While increasing focus on organic cultivation (certain agroforestry systems), and gradually easing regulations, the sales for biocontrol will only grow. Microbials occupy 65% of the market share as of 2019, and will grow due to use in seed treatment, on field and postharvest applications. North American and European farming sectors have most eagerly adopted biological controls and are major markets (Mordor Intelligence, 2019). In order to avoid the hefty registration costs of BCAs, many companies register their products as biofertilizers. Also, the time required for BCA registration against a particular pathogen or crop combination nearly is twice as long, and the double the cost of registration, as required for a biofertilizer for general use in all agricultural crops. Thus, most authorized bacterial agents are registered as biofertilizers and not as BCAs. Global trade in forestry microbial biocontrol is yet to be quantified, and remains a research gap. Country-wise breakup of commercial microbials for forest disease management is also lacking, since the published literature primarily caters to agriculture (Velivelli et al., 2014).

6.6 Future directions

Coevolution of pathogens finally leads to the emergence of new races each year. Beneficial fungi can counteract the effects of emergence of newer races of pathogens by means of their multifaceted interactions. Sound knowledge of the ecology of such fungi is important for the effective management of severe epiphytotics. Future research on BCAs against forest pathogens should concentrate upon the development of microbial consortia composed of native and stable isolates, nontoxic carrier material, to increase their bioefficacy and shelf life. Introduction of endophytic microbes into this scenario should be encouraged. For instance, an Indonesia centric review stated that diseases afflict about 8,668,670 ha of forest land, compromising plantation forestry. Rapidly growing exotic tree crops like *Acacia* and *Eucalyptus* are most at risk with emergence of newer pathogens. The most widespread fungal incitants in Indonesia include *Ceratocystis manginecans* (Ceratocystis wilt and dieback), *Ganoderma philippii* (red root rot), *Phellinus noxius* (brown root rot and heart rot), and *Fusarium* spp. (Fusarium wilt), whereas the major bacterial pathogens are *Ralstonia* spp. (bacterial wilt) and *Xanthomonas* spp. (leaf streak). As one key component of integrated pest management, biocontrol measures play significant roles in managing major diseases of tropical plantation forests in Indonesia and other countries. Indonesian forestry companies dedicated research priority to biocontrol product development. To facilitate this, prospecting and isolation of different fungal antagonists, viz., *T. harzianum, T. viride, T. polysporum, T. hamatum, Gliocladium viride, Phlebiopsis* spp., and *Cerrena* sp. from various Indonesian tree plantation sites, was performed for future implementation in an integrated disease management program (Gafur, 2020). International cooperation, among countries opting for biocontrol as a viable commercial alternative, also is a prerequisite, before biocontrol ceases to exist as a niche market, especially for the even more obscure and niche forestry sector. The IOBC has made an appeal to signatories creating the legal framework for Access and Benefit Sharing guidelines under the Nagoya Protocol, to formulate regulations that include unambiguous guidelines to ease the exchange in biocontrol agents. The IOBC strongly recommended that BCAs should be excluded from financial obligations under ABS, especially since classical biocontrol is nonprofit, benefitting developed and developing nations alike. Further, the IOBC designed a best practices guide, including a draft ABS agreement, for collecting and studying BCAs for research purpose and noncommercial release into the environment by signatories of the Nagoya Protocol. Global implementation of the IOBC recommendations would bolster up the future biocontrol (Van Lenteren, 2021).

6.7 Conclusion

The UN Sustainable Development Goals support forest proliferation (FAO, 2020). The sustainability of forests relies upon tree protection. Climate smart practices include greater focus on forest health, where diseases and pathogen attacks are a prominent feature.

Concerns over chemical control of pathogen management in commercial forest businesses are tipping the odds in favor of biocontrol development. Biological control has been practiced for over a century, since Hartley (1921) used biocontrol application of fungi in forest nurseries, for *Pythium* damping off management. Biological control research for woody perennials crop diseases is scant when compared to the attention focused upon annuals. Tree breeding research for resistance remains an important disease management tool, but trees are long-duration crops, likely to be outrun by the rate of emergence of newer pathogens. Integrated forest disease management would strike the correct balance between tree health and pathogen risk. Developments of high-throughput technologies like next-generation sequencing (NGS) provide detailed scenarios of fungal/bacterial/viral community structure, thus easing the processes of bioprospecting for potential BCAs.

Increasing burden on forest resources makes it imperative to need to identify and market BCAs as forestry crop protectants to boost productivity. Besides procuring BCAs, reliable taxonomic identification, field assessment, and registration of BCAs require comprehensive and concerted efforts from governmental agencies, especially forest departments, private forest businesses, research, and industrial segments to uphold sustainable forestry. Geopolitical registration bottlenecks slacken testing and marketing of novel plant protection products. Faith in the efficacy of BCAs must be instilled in the forest farming community, since they are the end users, by providing ample proof regarding microbial product safety as well as cost-effectiveness.

References

Alenezi, F.N., Fraser, S., Belka, M., Dogmus, T.H., Heckova, Z., Oskay, F., Belbahri, L., Woodward, S., 2016. Biological control of Dothistroma needle blight on pine with *Aneurinibacillus migulanus*. For. Pathol. 46 (5), 555-558. https://doi.org/10.1111/efp.12237.

Alfiky, A., Weisskopf, L., 2021. Deciphering Trichoderma-plant-pathogen interactions for better development of biocontrol applications. J. Fungi 7 (1), 61. https://doi.org/10.3390/jof7010061.

Amorini, E., Annesi, T., Cutini, A., Farina, A., Manetti, M.C., Motta, E., et al., 2000. La Pineta di Castel Fusano. Problematichegestionali e Patologiche del Pinodomestico. Atti del II Congresso S.I.S.E.F. Bologna. Avenue Media, Italia, pp. 45-49 (In Italian).

Anagnostakis, S., 1987. Chestnut blight: the classical problem of an introduced pathogen. Mycologia, 23-37. https://doi.org/10.1080/00275514.1987.12025367.

Annesi, T., Curcio, G., D'Amico, L., Motta, E., 2005. Biological control of *Heterobasidion annosum* on *Pinus pinea* by *Phlebiopsis gigantea*. For. Pathol. 35, 127-134.

Askew, S.E., Shamoun, S.F., van der Kamp, B.J., 2011. Assessment of *Colletotrichum gloeosporioides* as a biological control agent for management of hemlock dwarf mistletoe (*Arceuthobium tsugense*). For. Pathol. 41 (6), 444-452. https://doi.org/10.1111/j.1439-0329.2010.00698.x.

Attiwill, P.M., 1994. Ecological disturbance and the conservative management of eucalypt forests in Australia. For. Ecol. Manag. 63 (2–3), 301–346.

Banerjee, S., Singh, S., Pandey, S., Bhandari, M.S., Pandey, A., Giri, K., 2020. Biocontrol potential of *Pseudomonas azotoformans, Serratia marcescens* and *Trichoderma virens* against Fusarium wilt of *Dalbergia sissoo*. For. Pathol. https://doi.org/10.1111/efp.12581.

Barea, J.M., 2015. Future challenges and perspectives for applying microbial biotechnology in sustainable agriculture based on a better understanding of plant-microbiome interactions. J. Soil Sci. Plant Nutr. 15 (2), 261–282.

Bigler, F., Babendreier, D., Kuhlmann, U., 2006. Environmental Impact of Invertebrates for Biological Control of Arthropods: Methods and Risk Assessment. CABI Publishing, Wallingford.

Booth, T.H., Jovanovic, T., Old, K.M., Dudzinski, M.J., 2000. Climatic mapping to identify high-risk areas for Cylindrocladium quinqueseptatum leaf blight on eucalypts in mainland South East Asia and around the world. Environ. Pollut. 108 (3), 365–372.

Bose, R., Pandey, S., Joshi, P., Banerjee, S., Pandey, A., Bhandari, M.S., 2020. First report of *Calonectria cerciana* causing leaf blight of *Eucalyptus* in northern India. For. Pathol. 2020. https://doi.org/10.1111/efp.12658, e12658.

Boyd, I.L., Freer-Smith, P.H., Gilligan, C.A., Godfray, H.C.J., 2013. The consequence of tree pests and diseases for ecosystem services. Science 342 (6160). https://doi.org/10.1126/science.1235773.

Butt, T.M., Jackson, C., Magan, N., 2001. Fungi as Biocontrol Agents: Progress Problems and Potential. CABI.

CAB International, 2000. *Pinus pinea* (original text by a. Cutini, IstitutoSperimentale per la Selvicoltura, Italy). In: Forestry Compendium. CAB International, Wallingford, UK.

Capretti, P., 1998. Italy. In: Woodward, S., Stenlid, J., Karjalainen, R., Huttermann, A., Wallingford, U.K. (Eds.), *Heterobasidion annosum*: Biology, Ecology, Impact and Control. CAB International, pp. 377–385.

Capretti, P., Korhonen, K., Mugnai, L., Romagnoli, C., 1990. An intersterility group of *Heterobasidion annosum* specialized to *Abies alba*. Eur. J. For. Pathol. 20, 231–240.

Català, S., Pérez-Sierra, A., Abad-Campos, P., 2015. The use of genus-specific amplicon pyrosequencing to assess *Phytophthora* species diversity using eDNA from soil and water in northern Spain. PLoS One 10 (3), e0119311.

Cazorla, F.M., Mercado-Blanco, J., 2016. Biological control of tree and woody plant diseases: an impossible task? BioControl 61 (3), 233–242.

Chakraborty, S., Tiedemann, A.V., Teng, P.S., 2000. Climate change: potential impact on plant diseases. Environ. Pollut. 108 (3), 317–326. https://doi.org/10.1016/s0269-7491(99)00210-9.

Chase, T.E., Ullrich, R.C., 1988. *Heterobasidion annosum* root and butt rot of trees. Adv. Plant Pathol. 6, 501–510.

Chen, L., Bóka, B., Kedves, O., Nagy, V.D., Szűcs, A., Champramary, S., et al., 2019. Towards the biological control of devastating forest pathogens from the genus *Armillaria*. Forests 10 (11), 1013.

Cheng, K., Yu, S., 2020. Neighboring trees regulate the root-associated pathogenic fungi on the host plant in a subtropical forest. Ecol. Evol. 10 (9), 3932–3943.

Chou, H., Xiao, Y.T., Tsai, J.N., Li, T.T., Wu, H.Y., Liu, L.Y.D., Tzeng, D.S, Chung, C.L., 2019. In vitro and in planta evaluation of *Trichoderma asperellum* TA as a biocontrol agent against *Phellinus noxius*, the cause of brown root rot disease of trees. Plant Dis. 103 (11), 2733–2741. https://doi.org/10.1094/PDIS-01-19-0179-RE.

Cleary, M., Oskay, F., Doğmuş, H.T., Lehtijärvi, A., Woodward, S., Vettraino, A.M., 2019. Cryptic risks to forest biosecurity associated with the global movement of commercial seed. Forests 10 (5), 459.

Collinge, D.B., Jørgensen, H.J.L., Latz, M.A.C., Manzotti, A., Ntana, F., Rojas, E.C., Jensen, B., 2019. Searching for novel fungal biological control agents for plant disease control among endophytes. Endophytes for a Growing World, pp. 25–51. https://doi.org/10.1017/9781108607667.003.

Contreras-Cornejo, H.A., Macías-Rodríguez, L., Beltrán-Peña, E., Herrera-Estrella, A., López-Bucio, J., 2011. *Trichoderma*-induced plant immunity likely involves both hormonal- and camalexin-dependent mechanisms in *Arabidopsis thaliana* and confers resistance against necrotrophic fungi *Botrytis cinerea*. Plant Signal. Behav. 6, 1554–1563. https://doi.org/10.4161/psb.6.10.17443.

Crous, P.W., Wingfield, M.J., Cheewangkoon, R., Carnegie, A.J., Burgess, T.I., Summerell, B.A., et al., 2019. Foliar pathogens of eucalypts. Stud. Mycol. 94, 125–298. https://doi.org/10.1016/j.simyco.2019.08.001.

Dai, Y.C., Yuan, H.S., Wei, Y.L., Korhonen, K., 2006. New records of *Heterobasidion parviporum* from China. For. Pathol. 36, 287–293.

Dai, Y.C., Korhonen, K., 1999. *Heterobasidion annosum* group S identified in northeastern China. Eur. J. For. Pathol. 29, 273–279.

Dai, Y.C., Vainio, E.J., Hantula, J., Niemela, T., Korhonen, K., 2003. Investigations on *Heterobasidion annosum* s. lat. in central and eastern Asia with the aid of mating tests and DNA fingerprinting. For. Pathol. 33, 269–286.

Dai, Y., Wu, X.Q., Wang, Y.H., Zhu, M.L., 2020. Biocontrol potential of *Bacillus pumilus* HR10 against Sphaeropsis shoot blight disease of pine. Biol. Control 152. https://doi.org/10.1016/j.biocontrol.2020.104458, 104458.

De Silva, N.I., Brooks, S., Lumyong, S., Hyde, K.D., 2019. Use of endophytes as biocontrol agents. Fungal Biol. Rev. 33 (2), 133–148. https://doi.org/10.1016/j.fbr.2018.10.001.

Delatour, C., Guillamin, J.J., 1985. Root and butt rots in temperate regions. Eur. J. For. Pathol. 15 (5–6), 258–263. https://doi.org/10.1111/j.1439-0329.1985.tb01097.x.

Dennis, C., Webster, J., 1971. Antagonistic properties of species-groups of *Trichoderma*: II. Production of volatile antibiotics. Trans. Br. Mycol. Soc. 57 (1), 41–48. https://doi.org/10.1016/S0007-1536(71)80078-5.

Drenkhan, R., Ganley, B., Martín-García, J., Vahalík, P., Adamson, K., Adamčíková, K., et al., 2020. Global geographic distribution and host range of *Fusarium circinatum*, the causal agent of pine pitch canker. Forests 11 (7), 724. https://doi.org/10.1094/PHYTO-10-20-0445-R.

Dumas, M.T., Laflamme, G., 2013. Efficacy of two *Phlebiopsis gigantea* formulations in preventing *Heterobasidion* irregulare colonization of red pine stumps in eastern Canada. Phytoprotection 93, 25–31. https://doi.org/10.7202/1018887ar.

Dyck, B., 2010. Global Forest Biosecurity Threats and the Risk to New Zealand. https://www.nzffa.org.nz/farm-forestry-model/the-essentials/forest-health-pests-and-diseases/biosecurity/forest-biosecurity-threats/.

Eilenberg, J., Hajek, A., Lomer, C., 2001. Suggestions for unifying the terminology in biological control. BioControl 46, 387–400. https://doi.org/10.1023/A:1014193329979.

Elad, Y., Pertot, I., Marina, A., Prado, A.M., Stewart, A., 2016. Plant Hosts of *Botrytis* spp. In: Fillinger, S., Elad, Y. (Eds.), *Botrytis*—The Fungus, the Pathogen and Its Management in Agricultural Systems. Springer, Cham, Switzerland, pp. 413–448.

Elvira-Recuenco, M., Pando, V., Berbegal, M., Manzano Muñoz, A., Iturritxa, E., Raposo, R., 2021. Influence of temperature and moisture duration on pathogenic life-history traits of predominant haplotypes of *Fusarium circinatum* on *Pinus* spp. in Spain. Phytopathology. https://doi.org/10.1094/PHYTO-10-20-0445-R.

Ennos, R.A., 2015. Resilience of forests to pathogens: an evolutionary ecology perspective. For.: Int. J. For. Res. 88 (1), 41–52. https://doi.org/10.1093/forestry/cpu048.

FAO, 2001. Protecting plantations from pests and diseases. Report based on the work of W.M. Ciesla. Forest Plantation Thematic Papers, Working Paper 10. In: Forest Resources Development Service, Forest Resources Division. FAO, Rome. http://www.fao.org/3/ac130e/ac130e.pdf.

FAO, UNEP, 2020. The State of the World's Forests 2020. In Brief. Forests, Biodiversity and People. FAO & UNEP, Rome, Italy. https://doi.org/10.4060/ca8985en.

Franklin, J.F., Johnson, K.N., Johnson, D.L., 2018. Ecological Forest Management. Waveland Press, pp. 1–593.

Fraser, S., McTaggart, A.R., Roux, J., Wingfield, M.J., 2021. Hyperparasitism by *Sphaerellopsis macroconidialis* may lower over-wintering survival of *Uromycladium acaciae*. For. Pathol. https://doi.org/10.1111/efp.12691, e12691.

Fujimori, T., 2001. Ecological and Silvicultural Strategies for Sustainable Forest Management. Elsevier, pp. 1–337.

Gafur, A., 2020. Development of biocontrol agents to manage major diseases of tropical plantation forests in Indonesia: A review. In: Zaļuma, T., Kenigsvalde, A., Brūna, K., Kļaviņa, L., Burņeviča, D., Vasaitis, R. (Eds.), Environmental Sciences Proceedings. vol. 3. Multidisciplinary Digital Publishing Institute, p. 11. No. 1.

Gaitnieks, T., 2020. Natural infection and colonization of pre-commercially cut stumps of *Picea abies* and *Pinus sylvestris* by Heterobasidion rot and its biocontrol fungus *Phlebiopsis gigantea*. Biol. Control 143, 104208. https://doi.org/10.1016/j.biocontrol.2020.104208.

Garbelotto, M., Gonthier, P., 2013. Biology, epidemiology, and control of *Heterobasidion* species worldwide. Ann. Rev. Phytopathol. 51, 39–59. https://doi.org/10.1146/annurev-phyto-082712-102225.

Garnas, J.R., Hurley, B.P., Slippers, B., Wingfield, M.J., 2012. Biological control of forest plantation pests in an interconnected world requires greater internationalfocus. Int. J. Pest Manage. 58 (3), 211–223. https://doi.org/10.1080/09670874.2012.698764.

Ghyselinck, J., Velivelli, S.L., Heylen, K., O'Herlihy, E., Franco, J., Rojas, M., et al., 2013. Bioprospecting in potato fields in the central Andean highlands: screening of rhizobacteria for plant growth-promoting properties. Syst. Appl. Microbiol. 36 (2), 116–127. https://doi.org/10.1016/j.syapm.2012.11.007.

Glaeser, S.P., Kämpfer, P., 2015. Multilocus sequence analysis (MLSA) in prokaryotic taxonomy. Syst. Appl. Microbiol. 38 (4), 237–245. https://doi.org/10.1016/j.syapm.2015.03.007.

Glick, B.R., 2012. Plant growth-promoting bacteria: mechanisms and applications. Scientifica. https://doi.org/10.6064/2012/963401.

Głowacka, B., Kolk, A., Janiszewski, W., Rosa-Gruszecka, A., Pudełko, M., Łukaszewicz, J., Krajewski, S., 2013. Środki ochrony roślin oraz produkty do rozkładu pni drzew leśnych zalecane do stosowania w leśnictwie w roku 2013. Instytut Badawczy Leśnictwa Analizy i Raporty-19. Wyd. IBL, 77 s.

Gupta, P., Rani, R., Usmani, Z., Chandra, A., Kumar, V., 2019. The role of plant-associated bacteria in phytoremediation of trace metals in contaminated soils. New and Future Developments in Microbial Biotechnology and Bioengineering. Elsevier, pp. 69–76.

Hajek, A.E., Eilenberg, J., 2018. Natural Enemies: An Introduction to Biological Control. Cambridge University Press, p. 426.

Halecker, S., Wennrich, J.-P., Rodrigo, S., Andrée, N., Rabsch, L., Baschien, C., Steinert, M., Stadler, M., Surup, F., Schulz, B., 2020. Fungal endophytes for biocontrol of ash dieback: the antagonistic potential of *Hypoxylon rubiginosum*. Fungal Ecol. 45, 1–12. https://doi.org/10.1016/j.funeco.2020.100918.

Hao, F., Wu, M., Li, G., 2021. Characterization of a novel genomovirus in the phytopathogenic fungus *Botrytis cinerea*. Virology 553, 111–116. https://doi.org/10.1016/j.virol.2020.11.007.

Hartley, C., 1921. Damping-off in Forest Nurseries. USDA Bureau of PlantIndustry, Washington (DC), Bulletin, p. 99.

Havenga, M., Wingfield, B.D., Wingfield, M.J., Marincowitz, S., Dreyer, L.L., Roets, F., Japarudin, Y., Aylward, J., 2021. Genetic recombination in *Teratosphaeria destructans* causing a new disease outbreak in Malaysia. For. Pathol. https://doi.org/10.1111/efp.12683, e12683.

Hepting, G.H., 1963. Climate and forest diseases. Annu. Rev. Phytopathol. 1 (1), 31–50.

Hessenauer, P., Feau, N., Gill, U., Schwessinger, B., Brar, G.S., Hamelin, R.C., 2021. Evolution and adaptation of forest and crop pathogens in the Anthropocene. Phytopathology 111 (1), 49–67. https://doi.org/10.1094/PHYTO-08-20-0358-FI.

Himaman, W., Thamchaipenet, A., Pathom-Aree, W., Duangmal, K., 2016. Actinomycetes from eucalyptus and their biological activities for controlling eucalyptus leaf and shoot blight. Microbiol. Res. 188, 42–52. https://doi.org/10.1016/j.micres.2016.04.011.

Hiratsuka, Y., Chakravarty, P., 1999. Role of Phialemonium curvatum as a potential biological control agent against a blue stain fungus on aspen. Eur. J. Plant Pathol. 29 (4), 305–310. https://doi.org/10.1046/J.1439-0329.1999.00160.X.

Holah, J.C., Wilson, M.V., Hansen, E.M., 1993. Effects of a native forest pathogen, *Phellinus weirii*, on Douglas-fir Forest composition in western Oregon. Can. J. For. Res. 23 (12), 2473–2480.

Holdenrieder, O., Greig, B.J.W., 1998. Biological methods of control. In: Woodward, S., Stenlid, J., Karjalainen, R., Huttermann, A., Wallingford, U.K. (Eds.), Heterobasidion Annosum: Biology, Ecology, Impact and Control. International, pp. 235–258. http://www.fao.org/3/ac130e/ac130e.pdf. Retrieved 25 April 2021.

Inagaki, M., Kamo, K., Titin, J., Jamalung, L., Lapongan, J., Miura, S., 2010. Nutrient dynamics through fine litterfall in three plantations in Sabah, Malaysia, in relation to nutrient supply to surface soil. Nutr. Cycl. Agroecosyst. 88 (3), 381–395.

Iturritxa, E., Trask, T., Mesanza, N., Raposo, R., Elvira-Recuenco, M., Patten, C.L., 2017. Biocontrol of *Fusarium circinatum* infection of Young *Pinus radiata* trees. Forests 8 (2), 32. https://doi.org/10.3390/f8020032.

Joseph, S., Forsythe, S., 2012. Insights into the emergent bacterial pathogen *Cronobacter* spp., generated by multilocus sequence typing and analysis. Front. Microbiol. 3, 397. https://doi.org/10.3389/fmicb.2012.00397.

Joshi, P., Saini, A., Banerjee, S., Bose, R., Bhandari, M.S., Pandey, A., Pandey, S., 2021. Agriculturally important microbes: challenges and opportunities. In: Soni, R., Suyal, D.C., Bhargava, P., Goel, R. (Eds.), Microbiological Activity for Soil and Plant Health Management. Springer, Singapore. https://doi.org/10.1007/978-981-16-2922-8_1.

Köhl, J., Postma, J., Nicot, P., Ruocco, M., Blum, B., 2011. Stepwise screening of microorganisms for commercial use in biological control of plant-pathogenic fungi and bacteria. Biol. Control 57 (1), 1–12. https://doi.org/10.1016/j.biocontrol.2010.12.004.

Korhonen, K., 1978. Intersterility groups of *Heterobasidion annosum*. Commun. Inst. For. Fenn. 94, 1–25.

Korhonen, K., Stenlid, J., 1998. Biology of *Heterobasidion annosum*. Heterobasidion annosum : Biology, Ecology,Impact and Control (Woodward S., Stenlid, J., Karjalainen, R., and Hutterman, A., eds). CAB International, London, pp. 43–71.

Korhonen, K., Lipponen, K., Bendz, M., Ryen, I., Venn, K., Seiskari, P., Niemi, M., 1994. Control of Heterobasidion annosum by stump treatment with Rotstop, a new commercial formulation of Phlebiopsis gigantea. In: Johansson, M., Stenlid, J. (Eds.), on Root and Butt Rots. Wik, Sweden and Haikko, Finland. Uppsala, pp. 675–683.

Korhonen, K., Capretti, P., Karjalainen, R., Stenlid, J., Hütterman, A., 1998. Distribution of *Heterobasidion Annosum* Intersterility Groups in Europe. Heterobasidion annosum: Biology, Ecology, Impact and Control. (Woodward, S.,Stenlid, J., Karjalainen, R. and HÜtterman, A., eds). CAB International, London, pp. 93–105.

Kutschera, U., Hossfeld, U., 2011. Physiological phytopathology: origin and evolution of a scientific discipline. J. Appl. Bot. Food Qual. 85, 1–5.

Malmierca, M.G., Cardoza, R.E., Alexander, N.J., McCormick, S.P., Hermosa, R., Monte, E., Gutierrez, S., 2012. Involvement of *Trichoderma trichothecenes* in the biocontrol activity and induction of plant defense-related genes. Appl. Environ. Microbiol. 78 (14), 4856e4868.

Manion, P.D., 1991. Tree disease concepts (No. 634.963 M278 1991). Prentice Hall.

Marfetán, J.A., Greslebin, A.G., Taccari, L.E., Vélez, M.L., 2020. Rhizospheric microorganisms as potential biocontrol agents against Phytophthora austrocedri. Eur. J. Plant Pathol. 158, 721–732. https://doi.org/10.1007/s10658-020-02113-7.

Martín, J.A., Sobrino-Plata, J., Rodríguez-Calcerrada, J., Collada, C., Gil, L., 2019. Breeding and scientific advances in the fight against Dutch elm disease: will they allow the use of elms in forest restoration? New For. 50 (2), 183–215. https://doi.org/10.1007/s11056-018-9640-x.

Martínez-Arias, C., Macaya-Sanz, D., Witzell, J., et al., 2019. Enhancement of *Populus alba* tolerance to *Venturia tremulae* upon inoculation with endophytes showing in vitro biocontrol potential. Eur. J. Plant Pathol. 153, 1031–1042. https://doi.org/10.1007/s10658-018-01618-6.

Martínez-Arias, C., Sobrino-Plata, J., Ormeño-Moncalvillo, S., Gil, L., Rodríguez-Calcerrada, J., Martín, J.A., 2021. Endophyte inoculation enhances *Ulmus minor* resistance to Dutch elm disease. Fungal Ecol. 50, 101024. https://doi.org/10.1016/j.funeco.2020.101024.

Merril, W, Shigo, A.L., 1979. An expanded concept of tree decay. Phytopathology 69, 1158–1160.

Messing, R., Brodeur, J., 2018. Current challenges to the implementation of classical biological control. BioControl 63 (1), 1–9. https://doi.org/10.1007/s10526-017-9862-4.

Mills, P., Dehnen-Schmutz, K., Ilbery, B., Jeger, M., Jones, G., Little, R., 2011. Integrating natural and social science perspectives on plant disease risk, management and policy formulation. Philos. Trans. R. Soc., BPhilosophical Transactions of the Royal Society B: BiologicalSciencesBiol Scie 366 (1573), 2035–2044.

Mordor Intelligence, 2019. Crop Protection Chemicals Market – Growth, Trends, and Forecast (2019-2024), and Biological Control Market – Growth, Trends, and Forecast (2019-2024). https://www.mordorintelligence.com/industry-reports/biological-control-market. (Accessed 1 June 2021).

Navarro, M.O.P., Simionato, A.S., Barazetti, A.R., dos Santos, I.M.O., Cely, M.V.T., Chryssafidis, A.L., Andrade, G., 2017. Disease-induced resistance and plant immunization using microbes. In: Singh, D.P., Singh, H.B., Prabha, R. (Eds.), Plant-Microbe Interactions in Agro-ecological Perspectives: Fundamental Mechanisms, Methods and Functions. Springer, Singapore, pp. 447–465. https://doi.org/10.1007/978-981-10-5813-4_22.

Odum, E.P., 1953. Fundamentals of Ecology. W. B. Saunders, Philadelphia / London.

Okorski, A., Pszczółkowska, A., Oszako, T., Nowakowska, J.A., 2015. Current possibilities and prospects of using fungicides in forestry. For. Res. Pap. 76 (2), 191–206. https://doi.org/10.1515/frp-2015-0019.

Oliveira, L.S.S., Jung, T., Milenković, I., Tarigan, M., Horta Jung, M., Lumbangaol, P.D.M., Durán, Á., 2021. Damping-off, root rot and wilting caused by *Pythium myriotylum* on *Acacia crassicarpa* in Sumatra, Indonesia. For. Pathol. 51 (3), e12687.

Olson, A., Aerts, A., Asiegbu, F., Belbahri, L., Bouzid, O., et al., 2012. Trade-off between wood decay and parasitism: insights from the genome of a fungal forest pathogen. New Phytol. 194, 1001–1013. https://doi.org/10.1111/j.1469-8137.2012.04128.x.

Oszako, T., Voitka, D., Stocki, M., Stocka, N., Nowakowska, J.A., Linkiewicz, A., Hsiang, T., Belbahri, L., Berezovska, D., Malewski, T., 2021. *Trichoderma asperellum* efficiently protects *Quercus robur* leaves against *Erysiphe alphitoides*. Eur. J. Plant Pathol. 159, 295–308. https://doi.org/10.1007/s10658-020-02162-y.

Otrosina, W.J., Garbelotto, M., 2010. *Heterobasidion occidentale* sp. nov. and *Heterobasidion irregulare* nom. nov.: a disposition of North American *Heterobasidion* biological species. Fungal Biol. 114, 16–25. https://doi.org/10.1016/j.mycres.2009.09.001.

Özkara, A., Dilek, A., Konuk, M., 2016. Pesticides, environmental pollution, and health. In: Environmental Health Risk-Hazardous Factors to Living Species., https://doi.org/10.5772/63094. https://www.intechopen.com/books/environmental-health-risk-hazardous-factors-to-living-species/pesticides-environmental-pollution-and-health.

Paquette, A., Messier, C., 2010. The role of plantations in managing the world's forests in the Anthropocene. Front. Ecol. Environ. 8 (1), 27–34.

Park, Y.S., Chung, Y.J., 2006. Hazard rating of pine trees from a forest insect pest using artificial neural networks. For. Ecol. Manag. 222 (1–3), 222–233. https://doi.org/10.1016/j.foreco.2005.10.009.

Park, Y.S., Chung, Y.J., Moon, Y.S., 2013. Hazard ratings of pine forests to a pine wilt disease at two spatial scales (individual trees and stands) using self-organizing map and random forest. Ecol. Inform. 13, 40–46. https://doi.org/10.1016/j.ecoinf.2012.10.008.

Pellicciaro, M., Lione, G., Giordano, L., Gonthier, P., 2021. Biocontrol potential of *Pseudomonas protegens* against *Heterobasidion* species attacking conifers in Europe. Biol. Control 157. https://doi.org/10.1016/j.biocontrol.2021.104583, 104583.

Piombo, E., Abdelfattah, A., Droby, S., Wisniewski, M., Spadaro, D., Schena, L., 2021. Metagenomics approaches for the detection and surveillance of emerging and recurrent plant pathogens. Microorganisms 9 (1), 188. https://doi.org/10.3390/microorganisms9010188.

Putri, M.H., Handayani, K., Setiawan, W.A., Damayanti, B., Ratih, C.L., Arifiyanto, A., 2021. Screening of extracellular enzymes on serratia marcescens strain MBC1. J. Ris. Biol. Apl., 23-29. https://doi.org/10.26740/jrba.v3n1.p23-29.

Reglinksi, T., Dick, M., 2005. Biocontrol of forest nursery pathogens. N. Z. J. For. 50 (3), 19–26.

Rigling, D., Prospero, S., 2018. *Cryphonectria parasitica*, the causal agent of chestnut blight: invasion history, population biology and disease control. Mol. Plant Pathol. 19 (1), 7–20. https://doi.org/10.1111/mpp.12542.

Roy, G., Laflamme, G., Bussie'res, G., Dessureault, M., 2003. Field tests on biological control of Heterobasidionannosum by *Phaeothecadimorphospora* in comparison with *Phlebiopsis gigantea*. For. Pathol. 33, 127–140.

Ruangwong, O.U., Wonglom, P., Suwannarach, N., Kumla, J., Thaochan, N., Chomnunti, P., et al., 2021. Volatile organic compound from *Trichoderma asperelloides* TSU1: impact on plant pathogenic fungi. J. Fungi 7 (3), 187. https://doi.org/10.3390/jof7030187.

Sales, M.D.C., Costa, H.B., Fernandes, P.M.B., Ventura, J.A., Meira, D.D., 2016. Antifungal activity of plant extracts with potential to control plant pathogens in pineapple. Asian Pac. J. Trop. Biomed. 6, 26–31. https://doi.org/10.1016/j.apjtb.2015.09.026.

Santiago, T.R., Grabowski, C., Rossato, M., Romeiro, R.S., Mizubuti, E.S., 2015. Biological control of eucalyptus bacterial wilt with rhizobacteria. Biol. Control 80, 14–22. https://doi.org/10.1016/j.biocontrol.2014.09.007.

Sharma, A., Shukla, A., Attri, K., Kumar, M., Kumar, P., Suttee, A., et al., 2020. Global trends in pesticides: a looming threat and viable alternatives. Ecotoxicol. Environ. Saf. 201. https://doi.org/10.1016/j.ecoenv.2020.110812, 110812.

Shaw, M.W., Osborne, T.M., 2011. Geographic distribution of plant pathogens in response to climate change. Plant Pathol. 60 (1), 31–43. https://doi.org/10.1111/j.1365-3059.2010.02407.x.

Singh, T., Chittenden, C., 2021. Synergistic ability of chitosan and *Trichoderma harzianum* to control the growth and discolouration of common Sapstain fungi of *Pinus radiata*. Forests 12 (5), 542. https://doi.org/10.3390/f12050542.

Skidmore, A.M., Dickinson, C.H., 1976. Colony interactions and hyphal interference between *Septoria nodorum* and *Phylloplane* fungi. Trans. Br. Mycol. Soc. 66 (1), 57–64. https://doi.org/10.1016/S0007-1536(76)80092-7.

Solla, A., Moreno, G., Malewski, T., Jung, T., Klisz, M., Tkaczyk, M., Siebyla, M., Pérez, A., Cubera, E., Hrynyk, H., Szulc, W., 2021. Phosphite spray for the control of oak decline induced by *Phytophthora* in Europe. For. Ecol. Manag. 485, 118938.

Soutrenon, A., Levy, A., Legrand, P., Lung-Escarmant, B., Guillaumin, J.J., Delatour, C., 1998. Comparison between three stump treatments to control *Heterobasidion annosum* (urea, *Disodium* octoborate tetrahydrate, *Phlebiopsis gigantea*). In: Delatour, C., Guillaumin, J.J., Lung-Escarmant, B., Marcais, B.L.C. (Eds.), Root and Butt Rots of Forest Trees. vol. 89. INRA, Paris, pp. 381–389.

Stenberg, J.A., Sundh, I., Becher, P.G., Björkman, C., Dubey, M., Egan, P.A., Friberg, H., Gil, J.F., Jensen, D.F., Jonsson, M., Carlsson, M., 2021. When is it biological control? A framework of definitions, mechanisms, and classifications. J. Pest. Sci. 1-12. https://doi.org/10.1007/s10340-021-01354-7.

Sturrock, R.N., Frankel, S.J., Brown, A.V., Hennon, P.E., Kliejunas, J.T., Lewis, K.J., Worrall, J.J., Woods, A.J., 2011. Climate change and forest diseases. Plant Pathol. 60 (1), 133–149. https://doi.org/10.1111/j.1365-3059.2010.02406.x.

Suzuki, N., Cornejo, C., Aulia, A., Shahi, S., Hillman, B.I., Rigling, D., 2021. In-tree behavior of diverse viruses harbored in the chestnut blight fungus, *Cryphonectria parasitica*. J. Virol. 95 (6), e01962-20. https://doi.org/10.1128/JVI.01962-20.

Tainter, F.H., Baker, F.A., 1996. Principles of Forest Pathology. John Wiley & Sons.

Tashi-Oshnoei, F., Harighi, B., Abdollahzadeh, J., 2017. Isolation and identification of endophytic bacteria with plant growth promoting and biocontrol potential from oak trees. For. Pathol. 47 (5). https://doi.org/10.1111/efp.12360, e12360.

Terhonen, E., Sun, H., Buee, M., Kasanen, R., Paulin, L., Asiegbu, F.O., 2013. Effects of the use of biocontrol agent (*Phlebiopsis gigantea*) on fungal communities on the surface of *Picea abies* stumps. For. Ecol. Manag. 310, 428–433. https://doi.org/10.1016/j.foreco.2013.08.044.

Tiwari, S., Prasad, V., Lata, C., 2019. *Bacillus*: plant growth promoting bacteria for sustainable agriculture and environment. In: New and Future Developments in Microbial Biotechnology and Bioengineering. Elsevier, pp. 43–55.

Trogisch, S., Liu, X., Rutten, G., Xue, K., Bauhus, J., Brose, U., et al., 2021. The significance of tree-tree interactions for forest ecosystem functioning. Basic Appl. Ecol. https://doi.org/10.1016/j.baae.2021.02.003.

The Global Forest Goals Report 2021. (Accessed 21 February 2022).

Umrao, P.D., Kumar, V., Kaistha, S.D., 2021. Biocontrol potential of bacteriophage ϕsp1 against bacterial wilt-causing *Ralstonia solanacearum* in Solanaceae crops. Egypt. J. Biol. Pest Control 31 (1), 1–12. https://doi.org/10.1186/s41938-021-00408-3.

Usta, C., 2013. Microorganisms in biological pest control—a review (bacterial toxin application and effect of environmental factors). In: Current Progress in Biological Research. InTech, Rijeka, pp. 287–317. https://doi.org/10.5772/55786.

Vallières, C., Raulo, R., Dickinson, M., Avery, S.V., 2018. Novel combinations of agents targeting translation that synergistically inhibit fungal pathogens. Front. Microbiol. 9, 2355. https://doi.org/10.3389/fmicb.2018.02355.

Van Lenteren, J.C., 2021. Will the "Nagoya protocol on access and benefit sharing" put an end to biological control? In: Area-Wide Integrated Pest Management. CRC Press, pp. 655–667.

van Lenteren, J.C., Bolckmans, K., Köhl, J., Ravensberg, W., Urbaneja, A., 2018. Biological control using invertebrates and microorganisms: plenty of new opportunities. BioControl 63, 3959. https://doi.org/10.1007/s10526-017-9801-4.

Velivelli, S.L., De Vos, P., Kromann, P., Declerck, S., Prestwich, B.D., 2014. Biological control agents: from field to market, problems, and challenges. Trends Biotechnol. 32 (10), 493–496. https://doi.org/10.1016/j.tibtech.2014.07.002.

Wingfield, M.J., Hammerbacher, A., Ganley, R.J., 2008. Pitch canker caused by *Fusarium circinatum*—a growing threat to pine plantations and forests worldwide. Australas. Plant Pathol. 37, 319–334. https://doi.org/10.1071/AP08036.

Wingfield, M.J., Brockerhoff, E.G., Wingfield, B.D., Slippers, B., 2015. Planted forest health: the need for a global strategy. Science 349 (6250), 832–836. https://doi.org/10.1126/science.aac6674.

Worrall, J.J., Harrington, T.C., Blodgett, J.T., Conklin, D.A., Fairweather, M.L., 2010. *Heterobasidion annosum* and *H. parviporum* in the southern Rocky Mountains and adjoining states. Plant Dis. 94, 115–118.

Xue, Q.Y., Ding, G.C., Li, S.M., Yang, Y., Lan, C.Z., Guo, J.H., et al., 2013. Rhizocompetence and antagonistic activity toward genetically diverse *Ralstonia solanacearum* strains-an improved strategy for selecting biocontrol agents. Appl. Microbiol. Biotechnol. 97 (3), 1361–1371.

Yin, C., Wang, Y., Zhang, Y., 2020. Hypothesized mechanism of biocontrol against pine wilt disease by the nematophagous fungus *Esteya vermicola*. Eur. J. Plant Pathol. 156, 811–818. https://doi.org/10.1007/s10658-019-01930-9.

Yu, Z., Wang, Z., Zhang, Y., Wang, Y., Liu, Z., 2021. Biocontrol and growth-promoting effect of *Trichoderma asperellum* TaspHu1 isolate from *Juglans mandshurica* rhizosphere soil. Microbiol. Res. 242, 126596. https://doi.org/10.1016/j.micres.2020.126596.

Cold-tolerant and cold-loving microorganisms and their applications

Gayan Abeysinghe[a,*], H.K.S. De Zoysa[b,†], T.C. Bamunuarachchige[b], and Mohamed Cassim Mohamed Zakeel[c,d]

[a]Department of Biological Sciences, Faculty of Applied Sciences, Rajarata University of Sri Lanka, Mihintale, Sri Lanka, [b]Department of Bioprocess Technology, Faculty of Technology, Rajarata University of Sri Lanka, Mihintale, Sri Lanka, [c]Department of Plant Sciences, Faculty of Agriculture, Rajarata University of Sri Lanka, Puliyankulama, Anuradhapura, Sri Lanka, [d]Centre for Horticultural Science, Queensland Alliance for Agriculture and Food Innovation, The University of Queensland, Ecosciences Precinct, Dutton Park, QLD, Australia

7.1 Introduction

Microorganisms that grow under cold conditions are of two types: cold-loving psychrophiles and cold-tolerant psychrotrophs or psychrotolerants. Psychrophiles are considered as microorganisms that have an optimal growth temperature at or below 15°C, with a minimum of 0°C or even below that and a maximum of 20°C, while psychrotrophs can grow at low temperatures with an optimal growth above 15°C and a maximum of 20°C or above (Morita, 2001). However, there have been instances where the isolated psychrophiles show survival at higher temperatures such as 25°C (Mykytczuk et al., 2013). Many psychrophilic microbes have been isolated from cold environments, and the majority of them belong to bacteria. But studies focusing on psychrophilic fungi and yeast have also begun recently (Acuña-Rodríguez et al., 2019).

Psychrophiles are important in the industry due to their unique proteins that can withstand low temperatures or have the ability to

---

[*] Current address: Faculty of Life and Environmental Sciences, University of Tsukuba, Japan
[†] Current address: Department of Biology, University of Naples Federico II, Naples, Italy

surviveat low temperatures, thus increasing their applications, for instance high-altitude farming. Psychrophiles are armed with a variety of mechanisms that enable them to tolerate cold temperatures. These include cryoprotectants and antifreeze proteins, genome structure, membrane adaptations, and presence of differential gene expression (De Maayer et al., 2014).

7.2 Diversity of cold-tolerant mutants in cold ecosystems

Water constitutes nearly 70% of the biosphere of the earth, which is mostly at a cold temperature of approximately 5°C due to larger areas of polar regions (Achberger et al., 2017; Antranikian et al., 2005). Microorganisms can grow at a wider range of temperatures; thus, they can be divided into two main groups: psychrophiles and psychrotolerants. Psychrophilic microorganisms can grow at a temperature ranging from 0°C to 20°C, while psychrotolerant can grow between 20°C and 30°C. However, psychrophiles can also be found in permanently cold environments like glaciers, liquid water that exists beneath polar ice sheets, mountain regions, and deep sea (Antranikian et al., 2005; Achberger et al., 2017). Nonetheless, many studies have suggested that some microbes can survive at extreme cold temperatures like − 60°C, which are known as extremophiles and used in industries (Al-Ghanayem and Joseph, 2020; Berry and Foegeding, 1997). But psychrophiles are the major group which are used for both industrial and biotechnological applications (Al-Ghanayem and Joseph, 2020; Berry and Foegeding, 1997).

7.3 Mechanisms of cold tolerance in microorganisms

Psychrophiles and psychrotrophs consist of organisms from the domains of bacteria, archaea, and eukarya, which include certain plants and animals, yet here we have focused on the variety of response of the microorganisms towards cold. Although microorganisms due to their omnipresence face a number of abiotic stress conditions, cold stress poses an enormous impact on the regulation of biological and physiological processes. Many mechanisms of cold-tolerant microorganisms that enable growth and performance of cellular processes while withstanding the cold stress are not fully understood (Subramanian et al., 2011; Weinstein et al., 2000; Snider et al., 2000). Due to the realization of the potential use of enzymes of these cold-tolerant microorganisms in agriculture and biotechnology, researchers have exploited the ability of cold tolerance of the psychrophilic and psychrotrophic microorganisms.

Sensing of the external environmental temperature is of vital importance to any living organism. Phosphorylation and dephosphorylation mechanisms mediate the bacterial sensing of the environmental temperature (Shivaji and Prakash, 2010). Furthermore, membrane rigidifying is important in cold temperature sensing, and this activates a two-component signal transduction system, where transmembrane sensory domain recognizes the drop in temperature, subsequently phosphorylating the histidine by a histidine kinase, and the phosphate is transferred to an aspartate on the response regulator in the cytoplasm that initiates transcription of a number of cold-expressed genes (Shivaji et al., 2007). Additionally, there have been reports on complex phosphorelay systems (Shi et al., 2015; Zhang and Shi, 2005).

7.3.1 Cell membrane response

Cold stress affects the cell walls and membranes of microorganisms, and membranes are altered to sustain their function (Hassan et al., 2020; Siliakus et al., 2017; Bajerski et al., 2017). Major changes in membranes include the alteration of the composition of membrane lipids and carotenoids, fatty acid chain length, branching, configuration, and unsaturation of fatty acids (Chintalapati et al., 2004; Hassan et al., 2016).

Fatty acids with diverse melting points aid in maintaining the cell membrane fluidity under cold stress condition (Mansilla et al., 2004). Synthesis of temperature-dependent carotenoids as the polar carotenoids can stabilize the cell membrane as compared with its nonpolar molecules which modulate the fluidity of the unsaturated fatty acids in cold temperatures (Mansilla et al., 2004). In low temperatures, the levels of polar carotenoids have been shown to increase in *Shingobacterium antarcticum* (MTCC 675) and *Micrococcus roseus* (MTCC 678) (Shivaji and Prakash, 2010; Jagannadham et al., 2000). Incorporation of monounsaturated fatty acids is an important modification to cope with low temperature-induced stress where the bacteria use desaturases to introduce *cis* or *trans* double bonds, thus modulating the fluidity of the membrane. While *cis*-unsaturated fatty acids elevate the fluidity of the cell membrane, *trans*-unsaturated fatty acids decrease it. This is evident in the psychrophilic bacterium *Pseudomonas syringae*, which has shown an increase in the proportion of saturated to *trans*-monosaturated fatty acids with the increase in temperature (Shivaji and Prakash, 2010). Cold-tolerant Gram-positive and Gram-negative bacterial strains isolated from nonpolar glacial habitats possess cell membranes with straight-chain monounsaturated fatty acids and branched-chain fatty acids as predominant constituents (Hassan et al., 2020). Although polyunsaturated fatty acids are less efficient in modulating the fluidity of membranes, omega-3 and omega-6 polyunsaturated fatty acids are found in marine microorganisms (Siliakus et al., 2017; Shivaji and Prakash, 2010). Moreover, fungal strains, including *Cadophora fastigiate, Mortierella alpine, M. antarctica,* and *M. elongate* isolated from Antarctic region, have shown to produce fatty acids such as arachidonic acid, linoleic acid, and stearidonic acid, which change the fatty acid composition in cell membrane, while some psychrotolerant yeasts such as *Rhizopodium diobovatum* have shown to modulate the fluidity of cell membrane (Hassan et al., 2016; Weinstein et al., 2000; Turk et al., 2011).

7.3.2 Cryoprotectants

Microorganisms can accumulate low molecular weight, hydrophilic, and nontoxic solutes such as trehalose, glycine betaine, and carnitine that may act as cryoprotectants (Mishra et al., 2010; Shivaji and Prakash, 2010; Chattopadhyay, 2002). The cryoprotectant is presumed to stabilize the cellular proteins and membranes at low temperatures. The melting point of the cellular contents decreases with the formation and existence of cryoprotectants such as trehalose, a nonreducing disaccharide (α-D-glucopyranosil-1, 1-α-D-glucopyranoside), due to its high water-retaining property and less solubility at low temperatures, that could form crystals of 90% calorific sucrose (Bhattacharya, 2018). Many studies in fungal species such as *H. marvini* and *M. elongata* have demonstrated that these microbes show response to cold stress by accumulating intracellular trehalose and other cryoprotective carbohydrates that enhance resistance to freezing temperatures (Hassan et al., 2016; Weinstein et al., 2000). Similarly, glycine betaine, which is a quaternary ammonium solute, is a widely known cryoprotectant that was initially found in the food-borne pathogen *Listeria monocytogenes*, a bacterium that can withstand high osmolarity and low temperature conditions (Andrews and Harris, 2000). This bacterium has shown that sigma B protein, a part of RNA polymerase that is involved in bacterial gene expression, is responsible for the accumulation of betaine and ATP-dependent transport system in the uptake of betaine under cold stress (Chattopadhyay, 2002). In low temperatures, cell damage occurs by excessive shrinkage of cells due to hypertonic environment or intracellular ice injury caused by mechanical damage from internal ice crystals. Betaine has a high hydrophilic property that strongly binds to water molecules, thus minimizing water crystallization and subsequent ice formation inside cells (Yang et al., 2016).

7.3.3 Antifreeze proteins

Antifreeze proteins (AFPs) are found in many prokaryotes and eukaryotes and were initially identified in the bacterium *Moraxella* sp. (Yamashita et al., 2002; Hoshino et al., 2003; Garnham et al., 2008). AFPs are ice-binding proteins that prevent the growth of ice below the melting point of a solution by reducing the freezing point (Muñoz et al., 2017). This process is known as thermal hysteresis (TH), which is facilitated by the adhesion of AFP on the crystal surface of ice (Muñoz et al., 2017). AFPs have been found in several fungal species, including *Flammulina velupites*, *Pleurotus ostratus*, and *Coriolus versicolor*, and many bacterial species, including *Rhodococcus erythropolis*, *Micrococcus cryophilus* and *Marinomonas primoryensis* (Singh et al., 2014; Gilbert et al., 2004; Duman and Olsen, 1993). The majority of AFPs of bacterial origin have a lower TH, providing them the ability to tolerate low temperatures rather than freeze avoidance.

7.3.4 Cold acclimation proteins and cold shock response

Cold acclimation proteins that are also known as "caps" improve protein synthesis of bacteria at low temperatures (Margesin et al., 2007). These proteins are usually found in bacteria that inhabit permanently in cold environments, and thus, they are used as markers for identifying psychrotrophs (Mishra et al., 2010; Piette et al., 2010). *Pantoea ananas* KUIN-3, an ice nucleating bacterium, produces a type of caps known as Hsc 25 that enables the refolding of denatured enzymes (Kawahara et al., 2000).

In natural environments, microorganisms do not necessarily face a sudden decrease of temperature from mesophilic range to cold temperatures that could cause stress, but they encounter such situations in laboratory-induced scenarios, such as food production and storage. Bacteria respond to rapid temperature downshifts by the induction of cold shock proteins (CSPs) (Phadtare, 2004). Production of CSPs is a vital response of bacteria to cold shock. CSPs are small nucleic acid-binding proteins that are found in a range of microorganisms, including psychrophiles, mesophiles, and thermophiles (Czapski and Trun, 2014; Jin et al., 2014). Cold shock induces *csp* gene expression subsequently synthesizing CSPs (Table 7.1). The CSPs are composed of histone like proteins, RNA binding proteins, transcription factors, acyl lipid desaturases, subunit of DNA gyrase, heat shock protein (Hsc 66), and γ-glutamyl transpeptidase, and their structure is highly

Table 7.1 Proteins induced under cold stress condition and their functions.

Protein	Function	References
AceE	Pyruvate dehydrogenase, decarboxylase	Jones and Inouye (1994)
AceF	Pyruvate dehydrogenase, dihydrolipoamide acetyl transferase	Jones and Inouye (1994)
CspA	Cold-inducible RNA chaperone, transcriptional enhancer	Goldstein et al. (1990), Gualerzi et al. (2003)
CspB	Function unknown	Etchegaray et al. (1996)
CspC	Regulation of expression of stress response proteins RpoS and UspA	Shenhar et al. (2012), Phadtare and Inouye (2001)
CspD	Biofilm development, inhibition of DNA replication	Kim et al. (2010), Yamanaka and Inouye (2001a)
CspE	Regulation of expression of stress response proteins RpoS and UspA	Czapski and Trun (2014), Shenhar et al. (2012), Phadtare and Inouye (2001)
CspG	Function unknown	Nakashima et al. (1996)
CspI	Function unknown	Wang et al. (1999)
PNP	Degradation of RNA	Yamanaka and Inouye (2001b)
OtsA	Trehalose phosphate synthase	Kandror et al. (2002)
OtsB	Trehalose phosphatase	Kandror et al. (2002)
InfA	Protein chain initiation factor IF1	Gualerzi et al. (2003)
InfB	Protein chain initiation factor IF2, binding of charged tRNA-fmet to 30S ribosomal subunit	Gualerzi et al. (2003)
InfC	Protein chain initiation factor IF3, mRNA translation stimulation	Gualerzi et al. (2003)
Tig	Protein-folding chaperone, ribosome binding	Kandror et al. (2002)

conserved, although their thermostability is variable (Jin et al., 2014; Chattopadhyay, 2002; Keto-Timonen et al., 2016). In *E. coli*, CspA family has nine homologous proteins, CspA through CspI (Keto-Timonen et al., 2016; Chattopadhyay, 2002). CSPs have a highly conserved nucleic acid binding domain named cold shock domain (CSD) that consists of two nucleic acid binding motifs, ribonucleoprotein 1 and 2, that direct the binding to target nucleic acids (Lee et al., 2013). There are variations in the type and level of CSP expression in response to cold shock. For example, in *E. coli*, cold shock induces CspA, CspB, CspE, CspG, and CspI only (Table 7.1) (Keto-Timonen et al., 2016). However, *cspA* is one of the most expressed genes, and it is responsible for mRNA chaperone, which in turn is involved in transcriptional regulation of other cold shock genes (Keto-Timonen et al., 2016).

Clostridium botulinum group I and III strains have *cspA, cspB,* and *cspC* genes, and group II strains have only one type of *csp* genes, while type B toxic strains possess none of them (Söderholm et al., 2013).

7.3.5 RNA degradosome

RNA degradosome is a highly structured protein complex, which is composed of ribonucleases, RNA helicase, and glycolytic enzymes that perform bulk RNA decay in bacteria (Cho, 2017). Unlike most bacterial stress responses that are controlled by transcriptional regulators, cold shock response is believed to be driven by posttranscriptional control. A recent study has demonstrated a two-member mRNA surveillance system that enables a recovery of translation during acclimation where RNase R drives mRNA degradation, while CspS protein modulates mRNA secondary structure to adjust the protein expression (Zhang et al., 2018). Moreover, the need for ATP to unwind the RNA duplex by RNA helicase is reduced, leading to saving energy under cold stress.

7.4 Aspects of cold-tolerant enzymes

Lives in cold environments require broad adaptation to ensure that their physiological functions and metabolism including all required cellular activities to function well. Thus, psychrophilic microbes show plasticity during these extreme conditions to survive successfully. However, a key determinant of these adaptations depends on protein functions, which are directly involved in metabolism and cell cycle. Therefore, psychrophilic enzymes are the key factors for the successful establishment of psychrophilic microbes in cold environments (Feller, 2013; Hébraud and Potier, 1999). Nevertheless, microbes have to use some strategies to maintain sustainable growth and ensure their survival at low temperatures because some adaptations are energy-wise expensive and well-evolved enzymes are quite rare (Hébraud and Potier, 1999). Cold-active enzymes show exceptionally high efficacy, thermostability and alkaline stability, higher specific activity, and ease of inactivation and require very low activation energy at low temperature as opposed to mesophilic and thermophilic enzymes. There are several cold-active and cold-adaptive enzymes, especially lipases, polygalacturonases, esterases, proteases, cellulases (Alcalase, Natalase, and Lipolase Ultra), exopolysaccharides, and amylases (Al-Ghanayem and Joseph, 2020; Antranikian et al., 2005; Birgisson et al., 2003; Brakstad et al., 2017; Cavicchioli et al., 2011; Feller, 2013).

Psychrophiles can produce stable enzymes with higher catalytic activities at extremely low temperature by adapting to numerous structural changes. These structural features facilitate proper functioning of cold-adaptive enzymes at extreme conditions compared to the mesophiles and thermophiles. In addition, the alteration of phospholipid and fatty acid cell wall structure increases the membrane fluidity when exposed to low temperatures (Al-Ghanayem and Joseph, 2020; Brakstad et al., 2017; Hébraud and Potier, 1999; Berry and Foegeding, 1997). Generally, every 10°C decrease in temperature decreases the biochemical reaction rate of enzymes by two- to threefolds in cold-adapted microorganism. Moreover, the rate of reaction decreases by 16–80 times at 0°C as compared to 37°C (Wintrode et al., 2000). As compared to mesophilic enzymes, cold-adaptive enzymes are 10 times more active at low to moderate temperatures (up to 20–30°C) (De Oliveira et al., 2020).

One of the main characteristic features of psychrophilic enzymes is their higher catalytic activity, with an increasing affinity toward substrates at low temperatures and reduced activation energy. These diverse adaptations of cold enzymes guarantee their usefulness in industrial applications even under unfavorable conditions (Al-Ghanayem and Joseph, 2020; Feller, 2013; Białkowska et al., 2009). Another important feature of psychrophile enzymes is the flexible structure, which helps their activation with low kinetic energy. This inherent flexibility shows that these enzymes have less active enthalpy and negative active entropy compared to their counterparts in thermophiles (Białkowska et al., 2009; Cavicchioli et al., 2011; Feller, 2013). Higher flexibility of psychrophile enzymes is attributed to various features such as increased surface hydrophobicity, increased number of ion pairs, less electrostatic interactions, lower arginine/lysine ratio, increase in the extent of secondary structure formation, decreased secondary structure content, weaker interdomain and intersubunit interactions, reduced oligomerization, and reduced ratio of surface area to volume (Cavicchioli et al., 2011; Demirjian et al., 2001; Dhaulaniya et al., 2019). Most of these features are helpful for the structural stability and thermal stability of cold-adapted enzymes.

Genomic studies have shown that there are differences in the amino acid composition of psychrophiles compared to thermophiles (Cavicchioli et al., 2011; Demirjian et al., 2001; Dhaulaniya et al., 2019; Hébraud and Potier, 1999). Heat-liable activity is one of the special characteristics exhibited by psychrophilic enzymes, which signifies the heat liability of active sites of proteins. Active sites of these enzymes have revealed heat-liable property than the whole protein structure. Together with this, flexibility of active sites and unstable property, enhance or increase the activity of cold-adapted enzymes. For example, some enzymes such as isocitrate dehydrogenase and psychrophilic carbonic anhydrase are considered more stable based on these characteristics (Dhaulaniya et al., 2019; Feller, 2013; Giovanella et al., 2020).

The main disadvantages of microbial enzymes are that they show low enzyme activity, flexibility of protein denaturation, membrane fluidity, alterations in the cellular transport, low reaction rate, less availability, and low stability when they are exposed to harsh conditions (Al-Ghanayem and Joseph, 2020; De Oliveira et al., 2020). However, many novel techniques are available to overcome these impediments by increasing both quality and quantity of cold-active enzymes using modern techniques, including rDNA technology, omics approaches, protein engineering approaches, and via genetically improved strains (Al-Ghanayem and Joseph, 2020).

Enzyme engineering is a key approach of producing cold-active enzymes. Use of some cold-tolerant microorganisms to produce these enzymes and their applications are listed in Table 7.2. In addition, some mutagenized microorganisms are used to produce these enzymes with specific characteristics, including compatibility, active at a wide range of temperatures and pH, and thermostability.

7.4.1 Industrial and medical aspects

The significance of cold-tolerant microorganisms is that they are excellent candidates for industrial use. Cold-active enzymes derived from psychrotolerant and other microorganisms have been used as efficient tools in many industries. Increasing usage of cold-active enzymes in the preservation of food under low temperature to reduce spoilage and protect attractive traits of food, including taste and nutritional value, is among the many advantages. In cold storage, β-galactosidases or lactases are being used to remove lactose in milk by hydrolysis. Moreover, this is used to produce lactose-free milk and its derivatives for lactose-intolerant population (Nam and Ahn, 2011). Cold-active enzymes could aid in reducing the manufacturing cost of products. An Antarctic marine bacterium, *P. haloplanktis*, derived β-galactosidase is used in tagatose production (Van De Voorde et al., 2014). Cold-active xylanases from *P. haloplanktis* TAH3A and *Flavobacterium* sp. MSY-2 are being used in the production of cold-active xylanases that could convert the insoluble hemicellulose of dough into soluble sugars in bread production. (Sarmiento et al., 2015; Wang et al., 2012). Pectinases catalyze the degradation of plant carbohydrate pectin, and the use of cold-adapted pectinases is integral due to the low-temperature food processing. Another prominent use of the cold-active pectinases is in wine production. Although the majority of pectinases used in the industry are of mesophilic origin, several studies have put forth psychrophile-derived pectinases, including from *Cystofilobasidium capitatum, Mrakia frigida,* and *Cryptococcus*

Table 7.2 Use of mutagenized microorganisms with their application and aspects.

Host species	Mutant's recombinant enzymes	Application	Purpose/aspect	References
Antarctic *Bacillus* TA39	*Bacillus* TA39 protease (S39)	Detergent	• To increase specificity for synthetic substrate, and wider substrate profile • To increase detergent enzyme activity at room temperature	Al-Ghanayem and Joseph (2020)
Bacillus sphaericus P3C9	Protease	–	• Increase catalytic rate	Wintrode et al. (2000)
Escherichia coli strain CF1946	(p)ppGpp (guanosine 5'-triphosphate- 3'-diphosphate and guanosine 5'-diphosphate-3'-diphosphate)	–	• Cold shock response and the heat shock response	Jones et al. (199 Vanbogelen and Neidhardt (1990)
E. coli	*Bacillus* sp. strain TA41 alkaline protease	–	• Examine of structural characteristics of proteins	Berry and Foegeding (1997) Davail et al. (199
Bacillus subtilis without CspB gene	–	Food preservation	• Compared the effects of freezing on cell viability	Berry and Foegeding (1997)
E. coli C600	*Pseudomonas* sp. strain B11-1 lipases	–	• Able to increase in thermolability at higher temperatures and increase in activity at lower temperatures	Choo et al. (1998
E. coli JM101	*Pseudomonas fragi* strain IFO 3458 (PFL) lipase	–	• Able to enhance the flexibility of the solvent interactions	Alquati et al. (200
E. coli	*Psychrobacter immobilis* B10 lipase	–	• Enhance the cold adaptability	Arpigny et al. (1993)
E. coli	*Shewanella* sp. psychrophilic phosphatase	–	• Ability to show high catalytic activity at low temperature and protein-tyrosine-phosphatase activity.	Demirjian et al. (2001)
E. coli/pYOK3	*Bacillus psychrosaccharolyticu* Psychrophilic alanine racemase	–	• Stable at the low temperature optimum (0°C)	Demirjian et al. (2001)
E. coli	*Moraxella* TA144 lipase	–	• Able to catalyzes lipolysis at temperatures close to 0°C	Feller et al. (1991
E. coli DE3 (BL21)	*Serratia* sp. esterase	Detergent and other industrial applications	• Stable catalytic activity at low temperature, extreme salt tolerance, and good pH stability.	Jiang et al. (2016

cylindricus, which were isolated from Abashiri, Hokkaido, Japan (Nakagawa et al., 2004).

The novel trend in detergent industry is the enzymes/detergents that work efficiently under low temperatures due to the requirements such as energy saving, fabric protection, and reduced emission of carbon dioxide. Among the many newly discovered and introduced enzymes, the cold-adapted lipases from *Pseudomonas stutzeri* PS59, proteases from several bacterial strains from Arctic and Antarctic regions (Chen et al., 2013; Alias et al., 2014), amylases derived from marine bacterium *Zunongwangia profunda* (Qin et al., 2014) have offered interesting alternatives to the existing detergents (Sarmiento et al., 2015). Although there is substantial advancement in the introduction of cold-active enzymes in the detergent industry, still there is an apparent need of cold-adapted enzymes, which could be derived from psychrophilic microorganisms that could efficiently perform in cold washing practices.

When it comes to the medical and pharmaceutical applications, there is definitely a growing interest in psychrophilic and psychrotolerant microorganisms. Antarctic bacterium *Pseudoalteromonas* produces an antifreeze glycoprotein, Antarticine-NF3, which is widely used in cosmetics, especially for scar treatments (Bisht, 2011). Polyunsaturated fatty acids, which are a major component in the psychrotolerant mechanisms of bacteria that improves the fluidity of the bacterial membrane enabling efficient permeability and transport of solutes, may constitute in the cholesterol and triglyceride transportation opted for nervous system and cardiovascular health (Hamdan, 2018, Jadhav et al., 2010). Several fungal species living in cold temperature habitats have proven to be of excellent metabolite factories, from point of view of pharmaceuticals, like *Penicillium antarcticum*, a psychrotolerant-to-halotolerant species, which are able to produce patulin and asperentins (McRae and Seppelt, 1999), and *P. algidum* that produce pschophilin D and cycloaspeptide A and D (Dalsgaard et al., 2005). It is a well-known fact that psychrophilic or psychrotolerant microorganisms have great potential in a variety of fields as they are capable of producing cold-active enzymes, pharmaceuticals/metabolites, and other biotechnologically important compounds.

7.4.2 Environmental aspects

At the very beginning of the waste management process, conventional way of disposal of waste was practiced by burying those waste materials. With time and increases of waste with conventional practices, people felt the requirement of rapid and proper waste management systems. The degradation process was also very slow with the conventional practices. Further, this conventional method has many issues due to its difficulty, lack of public recognition, and profligate nature. These factors have introduced the modern-day waste management methods with new practices (Sivaperumal et al., 2017). Nevertheless, one of the major issues suffered by many areas is bioremediation strategies. Microorganisms offer potential and effective clean-up approaches to clean the contaminated environments. As bioremediation (or bioaugmentation) attempts to enhance the accelerated biodegradation process through the optimal conditions like nutrients, temperature, and bioavailability of contaminants, cold-adapted microorganisms play a vital and economical role (Sivaperumal et al., 2017; Garcia-Descalzo et al., 2012). Further, the bioremediation can occur under both aerobic and anaerobic conditions with less impacts on the environment (Sivaperumal et al., 2017).

Commercial requirements with increased regulatory demands, and environmental sustainability have enhanced the use of cold-adapted enzymes as a cleaning agent because some psychrophiles produce stable enzymes that are capable of use as eco-friendly and cost-effective methods in bioremediation (Al-Ghanayem and Joseph, 2020). Hydrolysis of substrate by cold-adapted enzymes reveals its usefulness for cleaning applications within the laundry, dishwashers, dairy, brewing, food, water treatment, medical devices, membrane filtration, petroleum, and a wide range of other industries (Cavicchioli et al., 2011).

With the huge amount of drilling, petroleum exploration has been polluting many environments with oil spill events. Therefore, biodegradation is more essential, and some of the microbes are actively involved in biodegradation of oils even at the cold environments. However, most of the psychrophilic or psychrotolerant microbes are involved with the biodegradation of oils from marine environments (Brakstad et al., 2017; Garcia-Descalzo et al., 2012; Giovanella et al., 2020). Some studies have revealed the complete degradation of alkanes using *Pseudomonas* sp. ST41 species at 4°C and degradation of polyaromatic hydrocarbon using *Pseudomonas mandelii* JR-1, including other petroleum hydrocarbons (aliphatic/aromatic alkane degradation, polychlorinated biphenyl and gaseous hydrocarbon). Biodegradation using native cold-adapted microorganisms present in Alpine soils and Antarctica and Arctic has been observed (Garcia-Descalzo et al., 2012; Giovanella et al., 2020; Kavitha, 2016; Nichols et al., 1999).

In waste management systems, many cold-tolerant microorganisms are used due to the characteristic features of their enzymes. Usually, for these systems, a different kind of broad enzyme complexes are used. Because, most of the industrial (dye industry, metal smelting industrial waste, mining activities, paper and pulp industry, petrochemical industrial waste, chemical weapon-producing industry, and agricultural wastes) and domestic sewages are with many different kinds of chemicals, enzymes, toxins, heavy metals, and protein-rich wastes, which cause hazardous damages to both environment and humans (Furhan, 2020; Giovanella et al., 2020; Gupta et al., 2020; Jin et al., 2019; Maiangwa et al., 2015; Sivaperumal et al., 2017). Therefore, many combinations of enzymes are required to treat the waste management system as it presents different kind of chemicals such as phenols, nitriles, aromatic amines, chlorinated compounds, dodecane, hexadecane, naphthalene, toluene, radionuclides (uranium, technetium, and cobalt), biopolymers (cellulase, chitin, proteins, lignin, and triacylglycerols), and keratinous waste (from leather and poultry industries) like hazardous materials and other microbial enzymes in industrial waste. During the degradation, those toxic materials become nontoxic, nonharmful substances and hydrolyze products (Al-Ghanayem and Joseph, 2020; Demirjian et al., 2001; Dempsey, 2017; Furhan, 2020; Garcia-Descalzo et al., 2012; Gupta et al., 2020; Maiangwa et al., 2015; Mishra et al., 2020; Sivaperumal et al., 2017). Some treated waste products are also used to produce biofuel as alternative energy sources (Garcia-Descalzo et al., 2012; Gupta et al., 2020; Jin et al., 2019). Most of the studies have reported commonly used cold-adapted enzymes such as amyloglucosidases, amidases, amylases, cellulases, glucoamylases, pectinases, proteases, lipases, lignin peroxidase, manganese peroxidase, tyrosinase, and laccase in waste management systems. The majority of those enzymes are extracted from the microorganisms living in different kind of cold-adapted environments and habitats like deep seas, Antarctic and Arctic Ocean sediments, soils from cold mountains, and soil from Antarctica (Dempsey, 2017; Furhan, 2020; Giovanella et al., 2020; Gupta et al., 2020; Jin et al., 2019; Kavitha, 2016; Kuddus et al., 2013; Sivaperumal et al., 2017). In addition, municipal water treatments of temperate countries are also conducted using the psychrophilic microorganisms. During this process, organic biomass is converted into microbial biomass. This is one of the largest applications of cold-adapted microbes in the biotechnology industry (Dempsey, 2017).

7.4.3 Agricultural aspects

One of the most limiting factors of using available biofertilizers in high-altitude agriculture is the inability to colonize plant roots due to reduced metabolite activities under low temperatures (Rawat et al., 2020). Hence, the application of cold-tolerant microbes in agriculture shows much prominence in the promotion of plant growth either as plant growth-promoting rhizobacteria (PGPR) or as other microbes facilitating availability of vital plant nutrients such as phosphates and providing protection through defense molecules. While the conventional technologies have unearthed cold-tolerant (psychrotrophic) and cold-loving (psychrophilic) microbes at a slower phase, the new combined omics technologies have led to the identification of many candidates that can be used as PGP microbes in cold climates.

7.4.3.1 Plant growth-promoting microbes

Plant growth-promoting (PGP) microbes exhibit many different mechanisms that directly or indirectly contribute to growth and development of plants. However, in cold climates, the activities of PGP microbes may become limited due to the inhibitory effect of the low temperatures. Although genetic engineering offers an alternative where cold-tolerant or cold-loving genes can be transferred to PGP for enhanced activities, the multigene nature of these characters and the concerns of releasing GMOs to the natural environment make it an unviable option. Hence, identification and isolation of psychrophilic and psychrotrophic or psychrotolerant microbes offer a more realistic avenue of improving plant growth through biofertilizers. Although most of the studied PGP microbes are bacteria, there have been several cases of psychrophilic fungi which include both endophytic fungal psychrophiles (Acuña-Rodríguez et al., 2019) and psychrophilic yeasts (Tapia-Vázquez et al., 2020).

7.4.3.2 Isolation of psychrophilic PGP microbes

Isolation of psychrophiles from extreme environments depends mainly on the techniques and the controls used. For instance, every tool used in the isolation such as pipettes, and culture vessels, may require prior cold treatment as these microbes are sensitive to temperature changes. Although they may survive certain temperatures, their rapid growth can only be expected at a particular temperature (Moyer and Morita, 2007). The lowest temperature survival has been shown by *Planococcus halocryophilus* at − 15°C. However, it grows best at 25°C (Mykytczuk et al., 2013). Henceforth, it is very difficult to figure out the ecology of a psychrophile simply because it has been isolated at a lower temperature. While media enrichment is a common method of isolating psychrophiles, there is an argument that since psychrophiles represent all taxa, it may not be fruitful in a larger scale (Bowman, 2001). However, with the development of culture-independent metagenomic sequencing methods, a broader selection of psychrophiles has been possible. Once such strains are detected by these techniques, strategies can be adopted to isolate them for further work. Hence, in the current workflow for psychrophilic PGP, identification may precede the isolation.

7.4.3.3 Identification of psychrophilic PGP microbes

Arguably, as discussed in the previous section, it may become advantageous to start the whole workflow with the identification of possible psychrophilic PGPs using culture-independent metagenomic techniques such as 16s rDNA sequencing for bacteria (Soni and Goel, 2010) and ITS sequencing for fungi (Schoch et al., 2012).

7.4.3.4 Characterization of psychrophilic PGP microbes

Once identification and isolation are conducted, the next step would be to characterize the microbes for selection as PGPs. Conventionally, this could be carried out by screening for plant growth-promoting traits such as solubilization of phosphates and production of siderophores IAA and other PGP molecules (Govarthanan et al., 2020). However, whole-genome sequencing (WGS), RNA sequencing (RNA-Seq), proteomics, and metabolomic fingerprinting have opened a whole new era for the selection of microbes based on their traits. While WGS can identify traditionally expected PGP traits, it can also identify novel characters that may contribute to plant growth promotion. Moreover, WGS captures the entire arsenal of PGP traits than a few traits captured by conventional screening. Although WGS gives the entire playing field, other techniques are vital for the selection of microbes for enhancing plant growth as biofertilizer as they may give the activities at the functional level. This becomes critical as some organisms are known to be dominant in samples, yet functionally, they may not be very active (White et al., 2016).

7.4.3.5 Selection of psychrophilic PGP microbes

Selection of psychrophiles for PGP depends not only on their functional traits but also on the capability of establishing in the given geographical area, which is more ecological. While some microbes may be recognized as functionally very active, they can hardly contribute as biofertilizers, if they cannot establish and persist. Hence, when traits for selection are considered, it is vital to look at both functional and ecological aspects. Moreover, a culture that does well when grown on a laboratory medium may not perform the same under the given environmental condition losing out in the natural competition.

7.4.3.6 Formulation of psychrophilic microbes as biofertilizers

Formulation of a strain or strains of PGP microbes is the final step in the biofertilizer production. Usually, formulation is geared toward maintaining the survival of the microbes within the product keeping in mind the storage period before application. While it is difficult to produce an excellent formulation for microbes, it will certainly be difficult to do it for psychrophiles. However, many psychrophiles are known to withstand temperatures near 20°C (Gounot, 1986). Hence, the challenge to formulate them will not be that difficult to overcome. The art of formulating a mixture of microbes, including bacteria and fungi, may represent a bigger challenge. Moreover, formulating plant growth-promoting fungi, especially arbuscular mycorrhizae (AM), could be extremely difficult as they are strictly symbiotic. Hence, they require carrier material that allows them to survive in the formulation, and these include the use of colonized seeds or roots with material such as compost or peat (Malusá et al., 2012).

7.5 Future prospects

Cold-tolerant and cold-loving microorganisms provide a substantial contribution to various industries. Engineered enzymes with increased efficacy would allow industries to maximize resource-use efficiency leading to profit maximization. The use of cold-tolerant and cold-loving microorganisms with PGP ability is crucial for future agriculture in cold climates to augment food production in order to meet global food demand to feed increasing world population. These microbes produced in formulations would ensure their activity when they are applied to plants under cold conditions or in cold climates.

References

Achberger, A.M., Michaud, A.B., Vick-Majors, T.J., Christner, B.C., Skidmore, M.L., Priscu, J.C., Tranter, M., 2017. Microbiology of subglacial environments. In: Margesin, R. (Ed.), Psychrophiles: From Biodiversity to Biotechnology. Springer, Cham, Switzerland.

Acuña-Rodríguez, I.S., Hansen, H., Gallardo-Cerda, J., Atala, C., Molina-Montenegro, M.A., 2019. Antarctic extremophiles: biotechnological alternative to crop productivity in saline soils. Front. Bioeng. Biotechnol. 7, 22.

Al-Ghanayem, A.A., Joseph, B., 2020. Current prospective in using cold-active enzymes as eco-friendly detergent additive. Appl. Microbiol. Biotechnol. 104, 2871–2882.

Alias, N., Mazian, A., Salleh, A.B., Basri, M., Rahman, R.N.Z.R.A., 2014. Molecular cloning and optimization for high level expression of cold-adapted serine protease from Antarctic yeast *Glaciozyma antarctica* Pi12. Enzyme Res. 2014, 197938.

Alquati, C., De Gioia, L., Santarossa, G., Alberghina, L., Fantucci, P., Lotti, M., 2002. The cold-active lipase of *Pseudomonas fragi*: heterologous expression, biochemical characterization and molecular modeling. Eur. J. Biochem. 269, 3321–3328.

Andrews, J.H., Harris, R.F., 2000. The ecology and biogeography of microorganisms on plant surfaces. Annu. Rev. Phytopathol. 38, 145–180.

Antranikian, G., Vorgias, C.E., Bertoldo, C., 2005. Extreme environments as a resource for microorganisms and novel biocatalysts. In: Ulber, R., Le Gal, Y. (Eds.), Marine Biotechnology I: Advances in Biochemical Engineering/Biotechnology. Springer, Berlin, Heidelberg.

Arpigny, J.L., Feller, G., Gerday, C., 1993. Cloning, sequence and structural features of a lipase from the antarctic facultative psychrophile *Psychrobacter immobilis* B10. Biochim. Biophys. Acta 1171, 331–333.

Bajerski, F., Wagner, D., Mangelsdorf, K., 2017. Cell membrane fatty acid composition of *Chryseobacterium frigidisoli* PB4T, isolated from Antarctic glacier forefield soils, in response to changing temperature and pH conditions. Front. Microbiol. 8, 677.

Berry, E.D., Foegeding, P.M., 1997. Cold temperature adaptation and growth of microorganisms. J. Food Prot. 60, 1583–1594.

Bhattacharya, S., 2018. Cryoprotectants and their usage in cryopreservation process. In: Cryopreservation Biotechnology in Biomedical and Biological Sciences. Intechopen.

Białkowska, A.M., Cieśliński, H., Nowakowska, K.M., Kur, J., Turkiewicz, M., 2009. A new β-galactosidase with a low temperature optimum isolated from the Antarctic *Arthrobacter* sp. 20B: gene cloning, purification and characterization. Arch. Microbiol. 191, 825-835.

Birgisson, H., Delgado, O., Arroyo, L.G., Hatti-Kaul, R., Mattiasson, B., 2003. Cold-adapted yeasts as producers of cold-active polygalacturonases. Extremophiles 7, 185-193.

Bisht, S.C., 2011. Cold active proteins in food and pharmaceutical industry. Biotech Article.

Bowman, J., 2001. Methods for psychrophilic bacteria. Methods Microbiol. 30, 591-614.

Brakstad, O.G., Lofthus, S., Ribicic, D., Netzer, R., 2017. Biodegradation of petroleum oil in cold marine environments. In: Margesin, R. (Ed.), Psychrophiles: From Biodiversity to Biotechnology. Springer International Publishing, Cham.

Cavicchioli, R., Charlton, T., Ertan, H., Omar, S.M., Siddiqui, K., Williams, T., 2011. Biotechnological uses of enzymes from psychrophiles. Microb. Biotechnol. 4, 449-460.

Chattopadhyay, M., 2002. Bacterial cryoprotectants. Resonance 7, 59-63.

Chen, M., Li, H., Chen, W., Diao, W., Liu, C., Yuan, M., Li, X., 2013. Isolation, identification and characterization of 68 protease-producing bacterial strains from the Arctic. Wei Sheng Wu Xue Bao = Acta Microbiol. Sin. 53, 702-709.

Chintalapati, S., Kiran, M., Shivaji, S., 2004. Role of membrane lipid fatty acids in cold adaptation. Cell. Mol. Biol. 50, 631-642.

Cho, K.H., 2017. The structure and function of the gram-positive bacterial RNA degradosome. Front. Microbiol. 8, 00154.

Choo, D.-W., Kurihara, T., Suzuki, T., Soda, K., Esaki, N., 1998. A cold-adapted lipase of an Alaskan psychrotroph, *Pseudomonas* sp. strain B11-1: gene cloning and enzyme purification and characterization. Appl. Environ. Microbiol. 64, 486-491.

Czapski, T.R., Trun, N., 2014. Expression of csp genes in *E. coli* K-12 in defined rich and defined minimal media during normal growth, and after cold-shock. Gene 547, 91-97.

Dalsgaard, P.W., Larsen, T.O., Christophersen, C., 2005. Bioactive cyclic peptides from the psychrotolerant fungus *Penicillium algidum*. J. Antibiot. 58, 141-144.

Davail, S., Feller, G., Narinx, E., Gerday, C., 1994. Cold adaptation of proteins. Purification, characterization, and sequence of the heat-labile subtilisin from the antarctic psychrophile *Bacillus* TA41. J. Biol. Chem. 269, 17448-17453.

De Maayer, P., Anderson, D., Cary, C., Cowan, D.A., 2014. Some like it cold: understanding the survival strategies of psychrophiles. EMBO Rep. 15, 508-517.

De Oliveira, T.B., De Lucas, R.C., Scarcella, A.S.D.A., Pasin, T.M., Contato, A.G., Polizeli, M.D.L.T.D.M., 2020. Cold-active lytic enzymes and their applicability in the biocontrol of postharvest fungal pathogens. J. Agric. Food Chem. 68, 6461-6463.

Demirjian, D.C., Morís-Varas, F., Cassidy, C.S., 2001. Enzymes from extremophiles. Curr. Opin. Chem. Biol. 5, 144-151.

Dempsey, M.J., 2017. Nitrification at low temperature for purification of used water. In: Margesin, R. (Ed.), Psychrophiles: From Biodiversity to Biotechnology. Springer International Publishing, Cham.

Dhaulaniya, A.S., Balan, B., Agrawal, P.K., Singh, D.K., 2019. Cold survival strategies for bacteria, recent advancement and potential industrial applications. Arch. Microbiol. 201, 1-16.

Duman, J.G., Olsen, T.M., 1993. Thermal hysteresis protein activity in bacteria, fungi, and phylogenetically diverse plants. Cryobiology 30, 322-328.

Etchegaray, J.P., Jones, P.G., Inouye, M., 1996. Differential thermoregulation of two highly homologous cold-shock genes, cspA and cspB, of *Escherichia coli*. Genes Cells 1, 171-178.

Feller, G., 2013. Psychrophilic enzymes: from folding to function and biotechnology. Scientifica 2013, 512840.

Feller, G., Thiry, M., Arpigny, J.L., Gerday, C., 1991. Cloning and expression in *Escherichia coli* of three lipase-encoding genes from the psychrotrophic antarctic strain Moraxella TA144. Gene 102, 111–115.

Furhan, J., 2020. Adaptation, production, and biotechnological potential of cold-adapted proteases from psychrophiles and psychrotrophs: recent overview. J. Genet. Eng. Biotechnol. 18, 1–13.

Garcia-Descalzo, L., Alcazar, A., Baquero, F., Cid, C., 2012. Biotechnological applications of cold-adapted bacteria. In: Extremophiles. John Wiley & Sons, Inc., Hoboken, NJ.

Garnham, C.P., Gilbert, J.A., Hartman, C.P., Campbell, R.L., Laybourn-Parry, J., Davies, P.L., 2008. A Ca^{2+}-dependent bacterial antifreeze protein domain has a novel β-helical ice-binding fold. Biochem. J. 411, 171–180.

Gilbert, J.A., Hill, P.J., Dodd, C.E., Laybourn-Parry, J., 2004. Demonstration of antifreeze protein activity in Antarctic lake bacteria. Microbiology 150, 171–180.

Giovanella, P., Vieira, G.A., Otero, I.V.R., Pellizzer, E.P., De Jesus Fontes, B., Sette, L.D., 2020. Metal and organic pollutants bioremediation by extremophile microorganisms. J. Hazard. Mater. 382, 121024.

Goldstein, J., Pollitt, N.S., Inouye, M., 1990. Major cold shock protein of *Escherichia coli*. Proc. Natl. Acad. Sci. 87, 283–287.

Gounot, A.-M., 1986. Psychrophilic and psychrotrophic microorganisms. Experientia 42, 1192–1197.

Govarthanan, M., Ameen, F., Kamala-Kannan, S., Selvankumar, T., Almansob, A., Alwakeel, S., Kim, W., 2020. Rapid biodegradation of chlorpyrifos by plant growth-promoting psychrophilic *Shewanella* sp. BT05: an eco-friendly approach to clean up pesticide-contaminated environment. Chemosphere 247, 125948.

Gualerzi, C.O., Giuliodori, A.M., Pon, C.L., 2003. Transcriptional and post-transcriptional control of cold-shock genes. J. Mol. Biol. 331, 527–539.

Gupta, S.K., Kataki, S., Chatterjee, S., Prasad, R.K., Datta, S., Vairale, M.G., Sharma, S., Dwivedi, S.K., Gupta, D.K., 2020. Cold adaptation in bacteria with special focus on cellulase production and its potential application. J. Clean. Prod. 258, 120351.

Hamdan, A., 2018. Psychrophiles: ecological significance and potential industrial application. S. Afr. J. Sci. 114, 1–6.

Hassan, N., Rafiq, M., Hayat, M., Shah, A.A., Hasan, F., 2016. Psychrophilic and psychrotrophic fungi: a comprehensive review. Rev. Environ. Sci. Biotechnol. 15, 147–172.

Hassan, N., Anesio, A.M., Rafiq, M., Holtvoeth, J., Bull, I., Haleem, A., Shah, A.A., Hasan, F., 2020. Temperature driven membrane lipid adaptation in glacial psychrophilic bacteria. Front. Microbiol. 11, 00824.

Hébraud, M., Potier, P., 1999. Cold shock response and low temperature adaptation in psychrotrophic bacteria. J. Mol. Microbiol. Biotechnol. 1, 211–219.

Hoshino, T., Kiriaki, M., Ohgiya, S., Fujiwara, M., Kondo, H., Nishimiya, Y., Yumoto, I., Tsuda, S., 2003. Antifreeze proteins from snow mold fungi. Can. J. Bot. 81, 1175–1181.

Jadhav, V.V., Jamle, M.M., Pawar, P.D., Devare, M.N., Bhadekar, R.K., 2010. Fatty acid profiles of PUFA producing Antarctic bacteria: correlation with RAPD analysis. Ann. Microbiol. 60, 693–699.

Jagannadham, M.V., Chattopadhyay, M.K., Subbalakshmi, C., Vairamani, M., Narayanan, K., Rao, C.M., Shivaji, S., 2000. Carotenoids of an Antarctic psychrotolerant bacterium, *Sphingobacterium antarcticus*, and a mesophilic bacterium, *Sphingobacterium multivorum*. Arch. Microbiol. 173, 418–424.

Jiang, H., Zhang, S., Gao, H., Hu, N., 2016. Characterization of a cold-active esterase from *Serratia* sp. and improvement of thermostability by directed evolution. BMC Biotechnol. 16, 1–11.

Jin, B., Jeong, K.-W., Kim, Y., 2014. Structure and flexibility of the thermophilic cold-shock protein of *Thermus aquaticus*. Biochem. Biophys. Res. Commun. 451, 402–407.

Jin, M., Gai, Y., Guo, X., Hou, Y., Zeng, R., 2019. Properties and applications of extremozymes from deep-sea extremophilic microorganisms: a mini review. Mar. Drugs 17, 656.

Jones, P.G., Inouye, M., 1994. The cold-shock response—a hot topic. Mol. Microbiol. 11, 811–818.

Jones, P.G., Cashel, M., Glaser, G., Neidhardt, F., 1992. Function of a relaxed-like state following temperature downshifts in *Escherichia coli*. J. Bacteriol. 174, 3903–3914.

Kandror, O., Deleon, A., Goldberg, A.L., 2002. Trehalose synthesis is induced upon exposure of *Escherichia coli* to cold and is essential for viability at low temperatures. Proc. Natl. Acad. Sci. 99, 9727–9732.

Kavitha, M., 2016. Cold active lipases—an update. Front. Life Sci. 9, 226–238.

Kawahara, H., Koda, N., Oshio, M., Obata, H., 2000. A cold acclimation protein with refolding activity on frozen denatured enzymes. Biosci. Biotechnol. Biochem. 64, 2668–2674.

Keto-Timonen, R., Hietala, N., Palonen, E., Hakakorpi, A., Lindström, M., Korkeala, H., 2016. Cold shock proteins: a minireview with special emphasis on Csp-family of enteropathogenic *Yersinia*. Front. Microbiol. 7, 1151.

Kim, Y., Wang, X., Zhang, X.S., Grigoriu, S., Page, R., Peti, W., Wood, T.K., 2010. *Escherichia coli* toxin/antitoxin pair MqsR/MqsA regulate toxin CspD. Environ. Microbiol. 12, 1105–1121.

Kuddus, M., Singh, P., Thomas, G., Al-Hazimi, A., 2013. Recent developments in production and biotechnological applications of C-phycocyanin. Biomed. Res. Int. 2013, 742859.

Lee, J., Jeong, K.-W., Jin, B., Ryu, K.-S., Kim, E.-H., Ahn, J.-H., Kim, Y., 2013. Structural and dynamic features of cold-shock proteins of *Listeria monocytogenes*, a psychrophilic bacterium. Biochemistry 52, 2492–2504.

Maiangwa, J., Ali, M.S.M., Salleh, A.B., Abd Rahman, R.N.Z.R., Shariff, F.M., Leow, T.C., 2015. Adaptational properties and applications of cold-active lipases from psychrophilic bacteria. Extremophiles 19, 235–247.

Malusá, E., Sas-Paszt, L., Ciesielska, J., 2012. Technologies for beneficial microorganisms inocula used as biofertilizers. Sci. World J. 2012, 491206.

Mansilla, M.C., Cybulski, L.E., Albanesi, D., De Mendoza, D., 2004. Control of membrane lipid fluidity by molecular thermosensors. J. Bacteriol. 186, 6681–6688.

Margesin, R., Neuner, G., Storey, K.B., 2007. Cold-loving microbes, plants, and animals—fundamental and applied aspects. Naturwissenschaften 94, 77–99.

McRae, C.F., Seppelt, R., 1999. Filamentous fungi of the Windmill Islands, continental Antarctica. Effect of water content in moss turves on fungal diversity. Polar Biol. 22, 389–394.

Mishra, P.K., Joshi, P., Bisht, S.C., Bisht, J.K., Selvakumar, G., 2010. Cold-tolerant agriculturally important microorganisms. In: Mageswari, D. (Ed.), Plant Growth and Health Promoting Bacteria. Springer-Verlag, Berlin.

Mishra, P.K., Joshi, S., Gangola, S., Khati, P., Bisht, J., Pattanayak, A., 2020. Psychrotolerant microbes: characterization, conservation, strain improvements, mass production, and commercialization. In: Goel, R., Soni, R., Suyal, D.C. (Eds.), Microbiological Advancements for Higher Altitude Agro-Ecosystems & Sustainability. Springer, Singapore.

Morita, R.Y., 2001. Psychrophiles and psychrotrophs. In: Encyclopedia of Life Sciences. Wiley, Chichester, UK.

Moyer, C.L., Morita, R.Y., 2007. Psychrophiles and psychrotrophs. In: Encyclopedia of Life Sciences. vol. 1. John Wiley & Sons Ltd, Chichester, UK.

Muñoz, P.A., Márquez, S.L., González-Nilo, F.D., Márquez-Miranda, V., Blamey, J.M., 2017. Structure and application of antifreeze proteins from Antarctic bacteria. Microb. Cell Fact. 16, 1–13.

Mykytczuk, N.C., Foote, S.J., Omelon, C.R., Southam, G., Greer, C.W., Whyte, L.G., 2013. Bacterial growth at −15°C; molecular insights from the permafrost bacterium *Planococcus halocryophilus* Or1. ISME J. 7, 1211–1226.

Nakagawa, T., Nagaoka, T., Taniguchi, S., Miyaji, T., Tomizuka, N., 2004. Isolation and characterization of psychrophilic yeasts producing cold-adapted pectinolytic enzymes. Lett. Appl. Microbiol. 38, 383–387.

Nakashima, K., Kanamaru, K., Mizuno, T., Horikoshi, K., 1996. A novel member of the cspA family of genes that is induced by cold shock in *Escherichia coli*. J. Bacteriol. 178, 2994–2997.

Nam, E., Ahn, J., 2011. Antarctic marine bacterium *Pseudoalteromonas* sp. KNOUC808 as a source of cold-adapted lactose hydrolyzing enzyme. Braz. J. Microbiol. 42, 927–936.

Nichols, D., Bowman, J., Sanderson, K., Nichols, C.M., Lewis, T., Mcmeekin, T., Nichols, P.D., 1999. Developments with Antarctic microorganisms: culture collections, bioactivity screening, taxonomy, PUFA production and cold-adapted enzymes. Curr. Opin. Biotechnol. 10, 240–246.

Phadtare, S., 2004. Recent developments in bacterial cold-shock response. Curr. Issues Mol. Biol. 6, 125–136.

Phadtare, S., Inouye, M., 2001. Role of CspC and CspE in regulation of expression of RpoS and UspA, the stress response proteins in *Escherichia coli*. J. Bacteriol. 183, 1205–1214.

Piette, F., D'amico, S., Leprince, P., Feller, G., 2010. Life in cold: a proteomic study of cold-repressed proteins in the Antarctic bacterium *Pseudoalteromonas haloplanktis* TAC125. Appl. Environ. Microbiol. 77, 3881–3883.

Qin, Y., Huang, Z., Liu, Z., 2014. A novel cold-active and salt-tolerant α-amylase from marine bacterium *Zunongwangia profunda*: molecular cloning, heterologous expression and biochemical characterization. Extremophiles 18, 271–281.

Rawat, J., Yadav, N., Pande, V., 2020. Role of rhizospheric microbial diversity in plant growth promotion in maintaining the sustainable agrosystem at high altitude regions. In: Mandal, S.D., Bhatt, P. (Eds.), Recent Advancements in Microbial Diversity. Elsevier.

Sarmiento, F., Peralta, R., Blamey, J.M., 2015. Cold and hot extremozymes: industrial relevance and current trends. Front. Bioeng. Biotechnol. 3, 148.

Schoch, C.L., Seifert, K.A., Huhndorf, S., Robert, V., Spouge, J.L., Levesque, C.A., Chen, W., Fungal Barcoding Consortium, 2012. Nuclear ribosomal internal transcribed spacer (ITS) region as a universal DNA barcode marker for fungi. Proc. Natl. Acad. Sci. 109, 6241–6246.

Shenhar, Y., Biran, D., Ron, E.Z., 2012. Resistance to environmental stress requires the RNA chaperones CspC and CspE. Environ. Microbiol. Rep. 4, 532–539.

Shi, Y., Ding, Y., Yang, S., 2015. Cold signal transduction and its interplay with phytohormones during cold acclimation. Plant Cell Physiol. 56, 7–15.

Shivaji, S., Prakash, J.S., 2010. How do bacteria sense and respond to low temperature? Arch. Microbiol. 192, 85–95.

Shivaji, S., Kiran, M., Chintalapati, S., 2007. Perception and transduction of low temperature in bacteria. In: Gerday, C., Glansdordd, N. (Eds.), Physiology and Biochemistry of Extremophiles. AMS Press, Washington.

Siliakus, M.F., Van Der Oost, J., Kengen, S.W., 2017. Adaptations of archaeal and bacterial membranes to variations in temperature, pH and pressure. Extremophiles 21, 651–670.

Singh, P., Hanada, Y., Singh, S.M., Tsuda, S., 2014. Antifreeze protein activity in Arctic cryoconite bacteria. FEMS Microbiol. Lett. 351, 14–22.

Sivaperumal, P., Kamala, K., Rajaram, R., 2017. Bioremediation of industrial waste through enzyme producing marine microorganisms. In: Da Cruz, A.G., Prudencio, E.S., Esmerino, E.A., Da Silva, M.C. (Eds.), Advances in Food and Nutrition Research. Elsevier.

Snider, C.S., Hsiang, T., Zhao, G., Griffith, M., 2000. Role of ice nucleation and antifreeze activities in pathogenesis and growth of snow molds. Phytopathology 90, 354–361.

Söderholm, H., Jaakkola, K., Somervuo, P., Laine, P., Auvinen, P., Paulin, L., Lindström, M., Korkeala, H., 2013. Comparison of *Clostridium botulinum* genomes shows the absence of cold shock protein coding genes in type E neurotoxin producing strains. Botulinum J. 2, 189–207.

Soni, R., Goel, R., 2010. Triphasic approach for assessment of bacterial population in different soil systems. Ekologija 56, 99–104.

Subramanian, P., Joe, M.M., Yim, W.-J., Hong, B.-H., Tipayno, S.C., Saravanan, V.S., Yoo, J.-H., Chung, J.-B., Sultana, T., Sa, T.-M., 2011. Psychrotolerance mechanisms in cold-adapted bacteria and their perspectives as plant growth-promoting bacteria in temperate agriculture. Korean J. Soil Sci. Fert. 44, 625–636.

Tapia-Vázquez, I., Sánchez-Cruz, R., Arroyo-Domínguez, M., Lira-Ruan, V., Sánchez-Reyes, A., Del Rayo Sánchez-Carbente, M., Padilla-Chacón, D., Batista-García, R.A., Folch-Mallol, J.L., 2020. Isolation and characterization of psychrophilic and psychrotolerant plant-growth promoting microorganisms from a high-altitude volcano crater in Mexico. Microbiol. Res. 232, 126394.

Turk, M., Plemenitaš, A., Gunde-Cimerman, N., 2011. Extremophilic yeasts: plasma-membrane fluidity as determinant of stress tolerance. Fungal Biol. 115, 950–958.

Van De Voorde, I., Goiris, K., Syryn, E., Van Den Bussche, C., Aerts, G., 2014. Evaluation of the cold-active *Pseudoalteromonas haloplanktis* β-galactosidase enzyme for lactose hydrolysis in whey permeate as primary step of d-tagatose production. Process Biochem. 49, 2134–2140.

Vanbogelen, R.A., Neidhardt, F.C., 1990. Ribosomes as sensors of heat and cold shock in *Escherichia coli*. Proc. Natl. Acad. Sci. 87, 5589–5593.

Wang, N., Yamanaka, K., Inouye, M., 1999. CspI, the ninth member of the CspA family of *Escherichia coli*, is induced upon cold shock. J. Bacteriol. 181, 1603–1609.

Wang, S.-Y., Hu, W., Lin, X.-Y., Wu, Z.-H., Li, Y.-Z., 2012. A novel cold-active xylanase from the cellulolytic myxobacterium *Sorangium cellulosum* So9733-1: gene cloning, expression, and enzymatic characterization. Appl. Microbiol. Biotechnol. 93, 1503–1512.

Weinstein, R.N., Montiel, P.O., Johnstone, K., 2000. Influence of growth temperature on lipid and soluble carbohydrate synthesis by fungi isolated from fellfield soil in the maritime Antarctic. Mycologia 92, 222–229.

White III, R.A., Bottos, E.M., Roy Chowdhury, T., Zucker, J.D., Brislawn, C.J., Nicora, C.D., Fansler, S.J., Glaesemann, K.R., Glass, K., Jansson, J.K., 2016. Molecule long read sequencing facilitates assembly and genomic binning from complex soil metagenomes. Msystems 1, e00045-16.

Wintrode, P.L., Miyazaki, K., Arnold, F.H., 2000. Cold adaptation of a mesophilic subtilisin-like protease by laboratory evolution. J. Biol. Chem. 275, 31635–31640.

Yamanaka, K., Inouye, M., 2001a. Induction of CspA, an *E. coli* major cold-shock protein, upon nutritional upshift at 37°C. Genes Cells 6, 279–290.

Yamanaka, K., Inouye, M., 2001b. Selective mRNA degradation by polynucleotide phosphorylase in cold shock adaptation in *Escherichia coli*. J. Bacteriol. 183, 2808–2816.

Yamashita, Y., Nakamura, N., Omiya, K., Nishikawa, J., Kawahara, H., Obata, H., 2002. Identification of an antifreeze lipoprotein from *Moraxella* sp. of Antarctic origin. Biosci. Biotechnol. Biochem. 66, 239–247.

Yang, J., Cai, N., Zhai, H., Zhang, J., Zhu, Y., Zhang, L., 2016. Natural zwitterionic betaine enables cells to survive ultrarapid cryopreservation. Sci. Rep. 6, 1–9.

Zhang, W., Shi, L., 2005. Distribution and evolution of multiple-step phosphorelay in prokaryotes: lateral domain recruitment involved in the formation of hybrid-type histidine kinases. Microbiology 151, 2159–2173.

Zhang, Y., Burkhardt, D.H., Rouskin, S., Li, G.-W., Weissman, J.S., Gross, C.A., 2018. A stress response that monitors and regulates mRNA structure is central to cold shock adaptation. Mol. Cell 70, 274–286. e7.

Plant growth-promoting diazotrophs: Current research and advancements

Chanda Vikrant Berde[a], P. Veera Bramhachari[b], and Vikrant Balkrishna Berde[c]

[a]Marine Microbiology, School of Earth, Ocean and Atmospheric Sciences, Goa University, Taleigao Plateau, Goa, India, [b]Department of Biotechnology, Krishna University, Machilipatnam, Andhra Pradesh, India, [c]Department of Zoology, Arts Commerce and Science College, Lanja, Maharashtra, India

8.1 Introduction

Increasing demands for food supply run parallel with increasing population. To meet the demands of increasing food production, it is necessary to increase the agricultural yields. Application of biofertilizers in the form of diverse nitrogen-fixing microorganisms called diazotrophs will ensure the optimization of agricultural yields. The overall increase of crop plant growth is achieved through plant growth-promoting rhizobacteria (PGPR). The PGPR covers microorganisms such as IAA (indole acetic acid) producers, phosphate solubilizers, potassium mobilizers, etc. Biological nitrogen fixation carried out by the diazotrophic microorganisms contributes to more than 60% of the fixed nitrogen on our planet. Thus, isolating very efficient nitrogen-fixing microorganisms, studying the mechanisms involved in nitrogen fixation, using these microorganisms as formulations for agricultural use for not only the leguminous but also the nonleguminous crops, will help in achieving success in increasing the crop yield (Singh, 2018). Research over the last few years has been handling various approaches to extend nitrogen fixation to crops other than legumes, develop inoculums with diazotrophs for nitrogen-deficient soils, and application of nonleguminous diazotrophs (Saumare et al., 2020). Those studies seek to illustrate diazotrophic agronomic value for the improvement of soil fertility and crop production as a nonpolluting, cost-effective method. The biological fixation of nitrogen produces about 200 million tonnes (Graham, 1992; Peoples et al., 2009).

With agricultural practices such as mixed cropping patterns, the nitrogen fixed in the soil during one crop can be efficiently utilized by the next crop. Thus, mixed cropping with leguminous and nonleguminous crops such as the soybean-wheat system, or the next season crops in crop rotation, can maximally utilize the fixed nitrogen (Fustec et al., 2010). This chapter focuses on the recent advances in the diazotrophic research apropos to terrestrial as well as diazotrophs from the ocean, the research focused on nitrogen-fixing genes, the enzymes and proteins involved as well as the efforts taken to apply the nitrogen-fixing ability of diazotrophs to nonleguminous plants.

8.2 Nitrogen fixation in diazotrophs

Atmospheric nitrogen fixation by diazotrophs is carried out by the enzyme nitrogenase, which catalyzes the conversion of nitrogen to ammonia (Hoffman et al., 2014; Einsle and Rees, 2020). All diazotrophic microorganisms harbor this enzyme. It is made up of two metalloprotein components, both of which play a role in ammonia formation. Component I, also called dinitrogenase, i.e., molybdo-ferro-protein (Mo-Fe-protein), is 2.2×10^5 Da protein, which reduces nitrogen as well as several substrates such as acetylene, protons, cyanide, isocyanide, and azide, as reported in nonphototrophic bacteria *R. rubrum* (Munson and Burris, 1969). Component II, also called dinitrogenase reductase, is an electron-transfer Fe protein. Catalysis requires a reduction source and Mg-ATP. In a catalytic loop of single-electron transfer and Mg-ATP hydrolysis, two member proteins associate and dissociate. The active substrate binding and reduction site, involving electron transfer from the Fe protein to FeMo-Co, is provided by an iron-molybdenum cofactor. Alternative nitrogenases of the form –V and –Fe, where Mo from FeMo-Co is substituted with V or Fe, were discovered as well. The extension of nitrogenase substrates to include CO and CO_2 has been reported (Seefeldt et al., 2020).

Many diazotrophs have another hydrogenase enzyme that is involved with the elimination of the hydrogen that is formed as nitrogen is fixed. Hydrogenases reuse H_2 for ATP synthesis and increase N_2 fixation speed (Johansson et al., 1983). Hydrogenase also has an oxygen sensitivity as nitrogenase enzyme and contains four iron and four molecule-labile sulfur atoms. Hence, the removal or utilization of hydrogen is important, to prevent the inactivation of the nitrogenase enzyme. In some diazotrophs like *Azotobacter vinelandii* and *A. chroococcum*, nitrogen fixation and hydrogen production may be carried out by the same enzyme complex. Hydrogen evolved is quickly removed from the cells by diffusion.

Nitrogenase enzyme is also inhibited by the higher concentration of ammonia. Hence, ammonia formed during nitrogen fixation needs to be converted to organic nitrogen compounds, to protect the nitrogenase enzyme from getting inactivated. Aerobic diazotrophs, which lack special compartments for nitrogen fixation, demonstrate a high respiratory rate, which is an adaptation to prevent the oxygen from reaching the nitrogenase enzyme and inactivating it. In the case of cyanobacteria, reduction of nitrogen occurs in heterocysts, which are thick-walled cells (Haselkorn, 2003). The O_2-producing photo-system II, ribulose bisphosphate carboxylase, is missing and the photosynthetic biliproteins may be lacking or diminished. In nonheterocystous cyanobacterial species such as *Lyngbya*, the nitrogen fixation occurs in cells in the center of the colony where oxygen penetration is less. In leguminous plants, nitrogen fixation takes place in root nodules that contain leghemoglobin to regulate oxygen tension.

8.3 Terrestrial nitrogen-fixing diazotrophs

The need to satisfy the growing demand for food productivity requires isolation and the efficiency of diazotrophs. The efficient relationship between diazotrophs and the host plant is a major prerequisite for nitrogen fixation, given the phylogenic and ecological richness of diazotrophic bacteria and their hosts. Hye Jia et al. (2014) identified diazotrophs from 576 endophytic bacteria of leaves, stems, and roots, from 10 rice cultivars. Eighty-one percent of these isolates produced ammonia and were classified with the application of special *nif* gene primary set as the diazotrophic bacteria. Diazotroph species of *Bacillus, Penibacillus, Microbacteria,* and *Klebsiella* have been reported as belonging to the genes of *nif*H. This group focused on the ability to fix nitrogen and other properties of the diazotrophs, including the ability to produce auxins and siderophores, solubilize phosphate and induce fungal resistance in plants. Diazotrophic bacterial isolation from maize mucilage has been confirmed by Higdon et al. (2020a) with the potential to fix nitrogen and having other plant growth-promoting attributes. Three large groups of sequences referring to the nitrogen-fixing gene were included in the sequences of the isolates. The sequences were homologous to *nif* genes (*nifHDKENB*) in the Dos Santos model (Dos Santos et al., 2012; Higdon et al., 2020b). Half of the overall diazotrophic isolates revealed the *nif*H gene and 193 isolates among these were belonging to *Enterobacter, Klebsiella, Metakosakonia, Pseudomonas, Rahnella,* and *Raoultella* species (Higdon et al., 2020b).

Lateral transfer of chromosomal symbiosis islands was observed in species of the genus *Mesorhizobium*, bearing the ability to nodulate legumes like chickpeas (Laranjo et al., 2014). This trait helps in nodulating new hosts, which is beneficial for the development of bioinoculants with wider host ranges. Thus, understanding the mechanisms of adaptation to new hosts and the symbiosis between the gene bearers and the host plant will enable more effective development of mesorhizobium bioinoculants as biofertilizers.

Recent studies by Elhady et al. (2020) focused on the effect of nodule size on nitrogen-fixing efficiency of *Bradyrhizobium japonicum* and the host soybean crop. Smaller nodules were formed as a result of *P. penetrans* invasion of plant roots. It affected the nodule size as well as the number of bacteroids in the infested roots; however, the number of nodules was higher. Therefore, it can be seen that the successful establishment of the diazotroph in the host plant, and establishment of nitrogen-fixing mechanism in the host, is still affected by external factors, biotic as well as abiotic.

Numerous cyanobacteria contain pale and dense cells called heterocysts in *Anabaena, Nostoc,* and *Cylindrospermopsis* and are filamentous (Ogawa and Carr, 1969; Haselkorn, 2003; Willis et al., 2015; Aly and Andrews, 2016; Zulkefli and Hwang, 2020). These are the nitrogen fixation sites (Haselkorn, 2003; Videau et al., 2016). In the absence of available combined nitrogen, heterocysts are formed, as ammonia prevents the differentiation of heterocysts as well as inhibits the nitrogenase enzyme. Nitrogen fixation takes place in in-house structured cells under reduced conditions; cyanobacteria such as *Lyngbya, Oscillatoria, Plectonomas,* and in bacterium do not produce heterocyst. The nitrogen fixation occurs in root nodules produced by *Rhizobium* (in leguminous plants) and *Frankia* (in nonleguminous plants). The lichens (a symbiotic structure created by cyanobacteria and fungi) are the site of nitrogen fixation in lower community of microorganisms.

Diazotrophism is seen in endophytic microorganisms, especially endophytic bacteria; these intracell colonizers bring about nitrogen fixation in tissues of the plants and promote their growth. The endophytic diazotrophs can be exploited as biofertilizers for sustainable agriculture. A strain of *P. polymyxa* can colonize nonnative hosts and fix atmospheric nitrogen within, promoting plant growth. *Gluconacetobacter diazotrophicus* is another well-studied endophytic diazotroph isolated from sugarcane. It has the nitrogen-fixing ability as well as additional plant growth-promoting traits (Puri et al., 2017). Diazotrophs imparting drought stress resistance to the plants such as strains of genus *Herbaspirillum* offer great hope for agriculture in drought-prone areas (Aguiar et al., 2016).

8.4 Nitrogen fixation in the ocean

Diazotrophs were isolated from the rhizosphere and soil in general, but during recent years, the focus is shifted to endophytic and marine diazotrophic microorganisms. The occurrence of diazotrophs, both bacteria and archaea, has been detected in the Arctic Ocean and found to play a role in the conversion of atmospheric nitrogen to bioavailable ammonia in the marine ecosystem. Both symbiotic cyanobacterial nitrogen fixation and heterotrophic diazotrophs have been reported (von Friesen and Riemann, 2020). The research workers, however, point out some gaps such as the inability to sample larger regions and the nonavailability of enough quantitative data to come to specific conclusions. This calls for future and in-depth research on diazotroph distribution, composition, and their activity in pelagic and sea ice-associated environments of the Arctic Ocean.

Though previously, there were reports of studies being carried out measuring nitrogen fixation and denitrification at the seafloor and pelagic zone using stable isotope technique (Fan et al., 2015). The same group has also worked on the diversity, abundance, and activity of nitrogen-fixing and denitrifying microorganisms at three stations in the southern North Sea. Their genomic analysis studies indicated the presence of *nif*H genes in anaerobic sulfur/iron reducers and sulfate reducers. These results are concomitant with the reports of the discoveries of diazotrophic methanogenic archaea *Methanosarcina barkeri* by ^{15}N radiotracer technique (Murray and Zinder, 1984; Leigh Nitrogen, 2000) and *Methanococcus thermolithotrophicus* by acetylene reduction assay (ARA) technique (Belay et al., 1984; Leigh Nitrogen, 2000).

Noncyanobacterial diazotrophs or heterotrophic diazotrophs are distributed widely in marine waters, including the oxygenated zones; however, the mechanism of nitrogenase protection from oxygen is yet to be understood (Pedersen et al., 2018; Geisler et al., 2019). A major fraction of the aquatic biosphere such as eutrophic estuaries has high ambient nitrogen concentrations and oxidized aphotic water. The diazotrophs found in these zones are closely associated with bacterioplankton and are referred to as planktonic heterotrophic diazotrophs. The prime requirement for the colonization of the diazotrophs onto surfaces is the initial colonization by bacterioplankton (Pedersen et al., 2018). Putative diazotrophs appeared after 80 h of colonization initiation, after the plankton, following the colonization by bacterioplankton. The surfaces for the colonization of diazotrophs can be natural particles that act as the nitrogen fixation loci. The natural particles or aggregates comprise polysaccharides that offer a microenvironment with less oxygen for the activity of the nitrogenase enzyme. It was also pointed out by the authors that resuspension of sediment material can promote pelagic N_2-fixation. Thus, these heterotrophic diazotrophs are responsible for the nitrogen fixation taking place in the ocean waters.

Work on diazotrophs associated with the particulate matter related to their high concentrations of nitrogen fixation rates and the existence of *nif*H genes indicates that they are extremely unique and specific. This relationship was shown for the first time recently with a direct staining approach (Geisler et al., 2019). Earlier such research was conducted mostly through indirect relations and various methodological especially statistical approaches. This new staining technique incorporates fluorescent tagging of active diazotrophs by nitrogenase immunolabeling and Alcian blue or concanavalin-A polysaccharide stain. Nucleic acid staining was used for the total bacteria. This approach provides nitrogen fixing frequencies, bacterial activity, and specific location of heterotrophic diazotrophs on artificial and natural aggregates (Geisler et al., 2019).

Diazotrophs in the marine environment contribute to nitrogen fixation, but these studies have been focusing only a certain hotspots. There is a dearth of knowledge of diazotrophic activities in the open oceans. The reasons are very less volumes being sampled, fewer sampling sites as compared to the vastness of the oceans, less frequency of sampling, and the practical difficulties of having a good geographic coverage. These difficulties have hampered the studies on diazotrophic diversity and distribution; hence, getting global nitrogen budget becomes a failure or is inaccurate. A solution for these inadequacies requires leveraging high spatiotemporal resolution measurements, and failure to employ these has come in the way of measurement methods according to Benavides and Robidart (2020). Increasing the spatiotemporal resolution of diazotroph activity and diversity will provide more accurate quantifications of nitrogen fluxes in ocean waters. A very recent study based on the application of combined values from two established acetylene-based assays was used to study the nitrogen cycling in coral reefs (El-Khaled et al., 2020). This method makes possible studying two processes, i.e., nitrogen fixation and denitrification, simultaneously by analyzing the gases formed during the processes. Gas chromatography is used for ethylene and nitrous oxide analyses formed during nitrogen fixation and denitrification, respectively.

8.5 Genomic and transcriptomics of diazotrophs

Mahmud et al. (2020) in their review have elaborated the need for research focusing on transferring nitrogen-fixing mechanisms to nonlegumes, with emphasis on molecular techniques. This in turn necessitates the importance of genome and proteomic/transcriptomic studies. The last two decades has seen a slow and assuring increase in reports on work pertaining to these aspects of diazotrophic research. From diazotrophic rhizobial genome, numerous symbiotic genes (*nod* genes) encoding for nodulation, and nitrogen-fixing genes (*nif* genes), have been identified.

In symbiotic diazotrophs, especially *Rhizobium* sp., *nif* genes are located on a megaplasmid adjacent to *nod* genes. In the nonsymbiotic diazotrophs like cyanobacteria, these genes are localized on the chromosome. *Nif* genes comprises gene cluster of 24-Kb nucleotides, between the genes encoding for histidine (his) and shikimic acid (shi A). The cluster is organized in seven operons, i.e., transcription units (e.g., QB AL FM VSUX NE YKDH J). These operons transcribe for nitrogenase, Fe-protein, and Mo-Fe-protein. In nonsymbiotic diazotroph *Azospirillum* sp., m/HDK cluster megaplasmid and the sequence homologous to *nod* genes were reported (Acosta-Cruz et al., 2012). The presence of plasmids has also been described for other diazotrophs, including *Anabaena, Azotobacter*, and *Frankia* (Elmerich et al., 1987).

Diazotrophs also harbor hydrogen uptake (or *Hup*) genes that help in the removal of hydrogen formed during nitrogen fixation. The need for the removal of hydrogen is due to the reduced efficiency of nitrogenase in the presence of hydrogen. There are reports of genetically engineered diazotrophic strain by transferring the *Hup* genes of *R. leguminosarum* into *Rhizobium* strain (Lambert et al., 1985). This is the world's first report of interspecific transfer of *Hup* genes. The efficient transfer and expression of *Hup* genes has made it easier for the chick-pea-*Rhizobium* system to improve symbiotic energy efficiency. Transfer of nitrogen-fixing genes to nonleguminous plants is one of the strategies to overcome nitrogen deficiency and also to reduce the use of chemical fertilizers. The best way to improve nitrogen availability in crop plants would be to transfer the genes m/genes into chloroplasts; however, lack of chloroplast transferring techniques and protection of nitrogenase from O_2 evolved during photosynthesis are the drawbacks of the process that need to be addressed (Long, 1989; Báscones et al., 2000).

According to Gaby and Buckley (2011), the diazotrophs are a poorly described group and many more diazotrophic strains and nitrogen-fixing genomes need to be discovered and studied. The authors have reported 16,989 *nif*H sequences so far. Other sequences include nitrogenase genes *nif*D, *nif*K, *nif*E, *nif*N, etc., which make up the total of 32,954 sequences in the database. This database allowed for a comparative study of the symbiotic systems designed to identify core genetic networks that shape the root nodule and to define strategies to transfer the nitrogen-fixing capability of nonlegume crops (López-Torrejón et al., 2016; Wardhani et al., 2019; Mahmud et al., 2020; van Heerwaarden et al., 2018). In a recent study, homologous coding sequences for the *acdS* and *ipdC/ppdC* genes were identified for the diazotrophs grouped in Dos Santos Positive (DSP) (Higdon et al., 2020b). In the case of PQQ genes, approximately 28% of all isolates examined had homologous *pqq*BCDE sequences, whereas 12% had coding sequences equal to *pqq*F and 90% had *pqq*DH matches (Higdon et al., 2020a).

The simultaneous developments in transcriptomics of diazotrophs and the nitrogen-fixing mechanisms have opened up a new era in this field. Mergaert et al. (2003) reported the discovery of the NCR (module-specific cysteine-rich) peptides in nodules of *Medicago truncatula*. Using transcriptome analysis, it was found that these have a signal peptide with a conserved cysteine motif and 300 plus members have been discovered in galeloid legumes as well as other plants (Wojciechowsk et al., 2004; Kondorosi et al., 2013; Pan and Wang, 2017; Kereszt et al., 2018). Further characterization of these NCR peptides has been reported by some workers (Kereszt et al., 2018; Lindstrom and Mousavi, 2019).

8.6 Beneficial mechanisms other than N-fixation provided by diazotrophs

Apart from fixing atmospheric nitrogen for the plants, the diazotrophs have been reported to have multiple other abilities that add to overall growth-promoting attributes. Phosphate solubilization helps in making inorganic phosphorus available, potassium mobilization provides essential potassium for plant growth and functions, the production of plant growth hormones like IAA benefits the plant root and shoot development, protection against plant diseases is provided by the production of secondary metabolites, etc. These are some of the additional properties observed and studied in diazotrophic microorganisms (Unpublished data). Genome mining showed that isolates of all diazotrophic groups possessed marker genes for multiple mechanisms of direct plant growth promotion (PGP). These findings reveal a potential to confer the targeted PGP traits to the host organism and also revealed phenotypic variation among isolates. Diazotrophs belonging to *Rhizobia, Bradyrhizobia, Azotobacter, Azospirillum, Pseudomonas, Klebsiella*, and *Bacillus* genera that harbor the beneficial mechanisms in addition to nitrogen fixation, play a vital role in the overall growth and yield of crop plants.

There is a lacunae of information and research on the other beneficial properties present in diazotrophs forming legume nodules except for the knowledge about their nitrogen-fixing abilities. Table 8.1 summarizes some of the plant growth-promoting attributes reported in some diazotrophs in addition to nitrogen-fixing properties.

Table 8.1 Plant growth-promoting properties found in diazotrophs.

Properties	Diazotrophic microorganism	References
Acetic indole (IAA), hydrocyanic acid (HCN), antibiotics and/or mycolytic enzymes, organic acid, and siderophores	Rhizobia	Gopalakrishnan et al. (2015, 2018)
Inorganic phosphate solubilization, nitrogen fixation, IAA production	Mycorrhizae	Agnolucci et al. (2019)
ACC-deaminase activity and production of IAA, phosphorous solubilization	12 Diazotrophic strains, including *Sphingomonas azotifigens* (JN085438), *Pseudomonas putida* (JN222977), *Herbispirillum* sp. (JF990839)	Laskar et al. (2013)
IAA production, FePO$_4$ solubilization, AlPO$_4$ solubilization, siderophores production	91 Isolates belonging to the Proteobacteria phylum	Zuluaga et al. (2020)
Higher auxin-producing activity, high siderophore-producing activity, high phosphate-solubilizing activity	Two species of *Penibacillus*, three species of *Microbacterium*, three *Bacillus* species, and four species of *Klebsiella*	Hye Jia et al. (2014)

8.7 Application in global agriculture

A variety of diazotroph-based biofertilizers are available globally for agricultural applications. Table 8.2 summarizes the available diazotroph-containing biofertilizers used for different crop cultivars and the difference in yields obtained. Use of diazotrophic biofertilizers impacted the economy of several countries such as Brazil with an economic benefit in terms of N fertilizer saving over USDA 2.5 billion per year (Bruno et al., 2003). Countries like Canada, Germany, the United Kingdom, Spain, Italy, France China, Japan, Australia, New Zealand, India, and the rest of Asia produce large numbers of biofertilizers that are based on nitrogen-fixing bacteria and contributing to economic growth making around USD 0.284–0.45 billion dollars (Swarnalakshmi et al., 2016). China holds more than 511 biofertilizer products and accounted for 43.2% of the biofertilizer market share for the Asia-Pacific region in 2017 (Market Data Forecast, 2018).

As a result of the availability of biofertilizers, the nitrogen fixed by crops is around 55–60 million tons per year (Vitousek et al., 2013; Figueiredo et al., 2013; Rao and Balachandar, 2017) and the highest contributor is *Bradyrhizobium* species found in the legumes of soybean as microsymbiont (Hungria and Mendes, 2015; Gyoguu et al., 2018). Application of the biological nitrogen-fixing diazotrophs along

Table 8.2 Commercially available diazotrophic biofertilizers and their impact on agricultural yields.

Commercial biofertilizer	Microorganism used	Country	Benefits	Plant cultivar	References
Diazotroph formulated with perlite-biochar carriers	*Rhizobium leguminosarum* bv. *phaseoli* LCS0306	Spain	Increased yield	Common bean	Pastor-Bueis et al. (2019)
Biofix and Legumefix	Rhizobia inoculants	Ghana	Grain yield (12%–19%)	Soybean and cowpea	Ulzen et al. (2016)
Sympal and Legumefix	Rhizobia inoculants	Zambia	Rise in yields from 2000 to 4000 kg/ha	Soybean	Thuita et al. (2018), Mathenge et al. (2019)
Nitragin	Root-associated bacteria	Germany	Three- to fourfold increase in yield	Soybean	William and Akiko (2020)
Maize seeds coated with *Azospirillum*	*Azospirillum*	Mexico, Argentina	Yield increase	Maize	Reis (2007)
Microbin and Azottein	Phosphate solubilizing and nitrogen fixing	Egypt	Grain yield increased	Barley	El-Sayed et al. (2000)
BioGro	BNF	Southeast Asia	Replaced 23%–52% of N chemical fertilizers without the loss of yield	Rice	Rose et al. (2014)
—	Rhizobia	Northern Nigeria	Yield increase by 447 kg/ha	Soybean	Ronner et al. (2016)
Biofertilizer	*Azospirillum brasilense*, *Azotobacter chroccocum*, and *Trichoderma lignorum*		Contributed to 60% of the nitrogen	Sugarcane	Serna-Cock et al. (2011)
Biofertilizer	*Herbaspirillum seropedicae*, *Pseudomonas* sp., and *Bacillus megaterium*		Yield increase from 18% to 57.31%	Sugarcane	Antunes et al. (2019)

with vermicompost and other soil conditioners has been observed to further improve the soils, especially poor alkaline soils (Mathenge et al., 2019).

8.8 Future challenges in agriculture: Application of diazotrophs to nonlegumes

Research into the genetics of diazotrophic microorganisms has led to the transfer of the benefits of nitrogen fixation to nonleguminous plants such as wheat, rice, sorghum, or maize. One way of achieving this is by inducing symbiosis between the diazotrophs and the non-nitrogen-fixing plants, resulting in the development of root nodules (Mus et al., 2016; Burén and Rubio, 2018). The plant needs to produce and secrete nodulation signals for the initiation of nodulation by the diazotroph. Another approach is to introduce the nitrogen-fixing genes in the nonleguminous plant itself (Dent and Cocking, 2017; Vicente and Dean, 2017; Burén et al., 2018).

The difficulty of this biosynthesis and the enzyme's oxygen sensitivity are a major challenge for this strategy to be implemented. It is also uncertain if the power and energy required to support nitrogenase catalysis can be provided by the cereal host (van Velzen et al., 2018). There are reports of the successful transfer of *nif* genes to nondiazotrophs such as *E. coli, Saccharomyces cerevisiae, Paenibacillus* sp., tobacco plants (Dixon and Postgate, 1972; Han et al., 2015; Oldroyd and Dixon, 2014; Burén et al. 2017; Burén and Rubio, 2018), as well as research carried out on the nonlegumes, *Parasponi* legumes, and Actinorrhizae, which shows that such a transfer of genes leading to acquiring nitrogen-fixing ability is possible. Significant developments in developing strategies in this regard were reported mostly in transgenic plants, yeast, etc. Ivleva et al. (2016) reported the integration of *nif*H and *nif*M genes into the tobacco chloroplast genome, which could encode active Fe protein of nitrogenase in the transgenic plant. Burén and Rubio (2018) described the transfer of *nif* cassette of nine genes in transgenic yeasts (*Saccharomyces cerevisiae*), which is sufficient for nitrogen fixation. Another report shows that the transfer of the nitrogenase gene to mitochondria and root plastids in eukaryotes can result in nitrogenase expression under a low-oxygen environment (Wardhani et al., 2019; Ivleva et al., 2016).

According to Postgate (1992), the production of a "diazoplast" in plants would allow plants to resolve their genetic and physiological bacterial dependency for nitrogen fixation. The new organelle has diazotrophic properties incorporated in chloroplasts. By adding an endosymbiotic prokaryote to the plant's genome, diazoplast could also be obtained similarly to a chloroplast. Research on the establishment of *Gluconacetobacter diazotrophicus* in root meristem cells indicate that symbiosomal vesicular cytoplasmic compartments are possible locations for diazoplast formation.

A third approach can be making use of the endophytic microbial community that is associated with the plant system, which will enable aerobic nitrogen fixation by microsymbionts, for improved plant growth (Kennedy and Islam, 2001). For example, Yonebayashi et al. (2014) have reported the association of endophyte and tumorous growth on sweet potato. Oliveira et al. (2003) have studied the effect of diazotrophic endophytic inoculants on sugarcane growth. Similarly, the nonsymbiotic diazotrophs also contribute significantly to nitrogen fixation as pointed out by the work done by Pankievicz et al. (2015) and van Deynze et al. (2018).

Nitrogen fixation efficacy and plant growth benefits in rhizobia-legume symbiosis have been successfully demonstrated for *Gluconacetobacter diazotrophicus* (Dent and Cocking, 2017). The development of nitrogen-fixing endophytic bioinoculants that can be useful to all staple food crops and an apt replacement for chemical nitrogen fertilizers, having yield benefits, is significant progress leading to Greener Nitrogen Revolution (Dent and Cocking, 2017).

8.9 Conclusions and future perspectives

The extensive application of synthetic fertilizers to meet the increasing nitrogen demand of agriculture has been seen in the last decades. This has ultimately led to environmental pollution especially due to the runoff of excess nitrates and soils losing fertility. The increasing concern over the deteriorating environment necessitates lowering the dependence on chemical fertilizers. Scientists are investigating bio-inoculants, in particular diazotrophic bioinoculants, in different dimensions to enhance legume and nonlegume plant growth. This is crucial for the future of sustainable farming, in an effort to guarantee food security and also to reduce air-polluting emissions from chemical fertilizers. Research dimensions being targeted include diazotrophs for nonlegumes, endophytic diazotrophs as bioinoculants, transferring *nif* genes to nondiazotrophs as well as crops and organelles of plants, etc. Future in situ studies are needed to establish the identity, activity, and ecology of particle-associated noncyanobacterial diazotrophs (NCDs) and also focus future research on diazotrophs of the aquatic system as well as environments in which diazotrophic research lacks or has been overlooked. The PGP properties of diazotrophs offer promise as effective bioinoculants of the future. By appropriate application of the naturally occurring endophytic diazotrophs, biological nitrogen fixation for cereals and other nonlegumes can be achieved.

References

Acosta-Cruz, E., Wisniewski-Dyé, F., Rouy, Z., Barbe, V., Valdés, M., Mavingui, P., 2012. Insights into the 1.59-Mbp largest plasmid of *Azospirillum brasilense* CBG497. Arch. Microbiol. 194 (9), 725–736.

Agnolucci, M., Avio, L., Pepe, A., Turrini, A., Cristani, C., Bonini, P., Cirino, V., Colosimo, F., Ruzzi, M., Giovannetti, M., 2019. Bacteria associated with a commercial mycorrhizal inoculum: community composition and multifunctional activity as assessed by Illumina sequencing and culture-dependent tools. Front. Plant Sci. 9, 1956.

Aguiar, N.O., Medici, L.O., Olivares, F.L., Dobbss, L.B., Torres-Netto, A., Silva, S.F., Novotny, E.H., Canellas, L.P., 2016. Metabolic profile and antioxidant responses during drought stress recovery in sugarcane treated with humic acids and endophytic diazotrophic bacteria. Ann. Appl. Biol. 168, 203–213.

Aly, W., Andrews, S., 2016. Iron regulation of growth and heterocyst formation in the nitrogen fixing cyanobacterium *Nostoc* sp. PCC 7120. J. Ecol. Health Environ. 4, 103–109.

Antunes, J.E.L., De Freitas, A.D.S., Oliveira, L.M.S., De Lyra, M.D.C.C.P., Fonseca, M.A.C., Santos, C.E.R.S., Oliveira, J.P., De Araújo, A.S.F., Figueiredo, M.V.B., 2019. Sugarcane inoculated with endophytic diazotrophic bacteria: effects on yield, biological nitrogen fixation and industrial characteristics. An. Acad. Bras. Cienc. 91, e20180990.

Báscones, E., Imperial, J., Ruiz-Argueso, T., Palacios, J.M., 2000. Generation of new hydrogen recycling *Rhizobiaceae* strains by introduction of a novel *hup* minitransposon. Appl. Environ. Microbiol. 66, 4292–4299.

Belay, N., Sparling, R., Daniels, L., 1984. Dinitrogen fixation by a thermophilic methanogenic bacterium. Nature 312, 286–288.

Benavides, M., Robidart, J., 2020. Bridging the spatiotemporal gap in diazotroph activity and diversity with high-resolution measurements. Front. Mar. Sci. 7, 568876.

Bruno, J.R.A., Boddey, R.M., Urquiaga, S., 2003. The success of BNF in soybean in Brazil. Plant Soil 252, 1–9.

Burén, S., Rubio, L.M., 2018. State of the art in eukaryotic nitrogenase engineering. FEMS Microbiol. Lett. 365, fnx274.

Burén, S., Young, E.M., Sweeny, E.A., Lopez-Torrejón, G., Veldhuizen, M., Voigt, C.A., Rubio, L.M., 2017. Formation of nitrogenase *nif*DK tetramers in the mitochondria of *Saccharomyces cerevisiae*. ACS Synth. Biol. 6, 1043–1055.

Burén, S., López-Torrejón, G., Rubio, L.M., 2018. Extreme bioengineering to meet the nitrogen challenge. Proc. Natl. Acad. Sci. U. S. A. 115, 12247.

Dent, D., Cocking, E.C., 2017. Establishing symbiotic nitrogen fixation in cereals and other non-legume crops: the Greener Nitrogen Revolution. Agric. Food Secur. 6, 7.

Dixon, R.A., Postgate, J.R., 1972. Genetic transfer of nitrogen fixation from *Klebsiella pneumoniae* to *Escherichia coli*. Nature 237, 102–103.

Dos Santos, P.C., Fang, Z., Mason, S.W., Setubal, J.C., Dixon, R., 2012. Distribution of nitrogen fixation and nitrogenase-like sequences amongst microbial genomes. BMC Genomics 13 (1), 162.

Einsle, O., Rees, D.C., 2020. Structural enzymology of nitrogenase enzymes. Chem. Rev. 120, 4969–5004.

Elhady, A., Hallmann, J., Heuer, H., 2020. Symbiosis of soybean with nitrogen fixing bacteria affected by root lesion nematodes in a density dependent manner. Sci. Rep. 10, 1619–1621.

El-Khaled, Y.C., Roth, F., Rädecker, N., Kharbatia, N., Jones, B.H., Voolstra, C.R., Wild, C., 2020. Simultaneous measurements of dinitrogen fixation and denitrification associated with coral reef substrates: advantages and limitations of a combined acetylene assay. Front. Mar. Sci. 7, 411.

Elmerich, C., Bozouklian, H., Vieille, C., Fogher, C., Perroud, B., Perrin, A., Vanderleydens, J., 1987. *Azospirillum*: genetics of nitrogen fixation and interaction with plants. Philos. Trans. R. Soc. Lond. B Biol. Sci. 317, 183–192.

El-Sayed, A.A., Elenein, R.A., Shalaby, E.E., Shalan, M.A., Said, M.A., 2000. Response of barley to biofertilizer with N and P application under newly reclaimed areas in Egypt. In: Proceedings of the 3rd International Crop Science Congress (ICSC), Hamburg, Germany, pp. 17–22.

Fan, H., Bolhuis, H., Stal, L.J., 2015. Drivers of the dynamics of diazotrophs and denitrifiers in North Sea bottom waters and sediments. Front. Microbiol. 6, 738.

Figueiredo, M.V.B., Mergulhão, A.E.S., Sobral, J.K., Junio, M.A.L., Araújo, A.S.F., 2013. Biological nitrogen fixation: importance, associated diversity, and estimates. In: Arora, N.K. (Ed.), Plant Microbe Symbiosis: Fundamentals and Advances. Springer India, Berlin/Heidelberg, Germany, p. 267.

Fustec, J., Lesuffleur, F., Mahieu, S., Cliquet, J.-B., 2010. Nitrogen rhizodeposition of legumes. A review. Agron. Sustain. Dev. 30, 57–66.

Gaby, J.B., Buckley, D.H., 2011. A global census of nitrogenase diversity. Environ. Microbiol. 13, 1790–1799.

Geisler, E., Bogler, A., Rahav, E., Bar-Zeev, E., 2019. Direct detection of heterotrophic diazotrophs associated with planktonic aggregates. Sci. Rep. 9, 9288–9297.

Gopalakrishnan, S., Sathya, A., Vijayabharathi, R., Varshney, R.K., Gowda, C.L.L., Krishnamurthy, L., 2015. Plant growth promoting rhizobia: challenges and opportunities. 3 Biotech 5, 355.

Gopalakrishnan, S., Srinivas, V., Vemula, A., Samineni, S., Rathore, A., 2018. Influence of diazotrophic bacteria on nodulation, nitrogen fixation, growth promotion and yield traits in five cultivars of chickpea. Biocatal. Agric. Biotechnol. 15, 35–42.

Graham, P.H., 1992. Stress tolerance in *Rhizobium* and *Bradyrhizobium*, and nodulation under adverse soil conditions. Can. J. Microbiol. 38, 475–484.

Gyogluu, C., Jaiswal, S.K., Kyei-Boahen, S., Dakora, F.D., 2018. Identification and distribution of microsymbionts associated with soybean nodulation in Mozambican soils. Syst. Appl. Microbiol. 41, 506–515.

Han, Y., Lu, N., Chen, Q., Zhan, Y., Liu, W., Wei, L., Zhu, B., Lin, M., Yang, Z., Yan, Y., 2015. Interspecies transfer and regulation of *Pseudomonas stutzeri* A1501 nitrogen fixation island in *Escherichia coli*. J. Microbiol. Biotechnol. 25, 1339–1348.

Haselkorn, R., 2003. Heterocysts. Annu. Rev. Plant Physiol. 29, 319–344.

Higdon, S.M., Pozzo, T., Tibbett, E.J., Chiu, C., Jeannotte, R., Weimer, B.C., et al., 2020a. Diazotrophic bacteria from maize exhibit multifaceted plant growth promotion traits in multiple hosts. PLoS One 15 (9), e0239081.

Higdon, S.M., Pozzo, T., Kong, N., Huang, B., Yang, M.L., Jeannotte, R., et al., 2020b. Genomic characterization of a diazotrophic microbiota associated with maize aerial root mucilage. bioRxiv 2020, 1–52. 2020.04.27.064337.

Hoffman, B.M., Lukoyanov, D., Yang, Z.-Y., Dean, D.R., Seefeldt, L.C., 2014. Mechanism of nitrogen fixation by nitrogenase: the next stage. Chem. Rev. 114, 4041–4062.

Hungria, M., Mendes, I.C., 2015. Nitrogen Fixation With Soybean: The Perfect Symbiosis? Biological Nitrogen Fixation. vol. 2 John Wiley & Sons, Inc., Hoboken, NJ, USA, pp. 1009–1024.

Hye Jia, S., Gururanib, M.A., Chun, S.-C., 2014. Isolation and characterization of plant growth promoting endophytic diazotrophic bacteria from Korean rice cultivars. Microbiol. Res. 169, 83–98.

Ivleva, N.B., Groat, J., Staub, J.M., Stephens, M., 2016. Expression of active subunit of nitrogenase via integration into plant organelle genome. PLoS One 11, e0160951.

Johansson, B.C., Nordlund, S., Baltscheffsky, H., 1983. In: Ormerod, J.G. (Ed.), The Phototrophic Bacteria: Anaerobic Life in the Light. Blackwells Science Publications, Oxford, p. 120.

Kennedy, I.R., Islam, N., 2001. The current and potential contribution of asymbiotic nitrogen fixation to nitrogen requirements on farms: a review. Aust. J. Exp. Agric. 41, 447–457.

Kereszt, A., Mergaert, P., Montiel, J., Endre, G., Kondorosi, E., 2018. Impact of plant peptides on symbiotic nodule development and functioning. Front. Plant Sci. 9, 1026.

Kondorosi, E., Mergaert, P., Kereszt, A., 2013. A paradigm for endosymbiotic life: cell differentiation of *Rhizobium* bacteria provoked by host plant factors. Annu. Rev. Microbiol. 67, 611–628.

Lambert, G.R., Cantrell, M.A., Hanus, F.J., Russell, S.A., Haddad, K.R., Evans, H.J., 1985. Intra- and interspecies transfer and expression of *Rhizobium japonicum* hydrogen uptake genes and autotrophic growth capability. Proc. Natl. Acad. Sci. U. S. A. 82, 3232–3236.

Laranjo, M., Alexandrea, A., Oliveira, S., 2014. Legume growth-promoting rhizobia: an overview on the Mesorhizobium genus. Microbiol. Res. 169, 2–17.

Laskar, F., Sharma, G.D., Deb, B., 2013. Characterization of plant growth promoting traits of diazotrophic bacteria and their inoculating effects on growth and yield of rice crops. Global Res. Anal. 2, 3–5.

Leigh Nitrogen, J.A., 2000. Fixation in methanogens: the archaeal perspective. Curr. Issues Mol. Biol. 2, 125–131.

Lindstrom, K., Mousavi, S.A., 2019. Effectiveness of nitrogen fixation in rhizobia. Microb. Biotechnol. 13, 1314–1335.

Long, S.R., 1989. *Rhizobium* genetics. Annu. Rev. Genet. 23, 483–506.

López-Torrejón, G., Jiménez-Vicente, E., Buesa, J.M., Hernandez, J.A., Verma, H.K., Rubio, L.M., 2016. Expression of a functional oxygen-labile nitrogenase component in the mitochondrial matrix of aerobically grown yeast. Nat. Commun. 7, 11426.

Mahmud, K., Makaju, S., Ibrahim, R., Missaoui, A., 2020. Current progress in nitrogen fixing plants and microbiome research. Plants 9, 97.

Market Data Forecast, 2018. Available online: https://www.marketdataforecast.com/market-reports/asiapacific-biofertilizers-market. (Accessed 15 June 2020).

Mathenge, C., Thuita, M., Masso, C., Gweyi-Onyango, J., Vanlauwe, B., 2019. Variability of soybean response to rhizobia inoculant, vermicompost, and a legume-specific fertilizer blend in Siaya County of Kenya. Soil Tillage Res. 194, 104–290.

Mergaert, P., Nikovics, K., Kelemen, Z., Maunoury, N., Vaubert, D., Kondorosi, A., Kondorosi, E., 2003. A novel family in *Medicago truncatula* consisting of more than 300 nodule-specific genes coding for small, secreted polypeptides with conserved cysteine motifs. Plant Physiol. 132, 161–173.

Munson, T.O., Burris, R.H., 1969. Nitrogen fixation by *Rhodospirillum rubrum* grown in nitrogen limited continuous culture. J. Bacteriol. 97, 1093–1098.

Murray, P.A., Zinder, S.H., 1984. Nitrogen fixation by a methanogenic archaebacterium. Nature 312, 284–286.

Mus, F., Crook, M.B., Garcia, K., Garcia Costas, A., Geddes, B.A., Kouri, E.D., Paramasivan, P., Ryu, M.-H., Oldroyd, G.E., Poole, P.S., et al., 2016. Symbiotic nitrogen fixation and the challenges to its extension to nonlegumes. Appl. Environ. Microbiol. 82, 3698–3710.

Ogawa, R.E., Carr, J.F., 1969. The influence of nitrogen on heterocyst production in blue-green algae. Limnol. Oceanogr. 14, 342–351.

Oldroyd, G.E., Dixon, R., 2014. Biotechnological solutions to the nitrogen problem. Curr. Opin. Biotechnol. 26, 19–24.

Oliveira, A.L.M., Canuto, E.L., Reis, V.M., Baldani, J.I., 2003. Response of micropropagated sugarcane varieties to inoculation with endophytic diazotrophic bacteria. Braz. J. Microbiol. 34, 59–61.

Pan, H., Wang, D., 2017. Nodule cysteine-rich peptides maintain a working balance during nitrogen-fixing symbiosis. Nat. Plants 3, 17048.

Pankievicz, V.C.S., do Amaral, F.P., Santos, K.F.D.N., Agtuca, B., Xu, Y., Schueller, M.J., Arisi, A.C.M., Stevens, M.B.R., de Souza, E.M., Pedrosa, F.O., et al., 2015. Robust biological nitrogen fixation in a model grass-bacterial association. Plant J. 81, 907–919.

Pastor-Bueis, R., Sánchez-Cañizares, C., James, E.K., González-Andrés, F., 2019. Formulation of a highly effective inoculant for common bean based on an autochthonous elite strain of *Rhizobium leguminosarum* bv. *phaseoli*, and genomic-based insights into its agronomic performance. Front. Microbiol. 10, 2724.

Pedersen, J.N., Bombar, D., Pearl, R.W., Riemann, L., 2018. Diazotrophs and N2-fixation associated with particles in coastal estuarine waters. Front. Microbiol. 9, 2759.

Peoples, M.B., Brockwell, J., Herridge, D.F., Rochester, I.J., Alves, B.J.R., Urquiaga, S., Boddey, R.M., Dakora, F.D., Bhattarai, S., Maskey, S.L., et al., 2009. The contributions of nitrogen-fixing crop legumes to the productivity of agricultural systems. Symbiosis 48, 1–17.

Postgate, J., 1992. The Leeuwenhoek lecture 1992: bacterial evolution and the nitrogen-fixing plant. Philos. Trans. R. Soc. Lond. B Biol. Sci. 338 (1286), 409–416.

Puri, A., Padda, K.P., Chanway, C.P., 2017. Plant growth promotion by endophytic bacteria in non-native crop hosts. In: Maheshwari, D.K., Annapurna, K. (Eds.), Endophytes: Crop Productivity and Protection. Springer International Publishing, Switzerland, pp. 11–45.

Rao, D.L.N., Balachandar, D., 2017. Nitrogen inputs from biological nitrogen fixation in Indian agriculture. In: Abrol, Y.P. (Ed.), The Indian Nitrogen Assessment, Sources of Reactive Nitrogen, Environmental and Climate Effects, Management Options, and Policies. Elsevier, Amsterdam, The Netherlands, pp. 117–132.

Reis, V.M., 2007. Uso de Bactérias Fixadores de Nitrogênio omo Inoculante para Aplicação em Gramíneas. Seropédica: Embrapa Agrobiologia; Embrapa Agrobiologia: Rodovia, Brasilia, vol. 232, p. 22.

Ronner, E., Franke, A., Vanlauwe, B., Dianda, M., Edeh, E., Ukem, B., Giller, K., 2016. Understanding variability in soybean yield and response to P-fertilizer and *Rhizobium* inoculants on farmers' fields in northern Nigeria. Field Crop Res. 186, 133–145.

Rose, M.T., Phuong, T.L., Nhan, D.K., Cong, P.T., Hien, N.T., Kennedy, I.R., 2014. Up to 52% N fertilizer replaced by biofertilizer in lowland rice via farmer participatory research. Agron. Sustain. Dev. 34, 857–868.

Saumare, A., Diedhiou, A.D., Thuita, M., Hafidi, M., Ouhdouch, Y., Gopalakrishnan, S., Kouisni, L., 2020. Exploiting biological nitrogen fixation: a route towards a sustainable agriculture. Plants 9, 1011–1033.

Seefeldt, L.C., Yang, Z.Y., Lukoyanov, D.A., Harris, D.F., Dean, D.R., Raugei, S., Hoffman, B.M., 2020. Reduction of substrates by nitrogenases. Chem 120 (12), 5082–5106. https://doi.org/10.1021/acs.chemrev.9b00556.

Serna-Cock, L., Arias-García, C., Valencia Hernandez, L.J., 2011. Effect of biofertilization on the growth of potted sugarcane plants (*Saccharum ocinarum*). Rev. Biol. Agroind. 9, 85–95.

Singh, I., 2018. Plant growth promoting Rhizobacteria (PGPR) and their various mechanisms for plant growth enhancement in stressful conditions: a review. Eur. J Biol. Res. 8, 191–213.

Swarnalakshmi, K., Vandana, Y., Senthilkumar, M., Dolly, W.D., 2016. Biofertilizers for higher pulse production in India: scope, accessibility and challenges. Indian J. Agron. 61, 173–181.

Thuita, M., Vanlauwe, B., Mutegi, E., Masso, C., 2018. Reducing spatial variability of soybean response to rhizobia inoculants in farms of variable soil fertility in Siaya Country of West Kenya. Biol. Fertil. Soils 261, 153–160.

Ulzen, J., Abaidoo, R.C., Mensah, N.E., Masso, C., AbdelGadir, A.H., 2016. Bradyrhizobium inoculants enhance grain yields of soybean and cowpea in Northern Ghana. Front. Plant Sci. 7, 1770.

van Deynze, A., Zamora, P., Delaux, P.-M., Heitmann, C., Jayaraman, D., Rajasekar, S., Graham, D., Maeda, J., Gibson, D., Schwartz, K.D., et al., 2018. Nitrogen fixation in a landrace of maize is supported by a mucilage-associated diazotrophic microbiota. PLoS Biol. 16, e2006352.

van Heerwaarden, J., Baijukya, F., Kyei-Boahen, S., Adjei-Nsiah, S., Ebanyat, P., Kamai, N., Wolde-meskel, E., Kanampiu, F., Vanlauwe, B., Giller, K., 2018. Soyabean response to *Rhizobium* inoculation across sub-Saharan Africa: patterns of variation and the role of promiscuity. Agric. Ecosyst. Environ. 261, 211–218.

van Velzen, R., Holmer, R., Bu, F., Rutten, L., van Zeijl, A., Liu, W., Santuari, L., Cao, Q., Sharma, T., Shen, D., et al., 2018. Comparative genomics of the nonlegume *Parasponia* reveals insights into evolution of nitrogen-fixing rhizobium symbioses. Proc. Natl. Acad. Sci. U. S. A. 115, 4700–4709.

Vicente, E.J., Dean, D.R., 2017. Keeping the nitrogen-fixation dream alive. Proc. Natl. Acad. Sci. U. S. A. 114, 3009–3011.

Videau, P., Rivers, O.S., Hurd, K., Ushijima, B., Oshiro, R.T., Ende, R.J., O'Hanlon, S.M., Cozy, L.M., 2016. The heterocyst regulatory protein HetP and its homologs modulate heterocyst commitment in *Anabaena* sp. strain PCC 7120. Proc. Natl. Acad. Sci. U. S. A. 113, E6984–E6992.

Vitousek, P.M., Menge, D.N.L., Reed, S.C., Cleveland, C.C., 2013. Biological nitrogen fixation: rates, patterns and ecological controls in terrestrial ecosystems. Philos. Trans. R. Soc. Lond. B Biol. Sci. 368, 1621.

von Friesen, L.W., Riemann, L., 2020. Nitrogen fixation in a changing arctic ocean: an overlooked source of nitrogen? Front. Microbiol. 11, 596426.

Wardhani, T.A., Roswanjaya, Y.P., Dupin, S., Li, H., Linders, S., Hartog, M., Geurts, R., van Zeijl, A., 2019. Transforming, genome editing and phenotyping the nitrogen-fixing tropical cannabaceae tree *Parasponia andersonii*. J. Vis. Exp. 150, e59971.

William, S., Akiko, A., 2020. History of Soybeans and Soyfoods in Eastern Europe (Including All of Russia) (1783–2020): Extensively Annotated Bibliography and Sourcebook. Soyinfo Center, Lafayette, CA, USA. ISBN1 1948436175.

Willis, A., Adams, M.P., Chuang, A.W., Orr, P.T., O'Brien, K.R., Burford, M.A., 2015. Constitutive toxin production under various nitrogen and phosphorus regimes of three ecotypes of *Cylindrospermopsis raciborskii* ((Wołoszyńska) Seenayya et Subba Raju). Harmful Algae 47, 27–34.

Wojciechowsk, M.F., Matt Lavin, I., Sanderson, M.J., 2004. A phylogeny of legumes (Leguminosae) based on analysis of the plastid *mat*K gene resolves many well-supported subclades within the family. Am. J. Bot. 91 (11), 1846–1862.

Yonebayashi, K., Katsumi, N., Nishi, T., Okazaki, M., 2014. Activation of nitrogen-fixing endophytes is associated with the tuber growth of sweet potato. Mass Spectrom (Tokyo) 3, A0032.

Zulkefli, N.S., Hwang, S.-J., 2020. Heterocyst development and diazotrophic growth of *Anabaena variabilis* under different nitrogen availability. Life 10, 279.

Zuluaga, M.Y.A., Lima Milani, K.M., Azeredo Goncalves, L.S., Martinez de Oliveira, A.L., 2020. Diversity and plant growth-promoting functions of diazotrophic/N-scavenging bacteria isolated from the soils and rhizospheres of two species of *Solanum*. PLoS One 15 (1), e0227422.

Role of mycorrhizae in plant-parasitic nematodes management

H.K. Patel[a], Y.K. Jhala[a], B.L. Raghunandan[b], and J.P. Solanki[a]
[a]Department of Agricultural Microbiology, Anand Agricultural University, Anand, Gujarat, India, [b]AICRP-Biological Control of Crop Pests, Anand Agricultural University, Anand, Gujarat, India

9.1 Introduction

Sir A. B. Frank first time used the Greek word "Mycorrhiza" in 1885, which clearly means "fungus roots." The fungus establishes a symbiotic relationship with its partner plant root in a similar way to rhizobium in legumes, which is obligate biotrophic in nature. The fungi take carbohydrates from the plants and in return supply the plants with nutrients, hormones, etc. Arbuscular mycorrhizal fungi (AMF) is found to occur on roots of most of the food, medicinal and aromatic, horticultural crops, and tropical trees. It has also been estimated that about 95% of the world's present species of vascular plants belong to families that are characteristically mycorrhizal, which demonstrates the usefulness of such organisms if we could apply them as biological control agents (Trapp, 1977). The presence of arbuscules and some cases vesicles is a diagnostic criterion for identifying arbuscular mycorrhizal fungi in a root. AMF are the members of the phylum Glomeromycota (Schubler et al., 2001), which presently comprises approximately 299 described species distributed among 31 genera, most of which have been defined mainly by spore morphology (Table 9.1).

The term "nematode" comes from the Greek words: *nema*, means "thread," and *toid*, means "form." Nematodes are multicellular, invertebrate, unsegmented, vermiform, long, and slender, but some species are swollen. Most people know them as roundworms because their cross section is round. These nematodes occupy different ecological niches like soil, marine, fresh water form as well as predator and parasite of vertebrate, invertebrate, and plants. Several parasitic forms of nematode cause disease in plants, humans, and animals. These nematodes also serve food for different organisms like bacteria,

Table 9.1 Classification of AM fungi.

Phylum: Glomeromycota		
Class: Glomeromycetes	**Class: Archaeosporomycetes**	**Class: Paraglomeromycetes**
Order: Diversisporales	**Order: Archaeosporales**	**Order: Paraglomales**
Family: Diversisporaceae	Family: Ambisporaceae	Family: Paraglomeraceae
Genus:	Genus:	Genus:
1. *Diversispora*	1. *Ambispora*	1. *Paraglomus*
2. *Otospora*	Family: Geosiphonceae	
3. *Tricispora*	Genus:	
4. *Redeckra*	1. *Geosiphon*	
Family: Acaulosporaceae	Family: Archaesporacea	
Genus:	Genus:	
1. *Acaulospora*	1. *Archaespora*	
2. *Kuklospora*	2. *Intraspora*	
Family: Sacculosporaceae		
Genus:		
1. *Sacculospora*		
Family: Pascifloraceae		
Genus:		
1. *Pasciflora*		
Order: Gigasporales		
Family: Scetellosporaceae		
Genus:		
1. *Scetellospora*		
2. *Orbispora*		
Family: Gigasporaceae		
Genus:		
1. *Gigaspora*		
Family: Intaornatosporaceae		
Genus:		
1. *Intraoatospora*		
2. *Paradentisculata*		
Family: Dentisculataceae		
Genus:		
1. *Dentisculata*		
2. *Quatunica*		
3. *Fusculata*		
Family: Racocetraceae		
Genus:		
1. *Cetraspora*		
2. *Racocetra*		

Table 9.1 Classification of AM fungi—cont'd

Phylum: Glomeromycota		
Class: Glomeromycetes	**Class: Archaeosporomycetes**	**Class: Paraglomeromycetes**
Order: Glomerales Family: Entrophosporaceae Genus: 1. *Viscospora* 2. *Claroideoglomus* 3. *Entrophospora* 4. *Albahypha* Family: Glomeraceae Genus: 1. *Simiglomus* 2. *Funneliformis* 3. *Septoglomus* 4. *Glomus*		

Modified from Goto, B.T., Silva, G.A., Assis, D., Silva, D.K., Souza, R.G., Ferreira, A.C., Oehl, F., 2012. Intraornatosporaceae (Gigasporales), a new family with two new genera and two new species. Mycotaxon 119(1), 117–132.

fungi, insects, and others. Some nematodes are parasite to insects and used as biocontrol agents to manage insect problem in plants. Other nematodes played some beneficial uses like cycling of mineral nutrients, decomposition of organic waste, including biodegradation of toxic compounds, acting as indicator of soil and water quality. Plant-parasitic nematode generally falls into two major groups: nematodes that spend their most of life stages in roots and soil and other that spend at least part of their life stages in foliage of plants. The soil forms are root-knot nematode (RKN), *Meloidogyne* sp., which have broad host range, and other commonly root-feeding nematodes are Stunt nematode (*Tylenchorhynchus* sp.), Lance nematode (*Hoplolaimus* sp.), Spiral nematode (*Helicotylenchus* sp.), Lesion nematode (*Pratylenchus* sp.), Cyst nematode (*Heterodera* sp.), and Ring nematode (*Criconema* sp.). The aerial parasitic nematodes generally belong to the genera *Aphelenchus* or *Aphelenchoides or Ditylenchus,* etc.

The average loss caused by phytonematode to different agricultural crops, including vegetables, sugar, cotton, cereals, pulses, oilseed, tuber, medicinal and aromatic plants, is more than 100 billion dollars every year. Due to hazardous effect of chemical nematicides on different types of environment like soil, water, air, and human too, most of effective nematicides are banned since 2005. Nowadays, nematological researchers are engaged to develop some new nematicides through the use of plants (Chitwood, 2002; Pandey and Bhargava, 2004) and microorganism (Sikora, 1992), and a significant success was achieved. *Pasteuria penetrans*, parasitic soil bacteria that infect and kill the nematodes, is becoming popular in United States. This bacteria attack targets phytonematodes and is highly durable too. This bacterium is known to produce heat- and chemical-resistant structure known as spore (endo), which is also resistant to drying, heat, and freezing also. Like in United States, Pasteuria Bioscience Company developed the formulation and hope it will be soon available in Asian market too. The safe way of phytonematode management is using different other biological options like the use of nematode parasitic bacteria, PGPR, nematophagous fungi, egg parasitic fungi, and arbuscular mycorrhizal fungi (Fig. 9.1).

The AMF and PPN both are zing on plant root system: in the first case, the association is beneficial to plant, while in case of the PPN, the association is harmful for root and the plant. AMF not only fulfill nutrient requirement of the plant but also provide other benefits

Mycorhizosphere: Rhizosphere + Hyphosphere
- Nutrient Transfer via AM fungi
- Increase Drought & Salt Tolerance
- **Increase resistance to Heavy Metals**
- **Resistance to Root Pathogens**
  - Competition for food source and colonization sites
  - Morphological changes in root
  - SAR responses (Salicylate & Jasmonate)
  - Systemic Resistance to Root Pathogens
  - Secondary metabolites and defence enzymes synthesis

Rhizosphere Only
- Nutrient Transfer via Root Hair
- More susceptibility towards Root Pathogens

(Modified from source https://giantveggiegardener.com/2013/06/05/the-importance-of-mycorrihizal-fungi-in-the-garden/)

Fig. 9.1 Schematic representation of difference between mycorrhizal and nonmycorrhizal plant to counter damage caused by pathogens (nematode) in plants.

to the plant besides managing nematode population; collectively, the AMF increase the crop yield significantly. The positive effects of AMF association are increased uptake of macro- and micronutrients along with resistance toward different biotic and abiotic stresses, ultimately making plant and soil healthy. On the other hand, injudicious and unwanted use of agrochemicals and agricultural practices like tillage can reduce effectiveness of AMF on plant. In organic cultivation, these AMF provide sufficient nutrient to plants, well colonized the root system, and reduced several inputs in crop cultivation. There are sufficient evidences that AMF reduced nematode infection if employed with other management practices carefully.

Management of plant pathogen through biological means is widely accepted phenomena and need of the day as it plays a key role in sustainable agriculture due to reliance on the use of natural resources, i.e., plant growth-promoting microbes and certain biological means, which are known for their antagonistic effect against plant pathogens (fungi, bacteria, and nematode). The association of plant roots with arbuscular mycorrhizal fungi (AMF) has been found to reduce the damage caused by soil-borne plant pathogens. Although few AM isolates have been tested in this regard, some recorded to be more effective. Furthermore, the magnitude of protection varies with the type of pathogen involved and soil and other environmental conditions. This prophylactic ability of the mycorrhizal fungi could be utilized in combination with other microbial biocontrol agents to enhance plant growth and to sustain plant health. Mycorrhizal symbionts have been recognized for their role in nutrient cycling in the agro-ecosystem, additionally found to protect plants against different biotic and abiotic stresses (Barea and Jeffries, 1995). Here, the term "mycorrhizosphere" comes in to the picture, means the extended root area of the plant colonized by AMF effectively. The AMF infection leads to change the root morphology, which ultimately changes the plant physiology. This physiological alteration changes biochemistry of plant metabolites and its secretions, including root exudates. As an effect, other beneficial microorganisms are being colonized the plant root and improving plant health. As a result, the plant becomes more resistant to different environmental stresses and plant pathogens (Linderman, 1988). The effect of AMF on disease incidence and intensity is governed by type of pathogen and environmental conditions, including microclimate. Generally, the host plant intercedes the interactions between pathogen and the symbionts. The aim of this chapter is (1) to explore the potential of AMF in suppression of the PPN and to understand their interaction and behavior and (2) to explore the possibilities of large-scale production and commercialization of mycorrhizal fungi.

9.2 Role of arbuscular mycorrhiza in the suppression of plant-parasitic nematodes

AMF can effectively control the damage caused by soil-borne plant pathogens whether it is bacteria, fungi, or PPN, and in several meta-analysis-based studies, these facts have been proved that the mycorrhizal fungi can reduce the disease or damage intensity (Cardoso and Kuyper, 2006; Sharma et al., 2004). In case of plant pathogenic bacteria and fungi, a general pattern of infestation was observed but in case of PPN varies, which depends on plant species and type of nematode as well as its feeding pattern. The symbiotic association between plant and AMF leads to an increase in plant growth and also helps to reduce the loss of plant produce by PPN, which leads the scientific community to investigate the effect of AMF on PPN management. The effect of vesicular arbuscular mycorrhizal (VAM) fungi inoculated with nematodes in roots of *Luffa cylindrica* (L.) Roem was studied under greenhouse conditions by Hajra et al. (2015). They have reported that damage produced by nematodes observed in treatments do not received the application of VAM fungi. Moreover, damage was observed in treatments receiving the application of fungi after 1 week of nematode inoculation. However, VAM fungi suppressed the effects of root-knot nematodes in such plants. They have also reported the presence of VAM fungi spores in cortical as well as in pith region of root in treatments receiving the application of fungi. Finally, they came to conclusion that VAM fungi caused an increase in plant growth by developing resistance to nematode infection and development, served as an effective and alternative biocontrol agent against nematode pests. Ceustermans et al. (2018) investigated the effect of AMF on PPN, and apple seedlings were inoculated with different AMF species in a pot experiment in the presence of the PPN *P. penetrans*. During their investigation, they have observed apple seedlings growth promotion in the presence of nematodes when mycorrhiza was inoculated into the soil. Moreover, a positive correlation was also found between percentage root length colonization of apple seedlings, by AM fungi species, and simultaneous nematode reduction in the soil of the seedlings. They have observed that native mycorrhizal species could colonize the roots of apple seedlings most efficiently, resulting in an excellent biological control. They have also recorded a synergistic effect, when two AMF strains were co-inoculated, leading to a significant increase in growth response of the seedlings. Inoculation of tomato roots with increasing levels of *Glomus fasciculatum* progressively enhanced the plant growth along with its own reproduction in terms of root colonization and chlamydospore production. The highest percentage of VAM colonization, the maximum number of chlamydospores, and the maximum reduction in number and size of galls of the nematode were observed in the plants inoculated simultaneously with 500 second-stage larvae and 200 chlamydospores per pot. Similarly, uptake of nutrients, viz., nitrogen, phosphorus, and potassium, was also enhanced considerably with the simultaneous inoculation than with the plants inoculated with the nematode alone (Mishra and Shukla, 1995). Talavera et al. (2001) found that tomato seedlings were infected with the nematode 3 weeks after mycorrhizal application and colonization of tomato roots by *G. mosseae* compensated for the reduction of plant growth caused by *M. incognita* infection with up to 85% reduction in soil nematode population. Besides beneficial effects of mycorrhizal fungi alone, they also conferred greater beneficial effects when applied together with bacterium. Talavera et al. (2002) found that tomato shoot and fruit weight were recorded higher in *M. incognita*-infested plants receiving a combined inoculation of *Glomus* sp. and *P. penetrans* in comparison with untreated nematode-infested plants. Siddiqui and Mahmood (1995) also observed that simultaneous use of biocontrol agents against pathogens gave better control than individual application. The most effective management of PPN multiplication was recorded in case of combined application of *G. mosseae*, *Trichoderma harzianum*, and *Verticillium chlamydosporium*. Baghel et al. (1990) studied an inoculative effect of *G. mosseae* on *Citrus jambhiri* seedlings and the AMF inoculation found to stimulate the seedling growth. Co-inoculation of AMF and *Tylenchulus semipenetrans* (Citrus nematode) was found to counter the ill-effect of the nematode. Tomato plants inoculated with mycorrhizal fungi were recorded with a low population density of *Nacobbus aberrans* juveniles (false root-knot nematode) within the roots compared with nonmycorrhizal plants after 12 days of inoculation (Marroa et al., 2018). The use of two species of AM fungi *Rhizophagus intraradices* and *Funneliformis mosseae* (individually as well as combined) was found to reduce the entry of nematodes in tomato roots. Al-Raddad and Ahmad (1995) observed that the advanced application of *G. mosseae* was found to reduce galling significantly along with a reduction in population density due to a negative effect of AMF on nematode reproduction. The capability of mycorrhized plants to flourish instead of infested by PPN can be considered to be the main effect of AMF on the host plants (Hussey and Roncadori, 1982). Moreover, in comparison with AMF-inoculated plant, the P-fertilized noninoculated plants showed more susceptibility toward nematode infection and damage, clearly indicating the contribution of factors other than phosphorus nutrient in the interactions (Smith, 1987).

Mycorrhiza was used to infect *Coffea canephora* plants and grafted onto *Coffea liberica* rootstock (Pham et al., 2020). During this experimentation, it was observed that the combination of AMF application with grafting significantly reduced the nematode infestation in the replanted soil. Additionally, they have also observed that the survival rate and growth of the coffee plant in the replanted soil treated by the combined technique were increased from 60.5% to 93.7% than individual treatment techniques. Sharma and Bhargava (1993) studied preinoculations of *G. fasciculatum* in tomato and recorded that there was a significant reduction in number of galls and egg masses per plant as well as a reduction in eggs per egg mass due to root-knot nematode *M. incognita*. They have also worked out an interaction that the magnitude of AMF-mediated nematode suppression was different in pot and field experiments. In the field study, the suppression of nematode population was almost double (58%) in comparison with pot (36%).

There are certain studies carried out which demonstrating that the sequence of AMF inoculation and colonization to plant root greatly affect the severity of nematode infection. Sharma et al. (1994) studied preinoculation effect of AMF to tomato plants on RKN *M. incognita* infestation and establishment. They have reported that preinoculation of AMF in tomato and other transplanted crops makes this slow-growing fungi to be colonized effectively in plant root before infection/colonization of RKN. Moreover, the penetration of the infective juveniles' penetration to root tissues was also minimized as well as found to affect the maturity of the nematode in comparison with control plant (noncolonized). Moreover, it is also reported by many workers (Jain and Sethi, 1987; Suresh and Bagyaraj, 1984; Sharma and Johri, 2002) that the early or preinoculation of AMF was found to resist/suppress the reproductive cycle of the nematodes and their penetration and development in the plant root in comparison with simultaneous application of both the root parasite (the AMF) and the nematodes. Specially for the effect on reproductive cycle, Priestel (1980) demonstrated that the AMF was found to suppress the multiplication rate of RKN *Meloidogyne* and the reason they have find out was a drastic reduction in egg laying by the nematode. Habte et al. (1999) evaluated three species of arbuscular mycorrhizal fungi, *G. aggregatum*, *G. intraradices*, and *G. mosseae*, for their effectiveness to suppress the plant-parasitic nematode *Meloidogyne incognita* in white clover and found that the growth of plant was significantly stimulated by mycorrhizal colonization. The range to which mycorrhizal inoculum reduces nematode damage varied with the species to species of this fungus; the extent of damage reduction, sometime, ranged from 19% to 49.8% based on shoot mass. The mycorrhizal inoculation and oil cakes incorporation significantly improved plant growth characters of *Crossandra* in comparison with that of nematode-inoculated checks. However, the magnitude of increase in each character under report varied with the combinations of *G. mosseae* and oil cakes. *G. mosseae* promoted better plant growth than their individual applications. The root galls were least on the plants treated with *G. mosseae* plus neem cake followed by *G. mosseae* + karanj cake- and *G. mosseae* + castor cake-treated plants (Nagesh and Reddy, 1997). The inoculation of mulberry with appropriate mycorrhizal fungi yields a considerable improvement in the plant growth characters. The VAM colonization induces a short of resistance for the root-knot nematode infestation and multiplication (Chandrashekar et al., 1995).

9.3 Interaction of arbuscular mycorrhiza with soil-borne nematodes associated with plants

The protective ability of mycorrhizae is generally observed against plant-parasitic nematode and is often related to the nature of the host plant, mycorrhizal symbionts, nematode, and condition of the soil and environment. AMF and PPN are commonly found to share the same niche with the plant rhizosphere and colonized the host plant roots. But these two obligate microbiotrophs exert opposite effects on the plant growth (Hussey and Roncadori, 1982). Generally, as a result of these interactions, the magnitude of nematode infestation is found to be reduced in mycorrhizal plants (Table 9.2). Parasitism of PPN eggs with AMF has been studied but the magnitude of parasitism was not that much sufficient, which can adversely affect nematode population (Schouteden et al., 2015), although in most cases, AMF was found to colonize only stressed or physiologically weak nematode eggs. The nematode parasitism by AMF is opportunistic and depends on carbon pull from biotrophic symbionts, rather than actual host-parasitic

Table 9.2 Consequence of plant-microbe interaction with special emphasis to host, AMF, and PPN

Interaction	Partner	Consequence
Positive	Plant	Due to suppressed nematode activity, significant improvement in plant growth and yield
	Fungi	Increase in root infection/sporulation
	Nematode	Suppressed attraction toward roots, penetration, or subsequent reproduction and development
Neutral	Plant	There will be no effect in plant growth and yield despite AMF colonization
	Fungi	No effect on root infection or sporulation
	Nematode	There will be no effect on nematode colonization and reproduction
Negative	Plant	Yield response to mycorrhiza suppressed
	Fungi	Root infection or sporulation suppressed
	Nematode	There will be a significant increase in root penetration and colonization followed by the development of nematode population

Modified from Hussey, R.S., Roncadori, R.W., 1982. Vesicular-arbuscular mycorrhizae may limit nematode activity and improve plant growth. Plant Dis. 66, 9–14.

relationship like between *Trichoderma* and phytopathogenic fungi (Siddiqui and Mahmood, 1995). The negative effect of AMF on nematode activity and reproduction is majorly due to histopathological changes brought by AMF in cells like smaller syncytia and fewer giant cells, which confer resistance against *Meloidogyne* sp. on the host plants. The alteration of host physiology and biochemical changes brought by mycorrhizal plants has been found to be one of the factors conferring the resistance against nematodes (Poveda et al., 2020).

A number of recent reports are available on fruit and vegetable crops where AM fungi was found to reduce nematode infestations; however, the degree of reduction was found to be varied due to AM application combinations (Vos et al., 2012; Alban et al., 2013; Koffi et al., 2013). In a study, the banana plants were colonized by *G. mosseae* or *G. intraradices*, two AMF, and the PPN *Radopholus similis* was inoculated after the establishment of the mycorrhizal colonization (Vos et al., 2012). The results clearly indicated lower nematode penetration in mycorrhized plants and in vitro chemotaxis bioassay point toward a reduced attraction of the nematodes to the mycorrhizal plant roots. The results indicated that a water-soluble compound in mycorrhizal root exudates is probably responsible for the mycorrhiza-induced resistance at the preinfectional level of *R. similis*. Timing of the interactions between the AMF and the nematode can be of prime importance. Vaast et al. (1998) noted that simultaneous inoculation of AMF with the root lesion nematode *P. coffeae* on coffee did not enhance tolerance of coffee, but early inoculation (4 months before coffee plants were challenged with the nematode) significantly improved the tolerance of coffee. In early inoculation, the nematode did not affect AM colonization, while in simultaneous inoculation the nematode suppressed mycorrhizal colonization. An experiment involving banana and two nematode species (*P. coffeae* and *R. similis* Cobb) showed that if the AM fungus *G. mosseae* was inoculated 2 months before the nematodes, the mycorrhizal fungus affected the nematodes negatively. The effects were to a larger extent due to the improvement of the nutritional status of banana (Elsen et al., 2003).

9.4 Mycorrhizal mode of action to manage plant-parasitic nematode

There are evidences that an increase in capacity for nutrient acquisition resulting from mycorrhiza association could help the plant to become stronger to resist stress. However, AMF symbiosis may also improve plant health through a more specific increase in protection (Anli et al., 2020). In fact, an AM-induced increase in resistance or a decrease in susceptibility requires the preestablishment of AM and extensive development of the symbiosis before pathogen attack. It has been hypothesized that it has been either physical or physiological changes to explain the effect of AMF on PPN (Marroa et al., 2018). In reviews of biocontrol of plant diseases, AMF are thought to be a biocontrol agent against plant diseases primarily by means of stress reduction. The biocontrol of plant diseases may be strongly affected by AMF following more than one mechanism. Different mechanisms have been suggested to account for this effect of AMF (Schouteden et al., 2015). One mechanism is through the changes in composition of microbial communities residing in the mycorrhizosphere. There is strong evidence that shifts in microbial community structure and the resulting microbial changes can influence the growth and health of plants (Rouphael et al., 2015). A major mechanism is nutritional, because plants with a high phosphorus status are less sensitive to pathogen damage. Nonnutritional mechanisms are also important, because AMF-colonized and noncolonized plants with the same internal phosphorus concentration may still be infected by pathogen at different level (Cardoso and Kuyper, 2006). Such nonnutritional mechanisms include the activation of plant defense systems, changes in secretion and concentration of root exudates and associated changes in mycorrhizosphere, increased lignification of cell walls, and competition for space for colonization along with site of infection. Activation of plant defense mechanisms, including the development of systemic resistance, has also been proposed, but the occurrence of this mechanism, and its effect in biocontrol, needs to be evaluated further (Vos et al., 2012).

9.4.1 Physical interferences

9.4.1.1 Alteration in root and root-associated tissues morphology

Besides an increased nutrient uptake, AMF colonized plants frequently showing enhanced growth of roots and healthy branches. The root morphological responses resulting from the colonization of mycorrhizal fungi somehow depend on plant characteristics, like gaining more benefits from mycorrhizal fungi which colonized tap roots in comparison with fibrous roots for gaining biomass and acquisition of nutrient. It has been demonstrated that AM colonization induces remarkable changes in root system morphology, as well as in the meristematic and nuclear activities of root cell (Yang et al., 2014). This might affect rhizosphere interaction and particularly pathogen infection development. The most consequence of AM colonization is an increase in branching, resulting in a relatively larger proportion of higher-order roots in the root system (Hooker et al., 1994). Moreover, an increase in lignin concentration in plant tissues is reported to inhibit the entry of nematode in AMF-colonized plant. Additionally, mycorrhizae also confers strong vascular network to plant root, and as a result, the flow of nutrient is also increased, which resulted in greater mechanic strength to plant tissues and subsequently reduced pathogen attack. Smaller syncytia with fewer cells were found to increase the resistance power of the host toward the nematodes. AMF application positively resulted from an elevation in plant root vigor, due to a higher uptake of nutrient, can possibly compensate the suppressed root growth caused by nematode. A decrease in the number of root branches of banana due to endoparasitic nematodes *R. similis* and *P. poffeae* seems to be compensated by an increase in branching due to colonization by the AMF *Funneliformis mosseae*. However, an increase in root branching has also an adverse effect on the host plant; it will increase the number of active infection sites, which depends on the type of nematode and the plant type. PPN, such as *R. similis*, prefers primary roots of the plant. In case of RKN and cyst nematodes, the root elongation zones and lateral root formation site are preferred for infection, as there is more secretion of root exudates, which attract the nematodes and can provide nutrients to nematodes (Curtis et al., 2009).

9.4.2 Biochemical and physiological disturbances
9.4.2.1 Availability of nutrients

Mycorrhizal fungi can increase water and nutrients supply for host plant, not only phosphorus, potassium, and nitrogen but also micronutrients such as zinc, iron, and magnesium (Baum et al., 2015). As the AM fungi are obligate biotrophic in nature, it receives prepared photosynthates from their host plant. Into similar line for plant protection from various abiotic stresses, viz., water, low temperature, or heavy metal toxicity, these fungi can also recompense the harm caused by nematodes. Even though an increased supply of phosphorus is found one of the major mechanisms as the mycorrhizal fungi governed biological control; on the contrary, the addition of phosphorus to nonmycorrhized plants was not found to gain the same level of reduction of nematode infestation (Singh et al., 2011). Opposite to the above facts, Fritz et al. (2006) conducted a study demonstrating the effect of phosphorus supplementation in mycorrhized and nonmycorrhized tomato plant on disease severity caused by *A. solani*. They have concluded that tomato plants colonized by *Rhizophagus irregularis* showed significantly less symptoms of the disease in comparison with nonmycorrhized plants, without an increase in P uptake. On the contrary, an addition of P resulted in an increase in disease severity. On the other hand, there could not be a positive effect on host plant, as in few studies suppressed plant growth was also evident as a result of AMF colonization (Smith and Smith, 2011).

Hosts with higher nutrient content are able to resist greater nematode infection, as observed in cotton infested with nematode *Rotylenchulus reniformis*. Regression analysis of nematode number against the nutrient in rice has also shown a positive relationship between migratory ectoparasitic *Helicotylenchus* spp. and magnesium concentration; on the other hand, a negative relationship was also recorded between endoparasitic nematode *Pratylenchus zeae* and zinc or iron and between *M. incognita* and magnesium and calcium contents (Coyne et al., 2004). These type of experiments have shown that the nutrient contents of the host plant are found to influence PPN populations. However, till date, there are no solid evidences recorded, which can establish that the mycorrhizal fungi-mediated nutrient uptake is one of the mechanisms for higher resistance against parasitic nematodes in plant.

9.4.2.2 Struggle for plant photosynthates and place for infection

For AMF, neither antibiosis nor hyperparasitism to plant pathogen has been reported so far; only direct effect of mycorrhizal fungi on plant pathogen through competition for nutrient and niche has been studied. As the mycorrhizal fungi and pathogen have the same physiological requirement, both are living in the same microenvironment and rely on common host particularly for infection site to obtain essential nutrients such as carbon and other photosynthates (Vos et al., 2014). The excess carbon requirement of AMF sometime inhibits the growth of plant pathogen (Parihar et al., 2020). Smith (1987) suggested that PPNs require host nutrients for multiplication and colonization and direct battle with AMF has been hypothesized as a means of their inhibitors. Since AMF, soil-dwelling phytopathogens, and PPN colonize the same root tissues, direct competition for niche/space has been considered as a mode of pathogen inhibition by AMF (Hussey and Roncadori, 1982; Linderman, 1988). Both localized and nonlocalized mechanisms could be responsible for competition for colonization sites mainly depending on the pathogen (fungus, nematode) and pathogen development is reduced due to the lack of resources and food supply (Tylka et al., 1991).

Competition for space could also play a major role in AMF-PPN interaction as they both hosted at the same site (Jung et al., 2012). Negative effects due to limitation in space can be exerted on PPN as arbuscules of AMF are exclusively formed inside the cortex region of root, which is also feeding site for endoparasitic nematode. Competition for space between mycorrhizal fungi and PPN could be put into consideration where the feeding cells extend into the root cortex tissues. Following DNA sequencing, del Mar Alguacil et al. (2011) observed that root galls produced by *M. incognita* in *Prunus persica* could be colonized by mycorrhizal fungi. However, AMF arbuscules are short-lived entities (Javot et al., 2011), and it seems complicated to discriminate that mycorrhizal fungi or nematode infected the infection site first.

9.4.2.3 Alteration in biochemical and chemical composition of root tissues

A best example of biological control is through the conservation of function of plant root system by fungal hyphae growing out into soil with an increase in the surface area of the roots for absorption and by maintaining the activity of root cells through arbuscules formation (Schouteden et al., 2015), which might affect rhizosphere interactions, particularly the development of pathogen infection. Mycorrhizal fungi colonization has also conferred structural and biochemical changes in the cell walls of the plants. Synthesis of β-1,3-glucans has been reported in mycorrhizae of pea. β-1,3-Glucans have also been detected within the structural host wall material around the point of penetration of mycorrhizal hyphae into plant cell (Sharma and Sharma, 2017). Triggering of specific plant defense response due to mycorrhizal colonization can be a most obvious basis for the protective ability of mycorrhizal fungi. Current research using modern molecular techniques is being used to detect substances and/or reactions elicited by mycorrhizal fungi involved in plant protection. According to Pozo et al. (2002), compounds like β-1,3-glucanases, peroxidases, callose, phytoalexins, enzymes of the phenylpropanoid pathway, chitinases, pathogenesis-related (PR) proteins, hydroxyproline-rich glycoproteins (HRGP), and some phenolics are secreted as a part of plant defense in reaction to mycorrhizal fungi colonization in plant root. There is evidence for an increase in phytoalexins in mycorrhizal roots infected by *Rhizoctonia solani*. It was reported that phytoalexin medicarpin increases during early stages of colonization in *Medicago truncatula*, but decreases slightly during later stages (Kaur and Suseela, 2020). Ismail and Hijri (2020) studied that mycorrhizal fungi may induce defense responses in potato to protect against infection with *Fusarium sambucinum*. They observed that, in response to *F. sambucinum* infection, the mycorrhizal treatment upregulated the expression of all defense genes except OSM-8e in potato roots at 72 and 120h postinfection. Overall, they have observed that mycorrhizae are a systemic bio-inducer and their effect could extend into noninfected parts; thus, mycorrhizal fungi significantly suppressed disease severity of *F. sambucinum* on potato plants compared with those infected and nonmycorrhizal plants with a decrease in the negative effects of *F. sambucinum* on biomass and potato tuber production. Furthermore, Lambais (2000) showed that the elicitor derived from an extract of extraradical mycelium of *Glomus intraradices* was able to induce phytoalexin synthesis in soybean cotyledons. Further results have shown that *G. intraradices* induces the expression of chalcone synthase, the first enzyme in metabolism of flavonoid compound, such as phytoalexin, in *M. truncatula* (Bonanomi et al., 2001). Increased levels of phytoalexins in mature mycorrhizal soybean roots could be elicited by the release of fungal cell wall molecules in senescent arbuscules. Pathogenesis-related proteins (PR proteins) are also induced in plants during resistance to pathogen infection. PR proteins are synthesized locally and in small amounts. For example, PR-b_1 proteins are only synthesized in cells containing living arbuscules.

Measurements of specific enzymes allow selective measurements of different fungi in the same root system. Alterations in the isoenzymatic patterns and biochemical properties of some defense-related enzymes such as chitinases, chitosanases, and β-1,3-glucanases have previously been shown during mycorrhizal colonization of tomato roots, with the induction of new isoforms (Pozo et al., 2002). These hydrolytic enzymes are believed to have a role in defense against invading fungal pathogens because of their potential to hydrolyze fungal cell wall polysaccharides (Adavi et al., 2020). Plant chitinase and/or β-1,3 glucanase elicitation has been reported in AM roots and plays a role as antifungal against soil/root pathogens. Mycorrhiza-induced chitinases isoforms appear to be a general phenomenon in AM roots. These chitinases release oligosaccharide elicitors from chitinous AM fungal cell walls, which in turn stimulate the general defense responses of plants (Kaur and Suseela, 2020). Peroxidase activity has been associated with epidermal and hypodermal cells, and increased in mycorrhizal roots, a process that can contribute to higher resistance to certain root pathogens. Moreover, wall-bound peroxidase activity has been observed during initial stages of mycorrhizal colonization; during later stage, its activity decreased. However, the activity was not detectable from the cells colonized by intracellular hyphae (Blilou et al., 2000).

9.4.3 Alteration in microflora of mycorrhizal rhizosphere

Mycorrhizal fungi play a very important role in nutrient uptake for plant as well as protection from various biotic and abiotic factors. Moreover, the mycelia of mycorrhizae fungi also secrete exudates composed of carbon and other elements into its rhizosphere zone peculiarly known as "mycorrhizosphere," which in turn enhances the formation of beneficial microbial communities along with soil aggregation. The beneficial microbial communities are composed of bacteria, fungi, protozoa, nematodes, and arthropods. They inhabit the mycorrhizosphere and interact with each other as well as with the plant system through direct or indirect mechanisms. These activities include the promotion of the microbial activity and population with the change in pH, release of nutrient from organic matter and nutrient recycling, protection against plant pathogens, advanced mycorrhizal formation, and changes in soil physico-chemical structure (Priyadharsini et al., 2016).

Meyer and Linderman (1986) demonstrated that the exudates secreted from maize and chrysanthemum root can effectively reduce the formation of sporangium and zoospores by the phytopathogen *Phytophthora cinnamomi*. In the same way, Secilia and Bagyaraj (1987) conducted a pot experiment on bacterial and actinobacterial populations of different AMF and analyzed them quantitatively as well as qualitatively. During their experiment, they have observed that pot cultures of *G. fasciculatum* possess higher number of actinobacterial antagonist to *Fusarium solani* and *Pseudomonas solanacearum* in comparison with nonmycorrhizal plants or plants colonized with other mycorrhizae. Siddiqui and Akhtar (2008) demonstrated the effects of *G. intraradices* and *Pseudomonas putida* alone and in combination with fertilizers on growth of tomato and the reproduction of root-knot nematode *M. incognita*. They have observed that the application of *P. putida* increased plant growth in comparison with *G. intraradices,* while combined application of both along with compost manure leads to greater increase than individual with a reduction in galling and nematode multiplication.

9.5 Mycorrhizal approaches for the management of plant pathogens

9.5.1 AMF integration with other approaches

In case of AMF found to reduce disease incidences, the approaches probably responsible for the suppression of AMF populations and their biocontrol potential like injudicious application of agrochemicals as well as unorganized tillage should be followed in a judicious manner. As AMF can improve the nutrient uptake of host plant and in addition also provide resistance to phytopathogens, it seems possible to opt for need-based tillage practices, which could help in effective colonization and establishment to plant root. According to Brito et al. (2012), frequent tillage can adversely influence AMF colonization on plant root as the extraradical hyphae get disturbed along with disturbance to other fungi. On the other hand, few studies have reported the long-term effects of cultural practices on rhizosphere microbiota. Mader et al. (2000) studied AMF colonization with 7-year-old alike crop rotation and tillage management practices in pot experiment, which were varying in only quantity and kind of fertilizer used. They have observed that the root length of colonized plant with AMF was recorded 30%–60% high in plants cultivated in soil of low-input farming practices in comparison with the plant cultivated in traditionally fertilized soil. Buysens et al. (2015) demonstrated that fungicides azoxystrobin and pencycuron applied at the recommended dose for *Rhizoctonia solani* control did not have an adverse effect on germination of spores and potato root colonization by *Rhizophagus irregularis*, while the development of extra-radical mycelium and spore production was found to reduce at 10 times of the recommended dose. While fungicide flutolanil at the recommended dose not affected adversely on spore germination, extra-radical growth of mycelium, root colonization, or arbuscular development. Based on their observations, they have concluded that the listed antifungal agrochemicals and AMF can be potentially used to manage phytopathogen *Rhizoctonia* in potato. Co-inoculation of AMF along with plant growth-promoting bacteria (PGPB) can effectively increase plant growth with an effective management of phytopathogens (Singh, 1992). Several studies have demonstrated that AMF showed a biocontrol potential against phytopathogens. Whether AMF could be used for biological control practically or possibly can act as carrier/mentor for PGPB having antibiosis potential, remains to be explored. Hence, combined inoculation effect of AMF with other biological control agents needs to be studied through pot as well as field experiment (Nanjundappa et al., 2019).

9.5.2 Host-symbiont-soil nutrient status

More intensive research should be carried out to study how AMF affect the host-pathogen relationship in a different way. The magnitude of mycorrhizae colonization depends on the biotrophism of the strain and the nutrient/fertility status of the soil. Basic research should be carried out to understand whether a native single AMF species or a mixture of AMF colonize and stimulate plant growth of that particular area of fairly fertile soil. Moreover, it is also important to access the responses of mycorrhized plant to phytopathogens in particular soil type, condition, and fertility level. A smart move would be to use different phosphorus level in soil, which includes field recommendation to produce nonmycorrhized plants of equal size and the same phosphorus level as mycorrhized plant. These conditions are important to understand the actual mechanism of resistance against phytopathogen infection imparted by mycorrhized plant (Emmanuel and Babalola, 2020).

9.5.3 Initial inoculum density and inoculations sequence of pathogen and symbionts

It seems difficult to come to exact interpretation regarding the magnitude of biological control effect unless a series of phytopathogen concentrations are being tested. While changing the phytopathogen concentrations to generate varying levels of disease infection or yield reduction, one should not solely use phytopathogen concentrations, which can cause adversely affected plants; otherwise, AMF are biased for not giving an equal chance for colonization and growth promotion before the challenge of nematode attack (Gough et al., 2020). As the pathogenic effect on root by fungi or PPN precedes AMF colonization, the sequence by which plants are inoculated can make effect on the type of interaction. So, an application strategy has been followed in most of the experiments where plants colonized using AMF 2–4 weeks before exposure of them against the phytopathogen (Sharma et al., 1994). The time of AMF application and colonization, whether it is before, along with, or after application with PPN, affect the impact of AMF for the management of PPN to a great extent. Research carried out by several scientific groups across the world emphasized on preinoculations of AMF was found to suppress PPN reproduction, growth, and development on roots at a major extent, in comparison with plants applied both the organisms simultaneously (Anjos et al., 2010).

Following these techniques, the AMF has been allowed sufficient time for colonizing host roots before they have been attached by the PPN; however, this technique can be adopted for protected cultivation or transplanted crops only. Sharma and Bhargava (1993) reported that preinoculations of tomato with AMF *G. fasciculatum* were found to reduce gall number and egg mass/plant and number of egg/mass of RKN *M. incognita*. Moreover, they have also observed that for pot and field experiments, the magnitude of nematode management was different as the nematode population was reduced up to 58% in pot in comparison with field, i.e., 36%. As an evident from several experiments, it is considered that the AMF are slow colonizer to host plant root and a minimal degree of root colonization is mandatory for an effective management of ill effects of phytopathogens. Inoculum dose of 1.0–5.0 IP/gram of soil is recommended for optimum growth and to achieve effective root colonization. On the other hand, such threshold levels of colonization required to manage PPN activities are being reported but higher levels of AMF colonization have not been reported to manage the magnitude of root infections by phytopathogens (Elsen et al., 2003). Moreover, a comparison of mycorrhized and phosphorus-fertilized nonmycorrhized plants of the same stage showed that the nonmycorrhized one was found to be more susceptible to PPN attack, indicating there are factors other than phosphorus nutrition in the interactions (Wanjohi et al., 2002).

9.6 Production and commercialization of AMF

9.6.1 Conventional methods

Several technologies are available for mass production of AMF for commercial production as well as for on-farm production (Sharma and Sharma, 2006), pot culture techniques (Kokkoris and Hart, 2019), nutrient film technique, and aeroponics (Das et al., 2020). Pot culture technique using a suitable host in sterile soil is the most widely accepted technique for obtaining more numbers of infective propagules (IP) for the propagation of mycorrhizal fungi (Kokkoris and Hart, 2019). In pot and field multiplication, there are many factors, which govern AMF development besides the host plant, like temperature, light intensity, size of the container, soil nutrient status, and the particle size of the growth substrate (Danesh and Tufenkci, 2017). Any humid environment, which inhibits the initial root development, can negatively affect the AMF colonization. Similarly, soil temperature is very much crucial factor for the growth of fungi and subsequently affects the host growth (Kilpeläinen et al., 2020). Cultures reaching the desired number of infective propagules (IP) can be stored using appropriate methods followed by dehydration. Moreover, AMF can be cultivated/multiplied on host plant using several substrates, like pebbles, sand particles, peat, expanded clay, perlite, vermiculite, soilrite, rockwool, and glass beads (Coelho et al., 2015). Apart from the above-mentioned environmental and physical factors, biological factors also play an important role in rapid culturing of AM fungi in pots. Bhowmik and Singh (2004) demonstrated the impact of PGPB in enhancing the mycorrhizal colonization of *Glomus mosseae* in pots on *Chloris gayana*.

Mycorrhizal fungi can also be produced by adopting aeroponics using suitable host plant seedlings roots precolonized using AMF and the modified Hoagland's solution having very low phosphorus (Hoagland and Arnon, 1938). In the aeroponic system, AMF colonization and spore formation were found to be excellent over the soil-dependent pot cultivation. *Acculospora kentinensis* has been successfully mass-produced using *Paspalum notatum* and *Ipomoea batatas* through aeroponic technique (Lakhiar et al., 2018). Mass multiplication of *G. intraradices* through aeroponic technique was also reported by some workers, where they compared the conventional atomizing disk with the ultrasonic nebulizer technology as misting sources (Mohammad et al., 2000). Further, Das et al. (2020) suggested that nutrient film technique (NFT) can also be utilized for the AMF mass production. Cultured host plants are arranged on a tending tray, on which a flow of a layer of nutrient solution is observed. As in the aeroponic technique, host plant seedlings need to be precolonized in separate medium. A modified nutrient film technique has been proposed to mass-multiply the AMF with the help of improvement in airflow and in between nutrient supplementation, with minimal P supply and use of glass beads as supporting materials.

9.6.2 In vitro/axenic cultivation of AM fungi (root organ culture)

Mycorrhizal fungi first time multiplied through in vitro culture in the early 1960s; after that, several milestones were established for axenic culturing of fungi. Meanwhile, Fortin et al. (2002) demonstrated a well-planned evaluation of the root organ culture technology and explained the basic improvements needed following which the AMF can more vigorously colonize the host plant roots. The axenic cultivation is the most striking mass multiplication technology to obtain pure, viable, rapid, and contaminant-free mother culture in limited space with several advantages over pot culture technique. The monospecific strains available can be used directly as starting material for large-scale inoculum production; a sole Petri dish culture is enough to generate sufficient spores in number and meters of hyphae within 4 months (Tiwari et al., 2003; Cranenbrouck et al., 2005).

Monoxenic cultures are developed through extraradical spores and infected root fragments. Sporocarps of *G. mosseae* have also been attempted to establish in vitro cultures of AM fungi. Spore sterilization can also be achieved with a solution consisting of chloramine T (oxidizing agent), Tween 20 (surfactant) and rinsed in an antibiotic solution (streptomycin and gentamycin). While the success of spore germination is solely dependent on sterilization, the presence of root exudates and 2% CO_2 can stimulate germination and postgermination mycelia development (Lakhiar et al., 2018). The infected roots, which come from trap culture roots, can be used as a section of vesicles for in vitro establishment. The root sections can be disinfected in an ultrasonic processor under aseptic conditions and incubated in MSR (modified Strullu-Romand) medium (Declerck et al., 1998). Mycelial growth from root pieces is usually observed within 2–15 days. Using segments of vesicles has more advantages over spores as far as contamination is concerned but the recovery of propagules is low compared to spores. Therefore, the vesicles are merely used for day-to-day inoculation for in vitro establishment. Regular subculturing is required to maintain higher infectivity. Continuous cultures can be obtained following the transfer of mycorrhized roots to fresh medium either with or without spores. It is always preferable to use actively growing roots.

Large-scale production of spores is a prerequisite, and several available statistical models can be used as descriptive and predictive tools for sporulation dynamics study (Declerck et al., 1998). Several forms of inoculum may be used, and the substrate may influence the effect of the inoculum. Different inoculums based on inert or sterilized substrates such as peat, expanded, or calcined clays can be utilized on commercial scale and are less vulnerable to contaminants (Coelho et al., 2015). AMF are available on commercial scale in different formulations and being sold by different manufactures as described in Table 9.3.

Table 9.3 List of commercial products and sources of AM inoculants and products.

Company name and location	Product name
Agronomic (P) Ltd., India	Majestic
Agriland Biotech (P) Ltd., India	Mycozone
Cadila Agritech (P) Ltd., India	Josh
Zydex Industries, India	Zytonic-M
Anand Agro Care, India	Dr. Bacto's VAM
IPL Biologicals Ltd., India	VAM SHAKTI, VAMLET & VAM-HD
Mycorrhizal Application, United States	MycoApply Endomaxx
Helena Agri-Enterprises, LLC, United States	Myco-Vam
Valent USA. LLC Agricultural Products, United States	MycoApply
Tainio Biologicals, Inc., United States	MycoGenesis
JH Biotech, Inc., United States	MYCORMAX
AgroScience Solutions, LLC, United States	Organic Mycorrhizal Fungi
Tainio Biologicals, Inc., United States	Spectrum + Myco
Vegalab Inc., United States	MYCO BIOBOOST
Tainio Biologicals, Inc., United States	Rhizogenesis
Shemin Garden, LLC, United States	Ecofungi
Pathway BioLogic, LLC, United States	Managefungi
Symborg Inc., Spain	MycoUp Activ
Symborg, Spain	ResidHC
Koppert, The Netherlands	Panoramix
INIFAP, México	Micorriza INIFAP
Plant Health Care, Mexico	MycorTree
OBA, Mexico	HIPER-GLOM
Vergel de Occidente, Mexico	Tec-Myc 60
Biokrone, Mexico	Glumix
Biofabrica Siglo, Mexico	Micorrizafer plus
BIOMIC, Mexico	TM-73
Italpolina SpA/ATENS, Italy/Spain/Chile	Aegis Gel
GroundWork BioAG, Israel	Rootella F, Rootella X
Agronutrition, France	CONNECTIS
MYCOSYM TRITION SL/Biosim, Chile	Mycosim Tri-Ton
PremierTech, Canada	Activ Plus
Symborg Business Development, Chile and Spain	MycoUp
S.L./Symborg Chile SpA, Chile and Spain	Resid HC

9.7 Conclusion

Due to complexity of the micro-soil-plant agro-ecosystem and the influence of key prevailing environmental situation, it is difficult to reach practical conclusions. We can certainly find the appropriate amalgamation of different features to exploit the prophylactic capabilities of mycorrhizal fungi. From the above facts, it can be concluded that (1) mycorrhizal associations can reduce the incidence and damage caused by nematodes, (2) the capabilities of mycorrhizal fungi to impart resistance/tolerance to host plant roots through symbiosis vary with the different genus/species of mycorrhizae, and (3) it is not possible to protect the plants effectively from all pathogens as it is also governed by soil and other environmental factor. Hence, to achieve the high degree of control in sustainable agro-ecosystems predominantly with nursery and horticulture crops in the future, we must use indigenous and appropriate arbuscular mycorrhizal fungi preferably co-inoculated with other plant growth-enhancing microorganism in the soil. The soil and crop cultural techniques, which may sustain a level of effective AMF colonization and other antagonistic to suppress disease, must be taken into consideration. Besides the assessment of the level of inoculums required to suppress a particular pathogen, the techniques of large-scale arbuscular mycorrhizal fungi production and application in the field must be standardized.

References

Adavi, Z., Tadayon, M.R., Razmjoo, J., et al., 2020. Antioxidant enzyme responses in potato (*Solanum tuberosum*) cultivars colonized with arbuscular mycorrhizas. Potato Res. 63, 291–301. https://doi.org/10.1007/s11540-019-09440-1.

Alban, R., Guerrero, R., Toro, M., 2013. Interactions between a root-knot nematode (*Meloidogyne exigua*) and arbuscular mycorrhizae in coffee plant development (*Coffea arabica*). Am. J. Plant Sci. 4, 19–23. https://doi.org/10.4236/ajps.2013.47A2003.

Al-Raddad, A., Ahmad, M., 1995. Interaction of *Glomus mosseae* and *Paecilomyces lilacinus* on *Meloidogyne javanica* of tomato. Mycorrhiza 5 (3), 233–236.

Anjos, E., Cavalcante, U., Gonçalves, D., Pedrosa, E., Santos, V., Maia, F., 2010. Interactions between an arbuscular mycorrhizal fungus (*Scutellospora heterogama*) and the root-knot nematode (*Meloidogyne incognita*) on sweet passion fruit (*Passiflora alata*). Braz. Arch. Biol. Technol. 53 (4), 801–809. https://doi.org/10.1590/S1516-89132010000400008.

Anli, M., Kaoua, M., ait-el-Mokhtar, M., Boutasknit, A., ben-Laouane, R., Toubali, S., Baslam, M., Lyamlouli, K., Hafidi, M., Meddich, A., 2020. Seaweed extract application and arbuscular mycorrhizal fungal inoculation: a tool for promoting growth and development of date palm (*Phoenix dactylifera* L.) cv Boufgous. S. Afr. J. Bot. 132, 15–21.

Baghel, P.P.S., Bhatti, D.S., Jalali, B.L., 1990. Interaction of VA mycorrhizal fungus and *Tylenchulus semipenetrans* on citrus. In: Jalali, B.L., Chand, H. (Eds.), Proceedings of the National Conference on Mycorrhiza. Haryana Agricultural University, Hisar February 14–16, 1990. TERI, New Delhi, p. 210.

Barea, J.M., Jeffries, P., 1995. Arbuscular mycorrhizas in sustainable soil plant systems. In: Hock, B., Varma, A. (Eds.), Mycorrhiza Structure, Function, Molecular Biology and Biotechnology. Springer, Heidelberg, pp. 521–559.

Baum, C., El-Tohamy, W., Gruda, N., 2015. Increasing the productivity and product quality of vegetable crops using arbuscular mycorrhizal fungi: a review. Sci. Hortic. 187, 131–141. https://doi.org/10.1016/j.scienta.2015.03.002.

Bhowmik, S.N., Singh, C.S., 2004. Mass multiplication of AM inoculum: effect of plant growth-promoting rhizobacteria and yeast in rapid culturing of *Glomus mosseae*. Curr. Sci. 86, 705–709.

Blilou, I., Bueno, P., Ocampo, J.A., Garcıa-Garrido, J.M., 2000. Induction of catalase and ascorbate peroxidase activities in tobacco roots inoculated with the arbuscular mycorrhizal fungus *Glomus mosseae*. Mycol. Res. 104, 722–725.

Bonanomi, A., Oetiker, J.H., Guggenheim, R., Boller, T., Wiemken, A., Vögeli-Lange, R., 2001. Arbuscular mycorrhizas in mini-mycorrhizotrons: first contact of *Medicago truncatula* roots with *Glomus intraradices* induces chalcone synthase. New Phytol. 150, 573–582.

Brito, G.M., Carvalho, M., Chatagnier, O., Tuinen, D., 2012. Impact of tillage system on arbuscular mycorrhiza fungal communities in the soil under Mediterranean conditions. Soil Tillage Res. 121, 63–67. https://doi.org/10.1016/j.still.2012.01.012.

Buysens, C., Dupré de Boulois, H., Declerck, S., 2015. Do fungicides used to control *Rhizoctonia solani* impact the non-target arbuscular mycorrhizal fungus *Rhizophagus irregularis*. Mycorrhiza 25, 277–288. https://doi.org/10.1007/s00572-014-0610-7.

Cardoso, I.M., Kuyper, T.W., 2006. Mycorrhizas and tropical soil fertility agriculture. Ecosyst. Environ. 116, 72–84.

Ceustermans, A., Hemelrijck, W.V., Campenhout, J.V., Bylemans, D., 2018. Effect of arbuscular mycorrhizal fungi on *Pratylenchus penetrans* infestation in apple seedlings under greenhouse conditions. Pathogens 7, 76. https://doi.org/10.3390/pathogens7040076.

Chandrashekar, D.S., Shetty, H.S., Datta, R.K., 1995. Effect of VAM fungus, *Glomus fasciculatum* and *Acaulospora laevis* against root-knot nematode of mulberry. Mycorrhiza News 7 (2), 7–8.

Chitwood, D.J., 2002. Phytochemical based strategies for nematode control. Annu. Rev. Phytopathol. 40, 221–249.

Coelho, I.R., Pedone-Bonfim, M.V., Silva, F.S., Maia, L.C., 2015. Optimization of the production of mycorrhizal inoculum on substrate with organic fertilizer. Braz. J. Microbiol. 45 (4), 1173–1178. https://doi.org/10.1590/s1517-83822014000400007.

Coyne, D.L., Sahrawat, K.L., Plowright, R.A., 2004. The influence of mineral fertilizer application and plant nutrition on plant-parasitic nematodes in upland and lowland rice in Côte d'Ivoire and its implications in long term agricultural research trials. Exp. Agric. 40, 245–256. https://doi.org/10.1017/S0014479703001595.

Cranenbrouck, S., Voets, L., Bivort, C., 2005. Mthodologies for in vitro cultivation of AM fungi with root organs. In: Declerck, S., Strullu, D.G., Fortin, A. (Eds.), *In Vitro* Culture of Mycorrhizas. Springer Verlag Berlin Heidelberg, pp. 342–375.

Curtis, R., Robinson, A., Perry, R., 2009. Hatch and host location. In: Perry, R.N., Moens, M., Starr, J.L. (Eds.), Root-Knot Nematodes. CAB International, Wallingford, pp. 139–162.

Danesh, Y.R., Tufenkci, S., 2017. *In vitro* culturing of mycorrhiza and mycorrhiza like fungi. Int. J. Agric. Tech. 13 (7.2), 1675–1689.

Das, D., Torabi, S., Chapman, P., Gutjahr, C., 2020. A flexible, low-cost hydroponic co-cultivation system for studying arbuscular mycorrhiza symbiosis. Front. Plant Sci. 11, 63. https://doi.org/10.3389/fpls.2020.00063.

Declerck, S., Strullu, D.G., Plenchette, C., 1998. Monoxenic culture of the intraradical forms of *Glomus* sp. isolated from a tropical ecosystem: a proposed methodology for germplasm collection. Mycologia 99, 579–585.

del Mar Alguacil, M., Torrecillas, E., Lozano, Z., Roldán, A., 2011. Evidence of differences between the communities of arbuscular mycorrhizal fungi colonizing galls and roots of *Prunus* persica infected by the root-knot nematode *Meloidogyne incognita*. Appl. Environ. Microbiol. 77, 8656–8661. https://doi.org/10.1128/AEM.05577-11.

Elsen, A., Baimey, H., Swennen, R., De Waele, D., 2003. Relative mycorrhizal dependency and mycorrhiza-nematode interaction in banana cultivars (*Musa* spp.) differing in nematode susceptibility. Plant Soil 256, 303–313. https://doi.org/10.1023/A:1026150917522.

Emmanuel, O.C., Babalola, O.O., 2020. Productivity and quality of horticultural crops through co-inoculation of arbuscular mycorrhizal fungi and plant growth promoting bacteria. Microbiol. Res. 239. https://doi.org/10.1016/j.micres.2020.126569, 126569.

Fortin, J.A., Becard, G., Declerck, S., Dalpe, Y., Arnaud, M., Coughlan, A.P., Piche, Y., 2002. Arbuscular mycorrhiza on root-organ cultures. Can. J. Bot. 80, 1–20.

Fritz, M., Jakobsen, I., Lyngkjaer, M.F., Thordal-Christensen, H., Pons-Kühnemann, J., 2006. Arbuscular mycorrhiza reduces susceptibility of tomato to *Alternaria solani*. Mycorrhiza 16, 413–419. https://doi.org/10.1007/s00572-006-0051-z.

Gough, E.C., Owen, K.J., Zwart, R.S., Thompson, J.P., 2020. A systematic review of the effects of arbuscular mycorrhizal fungi on root-lesion nematodes, *Pratylenchus* spp. Front. Plant Sci. 11, 923. https://doi.org/10.3389/fpls.2020.00923.

Habte, M., Zhang, Y.C., Schmitt, D.P., 1999. Effectiveness of *Glomus* species in protecting white clover against nematode damage. Can. J. Bot. 77 (1), 135–139.

Hajra, N., Shahina, F., Firoza, K., Maria, R., 2015. Damage induced by root-knot nematodes and its alleviation by vesicular arbuscular mycorrhizal fungi in roots of *Luffa cylindrical*. Pak. J. Nematol. 33 (1), 71–78.

Hoagland, D.R., Arnon, D.I., 1938. The Water-Culture Method for Growing Plants Without Soil. University of California, College of Agriculture, Agriculture Experiment Station Circular 347, Berkeley, CA, USA.

Hooker, J.E., Gianinazzi, S., Vestberg, M., Barea, J.M., Atkinson, D., 1994. The application of arbuscular mycorrhizal fungi to micropropagation systems: an opportunity to reduce chemical inputs. Agric. Sci. Finl. 3, 227–232.

Hussey, R.S., Roncadori, R.W., 1982. Vesicular-arbuscular mycorrhizae may limit nematode activity and improve plant growth. Plant Dis. 66, 9–14.

Ismail, Y., Hijri, M., 2020. Arbuscular mycorrhisation with *Glomus irregulare* induces expression of potato PR homologues genes in response to infection by *fusarium sambucinum*. Funct. Plant Biol. 39 (3), 236–245. https://doi.org/10.1071/FP11218.

Jain, R.K., Sethi, C.L., 1987. Pathogenicity of *Heterodera cajani* on cowpea as influenced by the presence of VAM fungi *Glomus fasciculatum* or *G. epigaeus*. Ind. J. Nematol. 17 (2), 165–170.

Javot, H., Penmetsa, R.V., Breuillin, F., Bhattarai, K.K., Noar, R.D., Gomez, S.K., et al., 2011. *Medicago truncatula* mtpt4 mutants reveal a role for nitrogen in the regulation of arbuscule degeneration in arbuscular mycorrhizal symbiosis. Plant J. 68, 954–965. https://doi.org/10.1111/j.1365-313X.2011.04746.x.

Jung, S.C., Martinez-Medina, A., Lopez-Raez, J.A., Pozo, M.J., 2012. Mycorrhiza-induced resistance and priming of plant defenses. J. Chem. Ecol. 38, 651–664. https://doi.org/10.1007/s10886-012-0134-6.

Kaur, S., Suseela, V., 2020. Unraveling arbuscular mycorrhiza-induced changes in plant primary and secondary metabolome. Metabolites 10 (8), 335. https://doi.org/10.3390/metabo10080335.

Kilpeläinen, J., Aphalo, P.J., Lehto, T., 2020. Temperature affected the formation of arbuscular mycorrhizas and ectomycorrhizas in *Populus angustifolia* seedlings more than a mild drought. Soil Biol. Biochem. 146. https://doi.org/10.1016/j.soilbio.2020.107798, 107798.

Koffi, M.C., Vos, C., Draye, X., Declerck, S., 2013. Effects of Rhizophagus irregularis MUCL41833 on the reproduction of *Radopholus similis* in banana plantlets grown under in vitro culture conditions. Mycorrhiza 23, 279–288. https://doi.org/10.1007/s00572-012-0467-6.

Kokkoris, V., Hart, M., 2019. *In vitro* propagation of arbuscular mycorrhizal fungi may drive fungal evolution. Front. Microbiol. 10, 2420. https://doi.org/10.3389/fmicb.2019.02420.

Lakhiar, I.A., Gao, J., Syed, T.Z., Chandio, F.A., Buttar, N.A., 2018. Modern plant cultivation technologies in agriculture under controlled environment: a review on aeroponics. J. Plant Interact. 13 (1), 338–352. https://doi.org/10.1080/17429145.2018.1472308.

Lambais, M.R., 2000. Regulation of plant defence-related genes in arbuscular mycorrhizae. In: Podila, G.K., Douds, D.D. (Eds.), Current Advances in Mycorrhizae Research. The American Phytopathological Society, Minnesota, USA, pp. 45–59.

Linderman, R.G., 1988. Mycorrhizal interactions with the rhizosphere effect. Phytopathology 78, 366–371.

Mader, P., Edenhofer, S., Boller, T., Wiemken, A., Niggli, U., 2000. Arbuscular mycorrhizae in a long-term field trial comparing low-input (organic, biological) and high-input (conventional) farming systems in a crop rotation. Biol. Fertil. Soils 31, 150–156.

Marroa, N., Cacciab, M., Doucetb, M.E., Cabelloc, M., Becerraa, A., 2018. Mycorrhizas reduce tomato root penetration by false root-knot nematode *Nacobbus aberrans*. Appl. Soil Ecol. 124, 262–265.

Meyer, J.R., Linderman, R.G., 1986. Response of subterranean clover to dual inoculation with vesicular-arbuscular mycorrhizal fungi and a plant growth-promoting bacterium, *Pseudomonas putida*. Soil Biol. Biochem. 18, 185–190.

Mishra, A., Shukla, B.N., 1995. Studies on management of root-knot (*Meloidogyne incognita*) of tomato by *Glomus fasciculatum* and some pesticides. Mycorrhiza News 7 (4), 8–11.

Mohammad, A., Khan, A.G., Kuek, C., 2000. Improved aeroponic culture of inocula of arbuscular mycorrhizal fungi. Mycorrhiza 9, 337–339.

Nagesh, M., Reddy, P., 1997. Management of *Meloidogyne incognita* on *Crossandra undularefolia* using vesicular arbuscular mycorrhiza, *Glomus mossae* and oil cakes. Mycorrhiza News 9 (1), 12–14.

Nanjundappa, A., Bagyaraj, D.J., Saxena, A.K., et al., 2019. Interaction between arbuscular mycorrhizal fungi and *Bacillus* spp. In soil enhancing growth of crop plants. Fungal Biol. Biotechnol. 6, 23. https://doi.org/10.1186/s40694-019-0086-5.

Pandey, R., Bhargava, S., 2004. Phytochemicals in the management of phytoparasitic nematodes. Ann. Rev. Plant Pathol. 3, 309–329.

Parihar, M., Rakshit, A., Meena, V.S., et al., 2020. The potential of arbuscular mycorrhizal fungi in C cycling: a review. Arch. Microbiol. 202, 1581–1596. https://doi.org/10.1007/s00203-020-01915-x.

Pham, T., Giang, B.L., Nguyen, N.H., Yen, P., Hoang, M., Ha, B., Le, N., 2020. Combination of mycorrhizal symbiosis and root grafting effectively controls nematode in replanted coffee soil. Plants 9, 555.

Poveda, J., Abril-Urias, P., Escobar, C., 2020. Biological control of plant-parasitic nematodes by filamentous fungi inducers of resistance: *Trichoderma*, mycorrhizal and endophytic fungi. Front. Microbiol. 11, 992. https://doi.org/10.3389/fmicb.2020.00992.

Pozo, M.J., Slezack-Deschaumes, S., Dumas-Gaudot, E., Gianinazzi, S., Azcón-Aguilar, C., 2002. Plant defense responses induced by arbuscular mycorrhizal fungi. In: Gianinazzi, S., Schüepp, H., Barea, J.M., Haselwandter, K. (Eds.), Mycorrhizal Technology in Agriculture. Birkhäuser, Basel, https://doi.org/10.1007/978-3-0348-8117-3_8.

Priestel, G., 1980. Wechseibeziehung zwischen der endotropics Mycorrhiza und dem Wurzelgallennematoden Melotdogyne incogm. (Kofoid and White, 1919) Chitwood, 1949 an Gurke. Dissertation, Hannover, West Germany, 103 pp.

Priyadharsini, P., Rojamala, K., Ravi, R.K., Muthuraja, R., Nagaraj, K., Muthukumar, T., 2016. Mycorrhizosphere: the extended rhizosphere and its significance. In: Choudhary, D., Varma, A., Tuteja, N. (Eds.), Plant-Microbe Interaction: An Approach to Sustainable Agriculture. Springer, Singapore, https://doi.org/10.1007/978-981-10-2854-0_5.

Rouphael, Y., Franken, P., Schneider, C., Schwarz, D., Giovannetti, M., Agnolucci, M., Pascale, S., Bonini, P., Colla, G., 2015. Arbuscular mycorrhizal fungi act as biostimulants in horticultural crops. Sci. Hortic. 196, 91–108. https://doi.org/10.1016/j.scienta.2015.09.002.

Schouteden, N., De Waele, D., Panis, B., Vos, C.M., 2015. Arbuscular mycorrhizal fungi for the biocontrol of plant-parasitic nematodes: a review of the mechanisms involved. Front. Microbiol. 6, 1280. https://doi.org/10.3389/fmicb.2015.01280.

Schubler, A., Schwarzott, D., Walker, C., 2001. A new fungal phylum, the Glomeromycota, phylogeny and evolution. Mycol. Res. 105, 1413–1421.

Secilia, J., Bagyaraj, D.J., 1987. Bacteria and actinomycetes associated with pot cultures of vesicular-arbuscular mycorrhizas. Can. J. Microbiol. 33, 1069–1073.

Sharma, M.P., Bhargava, S., 1993. Potential of VAM fungus *G .fasciculatum* against the root-knot nematode *Meloidogyne incognita* on tomato. Mycorrhiza News 4 (4), 4–5.

Sharma, A.K., Johri, B.N., 2002. Arbuscular mycorrhiza and plant disease. In: Sharma, A.K., Johri, B.N. (Eds.), Arbuscular Mycorrhizae: Interactions in Plants, Rhizosphere and Soils. Oxford and IBH Publishing Co. Pvt. Ltd., New Delhi, pp. 69–96.

Sharma, M.P., Sharma, S.K., 2006. Arbuscular mycorrhizal fungi: an emerging bioinoculant for production of soybean. SOPA Digest 3 (9), 10–16.

Sharma, I.P., Sharma, A.K., 2017. Physiological and biochemical changes in tomato cultivar PT-3 with dual inoculation of mycorrhiza and PGPR against root-knot nematode. Symbiosis 71, 75–183. https://doi.org/10.1007/s13199-016-0423-x.

Sharma, M.P., Bhargava, S., Verma, M.K., Adholeya, A., 1994. Interaction between the endomycorrhizal fungus *Glomus fasciculatum* and the root-knot nematode, *Meloidogyne incognita* on tomato. Ind. J. Nematol. 24 (2), 34–39.

Sharma, M.P., Gaur, A., Tanu, Sharma, O.P., 2004. Prospects of arbuscular mycorrhiza in sustainable management of root and soil borne diseases of vegetable crops. In: Mukerji, K.G. (Ed.), Disease Management of Fruits and Vegetables. Fruit and Vegetable Diseases, vol. 1. Kluwer Academic Publishers, The Netherlands, pp. 501–539.

Siddiqui, Z., Akhtar, M., 2008. Effects of fertilizers, AM fungus and plant growth promoting rhizobacterium on the growth of tomato and on the reproduction of root-knot nematode *Meloidogyne incognita*. J. Plant Interact. 3 (4), 263–271. https://doi.org/10.1080/17429140802272717.

Siddiqui, Z.A., Mahmood, I., 1995. Some observations on the management of the wilt disease complex of pigeonpea by treatment with a vesicular arbuscular fungus and biocontrol agents for nematodes. Bioresour. Technol. 54, 227–230.

Sikora, R.A., 1992. Management of the antagonistic potential in agricultural ecosystem for the biological control of plant parasitic nematode. Annu. Rev. Phytopathol. 32, 245–270.

Singh, C.S., 1992. Mass inoculum production of vesiculararbuscular (VA) mycorrhizae. Impact of N2-fixing and Psolubilizing bacterial inoculation on VA-mycorrhiza. Zentralbl. Mikrobiol. 147, 503–508.

Singh, L.P., Gill, S.S., Tuteja, N., 2011. Unraveling the role of fungal symbionts in plant abiotic stress tolerance. Plant Signal. Behav. 6, 175–191. https://doi.org/10.4161/psb.6.2.14146.

Smith, G.S., 1987. Interaction of nematodes with mycorrhizal fungi. In: Veech, J.A., Dickson, D.W. (Eds.), Vistas on Nematology. Society of Nematologists, Hyattsville, MD, pp. 292–300.

Smith, S.E., Smith, F.A., 2011. Roles of arbuscular mycorrhizas in plant nutrition and growth: new paradigms from cellular to ecosystem scales. Annu. Rev. Plant Biol. 62, 227–250. https://doi.org/10.1146/annurev-arplant-042110-103846.

Suresh, C.K., Bagyaraj, D.J., 1984. Interaction between a vesicular arbuscular mycorrhiza and a root knot nematode arbuscular mycorrhiza and a root knot nematode and its effect on growth and chemical composition of tomato. Nematol. Mediterr. 12, 31–39.

Talavera, M., Itou, K., Mizukubo, T., 2001. Reduction of nematode damage by root colonization with arbuscular mycorrhiza (*Glomus* spp) in tomato—*Meloidogyne incognita* and carrot—*Pratylenchus penetrans* pathosystems. Appl. Entomol. Zool. 36, 387–392.

Talavera, M., Itou, M., Mizukubo, T., 2002. Combined application of *Glomus* sp. and *Pasturia penetrans* for reducing *Meloidogyne incognita* populations and improving tomato growth. Appl. Entomol. Zool. 37, 61–67.

Tiwari, P., Adholeya, A., Prakash, A., 2003. Commercialization of arbuscular mycorrhizal biofertilizer. In: Arora, D.K. (Ed.), Fungal Biotechnology in Agricultural, Food, and Environmental Applications. Marcel Dekker Inc., New York, NY, pp. 195–201.

Trapp, J.M., 1977. Selection of fungi for ectomycorrhizal inoculation in nurseries. Annu. Rev. Phytopathol. 15, 203–222.

Tylka, G.L., Hussey, R.S., Roncadori, R.W., 1991. Interactions of vesicular-arbuscular mycorrhizal fungi, phosphorus, and *Heterodera glycines* on soybean. J. Nematol. 23, 122–133.

Vaast, P., Caswell-Chen, E.P., Zasoski, R.J., 1998. Influence of a root-lesion nematode, Pratylenchus coffeae, and two arbuscular mycorrhizal fungi, *Acaulospora mellea* and *Glomus clarum* on coffee (*Coffea arabica* L.). Biol. Fertil. Soils 26, 130–135.

Vos, C., Van Den Broucke, D., Lombi, F.M., De Waele, D., Elsen, A., 2012. Mycorrhiza-induced resistance in banana acts on nematode host location and penetration. Soil Biol. Biochem. 47, 60–66. https://doi.org/10.1016/j.soilbio.2011.12.027.

Vos, C.M., Yang, Y., De Coninck, B., Cammue, B.P.A., 2014. Fungal (-like) biocontrol organisms in tomato disease control. Biol. Control 74, 65–81. https://doi.org/10.1016/j.biocontrol.2014.04.004.

Wanjohi, W.J., Waudo, S.W., Sikora, R., 2002. Effect of inorganic phosphatic fertilizers on the efficacy of an arbuscular mycorrhiza fungus against a root-knot nematode on pyrethrum. Int. J. Pest Manage. 48 (4), 307–313. https://doi.org/10.1080/09670870210149862.

Yang, H., Zhang, Q., Dai, Y., Liu, Q., Tang, J., Bian, X., et al., 2014. Effects of arbuscular mycorrhizal fungi on plant growth depend on root system: a meta-analysis. Plant Soil 389, 361–374. https://doi.org/10.1007/s11104-014-2370-8.

Plant growth-promoting and biocontrol potency of rhizospheric bacteria associated with halophytes

Kalpna D. Rakholiya, Mital J. Kaneria, Paragi R. Jadhav, and Satya P. Singh

Department of Biosciences (UGC-CAS), Saurashtra University, Rajkot, Gujarat, India

10.1 Introduction

Environmental stress adversely affects the agricultural crops by reducing growth and crop yield worldwide (Yang et al., 2009; Leontidou et al., 2020). Generally, the crops encounter two types of environmental stresses: biotic and abiotic. Agricultural crops are continuously exposed to abiotic stress that includes drought, nutrient deficiency, oxidative stress, salinity, metal toxicity, and extremes of temperatures. Biotic stress includes effects caused by living organisms, such as insects, bacteria, fungi, viruses, nematodes, and herbivores. Causative agents of the biotic stress directly deprive the host of nutritional availability, leading to the death of the host plant, thus accounting for enormous losses in agricultural crop production (Ma et al., 2020). On the other hand, abiotic stress factors, including drought, salinity, waterlog, temperature extremes, mineral nutrients, and heavy metals, also badly affect the growth as well as yield of crop plants worldwide.

10.2 Plant-microbe interactions and mitigation of abiotic stress

Plant rhizospheric microorganism interactions are complex, have a symbiotic relationship, and play a significant role as a biofertilizer to enhance the plant growth and control of pathogens. Application of rhizospheric bacteria in agricultural practices can be an effective strategy in abiotic and biotic stress management. The beneficial bacteria positively affect the physiology of the plants and improve soil fertility besides improving the overall health of the plants in harsh environmental conditions. Rhizospheric bacteria (PGPR) comprise diverse genera: *Pseudomonas, Bacillus, Arthrobacter, Halomonas, Serratia, Nitrinicola, Achromobacter, Polymyxa, Paenibacillus, Citrobacter, Enterobacter, Microbacterium, Lysinibacillus, Shigella, Klebsiella, Pantoea,* and *Burkholderia* (Rakholiya et al., 2017). However, there is an eminent need to explore, identify, and investigate many unidentified microorganisms present in the rhizospheric niche. In this context, applications of many newer strategies, such as metagenomics and metatranscriptomics need to be adapted toward developing sustainable agriculture.

10.2.1 Plant growth-promoting rhizospheric bacteria (PGPR) and their activities

PGPR promote plant growth directly by enhancing the uptake and thus the availability of the macro and micro essential nutrients, such as nitrogen, phosphorous, iron, and potassium and synthesis of plant growth hormones: cytokinin, auxin, and gibberellin. Additionally, the beneficial activities of PGPR are also reflected in the control of phytopathogens, production of various hydrolytic enzymes, hydrogen cyanide and siderophore production, besides inducing systemic resistance in plants.

10.2.2 Current scenario of soil salinity

Soil salinity is among the serious environmental issues on a global scale. It causes land degradation ultimately adversely affecting soil quality and crop productivity (Kaushal and Wani, 2016). It is a global threat to world agriculture reducing crop productivity. Salt stress induces several adverse effects on the morphological, biochemical, metabolic, and physiological traits of crops. In the Indian scenario, over 6.74 million hectares of land have been adversely affected by salinity (Mandal et al., 2018).

10.2.3 Effect of salt stress on the plant growth

High salinity in soils causes oxidative and water stress, ionic and osmotic imbalance, and nutrient deficiency. Besides, it affects other features of plant growth by inhibiting seed germination, vegetative growth, root development, photosynthetic machinery, and reproductive developments (Goyal et al., 2019; Ma et al., 2020; Jadav et al.,

2020). At high salt concentrations, the accessibility of Ca^{2+} and K^+ is adversely affected, while the accumulation of Na^+ and Cl^- is enhanced, which leads to ionic imbalance and reduced nutrient uptake in plants (Orozco-Mosqueda et al., 2020). Abiotic stress also generates oxidative stress through producing reactive oxygen species (ROS). ROS includes hydrogen peroxide, superoxide, hydroxyl radical, and singlet oxygen. They are highly reactive oxygen species and cause cellular damage through oxidation of lipids, proteins, and nucleic acids. Due to the ROS stress, spatial configurations of various membrane proteins or enzymes are disturbed, leading to enhanced membrane permeability and ion leakage, chlorophyll destruction, metabolism perturbations, and even severe injury or death of the plants (Lopez-Raez, 2016).

10.2.4 Mechanisms of halotolerant bacteria

Halotolerant plant growth-promoting bacteria are fast emerging as efficient biological tools to mitigate the adverse effects of abiotic stress and improve the growth of plants, concomitantly restoring the degraded saline soil. A wide range of bacteria are present in the rhizosphere regions exhibiting beneficial effects on plants (Leontidou et al., 2020; Rakholiya et al., 2018a,b). Certain bacteria possess specific features to alleviate stress due to their inherent genetic and metabolic capabilities (Rakholiya et al., 2017; Deng et al., 2019; Goswami and Deka, 2020; Patel et al., 2020).

Production of natural polysaccharides during unfavorable conditions is a well-known characteristic of HT-PGPR (halotolerant plant growth-promoting rhizospheric bacteria). Exopolysaccharides produced by HT-PGPR help in the formation of a watery-nutrient-rich layer around the root surface, known as rhizosheath (Arora et al., 2020). Rhizosheath serves as a physical barrier against deposition of ionic salts and also acts as an active site of the nutrient cycling, cation uptake, symbiotic association, nodulation, and maintaining osmotic equilibrium in the root region of the plants exposed to high salt concentrations. Halotolerant/halophilic bacteria produce antioxidant enzymes under stress conditions and protect the plant from ROS (Table 10.1).

Isolation and identification of rhizospheric bacteria from salty and arid ecosystems have received considerable attention, due to their potential use as an alternative and sustainable strategy to mitigate the ill effects of extreme environments. The results included in this chapter have focused to explore halophytes for the isolation, cultivation, and further chaptalization of bacteria for plant growth promotion and biocontrol efficacy.

Table 10.1 List of reported salt-tolerant halophytic rhizospheric bacteria and their role in plant growth promotion and biocontrol.

No.	Host plant	PGPR	Methods	References
1	Salsola stocksii and Atriplex amnicola	Bacillus sp., Halobacillus sp., Pseudomonas sp.	BI, PS, IAA, ACC, SP, HCN, AF, AMY, CEL, LIP, PE	Mukhtar et al. (2020)
2	Capparis decidua	Serratia marcescens	ACC, PS, IAA, SP, NF, HCN, BI, AF, PE	Singh and Jha (2016)
3	Salsola grandis	Arthrobacter sp., Bacillus sp.	PE, ST, IAA, PS, SP, BI	Kataoka et al. (2017)
4	Sesuvium verrucosum	Enterobacter cancerogenus, Vibrio cholerae, Bacillus subtilis, Escherichia coli	ST, BI, PS, NF, IAA, ACC	El-Awady et al. (2015)
5	Suaeda fruticosa	B. licheniformis	IAA, PS, NH3, SP, HCN, CHI, PE, BI	Goswami et al. (2014)
6	Salicornia brachiate	Brachybacterium saurashtrense sp. nov., Zhihengliuella sp., Brevibacterium casei, Haererehalobacter sp., Halomonas sp., Vibrio sp., Cronobacter sakazakii, Pseudomonas sp., Rhizobium radiobacter, and Mesorhizobium sp.	ST, NF, IAA,SP,ACC, BI, SG	Jha et al. (2012a)
7	Sonchus arvensis L., Solanum surattense Burm. F., Lactuca dissecta D. Don., and Chrysopogon aucheri	Rhizobium sp.	MP, BC, MC	Naz et al. (2009)
8	Atriplex leucoclada, Haloxylon salicornicum, Lespedeza bicolor, Suaeda fruticosa, and Salicornia virginica	Bacillus sp. and Arthrobacter pascens	CAT, PS, BI, SG	Ullah and Bano (2015)
9	Salicornia hispanica	Staphylococcus equorum	BP	Vega et al. (2020)
10	Cressa cretica	Planctomyces sp., Halomonas sp., and Jeotgalibacillus sp.	SG	Etemadi et al. (2020)
11	Limonium sinense	Glutamicibacter halophytocola KLBMP 5180	ST, SG, SP	Xiong et al. (2019)
12	Elaeagnus angustifolia L.	Bacillus amyloliquefaciens	SG	Pan et al. (2020)
13	Cenchrus ciliaris L.	Pseudomonas moraviensis and Bacillus cereus	MC, BI, SG	Ul Hassan and Bano (2015)
14	Suaeda maritime L.	Bacillus amyloliquefaciens	PS, NF, SG	Rueda-Puente et al. (2019)

Table 10.1 List of reported salt-tolerant halophytic rhizospheric bacteria and their role in plant growth promotion and biocontrol—cont'd

No.	Host plant	PGPR	Methods	References
15	Bacopa monnieri L.	Serratia sp.	AF	Jimtha et al. (2017)
16	Prosopis strombulifera	Bacillus sp., Lysinibacillus sp., Pseudomonas sp., Achromobacter sp., and Brevibacterium sp.	PS, SP, IAA, ACC, PRO, AF, NF	Sgroy et al. (2009)
17	Limonium sinense	Glutamicibacter halophytocola	BI, PS, IAA, NF, ACC, PE	Qin et al. (2018)
18	Ipomoea hederacea	Bradyrhizobium japonicum, Pseudomonas putida, and Bacillus megaterium	IAA, SG, SEM-TEM Microscopy	Kim and Kremer (2005)
19	Sesbania bispinosa	Escherichia coli, Pseudomonas fluorescens, and Burkholderia sp.	BI, SG, PE	Dasgupta et al. (2015)
20	Salsola tetrandra	Pseudomonas knackmussii MLR6	ST, PS, SP, IAA, CEL, CHI, PRO, HCN, NH_3, NF, MC	Rabhi et al. (2018)
21	Limonium sinense	Bacillus sp., Pseudomonas sp., Klebsiella sp., Serratia sp., Arthrobacter sp., Streptomyces sp., Isoptericola sp., and Microbacterium sp.	ACC, IAA, BI, SG	Qin et al. (2014)
22	Cenchrus ciliaris	Bacillus cereus, Pseudomonad moraviensis, and Stenotrophomonas maltophilia	ST, BC, MC	Hassan et al. (2018)
23	Artemisia princeps, Chenopodium ficifolium, Echinochloa crus-galli, and Oenothera biennis	Arthrobacter woluwensis, Microbacterium oxydans, Arthrobacter aurescens, Bacillus megaterium, and Bacillus aryabhattai	IAA, SP, PS, SG	Khan et al. (2019)
24	Suaeda maritime	Zhihengliuella halotolerans, and Brachybacterium sp.	NF	Alishahi et al. (2020)
25	Suaeda fruticosa L.	Gracilibacillus sp., Staphylococcus sp., Virgibacillus sp., Salinicoccus sp., Bacillus sp., Zhihengliuella sp., Brevibacterium sp., Oceanobacillus sp., Exiguobacterium sp., Pseudomonas sp., Arthrobacter sp., and Halomonas genera	BI, ST, PS, ACC, SG, PE	Aslam and Ali (2018)

Continued

Table 10.1 List of reported salt-tolerant halophytic rhizospheric bacteria and their role in plant growth promotion and biocontrol—cont'd

No.	Host plant	PGPR	Methods	References
26	Cynodon dactylon	Bacillus cereus	MP, BC, MC, PS, SP, IAA, CHI, PRO, CAT	Chakraborty et al. (2011)
27	Salicornia brachiate	Paenibacillus polymyxa and Bacillus subtilis	PS, IAA, ACC, SP, HCN, NH_3, ST, PE, BI	Karnwal (2019)
28	Halimione portulacoides and Salicornia ramosissima	Vibrio spartinae, Bacillus siamensis, Marinobacter sediminum, Thalassospira australica, Halomonas taeanensis, and Pseudarthrobacter oxydans	ST, NF, PS, SP, IAA, ACC, BI, SG	Mesa-Marín et al. (2019)
29	Arthrocnemum indicum	Klebsiella sp., Pseudomonas sp., Agrobacterium sp., and Ochrobactrum sp.	BC, MC, NF, IAA, PS, ACC, SG, PE	Sharma et al. (2016)
30	Avicennia marina	B. subtilis	PS	Abhijith et al. (2017)
31	Spartina maritime	Bacillus methylotrophicus, Bacillus aryabhattai, B. aryabhattai, and Bacillus licheniformis	NF, PS, SP, IAA, BI, SG, PE	Mesa et al. (2015)
32	Jatropha curcas	Enterobacter cancerogenus	BI, ACC, PS, SP, NH_3, IAA, SG, PE	Jha et al. (2012a,b)
33	Salicornia rubra, Sarcocornia utahensis, and Allenrolfea occidentalis	Halomonas sp., Bacillus sp., and Kushneria sp.	SG, PE, BI	Kearl et al. (2019)
34	Distichlis spicata	Bacillus sp. and Pseudomonas sp.	SG, PS, IAA, SP, ACC, ST, BI	Palacio-Rodríguez et al. (2017)
35	Kochia indica	Klebsiella sp. and Vibrio sp.	IAA, PS, MP, BC	Yaqoob et al. (2013)
36	Arthrocnemum macrostachyum, Halocnemum strobilaceum, Limoniastrum monopetalum (L.), Mesembryanthemum forsskaolii Hochst, Mesembryanthemum crystallinum L., and Suaeda pruinosa Lange	Bacillus sp., Halomonas sp., and Kocuria sp.	IAA, PS, ST, MC BI	Saleh et al. (2017)
37	Salicornia brachiata L.	Bacillus aereus, Serratia nematodiphila, Pantoea agglomerans, Enterobacter sp., and Enterobacter sp.	BI, IAA, PS, SP, ACC, ST	Abbas et al. (2018)

Table 10.1 List of reported salt-tolerant halophytic rhizospheric bacteria and their role in plant growth promotion and biocontrol—cont'd

No.	Host plant	PGPR	Methods	References
38	Dichanthium annulatum	Bacillus, Exiguobacterium, Kocuria, Citricoccus, and Staphylococcus	MP, BC, MC, ST, CEL, AMY, PRO, LIP	Mehnaz et al. (2018)
39	Tamarix chinensis, Suaeda salsa, and Zoysia sinica	Arthrobacter and Bacillus megaterium	SG, BI, BC, PE	Fan et al. (2016)
40	Artemisia princeps, Chenopodium ficifolium, Oenothera biennis, and Echinochloa crus-galli	Arthrobacter woluwensis	IAA, PS, SP, BI, MC	Khan et al. (2019)
41	Salicornia sp.	Staphylococcus sp.	ST, IAA, ACC, PS, SG, MC, PE	Komaresofla et al. (2019)
42	Brachiaria mutica (Para grass)	Pseudomonas chlororaphis and P. aurantiaca	BC, MC, IAA, HCN, SP, PS, ZS. SG	Izzah et al. (2017)
43	Suaeda fruticosa	Pseudomonas sp.	PS, IAA, NH_3, SP, BI	Kayasth et al. (2013)
44	Atriplex halimus	Pseudomonas sp.	BC, PS, IAA, SP, HCN, AF, BI	Arif and Ghoul (2018)
45	Salicornia bigelovii	Streptomyces chartreusis, S. tritolerans, and S. rochei	ST, IAA, ACC, SP, NF, NH_3, PS	Mathew et al. (2020)

10.2.5 Analysis of PGP traits and biocontrol activity of halophytes

Rhizospheric bacteria were isolated from three halophytes (*Heliotropium indicum* L., *Trianthema portulacastrum* L., and *Ipomoea pes-caprae* L.) from Porbandar, Gujarat, India (21.63°N, 69.6°E.) during September 2019. A total of 23 isolates were selected for further study based on salt tolerance capacity. Phosphate solubilization was assessed using Pikovskaya agar plate method according to Pikovskaya (1948). The solubilization index (SI) was determined as described by the formula of Elias et al. (2016). Indole-3-acetic acid (IAA) production was an assay as described by Ehmann (1977). Ammonia (NH_3) production detection was determined by the method described by Cappuccino and Sherman (1992). Zinc solubilization was detected according to the method of Dhaked et al. (2017). The enzyme production was determined by spot plate assay as described for catalase (Marakana et al., 2018), protease (Vijayaraghavan and Vincent, 2013), amylase (Tsegaye et al., 2019), and lipase (Verma and Sharma, 2014). The antagonistic effects of rhizospheric isolates were assessed against eight fungal plant pathogens viz., *Fusarium oxysporum, Aspergillus niger, Aspergillus flavus, Rhizoctonia bataticola, Alternaria burnsii, Botryodiplodia theobromae, Sclerotium rolfsii*, and *Phomopsis vexans* by disc diffusion assay (Rakholiya et al., 2014). Percent inhibition was calculated after 120 hours of incubation. All experiments were performed individually in triplicate. Results are expressed as mean value ($n=3$).

10.2.6 Experimental outcomes

Screening of the selected rhizospheric isolates for in vitro plant growth-promoting traits and production of the hydrolytic enzymes is displayed in Table 10.2, while the biocontrol activities of the isolates

Table 10.2 Screening of selected rhizospheric isolates for in vitro plant growth-promoting traits and hydrolytic enzyme production.

No.	Halophytes	Strain code	Identified strain	Gram reaction
1	*Heliotropium indicum* L.	HI 4	*Bacillus sonorensis*	Positive
2		HI 18	–	Positive
3		HI 19	–	Negative
4		HI 20	–	Negative
5		HI 30	–	Positive
6		HI 32	–	Positive
7		HI 47	*Staphylococcus sciuri*	Positive
8		HI 49	–	Positive
9		HI 57	–	Positive
10		HI 58	–	Negative
11	*Trianthema portulacastrum* L.	TP 7	–	Positive
12		TP 23	–	Positive
13		TP 29	–	Positive
14		TP 36	*Staphylococcus sciuri*	Positive
15		TP 37	–	Positive
16		TP 51	–	Negative
17		TP 60	–	Positive
18	*Ipomoea pes-caprae* L.	IP 16	–	Positive
19		IP 31	*Staphylococcus sciuri*	Positive
20		IP 34	–	Positive
21		IP 35	–	Positive
22		IP 40	*Bacillus subtilis*	Positive
23		IP 64	–	Positive

Bold value indicates potential efficacy

against plant pathogens are shown in Fig. 10.1. Of 23 isolates studied, all produced IAA and ammonia and were able to solubilize phosphate. Maximum IAA production was observed in isolate IP40. Phosphate solubilization and ammonia were higher in the isolates HI20 and HI57, respectively. Isolate IP64 displayed greater efficacy to solubilize Zn. Besides possessing plant growth-promoting traits, the halophytic rhizospheric isolates produced hydrolytic enzymes to varying extents. With respect to the biocontrol activities, all isolates showed different percentages of inhibition against eight plant pathogens (Fig. 10.1). Isolate TP60 had the highest percentage of inhibition at 114.29% against *F. oxysporum (FO)*. However, further studies are necessary to assess the suitability of the selected PGPR strains at a larger scale in the salt-affected areas.

PGPR traits				Extracellular hydrolytic enzyme (enzyme index)			
IAA production ($\mu g\,mL^{-1}$)	Phosphate solubilization ($\mu g\,mL^{-1}$)	Ammonia production ($\mu mol\,mL^{-1}$)	Zn solubilization (%SE)	Amylase	Protease	Lipase	Catalase
49.00	14.71	17.13	0.00	2.64	2.38	0.00	Negative
37.27	29.43	20.66	0.00	1.12	0.00	0.00	Positive
39.36	4.29	9.13	15.38	4.14	0.00	1.13	Negative
40.50	**51.57**	7.74	120.00	1.58	**3.00**	0.00	**Positive (+++)**
33.55	11.00	4.13	0.00	1.00	2.10	0.00	Positive
38.27	29.00	8.74	0.00	2.38	0.00	1.33	Negative
17.82	1.71	9.27	36.36	1.47	0.00	1.13	Positive
30.00	5.14	12.44	–	1.89	0.00	0.00	Positive
43.89	2.86	**25.71**	150.00	1.13	0.00	0.00	Positive
26.27	6.57	19.49	137.50	**5.00**	0.00	0.00	Positive
17.82	11.57	5.26	62.50	2.14	1.56	1.50	Positive
34.82	19.00	17.32	50.00	1.68	1.38	1.08	Positive
59.27	10.29	14.79	0.00	3.00	0.00	1.45	Positive
48.00	5.29	5.74	75.00	2.78	2.11	0.00	Positive
29.18	2.29	15.16	0.00	2.33	1.33	0.00	Positive
41.09	0.43	11.36	–	2.17	0.00	0.00	Negative
39.45	0.14	5.43	50.00	2.38	2.22	1.13	Positive
43.64	22.86	4.11	0.00	2.00	0.00	0.00	Positive
44.73	22.43	18.89	150.00	2.25	0.00	0.00	Positive
46.73	5.43	10.93	33.33	3.67	2.80	**1.75**	Negative
28.91	0.86	18.36	0.00	2.00	1.88	0.00	Positive
62.55	1.29	13.36	60.00	1.88	0.00	0.00	Negative
18.55	12.57	24.23	**250.00**	2.36	1.75	1.11	Positive

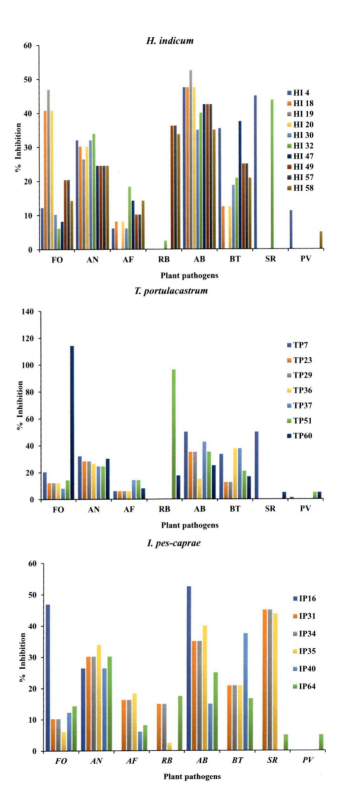

Fig. 10.1 Antifungal activity of halophytic rhizospheric bacteria.

10.3 Concluding remarks and future prospects

ST-PGPR offer eco-friendly biomes representing a sustainable approach for the enhancement of soil fertility and plant growth and protection against pathogens under high salinity. However, it requires further intensive research to understand plant-microbe interaction, signaling molecules and pathways and microbial population dynamics under varying levels of salinity.

Acknowledgments

The work described in this chapter from the SPS Laboratory at the Saurashtra University was supported under various programs: UGC-CAS, DST-FIST, DBT-Multi-Institutional Project, MoES (Government of India) Networking Project, and the Saurashtra University. SPS acknowledges DST-SERB International Travel Fellowships to present his work in Hamburg (Germany), Cape Town (South Africa), and Kyoto (Japan). SPS also acknowledge the award of the UGC-BSR Faculty Fellowship. MK acknowledges DST-SERB International Travel Fellowships to present his work in France. KR acknowledges DST National Postdoctoral Fellowship.

References

Abbas, H., Patel, R.M., Parekh, V., B., 2018. Culturable endophytic bacteria from halotolerant *Salicornia brachata* L.: isolation and plant growth promoting traits. Ind. J. Appl. Microbiol. 21, 10–21.

Abhijith, R., Vennila, A., Purushothaman, C.S., 2017. Occurrence of phosphate-solubilizing bacteria in rhizospheric and pneumatophoric sediment of *Avicennia marina*. Int J Fish Aqua Studies 5, 284–288.

Alishahi, F., Alikhani, H.A., Khoshkholgh-Sima, N.A., Etesami, H., 2020. Mining the roots of various species of the halophyte *Suaeda* for halotolerant nitrogen-fixing endophytic bacteria with the potential for promoting plant growth. Int. Microbiol. 2, 1–3.

Arif, F., Ghoul, M., 2018. Halotolerance of indigenous fluorescent *Pseudomonads* in the presence of natural osmoprotectants. Annu. Res. Rev. Biol. 24, 1–11.

Arora, N.K., Fatima, T., Mishra, J., Mishra, I., Verma, S., Verma, R., Verma, M., Bhattacharya, A., Verma, P., Mishra, P., Bharti, C., 2020. Halo-tolerant plant growth promoting rhizobacteria for improving productivity and remediation of saline soils. J. Adv. Res. 26, 69–82.

Aslam, F., Ali, B., 2018. Halotolerant bacterial diversity associated with *Suaeda fruticosa* (L.) Forssk. improved growth of maize under salinity stress. Agron 8, 131.

Cappuccino, J.C., Sherman, N., 1992. Microbiology: A Laboratory Manual, third ed. Benjamin/Cummings Publications Co., New York.

Chakraborty, U., Roy, S., Chakraborty, A.P., Dey, P., Chakraborty, B., 2011. Plant growth promotion and amelioration of salinity stress in crop plants by a salt-tolerant bacterium. Recent Res. Sci. Technol. 3, 61–70.

Dasgupta, D., Ghati, A., Sarkar, A., Sengupta, C., Paul, G., 2015. Application of plant growth promoting rhizobacteria (PGPR) isolated from the rhizosphere of *Sesbania bispinosa* on the growth of chickpea (*Cicer arietinum* L.). Int. J. Curr. Microbiol. App. Sci. 4, 1033–1042.

Deng, S., Wipf, H.M.L., Pierroz, G., Raab, T.K., Khanna, R., Coleman-Derr, D., 2019. A plant growth-promoting microbial soil amendment dynamically alters the strawberry root bacterial microbiome. Sci. Rep. 9, 17677.

Dhaked, B.S., Triveni, S., Reddy, R.S., Padmaja, G., 2017. Isolation and screening of potassium and zinc solubilizing bacteria from different rhizosphere soil. Int. J. Curr. Microbiol. App. Sci. 6, 1271–1281.

Ehmann, A., 1977. The Van Urk-Salkowski reagent-a sensitive and specific chromogenic reagent for silica gel thin-layer chromatographic detection and identification of indole derivatives. J. Chromatogr. 132, 267–276.

El-Awady, M.A., Hassan, M.M., Al-Sodany, Y.M., 2015. Isolation and characterization of salt tolerant endophytic and rhizospheric plant growth-promoting bacteria (PGPB) associated with the halophyte plant (*Sesuvium verrucosum*) grown in KSA. Int. J. Appl. Sci. Biotechnol. 3, 552–560.

Elias, F., Woyessa, D., Muleta, D., 2016. Phosphate solubilization potential of rhizosphere fungi isolated from plants in Jimma Zone, Southwest Ethiopia. Int. J. Microbiol. 2016, 5472601.

Etemadi, N., Müller, M., Etemadi, M., Brandón, M.G., Ascher-Jenull, J., Insam, H., 2020. Salt tolerance of *Cressa cretica* and its rhizosphere microbiota. Biologia 75, 355–366.

Fan, P., Chen, D., He, Y., Zhou, Q., Tian, Y., Gao, L., 2016. Alleviating salt stress in tomato seedlings using *Arthrobacter* and *Bacillus megaterium* isolated from the rhizosphere of wild plants grown on saline-alkaline lands. Int. J. Phytorem. 18, 1113–1121.

Goswami, M., Deka, S., 2020. Plant growth-promoting rhizobacteria—alleviators of abiotic stresses in soil: a review. Pedosphere 30 (1), 40–61.

Goswami, D., Dhandhukia, P., Patel, P., Thakker, J.N., 2014. Screening of PGPR from saline desert of Kutch: growth promotion in *Arachis hypogea* by *Bacillus licheniformis* A2. Microbiol. Res. 169, 66–75.

Goyal, D., Prakash, O., Pandey, J., 2019. Rhizospheric microbial diversity: an important component for abiotic stress management in crop plants toward sustainable agriculture. In: Singh, J.S., Singh, D.P. (Eds.), New and Future Developments in Microbial Biotechnology and Bioengineering. Elsevier, Duivendrecht, pp. 115–134.

Hassan, T.U., Bano, A., Naz, I., 2018. Halophyte root powder: an alternative biofertilizer and carrier for saline land. Soil Sci. Plant Nutr. 64, 653–661.

Izzah, S., Rizwan, M., Baig, D.N., Saleem, R.S., Malik, K.A., Mehnaz, S., 2017. Secondary metabolites production and plant growth promotion by *Pseudomonas chlororaphis* and *P. aurantiaca* strains isolated from cactus, cotton, and para grass. J. Microbiol. Biotechnol. 27, 480–491.

Jadav, P., Rakholiya, K., Kaneria, M., Singh, S.P., 2020. Isolation and characterization of plant growth promoting rhizospheric bacteria from *Limonium stocksii*. In: Proceedings of the National Conference on Innovations in Biological Sciences (NCIBS) 2020., https://doi.org/10.2139/ssrn.3568430.

Jha, B., Gontia, I., Hartmann, A., 2012a. The roots of the halophyte *Salicornia brachiata* are a source of new halotolerant diazotrophic bacteria with plant growth-promoting potential. Plant Soil 356, 265–277.

Jha, C.K., Patel, B., Saraf, M., 2012b. Stimulation of the growth of *Jatropha curcas* by the plant growth promoting bacterium *Enterobacter cancerogenus* MSA2. World J. Microbiol. Biotechnol. 28, 891–899.

Jimtha, C.J., Jishma, P., Sreelekha, S., Chithra, S., Radhakrishnan, E.K., 2017. Antifungal properties of prodigiosin producing rhizospheric *Serratia* sp. Rhizosphere 3, 105–108.

Karnwal, A., 2019. Screening, isolation and characterization of culturable stress-tolerant bacterial endophytes associated with *Salicornia brachiata* and their effect on wheat (*Triticum aestivum* L.) and maize (*Zea mays*) growth. J. Plant Prot. Res. 59, 293–303.

Kataoka, R., Güneri, E., Turgay, O.C., Yaprak, A.E., Sevilir, B., Başköse, I., 2017. Sodium-resistant plant growth-promoting rhizobacteria isolated from a halophyte, *Salsola grandis*, in saline-alkaline soils of Turkey. Eurasian J. Soil Sci. 6, 216–225.

Kaushal, M., Wani, S.P., 2016. Rhizobacterial-plant interactions: strategies ensuring plant growth promotion under drought and salinity stress. Agric. Ecosyst. Environ. 231, 68–78.

Kayasth, M., Kumar, V., Gera, R., Dudeja, S.S., 2013. Isolation and characterization of salt tolerant phosphate solubilizing strain of *Pseudomonas* sp. from rhizosphere soil of weed growing in saline field. Ann. Biol. 29, 224–227.

Kearl, J., McNary, C., Lowman, J.S., Mei, C., Aanderud, Z.T., Smith, S.T., West, J., Colton, E., Hamson, M., Nielsen, B.L., 2019. Salt-tolerant halophyte rhizosphere bacteria stimulate growth of alfalfa in salty soil. Front. Microbiol. 10, 1849.

Khan, M.A., Ullah, I., Waqas, M., Hamayun, M., Khan, A.L., Asaf, S., Kang, S.M., Kim, K.M., Jan, R., Lee, I.J., 2019. Halo-tolerant rhizospheric *Arthrobacter woluwensis* AK1 mitigates salt stress and induces physio-hormonal changes and expression of GmST1 and GmLAX3 in soybean. Symbiosis 77, 9–21.

Kim, S.J., Kremer, R.J., 2005. Scanning and transmission electron microscopy of root colonization of morning glory (*Ipomoea* sp.) seedlings by rhizobacteria. Symbiosis 39, 117–124.

Komaresofla, B.R., Alikhani, H.A., Etesami, H., Khoshkholgh-Sima, N.A., 2019. Improved growth and salinity tolerance of the halophyte *Salicornia* sp. by co-inoculation with endophytic and rhizosphere bacteria. Appl. Soil Ecol. 138, 160–170.

Leontidou, K., Genitsaris, S., Papadopoulou, A., Kamou, N., Bosmali, I., Matsi, T., Madesis, P., Vokou, D., Karamanoli, K., Mellidou, I., 2020. Plant growth promoting rhizobacteria isolated from halophytes and drought-tolerant plants: genomic characterisation and exploration of phyto-beneficial traits. Sci. Rep. 10, 14857.

Lopez-Raez, J.A., 2016. How drought and salinity affect Arbuscular mycorrhizal symbiosis and strigolactone biosynthesis? Planta 243, 1375–1385.

Ma, Y., Dias, M.C., Freitas, H., 2020. Drought and salinity stress responses and microbe-induced tolerance in plants. Front. Plant Sci. 13, 2020.

Mandal, S., Raju, R., Kumar, A., Kumar, P., Sharma, P.C., 2018. Current status of research, technology response and policy needs of salt-affected soils in India—a review. J. Indian Soc. Coast. Agric. Res. 36, 40–53.

Marakana, T., Sharma, M., Sangani, K., 2018. Isolation and characterization of halotolerant bacteria and it's effects on wheat plant as PGPR. J. Pharm. Innov. 7, 102–110.

Mathew, B.T., Torky, Y., Amin, A., Mourad, A.H., Ayyash, M.M., El-Keblawy, A., Hilal-Alnaqbi, A., AbuQamar, S.F., El-Tarabily, K.A., 2020. Halotolerant marine rhizosphere-competent actinobacteria promote *Salicornia bigelovii* growth and seed production using seawater irrigation. Front. Microbiol. 11, 552.

Mehnaz, D., Abdulla, K., Mukhtar, S., 2018. Isolation and characterization of haloalkaliphilic bacteria from the rhizosphere of *Dichanthium annulatum*. J. Adv. Res. Biotechnol. 3, 1–9.

Mesa, J., Mateos-Naranjo, E., Caviedes, M.A., Redondo-Gómez, S., Pajuelo, E., Rodríguez-Llorente, I.D., 2015. Scouting contaminated estuaries: heavy metal resistant and plant growth promoting rhizobacteria in the native metal rhizoaccumulator *Spartina maritima*. Mar. Pollut. Bull. 90, 150–159.

Mesa-Marín, J., Pérez-Romero, J.A., Mateos-Naranjo, E., Bernabeu-Meana, M., Pajuelo, E., Rodríguez-Llorente, I.D., Redondo-Gómez, S., 2019. Effect of plant growth-promoting rhizobacteria on *Salicornia ramosissima* seed germination under salinity, CO_2 and temperature stress. Agronomy 9, 655.

Mukhtar, S., Zareen, M., Khaliq, Z., Mehnaz, S., Malik, K.A., 2020. Phylogenetic analysis of halophyte associated rhizobacteria and effect of halotolerant and halophilic phosphate solubilizing biofertilizers on maize growth under salinity stress conditions. J. Appl. Microbiol. 128, 556–573.

Naz, I., Bano, A., UL-Hassan, T., 2009. Morphological, biochemical and molecular characterization of rhizobia from halophytes of Khewra Salt Range and Attock. Pak. J. Bot. 41, 3159–3168.

Orozco-Mosqueda, M., Glick, B.R., Santoyo, G., 2020. ACC deaminase in plant growth-promoting bacteria (PGPB): an efficient mechanism to counter salt stress in crops. Microbiol. Res. 235, 126439.

Palacio-Rodríguez, R., Coria-Arellano, J.L., López-Bucio, J., Sánchez-Salas, J., Muro-Pérez, G., Castañeda-Gaytán, G., Sáenz-Mata, J., 2017. Halophilic rhizobacteria from *Distichlis spicata* promote growth and improve salt tolerance in heterologous plant hosts. Symbiosis 73, 179–189.

Pan, J., Huang, C., Peng, F., Zhang, W., Luo, J., Ma, S., Xue, X., 2020. Effect of arbuscular mycorrhizal fungi (AMF) and plant growth-promoting bacteria (PGPR) inoculations on *Elaeagnus angustifolia* L. in saline soil. Appl. Sci. 10, 945.

Patel, J.S., Yadav, S.K., Bajpai, R., Teli, B., Rashid, M., 2020. PGPR secondary metabolites: an active syrup for improvement of plant health. In: Molecular Aspects of Plant Beneficial Microbes in Agriculture. Elsevier, pp. 195–208.

Pikovskaya, R.I., 1948. Mobilization of phosphorus in soil in connection with the vital activity of some microbial species. Mikrobiologiya 17, 362–370.

Qin, S., Zhang, Y.J., Yuan, B., Xu, P.Y., Xing, K., Wang, J., Jiang, J.H., 2014. Isolation of ACC deaminase-producing habitat-adapted symbiotic bacteria associated with halophyte *Limonium sinense* (Girard) Kuntze and evaluating their plant growth-promoting activity under salt stress. Plant Soil 374, 753–766.

Qin, S., Feng, W.W., Zhang, Y.J., Wang, T.T., Xiong, Y.W., Xing, K., 2018. Diversity of bacterial microbiota of coastal halophyte *Limonium sinense* and amelioration of salinity stress damage by symbiotic plant growth-promoting actinobacterium *Glutamicibacter halophytocola* KLBMP 5180. Appl. Environ. Microbiol. 84, e01533-18.

Rabhi, N.E., Silini, A., Cherif Silini, H., Yahiaoui, B., Lekired, A., Robineau, M., Esmaeel, Q., Jacquard, C., Vaillant Gaveau, N., Clément, C., Aït Barka, E., 2018. *Pseudomonas knackmussii* MLR6, a rhizospheric strain isolated from halophyte, enhances salt tolerance in *Arabidopsis thaliana*. J. Appl. Microbiol. 125, 1836–1851.

Rakholiya, K., Vaghela, P., Rathod, T., Chanda, S., 2014. Comparative study of hydroalcoholic extracts of *Momordica charantia* L. against foodborne pathogens. Indian J. Pharm. Sci. 76, 148.

Rakholiya, K.D., Kaneria, M.J., Singh, S.P., Vora, V.D., Suataria, G.S., 2017. Biochemical and proteomics analysis of the plant growth promoting rhizobacteria in stress conditions. In: Singh, R.P., Kothari, R., Koringa, P.G., Singh, S.P. (Eds.), Understanding Host-Microbiome Interactions—An Omics. Springer Nature, Singapore, pp. 227–245.

Rakholiya, K.D., Sureja, K., Chaniyara, R., Parmar, K., Mayatra, D., Kaneria, M.J., Singh, S.P., 2018a. Functional abilities of cultivable rhizospheric bacteria in *Phyllanthus fraternus* and their potential plant growth promotional and phyto-pathogen control. In: Proceedings of 10th National Symposium at Christ College, Rajkot, India, pp. 123–129.

Rakholiya, K.D., Parmar, K., Sureja, K., Chaniyara, R., Mayatra, D., Kaneria, M.J., Singh, S.P., 2018b. Cultivable plant growth promoting activities of rhizospheric bacteria isolated from indigenous medicinal plant *Cardiospermum halicacabum*. In: Proceedings of 10th National Symposium at Christ College, Rajkot, India, pp. 130–135.

Rueda-Puente, E.O., Bianciotto, O., Farmohammadi, S., Zakeri, O., Elías, J.L., Hernández-Montiel, L.G., Bernardo, M.A., 2019. Plant growth-promoting bacteria associated to the halophyte *Suaeda maritima* (L.) in Abbas, Iran. Sabkha Ecosys 1, 289–300.

Saleh, M.Y., Sarhan, M.S., Mourad, E.F., Hamza, M.A., Abbas, M.T., Othman, A.A., Youssef, H.H., Morsi, A.T., Youssef, G.H., El-Tahan, M., Amer, W.A., 2017. A novel plant-based-sea water culture media for *in vitro* cultivation and in situ recovery of the halophyte microbiome. J. Adv. Res. 8, 577–590.

Sgroy, V., Cassán, F., Masciarelli, O., Del Papa, M.F., Lagares, A., Luna, V., 2009. Isolation and characterization of endophytic plant growth-promoting (PGPB) or stress homeostasis-regulating (PSHB) bacteria associated to the halophyte *Prosopis strombulifera*. Appl. Microbiol. Biotechnol. 85, 371–381.

Sharma, S., Kulkarni, J., Jha, B., 2016. Halotolerant rhizobacteria promote growth and enhance salinity tolerance in peanut. Front. Microbiol. 7, 1600.

Singh, R.P., Jha, P.N., 2016. The multifarious PGPR Serratia marcescens CDP-13 augments induced systemic resistance and enhanced salinity tolerance of wheat (*Triticum aestivum* L.). PLoS One 11, 6.

Tsegaye, Z., Gizaw, B., Tefera, G., Feleke, A., Chaniyalew, S., Alemu, T., Assefa, F., 2019. Isolation and biochemical characterization of Plant Growth Promoting (PGP) bacteria colonizing the rhizosphere of Tef crop during the seedling stage. Biomed. J. Sci. Technol. Res. 14, 1586–1597.

Ul Hassan, T., Bano, A., 2015. The stimulatory effects of L-tryptophan and plant growth promoting rhizobacteria (PGPR) on soil health and physiology of wheat. J. Soil Sci. Plant Nutr. 15, 190–201.

Ullah, S., Bano, A., 2015. Isolation of plant-growth-promoting rhizobacteria from rhizospheric soil of halophytes and their impact on maize (*Zea mays* L.) under induced soil salinity. Can. J. Microbiol. 61, 307–313.

Vega, C., Rodríguez, M., Llamas, I., Béjar, V., Sampedro, I., 2020. Silencing of phytopathogen communication by the halotolerant PGPR *Staphylococcus equorum* strain EN21. Microorganisms 8, 42.

Verma, S.H., Sharma, K.P., 2014. Isolation, identification and characterization of lipase producing microorganisms from environment. Asian J. Pharm. Clin. Res. 7, 219–222.

Vijayaraghavan, P., Vincent, S.G., 2013. A simple method for the detection of protease activity on agar plates using bromocresolgreen dye. J. Biochem. Technol. 4, 628–630.

Xiong, Y.W., Gong, Y., Li, X.W., Chen, P., Ju, X.Y., Zhang, C.M., Yuan, B., Lv, Z.P., Xing, K., Qin, S., 2019. Enhancement of growth and salt tolerance of tomato seedlings by a natural halotolerant actinobacterium *Glutamicibacter halophytocola* KLBMP 5180 isolated from a coastal halophyte. Plant Soil 445, 307–322.

Yang, J., Kloepper, J.W., Ryu, C.M., 2009. Rhizosphere bacteria help plants tolerate abiotic stress. Trends Plant Sci. 14, 1–4.

Yaqoob, C., Awan, H.A., Maqbool, A., Malik, K.A., 2013. Microbial diversity of the rhizosphere of kochia (*Kochia indica*) growing under saline conditions. Pak. J. Bot. 45, 59–65.

Nanotechnology for plant growth promotion and stress management

Pooja Sharma[a], Ashutosh Shukla[b], Mamta Yadav[b], Anuj Kumar Tiwari[c], Ravindra Soni[d], Sudhir Kumar Srivastava[e], and Surendra Pratap Singh[b]

[a]*Department of Environmental Microbiology, School for Environmental Sciences, Babasaheb Bhimrao Ambedkar Central University, Lucknow, Uttar Pradesh, India,* [b]*Plant Molecular Biology Laboratory, Department of Botany, Dayanand Anglo-Vedic (PG) College, Chhatrapati Shahu Ji Maharaj University, Kanpur, Uttar Pradesh, India,* [c]*Department of Botany, Bhavan's Mehta Mahavidyalaya, Bharwari, Uttar Pradesh, India,* [d]*Department of Agricultural Microbiology, College of Agriculture, Indira Gandhi Krishi Vishwavidyalaya, Raipur, Chhattisgarh, India,* [e]*Chemical Research Laboratory, Department of Chemistry, Dayanand Anglo-Vedic (PG) College, Chhatrapati Shahu Ji Maharaj University, Kanpur, Uttar Pradesh, India*

11.1 Introduction

Nanoparticles (NPs) are organic, inorganic, or hybrid materials measuring at least 1–100 nm. NPs occurring in the environment of nature may be produced through photochemical reaction mechanisms, forest fires, volcanic eruptions, simple erosion, animals and plants, or even through microorganisms (Navarro García et al., 2006; Dahoumane et al., 2017). Production of plant-driven NPs has emerged as an effective biological source of green NPs, which, due to their green nature and the ease of production processes compared with others, has attracted special attention from scientists in recent years (Kitching et al., 2015; Iravani, 2011). A range of plant species and microbes are now used in the synthesis of NPs to exploit green nanotechnology, including bacteria, algae, and fungi. For instance, gold nanoparticles are formed from *Medicago sativa* and the plant species of Sesbania. Similarly, it can be generated inside live plants like *M. sativa*, *Brassica juncea*, and *Helianthus annus*, inorganic nanoparticles made up of cobalt, silver, nickel, zinc, and copper (Panpatte et al., 2016).

Three-dimensional (3D) objects are all NPs. One-dimensional (1D) NP is the NPs that have two nanoscale and one macroscale dimension (nanoscales, nanotubes). In contrast, 2D NPs have one nanoscale dimension and two macroscale dimensions (nanolayers, nanofilms) (Bernhardt et al., 2010). To synthesize or make zero-dimensional NPs with well-controlled dimensions, a wide range of chemical and physical methods were established (Tiwari et al., 2012). Zero-dimensional NPs, like quantum dots, are widely acceptable to light-emitting diodes, solar cells, and transistors for single electron as employed in lasers (Stouwdam and Janssen, 2008; Pradhan et al., 2010). Three-dimensional NPs have recently gained great study interest because of their huge surface area and other superior features, such as absorption sites for all molecules in a limited region, resulting in enhanced molecular transport. In terms of sustainable agricultural and environmental systems, the improvement and development of new technologies for the production of NPs with their prospective application have particular importance (Cheng et al., 2016).

Nanoparticles are a distinctive material category with special aspects (Gajjar et al., 2009; Thomas et al., 2018; Zargar et al., 2011; Zia et al., 2017). Particles of nanoscale metal are among the most significant product of nanotechnology. Due to their distinctive optical features, catalytic nature, and electromagnetic applications, these particles have gained interest. While several chemical and physical processes are extensively employed in nanoparticle manufacture, they have disadvantages, like the usage of harmful substances, generation of environmentally polluting waste, high energy consumption, and poor efficiency. Researchers also are exploring novel techniques of manufacturing nanoparticles that are of considerable importance to the biosynthesis of nanoparticles due to their environmental compatibility and financial advantages (Thakkar et al., 2010; Sana and Dogiparthi, 2018; Rasheed et al., 2017). Several plants and their products can be used in the biosynthesis of nanoparticles and contain alkaloids, steroids, flavonoids, saponins, tannins, and phenols, that helps decrease the silver ions by converting numerous functional groups, ketone, hydroxyl, and aldehydes into silver nanoparticles (Youdim et al., 2004; Gurunathan et al., 2014). Silver nanoparticles have various applications for study in medical technology, pharmacy, and many others. Therefore, biological nanoparticles are not ecologically toxic because they are made by plants (Patil et al., 2018). In addition, increased antibiotic resistance increases the expense of treatment, increases the time to recover, and leads to failure in antibiotic treatment. Consequently, new antibiotics must be produced, new and comprehensive treatment techniques used to combat infections should be used, and antibiotic-resistant organisms are prevented (Morones et al., 2005; Moghtader et al., 2016). In addition, new antimicrobial materials are necessary for some patients due to the presence of hazards, like anaphylactic liver and renal failure. Nanoparticles are applied as antibacterial materials by the advancement of nanoscience, the facility of manufacturing, and evidence of antimicrobial qualities of silver nanoparticles in medicine (Choi et al., 2008; Willets and Van Duyne, 2007; Dinesh et al., 2012; Babu Maddinedi et al., 2017). There are several reports on silver nanoparticles produced by the plant extract and green technique like carob leaf, banana peel, neem leaves, *Ocimum tenuiflorum* leaf extract, *Urtica dioica* Linn., *Azadirachta indica* leaf extract, olive leaf, *Solanum trilobatum*, *Datura stramonium*, green tea (*Camellia sinensis*), *Combretum erythrophyllum*, *Megaphrynium macrostachyum* Shikakai, *Sida cordifolia*, *Enicostemma axillare*, *Zataria multiflora*, coffee extract, *Givotia moluccana*, green *coffee bean*, *Embelia ribes*, *A. indica* leaf extract, and *Ocimum sanctum* leaf extract (Awwad et al., 2013; Khalil et al., 2014; Logeswari et al., 2015; Ibrahim, 2015; Manikandan et al., 2017; Torres et al., 2017; Ahmed et al., 2018).

11.2 Synthesis, absorption, and translocation of nanoparticles in plants

In translation research, nanotechnology is one of the most extensively employed technologies. The production of metal nanoparticles using biological materials has gained considerable attention with an environmentally friendly approach (Medvedeva et al., 2007). The soil or aqueous metal ions bind to the organism either through the cell wall or peptides synthesized by organisms, and it assembles into stable nanoparticle structure (Yong et al., 2002). The range of biological approaches for engineering and nanoparticle production depends on several variables. Nanoparticles made up of copper, cadmium, silver, zinc oxide, gold, palladium, cadmium, platinum, and titanium dioxide are the microbiological resources for most of the widely researched salts of metals and metals (Mousavi et al., 2018; Gahlawat and Choudhury, 2019).

Plants are considered to just be best suited compared to green nanoparticle synthesis microorganisms because they are not pathogenic and different pathways are fully investigated. A broad range of nanoparticles of metal were generated with various plants. In contrast to their counterpart massive substance with multiple uses in many domains of human concern, these nanoparticles have distinct optical, thermal, magnetic, physical, chemical, and electric qualities (Durán and Seabra, 2012; Husseiny et al., 2007). There are many different biological components employed for the synthesis of Ag NPs. The extract of *Jatropha curcas* results in homogeneous

Ag NPs of AgNO$_3$ salt produced within 4 h (10–20 nm). *Acalypha indica* leaf extracts were shown to be able to synthesize Ag NPs. Its resultant Ag NPs varied from 20 to 30 nm in length were extensively uniform (Krishnaraj et al., 2010). Ag NPs synthesis by *M. sativa* seed is significantly used in lab work. Ag$^+$ was lowered almost instantly as nanoparticles were observed to have exposed metal salt within a minute and 90% of Ag$^+$ had a decrease in < 50 min at a temperature of 30°C (Lukman et al., 2011). By such a part of the leaf broth, nanoparticles were spherical and stabilized. Fruit extract *Terminalia chebula* was used for producing NPs (Jebakumar Immanuel Edison and Sethuraman, 2012). Ag nanoparticle of cubic form in the size from 50 to 200 nm, generated by *Eucalyptus macrocarpa* leaf extract (Poinern et al., 2013); *Nyctanthes arbor-tristis* (night jasmine) NPs; Ag and Au nanoparticle secreted from leaf extract of *Coriandrum sativum*. Phyllanthine is derived from *Phyllanthus amarus* using generative nanoparticles of Ag and Au. Its investigation is unusual in using a single plant extract in the assessment of prior studies in which the complete plant was working for the combination of metallic NPs. The development of triangular and hexagonal Au NPs was caused by low concentrations of *Phyllanthus* while increasing absorptions formed enlarged round NPs (Kasthuri et al., 2008). Nanoparticle-soluble starch, cellulose, dextran, chitosan, alginic acid, and hyaluronic acid generated from plants and polysaccharides can be successfully bound and investigated for the synthesis of NPs (Park et al., 2011). Such molecules provide advantages by employing less harmful chemicals (Sangaru et al., 2004). The persistence of NPs was due to the leaf's compounds of terpenoid and flavanone (Sangaru et al., 2004). Phytochemically decreased *Staphylococcus aureus* NiOs address drug-tolerance difficulties to a certain extent. The creation of bimetallic nanoparticles was caused by alloying Ag and Au. Its synthesis involves a competitive reduction between two aquatic solutions which are employed together with a plant extract as a type of metallic ion precursor. Several plants were employed efficiently to produce AU and Ag NPs, such as *Western Anacardium, A. indica*, mahogany, and cross-breeding extracts *Swietenia* (Sangaru et al., 2004; Sheny et al., 2011; Jacob et al., 2012; Mondal et al., 2011).

The synthesis of copper oxide (CuO) and copper (Cu) NPs was based on extracts from different plants. Cu nanoparticles of 40–100 nm have been composed of extract from *Magnolia kobus* leaf and *Syzygium aromaticum* in the sphere to granular forms (Subhankari and Nayak, 2013; Lee et al., 2013). *Euphorbia nivulia* stem latex is used to produce Cu NPs which have been alleviated and covered by terpenes and latex peptides (Padil and Černík, 2013; Valodkar et al., 2011).

Diospyros kaki extract of leaf and amines, alcohols, carboxylic acids have demonstrated the creation of the first platinum nanoparticles (Song et al., 2009). *Cinnamomum zeylanicum* is used to manufacture palladium NPs (Roopan et al., 2012; Sathishkumar et al., 2009). The leaf extract of *Glycine max* produced NPs. *Coffea arabica* and *C. sinensis* extracts were used to manufacture palladium-like NPs (Petla et al., 2012). Also, antioxidants like chlorogenic acid, geniposide, crocetin, and crocin in the extracts work as stabilizing and reducing agents when an extract of *Gardenia jasminoides is* employed for the synthesis of palladium NPs (Jia et al., 2009). For the manufacture of platinum and palladium nanoparticles, other plants such as *O. sanctum* wooden nanomaterials of plants, and *Pinus resinosa* lignin were used (Coccia et al., 2012; Lin et al., 2011). The use of many plant extracts like *Annona squamosa* peel, *Cocos nucifera* coir, *Nyctanthes arbor-tristis* leaf extracts, *Psidium guajava*, *Eclipta prostrata*, and *Catharanthus roseus* was effectively produced with nanoparticles containing titanium (TiO_2) metal oxide (Roopan et al., 2012). The latex *Calotropis procera*, *Aloe vera*, *Physalis alkekengi*, and *Sedum alfredii* have been used for the production of ZnO. Spherical nanoparticles of biogenic indium oxide (In_2O_3) were produced utilizing leaf extracts from *Aloe vera* with a tunable size range of 5–50 nm (*Aloe barbadensis*).

Iron (Fe) NPs are produced, such as *Datura innoxia*, *A. indica*, *C. procera*, *Tinospora cordifolia*, *Tridax procumbens*, *Cymbopogon citratus*, and *Euphorbia milii* in aqueous *Sorghum bicolor* bran extract and leaf extracts. *J. curcas* latex is utilized for synthesizing spherical Pb nanoparticles of 10–12.5 nm in size (Joglekar et al., 2011). The use of extracts of plant components or entire vegetable extracts is a synthesis of metallic nanoparticles. In addition, in vivid plants, metallic nanoparticles may be formed and a new strategy for Pd NP synthesis action (Table 11.1).

11.3 Nanoparticles in plant growth and stress tolerance

Mechanisms for the absorption, transport, and accumulation of different NPs are elaborated in plant organs (Kaphle et al., 2018). NPs in the plant are based on the structure and physiology of the plant cells, their transport, accumulation, communication with the land, and the constancy of plant cells (Janmohammadi et al., 2016). The cell wall of plants functions as a special barrier to the entry of NPs into the cell and solubilization and transmission of NPs (Naruka et al., 2000; Mousavi et al., 2007). Most experiments show that the pores of the cell wall are the fundamental limit on the entry of NPs into cells. The structure of the NPs is a second limit element that impacts their penetration

Table 11.1 Effect of NPs on plant growth and development.

Nanomaterials	Plants	Response	Reference
ZnO	*Triticum aestivum*	Improved grain yield and biomass	Du et al. (2019)
CuO	*Spinacia oleracea*	Improved photosynthesis	Wang et al. (2019)
TiO$_2$	*Spinacia oleracea*	Increased biomass chlorophyll, nitrogen, and protein content	Yang et al. (2007)
SiO$_2$ NPs	*Oryza sativa*	Alleviated heavy metal toxicity and improved growth by decreasing bio-concentration and translocation in plants	Wang et al. (2016)
Ag NPs	*Vigna sinensis*	Enhanced growth and biomass by stimulating root nodulation and soil bacterial diversity	Mehta et al. (2016)
ZnO	*Coffea arabica*	Enhanced growth, biomass accumulation, and net photosynthesis	Rossi et al. (2019)
ZnO	*Cyamopsis tetragonoloba*	Improved plant growth, biomass	Raliya and Tarafdar (2013)
MWCNTs	*Hordeum vulgare*	Enhanced germination	Lahiani et al. (2013)
MWCNTs	*Zea mays, Arachis hypogaea, Allium sativum, Triticum aestivum*	Improved and rapid germination	Srivastava and Rao (2014)

into the membrane of cells or promotes a radical surface or radical exudate attachment (Ling and Silberbush, 2002). Positive loading of NPs can boost their cell wall adhesion. In their activity on rhizosphere and influence on plants, shape and NP coatings may play an important impact. The collected nanoparticles from plant roots are transferred to various tissues in the aerial section of the plant (Tripathi et al., 2017). All plant features and NPs have a key role to play in translocating NPs. In earlier investigations, gold NPs can collect in the shoot of *Oryza sativa* (Zhu et al., 2012).

By contrast, negative Au NPs are more resourcefully transferred from the roots to the plant. Earlier studies have shown that the most stable NPs are titanium dioxide (TiO$_2$) and silicon (SiO$_2$) and can be found in their pristine form in plants (Larue et al., 2011; Servin et al., 2012). Species of plants are accumulating and NPs are able to transform via variance. Zinc oxide (ZnO) NP changes were evaluated using X-ray (XAS) synchrotron spectroscopy while exposed to various plants (Castillo-Michel et al., 2017). In *Zea mays*, optimum deposition of Zn takes place in the roots and shoots in various forms under the aquatic treatment of ZnO NPs. This could be attributed to enhanced rhizosphere dissolution, plant uptake, and ionic Zn transport (Lv et al., 2015). Speciation of comparable Zn deposited in soil-grown wheat was also found (Dimkpa et al., 2012, 2013). CeO$_2$ NPs were found to be translocated into Zn-citrus within the plant tissue in the form of NPs and Zn biotransforming. Transportation of CuO NPs from the roots to the shooting via xylem was carried out in *Z. mays* and further transferred to the roots by phloem. Earlier investigations revealed that *Triticum aestivum* (*T. aestivum* sp.) has a role in NP TiO$_2$ translocation. NPs engage after entry with cellular and subcellular plants, producing changes in physiological and morphological circumstances. NP impacts on plant systems could be determined by chemical composition, reactivity, size, and in particular by NP concentration. Several NPs have shown the ability to boost the germination and growth of plants at levels below specified limit levels. Several experiments were conducted largely under artificial circumstances of treatment, including medium growth plate aquatic environments.

The benefit of nanoparticles in cultivated plants has shown enhanced germination of the seeds, greater shooting and root length, higher fruit output, improved metabolite content, and significant growth of the plant and seedling biomass in many plants. The influence of nanoparticles has also been demonstrated by boosting efficiency in nitrogen use and raising photosynthesis rates in many major crop plants, including peanuts, spinach, and soya, on several plant biochemicals relevant to growth and development in plants. The significance of nanoparticles in enhancing plant nutrient uptake and tolerance to various diseases and stress responses is well documented. NPs were able to influence the growth of the plant by changing the plant's certain physiological processes. Most research shows that NPs can have hazardous effects above a specific dosage, and plant toxicity-based researches have examined their germinating and biomass buildup impacts. Several investigations have revealed that plants may potentially be impacted by NPs (Fig. 11.1).

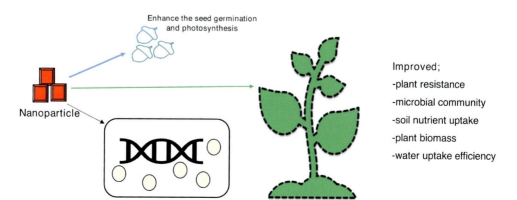

Fig. 11.1 The positive effect of nanoparticle on plant growth and development.

11.4 Green nanoparticle

Biochemically safe, economically efficient, and environmentally beneficial nanoparticle synthesis employs plants with green synthesizing technology. Plants have been granted the ability from their neighboring niche to eat and collect inorganic metal ions. The physiological entities are recognized for their extra and intracellular synthesis of nanoparticles. The capacity of each biological system to deliver metal nanoparticles is different. Successful implementation via the use of nanotechnology of medications and tissue engineering has already shown an essential contribution to translation research into and uses of pharmaceuticals. For the manufacture of silver nanoparticles (Ag NPs) with antibacterial characteristics, plant extracts of *Solanum trilobatum*, *Citrus sinensis*, *Centella asiatica*, *Syzygium cumini*, and *O. tenuiflorum* biomass have been employed (Shankar et al., 2003). *Lysiloma acapulcensis* is the endemic tree of the South of Mexico and is rich in traditional treatment for its therapeutic powers for respiratory, gastrointestinal, urinary, and skin illnesses, and has empirically been employed for the treatment of traditional medications (González-Cortazar et al., 2018). The extract contains plenty of tannins that provide access to antibacterial characteristics (Navarro García et al., 2006; García et al., 2003). Ag NPs can then be combined with *L. acapulcensis* to make a good option for low cytotoxicity infectious illnesses.

11.5 Genetic engineering in nanoparticle

Global warming and the growing world population as well as genetic plant changes have enhanced food production by providing crops with specific genetic characteristics. In the transportation of external molecules to plant cells, plant cell walls serve as a barrack. Globally, several ways for the transmission of DNA in plant cells are used to meet this issue and create plant genetic transformations based on biolistic techniques or Agrobacterium transformation. The varied and complicated approaches for the distribution of genes and protein discovered for animal systems have comparatively fewer plants. The current plant genetic modification generally involves two primary stages: the supply of genetic supply and regeneration of the processed plant, and the latter's requirements and complexity depend heavily on which method of supply is utilized and how stable the transformation is. In plant cell cultures, several early experiments were conducted in nanomaterial plant genetic engineering. Due to their capacity to enter plant cell walls without external force, the wide applicability of host range, and highly customizable physicochemical features, nanoparts are appealing materials for biomolecule transmission. For example, a good strategy for DNA delivery has been established as a silicon carbide-mediated transformation (soja, corn, rice, tobacco, and cotton) (Lau et al., 2017;

Sanzari et al., 2019). If development in animal systems was lagging, the latest form from plants indicated that NMs in plants grown can surpass the cell wall barriers and reduce the constraints of current transgene water supplies. The dsRNA loaded on clay nanosheet when sprayed on the tobacco leave provides protection from the Cauliflower Mosaic Virus (CaMV) infection, compared to unsprayed tobacco leaves, through RNAi (Mitter et al., 2017). In cotton plants, there is a successful, stable genetic transformation by magnetic nanoparticles (MNP). The β-glucuronidase (GUS) reporter gene-MNP complex has been magnetically infiltrated into cotton pollen grains without affecting the pollen's viability. Transgenic cotton plants have succeeded in producing magnetofected pollen and the genome has been successfully integrated with exogenous DNA which is effectively expressed and stabilized in the self-acquired descendants (Zhao et al., 2017). The infusion scaffolds of carbon nanotubes utilized for the transfer of linear and plasma DNA and siRNA to *T. aestivum*, *Eruca sativa*, *Gossypium hirsutum*, and *Nicotiana benthamiana* leaves resulted in the highly temporary production of green, fluorescent protection protein (GFP). In addition, the short-interfering RNA (siRNA) was provided to *N. benthamiana* plants, which produced GFP to allow the gene to be silenced by 95% (Demirer et al., 2019). MSNs are employed to transport Cre recombinase, which contains loxP sites integrated into the chromosomal DNA as carriers in *Z. mays* of immaterial embryos. After the biological application of designed MSNs into plant tissues, loxP was appropriately recombined to establish an efficient genome edition (Martin-Ortigosa et al., 2014). NP-based delivery of biomolecules to the plant hence enables a single-guide RNA (sgRNA), DNA, and RNP supply to be edited with enhanced performances of the plant genome (Fig. 11.2).

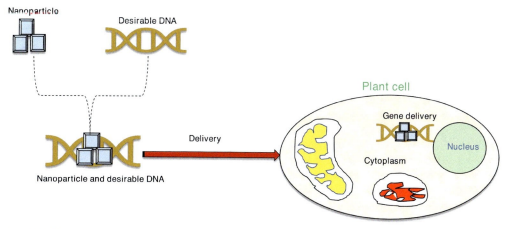

Fig. 11.2 Nanomaterials working as a substitute of conventional gene delivery systems for plant genetic engineering.

11.6 Conclusion

Studies on nanotechnology have recommended using NPs as a powerful way to alleviate existing difficulties caused in traditional agricultural systems by conventional fertilizers. Benefits of NP-containing nutrients have been shown to the considerable systematic suggestion of their effectiveness in enlightening the micronutrients of the plant, reflecting enhanced growing parameters and important physiological improvements, such as chlorophyll and carotenoids, photosynthesis, metabolism, and transpiration. For the development of an application strategy, the fundamental dosage of NPs, the period of exposure, translocation, and accumulation, and plant action mechanism are important. Furthermore, it is necessary to determine the precise effects and consequences on ecosystems of NPs on the accumulation of water, soil, and air.

References

Ahmed, S., Kaur, G., Sharma, P., Singh, S., Ikram, S., 2018. Fruit waste (peel) as bioreductant to synthesize silver nanoparticles with antimicrobial, antioxidant and cytotoxic activities. J. Appl. Biomed. 16 (3), 221–231.

Awwad, A.M., Salem, N.M., Abdeen, A.O., 2013. Green synthesis of silver nanoparticles using carob leaf extract and its antibacterial activity. Int. J. Ind. Chem. 4 (1), 1–6.

Babu Maddinedi, S., Mandal, B.K., Maddili, S.K., 2017. Biofabrication of size controllable silver nanoparticles—a green approach. J. Photochem. Photobiol. B Biol. 167, 236–241.

Bernhardt, E.S., Colman, B.P., Hochella, M.F., Cardinale, B.J., Nisbet, R.M., Richardson, C.J., Yin, L., 2010. An ecological perspective on nanomaterial impacts in the environment. J. Environ. Qual. 39 (6), 1954–1965.

Castillo-Michel, H.A., Larue, C., Del Real, A.E.P., Cotte, M., Sarret, G., 2017. Practical review on the use of synchrotron based micro-and nano-X-ray fluorescence mapping and X-ray absorption spectroscopy to investigate the interactions between plants and engineered nanomaterials. Plant Physiol. Biochem. 110, 13–32.

Cheng, H.N., Klasson, K.T., Asakura, T., Wu, Q., 2016. Nanotechnology in agriculture. In: Nanotechnology: Delivering on the Promise. vol. 2. American Chemical Society, pp. 233–242.

Choi, O., Deng, K.K., Kim Jr., N.J., Ross, L., Surampalli, R.Y., Hu, Z., 2008. The inhibitory effects of silver nanoparticles, silver ions, and silver chloride colloids on microbial growth. Water Res. 42 (12), 3066–3074.

Coccia, F., Tonucci, L., Bosco, D., Bressan, M., d'Alessandro, N., 2012. One-pot synthesis of lignin-stabilised platinum and palladium nanoparticles and their catalytic behaviour in oxidation and reduction reactions. Green Chem. 14 (4), 1073–1078.

Dahoumane, S.A., Jeffryes, C., Mechouet, M., Agathos, S.N., 2017. Biosynthesis of inorganic nanoparticles: a fresh look at the control of shape, size and composition. Bioengineering 4 (1), 14.

Demirer, G.S., Zhang, H., Matos, J.L., Goh, N.S., Cunningham, F.J., Sung, Y., Chang, R., Aditham, A.J., Chio, L., Cho, M.J., Staskawicz, B., 2019. High aspect ratio nanomaterials enable delivery of functional genetic material without DNA integration in mature plants. Nat. Nanotechnol. 14 (5), 456–464.

Dimkpa, C.O., McLean, J.E., Latta, D.E., Manangón, E., Britt, D.W., Johnson, W.P., Boyanov, M.I., Anderson, A.J., 2012. CuO and ZnO nanoparticles: phytotoxicity, metal speciation, and induction of oxidative stress in sand-grown wheat. J. Nanopart. Res. 14 (9), 1–15.

Dimkpa, C.O., Latta, D.E., McLean, J.E., Britt, D.W., Boyanov, M.I., Anderson, A.J., 2013. Fate of CuO and ZnO nano-and microparticles in the plant environment. Environ. Sci. Technol. 47 (9), 4734–4742.

Dinesh, S., Karthikeyan, S., Arumugam, P., 2012. Biosynthesis of silver nanoparticles from *Glycyrrhiza glabra* root extract. Arch. Appl. Sci. Res. 4 (1), 178–187.

Du, W., Yang, J., Peng, Q., Liang, X., Mao, H., 2019. Comparison study of zinc nanoparticles and zinc sulphate on wheat growth: from toxicity and zinc biofortification. Chemosphere 227, 109–116.

Durán, N., Seabra, A.B., 2012. Metallic oxide nanoparticles: state of the art in biogenic syntheses and their mechanisms. Appl. Microbiol. Biotechnol. 95 (2), 275–288. https://doi.org/10.1007/s00253-012-4118-9.

Gahlawat, G., Choudhury, A.R., 2019. A review on the biosynthesis of metal and metal salt nanoparticles by microbes. RSC Adv. 9 (23), 12944–12967.

Gajjar, P., Pettee, B., Britt, D.W., Huang, W., Johnson, W.P., Anderson, A.J., 2009. Antimicrobial activities of commercial nanoparticles against an environmental soil microbe, *Pseudomonas putida* KT2440. J. Biol. Eng. 3 (1), 1–13.

Garcia, V.N., Gonzalez, A., Fuentes, M., Aviles, M., Rios, M.Y., Zepeda, G., Rojas, M.G., 2003. Antifungal activities of nine traditional Mexican medicinal plants. J. Ethnopharmacol. 87 (1), 85–88.

González-Cortazar, M., Zamilpa, A., López-Arellano, M.E., Aguilar-Marcelino, L., Reyes-Guerrero, D.E., Olazarán-Jenkins, S., Ramírez-Vargas, G., Olmedo-Juárez, A., Mendoza-de-Gives, P., 2018. *Lysiloma acapulcensis* leaves contain anthelmintic metabolites that reduce the gastrointestinal nematode egg population in sheep faeces. Comp. Clin. Pathol. 27 (1), 189–197.

Gurunathan, S., Han, J.W., Kwon, D.N., Kim, J.H., 2014. Enhanced antibacterial and anti-biofilm activities of silver nanoparticles against Gram-negative and Gram-positive bacteria. Nanoscale Res. Lett. 9 (1), 1–17.

Husseiny, M.I., El-Aziz, M.A., Badr, Y., Mahmoud, M.A., 2007. Biosynthesis of gold nanoparticles using Pseudomonas aeruginosa. Spectrochim. Acta A Mol. Biomol. Spectrosc. 67 (3–4), 1003–1006. https://doi.org/10.1016/j.saa.2006.09.028.

Ibrahim, H.M., 2015. Green synthesis and characterization of silver nanoparticles using banana peel extract and their antimicrobial activity against representative microorganisms. J. Radiat. Res. Appl. Sci. 8 (3), 265–275.

Iravani, S., 2011. Green synthesis of metal nanoparticles using plants. Green Chem. 13 (10), 2638–2650.

Jacob, J., Mukherjee, T., Kapoor, S., 2012. A simple approach for facile synthesis of Ag, anisotropic Au and bimetallic (Ag/Au) nanoparticles using cruciferous vegetable extracts. Mater. Sci. Eng. C 32 (7), 1827–1834.

Janmohammadi, M., Amanzadeh, T., Sabaghnia, N., Ion, V., 2016. Effect of nano-silicon foliar application on safflower growth under organic and inorganic fertilizer regimes. Botanica 22 (1), 53–64.

Jebakumar Immanuel Edison, T., Sethuraman, M., 2012. Instant green synthesis of silver nanoparticles using *Terminalia chebula* fruit extract and evaluation of their catalytic activity on reduction of methylene blue. Process Biochem. 47, 1351–1357. https://doi.org/10.1016/j.procbio.2012.04.025.

Jia, L., Zhang, Q., Li, Q., Song, H., 2009. The biosynthesis of palladium nanoparticles by antioxidants in *Gardenia jasminoides* Ellis: long lifetime nanocatalysts for p-nitrotoluene hydrogenation. Nanotechnology 20 (38), 385601.

Joglekar, S., Kodam, K., Dhaygude, M., Hudlikar, M., 2011. Novel route for rapid biosynthesis of lead nanoparticles using aqueous extract of Jatropha curcas L. latex. Mater. Lett. 65 (19-20), 3170-3172.

Kaphle, A., Navya, P.N., Umapathi, A., Daima, H.K., 2018. Nanomaterials for agriculture, food and environment: applications, toxicity and regulation. Environ. Chem. Lett. 16 (1), 43-58.

Khalil, M.M., Ismail, E.H., El-Baghdady, K.Z., Mohamed, D., 2014. Green synthesis of silver nanoparticles using olive leaf extract and its antibacterial activity. Arab. J. Chem. 7 (6), 1131-1139.

Kitching, M., Ramani, M., Marsili, E., 2015. Fungal biosynthesis of gold nanoparticles: mechanism and scale up. J. Microbial. Biotechnol. 8 (6), 904-917.

Krishnaraj, C., Jagan, E.G., Rajasekar, S., Selvakumar, P., Kalaichelvan, P.T., Mohan, N., 2010. Synthesis of silver nanoparticles using Acalypha indica leaf extracts and its antibacterial activity against water borne pathogens. Colloids Surf. B Biointerfaces 76 (1), 50-56. https://doi.org/10.1016/j.colsurfb.2009.10.008.

Lahiani, M.H., Dervishi, E., Chen, J., Nima, Z., Gaume, A., Biris, A.S., Khodakovskaya, M.V., 2013. Impact of carbon nanotube exposure to seeds of valuable crops. ACS Appl. Mater. Interfaces 5 (16), 7965-7973.

Larue, C., Khodja, H., Herlin-Boime, N., Brisset, F., Flank, A.M., Fayard, B., Chaillou, S., Carrière, M., 2011. Investigation of titanium dioxide nanoparticles toxicity and uptake by plants. J. Phys. Conf. Ser. 304 (1), 012057. IOP Publishing.

Lau, H.Y., Wu, H., Wee, E.J., Trau, M., Wang, Y., Botella, J.R., 2017. Specific and sensitive isothermal electrochemical biosensor for plant pathogen DNA detection with colloidal gold nanoparticles as probes. Sci. Rep. 7 (1), 1-7.

Lee, H.J., Song, J.Y., Kim, B.S., 2013. Biological synthesis of copper nanoparticles using *Magnolia kobus* leaf extract and their antibacterial activity. J. Chem. Technol. Biotechnol. 88 (11), 1971-1977.

Lin, X., Wu, M., Wu, D., Kuga, S., Endo, T., Huang, Y., 2011. Platinum nanoparticles using wood nanomaterials: eco-friendly synthesis, shape control and catalytic activity for p-nitrophenol reduction. Green Chem. 13 (2), 283-287.

Ling, F., Silberbush, M., 2002. Response of maize to foliar vs. soil application of nitrogen-phosphorus-potassium fertilizers. J. Plant Nutr. 25 (11), 2333-2342.

Logeswari, P., Silambarasan, S., Abraham, J., 2015. Synthesis of silver nanoparticles using plants extract and analysis of their antimicrobial property. J. Saudi Chem. Soc. 19, 311-317.

Lukman, A.I., Gong, B., Marjo, C.E., Roessner, U., Harris, A.T., 2011. Facile synthesis, stabilization, and anti-bacterial performance of discrete Ag nanoparticles using Medicago sativa seed exudates. J. Colloid. Interface Sci. 353 (2), 433-444. https://doi.org/10.1016/j.jcis.2010.09.088.

Lv, J., Zhang, S., Luo, L., Zhang, J., Yang, K., Christie, P., 2015. Accumulation, speciation and uptake pathway of ZnO nanoparticles in maize. Environ. Sci. Nano 2 (1), 68-77.

Manikandan, V., Velmurugan, P., Park, J.H., Chang, W.S., Park, Y.J., Jayanthi, P., Cho, M., Oh, B.T., 2017. Green synthesis of silver oxide nanoparticles and its antibacterial activity against dental pathogens. 3 Biotech 7 (1), 72.

Martin-Ortigosa, S., Peterson, D.J., Valenstein, J.S., Lin, V.S.Y., Trewyn, B.G., Lyznik, L.A., Wang, K., 2014. Mesoporous silica nanoparticle-mediated intracellular Cre protein delivery for maize genome editing via loxP site excision. Plant Physiol. 164 (2), 537-547.

Medvedeva, N.V., Ipatova, O.M., Ivanov, Y.D., Drozhzhin, A.I., Archakov, A.I., 2007. Nanobiotechnology and nanomedicine. Biochem. (Mosc.) Suppl. Ser. B Biomed. Chem. 1 (2), 114-124.

Mehta, C.M., Srivastava, R., Arora, S., Sharma, A.K., 2016. Impact assessment of silver nanoparticles on plant growth and soil bacterial diversity. 3 Biotech 6 (2), 1-10.

Mitter, N., Worrall, E.A., Robinson, K.E., Li, P., Jain, R.G., Taochy, C., Fletcher, S.J., Carroll, B.J., Lu, G.M., Xu, Z.P., 2017. Clay nanosheets for topical delivery of RNAi for sustained protection against plant viruses. Nat. Plants 3 (2), 1–10.

Moghtader, M., Salari, H., Mozafari, H., Farahmand, A., 2016. Evaluation the qualitative and quantitative essential oil of *Calendula officinalis* and its antibacterial effects. Iran. J. Biotechnol. 29 (3), 331–339.

Mondal, S., Roy, N., Laskar, R.A., Sk, I., Basu, S., Mandal, D., Begum, N.A., 2011. Biogenic synthesis of Ag, Au and bimetallic Au/Ag alloy nanoparticles using aqueous extract of mahogany (*Swietenia mahogani* JACQ.) leaves. Colloids Surf. B Biointerfaces 82 (2), 497–504.

Morones, J.R., Elechiguerra, J.L., Camacho, A., Holt, K., Kouri, J.B., Ramírez, J.T., Yacaman, M.J., 2005. The bactericidal effect of silver nanoparticles. Nanotechnology 16 (10), 2346.

Mousavi, S.R., Galavi, M., Ahmadvand, G., 2007. Effect of zinc and manganese foliar application on yield, quality and enrichment on potato (*Solanum tuberosum* L.). Asian J. Plant Sci. 6 (8), 1256–1260.

Mousavi, S.M., Hashemi, S.A., Ghasemi, Y., Atapour, A., Amani, A.M., Savar Dashtaki, A., Babapoor, A., Arjmand, O., 2018. Green synthesis of silver nanoparticles toward bio and medical applications: review study. Artif. Cells Nanomed. Biotechnol. 46 (sup3), S855–S872.

Naruka, I.S., Gujar, K.D., Gopal, L., 2000. Effect of foliar application of zinc and molybdenum on growth and yield of okra (*Abelmoschus esculentus* L. Moench) cv. Pusa Sawani. Haryana J. Hort. Sci. 29 (3/4), 266–267.

Navarro García, V.M., Rojas, G., Gerardo Zepeda, L., Aviles, M., Fuentes, M., Herrera, A., Jiménez, E., 2006. Antifungal and antibacterial activity of four selected Mexican medicinal plants. Pharm. Biol. 44 (4), 297–300.

Padil, V.V.T., Černík, M., 2013. Green synthesis of copper oxide nanoparticles using gum karaya as a biotemplate and their antibacterial application. Int. J. Nanomedicine 8, 889.

Panpatte, D.G., Jhala, Y.K., Shelat, H.N., Vyas, R.V., 2016. Nanoparticles: the next generation technology for sustainable agriculture. In: Microbial Inoculants in Sustainable Agricultural Productivity. Springer, New Delhi, pp. 289–300.

Park, Y., Hong, Y.N., Weyers, A., Kim, Y.S., Linhardt, R.J., 2011. Polysaccharides and phytochemicals: a natural reservoir for the green synthesis of gold and silver nanoparticles. IET Nanobiotechnol. 5 (3), 69–78. https://doi.org/10.1049/iet-nbt.2010.0033.

Patil, M.P., Singh, R.D., Koli, P.B., Patil, K.T., Jagdale, B.S., Tipare, A.R., Kim, G.D., 2018. Antibacterial potential of silver nanoparticles synthesized using *Madhuca longifolia* flower extract as a green resource. Microb. Pathog. 121, 184–189.

Petla, R.K., Vivekanandhan, S., Misra, M., Mohanty, A.K., Satyanarayana, N., 2012. Soybean (*Glycine max*) leaf extract based green synthesis of palladium nanoparticles. J. Biomed. Nanotechnol. 3 (1), 14–19. https://doi.org/10.4236/jbnb.2012.31003.

Poinern, G.E.J., Chapman, P., Shah, M., Fawcett, D., 2013. Green biosynthesis of silver nanocubes using the leaf extracts from Eucalyptus macrocarpa. Nano Bull. 2 (1).

Pradhan, D., Sindhwani, S., Leung, K.T., 2010. Parametric study on dimensional control of ZnO nanowalls and nanowires by electrochemical deposition. Nanoscale Res. Lett. 5 (11), 1727–1736.

Raliya, R., Tarafdar, J.C., 2013. ZnO nanoparticle biosynthesis and its effect on phosphorous-mobilizing enzyme secretion and gum contents in Clusterbean (*Cyamopsis tetragonoloba* L.). Agric. Res. 2 (1), 48–57.

Rasheed, T., Bilal, M., Iqbal, H.M., Li, C., 2017. Green biosynthesis of silver nanoparticles using leaves extract of *Artemisia vulgaris* and their potential biomedical applications. Colloids Surf. B Biointerfaces 158, 408–415.

Roopan, S.M., Bharathi, A., Kumar, R., Khanna, V.G., Prabhakarn, A., 2012. Acaricidal, insecticidal, and larvicidal efficacy of aqueous extract of *Annona squamosa* L peel as biomaterial for the reduction of palladium salts into nanoparticles. Colloids Surf. B Biointerfaces 92, 209–212.

Rossi, L., Fedenia, L.N., Sharifan, H., Ma, X., Lombardini, L., 2019. Effects of foliar application of zinc sulfate and zinc nanoparticles in coffee (*Coffea arabica* L.) plants. Plant Physiol. Biochem. 135, 160–166.

Sana, S.S., Dogiparthi, L.K., 2018. Green synthesis of silver nanoparticles using *Givotia moluccana* leaf extract and evaluation of their antimicrobial activity. Mater. Lett. 226, 47–51.

Sangaru, S.S., Rai, A., Ahmad, A., Sastry, M., 2004. Rapid synthesis of Au, Ag, and bimetallic Au core-Ag shell nanoparticles using neem (*Azadirachta indica*) leaf broth. J. Colloid Interface Sci. 275, 496–502. https://doi.org/10.1016/j.jcis.2004.03.003.

Sanzari, I., Leone, A., Ambrosone, A., 2019. Nanotechnology in plant science: to make a long story short. Front. Bioeng. Biotechnol. 7, 120.

Sathishkumar, M., Sneha, K., Won, S.W., Cho, C.W., Kim, S., Yun, Y.S., 2009. Cinnamon zeylanicum bark extract and powder mediated green synthesis of nano-crystalline silver particles and its bactericidal activity. Colloids Surf. B Biointerfaces 73 (2), 332–338.

Servin, A.D., Castillo-Michel, H., Hernandez-Viezcas, J.A., Diaz, B.C., Peralta-Videa, J.R., Gardea-Torresdey, J.L., 2012. Synchrotron micro-XRF and micro-XANES confirmation of the uptake and translocation of TiO2 nanoparticles in cucumber (*Cucumis sativus*) plants. Environ. Sci. Technol. 46 (14), 7637–7643.

Shankar, S.S., Ahmad, A., Sastry, M., 2003. Geranium leaf assisted biosynthesis of silver nanoparticles. Biotechnol. Prog. 19 (6), 1627–1631.

Sheny, D.S., Mathew, J., Philip, D., 2011. Phytosynthesis of Au, Ag and Au-Ag bimetallic nanoparticles using aqueous extract and dried leaf of *Anacardium occidentale*. Spectrochim. Acta A Mol. Biomol. Spectrosc. 79 (1), 254–262.

Song, J.Y., Kwon, E.Y., Kim, B.S., 2009. Biological synthesis of platinum nanoparticles using *Diopyros kaki* leaf extract. Bioprocess Biosyst. Eng. 11 (1), 159.

Srivastava, A., Rao, D.P., 2014. Enhancement of seed germination and plant growth of wheat, maize, peanut and garlic using multiwalled carbon nanotubes. Eur. Chem. Bull. 3 (5), 502–504.

Stouwdam, J.W., Janssen, R.A., 2008. Red, green, and blue quantum dot LEDs with solution processable ZnO nanocrystal electron injection layers. J. Mater. Chem. 18 (16), 1889–1894.

Subhankari, I., Nayak, P.L., 2013. Synthesis of copper nanoparticles using *Syzygium aromaticum* (Cloves) aqueous extract by using green chemistry. World J. Nano Sci. Technol. 2 (1), 14–17.

Thakkar, K.N., Mhatre, S.S., Parikh, R.Y., 2010. Biological synthesis of metallic nanoparticles. Nanomedicine 6 (2), 257–262.

Thomas, B., Arul Prasad, A., Mary Vithiya, S., 2018. Evaluation of antioxidant, antibacterial and photo catalytic effect of silver nanoparticles from methanolic extract of Coleus vettiveroids—an endemic species. J. Nanostruct. 8 (2), 179–190.

Tiwari, J.N., Tiwari, R.N., Kim, K.S., 2012. Zero-dimensional, one-dimensional, two-dimensional and three-dimensional nanostructured materials for advanced electrochemical energy devices. Prog. Mater. Sci. 57 (4), 724–803.

Torres, I., Bustamante, J., Sierra, D.A. (Eds.), 2017. VII Latin American Congress on Biomedical Engineering CLAIB 2016, Bucaramanga, Santander, Colombia, October 26th-28th, 2016. vol. 60. Springer.

Tripathi, D.K., Singh, S., Singh, V.P., Prasad, S.M., Dubey, N.K., Chauhan, D.K., 2017. Silicon nanoparticles more effectively alleviated UV-B stress than silicon in wheat (*Triticum aestivum*) seedlings. Plant Physiol. Biochem. 110, 70–81.

Valodkar, M., Jadeja, R.N., Thounaojam, M.C., Devkar, R.V., Thakore, S., 2011. Biocompatible synthesis of peptide capped copper nanoparticles and their biological effect on tumor cells. Mater. Chem. Phys. 128 (1–2), 83–89.

Wang, S., Wang, F., Gao, S., Wang, X., 2016. Heavy metal accumulation in different rice cultivars as influenced by foliar application of nano-silicon. Water Air Soil Pollut. 227 (7), 1–13.

Wang, Y., Lin, Y., Xu, Y., Yin, Y., Guo, H., Du, W., 2019. Divergence in response of lettuce (var. ramosa Hort.) to copper oxide nanoparticles/microparticles as potential agricultural fertilizer. Environ. Pollut. Bioavail. 31 (1), 80–84.

Willets, K.A., Van Duyne, R.P., 2007. Localized surface plasmon resonance spectroscopy and sensing. Annu. Rev. Phys. Chem. 58, 267–297.

Yang, F., Liu, C., Gao, F., Su, M., Wu, X., Zheng, L., Hong, F., Yang, P., 2007. The improvement of spinach growth by nano-anatase TiO2 treatment is related to nitrogen photoreduction. Biol. Trace Elem. Res. 119 (1), 77–88.

Yong, P., Rowson, N.A., Farr, J.P.G., Harris, I.R., Macaskie, L.E., 2002. Bioaccumulation of palladium by *Desulfovibrio desulfuricans*. J. Chem. Technol. Biotechnol. 77 (5), 593–601.

Youdim, K.A., Qaiser, M.Z., Begley, D.J., Rice-Evans, C.A., Abbott, N.J., 2004. Corrigendum to Flavonoid permeability across an in situ model of the blood-brain barrier. Free Radic. Biol. Med. 36: 592-604; 2004. Free Radic. Biol. Med. 10 (36), 1342.

Zargar, M., Hamid, A.A., Bakar, F.A., Shamsudin, M.N., Shameli, K., Jahanshiri, F., Farahani, F., 2011. Green synthesis and antibacterial effect of silver nanoparticles using *Vitex negundo* L. Molecules 16 (8), 6667–6676.

Zhao, X., Meng, Z., Wang, Y., Chen, W., Sun, C., Cui, B., Cui, J., Yu, M., Zeng, Z., Guo, S., Luo, D., 2017. Pollen magnetofection for genetic modification with magnetic nanoparticles as gene carriers. Nat. Plants 3 (12), 956–964.

Zhu, Z.J., Wang, H., Yan, B., Zheng, H., Jiang, Y., Miranda, O.R., Rotello, V.M., Xing, B., Vachet, R.W., 2012. Effect of surface charge on the uptake and distribution of gold nanoparticles in four plant species. Environ. Sci. Technol. 46 (22), 12391–12398.

Zia, M., Gul, S., Akhtar, J., Haq, I.U., Abbasi, B.H., Hussain, A., Naz, S., Chaudhary, M.F., 2017. Green synthesis of silver nanoparticles from grape and tomato juices and evaluation of biological activities. IET Nanobiotechnol. 11 (2), 193–199.

Role of microbial biotechnology for strain improvement for agricultural sustainability

Akhila Pole[a], Anisha Srivastava[a], Mohamed Cassim Mohamed Zakeel[b,c], Vijay Kumar Sharma[d], Deep Chandra Suyal[e], Anup Kumar Singh[a], and Ravindra Soni[a]

[a]Department of Agricultural Microbiology, College of Agriculture, Indira Gandhi Krishi Vishwavidyalaya, Raipur, Chhattisgarh, India, [b]Department of Plant Sciences, Faculty of Agriculture, Rajarata University of Sri Lanka, Puliyankulama, Anuradhapura, Sri Lanka, [c]Centre for Horticultural Science, Queensland Alliance for Agriculture and Food Innovation, The University of Queensland, Ecosciences Precinct, Dutton Park, QLD, Australia, [d]Department of Post-harvest Science, Agricultural Research Organization-Volcani Centre, Rishon LeZion, Israel, [e]Department of Microbiology, Akal College of Basic Sciences, Eternal University, Sirmour, Himachal Pradesh, India

12.1 Introduction

The global demand for food products has risen significantly in recent decades (Elferink and Schierhorn, 2016). It is expected that global food demand will increase up to 59%–98% by 2050. Further, food demand is also growing in developing countries, where arable cropland supplies are scarce. In this situation, increasing food production from established land is difficult and fails to contribute to meet such a critical need (Bargaz et al., 2018). Therefore, to meet food production requirements for the growing human population, there has been an increase in dependency on chemical fertilizers and pesticides. These chemical fertilizers are highly soluble, quite high in nutrient content, and immediately available to plants but their overuse can cause negative effects such as leaching, acidification and alkalization of the soil, lessening of soil fertility, water resource pollution, destruction of microbes, destruction of beneficial insects and increased pest and disease attacks due to heightened susceptibility of crops, all of which form the basis of irreversible harm to the entire ecosystem (Chen, 2006). Furthermore, the use of chemical inputs for the replenishment of nutrients lost in crop removal is not considered acceptable advice going forward because research has found that their use, in the long run, slows down biological activity in the soil leading to soil health impairment (Aktar et al., 2009). Immediate attention should be given to the goal of agricultural sustainability, to carry out "safe cultivation" for food production without jeopardizing food quality and the wider environment. Thus, a huge challenge facing the current agriculture industry is to use the limited resources available while improving food production systems to satisfy present and future food needs without further deteriorating environmental quality. In this circumstance, the possible substitute for the productivity, consistency, and sustainability of the global food chain without chemicals can be microbial-based nutrient inputs. The choice of enhancing agricultural productivity and soil fertility with the help of superior eco-friendly management tools guarantees successful food security. Therefore, the substitute being suggested and practiced utilizes productive agrobioresources (microorganisms) that play a crucial role in mitigating many issues like insect pests, abiotic stress, diseases, and soil fertility (Tilman et al., 2011; Utuk and Daniel, 2015; Timmusk et al., 2017).

Nonetheless, there are various applications of biotechnology, and it is a rapidly expanding field in sustainable agriculture. The use of genetic engineering (GE) to modify living organisms or their components to produce functional products for different biological applications is highly utilized currently. Food and shelter will be in short supply by 2033 as the global population grows. This poses a major challenge to the agricultural sector in terms of meeting the growing demand for food and forestry products. It is expected that demand for agricultural products in 2050 is going to increase by at least 70%. At that moment, people are more conscious of agricultural demands and food insecurity, as well as the significance of sustainable agricultural practices in meeting potential agricultural demand (Barea, 2015). Microbial biotechnology aids value-added products, food protection, human nutrition, food safety, functional foods, plant and animal health, and overall fundamental agricultural research. The foundation for all biotechnology research is based on the genetic resources of microbes, plants, and animals, working on technical innovation, invention enhancement, and product development. All aspects of genetic resource management, including selection, conservation, evaluation, and use, have been significantly influenced by biotechnology. Biotechnology's molecular methods have sped up the precision of breeding by identifying, isolating, cloning, and moving desired genes between species (interspecies), rendering the Mendelian population theory obsolete. Biotechnology's ultimate goals are to identify single-nucleotide polymorphisms, define functions of specific genes, assign functions to unknown genes, and create improved transgenics with specific fruitful desired characteristics (Singh, 2000).

For decades, microbiological tools such as bioinoculants of bacteria, fungi, and algae have been created to efficiently promote plant growth and health. A bioinoculant is a substance that consists of live microorganisms that colonize the rhizosphere and phyllosphere of plant tissues and induce plant growth when applied to soil, seeds, or plant surfaces (Soni et al., 2008, 2017). Bacteria or fungi that are capable of one or a combination of beneficial processes such as nitrogen fixation, phosphate solubilization, oxidation of sulfur, production of plant hormones, and decomposition of organic compounds are commonly used as biofertilizers (Verma et al., 2019). Biofertilizers can increase crop yields while still being environmentally friendly (Mahanty et al., 2017). *Pseudomonas fluorescens* strain K-34, for example, releases organic acids that may be responsible for the plant's phosphate release. The plant hormone indole acetic acid (IAA) is produced by *P. fluorescens* K-34, *P. fluorescens* 1773/K, *Pseudomonas trivialis* strain BIHB 745, and *Bacillus circulans* (Parani and Saha, 2012). Biofertilizers, in general, perform nutrient cycling and ensure optimal crop growth and production (Bhardwaj et al., 2014).

12.2 Microbial inoculants in agriculture for sustainability

Microbial inoculants or bioinoculants are referred to as plant stimulators because of their favorable impacts on agricultural crop development. Since microbes can influence plant growth and development at several stages of life such as flower germination, the interaction of the host plant with microbes is critical for plant growth (Gaj et al., 2013). Microbial inoculants enhance natural processes in the rhizosphere soil, assist with uptake of nutrients, and enhance productivity, abiotic stress tolerance, and crop quality (Suyal et al., 2016; Packialakshmi et al., 2020; Kushwah et al., 2021). In agricultural management, these microbial inoculants are regularly aimed at increasing agricultural productivity, reducing chemical contributions for crop production and protection, and improving agriculture sustainability (Gallart et al., 2018). Soil inoculants are added to the soil by incorporating various groups of soil species into a single inoculant, allowing them to benefit from a variety of plant-growth-promoting mechanisms. It would be difficult to predict the mechanisms used by each type of organism that lives in the soil (Ganbaatar et al., 2017). However, while a certain group of rhizobacteria isolated from soil can have beneficial properties on plants, some of them fail to produce expected results when inoculated into the soil. *Azospirillum*, *Rhizobium*, and *Agrobacterium*, for example, when used as a seed inoculant in the field, sometimes increased crop yield (Sahu et al., 2020; Kumar et al.,

2020), while *Pseudomonas* species did not. Even after the introduction of genetically changed bioinoculants, it is impossible to anticipate bacterial life and permanence, as well as the expression of changes in their features.

Nowadays, it is possible to examine a newly altered bioinoculant and their interaction with their field ecological unit and to observe the effect on the local microflora. In the same manner, as bacteria are released in fields, local environmental factors play a significant role in determining their endurance and persistence. With some inoculant releases, temporary changes favor the new bacteria and disfavor some established populations of bacteria and fungi in the rhizosphere of the plant. However, the improvements were minor compared to what would be expected under normal agricultural practices as some studies have found intragenic and intergenic gene transfer among soil bacteria (García-Fraile et al., 2015). Microbial-based agricultural enhancements are becoming more popular around the world, particularly in poorer nations such as Asia and Africa, where a multistrain produced from rhizosphere soil increased grain output by 10% (García-Salamanca et al., 2013; Goold et al., 2018). Enabling the colonization or conservation of new inoculants in the rhizosphere is one of the most difficult aspects of microbial inoculant introduction in agriculture. Numerous studies have demonstrated effective microbial colonization in the soil, while in the agricultural context, yields are often inconsistent or variable, with a rapid reduction in the number and activity of inoculants in the soil (Gupta and Pandey, 2019). Competition with indigenous soil microorganisms, as well as changes in conditions of growth such as temperature, pH, humidity, and texture, are thought to be the key causes of the inoculant population decline. Tilling and the abundant use of agrochemicals affect the efficiency of microbial inoculants. The host plant's selection and interaction with the inoculant is another consistent factor in population success (Hardoim et al., 2015).

12.2.1 Microbial inoculants as biofertilizers

Maintaining soil fertility is critical for meeting the food needs of an ever-increasing population. Biofertilizers are one of the best options for meeting this need because they consist of beneficial microbes that can boost the productivity of plants and the production of food without compromising environmental stability. Biofertilizers contain microorganisms that provide nutrients to plants that have already started to sprout in the field. Nitrogen-fixing bacteria, mycorrhizal fungi, phosphorus, sulfur, and potassium solubilizers are examples of rhizosphere microorganisms used as biofertilizers that have been researched extensively (Harman et al., 2004; Hart et al., 2017; He et al.,

2015). Microbial inoculants may be applied as dry formulations or as liquids. For microbial inoculants to perform optimally in the field, a greater quantity of inoculum is required, which is practically impossible to obtain naturally (Horton et al., 2014). Microorganisms used as biofertilizers can enter and invade the plant's rhizosphere zone during inoculation of biofertilizer to cropland, establishing a plant-microbe relationship and encouraging plant growth through direct and indirect mechanisms. Microorganisms used as biofertilizers are known as symbiotic biofertilizers based on their mechanisms of action. Symbiotic microorganisms form specialized roots or structures in and around the roots of host plants and maintain a symbiotic relationship with them. Nitrogen-fixing rhizobacteria and arbuscular mycorrhizal fungi (AMF) are the primary symbiotic species (Jacoby et al., 2017). Nitrogen-fixing bacteria, such as rhizobia species, are commonly found in the nodules of legume roots. The relationship of mycorrhizal symbiosis with plant roots favors nitrogen, phosphorous, and water absorption by the plant (Dash et al., 2019a,b). The two major forms of mycorrhizal plant associations are ectomycorrhiza and endomycorrhiza. Plant roots are colonized by ectomycorrhizal fungi as a net in the outer cell wall layers of plant roots, without entering plant cells (for example, fungal species belonging to the phyla Ascomycota and Basidiomycota). Arbuscular mycorrhiza, one of the endomycorrhizal fungi, colonizes the root cortex and forms vesicles and arbuscules in root cells with adsorptive and storage functions. Glomeraceae species in the genera *Rhizophagus*, *Glomus*, and *Funneliformis* (previously all in the *Glomus* genus) are possibly the most widespread AMF in anthropogenic and natural environments (Jambhulkar et al., 2016). Plant-growth-promoting rhizobacteria (PGPR) such as *Azospirillum*, *Pseudomonas*, *Actinobacter*, *Azotobacter*, *Bacillus*, and others colonize the rhizosphere region and live outside the plant roots (Goel et al., 2017; Jones and Hinsinger, 2008). The most studied beneficial microbial group is the PGPR. Some species of PGPR are extremely selective, affecting just specific organisms, resulting in field inconsistencies in terms of quality and efficiency (Kloepper et al., 1980). Numerous writers have suggested various concepts to organize biofertilizers, such as plant-growth-promoting bacteria (PGPB) and plant stress homeostasis-regulating bacteria (PSHB). PGPB are organisms that increase mineral nutrient intake and synthesize phytohormones, both of which have a direct impact on plant growth whereas PSHB are microbes that help plants cope with stress (Jung et al., 2012). In addition, some biofertilizers have been shown to improve the nutritional value of vegetables by improving the metabolic contents of those plants, which lead to antioxidant capacity in human health. Inoculation with *Glomus fasciculatum* and *Azotobacter chroococcum*, for example, increased the quantity of total carotenoid material, anthocyanins, and phenolic compounds in lettuce (Kunin et al., 2007). Similarly, soybean seedlings inoculated with Rhizobacteria reported a 75% rise in phenolic acid biosynthesis (Liu et al., 2018). The viability of microbes used for inoculated seeds varies depending on treatment methods and storage conditions, and this affects inoculation performance in the field.

12.2.1.1 Plant-growth-promoting rhizobacteria (PGPR)

The rhizosphere of many plant species is colonized by PGPR, causing advantageous effects for the host, such as increased plant growth, and increased resistance against disease-causing plant pathogens like bacteria, viruses, fungi, and nematodes (Kloepper et al., 2004). PGPR is a group of bacteria that are currently used as biofertilizers and biocontrol agents (Pii et al., 2015; Paterson et al., 2017). *Azotobacter, Alcaligenes, Azospirillum, Arthrobacter, Bacillus, Burkholderia, Pseudomonas, Klebsiella, Serratia, Rhizobium*, and *Enterobacter* are the most well-known PGPR genera (Kloepper et al., 2004). Increased seed germination rate, leaf area, root development, chlorophyll content, nutrient absorption, hydraulic activity, shoot and root weights, protein content, abiotic stress tolerance, yield, and delayed senescence are some of the benefits of PGPR (Adesemoye and Kloepper, 2009). As a result, PGPR are often used as biofertilizers. *Azotobacter vinelandii* and *A. chroococcum*, for example, are used as nitrogen fertilizers all over the world. Phosphorus fertilization is carried out using *Bacillus megaterium, Bacillus amyloliquefaciens* IT45, and *P. fluorescens* strains. There are biofertilizer products based on *Frateuria aurantia* that provide potassium nutrition for crops. Strains like *Thiobacillus thiooxidans, Delftia acidovorans*, and *Bacillus* spp. are used as ingredients in Asia for zinc, sulfur, and silicate fertilization. Furthermore, there are > 10 products with an unknown mechanism (other than "bean growth promotion") that contain 1–30 PGPR strains (Mącik et al., 2020). Many *Pseudomonas* spp. contain antibiotics such as 2,4-diacetylphloroglucinol, pyrrolnitrin (PRN), pyoluteorin (PLT), and hydrogen cyanide, and the strain *P. fluorescens* ZX secretes lytic enzymes as a form of biocontrol (Sharma et al., 2017). Moreover, several products for biocontrol are PGPR-derived and are currently available based on *Streptomyces griseoviridis, Pseudomonas* strains, and *Bacillus* strains (Pii et al., 2015; Mącik et al., 2020). *P. fluorescens* A506, for example, competes with the pathogen *Erwinia amylovora* and prevents pome fruit fire blight by Cabrefiga et al. (2007). In addition, *Pseudomonas chlororaphis* is effective against *Drechslera graminae* (barley leaf stripe), *Drechslera teres* (barley net blotch), and *Fusarium* pathogens (Johnsson et al., 1998; Puopolo et al., 2011). *Bacillus subtilis* improves *Arabidopsis* growth and salt resistance (Han et al., 2014), and protects blueberries from the mummy berry fungus *Monilinia vaccinii-corymbosi* (Ngugi et al., 2005), while *B. amyloliquefaciens* controls lettuce bottom rot (Chowdhury et al., 2013). Similarly, *S. griseoviridis* K61 is effective against a variety of fungal pathogens, including *Fusarium oxysporum* f.sp. *lycopersici* and *Verticillium dahliae*, which cause tomato vascular wilt and verticillium wilt, respectively (Minuto et al., 2006).

12.2.1.2 Role of PGPR in plant growth and development

Plant-growth-promoting rhizobacteria (PGPR) use various direct and indirect mechanisms to enhance the growth and development of the plant. Different roles of PGPR for agricultural productivity and sustainability include:

1. Solubilization of phosphorus
2. Mobilization of Potash
3. Production of biofertilizer and biopesticide using microorganisms
4. Production of plant-growth regulators
5. Volatile organic compounds (VOCs) production
6. Microbes as biotic elicitors
7. In stress agriculture, microbial reactions are observed
8. Microbial antagonism
9. Environmental and natural resource conservation

12.2.2 Microbial inoculants as biocontrol agents

Biocontrol agents such as various bacteria, fungi, and actinobacteria are used to protect plants from pathogens and act as antibacterial and antifungal agents. A wide range of bioinoculants in different formulations are available on the market, with various benefits and applications in agriculture and horticulture crops. Microbial inoculants have various functions like competing for nutrients and secondary metabolites or releasing extracellular hydrolytic enzymes that are harmful to plant pathogens at low concentrations (Lugtenberg, 2015). Herbicidal activity has been identified for some of the microbial inoculants. Researchers have already recorded *Colletotrichum coccodes*, a velvetleaf mycoherbicide, and Striga mycoherbicides as examples (Salgarello et al., 2013; Mendes et al., 2013). Another well-known example is *Trichoderma harzianum*'s antibiotics, which inhibit wood decay and pathogenic fungi (Sachs, 2015). *Aspergillus niger, Aspergillus fumigatus, Pseudomonas funiculosum, Penicillium citrinum, Penicillium aurantiogriseum,* and *Trichoderma koningii* have all been found to be effective against the plant pathogenic fungus *Phytophthora infestans* (Muhammad et al., 2019). The efficacy of *Amphibacillus xylanus, B. amyloliquefaciens, Sporolactobacillus inulinus,* and *Microbacterium oleivorans* in inhibiting fungal pathogen growth has been documented by several researchers (Augé et al., 2015). *Mitsuaria* sp. had a strong biocontrol impact on bacterial leaf spot infections. *Pseudomonas* spp.

have shown promising biocontrol effect on Fusarium wilts (Müller et al., 2016). *Bacillus* spp., which have the ability to produce volatile inhibitory substances in many plants, have been discovered to be beneficial in the biological control of microbial infections (Müller et al., 2016). For Pythium disease, Rhizobia sp. has been shown as a promising biocontrol agent (Muñoz et al., 2019).

12.2.3 Different microbial groups in soil ecosystem

12.2.3.1 Rhizobia

Rhizobia are one of the oldest farming resources, with industrial development beginning at the end of the nineteenth century. The alpha-proteobacterial nitrogen-fixing genera such as *Rhizobium, Bradyrhizobium, Agrobacterium, Azorhizobium, Methylobacterium, Allorhizobium, Mesorhizobium, Devosia, Sinorhizobium, Ochrobactrum,* and *Phyllobacterium*, and beta-proteobacterial genera like *Cupriavidus* and *Burkholderia* with legumes as the host plant, can all develop nodules. Despite leguminous crop plants being used in limited amounts in agriculture, they are capable of fixing up to 200–300 kg of nitrogen per hectare per crop. Furthermore, in rice production, the green manure Azolla is generally used which shows symbiotic associations with cyanobacteria to fix nitrogen (Franche et al., 2009).

12.2.3.2 Mycorrhizal fungi

Apart from PGPR, one of the most widely used biofertilizers is mycorrhizal fungi (Vosátka et al., 2012). They help plants to absorb water and nutrients, especially phosphorus; they lessen abiotic stress; and they also act as biocontrol agents against pathogens that damage roots such as those belonging to the genera *Phytophthora, Fusarium*, and *Pythium*, and nematodes (Smith and Read, 2010; Smith and Smith, 2011; Benami et al., 2020). By improving the antioxidant and vitamin content of edible parts, mycorrhizal fungi can boost the food-quality properties of crop plants (Gianinazzi et al., 2010; Albrechtova et al., 2012). Furthermore, by altering the soil composition, they enhance the physical properties of the soil. For example, hyphae may cause soil particles to entangle with one another, forming macroaggregates (Miller and Jastrow, 2000). The most frequently employed variety to increase crop efficiency is arbuscular mycorrhizal fungi (AMF), which penetrate the root cortical cells of a host plant and create highly branching structures called arbuscules (Smith and Smith, 2011). Glomalin, an AMF glycoprotein, is an essential component of mycorrhizal soil that reduces macroaggregate disruption during wetting and drying and prevents expedited water movement through soil pores, preventing the development of gas-water interfaces (Miller and Jastrow, 2000). AMF are favored for improving crop production as they are naturally present in over 90% of plant species, although several agricultural activities, such as tillage and fertilization, lower the number of AMF in the fields (Rillig et al., 2016). Protecting host plants from root-damaging worms by AMF strains leads to an increase in the yields of tomato by 26% and yields of carrot by 300% (Affokpon et al., 2011). According to a metaanalysis, when the soil community is complex and phosphorus is limited instead of nitrogen, the nonlegume plants, woody plants, and C_4 grasses respond to AMF better in comparison with C_3 grasses and legumes (Pirttilä et al., 2021). Apart from the details of the plant-mycorrhiza relationship, Hoeksema et al. (2010) claimed that the outcome of mycorrhizal inoculation is influenced by soil fertility and biocomplexity, as well as environmental factors including salinity and pH. There is a MycoDB database that contains metaanalysis data on the response of plant productivity to the application of mycorrhiza spanning over 10 years allowing for better predictability of mycorrhizal inoculation (Chaudhary et al., 2016). Despite the positive outcomes of mycorrhizal inoculants, there are many technical difficulties in their manufacture. In axenic circumstances, some mutualistic symbiont strains are not cultivable, making up-scaling mass production of mycorrhizal fungi difficult. The mycorrhizal inocula may not be pure and may contain fungal pathogens, necessitating the use of molecular tests to determine the inocula composition (Vosátka et al., 2012). However, many mycorrhizal fungi-related biofertilizers and biocontrol products, primarily based on *Glomus iranicum* strains, are available to boost crop water and nutrient absorption as well as nematode tolerance.

12.2.3.3 Endophytic fungi

Despite their multiple beneficial roles for plants and capability for biocontrol and biofertilization, endophytic fungi have been rarely established as biotechnological tools in the field of agriculture (Kumar et al., 2019, 2020). *Piriformospora indica* is a promising and well-studied endophyte that colonizes the roots of barley and corn (Varma et al., 1999; Waller et al., 2005; Kumar et al., 2009). Similarly, the uptake of phosphorus and sulfur is increased by endophytic fungi, so as to increase in biomass production, production of seed, and early flowering is promoted (Oelmüller et al., 2009). *P. indica* aids the host plant in overcoming abiotic stresses including drought, heat, and salt, as well as inducing plant resistance to pathogens, insects, heavy metals, and toxins (Das et al., 2012). Recently, DNA and mRNA of *P. indica* were discovered in soil only for a maximum of fifteen months (Rabiey et al., 2017), indicating that the fungus thrives in the southern hemisphere. Overall, the effect of *P. indica* has been fully examined and evaluated in as many as 150 distinct plant species (Shrivastava and Varma, 2014). Fungi such as clavicipitaceous endophytes of grasses are utilized in agriculture as biocontrol agents (Kauppinen et al., 2016). *Epichloë coenophiala* has been inoculated into tall fescue to boost insect resistance (Bouton, 2007; Bouton et al., 2002), and in New Zealand and Australia, ryegrass cultivars treated with fungal endophytes have been employed to reduce pasture damage from insect herbivores (Johnson et al., 2013; Young, 2012). Yet, clavicipitaceous endophytes are currently used only in pasture grasses (Kauppinen et al., 2016).

12.2.3.4 Rhizospheric fungi

Numerous fungi present in the soil are also useful in agriculture to improve the health and fitness of the crops (Table 12.1). *Penicillium bilaiae*, a rhizospheric fungus, is an example which, like mycorrhizal fungi, aids the plant in phosphate acquisition (Aamir et al., 2020; Harvey et al., 2009). Many products contain members of the rhizospheric and epiphytic *Trichoderma* spp. because of their ability to minimize biotic and abiotic stress on host plant such as plant pathogen and nematode control (Vosátka et al., 2012). *Trichoderma viride* is antagonistic against the genera *Rhizoctonia, Pythium, Sclerotium*,

Table 12.1 Microorganisms used in agriculture for sustainability.

S. no.	Name	Species	Organism type
1.	Azotovit	*Azotobacter chroococcum*	PGPR
2.	AQ10 biofungicide	*Ampelomyces quisqualis* M-10	Soil fungi
3.	Bio-N	*Azotobacter* sp.	PGPR
4.	Bio-K	*Frateuria aurantia*	PGPR
5.	Biofox C	*Fusarium oxysporum*	Soil fungi
6.	Biogro	*Pseudomonas fluorescens, Candida tropicalis, Bacillus amyloliquefaciens, Bacillus subtilis*	mix
7.	Blight Ban A506	*Pseudomonas fluorescens* A506	PGPR
8.	Blossom-Protect, Boni-Protect	*Aureobasidium pullulans* DSM 14940 (*CF* 10), DSM 14941 (*CF* 40)	Epiphytic fungi
9.	Myc 800	*Rhizophagus intraradices*	Mycorrhiza
10.	Mykoflor	*Rhizophagus irregularis, Funneliformis mosseae, Claroideoglomus etunicatum*	Mycorrhiza
11.	Nitrofix	*Azospirillum* str. Az39	PGPR
12.	Nitragin Gold	*Rhizobium meliloti*	Rhizobia
13.	Novodor	*Bacillus thuringiensis morrisoni*	Entomopathogenic bacterium
14.	Phylazonit-M	*Pseudomonas putida, Azotobacter chroococcum, Bacillus circulans, Bacillus megaterium*	Mix
15.	Rhizocell	*Bacillus amyloliquefaciens* IT45	PGPR
16.	RhizoMyco, RhizoMyx, RhizoPlex	A mixture of 18 species of endo- and ectomycorrhizal fungi	Mycorrhiza

and *Fusarium* which are soilborne pathogens. Similar fungal pathogens are controlled by *T. harzianum;* instead, it also controls wood-rot fungi, *Gaeumannomyces, Sclerotinia, Botrytis,* and *Verticillium.* For treatment against fungal infections, *T. harzianum* and *Trichoderma polysporum* are used in combination to broaden the spectrum and *Trichoderma* spp. can also be employed with mycorrhizal fungi (Vosátka et al., 2012). The *Trichoderma atroviride* strain, when combined with *Glomus intraradices,* can produce siderophores and auxin-like compounds that can boost the growth of pepper, tomato, lettuce, zucchini, and melon by 56%–167% (Colla et al., 2015). *T. harzianum* may also cause host plants to develop, flower, and produce secondary metabolites cannabidiol in hemp (Kakabouki et al., 2021).

12.2.3.5 Entomopathogenic and mycoparasitic fungi

Aside from plant-associated fungi, there are other fungal products dependent on plant-pathogen mycoparasitism. Mycoparasitic fungi such as *Ampelomyces quisqualis* cause powdery mildew, and *Fusarium*

Application	Target	Reference
Biofertilizer	Improves nitrogen availability in maize and broccoli	Alafeea et al. (2019)
Biocontrol	Controls powdery mildews in grapes, apples, aubergine, tomatoes, strawberry, cucumber	Shishkoff and McGrath (2002)
Biofertilizer	Improves nitrogen acquisition in tomato	Youssef and Eissa (2017)
Biofertilizer	Improves potassium acquisition in tomato	Youssef and Eissa (2017)
Biocontrol	Against root pathogens tomato, cucumber, pepper, monocot, and dicot species	Minuto et al. (1995)
Biofertilizer	Nitrogen fixation, phosphorous solubilization, organic matter mineralization	Hien et al. (2014)
Biocontrol	Against blight in fruits, tomato, potato	Sharifazizi et al. (2017)
Biocontrol	Developed and registered to control postharvest diseases of apple as the product	Weiss et al. (2006)
Biofertilizer	Improves nutrient acquisition in strawberry	Mikiciuk et al. (2019)
Biofertilizer	Improves nutrient acquisition in strawberry	Mikiciuk et al. (2019)
Biofertilizer	Improves nitrogen acquisition	Okon et al. (2015)
Biofertilizer	Provides nitrogen fixation for alfalfa, sweet clover, or clover	Smith (1995)
Biocontrol	Controls potato beetle larvae	Wraight and Ramos (2017)
Biofertilizer	Improves phosphate availability in rice, maize	Ingle and Padole (2017)
Biofertilizer	Phosphorous solubilization and mineralization in vegetables, horticultural, and field crops	Xie et al. (2018)
Biofertilizer	Improves nutrient and water availability in several crops	Poddar et al. (2020)

proliferatum causes downy mildew on crop plants. Furthermore, pathogenic insects that infest a variety of crops like aphids, whiteflies, mites, and thrips, including their various developmental stages, are controlled by the entomopathogenic fungus *Beauveria bassiana* (Vosátka et al., 2012; Perdikis et al., 2008). *Chaetomium cupreum* also protects plants from fungi including early and late blight, rust, stem and tuber rot, and leaf blot (Soytong et al., 2001).

12.2.4 Microbial inoculants for biotic and abiotic stress tolerance

Crops are under more stress as a result of climate change, natural and anthropogenic causes, resulting in lower crop productivity. Even though several trials and experiments have been performed to establish methods to cope with abiotic stress, finding a long-term solution appears to be a difficult task. Microorganisms have been shown to favor plants in such situations and may be able to aid in the battle against stress through a variety of direct and indirect mechanisms. There are two types of stress: biotic and abiotic. Pests and plant pathogens (insects, bacteria, fungi, viruses, and nematodes) cause biotic stress, while abiotic stress may be caused by drought, floods, temperature, salinity, heavy metals, gases, and nutrient quantity (Dash et al., 2019a,b; Dubey et al., 2020). These stress conditions may result in a yield drop depending on plant characteristics and soil types, as well as the ecology and development of plant-soil microbial interactions (Nguyen et al., 2017). There would also be nutritional deficiency, hormonal imbalances, physiological disturbances (abscission, senescence, and epinasty), and vulnerability to illnesses (Nicolás et al., 2014). PGPRs are effective bacteria that can function in a variety of soils and aid plant growth and development through direct or indirect mechanisms. Biological nitrogen fixation is a direct process to stabilize nitrogen. Plants receive phytohormones such as indole-3-acetic acid (IAA), phosphates, and iron dissolved by bacterial siderophores. In the indirect process, physiological changes can be brought by PGPR at the molecular level by the release of enzymes such as bacterial 1-aminocyclopropane-1-carboxylic acid (ACC) deaminase. Plant ethylene production is directly regulated by ACC deaminase, resulting in altered plant growth and development. Bacterial strains that produce ACC deaminase help plants overcome the negative effects of stress. Plants treated with bacteria that produce ACC deaminase showed widespread root growth due to lower ethylene levels. Plants may be able to combat a variety of stressors in this way. To improve plant growth, PGPR is used with ACC deaminase activity to face a deficiency of nutrient, heavy metal stress, salinity, and drought (Nora et al., 2019; O'Callaghan, 2016; Öpik et al., 2010). Chickpea plants, for example, used the PGPR bacteria *Pseudomonas putida*-MTCC5279 to avoid drought stress by regulating the expression of genes involved in jasmonate (MYC2), salicylic acid (PR1), and ethylene biosynthesis (ACS and ACO) signaling. Drought tolerance was demonstrated by *P. putida* by altering membrane integrity, ROS scavenging capacity, and osmolyte accumulation (Owen et al., 2015). Similarly, a strain of *P. fluorescens* REN1 that produces ACC deaminase improved root elongation in rice plants under constant flooding conditions (Pandey et al., 2017). Inoculation of tomato plants exposed to low temperatures with *Pseudomonas frederiksbergensis* OS261 and *Pseudomonas vancouverensis* OB155 improved antioxidant activity and cold acclimation gene expression in leaf tissues (Pellegrino et al., 2015). AMF has also been found to increase resistance to stresses such as salinity and drought (Pellegrino et al., 2012; Purvis et al., 2019).

12.2.4.1 Mitigation of drought stress

Drought is the most detrimental environmental factor that decreases crop productivity by inhibiting the growth of the plant. Climate change, agronomic, and edaphic factors all contribute to drought stress (Rastegari et al., 2020). Drought stress is expected to escalate in the future if global freshwater supplies and climatic factors remain a problem (Nadeem et al., 2019). Drought would impede the production of feed, wood, and most significantly food due to changes in global temperature and precipitation. As a result, drought-tolerant crop production has become a prerequisite for a sustainable future to ensure food security. Most bioenergy crops that are used in biofuel generation, such as poplar and miscanthus, are drought tolerant. As a result, improving bioenergy crops' drought tolerance and significantly improving their water use efficiency (WUE) is critical for long-term biomass production in arid and semiarid regions (Von Cossel et al., 2019). Despite the widespread use of genetic engineering techniques to promote drought tolerance in plants, progress has been slow due to the existence of multiple genes and the complexity of the traits involved (Khan et al., 2019a,b; Rastegari et al., 2020). The rhizosphere and associated bacteria have been found to play an essential role in restricting a plant's ability to cope with drought stress (Kour et al., 2019). PGPR provides them the ability to withstand drought by assisting in the production of volatile organic compounds (VOCs), phytohormones, and exopolysaccharides (EPSs) (Naseem et al., 2018; Tiwari et al., 2020). They also aid in the absorption of osmolytes and antioxidants. Furthermore, they can change the morphology of roots to control stress-responsive genes and plant's overall response to stress (Sharma et al., 2019). For example, inoculation of *Azospirillum* species that produce indole acetic acid (IAA) enhanced the growth of roots and triggered the development of lateral roots which improved the drought tolerance of wheat plants (Vurukonda et al., 2018). Similarly, *Bacillus thuringiensis*, an IAA-producing plant-growth-promoting microbe, increased nutrient availability and boosted the plant's metabolic processes and promoted the growth of *Lavandula dentata* during drought (Armada et al., 2016). Green fluorescent protein (GFP)-labeled *Pseudomonas* and *Acinetobacter*, that were used for inoculation with bacteria, generated a drought-stress response in grapevine and *Arabidopsis* plants so that they were able to withstand drought conditions (Rolli et al., 2015). An increase in abscisic acid (ABA) concentration in shoots and stomatal conductance was reported following *B. subtilis* inoculation in *Platycladus orientalis* leaves, indicating that the plant was drought resistant. Water content in leaves increased, cytokinin levels increased, and water potential improved dramatically as ABA levels increased (Liu et al., 2013). In another study, inoculating *Arabidopsis* plants with *Phyllobacterium brassicacearum* strain STM196, an isolate from the rhizosphere of *Brassica napus*, aided drought adaptation by boosting ABA concentrations, lowering transpiration in leaves, and enhancing resistance to osmotic stress (Ahkami et al., 2017). Also, when soybean plants were inoculated with *P. putida* strain H-2-3, the gibberellin-producing rhizobacterium, length of shoots, and fresh weight were enhanced under drought conditions (Kang et al., 2014). In contrast with control plants under drought stress, they produced more chlorophyll, abscisic acid, and salicylic acid (Radhakrishnan et al., 2017).

12.2.4.2 Salinity stress mitigation

Salinity is the other significant environmental factor that has a negative impact on plant production around the world. Excessive salt in the soil causes toxicity and ionic disparity in plants, resulting in water scarcity and metabolic activity imbalances owing to hyperosmotic stress (Rajawat et al., 2020). Plants respond to salinity stress in different ways, such as the production of polyamines and osmolytes, the activation of defense mechanisms, the prevention of reactive oxygen species deposition, and the regulation of ion transport (Khan et al., 2019a, 2019b; Yadav et al., 2020). When wheat seedlings were exposed to PGPR such as *Bacillus*, *Enterobacter*, *Paenibacillus*, and others, exopolysaccharides (EPSs) were generated under extreme saline circumstances, and the plant's absorption of Na^+ ions was greatly reduced while biomass output increased (Egamberdieva et al., 2019). In a recent study, the carotenoid producing and halotolerant *Dietzia natronolimnaea* strain STR1 was used to counteract the effects of salinity in wheat plants. When halotolerant PGPR was inoculated in wheat plants, they had enhanced proline levels and produced more antioxidants, indicating that they were more saline responsive. Another study found that tomato plants inoculated with PGPR decreased the negative effects of ethylene released under stress conditions on root growth by increasing the activity of the enzyme ACC deaminase. This resulted in enhanced plant growth in saline and water-deficient environments (Ilangumaran and Smith, 2017). Furthermore, PGPR activation triggered certain plant pathways such as ABA signaling, Fe transport, and SOS pathways, among others (Bharti et al., 2016). Inoculated peanut seedlings also had better ion homeostasis, less ROS accumulation, and improved growth under saline conditions than the uninoculated peanut seedlings (Shukla et al., 2012). Another study found that foliar application of silicon combined with *Enterobacter cloacae* and *Bacillus drentensis* helped mung beans to withstand salinity (Ahkami et al., 2017). Furthermore, when *Haerero halobacter*, *Brachybacterium saurashtrense*, and *Brevibacterium casei* inoculated peanut seedlings were exposed to extreme saline conditions by incorporating 100 M NaCl, overall improved growth was observed (Shukla et al., 2012).

12.2.4.3 Mitigation of heavy metals stress

At low concentrations, heavy metals including Ni, As, Cr, Cd, Cu, Pb, Zn, and others are necessary for microorganism and plant growth and metabolic activities, but they can represent a major threat if the concentration surpasses the tolerance limits (Singh et al., 2011). In soil, toxic heavy metals have a major impact on plant characteristics and phytoremediation potentials, whereas bacteria in the soil can considerably boost the plant's phytoremediation capability through synergistic action, hence the phrase "microbe-assisted phytoremediation" (Sharaff et al., 2020). According to findings, PGPR can also help protect the host plant from the harmful effects of heavy metal toxicity (Wani et al., 2008). *Azotobacter*, *Bacillus*, *Mesorhizobium*, *Sinorhizobium*, *Rhizobium*, *Bradyrhizobium*, and *Pseudomonas* are only a few of the genera that PGPR are known to protect (Rai et al., 2020). In a report, *Bacillus licheniformis* was found to dramatically boost rice plant seed germination and improve the biochemical characteristics of rice when it was stressed by Ni. As a result, the strain's potential for shielding rice plants from heavy metal toxicity has been highlighted (Jamil et al., 2014). PGPR, like most microbes, have developed specific mechanisms to withstand heavy metals: mobilization, immobilization, and conversion of heavy metals into inactive or less harmful forms that can be used (Tiwari and Lata, 2018). To improve heavy metal resistance, PGPR are known to use one of five mechanisms: (1) heavy metal extrusion via efflux pumps; (2) exclusion of heavy metal from target sites through direct removal; (3) complex formation, such as the creation of thiol-containing complex structures, can inactivate heavy metals; (4) the conversion of extremely hazardous Cr^{4+} to less toxic Cr^{6+} as an example of heavy metal biotransformation from a hazardous oxidation state to a less toxic oxidation state; and (5) methylation and demethylation as processes of adding or removing methyl from heavy metals

(Ma et al., 2016). Plants, including microorganisms, have a range of methods for coping with heavy metal resistance; however, the molecular mechanism by which microbes and plants cooperate to combat heavy metal toxicity remains unknown. Furthermore, by better understanding plant-microbe interactions, the genes involved, and regulatory processes, it may be feasible to develop plants to grow faster in places where heavy metals are present (Mishra et al., 2017).

12.2.4.4 Mitigation of heat stress

In abiotic stresses, temperature is one important limiting factor that harms the development, homeostasis, and metabolic activities of plants and microorganisms. Bioprospecting PGPR for the ability to promote plant growth at lower temperatures could potentially boost global crop productivity, which is particularly important given the current rate of global warming (Kour et al., 2020). There is less experimental evidence supporting the impact of PGPR isolates on crop production at high temperatures. While thermostable PGPR isolates have been published in the literature, they are incapable of providing the host plant with thermostability (Rodriguez et al., 2008). Nonetheless, several studies have shown that PGPR isolates can be used to mitigate the negative effects of low-temperature-induced stress (Dimkpa et al., 2009; Suyal et al., 2018; Suyal et al., 2019a,b). Low-temperature stress has resulted in increased synthesis of proline, sugar, anthocyanin, and other compounds (Dimkpa et al., 2009). With *Burkholderia phytofirmans* inoculated grapevine plants, enhanced levels of phenols, proline, and carbohydrates, as well as enhanced accumulation of starch, were observed. However, when exposed to low temperatures (4°C), grapevine plants inoculated with PGPR produced less biomass and had an electrolyte imbalance (Kumar et al., 2019).

12.2.4.5 Combating elevated carbon dioxide

Various levels of photosynthesis are important in the absorption of carbon dioxide (CO_2) from the atmosphere and its conversion to organic carbon in plant biomass. The photosynthetic process in C_3 plants is aided by rising CO_2 levels in the atmosphere, which aids the proliferation of rhizospheric bacteria and improves photosynthate localization in soil. Climate change has a significant impact on plant composition and diversity, posing a danger to soil microorganisms and soil edaphic characteristics, such as organic matter quantity and consistency. It also has an adverse effect on nutrient cycles such as the biomass, methane, and nitrogen cycles, as well as terrestrial ecosystem climates (Malyan et al., 2019). The use of PGPR has improved reforestation, grassland management technology (Van Der Heijden et al., 2006), and ecosystem regeneration (Requena et al., 2001). At reduced atmospheric CO_2 levels, PGPR have a remarkable ability to increase carbon accumulation in terrestrial systems by improving the productivity of crops and reducing the loss of carbon by microbial respiration (Nie et al., 2015). However, the probability of potential CO_2 concentrations in the atmosphere also expands the horizon of PGPR use. Through plant-microbe interactions, the influence of microorganisms on the host plant is clearly understood, but the molecular mechanisms involved are still unknown. To fully exploit the ability of PGPR, further research into rhizobacteria colonization mechanisms and plant growth dynamics is needed.

12.3 Microbial biotechnology for sustainable agriculture

Nature is a rich storehouse of everything that all living organisms require to meet their basic needs. As a result of natural selection, nature has occasionally altered the genetic makeup of populations. Important microorganisms such as diazotrophic bacteria, fungi, biological control agents, and plant-growth-promoting rhizobacteria (PGPR) are important in reducing the toxicity of chemical use and maintaining essential ecosystem functioning involving plants and soil. Microbial biotechnology plays a significant role in sustainable agriculture by lowering reliance on agrochemicals and managing biotic and abiotic challenges. This management entails several processes, encompassing microbe selection through target and gene selection from related or unrelated genetic resources.

Microbial biotechnology and genetic engineering, when combined, lead to decline in the virulence of plant and animal pathogens, improved disease-diagnostic tools, and enhanced microbial agents for biological control of plant and animal pests. Augmented bioremediation bacteria present in soil are one of the most important components for the long-term production of nutritious, organic food (Lugtenberg, 2015). Rhizospheres serve as microhabitats and are rich in microbial variety (Hirsch et al., 2013), and so agriculture biologists are focusing their research on bacteria, fungi, and archaea in this important soil zone (Spence and Bais, 2013).

Microbes connected with plants have a wide range of trophic/living behaviors, including saprophytic and symbiotic connections. These connections could be advantageous or destructive. Most soil microorganisms live in the rhizosphere, although certain fungi and bacteria live inside plant tissue and are endophytic in nature (Brader et al., 2014; Mercado-Blanco, 2015). These endophytes do not cause disease in plant, but they do create crucial secondary metabolites in response to pathogens and environmental changes which affect plant growth. Beneficial microorganisms contribute to plant growth by performing vital functions such as (i) PGPR, (ii) decomposer of organic substances, and (iii) protection against plant pathogens. Useful plant mutualistic symbionts include bacteria that fix N_2 and multifunctional arbuscular mycorrhizal (AM) fungi. N_2-fixing bacteria in mutualistic symbiosis with legume plants belong to numerous genera, yet they are all grouped and referred to as "rhizobia" (de Bruijn, 2015; Olivares et al., 2013). The actinomycetes, of which the genus *Frankia* is a member, create N_2-fixing nodules on the roots of so-called actinorhizal plant species, which are ecologically significant (Normand et al., 2007). Biotechnology provides excellent prospects to boost global agricultural productivity while also reducing the usage of agrochemicals such as inorganic fertilizers, insecticides, and weedicides rodenticides. Modern biotechnology has played a critical part in achieving environmental sustainability through the use of environmentally friendly crops such as herbicide-tolerant, insect-resistant varieties, and crops that can fix atmospheric nitrogen, resulting in environmental cleansing. In 2005, the United Nations General Assembly recognized the importance of innovative tools such as biotechnology in food security, agricultural sustainability, and economic development (Ene-Obong, 2007).

The agricultural sector's incredible ability to increase crop output and production without jeopardizing bioresource assets will also be critical. Biotechnology has a lot of promise when it comes to developing sustainable agriculture (Anderson et al., 2016). Traits are designed to help increase yield, decrease inputs, and protect crops from viral, bacterial, fungal, and insect pests, as well as improve crop performance and productivity under biotic and abiotic stress conditions.

12.3.1 Strain improvement

An industrial microbiology company with a desire to increase profits despite rivals vying for a similar market has many choices. Although the organization's technological practices remain unchanged, it can pursue more aggressive marketing strategies, such as more appealing packaging. It could make better use of its human capital, lowering prices, or it could use a more effective extraction method to remove the material from the fermentation broth. The fermenter's operations could be enhanced by using an extra efficient medium, an improved brewing environment, better ways of monitoring fermentation process engineering, or use of a genetically improved strain of microorganism. Of the above possibilities, the greatest potential is attributed to strain improvement which gives higher profitability.

While it is crucial to recognize the importance of strain improvement, it is also essential to remember that an improved strain can introduce previously unnoticed problems. A higher-yielding strain, for example, may require further aeration or foam control; the products may present new extraction challenges or even necessitate the use of an entirely new fermentation medium. The benefits of using a more active strain must be balanced against the potential for higher costs due to higher extraction costs, richer media, more costly fermenting operations, and other previously unknown issues. The ability of a microorganism to produce a specific product is determined by its ability to secrete a specific set of enzymes. The synthesis of enzymes is largely determined by the genetic makeup of the species. As a result, the improvement of strains can be summarized as follows:

(i) controlling the function of the enzymes secreted by the organisms;
(ii) in the case of extracellular produced metabolites, increasing the easy absorption of these products to the exterior of the cell;
(iii) from a natural population choosing appropriate generating strains;
(iv) manipulating the current genetic mechanism in producing organism; and
(v) using rDNA technology or gene modification to introduce modified genetic properties into the organism.

12.3.2 Selection from naturally occurring variants

Naturally occurring variants that overproduce the desired product are sought in this form of selection. Strains that were discovered, but not chosen, should not be discarded; the better ones are normally retained as stock cultures in the organization's culture set for potential genetic manipulations. Natural variant selection is a common aspect in biotechnology and industrial microbiology. In antibiotic development for example, instead of surface culturing of fungi, it was discovered that the submerged conditions of deep-tank fermentation produced a higher yield of natural variants which boost yields in both *Penicillium rubens* and *Penicillium griseofulvin* (Okafor and Okeke, 2017). Another example is the production of lager beer, where the endless selection of yeasts that flocculate gradually gave rise to strains that are still used in the beverage's production. Similarly, yeasts from the best vats were used in wine fermentation over and over before suitable yeasts were found. This form of selection is not only slow but also out of the biotechnologist's control, which is an unacceptable state in the very competitive world of modern industry.

12.3.3 Manipulation of the genome of organisms

Genetic engineering, also known as gene cloning, recombinant DNA technology, or molecular cloning, is the process of creating new combinations of genetic material. By inserting nucleic acid molecules formed outside the cell into a virus, bacterial plasmid, or other vector system, the desired trait can be incorporated into the host species where they propagate easily but do not occur naturally.

The desired DNA that will be injected into the host bacterium could come from a eukaryotic cell, a prokaryotic cell, or even be chemically synthesized. The vector-foreign DNA complex that is inserted into the host DNA is often referred to as a DNA chimera. A community of organisms that can mate and produce fertile offspring is referred to as a genus. Genetic engineering has made it possible to cross the species barrier whereby artificially modified DNA from one organism can be inserted into another species that would not be possible naturally. Metabolic products can be produced by engineered cells that are significantly distinct from those of unchanged natural species' cells, thanks to this technology. The stepwise process involved in in vitro recombination or genetic engineering is briefly described below. The majority of research until now has used *Escherichia coli* as the model microorganism in molecular genetics.

1. Extract a particular portion of the donor organism's DNA.
2. The spliced DNA piece is attached to a self-replicating piece of DNA vector, which may be from a phage or plasmid.
3. Insertion of the phage or plasmid vector carrying desired DNA into the host cell (i.e., the DNA chimera).
4. Identification (or isolation) of cells that have successfully received and maintained the vector and its DNA.

12.3.4 Role of genetic engineering in strain improvement

The application of microbial inoculants has the potential to increase agricultural productivity without compromising the quality of agricultural land. They have the potential to reduce the negative effects of chemical inputs while still increasing the quality and quantity of farm products. Chemical fertilizers are used less often than microbial inoculants. Microbial inoculants, which are closely linked to plant development, health, and productivity, would be a promising environmentally friendly solution for the future green revolution (Hardoim et al., 2015). Genetic engineering has progressed to the point that modified microbes can be used for anything from bio-synthesis to bio-remediation. The tools for the development and application of microbial inoculum are constantly evolving and improving. Several growth-promoting bacterial species are already used as bio-fertilizers around the world, helping to promote plant growth, and as a result are crucial in forestry and agriculture production (Sapkota et al., 2015.). Furthermore, since soil microbial community is multifaceted, active, and varies in structure between parts and levels, researchers face a real challenge. The sampling of entities is a major issue in such biotechnology research. Replication number, size of the sample, sampling form (randomized or normal intervals), microsite variance, and spatial scaling are all major concerns. In addition, rhizospheric soil, which has received a lot of research attention, is difficult to describe precisely. However, the side distance outcome on bulk soil is more reliable (Sudheer et al., 2020). Timecourse reports are also important in monitoring application results according to the buffering capacity of the agroecosystem (Sudheer et al., 2020). The process used to analyze microbial groups in soil at taxonomic and functional stages, on the other hand, is difficult to use and limit the number of samples taken. The analysis of culture-dependent approaches is usually limited to a small number of samples whereas culture-independent methods seldom allow for the definitive description of taxonomic groups. Furthermore, due to the bias introduced by DNA extraction and PCR amplification, culture independent (metagenomics) methods have some inherent limitations (Savka et al., 2013). A variety of advance molecular methodologies have been developed to investigate the unknown of microbial isolates that live in the rhizosphere zone and how they communicate with one another, as well as the impact of climate change on soil microbiome (Sudheer et al., 2020).

Another point to note is that molecular markers can be used to detect the complete soil microbiome as well as unculturable microbes using these molecular techniques. DNA can be used to classify the functional capacity and phylogenetic identity of microbes, and RNA can be used to investigate gene expression in a given state. For several years, environmental samples have been studied using metagenomic experiments (Soni and Goel, 2011; Soni et al., 2016, 2017, 2021; Suyal et al., 2019a,b; Goel et al., 2017). DNA isolation and cloning are used in metagenomic research, and certain genes and operons are subjected to various techniques such as cloning, high-throughput sequencing, PCR amplification, and microarray hybridization. The construction of metagenomic soil libraries or e-libraries that could be screened for functional and structural genes, as well as phenotypic characters associated with proteins, enzymes, and secondary metabolites, will enable advanced profiling of microbial communities (Etesami et al., 2014). Different PCR-based quantitative methods have made it easier to amplify microbial DNA extracted from soil samples, because the majority of currently available microbiological tools used to decode the microbial diversity of the host plant's rhizosphere have often led to inconsistencies in identification (Marschner and Rumberger, 2004).

These differences can be linked to inoculant competence in the field, plant genotype, interactions with the current plant-host microbiome, and environmental factors. Moreover, the soil ecosystem is highly complex, with localized microenvironments influenced by temperature and humidity fluctuations. Microbial populations in the soil may play an important role in colonizing the new microbes. Scientists are searching continuously for new potent and stable microbial strains, as well as gaining a deeper understanding of biofertilizers and biocontrol agents and their mechanisms of action. This knowledge will be important in producing dependable microbial products for improved and sustainable crop production.

12.3.5 Modification of microbial genomes

Microorganisms that have had their genomes altered will produce more effective biofertilizers or biocontrol agents. Biocontrol agents that have been genetically enhanced have already been cultivated and evaluated (Spadaro and Gullino, 2005; Leger and Wang, 2010). *Pseudomonas syringae*, for example, was changed early on to improve strawberry and potato frost resistance by eliminating the ice nucleation protein from the bacterial genome (Skirvin et al., 2000). The antibiotic compounds 4-diacetylphloroglucinol (PHL) and pyoluteorin have been genetically engineered in *P. fluorescens* strains F113 and CHA0, as well as *P. putida* WCS358 (PLT). Furthermore, a *P. fluorescens* CHA0 strain has been genetically engineered to develop excessive amounts of phytohormone indole acetic acid (IAA). In a microcosm, it is observed that alpha modulation is reduced due to the overproduction of PHL and PLT by *P. fluorescens* strain CHA0 which also reduces the number of rhizobia, *Sinorhizobium meliloti*. Overproduction of the IAA-CHA0 strain leads to increased root yield in natural soil instead of decreasing in autoclaved soil (Mark et al., 2006). The invention of the CRISPR/Cas method for genome editing has resurrected the use of genetically modified microorganisms to combat plant diseases (Glandorf, 2019; Shelake et al., 2019). CRISPR/Cas-mediated production of nonpathogenic fungal strains has recently been suggested as a biocontrol strategy (Muñoz et al., 2019) as closely related nonpathogenic plant microbes, like fungal endophytes, are an important part of plant microbiomes (Ganley et al., 2004). Consequently, changes in the microbial population composition of the rhizosphere have been observed in several experiments with genetically enhanced biocontrol agents (Mark et al., 2006). More crucially, the genetically enhanced microorganisms when released into the ecosystem as biocontrol agents would most likely spread through microbiomes, further enhancing the benefits.

12.4 Advantages and limitation

Biofertilizers and biopesticides made from natural or microbiological sources offer significant advantages over synthetic fertilizers. The following outlines the benefits and drawbacks:
   I. They are less expensive than synthetic pesticides; therefore, farming costs could be decreased with good management.

II. They are less harmful to the environment than traditional herbicides and will not harm nontarget organisms.
III. Microbial pesticides must be regularly monitored to ensure that they do not develop the ability to damage nontarget organisms, including humans.
IV. When compared to conventional and synthetic fertilizers, organic biofertilizers are used in much smaller quantities (Suyal et al., 2021).
V. Skilled personnel are required for the manufacture and maintenance of biofertilizers.

12.5 Concluding remarks

Current issues with the use of chemical pesticides, as well as a focus on low-input, sustainable agriculture, have pushed the use of microbial agents in integrated pest management systems to the forefront. Because of the widespread use of artificial fertilizers, agricultural land has become polluted. Microorganisms provide a number of significant advantages over other types of control agents and approaches. The main advantage of these microorganisms for pest control is their eco-friendliness due to the pathogens' host specificity. Bacteria, algae, fungi, and other microorganisms have the capacity to control pests without harming other beneficial populations. Furthermore, a variety of microorganisms have the ability to promote nitrogen fixation and have been successfully used as biofertilizers. The application of microbial technology to the development of cost-effective and long-term agriculture would be beneficial for future agricultural prosperity. In order to achieve cost-effective and sustainable agriculture, a mixture of traditional and modern technology will be necessary going forward.

References

Aamir, M., Rai, K.K., Zehra, A., Dubey, M.K., Kumar, S., Shukla, V., Upadhyay, R.S., 2020. Microbial bioformulation-based plant biostimulants: a plausible approach toward next generation of sustainable agriculture. In: Microbial Endophytes. Woodhead Publishing, pp. 195–225.

Adesemoye, A.O., Kloepper, J.W., 2009. Plant-microbes interactions in enhanced fertilizer-use efficiency. Appl. Microbiol. Biotechnol. 85 (1), 1–12.

Affokpon, A., Coyne, D.L., Lawouin, L., Tossou, C., Agbèdè, R.D., Coosemans, J., 2011. Effectiveness of native West African arbuscular mycorrhizal fungi in protecting vegetable crops against root-knot nematodes. Biol. Fertil. Soils 47 (2), 207–217.

Ahkami, A.H., White III, R.A., Handakumbura, P.P., Jansson, C., 2017. Rhizosphere engineering: enhancing sustainable plant ecosystem productivity. Rhizosphere 3, 233–243.

Aktar, W., Sengupta, D., Chowdhury, A., 2009. Impact of pesticides use in agriculture: their benefits and hazards. Interdiscip. Toxicol. 2 (1), 1–12.

Alafeea, R.A.A., Alamery, A.A., Kalaf, I.T., 2019. Effect of bio fertilizers on increasing the efficiency of using chemical fertilizers on the yield component of maize (*Zea mays* L.). Plant Arch. 19 (2), 303–306.

Albrechtova, J., Latr, A., Nedorost, L., Pokluda, R., Posta, K., Vosatka, M., 2012. Dual inoculation with mycorrhizal and saprotrophic fungi applicable in sustainable cultivation improves the yield and nutritive value of onion. Sci. World J. 2012.

Anderson, J.A., Gipmans, M., Hurst, S., Layton, R., Nehra, N., Pickett, J., Shah, D.M., Souza, T.L.P., Tripathi, L., 2016. Emerging agricultural biotechnologies for sustainable agriculture and food security. Journal of Agricultural and Food Chemistry 64 (2), 383–393.

Armada, E., Probanza, A., Roldán, A., Azcón, R., 2016. Native plant growth promoting bacteria *Bacillus thuringiensis* and mixed or individual mycorrhizal species improved drought tolerance and oxidative metabolism in *Lavandula dentata* plants. J. Plant Physiol. 192, 1–12.

Augé, R.M., Toler, H.D., Saxton, A.M., 2015. Arbuscular mycorrhizal symbiosis alters stomatal conductance of host plants more under drought than under amply watered conditions: a meta-analysis. Mycorrhiza 25 (1), 13–24.

Barea, J.M., 2015. Future challenges and perspectives for applying microbial biotechnology in sustainable agriculture based on a better understanding of plant-microbiome interactions. Journal of Soil Science and Plant Nutrition 15 (2), 261–282.

Bargaz, A., Lyamlouli, K., Chtouki, M., Zeroual, Y., Dhiba, D., 2018. Soil microbial resources for improving fertilizers efficiency in an integrated plant nutrient management system. Front. Microbiol. 9, 1606.

Benami, M., Isack, Y., Grotsky, D., Levy, D., Kofman, Y., 2020. The economic potential of arbuscular mycorrhizal fungi in agriculture. In: Grand Challenges in Fungal Biotechnology. Springer, Cham, pp. 239–279.

Bhardwaj, D., Ansari, M.W., Sahoo, R.K., Tuteja, N., 2014. Biofertilizers function as key player in sustainable agriculture by improving soil fertility, plant tolerance and crop productivity. Microb. Cell Factories 13 (1), 1–10.

Bharti, N., Pandey, S.S., Barnawal, D., Patel, V.K., Kalra, A., 2016. Plant growth promoting rhizobacteria *Dietzia natronolimnaea* modulates the expression of stress responsive genes providing protection of wheat from salinity stress. Sci. Rep. 6 (1), 1–16.

Bouton, J., 2007. The economic benefits of forage improvement in the United States. Euphytica 154 (3), 263–270.

Bouton, J.H., Latch, G.C., Hill, N.S., Hoveland, C.S., McCann, M.A., Watson, R.H., Parish, J.A., Hawkins, L.L., Thompson, F.N., 2002. Reinfection of tall fescue cultivars with non-ergot alkaloid–producing endophytes. Agron. J. 94 (3), 567–574.

Brader, G., Compant, S., Mitter, B., Trognitz, F., Sessitsch, A., 2014. Metabolic potential of endophytic bacteria. Current Opinion in Biotechnology 27, 30–37.

Cabrefiga, J., Bonaterra, A., Montesinos, E., 2007. Mechanisms of antagonism of *Pseudomonas fluorescens* EPS62e against *Erwinia amylovora*, the causal agent of fire blight. Int. Microbiol. 10 (2), 123.

Chaudhary, V.B., Rúa, M.A., Antoninka, A., Bever, J.D., Cannon, J., Craig, A., Duchicela, J., Frame, A., Gardes, M., Gehring, C., Ha, M., 2016. MycoDB, a global database of plant response to mycorrhizal fungi. Sci. Data 3 (1), 1–10.

Chen, J.H., 2006. The combined use of chemical and organic fertilizers and/or biofertilizer for crop growth and soil fertility. In: International Workshop on Sustained Management of the Soil-Rhizosphere System for Efficient Crop Production and Fertilizer Use. vol. 16(20), pp. 1–11.

Chowdhury, S.P., Dietel, K., Rändler, M., Schmid, M., Junge, H., Borriss, R., Hartmann, A., Grosch, R., 2013. Effects of *Bacillus amyloliquefaciens* FZB42 on lettuce growth and health under pathogen pressure and its impact on the rhizosphere bacterial community. PLoS One 8 (7), e68818.

Colla, G., Rouphael, Y., Di Mattia, E., El-Nakhel, C., Cardarelli, M., 2015. Co-inoculation of *Glomus intraradices* and *Trichoderma atroviride* acts as a biostimulant to promote growth, yield and nutrient uptake of vegetable crops. J. Sci. Food Agric. 95 (8), 1706–1715.

Das, A., Sherameti, I., Varma, A., 2012. Contaminated soil: physical, chemical and biological components. In: Bio-Geo Interactions in Metal-Contaminated Soils. Springer, Cham, pp. 1–15.

Dash, B., Soni, R., Goel, R., 2019a. Rhizobacteria for reducing heavy metal stress in plant and soil. In: Sayyed, R.Z., et al. (Eds.), Plant Growth Promoting Rhizobacteria for Sustainable Stress Management, Microorganisms for Sustainability. Springer Nature Singapore Pte Ltd, p. 12.

Dash, B., Soni, R., Kumar, V., Suyal, D.C., Dash, D., Goel, R., 2019b. Mycorrhizosphere: microbial interactions for sustainable agricultural production. In: Varma, A., Choudhary, D.K. (Eds.), Mycorrhizosphere and Pedogenesis. Springer Nature Singapore Pte Ltd, p. 2019.

de Bruijn, F.J., 2015. The quest for biological nitrogen fixation in cereals: A perspective and prospective. In: de Bruijn, F.J. (Ed.), Biological Nitrogen Fixation. Wiley-Blackwell Publishers, Hoboken, pp. 1089–1101.

Dimkpa, C., Weinand, T., Asch, F., 2009. Plant–rhizobacteria interactions alleviate abiotic stress conditions. Plant Cell Environ. 32 (12), 1682–1694.

Dubey, P., Kumar, V., Karthika, P., Sonwani, R., Singh, A.K., Suyal, D.C., Soni, R., 2020. Microbe assisted plant stress management. In: de Mandal, S., Bhatt, P. (Eds.), Recent Advancements in Microbial Diversity. Elsevier Inc., https://doi.org/10.1016/B978-0-12-821265-3.00015-3.

Egamberdieva, D., Wirth, S., Bellingrath-Kimura, S.D., et al., 2019. Salt-tolerant plant growth promoting rhizobacteria for enhancing crop productivity of saline soils. Front. Microbiol. 10, 2791.

Elferink, M., Schierhorn, F., 2016. Global demand for food is rising. Can we meet it. Harv. Bus. Rev. 7 (4), 2016.

Ene-Obong, E.E., 2007. Achieving the millennium development goals (MDGS) in Nigeria: The role of agricultural biotechnology. In: Proc. Of the 20th annl. conf. biotechnology society of Nigeria (BSN), November. Ebonyi State University, Abakaliki, Nigeria.

Etesami, H., Hosseini, H.M., Alikhani, H.A., 2014. Bacterial biosynthesis of 1-aminocyclopropane-1-caboxylate (ACC) deaminase, a useful trait to elongation and endophytic colonization of the roots of rice under constant flooded conditions. Physiol. Mol. Biol. Plants 20 (4), 425–434.

Franche, C., Lindström, K., Elmerich, C., 2009. Nitrogen-fixing bacteria associated with leguminous and non-leguminous plants. Plant Soil 321 (1), 35–59.

Gaj, T., Gersbach, C.A., Barbas III, C.F., 2013. ZFN, TALEN, and CRISPR/Cas-based methods for genome engineering. Trends Biotechnol. 31 (7), 397–405.

Gallart, M., Adair, K.L., Love, J., Meason, D.F., Clinton, P.W., Xue, J., Turnbull, M.H., 2018. Host genotype and nitrogen form shape the root microbiome of *Pinus radiata*. Microb. Ecol. 75 (2), 419–433.

Ganbaatar, O., Cao, B., Zhang, Y., Bao, D., Bao, W., Wuriyanghan, H., 2017. Knockdown of *Mythimna separata* chitinase genes via bacterial expression and oral delivery of RNAi effectors. BMC Biotechnol. 17 (1), 1–11.

Ganley, R.J., Brunsfeld, S.J., Newcombe, G., 2004. A community of unknown, endophytic fungi in western white pine. Proc. Natl. Acad. Sci. 101 (27), 10107–10112.

García-Fraile, P., Menéndez, E., Rivas, R., 2015. Role of bacterial biofertilizers in agriculture and forestry. AIMS Bioeng. 2 (3), 183–205.

García-Salamanca, A., Molina-Henares, M.A., van Dillewijn, P., Solano, J., Pizarro-Tobías, P., Roca, A., Duque, E., Ramos, J.L., 2013. Biodiversity in adjacent niches. Microb. Biotechnol. 6 (1), 36–44.

Gianinazzi, S., Gollotte, A., Binet, M.N., van Tuinen, D., Redecker, D., Wipf, D., 2010. Agroecology: the key role of arbuscular mycorrhizas in ecosystem services. Mycorrhiza 20 (8), 519–530.

Glandorf, D.C., 2019. Re-evaluation of biosafety questions on genetically modified biocontrol bacteria. Eur. J. Plant Pathol. 154 (1), 43–51.

Goel, R., Kumar, V., Suyal, D.C., Dash, B., Kumar, P., Soni, R., 2017. Root-associated bacteria: rhizoplane and endosphere. In: Singh, D.P., et al. (Eds.), Plant-Microbe Interactions in Agro-Ecological Perspectives. Springer Nature Singapore Pte Ltd, https://doi.org/10.1007/978-981-10-5813-4_9.

Goold, H.D., Wright, P., Hailstones, D., 2018. Emerging opportunities for synthetic biology in agriculture. Genes 9 (7), 341.

Gupta, S., Pandey, S., 2019. ACC deaminase producing bacteria with multifarious plant growth promoting traits alleviates salinity stress in French bean (*Phaseolus vulgaris*) plants. Front. Microbiol. 10, 1506.

Han, Q.Q., Lü, X.P., Bai, J.P., Qiao, Y., Paré, P.W., Wang, S.M., Zhang, J.L., Wu, Y.N., Pang, X.P., Xu, W.B., Wang, Z.L., 2014. Beneficial soil bacterium *Bacillus subtilis* (GB03) augments salt tolerance of white clover. Front. Plant Sci. 5, 525.

Hardoim, P.R., Van Overbeek, L.S., Berg, G., Pirttilä, A.M., Compant, S., Campisano, A., Döring, M., Sessitsch, A., 2015. The hidden world within plants: ecological and evolutionary considerations for defining functioning of microbial endophytes. Microbiol. Mol. Biol. Rev. 79 (3), 293–320.

Harman, G.E., Howell, C.R., Viterbo, A., Chet, I., Lorito, M., 2004. Trichoderma species—opportunistic, avirulent plant symbionts. Nat. Rev. Microbiol. 2 (1), 43–56.

Hart, M.M., Antunes, P.M., Abbott, L.K., 2017. Unknown risks to soil biodiversity from commercial fungal inoculants. Nat. Ecol. Evol. 1 (4), 1.

Harvey, P.R., Warren, R.A., Wakelin, S., 2009. Potential to improve root access to phosphorus: the role of non-symbiotic microbial inoculants in the rhizosphere. Crop Pasture Sci. 60 (2), 144–151.

He, Y., Wu, Z., Tu, L., Han, Y., Zhang, G., Li, C., 2015. Encapsulation and characterization of slow-release microbial fertilizer from the composites of bentonite and alginate. Appl. Clay Sci. 109, 68–75.

Hien, N.T., Toan, P.V., Choudhury, A.T., Rose, M.T., Roughley, R.J., Kennedy, I.R., 2014. Field application strategies for the inoculant biofertilizer BioGro supplementing fertilizer nitrogen application in rice production. J. Plant Nutr. 37 (11), 1837–1858.

Hirsch, P.R., Miller, A.J., Dennis, P.G., 2013. Do root exudates exert more influence on rhizosphere bacterial community structure than other rhizodeposits? Molecular Microbial Ecology of the Rhizosphere 1, 229–242.

Hoeksema, J.D., Chaudhary, V.B., Gehring, C.A., Johnson, N.C., Karst, J., Koide, R.T., Pringle, A., Zabinski, C., Bever, J.D., Moore, J.C., Wilson, G.W., 2010. A meta-analysis of context-dependency in plant response to inoculation with mycorrhizal fungi. Ecol. Lett. 13 (3), 394–407.

Horton, M.W., Bodenhausen, N., Beilsmith, K., Meng, D., Muegge, B.D., Subramanian, S., Vetter, M.M., Vilhjálmsson, B.J., Nordborg, M., Gordon, J.I., Bergelson, J., 2014. Genome-wide association study of Arabidopsis thaliana leaf microbial community. Nat. Commun. 5 (1), 1–7.

Ilangumaran, G., Smith, D.L., 2017. Plant growth promoting rhizobacteria in amelioration of salinity stress: a systems biology perspective. Front. Plant Sci. 8, 1768.

Ingle, K.P., Padole, D.A., 2017. Phosphate solubilizing microbes: an overview. Int. J. Curr. Microbiol. App. Sci. 6 (1), 844–852.

Jacoby, R., Peukert, M., Succurro, A., Koprivova, A., Kopriva, S., 2017. The role of soil microorganisms in plant mineral nutrition—current knowledge and future directions. Front. Plant Sci. 8, 1617.

Jambhulkar, P.P., Sharma, P., Yadav, R., 2016. Delivery systems for introduction of microbial inoculants in the field. In: Microbial Inoculants in Sustainable Agricultural Productivity. Springer, New Delhi, pp. 199-218.

Jamil, M., Zeb, S., Anees, M., Roohi, A., Ahmed, I., Ur Rehman, S., Rha, E.S., 2014. Role of *Bacillus licheniformis* in phytoremediation of nickel contaminated soil cultivated with rice. Int. J. Phytorem. 16 (6), 554-571.

Johnson, L.J., de Bonth, A.C., Briggs, L.R., Caradus, J.R., Finch, S.C., Fleetwood, D.J., Fletcher, L.R., Hume, D.E., Johnson, R.D., Popay, A.J., Tapper, B.A., 2013. The exploitation of epichloae endophytes for agricultural benefit. Fungal Divers. 60 (1), 171-188.

Johnsson, L., Hökeberg, M., Gerhardson, B., 1998. Performance of the *Pseudomonas chlororaphis* biocontrol agent MA 342 against cereal seed-borne diseases in field experiments. Eur. J. Plant Pathol. 104 (7), 701-711.

Jones, D.L., Hinsinger, P., 2008. The rhizosphere: Complex by design. Plant and Soil 312, 1-6.

Jung, S.C., Martinez-Medina, A., Lopez-Raez, J.A., Pozo, M.J., 2012. Mycorrhiza-induced resistance and priming of plant defenses. J. Chem. Ecol. 38 (6), 651-664.

Kakabouki, I., Tataridas, A., Mavroeidis, A., Kousta, A., Karydogianni, S., Zisi, C., Kouneli, V., Konstantinou, A., Folina, A., Konstantas, A., Papastylianou, P., 2021. Effect of colonization of *Trichoderma harzianum* on growth development and CBD content of hemp (*Cannabis sativa* L.). Microorganisms 9 (3), 518.

Kang, S.M., Radhakrishnan, R., Khan, A.L., Kim, M.J., Park, J.M., Kim, B.R., Shin, D.H., Lee, I.J., 2014. Gibberellin secreting rhizobacterium, *Pseudomonas putida* H-2-3 modulates the hormonal and stress physiology of soybean to improve the plant growth under saline and drought conditions. Plant Physiol. Biochem. 84, 115-124.

Kauppinen, M., Saikkonen, K., Helander, M., Pirttilä, A.M., Wäli, P.R., 2016. Epichloë grass endophytes in sustainable agriculture. Nat. Plants 2 (2), 1-7.

Khan, A., Khan, A.L., Muneer, S., Kim, Y.H., Al-Rawahi, A., Al-Harrasi, A., 2019a. Silicon and salinity: crosstalk in crop-mediated stress tolerance mechanisms. Front. Plant Sci. 10, 1429.

Khan, S., Anwar, S., Yu, S., Sun, M., Yang, Z., Gao, Z.Q., 2019b. Development of drought-tolerant transgenic wheat: achievements and limitations. Int. J. Mol. Sci. 20 (13), 3350.

Kloepper, J.W., Leong, J., Teintze, M., Schroth, M.N., 1980. Enhanced plant growth by siderophores produced by plant growth-promoting rhizobacteria. Nature 286 (5776), 885-886.

Kloepper, J.W., Ryu, C.M., Zhang, S., 2004. Induced systemic resistance and promotion of plant growth by *Bacillus* spp. Phytopathology 94 (11), 1259-1266.

Kour, D., Kaur, T., Devi, R., Rana, K.L., Yadav, N., Rastegari, A.A., Yadav, A.N., 2020. Biotechnological applications of beneficial microbiomes for evergreen agriculture and human health. In: Trends of Microbial Biotechnology for Sustainable Agriculture and Biomedicine Systems: Perspectives for Human Health. Elsevier, Amsterdam, pp. 255-279.

Kumar, M., Yadav, V., Tuteja, N., Johri, A.K., 2009. Antioxidant enzyme activities in maize plants colonized with *Piriformospora indica*. Microbiology 155 (3), 780-790.

Kumar, M., Kour, D., Yadav, A.N., Saxena, R., Rai, P.K., Jyoti, A., Tomar, R.S., 2019. Biodiversity of methylotrophic microbial communities and their potential role in mitigation of abiotic stresses in plants. Biologia 74 (3), 287-308.

Kour, D., Rana, K.L., Yadav, A.N., Yadav, N., Kumar, V., Kumar, A., Sayyed, R.Z., Hesham, A.E.L., Dhaliwal, H.S., Saxena, A.K., 2019. Drought-tolerant phosphorus-solubilizing microbes: Biodiversity and biotechnological applications for alleviation of drought stress in plants. In: Plant growth promoting rhizobacteria for sustainable stress management. Springer, Singapore, pp. 255-308.

Kumar, V., Jain, L., Soni, R., Kaushal, P., Goel, R., 2020. Bioprospecting of endophytic microbes from higher altitude plants: recent advances and their biotechnological applications. In: Goel, R., et al. (Eds.), Microbiological Advancements for Higher Altitude Agroecosystems & Sustainability, Rhizosphere Biology., https://doi.org/10.1007/978-981-15-1902-4_18.

Kunin, V., Sorek, R., Hugenholtz, P., 2007. Evolutionary conservation of sequence and secondary structures in CRISPR repeats. Genome Biol. 8 (4), 1–7.

Kushwah, S., Singh, A.K., Chowdhury, T., Soni, R., 2021. Comparative effect of selected bacterial consortia on yield and yield attributes of chickpea (Cicer arietinum L.). Asian J. Microbiol. Biotechnol. Environ. Sci. 23 (1), 72–74.

Leger, R.J.S., Wang, C., 2010. Genetic engineering of fungal biocontrol agents to achieve greater efficacy against insect pests. Appl. Microbiol. Biotechnol. 85 (4), 901–907.

Liu, F., Xing, S., Ma, H., Du, Z., Ma, B., 2013. Cytokinin-producing, plant growth-promoting rhizobacteria that confer resistance to drought stress in Platycladus orientalis container seedlings. Appl. Microbiol. Biotechnol. 97 (20), 9155–9164.

Liu, J., Abdelfattah, A., Norelli, J., Burchard, E., Schena, L., Droby, S., Wisniewski, M., 2018. Apple endophytic microbiota of different rootstock/scion combinations suggests a genotype-specific influence. Microbiome 6 (1), 1–11.

Lugtenberg, B., 2015. Life of microbes in the rhizosphere. In: Principles of Plant-Microbe Interactions. Springer, Cham, pp. 7–15.

Ma, Y., Oliveira, R.S., Freitas, H., Zhang, C., 2016. Biochemical and molecular mechanisms of plant-microbe-metal interactions: relevance for phytoremediation. Front. Plant Sci. 7, 918.

Mącik, M., Gryta, A., Frąc, M., 2020. Biofertilizers in agriculture: an overview on concepts, strategies and effects on soil microorganisms. Adv. Agron. 162, 31–87.

Mahanty, T., Bhattacharjee, S., Goswami, M., Bhattacharyya, P., Das, B., Ghosh, A., Tribedi, P., 2017. Biofertilizers: a potential approach for sustainable agriculture development. Environ. Sci. Pollut. Res. 24 (4), 3315–3335.

Malyan, S.K., Kumar, A., Baram, S., Kumar, J., Singh, S., Kumar, S.S., Yadav, A.N., 2019. Role of fungi in climate change abatement through carbon sequestration. In: Recent Advancement in White Biotechnology Through Fungi. Springer, Cham, pp. 283–295.

Mark, G.L., Morrissey, J.P., Higgins, P., O'gara, F., 2006. Molecular-based strategies to exploit Pseudomonas biocontrol strains for environmental biotechnology applications. FEMS Microbiol. Ecol. 56 (2), 167–177.

Marschner, P., Rumberger, A., 2004. Rapid changes in the rhizosphere bacterial community structure during re-colonization of sterilized soil. Biol. Fertil. Soils 40 (1), 1–6.

Mendes, R., Garbeva, P., Raaijmakers, J.M., 2013. The rhizosphere microbiome: significance of plant beneficial, plant pathogenic, and human pathogenic microorganisms. FEMS Microbiol. Rev. 37 (5), 634–663.

Mercado-Blanco, J., 2015. Life of microbes inside the plant. In: Principles of plant-microbe interactions. Springer, Cham, pp. 25–32.

Mikiciuk, G., Sas-Paszt, L., Mikiciuk, M., Derkowska, E., Trzciński, P., Głuszek, S., Lisek, A., Wera-Bryl, S., Rudnicka, J., 2019. Mycorrhizal frequency, physiological parameters, and yield of strawberry plants inoculated with endomycorrhizal fungi and rhizosphere bacteria. Mycorrhiza 29 (5), 489–501.

Miller, R.M., Jastrow, J.D., 2000. Mycorrhizal fungi influence soil structure. In: Arbuscular Mycorrhizas: Physiology and Function. Springer, Dordrecht, pp. 3–18.

Minuto, A., Migheli, Q., Garibaldi, A., 1995. Evaluation of antagonistic strains of Fusarium spp. in the biological and integrated control of Fusarium wilt of cyclamen. Crop Prot. 14 (3), 221–226.

Minuto, A., Spadaro, D., Garibaldi, A., Gullino, M.L., 2006. Control of soilborne pathogens of tomato using a commercial formulation of Streptomyces griseoviridis and solarization. Crop Prot. 25 (5), 468–475.

Mishra, J., Singh, R., Arora, N.K., 2017. Alleviation of heavy metal stress in plants and remediation of soil by rhizosphere microorganisms. Frontiers in Microbiology 8, 1706.

Muhammad, T., Zhang, F., Zhang, Y., Liang, Y., 2019. RNA interference: a natural immune system of plants to counteract biotic stressors. Cells 8 (1), 38.

Müller, D.B., Vogel, C., Bai, Y., Vorholt, J.A., 2016. The plant microbiota: systems-level insights and perspectives. Annu. Rev. Genet. 50, 211–234.

Muñoz, I.V., Sarrocco, S., Malfatti, L., Baroncelli, R., Vannacci, G., 2019. CRISPR-Cas for fungal genome editing: a new tool for the management of plant diseases. Front. Plant Sci. 10, 135.

Nadeem, M., Li, J., Yahya, M., Sher, A., Ma, C., Wang, X., Qiu, L., 2019. Research progress and perspective on drought stress in legumes: a review. Int. J. Mol. Sci. 20 (10), 2541.

Naseem, H., Ahsan, M., Shahid, M.A., Khan, N., 2018. Exopolysaccharides producing rhizobacteria and their role in plant growth and drought tolerance. J. Basic Microbiol. 58 (12), 1009–1022.

Ngugi, H.K., Dedej, S., Delaplane, K.S., Savelle, A.T., Scherm, H., 2005. Effect of flower-applied Serenade biofungicide (*Bacillus subtilis*) on pollination-related variables in rabbiteye blueberry. Biol. Control 33 (1), 32–38.

Nguyen, T.H., Phan, T.C., Choudhury, A.T., Rose, M.T., Deaker, R.J., Kennedy, I.R., 2017. BioGro: a plant growth-promoting biofertilizer validated by 15 years' research from laboratory selection to rice farmer's fields of the Mekong Delta. In: Agro-Environmental Sustainability. Springer, Cham, pp. 237–254.

Nicolás, C., Hermosa, R., Rubio, B., Mukherjee, P.K., Monte, E., 2014. Trichoderma genes in plants for stress tolerance-status and prospects. Plant Sci. 228, 71–78.

Nie, M., Bell, C., Wallenstein, M.D., Pendall, E., 2015. Increased plant productivity and decreased microbial respiratory C loss by plant growth-promoting rhizobacteria under elevated CO2. Sci. Rep. 5 (1), 1–6.

Nora, L.C., Westmann, C.A., Guazzaroni, M.E., Siddaiah, C., Gupta, V.K., Silva-Rocha, R., 2019. Recent advances in plasmid-based tools for establishing novel microbial chassis. Biotechnol. Adv. 37 (8), 107433.

Normand, P., Lapierre, P., Tisa, L.S., Gogarten, J.P., Alloisio, N., Bagnarol, E., Bassi, C.A., Berry, A.M., Bickhart, D.M., Choisne, N., Couloux, A., 2007. Genome characteristics of facultatively symbiotic Frankia sp. strains reflect host range and host plant biogeography. Genome Research 17 (1), 7–15.

O'Callaghan, M., 2016. Microbial inoculation of seed for improved crop performance: issues and opportunities. Appl. Microbiol. Biotechnol. 100 (13), 5729–5746.

Oelmüller, R., Sherameti, I., Tripathi, S., Varma, A., 2009. *Piriformospora indica*, a cultivable root endophyte with multiple biotechnological applications. Symbiosis 49 (1), 1–17.

Okafor, N., Okeke, B.C., 2017. Modern Industrial Microbiology and Biotechnology. CRC Press.

Okon, Y., Labandera-Gonzales, C., Lage, M., Lage, P., 2015. Agronomic applications of *Azospirillum* and other PGPR. In: Biological Nitrogen Fixation. John Wiley & Sons Inc, New Jersey, pp. 921–933.

Olivares, J., Bedmar, E.J., Sanjuán, J., 2013. Biological nitrogen fixation in the context of global change. Molecular Plant-Microbe Interactions 26 (5), 486–494.

Öpik, M., Vanatoa, A., Vanatoa, E., Moora, M., Davison, J., Kalwij, J.M., Reier, Ü., Zobel, M., 2010. The online database MaarjAM reveals global and ecosystemic distribution patterns in arbuscular mycorrhizal fungi (Glomeromycota). New Phytol. 188 (1), 223–241.

Owen, D., Williams, A.P., Griffith, G.W., Withers, P.J., 2015. Use of commercial bio-inoculants to increase agricultural production through improved phosphrous acquisition. Appl. Soil Ecol. 86, 41–54.

Packialakshmi, J.S., Dash, B., Singh, A.K., Suyal, D.C., Soni, R., 2020. Nutrient cycling at higher altitudes. In: Goel, R., et al. (Eds.), Microbiological advancements for higher altitude agroecosystems & sustainability, rhizosphere biology. Springer Nature Singapore Pte Ltd. https://doi.org/10.1007/978-981-15-1902-4_18.

Pandey, P., Irulappan, V., Bagavathiannan, M.V., Senthil-Kumar, M., 2017. Impact of combined abiotic and biotic stresses on plant growth and avenues for crop improvement by exploiting physio-morphological traits. Front. Plant Sci. 8, 537.

Parani, K., Saha, B.K., 2012. Prospects of using phosphate solubilizing *Pseudomonas* as bio fertilizer. Eur. J. Biol. Sci. 4 (2), 40–44.

Paterson, J., Jahanshah, G., Li, Y., Wang, Q., Mehnaz, S., Gross, H., 2017. The contribution of genome mining strategies to the understanding of active principles of PGPR strains. FEMS Microbiol. Ecol. 93 (3), fiw249.

Pellegrino, E., Turrini, A., Gamper, H.A., Cafà, G., Bonari, E., Young, J.P.W., Giovannetti, M., 2012. Establishment, persistence and effectiveness of arbuscular mycorrhizal fungal inoculants in the field revealed using molecular genetic tracing and measurement of yield components. New Phytol. 194 (3), 810–822.

Pellegrino, E., Öpik, M., Bonari, E., Ercoli, L., 2015. Responses of wheat to arbuscular mycorrhizal fungi: a meta-analysis of field studies from 1975 to 2013. Soil Biol. Biochem. 84, 210–217.

Perdikis, D., Kapaxidi, E., Papadoulis, G., 2008. Biological control of insect and mite pests in greenhouse solanaceous crops. Eur. J. Plant Sci. Biotechnol. 2 (1), 125–144.

Pii, Y., Mimmo, T., Tomasi, N., Terzano, R., Cesco, S., Crecchio, C., 2015. Microbial interactions in the rhizosphere: beneficial influences of plant growth-promoting rhizobacteria on nutrient acquisition process. A review. Biol. Fertil. Soils 51 (4), 403–415.

Pirttilä, A.M., Mohammad Parast Tabas, H., Baruah, N., Koskimäki, J.J., 2021. Biofertilizers and biocontrol agents for agriculture: how to identify and develop new potent microbial strains and traits. Microorganisms 9 (4), 817.

Poddar, R., Sen, A., Kundu, R., Das, H., Bandopadhyay, P., 2020. Response of various mycorrhizal inoculants on rice growth, productivity and nutrient uptake. Int. J. Bioresour. Stress Manag. 11 (2), 171–177.

Puopolo, G., Raio, A., Pierson III, L.S., Zoina, A., 2011. Selection of a new *Pseudomonas chlororaphis* strain for the biological control of *Fusarium oxysporum* f. sp. *radicis-lycopersici*. Phytopathol. Mediterr. 50 (2), 228–235.

Purvis, B., Mao, Y., Robinson, D., 2019. Three pillars of sustainability: in search of conceptual origins. Sustain. Sci. 14 (3), 681–695.

Rabiey, M., Ullah, I., Shaw, L.J., Shaw, M.W., 2017. Potential ecological effects of *Piriformospora indica*, a possible biocontrol agent, in UK agricultural systems. Biol. Control 104, 1–9.

Radhakrishnan, R., Hashem, A., Abd Allah, E.F., 2017. *Bacillus*: a biological tool for crop improvement through bio-molecular changes in adverse environments. Front. Physiol. 8, 667.

Rai, P.K., Singh, M., Anand, K., Saurabh, S., Kaur, T., Kour, D., Yadav, A.N., Kumar, M., 2020. Role and potential applications of plant growth-promoting rhizobacteria for sustainable agriculture. In: New and Future Developments in Microbial Biotechnology and Bioengineering. Elsevier, pp. 49–60.

Rajawat, M.V.S., Singh, R., Singh, D., Yadav, A.N., Singh, S., Kumar, M., Saxena, A.K., 2020. Spatial distribution and identification of bacteria in stressed environments capable to weather potassium aluminosilicate mineral. Braz. J. Microbiol. 51 (2), 751–764.

Rastegari, A.A., Yadav, A.N., Yadav, N. (Eds.), 2020. New and Future Developments in Microbial Biotechnology and Bioengineering: Trends of Microbial Biotechnology for Sustainable Agriculture and Biomedicine Systems: Perspectives for Human Health. Elsevier.

Requena, N., Perez-Solis, E., Azcón-Aguilar, C., Jeffries, P., Barea, J.M., 2001. Management of indigenous plant-microbe symbioses aids restoration of desertified ecosystems. Appl. Environ. Microbiol. 67 (2), 495.

Rillig, M.C., Sosa-Hernández, M.A., Roy, J., Aguilar-Trigueros, C.A., Vályi, K., Lehmann, A., 2016. Towards an integrated mycorrhizal technology: harnessing mycorrhiza for sustainable intensification in agriculture. Front. Plant Sci. 7, 1625.

Rodriguez, R.J., Henson, J., Van Volkenburgh, E., Hoy, M., Wright, L., Beckwith, F., Kim, Y.O., Redman, R.S., 2008. Stress tolerance in plants via habitat-adapted symbiosis. ISME J. 2 (4), 404–416.

Rolli, E., Marasco, R., Vigani, G., Ettoumi, B., Mapelli, F., Deangelis, M.L., Gandolfi, C., Casati, E., Previtali, F., Gerbino, R., Pierotti Cei, F., 2015. Improved plant resistance to drought is promoted by the root-associated microbiome as a water stress-dependent trait. Environmental Microbiology 17 (2), 316–331.

Sachs, J.L., 2015. Engineering microbiomes to improve plant and animal health. Trends Microbiol. 23, 606–617.

Sahu, B., Singh, A.K., Chaubey, A.K., Soni, R., 2020. Effect of rhizobium and phosphate solubilizing bacteria inoculation on growth and yield performance of Lathyrus (*Lathyrus sativus* L.) in Chhattisgarh plains. Curr. J. Appl. Sci. Technol. 39 (48), 237–247. https://doi.org/10.9734/cjast/2020/v39i4831226.

Salgarello, M., Visconti, G., Barone-Adesi, L., 2013. Interlocking circumareolar suture with undyed polyamide thread: a personal experience. Aesthet. Plast. Surg. 37 (5), 1061–1062.

Sapkota, R., Knorr, K., Jørgensen, L.N., O'Hanlon, K.A., Nicolaisen, M., 2015. Host genotype is an important determinant of the cereal phyllosphere mycobiome. New Phytol. 207 (4), 1134–1144.

Savka, M.A., Dessaux, Y., McSpadden Gardener, B.B., Mondy, S., Kohler, P.R., de Bruijn, F.J., Rossbach, S., 2013. The "biased rhizosphere" concept and advances in the omics era to study bacterial competitiveness and persistence in the phytosphere. Mol. Microb. Ecol. Rhizosphere 1, 1145–1161.

Sharaff, M.S., Subrahmanyam, G., Kumar, A., Yadav, A.N., 2020. Mechanistic understanding of root microbiome interaction for sustainable agriculture in polluted soils. In: Trends of Microbial Biotechnology for Sustainable Agriculture and Biomedicine Systems: Diversity and Functional Perspectives. Elsevier, Amsterdam, pp. 61–84.

Sharifazizi, M., Harighi, B., Sadeghi, A., 2017. Evaluation of biological control of *Erwinia amylovora*, causal agent of fire blight disease of pear by antagonistic bacteria. Biol. Control 104, 28–34.

Sharma, P., Bora, L.C., Puzari, K.C., Baruah, A.M., Baruah, R., Talukdar, K., Kataky, L., Phukan, A., 2017. Review on bacterial blight of rice caused by *Xanthomonas oryzae* pv. *oryzae*: different management approaches and role of *Pseudomonas fluorescens* as a potential biocontrol agent. Int. J. Curr. Microbiol. App. Sci. 6, 982–1005.

Sharma, A., Shahzad, B., Kumar, V., et al., 2019. Phytohormones regulate accumulation of osmolytes under abiotic stress. Biomol. Ther. 9, 285.

Shelake, R.M., Pramanik, D., Kim, J.Y., 2019. Exploration of plant-microbe interactions for sustainable agriculture in CRISPR era. Microorganisms 7 (8), 269.

Shishkoff, N., McGrath, M.T., 2002. AQ10 biofungicide combined with chemical fungicides or AddQ spray adjuvant for control of cucurbit powdery mildew in detached leaf culture. Plant Disease 86 (8), 915–918.

Shrivastava, S., Varma, A., 2014. From *Piriformospora indica* to rootonic: a review. Afr. J. Microbiol. Res. 8 (32), 2984–2992.

Shukla, P.S., Agarwal, P.K., Jha, B., 2012. Improved salinity tolerance of *Arachis hypogaea* (L.) by the interaction of halotolerant plant-growth-promoting rhizobacteria. J. Plant Growth Regul. 31 (2), 195–206.

Singh, R.B., 2000. Biotechnology, biodiversity and sustainable agriculture-A contradiction. In: Regional conference in agricultural biotechnology proceedings: Biotechnology research and policy-needs and priorities in the context of southeast asia's agricultural activities. SEARCA (SEAMEO)/FAO/APSA, Bangkok.

Singh, R., Gautam, N., Mishra, A., Gupta, R., 2011. Heavy metals and living systems: an overview. Indian J. Pharmacol. 43 (3), 246.

Skirvin, R.M., Kohler, E., Steiner, H., Ayers, D., Laughnan, A., Norton, M.A., Warmund, M., 2000. The use of genetically engineered bacteria to control frost on strawberries and potatoes. Whatever happened to all of that research? Sci. Hortic. 84 (1–2), 179–189.

Smith, R.S., 1995. Inoculant formulations and applications to meet changing needs. In: Nitrogen Fixation: Fundamentals and Applications. Springer, Dordrecht, pp. 653–657.

Smith, S.E., Read, D.J., 2010. Mycorrhizal symbiosis. Academic Press.

Smith, S.E., Smith, F.A., 2011. Roles of arbuscular mycorrhizas in plant nutrition and growth: new paradigms from cellular to ecosystem scales. Annu. Rev. Plant Biol. 62, 227–250.

Soni, R., Goel, R., 2011. nifH homologs from soil metagenome. Ekologija 57 (3), 87–95.

Soni, R., Kumari, S., Zaidi, M.G.H., Goel, R., 2008. Practical applications of rhizospheric bacteria in biodegradation of polymers from plastic wastes. In: Ahmad, I., Oichtel, H. (Eds.), Plant-Bacteria Interaction Strategies and Techniques to Promote Plant Growth. WILEY-VCH Weinheim, Germany, pp. 235–243.

Soni, R., Suyal, D.C., Sai, S., Goel, R., 2016. Exploration of nifH gene through soil metagenomes of the western Indian Himalayas. 3 Biotech 6, 25.

Soni, R., Kumar, V., Suyal, D.C., Jain, L., Goel, R., 2017. Metagenomics of plant rhizosphere microbiome. In: Singh, R.P., et al. (Eds.), Understanding Host-Microbiome Interactions—An Omics Approach. Springer Nature Singapore Pte Ltd, pp. 193–206, https://doi.org/10.1007/978-981-10-5050-3_12.

Soni, R., Suyal, D.C., Sahu, B., Phulara, S.C., 2021. Metagenomics: an approach to unravel the plant microbiome and its function. In: Verma, A., Saini, J.K., Singh, H.B., Hesham, A.E.-L. (Eds.), Phytomicrobiome Interactions and Sustainable Agriculture. John Wiley & Sons Ltd., UK.

Soytong, K., Kanokmedhakul, S., Kukongviriyapa, V., Isobe, M., 2001. Application of *Chaetomium* species (Ketomium) as a new broad spectrum biological fungicide for plant disease control. Fungal Divers. 7, 1–15.

Spadaro, D., Gullino, M.L., 2005. Improving the efficacy of biocontrol agents against soilborne pathogens. Crop Prot. 24 (7), 601–613.

Spence, C., Bais, H., 2013. Probiotics for plants: Rhizospheric microbiome and plant fitness. Molecular microbial ecology of the rhizosphere 1, 713–721.

Sudheer, S., Bai, R.G., Usmani, Z., Sharma, M., 2020. Insights on engineered microbes in sustainable agriculture: biotechnological developments and future prospects. Curr. Genomics 21 (5), 321–333.

Suyal, D.C., Soni, R., Sai, S., Goel, R., 2016. Microbial inoculants as biofertilizer. In: Singh, D.P. (Ed.), Microbial Inoculants in Sustainable Agricultural Productivity. Springer, India, pp. 311–318, https://doi.org/10.1007/978-81-322-2647-5_19.

Suyal, D.C., Kumar, S., Joshi, D., Soni, R., Goel, R., 2018. Quantitative proteomics of psychotrophic diazotroph in response to nitrogen deficiency and cold stress. J. Proteome 187, 235–242. https://doi.org/10.1016/j.jprot.2018.08.005.

Suyal, D.C., Kumar, S., Joshi, D., Soni, R., Goel, R., 2019a. Differential protein profiling of soil diazotroph *Rhodococcus qingshengii* S10107 towards low-temperature and nitrogen deficiency. Sci. Rep. https://doi.org/10.1038/s41598-019-56592-8.

Suyal, D.C., Joshi, D., Debbarma, P., Soni, R., Das, B., Goel, R., 2019b. Soil metagenomics: unculturable microbial diversity and its function. In: Varma, A., Choudhary, D.K. (Eds.), Mycorrhizosphere and Pedogenesis. Springer Nature Singapore Pte Ltd.

Suyal, D.C., Soni, R., et al., 2021. Microbiome change of agricultural soil under organic farming practices. Biologia 76, 1315–1325.

Tilman, D., Balzer, C., Hill, J., Befort, B.L., 2011. Global food demand and the sustainable intensification of agriculture. Proc. Natl. Acad. Sci. 108 (50), 20260–20264.

Timmusk, S., Behers, L., Muthoni, J., Muraya, A., Aronsson, A.C., 2017. Perspectives and challenges of microbial application for crop improvement. Front. Plant Sci. 8, 49.

Tiwari, S., Lata, C., 2018. Heavy metal stress, signaling, and tolerance due to plant-associated microbes: an overview. Front. Plant Sci. 9, 452.

Tiwari, P., Bajpai, M., Singh, L.K., Mishra, S., Yadav, A.N., 2020. Phytohormones producing fungal communities: metabolic engineering for abiotic stress tolerance in crops. In: Agriculturally Important Fungi for Sustainable Agriculture. Springer, Cham, pp. 171–197.

Utuk, I.O., Daniel, E.E., 2015. Land degradation: a threat to food security: a global assessment. J. Environ. Earth Sci. 5 (8), 13–21.

Van Der Heijden, M.G., Bakker, R., Verwaal, J., Scheublin, T.R., Rutten, M., Van Logtestijn, R., Staehelin, C., 2006. Symbiotic bacteria as a determinant of plant community structure and plant productivity in dune grassland. FEMS Microbiol. Ecol. 56 (2), 178–187.

Varma, A., Verma, S., Sudha, Sahay, N., Bütehorn, B., Franken, P., 1999. *Piriformospora indica*, a cultivable plant-growth-promoting root endophyte. Appl. Environ. Microbiol. 65 (6), 2741–2744.

Verma, M., Mishra, J., Arora, N.K., 2019. Plant growth-promoting rhizobacteria: diversity and applications. Environ. Biotechnol. Sustain. Future 2018, 129–173.

Von Cossel, M., Wagner, M., Lask, J., Magenau, E., Bauerle, A., Von Cossel, V., Warrach-Sagi, K., Elbersen, B., Staritsky, I., Van Eupen, M., Iqbal, Y., 2019. Prospects of bioenergy cropping systems for a more social-ecologically sound bioeconomy. Agronomy 9 (10), 605.

Vosátka, M., Látr, A., Gianinazzi, S., Albrechtová, J., 2012. Development of arbuscular mycorrhizal biotechnology and industry: current achievements and bottlenecks. Symbiosis 58 (1), 29–37.

Vurukonda, S.S.K.P., Giovanardi, D., Stefani, E., 2018. Plant growth promoting and biocontrol activity of *Streptomyces* spp. as endophytes. Int. J. Mol. Sci. 19 (4), 952.

Waller, F., Achatz, B., Baltruschat, H., Fodor, J., Becker, K., Fischer, M., Heier, T., Hückelhoven, R., Neumann, C., von Wettstein, D., Franken, P., 2005. The endophytic fungus *Piriformospora indica* reprograms barley to salt-stress tolerance, disease resistance, and higher yield. Proc. Natl. Acad. Sci. 102 (38), 13386–13391.

Wani, P.A., Khan, M.S., Zaidi, A., 2008. Effects of heavy metal toxicity on growth, symbiosis, seed yield and metal uptake in pea grown in metal amended soil. Bulletin of Environmental Contamination and Toxicology 81 (2), 152–158.

Weiss, A., Mögel, G., Kunz, S., 2006. Development of "Boni-Protect"—a yeast preparation for use in the control of postharvest diseases of apples. In: Ecofruit-12th International Conference on Cultivation Technique and Phytopathological Problems in Organic Fruit-Growing: Proceedings to the Conference from 31st January to 2nd February 2006 at Weinsberg/Germany. Fördergemeinschaft Ökologischer Obstbau eV (FÖKO), pp. 113–117.

Wraight, S.P., Ramos, M.E., 2017. Characterization of the synergistic interaction between *Beauveria bassiana* strain GHA and *Bacillus thuringiensis morrisoni* strain tenebrionis applied against Colorado potato beetle larvae. J. Invertebr. Pathol. 144, 47–57.

Xie, L., Lehvävirta, S., Timonen, S., Kasurinen, J., Niemikapee, J., Valkonen, J.P., 2018. Species-specific synergistic effects of two plant growth—promoting microbes on green roof plant biomass and photosynthetic efficiency. PloS One 13 (12), e0209432.

Yadav, T., Kumar, A., Yadav, R.K., Yadav, G., Kumar, R., Kushwaha, M., 2020. Salicylic acid and thiourea mitigate the salinity and drought stress on physiological traits governing yield in pearl millet-wheat. Saudi Journal of Biological Sciences 27 (8), 2010–2017.

Young, C.A. (Ed.), 2012. Epichloae, Endophytes of Cool Season Grasses: Implications, Utilization and Biology. Samuel Roberts Noble Foundation.

Youssef, M.A., Eissa, M.A., 2017. Comparison between organic and inorganic nutrition for tomato. J. Plant Nutr. 40 (13), 1900–1907.

Harnessing the potential of genetically improved bioinoculants for sustainable agriculture: Recent advances and perspectives

Vinay Kumar[a], Anisha Srivastava[b], Lata Jain[a], Sorabh Chaudhary[c], Pankaj Kaushal[a], and Ravindra Soni[b]

[a]*ICAR—National Institute of Biotic Stress Management, Raipur, India,* [b]*Department of Agricultural Microbiology, College of Agriculture, Indira Gandhi Krishi Vishwavidyalaya, Raipur, Chhattisgarh, India,* [c]*Department of Crop Protection, Central Potato Research Institute, Meerut, India*

13.1 Introduction

Microbial communities associated with plants are known as plant microbiome, and this plant microbiome is referred to as a key determinant of plant growth, health, and development (Berg et al., 2017). Microbial communities associated with plants can be broadly categorized in two classes, namely phylospheric and rhizospheric microbes. The microbial communities are highly responsive to environmental factors; any changes in the environment may change the structure, composition, and biomass of a microbial community. Soil/rhizospheric microbes play key a crucial role in the cycling of nutrients, namely nitrogen, phosphorus, and providing protection to the plants against biotic and abiotic stress (Goel et al., 2017a, 2018b; Kumar et al., 2021c). Microorganisms could play an important role in adaptation strategies and increase of tolerance to various abiotic (high and low temperatures, drought, salinity, and metal toxicity) and biotic (Insect, pest, and disease) stresses in agricultural plants (Goel et al., 2018a; Kumar et al., 2020a,b).

Microbial inoculants are a group of beneficiary microorganisms applied to either in the soil or in the plant in order to improve productivity and crop health (Alori and Babalola, 2018). These microbes are known as soil inoculants or bioinoculants which are agricultural amendments that use beneficial rhizospheric or endophytic microbes to promote plant health (Babalola and Glick, 2012; Kumar et al., 2020a). In recent years, application of plant growth promoting rhizobacteria (PGPR) as microbial inoculants is attaining more awareness as an eco-friendly method of crop improvement as potential supplements of chemical fertilizers, which harms the environment. These microbial inoculants are widely used as a biocontrol agent (biopesticide, bioherbicide, biofungicide) and biofertilizer. The microbes present near to the plant roots (rhizosphere) and inside the plants (endosphere) trigger different mechanisms such as production of indole acetic acid, gibberellins and other substances which promote root hairs formation resulting facilitates nutrients uptake by plants and plant tolerance to stress (Soni et al., 2017). Microbial bioinoculants contribute to plants in multiple aspects, including essential functions such as improving germination of seeds, nutrient supply by fixation of nitrogen, solubilization and mobilization of phosphorus and minerals, providing resistance to various biotic and abiotic factors, and physiology and production of bioactive metabolites. Keeping in view the importance of microbial inoculants, exploitation and prospecting of beneficial microbes and their formulations may be viable option for enhancing crop productivity for a rapidly growing human population (Table 13.1). The role of microbial bioinoculants as biofertilizers and biocontrol agents for improving crop productivity is depicted in Fig. 13.1. It has been reported that different species of *Anabaena, Phormidium, Oscillatoria,* and *Chroococcidiopsis* establish loose association with wheat roots and stimulate growth by releasing cytokinin and IAA (Hussain and Hasnain, 2011).

13.2 Importance of manipulation of microbial bioinoculants

Microbial communities naturally found in the soil, rhizosphere, and inside the plant tissues are known to have various plant-growth-promoting activities. These microorganisms can influence both the efficiency of nutrient availability to crop plants and the soil biodiversity, and they are also involved in regulating interactions between plants and other pathogenic microflora. They may have certain limitations or low level of gene expression for desired traits which can be improved by using recent molecular biology tools, for example, either genetic or genome-editing techniques. The following are the possible trait effects

Table 13.1 List of potential bioinoculants used as biofertilizers and biocontrol agents.

S. No.	Functions of microbial inoculants	Name of the bacterial/fungal strains	Type of association and/or properties
1.	Nitrogen-fixing microbes	*Rhizobium, Anabaena*, and *Frankia* *Azotobacter, Clostridium, Beijerinckia, Klebsiella*, and *Nostoc* *Azospirillum*	Symbiotic association Free living microbe Associate symbiotic
2.	Phosphate-mobilizing microbes	*Pezizella ericae* *Glomus* sp., *Gigaspora* sp., *Acaulospora* sp., *Scutellospora* sp., and *Sclerocystis* sp. *Laccaria* sp., *Pisolithus* sp., and *Boletus* sp.	Ericoid mycorrhizae Arbuscular mycorrhiza Ectomycorrhiza
3.	Phosphate-solubilizing microbes	*Aspergillus awamori* and *Penicillium* sp. *Bacillus circulans, Bacillus subtilis, Bacillus megaterium var phosphaticum*, and *Pseudomonas striata*	Fungi Bacteria
4.	Plant-growth-promoting rhizobacteria	*Pseudomonas fluroscence*	Bacteria
5.	Biofertilizers for micronutrients	*Bacillus* sp.	Silicate and zinc
6.	Phytostimulators	*Azospirillum*	Microbes expressing phytohormones production
7.	Biological control agents	*Trichoderma, Pseudomonas*, and *Bacillus*	Microbes have antimicrobial properties against phytopathogens
8.	K-solubilizer	*Frateuria aurantia*	Bacteria help in potassium solubilization
9.	Silicate solubilizer	*Thiobacillus thiooxidans*	Silicate solubilization

stemming from functional genes: (i) increased competitiveness in the environment; (ii) enhanced interaction with the target plants; (iii) increased resistance to adverse conditions; and (iv) improved abilities to produce substances that promote plant growth and health. Majority of the commercially available bioinoculants containing single organism are based on combination of two or more microbial beneficial strains, which is called consortium. A consortium of beneficial microbes may have increased spectrum of beneficial effect of inoculum on plants. The plant microbiome is a complex web of species interactions governed, to a large extent, by chemical communication between plants and microbes as well as microbe-microbe communication. The probable impacts of microbial inoculants on pathogenic microbes, plants, and plant microbiome composition and diversity are depicted in Fig. 13.2.

Fig. 13.1 Role of microbial bioinoculants as biofertilizers and biocontrol agents for improving crop productivity.

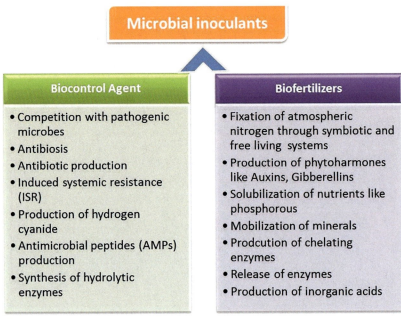

13.3 Genetic traits need to be manipulated for enhancing efficacy of microbes

Microorganisms are considered as rich biosynthetic repositories and have potential to produce large number of biomolecules and natural products of interest (Kumar et al., 2021b). Beneficial microbes which are good for all the activities do not have any adverse effect on growth and development of plants. But they are not able to produce the desired quality of particular trait/molecule of interest; such isolates need genetic manipulation to enhance the activity of gene or production of biomolecules in higher quantities. For instance, *Azotobacters* have ability to fix atmospheric nitrogen and are found naturally in many crops, namely cereals, pulses, legumes, and vegetable crops. The microbes have been engineered for better solubilization of phosphate, better excretion of fixed nitrogen, lesser consumption of ammonia, enhanced fixation of atmospheric nitrogen, and for sustained nitrogen fixation even in the presence of chemically synthesized nitrogenous fertilizers. With the advances in genetic engineering, multiomics technologies allow to get much deeper insights into the microbial communities associated with plants (Berg et al., 2016). PGPR-mediated innovative approaches targeting the mode of action of compound may contribute to enhance plant growth and disease resistance, and serve as an efficient bioinoculant for sustainable agriculture (Meena et al., 2020).

13.4 Genetic engineering techniques used for genetic manipulation of microbes

Microbial adaptation to stress is a complex regulatory process which involved a number of genes and regulatory mechanisms. Metagenomics approach is widely used to understand the structure and composition of microbial diversity in soil and plant tissues, and this technique offers an added advantage to get information related to entire microbial communities including cultural and nonculturable microbes (Soni et al., 2017; Goel et al., 2017b). The integrated approach of system biology including the genomics, metagenomics, metabolomics, transcriptomics, and proteomics is wieldy applied to engineer or manipulate the particular traits or genes in the microorganism to enhance or reduce the gene expression of specific or product of interest. Gene-editing tools, namely ZFNs, TALEN, and CRISPR Cas9, are usually applied to get the specific function by editing of particular genes and enzymes in microorganisms for production of various enzymes, hormones, molecular, and compounds (Sudheer et al., 2020).

13.5 Genetic manipulations/improvement of bioinoculants for different traits with examples

13.5.1 Microbial inoculants as biocontrol agents

Biological control can be as any condition or practice by which survivability of microbial phytopathogens is reduced through the application of living microbial inoculants, resulting in reduction in the disease incidence and yield losses. On the basis of previous studies, about 300 species belonging to 113 fungal genera are identified as biocontrol agents against various phytopathogens. Nine genera, namely *Trichoderma, Verticillium, Talaromyces, Pichia, Fusarium, Penicillium, Candida, Alternaria,* and *Aspergillus,* are considered as potential genera (Thambugala et al., 2020). *Trichoderma* are recognized as the most important genus containing 25 different biocontrol agents and are widely used for controlling the plant diseases. Besides, bacteria genera such as *Azotobacter, Azospirillum, Arthrobacter, Bacillus, Burkholderia, Enterobacter, Klebsiella,* and *Pseudomonas* have been reported for the control and management of diseases in plants. These biocontrol agents play a crucial role in controlling the growth of phytopathogens under field conditions through various processes like

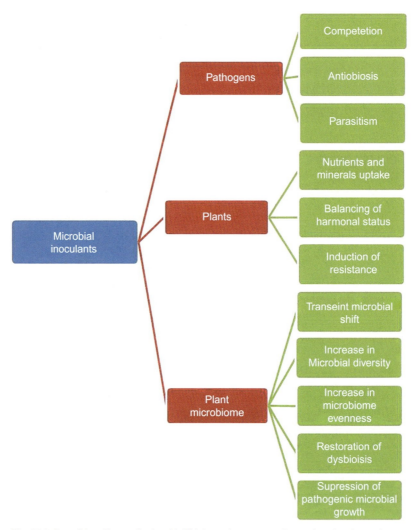

Fig. 13.2 Possible effects of microbial bioinoculants on pathogenic microbes, plants, and plant microbiome composition and diversity.

competition and antibiosis. The success of biological control by antagonistic microbes depends on the complex interactions among the host-plant pathogen and microbial inoculant (Tritrophic interaction) occurred under the field conditions (Lal et al., 2014; Chaudhary et al., 2020b). The collective outcome of these interactions may lead to either suppress or enhance the activity of microbial inoculants for biological control. In order to access the biological potential and efficacy of microbial inoculants, knowledge about the mode of action of microbial inoculants is the key factor for attaining an effective control of plant pathogens in their hosts. The understanding of the mechanisms of biocontrol will be helpful in establishing favorable conditions for interactions between pathogen and microbial inoculant which is very crucial for the success of biocontrol strategies in a particular pathosystem (Alizadeh et al., 2020). The most common mechanisms employed by the biocontrol agents for biocontrol processes include indirect and direct antagonism (Parveen et al., 2016).

The mechanism of biocontrol or antagonistic involves competition, antagonism, and other interactions with host and associated microbes (Kumar et al., 2020a; Dubey et al., 2020). Microbes are involved in providing protection to the plants against pathogenic microbes through direct and indirect mechanism. Microbial inoculants may induce or prime resistance against phytopathogens in host plant without having any direct interaction with the phytopathogen (Pieterse et al., 2014; Dubey et al., 2020). In this complex process, bioinoculants are initially established on the host plants followed by release of specific inducers (MAMPs) that are recognized by particular receptors (PRRs) by the plant and subsequently triggering of induced systemic resistance (ISR) within the host plant. The induction of ISR makes the host plant ready to defend themselves against phytopathogens (Kohl et al., 2019). Induction of ISR in plants may cause alteration or changes in various processes including enhancing protection of cell wall, modification in metabolic and physiological pathways in plants to produce active metabolites with antimicrobial activities, and phytoalexins. In addition, ISR also involved in deposition of callose restricts pathogens entry by creating structural barriers, higher expression of pathogenesis-related proteins (PR proteins), hydrolytic enzymes, and other defense-related compounds (Compant et al., 2005). Another possible mechanism of indirect interaction with pathogens is the competition for available nutrients (Spadaro and Droby, 2016). The competition for nutrient occurs through peculiar enzymatic activities and other mechanisms which bind limiting nutrients. Microbial inoculants have the ability to synthesize high amount of siderophores with high affinity for iron playing an important role in disease reduction (Lugtenberg and Kamilova, 2009; Kumar et al., 2020a, b). Raaijmakers et al. (2006) demonstrated that some isolates of *Pseudomonas* spp. reduce the pathogen populations in rhizosphere through siderophores-mediated iron competition. Iron-binding siderophores producing *Trichoderma asperellum* are found effective for controlling the *Fusarium* wilt in tomato (Segarra et al., 2010). Niche exclusion is an another control strategy used by microbial inoculants in excluding the pathogens from rhizosphere and inhibiting their pathogenesis (Pathak et al., 2017).

The direct antagonistic mechanisms employed by the microbial inoculants are hyperparasitism (Prasad et al., 2016; Ghorbanpour et al., 2018) or antibiosis (Chaudhary et al., 2020a, b; Alfiky and

Weisskopf, 2021). Antibiotic metabolites produced by bacterial antagonists include pyoluteorin, tensin, pyrrolnitrin, tropolone, phenazine-1-carboxylic acid (PCA), 2,4-diacetylphloroglucinol (DAPG), kanosamine, oomycin-A, cyclic lipopeptides, DDR, butyrolactones, and viscosinamide (Bhattacharyya and Jha, 2012). In addition to the antibiotic produced by the bacterial inoculants, bacterial endophytes have been reported to produce various novel antimicrobial peptides (AMPs) which have bactericidal and fungicidal activities (Kumar et al., 2020a).

Fungal bioinoculants are known to synthesize antimicrobial metabolites for restricting the growth of pathogenic microbes. *Trichoderma* and *Clonostachys* have been reported to produce a variety of antibiotics, namely 6-PAP, gliovirin, gliotoxin, trichodermin, trichodermol, harzianolide, viridian, and many more compounds having antimicrobial activities (Ghorbanpour et al., 2018). Parasitism referred to as the direct competitive interaction between two organisms where one organism is gaining nutrients from the other (host), and if the host is a parasite (plant pathogen), such interaction is called hyperparasitism. The parasitism and hyperparasitism interaction are often observed in fungi, whereas in bacteria, it is very rarely reported (Kohl et al., 2019). Mycoparasitism is mediated through the synthesis of cell wall-degrading enzymes (CWDEs) such as chitinase, cellulase, laminarinase, glucanase, and protease or proteinase (Jadhav and Sayyed, 2016). These enzymes showed hydrolytic activity against oomycetes and restrict or inhibit fungal sporulation and mycelial growth (Saraf et al., 2014). In *Trichoderma*, ech42 and prb1 genes are overexpressed during mycoparasitism as these genes belong to the mycoparasitism-related gene families. The synergistic transcription of various genes involved in cell wall degradation in *Trichoderma atroviride* in interaction with *Phytophthora capsici* and *Botrytis cinerea* was reported (Reithner et al., 2011).

Apart from these, microbial inoculants inhibit the growth of pathogens by producing antimicrobial metabolites, hydrogen cyanide (HCN), endotoxins, exopolysaccharides, biosurfactants, and volatile compounds like acetoin and 2,3-butanediol (Arora et al., 2020). Endotoxins (Cry proteins) exhibit lethal reaction upon ingestion by several classes of insects and nematodes (Schünemann et al., 2014). Additionally, *exo*-polysaccharides produced by microbial inoculants provide protection from pathogens by the formation of biofilm, protection in a hydrated and nutrient-rich environment. Biosurfactants are highly important for increasing the surface area, bacterial pathogenesis, biofilm formation, and quorum sensing. They also form channels in pathogens cell wall, disturb their cell surface properties, and in turn reduce pathogenesis property (Raaijmakers et al., 2006; Afsharmanesh et al., 2013).

The biocontrol efficacy of microbial inoculants can be influenced by various abiotic, biotic, and other environmental factors, resulting in less disease suppression and insufficient reduction of phytopathogen populations. During last decade, application of genetically improved microbial inoculants as bicontrol agents has been reported (Marian and Shimizu, 2019). Genetically manipulated bioagents with altered expression of biocontrol traits such as production of antimicrobial compounds and colonization ability have been developed to improve biocontrol efficacy (Table 13.2). Genetically improved entomopathogenic bacteria displayed many advantages over their wild-type strains, that is, increased pathogenicity, reduced spraying requirements, and long-time persistence. Genetic alterations have been applied to strains of the *genus Bacillus, Lysinibacillus, Pseudomonas, Serratia, Photorhabdus,* and *Xenorhabdus* (Azizoglu et al., 2020). Biocontrol potential of bioagents can be increased by improving their ability to produce antimicrobial compounds, that is, antibiotics, hydrolytic enzymes, and bacteriocins (Marian and Shimizu, 2019). Introduction of foreign gene(s) responsible for the synthesis of antibiotics and hydrolytic enzymes enhanced the biocontrol performance of microbial inoculants against various phytopathogens. In an experiment, Jing et al. (2018) developed a *retS* mutant of *Pseudomonas protegens* Pf-5 to improve the production of DAPG, which showed the greater suppression activity against *Rhizoctonia solani*. Similarly, Shi et al. (2019) engineered *P. protegens* H78 using multiple gene knockout strategies to increase the synthesis of antimicrobial compound pyolutcorin (Plt). Phenazine-1-carboxylic acid (PCA), synthesized by *Pseudomonas* and *Streptomyces*, was reported as effective antifungal compound against *Phytophthora capsici, P. infestans, Rhizoctonia solani,* and *Botrytis cinerea* (Jin et al., 2015, Liu et al., 2016; Morrison et al., 2017). A seven-gene operon from *Pseudomonas synxantha* was introduced into *Pseudomonas fluoresecns* to enhance the synthesis of PCA, and the recombinant strain showed inhibition activity against *Gaeumannomyces graminis* var. *tritici*, by producing antifungal cyclic lipopeptide (Yang et al., 2017).

Integration of hydrolytic enzyme-related gene(s) into certain microbial inoculants can enhance their biocontrol potential against various phytopathogens. Huang et al. (2006) reported that the engineered *Burkholderia vietnamiensis* P418 by introducing a chitinase-encoding gene chi113 from *Bacillus subtilis* Ap113 and transformed strain showed greater inhibition potential against *Rhizoctonia solani* and *Verticillium dahliae*. Similarly, a chitinase gene from *Bacillus subtilis* was cloned and integrated into the chromosome of *B. vietnamiensis* P418 using transposon vector. The transformed strain P418-37 showed enhanced in-vitro inhibition activity against *Rhizoctonia solani, Fusarium oxysporum, G. graminis, Verticillium dahlia,* and

Table 13.2 Genetically improved microbial bioinoculants with biocontrol properties against various phytopathogens.

Source/name of microorganisms/gene	Improved strain	Host plant/pathogen	Function	References
Pseudomonas fluorescens CHA0/pyoluteorin & *phl* gene	*P. fluorescens* CHA0-1	Cucumber/*Pythium ultimum*	Enhanced pyoluteorin & phloroglucinol synthesis	Haas et al. (1990)
Pseudomonas fluorescens 2–79/phenazine-1-carboxylic acid	*P. putida* & *P. fluorescens*	Wheat/*Gaeumannomyces graminis* var. *tritici*	Enhanced antibiosis	Bull et al. (1991)
Pseudomonas strain/*phl*	*P. aureofaciens* Q2-87	Wheat/*G. graminis* var. *tritici*, *Phythium ultimum*, *Rhizoctonia solani*	Enhanced phloroglucinol activity	Vincent et al. (1991)
Trichoderma atroviride/*prb1* proteinase gene	*T. atroviride* strain	*Rhizoctonia solani*	Enhanced lytic activity	Flores et al. (1997)
Serratia marcescens/*chit33* chitinase gene	*T. harzianum*	Cotton/*Rhizoctonia solani*	Enhanced lytic activity	Limon et al. (1999)
Trichoderma harzianum/*ech42* chitinase gene	*T. virens* strain	*Rhizoctonia solani*	Enhanced lytic activity	Baek et al. (1999)
Pseudomonas fluorescens WCS365/*sss*	*P. fluorescens* WCS365	Tomato/*Fusarium oxysporum* f. sp. *lycopersici*	Enhanced colonization	Dekkers et al. (2000)
Enterobacter cloacae UW4/*acdS*	*P. fluorescens* CHA0	Cucumber/*Phythium ultimum*, Potato/*Erwinia carotovora*	Enhanced biocontrol activity	Wang et al. (2000)
Serraetia marcescens/*chiA* chitinase gene	*P. fluorescens* strain	Bean/*Rhizoctonia solani*	Enhanced biocontrol activity	Downing and Thornson (2000)
Pseudomonas fluorescens F113	*P. fluorescens* F113Rif	Sugarbeet/*Phythium ultimum*	Enhanced 2,4-diacetylphloroglucinol production	Delany et al. (2002)
Pseudomonas fluorescens P5/*chiB* chitinase gene	*P. fluorescens* P5-1	Wheat/*G. graminis* var. *tritici*, Cotton/*R. solani*, Rice/*R. solani*	Enhanced biocontrol activity	Xiao-Jing et al. (2005)

Gene/source	Strain	Target/host	Outcome	Reference
Aspergillus niger/goxA	T. atroviride SJ3-4	Bean/Rhizoctonia solani, Phythium ultimum, Botrytis cinerea	Enhanced ISR activity	Brunner et al. (2005)
Serratia marcescens B2/ endochitinase chiA	Erwinia ananas NR1	Rice/Pyricularia oryzae	Enhanced lytic activity	Someya and Akutsu (2005)
Trichoderma brevicompatum/tri-5-trichodiene synthase gene	T. brevicompatum Tb41tri5	Aspergillus fumigates, Fusarium spp.	Enhanced biocontrol activity	Tijerino et al. (2011)
Metarhizium anisopliae/ chitinase chit42 gene	T. koningii	Maize/Bombyx mori, Fusarium verticillioides	Enhanced lytic activity	Li et al. (2012)
Trichoderma atroviride/ chitinase chit42 gene	T. harzianum Chit 42-9	Canola/Sclerotinia sclerotiorum	Enhanced chitinase activity	Kowsari et al. (2014)
Clonostachys rosea 67-1/ endochitinase chi67-1 gene	C. rosea Rc4-4	Soybean/Sclerotinia sclerotiorum	Enhanced biocontrol activity	Sun et al. (2017)
Pseudomonas protegens Pf-5/inactivate retS gene	P. protegens Pf-5	Rice/Rhizoctonia solani	Enhanced 2,4-diacetylphloroglucinol production and colonization	Jing et al. (2018)
Trichoderma harzianum/ chit42 & qid74	T. harzianum T13 & T15	Bean/Rhizoctonia solani	Induced plant defensive responses	Eslahi et al. (2021)

Bipolaris sorokiniana (Zhang et al., 2012). A chitinase gene (PtChi19) from *Pseudoalteromonas tunicate* CCUG44952 T was introduced into the genome of *Escherichia coli* (*E. coli*) BL21. The recombinant strain showed prominent growth inhibition against *Fusarium oxysporum, Armillaria mellea,* and *Aspergillus niger* (Garcia-Fraga et al., 2015). Recently, Okay and Alshehri (2020) introduced chitinase A (chiA) gene from *Serratia marcescens* into the genome of *Bacillus subtilis* 168, and recombinant strain showed 2.15-fold higher endochitinase activity than the parental strain. Similarly, other studies reported that the certain *Trichoderma* strains were engineered for insertion of chitinase-expressing genes for improving biocontrol potential against various fungal pathogens. In a study, Limon et al. (1999) cotransformed *Trichoderma harzianum* with *amdS* gene and its own *chit33* gene under the control of the pki constitutive promoter from *Trichoderma reesei*. The extracellular chitinase activity was increased up to 200-fold in transformants than the wild type and was highly effective in suppression of *R. solani* growth. Li et al. (2012) integrated a chitinase gene *chit42* from *Metarhizium anisopliae* CY1 into the genome of *Trichoderma koningii* through protoplast. The resulted transgenic *Trichoderma* strain harboring *chit42* gene exhibited lethal activities on the Asian corn borer larvae. Isolates of *Trichoderma harzianum* were transformed with constitutively expressed transgene of *Chit42*, displayed high level of chitinase activity than parental isolates, and showed higher growth suppression activity of *S. sclerotiorum* (Kowsari et al., 2014).

13.5.2 Microbial inoculants as biofertilizers

Microorganisms' bacteria, fungi, and blue green algae are largely used as biofertilizers. These microbes are added to the rhizosphere of the plant to enhance their activity in the soil. Biological fertilizer is a substance which contains living microorganisms which when applied to plants either on the surfaces or on the soil has the ability to colonize the rhizosphere or the interior of the plant. Biofertilizers composed of active strains of microorganisms like bacteria, fungi, and algae and are known to promote plant growth by increasing the accessibility and availability of primary nutrients to the host plant. In sustainable agriculture, microbial inoculants (biofertilizers) can be used as inoculants into the roots or soil, in seed, leave, and seedling alone or in combination, promoting plant growth without deleterious side effects for environment and enhancing crop yields (Mishra et al., 2017). Among these microbe's nitrogen-fixing soil bacteria (*Azotobacter, Rhizobium*), cyanobacteria (*Anabaena*), phosphate-solubilizing bacteria (*Pseudomonas* sp.), and AM fungi are most commonly used as biofertilizers. Similarly, phytohormone-producing and cellulolytic microorganisms are used in formulation of biostimulants. These inoculants can be used as an applicant on the soil, plant, or composting pits, and help in increasing different biological activity which mobilizes different nutrients which are required by the plant for their growth directly or indirectly and to stimulate microbial activity (Suyal et al., 2016). Recently, microbial inoculants are gaining momentum because of huge advantages as they play a crucial role in decomposition of organic matter, enhancing enzyme synthesis and production of plant hormones within plants. In addition, microbial inoculant helps in the maintenance of soil health and reduction of environmental pollution which occurs due to use of chemical fertilizers in agriculture.

Presently, a variety of biofertilizers are commercially available in the market, and these formulations were developed using different strategies to ensure the maximum viability of the microbe in the formulations. The strategy includes (i) optimization of formulation, (ii) application of/drought-tolerant/thermos-tolerant isolates, (iii) application of liquid biofertilizer, and (iv) genetically modified strains. The application of biofertilizers enhances soil properties and soil health, and reduces soil-borne diseases, which results in increases in crop yield and productivity.

13.5.3 Microbial inoculants for biotic and abiotic stresses

Microbes play a vital role in adaptation strategies and increased the tolerance to various stresses. Beneficial microbes help in combating various biotic and abiotic stresses as they can colonize the rhizosphere of the plant and promote growth and health through the plants by employing direct and indirect mechanisms (Raghu et al., 2021). Microorganisms belonging to different genera such as *Azospirillum, Achromobacter, Burkholderia, Bacillus, Enterobacter Pseudomonas, Pantoea, Paenibacillus, Rhizobium, Microbacterium,* and *Methylobacterium* have been reported to provide abiotic stress tolerance to host plants (Meena et al., 2017; Dubey et al., 2020).

PGPR has been reported to mitigate the impact of abiotic stresses, namely low and high temperature, drought stress, and salinity stress on plants by producing exopolysaccharides and formation of biofilm. They also involved in mitigating impact of drought stress through a process known as induced systemic tolerance (IST). Induced systemic tolerance includes bacterial production of cytokinins, antioxidants, and degradation of the ethylene precursor ACC by bacterial ACC deaminase (Dubey et al., 2020). Hahm et al. (2017) reported three PGPR isolates, namely *Brevibacterium iodinum, Microbacterium oleivorans,* and *Rhizobium massiliae,* isolated from pepper rhizosphere of saline conditions, and plants inoculated with PGPR showed significantly higher plant height, fresh weight, dry weight, and total chlorophyll content than uninoculated plants. The bacterial cytokinin production allows the abscisic acid (ABA) accumulation in leaves, which in turn promotes closure of stomata (Figueiredo et al., 2008; Yang et al., 2009). Arbuscular mycorrhizal fungi, mycorrhizae, rhizospheric microbes, and endophytic bacteria and fungi have been reported to mitigate the abiotic stress in plants, and these microbes also help plants to provide tolerance to stresses by producing plant-growth-promoting substances, molecule, antioxidants, and secondary metabolites (Chakrabort and Saha, 2019; Dash et al., 2019; Kumar et al., 2020b, 2021c).

13.5.4 Microbial inoculants with potential to produce phytohormones and metabolites

The microbial inoculants normally colonize on the root zone (rhizosphere) or inside the plant (endosphere) which leads to stimulation of various compounds, enzymes, hormones, and metabolites to speed up growth and enhance tolerance to various stresses in the plant. The direct role of plant-growth-promoting (PGP) microbes includes increase in availability and uptake of nutrients regulation of phytohormone synthesis and induction of systemic resistance via indirect mechanisms (Abhilash et al., 2016; Khoshru et al., 2020; Dubey et al., 2020). Furthermore, rhizospheric and endophytic microbes including bacterial and fungal isolates have potential to produce different types of phytohormones, metabolites, and antimicrobial peptides in the tissues of a plant which are involved in the modulation of important physiological processes including hormonal balance in the plants (Goel et al., 2017a; Chaudhary et al., 2020b; Kumar et al., 2021b). Microbial inoculants assist the plants in mitigating the adverse impact of various biotic and abiotic stresses through the synthesis of hormones such as auxin, gibberellin, cytokinin, abscisic acid, jasmonates, strigolactones, brassinosteroids, and secondary metabolites (Saravanakumar et al., 2011; Oosten et al., 2017; Kumar et al., 2020a, 2021a). PGPM helps the plant in the production of various hormones like auxin, gibberellin, cytokinin, ACC-deaminase concentrations, and volatile metabolites (VOC) which helps the plant for inducing disease resistance and tolerance to abiotic stresses. These microbes play a crucial role in alleviating stress by producing exopolysaccharides (EPSs), antioxidants, and osmoregulants, which reduces oxidative stress (Varma et al., 2019; Khan et al., 2020). Thus, promoting leaf area, increasing foliage, seedling vigor, plant height, chlorophyll content, seed germination, root development, photosynthetic rates, and biomass production ultimately increase plant growth, for example, *Sphingomonas* sp. and *Serratia marcescens*, which stimulates root and shoot growth of soybean and by producing higher amount of IAA as compared to the control plants (Asaf et al., 2017). Moreover, a beneficial microbe (bacteria and fungi) also acts as biocontrol agent by increasing resistance against plant pests and pathogens through competition for nutrients, antagonism, expression of genes for production of hydrolytic enzymes, chitinases, and induces systemic resistance and production of antimicrobial lipopeptides (Bhat et al., 2019; Chaudhary et al., 2020a; Kumar et al., 2019; Sahu et al., 2020). Naturally, a less quantity of phytohormones synthesized by microbes enhance the condition of the plant under stress conditions including temperature, salt, drought, flood, and heavy metal toxicity (Egamberdieva, 2011; Liu et al., 2013a, b; Ngumbi and Kloepper, 2014; Goel et al., 2017b; Goel et al., 2018a; Dubey et al., 2020; Kumar et al., 2021c).

Several bacterial species have potential to produce auxins, indole-3-acetic acid (IAA), which play a significant role in bacteria-plant interactions, as well as in pathogenesis to phytostimulation (Spaepen et al., 2007). Studies have reported the positive effects of bacteria associated with the plants enhance plant growth under abiotic stress conditions by producing IAA (Kumar et al., 2021c). In case of barley, plant biomass and resistance against salt stress are stimulated by *Ensifer garamanticus* and *Curtobacterium flaccumfaciens* bacterial isolates (Cardinale et al., 2015). Similarly, *Pseudomonas* strain recovered from high temperature environments was found to synthesize IAA under salt stress which stimulates the increases in maize root and shoot biomass (Egamberdieva, 2009; Mishra et al., 2017). Phytohormones produced by microbial inoculants also play a crucial role in metal-plant interaction and improving phytoextraction by plants. Ma et al. (2009) observed that the bacterium, *Achromobacter xylosoxidans*, improves the root system in *Brassica juncea* by production of IAA, which leads to the increase in copper phytoextraction.

Furthermore, ethylene is a gaseous phytohormone produced by most of the plants, acts at very low concentrations, and is involved in the regulation of numerous processes related to the plant growth, development, and senescence (Shaharoona et al., 2006; Saleem et al., 2007). Besides acting as a plant growth regulator, ethylene is also known as stress phytohormone. Ethylene production is substantially speeded and adversely affects the growth of the roots resulting in the effects on the plant growth. Several mechanisms have been identified to reduce ethylene concentration in plants. One of such mechanisms involves 1-aminocyclopropane-1-carboxylate (ACC) deaminase enzyme of bacterial origin (Jalili et al., 2009; Farajzadeh et al., 2012) as this enzyme regulates ethylene biosynthesis in plants by metabolizing ACC into α-ketobutyric acid and ammonia (Saleem et al., 2007; Arshad et al., 2007).

Khalid et al. (2017) reported that biofertilizers used for inoculation of spinach were found to have more phenolic content of 58.72 and 51.43% which is higher than the total uninoculated spinach. These phenolic metabolites take part in the prevention of cardiovascular disorders, cancer, and neurodegenerative diseases (Rodríguez-Morató et al., 2015). It was also reported that lettuce inoculated with *Glomus fasciculatum* and *Azotobacter chroococcum* led to enhance the level of carotenoids, phenolic compounds, and anthocyanins in vegetable crops (Baslam et al., 2011). Arbuscular mycorrhizal fungi (AMF) helps in biosynthesis of antioxidants (Carlsen et al., 2008; Nisha and RajeshKumar, 2010; Eftekhari et al., 2012). Inoculating soybean seedlings with rhizobacteria has shown phenolic acid increased to about 75% (Taie et al., 2008). It was also reported that the alkaloid content in *Catharanthus roseus* was increased upon inoculation with *Pseudomonas fluorescens* and *Bacillus megaterium* (Karthikeyan et al., 2010), and plant health-promoting compounds produced by biofertilizers include ferulic acid, flavonols, caffeic acid, flavones, coumaric acid, and chlorogenic acid (Alarcón-Flores et al., 2014).

13.6 Conclusion

Microorganisms used as bioinoculants such as rhizobacteria, fungi, and endophytes play a crucial role in the plant growth and development by providing the certain plant-growth-promoting substances like hormones, enzymes, and stress metabolites which confer resistance/tolerance to various biotic and abiotic stresses. These microorganisms operate through a variety of mechanisms like triggering osmotic pressure and inducing expression of novel genes in plants. Microbial inoculants play an important role in plant stress tolerance and adaptation to various stresses. Knowledge of specific genes and compounds produced by microorganism has opened a new area of research, and advanced molecular biology techniques allow to edit or modify the particular genes to increase or decrease the production of specific enzymes or compound of interest. In coming days, microbial inoculation will become a cost-effective and eco-friendly option to alleviate stresses in plants in a shorter time. Microbial-assisted selection approach will also be useful for breeding crops for crop improvement.

References

Abhilash, P.C., Dubey, R.K., Tripathi, V., Gupta, V.K., Singh, H.B., 2016. Plant growth-promoting microorganisms for environmental sustainability. Trends Biotechnol. 34, 847–850. https://doi.org/10.1016/j.tibtech.2016.05.005.

Afsharmanesh, H., Ahmadzadeh, M., Majdabadi, A., Motamedi, F., Behboudi, K., Javan-Nikkhah, M., 2013. Enhancement of biosurfactants and biofilm production after gamma irradiation-induced mutagenesis of *Bacillus subtilis* UTB1, a biocontrol agent of *Aspergillus flavus*. Arch. Phytopathol. Plant Protect. 46 (15), 1874–1884.

Alarcón-Flores, M.I., Romero-González, R., Vidal, J.L., Frenich, A., 2014. Determination of phenolic compounds in artichoke, garlic and spinach by ultra-high-performance liquid chromatography coupled to tandem mass spectrometry. Food Anal. Methods 7, 2095–2106. https://doi.org/10.1007/s12161-014-9852-4.

Alfiky, A., Weisskopf, L., 2021. Deciphering *Trichoderma*-plant-pathogen interactions for better development of biocontrol applications. J. Fungi. 7 (1), 61.

Alizadeh, M., Vasebi, Y., Safaie, N., 2020. Microbial antagonists against plant pathogens in Iran: a review. Open Agric. 5, 404–440.

Alori, E.T., Babalola, O.O., 2018. Microbial inoculants for improving crop quality and human health in Africa. Front. Microbiol. 9, 2213. https://doi.org/10.3389/fmicb.2018.02213.

Arora, N., Fatima, T., Mishra, I., Verma, S., 2020. Microbe-based inoculants: role in next green revolution. In: Shukla, V., Kumar, N. (Eds.), Environmental Concerns and Sustainable Development. Springer Nature, Singapore, pp. 191–246.

Arshad, M., Saleem, M., Hussain, S., 2007. Perspectives of bacterial ACC deaminase in phytoremediation. Trends Biotechnol. 25, 356–362.

Asaf, S., Khan, A.L., Khan, M.A., Imran, Q.M., Yun, B.W., Lee, I.J., 2017. Osmoprotective functions conferred to soybean plants via inoculation with *Sphingomonas sp.* LK11 and exogenous trehalose. Microbiol. Res. 205, 135–145. https://doi.org/10.1016/j.micres.2017.08.009.

Azizoglu, U., Jouzani, G.S., Yilmaz, N., Baz, E., Ozkok, D., 2020. Genetically modified entomopathogenic bacteria, recent developments, benefits and impacts: a review. Sci. Total Environ. 734, 139169.

Babalola, O.O., Glick, B.R., 2012. The use of microbial inoculants in African agriculture: current practice and future prospects. J. Food Agric. Environ. 10, 540–549.

Baek, J.M., Howell, C.R., Kenerley, C.M., 1999. The role of an extracellular chitinase from *Trichoderma virens* Gv29-8 in the biocontrol of *Rhizoctonia solani*. Curr. Genet. 35, 41–50.

Baslam, M., Garmendia, I., Goicoechea, N., 2011. Arbuscular mycorrhizal fungi (AMF) improved growth and nutritional quality of greenhouse-grown lettuce. J. Agric. Food Chem. 59, 5504–5515. https://doi.org/10.1021/jf200501c.

Berg, G., Rybakova, D., Grube, M., Köberl, M., 2016. The plant microbiome explored: implications for experimental botany. J. Exp. Bot. 67, 995–1002.

Berg, G., Köberl, M., Rybakova, D., Muller, H., Grosch, R., Smalla, K., 2017. Plant microbial diversity is suggested as the key to future biocontrol and health trends. FEMS Microbiol. Ecol. 93, 5.

Bhat, M.A., Rasool, R., Ramzan, S., 2019. Plant growth promoting rhizobacteria (PGPR) for sustainable and eco-friendly agriculture. Acta Sci. Agric. 3, 23–25.

Bhattacharyya, P.N., Jha, D.K., 2012. Plant growth-promoting rhizobacteria (PGPR): emergence in agriculture. World J. Microbiol. Biotechnol. 28, 1327–1350.

Brunner, K., Zeilinger, S., Ciliento, R., Woo, S.L., Lorito, M., Kubicek, C.P., Mach, R.L., 2005. Improvement of the fungal biocontrol agent *Trichoderma atroviride* to enhance both antagonism and induction of plant systemic disease resistance. Appl. Environ. Microbiol. 71, 3959–3965.

Bull, C.T., Weller, D.M., Thomashow, L.S., 1991. Relationship between root colonization and suppression of *Gaeumannomyces graminis* var. *tritici* by *Pseudomonas fluorescens* strain 2-79. Phytopathology 81, 954–959.

Cardinale, M., Ratering, S., Suarez, C., Montoya, A.M.Z., Geissler-Plaum, R., Schnell, S., 2015. Paradox of plant growth promotion potential of rhizobacteria and their actual promotion effect on growth of barley (*Hordeum vulgare* L.) under salt stress. Microbiol. Res. 181, 22–32. https://doi.org/10.1016/j.micres.2015.08.002.

Carlsen, S., Understrup, A., Fomsgaard, I., Mortensen, A., Ravnskov, S., 2008. Flavonoids in roots of white clover: interaction of arbuscular mycorrhizal fungi and a pathogenic fungus. Plant and Soil 302, 33–43. https://doi.org/10.1007/s11104-007-9452-9.

Chakrabort, S., Saha, S., 2019. Role of microorganisms in abiotic stress management. Int. J. Sci. Environ. Technol. 8 (5), 1028–1039.

Chaudhary, S., Sagar, S., Kumar, M., Lal, M., Kumar, V., Tomar, A., 2020a. Molecular cloning, characterization and semiquantitative expression of endochitinase gene from the mycoparasitic isolate of *Trichoderma harzianum*. Res. J. Biotechnol. 15 (4), 40–56.

Chaudhary, S., Sagar, S., Lal, M., Tomar, A., Kumar, V., Kumar, M., 2020b. Biocontrol and growth enhancement potential of *Trichoderma spp.* against *Rhizoctonia solani* causing sheath blight disease in Rice. J. Environ. Biol. 41, 1034–1045.

Compant, S., Duffy, B., Nowak, J., Clement, C., Barka, E.A., 2005. Use of plant growth-promoting bacteria for biocontrol of plant diseases: principles, mechanisms of action, and future prospects. Appl. Environ. Microbiol. 71, 4951–4959.

Dash, B., Soni, R., Kumar, V., Suyal, D.C., Dash, D., Goel, R., 2019. Mycorrhizosphere: microbial interactions for sustainable agricultural production. In: Varma, A., Choudhary, D. (Eds.), Mycorrhizosphere and Pedogenesis. Springer, Singapore, https://doi.org/10.1007/978-981-13-6480-8_18.

Dekkers, L.C., Mulders, I.H., Phoelich, C.C., Chin-A-Woeng, T.F., Wijfjes, A.H., Lugtenberg, B.J., 2000. The *sss* colonization gene of the *tomato-Fusarium oxysporum* f. sp. *radicis-lycopersici* biocontrol strain *Pseudomonas fluorescens* WCS365 can improve root colonization of other wild-type *Pseudomonas* spp. bacteria. Mol. Plant Microbe Interact. 13, 1177–1183.

Delany, I.R., Walsh, U.F., Ross, I., Fenton, A.M., Corkery, D.M., O'Gara, F., 2002. Enhancing the biocontrol efficacy of *Pseudomonas fluorescens* F113 by altering the regulation and production of 2,4-diacetylphloroglucinol. In: Powlson, D.S., Bateman, G.L., Davies, K.G., Gaunt, J.L., Hirsch, P.R. (Eds.), Interactions in the Root Environment: An Integrated Approach. Developments in Plant and Soil Sciences. vol. 96. Springer, Dordrecht, pp. 195–205.

Downing, K.J., Thornson, J.A., 2000. Introduction of the *Serratia marcescens* chiA gene into an endophytic *Pseudomonas fluorescens* for the biocontrol of phytopathogenic fungi. Can. J. Microbiol. 46, 363–369.

Dubey, P., Kumar, V., Ponnusamy, K., Sonwani, R., Singh, A.K., Suyal, D.C., Soni, R., 2020. Microbe assisted plant stress management. In: De Mandal, S., Bhatt, P. (Eds.), The Recent Advancements in Microbial Diversity 1st Edition. Elsevier (Academic), https://doi.org/10.1016/B978-0-12-821265-3.0001. Publisher.

Eftekhari, M., Alizadeh, M., Ebrahimi, P., 2012. Evaluation of the total phenolics and quercetin content of foliage in mycorrhizal grape (*Vitis vinifera* L.) varieties and effect of postharvest drying on quercetin yield. Ind. Crop Prod. 38, 160–165. https://doi.org/10.1016/j.indcrop.2012.01.022.

Egamberdieva, D., 2009. Alleviation of salt stress by plant growth regulators and IAA producing bacteria in wheat. Acta Physiol. Plant. 31, 861–864. https://doi.org/10.1007/s11738-009-0297-0.

Egamberdieva, D., 2011. Survival of *Pseudomonas extremorientalis* TSAU20 and *P. chlororaphis* TSAU13 in the rhizosphere of common bean (*Phaseolus vulgaris*) under saline conditions. Plant Soil Environ. 57, 122–127.

Eslahi, N., Kowsari, M., Zamani, M., Motallebi, M., 2021. The profile change of defense pathways in *Phaseolus vulgaris* L. by biochemical and molecular interactions of *Trichoderma harzianum* transformants overexpressing a chimeric chitinase. Biol. Control 152, 104304.

Farajzadeh, D., Yakhchali, B., Aliasgharzad, N., Sokhandan-Bashir, N., Farajzadeh, M., 2012. Plant growth promoting characterization of indigenous *Azotobacteria* isolated from soils in Iran. Curr. Microbiol. 64, 397–403.

Figueiredo, M.V.B., Burity, H.A., Martinez, C.R., Chanway, C.P., 2008. Alleviation of drought stress in the common bean (Phaseolus vulgaris L.) by co-inoculation with Paenibacillus polymyxa and Rhizobium tropici. Appl. Soil Ecol. 40, 182–188. https://doi.org/10.1016/j.apsoil.2008.04.005.

Flores, A., Chet, I., Herrera-Estrella, A., 1997. Improved biocontrol activity of the proteinase-encoding gene *prbl*. Curr. Genet. 31, 30–37.

Garcia-Fraga, B., de Silva, A.F., Lopez-Seijas, J., Sieiro, C., 2015. A novel family 19 chitinase from the marine-derived *Pseudoalteromonas tunicata* CCUG 44952T: heterologous expression characterization and antifungal activity. Biochem. Eng. J. 93, 84–93.

Ghorbanpour, M., Omidvari, M., Abbaszadeh-Dahaji, P., Omidvar, R., Kariman, K., 2018. Mechanisms underlying the protective effects of beneficial fungi against plant diseases. Biol. Control 117, 147–157.

Goel, R., Kumar, V., Suyal, D.K., Dash, B., Kumar, P., Soni, R., 2017a. Root-associated bacteria: rhizoplane and endosphere. In: Singh, D., Singh, H., Prabha, R. (Eds.), Plant-Microbe Interactions in Agro-Ecological Perspectives. Springer, Singapore, https://doi.org/10.1007/978-981-10-5813-4_9.

Goel, R., Kumar, V., Suyal, D.C., Narayan, S.R., 2018a. Toward the Unculturable microbes for sustainable agricultural production. In: Meena, V. (Ed.), Role of Rhizospheric Microbes in Soil. Springer, Singapore DOI, https://doi.org/10.1007/978-981-10-8402-7_4.

Goel, R., Suyal, D.C., Kumar, V., Jain, L., Soni, R., 2018b. Stress-tolerant beneficial microbes for sustainable agricultural production. In: Panpatte, D., Jhala, Y., Shelat, H., Vyas, R. (Eds.), Microorganisms for Green Revolution. Microorganisms for Sustainability. vol 7. Springer, Singapore, https://doi.org/10.1007/978-981-10-7146-1_8.

Goel, R., Suyal, D.C., Narayan, D.B., Soni, R., 2017b. Soil metagenomics: A tool for sustainable agriculture. In: Kalia, V., Shouche, Y., Purohit, H., Rahi, P. (Eds.), Mining of Microbial Wealth and Metagenomics. Springer Nature, Singapore, pp. 217–225.

Haas, D., Keel, C., Laville, J., Maurhofer, M., Oberhansli, T., Schnider, U., Voisard, C., Wuthrich, B., Defago, G., 1990. Secondary metabolites of *Pseudomonas puorescens* strain CHAO involved in the suppression of root diseases. In: Hennecke, H., Verma, D.P.S. (Eds.), Advances in Molecular Genetics of Plant-Microbe Interactions. vol. I. Kluwer Academic Publishers, Dordrecht, The Netherlands, pp. 450–456.

Hahm, M.S., Son, J.S., Hwang, Y.J., Kwon, D.K., Ghim, S.Y., 2017. Alleviation of salt stress in pepper (*Capsicum annum* L.) plants by plant growth-promoting rhizobacteria. J. Microbiol. Biotechnol. 27, 1790–1797. https://doi.org/10.4014/jmb.1609.09042.

Huang, Y.J., Yang, H.T., Zhou, H.Z., 2006. Chitinase gene from *Bacillus subtilis*: cloning sequencing and expression in Burkholderia B418. J. Biol. Control. 22, 72–77.

Hussain, A., Hasnain, S., 2011. Phytostimulation and biofertilization in wheat by cyanobacteria. J. Ind. Microbiol. Biotechnol. 38 (1), 85–92. https://doi.org/10.1007/s10295-010-0833-3.

Jadhav, H.P., Sayyed, R.Z., 2016. Hydrolytic enzymes of rhizospheric microbes in crop protection. MOJ Cell Sci. Rep. 3 (5), 135–136.

Jalili, F., Khavazi, K., Pazira, E., Nejati, A., Rahmani, H.A., Sadaghiani, H.R., Miransari, M., 2009. Isolation and characterization of ACC deaminase-producing fluorescent pseudomonads, to alleviate salinity stress on canola (*Brassica napus* L.) growth. J. Plant Physiol. 166, 667–674.

Jin, K., Zhou, L., Jiang, H., Sun, S., Fang, Y., Liu, J., Zhang, X., He, Y.W., 2015. Engineering the central biosynthetic and secondary metabolic pathways of *Pseudomonas aeruginosa* strain PA1201 to improve phenazine-1-carboxylic acid production. Metab. Eng. 32, 30–38.

Jing, X., Cui, Q., Li, X., Yin, J., Ravichandran, V., Pan, D., Fu, D., Fu, J., Tu, Q., Wang, H., Bian, X., Zhang, Y., 2018. Engineering *Pseudomonas protegens* Pf-5 to improve its antifungal activity and nitrogen fixation. Microbial. Biotechnol. 13 (1), 118–133.

Karthikeyan, B., Joe, M.M., Jaleel, C.A., Deiveekasundaram, M., 2010. Effect of root inoculation with plant growth promoting rhizobacteria (PGPR) on plant growth, alkaloid content and nutrient control of *Catharanthus roseus* (L.) G. Don. Nat. Croat. 1, 205–212. https://doi.org/10.1016/j.scienta.2007.04.013.

Khalid, M., Hassani, D., Bilal, M., Asad, F., Huang, D., 2017. Influence of bio-fertilizer containing beneficial fungi and rhizospheric bacteria on health promoting compounds and antioxidant activity of *Spinacia oleracea* L. Bot. Stud. 58, 35. https://doi.org/10.1186/s40529-017-0189-3.

Khan, N., Bano, A., Ali, S., Babar, M.A., 2020. Crosstalk amongst phytohormones from planta and PGPR under biotic and abiotic stresses. Plant Growth Regul. 90, 189–203. https://doi.org/10.1007/s10725-020-00571-x.

Khoshru, B., Mitra, D., Khoshmanzar, E., Myo, E.M., Uniyal, N., Mahakur, B., et al., 2020. Current scenario and future prospects of plant growth-promoting rhizobacteria: an economic valuable resource for the agriculture revival under stressful conditions. J. Plant Nutr. 43, 3062–3092. https://doi.org/10.1080/01904167.2020.1799004.

Kohl, J., Kolnarr, R., Ravensberg, W.J., 2019. Mode of action of microbial biological control agents against plant disease: relevance beyond efficacy. Front. Plant Sci. 10, 845.

Kowsari, M., Zamani, M.R., Motallebi, M., 2014. Enhancement of *Trichoderma harzianum* activity against *Sclerotinia sclerotiorum* by overexpression of *Chit42*. Iran. J. Biotechnol. 12 (2), 26–31.

Kumar, V., Jain, L., Jain, S.K., Chaturvedi, S., Kaushal, P., 2020a. Bacterial endophytes of Rice (*Oryza sativa* L.) and their potential for plant growth promotion and antagonistic activities. S. Afr. J. Bot. 34, 50–63.

Kumar, V., Jain, L., Kaushal, P., Soni, R., 2021b. Fungal endophytes and their applications as growth promoters and biological control agents. In: Sharma, V.K., Shah, M.P., Parmar, S., Kumar, A. (Eds.), Fungi Bio-prospects in Sustainable Agriculture, Environment and Nano-technology. vol. 1. Elsevier publishers, pp. 315–337, https://doi.org/10.1016/B978-0-12-821394-0.00012-3.

Kumar, V., Jain, L., Soni, R., Kaushal, P., Goel, R., 2020b. Bio-prospecting of endophytic microbes from higher altitude plants: recent advances and their biotechnological applications. In: Goel, R., Soni, R., Suyal, D. (Eds.), Microbiological Advancements for Higher Altitude Agro-Ecosystems & Sustainability. Rhizosphere Biology. Springer, Singapore, https://doi.org/10.1007/978-981-15-1902-4_18.

Kumar, J., MuraliBaskaran, R.K., Jain, S.K., Sivalingam, P.N., Mallikarjuna, J., Kumar, V., Sharma, K.C., Sridhar, J., Mooventhan, P., Dixit, A., Ghosh, P.K., 2021a. Emerging and re-emerging biotic stresses of agricultural crops in India and novel tools for their better management. Curr. Sci. 121 (1), 26–36. https://doi.org/10.18520/cs/v121/i1/26-36.

Kumar, V., Sahu, B., Suyal, D.C., Karthika, P., Singh, M., Singh, D., Kumar, S., Yadav, A.N., Soni, R., 2021c. Strategies for abiotic stress Management in Plants through Soil Rhizobacteria. In: Yadav, A.N. (Ed.), Soil Microbiomes for Sustainable Agriculture. Sustainable Development and Biodiversity. vol. 27. Springer, Cham, https://doi.org/10.1007/978-3-030-73507-4_11.

Kumar, V., Soni, R., Jain, L., Dash, B., Goel, R., 2019. Endophytic Fungi: Recent advances in identification and explorations. In: Singh, B. (Ed.), Advances in Endophytic Fungal Research. Fungal Biology. Springer, Cham, https://doi.org/10.1007/978-3-030-03589-1_13.

Lal, M., Singh, V., Kandhari, J., Sharma, P., Kumar, V., Murti, S., 2014. Diversity analysis of *Rhizoctonia solani* causing sheath blight of rice in India. Afr. J. Biotechnol. 13 (51), 4594–4605.

Li, Y.Y., Tang, J., Fu, K.H., Gao, S.G., Wu, Q., Chen, J., 2012. Construction of transgenic *Trichoderma koningii* with *chit42* of *Metarhizium anisopliae* and analysis of its activity against the Asian corn borer. J. Environ. Sci. Health B 47 (7), 622–630.

Limon, M.C., Pintor-Toro, J.A., Bcnitei, T.M., 1999. Increased antifungal activity of *Trichoderma harzianum* transformants that overexpress a 33-kDa chitinase. Phytopathology 89, 254–261.

Liu, K., Hu, H., Wang, W., Zhang, X., 2016. Genetic engineering of *Pseudomonas chlororaphis* GP72 for the enhanced production of 2-Hydroxyphenazine. Microb. Cell Fact. 15 (1), 131.

Liu, Y., Shi, Z., Yao, L., Yue, H., Li, H., Li, C., 2013b. Effect of IAA produced by *Klebsiella oxytoca* Rs-5 on cotton growth under salt stress. J. Gen. Appl. Microbiol. 59, 59–65. https://doi.org/10.2323/jgam.59.59.

Liu, F., Xing, S., Ma, H., Du, Z., Ma, B., 2013a. Cytokinin-producing, plant growth-promoting rhizobacteria that confer resistance to drought stress in *Platycladus orientalis* container seedlings. Appl. Microbiol. Biotechnol. 97, 9155–9164. https://doi.org/10.1007/s00253-013-5193-2.

Lugtenberg, B., Kamilova, F., 2009. Plant-growth-promoting rhizobacteria. Annu. Rev. Microbiol. 63, 541–556.

Ma, Y., Rajkumar M., Freitas H., 2009. Inoculation of plant growth promoting bacterium *Achromobacter xylosoxidans* strain Ax10 for the improvement of copper phytoextraction by *Brassica juncea*. J. Environ. Manage. 90 (2), 831–837. https://doi.org/10.1016/j.jenvman.2008.01.014.

Marian, M., Shimizu, M., 2019. Improving performance of microbial biocontrol agents against plant diseases. J. Gen. Plant Dis. 85, 329–336.

Meena, K.K., Sorty, A.M., Bitla, U.M., Choudhary, K., Gupta, P., Pareek, A., Singh, D.P., Prabha, R., Sahu, P.K., Gupta, V.K., Singh, H.B., Krishanani, K.K., Minhas, P.S., 2017. Abiotic stress responses and microbe-mediated mitigation in plants: the omics strategies. Front. Plant Sci. 8, 172. https://doi.org/10.3389/fpls.2017.00172.

Meena, M., Swapnil, P., Divyanshu, K., Kumar, S., Harish, Tripathi, Y.N., Zehra, A., Marwal, A., Upadhyay, R.S., 2020. PGPR-mediated induction of systemic resistance and physiochemical alterations in plants against the pathogens: current perspectives. J. Basic Microbiol. 60 (10), 828–861. https://doi.org/10.1002/jobm.202000370.

Mishra, S.K., Khan, M.H., Misra, S., Dixit, K.V., Khare, P., Srivastava, S., et al., 2017. Characterisation of *Pseudomonas* spp. and *Ochrobactrum* sp. isolated from volcanic soil. Antonie Van Leeuwenhoek 110, 253–270. https://doi.org/10.1007/s10482-016-0796-0.

Morrison, C.K., Arseneault, T., Novinscak, A., Fillion, M., 2017. Phenazine-1-carboxylic acid production by *Pseudomonas fluorescens* LBUM636 alters *Phytophthora infestans* growth and late blight development. Phytopathology 107 (3), 273–279.

Ngumbi, E., Kloepper, J., 2014. Bacterial-mediated drought tolerance: current and future prospects. Appl. Soil Ecol. 105, 109–125. https://doi.org/10.1016/j.apsoil.2016.04.009.

Nisha, M.C., RajeshKumar, S., 2010. Influence of arbuscular mycorrhizal fungi on biochemical changes in Wedilla chinensis (Osbeck) Merril. Anc. Sci. Life 29, 26.

Okay, S., Alshehri, W.A., 2020. Over expression of chitinase A gene from *Serratia marcescens* in *Bacillus subtilis* and characterization of enhanced chitinolytic activity. Braz. Arch. Biol. Technol. 63, 1–8.

Oosten, M.J.V., Pepe, O., Pascale, S., Silletti, S., Maggio, A., 2017. The role of biostimulants and bioeffectors as alleviators of abiotic stress in crop plants. Chem. Biol. Technol. Agric. 4, 5. https://doi.org/10.1186/s40538-017-0089-5.

Parveen, S., Wani, A.H., Bhat, M.Y., Koka, J.A., 2016. Biological control of postharvest fungal rots of rosaceous fruits using microbial antagonists and plant extracts. Czech Mycol. 98 (1), 41–66.

Pathak, D., Lone, R., Koul, K.K., 2017. Arbuscular mycorrhizal fungi (AMF) and plant growth promoting rhizobacteria (PGPR) association in potato (*Solanum tuberosum* L.): a brief review. In: Kumar, V., Kumar, M., Sharma, S., Prasad, R. (Eds.), Probiotics and Plant Health. Springer, Singapore, pp. 401–420.

Pieterse, C.M.J., Zamioudis, C., Berendsen, R.L., Weller, D.M., Van Wees, S.C.M., Bakker, P.A.H.M., 2014. Induced systemic resistance by beneficial microbes. Annu. Rev. Phytopathol. 52, 347–375.

Prasad, L., Chaudhary, S., Sagar, S., Tomar, A., 2016. Mycoparasitic capabilities of diverse native strains of *Trichoderma* spp. against *Fusarium oxysporum* f. sp. *lycopersici*. J. Appl. Nat. Sci. 8 (2), 769–776.

Raaijmakers, J.M., de Bruijn, I., de Kock, M.J., 2006. Cyclic lipopeptide production by plant-associated *Pseudomonas* spp.: diversity, activity, biosynthesis, and regulation. Mol. Plant Microbe Interact. 19, 699–710.

Raghu, S., Kumar, S., Suyal, D.C., Sahu, B., Kumar, V., Soni, R., 2021. Molecular tools to explore rhizosphere microbiome. In: Nath, M., Bhatt, D., Bhargava, P., Choudhary, D.K. (Eds.), Microbial Metatranscriptomics Belowground. Springer, Singapore, https://doi.org/10.1007/978-981-15-9758-9_2.

Reithner, B., Ibarra-Laclette, E., Mach, R.L., Herrera-Estrella, A., 2011. Identification of mycoparasitism-related genes in *Trichoderma atroviride*. Appl. Environ. Microbiol. 77, 4361–4370.

Rodríguez-Morató, J., Xicota, L., Fito, M., Farre, M., Dierssen, M., de la Torre, R., 2015. Potential role of olive oil phenolic compounds in the prevention of neurodegenerative diseases. Molecules 20, 4655–4680. https://doi.org/10.3390/molecules20034655.

Sahu, B., Suyal, D.C., Prasad, P., Kumar, V., Singh, A.K., Kushwaha, S., Karithika, P., Chaubey, A., Soni, R., 2020. Microbial diversity of chickpea rhizosphere. In: Sharma, S.K., Singh, U.B., Sahu, P.K., Singh, H.V., Sharma, P.K. (Eds.), Rhizosphere Microbes. Microorganisms for Sustainability. vol 23. Springer, Singapore, https://doi.org/10.1007/978-981-15-9154-9_20.

Saleem, M., Arshad, M., Hussain, S., Bhatti, A.S., 2007. Perspective of plant growth promoting rhizobacteria (PGPR) containing ACC deaminase in stress agriculture. J. Ind. Microbiol. Biotechnol. 34, 635–648.

Saraf, M., Pandya, U., Thakkar, A., 2014. Role of allelochemicals in plant growth promoting rhizobacteria for biocontrol of phytopathogens. Microbiol. Res. 169 (1), 18–29.

Saravanakumar, D., Kavino, M., Raguchander, T., Subbian, P., Samiyappan, R., 2011. Plant growth promoting bacteria enhance water stress resistance in green gram plants. Acta Physiol. Plant. 33, 203–209. https://doi.org/10.1007/s11738-010-0539-1.

Schünemann, R., Knaak, N., Fiuza, L.M., 2014. Mode of action and specificity of *Bacillus thuringiensis* toxins in the control of caterpillars and stink bugs in soybean culture. ISRN Microbiol. 2014, 135675. 12.

Segarra, G., Casanova, E., Avilés, M., Trillas, I., 2010. *Trichoderma asperellum* strain T34 controls *Fusarium* wilt disease in tomato plants in soilless culture through competition for iron. Microb. Ecol. 59, 141–149.

Shaharoona, B., Arshad, M., Zahir, Z.A., 2006. Effect of plant growth promoting rhizobacteria containing ACC-deaminase on maize (*Zea mays* L.) growth under axenic conditions and on nodulation in mung bean (*Vigna radiata* L.). Lett. Appl. Microbiol. 42, 155–159.

Shi, H., Huang, X., Wang, Z., Guan, Y., Zhang, X., 2019. Improvement of pyoluteorin production in *Pseudomonas protegens* H78 through engineering its biosynthetic and regulatory pathways. Appl. Microbiol. Biotechnol. 103 (8), 3465–3476.

Someya, N., Akutsu, K., 2005. Biocontrol of plant diseases by genetically modified microorganisms: Current status and future prospects. In: Siddiqui, Z.A. (Ed.), PGPR: Biocontrol and Biofertilization. Springer, The Netherlands, pp. 297–312.

Soni, R., Kumar, V., Suyal, D.C., Jain, L., Goel, R., 2017. Metagenomics of plant rhizosphere microbiome. In: Singh, R., Kothari, R., Koringa, P., Singh, S. (Eds.), Understanding Host-Microbiome Interactions—An Omics Approach. Springer, Singapore, https://doi.org/10.1007/978-981-10-5050-3_12.

Spadaro, D., Droby, S., 2016. Development of biocontrol products for postharvest diseases of fruit: the importance of elucidating the mechanisms of action of yeast antagonists. Trends Food Sci. Technol. 47, 39–49.

Spaepen, S., Vanderleyden, J., Remans, R., 2007. Indole-3-acetic acid in microbial and microorganism-plant signaling. FEMS Microbiol. Rev. 31, 425–448.

Sudheer, S., Bai, R.G., Usmani, Z., Sharma, M., 2020. Insights on engineered microbes in sustainable agriculture: biotechnological developments and future prospects. Curr. Genomics 21 (5), 321–333.

Sun, Z.B., Sun, M.H., Zhou, M., Li, S.D., 2017. Transformation of the endochitinase gene Chi67-1 in *Clonostachys rosea* 67-1 increases its biocontrol activity against *Sclerotinia sclerotiorum*. AMB Express 7, 1.

Suyal, D.C., Soni, R., Sai, S., Goel, R., 2016. Microbial inoculants as biofertilizer. In: Singh, D.P. (Ed.), Microbial Inoculants in Sustainable Agricultural Productivity. Springer India, New Delhi, pp. 311–318, https://doi.org/10.1007/978-81-322-2647-5_18.

Taie, H.A., El-Mergawi, R., Radwan, S., 2008. Isoflavonoids, flavonoids, phenolic acids profiles and antioxidant activity of soybean seeds as affected by organic and bioorganic fertilization. Am. Eur. J. Agric. Environ. Sci. 4, 207–213.

Thambugala, K.M., Daranagama, D.A., Phillips, A.J.L., Kannangara, S.D., Promputtha, I., 2020. Fungi vs. Fungi in biocontrol: an overview of fungal antagonists applied against fungal plant pathogens. Front. Cell. Infect. Microbiol. 10, 604923.

Tijerino, A., Cardoza, R.E., Moraga, J., Malmierca, M.G., Vicente, F., Aleu, J., Collado, I.G., Gutikrrez, S., Monte, E., Hermosa, R., 2011. Overexpression of the trichodiene synthase gene tri5 increases trichodermin production and antimicrobial activity in *Trichoderma brevicompactum*. Fungal Genet. Biol. 48, 285–296.

Varma, A., Tripathi, S., Prasad, R., 2019. Plant Biotic Interactions. Springer, Cham, https://doi.org/10.1007/978-3-030-26657-8.

Vincent, M.N., Harrison, J.M., Brackin, J.M., Kovacevich, Mukerji, P., Weller, D.M., Pierson, E.A., 1991. Genetic analysis of the antifungal activity of a soilborne *Pseudomonas aureofaciens* strain. Appl. Environ. Microbiol. 57, 2928–2934.

Wang, C., Knill, E., Glick, B.R., Defago, G., 2000. Effect of transferring 1-aminocyclopropane-1-carboxylic acid (ACC) deaminase genes into *Pseudomonas fluorescens* strain CHAO and its gacA derivative CHA96 on their growth-promoting and disease-suppressive capacities. Can. J. Microbiol. 46, 898–907.

Xiao-Jing, X., Li-Qun, Z., You-Yong, Z., Wen-Hua, T., 2005. Improving biocontrol effect of *Pseudomonas puorescens* P5 on plant diseases by genetic modification with chitinase gene. Chin. J. Agric. Biotechnol. 2, 23–27.

Yang, J., Kloepper, J.W., Ryu, C.M., 2009. Rhizosphere bacteria help plants tolerate abiotic stress. Trends Plant Sci. 14, 1–4. https://doi.org/10.1016/j.tplants.2008.10.004.

Yang, M., Mavrodi, D.V., Mavrodi, O.V., Thomashow, L.S., Weller, D.M., 2017. Construction of a recombinant strain of *Pseudomonas fluorescens* producing both phenazine-1-carboxylic acid and cyclic lipopeptide for the biocontrol of take-all disease of wheat. Eur. J. Plant Pathol. 149, 683–694.

Zhang, X., Huang, Y., Harvey, P.R., Ren, Y., Zhang, G., Zhou, H., Yang, H., 2012. Enhancing plant disease suppression by *Burkholderia vietnamiensis* through chromosomal integration of *Bacillus subtilis* chitinase gene chi113. Biotechnol. Lett. 34 (2), 287–293.

Omics technologies for agricultural microbiology research

Jagmohan Singh[a,b,*], Dinesh K. Saini[c,*], Ruchika Kashyap[d,*], Sandeep Kumar[e,*], Yuvraj Chopra[f,*], Karansher S. Sandhu[g,*], Mankanwal Goraya[h,*], and Rashmi Aggarwal[a,*]

[a]Division of Plant Pathology, ICAR-Indian Agricultural Research Institute, New Delhi, India, [b]Guru Angad Dev Veterinary and Animal Sciences University, Ludhiana, India, [c]Department of Genetics and Plant Breeding, Chaudhary Charan Singh University, Meerut, Uttar Pradesh, India, [d]Department of Agronomy, Horticulture, and Plant Sciences, South Dakota State University, Brookings, SD, United States, [e]ICAR - Indian Institute of Natural Resins and Gums, Ranchi, Jharkhand, India, [f]College of Agriculture, Punjab Agricultural University, Ludhiana, India, [g]Department of Crop and Soil Sciences, Washington State University, Pullman, WA, United States, [h]Department of Plant Pathology, North Dakota State University, Fargo, ND, United States

14.1 Introduction

In times of global climatic change, when challenges such as increasing nutritional food demand, changing diet habits, and increasing crop losses seem to haunt the agricultural production, omics technology is credited for the sweeping success in deciphering the numerous molecules benefiting the field (Li and Yan, 2020; Van Emon, 2016; Kaur et al., 2021). Many imminent large inter-related areas like ecotoxicogenomics, foodomics, genomics, metabolomics, proteomics, and transcriptomics share a common suffix, "Omics." This commonness has made it the tittle- tattle of the scientific world. This technology has allowed us to track all the natural and environmental changes in any organism when its environment, genetic, or nutritional state is altered (Righetti et al., 2014).

Understanding the complex mechanisms in all living beings involved in agriculture has introduced a component of certainty in agricultural science and expanded the science of food and feed to sustainable, analytical, and health-driven status (Van Emon, 2016). Plant

---

[*] All authors contributed equally.

health and performance are often associated with plant-microbial interactions (Bulgarelli et al., 2012). Bulk microbial communities are found near the root zone due to a phenomenon called "the rhizosphere effect" (Bakker et al., 2013), which helps enhance the P-solubilizing, N-mobilizing, and iron-chelating properties of plants (Li et al., 2019). These synergistic interactions have given rise to the term "holobiont" for the plants. These interactions are plants, the first line of defense that increases plant tolerance to abiotic and biotic stresses (Bakker et al., 2018; Mendes et al., 2018; Vandenkoornhuyse et al., 2015).

Nevertheless, diverse changes constantly occur in the crop cultivations and the microbiota associated with them. This dynamic scenario leaves us with a rather intriguing thought about the impact of these shifts on agricultural production (Cordovez et al., 2019). Meta-omics technologies are currently used to gain insights into the diverse mycobiome that could benefit and boost agricultural production (Pozo et al., 2021). Thus, omics technologies have paved ways to decipher the complex microbiota processes associated with plants, in turn affecting agriculture.

The evolution of agricultural genomics began with the sequencing of several plant genomes, the first being "*Arabidopsis thaliana*." After that, several agriculturally important genes and traits were discovered in various organisms using genetic and genomic technologies, whose data are collected and stored in public databases like GenBank (Kumar et al., 2015; Pathak et al., 2018). This genetic mining has been rewarding and has answered complex biological questions. This is true in tomatoes, where genetic changes resulted in larger and better-quality cultivars than their wild relatives (Lin et al., 2014). Gene expression studies have made it easier for breeders to add or knock down a particular gene or genes to produce desirable cultivars (Van Emon, 2016; Sandhu et al., 2021a,b). In a study by Mendes et al. (2018), it was found that several unknown plant traits causing an increased abundance of beneficial bacteria in the rhizosphere were coselected while breeding for *Fusarium oxysporum* resistance, and this discovery was possible through metagenome analysis. Recently, the use of comparative genomics helped in elucidating genetic determinants of symbiosis (Miyauchi et al., 2020).

To further understand the biological systems, transcriptomics comes in handy. It studies mRNAs expression level in the target cell populations, hence termed expression profiling or large-scale gene expression analysis. It also detects expression level changes due to biotic and abiotic stresses (Valdés et al., 2013). Next-generation sequencing (NGS) has restructured transcriptomics by paving economical and efficient ways for gene identification, which has led to the development of molecular markers (Muthamilarasan et al., 2015). The other approach commonly used is microarray technology that uses information from expressed sequence tags (ESTs) and genome sequences to generate information about biological parts under specific conditions (Pathak et al., 2018). Furthermore, metatranscriptomics plays a vital role in analyzing plant-microbial interactions and unraveling potentially beneficial unknown microflora (Schenk et al., 2012).

Earlier, most of the tools focused on nucleic acids and not proteins. However, the ubiquitous presence of proteins and their role in varied functional pathways in different crops and the inter-related organisms made us dig deeper and see whether mRNA expression results in functional proteins' formation. Proteomics helped us study this correlation between mRNA and proteins and further understand proteins' involvement in those functional pathways along with other protein-related activities by comprehending protein profiles and protein interactions (Deracinois et al., 2013). The three main proteomics components are expression proteomics, functional proteomics, and structural proteomics (Chandrasekhar et al., 2014). This technology helps us decipher the role of zillions of plant proteins in varied colors, flavor, texture, yield, and anything related (Roberts, 2002). Scientists are also studying the world of proteins in microbial communities and have referred to the study as "Metaproteomics." This exploration helps us unravel the metabolic processes behind selective colonization and adaptation of microbial communities in soil rhizospheres (Trivedi et al., 2021).

The final output in gene expression and protein interactions is the metabolites in general. To further fathom the biochemical systems in living organisms and understand the connections between DNA, RNA, and proteins, we need to take metabolites (by-products) into the picture. These molecules are studied by another branch of omics called "Metabolomics." Several analytical tools like mass spectrometry (MS), nuclear magnetic resonance (NMR) spectroscopy, gas chromatography (GC), liquid chromatography (LC), and capillary electrophoresis (CE) are nowadays used to sketch the impact of the various centennial of metabolites, leading to the production of large, complex data sets. This information, along with the one generated by transcriptomics and proteomics, can be used to create a clearer picture of what is happening inside the living world of both plants and microbes and can be then used to enhance production (Dixon et al., 2006; Witzel et al., 2015).

All the omics technologies are inter-related and woven with the same thread as the data generated from one could be used as input for the other owing to multiple layers of data. Also, a vast number of high-quality reference sequences have widened the horizons of omics and related technologies. At present, it has become effortless and affordable to generate omics data (Li and Yan, 2020) under different environmental conditions (Groen et al., 2020). The omics package gives us all, from accelerated QTL cloning to increased mapping resolution, rapid gene, and pathway identification (Li and Yan, 2020).

Talking about the advancements, the insertion of Bt gene producing toxic insect protein from *Bacillus thuringiensis* against in corn and cotton (James, 2014), development of high-quality vegetable oil, with higher oleic (78%) and stearic acid (40%) in cottonseed (Liu et al., 2002), and insertion of wild potato genes in cultivated potato for silencing the genes producing more acrylamide in fried potatoes (Van Emon, 2016) are all results of the omics technologies. Thus, omics technologies are our insurance against adverse situations and provide us with food, nutritional, and energy security, and scientists are aiming at plant-associated microbiomes to seek these. These technologies are the cradle that will drive agricultural research to a new level providing insights into both microbe and plant species related to agriculture that we thought we knew everything about.

14.2 Genomics

14.2.1 Genome sequencing

New sequencing technologies such as SOLiD, Solexa, and 454 pyrosequencing developed by Applied Biosystems, Illumina, and Roche, respectively, have revolutionized the field of microbiology by reducing the cost and increasing throughput of DNA sequencing. These new tools came up with new computational and technical challenges, in addition to new research opportunities. The utilization of these genomic sequencing techniques aided in the study of de novo genome assembly, polymorphism detection, sRNA discovery, metagenomics, epigenetics, and expression profiling. The large data sets generated from these technologies complement the freely available software for sequence alignment and assembly and various other downstream processing pipelines for deciphering biological conclusions (Sandhu et al., 2020). This section outlines various genome sequencing tools available and their use in microbial genome sequence with associated bioinformatics tools and challenges.

14.2.2 Sanger sequencing

In 1977, Sanger sequencing provided a breakthrough with the genome sequence of a bacteriophage using the dideoxy method on chain termination. This method required a mixture of four dNTPs, DNA polymerase, single-stranded DNA, and a dideoxynucleoside triphosphate. DNA polymerase performs the synthesis using template strand and dNTPs with chain termination by ddNTP (Samantra et al., 2021). The optimization of DNA fragment and sequence size is obtained by trying different ddNTP/dNTP ratios. The reaction products are separated over the gel with ddNTP information used for chain termination, and the complete DNA sequence is determined. Sanger sequencing provided a remarkable opportunity in life sciences and helped decipher various important genome sequences and shaped the understanding of various microbe's action and mechanism. *Pseudomonas protegens* was the first microbial biocontrol agent whose genome was sequenced in 2005 using the Sanger sequencing approach (Paulsen et al., 2005). In 2007, using the same approach, the *Bacillus amyloliquefaciens* genome was sequenced (Sandhu et al., 2022).

14.2.3 Next-generation sequencing

With the advent of next-generation sequencing (NGS) technologies in the mid-2000s, research in the biological sciences has reached a new hype due to extensive sequencing data production. NGS has several advantages over the traditional Sanger sequencing, namely array-based sequencing aid in multiplexing of DNA fragments, in-vitro clonal library construction, and amplification of DNA fragments. Several NGS tools can be distinguished from each other depending upon how DNA is immobilized over the solid substrate. The complete details of those are provided below.

14.2.3.1 High-throughput pyrosequencing on beads

Roche (454 pyrosequencing) platform was the first NGS platform available in 2005. For sequencing, DNA molecule is first sheared either with sonification or with enzyme-based digestion to subsequently ligate with the oligonucleotide adapters. The ligated molecule is later attached to the bead, and PCR product is amplified in the water-oil emulsion and pyro-sequenced. The addition of each unlabeled nucleotide in each pyrosequencing cycle results in the release of inorganic pyrophosphate (PPi), which is then detected. The iterations of these pyrogenic cycles generate the DNA molecule with a length of 350–500 bp (Margulies et al., 2006). Roche technology is prone to high error rates when reading the homopolymeric sequences, where sometimes n nucleotides are read as n-1. A soil inhabitant bacterium, *Myxococcus xanthus,* was the first one to be sequenced with pyrosequencing technique in 2006 (Vos and Velicer, 2006; Wu et al., 2011a,b), used the 454-pyrosequencing technique for complete genome sequence of *Pseudomonas aeruginosa,* and identified the core genome sequences which were conserved among different strains.

14.2.3.2 Sequencing by ligation on beads

SOLID technology involves converting an epifluorescence microscope to a rapid nonelectrophoretic DNA sequencer (Shendure et al., 2005). DNA fragment with attached adapters is attached over the beads and amplified in the water-oil emulsion. PCR amplicons are hybridized with PCR primer by immobilization over the solid planar substrate. Each cycle of sequencing proceeds with ligation of fluorescently labeled DNA octamer, which helps identify the nucleotide, and iteration of this step reveals the complete DNA sequence. The cost-per-base sequencing was one-tenth compared to conventional Sanger sequencing (Shendure et al., 2005).

14.2.3.3 Sequencing by synthesis on a glass solid phase surface

The Illumina genome analyzer involves fragmentation of template DNA with ligation of oligonucleotide adaptor. The amplification process is known as bridged PCR. The forward and reverse primers are attached to the glass surface with a flexible linker. Hybridization occurs between primers and the flanked DNA fragments over the glass surface (Turcatti et al., 2008). Amplification with bridged PCR results in a cluster of clonal amplicons. Each cluster over the array contains the amplification product from a single DNA fragment. Sequencing proceeds in the cycles with a modified polymerase and fluorescently labeled nucleotides. Each cycle enlarges the fragment by one base pair, followed by the fluorescent label's cleavage for reporting the associated nucleotide. This Illumina technology results in an error rate of 1%–1.5% due to signal decay and incomplete fluorescent reporter cleavage (Turcatti et al., 2008).

14.2.3.4 Third-generation sequencing

NGS mainly suffers bias introduced in reading distribution during the PCR amplification step, affecting the total coverage. However, third-generation sequencing avoids this bias with direct DNA sequence, thus avoiding bias during the amplification step and synchronization defects. Single-molecule real-time (SMRT) technology (Pacific Bioscience) was the first long-read sequencing technique (Rhoads and Au, 2015). The adapter molecule over both ends of the DNA sequence is ligated to construct the circular DNA molecule library, loaded into SMRT cell with wells. DNA polymerase is immobilized at the base of each well that binds to the adaptor on the circular DNA to initiate the replication. Fluorescently labeled nucleotides are added to each well. With each base's addition, a light pulse is generated to identify the base and iteration of this produce the complete DNA sequence. Read length up to 10 kbp can be obtained from the SMRT sequencing technology; however, it comes up with a high error rate (10%–15%) (Rhoads and Au, 2015; Aggarwal et al., 2019). Pac Bio sequencing is used for complete genome sequence and plasmids of the *Pantoea agglomerans* (Lim et al., 2014).

HeliScope is another example of the third-generation sequencing platform for single DNA molecule sequencing. Here a highly sensitive fluorescence detection system is used for single-DNA-molecule sequencing. DNA library is made by random shearing of the DNA fragment followed by poly-A tailing, which is then hybridized to surface tethered poly T oligomers. This results in creating an array of primer annealed single DNA molecules that are then extended with a single nucleotide. The fluorophore is attached to each nucleotide. The recorded image is used to analyze the nucleotide being incorporated into the growing strand. This cycle is repeated iteratively to obtain the complete DNA sequence (Braslavsky et al., 2003).

MinION technology uses electrophoresis to move the DNA molecule through a nanopore using a constant electric field and an electrolytic solution. The change in current pattern and magnitude is measured as nucleic acid passes through the nanopore (Lu et al., 2016). DNA molecule is shredded with Covaris g-TUBE and repaired with PreCR step for library preparation. Blunt end DNA molecule is created before addition of poly-A tail at the 3′-OH end. Later, two adapters, namely a hairpin and Y adapter, are added to the DNA. A motor protein is used to separate the double-stranded DNA at the Y adapter, and the single-stranded strand is fed to the nanopore. Base identification and read length up to a few hundred thousand bases occur in the nanopore with an accuracy ranging from 65% to 88%. In case a single DNA strand is used as a template, the system is called 1-dimensional; otherwise, it is a 2-D. This system is attracting pathogen diagnostic and surveillance genomics due to its small size, cost, and real-time performance (Lu et al., 2016). The complete detail of various sequencing platforms is provided in Table 14.1.

14.2.4 Diversity analysis

Historically, cell structure and metabolism were used to differentiate between different microorganisms or their respective strains. Diversity within a species could also be addressed with the help of opportune markers, like protein and capsular serotypes. In early 1970s, DNA-DNA based hybridization was introduced to differentiate among the species, with more than 70% homology led to the conclusion that both are the same species. With the advancement in sequencing technologies, several marker systems were introduced for microbes, among which 16S ribosomal RNA (rRNA) marker was the first one to be utilized. 16S rRNA is the molecule which is present in most of archaeal and bacterial genomes. Sequencing of the 16S rRNA molecules has been long used to differentiate different strains and separation of different clusters (Acinas et al., 2004). 16S rRNAs are being even used today with some exceptions where they do not work, like *Bacillus anthracis*, *Bacillus thuringiensis*, and *Bacillus cereus*, as they differ in only a few bases. Furthermore, polymorphism located within the primer design region is poorly located by the universal primer, and a useful information is lost (Huber et al., 2002; Aggarwal et al., 2017).

Moreover, 16S rRNA is least useful for differentiating the subpopulations with a family. To solve these issues, multilocus enzyme electrophoresis (MLEE) is used which targeted isoforms of fifteen different metabolic enzymes for classification. MLEE did not get much attraction owing to its low throughput and time-consuming laboratory experiments, and was quickly replaced by the multilocus sequence typing (MLST) (Gurjar et al., 2021a,b). MLST came up with the genomic era and uses the sequence of house-keeping genes for differentiating the strains and organisms. This is high-throughput technique and allows easy comparisons of organism, and several public repositories have been created for almost 60 microbial species (Medini et al., 2008; Aggarwal et al., 2018a,b). One major limitation of MLST is for species where house-keeping genes are so uniform or have little sequence variation, thus causing problem to provide sufficient species differentiation. Single-nucleotide polymorphism (SNP) is another commonly used approach for differentiating strains (Gurjar et al., 2021a). Till now, most of the studies use the SNP and MLST in

Table 14.1 Characteristics of most commonly used sequencing platforms for genome sequencing in microbes.

Sequencing platform	Maximum sequence length (bp)	Maximum output per run	Strengths	Limitations
454 pyrosequencing	600	700 Mb	Size of read length is long	Homopolymer errors and high cost
NextSeq 500	2×300	120 Gb	High throughput with low cost	Guanine cytokinin bias
HiSeq	2×130	1000 Gb	High accuracy and throughput	Short reads and guanine cytokinin bias
SOLiD	2×60	320 Gb	Better accuracy	Short reads
PacBio	11,000	400 Mb	Lack of amplification bias	Significant error rate
MinION	6000	500 Mb	Size of read length is long	High error rate
HeliScope	45	35 Gb	Limited amplification bias	Short reads and high error rate

combination for studying various variations within the microbial's genomes (Medini et al., 2008; Aggarwal et al., 2018b).

14.2.5 Application of genomics in agricultural microbiological research

Several genomics studies have been conducted to reveal the genome sequence, diversity, and gene annotations for different biocontrol agents, arbuscular mycorrhizal fungi, nitrogen-fixing, and plant-growth-promoting bacteria (Sandhu et al., 2021; Singh et al., 2019). Bashyal et al. (2021) studied the different expressions of *Trichoderma harzianum* treated drought-stressed rice plant using next-generation sequencing techniques and provided its potential for rice growth under drought stress. Pangenome sequence of *Pseudomonas fluorescens* reveals 76% of the protein-coding genes are conserved, and aids in denitrification, protein secretion, motility, chemotaxis, and diterpenoids catabolism in *Beta vulgaris* L (Redondo-Nieto et al., 2013). The complete details of various other studies using genomics for studying microorganisms are provided in Table 14.2.

Table 14.2 Application of genomic tools for studying the microorganism in agricultural microbiology

Microorganism	Host	Nature of organism	Purpose of study	Reference
Trichoderma harzianum	*Oryza sativa* l	Enhances rice growth in drought-stressed soils	Differential gene expression was obtained in *Trichoderma harzianum* treated drought-stressed plants using next-generation sequencing techniques	Bashyal et al. (2021)
Chaetomium globosum	*Cicer arietinum* L.	Biocontrol agent	*Chaetomium globosum* was observed effect antagonist against the Ascochyta blight of chickpea	Rajakumar et al. (2005)
Lysobacter spp.	*Solanum lycopersicum* L.	Biocontrol agent	Comparative genomic analysis showed that lytic enzymes produced are effective against various soil-borne pathogens	Lee et al. (2013)
Lysobacter capsici	*Capsicum chinense* L.	Biocontrol agent	Genome sequencing of different strains was performed to identify the core genome region for its role in biocontrol agent	De Bruijn et al. (2015)

Continued

Table 14.2 Application of genomic tools for studying the microorganism in agricultural microbiology—cont'd

Microorganism	Host	Nature of organism	Purpose of study	Reference
Pseudomonas fluorescens	Verticillium dahliae L.	Biocontrol agent	Genome sequencing reveals the gene predicting factors such as volatile components, detoxifying compounds, siderophores, and secretion system	Martínez-García et al. (2015)
Meliniomyces bicolor	Ericaceae family	Arbuscular mycorrhizal fungi	Comparative genomic analysis reveals the upregulation of lipases, cell wall degrading enzymes, proteases, and lipases	Martino et al. (2018)
Pseudomonas fluorescens	Beta vulgaris L.	Plant-growth-promoting rhizobacterium	Pangenome sequences reveal that 76% of the protein-coding genes are conserved, and aid in denitrification, protein secretion, motility, chemotaxis, and diterpenoid catabolism	Redondo-Nieto et al. (2013)
Rhizobia spp.	Acacia acuminata L.	Nitrogen-fixing bacteria	Genomic diversity was identified, which was driven by the tree size	Dinnage et al. (2018)
Medicago truncatula	Legume family	Nitrogen-fixing bacteria	It was observed that NCR211 peptide plays an important role in the function and survival of different rhizobia spp. in the host cell	Kim et al. (2015)
Rhizophagus clarus		Arbuscular mycorrhizal fungi	Genome sequencing reveals several genes which were absent from arbuscular mycorrhizal fungi and imply that these fungi have a high dependency on host plants than other biotrophic fungi	Kobayashi et al. (2018)

14.3 Transcriptomics

Transcriptomics term is coined by Charles Auffray for study of transcriptome, which is the entire set of transcripts expressed in a cell or tissue or organ or organism at a time point under specific conditions. Transcriptome of an organism varies with environmental conditions, life stages, and tissue in which it is observed. All the somatic cells in an organism except gametic cells have the same DNA constitution in their nucleus; even then, these possess different structure and function. It is governed by the set of genes that are transcribed as mRNA and further translated as proteins, which are functional units. Therefore, studying the transcriptome of an organism under different pathological or physiological conditions is important to understand the mechanism that operates in an organism under stress conditions.

14.3.1 Different transcriptomic technologies

The first effort in the field of transcriptomics was made in the late 1970s by preparing cDNA libraries of silk moths with help of reverse transcriptase (RT) enzyme (Sim et al., 1979). The Sanger sequencing, which was the 1st and still today used method of DNA sequencing, came into existence in 1977. In the 1980s, Sanger sequencing was implied for sequencing RNA transcripts popularly known as expressed sequence tags (ESTs) (Lowe et al., 2017).

In the 1990s, DDRT-PCR technique was the most commonly used method for gene expression studies (Liang and Pardee, 1992). In this technique, first total RNA is isolated from the sample tissue, then converted to cDNA using RT enzyme. The amplification is performed by using 3′ anchored primer and some arbitrary primers. The PCR product is separated by polyacrylamide gel electrophoresis (PAGE) for comparison. The challenges with this technique are that it is more prone to error due to deposition of different mRNA as a single band, amplifies only a fragment, not complete mRNA, thus does not cover the entire transcriptome (Liang and Pardee, 1992).

Later in the late 1990s, cDNA-AFLP, another PCR-based technique, was used for the detection of differentially expressed transcripts (Bachem et al., 1998). The cDNA is synthesized from total mRNA subjected to restriction digestion by using two restriction enzymes. The restricted digested product is subjected to ligation by known adaptors to the stick ends. The primers complementary to adopters are used for PCR amplification. Later, PCR product is resolved on the PAGE, cloned, and sequenced to identify differentially expressed genes (Bachem et al., 1998). As AFLP markers exhibited dominant traits only, there are chances of losing codominant traits.

In 1995, the first semi-high throughput method of sequencing called serial analysis of gene expression (SAGE) was developed at Johns Hopkins University (Velculescu et al., 1995). The short sequence tag (10–14 bp) is developed from each transcript by ligation and location. These sequence tags are further linked together to form long serial molecules known as concatemers, which are cloned and sequenced. The sequence tags are quantified for their recurrence by using computer programs that determine the expression level of the gene corresponding to a particular target (Velculescu et al., 1995).

Massively parallel signature sequencing (MPSS) uses a small signature sequence (17-20 bp) adjacent to the 3′-end of mRNA to identify mRNA. One type of DNA sequence is cloned on a microbead. The microbeads are arranged in a flow cell to sequence and quantify (Reinartz et al., 2002). The MPSS has higher specificity than SAGE due to a longer sequence tag (i.e., 17–20 bp) than the SAGE tag (i.e., 9–10 bp). Another advantage over SAGE is the presence of a larger library (~20 times) than SAGE.

Later, microarray well-refined technology came into existence which used a series of oligonucleotides arranged on a glass or silicon chip. Microarray quantifies a particular transcript by their hybridization to its complementary probe immobilized on chip. It is a high-throughput technology that assays 1000s of transcripts simultaneously and highly reduces the cost per transcript (Itzkovitz and van Oudenaarden, 2011). The drawback of microarray is that it can detect only known sequences, high background hybridization, and probe saturation.

Recently, RNA sequencing emerged as the most efficient high-throughput technology that does not require any prior information about the sequence. The RNA-seq is more widely adopted due to advances in the high-throughput sequencing technologies such as Illumina, PacBIO, and oxford nanopore (Singh et al., 2021). In this technique, total RNA is isolated and screened for mRNA using probes specific to the poly-A tail. The cDNA is synthesized from total mRNA and subjected to sequencing platforms (Knierim et al., 2011). When RNA sequencing is done for noncoding RNA, the library preparation protocol is modified. The siRNA or miRNA is separated out on the basis of their size using exclusion gel or size selection magnetic beads, and linkers are added for purification and then subjected to cDNA synthesis. Currently, nanopore technology is used to sequence RNA without synthesizing cDNA. It prevents the biases that occur during cDNA synthesis and also detects the modified bases. The RNA-seq data analysis is performed by using various bioinformatics tools (Garalde et al., 2018). The read quality is checked by using FastQC and FaQCs. The good quality reads are assembled by aligning to the reference genome or prepared with the help of tools such as Trinity, SPAdes, and RSEM (Hölzer and Marz, 2019). The quantification of DEDs is done by using tools such as cuffdiff2, EdgeR, and DEseq2 (Rajkumar et al., 2015). The advantage of RNA-seq over other methods is that it can be used for organisms whose genomic information is not available. It has high accuracy and high reproducibility in data (Sandhu et al., 2021a; Singh et al., 2020).

14.3.2 Application of transcriptomics in agricultural microbiological research

Several transcriptomic studies are conducted to understand the manipulations that a rhizobacterium does when it interacts with the plant. *Bacillus subtilis* FB17, a root-colonizing bacterium, upregulates 167 genes in Arabidopsis, and those were found related to metabolism, stress response, and plant defense. The mutant for upregulated genes showed decreased colonization, whereas mutants of downregulated genes enhanced the colonization of FB17 (Lakshmanan et al., 2013). The arbuscular mycorrhizal fungi play an essential role in enhancing plant nutrition and productivity. The transcriptome analysis of sunflower roots colonized by *Rhizoglomus irregulare* states the DEGs involved in membrane transport, mutualistic relationship, and also some upregulated putative genes that could be characterized in near future (Vangelisti et al., 2018). The nitrogen-fixing bacteria *Sinorhizobium meliloti* form root nodules in *Medicago truncatula* to fix nitrogen. The RNA-seq of *S. meliloti* and *M. truncatula* stated differential expression of 482 genes related to homeostasis, cell division, energy metabolism, stress response, and nitrogen fixation (Lang and Long, 2015). The transcriptome analysis of PGPR, *Bacillus amyloliquefaciens* SQR9, colonizing the maize roots reveals that root exudates stimulated the expression of metabolic-related proteins and later genes involved in biofilm formation for better establishment (Zhang et al., 2015).

The dual transcriptome of *Trichoderma harzianum* and tomato plants proposed involvement of salicylic acid, not jasmonate in orchestrating the defense response and provided evidence of the participation of cytosine methylation and alternative splicing in gene regulation during Trichoderma-tomato interaction (De Palma et al., 2019). The transcriptome analysis of biocontrol agent *Chaetomium globosum* Cg2 while interacting with *Bipolaris sorokiniana*, the causal agent of spot blotch of wheat, reveals the upregulation of genes related to catalytic activity, metabolic activity, and hydrolysis (Darshan et al., 2020). The comparative transcriptome analysis states that *Bacillus velezensis* F21 DEGs trigger induced systemic resistance (ISR) and upregulate various transcription factors and resistance genes in watermelon against fusarium wilt (Jiang et al., 2019). The take-all disease of wheat caused by *Gaeumannomyces graminis* var. *tritici* (Ggt) was controlled by pretreatment with *Bacillus velezensis*. The 4134 DEGs were expressed in Ggt-infected wheat, and 2011 DEGs were detected in Ggt-infected plants pretreated with *B. velezensis*. The Bv reduces the pathogenicity of Ggt by cell wall hydrolase, hyphobia formation, and inhibition of papain (Kang et al., 2019) (Table 14.3).

Table 14.3 Application of transcriptomics for studying the microorganism in agricultural microbiology.

Microorganism	Host	Nature of organism	Purpose of study	Reference
Sinorhizobium meliloti	Medicago truncatula	Nitrogen-fixing bacteria	Analysis of gene expression in nodules formed by wild-type bacteria on six plant mutants with defects in nitrogen fixation	Lang and Long (2015)
Rhizophagus irregularis Glomus aggregatum	Glycine max	Arbuscular mycorrhizal (AM) fungi	Transcriptomic analysis reveals the possible roles of sugar metabolism and export for positive mycorrhizal growth responses in soybean	Zhao et al. (2019a)
Meliniomyces bicolor, Meliniomyces variabilis, Oidiodendron maius, and Rhizoscyphus ericae	Ericaceae family	Arbuscular mycorrhizal (AM) fungi	Comparative genomics and transcriptomics depict ericoid mycorrhizal fungi as versatile saprotrophs and plant mutualists	Martino et al. (2018)
Pseudomonas putida	Zea mays	Plant-growth-promoting rhizobacteria	Transcriptomic profiling of maize seedlings reveals mechanism of drought resistance offered by Pseudomonas putida stain FBKV2	SkZ et al. (2018)
Trichoderma virens	Lycopersicum esculatum and Zea mays	Biocontrol agent	Host-specific gene expression may contribute to the ability of T. virens to colonize the roots of a wide range of plant species	Morán-Diez et al. (2015)
Azospirillum brasilense	Oryza sativa	Nitrogen-fixing bacteria	RNA-seq reveals the interactions between nonlegumes and beneficial bacteria that have long-term implications toward sustainable agriculture	Thomas et al. (2019)
Chaetomium globosum and Bipolaris sorokiniana	–	Biocontrol agent	First effort to unravel the biocontrol mechanism of C. globosum against B. sorokiniana	Darshan et al. (2020)
Rhizoglomus irregulare	Helianthus annuus	Arbuscular mycorrhizal (AM) fungi	Reveals important DEGs having roles in AM symbiosis	Vangelisti et al. (2018)
Paenibacillus polymyxa	Capsicum annuum L.	Biocontrol agent	P. polymyxa SC2 induces systemic resistance in pepper by stimulating expression of corresponding transcription regulators	Liu et al. (2021)
Bacillus amyloliquefaciens and Pseudomonas fluorescens	Musa acuminata	Plant-growth-promoting rhizobacteria	Evaluation of differences in banana gene expression profiles in response to the rhizobacteria	Gamez et al. (2019)
Bacillus velezensis and Gaeumannomyces graminis var. tritici	Triticum aestivum	Biocontrol agent	To explore potential biocontrol mechanisms involved in the interference of antagonistic bacteria with fungal pathogenicity	Kang et al. (2019)

14.4 Proteomics

Proteomics provides a global outlook of the protein complement of biological systems, facilitating the dissection of both protein expression and function. The technique uses two-dimensional electrophoresis and protein characterization for gene expression analysis under physiological and pathophysiological conditions. The broadly available whole-genome sequences for several microbial groups have augmented large-scale proteomic studies in microbiology. A detailed list of plant pathogenic fungal genome sequences accessible to the public is provided by González-Fernández et al. (2010). Nowadays, proteomics is an important research in the field of life science and is developed as a fundamental discipline in the era of postgenomics (Fern´andez et al., 2010). "Proteomics" term was given by Marc Wilkins, explaining the "PROTein complement of a genOME" (Wilkins et al., 1996). Through proteomics, we aim to understand the expression/regulation and function of the many individual protein types being created in a living organism, their inter- and intra-molecular interactions, and how they respond to changes in biotic and abiotic environments. This makes proteomics suitable to study metabolic changes in response to eclectic stress conditions. Stress conditions like biotic and abiotic stress make crop production systems less efficient and less productive. Proteomic studies can be an important tool in agricultural research to better understand the molecular talks in the plant biological systems during stress.

14.4.1 Types of proteomics

On the basis of protein response to stress, proteomics is classified into three main types: (i) expression proteomics, (ii) structural proteomics, and (iii) functional proteomics.

14.4.1.1 Expression proteomics

Expression proteomics studies are useful in comparing the quantitative and qualitative expression profiles of proteomes (total proteins) under contrasting conditions. These studies help better understand the protein expression patterns in abnormal cells or cells under stress (biotic/abiotic), for example, comparing tissue samples from a diseased plant and a healthy plant to analyze differential protein expression. Protein expression/regulation data can be recorded, and protein activities in protein complexes and signaling pathways can also be analyzed using 2-D gel electrophoresis or mass spectrometry. Analysis of this information will give valuable insights into the molecular biology of plant disease development and develop disease-specific diagnostic markers or therapeutic targets.

14.4.1.2 Structural proteomics

Structural proteomics provides insight about the three-dimensional shape and structural complexities of proteins. Structural proteomics can also be helpful in organellar studies, as it provides details about the structure and function of the proteins present in any particular cell organelle, for example, it can be used to identify proteins present in a complex system like mitochondria and score all the different protein interactions possible between the scored proteins and protein complexes. And for determining these complex three-dimensional protein structures and shapes, technologies like X-ray crystallography and NMR spectroscopy are used.

14.4.1.3 Functional proteomics

Functional proteomics studies address two main targets: (a) explaining the biological function of the unknown protein by associating them to a partner protein with known biological function, and (b) unrevealing molecular mechanisms within the cell from the interpretation of in-vivo protein-protein interactions (PPIs).

14.4.2 Techniques of proteomics

Proteomics studies are multidisciplinary and involve both analytical and bioinformatics tools to define protein structure and functions. Proteomic analysis includes three main steps: (1) protein extraction and separation, (2) protein identification, and (3) verification of identified protein. These steps are then followed by bioinformatics analysis that includes: (1) searching databases, (2) protein-protein interactions (PPIs) analysis, and (3) statistical analysis. A flowchart of steps in proteomics study is depicted in Fig. 14.1. The initial three steps involve the use of analytical techniques, and the final three steps use bioinformatics tools. Here we intend to explain the techniques first, to be followed by an overall view of the methodology. Various proteomics papers have been published in plant-pathogenic fungi interactions and fungi for biotechnological or agricultural applications; a detailed list is provided by Fern´andez et al. (2010).

14.4.2.1 Analytical techniques involved in proteomics

Analytical techniques are used for the identification and verification of identified proteins. Broadly, there are two types of analytical techniques in proteomics: (i) 2DE-gel-based proteomics or classical proteomics, and the more advanced (ii) gel free-MS-based proteomics.

Classical 2DE-based proteomics

The invention of 2-dimensional electrophoresis in 1975 by O'Farrel and Klose led to development of 2DE-based proteomics. Here, the protein separation is based on two principles, first based on molecular

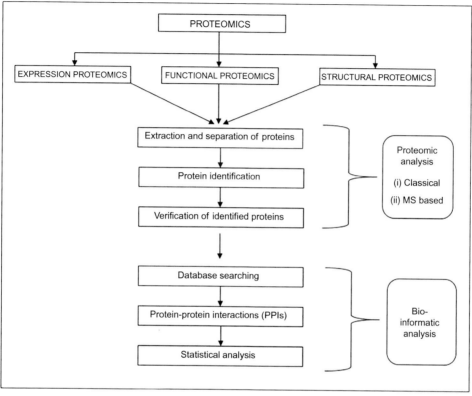

Fig. 14.1 An overview of proteomic studies.

charge (p*I*) in first dimension (isoelectric focusing, IEF) and in the second dimension (SDS-PAGE) by mass/molecular weight (*M*r) (O'Farrell, 1975). Based on these two orthogonal physicochemical properties (p*I* and *M*r) of the proteins, hundreds of proteins are investigated simultaneously on a 2D gel. This technique can compare the quantity of proteins and show their isoforms on the same gel. After separation, proteins are stained for visualization, cut out as pieces of gel, and then followed by an in-gel digestion with trypsin. Now, for identification of proteins by mass spectrometry (MS), the digested tryptic peptides are excised from the slice of gel and subsequently sequenced (Fig. 14.2). As the technique combines 2-DE with MS, it is called 2DE-MS. Use of 2DE is limited to qualitative experiments as it has its limitations because of low reliability as the results are not reproducible, ineffectiveness in detecting low abundant, hydrophobic proteins, low responsiveness in identifying proteins with low molecular masses (*M*r < 10 kD) or larger (*M*r > 150 kD), and very low pH values (pH < 3) or very high (pH > 10). So, 2DE-MS is incapable of mapping or identifying a vast majority of proteins within an organism. The method is also very labor-intensive because 2DE gels are not adapted

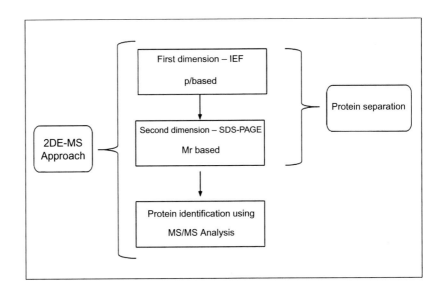

Fig. 14.2 Classical proteomics methods—2DE-MS.

to automation. A modification of 2DE is 2D-DIGE (Tonge et al., 2001) which uses fluorescent dyes in order to label proteins before running them on the gel. Because of proteins being labeled, the results from 2D-DIGE are reliable. The fluorescence signals from different fluorescent dyes are also specific, which shows the presence of specific type of proteins (Ye et al., 2007).

Gel free-MS-based proteomics

MS-centered proteomics is also known as bottom-up or shotgun proteomics, this technique targets proteolytic peptides for analysis instead of proteins themselves as was the case in 2DE (Duncan et al., 2010). MS-centered proteomics provides comprehensiveness, sensitivity, and versatility thus has been widely adopted. MS can be used in protein quantitative studies, but it is not innately quantitative. Because of differences in physicochemical properties for different peptide species, there are variations in ionization efficiencies and variation in signal intensities, and this can in turn alter the mass spectrometric response to a huge extent. So, labeling is used to distinguish between peptide species and for the determination of any quantitative changes in samples. There are two types of labeling techniques: (1) in-vivo labeling techniques which require the metabolic incorporation of an isotopic label like SILAC (Ong et al., 2002), N15 labeling (Kigawa et al., 1995), and stable isotope probing (SIP) (Jehmlich et al., 2008), and (2) in-vitro/chemical labeling techniques where proteins are characterized based on their chemical or enzymatic properties regardless of their origin, like ICAT (Gygi et al., 1999; Li et al., 2003; Schmidt et al., 2004) and its derivatives regarded as the first generation chemical tagging methods and the commercial isobaric tags including both iTRAQ (Wiese et al., 2007) and isobaric mass tags (TMT) (Ross et al., 2004; Thompson et al., 2003). Although some label-free approaches have also been developed recently, they are not as accurate as isotope labeling strategies. Label-free approaches involve two major concepts: (i) quantitation based on spectral counting (Collier et al., 2010; Choi et al., 2008); and (ii) quantitation on the basis of AUC determination at MS1 level, where the total signal of the monoisotopic peak is counted (Silva et al., 2005, 2006; Otto et al., 2014). These techniques make MS-centered proteomics suitable for quantitative studies. Along with labeling techniques, technologies like multidimensional liquid chromatography have been combined with MS/MS analysis (MuDPIT) for the simplification of the proteome (Washburn et al., 2001; Wolters et al., 2001). Although challenging, MuDPIT is a powerful technique as it extends the fraction of proteome that can be analyzed. MuDPIT has been extensively used in structural proteomics of subcellular organelles.

14.4.2.2 Bioinformatic tools involved in proteomics

Bioinformatics tools simplify the complex multistep process of processing and analysis of data from proteomics studies. Bioinformatics provides essential analytical tools which help dissect data from proteomics studies. For protein identification and validation, data from analytical studies of proteomics are subjected to search at several databases, and available search engines include Sequest, Mascot, and X!tandem. Identifying proteins from database searches is a very time-consuming and a computationally intensive task. To obviate this, good-quality data should be used which provides effective searches because of the tighter constraints and thus takes less time. As many proteins are common between multiple biochemical processes, so in shotgun proteomics strategies, it becomes really difficult to associate identified peptides with their precursor proteins. To resolve this, protein prophet database tool is used for peptide validation. By combining the probabilities associated with peptides identified by MS/MS, this tool computes precise probabilities for the proteins present. The protein inference drawn from identification and validation is then annotated to be deposited in an organized repository. These data repositories act as a storage, and help in the retrieval, and exchange of data and results (Chandramouli and Qian, 2009).

14.4.3 Steps in any proteomic analysis

Involving two kinds of aforementioned analysis, proteomics studies include six brief steps in total. A detailed summary of which is provided in this article (Alharbi, 2020).

14.4.3.1 Preparation of the samples and extraction of proteins

Firstly, different types of samples are scored. These samples can be the diseased tissues to conduct the expression proteomics or a microbial sample from the agricultural system for identification of changes in the proteome as a response to change biological conditions. Pretreatment and preparation depend on the nature of the sample, and specific lysis buffers and varying digestive processes are needed according to the type of sample. A general procedure follows the action with lysis buffers for protein isolation, which is then followed in series by a fractionation process for reducing the size of proteins, abstraction of the protein that have high expression, a depletion step to separate proteins with high abundance, and then regulate the protein concentrations enrichment and dialysis. Further, the proteins are extracted and separated from the sample using conventional 1-DE and 2-DE techniques. 2D-DIGE can be used for efficient separation of different proteins because of the fluorescent labeling (Ye et al., 2007). After running on the gel, proteins can be extracted by cutting the bands (gel plug) and putting them in a tube for digestion to cut the proteins as well as for reduction and alkylation to prevent proteins folding due to the broken bond of proteins.

14.4.3.2 Protein identification

After all the extractions and separation of proteins, MS analysis is performed for protein identification. MS is the most commonly used technique for protein identification because of its accuracy and sensitivity. There are many MS types, such as liquid chromatography-tandem mass spectrometry (LC-MS/MS), surface-enhanced laser desorption/ionization time-of-flight mass spectrometry (SELDI-TOF/MS), two-dimensional gel electrophoresis–mass spectrometry (2-DE/MS), and matrix-assisted laser desorption/ionization time-of-flight mass spectrometry (MALDI-TOF/MS), which are some of the common techniques used in proteomic studies.

LC-MS/MS (Sechi, 2016; Chen and Pramanik, 2009) is also called high-throughput technique as it can identify many proteins at the same time. And because of its high selectivity as well as high sensitivity, LC-MS/MS can detect very low amounts of biological substances. SELDI-TOF/MS (Tang et al., 2004; Li et al., 2002) involves the use of a chromatographic surface which is bound to a specific part of the desired protein, and the rest will be separated. In this technique, various chips are used to segregate proteins from the sample. The type of chip depends on absorption, electrostatic interaction, or biochemical affinity between the specific part of the desired protein and chromatographic surface. 2-DE/MS (Alharbi, 2020) combines the conventional 2-DE technique with MS although it has the same limitations as the

2-DE technique, and it provides an advantage of separating a myriad of proteins in one time on the same gel. MALDI-TOF/MS (Neville et al., 2011) is the most powerful technique in proteomics studies for protein identification. It is an ideal proteomics-based MS technique for the identification of proteins with a high molecular weight.

14.4.3.3 Verification of proteins

Following the identification of proteins, ELISA (He, 2013) and Western blot (Ghaemmaghami et al., 2003; Taylor and Posch, 2014) are the techniques used for verification of identified proteins and PPIs analysis to detect proteins. SDS-PAGE is used to separate proteins in Western blot technique. Using SDS-PAGE gives an estimate of the molecular weight of separated proteins and their isoforms. ELISA includes the use of enzymes and can be used in quantifying the level of targeted proteins. Being more accurate and sensitive, ELISA is considered better than the Western blot. This is the final step of proteomics analysis; all the subsequent steps are conducted to augment the understanding of results from the proteomics analysis by using bioinformatics tools.

14.4.3.4 Database searching

Mascot, MS-Tag, Phenyx, and Sonar are some of the databases that can be used to search for proteins using their MS/MS data. They help in peptide identification via database searches and also evaluate the quality of data. Tools like "Protein Prophet" are used to judge the validity of the protein inference and link a probability to it. The peptide data from MS/MS analysis are assigned probabilities using this software. It also computes probabilities for the proteins present (Cottrell, 2011; Alharbi, 2020; Chandramouli and Qian, 2009). A detailed list of fungal genome and proteome databases with some other useful resources is provided by Fern´andez et al. (2010).

14.4.3.5 Protein-protein interactions (PPIs) analysis

PPI can be conducted using STRING software. Its purpose is to establish a physical connection between identified proteins and score their interactions. This analysis is very important for functional proteomics studies (Cottrell, 2011; Alharbi, 2020).

14.4.3.6 Statistical analysis

Statistical Analysis System (SAS) software (Delwiche and Slaughter, 2019; Lafler, 2001) or Statistical package for social sciences (SPSS) (Landau and Everitt, 2003) is useful for statistical analysis in proteomics. By finding novel insights in MS/MS data, they save a lot of time (Lualdi and Fasano, 2019).

14.4.4 Applications of proteomics in agricultural research

Proteomics is an excellent way to study changes in metabolism in response to different stress conditions. In context to agricultural research, this technique can be used both for the plants and for the microorganisms interacting with them. Various studies have been conducted to better under the mechanism and biology of different stress and response of different organisms. A similar study was conducted (Godfrey et al., 2009) to better understand the infection process by barley powdery mildew fungus (*Blumeria graminis* f.sp. *hordei*), an effort to elucidate the underlying molecular mechanisms of haustoria (nutrient absorbing outgrowth) through proteomics. So, in study of biotic stresses, the plants and microbes are interconnected. Just as the aforementioned study is particularly about the barley powdery mildew fungus but the pathogen causes a biotic stress (powdery mildew) in plants and in order to better understand the plant diseases development, it is important to know the infection mechanism of the pathogen. Thus, proteomics also helps better understand plant-microbe interactions.

For abiotic stress, factors like nutrient stress or effect of high temperature can be studied. In a similar study, Lidbury et al. (2016) conducted a proteome analysis of *Pseudomonas* strains to better understand their global proteomic response to phosphorus stress. The study provides an in-depth assessment about phosphorus scavenging capabilities of these plant-growth-promoting rhizobacteria (PGPR). Another proteomic study to better understand the response of microorganisms under abiotic stresses was conducted for *Bacillus thuringiensis* strain YBT-1520 (Wu et al., 2011a,b). The study explains strategies used by the bacteria for its survival under long-term heat stress conditions. These strategies can be replicated in other microorganisms to increase their heat stress tolerance. Proteomic studies of microbes can elucidate complex biochemical pathways. This ensures a better understanding of processes microbial organisms are involved with. Guan et al. (2014) conducted proteomic analysis of *Propionibacterium acidipropionici* to better understand its mechanism of acid tolerance. The insights acquired from the study can then be used to enhance production of propionic acid. Such studies can be helpful for the agricultural industry and postharvest processing of agricultural products. A better understanding of these complex biochemical pathways working during stress or characteristic microbial function ensures a more calculated approach toward their management and aims at the sustainable agriculture of tomorrow. Some other recently conducted proteomics studies are presented in Table 14.4.

Table 14.4 Application of proteomics for studying the microorganism in agricultural microbiology.

Microorganism	Host	Nature of microorganism	Purpose of study	Reference
Enterobacter cloacae SBP-8	Triticum aestivum	PGPR	Evaluated adaptation mechanisms of wheat seedlings when exposed to high concentration of NaCl in response to Enterobacter cloacae SBP-8 inoculation	Singh et al. (2017)
Pseudomonas spp.	–	PGPR	The study provides an in-depth assessment about phosphorus scavenging capabilities of these soil bacteria	Lidbury et al. (2016)
Propionibacterium acidipropionici	–	Fermentation bacteria	To better understand the mechanism of acid tolerance of P. acidipropionici and enhance the production of propionic acid through fermentation by propionibacteria	Guan et al. (2014)
Bacillus thuringiensis	–	Microbial pesticide	The study explains various survival strategies used by Bacillus thuringiensis under long-term heat stress	Wu et al. (2011a,b)
Thielaviopsis basicola	Gossypium hirsutum	Plant pathogenic	To understand the molecular basis of defense mechanisms in cotton roots during interaction with black root rot fungus	Coumans et al. (2009)
Blumeria graminis f.sp. hordei	Hordeum vulgare	Plant pathogenic	Understand the molecular mechanisms of haustoria and acquire more insight about infection process of plant pathogenic fungus	Godfrey et al. (2009)
Uromyces appendiculatus	Phaseolus vulgaris	Plant pathogenic	Conduct a quantitative expression analysis to compare the proteome of plant infected by a virulent and avirulent obligate rust fungus	Lee et al. (2009)
Plasmodiophora brassicae	Brassica napus	Plant pathogenic	Understanding the metabolic changes after pathogen infection, which may result in susceptibility of the host	Cao et al. (2008)
Sclerotinia sclerotiorum	Brassica napus	Plant pathogenic	The study gives novel insight into B. napus – S. sclerotiorum interaction. Role of many proteins involved in this interaction is explained	Liang et al. (2008)
Altemaria brassicicola	Arabidopsis thaliana	Plant pathogenic	Proteomic approach was used to analyze change in secretome in response to salicylic acid and identify critical components involved in resistance to A. brassicicola	Oh et al. (2005)

14.5 Metabolomics

Metabolomics, one of the latest member of "omics" techniques, is generally considered the eventual level of postgenomic analysis as it may reveal alterations in metabolite fluxes that are regulated by only small changes within gene expression quantified using transcriptomics and/or by examining the proteome that elucidates posttranslational regulation over enzymatic activities. Metabolites are low-molecular-weight chemical compounds (with less than 1000 Da molecular weight) that play a vital role in microbial metabolism (Baidoo, 2019). Together, all the metabolites of a particular microorganism are referred to as its metabolome (Baidoo, 2019). The word "metabolome" was introduced to the literature in 1998 (Kell and Oliver, 2016), and concerning the micro-organisms, it can be grouped into three different matrices: in the culture headspace, in the extracellular medium, and within the cell.

In general, there are three types of metabolites present in biological samples (or microorganisms): (i) water-insoluble or nonpolar, (ii) water-soluble or polar, and (iii) volatiles (Horak et al., 2019). Further, metabolites can also be classified based on their function as either primary metabolites or secondary metabolites and their origin as either exogenous metabolites or endogenous metabolites (Pinu and Villas-Boas, 2017). Exogenous metabolites originate from outside the cells, while endogenous metabolites arise from within a cell (Pinu and Villas-Boas, 2017). Primary or secondary metabolites are produced based on the growth phase of microbial cells. Primary metabolites are directly involved in normal development, growth, and reproduction. In contrast, secondary metabolites are not directly involved in those processes but generally have crucial ecological functions (such as response to stress, etc.) (Olivoto et al., 2017).

In general, metabolomic studies are conducted either in a targeted manner or in a nontargeted way. Targeted metabolomics deals with the detection and measurement of a particular set of already known metabolites belonging to certain biochemical pathways. It allows the precise establishment of relations and interactions between known metabolites, and helps in placing them in a specific biological context (Bingol, 2018). Data generated from a targeted metabolomics study cannot be further investigated for examining the other hypothesis regarding the compounds absent in the starting panel. On the other hand, nontargeted metabolomics is an unbiased approach that allows the simultaneous identification of different novel metabolites in a given biological sample (Bingol, 2018).

Overall, metabolomics is considered a technically demanding multidisciplinary research that requires proficiency in the fields of biology, organic chemistry, analytical chemistry, chemometrics, and also in informatics sciences. The metabolomic analysis comprises three different experimental parts: (i) sampling or preparation of sample and metabolite extraction, (ii) measurement or acquisition of data using any of the analytical approaches, and (iii) data processing (data analyses and data interpretation). Fundamentally, all these steps are strongly interdependent and interrelated (Fig. 14.3) (Dettmer et al., 2007).

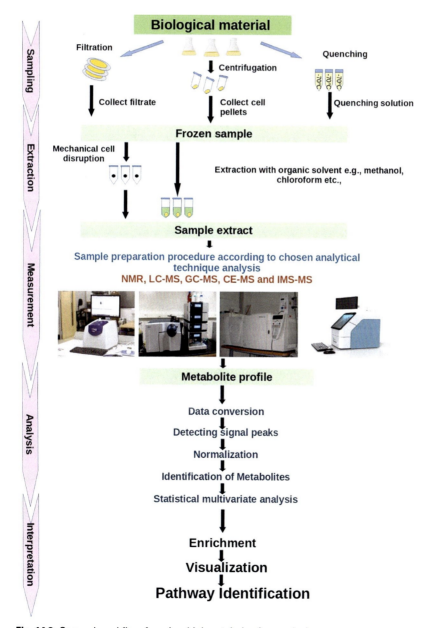

Fig. 14.3 General workflow for microbial metabolomics analysis.

14.5.1 Preparation of samples and extraction of metabolites

Sample preparation or sampling is of the uppermost significance to analytical quantification as it allows conservation of the metabolic solidarity of sampled biological material. Further, efficient metabolite quenching (ceasing all cellular activities) and extraction of metabolites are crucial to capture precise metabolite information (Baidoo, 2019). Different quenching agents such as pH, a chemical agent (e.g., organic solvent), or temperature (generally very low temperature) are used for metabolite quenching which allows a fast and sudden ceasing of the metabolism in a given sample (Pinu et al., 2017). In some cases, microbial culture is directly transferred to a quenching agent, and then either filtration or centrifugation is used to separate the microbial biomass from the medium. Organic solvents show an affinity for a wide range of cell membrane structures; hence, these are used extensively as extraction agents. Furthermore, information on the physical and chemical characteristics of the target analytes may assist in ameliorating extraction efficiency as approaches can be customized to concerned metabolites (Prasannan et al., 2018). Among others, methanol-based extraction strategies are quite popular for the extraction of intracellular metabolites.

14.5.2 Analytical techniques

Once the biological sample is prepared, available analytical approaches can be utilized to discriminate and characterize different chemical compounds present in the sample. Depending upon physical and chemical characteristics of the target metabolites, either of the following separation techniques integrated with compound detection techniques (such as nuclear magnetic resonance or mass spectrometry) is used for measurements of the metabolites: gas chromatography (GS), capillary electrophoresis (CE), liquid chromatography (LC), and ion mobility spectrometry (IMS) (Bingol, 2018; David and Rostkowski, 2020). Each separation and detection technique has a different resolution, sensitivity, and technological restrictions in identifying and characterizing the metabolites. In addition to the physical and chemical characteristics of target metabolites, the selection of an appropriate method also depends upon the type of analysis (targeted or nontargeted) to be performed (see Table 14.5 for more details).

An investigation of literature recently carried out by Miggiels et al. (2019) exhibits the increasing use of metabolomics in different organisms; a similar trend was also reported earlier (Kuehnbaum and Britz-McKibbin, 2013). A fresh search conducted on PubMed

Table 14.5 Comparison of some of the major analytical techniques used for microbial metabolomic profiling.

Analytical technique (approach type)	Pros	Cons
NMR (targeted)	(i) Quantitative (ii) Minimal sample preparation (iii) Nondestructive (iv) Rapid analysis	(i) >1 peak per metabolite or component (ii) Low sensitivity (iii) Identification is tedious and laborious because of complex matrix
GC-MS (targeted and nontargeted)	(i) Robust and sensitive (ii) Greater chromatographic (iii) resolution (iv) Availability of in-house and commercial MS libraries (v) Concurrent analysis of various groups of metabolites	(i) Do not allow thermo-labile compounds (ii) Derivatization is needed for nonvolatile metabolites
LC-MS (targeted and nontargeted)	(i) Derivatization is generally not needed (ii) High sensitivity (iii) Thermo-stable metabolites can be analyzed (iv) Capacity to handle large sample	(i) Desalting may be needed (ii) Poor-to-average chromatographic resolution (iii) Restricted commercial libraries (iv) Matrix effects
IMS-MS (nontargeted)	(i) Improved peak capacity (ii) Generation of multidimensional data which facilitate precise identification of metabolite(s)	(i) Limited linear range (ii) Fast electronics are needed (iii) Not appropriate for nonvolatile analytes (iv) Complex spectra and interferences
CE-MS (targeted and nontargeted)	(i) Separation of metabolite isomers (i) Small volume of sample is needed (ii) High resolution (iii) Generally, no derivatization is needed	(i) Poor sensitivity and reproducibility (ii) Restricted commercial libraries (iii) Difficulty in interfacing with mass spectrometry

(https://pubmed.ncbi.nlm.nih.gov/) in April 2021 shows that 6,863 papers were mentioning the "metabolomics" technique published in 2020 and already 2,151 published as of April 03, 2021. The trends represented in Fig. 14.4 show that NMR is recently a dominant technique followed by LC-MS and GC-MS, while IMS-MS is still the least-utilized technique.

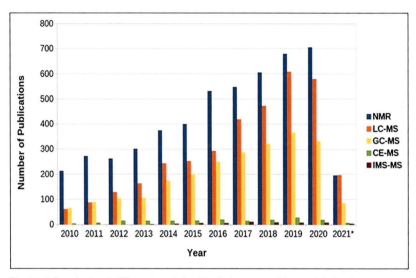

Fig. 14.4 Popularity of different analytical techniques used for metabolomics studies from 2010 to 2021. The popularity was based on the number of publications obtained from PubMed (https://pubmed.ncbi.nlm.nih.gov/). The utilized keywords were as follows: (1) "metabolomics" AND "NMR," (2) "metabolomics" AND "LC-MS," (3) "metabolomics" AND "GC-MS," (4) "metabolomics" AND "CE-MS," and (5) "metabolomics" AND "ion mobility mass spectrometry." Ion mobility mass spectrometry was used instead of IMS-MS, because the use of IMS-MS generated many unspecific publications (data retrieved on April 03, 2021).

14.5.2.1 Nuclear magnetic resonance (NMR)

NMR appeared as a used analytical method in more than 45% of the papers published in the last decade (https://pubmed.ncbi.nlm.nih.gov/). This technique is relatively cheap and more accepted for high-throughput metabolomics compared to mass spectrometry (MS). Moreover, this technique has comparatively higher reproducibility than MS. This technique can efficiently characterize the highly abundant, but hard to detect and analyze metabolic compounds and thereby complement other techniques such as LC and GC-MS. It has very little sensitivity compared to MS, which is considered its major limitation (Table 14.5). Some recently published papers have presented the future of the utilization of NMR in high-throughput metabolomics studies (Markley et al., 2017; David and Rostkowski, 2020; Letertre et al., 2020). Moreover, several different approaches intending to improve the sensitivity of this technique have also been published including techniques such as the use of higher field magnets, microcoil-NMR probes, cryogenically cooled probes, use of high-temperature superconducting oils, and hyperpolarization (Moser et al., 2017; Saggiomo and Velders, 2015; Jézéquel et al., 2015; Ramaswamy et al., 2013; Bucher et al., 2020).

14.5.2.2 Mass spectrometry

A single-generic method generally does not allow the thorough analysis of a complete set of metabolites present in a biological entity owing to the complexity of the metabolome. Therefore, it becomes imperative to use other complementary methods to achieve the goal. In most situations, MS, integrated with diverse high-throughput separation techniques, namely CE, LC, or GC, is a better alternative owing to its high sensitivity and selectivity (Table 14.5). Several different mass analyzers (a necessary part of the instruments required to differentiate mass peaks) have been placed in MS by various companies.

Resolving power and mass resolution are vital parameters representing to what amount the resolution of particular ions may be attained. Resolving power is the ability of an instrument to distinguish two adjacent ions having equal intensity, while mass resolution refers to the extent of separation between two mass spectral peaks of equal width and height observed in a particular spectrum. The working efficiency of different mass analyzers is generally determined by accuracy, mass resolving power, tandem analysis capabilities, and acquisition speed. Depending upon mass accuracy and resolving power, low-resolution and high-resolution mass spectrometry (LR-MS and HR-MS) may be distinguished (Dettmer et al., 2007; David and Rostkowski, 2020). A brief comparison of some of the major analytical techniques used for microbial metabolomic profiling is given in Table 14.5.

GC-MS is considered the most efficient platform for microbial metabolomics. It is mostly used to characterize low-molecular-weight volatile compounds (Table 14.5). Polar and nonvolatile compounds can also be detected with GC-MS after proper chemical derivatization. However, this requirement of chemical derivatization makes nontargeted metabolomics more challenging with GC-MS. In the case of complex samples, for achieving increased peak capacities and resolving power of separation, the utilization of two-dimensional (2D)-GC is recommended; it utilizes two columns that are connected serially and have different stationary-phase selectivities. The 2D-GC is integrated with fast detectors, for instance, electron impact time-of-flight MS or flame ionization detector (Allwood et al., 2009).

The utilization of LC-MS platforms, typically coupled with electrospray ionization (ESI), has recently increased in metabolomics (Fig. 14.4) (David and Rostkowski, 2020). LC-MS platform can be used for both targeted and nontargeted metabolomics. Besides, the utilization of ESI source in LC-MS also offers a soft ionization process that permits structural explanation (Miller et al., 2019). In contrast to 2D-GC-MS, the 2D-LC-MS has not been widely utilized for profiling of metabolites, owing to reduced sensitivity and complex experimental setup.

Ion mobility spectroscopy (IMS) separates ions based on their size, shape, and charge, which makes it quite applicable in the separation of isomeric structures (Eiceman et al., 2013). The implementation of IMS for utilization in nontargeted metabolomics studies has been recently reviewed elsewhere (David and Rostkowski, 2020). The researchers stated that IMS seems like an exceedingly attractive option for enhancing the dynamic range, coverage, and throughput of analytic quantifications in presently used analytical methods (David and Rostkowski, 2020). Various IMS-based platforms such as traveling wave IMS, drift tube IMS, trapped IMS, differential IMS, overtone IMS, transversal modulation IMS, and field asymmetric IMS are presently available (David and Rostkowski, 2020).

Similarly, CE-MS has also come up as a powerful analytical technique for the high-throughput profiling of charged and highly polar metabolites (Ramautar et al., 2019), but till now, it has not been much utilized for microbial metabolomics studies (Baidoo, 2019) (Table 14.5). It gives high-resolution separations with high sensitivity and detection selectivity, which permits the comprehensive characterization of many metabolites in a single analysis. Recent applications and developments have been widely discussed elsewhere (Ramautar et al., 2019; Stolz et al., 2019).

A general limitation with most of the above-discussed analytical techniques is that they require a significant amount of initial biological samples and do not support in-situ analysis. Therefore, especially when analyzing localized areas of microbes on any surface, for instance, localized areas of plant-growth-promoting bacteria colonization of roots, techniques that permit imaging of biological tissues through a high-throughput approach, and in-situ metabolic profiling have distinct advantages. Moreover, as the metabolome is continually changing, numerous biological queries may only be answered if the target samples are examined in real time, in vivo, and/or in situ. Ambient MS permits the surface sampling, direct soft ionization, and in-situ analysis in real time in the ambient condition, with no or little sample preparation. Some of the ambient MS techniques used in metabolomics are as follows: desorption electrospray ionization-MS (Garza et al., 2018), desorption sonic spray ionization (Haddad et al., 2006), liquid micro junction surface sampling probe (Van Berkel et al., 2009), direct analysis in real time (Cody et al., 2005), probe electrospray ionization (Hiraoka et al., 2007), laser ablation electrospray ionization (Stopka et al., 2017), rapid evaporative ionization-MS (Balog et al., 2015), and paper spray-MS (Liu et al., 2011).

14.5.3 Data processing and interpretation

Once obtained, raw signals (NMR data, spectra, or chromatographs) are preprocessed using ad hoc software tools for the identification of compounds. Normally, this preprocessing includes retention time correction, noise reduction, peak detection, etc. Various input data are devised to generate suitable data matrices for future analysis (Martínez-Arranz et al., 2015). Then, data normalization is carried out which helps in reducing technical variations or systematic biases and avoiding misidentification of metabolites. Followed by data normalization, significant differences between biological sample sets are identified utilizing suitable statistical methods. Generally, statistical analysis for any metabolomic data sets includes two phases: first, different multivariate and univariate methods are employed to generate an overview of the concerned data sets and to identify the compounds that exhibit significant alterations; then, data mining methods are employed to differentiate batches of functionally related compounds (Cambiaghi et al., 2017; Lamichhane et al., 2018).

Once identified, metabolites are further shortlisted based on corresponding P-values. In the final step, the shortlisted metabolites are associated with the biological context via metabolomic pathway enrichment analysis performed using ad hoc software tools. Following the identification of metabolic pathways, the generated information is generally merged with transcriptomic and proteomic data to get a detailed picture of all the biological processes involved (Cambiaghi et al., 2017). To represent complete interactions that the target data explain, different kinds of visualization tools (network-based) are generally utilized by researchers. Based on the type of interaction, a biological network may be constituted via using a different type of graph. A detailed description of different steps involved in the analysis of metabolomic data, associated tools, presently used strategies, etc., can be found elsewhere (Cambiaghi et al., 2017; Considine et al., 2018).

14.5.4 Applications of metabolomics in agricultural research

Although the identification of specific transcripts and proteins gives proof of the potential for a certain function or pathway to be active, it is the only metabolomic analysis that supplies mechanistic evidence that, indeed, specific metabolism is happening. Metabolomics has helped us in the elucidation of several complex metabolic pathways in microbes associated with plants. Studies using *Pennisetum giganteum and Glycine max,* for instance, identified and characterized specific metabolic pathways involved with nitrogen fixation capacities of the bacterial cells (Lin et al., 2020; Stopka et al., 2017). The ammonia-secreting performance is one of the major factors that determine the nitrogen-fixing capacity of the cell. Using GC-TOF-MS metabolomic technique, Lin et al. (2020) found that Gly-Pro, mannitol, fructose-6-phosphate, glycerol, diacylglycerol, and Glu are some of the important substances that regulate the secretion of ammonia from *K. variicola* GN02, an endophytic nitrogen-fixing bacterium generally found in the roots of *P. giganteum*.

In another study, LAESI-MS was used to investigate the symbiosis between soybean and its compatible symbiont bacteria. Results revealed several metabolic pathways including riboflavin, zeatin, and purine synthesis pathways within the nodule influencing the nitrogen-fixing capacity of the cells (Stopka et al., 2017). Similar efforts have also identified various metabolites or metabolic pathways involved in P-solubilization and the promotion of plant growth (Vinci et al., 2018; Xie et al., 2020). The effects of inoculation of *Bacillus amyloliquefaciens* on the growth, development, and metabolic pathways of maize plants were investigated using GC-MS in pot soils treated with different types of phosphorus fertilizers. Inoculated plants exhibited larger N and P content and significantly increased amount of fructose, glucose, GABA, and alanine metabolites in leaves which resulted in the augmented photosynthetic activity (Vinci et al., 2018). Most recently in 2020, a metabolomic study revealed the involvement of 23 types of organic acids mainly including D-(-)-quinic acid and gluconic acid in the regulation of phosphate-solubilizing ability of the bacteria, *Acinetobacter* sp. Ac-14. Moreover, few studies have also used metabolomics to investigate plant growth-promoting bacteria (PGPB)-plant interactions (Li et al., 2014; Agtuca et al., 2020). Some other recently conducted metabolomics studies with their significant findings are presented in Table 14.6.

Table 14.6 Some recently conducted microbial metabolomics studies.

Microorganism (nature of organism) [crop]	Purpose of the study	Major finding(s) of the study	References
Bacillus amyloliquefaciens Y1 (nematocidal activity) [*Solanum lycopersicum*]	To investigate the role and probable mechanism of *Bacillus amyloliquefaciens* Y1 against the root-knot nematode (*Meloidogyne incognita*)	One dipeptide cyclo (d-Pro-l-Leu) was identified in culture of *B. amyloliquefaciens* Y1 which was responsible for its nematocidal activity	Jamal et al. (2017)
Klebsiella variicola GN02 (nitrogen-fixing) [*Pennisetum giganteum*]	To reveal the major factors which affect the ammonia secretion in *K. variicola*	Gly-Pro, mannitol, and fructose-6-phosphate were found to have significant impact on ammonia-secreting performance of the bacteria which ultimately determines its nitrogen-fixing capacity	Lin et al. (2020)

Table 14.6 Some recently conducted microbial metabolomics studies—cont'd

Microorganism (nature of organism) [crop]	Purpose of the study	Major finding(s) of the study	References
Bacillus amyloliquefaciens (Phosphorus solubilizing) [*Zea mays*]	To explore the effects of the inoculation of *Bacillus amyloliquefaciens* on the metabolic processes and growth of maize plants treated with different P fertilizers	Inoculated plants exhibited larger P and N contents and differential metabolic pattern (significantly improved fructose, glucose, GABA, and alanine metabolites) thereby improved photosynthetic activity in maize plants	Vinci et al. (2018)
Arbuscular mycorrhizal fungi (AMF) (symbiotic association of microbes with plants) [*Triticum durum*]	To evaluate the effects of inoculation of AMF with plant-growth-promoting rhizobacteria on changes in metabolic pathways in durum wheat roots grown under P-rich and N-limited conditions	Inoculation with AMF reduced free amino acid content in durum wheat roots	Saia et al. (2015)
Bacillus subtilis Dcl1 (antifungal activity) [*Curcuma longa*]	To understand the mechanistic basis of plant-specific adaptations present in *Bacillus subtilis* Dcl1	Metabolites such as surfactin, iturin fengycin, griseofulvin and fusaricidin were detected in bacteria which showed antifungal activity and facilitated the adaptation of plants to various stresses such as drought and oxidative stress	Jayakumar et al. (2020)
Acinetobacter sp. Ac-14 (phosphorus solubilizing) [*Arabidopsis thaliana*]	To isolate novel phosphorus solubilizing bacteria (PSB) from the rhizosphere soil in the Karst rocky desertification regions in China and to understand the mechanisms of P-solubilization and promotion of plant growth	Analysis identified the *Acinetobacter* sp. Ac-14 as the most promising PSB which produced 23 types of organic acids, mainly including D-(-)-quinic acid and gluconic acid	Xie et al. (2020)
Herbaspirillum seropedicae SmR1 (plant growth–promoting bacteria-PGPB) [*Setaria viridis*]	To identify the metabolites associated with localized regions of *Setaria viridis* roots colonized by the bacteria	Metabolites involved in zeatin, purine, and riboflavin pathways were found significantly more abundant in inoculated plants	Agtuca et al. (2020)
Bradyrhizobium japonicum (nitrogen-fixing soil bacteria) [*Glycine max*]	To explore the well-characterized symbiosis between soybean and its compatible symbiont (*Bradyrhizobium japonicum*)	Several metabolic processes (e.g., zeatin, purine, and riboflavin synthesis) were revealed within the nodule	Stopka et al. (2017)
Pseudomonas sp. TLC 6-6.5-4 (PGPB) [*Zea mays*]	To investigate the influence of different ways of application of PGPB on PGPB – maize – soil interactions	Analysis revealed that PGPB inoculation upregulated hormone biosynthesis, photosynthesis, and tricarboxylic acid cycle metabolites in plants	Li et al. (2014)

14.6 Integrated omics

As explained above, many omics platforms, viz. genomics, transcriptomics, proteomics, ionomics, metabolomics/fluxomics, have emerged in the past few years (Whon et al., 2021). These techniques have provided biologists with powerful means to generate an enormous amount of data. However, a systems biology approach is required to integrate data from all these tools and reveal the cross-talk between different omics layers (Pinu et al., 2019), where, systems biology is a holistic method for generating understanding of biology at a systems level rather than a reductionist point of view which tries to understand the system by focusing on one aspect at a time (Chuang et al., 2010). We analyzed the trend of research publications in various omics approaches and integrated/multi-omics using the web of science database (Fig. 14.5). The search revealed a recent focus on the integrated omics vis-a-vis individual omics method.

14.6.1 Why integrate different omics data?

Transcriptomics and proteomics: There are a loose correlation between the mRNA expression and the corresponding proteins. This happens because of the posttranslational modifications involving (in)activation/degradation, posttranscriptional control, varying half-lives of proteins, mRNAs, etc. The collection of data from both these omics methods complements each other to achieve better coverage of metabolic changes and nullifying collection methods. Thus, it is the most commonly used multi-omics approach (Jamil et al., 2020). Concluding expression results with confidence become easier by using proteomic and transcriptomic together for cross-validation and utilization of data generated from both approaches which can give

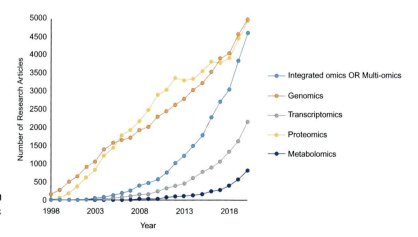

Fig. 14.5 Publication trend in different omics techniques from the year 1998 to 2020 (keywords as indicated).

new insights. Similarly, integration of transcriptomics and metabolomics may generate a link between the informational pathways and end products, i.e., metabolites. Increased interest in metabolite profiling has fuelled interest in this field (Sahoo et al., 2020). Genomics information can also predict the potential of microbes present in the soil or other niches but cannot reveal the actual activity carried out by the microbial community. In recent years, several examples have emerged where over two omics techniques have been used simultaneously (Zhang et al., 2020; Zolti et al., 2020; Vílchez et al., 2020).

14.6.2 Levels of multi-omics data integration

There is a classification framework where the integration approaches have been divided into integration postanalysis, integrated analysis, and system modeling (Pinu et al., 2019). The postanalysis integration examines each omics data separately and then integrates them. Integrated analysis combines data from different omics before construal. Here, we have followed the classification given by Jamil et al. (2020) for the data integration workflow approaches.

14.6.2.1 Element based

It is an unbiased approach and doesn't rely on prior information. We can use three different techniques, as explained below:

Correlation

It involves the use of the correlative relationship between different omics data. Pearson's correlation coefficient calculates the linear relation, and Spearman's correlation coefficient is performed to establish a ranked relationship (Zhang et al., 2010). However, some studies also use Fisher's method to transform skewed data into a normal distribution. However, in many studies, the relationship between levels of transcripts and their corresponding proteins was quite weak. These studies can be useful when the prior biochemical information is not available (Graw et al., 2020).

Clustering

Here, omics data are grouped as per the similarity among them, for example similarity in expression. The clustering methods involve the use of modern tools like machine learning like k-means and random forest (Jamil et al., 2020).

Multivariate

This approach allows the prediction of various aspects, viz. variance or covariance. The analysis is done using techniques like principal component analysis and partial least squares (Misra et al., 2018).

14.6.2.2 Pathway based

In this approach, the prior available information is used for integration.

Coexpression

It uses statistics for analyzing the strength of coexpression. A weighted network is then made from these data using tools like Weighted Gene Coexpression Network Analysis (WGCNA). The network data are then fed into the cytoscape for network visualization. Further, it is possible to integrate these networks with available pathway databases.

Pathway mapping

As the name suggests, it includes mapping the omics generated data to the available pathway databases. These pathway databases contain the pathway annotation information, thus linking genes and gene products to their metabolic role. However, the pathway annotation in related species needs to be performed carefully as important regions in BLAST results are sometimes not considered, thus decreasing reliance on such correlations. Thus, such analysis requires further confirmation.

14.6.2.3 Mathematical based

Models are generated in this approach with the help of mathematical analysis to test biological hypothesis.

Differential

It involves the use of modeling and differential equations for the integration of biological data. However, very high coverage and the well-characterized organism is a prerequisite for this analysis. Due to the relative simplicity of microbial metabolism and less compartmentalization, their model generation is relatively easier (Rai et al., 2017).

Genome-scale analysis

As opposed to the differential analysis which is targeted at a particular purpose, for example one pathway's metabolic flux or protein synthesis rate. Also, the differential method is based on the experimental measurements for model development. Genome-scale analysis, on the other hand, involves model development first followed by its validation.

14.6.3 Integrated omics in agricultural microbiology

Relatively few integrative biology studies exist where omics integration has been used to understand plant-microbe interaction or soil microbiology. Larsen et al. (2015) used an integrated approach for plant-mycorrhizal interaction. In another study, the microbial community of permafrost soils was studied to identify its various characteristics. For this, a combination of transcriptomics, metabolomics, and proteomics was used where the data were analyzed separately for each technique and later correlated with each other (Hultman et al., 2015). In an interesting comparison study, Karp et al. (2019) compared the widely used microbial genome web portals, namely BioCyc, KEGG, KBase, PATRIC, Ensembl Bacteria, and IMG. BioCyc was reported to have maximum capabilities to aid omics data analysis, including regulatory network analysis tools. The microbial interactions and metabolic pathways were also studied in an anammox reactor using an integrative approach (Wang et al., 2019). Zhang et al. (2020) used metagenomics and metaproteomics to study antibiotic resistance spread via cow manures and found the reduced spread via composting as compared to the mechanical drying and precipitation. In an interesting study, metabarcoding, metaproteomics, and metabolomics were used together to analyze the fermentation in pu-erh Chinese tea and the change in its quality and metabolic composition (Zhao et al., 2019a,b).

14.6.4 Opportunities and challenges in integrative omics

There can be difficulties in data integration due to variation in output as the number of molecules detected can vary in different platforms, the type of generated information, and the technological limitations of the omics platform (Pinu et al., 2019). Also, only qualitative data cannot reveal full information as there is limited information to explain phenotypic changes. A system biology experiment needs to consider the omics platforms, which can give maximum value in the hypothesis testing, the scope of the work to be done, the number of replications required, etc. The use of metabolomics at initial steps and the integration upward in the systems biology (followed by proteomics, transcriptomics, and genomics) can be a good alternative to the traditional top-down approach from a cost-benefit perspective as metabolites are closer representative of phenotype (Pinu et al., 2019). Such study, though, may have lesser coverage and relatively lesser information.

Although intuitiveness and simplicity is there in the element-based approach, Pearson's correlation coefficients can be biased toward the outliers needing alternatives like bi-weight mid-correlation. Clustering and multivariate analysis need familiarity with machine learning and model selection, respectively. Pathway-based integrations are also intuitive and have multiple resources available. It can decipher the metabolic regulations and interactions. Lack of pathway annotation in the less studied organisms and requirement of programming knowledge are major limitations in this approach. Mathematical methods also suffer from complexity and require good knowledge of programming and mathematics. A few of the software tools currently being employed in omics data integration are listed below in Table 14.7 which can aid agricultural microbiology researchers in the data interpretation.

Table 14.7 Major software currently being employed in the multi-omics data integration.

Name of omics data integration tool	Platform	Function	Source/website	Reference
BioCyc/MetaCyc	Web-based	Prediction of metabolic pathways in a sequenced genome	https://biocyc.org https://metacyc.org	Caspi et al. (2020)
IMPaLA	Web-based	Pathway analysis of multi-omics data along with enrichment analysis	http://impala.molgen.mpg.de	Kamburov et al. (2011)
IntegrOmics	R	Analyses of two types of 'omics' of identical samples	http://math.univ-toulouse.fr/biostat	Lê Cao et al. (2009)
MapMan4	Java	Visualization of gene expression data with support for metabolomic and proteomic analysis	https://mapman.gabipd.org/	Schwacke et al. (2019)
mixOmics	R	Statistical integration and data exploration, dimension reduction, and visualization	http://mixomics.org/	Rohart et al. (2017)
Omics Integrator	Web-based	Integration based on the protein-protein interaction	http://fraenkel.mit.edu/omicsintegrator	Tuncbag et al. (2016)
PaintOmics 3	Web-based	Visualization of integrated transcriptomics and metabolomics data	http://www.paintomics.org	Hernández-de-Diego et al. (2018)
Weighted Gene Coexpression Network Analysis (WGCNA)	R	Weighted network analysis based on correlation studies	https://labs.genetics.ucla.edu/horvath/CoexpressionNetwork/Rpackages/WGCNA/	Langfelder and Horvath (2008)

14.7 Conclusion

Mapping all the metabolites and informational molecules in a cell has been a dream of biologists. With the emergence of omics tools like transcriptomics, proteomics, and metabolomics, coverage of various molecules in a single experiment has increased, giving a snapshot of a particular stage, tissue, or condition. As more of these techniques become affordable and abundant, the way researchers approach biological problems is also changing. There is an increasing shift in seeing organisms and their processes as a system rather than an isolated reaction or a pathway. Considering the amount of attention being given to various omics platforms, we expect an explosion in agricultural microbiology data. Some problems often associated with omics experiments are related to experimental designs and quality control. These problems can be addressed by using a standardized set of procedures ensuring sufficient reproducibility. Use of a combination of these omics' techniques can provide valuable insights into the plant-microbe interactions and soil microbial community analysis.

References

Acinas, S.G., Klepac-Ceraj, V., Hunt, D.E., Pharino, C., Ceraj, I., Distel, D.L., Polz, M.F., 2004. Fine-scale phylogenetic architecture of a complex bacterial community. Nature 430, 551–554. https://doi.org/10.1038/nature02649.

Aggarwal, R., Sharma, S., Gupta, S., Manjunatha, C., Singh, V.K., Kulshreshtha, D., 2017. Gene-based analysis of Puccinia species and development of PCR-based marker to detect *Puccinia striiformis* f. sp. *tritici* causing yellow rust of wheat. J. Gen. Plant Pathol. 83 (4), 205–215. https://doi.org/10.1007/s10327-017-0723-x.

Aggarwal, R., Kulshreshtha, D., Sharma, S., Singh, V.K., Manjunatha, C., Bhardwaj, S.C., Saharan, M.S., 2018a. Molecular characterization of Indian pathotypes of *Puccinia striiformis* f. sp. *tritici* and multigene phylogenetic analysis to establish inter-and intraspecific relationships. Genet. Mol. Biol. 41 (4), 834–842. https://doi.org/10.1590/1678-4685-gmb-2017-0171.

Aggarwal, R., Sharma, S., Gupta, S., Banerjee, S., Bashyal, B.M., Bhardwaj, S.C., 2018b. Molecular characterization of predominant Indian wheat rust pathotypes using URP and RAPD markers. Indian J. Biotechnol. 17, 327–336.

Aggarwal, R., Sharma, S., Singh, K., Gurjar, M.S., Saharan, M.S., Gupta, S., Bashyal, B.M., Gaikwad, K., 2019. First draft genome sequence of wheat spot blotch pathogen *Bipolaris sorokiniana* BS_112 from India, obtained using hybrid assembly. Microbiol. Resour. Announc. 8 (38). https://doi.org/10.1128/MRA.00308-19.

Agtuca, B.J., Stopka, S.A., Tuleski, T.R., do Amaral, F.P., Evans, S., Liu, Y., Xu, D., Monteiro, R.A., Koppenaal, D.W., Paša-Tolić, L., Anderton, C.R., 2020. In-situ metabolomic analysis of Setaria viridis roots colonized by beneficial endophytic bacteria. Mol. Plant-Microbe Interact. 33, 272–283. https://doi.org/10.1094/MPMI-06-19-0174-R.

Alharbi, R.A., 2020. Proteomics approach and techniques in identification of reliable biomarkers for diseases. Saudi J. Biol. Sci. 27, 968–974. https://doi.org/10.1016/j.sjbs.2020.01.020.

Allwood, J.W., Erban, A., de Koning, S., Dunn, W.B., Luedemann, A., Lommen, A., Kay, L., Löscher, R., Kopka, J., Goodacre, R., 2009. Inter-laboratory reproducibility of fast gas chromatography–electron impact–time of flight mass spectrometry (GC-EI-TOF/MS) based plant metabolomics. Metabolomics 5, 479–496. https://doi.org/10.1007/s11306-009-0169-z.

Bachem, C.W.B., Oomen, R.J.F.J., Visser, R.G.F., 1998. unav. Springer Science and Business Media LLC, https://doi.org/10.1023/a:1007468801806.

Baidoo, E.E., 2019. Microbial metabolomics: a general overview. Microb. Metabolom., 1–8. https://doi.org/10.1007/978-1-4939-8757-3_1.

Bakker, P.A.H.M., Berendsen, R.L., Doornbos, R.F., Wintermans, P.C.A., Pieterse, C.M.J., 2013. The rhizosphere revisited: root microbiomics. Front. Plant Sci. 4, 1–7. https://doi.org/10.3389/fpls.2013.00165.

Bakker, P.A.H.M., Pieterse, C.M.J., de Jonge, R., Berendsen, R.L., 2018. The soil-borne legacy. Cell 172, 1178–1180. https://doi.org/10.1016/j.cell.2018.02.024.

Balog, J., Kumar, S., Alexander, J., Golf, O., Huang, J., Wiggins, T., Abbassi-Ghadi, N., Enyedi, A., Kacska, S., Kinross, J., Hanna, G.B., 2015. In vivo endoscopic tissue identification by rapid evaporative ionization mass spectrometry (REIMS). Angew. Chem. 127, 11211–11214. https://doi.org/10.1002/ange.201502770.

Bashyal, B.M., Parmar, P., Zaidi, N.W., Aggarwal, R., 2021. Molecular programming of drought-challenged *Trichoderma harzianum*-bioprimed rice (*Oryza sativa* L.). Front. Microbiol. 12. https://doi.org/10.3389/fmicb.2021.655165.

Bingol, K., 2018. Recent advances in targeted and untargeted metabolomics by NMR and MS/NMR methods. High-throughput 7, 9. https://doi.org/10.3390/ht7020009.

Braslavsky, I., Hebert, B., Kartalov, E., Quake, S.R., 2003. Sequence information can be obtained from single DNA molecules. Proc. Natl. Acad. Sci. U. S. A. 100, 3960–3964. https://doi.org/10.1073/pnas.0230489100.

Bucher, D.B., Glenn, D.R., Park, H., Lukin, M.D., Walsworth, R.L., 2020. Hyperpolarization-enhanced NMR spectroscopy with femtomole sensitivity using quantum defects in diamond. Phys. Rev. X 10, 021053. https://doi.org/10.1103/PhysRevX.10.021053.

Bulgarelli, D., Rott, M., Schlaeppi, K., Ver Loren van Themaat, E., Ahmadinejad, N., Assenza, F., Rauf, P., Huettel, B., Reinhardt, R., Schmelzer, E., Peplies, J., Gloeckner, F.O., Amann, R., Eickhorst, T., Schulze-Lefert, P., 2012. Revealing structure and assembly cues for Arabidopsis root-inhabiting bacterial microbiota. Nature 488, 91–95. https://doi.org/10.1038/nature11336.

Cambiaghi, A., Ferrario, M., Masseroli, M., 2017. Analysis of metabolomic data: tools, current strategies and future challenges for omics data integration. Brief. Bioinform. 18, 498–510. https://doi.org/10.1093/bib/bbw031.

Cao, T., Srivastava, S., Rahman, M.H., Kav, N.N.V., Hotte, N., Deyholos, M.K., Strelkov, S.E., 2008. Proteome-level changes in the roots of Brassica napus as a result of *Plasmodiophora brassicae* infection. Plant Sci. 174, 97–115. https://doi.org/10.1016/j.plantsci.2007.10.002.

Caspi, R., Billington, R., Keseler, I.M., Kothari, A., Krummenacker, M., Midford, P.E., Ong, W.K., Paley, S., Subhraveti, P., Karp, P.D., 2020. The MetaCyc database of metabolic pathways and enzymes - a 2019 update. Nucleic Acids Res. 48, D445–D453. https://doi.org/10.1093/nar/gkz862.

Chandramouli, K., Qian, P.-Y., 2009. Proteomics: challenges, techniques and possibilities to overcome biological sample complexity. Hum. Genomics Proteomics 2009. https://doi.org/10.4061/2009/239204.

Chandrasekhar, K., Dileep, A., Lebonah, D.E., Kumari, J.P., 2014. A short review on proteomics and its applications. Int. Lett. Nat. Sci. 17, 77–84. https://doi.org/10.18052/www.scipress.com/ilns.17.77.

Chen, G., Pramanik, B.N., 2009. Application of LC/MS to proteomics studies: current status and future prospects. Drug Discov. Today 14, 465–471. https://doi.org/10.1016/j.drudis.2009.02.007.

Choi, H., Fermin, D., Nesvizhskii, A.I., 2008. Significance analysis of spectral count data in label-free shotgun proteomics. Mol. Cell. Proteomics 7, 2373-2385. https://doi.org/10.1074/mcp.M800203-MCP200.

Chuang, H.-Y., Hofree, M., Ideker, T., 2010. A decade of systems biology. Annu. Rev. Cell Dev. Biol. 26, 721-744. https://doi.org/10.1146/annurev-cellbio-100109-104122.

Cody, R.B., Laramée, J.A., Durst, H.D., 2005. Versatile new ion source for the analysis of materials in open air under ambient conditions. Anal. Chem. 77, 2297-2302. https://doi.org/10.1021/ac050162j.

Collier, T.S., Sarkar, P., Franck, W.L., Rao, B.M., Dean, R.A., Muddiman, D.C., 2010. Direct comparison of stable isotope labeling by amino acids in cell culture and spectral counting for quantitative proteomics. Anal. Chem. 82, 8696-8702. https://doi.org/10.1021/ac101978b.

Considine, E.C., Thomas, G., Boulesteix, A.L., Khashan, A.S., Kenny, L.C., 2018. Critical review of reporting of the data analysis step in metabolomics. Metabolomics 14, 1-16. https://doi.org/10.1007/s11306-017-1299-3.

Cordovez, V., Dini-Andreote, F., Carrión, V.J., Raaijmakers, J.M., 2019. Ecology and evolution of plant microbiomes. Annu. Rev. Microbiol. 73, 69-88. https://doi.org/10.1146/annurev-micro-090817-062524.

Cottrell, J.S., 2011. Protein identification using MS/MS data. J. Proteome 74, 1842-1851. https://doi.org/10.1016/j.jprot.2011.05.014.

Coumans, J.V.F., Poljak, A., Raftery, M.J., Backhouse, D., Pereg-Gerk, L., 2009. Analysis of cotton (*Gossypium hirsutum*) root proteomes during a compatible interaction with the black root rot fungus *Thielaviopsis basicola*. Proteomics 9, 335-349. https://doi.org/10.1002/pmic.200800251.

Darshan, K., Aggarwal, R., Bashyal, B.M., Singh, J., Shanmugam, V., Gurjar, M.S., Solanke, A.U., 2020. Transcriptome profiling provides insights into potential antagonistic mechanisms involved in *Chaetomium globosum* against *Bipolaris sorokiniana*. Front. Microbiol. 11, 578115. https://doi.org/10.3389/fmicb.2020.578115.

David, A., Rostkowski, P., 2020. Analytical techniques in metabolomics. In: Environmental Metabolomics. Elsevier, pp. 35-64, https://doi.org/10.1016/B978-0-12-818196-6.00002-9.

De Bruijn, I., Cheng, X., de Jager, V., Expósito, R.G., Watrous, J., Patel, N., Postma, J., Dorrestein, P.C., Kobayashi, D., Raaijmakers, J.M., 2015. Comparative genomics and metabolic profiling of the genus Lysobacter. BMC Genomics 16, 991. https://doi.org/10.1186/s12864-015-2191-z

De Palma, M., Salzano, M., Villano, C., Aversano, R., Lorito, M., Ruocco, M., Docimo, T., Piccinelli, A.L., D'Agostino, N., Tucci, M., 2019. Transcriptome reprogramming, epigenetic modifications and alternative splicing orchestrate the tomato root response to the beneficial fungus *Trichoderma harzianum*. Hortic. Res. 6, 5. https://doi.org/10.1038/s41438-018-0079-1.

Delwiche, L.D., Slaughter, S.J., 2019. The Little SAS Book: A Primer. SAS Institute.

Deracinois, B., Flahaut, C., Duban-Deweer, S., Karamanos, Y., 2013. Comparative and quantitative global proteomics approaches: an overview. Proteomes 1, 180-218. https://doi.org/10.3390/proteomes1030180.

Dettmer, K., Aronov, P.A., Hammock, B.D., 2007. Mass spectrometry-based metabolomics. Mass Spectrom. Rev. 26, 51-78. https://doi.org/10.1002/mas.20108.

Dinnage, R., Simonsen, A.K., Barrett, L.G., Cardillo, M., Raisbeck-Brown, N., Thrall, P.H., Prober, S.M., 2018. Larger plants promote a greater diversity of symbiotic nitrogen-fixing soil bacteria associated with an Australian endemic legume. J. Ecol. https://doi.org/10.1111/1365-2745.13083.

Dixon, R.A., Gang, D.R., Charlton, A.J., Fiehn, O., Kuiper, H.A., Reynolds, T.L., Tjeerdema, R.S., Jeffery, E.H., German, J.B., Ridley, W.P., Seiber, J.N., 2006. Applications of metabolomics in agriculture. J. Agric. Food Chem. 54, 8984-8994. https://doi.org/10.1021/jf061218t.

Duncan, M.W., Aebersold, R., Caprioli, R.M., 2010. The pros and cons of peptide-centric proteomics. Nat. Biotechnol. 28, 659-664. https://doi.org/10.1038/nbt0710-659.

Eiceman, G.A., Karpas, Z., Hill Jr., H.H., 2013. Ion Mobility Spectrometry. CRC Press.

Gamez, R.M., Rodríguez, F., Vidal, N.M., Ramirez, S., Vera Alvarez, R., Landsman, D., Mariño-Ramírez, L., 2019. Banana (*Musa acuminata*) transcriptome profiling in response to rhizobacteria: *Bacillus amyloliquefaciens* Bs006 and *Pseudomonas fluorescens* Ps006. BMC Genomics 20, 378. https://doi.org/10.1186/s12864-019-5763-5.

Garalde, D.R., Snell, E.A., Jachimowicz, D., Sipos, B., Lloyd, J.H., Bruce, M., Pantic, N., Admassu, T., James, P., Warland, A., Jordan, M., Ciccone, J., Serra, S., Keenan, J., Martin, S., McNeill, L., Wallace, E.J., Jayasinghe, L., Wright, C., Blasco, J., Turner, D.J., 2018. Highly parallel direct RNA sequencing on an array of nanopores. Nat. Methods 15, 201-206. https://doi.org/10.1038/nmeth.4577.

Garza, K.Y., Feider, C.L., Klein, D.R., Rosenberg, J.A., Brodbelt, J.S., Eberlin, L.S., 2018. Desorption electrospray ionization mass spectrometry imaging of proteins directly from biological tissue sections. Anal. Chem. 90, 7785-7789. https://doi.org/10.1021/acs.analchem.8b00967.

Ghaemmaghami, S., Huh, W.-K., Bower, K., Howson, R.W., Belle, A., Dephoure, N., O'Shea, E.K., Weissman, J.S., 2003. Global analysis of protein expression in yeast. Nature 425, 737-741. https://doi.org/10.1038/nature02046.

Godfrey, D., Zhang, Z., Saalbach, G., Thordal-Christensen, H., 2009. A proteomics study of barley powdery mildew haustoria. Proteomics 9, 3222-3232. https://doi.org/10.1002/pmic.200800645.

González-Fernández, R., Prats, E., Jorrín-Novo, J.V., 2010. Proteomics of plant pathogenic fungi. J. Biomed. Biotechnol., 2010.

Graw, S., Chappell, K., Washam, C.L., Gies, A., Bird, J., Robeson, M.S., Byrum, S.D., 2020. Multi-omics data integration considerations and study design for biological systems and disease. Mol. Omics. https://doi.org/10.1039/d0mo00041h.

Groen, S.C., Ćalić, I., Joly-Lopez, Z., Platts, A.E., Choi, J.Y., Natividad, M., Dorph, K., Mauck, W.M., Bracken, B., Cabral, C.L.U., Kumar, A., Torres, R.O., Satija, R., Vergara, G., Henry, A., Franks, S.J., Purugganan, M.D., 2020. The strength and pattern of natural selection on gene expression in rice. Nature 578, 572-576. https://doi.org/10.1038/s41586-020-1997-2.

Guan, N., Shin, H., Chen, R.R., Li, J., Liu, L., Du, G., Chen, J., 2014. Understanding of how *Propionibacterium acidipropionici* respond to propionic acid stress at the level of proteomics. Sci. Rep. 4, 6951. https://doi.org/10.1038/srep06951.

Gurjar, M.S., Singh, J., Saharan, M.S., Aggarwal, R., 2021a. *Tilletia indica*: biology, variability, detection, genomics and future perspective. Indian Phytopathol., 1-11. https://doi.org/10.1007/s42360-021-00319-1.

Gurjar, M.S., Aggarwal, R., Jain, S., Sharma, S., Singh, J., Gupta, S., Agarwal, S., Saharan, M.S., 2021b. Multilocus sequence typing and single nucleotide polymorphism analysis in *Tilletia indica* isolates inciting Karnal bunt of wheat. J. Fungi 7 (2), 103. https://doi.org/10.3390/jof7020103.

Gygi, S.P., Rist, B., Gerber, S.A., Turecek, F., Gelb, M.H., Aebersold, R., 1999. Quantitative analysis of complex protein mixtures using isotope-coded affinity tags. Nat. Biotechnol. 17, 994-999. https://doi.org/10.1038/13690.

Haddad, R., Sparrapan, R., Eberlin, M.N., 2006. Desorption sonic spray ionization for (high) voltage-free ambient mass spectrometry. Rapid Commun. Mass Spectrom. 20, 2901-2905. https://doi.org/10.1002/rcm.2680.

He, J., 2013. Practical guide to ELISA development. In: The Immunoassay Handbook. Elsevier, pp. 381-393, https://doi.org/10.1016/B978-0-08-097037-0.00025-7.

Hernández-de-Diego, R., Tarazona, S., Martínez-Mira, C., Balzano-Nogueira, L., Furió-Tarí, P., Pappas, G.J., Conesa, A., 2018. PaintOmics 3: a web resource for the pathway analysis and visualization of multi-omics data. Nucleic Acids Res. 46, W503-W509. https://doi.org/10.1093/nar/gky466.

Hiraoka, K., Nishidate, K., Mori, K., Asakawa, D., Suzuki, S., 2007. Development of probe electrospray using a solid needle. Rapid Commun. Mass Spectrom. 21, 3139–3144. https://doi.org/10.1002/rcm.3201.

Hölzer, M., Marz, M., 2019. De novo transcriptome assembly: A comprehensive cross-species comparison of short-read RNA-Seq assemblers. Gigascience 8. https://doi.org/10.1093/gigascience/giz039.

Horak, I., Engelbrecht, G., van Rensburg, P.J., Claassens, S., 2019. Microbial metabolomics: essential definitions and the importance of cultivation conditions for utilizing Bacillus species as bionematicides. J. Appl. Microbiol. 127, 326–343. https://doi.org/10.1111/jam.14218.

Huber, H., Hohn, M.J., Rachel, R., Fuchs, T., Wimmer, V.C., Stetter, K.O., 2002. A new phylum of Archaea represented by a nanosized hyperthermophilic symbiont. Nature 417, 63–67. https://doi.org/10.1038/417063a.

Hultman, J., Waldrop, M.P., Mackelprang, R., David, M.M., McFarland, J., Blazewicz, S.J., Harden, J., Turetsky, M.R., McGuire, A.D., Shah, M.B., VerBerkmoes, N.C., Lee, L.H., Mavrommatis, K., Jansson, J.K., 2015. Multi-omics of permafrost, active layer and thermokarst bog soil microbiomes. Nature 521, 208–212. https://doi.org/10.1038/nature14238.

Itzkovitz, S., van Oudenaarden, A., 2011. Validating transcripts with probes and imaging technology. Nat. Methods 8, S12–S19. https://doi.org/10.1038/nmeth.1573.

Jamal, Q., Cho, J.Y., Moon, J.H., Munir, S., Anees, M., Kim, K.Y., 2017. Identification for the first time of cyclo (d-Pro-l-Leu) produced by *Bacillus amyloliquefaciens* Y1 as a nematocide for control of *Meloidogyne incognita*. Molecules 22, 1839. https://doi.org/10.3390/molecules22111839.

James, C., 2014. Global status of commercialized biotech/GM crops: 2010 global area of biotech crops million hectares (1996-2010). ISAAA Br. 49, 16–27.

Jamil, I.N., Remali, J., Azizan, K.A., Nor Muhammad, N.A., Arita, M., Goh, H.-H., Aizat, W.M., 2020. Systematic multi-omics integration (MOI) approach in plant systems biology. Front. Plant Sci. 11, 944. https://doi.org/10.3389/fpls.2020.00944.

Jayakumar, A., Nair, I.C., Radhakrishnan, E.K., 2020. Environmental adaptations of an extremely plant beneficial *Bacillus subtilis* Dcl1 identified through the genomic and metabolomic analysis. Microb. Ecol. 1–16. https://doi.org/10.1007/s00248-020-01605-7.

Jehmlich, N., Schmidt, F., von Bergen, M., Richnow, H.-H., Vogt, C., 2008. Protein-based stable isotope probing (protein-SIP) reveals active species within anoxic mixed cultures. ISME J. 2, 1122–1133. https://doi.org/10.1038/ismej.2008.64.

Jézéquel, T., Deborde, C., Maucourt, M., Zhendre, V., Moing, A., Giraudeau, P., 2015. Absolute quantification of metabolites in tomato fruit extracts by fast 2D NMR. Metabolomics 11, 1231–1242. https://doi.org/10.1007/s11306-015-0780-0.

Jiang, C.-H., Yao, X.-F., Mi, D.-D., Li, Z.-J., Yang, B.-Y., Zheng, Y., Qi, Y.-J., Guo, J.-H., 2019. Comparative transcriptome analysis reveals the biocontrol mechanism of *Bacillus velezensis* F21 against fusarium wilt on watermelon. Front. Microbiol. 10, 652. https://doi.org/10.3389/fmicb.2019.00652.

Kamburov, A., Cavill, R., Ebbels, T.M.D., Herwig, R., Keun, H.C., 2011. Integrated pathway-level analysis of transcriptomics and metabolomics data with IMPaLA. Bioinformatics 27, 2917–2918. https://doi.org/10.1093/bioinformatics/btr499.

Kang, X., Guo, Y., Leng, S., Xiao, L., Wang, L., Xue, Y., Liu, C., 2019. Comparative transcriptome profiling of *Gaeumannomyces graminis* var. *tritici* in wheat roots in the absence and presence of biocontrol *Bacillus velezensis* CC09. Front. Microbiol. 10, 1474. https://doi.org/10.3389/fmicb.2019.01474.

Karp, P.D., Ivanova, N., Krummenacker, M., Kyrpides, N., Latendresse, M., Midford, P., Ong, W.K., Paley, S., Seshadri, R., 2019. A comparison of microbial genome web portals. Front. Microbiol. 10, 208. https://doi.org/10.3389/fmicb.2019.00208.

Kaur, B., Sandhu, K.S., Kamal, R., Kaur, K., Singh, J., Röder, M.S., Muqaddasi, Q.H., 2021. Omics for the improvement of abiotic, biotic, and agronomic traits in major cereal crops: applications, challenges, and prospects. Plants. https://doi.org/10.3390/plants10101989.

Kell, D.B., Oliver, S.G., 2016. The metabolome 18 years on: a concept comes of age. Metabolomics 12, 1–8. https://doi.org/10.1007/s11306-016-1108-4.

Kigawa, T., Muto, Y., Yokoyama, S., 1995. Cell-free synthesis and amino acid-selective stable isotope labeling of proteins for NMR analysis. J. Biomol. NMR 6, 129–134. https://doi.org/10.1007/BF00211776.

Kim, M., Chen, Y., Xi, J., Waters, C., Chen, R., Wang, D., 2015. An antimicrobial peptide essential for bacterial survival in the nitrogen-fixing symbiosis. Proc. Natl. Acad. Sci. U. S. A. 112, 15238–15243. https://doi.org/10.1073/pnas.1500123112.

Knierim, E., Lucke, B., Schwarz, J.M., Schuelke, M., Seelow, D., 2011. Systematic comparison of three methods for fragmentation of long-range PCR products for next generation sequencing. PLoS One 6. https://doi.org/10.1371/journal.pone.0028240, e28240.

Kobayashi, Y., Maeda, T., Yamaguchi, K., Kameoka, H., Tanaka, S., Ezawa, T., Shigenobu, S., Kawaguchi, M., 2018. The genome of *Rhizophagus clarus* HR1 reveals a common genetic basis for auxotrophy among arbuscular mycorrhizal fungi. BMC Genomics 19, 465. https://doi.org/10.1186/s12864-018-4853-0.

Kuehnbaum, N.L., Britz-McKibbin, P., 2013. New advances in separation science for metabolomics: resolving chemical diversity in a post-genomic era. Chem. Rev. 113, 2437–2468. https://doi.org/10.1021/cr300484s.

Kumar, A., Pathak, R.K., Gupta, S.M., Gaur, V.S., Pandey, D., 2015. Systems biology for smart crops and agricultural innovation: filling the gaps between genotype and phenotype for complex traits linked with robust agricultural productivity and sustainability. OMICS 19, 581–601.

Lafler, K.P., 2001. Basic SAS® PROCedures for Generating Quick Results.

Lakshmanan, V., Castaneda, R., Rudrappa, T., Bais, H.P., 2013. Root transcriptome analysis of *Arabidopsis thaliana* exposed to beneficial *Bacillus subtilis* FB17 rhizobacteria revealed genes for bacterial recruitment and plant defense independent of malate efflux. Planta 238, 657–668. https://doi.org/10.1007/s00425-013-1920-2.

Lamichhane, S., Sen, P., Dickens, A.M., Hyötyläinen, T., Orešič, M., 2018. An overview of metabolomics data analysis: current tools and future perspectives. Compr. Anal. Chem. 82, 387–413.

Landau, S., Everitt, B.S., 2003. A Handbook of Statistical Analyses using SPSS. Chapman and Hall/CRC, https://doi.org/10.1201/9780203009765.

Lang, C., Long, S.R., 2015. Transcriptomic analysis of *Sinorhizobium meliloti* and *Medicago truncatula* symbiosis using nitrogen fixation-deficient nodules. Mol. Plant-Microbe Interact. 28, 856–868. https://doi.org/10.1094/MPMI-12-14-0407-R.

Langfelder, P., Horvath, S., 2008. WGCNA: an R package for weighted correlation network analysis. BMC Bioinform. 9, 559. https://doi.org/10.1186/1471-2105-9-559.

Larsen, P.E., Sreedasyam, A., Trivedi, G., Desai, S., Dai, Y., Cseke, L.J., Collart, F.R., 2015. Multi-omics approach identifies molecular mechanisms of plant-fungus mycorrhizal interaction. Front. Plant Sci. 6, 1061. https://doi.org/10.3389/fpls.2015.01061.

Lê Cao, K.-A., González, I., Déjean, S., 2009. integrOmics: an R package to unravel relationships between two omics datasets. Bioinformatics 25, 2855–2856. https://doi.org/10.1093/bioinformatics/btp515.

Lee, J., Feng, J., Campbell, K.B., Scheffler, B.E., Garrett, W.M., Thibivilliers, S., Stacey, G., Naiman, D.Q., Tucker, M.L., Pastor-Corrales, M.A., Cooper, B., 2009. Quantitative proteomic analysis of bean plants infected by a virulent and avirulent obligate rust fungus. Mol. Cell. Proteomics 8, 19–31. https://doi.org/10.1074/mcp.M800156-MCP200.

Lee, Y.S., Park, Y.S., Anees, M., Kim, Y.C., Kim, Y.H., Kim, K.Y., 2013. Nematicidal activity of *Lysobacter capsici* YS1215 and the role of gelatinolytic proteins against root-knot nematodes. Biocontrol Sci. Tech. 23, 1427-1441. https://doi.org/10.1080/09583157.2013.840359.

Letertre, M.P., Dervilly, G., Giraudeau, P., 2020. Combined nuclear magnetic resonance spectroscopy and mass spectrometry approaches for metabolomics. Anal. Chem. https://doi.org/10.1016/bs.coac.2018.07.001.

Li, Q., Yan, J., 2020. Sustainable agriculture in the era of omics: knowledge-driven crop breeding. Genome Biol. 21, 5-9. https://doi.org/10.1186/s13059-020-02073-5.

Li, J., Zhang, Z., Rosenzweig, J., Wang, Y.Y., Chan, D.W., 2002. Proteomics and bioinformatics approaches for identification of serum biomarkers to detect breast cancer. Clin. Chem. 48, 1296-1304. https://doi.org/10.1093/clinchem/48.8.1296.

Li, J., Steen, H., Gygi, S.P., 2003. Protein profiling with cleavable isotope-coded affinity tag (cICAT) reagents: the yeast salinity stress response. Mol. Cell. Proteomics 2, 1198-1204. https://doi.org/10.1074/mcp.M300070-MCP200.

Li, K., Pidatala, V.R., Shaik, R., Datta, R., Ramakrishna, W., 2014. Integrated metabolomic and proteomic approaches dissect the effect of metal-resistant bacteria on maize biomass and copper uptake. Environ. Sci. Technol. 48, 1184-1193. https://doi.org/10.1021/es4047395.

Li, X., Jousset, A., de Boer, W., Carrión, V.J., Zhang, T., Wang, X., Kuramae, E.E., 2019. Legacy of land use history determines reprogramming of plant physiology by soil microbiome. ISME J. 13, 738-751. https://doi.org/10.1038/s41396-018-0300-0.

Liang, P., Pardee, A.B., 1992. Differential display of eukaryotic messenger RNA by means of the polymerase chain reaction. Science 257, 967-971. https://doi.org/10.1126/science.1354393.

Liang, Y., Srivastava, S., Rahman, M.H., Strelkov, S.E., Kav, N.N.V., 2008. Proteome changes in leaves of Brassica napus L. as a result of *Sclerotinia sclerotiorum* challenge. J. Agric. Food Chem. 56, 1963-1976. https://doi.org/10.1021/jf073012d.

Lidbury, I.D.E.A., Murphy, A.R.J., Scanlan, D.J., Bending, G.D., Jones, A.M.E., Moore, J.D., Goodall, A., Hammond, J.P., Wellington, E.M.H., 2016. Comparative genomic, proteomic and exoproteomic analyses of three Pseudomonas strains reveals novel insights into the phosphorus scavenging capabilities of soil bacteria. Environ. Microbiol. 18, 3535-3549. https://doi.org/10.1111/1462-2920.13390.

Lim, J.-A., Lee, D.H., Kim, B.-Y., Heu, S., 2014. Draft genome sequence of *Pantoea agglomerans* R190, a producer of antibiotics against phytopathogens and foodborne pathogens. J. Biotechnol. 188, 7-8. https://doi.org/10.1016/j.jbiotec.2014.07.440.

Lin, T., Zhu, G., Zhang, J., Xu, X., Yu, Q., Zheng, Z., Zhang, Z., Lun, Y., Li, S., Wang, X., Huang, Z., Li, J., Zhang, C., Wang, T., Zhang, Y., Wang, A., Zhang, Y., Lin, K., Li, C., Xiong, G., Xue, Y., Mazzucato, A., Causse, M., Fei, Z., Giovannoni, J.J., Chetelat, R.T., Zamir, D., Städler, T., Li, J., Ye, Z., Du, Y., Hurang, S., 2014. Genomic analyses provide insights into the history of tomato breeding. Nat. Genet. 46 (11), 1220-1226.

Lin, B., Zheng, X., Zheng, S., Luo, M., Lin, Z., 2020. Metabolomics analysis of ammonia secretion during the fermentation of *Klebsiella variicola* GN02 with highly efficient endophytic nitrogen-fixing bacteria. Appl. Biochem. Microbiol. 56, 400-411. https://doi.org/10.1134/S0003683820040109.

Liu, Q., Singh, S., Green, A., 2002. High-oleic and high-stearic cottonseed oils: nutritionally improved cooking oils developed using gene silencing. J. Am. Coll. Nutr. 21 (Suppl. 3), 205S-211S. https://doi.org/10.1080/07315724.2002.10719267.

Liu, J., Wang, H., Cooks, R.G., Ouyang, Z., 2011. Leaf spray: direct chemical analysis of plant material and living plants by mass spectrometry. Anal. Chem. 83, 7608-7613. https://doi.org/10.1021/ac2020273.

Liu, H., Li, Y., Ge, K., Du, B., Liu, K., Wang, C., Ding, Y., 2021. Interactional mechanisms of *Paenibacillus polymyxa* SC2 and pepper (Capsicum annuum L.) suggested by transcriptomics. BMC Microbiol. 21, 70. https://doi.org/10.1186/s12866-021-02132-2.

Lowe, R., Shirley, N., Bleackley, M., Dolan, S., Shafee, T., 2017. Transcriptomics technologies. PLoS Comput. Biol. 13. https://doi.org/10.1371/journal.pcbi.1005457, e1005457.

Lu, H., Giordano, F., Ning, Z., 2016. Oxford nanopore minion sequencing and genome assembly. Genom. Proteom. Bioinform. 14, 265–279. https://doi.org/10.1016/j.gpb.2016.05.004.

Lualdi, M., Fasano, M., 2019. Statistical analysis of proteomics data: a review on feature selection. J. Proteome 198, 18–26. https://doi.org/10.1016/j.jprot.2018.12.004.

Margulies, M., Egholm, M., Altman, W.E., Attiya, S., Bader, J.S., Bemben, L.A., Berka, J., Braverman, M.S., Chen, Y.-J., Chen, Z., Dewell, S.B., de Winter, A., Drake, J., Du, L., Fierro, J.M., Forte, R., Gomes, X.V., Godwin, B.C., He, W., Helgesen, S., Rothberg, J.M., 2006. Erratum: corrigendum: genome sequencing in microfabricated high-density picolitre reactors. Nature 441, 120. https://doi.org/10.1038/nature04726.

Markley, J.L., Brüschweiler, R., Edison, A.S., Eghbalnia, H.R., Powers, R., Raftery, D., Wishart, D.S., 2017. The future of NMR-based metabolomics. Curr. Opin. Biotechnol. 43, 34–40. https://doi.org/10.1016/j.copbio.2016.08.001.

Martínez-Arranz, I., Mayo, R., Pérez-Cormenzana, M., Mincholé, I., Salazar, L., Alonso, C., Mato, J.M., 2015. Enhancing metabolomics research through data mining. J. Proteome 127, 275–288. https://doi.org/10.1016/j.jprot.2015.01.019.

Martínez-García, P.M., Ruano-Rosa, D., Schilirò, E., Prieto, P., Ramos, C., Rodríguez-Palenzuela, P., Mercado-Blanco, J., 2015. Complete genome sequence of *Pseudomonas fluorescens* strain PICF7, an indigenous root endophyte from olive (Olea europaea L.) and effective biocontrol agent against *Verticillium dahliae*. Stand. Genomic Sci. 10, 10. https://doi.org/10.1186/1944-3277-10-10.

Martino, E., Morin, E., Grelet, G.-A., Kuo, A., Kohler, A., Daghino, S., Barry, K.W., Cichocki, N., Clum, A., Dockter, R.B., Hainaut, M., Kuo, R.C., LaButti, K., Lindahl, B.D., Lindquist, E.A., Lipzen, A., Khouja, H.-R., Magnuson, J., Murat, C., Ohm, R.A., Perotto, S., 2018. Comparative genomics and transcriptomics depict ericoid mycorrhizal fungi as versatile saprotrophs and plant mutualists. New Phytol. 217, 1213–1229. https://doi.org/10.1111/nph.14974.

Medini, D., Serruto, D., Parkhill, J., Relman, D.A., Donati, C., Moxon, R., Falkow, S., Rappuoli, R., 2008. Microbiology in the post-genomic era. Nat. Rev. Microbiol. 6, 419–430. https://doi.org/10.1038/nrmicro1901.

Mendes, L.W., Raaijmakers, J.M., De Hollander, M., Mendes, R., Tsai, S.M., 2018. Influence of resistance breeding in common bean on rhizosphere microbiome composition and function. ISME J. 12, 212–224. https://doi.org/10.1038/ismej.2017.158.

Miggiels, P., Wouters, B., van Westen, G.J., Dubbelman, A.C., Hankemeier, T., 2019. Novel technologies for metabolomics: more for less. Trends Anal. Chem. 120, 115323. https://doi.org/10.1016/j.trac.2018.11.021.

Miller, S.B., Heuberger, A.L., Broeckling, C.D., Jahn, C.E., 2019. Non-targeted metabolomics reveals sorghum rhizosphere-associated exudates are influenced by the belowground interaction of substrate and sorghum genotype. Int. J. Mol. Sci. 20, 431. https://doi.org/10.3390/ijms20020431.

Misra, B.B., Langefeld, C.D., Olivier, M., Cox, L.A., 2018. Integrated omics: tools, advances, and future approaches. J. Mol. Endocrinol. https://doi.org/10.1530/JME-18-0055.

Miyauchi, S., Kiss, E., Kuo, A., Drula, E., Kohler, A., Sánchez-García, M., Morin, E., Andreopoulos, B., Barry, K.W., Bonito, G., Buée, M., Carver, A., Chen, C., Cichocki, N., Clum, A., Culley, D., Crous, P.W., Fauchery, L., Girlanda, M., Hayes, R.D., Kéri, Z., LaButti, K., Lipzen, A., Lombard, V., Magnuson, J., Maillard, F., Murat, C., Nolan, M., Ohm, R.A., Pangilinan, J., de Pereira, F.M., Perotto, S., Peter, M., Pfister, S., Riley, R., Sitrit, Y., Stielow, J.B., Szöllősi, G., Žifčáková, L., Štursová, M., Spatafora, J.W., Tedersoo, L., Vaario, L.M., Yamada, A., Yan, M., Wang, P., Xu, J., Bruns, T., Baldrian, P., Vilgalys, R., Dunand, C., Henrissat, B., Grigoriev, I.V., Hibbett, D., Nagy, L.G., Martin, F.M., 2020. Large-scale genome sequencing of mycorrhizal fungi provides insights into the early evolution of symbiotic traits. Nat. Commun. 11, 1–17. https://doi.org/10.1038/s41467-020-18795-w.

Morán-Diez, M.E., Trushina, N., Lamdan, N.L., Rosenfelder, L., Mukherjee, P.K., Kenerley, C.M., Horwitz, B.A., 2015. Host-specific transcriptomic pattern of *Trichoderma virens* during interaction with maize or tomato roots. BMC Genomics 16, 8. https://doi.org/10.1186/s12864-014-1208-3.

Moser, E., Laistler, E., Schmitt, F., Kontaxis, G., 2017. Ultra-high field NMR and MRI—the role of magnet technology to increase sensitivity and specificity. Front. Phys. 5, 33. https://doi.org/10.3390/ijms20020431.

Muthamilarasan, M., Dhaka, A., Yadav, R., Prasad, M., 2015. Exploration of millet models for developing nutrient rich graminaceous crops. Plant Sci. 242, 89–97. https://doi.org/10.1016/j.plantsci.2015.08.023.

Neville, S.A., Lecordier, A., Ziochos, H., Chater, M.J., Gosbell, I.B., Maley, M.W., van Hal, S.J., 2011. Utility of matrix-assisted laser desorption ionization-time of flight mass spectrometry following introduction for routine laboratory bacterial identification. J. Clin. Microbiol. 49, 2980–2984. https://doi.org/10.1128/JCM.00431-11.

O'Farrell, P.H., 1975. High resolution two-dimensional electrophoresis of proteins. J. Biol. Chem. 250, 4007–4021.

Oh, I.S., Park, A.R., Bae, M.S., Kwon, S.J., Kim, Y.S., Lee, J.E., Kang, N.Y., Lee, S., Cheong, H., Park, O.K., 2005. Secretome analysis reveals an Arabidopsis lipase involved in defense against *Alternaria brassicicola*. Plant Cell 17, 2832–2847. https://doi.org/10.1105/tpc.105.034819.

Olivoto, T., Nardino, M., Carvalho, I.R., Follmann, D.N., Szareski, V.I.J., Ferrari, M., de Pelegrin, A.J., de Souza, V.Q.O., 2017. Plant secondary metabolites and its dynamical systems of induction in response to environmental factors: a review. Afr. J. Agric. Res. 12, 71–84. https://doi.org/10.5897/AJAR2016.11677.

Ong, S.-E., Blagoev, B., Kratchmarova, I., Kristensen, D.B., Steen, H., Pandey, A., Mann, M., 2002. Stable isotope labeling by amino acids in cell culture, SILAC, as a simple and accurate approach to expression proteomics. Mol. Cell. Proteomics 1, 376–386. https://doi.org/10.1074/mcp.m200025-mcp200.

Otto, A., Becher, D., Schmidt, F., 2014. Quantitative proteomics in the field of microbiology. Proteomics 14, 547–565. https://doi.org/10.1002/pmic.201300403.

Pathak, R.K., Baunthiyal, M., Pandey, D., Kumar, A., 2018. Augmentation of crop productivity through interventions of omics technologies in India: challenges and opportunities. 3 Biotech 8, 1–28. https://doi.org/10.1007/s13205-018-1473-y.

Paulsen, I.T., Press, C.M., Ravel, J., Kobayashi, D.Y., Myers, G.S.A., Mavrodi, D.V., DeBoy, R.T., Seshadri, R., Ren, Q., Madupu, R., Dodson, R.J., Durkin, A.S., Brinkac, L.M., Daugherty, S.C., Sullivan, S.A., Rosovitz, M.J., Gwinn, M.L., Zhou, L., Schneider, D.J., Cartinhour, S.W., Loper, J.E., 2005. Complete genome sequence of the plant commensal *Pseudomonas fluorescens* Pf-5. Nat. Biotechnol. 23, 873–878. https://doi.org/10.1038/nbt1110.

Pinu, F.R., Villas-Boas, S.G., 2017. Extracellular microbial metabolomics: the state of the art. Metabolites 7, 43. https://doi.org/10.3390/metabo7030043.

Pinu, F.R., Villas-Boas, S.G., Aggio, R., 2017. Analysis of intracellular metabolites from microorganisms: quenching and extraction protocols. Metabolites 7, 53. https://doi.org/10.3390/metabo7040053.

Pinu, F.R., Beale, D.J., Paten, A.M., Kouremenos, K., Swarup, S., Schirra, H.J., Wishart, D., 2019. Systems biology and multi-omics integration: viewpoints from the metabolomics research community. Metabolites 9. https://doi.org/10.3390/metabo9040076.

Pozo, M.J., Zabalgogeazcoa, I., De Aldana, B.R.V., Martinez-medina, A., 2021. ScienceDirect Plant Biology Untapping the potential of plant mycobiomes for applications in agriculture. Curr. Opin. Plant Biol. 60, 102034. https://doi.org/10.1016/j.pbi.2021.102034.

Prasannan, C.B., Jaiswal, D., Davis, R., Wangikar, P.P., 2018. An improved method for extraction of polar and charged metabolites from cyanobacteria. PLoS One 13, e0204273. https://doi.org/10.1371/journal.pone.0204273.

Rai, A., Saito, K., Yamazaki, M., 2017. Integrated omics analysis of specialized metabolism in medicinal plants. Plant J. 90, 764–787. https://doi.org/10.1111/tpj.13485.

Rajakumar, E., Aggarwal, R., Singh, B., 2005. Fungal antagonists for the biological control of ascochyta blight of chickpea. Acta Phytopathol. Entomol. Hung. 40, 35–42. https://doi.org/10.1556/APhyt.40.2005.1-2.5.

Rajkumar, A.P., Qvist, P., Lazarus, R., Lescai, F., Ju, J., Nyegaard, M., Mors, O., Børglum, A.D., Li, Q., Christensen, J.H., 2015. Experimental validation of methods for differential gene expression analysis and sample pooling in RNA-seq. BMC Genomics 16, 548. https://doi.org/10.1186/s12864-015-1767-y.

Ramaswamy, V., Hooker, J.W., Withers, R.S., Nast, R.E., Brey, W.W., Edison, A.S., 2013. Development of a 13C-optimized 1.5-mm high temperature superconducting NMR probe. J. Magn. Reson. 235, 58–65. https://doi.org/10.1016/j.jmr.2013.07.012.

Ramautar, R., Somsen, G.W., de Jong, G.J., 2019. CE-MS for metabolomics: developments and applications in the period 2016–2018. Electrophoresis 40, 165–179. https://doi.org/10.1002/elps.201800323.

Redondo-Nieto, M., Barret, M., Morrissey, J., Germaine, K., Martínez-Granero, F., Barahona, E., Navazo, A., Sánchez-Contreras, M., Moynihan, J.A., Muriel, C., Dowling, D., O'Gara, F., Martín, M., Rivilla, R., 2013. Genome sequence reveals that *Pseudomonas fluorescens* F113 possesses a large and diverse array of systems for rhizosphere function and host interaction. BMC Genomics 14, 54. https://doi.org/10.1186/1471-2164-14-54.

Reinartz, J., Bruyns, E., Lin, J.-Z., Burcham, T., Brenner, S., Bowen, B., Kramer, M., Woychik, R., 2002. Massively parallel signature sequencing (MPSS) as a tool for in-depth quantitative gene expression profiling in all organisms. Brief. Funct. Genomic. Proteomic. 1, 95–104. https://doi.org/10.1093/bfgp/1.1.95.

Rhoads, A., Au, K.F., 2015. PacBio sequencing and its applications. Genom. Proteom. Bioinform. 13, 278–289. https://doi.org/10.1016/j.gpb.2015.08.002.

Righetti, P.G., Fasoli, E., D'Amato, A., Boschetti, E., 2014. Making progress in plant proteomics for improved food safety. In: Comprehensive Analytical Chemistry. vol. 64. Elsevier, pp. 131–155.

Roberts, J.K., 2002. Proteomics and a future generation of plant molecular biologists. Func. Genomics. 48, 143–154.

Rohart, F., Gautier, B., Singh, A., Lê Cao, K.-A., 2017. mixOmics: an R package for omics feature selection and multiple data integration. PLoS Comput. Biol. 13, e1005752. https://doi.org/10.1371/journal.pcbi.1005752.

Ross, P.L., Huang, Y.N., Marchese, J.N., Williamson, B., Parker, K., Hattan, S., Khainovski, N., Pillai, S., Dey, S., Daniels, S., Purkayastha, S., Juhasz, P., Martin, S., Bartlet-Jones, M., He, F., Jacobson, A., Pappin, D.J., 2004. Multiplexed protein quantitation in *Saccharomyces cerevisiae* using amine-reactive isobaric tagging reagents. Mol. Cell. Proteomics 3, 1154–1169. https://doi.org/10.1074/mcp.M400129-MCP200.

Saggiomo, V., Velders, A.H., 2015. Simple 3D printed scaffold-removal method for the fabrication of intricate microfluidic devices. Adv. Sci. 2, 1500125. https://doi.org/10.1002/advs.201500125.

Sahoo, J.P., Behera, L., Sharma, S.S., Praveena, J., Nayak, S.K., Samal, K.C., 2020. Omics studies and systems biology perspective towards abiotic stress response in plants. AJPS 11, 2172–2194. https://doi.org/10.4236/ajps.2020.1112152.

Saia, S., Ruisi, P., Fileccia, V., Di Miceli, G., Amato, G., Martinelli, F., 2015. Metabolomics suggests that soil inoculation with arbuscular mycorrhizal fungi decreased free amino acid content in roots of durum wheat grown under N-limited, P-rich field conditions. PLoS One 10, e0129591. https://doi.org/10.1371/journal.pone.0129591.

Samantra, K., Shiv, A., de Sousa, L.L., Sandhu, K.S., Priyadarshini, P., Mohapatra, S.R., 2021. A comprehensive review on epigenetic mechanisms and application of epigenetic modifications for crop improvement. Environ. Exp. Bot 188, 104479. https://doi.org/10.1016/j.envexpbot.2021.104479.

Sandhu, K.S., Lozada, D.N., Zhang, Z., Pumphrey, M.O., Carter, A.H., 2020. Deep learning for predicting complex traits in spring wheat breeding program. Front. Plant Sci. 11, 613325. https://doi.org/10.3389/fpls.2020.613325.

Sandhu, K.S., Merrick, L.F., Sankaran, S., Zhang, Z., Carter, A.H., 2022. Prospectus of genomic selection and phenomics in cereal, legume and oilseed breeding programs. Front. Genet. 12. https://doi.org/10.3389/fgene.2021.829131.

Sandhu, K.S., Mihalyov, P.D., Lewien, M.J., Pumphrey, M.O., Carter, A.H., 2021. Combining genomic and phenomic information for predicting grain protein content and grain yield in spring wheat. Front. Plant Sci. 12, 613300. https://doi.org/10.3389/fpls.2021.613300.

Sandhu, K.S., Mihalyov, P.D., Lewien, M.J., Pumphrey, M.O., Carter, A.H., 2021a. Genomic selection and genome-wide association studies for grain protein content stability in a nested association mapping population of wheat. Agronomy 11 (12), 2528.

Sandhu, K.S., Patil, S.S., Pumphrey, M., Carter, A., 2021b. Multitrait machine-and deep-learning models for genomic selection using spectral information in a wheat breeding program. Plant Genome 14 (3), e20119.

Schenk, P.M., Carvalhais, L.C., Kazan, K., 2012. Unraveling plant-microbe interactions: can multi-species transcriptomics help? Trends Biotechnol. 30, 177–184. https://doi.org/10.1016/j.tibtech.2011.11.002.

Schmidt, F., Donahoe, S., Hagens, K., Mattow, J., Schaible, U.E., Kaufmann, S.H.E., Aebersold, R., Jungblut, P.R., 2004. Complementary analysis of the *Mycobacterium tuberculosis* proteome by two-dimensional electrophoresis and isotope-coded affinity tag technology. Mol. Cell. Proteomics 3, 24–42. https://doi.org/10.1074/mcp.M300074-MCP200.

Schwacke, R., Ponce-Soto, G.Y., Krause, K., Bolger, A.M., Arsova, B., Hallab, A., Gruden, K., Stitt, M., Bolger, M.E., Usadel, B., 2019. MapMan4: a refined protein classification and annotation framework applicable to multi-omics data analysis. Mol. Plant 12, 879–892. https://doi.org/10.1016/j.molp.2019.01.003.

Sechi, S. (Ed.), 2016. Quantitative Proteomics by Mass Spectrometry, Methods in Molecular Biology. Springer New York, New York, NY, https://doi.org/10.1007/978-1-4939-3524-6.

Shendure, J., Porreca, G.J., Reppas, N.B., Lin, X., McCutcheon, J.P., Rosenbaum, A.M., Wang, M.D., Zhang, K., Mitra, R.D., Church, G.M., 2005. Accurate multiplex polony sequencing of an evolved bacterial genome. Science 309, 1728–1732. https://doi.org/10.1126/science.1117389.

Silva, J.C., Denny, R., Dorschel, C.A., Gorenstein, M., Kass, I.J., Li, G.-Z., McKenna, T., Nold, M.J., Richardson, K., Young, P., Geromanos, S., 2005. Quantitative proteomic analysis by accurate mass retention time pairs. Anal. Chem. 77, 2187–2200. https://doi.org/10.1021/ac048455k.

Silva, J.C., Gorenstein, M.V., Li, G.-Z., Vissers, J.P.C., Geromanos, S.J., 2006. Absolute quantification of proteins by LCMSE: a virtue of parallel MS acquisition. Mol. Cell. Proteomics 5, 144–156. https://doi.org/10.1074/mcp.M500230-MCP200.

Sim, G.K., Kafatos, F.C., Jones, C.W., Koehler, M.D., Efstratiadis, A., Maniatis, T., 1979. Use of a cDNA library for studies on evolution and developmental expression of the chorion multigene families. Cell 18, 1303–1316. https://doi.org/10.1016/0092-8674(79)90241-1.

Singh, R.P., Runthala, A., Khan, S., Jha, P.N., 2017. Quantitative proteomics analysis reveals the tolerance of wheat to salt stress in response to *Enterobacter cloacae* SBP-8. PLoS One 12. https://doi.org/10.1371/journal.pone.0183513, e0183513.

Singh, J., Aggarwal, R., Bashyal, B.M., Darshan, K., Parmar, P., Saharan, M.S., Hussain, Z., Solanke, A.U., 2021. Transcriptome reprogramming of tomato orchestrate the hormone signaling network of systemic resistance induced by *Chaetomium globosum*. Front. Plant Sci. https://doi.org/10.3389/fpls.2021.721193. In this issue.

Singh, J., Aggarwal, R., Gujjar, M.S., Sharma, S., Saharan, M.S., 2019. Identification of carbohydrate active enzymes from whole genome sequence of *Tilletia indica* and sporulation analysis. Indian J. Agric. Sci. 89 (6), 1023–1026. In this issue.

Singh, J., Aggarwal, R., Gurjar, M.S., Sharma, S., Jain, S., Saharan, M.S., 2020. Identification and expression analysis of pathogenicity-related genes in *Tilletia indica* inciting Karnal bunt of wheat. Australas. Plant Pathol. 49, 393–402.

SkZ, A., Vardharajula, S., Vurukonda, S.S.K.P., 2018. Transcriptomic profiling of maize (Zea mays L.) seedlings in response to *Pseudomonas putida* stain FBKV2 inoculation under drought stress. Ann. Microbiol. 68, 331–349. https://doi.org/10.1007/s13213-018-1341-3.

Stolz, A., Jooß, K., Höcker, O., Römer, J., Schlecht, J., Neusüß, C., 2019. Recent advances in capillary electrophoresis-mass spectrometry: instrumentation, methodology and applications. Electrophoresis 40, 79–112. https://doi.org/10.1002/elps.201800331.

Stopka, S.A., Agtuca, B.J., Koppenaal, D.W., Paša-Tolić, L., Stacey, G., Vertes, A., Anderton, C.R., 2017. Laser-ablation electrospray ionization mass spectrometry with ion mobility separation reveals metabolites in the symbiotic interactions of soybean roots and rhizobia. Plant J. 91, 340–354. https://doi.org/10.1111/tpj.13569.

Tang, N., Tornatore, P., Weinberger, S.R., 2004. Current developments in SELDI affinity technology. Mass Spectrom. Rev. 23, 34–44. https://doi.org/10.1002/mas.10066.

Taylor, S.C., Posch, A., 2014. The design of a quantitative western blot experiment. Biomed. Res. Int. 2014, 361590. https://doi.org/10.1155/2014/361590.

Thomas, J., Kim, H.R., Rahmatallah, Y., Wiggins, G., Yang, Q., Singh, R., Glazko, G., Mukherjee, A., 2019. RNA-seq reveals differentially expressed genes in rice (*Oryza sativa*) roots during interactions with plant-growth promoting bacteria, *Azospirillum brasilense*. PLoS One 14. https://doi.org/10.1371/journal.pone.0217309, e0217309.

Thompson, A., Schäfer, J., Kuhn, K., Kienle, S., Schwarz, J., Schmidt, G., Neumann, T., Johnstone, R., Mohammed, A.K.A., Hamon, C., 2003. Tandem mass tags: a novel quantification strategy for comparative analysis of complex protein mixtures by MS/MS. Anal. Chem. 75, 1895–1904. https://doi.org/10.1021/ac0262560.

Tonge, R., Shaw, J., Middleton, B., Rowlinson, R., Rayner, S., Young, J., Pognan, F., Hawkins, E., Currie, I., Davison, M., 2001. Validation and development of fluorescence two-dimensional differential gel electrophoresis proteomics technology. Proteomics 1, 377–396. https://doi.org/10.1002/1615-9861(200103)1:3<377::AID-PROT377>3.0.CO;2-6.

Trivedi, P., Mattupalli, C., Eversole, K., Leach, J.E., 2021. Enabling sustainable agriculture through understanding and enhancement of microbiomes. New Phytologist 230 (6), 2129–2147. https://doi.org/10.1111/nph.17319.

Tuncbag, N., Gosline, S.J.C., Kedaigle, A., Soltis, A.R., Gitter, A., Fraenkel, E., 2016. Network-based interpretation of diverse high-throughput datasets through the omics integrator software package. PLoS Comput. Biol. 12. https://doi.org/10.1371/journal.pcbi.1004879, e1004879.

Turcatti, G., Romieu, A., Fedurco, M., Tairi, A.-P., 2008. A new class of cleavable fluorescent nucleotides: synthesis and optimization as reversible terminators for DNA sequencing by synthesis. Nucleic Acids Res. 36. https://doi.org/10.1093/nar/gkn021, e25.

Valdés, A., Ibáñez, C., Simó, C., García-Cañas, V., 2013. Recent transcriptomics advances and emerging applications in food science. TrAC - Trends Anal. Chem. 52, 142–154. https://doi.org/10.1016/j.trac.2013.06.014.

Van Berkel, G.J., Kertesz, V., King, R.C., 2009. High-throughput mode liquid microjunction surface sampling probe. Anal. Chem. 81, 7096–7101. https://doi.org/10.1021/ac901098d.

Van Emon, J.M., 2016. The omics revolution in agricultural research. J. Agric. Food Chem., 36–44.

Vandenkoornhuyse, P., Quaiser, A., Duhamel, M., Le Van, A., Dufresne, A., 2015. The importance of the microbiome of the plant holobiont. New Phytol. 206, 1196–1206. https://doi.org/10.1111/nph.13312.

Vangelisti, A., Natali, L., Bernardi, R., Sbrana, C., Turrini, A., Hassani-Pak, K., Hughes, D., Cavallini, A., Giovannetti, M., Giordani, T., 2018. Transcriptome changes induced by arbuscular mycorrhizal fungi in sunflower (*Helianthus annuus* L.) roots. Sci. Rep. 8, 4. https://doi.org/10.1038/s41598-017-18445-0.

Velculescu, V.E., Zhang, L., Vogelstein, B., Kinzler, K.W., 1995. Serial analysis of gene expression. Science 270, 484–487. https://doi.org/10.1126/science.270.5235.484.

Vílchez, J.I., Yang, Y., He, D., Zi, H., Peng, L., Lv, S., Kaushal, R., Wang, W., Huang, W., Liu, R., Lang, Z., Miki, D., Tang, K., Paré, P.W., Song, C.-P., Zhu, J.-K., Zhang, H., 2020. DNA demethylases are required for myo-inositol-mediated mutualism between plants and beneficial rhizobacteria. Nat. Plants 6, 983–995. https://doi.org/10.1038/s41477-020-0707-2.

Vinci, G., Cozzolino, V., Mazzei, P., Monda, H., Savy, D., Drosos, M., Piccolo, A., 2018. Effects of *Bacillus amyloliquefaciens* and different phosphorus sources on Maize plants as revealed by NMR and GC-MS based metabolomics. Plant Soil 429, 437–450. https://doi.org/10.1007/s11104-018-3701-y.

Vos, M., Velicer, G.J., 2006. Genetic population structure of the soil bacterium *Myxococcus xanthus* at the centimeter scale. Appl. Environ. Microbiol. 72, 3615–3625. https://doi.org/10.1128/AEM.72.5.3615-3625.2006.

Wang, Y., Niu, Q., Zhang, X., Liu, L., Wang, Y., Chen, Y., Negi, M., Figeys, D., Li, Y.-Y., Zhang, T., 2019. Exploring the effects of operational mode and microbial interactions on bacterial community assembly in a one-stage partial-nitritation anammox reactor using integrated multi-omics. Microbiome 7, 122. https://doi.org/10.1186/s40168-019-0730-6.

Washburn, M.P., Wolters, D., Yates, J.R., 2001. Large-scale analysis of the yeast proteome by multidimensional protein identification technology. Nat. Biotechnol. 19, 242–247. https://doi.org/10.1038/85686.

Whon, T.W., Shin, N.-R., Kim, J.Y., Roh, S.W., 2021. Omics in gut microbiome analysis. J. Microbiol. 59, 292–297. https://doi.org/10.1007/s12275-021-1004-0.

Wiese, S., Reidegeld, K.A., Meyer, H.E., Warscheid, B., 2007. Protein labeling by iTRAQ: a new tool for quantitative mass spectrometry in proteome research. Proteomics 7, 340–350. https://doi.org/10.1002/pmic.200600422.

Wilkins, M.R., Sanchez, J.C., Gooley, A.A., Appel, R.D., Humphery-Smith, I., Hochstrasser, D.F., Williams, K.L., 1996. Progress with proteome projects: why all proteins expressed by a genome should be identified and how to do it. Biotechnol. Genet. Eng. Rev. 13, 19–50. https://doi.org/10.1080/02648725.1996.10647923.

Witzel, K., Neugart, S., Ruppel, S., Schreiner, M., Wiesner, M., Baldermann, S., 2015. Recent progress in the use of omics technologies in brassicaceous vegetables. Front. Plant Sci. 6, 1–14. https://doi.org/10.3389/fpls.2015.00244.

Wolters, D.A., Washburn, M.P., Yates, J.R., 2001. An automated multidimensional protein identification technology for shotgun proteomics. Anal. Chem. 73, 5683–5690. https://doi.org/10.1021/ac010617e.

Wu, D., He, J., Gong, Y., Chen, D., Zhu, X., Qiu, N., Sun, M., Li, M., Yu, Z., 2011a. Proteomic analysis reveals the strategies of *Bacillus thuringiensis* YBT-1520 for survival under long-term heat stress. Proteomics 11, 2580–2591. https://doi.org/10.1002/pmic.201000392.

Wu, D.-Q., Ye, J., Ou, H.-Y., Wei, X., Huang, X., He, Y.-W., Xu, Y., 2011b. Genomic analysis and temperature-dependent transcriptome profiles of the rhizosphere originating strain *Pseudomonas aeruginosa* M18. BMC Genomics 12, 438. https://doi.org/10.1186/1471-2164-12-438.

Xie, J., Yan, Z., Wang, G., Xue, W., Li, C., Che, X., Chen, D., 2020. A bacterium isolated from soil in karst rocky desertification region has efficient phosphate-solubilizing and plant growth-promoting ability. Front. Microbiol. 11, 3612. https://doi.org/10.3389/fmicb.2020.625450.

Ye, M., Jiang, X., Feng, S., Tian, R., Zou, H., 2007. Advances in chromatographic techniques and methods in shotgun proteome analysis. TrAC - Trends Anal. Chem. 26, 80–84. https://doi.org/10.1016/j.trac.2006.10.012.

Zhang, W., Li, F., Nie, L., 2010. Integrating multiple "omics" analysis for microbial biology: application and methodologies. Microbiology (Reading, Engl.) 156, 287–301. https://doi.org/10.1099/mic.0.034793-0.

Zhang, N., Yang, D., Wang, D., Miao, Y., Shao, J., Zhou, X., Xu, Z., Li, Q., Feng, H., Li, S., Shen, Q., Zhang, R., 2015. Whole transcriptomic analysis of the plant-beneficial rhizobacterium *Bacillus amyloliquefaciens* SQR9 during enhanced biofilm formation regulated by maize root exudates. BMC Genomics 16, 685. https://doi.org/10.1186/s12864-015-1825-5.

Zhang, L., Li, L., Sha, G., Liu, C., Wang, Z., Wang, L., 2020. Aerobic composting as an effective cow manure management strategy for reducing the dissemination of antibiotic resistance genes: an integrated meta-omics study. J. Hazard. Mater. 386, 121895. https://doi.org/10.1016/j.jhazmat.2019.121895.

Zhao, M., Su, X.Q., Nian, B., Chen, L.J., Zhang, D.L., Duan, S.M., Wang, L.Y., Shi, X.Y., Jiang, B., Jiang, W.W., Lv, C.Y., Wang, D.P., Shi, Y., Xiao, Y., Wu, J.-L., Pan, Y.H., Ma, Y., 2019a. Integrated meta-omics approaches to understand the microbiome of spontaneous fermentation of traditional Chinese Pu-erh Tea. mSystems 4. https://doi.org/10.1128/mSystems.00680-19.

Zhao, S., Chen, A., Chen, C., Li, C., Xia, R., Wang, X., 2019b. Transcriptomic analysis reveals the possible roles of sugar metabolism and export for positive mycorrhizal growth responses in soybean. Physiol. Plant. 166, 712–728. https://doi.org/10.1111/ppl.12847.

Zolti, A., Green, S.J., Sela, N., Hadar, Y., Minz, D., 2020. The microbiome as a biosensor: functional profiles elucidate hidden stress in hosts. Microbiome 8, 71. https://doi.org/10.1186/s40168-020-00850-9.

Plant growth-promoting microorganism-mediated abiotic stress resilience in crop plants

Sonth Bandeppa[a], Priyanka Chandra[b], Savitha Santosh[c], Saritha M[d], Seema Sangwan[e], and Samadhan Yuvraj Bagul[f]

[a]ICAR-Indian Institute of Rice Research, Hyderabad, Telangana, India, [b]ICAR-Central Soil Salinity Research Institute, Karnal, Haryana, India, [c]ICAR-Central Institute for Cotton Research, Nagpur, Maharashtra, India, [d]ICAR-Central Arid Zone Research Institute, Jodhpur, Rajasthan, India, [e]Division of Microbiology, Indian Agricultural Research Institute, New Delhi, Delhi, India, [f]ICAR-Directorate of Medicinal and Aromatic Plant Research, Boriavi, Gujarat, India

15.1 Introduction

A major deterrent in crop productivity the world over is the various biotic and abiotic stresses that cause enormous crop losses. Suboptimal environmental factors that act singly or simultaneously, such as temperature extremes, drought, waterlogging, salinity, and pollutants such as heavy metals, are the major stressors responsible for major reduction in agricultural production. "Abiotic stresses cause more than 50% losses in crop productivity and are the major concerns for food and nutritional security of additional 0.4 billion Indians by 2050" (ICAR-NIASM, 2015). Ashraf et al. (2008) attributed 15%–40% of the yield losses to temperature extremes, 20% to salinity, and 17% to drought besides other forms of stresses. Climate change has increased the intensity of heat stress which could reduce crop yields by 15%–35% for every temperature increase of 3–4°C, leading to economic losses (Bita and Gerats, 2013; Ortiz et al., 2008). In India, soil moisture deficit or drought alone affects nearly two-thirds of areas comprising the arid and semiarid ecosystems (Dutta et al., 2015). Many of the major rainfed crops, during their growth period, are very often exposed to drought, resulting in significant yield losses. Besides, around 6.74 million ha area in the country is affected by salinity and estimates suggest that nearly 10% additional area is getting salinized every year (Kumar and Sharma, 2020). On the whole, nearly 147 million ha of land is subjected to soil degradation from erosion, salinity, alkalinity, acidification, waterlogging, and a combination of factors due to different forces (Kumar and Sharma, 2020; ICAR-NIASM, 2015). To overcome the adverse impacts of such diverse stresses, plants have evolved adaptive strategies, which are more or less analogous in nature, through the initiation of a stress-specific signaling cascade (Andreasson and Ellis, 2010). For example, plants try to avoid salinity stress by keeping salt-sensitive tissues away from the salinity zone and by immediately maintaining the osmotic equilibrium (Silva et al., 2010). Mitigation strategies in plants against heat stress comprise activation of mechanisms that aid in the maintenance of membrane stability and accumulation of antioxidant metabolites (Meena et al., 2017). Similarly, plants in response to freezing temperatures increase their antifreezing response, leading to cold acclimation (Thomashow, 2010). However, these adaptation strategies come with a cost. In most cases, the growth and yield are reduced under stressed environments due to channeling of resources toward maintenance of plant homeostasis. So, for sustainable crop production under field conditions where combined stresses prevail, efficient resource management and adaptation strategies are needed. Resource management practices such as conservation agriculture, water harvesting, shifting the crop calendars, and development of stress-tolerant varieties are in place (Venkateswarlu and Grover, 2009). However, most of these are high-cost technologies that are inaccessible by the small and marginal farmers who make up the majority of the agrarian communities in India. Also, as stated by Soumyabrata et al. (2017), "these strategies being long drawn and cost intensive, there is a need to develop simple and low cost biological methods for the management of abiotic stress, which can be used on short term basis."

Soil microorganisms are potential means to help crop plants adapt to and achieve increased tolerance to abiotic stresses. Soil microorganisms promote plant growth and productivity by establishing beneficial relationships with them, producing phytohormones, and solubilizing nutrients and making them available for uptake by plants, and by surviving under extreme conditions. These plant growth-promoting rhizobacteria (PGPRs) relieve the impact of abiotic stresses on plants effectively through the production of exopolysaccharides, biofilm formation, and induced systemic tolerance (IST) (Khanna et al., 2019a,b; Yang et al., 2009). Interactions between plant and microorganisms cause alteration of cellular and molecular mechanisms connected with stress tolerance, thereby evoking various kinds of local and systemic responses that improve metabolic capability of the plants to fight against abiotic stresses (Meena et al., 2017; Nguyen et al., 2016; Onaga and Wydra, 2016). The role of rhizobacteria such as *Pseudomonas, Azotobacter, Azospirillum, Rhizobium, Pantoea,*

Bacillus, *Enterobacter*, and *Burkholderia*, cyanobacteria and fungi such as *Trichoderma, Aspergillus, Chaetomium,* and *Phoma*, and arbuscular mycorrhiza in plant growth promotion and abiotic stress tolerance has been reported (Kapoor et al., 2013; Khan et al., 2012a,b; Nagaraju et al., 2012; Oliveira et al., 2009; Sahoo et al., 2014; Singh et al., 2011; Sorty et al., 2016; Vardharajula et al., 2011a,b). Rhizobacteria-induced stress resilience has been associated with changes in the levels of phytohormones, defense-related proteins and enzymes, antioxidants, and epoxypolysaccharides, along with upregulation of aquaporin, dehydrin, and malondialdehyde genes (Pandey et al., 2016). Further, the studies on microbe-mediated stress resilience strategies have demonstrated enhanced germination, superior sustenance, enhanced ability to combat adverse conditions of environment, and superior yield in plants (Meena et al., 2017). The bioprospecting and selection of stress-tolerant microorganisms for their application in crop production could therefore be a practical possibility in overcoming productivity limitations of crop plants in stress-prone areas.

15.2 Diverse abiotic stress affecting plants

One of the main causes of fiasco in crops worldwide is the diverse abiotic stress compromising more than half of the production of significant crops (Hirayama and Shinozaki, 2010). Plants are affected from morphological to molecular levels having visible signs at the stages during development (Fahad et al., 2017), e.g., leaf wilting and abscission, shrinking leaf area, and water depletion through transpiration. Major abiotic stresses are water scarcity, salinity, cold, and heavy metal pollution, which affect the crop development and its yield due to constantly changing environmental conditions.

15.2.1 Drought

A stress where there is deficiency of water gives rise to drought, which usually arises with a period with scanty rainfall. Water is an important crucial element for survival and transportation of other nutrients in plants, leading to stress when plants are not able to absorb sufficient water from soil because of the decreased quantity and constant loss of water through evaporation or transpiration. Hence, the result comes out as reduced vitality of plants because of the deficiency of water which leads to water stress or drought (Ashkavand et al., 2018). High salinity in plants may cause water deficit and stress in plants as its leads to a decrease in soil water potential. Soil having smaller osmotic potential than water makes plant roots too hard that it becomes difficult for water uptake from soil. Besides salinity, extreme temperature can also trigger the water stress because of the enhanced rate of transpiration or evaporation leading to water loss when the temperature is high and formation of ice crystals in the extracellular spaces in plant cells when the temperature is low. Ice crystal formation leads to water efflux intracellularly, thus causing plasmolysis and death of the cell. Inhibition of photosynthesis affects thylakoid membrane, thus ultimately damaging the plant growth. Water stress also causes amassing toxic ions in plant cells (Yadav et al., 2020). Thus, water deficiency depreciates the plant growth and advancement, making drought stress a complicated abiotic stress that ultimately results in decreased plant yield.

15.2.2 Salinity

Soil salinity can be described as gradual build of salt in the soil and causes a visible drop in crop yield as it causes an increase in the concentration of sodium ions and sequestration in plant parts (Bockheim and Gennadiyev, 2000). The main areas posed by salinity are those environments where the volume of precipitation is less as compared to the rates of evaporation and transpiration. The salinity can be divided into two categories: primary soil salinity, where the salt concentration is increased in the subsoil part; and secondary soil salinity, where anthropogenic activities cause environmental pollution and modify the soil content, e.g., irrigation using salty water or enhanced use of fertilizers in soil ultimately causing salt to concentrate in soil (Carillo et al., 2011). Salinization leading to decreased agriculture output and a severe risk to this sector has now become a major global problem. Salinity has a direct impact on the growth and development by increasing ionic toxicity and osmotic stress, thus limiting the yield of the crop. Plant cells are affected by osmotic stress in ordinary conditions than the soil because they use the osmotic pressure in absorbing water and the necessary nutrients into root cells from the soil. But under salinity stress, salt content in the soil rises because osmotic pressure of soil increases more than that of plant cells. Therefore, the absorption of water and minerals such as K^+ and Ca^{2+} is decreased, whereas that of Na^+ and Cl^- ions is increased in plant cells, which ultimately damages the cell membrane leading to plasmolysis. Some other adverse consequences of salinity on the plants are shrinkage of membrane affecting its function and cell growth and low metabolic activity in cytosol and excessive production of reactive oxygen species (ROS) which ultimately leads to cell death by damaging the cell. Quality and quantity of plant production are adversely affected by high salinity, particularly the ion toxicity which inhibits the germination of seeds, and as a consequence, the maturation and development phases of crops (Akbari et al., 2007).

15.2.3 Temperature

Like drought and salinity, another important abiotic stress is extreme climatic temperature. Plant development and its growth are affected by undesirable fluctuations and extreme changes in temperature, e.g., extreme hot and extreme cold. Very low temperature during the growth of plants exposes them to cold stress, which reduces their productivity and yield. Quality of crop development and life after harvesting get reduced drastically due to this major abiotic stress. All functional characteristics at the cellular level are affected badly by chilling. The impact of cold stress in plants can be seen in three ways: Crops when grown above 0°C and below 15°C get highly damaged and are not resistant to chilling, crops get wounded when exposed to low temperatures due to ice formation in tissues but they can tolerate low temperatures, and plants can tolerate very low-temperature conditions. This abiotic tress cause changes in membrane functions and decrease the protoplasmic streaming, leading to leakage of electrolytes and plasmolysis which cause cellular death at the end. The respiration rate of the cell gets fluctuate due to metabolism impairment and abnormal anaerobic respiration cause accumulation of abnormal metabolites in the cell. All these consequences limit the plant growth, further causing delayed and abnormal ripening of fruits along with internal discoloration, which is also called vascular browning, ultimately resulting in decay and the death of the plant (Devasirvatham and Tan, 2018). On the other side, heat stress arises when crops are subjected to high-temperature conditions that affect the functioning or development of plant directly causing permanent injury to the plant. High temperature stress reduces the duration of photosynthesis, which is needed for complete production and ripening of fruits as rate of sexual development is enhanced in plants. The rate of transpiration and evaporation increases, which causes more water loss. Seed germination, emergence of plants, their growth, and development all get weakened. Irreversible drought stress is also triggered when plants are exposed to high temperature, ultimately causing plant death (Takahashi et al., 2013).

15.2.4 Metal toxicity

The presence of toxic ions in excessive amount in the soil which belong to a group of inorganic chemicals and are nonbiodegradable causes heavy metal stress. These inorganic chemicals having density more than $5\,g\,cm^{-3}$ enter the food chain through irrigation, soil, and potable water and have a detrimental toxic effect on plant cells. They also cause various mutagenic disorders in crops by affecting the gene expression (Flora et al., 2008; Wuana and Okieimen, 2011). Although some of the metals are required as vital nutrients in minute quantities for the metabolism and growth of plant, which can be categorized into essential (Fe, Mg, Mo, Zn, Mn, Cu, and Ni) and nonessential (Ag, Cr, Cd, Co, As, Sb, Pb, Se, and Hg) (Schutzendubel, 2002), plants have the ability to absorb these metal ions that travel through water channel from the soil and play a vital role in protein and enzyme structure. However, their excessive amount can hinder plant growth and development by causing toxicity. Their excessive concentration can disturb the protein structure, which indirectly interferes with other significant cellular molecules. It results in the inhibition of ordinary metabolic procedure and acts as a barrier in plant development (Hossain et al., 2015).

15.3 Microbe-mediated abiotic stress alleviation

Global climate change is the worldwide concern due to frequency, intensity, and duration of the environmental stresses that possessed a significant threat to the agriculture and food security worldwide (Jones et al., 2014), and only 3.5% of areas are safe from environmental stress (Van Velthuizen, 2007). Microorganisms could play a vital role in alleviating abiotic stress (Cooper et al., 2014) and provide basic support to the plants in tolerating abiotic stresses (Turner et al., 2013). Therefore, microbe-mediated approaches could be the sustainable option to mitigate the environmental stress. Water stress for agriculture is enormous, as it affects all the growth stages of crop and causes huge economic losses in many parts of the world (Borsani et al., 2001). Under drought stress, plants are adversely affected by accumulation of reactive oxygen species (ROS) (Sairam et al., 2004). Production of antioxidant enzymes such as superoxide dismutase (SOD), catalase (CAT), and ascorbate peroxidase could prevent oxidative damage by eliminating ROS. Production of exopolysaccharides by *Pseudomonas* sp. has been documented, and they possess the water-holding as well as cementing properties, thus playing a key role in soil structure formation (Sandhya et al., 2009). *Trichoderma* sp. has been reported to increase plant growth, enhanced yield, and decrease in harmful root microflora. It has also been characterized for nutrient solubilization and production of phytohormones and secondary metabolites in response to water stress (Shukla et al., 2012). Saline soil also greatly affect the crop yield, inoculation with rhizobacteria, and arbuscular fungi (Rabab and Reda, 2018) improves plant growth under salt stress.

Heat stress is a challenging factor as the global temperature is rising gradually, which has a negative impact on agriculture. Under heat stress, plants could experience impaired cellular processes (Kumar et al., 2001).

ROS are produced at high temperatures, causing oxidative stress, resulting in damage to the cell membrane and building blocks of the cell. Khan et al. (2013) reported endophytic fungus *Penicillium resedanum* LK6 produced flavonoids (daidzein and m-glycitin) and high proline content, which conferred heat stress in the *Capsicum annuum* plant. ACC deaminase activity has been shown to have a significant effect on root elongation through breakdown of ethylene by bacteria and enhanced plant height and chlorophyll content (Meena et al., 2015). Ali et al. (2017) reported endophytic fungi in cucumber could alleviate heat stress by producing total sugars, antioxidant enzymes, and secondary metabolites, which increased photosynthesis. Park et al. (2017) ameliorated heat stress by application of *Bacillus aryabhattai* SRB02 in soybean plants, and the isolate could produce an elevated level of ABA, which mediated stomatal closure.

Metal toxicity: Industrial revolution has made significant impacts on human life; however, it also raised issues regarding heavy metal contamination in soils which is toxic to plants, ultimately declining the crop productivity and nutrient availability (Etesami, 2018). Inoculation of endophytic fungus with soybean could secret gibberellins and downregulated ABA under copper and cadmium toxicity. Ali et al. (2017) reported the use of actinomyces to alleviate heavy metal toxicity from *Brassica juncea* and found that the plant showed tolerance to Zn, Pb, Cd, and Cu. Increased root and shoot were recorded by producing different antioxidant enzymes, thus alleviating the heavy metal stress. Su et al. (2017) reported improved oil quality by inoculating *Piriformospora indica* with *Brassica napus* under different heavy metal stresses. The inoculated plant showed downregulation of erucic acid and glucosinolate genes and showed enhanced plant growth and crop yield and improved oil quality (some important plant growth-promoting microorganisms involved in alleviation of abiotic stress are listed in Table 15.1).

15.4 Mechanisms of microbe-mediated abiotic stress tolerance

The survival and adaptation of a plant to various abiotic stresses depend on its inherent defense mechanism and also interaction with plant-associated beneficial microflora. The plant growth-promoting microbes have innate metabolic and genetic ability to alleviate abiotic stress in crop plants (Gopalakrishnan et al., 2014). The mechanisms of PGPR- or PGPM-mediated abiotic stress tolerance in plants include phytohormonal activity; 1-aminocyclopropane-1-carboxylate (ACC) deaminase activity; production of osmolytes, volatile compounds, and EPSs; antioxidant defense; and nutrient uptake.

Table 15.1 Some plant growth-promoting microorganisms involved in alleviation of abiotic stress.

Stress	Microorganism	Host crop	Mechanism	References
Drought	*Achromobacter piechaudii* ARV8	Pepper, tomato	ACC deaminase production	Mayak et al. (2004)
	Pseudomonas mendocina	Lettuce	Osmolyte production	Kohler (2008)
	Pseudomonas aeruginosa GGRJ21	Mung bean	ROS scavenging enzyme and osmolyte production	Sarma and Saikia (2013)
	Pseudomonas azotoformans ASS1	*Trifolium arvense*	Bioaccumulation, PGP	Ma et al. (2016)
	Paenibacillus polymyxa and *Rhizobium tropici*	Common bean	Phytohormone production	Figueiredo et al. (2008)
	Alcaligenes faecalis AF3	Maize	Exopolysaccharide	Naseem and Bano (2014)
	Klebsiella variicola F2, *Pseudomonas fluorescens* YX2, *Raoultella planticola* YL2	Maize	Accumulation of glycine betaine	Gou et al. (2015a,b)
	Pantoea agglomerans	Wheat	Exopolysaccharide production	Amellal et al. (1998)
	Phyllobacterium brassicacearum	*Arabidopsis thaliana*	Increased level of ABA	Bresson et al. (2013)
Salinity	*Azospirillum* sp.	Wheat	Phytohormone production	Pereyra et al. (2012)
	Pseudomonas pseudoalcaligenes	Rice	Accumulation of glycine betaine	Jha et al. (2011)
	Bacillus sp., *Zhihengliuella halotolerance*	Wheat	PGP	Orhan (2016)
	Arbuscular mycorrhizal fungi (AMF)	Fenugreek	Increased levels of proline	Rabab and Reda (2018)
	AMF	Tomato	High antioxidant enzyme activity	Ebrahim and Saleem (2017)
	Curtobacterium albidum	Rice	EPS production, ACC deaminase, antioxidant enzyme activity	Vimal et al. (2018)
	Bacillus amyloliquefaciens RWL1	Rice	ABA production	Shahzad et al. (2017)
	Penicillium sp.	Sesame	High sugar and reduced oxidative damage	Radhakrishnan and Lee (2015)
	Enterobacter cloacae HSNJ4	Canola	IAA, ACC deaminase activity	Li et al. (2017)
	Methylobacterium oryzae CBMB20 and *Glomus etunicatum*	Maize	Reduced Na uptake	Lee et al. (2015)

Stress	Microorganism	Plant	Effect	Reference
Heavy metal toxicity	*Mucor circinelloides* and *Trichoderma aspereHum*	*Arabidopsis thaliana*	Higher antioxidant enzyme activity	Zhang et al. (2018)
	Proteus mirabilis T2Cr, CrP450	Maize	Decreased Cr uptake, elevated antioxidant enzyme	Islam et al. (2016)
	Cellulosimicrobium funkei KM032184	*Phaseolus vulgaris*	Increased antioxidant enzyme activity	Karthik et al. (2016)
	Streptomyces pactum	*Brassica juncea*	Increased antioxidant enzyme activity	Ali et al. (2017)
	Funneliformis mosseae, *F. caledonium*	Sunflower	Increased leaf catalase and reduced HM toxicity	Zhang et al. (2018)
	Trichoderma sp.	Chickpea	Organic acid production, biotransformation	Tripathi et al. (2017)
	Enterobacter ludwigii CDP-14	Wheat	Increased compatible solutes, total sugars, IAA, ACC deaminase activity	Singh et al. (2018)
	Piriformospora indica	Sunflower	Accumulation of proline content	Shahabivand et al. (2017)
	Pseudomonas aeruginosa CPSB1	Wheat	Phytohormone, antioxidant enzyme	Rizvi and Khan (2017)
	Enterobacter aerogenes K6	Rice	Bioaccumulation, reduced oxidative stress	Pramanik et al. (2018)
	Pseudomonas libanensis TR1, *P. reactans* Ph3R3	*Brassica oxyrrhina*	Reduced MDA, proline content	Ma et al. (2016)
	Pseudomonas putida	*Eruca sativa*	IAA, ACC deaminase production	Kamran et al. (2016)
	Septoglomus deserticola, *S. constrictum*	Tomato	Modulation of oxidative stress	Duc et al. (2018)
Heat	*Pseudomonas putida* AKMP7	Wheat	Improved level of cellular metabolites, antioxidant enzymes	Ali et al. (2011)
	Thermomyces sp.	Cucumber	Accumulation of total sugars and antioxidant enzymes	Ali et al. (2017)
	Glomus mosseae, *G. fasciculatum*	Cyclamen	Increased antioxidants activity	Matsubara and Ishioka (2013)
	Penicillium resedanum LK6	*Capsicum annuum*	Increased level of flavonoids, proline accumulation	Khan et al. (2013)
	Exophiala sp.	Cucumber	Increased fatty acid content, total polyphenol, flavonoids	Khan et al. (2012a,b)
	Pseudomonas aeruginosa 2CpS1	Wheat	ACC deaminase activity, nutrient uptake	Meena et al. (2015)
	Bacillus safensis, *Ochrobactrum pseudogrignonense*	Wheat	Increased level of redox enzyme and accumulation of osmolytes	Sarkar et al. (2018a,b)
	Bacillus aryabhattai SRB02	Soybean	ABA-induced stomatal closure	Park et al. (2017)

15.4.1 Phytohormone production

The growth and development of plants depends on their response to various internal and external stimuli (Wolters and Jürgens, 2009). These responses are mediated by small signaling chemicals called phytohormones, mainly including auxins, gibberellins (GA), and abscisic acid (ABA). Phytohormones are the most important plant growth regulators synthesized in small quantities with a well-documented impact on plant growth (Kazan, 2013) and play an immense role in alleviation of abiotic stresses (Hu et al., 2013). Phytohormones, when applied exogenously, have improved plant growth and metabolism under stress conditions. However, abiotic stresses affect the production of endogenous phytohormones (Du et al., 2013). The beneficial rhizobacteria can modulate endogenous production of phytohormones (Kang et al., 2012) and have beneficial effects similar to exogenous application of phytohormones (Turan et al., 2014; Shahzad et al., 2016). Nitric oxide produced by *Azospirillum brasilense* acts as a signaling molecule in the IAA induced pathway leading to enhanced lateral root and root hair development in tomato plants (Creus et al., 2005; Molina-Favero et al., 2008) is a good example of endogenous phytohormone modulation. The PGPRs *Arthrobacter protophormiae* (SA3) and *Dietzia natronolimnaea* (STR1) and *Bacillus subtilis* (LDR2) confer abiotic stress tolerance in wheat by enhancing IAA content, reducing ABA/ACC content, and modulating the expression of a regulatory component (CTR1) of ethylene signaling pathway and DREB2 transcription factor (Barnawal et al., 2017). Various rhizobacteria were found to be involved in the synthesis of stress hormone ABA. Shahzad et al. (2017) reported *Bacillus amyloliquefaciens* as a synthesizer of ABA in salinity-stressed rice crop. Similarly, *Phyllobacterium brassicacearum* increased the level of ABA in *Arabidopsis thaliana* (Bresson et al., 2013). Recently, Li et al. (2020) identified a novel rhizobacterium, *Kocuria rhizophila* Y1, for mitigating the deleterious effects of salinity on maize growth mainly by regulating plant growth hormones. Plants adapt to various environmental stresses mainly by changing root morphology, a process in which phytohormones play a major role (Potters et al., 2007). Various plant species inoculated with phytohormone producing rhizobacteria (especially IAA and CK) increased root growth and/or enhanced formation of lateral roots (Table 15.1).

15.4.2 ACC deaminase activity

The production of ACC deaminase enzyme is another widespread mechanism of plant growth promotion exerted by PGPRs under abiotic stress (Saleem et al., 2007). The plant metabolism is regulated by ethylene; this in turn is regulated by biotic and abiotic stresses (Hardoim et al., 2008). Under stress conditions, ethylene alters plant metabolism, leading to a reduction in root and shoot growth. The production of ACC, which is an immediate precursor of the ethylene biosynthetic pathway, can be regulated by ACC deaminase activity. Rhizobacteria possessing this enzyme will hydrolyze ACC to nitrogen and energy for plant utilization, which in turn reduces the deleterious effects of ethylene under stress conditions (Glick, 2005). The PGPRs possess ACC deaminase in very low concentrations till stress induction. The concentration of the enzyme increases gradually upon stress induction. Inoculation of rice plants with ACC deaminase-producing *Enterobacter* sp. P23 under salt stress promoted seedling growth along with a decreased concentration of ethylene (Sarkar et al., 2018a). The seed bacterization of French bean with ACC deaminase-producing *Aneurinibacillus aneurinilyticus* and *Paenibacillus* sp. significantly reduced (~60%) stress-stimulated ethylene levels under salinity conditions (Gupta and Pandey, 2019). ACC deaminase containing *Klebsiella* sp. SBP-8 significantly increased the salinity tolerance of wheat (Singh et al., 2015). The effect of drought stress on plant growth and yield of pea under pot and field conditions was eliminated by ACC deaminase-producing bacteria (Arshad et al., 2008). Lim and Kim (2013) reported *Bacillus licheniformis* K11 as ACC deaminase-producing rhizobacteria capable of surviving under drought stress and influencing plant growth in pepper. Okra plants inoculated with ACC deaminase-producing bacteria *Enterobacter* sp. UPMR18 showed higher germination percentage, growth parameters, and chlorophyll content than the control plants under salinity stress (Habib et al., 2016a,b).

15.4.3 Production of osmolytes

Plants adapt to adverse environmental stresses by metabolic adjustment leading to accumulation of osmolytes/compatible solutes. These include proline, sugars, polyamines, betaines, quaternary ammonium compounds, amino acids, and other low molecular weight metabolites (Chen and Jiang, 2010; Slama et al., 2015). Sometimes root exudates also include the osmolytes. The osmotic differences linking cells' surroundings and cytosol are stabilized by the osmolytes (Hussain Wani et al., 2013), thus protecting plants from oxidative damage by inhibiting the production of ROS (Alia Saradhi and Mohanty, 1993). The osmolytes produced by PGPRs and plant-produced osmolytes under abiotic stress collectively act and stimulate plant growth (Paul et al., 2008). PGPRs can produce various osmoprotectant compounds along with phytohormones, some include drought-tolerant *Bacillus* spp. that are known to produce phytohormone and osmolytes such as proline, sugars, free amino acids could alleviate the negative impact of drought in maize (Vardharajula et al., 2011a,b). In salinity-induced capsicum plants, the multifunctional *Bacillus fortis* SSB21 increased the biosynthesis of proline and decreased the ethylene level and lipid peroxidation (Yasin et al., 2018). This contributed

Table 15.2 Plant growth-promoting bacteria involved in enhancing synthesis of osmolytes under different abiotic stresses.

PGPR strain	Host plant	Osmolytes involved	Stress	References
Pseudomonas putida	*Zea mays* L.	Proline	Drought	Sandhya et al. (2010)
Bacillus polymyxa	*Solanum lycopersicum* L.	Proline	Drought	Shintu and Jayaram (2015)
Rhizobium etli	*Phaseolus vulgaris* L.	Trehalose	Drought	Suarez et al. (2008)
Burkholderia phytofirmans	*Vitis vinifera* L.	Carbohydrates, proline, and phenol	Drought	Barka et al. (2006)
Pseudomonas putida MTCC5279	*Cicer arietinum* L.	Proline and sugars	Drought	Tiwari et al. (2016)
A. brasilense	*Zea mays* L.	Trehalose	Drought	Rodriguez et al. (2009)
Klebsiella variicola F2, *Pseudomonas fluorescens* YX2, and *Raoultella planticola* YL2	*Zea mays* L.	Choline and glycine betaine	Drought	Gou et al. (2015a,b)
Azospirillum brasilense	*Zea mays* L.	Proline	Drought	Casanovas et al. (2002)
A. lipoferum	*Zea mays* L.	Free amino acids, pralines, soluble sugars	Drought	Qudsaia et al. (2013)
Pseudomonas aeruginosa and *Burkholderia gladioli*	*Solanum lycopersicum* L.	Trehalose, proline, and glycine betaine	Cadmium toxicity	Khanna et al. (2019a,b)
A. calcoaceticus EU-LRNA-72 and *Penicillium* sp. EU-FTF-6	*Setaria italica* L.	Glycine betaine, proline, sugars	Drought	Kour et al. (2020a,b)
Bacillus safensis and *Ochrobactrum pseudogrignonense*	*Triticum aestivum* L.	Glycine betaine, proline	Heat	Sarkar et al. (2018b)
Bacillus firmus SW5	*Glycine max* L.	Glycine betaine, proline	Salinity	El-Esawi et al. (2018)

to improved plant growth by activating physiological and biochemical processes involved in the mitigation of the salinity. They are many PGPRs producing different types of osmolytes to overcome different abiotic stresses in crop plants (Table 15.2).

15.4.4 Biosynthesis of antioxidant enzymes

Free radicals, also known as reactive oxygen species (ROS), are generated by plant cells under normal metabolism and get neutralized by antioxidants present in cells (Bowler et al., 1992). During abiotic stresses, this mechanism gets disrupted. These ROS, including superoxide radicals $\left(O_2^{\cdot-}\right)$, hydrogen peroxide ($H_2O_2$), hydroxyl radicals (OH$^\bullet$), and singlet oxygen ($^1O_2$), cause damage to DNA, photosynthetic pigments, proteins, and lipids of cells under stressed conditions (Dewir et al., 2006). Several plant antioxidant enzymes, such as peroxidase (POX), superoxide dismutase (SOD), polyphenol oxidase (PPO), and catalase (CAT), protect plant cells from these ROS under stressed conditions (Sen and Alikamanoglu, 2013). The application of PGPRs to plants enhances their tolerance to such environmental stresses (Enebe and Babalola, 2018). There are several mechanisms such as improving the availability of nutrients required by plants, production of plant growth hormones such as gibberellin, indole acetic acid, 1-aminocyclopropane-1-carboxylate (ACC) deaminase, and siderophores, and biocontrol of plant pathogens through which PGPRs improve plant growth and enhance tolerance against abiotic stresses (Batool et al., 2020). Production and modulation of antioxidant enzymes by PGPRs are one of the predominant mechanisms that improve IST in plants (Bhat et al., 2020). Tiwari et al. (2016) reported modulation of differential expression of genes supporting membrane integrity maintenance, accumulation of osmolytes (proline, glycine betaine), and ROS neutralization by *Pseudomonas putida* in *Cicer arietinum* under drought conditions. The genes responsible for ethylene biosynthesis, salicylic acid, jasmonate transcription activation were modulated along with them code for antioxidant enzymes; dehydrins and transcription factors expression were activated. Several researchers reported the improvement of SOD and POX enzymatic activity in many crops when they were co-inoculated with PGPRs in salt stress (Azarmi et al., 2016; Jha et al., 2011; Han and Lee, 2005). *Serratia nematodiphila* produces plant hormones gibberellin enhances *Capsicum annum* growth by producing more gibberellic acid and ABA consequently reduces the concentration of salicylate and jasmonate under low-temperature-stress conditions (Kang et al., 2015). At low temperature stress, *Burkholderia phytofirmans* modulated starch and concentrations of soluble sugars and also adjusted photosynthesis parameters in *Vitis vinifera* which enhanced the tolerance against stress (Fernandez et al., 2012). *Pseudomonas vancouverensis* and *Pseudomonas frederiksbergensis* regulate the pronouncement of antioxidant enzymes and old acclimation genes in *Solanum lycopersicum* (Subramanian et al., 2015). Habib et al. (2016a,b) reported the PGPR increased antioxidant enzyme activities (superoxide dismutase, ascorbate, and catalase) and modulated the ROS pathway in okra enhancing salt tolerance. *Bacillus amyloliquefaciens* mitigates several abiotic stresses by modulating genes responsible for the regulation of antioxidant enzymes in rice (Tiwari et al., 2017). This PGPR also improves antioxidant enzymes (peroxidase, catalase activity, and glutathione) in maize which scavenge ROS under salt stress. The PGPR also modulates genes involved in pumping pyrophosphatase, ion homeostasis, and proline biosynthesis (Chen et al., 2016). Halotolerant *Dietzia natronolimnaea* upregulates TaABARE and TaOPR1, leading to activation of ABA-signaling cascade. It also activates antioxidant enzymes and ion transporters through regulation of the salt overly sensitive (SOS) pathway under salt-stressed conditions (Bharti et al., 2016). Similarly, *Bacillus pumilus* improved rice growth by supporting higher expression of antioxidant enzyme, leading to cell protection under salinity and high boron stresses (Khan et al., 2016).

15.4.5 Nutrient uptake

Under abiotic stresses, plant growth is reduced mainly due to nutritional unavailability. PGPRs are known to fix, mineralize, and/or solubilize plant nutrients and provide to them. Nitrogen fixation potential of microbes provides nitrogen to plants under abiotic stress conditions. *Sinorhizobium meliloti* and *Sinorhizobium medicae* are symbiotic N_2 fixers and have the potential to mitigate drought stress in *Medicago truncatula* through accumulation of K^+ and ethylene, cytokinin production, and accumulation of sugars and amino acids (Staudinger et al., 2016). *Bradyrhizobium* under drought stress converts atmospheric nitrogen into $NH4^+$ and provides essential amino acids for protein formation in groundnut (Delfini et al., 2010). *Bradyrhizobium* also aided tolerance in cowpea against water scarcity stress and supplied nitrate and proline (Barbosa et al., 2013).

PGPRs also mobilize and/or solubilize phosphorus to plants under several stress conditions. Egamberdiyeva and Hoflich (2003) reported that in wheat, PGPRs enhance the uptake of nutrients under drought and salinity stress. *Bacillus pumilus*, *Arthrobacter* sp., *Bacillus cereus*, and *Bacillus aquimaris* have phosphorus solubilization ability and enhance P content in wheat plants (Upadhyay et al., 2009). In wheat, halophilic bacteria *Planococcus rifietoensis* possess several plant growth-promoting potentials that improve plant growth and yield under salinity stress (Rajput et al., 2013). Phosphate solubilizers, *Arthrobacter woluwensis*, *Microbacterium oxydans*, *Arthrobacter aurescens*, *Bacillus megaterium*, and *Bacillus aryabhattai* unveiling production of indole-3-acetic acid, gibberellin, and siderophores also increases plant growth attributes and chlorophyll content in soybean plants under salt stress conditions (Khan et al., 2019). Drought-tolerant and P-solubilizing bacteria, *Pseudomonas libanensis*, substantially improve physiological properties and growth parameters in wheat under water-deficit conditions (Kour et al., 2020a,b). *Pseudomonas putida* and *Bacillus amyloliquefaciens* possess mineral solubilization, ACC deaminase activity, biofilm formation, IAA and siderophore potential ameliorates drought stress in chickpea through various mechanisms (Kumar et al., 2016).

15.4.6 Other mechanisms

Beneficial microbes effectively colonize the roots of plants by forming a biofilm, which results in successful plant-microbe interactions (Seneviratne et al., 2010). These biofilms possess the potential to provide protection from several abiotic stresses. PGPRs present in the rhizosphere, generally biofilm around the roots which protect plants against multiple stresses (Rosier et al., 2018). *Pseudomonas putida* provide protection from drought by colonizing root surfaces and forming a biofilm (Ansari and Ahmad, 2018). Similarly, *Bacillus amyloliquefaciens* improves salt stress tolerance in barley by forming a biofilm around the roots (Kasim et al., 2016) and *Bacillus subtilis* provides protection against tomato wilt disease by the same mechanism (Chen et al., 2013). Along with biofilm formation, extracellular exopolysaccharides (EPSs) are also secreted by PGPRs (Qurashi and Sabri, 2012). EPSs are the complex of different polymers of sugars which protect plants from different types of stresses (Costa et al., 2018). EPS production by PGPRs in salt-affected soils reduces salinity and osmotic stress around the roots (Tewari and Arora, 2014). EPSs act like cement and bind the particles of soils, leading to aggregate formation which improves water retention capacity and sustains fertility of the soil, thus improving plant growth and productivity (Costa et al., 2018). EPSs promote bacterial colonization around plant roots and in soil particles, which eventually improves the structure of the soil. EPS and biofilm complex association with PGPRs helps to retain moisture, which is effective in drought stress and also protects plant roots from various pathogens (Santaella et al., 2008). Under drought stress, *Bacillus subtilis* and *Azospirillum brasilense* have the ability to produce EPSs and aid wheat plants (Kasim et al., 2013). *Proteus penneri*, *Pseudomonas aeruginosa*, and *Alcaligenes faecalis* are EPS-producing bacteria that improve soil physiochemical properties and plant growth under drought stress in maize (Naseem and Bano, 2014). EPS-producing bacterial strains *Planomicrobium chinense* and *Bacillus cereus* demonstrated a positive impact on the growth of wheat grown under rainfed conditions (Khan and Bano, 2019).

15.5 Conclusion

In the era of ecological issues, environmental challenges, and climate change along with depletion of nutrients in soils, such rhizospheric plant growth-promoting microbes are very much beneficial in mediating biotic resistance while providing tolerance to plants against various abiotic stresses. Therefore, the screening and identification of such potential microbes are very much required to enhance resistance to several abiotic stresses. Induction of IST and ISR in plants by PGPRs is the predominant step in the mitigation of biotic and abiotic stresses. Hence, investigation of mechanisms of triggering IST and decoding of signaling cascades induced by PGPRs are very much important to reveal the original mechanisms. The most important mechanisms are root and soil interactions with microbes and the physiological changes induced by them. PGPRs through release of plant hormones including IAA and nitric oxide, production of ACC deaminase, and modifications in cell wall/cell membrane mitigate several stresses. Based on the earlier understanding, our suggestions for the future avenues of research approaches are to (1) investigate the genetic mechanisms for the plant-soil-microbial interactions; (2) characterize active metabolites/compounds responsible for providing protection against stresses and also elucidate their modes of action; and (3) identify the genetic traits that are responsive to PGPRs.

The successful application through colonization and establishment of PGPRs are also required for that several experimentation needs to be conducted under controlled conditions and comparing with that of natural environment. Field trails for the evaluation of PGPRs along with the effect of soil type, management practices such as tillage, crop rotation, and application of pesticides as these are important factors influencing persistence and establishment of PGPRs. Studies on the practical use and commercialization of PGPRs focusing on agricultural systems and their sustainability are, thus, urgently needed.

References

Akbari, G., Modarres-Sanavy, S.A.M., Yousefzadeh, S., 2007. Effect of auxin and salt stress (NaCl) on seed germination of wheat cultivars (*Triticum aestivum* L.). Pak. J. Biol. Sci. 10 (15), 2557–2561.

Ali, S.Z., Sandhya, V., Grover, M., Linga, V.R., Bandi, V., 2011. Effect of inoculation with a thermotolerant plant growth promoting *Pseudomonas putida* strain AKMP7 on growth of wheat under heat stress. J. Plant Interact. 6 (4), 239–246.

Ali, A.H., Abdelrahman, M., Radwan, U., et al., 2017. Effect of thermomyces fungal endophyte isolated from extreme hot desert adapted plant on heat stress tolerance of cucumber. Appl. Soil Ecol. 124, 155–162.

Alia Saradhi, P.P., Mohanty, P., 1993. Proline in relation to free radical production in seedlings of Brassica juncea raised under sodium chloride stress. Plant Soil 156 (1), 497–500.

Amellal, N., Burtin, G., Bartoli, F., Heulin, T., 1998. Colonization of wheat roots by an exopolysaccharide producing *Pantoea agglomerans* strain and its effect on rhizosphere soil aggregation. Appl. Environ. Microbiol. 64, 3740–3747.

Andreasson, E., Ellis, B., 2010. Convergence and specificity in the *Arabidopsis* MAPK nexus. Trends Plant Sci. 15, 106–113. https://doi.org/10.1016/j.tplants.2009.12.001.

Ansari, F.A., Ahmad, I., 2018. Biofilm development, plant growth promoting traits and rhizosphere colonization by *Pseudomonas entomophila* FAP1: a promising PGPR. Adv. Microbiol. 8, 235–251.

Arshad, M., Sharoona, B., Mahmood, T., 2008. Inoculation with *Pseudomonas sp.* containing ACC deaminase partially eliminate the effects of drought stress on growth, yield and ripening of pea (*P. sativum* L.). Pedosphere 18, 611–620.

Ashkavand, P., Zarafshar, M., Tabari, M., Mirzaie, J., Nikpour, A., Kazem-Bordbar, S., 2018. Application of SIO2 nanoparticles AS pretreatment alleviates the impact of drought on the physiological performance of *Prunus mahaleb* (Rosaceae). Bol. Soc. Argent. Bot. 53, 207.

Ashraf, M., Athar, H.R., Harris, P.J.C., Kwon, T.R., 2008. Some prospective strategies for improving crop salt tolerance. Adv. Agron. 97, 45–110.

Azarmi, F., Mozafari, V., Abbaszadeh Dahaji, P., et al., 2016. Biochemical, physiological and antioxidant enzymatic activity responses of pistachio seedlings treated with plant growth promoting rhizobacteria and Zn to salinity stress. Acta Physiol. Plant. 38, 21.

Barbosa, M.A.M., da Silva Lobato, A.K., et al., 2013. *Bradyrhizobium* improves nitrogen assimilation, osmotic adjustment and growth in contrasting cowpea cultivars under drought. Aust. J. Crop. Sci. 7 (13), 1983–1989.

Barka, E.A., Nowak, J., Clement, C., 2006. Enhancement of chilling resistance of inoculated grapevine plantlets with a plant growth-promoting rhizobacterium *Burkholderia phytofirmans* strain PsJN. Appl. Environ. Microbiol. 72 (11), 7246–7252.

Barnawal, D., Bharti, N., Pandey, S.S., Pandey, A., Chanotiya, C.S., Kalra, A., 2017. Plant growth-promoting rhizobacteria enhance wheat salt and drought stress tolerance by altering endogenous phytohormone levels and TaCTR1/TaDREB2 expression. Physiol. Plant. 161 (4), 502–514.

Batool, T., Ali, S., Seleiman, M.F., et al., 2020. Plant growth promoting rhizobacteria alleviates drought stress in potato in response to suppressive oxidative stress and antioxidant enzymes activities. Sci. Rep. 10, 16975.

Bharti, N., Pandey, S.S., Barnawal, D., Patel, V.K., Kalra, A., 2016. Plant growth promoting rhizobacteria *Dietzian atronolimnaea* modulates the expression of stress responsive genes providing protection of wheat from salinity stress. Sci. Rep. 6, 34768.

Bhat, M.A., Kumar, V., Bhat, M.A., et al., 2020. Mechanistic insights of the interaction of plant growth-promoting rhizobacteria (PGPR) with plant roots toward enhancing plant productivity by alleviating salinity stress. Front. Microbiol. 11, 1952.

Bita, C.E., Gerats, T., 2013. Plant tolerance to high temperature in a changing environment: scientific fundamentals and production of heat stress-tolerant crops. Front. Plant Sci. 4, 273. https://doi.org/10.3389/fpls.2013.00273.

Bockheim, J.G., Gennadiyev, A.N., 2000. The role of soil-forming processes in the definition of taxa in soil taxonomy and the world soil reference base. Geoderma 95 (1–2), 53–72.

Borsani, O., Valpuesta, V., Botella, M.A., 2001. Evidence for a role of salicylic acid in the oxidative damage generated by NaCl and osmotic stress in Arabidopsis seedlings. Plant Physiol. 126, 1024–1030.

Bowler, C., Montagu, M.V., Inzé, D., 1992. Superoxide dismutase and stress tolerance. Annu. Rev. Plant Physiol. Plant Mol. Biol. 43, 83–116.

Bresson, J., Varoquaux, F., Bontpart, T., Touraine, B., Vile, D., 2013. The PGPR strain *Phyllobacterium brassicacearum* STM196 induces a reproductive delay and physiological changes that result in improved drought tolerance in Arabidopsis. New Phytol. 200 (2), 558–569.

Carillo, P., Grazia, M., Pontecorvo, G., Fuggi, A., Woodrow, P., 2011. Salinity stress and salt tolerance. In: Shanker, A. (Ed.), Abiotic Stress in Plants-Mechanisms and Adaptations. Intech Open, Rijeka. http://www.intechopen.com/books/abiotic-stress-in-plants-mechanisms-and-adaptations/salinity-stress-and-salt-tolerance.

Casanovas, E.M., Barassi, C.A., Sueldo, R.J., 2002. Azospirillum inoculation mitigates water stress effects in maize seedlings. Cereal Res. Commun. 30, 343–350.

Chen, H., Jiang, J.G., 2010. Osmotic adjustment and plant adaptation to environmental changes related to drought and salinity. Environ. Rev. 18, 309–319.

Chen, Y., Yan, F., Chai, Y., et al., 2013. Biocontrol of tomato wilt disease by *Bacillus subtilis* isolates from natural environments depends on conserved genes mediating biofilm formation. Environ. Microbiol. 15 (3), 848–864.

Chen, L., Liu, Y., Wu, G., Veronican Njeri, K., Shen, Q., Zhang, N., et al., 2016. Induced maize salt tolerance by rhizosphere inoculation of *Bacillus amyloliquefaciens* SQR9. Physiol. Plant. 158, 34–44.

Cooper, M., Gho, C., Leafgren, R., Tang, T., Messina, C., 2014. Breeding drought-tolerant maize hybrids for the US corn-belt: discovery to product. J. Exp. Bot. 65, 6191–6204.

Costa, O.Y.A., Raaijmakers, J.M., Kuramae, E.E., 2018. Microbial extracellular polymeric substances: ecological function and impact on soil aggregation. Front. Microbiol. 9, 1636.

Creus, C.M., Graziano, M., Casanovas, E.M., Pereyra, M.A., Simontacchi, M., Puntarulo, S., Barassi, C.A., Lamattina, L., 2005. Nitric oxide is involved in the *Azospirillum brasilense*-induced lateral root formation in tomato. Planta 221 (2), 297–303.

Delfini, R., Belgoff, C., Fernández, E., Fabra, A., Castro, S., 2010. Symbiotic nitrogen fixation and nitrate reduction in two peanut cultivars with different growth habit and branching pattern structures. Plant Growth Regul. 61, 153–159.

Devasirvatham, V., Tan, D., 2018. Impact of high temperature and drought stresses on chickpea production. Agronomy 8 (8), 145.

Dewir, Y.H., Chakrabarty, D., Ali, B.M., Hahna, E.J., Paek, K.Y., 2006. Lipid peroxidation and antioxidant enzyme activities of *Euphorbia millii* hyperhydric shoots. Environ. Exp. Bot. 58, 93–99.

Du, H., Liu, H., Xiong, L., 2013. Endogenous auxin and jasmonic acid levels are differentially modulated by abiotic stresses in rice. Front. Plant Sci. 4, 397.

Duc, N.H., Csintalan, Z., Posta, K., 2018. Arbuscular mycorrhizal fungi mitigate negative effects of combined drought and heat stress on tomato plants. Plant Physiol. Biochem. 132, 297–307.

Dutta, D., Kundu, A., Patel, N.R., Saha, S.K., Siddiqui, A.R., 2015. Assessment of agricultural drought in Rajasthan (India) using remote sensing derived vegetation condition index (VCI) and standardized precipitation index (SPI). Egypt. J. Remote Sens. Space Sci. 18 (1), 53–63. https://doi.org/10.1016/j.ejrs.2015.03.006.

Ebrahim, M.K.H., Saleem, A., 2017. Alleviating salt stress in tomato inoculated with mycorrhizae: photosynthetic performance and enzymatic antioxidants. J. Taibah. Univ. Sci. 11 (6), 850–860.

Egamberdiyeva, D., Hoflich, G., 2003. Influence of growth-promoting bacteria on the growth of wheat in different soils and temperatures. Soil Biol. Biochem. 35, 973–978.

El-Esawi, M.A., Alaraidh, I.A., Alsahli, A.A., Alamri, S.A., Ali, H.M., Alayafi, A.A., 2018. *Bacillus firmus* (SW5) augments salt tolerance in soybean (*Glycine max* L.) by modulating root system architecture, antioxidant defense systems and stress-responsive genes expression. Plant Physiol. Biochem. 132, 375–384.

Enebe, M.C., Babalola, O.O., 2018. The influence of plant growth-promoting rhizobacteria in plant tolerance to abiotic stress: a survival strategy. Appl. Microbiol. Biotechnol. 102 (18), 7821–7835.

Etesami, H., 2018. Can interaction between silicon and plant growth promoting rhizobacteria benefit in alleviating abiotic and biotic stresses in crop plants? Agric. Ecosyst. Environ. 253, 98–112.

Fahad, S., Bajwa, A.A., Nazir, U., Anjum, S.A., Farooq, A., Zohaib, A., 2017. Crop production under drought and heat stress: plant responses and management options. Front. Plant Sci. 29, 8.

Fernandez, O., Theocharis, A., Bordiec, S., Feil, R., Jacquens, L., Clément, C., Florence, F., AitBarka, E., 2012. *Burkholderi aphytofirmans* PsJN acclimates grapevine to cold by modulating carbohydrate metabolism. Mol. Plant-Microbe Interact. 25, 496–504.

Figueiredo, M.V.B., Burity, H.A., Martinez, C.R., Chanway, C.P., 2008. Alleviation of drought stress in the common bean by co-inoculation with *Paenibacillus polymyxa* and *Rhizobium tropici*. Appl. Soil Ecol. 40, 182–188.

Flora, S.J.S., Mittal, M., Mehta, A., 2008. Heavy metal induced oxidative stress and its possible reversal by chelation therapy. Indian J. Med. Res. 128 (4), 501–523.

Glick, B.R., 2005. Modulation of plant ethylene levels by the bacterial enzyme ACC deaminase. FEMS Microbiol. Lett. 251 (1), 1–7.

Gopalakrishnan, S., Sathya, A., Vijayabharathi, R., Varshney, R.K., Gowda, C.L.L., Krishnamurthy, L., 2014. Plant growth promoting rhizobia: challenges and opportunities. 3 Biotech 5 (4), 355–377.

Gou, W., Tian, L., Ruan, Z., Zheng, P., Chen, F., Zhang, L., et al., 2015a. Accumulation of choline and glycine betaine and drought stress tolerance induced in maize (*Zea mays*) by three plant growth promoting rhizobacteria (PGPR) strains. Pak. J. Bot. 47 (2), 581–586.

Gou, W., Tian, L., Ruan, Z., Zheng, P., Chen, F., et al., 2015b. Accumulation of choline and glycine betaine and drought stress tolerance induced in maize by three plant growth promoting rhizobacteria strains. Pak. J. Bot. 47, 581–586.

Gupta, S., Pandey, S., 2019. ACC deaminase producing bacteria with multifarious plant growth promoting traits alleviates salinity stress in French bean (*Phaseolus vulgaris*. L) plants. Front. Microbiol. 10, 1506.

Habib, S., Kausar, H., Halimi, M., 2016a. Plant growth-promoting rhizobacteria enhance salinity stress tolerance in okra through ROS-scavenging enzymes. Biomed. Res. Int. 2016, 6284547.

Habib, S.H., Kausar, H., Saud, H.M., 2016b. Plant growth-promoting rhizobacteria enhance salinity stress tolerance in okra through ROS-scavenging enzymes. Biomed. Res. Int. 2016, 1–10.

Han, H.S., Lee, K.D., 2005. Plant growth promoting rhizobacteria effect on antioxidant status, photosynthesis, mineral uptake and growth of lettuce under soil salinity. Res. J. Agric. Biol. Sci. 1, 210–215.

Hardoim, P.R., van Overbeek, L.S., Elsas, J.D.V., 2008. Properties of bacterial endophytes and their proposed role in plant growth. Trends Microbiol. 16 (10), 463–471.

Hirayama, T., Shinozaki, K., 2010. Research on plant abiotic stress responses in the post-genome era: past, present and future. Plant J. 61 (6), 1041–1052.

Hossain, Z., Mustafa, G., Komatsu, S., 2015. Plant responses to nanoparticle stress. Int. J. Mol. Sci. 16 (11), 26644–26653.

Hu, Y.F., Zhou, G., Na, X.F., Yang, L., Nan, W.B., Liu, X., et al., 2013. Cadmium interferes with maintenance of auxin homeostasis in Arabidopsis seedlings. J. Plant Physiol. 170, 965–975.

Hussain Wani, S., Brajendra Singh, N., Haribhushan, A., Iqbal Mir, J., 2013. Compatible solute engineering in plants for abiotic stress tolerance—role of glycine betaine. Curr. Genomics 14 (3), 157–165.

ICAR-NIASM, 2015. Vision 2050. National Institute of Abiotic Stress Management, Indian Council of Agricultural Research, Baramati, Pune, p. 7.

Islam, F., Yasmeen, T., Arif, M.S., Riaz, M., Shahzad, S.M., Imran, Q., Ali, I., 2016. Combined ability of chromium (Cr) tolerant plant growth promoting bacteria (PGPB) and salicylic acid (SA) in attenuation of chromium stress in maize plants. Plant Physiol. Biochem. 108, 456–467.

Jha, Y., Subramanian, R.B., Patel, S., 2011. Combination of endophytic and rhizospheric plant growth promoting rhizobacteria in *Oryza sativa* shows higher accumulation of osmoprotectant against saline stress. Acta Physiol. Plant. 33, 797–802.

Jones, L., Provins, A., Holland, M., Mills, G., Hayes, F., et al., 2014. A review and application of the evidence for nitrogen impacts on ecosystem services. Ecosyst. Serv. 7, 76–88.

Kamran, M.A., Eqani, S.A., Bibi, S., Xu, R.K., Amna, M.M.F.H., Katsoyiannis, A., Bokhari, H., Cahudhary, H.J., 2016. Bioaccumulation of nickel by *E. sativa* and role of plant growth promoting rhizobacteria under nickel stress. Ecotoxicol. Environ. Saf. 126, 256–263.

Kang, S.M., Khan, A.L., Hamayun, M., Hussain, J., Joo, G.J., You, Y.H., Kim, J.G., Lee, I.J., 2012. Gibberellin-producing *Promicromonospora* sp. SE188 improves *Solanum lycopersicum* plant growth and influences endogenous plant hormones. J. Microbiol. 50 (6), 902–909.

Kang, S.M., Khan, A., Waqas, M., You, Y.H., Hamayun, M., Joo, G.J., Shahzad, R., Choi, K.S., Lee, I.J., 2015. Gibberellin-producing *Serratia nematodiphila* PEJ1011 ameliorates low temperature stress in *Capsicum annuum* L. Eur. J. Soil Biol. 68, 85–93.

Kapoor, R., Evelin, H., Mathur, P., Giri, B., 2013. Arbuscular mycorrhiza: approaches for abiotic stress tolerance in crop plants for sustainable agriculture. In: Tuteja, N., Singh Gill, S. (Eds.), Plant Acclimation to Environmental Stress. Springer, New York, pp. 359–401, https://doi.org/10.1007/978-1-4614-5001-6_14.

Karthik, C., Oves, M., Thangabalu, R., Sharma, R., Santosh, S.B., Arulselvi, P.I., 2016. *Cellulosimicrobium funkei*-like enhances the growth of *phaseolus vulgaris* by modulating oxidative damage under chromium (VI) toxicity. J. Adv. Res. 7, 839–850.

Kasim, W.A., Osman, M.E., Omar, M.N., Abd El-Daim, I.A., Bejai, S., Meijer, J., 2013. Control of drought stress in wheat using plant growth promoting bacteria. J. Plant Growth Regul. 32, 122–130.

Kasim, W.A., Gaafar, R.M., Abou-Ali, R.M., Omar, M.N., Hewait, H.M., 2016. Effect of biofilm forming plant growth promoting rhizobacteria on salinity tolerance in barley. Ann. Agric. Sci. 61, 217–227.

Kazan, K., 2013. Auxin and the integration of environmental signals into plant root development. Ann. Bot. 112, 1655–1665.

Khan, N., Bano, A., 2019. Exopolysaccharide producing rhizobacteria and their impact on growth and drought tolerance of wheat grown under rainfed conditions. PLoS One 14, e0222302.

Khan, A.l., Hamayun, M., Waqas, M., Kang, S.M., Kim, Y.H., Kim, D.H., Lee, I.J., 2012a. *Exophiala* sp. LHL08 association gives heat stress tolerance by avoiding oxidative damage to cucumber plants. Boil. Fertil. Soils 48, 519–529.

Khan, A.L., Shinwari, Z.K., Kim, Y.H., Waqas, M., Hamayun, M., Kamran, M., Lee, I.J., 2012b. Role of endophyte *Chaetomium globosum* LK4 in growth of *Capsicum annum* by production of gibberellins and indole acetic acid. Pak. J. Bot. 44, 1601–1607.

Khan, A.L., Kang, S.M., Dhakal, K.H., Hussain, J., Adnan, M., Kim, J.G., Lee, I.J., 2013. Flavonoides and amino acid regulation in capsicum annum by endophytic fungi under different heat stress regimes. Sci. Hortic. 155, 1–7.

Khan, A., Zhao, X.Q., Javed, M.T., Khan, K.S., Bano, A., Shen, R.F., et al., 2016. *Bacillus pumilus* enhances tolerance in rice (*Oryza sativa* L.) to combined stresses of NaCl and high boron due to limited uptake of Na+. Environ. Exp. Bot. 124, 120–129.

Khan, M.A., Asaf, S., Khan, A.L., Arjun, R., Rahmatullah, J., Sajid, A., et al., 2019. Halotolerant rhizobacterial strains mitigate the adverse effects of NaCl stress in soybean seedlings. Biomed. Res. Int. 2019, 9530963.

Khanna, K., Jamwal, V.L., Sharma, A., Gandhi, S.G., Ohri, P., Bhardwaj, R., Al-Huqail, A.A., Siddiqui, M.H., Ali, H.M., Ahmad, P., 2019a. Supplementation with plant growth promoting rhizobacteria (PGPR) alleviates cadmium toxicity in *Solanum lycopersicum* by modulating the expression of secondary metabolites. Chemosphere 230, 628–639.

Khanna, K., Kaur, R., Bali, S., Sharma, A., Bakshi, P., Saini, P., Thukral, A.K., Ohri, P., Mir, B.A., Choudhary, S.P., Bhardwaj, R., 2019b. Role of beneficial microorganisms in abiotic stress tolerance in plants. In: Hasanuzzaman, M., Fujita, M., Oku, H., Islam, M.T. (Eds.), Plant Tolerance to Environmental Stress: Role of Phytoprotectants. CRC Press, Boca Raton. 22 pages.

Kohler, J., 2008. Plant-growth-promoting rhizobacteria and arbuscular mycorrhizal fungi modify alleviation biochemical mechanisms in water-stressed plants. Funct. Plant Biol. 35, 141–151.

Kour, D., Rana, K.L., Sheikh, I., et al., 2020a. Alleviation of drought stress and plant growth promotion by *Pseudomonas libanensis* EU-LWNA-33, a drought-adaptive phosphorus-solubilizing bacterium. Proc. Natl. Acad. Sci. India, Sect. B. Biol. Sci. 90, 785–795.

Kour, D., Rana, K.L., Yadav, A.N., Sheikh, I., Kumar, V., Dhaliwal, H.S., Saxena, A.K., 2020b. Amelioration of drought stress in foxtail millet (*Setaria italica* L.) by P-solubilizing drought-tolerant microbes with multifarious plant growth promoting attributes. Environ. Sustain. 3 (1), 23–34.

Kumar, P., Sharma, P.K., 2020. Soil salinity and food security in India. Front. Sustain. Food Syst. 4, 533781. https://doi.org/10.3389/fsufs.2020.533781.

Kumar, B., Pandey, D.M., Goswami, C.L., Jain, S., 2001. Effect of growth regulators on photosynthesis, transpiration and related parameters in water stressed cotton. Biol. Plant. 44, 475–478.

Kumar, M., Mishra, S., Dixit, V., Kumar, M., Agarwal, L., Chauhan, P.S., Nautiyal, C.S., 2016. Synergistic effect of *Pseudomonas putida* and *Bacillus amyloliquefaciens* ameliorates drought stress in chickpea (*Cicer arietinum* L.). Plant Signal. Behav. 11, e1071004.

Lee, Y., Krishnamoorthy, R., Selvakumar, G., Kim, K., Sa, T., 2015. Alleviation of salt stress in maize plant by co-inoculation of arbuscular mycorrrhizal fungi and *Methylibacterium oryzae* CBMB20. J. Korean Soc. Appl. Biol. Chem. 58, 533–540.

Li, H., Lei, P., Xiao, P., Li, S., Xu, H., Xu, Z., Feng, X., 2017. Enhanced tolerance to salt stress in canola seedlings inoculated with the halotolerant *Enterobacter clocae* HSNJ4. Appl. Soil Ecol. 119, 26–34.

Li, X., Sun, P., Zhang, Y., Jin, C., Guan, C., 2020. A novel PGPR strain *Kocuria rhizophila* Y1 enhances salt stress tolerance in maize by regulating phytohormone levels, nutrient acquisition, redox potential, ion homeostasis, photosynthetic capacity and stress-responsive genes expression. Environ. Exp. Bot. 174, 104023.

Lim, J.H., Kim, S.D., 2013. Induction of drought stress resistance by multi-functional PGPR *Bacillus licheniformis* K11 in pepper. Plant Pathol. J. 29 (2), 201–208.

Ma, Y., Rajkumar, M., Zhang, C., Freitas, H., 2016. Inoculation of *Brassica oxyrrhina* with plant growth promoting bacteria for the improvement of heavy metal phytoremediation under drought conditions. J. Hazard. Mater. 320, 36–44.

Matsubara, Y., Ishioka, C., 2013. Bioregulation potential of arbuscular mycorrhizal fungi on heat stress and anthracnose tolerance in Cyclamen. Acta Hortic. 1037, 813–818.

Mayak, S., Tirosh, T., Glick, B.R., 2004. Plant growth promoting bacteria confer resistance in tomato plants to salt stress. Plant Physiol. Biochem. 42, 565–572.

Meena, H., Ahmed, M.A., Prakash, P., 2015. Amelioration of heat stress in wheat, *Triticum aestivum* by PGPR (*Pseudomonas aeruginosa* strain 2CpS1). Biosci. Biotech. Res. Comm. 8 (2), 171–174.

Meena, K.K., Sorty, A.M., Bitla, U.M., Choudhary, K., Gupta, P., Pareek, A., Singh, D.P., Prabha, R., Sahu, P.K., Gupta, V.K., Singh, H.B., Krishanani, K.K., Minhas, P.S., 2017. Abiotic stress responses and microbe-mediated mitigation in plants: the omics strategies. Front. Plant Sci. 8, 172. https://doi.org/10.3389/fpls.2017.00172.

Molina-Favero, C., Creus, C.M., Simontacchi, M., Puntarulo, S., Lamattina, L., 2008. Aerobic nitric oxide production by *Azospirillum brasilense* Sp245 and its influence on root architecture in tomato. Mol. Plant-Microbe Interact. 21 (7), 1001–1009.

Nagaraju, A., Murali, M., Sudisha, J., Amruthesh, K.N., Murthy, S.M., 2012. Beneficial microbes promote plant growth and induce systemic resistance in sunflower against downy mildew disease caused by *Plasmopara halstedii*. Curr. Bot. 3 (5), 12–18.

Naseem, H., Bano, A., 2014. Role of plant growth-promoting rhizobacteria and their exopolysaccharide in drought tolerance of maize. J. Plant Interact. 9 (1), 689–701.

Nguyen, D., Rieu, I., Mariani, C., van Dam, N.M., 2016. How plants handle multiple stresses: hormonal interactions underlying responses to abiotic stress and insect herbivory. Plant Mol. Biol. 91, 727–740. https://doi.org/10.1007/s11103-016-0481-8.

Oliveira, C.A., Alves, V.M.C., Marriel, I.E., Gomes, E.A., Scotti, M.R., Carneiro, N.P., Guimarães, C.T., Schaffert, R.E., Sá, N.M.H., 2009. Phosphate solubilizing microorganisms isolated from rhizosphere of maize cultivated in an oxisol of the Brazilian Cerrado biome. Soil Biol. Biochem. 41, 1782–1787. https://doi.org/10.1016/j.soilbio.2008.01.012.

Onaga, G., Wydra, K., 2016. Advances in plant tolerance to abiotic stresses. In: Abdurakhmonov, I.Y. (Ed.), Plant Genomics. InTech, Rijeka, https://doi.org/10.5772/64350.

Orhan, F., 2016. Alleviation of salt stress by halotolerant and halophilic plant growth promoting bacteria in wheat. Braz. J. Microbiol. 47, 621–627.

Ortiz, R., Braun, H.J., Crossa, J., Crouch, J.H., Davenport, G., Dixon, J., et al., 2008. Wheat genetic resources enhancement by the international maize and wheat improvement center (CIMMYT). Genet. Resour. Crop. Evol. 55, 1095–1140. https://doi.org/10.1007/s10722-008-9372-4.

Pandey, V., Ansari, M.W., Tula, S., Yadav, S., Sahoo, R.K., Shukla, N., et al., 2016. Dose-dependent response of *Trichoderma harzianum* in improving drought tolerance in rice genotypes. Planta 243, 1251–1264. https://doi.org/10.1007/s00425-016-2482-x.

Park, Y.G., Mun, B.G., Kang, S.M., Hussain, A., Shahzad, R., Seo, C.W., et al., 2017. *Bacillus aryabhattai* SRB02 tolerates oxidative and nitrosative stress and promotes the growth of soybean by modulating the production of phytohormones. PLoS One 12, e0173203.

Paul, M.J., Primavesi, L.F., Jhurreea, D., Zhang, Y., 2008. Trehalose metabolism and signaling. Annu. Rev. Plant Biol. 59, 417–441.

Pereyra, M.A., Garcia, P., Colabelli, M.N., Barassi, C.A., Creus, C.M., 2012. A better water status in wheat seedlings induced by *Azospirilum* under osmotic stress is related to morphological changes in xylem vessels of the coleoptiles. Appl. Soil Ecol. 53, 94–97.

Potters, G., Pasternak, T.P., Guisez, Y., Palme, K.J., Jansen, M.A., 2007. Stress-induced morphogenic responses: growing out of trouble? Trends Plant Sci. 12 (3), 98–105.

Pramanik, K., Mitra, S., Sarkar, A., Maiti, T.K., 2018. Alleviation of phytotoxic effects of cadmium on rice seedlings by cadmium resistant PGPR strain *Enterobacter aerogens* MCC 3090. J. Hazard. Mater. 351, 317–329.

Qudsaia, B., Noshinil, Y., Asghari, B., Nadia, Z., Abida, A., Fayazul, H., 2013. Effect of *Azospirillum* inoculation on maize (*Zea mays* L.) under drought stress. Pak. J. Bot. 45, 13–20.

Qurashi, A.W., Sabri, A.N., 2012. Bacterial exopolysaccharide and biofilm formation stimulate chickpea growth and soil aggregation under salt stress. Braz. J. Microbiol. 43 (3), 1183–1191.

Rabab, A.M., Reda, E.A., 2018. Synergistic effect of arbuscular mycorrhizal fungi on growth and physiology of salt stressed *Trigonella foenum graecum* plants. Biocatal. Agric. Biotechnol. 16, 538–544.

Radhakrishnan, R., Lee, I.J., 2015. *Penicillium* sesame interactions: a remedy for mitigating high salinity stress effects on primary and defence metabolites in plants. Environ. Exp. Bot. 116, 47–60.

Rajput, L., Imran, A., Mubeen, F., Hafeez, F., 2013. Salt-tolerant PGPR strain *Planococcus rifietoensis* promotes the growth and yield of wheat (*Triticum aestivum* L.) cultivated in saline soil. Pak. J. Bot. 45, 1955–1962.

Rizvi, A., Khan, M.S., 2017. Biotoxic impact of heavy metals on growth, oxidative stress and morphological changes in root structure of wheat and stress alleviation by *Pseudomonas aeruginosa* strain CPSB1. Chem. Aust. 185, 942–952.

Rodriguez, S.J., Suarez, R., Caballero, M.J., Itturiaga, G., 2009. Trehalose accumulation in *Azospirillum brasilense* improves drought tolerance and biomass in maize plants. FEMS Microbiol. Lett. 296, 52–59.

Rosier, A., Medeiros, F.H.V., Bais, H.P., 2018. Defining plant growth promoting rhizobacteria molecular and biochemical networks in beneficial plant-microbe interactions. Plant Soil 428, 35–55.

Sahoo, R.K., Ansari, M.W., Dangar, T.K., Mohanty, S., Tuteja, N., 2014. Phenotypic and molecular characterisation of efficient nitrogen-fixing *Azotobacter* strains from rice fields for crop improvement. Protoplasma 251, 511–523. https://doi.org/10.1007/s00709-013-0547-2.

Sairam, R., Srivastava, G., Agarwal, S., Meena, R., 2004. Differences in antioxidant activity in response to salinity stress in tolerant and susceptible wheat genotypes. Biol. Plant. 49, 85–91.

Saleem, M., Arshad, M., Hussain, S., Bhatti, A.S., 2007. Perspective of plant growth promoting rhizobacteria (PGPR) containing ACC deaminase in stress agriculture. J. Ind. Microbiol. Biotechnol. 34 (10), 635–648.

Sandhya, V., Ali, S.Z., Grover, M., Reddy, G., Venkateswarlu, B., 2009. Alleviation of drought stress effects in sunflower seedlings by exopolysaccharides producing *Pseudomonas putida* strain P45. Biol. Fertil. Soils 46, 17–26.

Sandhya, V., Ali, S.Z., Grover, M., Reddy, G., Venkateswarlu, B., 2010. Effect of plant growth promoting *Pseudomonas sp.* on compatible solutes, antioxidant status and plant growth of maize under drought stress. Plant Growth Regul. 62 (1), 21–30.

Santaella, C., Schue, M., Berge, O., Heulin, T., Achouak, W., 2008. The exopolysaccharide of *Rhizobium* sp. YAS34 is not necessary for biofilm formation on *Arabidopsis thaliana* and *Brassica napus* roots but contributes to root colonization. Environ. Microbiol. 10 (8), 2150–2163.

Sarkar, A., Ghosh, P.K., Pramanik, K., Mitra, S., Soren, T., Pandey, S., Mondal, M.H., Maiti, T.K., 2018a. A halotolerant *Enterobacter sp.* displaying ACC deaminase activity promotes rice seedling growth under salt stress. Res. Microbiol. 169 (1), 20–32.

Sarkar, J., Chakraborty, B., Chakraborty, U., 2018b. Plant growth promoting rhizobacteria protect wheat plants against temperature stress through antioxidant signaling and reducing chloroplast and membrane injury. J. Plant Growth Regul. 37 (4), 1396–1412.

Sarma, R., Saikia, R., 2013. Alleviation of drought stress in mung bean by strain *Pseudomonas aeruginosa* GGRJ21. Plant Soil 377, 111–126.

Schutzendubel, A., 2002. Plant responses to abiotic stresses: heavy metal-induced oxidative stress and protection by mycorrhization. J. Exp. Bot. 53 (372), 1351–1365.

Sen, A., Alikamanoglu, S., 2013. Antioxidant enzyme activities, malondialdehyde, and total phenolic content of PEG-induced hyperhydric leaves in sugar beet tissue culture. In Vitro Cell Dev. Biol. Plant 49, 396–404.

Seneviratne, G., Weerasekara, M.L.M.A.W., Seneviratne, K.A.C.N., Zavahir, J.S., Kecskés, M.L., Kennedy, I.R., 2010. Importance of biofilm formation in plant growth promoting rhizobacterial action. In: Maheshwari, D. (Ed.), Plant Growth and Health Promoting Bacteria. Microbiology Monographs. vol. 18. Springer, Berlin, Heidelberg.

Shahabivand, S., Parvaneh, A., Aliloo, A.A., 2017. Root endophytic fungus *Piriformospora indica* affected growth, cadmium portioning and chlorophyll fluorescence of sunflower under cadmium toxicity. Ecotoxicol. Environ. Saf. 145, 496–502.

Shahzad, R., Waqas, M., Khan, A.L., Asaf, S., Khan, M.A., Kang, S.M., et al., 2016. Seed-borne endophytic *Bacillus amyloliquefaciens* RWL-1 produces gibberellins and regulates endogenous phytohormones of *Oryza sativa*. Plant Physiol. Biochem. 106, 236–243.

Shahzad, R., Khan, A.L., Bilal, S., Waqas, M., Kang, S.M., Lee, I.J., 2017. Inoculation of abscisic acid-producing endophytic bacteria enhances salinity stress tolerance in *Oryza sativa*. Environ. Exp. Bot. 136, 68–77.

Shintu, P.V., Jayaram, K.M., 2015. Phosphate solubilising bacteria (*Bacillus polymyxa*)—an effective approach to mitigate drought in tomato (*Lycopersicon esculentum* Mill.). Trop. Plant Res. 2 (1), 17–22.

Shukla, N., Awasthi, R.P., Rawat, L., Kumar, J., 2012. Biochemical and physiological responses of rice as influenced by *Trichoderma harziannum* under drought stress. Plant Physiol. Biochem. 54, 78–88.

Silva, E.N., Ribeiro, R.V., Ferreira-Silva, S.L., Viégas, R.A., Silveira, J.A.G., 2010. Comparative effects of salinity and water stress on photosynthesis, water relations and growth of *Jatropha curcas* plants. J. Arid Environ. 74, 1130–1137. https://doi.org/10.1016/j.jaridenv.2010.05.036.

Singh, D.P., Prabha, R., Yandigeri, M.S., Arora, D.K., 2011. Cyanobacteria-mediated phenylpropanoids and phytohormones in rice (*Oryza sativa*) enhance plant growth and stress tolerance. Antonie Van Leeuwenhoek 100, 557–568. https://doi.org/10.1007/s10482-011-9611-0.

Singh, R.P., Jha, P., Jha, P.N., 2015. The plant-growth-promoting bacterium *Klebsiella sp.* SBP-8 confers induced systemic tolerance in wheat (*Triticum aestivum*) under salt stress. J. Plant Physiol. 184, 57–67.

Singh, R.P., Mishra, S., Jha, P., Raghuvanshi, S., Jha, P., 2018. Effect of inoculation of zinc resistant bacterium *Enterobacter ludwigii* CDP-14 on growth, biochemical parameters and zinc uptake in wheat plant. Ecol. Eng. 116, 163–173.

Slama, I., Abdelly, C., Bouchereau, A., Flowers, T., Savouré, A., 2015. Diversity, distribution and roles of osmoprotective compounds accumulated in halophytes under abiotic stress. Ann. Bot. 115 (3), 433–447.

Sorty, A.M., Meena, K.K., Choudhary, K., Bitla, U.M., Minhas, P.S., Krishnani, K.K., 2016. Effect of plant growth promoting bacteria associated with halophytic weed (*Psoralea corylifolia* L.) on germination and seedling growth of wheat under saline conditions. Appl. Biochem. Biotechnol. 180, 872–882. https://doi.org/10.1007/s12010-016-2139-z.

Soumyabrata, C., Abhijit, S., Adyant, K., Sahar, M., 2017. Role of microorganisms in abiotic stress management. Curr. Adv. Agric. Sci. 9 (1), 98–100.

Staudinger, C., Mehmeti-Tershani, V., Gil-Quintana, E., Gonzalez, E.M., Hofhans, F., Bachmann, G., Wienkoop, S., 2016. Evidence for a rhizobia-induced drought stress response strategy in *Medicago truncatula*. J. Proteome 136, 202–213.

Su, Z.Z., Wang, T., Shrivastava, N., Chen, Y.Y., Liu, X., Sun, C., Yin, Y., Gao, Q.K., Lou, B.G., 2017. Piriformospora indica promotes growth, seed yield and quality of *Brassica napus* L. Microbiol. Res. 199, 29–39.

Suarez, R., Wong, A., Ramirez, M., Barraza, A., Orozco Mdel, C., Cevallos, M.A., et al., 2008. Improvement of drought tolerance and grain yield in common bean by over expressing trehalose-6-phosphate synthase in rhizobia. Mol. Plant-Microbe Interact. 21, 958–966.

Subramanian, P., Mageswari, A., Kim, K., Lee, Y., Sa, T., 2015. Psychrotolerant endophytic *Pseudomonas* sp. strains OB155 and OS261 induced chilling resistance in tomato plants (*Solanum lycopersicum* Mill.) by activation of their antioxidant capacity. Mol. Plant-Microbe Interact. 28, 1073–1081.

Takahashi, D., Li, B., Nakayama, T., Kawamura, Y., Uemura, M., 2013. Plant plasma membrane proteomics for improving cold tolerance. Front. Plant Sci. 4, 90.

Tewari, S., Arora, N.K., 2014. Multifunctional exopolysaccharides from *Pseudomonas aeruginosa* PF23 involved in plant growth stimulation, biocontrol and stress amelioration in sunflower under saline conditions. Curr. Microbiol. 69 (4), 484–494.

Thomashow, M.F., 2010. Molecular basis of plant cold acclimation: insights gained from studying the CBF cold response pathway. Plant Physiol. 154, 571–577. https://doi.org/10.1104/pp.110.161794.

Tiwari, S., Lata, C., Chauhan, P.S., Nautiyal, C.S., 2016. *Pseudomonas putida* attunes morpho-physiological, biochemical and molecular responses in *Cicer arietinum* L. during drought stress and recovery. Plant Physiol. Biochem. 99, 108–117.

Tiwari, S., Prasad, V., Chauhan, P.S., Lata, C., 2017. *Bacillus amyloliquefaciens* confers tolerance to various abiotic stresses and modulates plant response to phytohormones through osmoprotection and gene expression regulation in rice. Front. Plant Sci. 8, 1510.

Tripathi, P., Singh, P.C., Mishra, A., Srivastava, S., Chauhan, R., Awasthi, S., Mishra, A., Dwivedi, S., Tripathi, P., Kalra, A., Tripathi, R.D., Nautiyal, C.S., 2017. Arsenic tolerant *Trichoderma sp.* reduces arsenic induced stress in chickpea (*Cicer arietinum*). Environ. Pollut. 223, 137–145.

Turan, M., Ekinci, M., Yildirim, E., Güneş, A., Karagz, K., Kotan, R., et al., 2014. Plant growth-promoting rhizobacteria improved growth, nutrient, and hormone content of cabbage (*Brassica oleracea*) seedlings. Turk. J. Agric. For. 38, 327–333.

Turner, T.R., James, E.K., Poole, P.S., 2013. The plant microbiome. Genome Biol. 14, 209.

Upadhyay, S.K., Singh, D.P., Saikia, R., 2009. Genetic diversity of plant growth promoting rhizobacteria isolated from rhizospheric soil of wheat under saline conditions. Curr. Microbiol. 59, 489–496.

Vardharajula, S., Ali, S.Z., Grover, M., Reddy, G., Bandi, V., 2011a. Drought-tolerant plant growth promoting *Bacillus* spp.: effect on growth, osmolytes, and antioxidant status of maize under drought stress. J. Plant Interact. 6, 1–14.

Vardharajula, S., Zulfikar Ali, S., Grover, M., Reddy, G., Bandi, V., 2011b. Drought-tolerant plant growth promoting Bacillus spp.: effect on growth, osmolytes, and antioxidant status of maize under drought stress. J. Plant Interact. 6 (1), 1–14.

Van Velthuizen, H., 2007. Mapping Biophysical Factors That Influence Agricultural Production and Rural Vulnerability. Food & Agriculture Organization, Rome, Italy.

Venkateswarlu, B., Grover, M., 2009. Can microbes help crops cope with climate change? Indian J. Microbiol. 49, 297–298.

Vimal, S.J., Patel, V.K., Singh, J.S., 2018. Plant growth promoting *Curtobacterium albidum* strain SR4: an agriculturally important microbe to alleviate salinity stress in paddy plants. Ecol. Indic. 105 (2019), 553–562.

Wolters, H., Jürgens, G., 2009. Survival of the flexible: hormonal growth control and adaptation in plant development. Nat. Rev. Genet. 10 (5), 305–317.

Wuana, R.A., Okieimen, F.E., 2011. Heavy metals in contaminated soils: a review of sources, chemistry, risks and best available strategies for remediation. ISRN Ecol. 2011, 1–20.

Yadav, S., Modi, P., Dave, A., Vijapura, A., Patel, D., Patel, M., 2020. Effect of Abiotic Stress on Crops. Sustainable Crop Production. Intech Open, pp. 1–21.

Yang, J., Kloepper, J.W., Ryu, C.-M., 2009. Rhizosphere bacteria help plants tolerate abiotic stress. Trends Plant Sci. 14 (1), 1–4.

Yasin, N.A., Akram, W., Khan, W.U., Ahmad, S.R., Ahmad, A., Ali, A., 2018. Halotolerant plant-growth promoting rhizobacteria modulate gene expression and osmolyte production to improve salinity tolerance and growth in Capsicum annum L. Environ. Sci. Pollut. Res. 25 (23), 23236–23250.

Zhang, Y., Hu, J., Bai, J., Wang, J., Yin, R., Wang, J., Lin, X., 2018. Arbuscular mycorrhizal fungi alleviate the heavy metal toxicity on sunflower (*Helianthus annus*) plants cultivated on a heavily contaminated field soil at a WEEE-recycling site. Sci. Total Environ. 628, 282–290.

Phosphate biofertilizers: Recent trends and new perspectives

Mohammad Saghir Khan[a], Asfa Rizvi[b], Bilal Ahmed[c], and Jintae Lee[c]

[a]Department of Agricultural Microbiology, Faculty of Agricultural Sciences, Aligarh Muslim University, Aligarh, Uttar Pradesh, India, [b]Department of Botany, School of Chemical and Life Sciences, Jamia Hamdard, New Delhi, India, [c]School of Chemical Engineering, Yeungnam University, Gyeongsan, Republic of Korea

16.1 Introduction

The consistent and terrifying increase in the human populations together with declining cultivable lands has threatened the world's food security, leading to acute problems in feeding humans (Salcedo Gastelum et al., 2020). Chemical fertilizers, though, have played a fundamental role in maintaining global crop productivity at current levels and will be even more crucial if yields are to be optimized worldwide. The continuous use of conventional chemical fertilizers in modern agronomic practices, however, causes surface water and groundwater pollution, waterway eutrophication, soil fertility depletion, and accumulation of toxic elements (Alori et al., 2017; Nyalemegbe et al., 2010). Consequently, they are integrated into the food chain causing human/animal health hazards. Conclusively, due to problems such as (i) leaching, (ii) photolysis, and (iii) microbial degradation associated with the use of chemical fertilizers, the attention is given to finding a viable and ecologically sustainable alternative strategy that can ensure competitive yields without damaging the health of soils. In this regard, the current trends of viable and agronomically feasible fertilizers generally referred to as "biofertilizers" have provided solutions to the expensive and hazardous chemical formulations. Indeed, biofertilizers, defined as the "preparation/formulations containing live microbes which enhances the soil fertility/nutrient availability either through BNF/SNF, solubilizing/mineralizing P or decomposing organic wastes or by accelerating plant growth by producing growth hormones with their biological activities" (Okur, 2018; Chaparro et al., 2014), are a breakthrough technology that promises a significant increase in food productivity at minimal cost. Among biofertilizers (Nafis et al., 2019; Umesha et al., 2018), a group of microorganisms capable of solubilizing/mineralizing insoluble inorganic (mineral) P and mineralizing insoluble organic P into soluble and available forms, commonly referred to as phosphate-solubilizing microorganisms (PSM), supply essential master key nutrients, especially P, to plants and thus increase the bioavailability of P for plant use (Abawari et al., 2020). The microbial intervention of P solubilization seems to be an effective way to solve the P availability in soils. However, the effectiveness of phosphate solubilizers in soils depends on factors such as the soil temperature, moisture, pH, salinity, and source of insoluble P, method of inoculation, the energy sources, and the strain of microorganism used (Soumare et al., 2020; Khan et al., 2007).

Microbiome-mediated management of soil P is the only eco-friendly and low-cost approach to enhance plant available P and seems a realistic alternative to reduce environmental risks of conventional phosphatic fertilizers. Soil/rhizosphere indeed is the largest reservoir that harbors many heterogeneous microbes capable of effectively releasing P from total soil P pool through solubilization and mineralization (Chawngthu et al., 2020; Manzoor et al., 2017). Microbial communities involving species of fungi (Doilom et al., 2020; Sarr et al., 2020), bacteria (Mohamed et al., 2019; Behera et al., 2017), and actinomycetes (Farhat et al., 2015; Saif et al., 2014) have been reported to solubilize P, and some of them can mobilize P in plants (Zhu et al., 2012). The phosphate solubilizers when applied as microbiological formulations have been reported to enhance the growth and yield of many food crops including legumes (Singh et al., 2018; Ditta et al., 2018), cereals (Batool and Iqbal, 2019; Tahir et al., 2018), and vegetables (Menéndez et al., 2020; Rizvi et al., 2014). The stress-tolerant P-solubilizing microbiomes such as salt-tolerant or halophilic organisms (Mahadevaswamy and Nagaraju, 2018; Marakana et al., 2018) and metal-tolerant P solubilizers (Yang et al., 2020; Biswas et al., 2018) have been found to facilitate the growth and development of food crops even under problem soils. The inoculation of seed/seedling/soil with P-solubilizing/mineralizing organisms either alone or in combination is, therefore, a promising strategy for the improvement of P absorption by plants, thereby reducing the use of lethal phosphatic fertilizers. Here, we intend to provide a recent perspective on the composition and functional diversity of PSM, mechanisms of P solubilization/mineralization, how PSM induce plant growth under conventional and stressful conditions, and their possible role as phosphate biofertilizer in food crop production.

16.2 Importance of P and rationale for using phosphate biofertilizers in agrosystems

Phosphorus, one of the major plant nutrients, is next only to nitrogen, accounting for 0.2% (w/w) of plant dry weight (Maharajan et al., 2018), and is the most restrictive factor for plant growth. Phosphorus, a light-bearing, nonmetallic plant nutrient, once absorbed from the soil solution as monovalent (H_2PO_4) and divalent (HPO_4) orthophosphate anions, facilitates the root morphogenesis including development, root anatomy modifications, and root hair density, leading to yield increase of crops and plant resistance against multiple diseases (Kondracka and Rychter, 1997). At cellular levels, P participates in cell division, growth of new tissues and nucleic acid structure, protein synthesis regulation, respiration, signal transduction, macromolecular biosynthesis, phospholipids, carbon metabolism and a wide range of enzymes, energy transfer, and photosynthesis (Khan et al., 2014; Hameeda et al., 2008; Saber et al., 2005). P-deficient plants, on the other hand, exhibit retarded growth (reduced cell and leaf expansion, respiration, and photosynthesis) and are often characterized by symptoms such as reddish leaves (enhanced anthocyanin formation) and necrosis in old leaf tips (Luiz et al., 2018) and dark green color (higher chlorophyll concentration).

Globally, 5.7 billion hectares of land contains a lower quantity of available P that can facilitate crop production (Hinsinger, 2001). It is reported that many soils have high reserves of total P accounting for approximately 0.05% of the soil dry weight on average, but only a small proportion (1%–5%) of the total soil P is available in a soluble form that can be taken up by plants (Mahdi et al., 2011; Takahashi and Anwar, 2007). In contrast, plant requires from 1 to 5 µM for optimal growth, for instance, in the case of grasses, and 5–60 µM for high-P-demanding crops, such as tomato and pea (Raghothama, 1999). However, the low P availability in agricultural soils is an important issue that affects over 2 billion hectares worldwide (Oberson et al., 2001). For instance, P deficiency has been reported to significantly reduce the crop yields by 5%–15% (Shenoy and Kalagudi, 2005). Most natural ecosystems in tropical and subtropical areas are predominantly acidic, rich in iron, and extremely P-deficient (Ramachandran et al., 2007) due to the strong fixation of P as insoluble P of Fe and Al resulting in P deficiency in the soil (Tariq et al., 2014). Thus, the majority of P remains unavailable for absorption by plants due to its rapid rate of fixation/complex formation with other soil constituents. Therefore, to counterbalance the low P availability (deficiency) and to allow plants to grow and functions normally in these regions, phosphatic fertilizers are frequently used on a large scale to maintain soil fertility in crop cultivation practices. Among synthetic fertilizers, approximately 52.3 billion tons of P-based fertilizers are applied annually worldwide to maintain available P levels in soil-plant systems (FAO, 2017). Of these, only 0.2% (< 10 μM) of applied P is used by plants (Islam et al., 2019; Alori et al., 2017) and the remaining is lost due to complex formation in the soils (Gyaneshwar et al., 2002), needing frequent P enrichment of the soil. The production of conventional phosphatic fertilizers is, however, a highly energy-intensive process requiring energy worth US$ 4 billion per annum in order to meet the global P need (Goldstein et al., 1993). However, the excessive and injudicious application of P fertilizers and their deposition in soils can pose serious environmental problems such as water eutrophication (Kang et al., 2011; Chang and Yang, 2009). Enhancement in crop productivity through uncontrolled fertilization to fulfill human food demands has also resulted in intense mining of P-containing minerals around the world. The world reserves of rock phosphate (RP), the only significant global resources of P, are declining consistently due to its unregulated use (Leghari et al., 2016), resulting in a massive increase in the cost of P fertilizers. It has been estimated that the P mines could be depleted by 2060 (Gilbert, 2009), while the amount of P in agricultural soils is sufficient to sustain maximum crop yields worldwide for only about 100 years (Walpola and Yoon, 2012). Additionally, the overuse and abuse of chemical fertilizers can cause unpredicted environmental hazards, damage human health, and pollute the surrounding environment, thereby adversely impacting the sustainability of soil-plant systems (Adesemoye and Kloepper, 2009). These factors together provide enough reasons to search for eco-friendly and inexpensively realistic alternative strategies for improving crop production in low-P or P-deficient soils (Vaxevanidou et al., 2015). In this context, organisms endowed with P-solubilizing (PS) activity, often termed phosphate-solubilizing microorganisms (PSM), a group of beneficial microorganisms capable of hydrolyzing organic and inorganic P compounds from insoluble P sources (Kalayu, 2019; Iwuagwu et al., 2013), are regarded as a promising substitute to chemical P fertilizers in improving soil conditions for plant growth under P-deficient soils. The PSM-plant interactions could, therefore, be of great practical value in maintaining the P pool of soils and consequently supplying enough P to plants.

16.3 Current status of phosphate biofertilizers

Soil-plant-microbe interactions in general and phosphate biofertilizers in particular have received greater attention in recent times, though the role of soil microbiota in mineral phosphate solubilization (mps) was known as early as 1903 but they were used as biofertilizers since the 1950s (Krasilinikov, 1957). In general, the biofertilizer technology is considered safe and preferred over the chemical fertilizers due to their ability to (i) release nutrients slowly from a source and according to the need of plants, (ii) complement other minerals, and (iii) provide growth factors that cannot be achieved through the use of chemical fertilizers (Yadav and Sarkar, 2019; García-Fraile et al., 2015). Broadly, biofertilizers that are commercially available to the farmers have been categorized as follows: (i) nitrogen-fixing biofertilizers (*Rhizobium*, *Bradyrhizobium*, *Azospirillum*, and *Azotobacter*), (ii) phosphate biofertilizers (*Bacillus*, *Pseudomonas*, and *Aspergillus*), (iii) phosphate-mobilizing biofertilizer (mycorrhiza), and (iv) plant growth-promoting biofertilizers (*Pseudomonas*). Currently, the biofertilizer market is valued around $1.57 billion and is expected to reach $1.88 billion very soon, while the global fertilizer market need will be around $245 billion. Biofertilizers based on P-solubilizing bacteria (phosphate biofertilizers) account for only 14% of this market (Market Data Forecast, 2018). The major phosphate biofertilizer-producing units in India are presented in Table 16.1. Among microbiological biofertilizers, P biofertilizers are gaining momentum due to their significant role in the nutrient management and nonhazardous nature while reducing chemical use for quality food production, especially in developing countries such as Africa and in Asia, which together account for 50% and 74%, respectively, of the global land mass and population (Ogbo, 2010). A commercial biofertilizer named "phosphobacterin" was first prepared and used in the former Soviet Union from *B. megaterium* var. *phosphaticum* adsorbed on kaolin. Later on, this microphos technology was adopted by East European countries and many Asian countries including India. For instance, a carrier-based formulation commonly called IARI microphos was developed by the Indian Agricultural Research Institute (IARI), New Delhi, India, using efficient P-dissolving strains of *P. striata*, *B. polymyxa* and *A. awamori*, *A. niger*, and *P. digitatum* (Gaur, 1990) packed in a wood charcoal and soil mixture. Some of the potential P-solubilizing microbes are *Bacillus* and *Pseudomonas* (Illmer and Schinner, 1992), while *Aspergillus* and *Penicillium* form the important fungal genera (Motsara et al., 1995), rhizobia (Chabot et al., 1996; Abd-Alla, 1994), actinomycetes (Faried et al., 2019; Poomthongdee et al., 2015), etc. Among PSM, phosphate-solubilizing bacteria (PSB) outnumber phosphate-solubilizing fungi (PSF) in soils by 2–150 times (Kucey, 1983).

Table 16.1 Main manufacturers of biofertilizers based on P-solubilizing microbes.

Product unit/country	Product name	Component	Form(s)	Performance declared by the manufacturer	References (Web ID)
TNAU Agritech Portal/India	Phosphobacteria and Phosphatika	Two bacteria and two fungal species	Powder/liquid	Increase yield 5%–30%	http://agritech.tnau.ac.in
Monarch Bio-Fertilisers and Research Centre/India	Phosphobacteria	Bacteria	Powder	Dissolve 30–50 kg of phosphorus/hectare	http://www.monarchbio.co.in/bio_fertilizers.html
SAFS Organic Enterprises/India	Phosphobacterium	Bacteria	Powder/liquid	No indication	https://www.indiamart.com/safsorganicenterprises/bio-fertilizer.html#bio-fertilizer-phosphobacterium
Agro Bio-Tech Research Centre Ltd/India	Phosphobacteria	*Bacillus megaterium* var. *phosphaticum*	Powder/liquid	No indication	http://www.abtecbiofert.com/products.htm
Ajay Bio-Tech (India) Ltd/India	Biophos	Bacteria and fungi	Powder (spore)	No indication	https://www.linkedin.com/company/ajay-biotech-india-ltd/?originalSubdomain=fr
International Panaacea Limited/India	Phosphofix	Bacteria	Liquid	Reduce 25%–30% phosphatic fertilizer requirement	https://www.iplbiologicals.com/
Varsha Bioscience and Technology India Private Limited/India	Phosphomax	*Bacillus megaterium*	Powder	Increase crop yield by 15%–25%	http://www.varshabioscience.com/products/phosphomax.html
T. Stanes Company Limited/India	Symbion-P	*Bacillus megaterium* var. *phosphaticum*	Liquid	Save up to 50% over the cost of phosphorus chemical fertilizer	http://www.tstanes.com/products-symbion-p.html
Novozymes Biologicals Limited/Canada	Jumpstart LCO	*Penicillium bilaii*	Powder/granular	Solubilize 8.25 kg/ha	https://www.novozymes.com/en/advance-your-business/agriculture/crop-production/jumpstart
AgriLife/India	P Sol B-BM	*Bacillus megaterium*	Powder (spore)	No indication	http://www.agrilife.in/bioferti_psolb_bm.htm

16.4 Development of phosphate biofertilizers: An overview

Phosphate solubilizers form an integral and essential component of soil-plant systems and significantly influence the soil P pool by transforming insoluble P to plant accessible P through solubilization/mineralization. However, the P-solubilizing/mineralizing efficiency of microbial P biofertilizers varies with soil hydrology, physicochemical features, and photosynthates besides age and genotypes of plants. Despite these, the abundance of PSM involving bacteria including conventional phosphate solubilizers (Table 16.2), nitrogen fixers (Table 16.3), and fungi and actinomycetes (Table 16.4) inhabit the conventional/stressed bulk soil/rhizospheres. The identification and functional analysis of such microbiomes are important in order to develop and commercialize phosphate biofertilizers. Generally, such organisms are recovered from a wide array of habitat including rhizosphere (Chouyia et al., 2020; Emami et al., 2019), nonrhizosphere soils (Mendoza-Arroyo et al., 2020; Doilom et al., 2020), rhizoplane (Selvi et al., 2017; Sarkar et al., 2012), phyllosphere/phylloplane (Batool et al., 2016), and rock phosphate deposit area soil (Aliyat et al., 2020; Azaroual et al., 2020) and even from problem soils (Teng et al., 2019; Fitriyanti et al., 2017) using standard microbiological methods (serial plate dilution method or enrichment culture technique). Since 1948, when Gerretsen proposed a method for the detection of P solubilizers, there have been several attempts to discover methods/media for cultivating PSM. In this regard, Pikovskaya (1948), bromophenol blue dye method (Gupta et al., 1994), and National Botanical Research Institute P (NBRIP) medium (Nautiyal, 1999) have been designed and commercialized for identifying PSM. Broadly, the PS efficiency of any organism is revealed through the formation of clear halo (a sign of solubilization) around microbial colonies (Fig. 16.1). The colonies showing P solubilization are purified by regular subculturing, and the persistence of PS activity is checked. Following selection, the PS activity of microbes is quantitatively assayed, after which they are produced at mass scale (microphos) for onward transfer to farm practitioners (farmers). The production of microphos, thus, involves three phases: (i) identification, selection, and testing of functional quality of P solubilizers; (ii) carrier-based inoculant preparation and polythene-based packaging following ISI regulation; and (iii) quality assessment of prepared P biofertilizers and distribution to consumers. For microphos production, peat, farmyard manure (FYM), soil, and cow dung cake powder have been suggested as suitable carriers. Finally, the cultures are packed in polybags and can safely be stored for about 3 months at $30 \pm 2°C$.

Table 16.2 Conventional and stress-tolerant phosphate-solubilizing bacterial biofertilizers.

Phosphate solubilizers

Conventional bacterial genera	Source/origin	Media/method used	References
Enterobacter	Rhizosphere of *Capsicum chinense*	Pikovskaya (PVK)	Mendoza-Arroyo et al. (2020)
Pseudomonas spp.	Wheat, barley, maize, oat, faba beans, peas	NBRIP	Elhaissoufi et al. (2020)
Acinetobacter, Pseudomonas, Massilia, Bacillus, Arthrobacter, Stenotrophomonas, Ochrobactrum, and Cupriavidus	Bulk soil	NBRIP medium	Wan et al. (2020)
Pseudomonas	Wheat rhizosphere	NBRIP	Liu et al. (2019)
Pseudomonas aeruginosa	Chili rhizosphere	Vanadomolybdate phosphoric yellow color method	Linu et al. (2019)
Bacillus subtilis, Serratia marcescens Pantoea, Pseudomonas, Serratia, and Enterobacter	Tomato rhizosphere Wheat rhizosphere	PVK NBRIP	Mohamed et al. (2018) Rfaki et al. (2020)
Burkholderia cepacia, B. contaminans	Sweet corn rhizosphere	PVK	Pande et al. (2020)

Endophytes

Aneurinibacillus sp. and *Lysinibacillus* sp.	Banana tree roots	NBRIP	Matos et al. (2017)

Stress–tolerant

Halophiles

Haloarcula, Halobacterium, Halococcus, Haloferax, Halolamina, Halosarcina, Halostagnicola, Haloterrigena, Natrialba, Natrinema, and Natronoarchaeum	Sediment, water, and rhizospheric soil	Haloarchaea P Solubilization (HPS) medium	Yadav et al. (2015a)

Alkaliphiles

Bacillus marisflavi, Chromohalobacter israelensis	Mangrove Merces, Batim salt pan	PVK	Prabhu et al. (2017)

Metal-tolerant

Enterobacter cancerogenus	NBRIP medium	Agricultural fields	Walpola and Hettiarachchi (2018)

Pesticide-tolerant

Insecticide-tolerant

Pseudomonas sp.		Pesticide-treated *Achillea clavennae* rhizosphere	PVK	Rajasankar et al. (2013)
Pestalotiopsis microspora, Aquabacterium commune, Bacillus spp.		Rhizospheres of groundnut, tomato, broad beans, and taro root	PVK	Thiruvengadam et al. (2020)

Fungicide-tolerant

Pseudomonas spp.		Cabbage and mustard rhizospheres	PVK	Khan et al. (2020)

The phosphate growth NBRIP (National Botanical Research Institute Phosphate) medium (Nautiyal, 1999) contained the following (in g L^{-1}): glucose, 10.0; $Ca_3(PO_4)_2$, 5.0; $MgCl_2·6H_2O$, 5.0; $MgSO_4·7H_2O$, 0.25; KCl, 0.2; and $(NH_4)_2SO_4$, 0.1 (pH 7.0). Pikovskaya (1948) contained the following (in g/L): yeast extract, 0.5; dextrose, 10.0; calcium phosphate, 5.0; ammonium sulfate, 0.5; potassium chloride, 0.2; magnesium sulfate, 0.1; manganese sulfate, 0.0001; ferrous sulfate, 0.0001; Sperber's basal medium (glucose: 10 g; $MgSO_4·7H_2O$: 0.25 g; $CaCl_2$: 0.1 g; agar: 15 g) supplemented with 2.5 g of tricalcium phosphate ($Ca_3(PO_4)_2$); and Haloarchaea P Solubilization (HPS) medium (in g/L): 10.0 glucose, 1.0 yeast extract, 5.0 tricalcium phosphate (TCP) or hydroxyapatite (HA) or rock phosphate (RP, a nondetrital sedimentary rock which contains high amounts of phosphate-bearing minerals (P_2O_5: 32%)), 195.0 NaCl, 35.0 $MgCl_2·6H_2O$, 50.0 $MgSO_4·7H_2O$, 5.0 KCl, 0.5 $(NH_4)_2SO_4$, 1.0 $NaNO_3$, $CaCl_2·2H_2O$, 0.05 KH_2PO_4, 0.03 NH_4Cl, traces $FeSO_4·7H_2O$, traces $MnSO_4·7H_2O$, and 20 agar. pH was adjusted to 7.4 with 1M Tris base and autoclaved. Filter-sterilized 8% (w/v) $NaHCO_3$ and 25% (w/v) sodium pyruvate solutions were added aseptically to the autoclaved medium.

Table 16.3 Phosphate-solubilizing nitrogen-fixing bacteria recovered from variable regions.

Nodule bacteria	Source/origin	Media/method used	References
Rhizobium, Agrobacterium, Phyllobacterium	Root nodules of Acacia cyanophylla	PVK	Lebrazi et al. (2020)
Mesorhizobium ciceri, M. tamadayense	Cicer canariense nodules	NBRIP	Menéndez et al. (2020)
A. tumefaciens syn. Rhizobium radiobacter	Nodules of Leucaena leucocephala growing in saline soil	NBRIP	Verma et al. (2020)
Rhizobium (pea)	Root nodules of field pea	PVK	Gebremedhin et al. (2019)
R. leguminosarum	Root nodule of Vicia faba	Sperber's basal medium	Shravanthi and Panchatcharam (2017)
Azotobacter	Rhizosphere of agricultural crops, maize rhizospheres	Sperber medium, PVK, and NBRIP	Bjelić et al. (2015), Nosrati et al. (2014)
Azospirillum strain	Wheat rhizospheres	PVK, MPVK, and LB	Ayyaz et al. (2016)
Sinorhizobium meliloti	Phaseolus lunatus nodules	YED-P	Ormeño et al. (2007)

Sperber medium composed of 10 g glucose, 0.5 g yeast extract, 0.1 g $CaCl_2$, and 0.25 g $MgSO_4 \cdot 7H_2O$ was supplemented with 2.5 g $Ca_3(PO_4)_2$ (TCP) and 15 g agar (in solid medium) per liter at pH 7.2 (Malboobi et al., 2009); in modified Pikovskaya (MPVK) medium, sucrose was replaced by sodium malate (10 g L^{-1}); and Luria-Bertani (LB) was supplemented with TCP.

Table 16.4 Fungal and actinomycetal phosphate solubilizers recovered from different sources.

Phosphate solubilizers	Source/origin	Media/method	References
Penicillium guanacastense	Pinus massoniana rhizosphere	NBRIP	Qiao et al. (2019)
Aspergillus hydei, Gongronella hydei, P. soli, and Talaromyces yunnanensis	Rhizosphere of Quercus rubra	PVK	Doilom et al. (2020)
Rhizopus stolonifer and R. oryzae	Rhizosphere of Prosopis juliflora and Solanum xanthocarpum	Tris Minimal agar media supplemented with rock phosphate	Patel et al. (2015)
Streptomyces roseocinereus and S. natalensis	Moroccan oat rhizosphere	MPVK	Chouyia et al. (2020)

16.4.1 How are phosphate biofertilizers applied?

Phosphate biofertilizers can be used for all crops including cereals (Wang et al., 2020; Elhaissoufi et al., 2020), vegetables (Rizvi et al., 2014), oilseeds (Santana et al., 2016), and pulses (Zafar et al., 2020; Ahmad et al., 2016). The most widely applied methods recommended

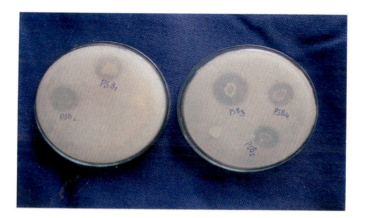

Fig. 16.1 Zone of insoluble phosphate (TCP) solubilization by bacteria on Pikovskaya agar plate.

for P-biofertilizer application are seed treatment, seedling dipping, and soil application. The seed surface inoculation of P biofertilizers is traditionally the most commonly used and easiest means of application. When appropriately applied, this method guarantees that each seed receives the introduced mps inoculum. However, the quantity of inoculum that can stick onto the seed surface, direct contact of bacterized seeds with seeds dressed with pesticides, movement of phosphatic biofertilizer away from the rooting/target zone, and exposure to natural biotic and abiotic stresses after planting are some of the limitations of P-biofertilizer application under field conditions. The use of a sticker solution, for example, gum arabic, however, may improve the attachment of the P inoculants onto the seed surface. Soil application is yet another method that has the advantage over other modes of PSM application. For instance, through soil application, there occurs greater aggregation of PSM per unit area, the direct contact of P biofertilizers with chemically treated seeds is minimized, the process of seed mixing is avoided, and the ability to withstand low moisture in the soil is better than the powdered form of P biofertilizers. Conclusively, the P biofertilizers, either alone wherein only one P-solubilizing organism is used and/or the coculture or mixed culture can be applied in varying cultivation practices (Zaidi et al., 2009).

16.4.2 Factors affecting the survival of phosphate-solubilizing microorganisms

The performance of phosphate biofertilizers under field conditions depends on the intrinsic competence, colonizing efficiency, and their survivability, which, however, is influenced greatly by changing environment of a specific habitat. The biotic factors that influence the survival of the inoculated PSM include competition and predation. The survival of the applied bacterial strains is also impacted by soil composition (Bashan et al., 1995), soil photosynthates, nutrient status, temperature, pH, and moisture content (Van Elsas et al., 1991). The density of the introduced PSM, however, declines rapidly after their application into soils (Ho and Ko, 1985).

16.5 Overview of P solubilization mechanisms

Broadly, the phosphate-solubilizing microbes transform the inaccessible P such as $Ca_3(PO_4)_2$, Fe_3PO_4, and Al_3PO_4 to plant accessible forms by acidification, chelation, exchange reactions, and polymeric substance formation (Chang and Yang, 2009; Delvasto et al., 2006) while they mineralize organic P into bioavailable form through enzymes and facilitate the P uptake by plant roots (Khan et al., 2007). The organic acid (OA) theory of P solubilization by phosphate biofertilizers, for instance, bacteria (Table 16.5) and fungi and actinomycetes (Table 16.6), is the most widely accepted mechanism of P

Table 16.5 Organic acids secreted by phosphate-solubilizing bacterial biofertilizers.

P-solubilizing bacteria	Organic acids	References
Bacillus, Burkholderia, Paenibacillus sp.	Gluconic, oxalic, citric, tartaric, succinic, formic, and acetic acids	Chawngthu et al. (2020)
Pantoea, Pseudomonas, Serratia, and *Enterobacter*	Oxalic, citric, gluconic, succinic, and fumaric acids	Rfaki et al. (2020)
Pseudomonas sp. strain AZ5	Acetic, oxalic, and gluconic acids	Zaheer et al. (2019)
Bacillus sp. strain AZ17	Acetic, citric, and lactic acids	Zaheer et al. (2019)
B. megaterium Y95, Y99, Y924, Y1412	Succinic acid, oxalic acid, citric acid	Zheng et al. (2018)
B. megaterium, Bacillus sp.	Gluconic acid, lactic acid, acetic acid, succinic acid, propionic acid	Saeid et al. (2018)
B. megaterium P17	Formic acid, acetic acid, propionic acid, butyric acid	Zhong et al. (2017)
Bacillus sp., *Pseudomonas* sp., *Proteus* sp., *Azospirillum* sp.	Citric acid, glutaric acid, glyoxylic acid, ketoglutaric acid, ketobutyric acid, malic acid, malonic acid, succinic acid, fumaric acid, tartaric acid, gluconic acid	Selvi et al. (2017)
Serratia sp.	Malic acid, lactic acid, acetic acid	Behera et al. (2017)
Pantoea sp.	Acetic acid, gluconic acid, formic acid, propionic acid	Sharon et al. (2016)
Pseudomonas veronii PSB12	Gluconic acid, propanedioic acid, propionate, butyrate, succinate, formate, citrate, and pentanoic acid	Chen et al. (2016)
Azospirillum strains	Acetic, citric, lactic, malic, and succinic acids	Ayyaz et al. (2016)
Rhizobium tropici UFLA03-08, *Acinetobacter* sp. UFLA03-09, *Paenibacillus kribbensis* UFLA03-10, *P. kribbensis* UFLA03-106, *Paenibacillus* sp. UFLA03-116	Propionic acid, gluconic acid, tartaric acid, malic acid	Marra et al. (2015)

Table 16.6 Organic acids secreted by phosphate-solubilizing fungi and actinomycetes.

	Source	Organic acids	References
Phosphate-solubilizing fungi			
Trichoderma, Aspergillus	Rhizosphere and rhizoplane of beans, corn, chilli, and tomato	Pyruvic, succinic, fumaric, malic, tartaric, oxalic acids	Zúñiga-Silgado et al. (2020)
Penicillium oxalicum and A. niger	Soil	Oxalic, citric, formic, tartaric, malic, acetic, and citric acids	Li et al. (2016)
Rhizopus stolonifer and R. oryzae	Prosopis juliflora and Solanum xanthocarpum rhizosphere	Gluconic, oxalic, propionic, and malic acids	Patel et al. (2015)
A. niger, P. canescens, Eupenicillium ludwigii, and P. islandicum	Eucalyptus rhizosphere	Oxalic acid, citric acid, gluconic acids	de Oliveira Mendes et al. (2014)
Phosphate-solubilizing actinomycetes			
Streptomyces sp. KP109810	–	Oxalic, lactic, citric, succinic, gluconic, malic, fumaric, acetic acid, propionic acids	Mohammed (2020)
Streptomyces sp. CTM396	Agricultural soils	Gluconic acid	Farhat et al. (2015)
Streptomyces mhce0811 and mhce0816	Wheat rhizosphere	Gluconic and malic acids	Jog et al. (2014)

availability in soils. The role of OA in solubilizing insoluble P may be due to the lowering of pH and chelation of cations and by competing with P for adsorption sites in soils (Nahas, 1996). The OAs secreted by PSM chelate mineral ions or reduce the pH to bring P into solution (Cunningham and Kuiack, 1992). As a consequence, the microbial cells and their surroundings are acidified, leading to the release of P ions from the P mineral by H^+ substitution for Ca^{2+} (Goldstein, 1994). The OAs, for example, oxalic acid, citric acid, and lactic acid, released into the liquid medium by PSM can be assayed by paper or thin-layer chromatography (TLC) (Mohammed, 2020) or by high-performance liquid chromatography (HPLC) (Rashid et al., 2004). In other studies, a significant portion of P was reported to be fixed in acidic soil (such as red soil) accumulating Fe or Al ions, but no correlation was observed between pH and the amount of solubilized P (Asea et al., 1988), suggesting the involvement of mechanisms other than organic acid secretion by PSM. So, the release of H^+, production of chelating substances and inorganic acids has been suggested as an alternative mechanism of P solubilization by PS microbes. As an example, 0–5000 µM of gluconic acid at pH 4–7 showed no effect on Ca-P solubility at pH > 6 (Illmer and Schinner, 1992). Also, the quantity of P solubilized under in vitro conditions does not correlate with the amounts of OA released into liquid culture medium highlighting the role of factors other than OA in microbe mediated solubilization of insoluble P. Inorganic acids like, HCl can also solubilize P but they are less effective compared with OA at the same pH (Kim et al., 1997). Moreover, pH, humidity and cation contents of soil and microbial compositions influence the P solubilization (Mohamed et al., 2018; Liu et al., 2015). Enzymes such as acid phosphatases (Behera et al., 2017; Swetha and Padmavathi, 2016), phytases (Liu et al., 2018; Menezes-Blackburn et al., 2016), and phospholipases (Zavaleta-Pastor et al., 2010) produced by many PSM induce the release of P from organic matters by the mineralization process in the rhizosphere soil.

16.6 Phosphate biofertilizers: Phyto-beneficial and eco-physiological perspective

16.6.1 Phyto-beneficial features

The phosphate biofertilizers when applied as microbial formulation can improve the P use efficiency (PUE) in agricultural soils. Broadly, plant inoculation with P biofertilizers seems to satisfy two major objectives of the modern agriculture: (i) when used either alone or in synergism, it increases the plant biomass (Sirinapa et al., 2021; Ahmad et al., 2021; Wang et al., 2020); and (ii) it improves the nutritional value of seeds and fruits of food crops (Tchakounté et al., 2020; Kang et al., 2019; Sharma et al., 2018). Together, they facilitate the growth and development, leading to an increase in plant yields by secreting various phyto-beneficial compounds (Table 16.7). Such plant growth promotory traits exhibited by valuable P biofertilizers include (i) excretion of phytohormones, such as indoleacetic acid (Zhang et al., 2017; Dasgupta et al., 2015) and gibberellin (Kang et al., 2019): together, these plant hormones improve root morphogenesis for better P acquisition; (ii) asymbiotic (Nosrati et al., 2014) or symbiotic N_2 fixation (Zaidi et al., 2017); (iii) management of plant pathogens through antibiotics (Mitra et al., 2020) or by secreting hydrolytic enzymes (Hamane et al., 2020); and (iv) production of low molecular weight iron-chelating compounds, siderophores (Mendoza-Arroyo et al., 2020; Abbas et al., 2019), and cyanogenic compounds (Boubekri et al., 2021): microbial excretion of HCN has been observed as an important antifungal feature to manage root-infecting fungi (Ramette et al., 2003) and is considered an environmentally sustainable strategy for biological control of weeds (Heydari et al., 2008). In addition, many P biofertilizers also produce the enzyme 1-aminocyclopropane-1-carboxylate (ACC) deaminase

Table 16.7 Plant growth-promoting active biomolecules released by phosphate biofertilizers affecting food crops.

Soil microbiota	Source	PGP activities	References
Streptomyces alboviridis P18-S. griseorubens BC3-S. griseorubens BC10 and Nocardiopsis alba BC11	Desert soils of Morocco	IAA, siderophore, HCN, and ammonia	Boubekri et al. (2021)
Bacillus strains		IAA	de Sousa et al. (2020)
Agrobacterium tumefaciens syn. Rhizobium radiobacter	Nodules of Leucaena leucocephala	Zinc solubilization, IAA, N_2 fixation, siderophores, EPS, salt tolerance	Verma et al. (2020)
S. roseocinereus, S. natalensis	Oat rhizospheres	Siderophores, IAA, ACC deaminase, and antimicrobial activity against Fusarium oxysporum, Botrytis cinerea, Phytophthora cactorum, and Phytophthora cryptogea	Chouyia et al. (2020)
Agrobacterium sp. NA11001, Phyllobacterium sp. C65, Bacillus sp. CS14, and Rhizobium sp. V3E1	Root nodules of Acacia cyanophylla	IAA	Lebrazi et al. (2020)
Acinetobacter sp., PGP27, Ensifer meliloti	Faba bean and wheat rhizosphere	K solubilization, IAA, EPS	Bechtaoui et al. (2019)
Bacillus sp.	Stevia rebaudiana rhizosphere	IAA, siderophores	Prakash and Arora (2019)
Lysinibacillus fusiformis, Bacillus sp., Paenibacillus sp.	Roots of wheat	IAA, siderophore production, protease activity, and antibacterial and antifungal inhibition	Akınrınıola et al. (2018)
Mesorhizobium ciceri and M. mediterraneum	Chickpea nodules	N_2 fixation, IAA	Zafar et al. (2017)
Klebsiella sp. Br1, K. pneumoniae Fr1, B. pumilus S1r1, Acinetobacter sp. S3r2	Maize roots	N_2 fixation, auxin production	Kuan et al. (2016)
A. chroococcum, P. polymyxa, Burkholderia cepacia	Maize rhizosphere Maize rhizosphere	IAA production, N_2 fixation Biocontrol activity	Kalaiarasi and Dinakar (2015) Zhao et al. (2014)
A. lipoferum, A. brasilense, A. chroococcum	–	Phytohormone	Gholami et al. (2012)

(Zhao et al., 2021; Kalam et al., 2020), which lowers the plant ethylene levels (Gupta and Pandey, 2019). Physiologically, ethylene (C_2H_4 or $H_2C=CH_2$), one of the phytohormones produced under stress (Sharma et al., 2019), can induce senescence, chlorosis, and abscission in plants, aggravating the lethal impact caused by the pathogens (Etesami and Glick, 2020; Dubois et al., 2018). Mechanistically, the enzyme ACC deaminase cleaves ACC, a precursor of ethylene synthesis, into α-ketobutyrate and ammonia (del Carmen Orozco-Mosqueda et al., 2020; Khan et al., 2014) and thus reduces the levels of the precursor and consequently the ethylene production. The decreased levels of ethylene allow the plants to grow normally under stress and, therefore, can improve the growth.

16.6.2 Eco-physiological traits

Globally, arable soils do not have optimal cultivation conditions, and therefore, they present additional agronomic challenges due to which plants fail to flourish under abiotic and biotic stresses. Among stresses, abiotic stresses include soil salinity, drought, flooding, extremes of temperature, and organic pollutants and metals, while plant pathogens, such as insects, fungi, bacteria, viruses, and nematodes, among biotic factors, affect the agricultural production (Gamalero and Glick, 2020; Gimenez et al., 2018; Santoyo et al., 2017). As an example, the salinization of soils greatly inhibits the growth of plants (Etesami and Beattie, 2017), and according to some reports, approximately 20% of the world's total arable land is impacted by salinity, especially in arid and semiarid regions (Gamalero et al., 2020; Hanin et al., 2016). In these regions, high temperatures (Narayanan, 2018; Fahad et al., 2017) and water scarcity (Zhao et al., 2020) also inversely affect the crop production, which leads to the enhancement in the cost of crop production (Rodríguez-Flores et al., 2019). Interestingly, the same variously stressed soils also harbor PSM, which, through one or simultaneous growth-promoting mechanisms, facilitate plant growth. The stress tolerance by P biofertilizers toward high salinity (Thant et al., 2018), drought (Kour et al., 2019), temperature (Yadav et al., 2015b), pesticides (Romauld, 2020), and heavy metals (Walpola and Hettiarachchi, 2018) may be an interesting option for their application under variously contaminated soils. To this end, many stress-tolerant bacteria have been isolated from contaminated soils with active physiological metabolisms. For example, heavy metals that are needed in trace amounts for both animals and plants persist in the environment and may alter the composition and functions of soil microbial populations (Rizvi and Khan, 2018). In contrast, the ability to withstand high metal concentrations by P solubilizers is advantageous in agriculture and serves as a good option to develop bioinoculants for augmenting crop production in contaminated soils (Rizvi et al., 2020). The abil-

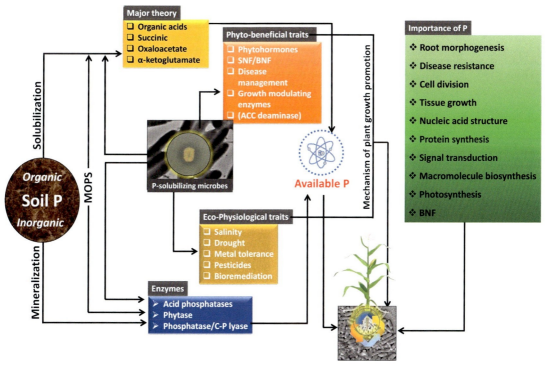

Fig. 16.2 Description of the importance of phosphate biofertilizers in augmenting crop production in different agrosystems. *MOPS*, mechanism of phosphate solubilization; *SNF*, symbiotic nitrogen fixation; *BNF*, biological nitrogen fixation.

ity to withstand stress could be chromosomal or plasmid mediated. The metal-tolerant phosphate-solubilizing bacteria, for example, *Pseudomonas putida* Ws3, have been reported to promote the growth of maize through the synthesis of IAA while decreasing the metal concentration in plants exhibiting largest effects on metal accumulation in roots and shoots (Marzban et al., 2016). A detailed description of importance of P to crops, mechanism of P solubilization, and how different active biomolecules secreted by phosphate biofertilizers augment crop production in different agrosystems is given in Fig. 16.2.

16.7 Trends of phosphate biofertilizers use: A key for sustainable agriculture

The sustainable agriculture system implies an integrated system of plant and animal production practices with a site-specific application which, over a long period of time, satisfy human needs, improve environmental quality, and maintain the economic viability of farm operations. Broadly, the benefits of sustainable agrosystems are to minimize the adverse impacts of different biotic and abiotic stressors/formulations onto plants employing inexpensive natural resources. Among many microbiological options, the application of phosphate biofertilizers in the food production system is promoted for many reasons. Chief among them is their ability to improve soil nutrient pool and supply macronutrients (especially P) to plants (Zhu et al., 2012), and they do not cause environmental pollution and protect plants from numerous stresses. The recent trends suggest that the mono- or coculture application of phosphate biofertilizers can now be used to enhance the growth and yield of many food crops such as vegetables and pulses (Table 16.8) and a variety of cereals (Table 16.9) both in pot experiments and under field conditions.

Table 16.8 Inoculation effect of phosphate biofertilizers on vegetables and pulse crops.

Bacterized crops	Bacteria	Conditions	Response	References
Vegetables				
Potato	*Bacillus pumilus*	Greenhouse	Increased length and weight of both roots and shoots	Yanez-Ocampo et al. (2020)
Cabbage	*B. subtilis*	Greenhouse	Significantly increased plant height, biomass, chlorophyll contents, and nutrient uptake	Kang et al. (2019)
Tomato	*Bacillus* PSB 24	Greenhouse	Better shoot and root growth, and dry and fresh weight	Sarsan (2016)
Pulses				
Chickpea Desi-type (Bittal-2016) and Kabuli-type (Punjab-Noor-2013)	*Pseudomonas* sp. strain AZ5, *Bacillus* sp. strain AZ17	Field	Increased grain yield, straw weight, nodule number, dry weight of nodules, Zn uptake, and P uptake	Zaheer et al. (2019)
Pea	*Rhizobium* sp.	Greenhouse	Significantly higher nodule number, nodule dry weight, shoot dry weight, and total plant N	Gebremedhin et al. (2019)
Greengram	*Pseudomonas striata* and *B. polymyxa*	Greenhouse	Increased plant height, branch/plant, leaves/plant, pods/plant, seeds/plant, pod length and 1000-seed weight, number of effective nodules, nodule density, chlorophylls a and b, carotenoids, protein and proline, and enhanced availability of P and N	Hassan et al. (2017)

Table 16.9 Inoculation effects of phosphate-solubilizing bacteria on the performance of selected cereal crops.

PSB inoculants	Growth parameters of cereals	References
Maize		
Azospirillum brasilense, B. subtilis, P. fluorescens	Improved P uptake efficiency and greater yield	Pereira et al. (2020)
B. subtilis strain 18MZR, *L. fusiformis* strain 31MZR	Improved plant height and nutrient uptake	Rafique et al. (2017)
B. flexus, B. megaterium, Sinorhizobium meliloti	Plant height, shoot and root dry weight, and P nutrition	Ibarra-Galeana et al. (2017)
Bacillus sp., *Azotobacter* sp., *Pseudomonas* sp.	Reduction in heavy metal uptake and overall growth improvement	Mohamed and Almaroai (2017)
Wheat		
Streptomyces alboviridis P18-*S. griseorubens* BC3-*S. griseorubens* BC10 and *Nocardiopsis alba* BC11	Improved root length, root volume, root dry weight, shoot length, and shoot dry weight	Boubekri et al. (2021)
Paenibacillus sp.	Significantly increased the plant height, biomass, root growth, and P uptake	Chen and Liu (2019)
Rice		
Pantoea sp.	Significantly increased the plant height, biomass, root growth, and P uptake	Chen and Liu (2019)
Millet		
Bacillus sp. (C2) and *Pseudomonas* sp.	Increased height, total chlorophyll, IAA, starch, and fresh and dry weight	Harinathan et al. (2016)

The application of phosphate biofertilizers is considered indeed a promising approach to enhancing food production while challenging the soil problems. Broadly, the benefits of adopting microbial management include enriching the P pool of soils, stimulated morphophysiological activities, and significant improvement in the quality and quantity of different food crops. Table 16.10 shows how different plant growth regulators released by phosphate biofertilizers after inoculation enhance the overall performance of maize plant growing under different conditions. Briefly, the growth of maize that was inoculated with phosphate biofertilizers was significantly improved

Table 16.10 Plant growth regulators, experimental conditions, and agronomic response of maize crop to phosphate biofertilizers.

Root-associated phosphate solubilizers	Plant growth modulators	Experimental conditions	Agronomic responses	References
Achromobacter xylosoxidans, *Leclercia adecarboxylata*	IAA, K-solubilizing activity	Greenhouse	Increased photosynthetic rate, stomatal conductance, chlorophyll, carotenoids, and grain yield	Danish et al. (2020)
Enterobacter cloacae	Siderophore, IAA, and ACC deaminase	Growth chamber	Increased root hairs and root length	Abedinzadeh et al. (2019)
B. mojavensis, *P. aeruginosa*, *Alcaligenes faecalis*, *P. syringae* pv. *syringae*, *B. cereus*	IAA, GA, K solubilization, N_2 fixation	Controlled growth chamber	Shoot and root length, root and shoot dry weight	Akintokun et al. (2019)
B. subtilis	Phytohormone	Field	Increased productivity and shoots P	Lobo et al. (2019)
P. kilonensis, *P. protegens*	Biocontrol	Field	Increased leaf yield, height, and length	Alori et al. (2019)
Chryseobacterium sp.	BNF and IAA	Greenhouse	Increased fresh and dry biomass of root and shoot, growth parameter	Youseif (2018)
Burkholderia cepacia	Biocontrol	Greenhouse	Increased leaf area, length, and shoot and root dry weight	Zhao et al. (2014)
Achromobacter sp., *Rhodococcus*	IAA	Controlled condition	Increased plant dry weight, grain yield	Qaisrani et al. (2014)

IAA, GA, BNF, P, K, and NA indicate indoleacetic acid, gibberellic acid, biological nitrogen fixation, phosphorus, potassium, and not available, respectively.

compared to the un-inoculated plants by one or more plant growth modulators secreted by them.

Phosphate biofertilizers play an important role in improving the P acquisition efficiency (PAE) of plants growing in P-deficient agricultural soils (Arif et al., 2017). For example, inoculation of P biofertilizers, *Pseudomonas* (rhizosphere P solubilizer), and *Bacillus* (endophyte P solubilizer) increased the P-efficiency (PE) index by 29.5% and 18.7% in two wheat cultivars (Marvdasht and Roshan, respectively) grown under greenhouse conditions (Emami et al., 2020). The co-inoculation of rhizospheric bacteria with endophytic bacteria increased the PAE of wheat cultivars relative to inoculation with single bacterial culture or un-inoculated plants. Moreover, the rhizospheric and endophytic bacterial formulations applied alone or in consortium synergistically improved the root and shoot dry matter, plant height, grain yield, length, surface area, and volume of roots of the two cultivars. A notable effect was for coculture application suggesting that the PGPR acted additively with plant growth-promoting endophytes. A similar increase in P uptake and grain yield of wheat due to inoculation with P-solubilizing *Pseudomonas* and *Bacillus* species has been reported (Walpola and Yoon, 2012). The studies by Mattos et al. (2020) revealed that the inoculation effect of PSB (*Bacillus* strains) on two sorghum (*Sorghum bicolor*) genotypes with different P responses (BR007-efficient and responsive and SC283-efficient and nonresponsive) grown in soil fertilized with RP and triple superphosphate (TSP) was variable. The results showed that the inoculation response was influenced by sorghum genotype, P source and microbial strain. Inoculation of the genotype BR007 significantly increased root biomass and P content under greenhouse while yield and grain P content were found maximum in the field but no effect was observed for genotype SC283. The application of PSB in combination with RP is therefore, a promising alternative to reduce the use of synthetic fertilizers in agro-ecosystems contributing toward the sustainable sorghum production. Barley plants bacterized with P-solubilizing *Enterobacter ludwigii* and *A. brasilense* and grown in open field conditions showed a beneficial effect on the dry weight, P assimilation and yield, especially in *E. ludwigii*-inoculated plants (Ribaudo et al., 2020). The plant P estimated at 60 DAS was 38%–56% greater in *E. ludwigii*-inoculated barley plants relative to noninoculated plants. The application of bacteria in the absence of fertilizer caused a substantial increase in biological yields (3795 kg/ha) and 1000-seed weight which was comparable to those recorded for the maximum dose of chemical fertilizer. The inoculation of bacteria with the intermediate fertilizer dose significantly increased the grain size. These results clearly demonstrated a promising increase in P due to *E. ludwigii* inoculation. In a field experiment, the P solubilizers *Kosakonia radicincitans* (PSB1) and *B. subtilis* (PSB2) applied with DAP and RP had variable impact on sugarcane

(Pyone, 2021). The inoculation of *B. subtilis* with DAP enhanced the soil available P, stalk population and yield of sugarcane. The PSB with RP further increased the sugarcane weight at harvest. From yield perspective, maximum increase was recorded with *B. subtilis* applied either with DAP or RP. The siderophore and IAA-positive P-solubilizing *Bacillus* sp. isolated from *Stevia rebaudiana* rhizosphere significantly increased the growth parameters, oil yield and P uptake of *M. arvensis* under greenhouse experiment (Prakash and Arora, 2019). Moreover, the highest oil yield and menthol content were recorded when *S. rebaudiana* was inoculated with *Bacillus* sp. and TCP. This approach suggested that the P biofertilizer in association with inorganic P can serve as a good choice to increase the production of menthol and oil yield of *M. arvensis*.

The integrated effect of co-inoculation of organisms belonging to two or more physiologically divergent groups have been found better in promoting crop performance compared to single or mono culture applications (Saleemi et al., 2017; Me Carty et al., 2017). For example, co-inoculation of a diazotrophic bacterium (*Paenibacillus beijingensis* BJ-18) and a P-solubilizing bacterium (*Paenibacillus* sp. B1) significantly increased plant biomass (length, fresh, and dry weight) and N content (root: 27%, shoot: 30%) and P content in root and shoots of wheat by 63% and 30%, respectively (Li et al., 2020). Coculture also significantly increased the total N (12%), available P (9%), and nitrogenase activity (69%) compared to sole application of *P. beijingensis* BJ-18. Conclusively, P-solubilizing bacteria improved soil available P and plant P uptake, and considerably stimulated BNF in the rhizosphere and endosphere of wheat seedlings. In a similar study, the dual application of P biofertilizers and potassium-solubilizing biofertilizers (KSB) in a field experiment carried out during rabi season demonstrated variable effect on nutrient uptake, quality parameter and economics of popcorn (Ghetiya et al., 2019). Application of 45 kg P_2O_5/ha with PSB (bioprimed seed) and in soil enhanced the biochemical characteristics such as leaf chlorophyll content, protein yield and nutrient content, grain P content, fodder N and P content in grain and fodder. Also, enhancement in N, P and K uptake, P and K use efficiency and higher grain yield (3452 kg/ha) and straw yield (5300 kg/ha) with higher net return (₹ 40,018 ha) and B:C ratio (2.11) were recorded. It could be concluded that seed inoculation and soil application of PSB and KSB enhanced microbial counts in soil which is capable for solubilizing insoluble form of nutrient and enhanced nutrient uptake and saving of 25% fertilizer dose of P and K. Phosphorus and nitrogen are the major plant nutrients that affect many important cellular processes such as root elongation, proliferation and changes of root architecture, seed development, and normal maturity of plants growing both under conventional and stressful conditions. Considering these, Patel and Panchal (2020) assessed the inoculation effects of

P-solubilizing and BNF and found a significant improvement in the root and shoot height, number of root hairs, length, width and weight of leaves, and chlorophyll content of cotton plants grown under greenhouse conditions.

16.7.1 Performance of phosphate biofertilizers under stress environment

Phosphate biofertilizers technology apart from playing game changer role in conventional agrosystems has also been found highly successful in enhancing the yield and quality of food crops under stressful conditions without deteriorating the soil health (Rizvi et al., 2020; Rizvi and Khan, 2017a). Broadly, the use of stress-tolerant P-solubilizing microbiomes in agrosystems is recommended (Li et al., 2017; Rafique et al., 2017) due to many reasons: (i) production methods are easy and cheap, (ii) they exhibit tolerance to single or multiple stressor molecules and still remain physiologically active; (iii) the ability to supply growth regulators is maintained even under a stressful environment; and (iv) they are harmless to the environment (Table 16.11). Some notable phosphate-solubilizing bacterial and fungal genera possessing stress-tolerating ability include *Cronobacter* (Saranya et al., 2018), *Azotobacter* (Rizvi et al., 2019), *Aerococcus*, *Pseudomonas* and *Pantoea* sp. (Chen and Liu, 2019), *Achromobacter* (Oves et al., 2019), *Bacillus* sp. (Shao et al., 2019; Paul and Sinha, 2015), and many fungi (Liaquat et al., 2020). For example, the PSB *B. cepacia* have been reported to magnify the growth of maize plants in the presence of NaCl up to 5% (Zhao et al., 2014). Phosphate-solubilizing bacteria belonging to both Gram-negative and Gram-positive groups trap the mobile metals in their polymeric (EPS) layers and cell wall (biosorption) and, therefore, protect the crops from metal toxicity (Gupta and Diwan, 2017; Shameer, 2016). For example, *P. putida* strain Ws3, an IAA-positive PSB when used as inoculant, alleviated the conditions for growth and increased the plant dry matter while declining the metal concentrations in maize plant with maximum impact on the shoot and root (Marzban et al., 2016). In a similar manner, *B. subtilis* strain BM2 applied with 195 mg Pb kg^{-1} in a greenhouse experiment enhanced the length and dry phyto-mass of shoots by 14% and 23%, respectively, over control. Also, strain BM2 improved the grain yield significantly by 49% at 870 mg Ni kg^{-1} and by 50% at 585 mg Pb kg^{-1} relative to nonbacterized plants. Moreover, *B. subtilis* BM2 relieved the metal stress on wheat and caused a significant drop in proline and malondialdehyde content and the activities of antioxidant enzymes, such as catalase (CAT), superoxide dismutase (SOD), and glutathione reductase (GR). This study, therefore, provided solutions to the metal toxicity problems faced by winter wheat and clearly suggests that the metal detoxification potential of microbiome could be greatly useful

Table 16.11 Metal tolerance ability and plant growth-promoting characteristics of phosphate-solubilizing root-associated microbiome.

Phosphate solubilizers	Source	HMT* M	HMT* TL	PGP traits	References
Serratia proteamaculans (Ai), Pseudomonas sp., P. veronii, P. psychrophile, Bacillus cereus	Roots of Acacia mangium, Eucalyptus camaldulensis, Pityrogramma calomelanos[a]	Pb	1875	Siderophores Nil	Yongpisanphop and Babel (2020)
Lysinibacillus varians strain KUBM17 P. putida strain KUBM18	Rhizospheric soil contaminated with industrial, sewage, or agrochemical wastes	High degree of Cd- and Pb-tolerant ability		Ammonia, IAA, N_2 fixation	Pal and Sengupta (2019)
Achromobacter xylosoxidans OS2	Metal-polluted soil[b]	Cd Cr Ni Cu Zn	300 1000 1200 1400 1500	Ammonia, IAA, HCN	Oves et al. (2019)
Cellulosimicrobium sp. (NF2)	Rhizosphere[a]	Cr Zn Cu Ni Pb Co	800 1500 400 500 1500 500	IAA, siderophores	Tirry et al. (2018)
C. funkei (AR6)	Phaseolus vulgaris rhizosphere[b]	Cr Pb Mn Cu Zn	1200 750 700 300 150	IAA, ammonia, catalase, hydrolytic enzymes	Karthik et al. (2017)
P. aeruginosa (CPSB1)	Chilli rhizosphere[a]	Cu Cd Cr	1400 1000 1000	IAA, siderophores, NH_3, ACC deaminase, HCN	Rizvi and Khan (2017a)
P. aeruginosa	Metal-polluted chilli rhizosphere[b]	Ni Zn Pb Cr	400 800 1000 400	IAA, siderophores, HCN, NH_3, ACC deaminase	Rizvi and Khan (2017b)
Ensifer adhaerens (OS3)	Chickpea root surface[b]	Cd Cr Cu Zn Ni	250 500 800 800 1000	IAA, siderophore, HCN, NH_3	Oves et al. (2017)

Table 16.11 Metal tolerance ability and plant growth-promoting characteristics of phosphate-solubilizing root-associated microbiome—cont'd

Phosphate solubilizers	Source	HMT* M	HMT* TL	PGP traits	References
P. aeruginosa (KUJM)	Sewage treatment plant[c]	As (III)	50	IAA	Biswas et al. (2018)
		As (V)	800		
		Cd	8		
		Co	18		
		Cu	7		
		Cr	2.5		
		Ni	3		
		Zn	14		
Achromobacter piechaudii	Stems of the Zn/Cd hyperaccumulator plant Sedum plumbizincicola[a]	Cd, Zn, Pb	100 of each metal	IAA, ACC deaminase	Ma et al. (2016)
P. putida CG29, B. safensis KM39	Rhizosphere of Phyllanthus urinaria[a]	Cd Pb	50 110	IAA, NH_3 siderophores, nitrification, nitrate reduction	Singh and Lal (2015)

*HMT indicates heavy metal tolerance and is expressed as [a]mg/L, [b]μg/L, and [c]mM, respectively; M and TL represent metal and tolerance level, respectively; IAA, indole-3-acetic acid; HCN, hydrogen cyanide; NH3, ammonia; ACC deaminase, 1-aminocyclopropane-1-carboxylate deaminase.

in the management of metal-polluted soils. Conclusively, the mixture of stress-tolerant PSM has been found more effective than sole bacterial application to promote food crops under stressed soils. Therefore, the soil-plant-PSM interactions, has great agronomic potential in future and promises to be an effective option for potentiating the bioremediation of metals and promotion of food crops even under metal stressed environment.

16.8 Molecular engineering of phosphate biofertilizers

Introduction or overexpression of genes associated with P solubilization/mineralization in beneficial soil microbiome is an agronomically attractive strategy to enhance the P-solubilizing potential among the nonphosphate-solubilizing microbes. Some recent developments in the manipulation of genes related to microbial P solubilization/mineralization and its relationship with the use of rhizobacteria as a new or improved phosphate biofertilizers are discussed.

Molecular engineering of phosphate biofertilizers and use of genetically engineered microphos under P-deficient soils have rarely been endeavored. Despite these, attempts have been made to isolate genes associated with mineral P solubilization (mps) in bacterial species. Broadly, the genes directly involved in both mps and organic P solubilization have been cloned and allowed to express in selected non-P solubilizers rhizobacteria in order to enhance the host range of phosphate biofertilizers (Fraga-Vidal et al., 2007; Krishnaraj and Goldstein, 2001). Besides these, chromosomal insertion of such genes under appropriate promoters has been reported (Rodríguez and Fraga, 1999). As an example, a genetic construct using the broad host range vector pKT230 and plasmid pMCG898, encoding for the *Erwinia herbicola* pyrroloquinoline quinone (PQQ) synthase, a gene involved in mps was developed (Rodrıguez et al., 2000). The resulting construct was transformed and expressed in *Escherichia coli* MC1061 and the recombinant plasmids were inserted into two recipient cells of *B. cepacia* IS-16 and *Pseudomonas* sp. PSS by conjugation. The clones possessing recombinant plasmids showed zone of solubilization on PVK plates supplemented with TCP suggesting a successful expression of mps activity of the *E. herbicola* gene in the recipient strains. Apart from inorganic P, phosphorus can also be released form organic compounds activated by enzymes, phosphatases, phytases, and phosphonatases. Of the three enzymes, acid phosphatases and phytases are important ones, and several acid phosphatase genes from Gram-negative bacteria have been isolated and characterized. For instance, the subcloning of the gene encoding the PhoC acid phosphatase from *M. morganii* (phoC gene) in a vector which allows the chromosomal integration of phoC gene in other PGPR such as *Azospirillum* spp. and *B. cepacia* strains is reported (Rodríguez et al., 2006). Also, a gene from *B. cepacia* that showed phosphatase activity was isolated. This gene codes for an outer membrane protein which improves its synthesis in the absence of soluble P and was suggested to be involved in P transport to the cell. The heterologous expression of these genes in agriculturally important bacterial strains is the next step in this approach. The *napA* phosphatase gene from the soil bacteria *M. morganii* was transferred to *B. cepacia* IS-16 using a broad-host range vector [pRK293]. An increase in the extracellular phosphatase activity of the recombinant strain was achieved. Insertion of the transferred genes into the bacterial chromosome is advantageous for stability and ecological safety. The preliminary success in the genetic manipulation of P-solubilizing microbes, therefore, provides a promising and exciting perspective for obtaining bacterial strains with enhanced P solubilization/mineralization ability and broader host range, which could serve as efficient microbial inoculants for furthering agricultural production.

16.9 Challenges and future prospects of phosphate biofertilizers

The excessive and injudicious applications of fertilizers have led to the advent of inexpensive phosphate biofertilizers, which represent a potential substitute for chemical phosphatic fertilizers to fulfill the P demands of plants and enhance yields in sustainable agrosystems. The discovery of phosphate biofertilizers, an ecologically attractive and economically sound biotechnological approach, opens up a new horizon for better plant productivity while preserving the deteriorating agro-ecosystems. Recently, the exploration and application of stress-tolerant (drought/saline/alkaline/metals/pesticides) phosphate solubilizers in optimizing food production have provided some solutions for crops growing in the polluted soils where application of conventional phosphate solubilizers is almost impossible. The molecular engineering of phosphate biofertilizers has given a new direction for broadening the host range and consequently enhancing the crop production using genetically engineered P microbiome. Despite the progress made so far in the phosphate biofertilizer technology, there is still a need to discover and foster region-specific technologies for producing phosphate biofertilizers and transferring the product to the farmers in a relatively short time. To achieve these, it is essential that researchers continue to learn more about phosphate biofertilizers and, urgently, translate this knowledge into a form that can readily be used by farm practitioners (farmers) for accelerating the food production inexpensively. However, the funding to conduct more finer researches in this area needs to be improved while the poor quality of such products and low level of adaptability to the farmers are some other issues that warrant urgent attention despite the fact that the phosphate biofertilizers are easy to use, nontoxic, eco-friendly, and sustainable. Collectively, the global scientists, institutions, and phosphate biofertilizer manufacturers need to work together so that the bottlenecks associated with the production of phosphate biofertilizers and their delivery to farmers are resolved swiftly.

References

Abawari, R.A., Tuji, F.A., Yadete, D.M., 2020. Phosphate solubilizing bio-fertilizers and their role in bio-available P nutrient: an overview. Int. J. Appl. Agric. Sci. 6 (6), 162.

Abbas, Z.R., Al-Ezee, A.M.M., Authman, S.H., 2019. Sidrophore production and phosphate solubilization by bacillus cereus and pseudomonas fluorescens isolated from iraqi soils and soil characterization. Int. J. Pharm. Clin. Res. 10 (01), 74–79.

Abd-Alla, M.H., 1994. Solubilization of rock phosphates by *Rhizobium* and *Bradyrhizobium*. Folia Microbiol. 39 (1), 53-56.

Abedinzadeh, M., Etesami, H., Alikhani, H.A., 2019. Characterization of rhizosphere and endophytic bacteria from roots of maize (*Zea mays* L.) plant irrigated with wastewater with biotechnological potential in agriculture. Biotechnol. Rep. 21, e00305.

Adesemoye, A.O., Kloepper, J.W., 2009. Plant–microbes interactions in enhanced fertilizer-use efficiency. Appl. Microbiol. Biotechnol. 85 (1), 1-12.

Ahmad, E., Zaidi, A., Khan, M.S., 2016. Effects of plant growth promoting rhizobacteria on the performance of greengram under field conditions. Jordan J. Biol. Sci. 9 (2), 79-88.

Ahmad, I., Ahmad, M., Hussain, A., Jamil, M., 2021. Integrated use of phosphate-solubilizing *Bacillus subtilis* strain IA6 and zinc-solubilizing *Bacillus* sp. strain IA16: a promising approach for improving cotton growth. Folia Microbiol. 66 (1), 115-125.

Akinrinlola, R.J., Yuen, G.Y., Drijber, R.A., Adesemoye, A.O., 2018. Evaluation of *Bacillus* strains for plant growth promotion and predictability of efficacy by in vitro physiological traits. Int. J. Microbiol. 2018. https://doi.org/10.1155/2018/5686874, 5686874.

Akintokun, A.K., Ezaka, E., Akintokun, P.O., Shittu, O.B., Taiwo, L.B., 2019. Isolation, screening and response of maize to plant growth promoting Rhizobacteria inoculants. Sci. Agric. Biohem. 50 (3), 181-190.

Aliyat, F.Z., Maldani, M., El Guilli, M., Nassiri, L., Ibijbijen, J., 2020. Isolation and characterization of phosphate solubilizing bacteria from phosphate solid sludge of the Moroccan phosphate mines. Open Agric. J. 14 (1), 16-24. https://doi.org/10.2174/1874331502014010016.

Alori, E.T., Glick, B.R., Babalola, O.O., 2017. Microbial phosphorus solubilization and its potential for use in sustainable agriculture. Front. Microbiol. 8, 971.

Alori, E.T., Babalola, O.O., Prigent-Combaret, C., 2019. Impacts of microbial inoculants on the growth and yield of maize plant. Open Agric. J. 13 (1), 1-8.

Arif, M.S., Shahzad, S.M., Yasmeen, T., Riaz, M., Ashraf, M., Ashraf, M.A., Mubarik, M.S., Kausar, R., 2017. Improving plant phosphorus (P) acquisition by phosphate-solubilizing bacteria. In: Essential Plant Nutrients. Springer, Cham, pp. 513-556.

Asea, P.E.A., Kucey, R.M.N., Stewart, J.W.B., 1988. Inorganic phosphate solubilization by two *Penicillium* species in solution culture and soil. Soil Biol. Biochem. 20 (4), 459-464.

Ayyaz, K., Zaheer, A., Rasul, G., Mirza, M.S., 2016. Isolation and identification by 16S rRNA sequence analysis of plant growth-promoting azospirilla from the rhizosphere of wheat. Braz. J. Microbiol. 47 (3), 542-550.

Azaroual, S.E., Hazzoumi, Z., Mernissi, N.E., Aasfar, A., Meftah Kadmiri, I., Bouizgarne, B., 2020. Role of inorganic phosphate solubilizing bacilli isolated from moroccan phosphate rock mine and rhizosphere soils in wheat (*Triticum aestivum* L) phosphorus uptake. Curr. Microbiol. 77, 2391-2404.

Bashan, Y., Puente, M.E., Rodriguez-Mendoza, M.N., Toledo, G., Holguin, G., Ferrera-Cerrato, R., Pedrin, S., 1995. Survival of *Azospirillum brasilense* in the bulk soil and rhizosphere of 23 soil types. Appl. Environ. Microbiol. 61 (5), 1938-1945.

Batool, S., Iqbal, A., 2019. Phosphate solubilizing rhizobacteria as alternative of chemical fertilizer for growth and yield of *Triticum aestivum* (Var. Galaxy 2013). Saudi J. Biol. Sci. 26 (7), 1400-1410.

Batool, F., Rehman, Y., Hasnain, S., 2016. Phylloplane associated plant bacteria of commercially superior wheat varieties exhibit superior plant growth promoting abilities. Front. Life Sci. 9 (4), 313-322.

Bechtaoui, N., Raklami, A., Tahiri, A.I., Benidire, L., El Alaoui, A., Meddich, A., Göttfert, M., Oufdou, K., 2019. Characterization of plant growth promoting rhizobacteria and their benefits on growth and phosphate nutrition of faba bean and wheat. Biol. Open 8 (7), bio043968.

Behera, B.C., Yadav, H., Singh, S.K., Mishra, R.R., Sethi, B.K., Dutta, S.K., Thatoi, H.N., 2017. Phosphate solubilization and acid phosphatase activity of *Serratia* sp. isolated from mangrove soil of Mahanadi river delta, Odisha, India. J. Genet. Eng. Biotechnol. 15 (1), 169–178.

Biswas, J.K., Banerjee, A., Rai, M., Naidu, R., Biswas, B., Vithanage, M., Dash, M.C., Sarkar, S.K., Meers, E., 2018. Potential application of selected metal resistant phosphate solubilizing bacteria isolated from the gut of earthworm (*Metaphire posthuma*) in plant growth promotion. Geoderma 330, 117–124.

Bjelić, D.Đ., Marinković, J.B., Tintor, B.B., Tančić, S.L., Nastasić, A.M., Mrkovački, N.B., 2015. Screening of *Azotobacter* isolates for PGP properties and antifungal activity. Zbornik Matice srpske za prirodne nauke 129, 65–72.

Boubekri, K., Soumare, A., Mardad, I., Lyamlouli, K., Hafidi, M., Ouhdouch, Y., Kouisni, L., 2021. The screening of potassium-and phosphate-solubilizing actinobacteria and the assessment of their ability to promote wheat growth parameters. Microorganisms 9 (3), 470.

Chabot, R., Antoun, H., Cescas, M.P., 1996. Growth promotion of maize and lettuce by phosphate-solubilizing *Rhizobium leguminosarum* biovar. phaseoli. Plant Soil 184 (2), 311–321.

Chang, C.H., Yang, S.S., 2009. Thermo-tolerant phosphate-solubilizing microbes for multi-functional biofertilizer preparation. Bioresour. Technol. 100 (4), 1648–1658.

Chaparro, J.M., Badri, D.V., Vivanco, J.M., 2014. Rhizosphere microbiome assemblage is affected by plant development. ISME J. 8 (4), 790–803.

Chawngthu, L., Hnamte, R., Lalfakzuala, R., 2020. Isolation and characterization of rhizospheric phosphate solubilizing bacteria from wetland paddy field of Mizoram, India. Geomicrobiol. J. 37 (4), 366–375.

Chen, Q., Liu, S., 2019. Identification and characterization of the phosphate-solubilizing bacterium Pantoea sp. S32 in reclamation soil in Shanxi, China. Front. Microbiol. 10, 2171.

Chen, W., Yang, F., Zhang, L., Wang, J., 2016. Organic acid secretion and phosphate solubilizing efficiency of *Pseudomonas* sp. PSB12: effects of phosphorus forms and carbon sources. Geomicrobiol. J. 33 (10), 870–877.

Chouyia, F.E., Romano, I., Fechtali, T., Fagnano, M., Fiorentino, N., Visconti, D., Idbella, M., Ventorino, V., Pepe, O., 2020. P-Solubilizing *Streptomyces roseocinereus* MS1B15 with multiple plant growth-promoting traits enhance barley development and regulate rhizosphere microbial population. Front. Plant Sci. 7, 1137.

Cunningham, J.E., Kuiack, C., 1992. Production of citric and oxalic acids and solubilization of calcium phosphate by *Penicillium bilaii*. Appl. Environ. Microbiol. 58 (5), 1451–1458.

Danish, S., Zafar-ul-Hye, M., Mohsin, F., Hussain, M., 2020. ACC-deaminase producing plant growth promoting rhizobacteria and biochar mitigate adverse effects of drought stress on maize growth. PLoS One 15 (4), e0230615.

Dasgupta, D., Sengupta, C., Paul, G., 2015. Screening and identification of best three phosphate solubilizing and IAA producing PGPR inhabiting the rhizosphere of *Sesbania bispinosa*. Screening 4 (6), 3968–3979.

de Oliveira Mendes, G., Moreira de Freitas, A.L., Liparini Pereira, O., Ribeiro da Silva, I., Bojkov Vassilev, N., Dutra Costa, M., 2014. Mechanisms of phosphate solubilization by fungal isolates when exposed to different P sources. Ann. Microbiol. 64, 239–249. https://doi.org/10.1007/s13213-013-0656-3.

de Sousa, S.M., de Oliveira, C.A., Andrade, D.L., de Carvalho, C.G., Ribeiro, V.P., Pastina, M.M., Marriel, I.E., de Paula Lana, U.G., Gomes, E.A., 2020. Tropical *Bacillus* strains inoculation enhances maize root surface area, dry weight, nutrient uptake and grain yield. J. Plant Growth Regul. 40, 1–11. https://doi.org/10.1007/s00344-020-10146-9.

del Carmen Orozco-Mosqueda, M., Glick, B.R., Santoyo, G., 2020. ACC deaminase in plant growth-promoting bacteria (PGPB): an efficient mechanism to counter salt stress in crops. Microbiol. Res. 235, 126439.

Delvasto, P., Valverde, A., Ballester, A., Igual, J.M., Muñoz, J.A., González, F., Blázquez, M.L., García, C., 2006. Characterization of brushite as a re-crystallization product formed during bacterial solubilization of hydroxyapatite in batch cultures. Soil Biol. Biochem. 38 (9), 2645–2654.

Ditta, A., Imtiaz, M., Mehmood, S., Rizwan, M.S., Mubeen, F., Aziz, O., Qian, Z., Ijaz, R., Tu, S., 2018. Rock phosphate-enriched organic fertilizer with phosphate-solubilizing microorganisms improves nodulation, growth, and yield of legumes. Commun. Soil Sci. Plant Anal. 49 (21), 2715–2725.

Doilom, M., Guo, J.W., Phookamsak, R., Mortimer, P.E., Karunarathna, S.C., Dong, W., Liao, C.F., Doilom, K., Pem, D., Suwannarach, N., Promputtha, I., 2020. Screening of phosphate-solubilizing fungi from air and soil in Yunnan, China: four novel species in *Aspergillus, Gongronella, Penicillium* and *Talaromyces*. Front. Microbiol. 11, 2443.

Dubois, M., Van den Broeck, L., Inzé, D., 2018. The pivotal role of ethylene in plant growth. Trends Plant Sci. 23 (4), 311–323.

Elhaissoufi, W., Khourchi, S., Ibnyasser, A., Ghoulam, C., Rchiad, Z., Zeroual, Y., Lyamlouli, K., Bargaz, A., 2020. Phosphate solubilizing rhizobacteria could have a stronger influence on wheat root traits and aboveground physiology than rhizosphere P solubilization. Front. Plant Sci. 11, 979.

Emami, S., Alikhani, H.A., Pourbabaei, A.A., Etesami, H., Sarmadian, F., Motessharezadeh, B., 2019. Effect of rhizospheric and endophytic bacteria with multiple plant growth promoting traits on wheat growth. Environ. Sci. Pollut. Res. 26 (19), 19804–19813.

Emami, S., Alikhani, H.A., Pourbabaee, A.A., Etesami, H., Motasharezadeh, B., Sarmadian, F., 2020. Consortium of endophyte and rhizosphere phosphate solubilizing bacteria improves phosphorous use efficiency in wheat cultivars in phosphorus deficient soils. Rhizosphere 14, 100196.

Etesami, H., Beattie, G.A., 2017. Plant-microbe interactions in adaptation of agricultural crops to abiotic stress conditions. In: Probiotics and Plant Health. Springer, Singapore, pp. 163–200.

Etesami, H., Glick, B.R., 2020. Halotolerant plant growth–promoting bacteria: prospects for alleviating salinity stress in plants. Environ. Exp. Bot. 178, 104124.

Fahad, S., Bajwa, A.A., Nazir, U., Anjum, S.A., Farooq, A., Zohaib, A., Sadia, S., Nasim, W., Adkins, S., Saud, S., Ihsan, M.Z., 2017. Crop production under drought and heat stress: plant responses and management options. Front. Plant Sci. 8, 1147.

FAO, 2017. The Future of Food and Agriculture – Trends and Challenges. Rome. Available from: www.fao.org/publications.

Farhat, M.B., Boukhris, I., Chouayekh, H., 2015. Mineral phosphate solubilization by *Streptomyces* sp. CTM396 involves the excretion of gluconic acid and is stimulated by humic acids. FEMS Microbiol. Lett. 362 (5). fnv008.

Faried, A.S.M., Mohamed, H.M., El-Dsouky, M.M., El-Rewainy, H.M., 2019. Isolation and Characterization of Phosphate Solubilizing Actinomycetes From Rhizosphere Soil. Doctoral dissertation, Ph.D. thesis, Fac of Agric, Assiut University, Assiut, Egypt.

Fitriyanti, D., Mubarik, N.R., Tjahjoleksono, A., 2017. Characterization and identification of phosphate solubilizing bacteria isolate GPC3. 7 from limestone mining region. IOP Conf. Ser. Earth Environ. Sci. 58 (1), 012016. IOP Publishing.

Fraga-Vidal, R., Mesa, H.R., de Villegas, T.G.D., 2007. Vector for chromosomal integration of the phoC gene in plant growth-promoting bacteria. In: First International Meeting on Microbial Phosphate Solubilization. Springer, Dordrecht, pp. 239–244.

Gamalero, E., Glick, B.R., 2020. The use of plant growth-promoting bacteria to prevent nematode damage to plants. Biology 9 (11), 381.

Gamalero, E., Bona, E., Todeschini, V., Lingua, G., 2020. Saline and arid soils: impact on bacteria, plants, and their interaction. Biology 9 (6), 116.

García-Fraile, P., Menéndez, E., Rivas, R., 2015. Role of bacterial biofertilizers in agriculture and forestry. AIMS Bioeng. 2 (3), 183–205.

Gaur, A.C., 1990. Phosphate Solubilizing Micro-organisms as Biofertilizer. Omega Scientific Publishers, New Delhi, India, p. 176.

Gebremedhin, A., Assefa, F., Argaw, A., 2019. Isolation and characterization of phosphate solubilising rhizobia nodulating wild field pea (*Pisum sativum* var. abyssinicum) from Southern Tigray, Ethiopia. J. Adv. Microbiol. 17, 1–11.

Ghetiya, K.P., Bhalu, V.B., Mathukia, R.K., Chovatia, P.K., Hadavani, J.K., 2019. Effect of phosphate and potash solubilizing bacteria on nutrient uptake, quality parameter and economics of popcorn (*Zea mays* L. Var. Everta). Int. J. Pure Appl. Biosci. 7 (1), 216–223.

Gholami, A., Biyari, A., Gholipoor, M., Asadi Rahmani, H., 2012. Growth promotion of maize (*Zea mays* L.) by plant-growth-promoting rhizobacteria under field conditions. Commun. Soil Sci. Plant Anal. 43 (9), 1263–1272.

Gilbert, N., 2009. Environment: the disappearing nutrient. Nature News 461 (7265), 716–718.

Gimenez, E., Salinas, M., Manzano-Agugliaro, F., 2018. Worldwide research on plant defense against biotic stresses as improvement for sustainable agriculture. Sustainability 10 (2), 391.

Goldstein, A.H., 1994. Recent progress in understanding the molecular genetics and biochemistry of calcium phosphate solubilization by gram negative bacteria. Biol. Agric. Hortic. 12, 185–193. https://doi.org/10.1080/01448765.1995.9754736.

Goldstein, A.H., Rogers, R.D., Mead, G., 1993. Mining by microbe. Bio/Technology 11, 1250–1254.

Gupta, P., Diwan, B., 2017. Bacterial exopolysaccharide mediated heavy metal removal: a review on biosynthesis, mechanism and remediation strategies. Biotechnol. Rep. 13, 58–71.

Gupta, S., Pandey, S., 2019. ACC deaminase producing bacteria with multifarious plant growth promoting traits alleviates salinity stress in French bean (*Phaseolus vulgaris*) plants. Front. Microbiol. 10, 1506.

Gupta, R., Singal, R., Shankar, A., Kuhad, R.C., Saxena, R.K., 1994. A modified plate assay for screening phosphate solubilizing microorganisms. J. Gen. Appl. Microbiol. 40 (3), 255–260.

Gyaneshwar, P., Kumar, G.N., Parekh, L.J., Poole, P.S., 2002. Role of soil microorganisms in improving P nutrition of plants. Plant Soil 245 (1), 83–93.

Hamane, S., Zerrouk, M.H., Lyemlahi, A.E., Aarab, S., Laglaoui, A., Bakkali, M., Arakrak, A., 2020. Screening and characterization of phosphate-solubilizing rhizobia isolated from *Hedysarum pallidum* in the northeast of Morocco. In: Phyto-Microbiome in Stress Regulation. Springer, Singapore, pp. 113–124.

Hameeda, B., Harini, G., Rupela, O.P., Wani, S.P., Reddy, G., 2008. Growth promotion of maize by phosphate-solubilizing bacteria isolated from composts and macrofauna. Microbiol. Res. 163 (2), 234–242.

Hanin, M., Ebel, C., Ngom, M., Laplaze, L., Masmoudi, K., 2016. New insights on plant salt tolerance mechanisms and their potential use for breeding. Front. Plant Sci. 7, 1787.

Harinathan, B., Sankaralingam, S., Palpperumal, S., Kathiresan, D., Shankar, T., Prabhu, D., 2016. Effect of phosphate solubilizing bacteria on growth and development of pearl millet and ragi. J. Adv. Biol. Biotechnol. 7, 1–7.

Hassan, W., Bashir, S., Hanif, S., Sher, A., Sattar, A., Wasaya, A., Atif, H., Hussain, M., 2017. Phosphorus solubilizing bacteria and growth and productivity of mung bean (*Vigna radiata*). Pak. J. Bot. 49 (3), 331–336.

Heydari, S., Rezvani-Moghadam, P., Arab, M., 2008. Hydrogen cyanide production ability by Pseudomonas fluorescence bacteria and their inhibition potential on weed germination. In: Proceedings "Competition for Resources in a Changing World: New Drive for Rural Development", Tropentag, Hohenheim.

Hinsinger, P., 2001. Bioavailability of soil inorganic P in the rhizosphere as affected by root-induced chemical changes: a review. Plant Soil 237 (2), 173–195.

Ho, W.C., Ko, W.H., 1985. Soil microbiostasis: effects of environmental and edaphic factors. Soil Biol. Biochem. 17 (2), 167–170.

Ibarra-Galeana, J.A., Castro-Martínez, C., Fierro-Coronado, R.A., Armenta-Bojórquez, A.D., Maldonado-Mendoza, I.E., 2017. Characterization of phosphate-solubilizing bacteria exhibiting the potential for growth promotion and phosphorus nutrition improvement in maize (*Zea mays* L.) in calcareous soils of Sinaloa, Mexico. Ann. Microbiol. 67 (12), 801–811.

Illmer, P., Schinner, F., 1992. Solubilization of inorganic phosphates by microorganisms isolated from forest soils. Soil Biol. Biochem. 24 (4), 389–395.

Islam, M.K., Sano, A., Majumder, M.S.I., Hossain, M.A., Sakagami, J.I., 2019. Isolation and molecular characterization of phosphate solubilizing filamentous fungi from subtropical soils in Okinawa. Appl. Ecol. Environ. Res. 17, 9145–9157.

Iwuagwu, M., Chukwuka, K.S., Uka, U.N., Amandianeze, M.C., 2013. Effects of biofertilizers on the growth of Zea mays L. Asian J. Microbiol. Biotechnol. Environ. Sci. 15 (2), 235–240.

Jog, R., Pandya, M., Nareshkumar, G., Rajkumar, S., 2014. Mechanism of phosphate solubilization and antifungal activity of *Streptomyces* spp. isolated from wheat roots and rhizosphere and their application in improving plant growth. Microbiology 160 (4), 778–788.

Kalaiarasi, R., Dinakar, S., 2015. Positive effect of different formulations of *Azotobacter* and *Paenibacillus* on the enhancement of growth and yield parameters in maize (*Zea mays* L.). Int. J. Curr. Microbiol. App. Sci. 4 (10), 190–196.

Kalam, S., Basu, A., Podile, A.R., 2020. Functional and molecular characterization of plant growth promoting *Bacillus* isolates from tomato rhizosphere. Heliyon 6 (8), e04734.

Kalayu, G., 2019. Phosphate solubilizing microorganisms: promising approach as biofertilizers. Int. J. Agron. 2019, 4917256.

Kang, J., Amoozegar, A., Hesterberg, D., Osmond, D.L., 2011. Phosphorus leaching in a sandy soil as affected by organic and inorganic fertilizer sources. Geoderma 161 (3–4), 194–201.

Kang, S.M., Hamayun, M., Khan, M.A., Iqbal, A., Lee, I.J., 2019. *Bacillus subtilis* JW1 enhances plant growth and nutrient uptake of Chinese cabbage through gibberellins secretion. J. Appl. Bot. Food Qual. 92, 172–178.

Karthik, C., Elangovan, N., Kumar, T.S., Govindharaju, S., Barathi, S., Oves, M., Arulselvi, P.I., 2017. Characterization of multifarious plant growth promoting traits of rhizobacterial strain AR6 under Chromium (VI) stress. Microbiol. Res. 204, 65–71.

Khan, M.S., Zaidi, A., Wani, P.A., 2007. Role of phosphate-solubilizing microorganisms in sustainable agriculture—a review. Agron. Sustain. Dev. 27 (1), 29–43.

Khan, M.S., Zaidi, A., Ahmad, E., 2014. Mechanism of phosphate solubilization and physiological functions of phosphate-solubilizing microorganisms. In: Phosphate Solubilizing Microorganisms. Springer, Cham, pp. 31–62.

Khan, S., Shahid, M., Khan, M.S., Syed, A., Bahkali, A.H., Elgorban, A.M., Pichtel, J., 2020. Fungicide-tolerant plant growth-promoting rhizobacteria mitigate physiological disruption of white radish caused by fungicides used in the field cultivation. Int. J. Environ. Res. Public Health 17 (19), 7251.

Kim, K.Y., McDonald, G.A., Jordan, D., 1997. Solubilization of hydroxyapatite by *Enterobacter agglomerans* and cloned *Escherichia coli* in culture medium. Biol. Fertil. Soils 24 (4), 347–352.

Kondracka, A., Rychter, A.M., 1997. The role of Pi recycling processes during photosynthesis in phosphate-deficient bean plants. J. Exp. Bot. 48 (7), 1461–1468.

Kour, D., Rana, K.L., Yadav, A.N., Yadav, N., Kumar, V., Kumar, A., Sayyed, R.Z., Hesham, A.E.L., Dhaliwal, H.S., Saxena, A.K., 2019. Drought-tolerant phosphorus-solubilizing microbes: biodiversity and biotechnological applications for alleviation of drought stress in plants. In: Plant Growth Promoting Rhizobacteria for Sustainable Stress Management. Springer, Singapore, pp. 255–308.

Krasilinikov, N.A., 1957. On the role of soil micro-organism in plant nutrition. Microbiologiya 26, 659–672.

Krishnaraj, P.U., Goldstein, A.H., 2001. Cloning of a Serratia marcescens DNA fragment that induces quinoprotein glucose dehydrogenase-mediated gluconic acid production in Escherichia coli in the presence of stationary phase Serratia marcescens. FEMS Microbiol. Lett. 205 (2), 215–220.

Kuan, K.B., Othman, R., Abdul Rahim, K., Shamsuddin, Z.H., 2016. Plant growth-promoting rhizobacteria inoculation to enhance vegetative growth, nitrogen fixation and nitrogen remobilisation of maize under greenhouse conditions. PLoS One 11 (3), e0152478.

Kucey, R.M.N., 1983. Phosphate-solubilizing bacteria and fungi in various cultivated and virgin Alberta soils. Can. J. Soil Sci. 63 (4), 671–678.

Lebrazi, S., Niehaus, K., Bednarz, H., Fadil, M., Chraibi, M., Fikri-Benbrahim, K., 2020. Screening and optimization of indole-3-acetic acid production and phosphate solubilization by rhizobacterial strains isolated from *Acacia cyanophylla* root nodules and their effects on its plant growth. J. Genet. Eng. Biotechnol. 18 (1), 1–12.

Leghari, S.J., Buriro, M., Jogi, Q., Kandhro, M.N., Leghari, A.J., 2016. Depletion of phosphorus reserves, a big threat to agriculture: challenges and opportunities. Sci. Int. 28 (3), 2697–2702.

Li, Z., Bai, T., Dai, L., Wang, F., Tao, J., Meng, S., Hu, Y., Wang, S., Hu, S., 2016. A study of organic acid production in contrasts between two phosphate solubilizing fungi: *Penicillium oxalicum* and *Aspergillus niger*. Sci. Rep. 6 (1), 1–8.

Li, Y., Li, Q., Guan, G., Chen, S., 2020. Phosphate solubilizing bacteria stimulate wheat rhizosphere and endosphere biological nitrogen fixation by improving phosphorus content. PeerJ 8. https://doi.org/10.7717/peerj.9062, e9062.

Liaquat, F., Munis, M.F.H., Haroon, U., Arif, S., Saqib, S., Zaman, W., Khan, A.R., Shi, J., Che, N., Liu, Q., 2020. Evaluation of metal tolerance of fungal strains isolated from contaminated mining soil of Nanjing, China. Biology 9 (12), 469.

Linu, M.S., Asok, A.K., Thampi, M., Sreekumar, J., Jisha, M.S., 2019. Plant growth promoting traits of indigenous phosphate solubilizing *Pseudomonas aeruginosa* isolates from Chilli (*Capsicum annuum* L.) Rhizosphere. Commun. Soil Sci. Plant Anal. 50 (4), 444–457.

Liu, Z., Li, Y.C., Zhang, S., Fu, Y., Fan, X., Patel, J.S., Zhang, M., 2015. Characterization of phosphate-solubilizing bacteria isolated from calcareous soils. Appl. Soil Ecol. 96, 217–224.

Liu, L., Li, A., Chen, J., Su, Y., Li, Y., Ma, S., 2018. Isolation of a phytase-producing bacterial strain from agricultural soil and its characterization and application as an effective eco-friendly phosphate solubilizing bioinoculant. Commun. Soil Sci. Plant Anal. 49 (8), 984–994.

Liu, X., Jiang, X., He, X., Zhao, W., Cao, Y., Guo, T., Li, T., Ni, H., Tang, X., 2019. Phosphate-solubilizing *Pseudomonas* sp. strain P34-L promotes wheat growth by colonizing the wheat rhizosphere and improving the wheat root system and soil phosphorus nutritional status. J. Plant Growth Regul. 38 (4), 1314–1324.

Lobo, L.L.B., Dos Santos, R.M., Rigobelo, E.C., 2019. Promotion of maize growth using endophytic bacteria under greenhouse and field conditions. Aust. J. Crop. Sci. 13 (12), 2067–2074.

Luiz, J., Young, M., Kanashiro, S., Jocys, T., Tavares, A.R., 2018. Silver vase bromeliad: plant growth and mineral nutrition under macronutrients omission. Sci. Hortic. 234, 318–322. https://doi.org/10.1016/j.scienta.2018.02.002.

Ma, Y., Zhang, C., Oliveira, R.S., Freitas, H., Luo, Y., 2016. Bioaugmentation with endophytic bacterium E6S homologous to *Achromobacter piechaudii* enhances metal rhizoaccumulation in host *Sedum plumbizincicola*. Front. Plant Sci. 7, 75.

Mahadevaswamy, Nagaraju, Y., 2018. Role of halophilic microorganisms in agriculture. J. Pharm. Phytochem. 7, 1063–1071.

Maharajan, T., Ceasar, S.A., Ajeesh Krishna, T.P., Ramakrishnan, M., Duraipandiyan, V., Naif Abdulla, A.D., Ignacimuthu, S., 2018. Utilization of molecular markers for improving the phosphorus efficiency in crop plants. Plant Breed. 137 (1), 10–26.

Mahdi, S.S., Hassan, G.I., Hussain, A., Rasool, F., 2011. Phosphorus availability issue-its fixation and role of phosphate solubilizing bacteria in phosphate solubilization. Res. J. Agric. Sci. 2 (1), 174–179.

Malboobi, M.A., Owlia, P., Behbahani, M., Sarokhani, E., Moradi, S., Yakhchali, B., Deljou, A., Heravi, K.M., 2009. Solubilization of organic and inorganic phosphates by three highly efficient soil bacterial isolates. World J. Microbiol. Biotechnol. 25 (8), 1471–1477.

Manzoor, M., Abbasi, M.K., Sultan, T., 2017. Isolation of phosphate solubilizing bacteria from maize rhizosphere and their potential for rock phosphate solubilization-mineralization and plant growth promotion. Geomicrobiol. J. 34 (1), 81–95.

Marakana, T., Sharma, M., Sangani, K., 2018. Isolation and characterization of halotolerant bacteria and it's effects on wheat plant as PGPR. Pharma Innov. J. 7, 102–110.

Market Data Forecast, 2018. Available from: https://www.marketdataforecast.com/market-reports/asia-pacific-biofertilizers-market. (Accessed August 2019).

Marra, L.M., Oliveira-Longatti, S.M.D., Soares, C.R., Lima, J.M.D., Olivares, F.L., Moreira, F., 2015. Initial pH of medium affects organic acids production but do not affect phosphate solubilization. Braz. J. Microbiol. 46 (2), 367–375.

Marzban, A., Ebrahimipour, G., Karkhane, M., Teymouri, M., 2016. Metal resistant and phosphate solubilizing bacterium improves maize (*Zea mays*) growth and mitigates metal accumulation in plant. Biocatal. Agric. Biotechnol. 8, 13–17.

Matos, A.D., Gomes, I.C., Nietsche, S., Xavier, A.A., Gomes, W.S., Dos Santos Neto, J.A., Pereira, M.C., 2017. Phosphate solubilization by endophytic bacteria isolated from banana trees. An. Acad. Bras. Cienc. 89 (4), 2945–2954.

Mattos, B.B., Marriel, I.E., De Sousa, S.M., Lana, U.G.D.P., Schaffert, R.E., Gomes, E.A., Paiva, C.A.D.O., 2020. Sorghum genotypes response to inoculation with phosphate solubilizing bacteria. Rev. Bras. Milho Sorgo 19, 14.

Me Carty, S.C., Chauhan, D.S., MeCarty, A.D., Tripathi, K.M., Selvan, T., 2017. Effect of Azotobacter and phosphobacteria on yield of wheat (*Triticum aestivum*). Vegetos Int. J. Plant Res. 30 (2). https://doi.org/10.5958/2229-4473.2017.00130.6.

Mendoza-Arroyo, G.E., Chan-Bacab, M.J., Aguila-Ramírez, R.N., Ortega-Morales, B.O., Canché Solís, R.E., Chab-Ruiz, A.O., Cob-Rivera, K.I., Dzib-Castillo, B., Tun-Che, R.E., Camacho-Chab, J.C., 2020. Inorganic phosphate solubilization by a novel isolated bacterial strain *Enterobacter* sp. ITCB-09 and its application potential as biofertilizer. Agriculture 10 (9), 383.

Menéndez, E., Pérez-Yépez, J., Hernández, M., Rodríguez-Pérez, A., Velázquez, E., León-Barrios, M., 2020. Plant growth promotion abilities of phylogenetically diverse *Mesorhizobium* strains: effect in the root colonization and development of tomato seedlings. Microorganisms 8 (3), 412.

Menezes-Blackburn, D., Inostroza, N.G., Gianfreda, L., Greiner, R., Mora, M.L., Jorquera, M.A., 2016. Phytase-producing *Bacillus* sp. inoculation increases phosphorus availability in cattle manure. J. Soil Sci. Plant Nutr. 16 (1), 200–210.

Mitra, D., Anđelković, S., Panneerselvam, P., Senapati, A., Vasić, T., Ganeshamurthy, A.N., Chauhan, M., Uniyal, N., Mahakur, B., Radha, T.K., 2020. Phosphate-solubilizing microbes and biocontrol agent for plant nutrition and protection: current perspective. Commun. Soil Sci. Plant Anal. 51 (5), 645–657.

Mohamed, H.M., Almaroai, Y.A., 2017. Effect of phosphate solubilizing bacteria on the uptake of heavy metals by corn plants in a long-term sewage wastewater treated soil. Int. J. Environ. Sci. Dev. 8 (5), 366–371.

Mohamed, E.A., Farag, A.G., Youssef, S.A., 2018. Phosphate solubilization by *Bacillus subtilis* and *Serratia marcescens* isolated from tomato plant rhizosphere. J. Environ. Prot. 9 (03), 266-277.

Mohamed, A.E., Nessim, M.G., Ibrahim Abou-el-seoud, I., Darwish, K.M., Shamseldin, A., 2019. Isolation and selection of highly effective phosphate solubilizing bacterial strains to promote wheat growth in Egyptian calcareous soils. Bull. Natl. Res. Cent. 43 (1), 1-13.

Mohammed, A.F., 2020. Influence of Streptomyces sp. Kp109810 on Solubilization of inorganic phosphate and growth of maize (*Zea mays* L.). J. Appl. Plant Protect. 9 (1), 17-24.

Motsara, M.R., Bhattacharyya, P.B., Srivastava, B., 1995. Biofertilizers their description and characteristics. In: Biofertiliser Technology. Marketing and Usage. A Source Book-Cum-Glossary. Fertiliser Development and Consultation Organization, New Delhi, pp. 9-18.

Nafis, A., Raklami, A., Bechtaoui, N., El Khalloufi, F., El Alaoui, A., Glick, B.R., Hafidi, M., Kouisni, L., Ouhdouch, Y., Hassani, L., 2019. Actinobacteria from extreme niches in morocco and their plant growth-promoting potentials. Diversity 11 (8), 139.

Nahas, E., 1996. Factors determining rock phosphate solubilization by microorganisms isolated from soil. World J. Microbiol. Biotechnol. 12 (6), 567-572.

Narayanan, S., 2018. Effects of high temperature stress and traits associated with tolerance in wheat. Open Access J. Sci. 2 (3), 177-186.

Nautiyal, C.S., 1999. An efficient microbiological growth medium for screening phosphate solubilizing microorganisms. FEMS Microbiol. Lett. 170 (1), 265-270.

Nosrati, R., Owlia, P., Saderi, H., Rasooli, I., Malboobi, M.A., 2014. Phosphate solubilization characteristics of efficient nitrogen fixing soil *Azotobacter* strains. Iran. J. Microbiol. 6 (4), 285-295.

Nyalemegbe, K.K., Oteng, J.W., Asuming-Brempong, S., 2010. Integrated organic-inorganic fertilizer management for rice production on the Vertisols of the Accra Plains of Ghana. West Afr. J. Appl. Ecol. 16 (1), 23-33.

Oberson, A., Friesen, D.K., Rao, I.M., Bühler, S., Frossard, E., 2001. Phosphorus transformations in an Oxisol under contrasting land-use systems: the role of the soil microbial biomass. Plant Soil 237 (2), 197-210.

Ogbo, F.C., 2010. Conversion of cassava wastes for biofertilizer production using phosphate solubilizing fungi. Bioresour. Technol. 101 (11), 4120-4124.

Okur, N., 2018. A review-bio-fertilizers-power of beneficial microorganisms in soils. Biomed. J. Sci. Tech. Res. 4 (4), 4028-4029.

Ormeño, E., Torres, R., Mayo, I., Rivas, R., Peix, A., Velázquez, E., Zuñiga, D., 2007. Phaseolus lunatus is nodulated by a phosphate solubilizing strain of *Sinorhizobium meliloti* in a Peruvian soil. In: First International Meeting on Microbial Phosphate Solubilization. Springer, Dordrecht, pp. 143-147.

Oves, M., Khan, M.S., Qari, H.A., 2017. *Ensifer adhaerens* for heavy metal bioaccumulation, biosorption, and phosphate solubilization under metal stress condition. J. Taiwan Inst. Chem. Eng. 80, 540-552.

Oves, M., Khan, M.S., Qari, H.A., 2019. Chromium-reducing and phosphate-solubilizing *Achromobacter xylosoxidans* bacteria from the heavy metal-contaminated soil of the Brass city, Moradabad, India. Int. J. Environ. Sci. Technol. 16 (11), 6967-6984.

Pal, A.K., Sengupta, C., 2019. Isolation of cadmium and lead tolerant plant growth promoting rhizobacteria: *Lysinibacillus* varians and *Pseudomonas putida* from Indian Agricultural Soil. Soil Sediment Contam. Int. J. 28 (7), 601-629.

Pande, A., Kaushik, S., Pandey, P., Negi, A., 2020. Isolation, characterization, and identification of phosphate-solubilizing *Burkholderia cepacia* from the sweet corn cv. Golden Bantam rhizosphere soil and effect on growth-promoting activities. Int. J. Veg. Sci. 26 (6), 591-607.

Patel, P., Panchal, K., 2020. Effect of free-living nitrogen fixing and phosphate solubilizing bacteria on growth of *Gossypium hirsutum* L. Asian J. Biol. Life Sci. 9 (2), 169.

Patel, S., Panchal, B., Karmakar, N., Rajkumar, J.S., 2015. Solubilization of rock phosphate by two Rhizopus species isolated from coastal areas of South Gujarat and its effect on chickpea. Ecol. Environ. Conserv. 21, 229–237.

Paul, D., Sinha, S.N., 2015. Isolation and characterization of a phosphate solubilizing heavy metal tolerant bacterium from River Ganga, West Bengal, India. Songklanakarin J. Sci. Technol. 37 (6), 651–657.

Pereira, S.I.A., Abreu, D., Moreira, H., Vega, A., Castro, P.M.L., 2020. Plant growth-promoting rhizobacteria (PGPR) improve the growth and nutrient use efficiency in maize (Zea mays L.) under water deficit conditions. Heliyon 6 (10), e05106.

Pikovskaya, R.I., 1948. Mobilization of phosphorus in soil in connection with vital activity of some microbial species. Mikrobiologiya 17, 362–370.

Poomthongdee, N., Duangmal, K., Pathom-aree, W., 2015. Acidophilic actinomycetes from rhizosphere soil: diversity and properties beneficial to plants. J. Antibiot. 68 (2), 106–114.

Prabhu, N., Borkar, S., Garg, S., 2017. Alkaliphilic and haloalkaliphilic phosphate solubilizing bacteria from coastal ecosystems of Goa. Int. J. Adv. Biotechnol. Res. 7 (4), 2015–2027.

Prakash, J., Arora, N.K., 2019. Phosphate-solubilizing *Bacillus* sp. enhances growth, phosphorus uptake and oil yield of Mentha arvensis L. 3 Biotech 9 (4), 1–9.

Pyone, A.P., 2021. Effect of phosphorus solubilizing bacteria on soil available phosphorus and growth and yield of sugarcane. Walailak J. Sci. Technol. 18, 1–9. https://doi.org/10.48048/wjst.2021.10754. 10754.

Qaisrani, M.M., Mirza, M.S., Zaheer, A., Malik, K.A., 2014. Isolation and identification by 16s rRNA sequence analysis of *Achromobacter*, *Azospirillum* and *Rhodococcus* strains from the rhizosphere of maize and screening for the beneficial effect on plant growth. Pak. J. Agric. Sci. 51 (1), 91–99.

Qiao, H., Sun, X.R., Wu, X.Q., Li, G.E., Wang, Z., Li, D.W., 2019. The phosphate-solubilizing ability of *Penicillium guanacastense* and its effects on the growth of Pinus massoniana in phosphate-limiting conditions. Biol. Open 8 (11), bio046797.

Rafique, M., Sultan, T., Ortas, I., Chaudhary, H.J., 2017. Enhancement of maize plant growth with inoculation of phosphate-solubilizing bacteria and biochar amendment in soil. Soil Sci. Plant Nutr. 63 (5), 460–469.

Raghothama, K.G., 1999. Phosphate acquisition. Annu. Rev. Plant Biol. 50 (1), 665–693.

Rajasankar, R., Gayathry, G.M., Sathiavelu, A., Ramalingam, C., Saravanan, V.S., 2013. Pesticide tolerant and phosphorus solubilizing *Pseudomonas* sp. strain SGRAJ09 isolated from pesticides treated *Achillea clavennae* rhizosphere soil. Ecotoxicology 22 (4), 707–717.

Ramachandran, K., Srinivasan, V., Hamza, S., Anandaraj, M., 2007. Phosphate solubilizing bacteria isolated from the rhizosphere soil and its growth promotion on black pepper (Piper nigrum L.) cuttings. In: First International Meeting on Microbial Phosphate Solubilization. Springer, Dordrecht, pp. 325–331.

Ramette, A., Frapolli, M., Défago, G., Moënne-Loccoz, Y., 2003. Phylogeny of HCN synthase-encoding hcnBC genes in biocontrol fluorescent pseudomonads and its relationship with host plant species and HCN synthesis ability. Mol. Plant-Microbe Interact. 16 (6), 525–535.

Rashid, M., Khalil, S., Ayub, N., Alam, S., Latif, F., 2004. Organic acids production and phosphate solubilization by phosphate solubilizing microorganisms (PSM) under in vitro conditions. Pak. J. Biol. Sci. 7 (2), 187–196.

Rfaki, A., Zennouhi, O., Aliyat, F.Z., Nassiri, L., Ibijbijen, J., 2020. Isolation, selection and characterization of root-associated rock phosphate solubilizing bacteria in Moroccan wheat (*Triticum aestivum* L.). Geomicrobiol. J. 37 (3), 230–241.

Ribaudo, C., Zaballa, J.I., Golluscio, R., 2020. Effect of the phosphorus-solubilizing bacterium *Enterobacter Ludwigii* on barley growth promotion. Am. Sci. Res. J. Eng. Technol. Sci. 63 (1), 144–157.

Rizvi, A., Khan, M.S., 2017a. Biotoxic impact of heavy metals on growth, oxidative stress and morphological changes in root structure of wheat (*Triticum aestivum* L.) and stress alleviation by *Pseudomonas aeruginosa* strain CPSB1. Chemosphere 185, 942–952.

Rizvi, A., Khan, M.S., 2017b. Cellular damage, plant growth promoting activity and chromium reducing ability of metal tolerant *Pseudomonas aeruginosa* CPSB1 recovered from metal polluted chilli (*Capsicum annuum*) rhizosphere. Acta Sci. Agric. 1, 36–46.

Rizvi, A., Khan, M.S., 2018. Heavy metal induced oxidative damage and root morphology alterations of maize (*Zea mays* L.) plants and stress mitigation by metal tolerant nitrogen fixing *Azotobacter chroococcum*. Ecotoxicol. Environ. Saf. 157, 9–20.

Rizvi, A., Khan, M.S., Ahmad, E., 2014. Inoculation impact of phosphate-solubilizing microorganisms on growth and development of vegetable crops. In: Phosphate Solubilizing Microorganisms. Springer, Cham, pp. 287–297.

Rizvi, A., Ahmed, B., Zaidi, A., Khan, M.S., 2019. Bioreduction of toxicity influenced by bioactive molecules secreted under metal stress by *Azotobacter chroococcum*. Ecotoxicology 28 (3), 302–322.

Rizvi, A., Zaidi, A., Ameen, F., Ahmed, B., AlKahtani, M.D., Khan, M.S., 2020. Heavy metal induced stress on wheat: phytotoxicity and microbiological management. RSC Adv. 10 (63), 38379–38403.

Rodríguez, H., Fraga, R., 1999. Phosphate solubilizing bacteria and their role in plant growth promotion. Biotechnol. Adv. 17 (4–5), 319–339.

Rodrıguez, H., Gonzalez, T., Selman, G., 2000. Expression of a mineral phosphate solubilizing gene from *Erwinia herbicola* in two rhizobacterial strains. J. Biotechnol. 84 (2), 155–161.

Rodríguez, H., Fraga, R., Gonzalez, T., Bashan, Y., 2006. Genetics of phosphate solubilization and its potential applications for improving plant growth-promoting bacteria. Plant Soil 287 (1), 15–21.

Rodríguez-Flores, J.M., Medellín-Azuara, J., Valdivia-Alcalá, R., Arana-Coronado, O.A., García-Sánchez, R.C., 2019. Insights from a calibrated optimization model for irrigated agriculture under drought in an irrigation district on the central Mexican high plains. Water 11 (4), 858.

Romauld, I., 2020. Isolation, screening and evaluation of multifunctional strains of high efficient phosphate solubilizing microbes from rhizosphere soil. Res. J. Pharma. Technol. 13 (4), 1823–1826.

Saber, K., Labidi, N., Debez, A., Abdelly, C., 2005. Effect of P on nodule formation and N fixation in bean. Agron. Sustain. Dev. 25 (3), 389–393.

Saeid, A., Prochownik, E., Dobrowolska-Iwanek, J., 2018. Phosphorus solubilization by *Bacillus* species. Molecules 23 (11), 2897.

Saif, S., Khan, M.S., Zaidi, A., Ahmad, E., 2014. Role of phosphate-solubilizing actinomycetes in plant growth promotion: current perspective. In: Phosphate Solubilizing Microorganisms. Springer, Cham, pp. 137–156.

Salcedo Gastelum, L.A., Díaz Rodríguez, A.M., Félix Pablos, C.M., Parra Cota, F.I., Santoyo, G., Puente, M.L., Bhattacharya, D., Mukherjee, J., de los Santos Villalobos, S., 2020. The current and future role of microbial culture collections in world food security. Front. Sustain. Food Syst. 4, 291.

Saleemi, M., Kiani, M.Z., Sultan, T., Khalid, A., Mahmood, S., 2017. Integrated effect of plant growth-promoting rhizobacteria and phosphate-solubilizing microorganisms on growth of wheat (*Triticum aestivum* L.) under rainfed condition. Agric. Food Secur. 6 (1), 1–8.

Santana, E.B., Marques, E.L.S., Dias, J.C.T., 2016. Effects of phosphate-solubilizing bacteria, native microorganisms, and rock dust on *Jatropha curcas* L. growth. Genet. Mol. Res. 15 (4), 15048729.

Santoyo, G., Pacheco, C.H., Salmerón, J.H., León, R.H., 2017. The role of abiotic factors modulating the plant-microbe-soil interactions: toward sustainable agriculture. A review. Span. J. Agric. Res. 15 (1), 13.

Saranya, K., Sundaramanickam, A., Shekhar, S., Meena, M., Sathishkumar, R.S., Balasubramanian, T., 2018. Biosorption of multi-heavy metals by coral associated phosphate solubilising bacteria *Cronobacter muytjensii* KSCAS2. J. Environ. Manage. 222, 396–401.

Sarkar, A., Islam, T., Biswas, G., Alam, S., Hossain, M., Talukder, N., 2012. Screening for phosphate solubilizing bacteria inhabiting the rhizoplane of rice grown in acidic soil in Bangladesh. Acta Microbiol. Immunol. Hung. 59 (2), 199–213.

Sarr, P.S., Tibiri, E.B., Fukuda, M., Zongo, A.N., Nakamura, S., 2020. Phosphate-solubilizing fungi and alkaline phosphatase trigger the P solubilization during the co-composting of sorghum straw residues with Burkina Faso phosphate rock. Front. Environ. Sci. 8, 174.

Sarsan, S., 2016. Effect of phosphate solubilising bacteria bacillus PSB24 on growth of tomato plants. Int. J. Curr. Microbiol. Appl. Sci. 5 (7), 311–320. https://doi.org/10.20546/ijcmas.2016.507.033.

Selvi, K.B., Paul, J.J.A., Vijaya, V., Saraswathi, K., 2017. Analyzing the efficacy of phosphate solubilizing microorganisms by enrichment culture techniques. Biochem. Mol. Biol. J. 3 (1), 1–7.

Shameer, S., 2016. Biosorption of lead, copper and cadmium using the extracellular polysaccharides (EPS) of *Bacillus* sp., from solar salterns. 3 Biotech 6 (2), 1–10.

Shao, W., Li, M., Teng, Z., Qiu, B., Huo, Y., Zhang, K., 2019. Effects of Pb (II) and Cr (VI) stress on phosphate-solubilizing bacteria (*Bacillus* sp. strain mrp-3): oxidative stress and bioaccumulation potential. Int. J. Environ. Res. Public Health 16 (12), 2172.

Sharma, T., Kumar, N., Rai, N., 2018. Inoculation effect of nitrogen-fixing and phosphate-solubilising bacteria on seed germination of brinjal (*Solanum melongena* L.). J. Graphic Era University, 7–19.

Sharma, A., Kumar, V., Sidhu, G.P.S., Kumar, R., Kohli, S.K., Yadav, P., Kapoor, D., Bali, A.S., Shahzad, B., Khanna, K., Kumar, S., 2019. Abiotic stress management in plants: role of ethylene. In: Roychoudhury, A., Tripathi, D.K. (Eds.), Molecular Plant Abiotic Stress: Biology and Biotechnology. John Wiley & Sons, pp. 185–208.

Sharon, J.A., Hathwaik, L.T., Glenn, G.M., Imam, S.H., Lee, C.C., 2016. Isolation of efficient phosphate solubilizing bacteria capable of enhancing tomato plant growth. J. Soil Sci. Plant Nutr. 16 (2), 525–536.

Shenoy, V.V., Kalagudi, G.M., 2005. Enhancing plant phosphorus use efficiency for sustainable cropping. Biotechnol. Adv. 23 (7–8), 501–513.

Shravanthi, G.V., Panchatcharam, P., 2017. Isolation and optimization studies of phosphate solubilizing *Rhizobium Leguminosarum*. Int. J. Recent Sci. Res. 8, 22209–22212.

Singh, Y., Lal, N., 2015. Investigations on the heavy metal resistant bacterial isolates in vitro from industrial effluents. World J. Pharm. Pharm. Sci. 4, 343–350.

Singh, R., Singh, V., Singh, P., Yadav, R.A., 2018. Effect of phosphorus and PSB on yield attributes, quality and economics of summer greengram (*Vigna radiata* L.). J. Pharmacogn. Phytochem. 7 (2), 404–408.

Sirinapa, C., Thongjoo, C., Islam, A.M., Yeasmin, S., 2021. Efficiency of phosphate-solubilizing bacteria to address phosphorus fixation in Takhli soil series: a case of sugarcane cultivation, Thailand. Plant Soil 460 (1), 347–357.

Soumare, A., Boubekri, K., Lyamlouli, K., Hafidi, M., Ouhdouch, Y., Kouisni, L., 2020. From isolation of phosphate solubilizing microbes to their formulation and use as biofertilizers: status and needs. Front. Bioeng. Biotechnol. 7, 425.

Swetha, S., Padmavathi, T., 2016. Study of acid phosphatase in solubilization of inorganic phosphates by *Piriformospora indica*. Pol. J. Microbiol. 65 (4), 407–412.

Tahir, M., Khalid, U., Ijaz, M., Shah, G.M., Naeem, M.A., Shahid, M., Mahmood, K., Ahmad, N., Kareem, F., 2018. Combined application of bio-organic phosphate and phosphorus solubilizing bacteria (*Bacillus* strain MWT 14) improve the performance of bread wheat with low fertilizer input under an arid climate. Braz. J. Microbiol. 49, 15–24.

Takahashi, S., Anwar, M.R., 2007. Wheat grain yield, phosphorus uptake and soil phosphorus fraction after 23 years of annual fertilizer application to an Andosol. Field Crop Res. 101 (2), 160–171.

Tariq, A., Sabir, M., Farooq, M., Maqsood, M.A., Ahmad, H.R., Warraich, E.A., 2014. Phosphorus deficiency in plants: responses, adaptive mechanisms, and signaling. In: Plant Signaling: Understanding the Molecular Crosstalk. Springer, New Delhi, pp. 133–148.

Tchakounté, G.V.T., Berger, B., Patz, S., Becker, M., Fankem, H., Taffouo, V.D., Ruppel, S., 2020. Selected rhizosphere bacteria help tomato plants cope with combined phosphorus and salt stresses. Microorganisms 8 (11), 1844.

Teng, Z., Shao, W., Zhang, K., Huo, Y., Li, M., 2019. Characterization of phosphate solubilizing bacteria isolated from heavy metal contaminated soils and their potential for lead immobilization. J. Environ. Manage. 231, 189–197.

Thant, S., Aung, N.N., Aye, O.M., Oo, N.N., Htun, T.M.M., Mon, A.A., Mar, K.T., Kyaing, K., Oo, K.K., Thywe, M., Phwe, P., 2018. Phosphate solubilization of *Bacillus megaterium* isolated from non-saline soils under salt stressed conditions. J. Bacteriol. Mycol. Open Access 6 (6), 335–341.

Thiruvengadam, S., Ramki, R., Rohini, S., Vanitha, R., Romauld, I., 2020. Isolation, screening and evaluation of multifunctional strains of high efficient phosphate solubilizing microbes from rhizosphere soil. Res. J. Pharm. Technol. 4, 1825–1828.

Tirry, N., Joutey, N.T., Sayel, H., Kouchou, A., Bahafid, W., Asri, M., El Ghachtouli, N., 2018. Screening of plant growth promoting traits in heavy metals resistant bacteria: prospects in phytoremediation. J. Genet. Eng. Biotechnol. 16 (2), 613–619.

Umesha, S., Singh, P.K., Singh, R.P., 2018. Microbial biotechnology and sustainable agriculture. In: Biotechnology for Sustainable Agriculture. Woodhead Publishing, pp. 185–205.

Van Elsas, J.D., Van Overbeek, L.S., Fouchier, R., 1991. A specific marker, pat, for studying the fate of introduced bacteria and their DNA in soil using a combination of detection techniques. Plant Soil 138 (1), 49–60.

Vaxevanidou, K., Christou, C., Kremmydas, G.F., Georgakopoulos, D.G., Papassiopi, N., 2015. Role of indigenous arsenate and iron (III) respiring microorganisms in controlling the mobilization of arsenic in a contaminated soil sample. Bull. Environ. Contam. Toxicol. 94 (3), 282–288.

Verma, M., Singh, A., Dwivedi, D.H., Arora, N.K., 2020. Zinc and phosphate solubilizing *Rhizobium radiobacter* (LB2) for enhancing quality and yield of loose leaf lettuce in saline soil. Environ. Sustain. 3, 209–218.

Walpola, B.C., Hettiarachchi, R.H.A.N, 2018. Isolation and characterization of phosphate-solubilizing and heavy-metal tolerant bacteria from agricultural fields in Matara District, Sri Lanka. Trop. Agric. Res. Ext. 21 (3–4), 51–58. https://doi.org/10.4038/tare.v21i3-4.5467.

Walpola, B.C., Yoon, M.H., 2012. Prospectus of phosphate solubilizing microorganisms and phosphorus availability in agricultural soils: a review. Afr. J. Microbiol. Res. 6 (37), 6600–6605.

Wan, W., Qin, Y., Wu, H., Zuo, W., He, H., Tan, J., Wang, Y., He, D., 2020. Isolation and characterization of phosphorus solubilizing bacteria with multiple phosphorus sources utilizing capability and their potential for lead immobilization in soil. Front. Microbiol. 11, 752.

Wang, J., Li, R., Zhang, H., Wei, G., Li, Z., 2020. Beneficial bacteria activate nutrients and promote wheat growth under conditions of reduced fertilizer application. BMC Microbiol. 20 (1), 1–12.

Yadav, K.K., Sarkar, S., 2019. Biofertilizers, impact on soil fertility and crop productivity under sustainable agriculture. Environ. Ecol. 37 (1), 89–93.

Yadav, A.N., Sharma, D., Gulati, S., Singh, S., Dey, R., Pal, K.K., Kaushik, R., Saxena, A.K., 2015a. Haloarchaea endowed with phosphorus solubilization attribute implicated in phosphorus cycle. Sci. Rep. 5 (1), 1–10.

Yadav, H., Gothwal, R.K., Solanki, P.S., Nehra, S., Sinha-Roy, S., Ghosh, P., 2015b. Isolation and characterization of thermo-tolerant phosphate-solubilizing bacteria from a phosphate mine and their rock phosphate solubilizing abilities. Geomicrobiol. J. 32 (6), 475–481.

Yanez-Ocampo, G., Mora-Herrera, M.E., Wong-Villarreal, A., De La Paz-Osorio, D.M., De La Portilla-Lopez, N., Lugo, J., Vaca-Paulin, R., Del Aguila, P., 2020. Isolated phosphate-solubilizing soil bacteria promotes in vitro growth of *Solanum tuberosum* L. Pol. J. Microbiol. 69 (3), 357–365.

Yang, Y., Ding, J., Chi, Y., Yuan, J., 2020. Characterization of bacterial communities associated with the exotic and heavy metal tolerant wetland plant Spartina alterniflora. Sci. Rep. 10 (1), 1–11.

Yongpisanphop, J., Babel, S., 2020. Characterization of Pb-tolerant plant-growth-promoting endophytic bacteria for biosorption potential, isolated from roots of Pb excluders grown in different habitats. Environ. Nat. Res. J. 18 (3), 268–274.

Youseif, S.H., 2018. Genetic diversity of plant growth promoting rhizobacteria and their effects on the growth of maize plants under greenhouse conditions. Ann. Agric. Sci. 63 (1), 25–35.

Zafar, M., Ahmed, N., Mustafa, G., Zahir, Z.A., Simms, E.L., 2017. Molecular and biochemical characterization of rhizobia from chickpea (*Cicer arietinum*). Pak. J. Agric. Sci. 54 (2), 373–381.

Zafar, N., Munir, M.K., Ahmed, S., Zafar, M., 2020. Phosphorus Solubilizing Bacteria (PSB) in combination with different Fertilizer sources to enhance yield performance of chickpea. Life Sci. J. 17 (8), 84–88.

Zaheer, A., Malik, A., Sher, A., Qaisrani, M.M., Mehmood, A., Khan, S.U., Ashraf, M., Mirza, Z., Karim, S., Rasool, M., 2019. Isolation, characterization, and effect of phosphate-zinc-solubilizing bacterial strains on chickpea (*Cicer arietinum* L.) growth. Saudi J. Biol. Sci. 26 (5), 1061–1067.

Zaidi, A., Khan, M.S., Ahemad, M., Oves, M., Wani, P.A., 2009. Recent advances in plant growth promotion by phosphate-solubilizing microbes. In: Khan, M.S., et al. (Eds.), Microbial Strategies for Crop Improvement. Springer-Verlag, Berlin Heidelberg, pp. 23–50.

Zaidi, A., Khan, M.S., Saif, S., Rizvi, A., Ahmed, B., Shahid, M., 2017. Role of nitrogen-fixing plant growth-promoting rhizobacteria in sustainable production of vegetables: current perspective. In: Microbial Strategies for Vegetable Production. Springer, Cham, pp. 49–79.

Zavaleta-Pastor, M., Sohlenkamp, C., Gao, J.L., Guan, Z., Zaheer, R., Finan, T.M., Raetz, C.R., López-Lara, I.M., Geiger, O., 2010. *Sinorhizobium meliloti* phospholipase C required for lipid remodeling during phosphorus limitation. Proc. Natl. Acad. Sci. 107 (1), 302–307.

Zhang, J., Wang, P., Fang, L., Zhang, Q.A., Yan, C., Chen, J., 2017. Isolation and characterization of phosphate-solubilizing bacteria from mushroom residues and their effect on tomato plant growth promotion. Pol. J. Microbiol. 66 (1), 57–65.

Zhao, K., Penttinen, P., Zhang, X., Ao, X., Liu, M., Yu, X., Chen, Q., 2014. Maize rhizosphere in Sichuan, China, hosts plant growth promoting *Burkholderia cepacia* with phosphate solubilizing and antifungal abilities. Microbiol. Res. 169 (1), 76–82.

Zhao, W., Liu, L., Shen, Q., Yang, J., Han, X., Tian, F., Wu, J., 2020. Effects of water stress on photosynthesis, yield, and water use efficiency in winter wheat. Water 12 (8), 2127.

Zhao, G., Wei, Y., Chen, J., Dong, Y., Hou, L., Jiao, R., 2021. Screening, identification and growth-promotion products of multifunctional bacteria in a Chinese Fir Plantation. Forests 12 (2), 120.

Zheng, B.X., Ibrahim, M., Zhang, D.P., Bi, Q.F., Li, H.Z., Zhou, G.W., Ding, K., Peñuelas, J., Zhu, Y.G., Yang, X.R., 2018. Identification and characterization of inorganic-phosphate-solubilizing bacteria from agricultural fields with a rapid isolation method. AMB Express 8 (1), 1–12.

Zhong, C., Jiang, A., Huang, W., Qi, X., Cao, G., 2017. Studies on the acid-production characteristics of Bacillus megaterium strain P17. AIP Conf. Proc. 1839 (1), 020054. AIP Publishing LLC.

Zhu, H.J., Sun, L.F., Zhang, Y.F., Zhang, X.L., Qiao, J.J., 2012. Conversion of spent mushroom substrate to biofertilizer using a stress-tolerant phosphate-solubilizing Pichia farinose FL7. Bioresour. Technol. 111, 410–416.

Zúñiga-Silgado, D., Rivera-Leyva, J.C., Coleman, J.J., Sánchez-Reyez, A., Valencia-Díaz, S., Serrano, M., de-Bashan, L.E., Folch-Mallol, J.L., 2020. Soil type affects organic acid production and phosphorus solubilization efficiency mediated by several native fungal strains from Mexico. Microorganisms 8 (9), 1337.

Plant-microbe interactions: Beneficial role of microbes for plant growth and soil health

Raghu Shivappa[a], Mathew Seikholen Baite[a], Prabhukarthikeyan S. Rathinam[a], Keerthana Umapathy[a], Prajna Pati[b], Anisha Srivastava[c], and Ravindra Soni[c]

[a]Crop Protection Division, ICAR-National Rice Research Institute (ICAR-NRRI), Cuttack, Odisha, India, [b]Department of Agricultural Entomology, Institute of Agricultural Sciences (IAS), Siksha 'O' Anusandhan Deemed to be University, Bhubaneswar, Odisha, India, [c]Department of Agricultural Microbiology, College of Agriculture, Indira Gandhi Krishi Viswavidyalaya, Raipur, Chhattisgarh, India

17.1 Introduction

Agriculture is the backbone of achieving global food security, and modern agriculture must provide sufficient food in terms of nutrition and calories to feed ever-growing population of the world. The population is projected to increase from 7.3 billion in 2015 to 9.8 billion by 2050. There are many challenges in achieving this goal such as biotic and abiotic stresses, increased temperature, erratic rainfall, floods, and drought. Among them, biotics are more important (Oerke and Dehne, 2004). Plants are counterattacked by many pests and diseases. These biotic stresses including fungal, bacterial, and viral diseases are responsible for losses ranging from 20% to 40% of global agricultural productivity (Savary et al., 2012). Plants not only attacked by harmful pathogens, but also benefited by a number of microorganisms, which directly or indirectly helps plants for their growth promotion, conferring host plant resistance to pathogens can reduce the impact of disease on crop development and yield. These entire processes take place in ecology through a complex networking called as plant-microbe interactions. Understanding these complex interactions lead us to new frontier areas of how they do interact in the system and their new potentialities, thereby addressing the issues related to plant-microbe interaction will solve many challenges of utilizing them for betterment of agricultural productivity which is an ultimate necessity of feeding the world's growing population.

17.2 Rhizobacteria

Rhizobacteria can be defined as those soil-borne bacteria which are found extensively in rhizosphere either as free living or in association with plant roots (Pielach et al., 2008). Among these, only 1%–2% bacteria in rhizosphere are involved in plant growth promotion (Beneduzi et al., 2012). Rhizosphere termed by Hiltner (1904) attributed as a nutrient-rich habitat for microorganisms and characterized by enhanced biomass productivity. This small area encircling the plant root system eventually becomes the interactive place among the plant, soil, and microfauna (Antoun and Prévost, 2005) which can affect plant growth and productivity. Among different types of dynamic microbial population residing in the rhizosphere (Goel et al., 2017; Kamaludeen and Ramasamy, 2008; Soni et al., 2017), bacteria have a staggering diversity which contributes the most to growth, physiology, and functional attributes of the plants (Babalola, 2010; Khan et al., 2009) and hence affect overall plant health through their secreted organic exudates (Sivasakthi et al., 2014). Microbial communities present in the rhizosphere, external and internal (the endosphere) plant tissues; all together constitute plant microbiota (Bai et al., 2015; Mendes et al., 2013; Mengoni et al., 2010; Pini et al., 2012), out of which rhizospheric population is referred as the rhizobiome. The beneficial soil bacteria are generally referred to as plant growth-promoting rhizobacteria (PGPR) which may either colonize the rhizosphere, rhizoplane, and intercellular spaces on root cells or they may inhabit inside the root cells through specialized structures. The former types of PGPR are designated as extracellular PGPR (ePGPR), while the latter ones are named as intracellular PGPR (iPGPR) (Gray and Smith, 2005). Plant growth-promoting rhizobacteria (PGPR) such as nitrogen-fixing legume endosymbiotic bacteria is known as rhizobia (Nadeem et al., 2014). Plant growth and yield are the net results of the interactions between the specific plant and its associated microbiome (Theis et al., 2016), as an example of which rhizobial "symbiotic nitrogen fixation" (SNF) (Sprent et al., 2013) found to be the true determinant of the evolutionary success of the leguminous plants (Sprent et al., 2013). In particular, As resistant microorganisms belonging to various rhizospheric bacterial genera, especially *Achromobacter, Arthrobacter, Azotobacter, Azospirillum, Bacillus, Brevundimonas, Comamonas, Enterobacter, Ensifer, Microbacterium, Ochrobactrum, Pseudomonas, Stenotrophomonas,* and *Serratia* and also documented for enhancing plant growth (Cavalca et al., 2013; Ghosh et al., 2011; Gray and Smith, 2005; Wang et al., 2011; Yang and Rosen, 2016), well adaptability to harsh soil conditions, plant growth-promoting activity, bioremediation ability of toxic heavy metals (Baum et al., 2006; Fomina et al., 2005; Schützendübel and Polle, 2002; Wenzel, 2009; Zimmer et al., 2009) made rhizobacteria as bioinoculants to promote plant growth under toxic metal stresses (Wani and Khan, 2010).

17.3 Ecological considerations for plant beneficial function of microbes in the field

Microorganisms do interact continuously with plant roots and soil constituents in the rhizosphere where root exudates and plant materials which are decaying provide sources of carbon compounds for these heterotrophic microbiota (Bisseling et al., 2009). The number of bacteria in the rhizosphere and rhizoplane is higher than in the soil devoid of plants. The soil devoid of plants does not have any attractive substances secreted by plant roots that are essential for microbes. When a seed starts to germinate, relatively large amount of carbon and nitrogenous compounds such as sugars, organic acids, amino acids, and vitamins excreted into the surrounding environment, and attracts large population of microorganisms inducing vigorous competition between the different species (Okon and Labandera-Gonzalez, 1994). The rhizosphere microorganisms typically differ between plant species (Bisseling et al., 2009).

17.4 Secondary metabolites in plant-microbe interaction

Numbers of antimicrobial compounds are produced in host plants when they are challenge inoculated with bacterial-fungal-viral plant pathogens. There are many antimicrobial compounds, including, but not limited to, antimicrobial peptides (AMPs), antifungal proteins, enzymes that break down pathogen cell wall and infection structure and phytoalexins (low molecular weight antimicrobial compounds) (Stuiver and Custers, 2001). In order to enhance disease resistance, genes encoding those antimicrobial compounds are expressed in host plant. Two of the most popular genes to express in transgenic plants were chitinase and glucanase as they can break down the fungal cell wall components chitin and glucan, respectively. Either of the two genes isolated from different plant or nonplant sources has been expressed in plants like rice (Datta et al., 2001), tomato (Jabeen et al., 2015), cotton (Emani et al., 2003), and brinjal (Simmonds et al., 2014), showing enhanced resistance to various pathogens. Antimicrobial peptides, positively charged, cysteine-rich, and thermostable small peptides having antimicrobial properties owing to their ability to damage pathogen membrane have also been used to develop transgenic crop plants with enhanced disease resistance. For example, transgenic rice (Iwai et al., 2002), tomato (Chan et al., 2005), papaya (Zhu et al., 2007), and wheat (Roy-Barman et al., 2006) plants expressing AMPs from different sources exhibited increased resistance to bacteria, *Burkholderia plantarii* and *Ralstonia solanacearum*, and fungus, *Phytophthora palmivora*, and *Neovossia indica*, respectively. In a similar note, transformation of phytoalexin biosynthetic gene like grapevine *Vst1* improved resistance of rice to *Pyricularia oryzae* (Stark-Lorenzen et al., 1997), of wheat to *Blumeria graminis* (Liang et al., 2000), and papaya to *Phytophthora palmivora* (Zhu et al., 2007).

17.4.1 Toxin production in host

This strategy is commonly followed to develop insect resistance in host plant species where an insecticidal toxin protein-coding gene is expressed. The development of transgenic plants expressing insecticidal proteins like CRY is considered as one of the significant achievements in pest management in the last three decades. As an alternative of chemical pesticides, the formulation of the gram-negative bacterium *Bacillus thuringiensis* (Bt) has been in use since 1938 as biopesticide owing to its ability to produce a number of insecticidal proteins, viz. cry (crystal), cyt (cytolytic), and vip (vegetative insecticidal proteins) (Schnepf et al., 1998). Scientists have isolated the genes coding for those toxin proteins from Bt and expressed in crop plants to develop host resistance against insect pests belonging to Lepidoptera, Diptera, and Coleoptera orders. These kinds of genetically engineered crops are commonly called as Bt-crops. Ingestion of CRY proteins can cause insect death as CRY proteins form pore in the insect mid-gut membrane that leads to loss of ions and cellular ATPs. Many countries including India are commercially cultivating different Bt-transgenic crops. Bt-cotton has been developed expressing different cry genes (*cry1Ac, cry2Ab, cry1C,* and *cry1Ab*) showing complete resistance against the cotton bollworm *Helicoverpa armigera* (Karihaloo and Kumar, 2009). Similarly, Bt-brinjal (Choudhary and Gaur, 2009) and Bt-poplar tree expressing *cry1Ac* (Hu et al., 2010), Bt-maize expressing *CryIAb* (Fearing et al., 1997), Bt-rice expressing *cry1Ab* and/or *cry1C* (Datta et al., 1998; Shu et al., 2000), Bt-potato expressing *cry3A* (Perlak et al., 1993), Bt-tomato expressing cry1Ab (Kumar and Kumar, 2004), and Bt-soybean expressing *cry1Ac* (Yu et al., 2013) have been developed to control major insect pest/s of the particular crop. Fusion of two different classes of *cry* genes is also used for effective control of insect pests. For chickpea pod borer resistance, Ganguly et al. (2014) developed transgenic plants expressing fused *cry1Ab/Ac* under the control of constitutive and pod-specific msg promoter. The insecticidal proteins from Bt have been the subject of intensive investigation for last two decades which documented their safety nature to nontarget organism (Schnepf et al., 1998). As a non-Bt-crop, corn has been genetically modified to produce the VIP3A protein to protect the crop from a range of secondary pests which are found to be resistant to CRY proteins (Syngenta). Another type of insecticidal protein is different classes of protease inhibitors which inhibit the protease enzymes required by insects for their normal digestion procedure. Trypsin inhibitor gene from cowpea and sweet potato was used to generate insect resistance in different crops (Xu et al., 1996; Yeh et al., 1997; Ding et al., 1998). CRY proteins have limitations against sap-sucking homopteran pest as they were found to be resistant to CRY protein. This limitation can be overcome by the use of plant lectin proteins which show entomotoxic activity as they interact with different glycoproteins or glycan structures in insects interfering with a number of physiological processes of the insect (Macedo et al., 2015). Plant growth-promoting rhizobacteria produce numerous secondary metabolites called toxic compounds that are lethal to pathogens. The compounds such as hydrogen cyanide, phenazines, pyrrolnitrin and pyoluteorins and several enzymes, antibiotics and metabolites, and phytohormones (Castro-Sowinski et al., 2007; Ramette et al., 2011; Jousset et al., 2011). When soil is limited with iron, PGPRs can produce low molecular weight compounds called siderophores that sequester iron in a competitive way, thus depriving pathogenic fungi of this essential and often scarcely bioavailable element (Pedraza et al., 2007).

17.5 Microbial-plant defense genes involved in interaction

The interaction between plant and microbe is intricate in a complex system. The interaction takes place at both cellular and molecular levels. For many years, these interactions are poorly understood at various levels such as cellular, molecular, and physiological levels. A plethora of genes (defense) is produced at different levels of plant-microbe interaction. With the advent of recombinant DNA technology and molecular markers, it is possible to understand the complex interactions. Very few of the total microbes are harmful to the plants when compared to vast benefits they provide. Some of the pathogens developed very sophisticated mechanisms to parasite and colonize the host plants. Best example of this is *Agrobacterium tumefaciens*, a soil bacterium, which causes crown gall of apple. A remarkable integrity has been developed between plant and *A. tumefaciens* for a successful disease establishment.

The entry of pathogens into the host cells is the first and most important step in disease response and successful establishment of the disease. The mode of entry depends on the type of pathogen invading the host. Bacteria enter into plant cells through trichomes, lenticels, stomata, hydathodes, and other openings. On the contrary, fungi use a specialized structure called hyphae and form penetration pegs that enter inside the cell and establish infection successfully. Viruses are obligate parasites that enter only through physical injuries and insect vectors (Layne, 1967; Fox et al., 1971; Getz et al., 1983; Mendgen et al., 1996). After breaking the primary barriers, it elicits immune response which ore of two types, one is microbial (or pathogen)-associated molecular patterns (MAMPs/PAMPs) triggered immunity (MTI/PTI) formerly called as basal or horizontal immunity and other is effector- triggered immunity (ETI) formerly called R-gene-based or vertical immunity (Padmanabhan et al., 2009; Sahu et al., 2012). Economically important concept for plant-pathogen interaction is gene-for-gene concept which is widely studied in different crops for disease resistance (Russell, 2013; Lucas, 1998). Plant defense through effector-triggered immunity (ETI) is based on the highly specific interaction between plant pathogen products called avirulent genes (Avr) and products from host-resistant genes (R) produced according to the gene-for-gene hypothesis (Flor, 1955), which suggests that plants have many R-genes and pathogens have many Avr genes and plant disease resistance is observed if the product of any particular R gene has recognition specificity for a compound produced due to a particular pathogen Avr gene. R-genes are structurally present in a central nucleotide-binding site (NBS) domain, a C-terminal LRR region to mediate pathogen recognition, and N-terminal variable domain mainly identified as TIR (Toll/Interleukin-1) or CC (Coiledcoil) (Elmore et al., 2011; Gururani et al., 2012).

17.5.1 Overexpression of host defense genes

Unlike animals, plants are unable to move from one place to other compelling them to endure any kind of stress at a standstill condition (Molla et al., 2015). They depend on their innate immune system that relies on the activation of an array of defense systems including production of phytoalexins, modification of cell wall, synthesis of PR genes, generation of reactive oxygen species (ROS), deposition of callose, and activation of different defense signal transduction pathways. In order to enhance resistance to pathogen by boosting plants own defense system, the endogenous defense genes or well-known defense genes from other plants are overexpressed in transgenic plants. Phenylalanine ammonia lyase (PAL) plays an important role in plant defense by mediating an important step in the shikimic acid pathway for the synthesis of many phenolic compounds, including lignin, which play a role in the defense mechanism (Molla et al., 2013). PAL gene overexpressing transgenic tobacco plants exhibited reduced susceptibility to the fungal pathogen *Cercospora nicotianae* (Shadle et al., 2003). Cao et al. (2008) showed that the overexpression of a rice defense-related F-box protein gene *OsDRF1* improves disease resistance through enhanced defense gene expression in transgenic tobacco. Elevated expression of PR genes in host plants in response to pathogen attack is a common phenomenon in many pathosystems. There are more than 10 different types of well-characterized PR proteins found in plants (Van Loon and Van Strien, 1999). Two well-known plant PR genes are β-1,3-glucanase (PR2) and chitinase (PR3) which have been widely utilized to generate transgenic plants for developing resistance against various pathogens. Some examples are given in Section 17.1. Thaumatin-like protein (TLP), a PR5 protein, exhibited an enhanced resistance to sheath blight disease when expressed in transgenic rice (Datta et al., 1999) and to fusarium wilt in transgenic banana (Mahdavi et al., 2012). Plant genome harbors different master controller genes that act in different defense signaling pathways. Arabidopsis *NPR1* (nonexpressor of pathogenesis-related genes 1) gene is a master controller of systemic acquired resistance (SAR) pathway that provides broad-spectrum resistance to pathogens. *NPR1* homologs were found to play an important role in defense response in many plant species like rice (Chern et al., 2005a,b), apple (Zhang et al., 2012), grapevine (Le Henanff et al., 2009), and gladiolus (Zhong et al., 2015). Overexpression of *NPR1* in transgenic plants exhibited increased resistance to diverse pathogens in rice, wheat, cotton, carrot, tomato, etc. (Kumar et al., 2013; Makandar et al., 2006; Quilis et al., 2008; Wally et al., 2009). However, constitutive expression of Arabidopsis *NPR1* gene resulted in the development of phenotypic abnormalities (Quilis et al., 2008; Fitzgerald et al., 2004). Constitutive expression of this kind of master switch gene can cause enhanced susceptibility to other pathogens and herbivores (Stuiver and Custers, 2001; Yuan et al., 2007). In order to avoid such kind of problems, recently two strategies have been demonstrated as successful (Molla et al., 2016; Xu et al., 2017). Restricting the expression of *AtNPR1* gene only in green tissues using the green tissue-specific promoter $P_{D540-544}$ has been demonstrated to enhance sheath blight resistance in rice without concomitant phenotypic cost (Molla et al., 2016). The enhanced tolerance to sheath blight pathogen was achieved by elevated expression of PR genes in transgenic plants (Molla et al., 2016). Another strategy was devised by Xu et al. (2017) where they developed a "TBF1-cassette" consisting of the immune-inducible promoter and two pathogen-responsive upstream open-reading frames (uORF-s_{TBF1}) of the TBF1 gene to control the expression of *AtNPR1* gene. The cassette acts in two different levels, viz. the transcriptional level control is mediated by TBF1 immune inducible promoter and translational level control is effected by uORFs$_{TBF1}$. Rice plant expressing *AtNPR1* gene controlled by TBF1-cassette exhibited broad-spectrum resistance to bacterial blight, bacterial leaf streak, and fungal blast pathogen (Xu et al., 2017). These two strategies could be effectively utilized in plant disease engineering with a controlled expression of master switch gene like *NPR1*.

17.6 RNAi and CRISPR/Cas9 technology to explore plant-microbe interaction

17.6.1 RNAi

Successful disease development or infestation is largely dependent on compatible interaction between pathogen-insect and plant host. There are genes in host plants that facilitate disease developments by different means and those genes are called as susceptibility (S) genes. If silenced or mutated, those S genes will no longer support a compatible interaction and ultimately can develop pathogen-specific or broad-spectrum resistance in the host (van Schie and Takken, 2014). Silencing or knockdown of S gene can be done using antisense-RNAi, hairpin-RNAi, or microRNA. The *mycoplasma-like organism* (*MLO*) and *eIF4E* represent two very good examples of S gene used in breeding powdery mildew-resistant barley and potyvirus-resistant pepper, respectively (Jørgensen, 1992; Ruffel et al., 2002). In a recent study, RNAi (RNA interference)-mediated silencing of six S genes in potato was demonstrated to develop reduced susceptibility to complete resistance against late blight pathogen *Phytophthora infestans* (Sun et al., 2016). Transgenic rice plants silencing a susceptibility gene, *Os8N3*, exhibited an enhanced resistance to PXO99^A strain of *Xanthomonas oryzae* pv. *oryzae* (Yang et al., 2006). In a similar note, tomato plant exhibited a reduction in susceptibility to Botrytis cinerea when the expression of tomato *phosphatidylinositol-phospholipase C* (*PI-PLC*) gene was knocked down by virus-induced gene silencing (Gonorazky et al., 2016). Mds1, a major susceptibility gene for infestation of wheat by the gall midge commonly known as Hessian fly, when silenced by RNA interference conferred immunity on normally susceptible wheat plants against all Hessian fly biotypes (Liu et al., 2013). Sometimes silencing of host S gene can cause the development of pleiotropic effects in unrelated tissues, which is a matter of concern. For example, silencing of *SlDMR1* and *SlPMR4* genes using RNAi resulted in resistance to the tomato powdery mildew fungus *Oidium neolycopersici*, but *SlDMR1* silenced plant exhibited reduced plant growth, whereas *SlPMR4* silenced plant was normal (Huibers et al., 2013).

17.6.2 CRISPR/Cas9

Understanding the interactions of plants and microbes such as bacteria, fungi, viruses, nematodes, and mycoplasmas has been the major area of investigation attempted for many years by many researchers. The high-throughput molecular technologies made these interesting but very imperative studies possible with newer and newer outcomes day by day. These technologies provided newer insights into the plants' interaction with microbes in relation to environmental factors and genotypes. The complex interplay between plant and pathogen, plant, and beneficial microbes involves several components at cellular and molecular levels (Boyd and O'Toole, 2012; Dracatos et al., 2018). Genome editing technology is the one and latest tool to study plant-microbe interactions. In recent years, we have witnessed the emergence of new technologies like use of meganucleases, zinc finger nucleases (ZENs), transcription activator-like effector nucleases (TALENs), and clustered regulatory interspaced palindrome repeats, popularly called as CRISPR/CRISPR-associated protein 9 (Cas9) (Borrelli et al., 2018). This technology has largely overtaken the other genome editing technologies because of the fact that it is easier to design and implement, more versatile, less expensive, and has a success rate. This technology has been developed based on the adaptive immune system of *Streptococcus pyogenes* (Jinek et al., 2012; Cong et al., 2013). This technology operates by creating DNA double-strand breaks (DSBs) at the target loci to stimulate genome editing via nonhomologous end joining (NHEJ) or homology-directed repair (HDR). For genome editing purposes, a specific gRNA (guide RNA) and a Cas9 expression cassette are cloned in transformation vector. The gRNA is a ~20-bp-long RNA sequence (complementary to target sequence of the genome to be edited) located within a longer RNA scaffold. The gRNA binds to DNA in the genome and the Cas9 nuclease creates nick (DSB) to the specific part of the genome. The DSB is then repaired by cellular repair mechanism that causes the mutation. The major advantage of this system is that CRISPR/Cas9 vector can be introduced in plants to create the targeted genome edits, and subsequently, selection can be done for progeny plants carrying the desired edits but have lost the CRISPR/Cas9 cassette through segregation. The edited plants will be free of any foreign DNA and therefore can be called as nontransgenic. The CRISPR/Cas9 system can be used efficiently to mutate the susceptibility genes and the repressor of defense genes in the host plant to confer resistance to pathogens. Wang et al. (2014) utilized the CRISPR/Cas9 genome editing system to mutate three homoeoalleles of *MLO* gene, a defense repressor in hexaploid bread wheat, to develop resistance against powdery mildew. Similarly, improved blast disease resistance has been obtained in rice plants when a transcription factor gene OsERF922 was knocked out via CRISPR/Cas9 system (Wang et al., 2016). In addition, the editing system can also be employed in developing resistance against DNA viruses. For example, the bean yellow dwarf virus (BeYDV) (geminivirus) genome has been targeted for destruction with the CRISPR/Cas system expressed in transgenic plants and a reduced virus load and disease symptoms were observed in the transgenic plants (Baltes et al., 2015). As virus enters into the host plant cell to propagate, the expression of CRISPR/Cas system targeted to the viral genome in plants is a good strategy to reduce the level of damage due to the particular viral infection. The application of CRISPR/Cas tools has mainly been explored against virus infection, followed by efforts to improve fungal and bacterial disease resistance. Rice blast, caused by *Magnaporthe oryzae*, is one of the most devastating diseases that affect rice production all over the world (Dean et al., 2012). Ethylene-responsive factors (ERFs) of the APETELA2/ERF (AP2/ERF) superfamily play an important role in rice adaptation to multiple biotic and abiotic stresses (Mizoi et al., 2012). The expression of OsERF922 is induced not only by abscisic acid (ABA) and salt stress, but also by *Magnaporthe oryzae*. OsERF922 knockdown by RNAi (RNA interference) leads to increased resistance to blast fungus. This clearly indicates that OsERF922 is a negative regulator of rice blast resistance (Liu et al., 2012). Using CRISPR/Cas9 technology-targeted modification of OsERF922 was made and generated Oserf922 knockout mutants (Wang et al., 2016). These null mutants showed an enhanced resistance to rice blast without affecting other major agronomic traits (Table 17.1).

Table 17.1 CRISPR/Cas9 applications in plant disease resistance.

Sl No	Plant species	Pathogen	Target gene	Gene function	References
1	*Triticum aestivum*	Powdery mildew (*Blumeria graminis* f. sp. *tritici*)	MLO-A1	Susceptibility (S) gene involved in powdery mildew disease	Wang et al. (2014)
2	*Oryza sativa, L. japonica*	Rice blast disease (*Magnaporthe oryzae*)	SEC3A	Subunit of the exocyst complex	Ma et al. (2018)
3	*Oryza sativa, L. japonica*	Rice blast disease (*Magnaporthe oryzae*)	ERF922	Transcription factor implicated in multiple stress responses	Wang et al. (2016)
4	*Solanum lycopersicum*	Powdery mildew (*Oidium neolycopersici*)	MLO1	Major responsible for powdery mildew vulnerability	Nekrasov et al. (2017)
5	*Vitis vinifera*	Gray mold (*Botrytis cinerea*)	WRKY52	Transcription factor involved in response to biotic stress	Wang et al. (2018)

Table 17.1 CRISPR/Cas9 applications in plant disease resistance—cont'd

Sl No	Plant species	Pathogen	Target gene	Gene function	References
6	Oryza sativa	Bacterial blight (Xanthomonas oryzae pv. oryzae)	SWEET13	Sucrose transporter gene	Li et al. (2012) and Zhou et al. (2015)
7	Citrus paradisi	Citrus canker (Xanthomonas citri subspecies citric)	LOB1	Susceptibility (S) gene-promoting pathogen growth and pustule formation	Jia et al. (2016)
8	Citrus sinensis Osbeck	Citrus canker (Xanthomonas citri subspecies citric)	LOB1	Susceptibility (S) gene-promoting pathogen growth and pustule formation	Peng et al. (2017)
9	Nicotiana benthamiana and Arabidopsis thaliana	BeYDV	CP, Rep, and IR	RCA mechanism	Ji et al. (2015)
10	Cucumis sativus	CVYV ZYMV PRSV-W	eIF4E	Host factor for RNA virus translation	Chandrasekaran et al. (2016)
11	Oryza sativa, L. japonica	RTSV	eIF4G	Host factor for RNA virus translation	Macovei et al. (2018)
12	Nicotiana benthamiana	TuMV	GFP1, GFP2, HC-Pro, and CP	Replication mechanism	Aman et al. (2018)
13	Nicotiana benthamiana	TYLCV BCTV MeMV	CP, Rep, and IR	RCA mechanism	Ali et al. (2015)
14	Nicotiana benthamiana	CLCuKoV MeMV TYLCV	CP, Rep, and IR	RCA mechanism	Ali et al. (2016)

References

Ali, Z., Abulfaraj, A., Idris, A., Ali, S., Tashkandi, M., Mahfouz, M.M., 2015. CRISPR/Cas9-mediated viral interference in plants. Genome Biol. 16 (1), 1–11.

Ali, Z., Ali, S., Tashkandi, M., Zaidi, S.S.E.A., Mahfouz, M.M., 2016. CRISPR/Cas9-mediated immunity to geminiviruses: differential interference and evasion. Sci. Rep. 6 (1), 1–13.

Aman, R., Ali, Z., Butt, H., Mahas, A., Aljedaani, F., Khan, M.Z., et al., 2018. RNA virus interference via CRISPR/Cas13a system in plants. Genome Biol. 19, 1. https://doi.org/10.1186/s13059-017-1381-1.

Antoun, H., Prévost, D., 2005. Ecology of plant growth promoting rhizobacteria. In: Siddiqui, Z.A. (Ed.), PGPR: Biocontrol and Biofertilization. Springer, Dordrecht. https://doi.org/10.1007/1-4020-4152-7_1.

Babalola, O.O., 2010. Beneficial bacteria of agricultural importance. Biotechnol Lett 32, 1559–1570. https://doi.org/10.1007/s10529-010-0347-0.

Bai, Y., Muller, D.B., Srinivas, G., Garrido-Oter, R., Potthoff, E., Rott, M., et al., 2015. Functional overlap of the Arabidopsis leaf and root microbiota. Nature 528, 364–369. https://doi.org/10.1038/nature16192.

Baltes, N.J., Hummel, A.W., Konecna, E., Cegan, R., Bruns, A.N., Bisaro, D.M., Voytas, D.F., 2015. Conferring resistance to geminiviruses with the CRISPR-Cas prokaryotic immune system. Nat. Plants 1 (10), 1–4.

Baum, C., Hrynkiewicz, K., Leinweber, P., Meißner, R., 2006. Heavy-metal mobilization and uptake by mycorrhizal and nonmycorrhizal willows (Salix×dasyclados). J. Plant. Nutr. Soil Sci. 169, 516–522.

Beneduzi, A., Ambrosini, A., Passaglia, L.M.P., 2012. Plant growth-promoting rhizobacteria (PGPR): their potential as antagonists and biocontrol agents. Genet. Mol. Biol. 35 (4 (Suppl)), 1044–1051.

Bisseling, T., Dangl, J.L., Schulze-Lefert, P., 2009. Next-generation communication. Science 324, 691.

Borrelli, V.M., Brambilla, V., Rogowsky, P., Marocco, A., Lanubile, A., 2018. The enhancement of plant disease resistance using CRISPR/Cas9 technology. Front. Plant Sci. 9, 1245.

Boyd, C.D., O'Toole, G.A., 2012. Second messenger regulation of biofilm formation: breakthroughs in understanding c-di-GMP effector systems. Annu. Rev. Cell Dev. Biol. 28, 439–462.

Cao, Y., Yang, Y., Zhang, H., Li, D., Zheng, Z., Song, F., 2008. Overexpression of a rice defense-related F-box protein gene OsDRF1 in tobacco improves disease resistance through potentiation of defense gene expression. Physiol. Plant 134, 440–452.

Castro-Sowinski, S., Herschkovitz, Y., Okon, Y., Jurkevitch, E., 2007. Effects of inoculation with plant growth-promoting rhizobacteria on resident rhizosphere microorganisms. FEMS Microbiol. Lett. 276 (1), 1–11.

Cavalca, L., Corsini, A., Bachate, S.P., Andreoni, V., 2013. Rhizosphere colonization and arsenic translocation in sunflower (Helianthusannuus L.) by arsenate reducing Alcaligenes sp. strain Dhal-L. World J. Microbiol. Biotechnol. 29, 1931–1940.

Chan, Y.L., Prasad, V., Chen, K.H., Liu, P.C., Chan, M.T., Cheng, C.P., 2005. Transgenic tomato plants expressing an Arabidopsis thionin (Thi2. 1) driven by fruit-inactive promoter battle against phytopathogenic attack. Planta 221 (3), 386–393.

Chandrasekaran, J., Brumin, M., Wolf, D., Leibman, D., Klap, C., Pearlsman, M., Sherman, A., Arazi, T., Gal-On, A., 2016. Development of broad virus resistance in non-transgenic cucumber using CRISPR/Cas9 technology. Mol. Plant Pathol. 17 (7), 1140–1153.

Chern, M., Canlas, P.E., Fitzgerald, H.A., Ronald, P.C., 2005a. Rice NRR, a negative regulator of disease resistance, interacts with Arabidopsis NPR1 and rice NH1. Plant J 43, 623–635. https://doi.org/10.1111/j.1365-313X.2005.02485.x.

Chern, M., Fitzgerald, H.A., Canlas, P.E., Navarre, D.A., Ronald, P.C., 2005b. Overexpression of a rice NPR1 homolog leads to constitutive activation of defense response and hypersensitivity to light. Mol. Plant-Microbe Interact. 18, 511–520. https://doi.org/10.1094/MPMI-18-0511.

Choudhary, B., Gaur, K., 2009. The Development and Regulation of Bt Brinjal in India (Eggplant/Aubergine). International Service for the Acquisition of Agri-biotech Applications.

Cong, L., Ran, F.A., Cox, D., Lin, S., Barretto, R., Habib, N., Hsu, P.D., Wu, X., Jiang, W., Marraffini, L.A., 2013. Multiplex genome engineering using CRISPR/Cas systems. Science 339 (6121), 819–823.

Datta, K., Vasquez, A., Tu, J., Torrizo, L., Alam, M.F., Oliva, N., Abrigo, E., Khush, G.S., Datta, S.K., 1998. Constitutive and tissue-specific differential expression of the cryIA (b) gene in transgenic rice plants conferring resistance to rice insect pest. Theor. Appl. Genet. 97 (1–2), 20–30.

Datta, K., Tu, J., Oliva, N., Ona, I., Velazahan, R., Mew, T.W., Muthukrishnan, S., Datta, S.K., 2001. Enhanced resistance to sheath blight by constitutive expression of infection-related rice chitinase in transgenic elite indica rice cultivars. Plant Sci. 160 (3), 405–414.

Datta, K., Velazhahan, R., Oliva, N., Ona, I., Mew, T., Khush, G., Muthukrishnan, S., et al., 1999. Overexpression of the cloned rice thaumatin-like protein (PR-5) gene in transgenic rice plants enhances environmental friendly resistance to Rhizoctonia solani causing sheath blight disease. Theor. Appl. Genet. 98, 1138–1145.

Dean, R., Van Kan, J.A.L., Pretorius, Z.A., et al., 2012. The top 10 fungal pathogens in molecular plant pathology. Mol. Plant Pathol. 13 (4), 414–430.

Ding, L.C., Hu, C.Y., Yeh, K.W., Wang, P.J., 1998. Development of insect-resistant transgenic cauliflower plants expressing the trypsin inhibitor gene isolated from local sweet potato. Plant Cell Rep. 17 (11), 854–860.

Dracatos, P.M., Haghdoust, R., Singh, D., Park, R.F., 2018. Exploring and exploiting the boundaries of host specificity using the cereal rust and mildew models. New Phytol. 218 (2), 453–462.

Elmore, J.M., Lin, Z.J.D., Coaker, G., 2011. Plant NB-LRR signaling: upstreams and downstreams. Curr. Opin. Plant Biol. 14 (4), 365–371.

Emani, C., Garcia, J.M., Lopata-Finch, E., Pozo, M.J., Uribe, P., Kim, D.J., Sunilkumar, G., Cook, D.R., Kenerley, C.M., Rathore, K.S., 2003. Enhanced fungal resistance in transgenic cotton expressing an endochitinase gene from Trichoderma virens. Plant Biotechnol. J. 1 (5), 321–336.

Fearing, P.L., Brown, D., Vlachos, D., Meghji, M., Privalle, L., 1997. Quantitative analysis of CryIA (b) expression in Bt maize plants, tissues, and silage and stability of expression over successive generations. Mol. Breed. 3 (3), 169–176.

Fitzgerald, H.A., Chern, M.S., Navarre, R., and Ronald, P.C., 2004. Overexpression of (At) NPR1 in rice leads to a BTH- and environment-induced lesion-mimic/cell death phenotype. Mol. Plant-Microbe Interact. 17, 140–151. https://doi.org/10.1094/MPMI.2004.17.2.140.

Flor, H.H., 1955. Host-parasite interactions in flax rust-its genetics and other implications. Phytopathology 45, 680–685.

Fomina, M.A., Alexander, I.J., Colpaert, J.V., Gadd, G.M., 2005. Solubilization of toxic metal minerals and metal tolerance of mycorrhizal fungi. Soil Biol. Biochem. 37, 851–866.

Fox, R.T.V., Manners, J.G., Myers, A., 1971. Ultrastructure of entry and spread of Erwinia carotovora var. atroseptica into potato tubers. Potato Res. 14 (2), 61–73.

Ganguly, M., Molla, K.A., Karmakar, S., Datta, K., Datta, S.K., 2014. Development of pod borer-resistant transgenic chickpea using a pod-specific and a constitutive promoter-driven fused cry1Ab/Ac gene. Theor. Appl. Genet. 127, 2555–2565. https://doi.org/10.1007/s00122-014-2397-5.

Getz, S., Fulbright, D.W., Stephens, C.T., 1983. Scanning electron microscopy of infection sites and lesion development on tomato fruit infected with Pseudomonas syringae pv. tomato. Phytopathology 73 (1), 39–43.

Ghosh, P., Rathinasabapathi, B., Ma, L.Q., 2011. Arsenic-resistant bacteria solubilized arsenic in the growth media and increased growth of arsenic hyperaccumulator Pteris vittata L. Bioresour. Technol. 102 (19), 8756–8761. https://doi.org/10.1016/j.biortech.2011.07.064.

Goel, R., Suyal, D.C., Narayan, Dash, B., Soni, R., 2017. Soil metagenomics: a tool for sustainable agriculture. In: Kalia, V., Shouche, Y., Purohit, H., Rahi, P. (Eds.), Mining of Microbial Wealth and MetaGenomics. Springer, Singapore. https://doi.org/10.1007/978-981-10-5708-3_13.

Gonorazky, G., Guzzo, M.C., Abd-El-Haliem, A.M., Joostern, M.H.A.J., Laxal, A.M.A., 2016. Silencing of the tomato phosphatidylinositol-phospholipase C2(SlPLC2) reduces plant susceptibility to Botrytis cinerea. Mol. Plant Pathol. 17 (9), 1354–1363. https://doi.org/10.1111/mpp.12365.

Gray, J., Smith, D.L., 2005. Intracellular and extracellular PGPR: commonalities and distinctions in the plant–bacterium signaling processes. Soil Biol. Biochem. 37, 3.

Gururani, M.A., Venkatesh, J., Upadhyaya, C.P., Nookaraju, A., Pandey, S.K., Park, S.W., 2012. Plant disease resistance genes: current status and future directions. Physiol. Mol. Plant Pathol. 78, 51–65.

Hiltner, L., 1904. Über neuere Erfahrungen und Probleme auf dem Gebiete der Bodenbakteriologie unter besonderer Berücksichtigung der Gründüngung und Brache. Arb. DLG 98, 59–78.

Hu, J.J., Yang, M.S., Lu, M.Z., 2010. Advances in biosafety studies on transgenic insect-resistant poplars in China. Biodivers. Sci. 18, 336–345. https://doi.org/10.3724/SP.J.1003.2010.336.

Huibers, R.P., Loonen, A.E.H.M., Gao, D., Van den Ackerveken, G., Visser, R.G.F., Bai, Y., 2013. Powdery mildew resistance in tomato by impairment of SlPMR4 and SlDMR1. PLoS ONE 8 (6). https://doi.org/10.1371/journal.pone.0067467, e67467.

Iwai, T., Kaku, H., Honkura, R., Nakamura, S., Ochiai, H., Sasaki, T., Ohashi, Y., 2002. Enhanced resistance to seed-transmitted bacterial diseases in transgenic rice plants overproducing an oat cell-wall-bound thionin. Mol. Plant-Microbe Interact. 15 (6), 515–521.

Jabeen, N., et al., 2015. Expression of rice chitinase gene in genetically engineered tomato confers enhanced resistance to fusarium wilt and early blight. Plant Pathol. J. 31, 252–258. https://doi.org/10.5423/PPJ.OA.03.2015.0026.

Ji, X., Zhang, H., Zhang, Y., Wang, Y., Gao, C., 2015. Establishing a CRISPR–Cas-like immune system conferring DNA virus resistance in plants. Nat. Plants 1 (10), 1–4.

Jia, H., Orbovic, V., Jones, J.B., Wang, N., 2016. Modification of the PthA4 effector binding elements in type I Cs LOB 1 promoter using Cas9/sg RNA to produce transgenic Duncan grapefruit alleviating XccΔpthA4: dCs LOB 1.3 infection. Plant Biotechnol. J. 14 (5), 1291–1301.

Jinek, M., Chylinski, K., Fonfara, I., Hauer, M., Doudna, J.A., Charpentier, E., 2012. A programmable dual-RNA–guided DNA endonuclease in adaptive bacterial immunity. Science 337 (6096), 816–821.

Jørgensen, J.H., 1992. Discovery, characterization and exploitation of Mlo powdery mildew resistance in barley. Euphytica 63, 141–152.

Jousset, A., Rochat, L., Lanoue, A., Bonkowski, M., Keel, C., Scheu, S., 2011. Plants respond to pathogen infection by enhancing the antifungal gene expression of root-associated bacteria. Mol. Plant-Microbe Interact. 24 (3), 352–358.

Kamaludeen, S.P.B., Ramasamy, K., 2008. Rhizoremediation of metals: harnessing microbial communities. Indian J. Microbiol. 48, 80–88.

Karihaloo, J.L., Kumar, P.A., 2009. Bt Cotton in India-A Status Report (Second Edition). Asia-Pacific Consortium on Agricultural Biotechnology (APCoAB), New Delhi, India, p. 56.

Khan, A.A., Jilani, G., Akhtar, M.S., Naqvi, S.S., Rasheed, M., 2009. Phosphorus solubilizing bacteria: occurrence, mechanisms and their role in crop production. J. Agric. Biol. Sci. 1, 48–58.

Kumar, V., Joshi, S.G., Bell, A.A., Rathore, K.S., 2013. Enhanced resistance against Thielaviopsis basicola in transgenic cotton plants expressing Arabidopsis NPR1 gene. Trans. Res. 22, 359–368. https://doi.org/10.1007/s11248-012-9652-9.

Kumar, H., Kumar, V., 2004. Tomato expressing Cry1A (b) insecticidal protein from Bacillus thuringiensis protected against tomato fruit borer, Helicoverpa armigera (Hübner)(Lepidoptera: Noctuidae) damage in the laboratory, greenhouse and field. Crop Prot. 23 (2), 135–139.

Layne, R.E.C., 1967. Foliar trichomes and their importance as infection sites for Corynebacterium michiganensé ontomato. Phytopathology 57, 981–985.

Le Henanff, G., Heitz, T., Mestre, P., Mutterer, J., Walter, B., Chong, J., 2009. Characterization of Vitis vinifera NPR1 homologs involved in the regulation of pathogenesis-related gene expression. BMC Plant Biol. 9, 54. https://doi.org/10.1186/1471-2229-9-54.

Li, T., Liu, B., Spalding, M.H., Weeks, D.P., Yang, B., 2012. High-efficiency TALEN-based gene editing produces disease-resistant rice. Nat. Biotechnol. 30 (5), 390–392.

Liang, H., Zheng, J., Shuange, J.I.A., Wang, D., Ouyang, J., Li, J., Li, L., Tian, W., Jia, X., Duan, X., Sheng, B., Hain, R., 2000. A transgenic wheat with a stilbene synthase gene resistant to powdery mildew obtained by biolistic method. Chin. Sci. Bull. 45, 634–638.

Liu, D., Chen, X., Liu, J., Ye, J., Guo, Z., 2012. The rice ERF transcription factor OsERF922 negatively regulates resistance to *Magnaportheoryzae* and salt tolerance. J. Exp. Bot. 63, 3899–3911. https://doi.org/10.1093/jxb/ers079.

Liu, X., Khajuria, C., Li, J., et al., 2013. Wheat *Mds-1* encodes a heat-shock protein and governs susceptibility towards the Hessian fly gall midge. Nat Commun 4, 2070. https://doi.org/10.1038/ncomms3070.

Lucas, J.A., 1998. Plant Pathology and Plant Pathogens. Blackwell Sci, Oxford, UK, p. 274.

Ma, J., Chen, J., Wang, M., Ren, Y., Wang, S., Lei, C., Cheng, Z., 2018. Disruption of OsSEC3A increases the content of salicylic acid and induces plant defense responses in rice. J. Exp. Bot. 69 (5), 1051–1064.

Macedo, M.L.R., Oliveira, C.F., Oliveira, C.T., 2015. Insecticidal activity of plant lectins and potential application in crop protection. Molecules 20 (2), 2014–2033.

Macovei, A., Sevilla, N.R., Cantos, C., Jonson, G.B., Slamet-Loedin, I., Čermák, T., Voytas, D.F., Choi, I.R., Chadha-Mohanty, P., 2018. Novel alleles of rice eIF4G generated by CRISPR/Cas9-targeted mutagenesis confer resistance to Rice tungro spherical virus. Plant Biotechnol. J. 16 (11), 1918–1927.

Mahdavi, F., Sariah, M., Mazih, M., 2012. Expression of rice thaumatin-like protein gene in transgenic banana plants enhances resistance to Fusarium wilt. Appl. Biochem. Biotechnol. 166 (4), 1008–1019.

Makandar, R., Essig, J.S., Schapaugh, M.A., Trick, H.N., Shah, J., 2006. Genetically engineered resistance to Fusarium head blight in wheat by expression of Arabidopsis NPR1. Mol. Plant-Microbe Interact. 19, 123–129. https://doi.org/10.1094/MPMI-19-0123.

Mendes, R., Garbeva, P., Raaijmakers, J.M., 2013. The rhizosphere microbiome: significance of plant beneficial, plant pathogenic, and human pathogenic microorganisms. FEMS Microbiol. Rev. 37 (Issue 5), 634–663. https://doi.org/10.1111/1574-6976.12028.

Mendgen, K., Hahn, M., Deising, H., 1996. Morphogenesis and mechanisms of penetration by plant pathogenic fungi. Annu. Rev. Phytopathol. 34 (1), 367–386.

Mengoni, A., Schat, H., Vangronsveld, J., 2010. Plants as extreme environments? Ni-resistant bacteria and Ni-hyperaccumulators of serpentine flora. Plant Soil 331, 5–16. https://doi.org/10.1007/s11104-009-0242-4.

Mizoi, J., Shinozaki, K., Yamaguchi-Shinozaki, K., 2012. AP2/ERF family transcription factors in plant abiotic stress responses. Biochim. Biophys. Acta 1819, 86–96. https://doi.org/10.1016/j.bbagrm.2011.08.004.

Molla, K.A., Karmakar, S., Chanda, P.K., Ghosh, S., Sarkar, S.N., Datta, S.K., Datta, K., 2013. Rice oxalate oxidase gene driven by green tissue-specific promoter increases tolerance to sheath blight pathogen (Rhizoctonia solani) in transgenic rice. Mol. Plant Pathol. 14, 910–922.

Molla, K.A., Debnath, A.B., Ganie, S.A., et al., 2015. Identification and analysis of novel salt responsive candidate gene based SSRs (cgSSRs) from rice (Oryza sativa L.). BMC Plant Biol 15, 122. https://doi.org/10.1186/s12870-015-0498-1.

Molla, K.A., Karmakar, S., Chanda, P.K., Sarkar, S.N., Datta, S.K., Datta, K., 2016. Tissue-specific expression of Arabidopsis NPR1 gene in rice for sheath blight resistance without compromising phenotypic cost. Plant Sci 250, 105–114. https://doi.org/10.1016/j.plantsci.2016.06.005.

Nadeem, S.M., Ahmad, M., Zahir, Z.A., Javaid, A., Ashraf, M., 2014. The role of mycorrhizae and plant growth promoting rhizobacteria (PGPR) in improving crop productivity under stressful environments. Biotechnol. Adv. 32 (2), 429–448.

Nekrasov, V., Wang, C., Win, J., Lanz, C., Weigel, D., Kamoun, S., 2017. Rapid generation of a transgene-free powdery mildew resistant tomato by genome deletion. Sci. Rep. 7 (1), 1–6.

Oerke, E.C., Dehne, H.W., 2004. Safeguarding production-pests, losses and crop protection in major crops. Crop Prot. 23, 275–285.

Okon, Y., Labandera-Gonzalez, C.A., 1994. Agronomic applications of Azospirillum: an evaluation of 20 years worldwide field inoculation. Soil Biol. Biochem. 26 (12), 1591–1601.

Padmanabhan, C., Zhang, X., Jin, H., 2009. Host small RNAs are big contributors to plant innate immunity. Curr. Opin. Plant Biol. 12 (4), 465–472.

Pedraza, R.O., Motok, J., Tortora, M.L., Salazar, S.M., Díaz-Ricci, J.C., 2007. Natural occurrence of Azospirillum brasilense in strawberry plants. Plant Soil 295 (1), 169–178.

Peng, A., Chen, S., Lei, T., Xu, L., He, Y., Wu, L., Yao, L., Zou, X., 2017. Engineering canker-resistant plants through CRISPR/Cas9-targeted editing of the susceptibility gene Cs LOB 1 promoter in citrus. Plant Biotechnol. J. 15 (12), 1509–1519.

Perlak, F.J., Stone, T.B., Muskopf, Y.M., Petersen, L.J., Parker, G.B., McPherson, S.A., Wyman, J., Love, S., Reed, G., Biever, D., Fischhoff, D.A., 1993. Genetically improved potatoes: protection from damage by Colorado potato beetles. Plant Mol. Biol. 22 (2), 313–321.

Pielach, C.A., Roberts, D.P., Kobayashi, D.Y., 2008. Metabolic behavior of bacterial biological control agents in soil and plant rhizospheres. Adv. Appl. Microbiol. 65, 199–215.

Pini, F., Frascella, A., Santopolo, L., Bazzicalupo, M., Biondi, E., Scotti, C., et al., 2012. Exploring the plant-associated bacterial communities in Medicago sativa L. BMC Microbiol 12, 78.

Quilis, J., Penas, G., Messeguer, J., Brugidou, C., San Segundo, B., 2008. The Arabidopsis AtNPR1 inversely modulates defense responses against fungal, bacterial, or viral pathogens while conferring hypersensitivity to abiotic stresses in transgenic rice. Mol. Plant-Microbe Interact. 21, 1215–1231. https://doi.org/10.1094/MPMI-21-9-1215.

Ramette, A., Frapolli, M., Fischer-Le Saux, M., Gruffaz, C., Meyer, J.M., Défago, G., Sutra, L., Moënne-Loccoz, Y., 2011. Pseudomonas protegens sp. nov., widespread plant-protecting bacteria producing the biocontrol compounds 2, 4-diacetylphloroglucinol and pyoluteorin. Syst. Appl. Microbiol. 34 (3), 180–188.

Roy-Barman, S., Sautter, C., Chattoo, B.B., 2006. Expression of the lipid transfer protein Ace-AMP1 in transgenic wheat enhances antifungal activity and defense responses. Transgenic Res. 15 (4), 435–446.

Ruffel, S., Dussault, M.H., Palloix, A., Moury, B., Bendahmane, A., Robaglia, C., Caranta, C., 2002. A natural recessive resistance gene against potato virus Y in pepper corresponds to the eukaryotic initiation factor 4E (eIF4E). Plant J 32, 1067–1075.

Russell, G.E., 2013. Plant Breeding for Pest and Disease Resistance: Studies in the Agricultural and Food Sciences. Elsevier Science, Burlington, https://doi.org/10.1016/C2013-0-06283-4.

Sahu, P.P., Puranik, S., Khan, M., Prasad, M., 2012. Recent advances in tomato functional genomics: utilization of VIGS. Protoplasma 249 (4), 1017–1027.

Savary, S., Ficke, A., Aubertot, J.N., Hollier, C., 2012. Crop losses due to diseases and their implications for global food production losses and food security. Food Sec. 4, 519–537.

Schnepf, E., Crickmore, N.V., Van Rie, J., Lereclus, D., Baum, J., Feitelson, J., Zeigler, D.R., Dean, D., 1998. Bacillus thuringiensis and its pesticidal crystal proteins. Microbiol. Mol. Biol. Rev. 62 (3), 775–806.

Schützendübel, A., Polle, A., 2002. Plant responses to abiotic stresses: heavy metal-induced oxidative stress and protection by mycorrhization. J. Exp. Bot. 53, 1351–1365.

Shadle, G.L., Wesley, S.V., Korth, K.L., Chen, F., Lamb, C., Dixon, R.A., 2003. Phenylpropanoid compounds and disease resistance in transgenic tobacco with altered expression of l-phenylalanine ammonia-lyase. Phtyochemistry 64, 156–161.

Shu, Q., Ye, G., Cui, H., Cheng, X., Xiang, Y., Wu, D., Gao, M., Xia, Y., Hu, C., Sardana, R., Altosaar, I., 2000. Transgenic rice plants with a synthetic cry1Ab gene from Bacillus thuringiensis were highly resistant to eight lepidopteran rice pest species. Mol. Breed. 6 (4), 433–439.

Simmonds, N.W., et al., 2014. Increased resistance to fungal wilts in transgenic eggplant expressing alfalfa glucanase gene. Physiol. Mol. Biol. Plants 20 (2), 143–150. https://doi.org/10.1007/s12298-014-0225-7.

Sivasakthi, S., Usharani, G., Saranraj, P., 2014. Biocontrol potentiality of plant growth promoting bacteria (PGPR)—pseudomonas fluorescens and Bacillus subtilis: a review. Afr. J. Agric. Res. 9 (16), 1265–1277.

Soni, R., Kumar, V., Suyal, D.C., Jain, L., Goel, R., 2017. Metagenomics of plant rhizosphere microbiome. In: Singh, R., Kothari, R., Koringa, P., Singh, S. (Eds.), Understanding Host-Microbiome Interactions—An Omics Approach. Springer, Singapore, https://doi.org/10.1007/978-981-10-5050-3_12.

Sprent, J.I., Ardley, J.K., James, E.K., 2013. From north to south: A latitudinal look at legume nodulation processes. S. Afr. J. Bot. 89, 31–41.

Stark-Lorenzen, P., Nelke, B., Hänßler, G., Mühlbach, H.P., Thomzik, J.E., 1997. Transfer of a grapevine stilbene synthase gene to rice (Oryza sativa L.). Plant Cell Rep. 16 (10), 668–673.

Stuiver, M.H., Custers, J.H., 2001. Engineering disease resistance in plants. Nature 411 (6839), 865–868.

Sun, K., Wolters, A.M., Vossen, J.H., Rouwet, M.E., Loonen, A.E., Jacobsen, E., Visser, R.G., Bai, Y., 2016. Silencing of six susceptibility genes results in potato late blight resistance. Transgen. Res. 25 (5), 731–742. https://doi.org/10.1007/s11248-016-9964-2.

Theis, K.R., Dheilly, N.M., Klassen, J.L., Brucker, R.M., Baines, J.F., Bosch, T.C.G., et al., 2016. Getting the hologenome concept right: an ecoevolutionary framework for hosts and their microbiomes. mSystems 1, e00028-16. https://doi.org/10.1128/mSystems.00028-16.

Van Loon, C., Van Strien, E.A., 1999. The families of pathogenesis-related proteins, their activities, and comparative analysis of PR-1 type proteins. Physiol. Mol. Plant Pathol. 55 (2), 85–97. https://doi.org/10.1006/pmpp.1999.0213.

van Schie, C.C.N., Takken, F.L.W., 2014. Host susceptibility genes 101: How to be a good. Ann. Rev. Phytopathol. 52 (1), 551–581.

Wally, O., Jayaraj, J., Punja, Z.K., 2009. Broad-spectrum disease resistance to necrotrophic and biotrophic pathogens in transgenic carrots (Daucus carota L.) expressing an Arabidopsis NPR1 gene. Planta 231, 131–141. https://doi.org/10.1007/s00425-009-1031-2.

Wang, Y., Cheng, X., Shan, Q., Zhang, Y., Liu, J., Gao, C., Qiu, J.L., 2014. Simultaneous editing of three homoeoalleles in hexaploid bread wheat confers heritable resistance to powdery mildew. Nat. Biotechnol. 32 (9), 947–951.

Wang, F., Wang, C., Liu, P., Lei, C., Hao, W., Gao, Y., Liu, Y.G., Zhao, K., 2016. Enhanced rice blast resistance by CRISPR/Cas9-targeted mutagenesis of the ERF transcription factor gene OsERF922. PLoS One 11 (4), e0154027.

Wang, W., Pan, Q., He, F., Akhunova, A., Chao, S., Trick, H., Akhunov, E., 2018. Transgenerational CRISPR-Cas9 activity facilitates multiplex gene editing in allopolyploid wheat. CRISPR J. 1 (1), 65–74.

Wang, Q., Xiong, D., Zhao, P., Yu, X., Tu, B., Wang, G., 2011. Effect of applying an arsenic-resistant and plant growth–promoting rhizobacterium to enhance soil arsenic phytoremediation by Populus deltoides LH05-17. J. Appl. Microbiol. 111 (5), 1065–1074. https://doi.org/10.1111/j.1365-2672.2011.05142.x.

Wani, P.A., Khan, M.S., 2010. Bacillus species enhance growth parameters of chickpea (Cicer arietinum L.) in chromium stressed soils. Food Chem. Toxicol. 48, 3262–3267.

Wenzel, W.W., 2009. Rhizosphere processes and management in plant-assisted bioremediation (phytoremediation) of soils. Plant Soil 321, 385–408.

Xu, D., Xue, Q., McElroy, D., Mawal, Y., Hilder, V.A., Wu, R., 1996. Constitutive expression of a cowpea trypsin inhibitor gene, CpTi, in transgenic rice plants confers resistance to two major rice insect pests. Mol. Breed. 2 (2), 167–173.

Xu, G., Yuan, M., Ai, C., Liu, L., Zhuang, E., Karapetyan, S., Wang, S., Dong, X., 2017. uORF-mediated translation allows engineered plant disease resistance without fitness costs. Nature. 545 (7655), 491–494. https://doi.org/10.1038/nature22372.

Yang, H.C., Rosen, B.P., 2016. New mechanisms of bacterial arsenic resistance. Biomed. J. 39 (1), 5–13. https://doi.org/10.1016/j.bj.2015.08.003.

Yang, B., Sugio, A., White, F.F., 2006. Os8N3 is a host disease-susceptibility gene for bacterial blight of rice. Proc. Natl. Acad. Sci. USA 103 (27), 10503–10508. https://doi.org/10.1073/pnas.0604088103.

Yeh, K.W., Lin, M.I., Tuan, S.J., Chen, Y.M., Lin, C.J., Kao, S.S., 1997. Sweet potato (Ipomoea batatas) trypsin inhibitors expressed in transgenic tobacco plants confer resistance against Spodoptera litura. Plant Cell Rep. 16 (10), 696–699.

Yu, H., Li, Y., Li, X., Romeis, J., Wu, K., 2013. Expression of Cry1Ac in transgenic Bt soybean lines and their efficiency in controlling lepidopteran pests. Pest Manag. Sci. 69 (12), 1326–1333. https://doi.org/10.1002/ps.

Yuan, Y., Zhong, S., Li, Q., Zhu, Z., Lou, Y., Wang, L., et al., 2007. Functional analysis of rice NPR1-like genes reveals that OsNPR1/NH1 is the rice orthologue conferring disease resistance with enhanced herbivore susceptibility. Plant Biotechnol. J. 5, 313–324. https://doi.org/10.1111/j.1467-7652.2007. 00243.x.

Zhang, J.Y., Qiao, Y.S., Lv, D., Gao, Z.H., Qu, S.C., Zhang, Z., 2012. Malus hupehensis NPR1 induces pathogenesis-related protein gene expression in transgenic tobacco. Plant Biol 14 (Suppl. 1), 46–56. https://doi.org/10.1111/j.1438-8677. 2011.00483.x.

Zhong, X., Xi, L., Lian, Q., Luo, X., Wu, Z., Seng, S., et al., 2015. The NPR1 homolog GhNPR1 plays an important role in the defense response of Gladiolus hybridus. Plant Cell Rep 34, 1063–1074. https://doi.org/10.1007/s00299-015-1765-1.

Zhou, J., Peng, Z., Long, J., Sosso, D., Liu, B., Eom, J.S., Huang, S., Liu, S., Vera Cruz, C., Frommer, W.B., White, F.F., 2015. Gene targeting by the TAL effector PthXo2 reveals cryptic resistance gene for bacterial blight of rice. Plant J. 82 (4), 632–643.

Zhu, Y.J., Agbayani, R., Moore, P.H., 2007. Ectopic expression of Dahlia merckii defensin DmAMP 1 improves papaya resistance to Phytophthora palmivora by reducing pathogen vigor. Planta 226 (1), 87–97.

Zimmer, D., Baum, C., Leinweber, P., Hrynkiewicz, K., Meissner, R., 2009. Associated bacteria increase the phytoextraction of cadmium and zinc from a metal-contaminated soil by mycorrhizal willows. Int. J. Phytorem. 11, 200–213.

Potash biofertilizers: Current development, formulation, and applications

Shiv Shanker Gautam[a], Manjul Gondwal[b], Ravindra Soni[c], and Bhanu Pratap Singh Gautam[b]

[a]Serve India Inter College, Roshanpur, Gadarpur, Udham Singh Nagar, Uttarakhand, India, [b]Department of Chemistry, Laxman Singh Mahar Govt. P.G. College, Pithoragarh, Uttarakhand, India, [c]Department of Agricultural Microbiology, College of Agriculture, Indira Gandhi Krishi Vishwavidyalaya, Raipur, Chhattisgarh, India

18.1 Introduction

A biofertilizer is a carrier-based bioinoculant that consists of living microorganisms such as bacteria, mycorrhizal fungi, and blue-green algae. These help enhance plant growth via increasing the availability of essential nutrients (Pandey and Singh, 2019). Moreover, biofertilizers may be defined as natural fertilizers that comprise an outsized population of a particular or a bunch of beneficial microorganisms to improve the productiveness of soil by multiple factors such as atmospheric nitrogen fixation, solubilization of soil phosphorus, stimulation of plant growth via synthesis of growth-promoting substances, facilitation of availability of nutrients for plants, etc. (Singh and Purohit, 2011; Bhattacharjee and Dey, 2014). Biofertilizers work in multiple directions such as enhancement of soil nutrients and their availability to plants, enrichment of soil microflora, protection of plants by bacterial and fungal pathogens and nematodes, balance soil pH, provide stress tolerance to plants, etc. (Nosheen et al., 2021). In comparison to chemical fertilizers, they are cost-effective and ecofriendly (Pandey and Singh, 2019). Based on nature and functions, biofertilizers can be divided into different groups such as nitrogen biofertilizers, phosphate solubilizing and mobilizing biofertilizers, potash biofertilizers, silicate and zinc solubilizing biofertilizers, plant growth promoting rhizobacteria (PGPR) biofertilizers, compost biofertilizers, etc. (Jha, 2017). Most organic fertilizers have nitrogen (N), phosphorus

482 Chapter 18 Potash biofertilizers

(P), potassium (K), iron (Fe), calcium (Ca), sulfur (S), hormones, vitamins, enzymes, and antibiotics as chief constituents. The major genera *Azotobacter, Clostridium, Rhizobium, Azospirillum, Bacillus, Pseudomonas, Glomus, Laccaria, Boletus,* and *Pezizella* have been reported as chief bioinoculants of biofertilizers (Nosheen et al., 2021).

The term "potash" refers to a gaggle of K-containing chemicals and minerals (Hocking, 2005). Potash may well be a K-rich salt that is extracted from subterranean deposits formed from evaporated sea beds numerous years ago. K could be a vital element for all plants and animals (Hasanuzzaman et al., 2018). Potassium chloride (KCl) is an important chemical within the potash. K is the seventh most plentiful element of our Earth's soil, yet only 1%–2% is available for plants (Manning, 2010; Jha, 2017). It is found in soil both in organic and inorganic forms, but plants can consume it only in soluble form. In soil, the highest concentration of K exists in the uppermost layer in three forms comprising fixed K, exchangeable K, and solution K (Fig. 18.1). The most plentiful elements in fertilizers are N, P, and K while copper (Cu), manganese (Mn), zinc (Zn), and boron (B) are trace elements necessary for normal plant growth. K is the third most significant plant

Potash Biofertilizers: Current Development, Formulation and Applications

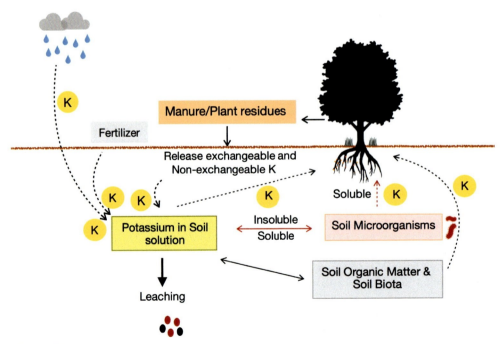

Fig. 18.1 Flow diagram of a potassium cycle.

nutrient. It plays an essential role in the expansion, metabolism, and development of plants. Without an appropriate supply of K, plants have inadequately developed roots while they are also slow growing, develop small seeds, and have lower yields (McAfee, 2008; White and Karley, 2010). They also have higher susceptibility to diseases (Amtmann et al., 2008; Armengaud et al., 2010) and pests (Amtmann et al., 2006; Troufflard et al., 2010). K assists plants in building starches, influencing root development, and controlling the opening and closing of the stomata, which is important for efficient water use. It activates plant enzymes to verify plants for using water efficiently. It is normally provided as K fertilizer in both intensive and extensive agricultural systems (Pettigrew, 2008; Dasan, 2012; Phua et al., 2012; Yadegari et al., 2012; Zhang et al., 2013).

Soil microorganisms have the ability to solubilize soil K and make it available for plants (Diep and Hieu, 2013). Moreover, such microorganisms improve nutrient and water retention, the soil nutrient cycle, stress management, yield, and disease resistance of crops. Hence, potash-based biofertilizers enhance the plant soil microflora. Fertilizers with K boost crop yields, improve crop quality, protect plants from extreme temperatures, fight stress and diseases, dissuade pests such as weeds and insects, prevent wilting, nourish roots and stems, and serve in moving food. In this chapter, we broadly discuss potash biofertilizers, their formulation, and applications for sustainable agriculture.

18.2 Potash biofertilizers and potassium solubilizing microorganisms (KSMs)

Potassic fertilizers are conveyed as potassium oxide (K_2O), named potash. The two most common forms of potash are sulfate of potash (SOP) and muriate of potash (MOP). First, SOP is a high-grade potash comprising two essential nutrients, K and S, that are helpful for growing crops. It can improve both the quality and yield of plants and make them more flexible to environmental stress, insects, and some diseases. The normal approach used for SOP production is called the Mannheim process. Second, potassium chloride (MOP) is largely used potash fertilizer for various fruits, crops, and chloride-loving vegetables such as corn, sugar beets, celery, and Swiss chard. It is the most common commercially used fertilizer, with more than 55 million tonnes sold and used every year. The application of potash fertilizers has not always enhanced crop yield because enriched soil with higher K is evenly spread over soil layers where most plant roots develop. The absorption of such depleted K in the soil solution may get rapidly restored from the reserves.

Potash biofertilizers are a selective microbial consortium that helps in the mobilization of unavailable soil K into the utilized form. In soil, four different forms of K occur: soil solution, exchangeable K, fixed K, and lattice K (Kumar et al., 2016). Solution K is available for plants in the form of exchangeable or nonexchangeable minerals. Some soils such as sandy, waterlogged, saline, and acidic soils are quite deficient in K. Various reports determined the importance of K in the growth, metabolism, and development of plants (Diep and Hieu, 2013; Hasanuzzaman et al., 2018; Kalayu, 2019; Liu et al., 2020). It improves crop quality via grain filling and kernel weight while strengthening straw, aiding in stress management, and increasing resistance against pests, microbial infections, and diseases (Amtmann et al., 2008; Armengaud et al., 2010). K activates plant enzymes, boosts photosynthesis, reduces respiration, maintains cell turgor, helps transport sugars and starches, facilitates nitrogen uptake, and is also important for protein synthesis. In contrast to biofertilizers, chemical-based fertilizers cause serious human health and environmental issues (Nicolopoulou-Stamati et al., 2016). The application of potash biofertilizer can overcome the use of chemical fertilizer and enhance sustainable agriculture.

A wide range of potassium-solubilizing microorganisms (KSMs) has been reported, including bacteria, fungi, and actinomycetes. The major KSB genera are *Achromobacter, Azospirillum, Azotobacter, Bacillus, Burkhoderia, Frateuria, Klebsiella, Enterobacter, Pantoea, Agrobacterium, Microbacterium, Myroides, Paenibacillus, Pseudomonas, Leclercia, Ralstonia, Raoultella*, etc. (Verma et al., 2015; Singh et al., 2010; Xue et al., 2016; Liu et al., 2020; Baghel et al., 2020; Zhang and Kong, 2014; Sangeeth et al., 2012; Bagyalakshmi et al., 2017). They are responsible for releasing certain enzymes that cause the solubilization of soil K. The fungal genera are *Aspergillus, Cladosporium, Glomus, Penicillium,* and *Rhizophagus* (Wu et al., 2005; Lopes-Assad et al., 2010; Baquall and Das, 2006; Islam et al., 2007; Singh and Kapoor, 1999). The actinomycetes with K solubilizing ability belong to the genera *Actinomyces, Streptomyces,* and *Streptoverticillium* (Han et al., 2018; Kumar et al., 2018). Moreover, KSB has also been described as silicate-solubilizing bacteria in the rhizosphere and nonrhizosphere soil of various crops (Lin et al., 2002; Liu et al., 2020).

18.2.1 Potassium-solubilizing bacteria (KSB)

Based on K-solubilizing capability, a wide range of bacteria has been studied and characterized in the last few decades. These are applicable on various crops, including *Achromobacter piechaudii, Azospirillum* spp.*, Azotobacter chroococcum, Bacillus mucilaginous, Bacillus edaphicus, B. cirulans,* B. subtilis, *Bacillus megaterium, Bacillus methylotrophicus, Burkholderia pyrrocinia, Burkholderia ubonensis, Rhizobium* sp.*, Frateuria, Klebsiella* spp., Enterobacter *cloacae,* Enterobacter *asburiae,*

Enterobacter aerogenes, Pantoea agglomerans, Agrobacterium spp., *Microbacterium* spp., *Myroides* spp., *Paenibacillus* spp., *Pseudomonas* spp., *Leclercia* spp., *Ralstonia* spp., *Raoultella* spp., etc. (Verma et al., 2015; Singh et al., 2010; Xue et al., 2016; Liu et al., 2020; Baghel et al., 2020; Zhang and Kong, 2014; Sangeeth et al., 2012; Bagyalakshmi et al., 2017). These bacteria have been studied on various crops such as wheat, maize, sorghum, brinjal, cotton and rape seed, elephant foot yam, rice, sorghum, groundnut, tomato, brinjal, pepper, cucumber, tobacco, legume, etc. In these crops, higher biomass yields have been observed due to higher K uptake with *B. methylotrophicus* and *Bacillus* spp. for rice (Liu et al., 2020; Raj, 2004); *B. edaphicus* for wheat, maize, brinjal, cotton and rape seed (Sheng and He, 2006; Wu et al., 2005; Ramarethinam and Chandra, 2005; Sheng, 2005); *Paenibacillus glucanolyticus* for black pepper (Sangeeth et al., 2012); *Bacillus mucilaginosus* for tomato (Lin et al., 2002); *Klebsiella variicola* and *E. cloacae* for tobacco (Zhang and Kong, 2014); and *B. subtilis* and *B. megaterium* for elephant foot yam (Anjanadevi et al., 2016). Moreover, PSB also activate plant enzymes along with PGP activities. Kumar et al. (2012) showed the solubilization capabilities of two strains of *Bacillus* spp. including the production of siderophores, IAA, phytases, and β-1,3-Glucanase. *A. chroococcum* is a dominant species of the *Azotobacter* genera of Indian soils, exhibiting PGP activity and nitrogen fixation (Jain et al., 2021). Similarly, *Bacillus mucilaginous, B. subtilis, B. megaterium, B. methylotrophicus, Burkholderia* spp., *Rhizobium* sp., *Klebsiella* spp., *Enterobacter* spp., *Agrobacterium* spp., *Microbacterium* spp., *Myroides* spp., *Paenibacillus* spp., and *Pseudomonas* spp. have PGP and nitrogen-fixing activities (Verma et al., 2014; Kour et al., 2020; Baghel et al., 2020; Singh et al., 2020a, b) (Fig. 18.1).

18.2.2 Potassium-solubilizing fungi (KSF)

A few studies have reported on potassium-solubilizing fungi (KSF). The major group of fungi are filamentous. While, yeasts include the genera *Aspergillus, Cladosporium, Glomus, Penicillium,* and *Rhizophagus* (Wu et al., 2005; Lopes-Assad et al., 2010; Baquall and Das, 2006; Islam et al., 2007). These fungi are applicable to various crops, including maize, sugarcane, wheat, mulberry, and legumes, due to their K-solubilizing abilities (Table 18.1). Prajapati et al. (2012) reported that A. niger has the highest K solubilization and acid production by utilizing feldspar and potassium aluminum silicate as insoluble sources of K. According to Kour et al. (2020), among the K-solubilizing fungi Ascomycota was reported as the dominant phylum, followed by Mucoromycota and Basidiomycota. Ascomycota included 13 species belongs to the genera *Aspergillus, Cladosporium, Fusarium, Macrophomina, Penicillium, Sclerotinia,* and *Trichoderma*. Mucoromycota consists of seven species of the genera *Glomus* and

Table 18.1 Potassium-solubilizing microorganisms used as potash biofertilizer in various crops.

Microorganism	Applicable crop/Niche	Ref.
BACTERIA		
Achromobacter piechaudii	Wheat	Verma et al. (2015)
Azospirillum spp.	Maize, sorghum, wheat	Rodriguez and Fraga (1999)
Azotobacter chroococcum, Bacillus mucilaginous, Rhizobium sp.	Maize, wheat	Singh et al. (2010)
Bacillus edaphicus	Wheat, maize, brinjal, cotton and rape seed	Sheng and He (2006), Wu et al. (2005), Ramarethinam and Chandra (2005), and Sheng (2005)
Bacillus cirulans	Maize	Xue et al. (2016)
Bacillus subtilis, Bacillus megaterium	Elephant foot yam	Anjanadevi et al. (2016)
Bacillus methylotrophicus, Bacillus spp.	Rice	Liu et al. (2020) and Raj (2004)
Bacillus mucilaginous	Sorghum, groundnut	Basak and Biswas (2009) and Sugumaran and Janarthanam (2007)
Bacillus mucilaginous	Tomato	Lin et al. (2002)
B. megaterium, B. mucilaginous	Brinjal, pepper, and cucumber	Han and Lee (2005) and Han et al. (2006)
Bacillus spp.	Chilly	Supanjani et al. (2006)
Burkholderia pyrrocinia, Burkholderia ubonensis	Bamboo	Ruangsanka (2014)
Burkholderia sp. FDN2-1	Corn	Baghel et al. (2020)
Frateuria aurantia	Brinjal, tobacco	Ramarethinam and Chandra (2005) and Subhashini (2015)
Klebsiella variicola, Enterobacter cloacae, Enterobacter asburiae, Enterobacter aerogenes, Pantoea agglomerans, Agrobacterium tumefaciens, Microbacterium foliorum, Myroides odoratimimus, and *Burkholderia cepacia*	Tobacco	Zhang and Kong (2014)
Paenibacillus glucanolyticus	Black pepper	Sangeeth et al. (2012)
Pseudomonas spp., *Burkholderia* spp., *Leclercia* spp., *Enterobacter* spp., *Ralstonia* sp., *Klebsiella* sp., *Kosakonia* sp., *Raoultella* sp.	*Mikania micrantha*	Sun et al. (2020)
Pseudomonas spp., *Bacillus* spp. and *Burkholderia* spp.	Tea	Bagyalakshmi et al. (2017)
FUNGI		
Rhizophagus irregularis, Glomus mosseae	Maize	Wu et al. (2005)
Aspergillus niger, Penicillium spp.	Soil	Sperberg (1958)
Aspergillus niger	Sugarcane	Lopes-Assad et al. (2010)
Cladosporium herbarum	Wheat	Islam et al. (2007) and Singh and Kapoor (1999)
Glomus fasciculatum	Mulberry	Baquall and Das (2006)
Penicillium lilacinum, Aspergillus sp., *A. flavus, A. niger, A. terrus, A. nidulans*	Legumes	Chhonkar and Subba Rao (1967)
ACTINOMYCETES		
Streptomyces rochei, Streptomyces sundarbansensis	*M. micrantha*	Han et al. (2018)
Streptoverticillium album	Soil	Kumar et al. (2018)

Rhizophagus, and Basidiomycota consists of two distinct species of genera the *Rhizoctonia* and *Schizophyllum*.

18.3 Current development of potash biofertilizers

Potash biofertilizers are a formulation of specific KSMs such as bacteria, fungi, and actinomycetes. Thus, they indirectly enrich the plant rhizospheric microflora and provide them a necessary supply of K with or without other essential minerals. In soil, four different forms of K occur such as soil solution, exchangeable K, fixed K, and lattice K. In that, solution K is available for plants. Applying potash biofertilizer in K-deficient soil may cause a change in insoluble K to soluble K via KSMs. India is an agricultural country with diverse geography, soil textures, and climate. Thus, there are major variations in crop production and yield. These variations may be fulfilled by K demand based on the distribution of agricultural soil K, which is especially found in sandy and lateritic soils because of leaching (Rengel and Damon, 2008; Mengel and Kirkby, 2001). The application of K-based biofertilizers to agriculture soils can enhance plant yield equal to other enriched soil. Thus, K biofertilizers still need to boost up in various developing countries like India because of major economic challenges and higher use of K fertilizers alone (Meena et al., 2016).

Biofertilizers are produced by applying simple technology with a low installation cost compared to chemical fertilizers (Singh et al., 2014). In step with Hasan (2002), the K productivity status of Indian agricultural soils is categorized as low (21%), medium (51%), and high (28%). Studies have also demonstrated that maintaining the available K reserves in soil guarantees a better opportunity for plants to gain their desirable economic yield. Currently, various commercial potash-based biofertilizers available in the market are called Bio Potash (Abtec), Biocedar-VAM (Mycorrhiza) (ESAF Swasraya Producers Company Pvt. Ltd.), Bio-fertilizer Shakti-Potash bacteria (Multiplex Biotech Pvt. Ltd.), Dr. Soil Slurry Enricher (Potash Mobilizing Bacteria) (Microbi Agrotech Pvt. Ltd.), Dr. Soil Slurry Enricher (Liquid Consortia) (Microbi Agrotech Pvt. Ltd.), Greenfert Potash Mobilizing Bacteria Biofertilizer (Greenfert Green Vision Technical Services), Krishi-Potash Mobilizer (Krishi Gokulam Biotech), LEGEND-Potash derived from Rhodophytes (FMC India Pvt. Ltd.), MICROBES (Utkarsh Agrochem Pvt. Ltd.), NPK Liquid Consortia (IFFCO), Potassium Mobilizing Biofertilizer (KMB) (IFFCO), PrimAgro K (Agroliquid), Potaz (Potassium Solubilizing Bacteria-KSB) (Utkarsh Natural Organics and Biotech), Samridhi Bio Potash (Samridhi Pvt. Ltd.), Tag Nano Potash (Tropical Agrosystem India (P) Ltd.), etc. (Table 18.2).

Table 18.2 List of potash biofertilizer-based commercially available products.

S. no.	Name of commercial product	Company name	Composition	Properties	Ref.
1	Bio Potash	Abtec	Potash mobilizing bacteria, that is, Frateuria aurantia	Useful for cereals, vegetables, plantation crops, and ornamental plants	www.abtecbiofert.com
2	Biocedar-VAM (Mycorrhiza)	ESAF Swasraya Producers Company Pvt. Ltd.	Mycorrhiza, fulvic acid, macro- and micronutrients, and proprietary constituents	Increases the microbial and mycorrhizal activities, promotes nutrient uptake, helps with rapid and extensive root growth, the uptake of phosphorus and other nutrients, and stress management	https://dominica.desertcart.com/products/
3	Bio-fertilizer Shakti (Potash bacteria)	Multiplex Bio-tech Pvt. Ltd.	*F. aurantia*	Helpful to solubilize soil K and make it available for plants	www.indiamart.com
4	Dr. Soil Slurry Enricher (Potash Mobilizing Bacteria)	Microbi Agrotech Pvt. Ltd.	Potash-mobilizing bacteria	Enhances organic waste and recycles and enriches slurry. Mobilizes the potash near the roots of plants	http://microbiagro.com/products
5	Dr. Soil Slurry Enricher (Liquid Consortia)	Microbi Agrotech Pvt. Ltd.	Nitrogen fixers (*Azotobacter* and *Azospirillum*), phosphate solubilizers, and potash mobilizers	Helps in nitrogen fixation, solubilize and mobilize phosphorus and potassium	http://microbiagro.com/products
6	Greenfert Potash Mobilizing Bacteria Biofertilizer	Greenfert Green Vision Technical Services	Potash-mobilizing bacteria	Regulates the formation of amino acids and proteins in the roots, increases the resistance of the crops to hot and dry conditions	https://www.indiamart.com/proddetail/greenfert-potash-mobilizing-bacteria-biofertilizer-21318630688.html
7	Krishi-Potash Mobilizer	Krishi Gokulam Biotech	Potash-mobilizing bacteria	Mobilize insoluble potassium in soil to make it easily available for plants	www.gokulambiotech.com
8	LEGEND—Potash derived from Rhodophytes	FMC India Pvt. Ltd.	Potash derived from rhodophytes	Affects the gene expression of plants and enhances inner plant health	https://www.kissanhut.com/product/fmc-legend/

9	MICROBES	Utkarsh Agrochem Pvt. Ltd.	Consortium of nitrogen-fixing bacteria, phosphorus, potash, sulfur, magnesium, calcium, boron, iron, copper, zinc, silicon-solubilizing bacteria	Provides crop protection from various bacterial and fungal pathogens and nematodes, provides essential elements in soluble form to soil, improves soil pH	https://www.utkarshagro.com/product/microbes/
10	NPK Liquid Consortia	IFFCO	Consortium of rhizobium, azotobacter, and acetobacter	Provides nitrogen, phosphorus and potassium to crops	https://iffco.in/index.php/ourproducts/index/bio-fertiliser
11	Potassium Mobilizing Biofertilizer (KMB)	IFFCO	Silicate bacteria, that is, *Bacillus mucilaginous, B. edaphicus, B. glucanolyticus, B. cirulans*	Releases K in the soil and makes soil potassium available for plants	https://iffco.in/index.php/ourproducts/index/bio-fertiliser
12	PrimAgro K	Agroliquid	Consists of ammonium nitrate, potassium carbonate, and potassium with beneficial soil bacteria, that is, *Bacillus subtilis, B. methylotrophicus*	Provides nitrogen, potassium, and sulfur to plants and soil. Improves nutrient retention and nutrient cycle of soil	https://www.agroliquid.com/products/potassium/primagro-k/
13	Potaz (potassium Solubilizing Bacteria—KSB)	Utkarsh Natural Organics and Biotech	Potassium-solubilizing bacteria	Solubilizes insoluble potash in the soil in soluble form and makes available to plants, improves water retention, color, texture, yield, and disease resistance of the crop	www.utkarshagrotech.com
14	Samridhi Bio Potash	Samridhi Pvt. Ltd.	Potash-mobilizing bacteria	Improves resistance of crop plants to disease and stress, gives higher yield (20%–30%), improves sucrose content in crop and shelf life	https://www.jaipurbiofertilizer.com/product-supplier/samridhi-bio-potash
15	Tag Nano Potash	Tropical Agrosystem India (P) Ltd.	Proteino-lacto-gluconate formulation, formulated with organic acid-based chelated potash, vitamins, and probiotics	Accelerates photosynthesis, reduces the intensity of chlorosis, provides stress tolerance to plants	https://product.statnano.com/product/6679/tag-nano-potash

These products have various field applications and properties, including the maintenance of mycorrhizal activities, promotion of nutrient uptake, rapid and extensive root growth, regulation of the uptake of phosphorus and other nutrients, stress management, helping in the solubilization of soil K and making it available for plants, helping with nitrogen fixation, regulating the formation of amino acids and proteins in the roots, increasing crop resistance to hot and dry conditions, influencing the gene expression of plants, and improving the inner health of plants. Moreover, these products provide crop protection from various bacterial and fungal pathogens as well as nematodes. They provide essential elements in soluble form to the soil, improve soil pH, enhance the shelf life of crops, accelerate photosynthesis, and reduce the intensity of chlorosis to plants.

18.4 Formulation of potash biofertilizers

Potassium is an essential macronutrient for plants. Many KMBs such as *Acidithiobacillus ferrooxidans, Paenibacillus* spp., *B. mucilaginosus, B. edaphicus*, and *B. circulans* have the capacity to solubilize soil K (Etesami et al., 2017). The effectiveness of the formulation of potash biofertilizers in field conditions mostly relies on collective phases such as microbial selection and development of a consortium, delivery system, application strategies, and the persistence of microbial strains with the rhizospheric soil and plant system (Bargaz et al., 2018). These steps are briefly discussed here.

18.4.1 Selection and development of consortium

A microbial consortium is a group of two or more symbiotic microorganisms capable of surviving under diverse conditions by the formation of synergistic population structures such as stromatolites, microbial mats, biofilms, etc. (Zhang et al., 2018). For the selection of microorganisms, the desirable characteristics must be met with inoculant strains such as extended shelf life, survival capability under harsh conditions, cost effectiveness, efficient delivery with the host plant, genetic stability, physiologically adaptable with the host environment, etc. For potash biofertilizer, the selection of isolates should be based on their potassium-solubilizing capability with some other plant growth promoting (PGP) activities (Wong et al., 2019; Singh et al., 2020a, b). In the laboratory, for isolation and detection, Aleksandrow agar medium is routinely used. The chief bacteria, that is, *Pseudomonas* spp., Frateuria aurantia, *Bacillus mucilaginous, B. edaphicus, Bacillus glucanolyticus, B. cirulans*, B. subtilis, *B. methylotrophicus, Burkholderia, A. ferrooxidans*, and *Paenibacillus* sp., and fungal organisms, that is, *Aspergillus* spp., are used in the development of the consortium for potash biofertilizers (Gore and Navale, 2017; Etesami et al., 2017; Ahmad et al., 2016; Sun et al., 2020). The bacterial consortium is generally prepared in a liquid medium to reach a high biomass yield. For this, a single colony of selected bacterial strains is inoculated in 100 mL of culture broth, that is, nutrient broth or Luria-Bertani (LB) broth, and incubated at 120 rpm for 24 h at room temperature ($28 \pm 2°C$) (Namsena et al., 2016; Vishwakarma et al., 2018; Singh et al., 2020a, b). The media composition and growth conditions such as temperature, pH, aeration, agitation, etc., should be optimum for the desired inoculants. Further, the grown active culture is used to prepare the bioformulation.

18.4.2 Bioformulation processes

In bioformulation preparation, the 24 h old microbial consortium is centrifuged at 10,000 rpm for 10 min before mixing with a suitable carrier (Singh et al., 2020a, b). The carrier acts as a delivery material for live inoculants during processing from the laboratory to the field. Individually or together, a suitable carrier such as rock power, talc, or charcoal can be used for the survival and effective delivery of the desired microbes into the soil (Arora et al., 2010; Samavat et al., 2014; Maheshwari et al., 2015; Vishwakarma et al., 2018; Singh et al., 2020a, b). During preparation stages, sterilization of the carriers is an essential step. For this, gamma irradiation at a dose rate of 4.0 kGy for 1 h and autoclaving at 121°C for 20 min, the most suitable way of carrier sterilization, have been used for selected carriers (El-Fattah et al., 2013; Sahai et al., 2019). The sterilized carrier is mixed with the microbial consortium and air dried overnight at up to 20% moisture content at room temperature ($28 \pm 2°C$) (Samavat et al., 2014). Furthermore, the mixture is packed and sealed in polypropylene bags presterilized by autoclaving or gamma irradiation, and stored at room temperature with 80% relative humidity (Arora et al., 2010; Suryadi et al., 2013; Namsena et al., 2016; Bargaz et al., 2018). In bioformulation, the bacterial number should be 5×10^8 cfu/g (Samavat et al., 2014; Jambhulkar and Sharma, 2014).

18.4.3 Desiccation tolerance testing

Bacterial desiccation is a natural abiotic stress condition that happens by freezing, heating, drying, and rewetting condition of soil due to low precipitation or irrigation. This condition becomes more lethal, especially for nonspore-forming bacteria. Multiple physiological mechanisms have been observed behind desiccation tolerance, including the synthesis of compatible solute disaccharide trehalose or hydroxypyrimidine hydroxyectoine (Roder et al., 2005; Narvaez-Reinaldo et al., 2010), the production of heat-shock proteins and enzymes, and exopolysaccharide modification or DNA repair (Berninger et al., 2018). Desiccation tolerance has great biotechnological interest through microbial cellular stabilization, which allows the long-term storage of formulated products for commercial use. Bacterial desiccation tolerance may be improved during the bioformulation process by applying some strategies, including drying methods such as freeze drying, vacuum drying, spray drying, fluidized bed drying, and air drying. Other strategies include the addition of external protectants, triggering of stress adaptation, triggering of exopolysaccharide secretion, and indirect protection by "helper" microbial strains (Berninger et al., 2018). Some sublethal stress has been suggested before the desiccation for cellular protective mechanisms and as external protectants (Liu et al., 2014). It includes the variation of pH, temperature, depletion of nutrients, anoxic conditions, salt stress, etc. Helper microbial strains do not have any biocontrol properties and have a supportive role in enhancing biocontrol properties with other potential strains in bioformulations (Garcia, 2011; Molina-Romero et al., 2017).

18.4.4 Storage stability testing

The determination of the storage stability of the bioformulation is an essential and critical factor in terms of microbial inoculants and product efficacy. Usually, the shelf life of a product and its microbial stability are 6–12 months (Berninger et al., 2018). The additives and low-temperature storage are essential factors for the survival and stability of microbial inoculants. The long-term storage (at 25°C (room temperature) and 4°C up to 180 days) of bioformulations need a periodic check for efficiency and stability. The stability test is performed by using serial dilution plating. In this method, 1 g of dry formulation is added to 10 mL of sterile saline solution. The serial dilution is prepared up to 10^5 and the microbial population can be estimated by pouring 100 μL of dilution on the nutrient agar. The microbial plates are incubated for 24–48 h at 28°C. The CFU estimation is done, which should be 10^8 CFU per gram sample. The above procedure is performed on a monthly basis (Wong et al., 2019).

18.4.5 Field efficacy

Some other factors such as the effectiveness, soil adaptability, and synergistic relationship with the available microflora also create challenges for introducing biofertilizers. For field efficacy, it is necessary to routinely check the soil parameters and laboratory tests prior to field trials. Sometimes, field trials give opposite results compared to laboratory testing. The determination of the availability of potassium in tested soil has its own importance. Abiotic and biotic factors also influence the long-term growth and survivability of applicable

KSB biofertilizers (Mishra and Arora, 2016; Jeyakumar et al., 2020). Sometimes, commercial products cannot work accurately under field conditions which may effective in laboratory and greenhouse conditions (Vassilev et al., 2015; Mishra and Arora, 2016) because of poor compatibility or stability of carriers (Bashan et al., 2016; Jampilek and Kralova, 2017; Stamenkovic et al., 2018; Aamir et al., 2020). The cell viability has been observed to decline during field trials while other introduced biofertilizers, that is, talc carrier-based ones, were stable during storage (Suryadi et al., 2013).

18.5 Field applications and crop improvement

Agriculture heavily depends on the use of herbicides, pesticides, and chemical fertilizers. In such a scenario, biofertilizer-based agriculture can be an alternative approach to improve soil health, crop yield, and the development of sustainable agriculture (Bashan et al., 2016; Ojuederie et al., 2019; Ijaz et al., 2019). Thus, introducing potential microorganisms via biofertilizers can play an essential role in such a context. Currently, molecular and proteomic approaches can be a valuable tool for exploring such environmental microbial communities (Wang et al., 2016; Eldakak et al., 2013); this is popularly known as metaproteomics. Metaproteomics can explain the taxonomic composition and microbial functioning including nutrient cycles, mutualistic relationships, metal utilization, eutrophication, nutrient mobilization and uptake, suppression of diseases, etc. (Baldrian and Lopez-Mondejar, 2014; Heyer et al., 2019). Various reports showed the microbial rhizospheric inoculation in legume and nonlegume crops for improving the uptake of macro- and microelements, including N, P, K, Mg, Zn, Ca, Fe, etc. (Sharma et al., 2005; Bargaz et al., 2018; Liu et al., 2019; Jeyakumar et al., 2020). K is the third most essential element for plant growth. It assists plants to resist drought and excessive temperatures and provides resistance to various diseases. Potash biofertilizers are applicable in a wide range of crops such as grapes, oranges, mangos, apples, sugarcane, pineapples, paddy, muskmelons, tomatoes, beans, wheat, watermelons, capsicum, pomegranates, gerbera, etc.

KSBs have multiple roles such as the protection of plant ions from salinity via intensifying growth-related physiological activities such as stomatal conductance, electrolyte leakage, and lipid peroxidation (Jha, 2017). It is also proficient at decomposing organic matter, and consequently has a major role in nutrient cycling. The KSB-rich soil of invasive plants shows better adaptation than other native plants (Sun et al., 2020). Similarly, mycorrhizae fungi are capable of absorbing soil macro- and micronutrients, including N, P, K, Mg, Zn, Ca, Fe, Cu, Mn, etc. This association is the mutualistic relationship between plant roots and fungi (Sylvia et al., 2005). Among these associations, the two most ecologically important are arbuscular mycorrhizae (AM) and ectomycorrhizae (ECM). The KSF-based potash biofertilizers work simultaneously to enhance and maintain the mycorrhizal association of their respective/applicable crops. Many studies have demonstrated the improvement of K uptake due to the application of such biofertilizers. Nakmee et al. (2016) reported the enhanced growth of *Sorghum bicolor* in plant height, number of leaves, biomass, total nitrogen, and phosphorus and potassium uptake. In this study, 10 AM fungi were selected, including *Glomus* spp., *Glomus aggregatum*, *G. fasciculatum*, *G. occultum*, *Acaulospora longula*, *A. scrobiculata*, *A.* spinosa, and *Scutellospora* sp. *A. scrobiculata* produced the highest biomass, grain dry weight, and total N uptake in shoots, followed by *Glomus* sp. and *A. scrobiculata* and *Scutellospora* sp. with the highest K uptake in shoots.

Potash biofertilizers are significantly capable of promoting nutrient uptake and maintaining soil pH. They are also helpful in stress management, supportive in the enhancement of the shelf life of crops via multiple roles, improve plant physiological responses, etc. Thus, potash biofertilizers work simultaneously at improving the soil texture, soil K, plant health, crop productivity, and rhizospheric environment.

18.6 Conclusions and future perspectives

Potash biofertilizers are an alternative way to replace chemical fertilizers via natural means. They can fulfill public demands for the development of sustainable agriculture via lowering the production cost, easing field applications, and decreasing the use of chemical fertilizers. Potash biofertilizers work equally in the development of plant absorption of soil K and the improvement of plant health and productivity by multiple means such as nutrient uptake, maintaining soil pH, helping in stress management, enhancing the shelf life of crops, better plant physiological responses, etc. The studies on KSB, KSF, and actinomycetes have provided broad applicability on a wider range of crops. But, we still need more studies on the ecological niches of other crops and their synergistic relationships with microbial communities. This could result in the development of more K-efficient biofertilizers and sustainable plant productivity in addition to improving our understanding of microbial interactions.

References

Aamir, M., Rai, K.K., Zehra, A., Dubey, M.K., 2020. Microbial bioformulation-based plant biostimulants: a plausible approach toward next generation of sustainable agriculture. In: Kumar, A., Radhakrishnan, E.K. (Eds.), Microbial Endophytes, pp. 195-225, https://doi.org/10.1016/B978-0-12-819654-0.00008-9.

Ahmad, M., Nadeem, S.M., Naveed, M., Zahir, Z.A., 2016. Potassium-solubilizing bacteria and their application in agriculture. In: Meena, V., Maurya, B., Verma, J., Meena, R. (Eds.), Potassium Solubilizing Microorganisms for Sustainable Agriculture. Springer, New Delhi, https://doi.org/10.1007/978-81-322-2776-2_21.

Amtmann, A., Hammond, J.P., Armengaud, P., White, P.J., 2006. Nutrient sensing and signaling in plants: potassium and phosphorus. Adv. Bot. Res. 43, 209–257.

Amtmann, A., Troufflard, S., Armengaud, P., 2008. The effect of potassium nutrition on pest and disease resistance in plants. Physiol. Plant. 133, 682–691.

Anjanadevi, I.P., John, N.S., John, K.S., Jeeva, M.L., Misra, R.S., 2016. Rock inhabiting potassium solubilizing bacteria from Kerala, India: characterization and possibility in chemical K fertilizer substitution. J. Basic Microbiol. 56, 67–77. https://doi.org/10.1002/jobm.201500139.

Armengaud, P., Breitling, R., Amtmann, A., 2010. Coronatine-intensive 1 (COI1)mediates transcriptional responses of Arabidopsis thaliana to external potassium supply. Mol. Plant 3 (2), 390–405.

Arora, N., Khare, E., Maheshwari, D.K., 2010. Plant growth promoting rhizobacteria: Constraints in bioformulation, commercialization, and future strategies. In: Maheshwari, D.K. (Ed.), Plant Growth and Health Promoting Bacteria. Microbiology Monographs, vol. 18, pp. 97–116, https://doi.org/10.1007/978-3-642-13612-2_5.

Baghel, V., Thakur, J.K., Yadav, S.S., Manna, M.C., et al., 2020. Phosphorus and potassium solubilization from rock minerals by endophytic *Burkholderia* sp. strain FDN2-1 in soil and shift in diversity of bacterial endophytes of corn root tissue with crop growth stage. Geomicrobiol J. 37 (6), 550–563. https://doi.org/10.1080/01490451.2020.1734691.

Bagyalakshmi, B., Ponmurugan, P., Balamurugan, A., 2017. Potassium solubilization, plant growth promoting substances by potassium solubilizing bacteria (KSB) from southern Indian tea plantation soil. Biocatal. Agric. Biotechnol. 12, 116–124.

Baldrian, P., Lopez-Mondejar, R., 2014. Microbial genomics, transcriptomics and proteomics: new discoveries in decomposition research using complementary methods. Appl. Microbiol. Biotechnol. 98, 1531–1537.

Baquall, M.F., Das, M.F., 2006. Influence of biofertilizers on macronutrient uptake by the mulberry plant and its impact on silkworm bioassay. Caspian J. Environ. Sci. 4, 98–109.

Bargaz, A., Lyamlouli, K., Chtouki, M., Zeroual, Y., Dhiba, D., 2018. Soil microbial resources for improving fertilizers efficiency in an integrated plant nutrient management system. Front. Microbiol. https://doi.org/10.3389/fmicb.2018.01606.

Basak, B.B., Biswas, D.R., 2009. Influence of potassium solubilizing microorganism (*Bacillus mucilaginosus*) and waste mica on potassium uptake dynamics by Sudan grass (*Sorghum vulgare* Pers.) grown under two Alfisols. Plant Soil 317, 235–255.

Bashan, Y., de Bashan, L.E., Prabhu, S.R., 2016. Superior polymeric formulations and emerging innovative products of bacterial inoculants for sustainable agriculture and the environment. In: Singh, H.B., Sarma, B.K., Keswani, C. (Eds.), Agriculturally Important Microorganisms—Commercialization and Regulatory Requirements in Asia. Springer Pvt. Ltd., Berlin, pp. 15–46, https://doi.org/10.1007/978-981-10-2576-1_2.

Berninger, T., Lopez, O.G., Bejarano, A., Preininger, C., Sessitsch, A., 2018. Maintenance and assessment of cell viability in formulation of non-sporulating bacteria inoculants. Microb. Biotechnol. https://doi.org/10.1111/1751-7915.12880.

Bhattacharjee, R., Dey, U., 2014. Biofertilizer, a way towards organic agriculture: a review. Afr. J. Microbiol. Res. 8 (24), 2332–2342. https://doi.org/10.5897/AJMR2013.6374.

Chhonkar, P.K., Subba Rao, N.S., 1967. Phosphate solubilisation by fungi associated with legume root nodules. Can. J. Microbiol. 13, 749–753.

Dasan, A.S., 2012. Compatibility of agrochemicals on the growth of phosphorous mobilizing bacteria Bacillus megaterium var. phosphaticum potassium mobilizing bacteria *Frateuria aurantia*. Appl. Res. Dev. Inst. J. 6 (13), 118–134.

Diep, C.N., Hieu, T.N., 2013. Phosphate and potassium solubilizing bacteria from weathered materials of denatured rock mountain, Ha Tien, Kiên Giang province Vietnam. Am. J. Life Sci. 1 (3), 88–92.

Eldakak, M., Milad, S.I.M., Nawar, A.I., Rohila, J.S., 2013. Proteomics: a biotechnology tool for crop improvement. Front. Plant Sci. 4, 35. https://doi.org/10.3389/fpis.2013.00035.

El-Fattah, D.A.A., Eweda, W.E., Zayed, M.S., Hassanein, M.K., 2013. Effect of carrier materials, sterilization method, and storage temperature on survival and biological activities of *Azotobacter chroococcum* inoculant. Ann. Agric. Sci. 58 (2), 111–118. https://doi.org/10.1016/j.aoas.2013.07.001.

Etesami, H., Emami, S., Alikhani, H.A., 2017. Potassium solubilizing bacteria (KSB): mechanisms, promotion of plant growth, and future prospects—a review. J. Soil Sci. Plant Nutr. 17 (4), 897–911. https://doi.org/10.4067/S0718-95162017000400005.

Garcia, A.H., 2011. Anhydrobiosis in bacteria: from physiology to applications. J. Biosci. 36, 939–950.

Gore, N.S., Navale, A.M., 2017. Effect of consortia of potassium solubilizing bacteria and fungi on growth, nutrient uptake and yield of banana. Indian J. Hortic. 74 (2), 189–197. https://doi.org/10.5958/0974-0112.2017.00041.X.

Han, H.S., Lee, K.D., 2005. Phosphate and potassium solubilizing bacteria effect on mineral uptake, soil availability and growth of eggplant. Res. J. Agric. Biol. Sci. 1 (2), 176–180.

Han, H.S., Supanjani, Lee, K.D., 2006. Effect of co-inoculation with phosphate and potassium solubilizing bacteria on mineral uptake and growth of pepper and cucumber. Plant Soil Environ. 52, 130–136.

Han, D., Wang, L., Luo, Y., 2018. Isolation, identification, and the growth promoting effects of two antagonistic actinomycete strains from the rhizosphere of *Mikania micrantha* Kunth. Microbiol. Res. 208, 1–11. https://doi.org/10.1016/j.micres.2018.01.003.

Hasan, R., 2002. Potassium status of soils in India. Better Crops Int. 16, 3–5.

Hasanuzzaman, M., Bhuyan, M.H.M.B., Nahar, K., Hossain, M.S., et al., 2018. Potassium: a vital regulator of plant responses and tolerance to abiotic stresses. Agronomy 8, 31. https://doi.org/10.3390/agronomy8030031.

Heyer, R., Schallert, K., Budel, A., Zoun, R., Dorl, S., Behne, A., et al., 2019. A robust and universal metaproteomics workflow for research studies and routine diagnostics within 24 h using phenol extraction, FASP digest, and the metaproteomeanalyzer. Front. Microbiol. 10, 1883. https://doi.org/10.3389/fmicb.2019.01883.

Hocking, M.B., 2005. Natural and derived sodium and potassium salts. In: Hocking, M.B. (Ed.), Handbook of Chemical Technology and Pollution Control, third ed. Academic Press, pp. 175–199, https://doi.org/10.1016/B978-012088796-5/50009-0.

Ijaz, M., Ali, Q., Ashraf, S., Kamran, M., Rehman, A., 2019. Development of future bioformulations for sustainable agriculture. In: Kumar, V., Prasad, R., Kumar, M., Choudhary, D. (Eds.), Microbiome in Plant Health and Disease. Springer, Singapore.

Islam, M.T., Deora, A., Hashidoko, Y., Rahman, A., et al., 2007. Isolation and identification of potential phosphate solubilizing bacteria from the rhizoplane of *Oryza sativa* L. cv. BR29 of Bangladesh. Verlag der. Z. Naturforsch. 62c, 103–110.

Jain, D., Sharma, J., Kaur, G., Bhojiya, A.A., et al., 2021. Phenolic and molecular diversity of nitrogen fixating plant growth promoting Azotobacter isolated from semiarid region of India. Biomed. Res. Int. https://doi.org/10.1155/2021/6686283, 6686283.

Jambhulkar, P.P., Sharma, P., 2014. Development of bioformulation and delivery system of Pseudomonas fluorescens against bacterial leaf blight of rice (*Xanthomonas oryzae* Pv. oryzae). J. Environ. Biol. 35 (5), 843–849.

Jampilek, J., Kralova, K., 2017. Nanomaterials for delivery of nutrients and growth-promoting compounds to plants. In: Prasad, R., Kumar, M., Kumar, V. (Eds.), Nanotechnology: An Agricultural Paradigm. Springer Pvt. Ltd., Singapore, pp. 177–226, https://doi.org/10.1007/978-981-10-4573-8_9.

Jeyakumar, S.P., Dash, B., Singh, A.K., Suyal, D.C., Soni, R., 2020. Nutrient cycling at higher altitudes. In: Goel, R., Soni, R., Suyal, D.C. (Eds.), Microbiological Advancements for Higher Altitude Agro-Ecosystems & Sustainability. Springer Nature Singapore Pvt. Ltd., Singapore, pp. 293–305.

Jha, Y., 2017. Potassium mobilizing bacteria: enhance potassium intake in paddy to regulates membrane permeability and accumulate carbohydrates under salinity stress. Braz. J. Biol. Sci. 4, 333–344.

Kalayu, G., 2019. Phosphate solubilizing microorganisms: promising approach as biofertilizers. Int. J. Agronomy. https://doi.org/10.1155/2019/4917256, 4917256.

Kour, D., Rana, K.L., Kaur, T., Yadav, N., et al., 2020. Potassium solubilizing and mobilizing microbes: biodiversity, mechanisms of solubilization, and biotechnological implication for alleviations of abiotic stress. In: Rastegari, A.A., Yadav, A.N., Yadav, N. (Eds.), New and Future Developments in Microbial Biotechnology and Bioengineering. Elsevier, pp. 177–202, https://doi.org/10.1016/B978-0-12-820526-6.00012-9.

Kumar, P., Dubey, R.C., Maheshwari, D.K., 2012. Bacillus strains isolated from rhizosphere showed plant growth promoting and antagonistic activity against phytopathogens. Microbiol. Res. 167 (8), 493–499.

Kumar, A., Patel, J.S., Bahadur, I., Meena, V.S., 2016. The molecular mechanisms of KSMs for enhancement of crop production under organic farming. In: Meena, V., Maurya, B., Verma, J., Meena, R. (Eds.), Potassium Solubilizing Microorganisms for Sustainable Agriculture. Springer, New Delhi, https://doi.org/10.1007/978-81-322-2776-2_5.

Kumar, A., Kumar, A., Patel, H., 2018. Role of microbes in phosphorus availability and acquisition by plants. Int. J. Curr. Microbiol. App. Sci. 7 (5), 1344–1347.

Lin, Q.M., Rao, Z.H., Sun, Y.X., Yao, J., Xing, L.J., 2002. Identification and practical application of silicate-dissolving bacteria. Agric. Sci. China 1, 81–85.

Liu, X.T., Hou, C.L., Zhang, J., Zeng, X.F., Qiao, S.Y., 2014. Fermentation conditions influence the fatty acid composition of the membranes of *Lactobacillus reuteri* I5007 and its survival following freeze-drying. Lett. Appl. Microbiol. 59, 398–403.

Liu, F., Hewezi, T., Lebeis, S.L., Pantalone, V., et al., 2019. Soil indigenous microbiome and plant genotypes cooperatively modify soybean rhizosphere microbiome assembly. BMC Microbiol. 19, 201. https://doi.org/10.1186/s12866-019-1572-x.

Liu, Z., Wang, H., Xu, W., Wang, Z., 2020. Isolation and evaluation of the plant growth promoting rhizobacterium *Bacillus methylotrophicus* (DD-1) for growth enhancement of rice seedling. Arch. Microbiol. 202, 2169–2179. https://doi.org/10.1007/s00203-020-01934-8.

Lopes-Assad, M.L., Avansini, S.H., Rosa, M.M., de Carvalho, J.R.P., Ceccato-Antonini, S.R., 2010. The solubilization of potassium-bearing rock powder by *Aspergillus niger* in small-scale batch fermentations. Can. J. Microbiol. 56, 598–605.

Maheshwari, D.K., Dubey, R.C., Agarwal, M., Dheeman, S., Aeron, A., Bajpai, V.K., 2015. Carrier based formulations of biocoenotic consortia of disease suppressive *Pseudomonas aeruginosa* KRP1 and *Bacillus licheniformis* KRB1. Ecol. Eng. 81, 272–277. https://doi.org/10.1016/j.ecoleng.2015.04.066.

Manning, D.A.C., 2010. Mineral sources of potassium for plant nutrition. A review. Agron. Sustain. Dev. 30, 281–294. https://doi.org/10.1051/agro/2009023.

McAfee, J., 2008. Potassium, a Key Nutrient for Plant Growth. Department of Soil and Crop Sciences.

Meena, V.S., Bahadur, I., Maurya, B.R., Kumar, A., et al., 2016. Potassium-solubilizing microorganism in evergreen agriculture: an overview. In: Meena, V., Maurya, B., Verma, J., Meena, R. (Eds.), Potassium Solubilizing Microorganisms for Sustainable Agriculture. Springer, New Delhi, https://doi.org/10.1007/978-81-322-2776-2_1.

Mengel, K., Kirkby, E.A., 2001. Principles of Plant Nutrition, fifth ed. Kluwer Acad. Publishers, Dordrecht.

Mishra, J., Arora, N.K., 2016. Bioformulations for plant growth promotion and combating phytopathogens: a sustainable approach. In: Arora, N.K., et al. (Eds.),

Bioformulations: For Sustainable Agriculture. Springer India Ltd., https://doi.org/10.1007/978-81-322-2779-3_1.

Molina-Romero, D., Baez, A., Quintero-Hernandez, V., Castaneda-Lucio, M., et al., 2017. Compatible bacterial mixture, tolerant to desiccation, improves maize plant growth. PLoS ONE 12 (11). https://doi.org/10.1371/journal.pone.0187913, e0187913.

Nakmee, P.S., Techapinyawat, S., Ngamprasit, S., 2016. Comparative potentials of native arbuscular mycorrhizal fungi to improve nutrient uptake and biomass of *Sorghum bicolor* Linn. Agric. Nat. Resour. 50 (3), 173–178. https://doi.org/10.1016/j.anres.2016.06.004.

Namsena, P., Bussaman, P., Rattanasena, P., 2016. Bioformulation of *Xenorhabdus stockiae* PB09 for controlling mushroom mite, *Luciaphorus perniciosus* Rack. Bioresour. Bioprocess 3, 19. https://doi.org/10.1186/s40643-016-0097-5.

Narvaez-Reinaldo, J.J., Barba, I., Gonzalez-Lopez, J., Tunnacliffe, A., Manzanera, M., 2010. Rapid method for isolation of desiccation-tolerant strains and xeroprotectants. Appl. Environ. Microbiol. 76 (15), 5254–5262. https://doi.org/10.1128/AEM.00855-10.

Nicolopoulou-Stamati, P., Maipas, S., Kotampasi, C., Stamatis, P., Hens, L., 2016. Chemical pesticides and human health: the urgent need for a new concept in agriculture. Front. Public Health 4, 148. https://doi.org/10.3389/fpubh.2016.00148.

Nosheen, S., Ajmal, I., Song, Y., 2021. Microbes as biofertilizers, a potential approach for sustainable crop production. Sustainability 13, 1868. https://doi.org/10.3390/su13041868.

Ojuederie, O.B., Olanrewaju, O.S., Babalola, O.O., 2019. Plant growth promoting rhizobacterial mitigation of drought stress in crop plants: implications for sustainable agriculture. Agronomy 9 (11), 712. https://doi.org/10.3390/agronomy9110712.

Pandey, V.C., Singh, V., 2019. Exploring the potential and opportunities of current tools for removal of hazardous materials from environments. In: Pandey, V.C., Bauddh, K. (Eds.), Phytomanagement of Polluted Sites. Elsevier, pp. 501–516, https://doi.org/10.1016/B978-0-12-813912-7.00020-X.

Pettigrew, W.T., 2008. Potassium influences on yield and quality production for maize, wheat, soybean and cotton. Physiol. Plant. 133, 670–681.

Phua, C.K.H., Abdul Wahid, A.N., Abdul Rahim, K., 2012. Development of multifunctional biofertilizer formulation from indigenous microorganisms and evaluation of their N_2-fixing capabilities on Chinese cabbage using 15 N tracer technique. Pak. J. Trop. Agric. Sci. 35 (3), 673–679.

Prajapati, K.B., Sharma, M.C., Modi, H.A., 2012. Isolation of two potassium solubilizing fungi from ceramic industry soils. Life Sci. Leafl. 5, 71–75.

Raj, S.A., 2004. Solubilization on a silicate and concurrent release of phosphorus and potassium in rice ecosystem. In: Biofertilizer Technology for Rice Based Cropping System. Scientific Publishers, Jodhpur, pp. 372–378.

Ramarethinam, S., Chandra, K., 2005. Studies on the effect of potash solubilizing/mobilizing bacteria *Frateuria aurantia* on Brinjal growth and yield. Pestology 11, 35–39.

Rengel, Z., Damon, P.M., 2008. Crops and genotypes differ in efficiency of potassium uptake and use. Physiol. Plant. 133, 624–636.

Roder, A., Hoffmann, E., Hagemann, M., Berg, G., 2005. Synthesis of the compatible solutes glucosylglycerol and trehalose by salt-stressed cells of *Stenotrophomonas* strains. FEMS Microbiol. Lett. 243 (1), 219–226. https://doi.org/10.1016/j.femsle.2004.12.005.

Rodriguez, H., Fraga, R., 1999. Phosphate solubilizing bacteria and their role in plant growth promotion. Biotechnol. Adv. 17, 319–339.

Ruangsanka, S., 2014. Identification of phosphate-solubilizing bacteria from the bamboo rhizosphere. Sci. Asia 40, 204–211.

Sahai, P., Sinha, V.B., Dutta, R., 2019. Bioformulation and nanotechnology in pesticide and fertilizer delivery system for eco-friendly agriculture: a review. Acta Sci. Agron. 3 (11), 2–10. https://doi.org/10.31080/ASAG.2019.03.0675.

Samavat, S., Heydari, A., Zamanizadeh, H.R., Rezaee, S., Aliabadi, A.A., 2014. Application of new bioformulations of *Pseudomonas aureofaciens* for biocontrol of cotton seedling sampling-off. J. Plant Prot. Res. 54 (4), 334–339.

Sangeeth, K.P., Bhai, R.S., Srinivasan, V., 2012. *Paenibacillus glucanolyticus*, a promising potassium solubilizing bacterium isolated from black pepper (*Piper nigrum* L.) rhizosphere. J. Spices Aromat. Crops 21 (2), 118–124.

Sharma, S., Aneja, M.K., Mayer, J., Munch, J.C., Schloter, M., 2005. Characterization of bacterial community structure in rhizosphere soil of grain legumes. Microb. Ecol. 49 (3), 407–415. https://doi.org/10.1007/s00248-004-0041-7.

Sheng, X.F., 2005. Growth promotion and increased potassium uptake of cotton and rape by a potassium releasing strain of Bacillus edaphicus. Soil Biol. Biochem. 37 (1), 1918–1922.

Sheng, X.F., He, L.Y., 2006. Solubilization of potassium-bearing minerals by a wild-type strain of *Bacillus edaphicus* and its mutants and increased potassium uptake by wheat. Can. J. Microbiol. 52 (1), 66–72. https://doi.org/10.1139/w05-117.

Singh, S., Kapoor, K.K., 1999. Inoculation with phosphate solubilizing microorganisms and a vesicular arbuscular mycorrhizal fungus improves dry matter yield and nutrient uptake by wheat grown in a sandy soil. Biol. Fertil. Soils 28, 139–144.

Singh, T., Purohit, S.S., 2011. Biofertilizers Technology. Agrobios, New Delhi.

Singh, G., Biswas, D.R., Marwah, T.S., 2010. Mobilization of potassium from waste mica by plant growth promoting rhizobacteria and its assimilation by maize (*Zea mays*) and wheat (*Triticum aestivum* L.). J. Plant Nutr. 33, 1236–1251.

Singh, S., Singh, B.K., Yadav, S.M., Gupta, A.K., 2014. Potential of biofertilizers in crop production in Indian agriculture. Am. J. Plant Nutr. Fert. Technol. 4, 33–40. https://doi.org/10.3923/ajpnft.2014.33.40.

Singh, R.K., Singh, P., Li, H.B., Song, Q.Q., et al., 2020a. Diversity of nitrogen-fixing rhizobacteria associated with sugarcane: a comprehensive study of plant-microbe interactions for growth enhancement in *Saccharum* spp. BMC Plant Biol. 20, 220. https://doi.org/10.1186/s12870-020-02400-9.

Singh, J., Singh, A.V., Upadhayay, V.K., Khan, A., 2020b. Comparative evaluation of developed carrier based bioformulations bearing multifarious PGP properties and their effect on shelf life under different storage conditions. Environ. Ecol. 38 (1), 96–103.

Sperberg, J.I., 1958. The incidence of apatite-solubilizing organisms in the rhizosphere and soil. Aust. J. Agric. Res. Econ. 9, 778.

Stamenkovic, S., Beskoski, V., Karabegovic, I., Lazic, M., Nikolic, N., 2018. Microbial fertilizers: a comprehensive review on current findings and future perspectives. Span. J. Agr. Res. 16. https://doi.org/10.5424/sjar/2018161-2012117, e09R01.

Subhashini, D.V., 2015. Growth promotion and increased potassium uptake of tobacco by potassium-mobilizing bacterium *Frateuria aurantia* grown at different potassium levels in Vertisols. Commun. Soil Sci. Plant Anal. 46 (2), 210–220. https://doi.org/10.1080/00103624.2014.967860.

Sugumaran, P., Janarthanam, B., 2007. Solubilization of potassium containing minerals by bacteria and their effect on plant growth. World J. Agric. Sci. 3, 350–355.

Sun, F., Ou, Q., Wang, N., Xuan Guo, Z., et al., 2020. Isolation and identification of potassium-solubilizing bacteria from *Mikania micrantha* rhizospheric soil and their effect on *M. micrantha* plants. Global Ecol. Conserv. 23. https://doi.org/10.1016/j.gecco.2020.e01141, e01141.

Supanjani, H.H.S., Jung, S.J., Lee, K.D., 2006. Rock phosphate potassium and rock solubilizing bacteria as alternative sustainable fertilizers. Agron. Sustain. Dev. 26, 233–240.

Suryadi, Y., Susilowati, D.N., Kadir, T.S., Zaffan, Z.R., Hikmawati, N., Mubarik, N.R., 2013. Bioformulation of antagonistic bacterial consortium for controlling blast, sheath blight and bacterial blight disease of rice. Asian J. Plant Pathol. 7 (3), 92–108. https://doi.org/10.3923/ajppaj.2013.92.108.

Sylvia, D., Fuhrmann, J., Hartel, P., Zuberer, D., 2005. Principles and Applications of Soil Microbiology. Pearson, Upper Saddle River, NJ.

Troufflard, S., Mullen, W., Larson, T.R., Graham, I.A., Crozier, A., Amtmann, A., Armengaud, P., 2010. Potassium deficiency induced the biosynthesis of oxylipins and glucosinolates in *Arabiodopsis thaliana*. Plant Biol. 10 (1), 172.

Vassilev, N., Vassileva, M., Lopez, A., Martos, V., Reyes, A., Maksimovic, I., Eichler-Lobermann, B., Malusa, E., 2015. Unexploited potential of some biotechnological techniques for biofertilizer production and formulation. Appl. Microbiol. Biotechnol. 99 (12), 4983–4996.

Verma, P., Yadav, A.N., Kazy, S.K., Saxena, A.K., Suman, A., 2014. Evaluating the diversity and phylogeny of plant growth promoting bacteria associated with wheat (*Triticum aestivum*) growing in central zone of India. Int. J. Curr. Microbiol. Appl. Sci. 3, 432–447.

Verma, P., Yadav, A.N., Khannam, K.S., Panjiar, N., et al., 2015. Assessment of genetic diversity and plant growth promoting attributes of psychrotolerant bacteria allied with wheat (*Triticum aestivum*) from the northern hills zone of India. Ann. Microbiol. https://doi.org/10.1007/s13213-014-1027-4.

Vishwakarma, K., Kumar, V., Tripathi, D.K., Sharma, S., 2018. Characterization of rhizobacterial isolates from *Brassica juncea* for multi trait plant growth promotion and their viability studies on carriers. Environ. Sustain. 1, 253–265. https://doi.org/10.1007/s42398-018-0026-y.

Wang, D.Z., Kong, L.F., Li, Y.Y., Xie, Z.X., 2016. Environmental microbial community proteomics: status, challenges and perspectives. Int. J. Mol. Sci. 17 (8), 1275. https://doi.org/10.3390/ijms17081275.

White, P.J., Karley, A.J., 2010. Potassium. In: Hell, R., Mendel, R.R. (Eds.), Cell Biology of Metals and Nutrients, Plant Cell Monographs. vol. 17. Springer, Berlin, pp. 199–224.

Wong, C.K.F., Saidi, N.B., Vadamalai, G., Teh, C.Y., Zulperi, D., 2019. Effect of bioformulations on the biocontrol efficacy, microbial viability and storage stability of a consortium of biocontrol agents against *Fusarium* wilt of Banana. J. Appl. Microbiol. 127 (2), 544–555. https://doi.org/10.1111/jam.14310.

Wu, S.C., Cao, Z.H., Li, Z.G., Cheung, K.C., Wong, M.H., 2005. Effects of biofertilizer containing N-fixer, P and K solubilizers and AM fungi on maize growth: a greenhouse trial. Geoderma 125, 155–166.

Xue, S., Miao, L., Ma, Y., Du, Y., Yan, H., 2016. Optimizing *Bacillus circulans* Xue-113168 for biofertilizer production and its effects on crops. Afr. J. Biotechnol. 15 (52), 2845–2853. https://doi.org/10.5897/AJB2016.15255.

Yadegari, M., Farahani, G.H.N., Mosadeghzad, Z., 2012. Biofertilizers effects on quantitative and qualitative yield of thyme (*Thymus vulgaris*). Afr. J. Agric. Res. 7 (34), 4716–4723.

Zhang, C., Kong, F., 2014. Isolation and identification of potassium-solubilizing bacteria from tobacco rhizospheric soil and their effect on tobacco plants. Appl. Soil Ecol. 82, 18–25.

Zhang, A., Zhao, G., Gao, T., Wang, W., et al., 2013. Solubilization of insoluble potassium and phosphate by *Paenibacillus kribensis* CX-7: a soil microorganism with biological control potential. Afr. J. Microbiol. Res. 7 (1), 41–47.

Zhang, S., Merino, N., Okamoto, A., Gedalanga, P., 2018. Interkingdom microbial consortia mechanisms to guide biotechnological applications. Microb. Biotechnol. 11 (5), 833–847. https://doi.org/10.1111/1751-7915.13300.

Phosphate-solubilizing microbial inoculants for sustainable agriculture

Sonth Bandeppa[a], Kiran Kumar[b], P.C. Latha[a], P.G.S. Manjusha[a], Amol Phule[a], and C. Chandrakala[a]

[a]*ICAR-Indian Institute of Rice Research, Hyderabad, Telangana, India,* [b]*Crop Production Unit, ICAR-Directorate of Groundnut Research, Junagadh, India*

19.1 Introduction

Phosphorus (P) is contained in every living plant cell and is indispensable for plant growth. It is a prominent limiting nutrient for the crop growth where it helps in nitrogen fixation, signal transduction, photosynthesis, energy transfer, transformation of sugars and starches, transfer of genetic characteristics, respiration, and biomolecular biosynthesis (Fernandez et al., 2007). The phosphorus deficiency leads to curtailment of the leaf size and retarded growth (Yeh et al., 2000). According to the Directorate of Economics and Statistics (DES), Ministry of Agriculture, India, holds the second-largest and most diverse sector of agricultural land with 159.7 million hectares under the cultivation (Directorate of Economics and Statistics (DES), 2019). To maintain sustainable food production from agricultural lands, soil plays a crucial role because it is the original source of nutrients which are categorized into macronutrients (nitrogen, phosphorus, potassium, most commonly known as NPK), secondary nutrient elements (Ca, Mg, and S), and micronutrients (B, Cl, Cu, Fe, Ni, Mn, Mo, and Zn) which help in the growth of crops. The average soil contains roughly 0.05% (w/w) P, but due to poor solubility and soil fixation, only 0.1% of the total P is accessible to plants and, on dry weight basis, it ranges from 0.2% to 0.8% in different crop plants.

Phosphorus is a limited resource widely used for agricultural purposes to maintain healthy soil and growth of plants. Mining rock phosphate and transporting it from production site to farmer fields are not environmentally benign, economically practical, or sustainable. At the present mining rate, global reserves will be depleted in about 500–600 years. The comprehensive P management entails changes of topsoil and rhizospheric activities, building of P efficient crops, and enhancing P recycling efficiency (Sharma et al., 2013). For minimizing P deficiency in soil as per synthetic, P fertilizers are applied, whereas it creates harmful effects on plants, and it depleted the soil health condition which causes the deprivation of soil fertility, disturbances in microbial biodiversity, and other metabolic activities, plummeted yield of crops (Toro, 2007). It is obligatory to search for low-cost and viable replacement to synthetic P fertilizers. In these circumstances, biofertilizers are requisite to make sure healthy soils. According to recent trends, biofertilizer is a potent fertilizer as it is a dense and stable organic fertilizer, which can lessen the use of chemical fertilizers.

Biofertilizers are the inoculant that consists of one or more strains/species of microorganisms that have potential of altering the nutrients from unusable to usable form by biological process such as P solubilization and/or mobilization. Microorganisms that are able to solubilize phosphorus-bearing nonsoluble inorganic and organic compounds are known as phosphate-solubilizing microorganisms (PSMs). PSMs have provided solution to the P-deficiency problems (Trimurtulu and Rao, 2014; Pal et al., 2015). Some of the important PSMs like bacteria (*Bacillus megaterium, Bacillus sp., Mycobacterium, Pseudomonas, Thiobacillus, Flavobacterium, Micrococcus*, etc.) are produced in a large scale as commercialized biofertilizers to promote the plant growth; some fungi such as *Aspergillus, Cladosporium, Fusarium, Paecilomyces, Penicillium*, and *Rhizoctonia* appear more advantageous agents in the solubilization of phosphates, whereas another type of fungi *Glomus sp.* form association with plant roots (Sinha et al., 2014). Since the 1950s, phosphate-solubilizing bacteria are being used as a biofertilizer (Kudashev, 1956). In this context, we discuss more in detail the role of PSM in aspects of their diversity, genetics, and mechanisms of P solubilization in soil for maintaining sustainable agriculture.

According to Vessey (2003), "a product containing living microorganisms, when applied to seed, plant surfaces, or soil, colonizes the rhizosphere or the inside of the plant and invigorates development by improving the supply or availability of primary nutrients to the host plant." A biofertilizer can be defined, according to Fuentes-Ramirez and Caballero-Mellado (2005), as "a product containing living microorganisms that exert direct or indirect favorable effects on plant development and crop output through several processes." Therefore, these microorganisms in biofertilizers promote plant growth by replenishing soil rhizosphere nutrient elements (i.e., biological N_2 fixation), making nutrients largely available to crop plants (i.e., nutrient solubilization), or expanding crop plant accessibility to nutrients (i.e., expanding the volume of soil reached by the root system), as long as the plants' nutritional content is improved.

19.2 Status and availability of soil phosphorus

In soil, phosphorus can be found in two different forms one is organic and another is inorganic which vary greatly in their bioavailability. By combining these two forms together makes up the total soil phosphorus. However, 95%–99% of insoluble phosphates are present in the soil. Indeed, most of the phosphorus is in immobile stage and not available for plant uptake from soil. These P forms differ in their behavior and responsibility for changes in soil (Turner et al., 2007). The inorganic phosphorus availability in soil is reduced and use of due biofertilizers promotes plant growth by making more nutrients access to root and maximum nutrients available to plants (Ranjan et al., 2013). The organically bound phosphorus fraction in soils is more compared to the inorganic. Organic phosphorus is categorized into three classes, phosphate esters, phosphonates (C–P bonds), and phosphoric acid anhydrides (Huang et al., 2017).

Inositol phosphate (soil phytate) is an important form of organic phosphorus in the soil and also other forms of organic P compounds (phosphomonoesters, phosphodiesters, and phosphotriesters) are present (Asea et al., 1988). Plants take phosphorus only in inorganic form, and for the conversion of organic phosphorus into inorganic, mineralization plays a pivot role by means of phosphatase enzymes. Minerals that bound to phosphates like calcium phosphate, iron phosphate, and aluminum phosphate are the constituents of inorganic phosphorus, used as a predominant source of the plants (Boitt et al., 2018). Mineralization process is done by soil microorganisms and plant roots in association with phosphatase secretion (Turner et al., 2007). Prominent forms of phosphorus in acidic soils are iron phosphates and aluminum phosphates, whereas calcium phosphates are in alkaline soils (Kumar and Shastri, 2017).

19.3 Importance of phosphate-solubilizing microorganism in agriculture

Soil microorganisms play a very crucial role in the development of soil healthy structure. Significantly, higher populations of PSMs are found in range and agricultural soils (Yahya and Azawi, 1998). They tend to facilitate the growth of plants by providing the essentials as shown in Fig. 19.1 and help in solubilization of bound phosphates to make them easily available (Sujatha et al., 2004). PSMs are important contributors to soil P pools by decomposing the organic residue by mineralization and immobilization. PSM assimilates soil microbial biomass into soluble P and prevents it from fixation or adsorption (Khan and Joergensen, 2009). Phosphate-solubilizing

Fig. 19.1 Role of phosphate solubilizing bacteria in plant growth and development.

microorganisms enhance the plant growth by increasing the efficiency of biological nitrogen fixation and production of plant growth hormones (gibberellins, auxins, cytokinin, ACC deaminase, and IAA) and increase the accessibility of other trace elements like zinc, boron, iron, copper, and manganese in agricultural fields. PSMs can be classified into two types, i.e., phosphate-solubilizing and phosphate-mobilizing microbes. Phosphate solubilizers are bacterial and fungal strains that secrete low molecular weight organic acids (e.g., malic, succinic, fumaric, and citric acid) to dissolve insoluble phosphorus compounds like tricalcium phosphate, hydroxyapatite, dicalcium phosphate, and rock phosphate and also mineralize organic P compounds (nucleic acids, phospholipids, sugar phosphates, phytic acid, phosphonates, and polyphosphates) through the production of phosphatases and phytases (Shrivastava et al., 2018). Mycorrhizal fungi are known as phosphorus-mobilizing microorganisms because they improve phosphorus intake by mobilizing them from the soil rather than solubilizing P molecules. Mycorrhizal fungi are classified into two categories: endomycorrhizal and ectomycorrhizal. In endomycorrhiza (AMF), the hypha of fungi inserts into root cortical cells and folds inward the cell membrane. AMF fabricates long external hyphae, which improves increased contact with soil phosphates, because of which increase P uptake by crop plants (Kumar et al., 2018).

19.4 Diversity of phosphate-solubilizing microorganisms

Phosphate-solubilizing microorganisms (PSMs) have a vast diversity in soils and they can be isolated directly from plants' rhizosphere. PSMs are categorized into different groups like bacteria, fungi, actinomycetes, vesicular-arbuscular mycorrhizae, and cyanobacteria as shown in Table 19.2 (Sperber, 1958). In soil, phosphate-solubilizing bacteria (PSB) like *Pseudomonas, Bacillus, Rhizobium,* and *Enterobacter* contribute 1% to 50%, whereas phosphate-solubilizing fungi (PSF) like *Aspergillus* and *Penicillium* are 0.1% to 0.5%. In comparison, phosphate-solubilizing activity is more for fungi than bacteria (Kucey, 1983). In detail, PSMs can solubilize insoluble phosphates to release soluble phosphates; it undergoes through various mechanisms such as chelation, ion-exchange reactions, and production of organic acids (Servin et al., 2020). Some groups of organisms like actinomycetes are attracted by recent trends, showing that they are capable of surviving in extreme conditions but also possess the production of phytohormones (Hamdali et al., 2008). When vesicular-arbuscular mycorrhizae (VAM; *Enterobacter agglomerans*) and PSB (*Glomus etunicatum*) were used together, the soil rhizosphere colonization and phosphatase activity were greater as compared to that when used individually (Kim et al., 1997a, b) (Table 19.1).

19.5 Mechanisms of P solubilization by PSMs

To solubilize inaccessible phosphorus compounds, PSMs play a major role to make them easily available for plants. The main mechanisms that are mediated by PSMs are organic/inorganic acid production, release of proton from NH_4 assimilation, chelation of cation bound to P, by release of protons, reduction of pH level, production of plant growth hormones, and release of biocontrol agents. There are four bacterial pathways for the oxidation of the reduced inorganic P compound, phosphite: C–P lyase, bacterial alkaline phosphatase (BAP), NAD:phosphite oxidoreductase, and an undetermined pathway for deriving energy from phosphite oxidation

Table 19.1 Diversity of phosphate-solubilizing microorganisms (PSMs) in soils.

Type	General solubilizing phosphate	References
Bacteria	*Pseudomonas* spp., *Agrobacterium* spp., and *Bacillus circulans*. *Azotobacter, Bacillus, Burkholderia, Enterobacter, Erwinia, Kushneria, Paenibacillus, Ralstonia, Rhizobium, Rhodococcus, Serratia, Bradyrhizobium, Salmonella, Sinomonas,* and *Thiobacillus*	Elias et al. (2016)
Fungi	*Achrothcium, Alternaria, Arthrobotrys, Aspergillus, Cephalosporium, Cladosporium, Curvularia, Cunninghamella, Chaetomium, Fusarium, Glomus, Helminthosporium, Micromonospora, Mortierella, Myrothecium, Pichia fermentans, Oidiodendron, Paecilomyces, Penicillium, Phoma, Populospora, Pythium, Rhizoctonia, Rhizopus, Saccharomyces, Schizosaccharomyces, Sclerotium, Torula, Trichoderma,* and *Yarrowia*	Alori et al. (2017)
Actinomycetes	Actinomyces, Streptomyces	Sperber (1958)
Cyanobacteria	*Anabaena, Calothrix braunii, Nostoc* sp., *Syctonema* sp.	
Vesicular-arbuscular mycorrhiza (VAM)	*Glomus fasciculatum, Gigaspora* sp., *Pisolithus* sp., *Rhizopogon* sp.	

in *Desulfotignum phosphitoxidans*. Two distinct enzymes for hypophosphite oxidation in bacteria have been found: hypophosphite:2-oxoglutarate dioxygenase and an enzyme homologous to formate dehydrogenase (White and Metcalf, 2007).

19.5.1 Organic acids

PSMs solubilize inorganic phosphates through the action of organic acids like citric acid, gluconic acid, oxalic acid, 2-ketogluconic acid, propionic acid, and tartaric acid (Marra et al., 2019) by chelation of cations linked to phosphate, minimizing the pH, metal ion complexation, and P for adsorption site (Kishore et al., 2015). Gluconic acid and 2-ketogluconic acid are most frequently excreted by PSB (Suleman et al., 2018; Duebel et al., 2000). PSBs' production of gluconic acid is aided by the conversion of glucose to gluconic acid, i.e., the quinoprotein pyrroloquinoline quinone-dependent periplasmic glucose dehydrogenase (PQQ-GDH) is involved in the direct oxidation of glucose (Rawat et al., 2020). Glucose dehydrogenase enzyme is encoded by the gene (gcd) and PQQ as a cofactor. Wan et al. (2020) reported that PQQ is a redox-active molecule enciphered by *pqq* operon that consists of core genes *pqq* A, B, C, D, E, and F which are behind the dehydrogenase activity and mineral phosphate solubilization.

19.5.2 Phosphatases and phytases

In terms of mineralization, the microbes mineralize organic phosphorus compounds (e.g., phytin, inositol phosphates, nucleic acids, and phospholipids) to orthophosphates by versatile microorganisms that include bacteria (e.g., *Bacillus subtilis* and *Arthrobacter*), fungi (e.g., *Aspergillus* and *Penicillium*), and actinomycetes (e.g., *Streptomyces*). The whole process of phosphorus mineralization can be defined in two intervention methods like biological and biochemical processes, in which biological mineralization means the release of Pi from organic matrix during the oxidation of carbon by soil rhizosphere microbes (Gressel et al., 1996), whereas biochemical mineralization is defined as the liberation of inorganic phosphorus from organic matrices by enzymatic hydrolysis exterior to the cell.

The insoluble forms of phosphorous such as iron phosphate (Fe_3PO_4), aluminum phosphate (Al_3PO_4), and tricalcium phosphate ($Ca_3PO_4)_2$ are converted to soluble phosphorous by phosphorous-solubilizing organisms inhabiting different soil ecosystems (Sharma et al., 2013a). Most importantly, microbes imbibe phosphorus that forms the composition of many macromolecules in the cell. Especially some microbes have the capability to store phosphorus as polyphosphates in special granules.

Polyphosphate-accumulating microorganisms (proteobacteria, actinobacteria, algae, and fungi) have the capability to accumulate inorganic polyphosphates and orthophosphates, whereas they act as phosphate and energy storage reservoirs (Chaudhry and Nautiyal, 2011). Consequently, phytate (IHP) is a phosphorus storage molecule and constitutes in grains. About 20%–80% of phosphorus in soils is found as an organic form. In alkaline soils, phytate is often precipitated as an unavailable calcium phytate, and for degradation/mineralization of this calcium phytate, the microorganisms play a crucial role in resolving precipitation by degrading it. To measure a greater abundance of phytate mineralizing bacteria via 16S rRNA gene copy number in alkaline soils, they analyzed three genes that are related to organic phosphorus degradation {*phoX* phosphatases, *phoD* phosphatases, and β-propeller phytase (*BPP*)} genes (Pittroff et al., 2020).

19.5.3 Exopolysaccharides (EPSs)

Exopolysaccharide (EPS) helps microorganisms to survive in unfavorable and stressful conditions. The plant growth-promoting *Rhodotorula sp.* yeast strain CAH2 may be protected by EPS production from unfavorable environmental conditions (Silambarasan et al., 2019). In soil, some metals like aluminum, copper, zinc, iron, potassium, and magnesium form complexes with EPSs (Ochoa-Loza et al., 2001).

19.5.4 Siderophores

Siderophore is an iron-binding ligand cum uptake protein that is needed to transport iron inside the cell. The ability to manufacture specific siderophore and to use a broad spectrum of siderophore has been identified as a factor in rhizobacteria's ability to colonize roots. Siderophores are growth inhibitors of many phytopathogenic fungi. For maintaining a strategy to chelate iron from P complexes, PSMs secrete siderophores as a strategy to chelate iron from P complexes in the soil (Collavino et al., 2010). Under alkaline conditions, many phosphate solubilizers like *B. megaterium, B. subtilis, Rhizobium radiobacter*, and *Pantoea allii* produce siderophores that invigorate the survival of organisms under stressful environments and improve phosphorus solubilization (Ferreira et al., 2019).

19.6 Types of phosphate biofertilizers

In market, chemically available phosphorus fertilizers are phosphoric acid, calcium orthophosphate, ammonium phosphate, ammonium polyphosphate, rock phosphate, and nitric phosphate which are commercially used. These fertilizers are not completely used by plants, but instead, it gets accumulated in the soil as organic phosphorus; the plants only utilize inorganic forms of phosphorus. For the conversion of organic form to inorganic form, PSM plays an important role. The microorganisms widely used for manufacturing phosphate-solubilizing biofertilizers are *Bacillus, Pseudomonas*, and *Aspergillus* in either liquid or solid form. For promoting plant growth, *Pseudomonas* species is used commercially as phosphate-solubilizing biofertilizer and mycorrhiza is used as phosphate-mobilizing biofertilizer. The popularity of phosphate-solubilizing biofertilizers is raised due to their ecofriendly, nontoxic, and nonhazardous properties in nature (Rai, 2006). Different types of phosphate biofertilizers available in different continents are listed in Table 19.2.

Microbial products can be either in solid or liquid form. Most widely used phosphate biofertilizers are in liquid forms because of longer shelf life, survival on soil, having high potential of PSMs, easy to produce, high commercial values, temperature tolerance, high cost, no contamination with 100% compared to carrier-based biofertilizers, and higher investment for production unit, whereas carrier-based biofertilizer is cheap, easier to produce with less investment, low shelf life, prone to contamination, temperature sensitive, low cell counts, difficult to automation, and less effective (Ngampimol and

Table 19.2 List of phosphorus biofertilizers commercially available in different continents of the world.

Strain(s)	Name of biofertilizer	Continent	References
Bacillus megaterium	Rhizosum P	Europe	Mehnaz (2016)
Azotobacter chroococcum and B. megaterium	Phylazonit M		Garcı́a-Fraile et al. (2017)
Penicillium bilaii	JumpStart		Saxena (2015)
B. megaterium	Bio-Phos	Asia	Dash et al. (2017)
Bacillus mucilaginosus and Bacillus subtilis	CBF		Celador-Lera et al. (2018)
Pseudomonas striata, and B. megaterium	P Sol B		Mehnaz (2016)
Pseudomonas fluorescens	FOSFORINA	North America	Uribe et al. (2010)
Penicillium janthinellum	FOSFOSOL	South America	Uribe et al. (2010)
Azorhizobium, Azoarcus, Azospirillum	Twin N	Australia	Adeleke et al. (2019)

Kunathigan, 2008). There are different types of phosphate biofertilizers available in the Indian market which are widely used as listed in Table 19.3 with Web ID references.

19.7 Production, quality standards, evaluation, and marketing of phosphate biofertilizers

To produce phosphate biofertilizers, the main step is to opt for desired microorganism that plays a crucial role as seed in crop cultivation. For the production of potential phosphate biofertilizer, there are many steps involved as shown in Fig. 19.2. Steps for initial screening are: (a) collection of soil samples from agricultural fields, (b) soil samples are subjected to the serial dilutions 10^{-1} to 10^{-9}, (c) isolation of phosphate-solubilizing microorganisms on selective media (PKV) by spread plate, (d) confirmation by halo zone formation on plates (capability of solubilizing TCP) (Panhwar et al., 2012), (e) study of morphological characteristics (on EMB agar to know the Gram nature of particular bacteria, pH test) and colony characteristics (Collee et al., 1996; Cheesbrough, 2000), and (f) study of biochemical tests (MR-VP test, starch hydrolysis test, catalase test, and hydrogen sulfide test) as per the instructions of

Table 19.3 Important phosphate biofertilizers available in Indian market.

Producer	Biofertilizer	Microorganism	Forms	Shelf life	Mode of action	Reference (Website Reference IDs)
T Stanes & Company Ltd.	SYMBION-P	*Bacillus* sp	Powder/liquid	1 year	Through biological process, it solubilizes the fixed phosphorous in the soil and makes it available to the plant	https://tstanes.com/nutrientmanagement/#1598704261617-ca39b313-3264
SOM Phytopharma (India) Ltd./Agri Life	P Sol B-3 variants (a) P Sol B-PS (b) P Sol B-BP (c) P Sol B-BM	(a) Vegetative cells of *Pseudomonas striata* (b) Endospores of *Bacillus polymyxa* (c) Endospores of *Bacillus megaterium*	(a) Powder/liquid (b) Powder/liquid (c) Powder/liquid	1 year 1 year 1 year	Transformation of soil phosphorus is the integral part of soil phosphorus cycle as they aid in releasing phosphorus from inorganic and organic pools through solubilization and mineralization	https://www.indiamart.com/somphytopharma-india/bio-fertilizer.html#agri-life-p-sol-b-bio-fertilizer
International Panaacea Limited	(a) Premium Phoster (b) Premium Phosphofix	(a) Phosphobacteria (b) Phosphobacteria	(a) Granules (b) Liquid	1 year 1 year	Converts chemically fixed soil phosphorus into available form and reduces 25–30% phosphatic fertilizer requirement	https://www.iplbiologicals.com/phosphate-solubilizing-bacteria/
Kan Biosys	Phosfert R	*Azotobacter chroococcum* and *B. polymyxa*	Liquid	1 year	Sustainable supply of phosphorus by PGPR action and 20 to 30% reduction in the use of phosphatic fertilizers	https://www.kanbiosys.com/nutrientmanagement/phosfert.php
Madras Fertilizers Limited	PSB	Phosphobacteria	Powder	1 year	Maintains soil fertility	http://madrasfert.co.in/marketing/bio-fertilizers/
Gujarat State Fertilizers & Chemicals Ltd.	PSB	Phosphobacteria	Powder	1 year	Suppresses plant pathogens, maintains soil fertility, and increases yield	https://www.gsfclimited.com/organic-products
National Fertilizers Limited	PSB	Phosphobacteria	Liquid	1 year	Increases yield by 10%–30% Secretes enzymes that mineralize organic phosphorous to a soluble form	https://www.nationalfertilizers.com/
Rashtriya Chemicals & Fertilizers Ltd	Biola	Phosphobacteria	Liquid	1 year	Releases organic acids which aid in decreasing the pH of the soil and also dissolves the fixed phosphorous and makes it available to the crop plants	https://www.rcfltd.com/products/products/14

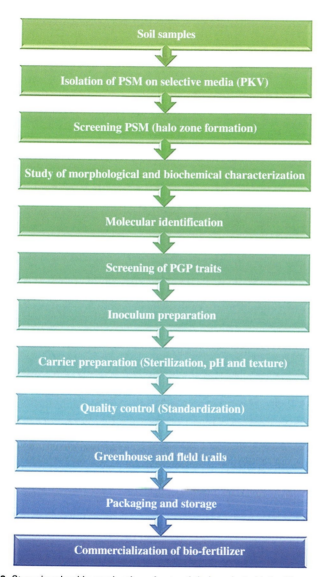

Fig. 19.2 Steps involved in production of potential phosphate biofertilizers.

Bergey's manual of systematic bacteriology (Claus and Berkery, 1986). Therefore, potential phosphate strains are obtained by performing initial screening.

The selected strains are subjected to production media of PKV broth. This starter culture is incubated for a week in BOD shaker for rapid production of phosphate-solubilizing bacteria. Growth rate was measured at regular intervals of starter culture. A loopful of culture is inoculated into small scale and tested on pot cultures. Again, the same process is done at a large scale incubated for more than 2 weeks, tested on greenhouse and field trials. Quality refers to the density, viability, and preservation of the available microorganisms (Somasegaran and Hoben, 1994). Quality control plays a major role to ensure that the product is safe to use in agricultural fields. The most important aspect is the maintenance of the phosphate-solubilizing biofertilizer for the success/failure. Quality results in active form of desired microorganisms per gram biofertilizer. There are many specifications to be taken into consideration to maintain biofertilizer quality like microbial density (CFU) at the time of manufacturing/expiry, shelf life, moisture, pH, allowable contamination, carrier material, and microbial strain (genus and species). Quality checking should be done at different stages of production (starter culture stage, broth culture, selection of the carrier, packaging, and storage) and it should be standardized (Machi, 2006). After that, evaluation for phosphate-solubilizing bacteria, for example, *B. megaterium*, was identified by PCR method, etc., CFU on selective medium for density of bacterial strain, and assessment of main activity by solubilizing of tricalcium phosphate as an indicator; these aspects come under regular inspection of evaluation by authorities under certain acts. Coming to the field trials, the sterilized biofertilizer is applied in either liquid form or solid form with carrier to know the growth rate and nutrient absorption of targeted crops. It is essential to evaluate precisely and register the biofertilizers under the regulation for marketing purposes.

In India, the Ministry of Agriculture's order on the control of fertilizers, as amended, regulates manufacturing and selling criteria for the various types of microorganisms that make up biofertilizers. In the case of bacteria, the minimum count of viable cells is 5×10^7 cells per gram of solid carrier, or 1×10^8 cells per mL of liquid carrier. The PSB activity can be measured spectrophotometrically (30% P solubilization) or by forming a solubilization zone of minimum 5 mm in a media with a thickness of at least 3 mm. The farmer instruction manual on biofertilizer should contain a reference on the soil best management practices, particularly on the reduction of the quantity of chemical fertilizers used. Under low nutritional conditions, PGPM colonization and activity are increased (e.g., a decrease of chemical fertilizers by 20%–50% has been demonstrated with numerous crops), and an efficient mycorrhizal symbiosis can replace up to 220 kg P_2O_5 ha^{-1} (Kelly et al., 2001). If under any circumstances, strains of PGPM species used in the formulation are potentially or opportunistic pathogens for mammals (e.g., *Enterobacter, Pantoea Klebsiella, and Burkholderia*), such data must be supplied and examined before the marketing authorization is granted (Guo et al., 2002).

19.8 Plant growth-promoting activities of PSMs

Inoculation of PGPR (PSM) improved seedling vigor, emergence, stalk growth, root growth, total biomass of the crop plants, early blooming, seed weight, increased grain, fodder, and fruit yields. PSMs promote growth directly through the generation of phytohormones, as well as the solubilization and uptake of nutrients. Induced systemic resistance (ISR), antibiotic protection against infections, formation of HCN, siderophore production, and synthesis of antifungal enzymes or lytic enzymes are indirect growth promotion mechanisms (Lucy et al., 2004).

19.9 Influence of PSMs on plant growth and yield

Interaction between PSMs and plants is generally synergistic in nature as they are involved in direct release of Pi for plant use and the plants provide the easily utilizable sugars for microbes (Perez et al., 2007) and root exudates in exchange that support PSM's growth. The PSMs either singly or in consortia as bioinoculants aid in improving plant growth while keeping the soil health as it is. The efficiency and economy of P fertilizers can be improved by deploying PSMs in legumes, cereals, and other crops. Beneficial effects of the inoculation with PSM in crop plants are listed in Table 19.4. Dry matter production and P uptake were significantly increased by the usage of PSMs. Adding rock phosphate and inoculating with PSMs applied to cereals, legumes, potatoes, and other field crops increased yield by 12%–15% and replaced 25%–27% of phosphate fertilizers in cereals, legumes, potatoes, and other agronomically important crops (Arun, 2007). The combined inoculation of arbuscular mycorrhiza and PSB improved uptake of native P from the soil and P from the phosphatic rock (Cabello et al., 2005). Inoculation with PSMs improved the sugarcane yield by 12.6% (Sundara et al., 2002). Growth and phosphorus content in two alpine Carex species improved by inoculation with *Pseudomonas fortinii* (Jilani et al., 2007).

19.10 Genetics of phosphate solubilization by PSMs

Organic P pool consists of phospholipids and phytate, which are hydrolyzed by phosphatase and phytase, respectively (Irshad and Yergeau, 2018). Acid phosphatase is mostly synthesized and secreted

Table 19.4 Plant growth-promoting effects of PSMs.

Host crop	Microorganism	Effect on host plant	References
Chickpea	*Mesorhizobium sp.*	Improved grain and straw yield, uptake of P and N, nodulation, shoot and root dry weight	Verma et al. (2013)
Tomato	*Pseudomonas aeruginosa*		Walpola and Yoon (2013)
	Pantoea agglomerans, Burkholderia anthina	Improved plant height, root length, phosphorus uptake, and available phosphorus content	
Moth bean	*Rhizobium* and *Pseudomonas sp.*	Increase in shoot and root length	Sharma et al. (2013b)
Cotton	*Bacillus sp.*	Increase in plant height, number of bolls per plant and boll weight and soil available P	Qureshi et al. (2012)
Rice	*Pseudomonas sp., Serratia sp., Azospirillum sp.*	Increase in plant growth and P uptake	Nico et al. (2012)
Glycine max	Fluorescent pseudomonads	Increased nutrient uptake, tolerance to stress, salinity, metal toxicity, and pesticide	Malviya and Singh (2012)
Cowpea	*Micrococcus sp.*	Increased root and shoot lengths, increase in dry biomass as well as a number of roots	Dastager et al. (2010)
Sunflower	*Bacillus sp.*	Increase in growth, yield, and quality of plant, and oil yield	Ekin (2010)
Soybean	*Pseudomonas sp.*	Increase in the number of nodules, dry weight of nodules, and grain yield, nutrient availability, and uptake	Son et al. (2006)
Gram	*Pseudomonas striata* and *Bacillus polymyxa*	Increase in nodulation, nitrogenase activity, dry matter content	Alagawadi and Gaur (1988)
Maize	*Rhizobium leguminosarum biovar phaseoli* (MPS)	Increase colonization and dry matter	Chabot et al. (1998)

into rhizosphere by plants, whereas alkaline phosphatase is entirely derived from soil microbes (Fraser et al., 2015). Three genes *phoX, phoA*, and *phoD* of prokaryotic origin encode for alkaline phosphatase (Huang et al., 2009); In prokaryotes, three genes *bpp, ptp, and hap* encode for β-propeller phytase, protein tyrosine phosphatase, and histidine acid phosphatase, respectively (Neal et al., 2017). In terrestrial ecosystems, *phoD* and *bpp*-harboring bacteria are more widely distributed (Hu et al., 2018). Inorganic P pool consisting of insoluble calcium, iron, and aluminum phosphates are generally dissolved by low molecular organic acid (e.g., gluconic acid, malic, and citric acid), which are released from gcd-harboring bacteria (Hanif et al., 2015; Rasul et al., 2019). For these reasons, *phoD, bpp*, and *gcd* genes are used as good biomarkers in studying soil inorganic and organic P transformation. Thirty-nine near-complete genomes involved in soil P cycling from 18 metagenomes were used to deduce the soil microbial phosphorus (P) cycling in terrestrial ecosystems. Relative abundance of key genes involved in microbial P cycling like *gcd* gene (glucose dehydrogenase gene involved in the production of gluconic acid), producing an enzyme that oversees inorganic P solubilization, was a prominent deciding factor of bioavailable soil P (Liang et al., 2020). Many PSB are Gram-negative bacteria and belong to *Pseudomonas Acinetobacter, Pantoea,* and *Enterobacter* and very few PSB are Gram-positive bacteria belonging to *Bacillus* (Wang et al., 2018).

Gluconic acid is formed by oxidation of glucose arbitrated by a membrane-bound glucose dehydrogenase (GDH), in which pyrroloquinoline quinone (PQQ) as a cofactor. PQQ are a category of quinoproteins (Wagh et al., 2014). PQQ synthesis pathway involves several genes presenting in a cluster, whose composition varies in different species of bacteria (Choi et al., 2008). Polyphosphate kinases (PPKs) encoded by *ppk* gene synthesize polyphosphate, a high-energy compound, thought to be involved in the process of inorganic P solubilization (Ishige and Noguchi, 2000). Conversion of insoluble tricalcium phosphate into soluble phosphorus by *Acinetobacter pittii* gp-1 is due to the strong expression of *pqq* gene. So, the expression of *pqq* gene is highly correlated with inorganic phosphorus solubilization (Oteino et al., 2015; Wan et al., 2020). Polyphosphate is hydrolyzed when phosphorus-accumulating bacteria undergo undernourished conditions (Li and Dittrich, 2019). *Burkholderia multivorans* WS-FJ9 strain has *AP-2, GspE,* and *GspF genes* involved in organic phosphate breakdown, *HlyB* gene was only related to inorganic phosphate solubilization, and *PhoR, PhoA, AP-1,* and *AP-3* genes work on both organic and inorganic phosphates (Liu et al., 2020). Expression levels of the *PhoA, PhoC,* and *PhoD* genes oversee the level of phosphate solubilization in *Streptomyces coelicolor*. Under the circumstances of low-soluble

Table 19.5 List of some cloned genes used in mineral phosphate solubilization.

Microbe	Gene/plasmid	Features	References
Erwinia herbicola	mps	Produces gluconic acid and solubilizes mineral P in *E. coli* HB101. Probably involved in PQQ1 synthesis	Goldstein and Liu (1987)
Enterobacter agglomerans	pKKY	Solubilizes P in *E. coli* JM109 and does not lower pH	Kim et al. (1997a, b)
Serratia marcescens	pKG3791	Makes gluconic acid and solubilizes mineral P	Krishnaraj and Goldstein (2001)

PQQ: Pyrroloquinoline quinone.

phosphate, *PhoA* and *PhoD* were upregulated, whereas *PhoC* exhibited the opposite expression phenomena (Apel et al., 2007). Yang et al. (2016) found various expression signatures under variable soluble phosphate levels by *GDH* of *Pseudomonas sp.* Wj1 and *Enterobacter sp.* Wj3 (Few genes involved in mineral phosphate solubilization are listed in Table 19.5).

19.11 Impact of application of phosphate biofertilizers on native soil microorganisms

The use of phosphate biofertilizer affects the physicochemical soil properties, structure, and functions of soil microcosm (Javorekova et al., 2015). The results of the application of biofertilizer in the plant rhizosphere vary based on native population dynamics. Some groups of microbes are enhanced, others may be inhibited, and some may show no changes in native microbial richness (Castro-Sowinski et al., 2007). Different techniques used to study shifts in microbial communities are denaturing gradient gel electrophoresis (DGGE), amplified ribosomal DNA restriction analysis (ARDRA), single-strand conformation polymorphism (SSCP), and the community-level physiological profiling (CLPP) with the use of BIOLOG plates (Javorekova et al., 2015). Trabelsi et al. (2012) used the t-RFLP technique, to study changes in rhizosphere after inoculation with rhizobial strains (*Rhizobium gallicum* and *Sinorhizobium meliloti*) and observed that the structure and diversity of α- and γ-proteobacteria along with Firmicutes and Actinobacteria were significantly affected. Inoculation with two *Azospirillum brasilense* strains has found to change the CLPP profiles of the microbes associated with rice (de Salamone et al., 2010). Modifications in CLPP profiles have been found in faba bean inoculated with *Rhizobium leguminosarum* (Siczek and Lipiec, 2016). Inoculation of *Medicago sativa* with *S. meliloti* improved the population of α-proteobacteria and decreased γ-proteobacteria in the plant rhizosphere, which is detected through SSCP technique (Wang et al., 2018). Inoculation of probiotic strain *Stenotrophomonas acidaminiphila* improved the bacterial population in the chlorothalonil contaminated plant rhizosphere of *Vicia faba* (Zhang et al., 2017).

19.12 Constraints in using phosphate biofertilizers

Constraints faced in spite of a glorious growth of biofertilizer industry over the last two and a half decades in India, and they are still far from their expected potential. The main limitations are limited nutrient mobilization capability compared to chemical analogues and a slow influence on crop development. Farmers' skepticism arises from inconsistent reactions in the field across a variety of agro-ecological niches and cropping systems. Still much more is needed for achieving success as the production and application of biofertilizers have several constraints (Nosheen et al., 2021) as follows.

Raw material constraints: Biofertilizers are typically made up of carrier-based inoculants containing beneficial microbes. Granular carrier material like peat, perlite, and charcoal is mostly recommended for soil inoculation of the biofertilizers (Mazid and Khan, 2014). These carrier materials for seed and soil treatment are not easily accessible to the small and marginal farmers. In India, these carriers are available not in right quantities nor in desirable quality, which is one of the main reasons for the lack of popularity of biofertilizers among the Indian farmers (Mahdi et al., 2010).

Field-level constraints: Competition between bioinoculant and natural flora of soil ecology. Many of the biofertilizers are not only crop specific but are also soil and agroclimate specific. The absence of region-specific strains is one of the main constraints associated with biofertilizer use. This limits their extensive and optimum use with anticipated performance (Mazid and Khan, 2014; Motghare and Gauraha, 2012).

Technical constraints: Unavailability of the experienced and professional workforce in manufacturing unit and biofertilizers possesses the tendency to mutate throughout the fermentation that increases the fee of production and first-rate manipulate. A wide variety of studies is needed to lessen such undesired adjustments (Singh et al., 2016).

Financial constraints: Unavailability of high-tech instruments and equipment for the production of quality product is there. In the absence of these facilities, the production of contamination-free product is uncertain (Pathak and Christopher, 2019).

Market constraints: Lack of facilities for storage and the shelf life of biofertilizers prepared with carriers like burn peat or lignite is usually less than 6 months. It has been suggested that the best effects from biofertilizers may only be obtained if they are employed within 3–4 months of manufacturing. But commonly the biofertilizers are subjected to very high temperature during transportation and storage which reduces their efficiency and leads to a lack of interest among the dealers due to a very low profit margin (Mathur et al., 2010; Das et al., 2015).

Low demand: Farmers avoid using biofertilizers due to a lack of proper information and advertising about their benefits from using this sustainable practice due to different methods of inoculation and no visual changes in the crop growth immediately as in the case of inorganic fertilizers (Mazid and Khan, 2014).

19.13 Future prospects

In spite of different ecological niches and multiple functional properties, development of commercial bioinoculants from PSMs faces a lot of challenges. Recent developments within the areas of purposeful diversity, rhizosphere colonizing ability, mode of actions, and considered application can aid in their use in sustainable agricultural systems. Thus, for obtaining high yields to feed the burgeoning population, the use of PSMs is must. Nowadays, consumers are much aware about the quality aspects of food-like organoleptic and nutritional properties. Future studies must aware of dealing with plant-microbe interactions, mode of actions, and flexibility to situations below excessive environments; enhancing the efficacy of biofertilizers; stabilizing microbes in the soil system; and reducing the pesticide applications. Development of the most efficient microbial inoculants, through the genetic manipulation, needs to be studied for the sustainable fastening the P cycle.

19.14 Conclusion

Application of phosphorus biofertilizers in agricultural ecosystems serves as supplementary, renewable, and ecofriendly source of supply, or making available the limited/finite P resource. Since they need the flexibility to remodel nutritionally necessary parts from nonusable to extremely digestible forms without any hazardous consequences on natural environment, they are imperative for an integrated plant nutrient system (IPNS). They aid in sustaining soil fertility and crop productivity on a higher scale, which is compulsory to achieve farming sustainability. They also overcome stumbling blocks arising from the increasing demand of the global population for food and fiber and also from the rampant chemical utilization in agro-ecosystems. The appliance of phosphatic biofertilizers shall therefore be considered within the ambit of the farming system during which they are applied, supporting synergistic interactions with totally different scientific discipline practices affecting soil's physiochemical conditions (pH, water availability, salinity, and organic matter), minimum tillage or precision agriculture irrigation, and pest and disease control.

References

Adeleke, R.A., Raimi, A.R., Roopnarain, A., Mokubedi, S.M., 2019. Status and prospectsof bacterial inoculants for sustainable management of agroecosystems. In: Giri, B., Prasad, R., Wu, Q.-S., Varma, A. (Eds.), Biofertilizers for Sustainable Agricultureand Environment. Soil Biol, vol. 55, pp. 137–172.

Alagawadi, A.R., Gaur, A.C., 1988. Associative effect of isoluble phosphates by some soil fungi isolated from nursery seed beds. Can. J. Microbiol. 16, 877–880.

Alori, E.T., Glick, B.R., Babalola, O.O., 2017. Microbial phosphorus solubilization and itspotential for use in sustainable agriculture. Front. Microbiol. 8, 971. 1–8.

Apel, A.K., Alberto Sola-Landa, A., Rodríguez-García, A., Martín, J.F., 2007. Phosphate control of phoA, phoC and phoD gene expression in *Streptomyces coelicolor* reveals significant differences in binding of PhoP to their promoter regions. Microbiology 153 (10), 3527–3537.

Arun, K.S., 2007. Bio-Fertilizers for Sustainable Agriculture. Mechanism of P-Solubilization. vol. 196 Agribios Publishers, p. 197.

Asea, P.E.A., Kucey, R.M.N., Stewart, J.W.B., 1988. Inorganic phosphate solubilisation by two *Penicillium* species in solution culture and soil. Soil Biol. Biochem. 20, 459–464.

Boitt, G., Simpson, Z.P., Tian, J., Black, A., Wakelin, S.A., Condron, L.M., 2018. Plant biomass management impacts on short-term soil phosphorus dynamics in a temperate grassland. Biol. Fertil. Soils 54, 397–409.

Cabello, M., Irrazabal, G., Bucsinszky, A.M., Saparrat, M., Schalamuk, S., 2005. Effect of an arbuscular mycorrhizal fungus, *Glomus mosseae*, and a rock-phosphate-solubilizing fungus, *Penicillium thomii*, on *Mentha piperita* growth in a soilless medium. J. Basic Microbiol. 45 (3), 182–189.

Castro-Sowinski, S., Herschkovitz, Y., Okon, Y., Jurkevitch, E., 2007. Effects of inoculationwith plant growth-promoting rhizobacteria on resident rhizosphere microorganisms. FEMS Microbiol. Lett. 276, 1–11.

Celador-Lera, L., Jimenez-Go'mez, A., Menendez, E., Rivas, R., 2018. Biofertilizers basedon bacterial endophytes isolated from cereals: potential solution to enhance these crops. In: Meena, V.S. (Ed.), Role of Rhizospheric Microbes in Soil Volume 1 Stress Management and Agricultural Sustainability, pp. 175–203.

Chabot, R., Beauchamp, C.J., Kloepper, J.W., Antoun, H., 1998. Effect of phosphorus on root colonization and growth promotion of maize by bioluminescent mutants of phosphate-solubilizing *Rhizobium leguminosarumbiovar phaseoli*. Soil Biol. Biochem. 30 (12), 1615–1618.

Chaudhry, V., Nautiyal, S.C., 2011. A high throughput method and culture medium for rapid screening of phosphate accumulating microorganisms. Bioresour. Technol. 102, 57–62.

Cheesbrough, M., 2000. District Laboratory Practise in Tropical Countries. Cambridge University Press, Cambridge.

Choi, O., Kim, J., Kim, J.G., Jeong, Y., Moon, J.S., Park, C.S., 2008. Pyrroloquinoline quinone is a plant growth promotion factor produced by *Pseudomonas fluorescens*. Plant Physiol. 146, 657–668.

Claus, D., Berkery, R.C.W., 1986. Genus pseudomonas. In: PHA, S., Mair, N.S., Sharpe, M.E. (Eds.), Bergeys Manual of Systematic Bacteriology. vol 1, pp. 140–219.

Collavino, M.M., Sansberro, P.A., Mroginski, L.A., Aguilar, O.M., 2010. Comparison of in vitro solubilization activity of diverse phosphate-solubilizing bacteria native to acid soil and their ability to promote Phaseolus vulgaris growth. Biol. Fertil. Soils 46, 727–738.

Collee, J.G., Miles, R.S., Watt, B., 1996. Test for the identification of bacteria. In: Collee, J.G., Faser, A.G., Marmion, B.P., Simmons, A. (Eds.), Mackie and McCartney practical medical microbiology, fourteenth ed. Churchill Livingstone, London, pp. 131–145.

Das, D., Dwivedi, B.S., Meena, M.C., Singh, V.K., Tiwari, K.N., 2015. Integrated nutrient management for improving soil health and crop productivity. Indian J. Fert. 11, 64–83.

Dash, N., Pahari, A., Dangar, T.K., 2017. Functionalities of phosphate-solubilizing bacteria of rice rhizosphere: techniques and perspectives. In: Shukla, P. (Ed.), Recent Advances in Applied Microbiology, pp. 151–163.

Dastager, S.G., Deepa, C.K., Pandey, A., 2010. Isolation and characterization of novel plant growth promoting Micrococcus sp NII-0909 and its interaction with cowpea. Plant Physiol. Biochemist 48 (12), 987–992.

Directorate of Economics and Statistics (DES), 2019. Ministry of Agriculture, India.

Duebel, A., Gransee, A., Merbach, W., 2000. Transformation of organic rhizo deposits by rhizoplane bacteria and its influence on the availability of tertiary calcium phosphate. J. Plant Nutr. Soil Sci. 163, 387–392.

Ekin, Z., 2010. Performance of phosphate solubilizing bacteria for improving growth and yield of sunflower (*Helianthus annuus L.*) in the presence of phosphorus fertilizer. Afr. J. Biotechnol. 9 (25), 3794–3800.

Elias, F., Woyessa, D., Muleta, D., 2016. Phosphate solubilization potential of rhizospherefungi isolated from plants in Jimma zone, Southwest Ethiopia. Int. J. Microbiol. 54726, 1–11.

Fernandez, L.A., Zalba, P., Gomez, M.A., Sagardoy, M.A., 2007. Phosphate-solubilization activity of bacterial strains in soil and their effect on soybean growth under greenhouse conditions. Biol. Fertil. Soils 43, 805–809.

Ferreira, C.M., Vilas-Boas, A., Sousa, C.A., Soares, H.M., Soares, E.V., 2019. Comparison of five bacterial strains producing siderophores with ability to chelate iron under alkaline conditions. AMB Express 9, 1–12.

Fraser, T.D., Lynch, D.H., Bent, E., Entz, M.H., Dunfield, K.E., 2015. Soil bacterial *phoD* gene abundance and expression and long-term management. Soil Biol. Biochem. 88, 137–147.

Fuentes-Ramirez, L.E., Caballero-Mellado, J., 2005. Bacterial biofertilizers. In: Siddiqui, Z.A. (Ed.), PGPR: Biocontrol and Biofertilization. Springer, pp. 143–172.

Garcı́a-Fraile, P., Menendez, E., Lera, L.C., Dı́ez-Mendez, A., Jimenez-Go'mez, A., Marcos-Garcı́a, M., Cruz-Gonza'lez, X.A., Martínez-Hidalgo, P., Mateos, P.F., Rivas, R., 2017. Bacterial probiotics: a truly green revolution. In: Kumar, V. (Ed.), Probiotics and Plant Health, pp. 131–162.

Goldstein, A.H., Liu, S.T., 1987. Molecular cloning and regulation of a mineral phosphate solubilizing gene from *Erwinia herbicola*. Biotechnology 5, 72–74.

Gressel, N., McColl, J.G., Preston, C.M., Newman, R.H., Powers, R., 1996. Linkages between phosphorus transformations and carbon decomposition in a forest soil. Biogeochemistry 33, 97–123.

Guo, X., Van Iersel, M.W., Chen, J., Brackett, R.E., Beuchat, L.R., 2002. Evidence of association of salmonellae with tomato plants grownhydroponically in inoculated nutrient solution. Appl. Environ. Microbiol. 68, 3639–3643.

Hamdali, H., Hafidi, M., Virolle, M.J., Ouhdouch, Y., 2008. Rock phosphate-solubilizing Actinomycetes: screening for plant growth-promoting activities. World J. Microbiol. Biotechnol. 24, 2565–2575.

Hanif, M.K., Hameed, S., Imran, A., Naqqash, T., Shahid, M., Van Elsas, J.D., 2015. Isolation and characterization of a β-propeller gene containing phosphor bacterium *Bacillus subtilis* strain KPS-11 for growth promotion of potato (*Solanum tuberosum* L.). Front. Microbiol. 6, 583.

Hu, Y., Xia, Y., Sun, Q., Liu, K., Chen, X., Ge, T., 2018. Effects of long-term fertilization on *phoD*-harboring bacterial community in karst soils. Sci. Total Environ. 62, 53–63.

Huang, L.M., Jia, X.X., Zhang, G.L., Shao, M.A., 2017. Soil organic phosphorus transformation during ecosystem development: a review. Plant Soil 417, 17–42.

Huang, H., Shi, P., Wang, Y., Luo, H., Shao, N., Wang, G., 2009. Diversity of beta-propeller phytase genes in the intestinal content of grass crap provides insight into the release of major phosphorus from phytate in nature. Appl. Environ. Microbiol. 75, 1508–1516.

Irshad, U., Yergeau, E., 2018. Bacterial subspecies variation and nematode grazing change P dynamics in the wheat rhizosphere. Front. Microbiol. 9, 1990.

Ishige, K., Noguchi, T., 2000. Inorganic polyphosphate kinase and adenylate kinase participate in the polyphosphate: AMP phosphotransferase activity of *Escherichia coli*. Proc. Natl. Acad. Sci. U. S. A. 97, 14168–14171.

Javorekova, S., Makova, J., Medo, J., Kovacsova, S., Charousova, I., Horak, J., 2015. Effect of bio-fertilizers application on microbial diversity and physiological profiling of microorganisms in arable soil. Euro. J. Soil Sci. 4, 54–61.

Jilani, G., Akram, A., Ali, R.M., Hafeez, F.Y., Shamsi, I.H., Chaudhry, A.N., Chaudhry, A.G., 2007. Enhancing crop growth, nutrients availability, economics and beneficial rhizosphere microflora through organic and biofertilizers. Ann. Microbiol. 57 (2), 177–184.

Kelly, R.M., Edwards, D.G., Thompson, J.P., Magarey, R.C., 2001. Responses of sugarcane, maize, and soybean to phosphorus and vesicular–arbuscular mycorrhizal fungi. Aust. J. Agric. Res. 52 (7), 731–743.

Khan, K.S., Joergensen, R.G., 2009. Changes in microbial biomass and P fractions in biogenic household waste compost amended with inorganic P fertilizers. Bioresour. Technol. 100, 303–309.

Kim, Y.K., Jordan, D., McDonald, A.G., 1997a. Effect of phosphate-solubilizing bacteria and vesicular-arbuscular mycorrhizae on tomato growth and soil microbial activity. Biol. Fertil. Soils 26, 79–87.

Kim, K.Y., McDonald, G.A., Jordan, D., 1997b. Solubilization of hydroxypatite by Enterobacter agglomerans and cloned Escherichia coli in culture medium. Biol. Fertil. Soils 24, 347–352.

Kishore, N., Pindi, P.K., Reddy, S.R., 2015. Phosphate solubilizing microorganisms: a critical review. In: Bahadur, B., Venkat Rajam, M., Sahijram, L., Krishnamurthy, K. (Eds.), Plant Biology and Biotechnology, pp. 307–333.

Krishnaraj, P.U., Goldstein, A.H., 2001. Cloning of a Serratia marcescens DNA fragment that induces quino protein glucose dehydrogenase-mediated gluconic acid production in Escherichia coli in the presence of stationary phase Serratia marcescens. FEMS Microbiol. Lett. 205, 215–220.

Kucey, R.M.N., 1983. Phosphate-solubilizing bacteria and fungi in various cultivated and virgin Alberta soils. Can. J. Soil Sci. 63, 671–678.

Kudashev, I.S., 1956. The effect of phosphor bacterin on the yield and protein content in grains of autumn wheat, maize and soybean. Doki. Akad. Skh. Nauk. 8, 20–23.

Kumar, M.S., Reddy, G.C., Phogat, M., Korav, S., 2018. Role of bio-fertilizers towards sustainable agricultural development: a review. J. Pharmacogn. Phytochem. 7, 1915–1921.

Kumar, R., Shastri, B., 2017. Role of phosphate-solubilising microorganisms in sustainable agricultural development. In: Singh, J., Seneviratne, G. (Eds.), Agro-Environmental Sustainability, pp. 271–303.

Li, J., Dittrich, M., 2019. Dynamic polyphosphate metabolism in cyanobacteria responding to phosphorus availability. Environ. Microbiol. 21, 572–583.

Liang, J.L., Liu, J., Jia, P., Yang, T.T., Zeng, Q.W., Zhang, S.C., Li, J.T., 2020. Novel phosphate-solubilizing bacteria enhance soil phosphorus cycling following ecological restoration of land degraded by mining. ISME J. 14 (6), 1600–1613.

Liu, Y.Q., Wang, Y.H., Kong, W.L., Liu, W.H., Xie, X.L., Wu, X.Q., 2020. Identification, cloning and expression patterns of the genes related to phosphate solubilization in Burkholderiamultivorans WS-FJ9 under different soluble phosphate levels. AMB Express 10, 1–11.

Lucy, M., Reed, E., Glick, B.R., 2004. Applications of free living plant growth-promoting rhizobacteria. Antonie Van Leeuwenhoek 86, 1–25.

Machi, S., 2006. Manual on Biofertilizer Production and Application. J.A.I.F, pp. 1–138.

Mahdi, S.S., Hassan, G.I., Samoon, S.A., Rather, H.A., Dar, S.A., Zehra, B., 2010. Biofertilizers in organic agriculture. J. Phytol. 2, 42–54.

Malviya, J., Singh, K., 2012. Characterization of novel plant growth promoting and biocontrol strains of fluorescent pseudomonads for crop. Int. J. Med. Res. 1, 235–244.

Marra, L.M., de Oliveira-Longatti, S.M., Soares, C.R.F.S., Olivares, F.L., de Souza Moreira, F.M., 2019. The amount of phosphate solubilization depends on the strain, C-source, organic acids and type of phosphate. Geomicrobiol J. 36, 232–242.

Mathur, N., Singh, J., Bohra, S., Bohra, A., Vyas, A., 2010. Microbes as biofertilizers. In: Tripathi, G. (Ed.), Cellular and Biochemical Science, pp. 1089–1113.

Mazid, M., Khan, T.A., 2014. Future of bio-fertilizers in Indian agriculture: an overview. Int. J. Agric. Food Res. 3, 10–23.

Mehnaz, S., 2016. An overview of globally available bioformulations. In: Arora, N.K. (Ed.), Bioformulations: For Sustainable Agriculture, pp. 267–281.

Motghare, H., Gauraha, R., 2012. Biofertilizers- Types and their Application. Krishi Sewa.

Neal, A.L., Rossmann, M., Brearley, C., Akkari, E., Guyomar, C., Clark, I.M., 2017. Land-use influences phosphatase gene micro diversity in soils. Environ. Microbiol. 19, 2740–2753.

Ngampimol, H., Kunathigan, V., 2008. The study of shelf life for liquid biofertilizer from vegetable waste. AU J.T. 11 (4), 204–208.

Nico, M., Ribaudo, C.M., Gori, J.I., Cantore, M.L., Curá, J.A., 2012. Uptake of phosphate and promotion of vegetative growth in glucose-exuding rice plants (Oryza sativa) inoculated with plant growth-promoting bacteria. Appl. Soil Ecol. 61, 190–195.

Nosheen, S., Ajmal, I., Song, Y., 2021. Microbes as biofertilizers, a potential approach for sustainable crop production. Sustainability 13, 1868.

Ochoa-Loza, F.J., Artiola, J.F., Maier, R.M., 2001. Stability constants for the complexation of various metals with a rhamnolipid biosurfactant. J. Environ. Qual. 30, 479–485.

Oteino, N., Lally, R.D., Kiwanuka, S., Lloyd, A., Ryan, D., Germaine, K.J., 2015. Plant growth promotion induced by phosphate solubilizing endophytic *Pseudomonas* isolates. Front. Microbiol. 6, 745.

Pal, S., Singh, H.B., Farooqui, A., Rakshit, A., 2015. Fungal biofertilizers in Indian agriculture: perception, demand and promotion. J. Eco-Friend Agric. 10, 101–113.

Panhwar, Q.A., Othman, R., Rahman, Z.A., Meon, S., 2012. Isolation and characterization of phosphate-solubilizing bacteria from aerobic rice. Afr. J. Biotechnol. 11, 2711–2719.

Pathak, A.K., Christopher, K., 2019. Study of socio-economic condition and constraints faced by the farmers in adoption of biofertilizer in Bhadohi district (Uttar Pradesh). J. Pharmacogn. Phytochem. 8, 1916–1917.

Perez, E., Sulbaran, M., Ball, M.M., Yarzabal, L.A., 2007. Isolation and characterization of mineral phosphate-solubilizing bacteria naturally colonizing a limonitic crust in the south-eastern Venezuelan region. Soil Biol. Biochem. 39 (11), 2905–2914.

Pittroff, S., Olsson, S., Ashlea Doolettee, R., Greinerd, Richardson, A.E., Nicolaisen, M., 2020. A novel microcosm for recruiting inherently competitive biofertilizer candidate microorganisms from soil environments. biorxiv, 1–38.

Qureshi, M.A., Ahmad, Z.A., Akhtar, N., Iqbal, A., Mujeeb, F., Shakir, M.A., 2012. Role of phosphate solubilizing bacteria (PSB) in enhancing P availability and promoting cotton growth. J. Anim. Plant Sci. 22 (1), 204–210.

Rai, M.K. (Ed.), 2006. Handbook of Microbial Biofertilizers. The Haworth Press, Inc. 13904–1580.

Ranjan, A., Mahalakshmi, M.R., Sridevi, M., 2013. Isolation and characterization ofphosphate-solubilizing bacterial species from different crop fields of Salem, TamilNadu, India. Int. J. Nutr., Pharmacol., Neurol. Dis. 3, 29–33.

Rasul, M., Yasmin, S., Suleman, M., Zaheer, A., Reitz, T., Tarkka, M.T., 2019. Glucose dehydrogenase gene containing phosphobacteria for biofortification of phosphorus with growth promotion of rice. Microbiol. Res. 223–225, 1–12.

Rawat, P., Das, S., Shankhdhar, D., Shankhdhar, S., 2020. Phosphate-solubilizing microorganisms: mechanism and their role in phosphate solubilization and uptake. Soil Sci. Plant Nutr., 1–20.

de Salamone, I.E.G., Di Salvo, L.P., Ortega, J.S.E., Sorte, P.M.F.B., Urquiaga, S., Teixeira, K.R.S., 2010. Field response of rice paddy crop to Azospirillum inoculation: physiology of rhizosphere bacterial communities and the genetic diversity of endophyticbacteria in different parts of the plants. Plant Soil 336, 351–362.

Saxena, S., 2015. Agricultural applications of microbes biofertilisers and biopesticides. Appl. Microbiol., 37–54.

Servin, P., Antoun, H., Taktek, S., de-Bashan, L., 2020. Designing a multi-species inoculant of phosphate rock-solubilizing bacteria compatible with arbuscular mycorrhizae for plant growth promotion in low-P soil amended with PR. Biol. Fertil. Soils 56, 521–536.

Sharma, S., Gaur, R.K., Choudhary, D.K., 2013. Solubilization of inorganic phosphate (Pi) and plant growth-promotion (PGP) activities by root-nodule bacteria isolated from cultivated legume, mothbean (*Vigna aconitifolia L.*) of the great Indian Thar desert. Res. J. Biotechnol. 8, 4–10.

Sharma, S.B., Sayyed, R.Z., Trivedi, M.H., Gobi, T.A., 2013a. Phosphate solubilizing microbes: sustainable approach for managing phosphorus deficiency in agricultural soils. SpringerPlus 2, 587–600.

Sharma, S.B., Sayyed, R.Z., Trivedi, M.H., Gobi, T.A., 2013b. Phosphate solubilizing microbes: sustainable approach for managing phosphorus deficiency in agricultural soils. SpringerPlus 2 (1), 1–14.

Shrivastava, M., Srivastava, P.C., D'Souza, S.F., 2018. Phosphate-solubilizing microbes: diversity and phosphates solubilization mechanism. In: Meena, V.S. (Ed.), Role of Rhizospheric Microbes in Soil Volume 2: Nutrient Management and Crop Improvement. Springer Nature, pp. 137–165.

Siczek, A., Lipiec, J., 2016. Impact of faba bean-seed rhizobial inoculation on microbialactivity in the rhizosphere soil during growing season. Int. J. Mol. Sci. 17, 1–9.

Silambarasan, S., Logeswari, P., Cornejo, P., Kannan, V.R., 2019. Evaluation of the production of exopolysaccharide by plant growth promoting yeast *Rhodotorula* sp. strain CAH2 under abiotic stress conditions. Int. J. Biol. Macromol. 121, 55–62.

Singh, M., Dotaniya, M.L., Mishra, A., Dotaniya, C.K., Regar, K.L., Lata, M., 2016. Role of biofertilizers in conservation agriculture. In: Bisht, J.K., et al. (Eds.), Conservation Agriculture, pp. 113–134.

Sinha, R.K., Valani, D., Chauhan, K., Agarwal, S., 2014. Embarking on a second green revolution for sustainable agriculture by vermiculture biotechnology using earthworms: reviving the dreams of sir Charles Darwin. J. Agric. Biotechnol. Sustain. Dev. 1, 113–128.

Somasegaran, P., Hoben, J., 1994. Hand Book for Rhizobia. Methods in legume—Rhizobium Technology for Research and Application. Springer, Verlag, New York.

Son, H.J., Park, G.T., Cha, M.S., Heo, M.S., 2006. Solubilization of insoluble inorganic phosphates by a novel salt-and pH-tolerant *Pantoeaagglomerans* R-42 isolated from soybean rhizosphere. Bioresour. Technol. 97 (2), 204–210.

Sperber, J.I., 1958. The incidence of apatite-solubilizing organisms in the rhizosphere and soil. Aust. J. Agric. Res. 9, 778–781.

Sujatha, S., Sirisham, S., Reddy, S.M., 2004. Phosphate solubilization by thermophilic microorganisms. Indian J. Microbiol. 44 (2), 101–104.

Suleman, M., Yasmin, S., Rasul, M., Yahya, M., Atta, B.M., Mirza, M.S., 2018. Phosphate solubilizing bacteria with glucose dehydrogenase gene for phosphorus uptake and beneficial effects on wheat. PLoS One 13.

Sundara, B., Natarajan, V., Hari, K., 2002. Influence of phosphorus solubilizing bacteria on the changes in soil available phosphorus and sugarcane and sugar yields. Field Crop Res. 77 (1), 43–49.

Toro, M., 2007. Phosphate solubilizing microorganisms in the rhizosphere of native plants from tropical savannas: An adaptive strategy to acid soils? In: Velazquez, C., Rodriguez-Barrueco, E. (Eds.), Developments in Plant and Soil Sciences. Springer, pp. 49–252.

Trabelsi, D., Ben Ammar, H., Mengoni, A., Mhamdi, R., 2012. Appraisal of the croprotation effect of rhizobial inoculation on potato cropping systems in relation to soil bacterial communities. Soil Biol. Biochem. 54, 1–6.

Trimurtulu, N., Rao, D.L.N., 2014. Liquid Microbial Inoculants and Their Efficacy on Field Crops. Agricultural Research Station., p. 54.

Turner, B.L., Richardson, A.E., Mullaney, E.J., 2007. Inositol Phosphates: Linking Agriculture and the Environment. CAB International, Wallingford, p. 304.

Uribe, D., Sanchez-Nieves, J., Vanegas, J., 2010. Role of microbial biofertilizers in thedevelopment of a sustainable agriculture in the tropics. In: Dion, P. (Ed.), Soil Biology and Agriculture in the Tropics, pp. 235–250.

Verma, J.P., Yadav, J., Tiwari, K.N., Kumar, A., 2013. Effect of indigenous *Mesorhizobium* spp. and plant growth promoting rhizobacteria on yields and nutrients uptake of chickpea (*Cicer arietinum L.*) under sustainable agriculture. Ecol. Eng. 51, 282–286.

Vessey, J.K., 2003. Plant growth promoting rhizobacteria as biofertilizers. Plant Soil 255, 571–586.

Wagh, J., Shah, S., Bhandari, P., Archana, G., Kumar, G.N., 2014. Heterologous expression of pyrroloquinoline quinone (*pqq*) gene cluster confers mineral phosphate solubilization ability to *Herbaspirillumseropedicae* Z67. Appl. Microbiol. Biotechnol. 98, 5117–5129.

Walpola, B.C., Yoon, M.H., 2013. Isolation and characterization of phosphate solubilizing bacteria and their co-inoculation efficiency on tomato plant growth and phosphorous uptake. Afr. J. Microbiol. Res. 7 (3), 266–275.

Wan, W., Qin, Y., Wu, H., Zuo, W., He, H., Tan, J., He, D., 2020. Isolation and characterization of phosphorus solubilizing bacteria with multiple phosphorus sources utilizing capability and their potential for lead immobilization in soil. Front. Microbiol. 11.

Wang, J., Li, Q., Xu, S., Zhao, W., Lei, Y., Song, C., Huang, Z., 2018. Traits-based integration of multi-species inoculants facilitates shifts of indigenous soil bacterial community. Front. Microbiol. 9, 1–13.

White, A.K., Metcalf, W.W., 2007. Microbial metabolism of reduced phosphorus compounds. Annu. Rev. Microbiol. 61, 379–400.

Yahya, A., Azawi, S.K.A., 1998. Occurrence of phosphate solubilizing bacteria in some Iranian soils. Plant Soil 117, 135–141.

Yang, M.Y., Wang, C.H., Wu, Z.H., Yu, T., Sun, M.H., Liu, J.J., 2016. Phosphorus dissolving capability, glucose dehydrogenase gene expression and activity of two phosphate solubilizing bacteria. Acta Microbiol Sin. 56 (04), 651–663.

Yeh, D.M., Lin, L., Wright, C.J., 2000. Effects of mineral nutrient deficiencies on leafdevelopment, visual symptoms and shoot-root ratio of *Spathiphyllum*. Sci. Hortic. 86, 223–233.

Zhang, Q., Saleem, M., Wang, C., 2017. Probiotic strain *Stenotrophomonas acidaminiphila* BJ1 degrades and reduces chlorothalonil toxicity to soil enzymes, microbial communities and plant roots. AMB Express 7, 1–8.

Trichoderma: Improving growth and tolerance to biotic and abiotic stresses in plants

Bahman Fazeli-Nasab[a], Laleh Shahraki-Mojahed[b], Ramin Piri[c], and Ali Sobhanizadeh[d]

[a]Research Department of Agronomy and Plant Breeding, Agricultural Research Institute, University of Zabol, Zabol, Iran, [b]Department of Biochemistry, School of Medicine, Zabol University of Medical Sciences, Zabol, Iran, [c]Department of Agronomy and Plant Breeding, Faculty of Agriculture, University of Tehran, Tehran, Iran, [d]Department of Horticultural Science, Faculty of Agriculture, University of Zabol, Iran

20.1 Introduction

Among the latest strategies used for improvement of growth and increase in seed germination as well as control of seed surface pathogens is seed priming with beneficial biotic organisms. Recently, the process of biosorption with string fungi has been widely considered for its easy cultivation and nonpathogenicity for humans and animals (Kavamura and Esposito, 2010; Rangel et al., 2018). Among them, different species of *Trichoderma* spp., commonly present in all soils and around plant roots, are among the most common cultivable fungi of special importance and are produced as soil modifiers on commercial scales (Sun et al., 2010; Tashakori Fard et al., 2017). Their main activities include biological control against soil pathogens, production of growth hormones (Sun et al., 2010; Tashakori Fard et al., 2017), soluble elements, increased absorption and transport of nutrients, detoxification, increased transfer of sugar and amino acids in plant roots, and enhanced induction resistance to environmental stresses and can increase the growth and development of plants (Mazhabi et al., 2011).

Increase in production can result from minimizing losses caused by weak seedlings, poor growing conditions, plant diseases, and chemical cost of disease control. Plant diseases, especially root diseases, cause significant losses in the production of plants such as tomatoes (Anhar et al., 2019; Yu et al., 2021). The fungus that promotes plant growth and *Trichoderma* antagonism has many strains with high value in agriculture since they act as biological control agents (BCA). It is well proved that due to their biological control capacity, they can be used for a wide range of plant pathogens (Harman et al., 2004b; Sood et al., 2020) as well as for plant growth in greenhouse and field conditions (Mayo-Prieto et al., 2020). Plant inoculation with *Trichoderma* species can provide conditions for refining soils with multiple contaminants, increasing plant growth, and improving soil fertility (Cao et al., 2008). Increased absorption of nutrients caused by the activity of *Trichoderma* fungi can also increase the growth and vigor of plants and make them resistant to pathogens (Singh et al., 2007; Naeimi et al., 2020).

Considering the great significance of using biological fertilizers contributing to sustainable agriculture, in this study, it is tried to evaluate the potential of *Trichoderma* in improving the growth and productivity of plants and its role in lowering biotic and abiotic stresses.

20.2 Trichoderma

The program of using various species of *Trichoderma* spp. as a biocontrol factors was initially introduced and analyzed in 1932 (Weindling, 1932). Since then, numerous researches have been done on this field and different commercial compounds have been produced in various forms and ways such as propagation in planting furrows, seed cover, and use in root area (Worrall et al., 2018; Singh et al., 2019b). Antagonistic fungi, *Trichoderma* spp., was first introduced in 1974, and based on Alexopoulos, classification was categorized in order of Moniliales and Hyphomycetes family (Alexopoulos et al., 1996). The soil fungus is one of the main components of soil mycoflora. On the contrary, a vast range of infectious factors which can produce extracellular enzymes such as amylolytic, pectolytic, proteolytic, lipolytic, and cellulitic have a great competition potential (Loc et al., 2020).

Trichoderma species are in fact free-biotic fungi that mostly live as saprophytes on soil debris. By applying various mechanisms such as mycoparasitism, antibiosis, competition for food and space, stimulation of plant resistance mechanisms, stimulation of plant growth and development, change of environmental conditions, especially the rhizosphere, and increase in solubility of mineral elements as the plants absorb them, they can exert their biocontrol activity (Negi et al., 2019; Swain et al., 2021). *Trichoderma* species can also live as parasites on other fungi and colonize the plant's roots and rhizospheres. They also produce a large number of hydrolytic enzymes which are of great significance in industries (Konappa et al., 2020b). However,

Trichoderma are among the microorganisms that are highly resistant to toxins and chemicals (natural and synthetic) and can also catalyze some of these compounds like hydrocarbons, chlorophenolics, polysaccharides, and xenobiotic pesticides. Some strains of *Trichoderma* are both strong producers of antibiotics and enzymes considered as opportunistic invaders with rapid growth and high spore production. All these characteristics have made them very ecologically successful, so they are significantly present in agricultural soils and meadows, forests, swamps, deserts, and in climatic regions such as the tundra, Antarctica, equator, as well as lakes and dead plant remains (Monte, 2001; Sasidharan et al., 2020).

Trichoderma fungi are a group of soil microorganisms which has a major role in conversion of organic residues and compost production, can easily and quickly ferment, can decompose hemicellulose and lignin, and are also useful in compost production. These fungi have increased the expression of auxin-regulating genes. In addition, the number of lateral roots in inoculated plants was more than four times that of inoculated plants. . Mutations in genes involved in auxin transport or messaging can further reduce the effects of *Trichoderma* on root growth and development (Gad et al., 2020; Yu et al., 2021).

Moreover, *Trichoderma* fungi are involved in both regulating plant growth and activating its immune response. For example, harzianolide, 6-phenylphaprone, has auxin-like activity and besides activating the plant defense response; it stimulates plant growth, so *Trichoderma* strains should be applied to colonize plant roots prior to stimulating plant growth and protection against pollutants. Such colonization implies the ability to adhere to and detect plant roots, to penetrate the plant, and to tolerate toxic metabolites produced by the plant in response to attacks made by a foreign pathogenic or nonpathogenic organism (Speckbacher et al., 2020)

Some strains of *Trichoderma* can colonize plant roots and penetrate the epidermis in long runs. The best strains can grow along with the plant roots to provide long-term beneficial effects. As an example, treatment of maize seeds with *T. harzianum* (T22) leads to growth and yield which in turn can provide more plant growth and yield, more green leaves, increased rooting in mature plants, and more efficient use of nitropene fertilizers. *Trichoderma* seed treatments can further affect plant yield in long term, since the fungus is considerably compatible with the fugitive-root space and is able to survive as a root symbiosis for a long time (Harman et al., 2021; Küçük et al., 2019). Infection of plant tissues with these microorganisms can also induce cell division by cytokines and facilitate formation of special structures like mycorrhiza root fungi, *Trichoderma*, and *Pseudomonas* fluorescence that has a bilateral coexistence with the plant. Application of *Trichoderma* fungi in agriculture has four advantages: (1) it can settle the plant roots, (2) *Trichoderma* fungi can control plant pathogens, (3) it improves plant health by increasing plant growth, and (4) it may stimulate root and plant growth (Harman et al., 2008, 2021; Mastouri et al., 2010).

Although *Trichoderma* can normally grow fast in different culture media, this growth rate varies from species to species (Loc et al., 2020; Gad et al., 2020). Among *Trichoderma* species, *T. harzianum* is the most universally known biocontrol agents mainly due to its rapid growth; considerable reproductive speed; high tolerance to adverse conditions; potential to use different food sources; capability to grow and colonize in plant roots; high invasion against pathogenic factors; utilization of various antagonistic mechanisms such as competition, parasitism, and antibiosis; ability to modify rhizosphere; efficiency in stimulating growth; and ultimately induction of resistance in plants (Gad et al., 2020; Yan and Khan, 2021). As mentioned earlier, activity of such fungi is potentially antagonistic. In other words, antagonists can induce resistance in plants in two paths SAR and ISR which are, respectively, affected by salicylic acid and ethylene and jasmonate. These antioxidant enzymes are considered as signals used to activate the pathways (Maithani et al., 2021; Kamle et al., 2020).

Some biochemical substances produced by *Trichoderma harzianum* strains are pyrone isonitrile compounds and their various derivatives, Alamethicin, dermadine, trichodermin, and trichotoxin (Reino et al., 2008). An important substance made by these fungi is 6-pentyl-alphapyrone, commonly acting as a growth stimulant for plants in low concentrations. In addition, interpreting the action of plant growth stimulants, researchers concluded that by producing biochemicals, different strains of *Trichoderma* are able to stimulate plant growth or reduce inhibitory effect of some compounds such as biological and chemical toxins in oil and even change contents of soluble elements in soil (Vinale et al., 2008). *T. harzianum* can also build IAA hormone that is involved in plant growth (Anhar et al., 2019). Moreover, the use of *T. harzianum* WKY1 as a dual-purpose biological agent for biological control of anthracnose disease and enhancing plant growth is reported (Chen et al., 2012).

IAA is the most natural and well-known auxin in vascular plants, which plays a significant role in the onset and emergence of lateral roots and stem growth (Sofo et al., 2012). Increased growth by different species of *Trichoderma* depends on the strains used. In fact, the potential of promoting plant growth varies greatly in strains and *Trichoderma* species (Anhar et al., 2019).

Some features of *Trichoderma* fungi such as high food and spatial competitiveness, abundant establishment, and sporulation in soil, especially around the roots of most crops and noncrops, and ability to induce resistance in the plant, can both reduce pathogens in the soil and along with biochemical mechanisms, stimulate the growth of underground or aerial organs of some of these plants. Thus, it is recommended to use their strains in the form of inoculants to improve the plant yield in arid regions, since these bacteria can increase the weight of plants and roots (Harman et al., 2004a; Konappa et al., 2020b).

Usage of *Trichoderma* spp. mixture with soil instead of its spore suspension provided a greater increase in dry weight of biomass (Baker et al., 1984; Raymaekers et al., 2020). Some isolates of *T. harzianum*, used just in 1% concentration in soil, increased the growth of shoots and roots in lettuce (Kamal et al., 2018; Zhang et al., 2020).

20.3 Role of *Trichoderma* in stimulating plant growth

Minimized losses caused by weak seedlings, poor growth condition, plant pathogens, and chemical costs of disease control can increase production rates. Increasing plant growth can be achieved by feeding artificial hormones or by using microbial potential, especially when they are classified as plant growth-promoting fungi (PGPF) (Anhar et al., 2019). Some plant growth-promoting and antagonistic fungi called *Trichoderma* are commonly used, since they are able to biologically control a wide range of plant pathogens. Also, they are useful for plant growth under in vivo and in vitro conditions. However, their efficiency largely depends on the physical, chemical, and biological conditions of the soil. *Trichoderma* species can colonize many plant roots, decompose plant debris, and are significantly involved in biodegradation (Islam et al., 2020).

Special mechanisms of *Trichoderma* make them beneficial fungi which can improve the plant growth, and, on the contrary, inhibit growth in plant pathogenic fungi (Fazeli-Nasab et al., 2019b). Several popular mechanisms useful for plant growth are biological control of soil diseases, production of antibiotics (Saber et al., 2017), penetration into the body of pathogenic fungi, detoxification and increased transfer of sugar and amino acids in plant roots, induction of resistance to environmental stresses, increased nutrient uptake by improving the solubility of the element, secretion of growth hormones and hormone-like, and finally, formation of enzyme cellulase which can directly stimulate ethylene synthesis in the plant as a response to the presence of pathogens (Sriwati et al., 2019; Gravel et al., 2007). Some *Trichoderma*, like *T. harzianum*, can build indole acetic acid (IAA) hormone which is also involved in plant growth (Anhar et al., 2019).

Any increase of plant roots caused by *Trichoderma* fungi can be related to secretion of indole acetic acid. In addition, the in vitro stimulating effect of root growth by these fungi is reported (Casimiro et al., 2001).

Trichoderma fungi are able to accelerate plant growth by raising the level of indole acetic acid (Martínez-Medina et al., 2014). *Trichoderma* fungi enhance the growth and germination indices in radish. Likewise, seed treatment with *Trichoderma* may improve seed condition and plant quality in long term (Anjum et al., 2020).

Trichoderma has also several benefits for agriculture. It is commonly used as a biological control agent against a wide range of plant pathogens and can improve plant growth capacity. Frequently, its stimulating effect on plant growth is observed by modifying soil conditions. Based on reports, increased seedling growth response is caused by the fact that *Trichoderma* strains mostly depend on the ability of survival and establishment in rhizosphere. Besides, it can be inferred from the findings that *Trichoderma*'s effects on seedling growth and vigor mainly contribute to the type/species of treated *Trichoderma* (Islam et al., 2020).

Applying *Trichoderma* strains to soil or using it directly to seedling roots is an economical and effective way to obtain strong tomato seedlings that can be later cultivated on the main fields. Besides its ability to provide an antimicrobial effect, *Trichoderma* can stimulate the biological activity of the resident microbial antagonist population and, consequently, promote plant growth. The development of tomato root system along with building several organic acids (e.g., gluconic, citric, or fumaric acids) in rhizosphere by *Trichoderma* (which reduces soil pH) can increase solubility of insoluble compounds, availability of micronutrients, and uptake of plant nutrients. An increase in absorption of plant nutrients and its transfer from the roots to the vegetative parts accompanied by the resulted plant stimulants can also enhance the agronomic traits of tomato seedlings (Islam et al., 2020).

The mechanisms (both direct and indirect) used by *Trichoderma* strains can further be used to affect seed germination and seedling vigor. In an experiment conducted to evaluate the effects of using three strains of *Trichoderma* including *T. harzianum* TR05, *T. virens* TR06, and *T. asperellum* TR08, the significant effect of these strains on germination, number of true leaves, branch length, root length, and seedling vigor, fresh and dry weights of tomato seedlings was observed compared to control (Islam et al., 2016, 2017).

Moreover, effectiveness of *Trichoderma* varies based on the physical, chemical, and biological conditions of the soil. *Trichoderma* species colonize many plant roots, break down plant debris, and play a significant role in biodegradation. Numerous studies were conducted on the positive and enhancing effects of *Trichoderma* on growth in various plants (Konappa et al., 2020a). Inoculation of maize seeds with *Trichoderma asperellum* for 1.5 h increased seedling length and H^+ ATPase activity (López-Coria et al., 2016). Treatment of wheat seeds with *Trichoderma harzianum* for 30 min also increased plant height, chlorophyll content, and the length of root and tillers (Meena et al., 2016). Treatment of chickpea seeds with *Trichoderma asperellum* in potted conditions improved plant growth indices as well (Singh et al., 2016a).

Treatment with *Trichoderma asperellum* for 24 h on a variety of eggplant, red pepper, guava, okra, squash, and tomato increased seed germination, plant growth indices, chlorophyll, phenylpropanoid, and lignin activity (Singh et al., 2016b). On the contrary, chlorophyll depletion is a negative result of stress on plants, yet such reduction significantly interferes with prevention of light inhibitory damage and reduces the light received by the leaves (Yu et al., 2021; Ahooi et al., 2021). *Trichoderma* can also stimulate vegetative growth and simultaneously transfer carbohydrates to vegetative reservoirs to use them for building chlorophylls (Singh et al., 2019a).

Inoculation of seeds with *Bacillus subtilis* and *Trichoderma harzianum* for 24 h increased germination, number of leaves, stem length, root length, plant height, number of spikes, number of florets per spike, and flowering time in *Antirrhinum majus* (Bhargava et al., 2015). Treatment of soybean seeds with *Trichoderma* fungi can also increase the rate of cumulative emergence of seedlings (Yazdani et al., 2012). Dual application of beneficial microorganisms using a combination of a bacterium (*Pseudomonas chlororaphis* MA342 or *Pseudomonas fluorescens* CHA0) and a fungal strain (*Clonostachys rosea* IK726d11 or *Trichoderma harzianum* T22) on onion and carrot seeds significantly affected the number of the recycled microorganisms on rhizosphere and ultimately increased root and plant growth (Bennett and Whipps, 2008).

Likewise, inoculation of red pepper plant with *Trichoderma viride* for 3 and 12 h increased germination, root and stem length, biomass, and seedling vigor index (Ananthi et al., 2014), but in no research, *Trichoderma* fungi did not increase the dry weight of millet seedlings under drought stress compared to prime treatment (Chepsergon et al., 2014). Seed inoculation with *Trichoderma viride* showed a significant increase in root and shoot growth of chickpea (Dubey et al., 2007). Different species of *Trichoderma* fungi improved plant growth by producing growth hormones such as gibberellin, auxin, and cytokinin. This increased seedling growth of tomato plants was related to synthesis of growth hormones such as indole acetic acid (Gravel et al., 2007).

Infections of plant tissues with *Trichoderma* which induce cell division through cytokinin and formation of special structures represent a bilateral symbiotic relationship with the plant. Meantime, synthesis of cytokinin in the root increases the lateral roots. It can also build chloroplasts with expanded viscosity in leaves so that chlorophyll and photosynthetic enzymes are synthesized more quickly (Sakakibara, 2006). Under drought stress, *Trichoderma* growth-stimulating fungi increased root length in cucumber and bitter gourd more significantly than control (Lo and Lin, 2002). These fungi can produce an auxin-like activity by building harzianolide and 6-phenyl alphapyrone (Vinale et al., 2008). Hence, seemingly, the main reason for increasing root growth in fungal treatments compared to bacterial treatments is an increase in auxin/cytokinin ratio. Similar results were also reported on increased root and shoot length (Dubey et al., 2007) and increased yield (Rojo et al., 2007) made by *Trichoderma harzianum*. As a result, biological treatments promote and activate plant growth and resistance mechanisms, respectively (Harman and Shoresh, 2007).

20.4 Role of *Trichoderma* in increasing germination indices

Rapid seed germination, strong seedling production, and rapid establishment are among the most important parameters in having a plant with high growth ability and remarkable resistance to natural conditions. Such rapid germination and considerable growth and establishment of seedlings in soil expose seeds and seedlings to soil pests and pathogens for a while, yet this critical stage is quickly passed (Anjum et al., 2020; Rezaloo et al., 2020).

Mean germination time and uniformity of germination in primed seeds of rapeseed, wheat, chickpeas, soybeans, alfalfa, corn, sorghum, watermelon, rice, lettuce, and beans which can significantly improve represent accelerated germination and increased seeding, mainly caused by the application of preplanting treatments (Duman, 2006). The secretions of these fungi include a growth-regulating factor which consequently will increase seed germination, plant growth, and nutrient uptake (Khoshmanzar et al., 2020). Inoculation of maize seeds with *Trichoderma* cultivated in soil with no fertilizer had the highest germination rate (82.7%) compared to seeds inoculated with *Trichoderma* in soil containing fertilizer (82.2%) (Okoth et al., 2011; Mehetre and Mukherjee, 2015). Treatment of soybean seeds with *Trichoderma* strains resulted in an increase in germination indices of soybean seedlings compared to control (Tancic, 2013). An increase of germination indices and seedling indices compared to control was also reported (Rahman et al., 2012).

Different species of *Trichoderma* have various mechanisms like solubility of some nutrients, reduced ethylene synthesis, auxin production, and successful competition to promote the efficiency of the growth factors in plants (Contreras-Cornejo et al., 2009, 2016). Impregnation of seeds with *Trichoderma* also makes several metabolic and biochemical changes on behalf of germination. For instance, in the case of these seeds, parts of proteins and carbohydrates are broken down by enzymes and hydrolyzing reactions, so they are ready to act in germination process. This can accelerate germination and reduce mean time of germination (Taylor, 2020; Saddiq et al., 2019).

Combining seed priming with the use of beneficial plant fungi can significantly improve seed germination and emergence, seedling establishment, crop growth, and yield parameters under stress. On the contrary, inoculation of seeds with *Trichoderma* increases the activity of antioxidants and in turn reduces the content of reactive oxygen species. In other words, it is the property of inducing resistance that occurs by this antagonist (Mastouri et al., 2010). Such increase and acceleration of plant germination are attributed to increased levels of indole acetic acid which is caused by *Trichoderma* fungi (Martínez-Medina et al., 2014). Possible reasons for this accelerated germination of treated seeds are increased activity of degrading enzymes such as alpha-amylase, increased bioenergy charge level in the form of increased ATP, increased RNA and DNA synthesis, increased number, and promotion in mitochondria function (Afzal et al., 2002; Li et al., 2017). In addition, significant effects on seed germination and seedling vigor were observed in seeds inoculated by *Trichoderma* spores (Hajieghrari and Mohammadi, 2016).

Despite numerous reports published on positive effects of *Trichoderma* on improvement and promotion of germination and growth indices, there remain conflicts on *Trichoderma* ineffectiveness proposed by some researchers.

For example, using 3-strain *Trichoderma* originated from three different sites of corn medium had no effect on seed germination (Mantja et al., 2015). The germination rate of tomato seeds was not affected by *Trichoderma* application (Azarmi et al., 2011), whereas germination indices of rice seeds inoculated with *Trichoderma* were higher compared to control (Doni et al., 2014), yet, germination of red pepper increased by using *Trichoderma* (Rahman et al., 2012). Among 5 strains of *Trichoderma*, *T. harzianum* IMI 392432 showed a significant content of germination rate in red pepper seeds in both laboratory and field conditions. Next, using *T. harzianum* IMI 392433, *T. harzianum* IMI 392434, *T. virens* IMI 392430, and *T. pseudokoningii* IMI 392431 can increase the germination of rice seeds (Khan et al., 2005).

Paradoxical effect of *Trichoderma* on increasing plant growth, reported even in strains of one species, had a different effect on germination rate of rice seeds (Anhar et al., 2019). Three strains of *Trichoderma* were also able to increase the germination of rice seeds and achieve the desirable standards in high-quality seeds. On the contrary, two-way germination rates were still below standard quality seed standards. Considering 101 types of *Trichoderma* strains and based on their relationship with improvement in the early growth stages of bean seedlings, it was inferred that 60% of the strains could produce IAA or auxin. In fact, the synthesis of each of these metabolites is a characteristic of certain strains, since the potential of producing such metabolites in species varies greatly. Although seven strains of *Trichoderma* significantly improved bean seedling growth, metabolite production varied widely among them, and even in some cases, none of the growth-promoting metabolites was produced (Hoyos-Carvajal et al., 2009).

According to some researchers, the difference in germination percentage of rice seeds is the difference in the potential of each strain to produce phytohormonal compounds like IAA (indole acetic acid). These auxin-like hormones are involved in promoting A-amylase activity. Alpha-amylase plays an essential role in starch motility. It also has a crucial function in seedling growth (Smith et al., 2005). It was found that the type of strain affected the germination rate of rice seeds so significantly that the highest and lowest germination rates in the strain were recorded in rhizosphere of Sizokan Balang cultivar (94%) and in control (74%), respectively (Anhar et al., 2019).

20.5 Significance of seed germination and seedling establishment in seed production

Seed germination is considered as a main factor in modern agriculture, since seed is a reproductive unit that as a string of life ensures the survival of species. In addition, due to the role seeds play in plant establishment, germination is an important stage in plant life. Accordingly, there are different proposed definitions for germination. Seed physiologists defined germination as the removal of root from seed coat (Azad et al., 2017). Seed analysis specialists also explained germination as the emergence and development of essential structures (based on seed type) from the embryo to signify seed ability to produce a natural plant under favorable conditions (Afrouz, 2018; Mehrabi et al., 2011). Others consider germination as the resumption of active fetal growth, which mostly leads to cleft palate and seedling emergence; however, noteworthy that it is impossible to clearly observe germination process. In all aforementioned definitions, a sort of measurement is implicit for seedling development, even though such events occur after germination (Fazeli-Nasab et al., 2016).

As reported in inoculation of different wheat seeds with various strains of *Trichoderma*, such natural stimulants are able to increase longitudinal growth and root growth density through producing a variety of plant hormones and, in turn, can make considerable infection in roots with infectious fungi. As a result, this could improve the efficiency of nitrogen and phosphorus consumption and increase the crop yield (Manske et al., 1995).

20.6 Factors affecting germination

Factors that generally affect germination are subdivided into two groups.

20.6.1 Seed factors

These factors deal with seed properties and, regardless of environment, affect seed germination. Among them, the most important ones are vigor, seed longevity, and dormancy (Afrouz, 2018; Fazeli-Nasab and Amozadeh, 2012).

20.6.2 Environmental factors

These factors include humidity, oxygen, temperature, and light.

20.7 Seedling establishment

Agriculture is a great challenge in most areas suffering issues such as low fertility and numerous environmental stresses such as drought, heavy metal stress, salinity, and high and low temperatures. The first problem to face in process of crop production under such conditions is the concern of proper germination and placement of crop in fields. Obviously, optimal germination followed by suitable establishment in the field can provide the production of acceptable products, both quantitatively and qualitatively (Finch-Savage et al., 2004; Amozadeh and Fazeli-Nasab, 2012). Increasing the speed of emergence and better establishment of the plant causes the plant to make better use of soil moisture, nutrients, and sunlight and ultimately improves the plant's growth potential and yield (Azad et al., 2017).

A great number of factors such as spread of mechanized crops, growth of huge industrial plants, and presence of adverse environmental conditions such as drought, soil salinity, and heavy metal stress necessitate the need for germination and rapid and uniform establishment. Further researches indicate improvement of germination behavior and its related indicators such as mean germination time, seed vigor, root length, stem length, germination rate, and initial establishment in primed seeds (Afrouz, 2018).

20.8 Effect of environmental stresses on seedling establishment

Low crop yield is frequently caused by environmental stresses. In most regions of the world, plants are often exposed to abiotic stresses such as salinity, drought, too high or too low temperatures, heavy metal poisoning, UV rays, and herbicides which can threaten plant production (Ahmad and Prasad, 2011; Shirazi et al., 2016). Due to imitations in land resources, providing healthy food security for the ever-growing population of the world is among the most critical issues that cause the least damage to the environment. Today, increasing industrial activities that result in production of various pollutants such as heavy metals is one of the most serious concerns facing modern man (Fazeli-Nasab et al., 2021; Fazeli-Nasab, 2021).

Stress, defined as an abnormal trend of physiological processes, is caused by the influence of one or more biological and environmental factors. In other words, stress is an organism's exposure to a strong influence of an environmental factor which leads to a failure in its appearance, efficiency, or value (Rasouli et al., 2020). Normally, stress is the excessive pressure of some opposing forces that inhibit a proper activity of natural systems. Stress can also be a factor that affects responses (Shahi et al., 2019). Environmental stresses (e.g., salinity, drought, cold, heat, and heavy elements) are among the most important factors that leave adverse effects on crops such as rapeseed. Environmental stresses may also include biological and physicochemical stresses. Biological stresses are of two kinds: attacks by pests and diseases and competition. Physicochemical stresses are physical, chemical, radiative, temporal, and hydro. Here, stress caused by heavy metals is considered as chemical stress (Forouzandeh et al., 2019; Shabala, 2017; Ahanger et al., 2017).

Heavy metals directly increase the harmful effects of reactive oxygen species by causing rises in their cellular concentration and, at the same time, promote the capacity of cellular antioxidants. Most environmental factors can induce oxidative stress in cells by producing superoxide anions. Moreover, by breaking the electron transfer chain, they contribute to the production of superoxide anions (Fazelienasab et al., 2004; Amozadeh and Fazeli-Nasab, 2012; Fazeli-Nasab et al., 2012).

These elements can affect the activity of NADPH oxidase and cause an increase in O_2^- and H_2O_2, respectively. In turn, increased peroxidase enhances root death, but it can decrease H_2O_2 as well. Finally, root death prevents the uncontrolled entry of elements into the cell (Fazeli-Nasab and Sayyed, 2019).

20.9 Significance of regarding stresses of heavy metals

The industrialization of societies has left numerous consequences and consequences on people's lives, especially in terms of environment. Disregarding the environment and the harmful effects of pollution which were introduced in the mid-20th century have recently brought up so serious issues and concerns in all aspects that special environmental standards have been developed in various fields for industries to be implemented (Rasouli et al., 2020).

Burial and incineration of pollutants have dangerous effects both on the environment and on human health. Physical removal of contaminants and their removal from soil with solvents is very effective in reducing pollutants, yet this method is quite extravagant. Although using a bioreactor can be effective, it requires to transfer contaminated soil from the given site to a reactor which can also cost a lot. However, logically, using plant techniques or plant resilience for greater resistance or higher confrontation with heavy metals is regarded as an effective technology for cleansing such contaminated sites (Afrouz, 2018).

20.10 Application of seed biological treatments (biopriming) in tolerance induction to environmental stresses

Nowadays, various technologies are evaluated by researchers to improve seed quality, increase the percentage, speed, and uniformity of germination, and provide better seedling establishment under adverse environmental conditions. Among these technologies is pretreatment or seed priming. Seed pretreatment is a technique by which seeds are physiologically and biochemically prepared for germination prior to being placed in a culture medium (Ashraf and Foolad, 2005).

The most important kinds of seed priming are osmopriming, hydropriming, halopriming, biopriming, and hormone priming. Hence, in modern sustainable and organic farming systems, using biopriming and biofertilizers is important in increasing crop production and maintaining sustainable soil fertility. The term biofertilizers do not exclusively refer to organic matter derived from animal manure, plant fertilizers, and green manure. It also covers bacterial and fungal microorganisms, especially plant growth-promoting bacteria (PGPR) and materials derived from their activity which are regarded as the most important biofertilizers (Fazeli-Nasab and Sayyed, 2019; Shaban et al., 2015).

The mechanisms which are used by these bacteria and can promote growth are not fully understood; however, among their features, one can generally refer to the ability in producing some growth-promoting hormones, especially auxin, gibberellin, and cytokinin, contribution to nitrogen fixation, control of plant pathogens through production of antibiotics, enzymes and fungicides, mineral phosphorus solubility and organic phosphate mineralization, production of phytohormones and vitamins, and development of root system in plants. These bacteria are able to increase the quantity and quality of different plants by enhancing germination rate, increasing the length and weight of the roots, and accelerating the elongation of the embryonic and lateral roots (Khan et al., 2020).

Seed biopriming is mainly used for legumes that can coexist with nitrogen-fixing bacteria. Researchers are trying to apply this feature to other crops, especially cereals. Experiments have shown that seed biopriming cannot only play a key role in providing some plant needs but it can also leave positive effects on improving plant behavior against pathogens. The application of biofertilizers, especially PGPR, is considered as the most important strategy in integrated plant nutrition management with a sustainable agricultural system and sufficient input is using a combination of chemical fertilizers with these bacteria (Fathollahy and Mozaffari, 2020).

Significant interactions between biopriming and heavy metals were observed in terms of leaf osmotic potential, leaf carotenoids, leaf chlorophyll *b*, chlorophyll *a*, relative leaf water content, and leaf greenness index at a probability level of 1%. The interaction of different concentrations of biopriming and heavy metals in terms of quantum yield, leaf catalase activity, root catalase activity, leaf peroxidase activity, and root peroxidase activity of seedlings was also significant at the probability level of 1% (Afrouz, 2018).

Under no stress, *Trichoderma* and *Azospirillum* caused a further increase in leaf carotenoid content. *Trichoderma* and *Pseudomonas* under the combined toxicity of aluminum and copper caused a more significant increase in leaf carotenoid activity than other priming treatments (Afrouz, 2018). An increase in carotenoids was reported which was made by inoculation of seeds with fungi (Guler et al., 2016); however, in another study, it was reported that *Azospirillum* inoculation on carotenoid content was insignificant (Zaied et al., 2003). Moreover, it was reported that natural growth stimulants are not only effective in increasing growth and chlorophyll content, but also have a positive effect on absorption and especially the transport of elements (Hajeeboland et al., 2004). In fact, this work which is seemingly not exclusive can be significant for low-consumption elements. In other words, coexistence with fungi enables the plant to be less exposed to light inhibition and degradation of pigments although it occurs mostly in the presence of carotenoids that act as an effective light isolator. Increased uptake of nitrogen and magnesium under environmental stress conditions also increases the content of photosynthetic pigments in plants (Seyed Sharifi, 2016).

Seeds that use different biopriming methods differ so significantly in terms of germination and seedling establishment that seeds with no biopriming had lower germination and establishment than bioprimed seeds. Adding various concentrations of heavy metal stress to nutrient solution caused adverse effects of stress on seeds and physiological and biochemical conditions of seedlings. Regarding seed germination rate and seedling emergence and compared to control (base concentrations of nutrient solution), they showed a greater effect than different doses of stress (Afrouz, 2018).

The fact that all measured parameters had a significant effect on interaction of heavy metal stress and biopriming shows the significant effect of biopriming on corn seeds. Noteworthy, biopriming with *Trichoderma* had a better effect than the other three bacteria (*Trichoderma*, *Azotobacter*, and *Mycorrhiza*). It can also be due to presence of a wide range of free fungus activities in different pH conditions in terms of alkalinity or acidity of the medium, since at low pH, they have a high activity against bacteria (Afrouz, 2018).

20.11 Biopriming steps based on fungi microstructure

In step one, 5.25 g of PDA and 0.62 g of agar in 125 mL of distilled water were mixed in Erlenmeyer flask to prepare the culture medium (Fig. 20.1). To disinfect the culture medium, Erlenmeyer was placed in an autoclave at 121°C for 2 h. The disinfected culture medium was poured into 90-mm Petri dishes to cool.

Fig. 20.1 Preparation of Nutrient Agar (NA) culture medium for bacterial culture and Potato Dextrose Agar (PDA) for fungal culture. Photo: Piri.

Using a loop (laboratory tube), the colonies of fungal strains were cultured on PDA medium and placed in an incubator (Parsian TEB model) at 27°C. After growing the fungi and abundant sporulation for 10 days, the spores were poured into the Erlenmeyer sterol with the culture medium (Fig. 20.2). In the next step, the fungal colonies along with the culture medium were removed in a square shape (Fig. 20.3A) and placed in 20 mL of distilled water (Fig. 20.3B). It should be noted that a shaker can be used to separate the fungal colonies from the culture medium.

To study the concentration of the fungal suspension inoculum using a homocytometer slide (blood cell count), the spores were counted under a microscope (suspension of 10^7 spores per milliliter suitable for seed inoculation) (Fig. 20.4). After achieving the appropriate density, the fungicides of the seeds were placed in this suspension for 2 h (Fig. 20.5). After inoculation, the culture was performed in a Petri dish on top of the paper (Fig. 20.6).

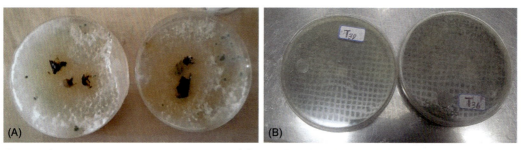

Fig. 20.2 Growth of fungal colonies on the 2nd day (A), growth of colonies of *Trichoderma harzianum* on the 10th day (B). Photo: Piri, Bakhit.

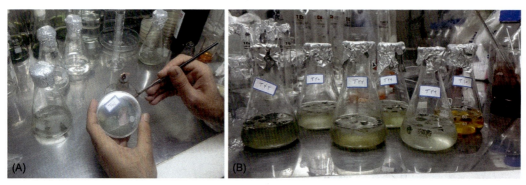

Fig. 20.3 Place the fungal colonies in distilled water. Photo: Piri, Bakhit.

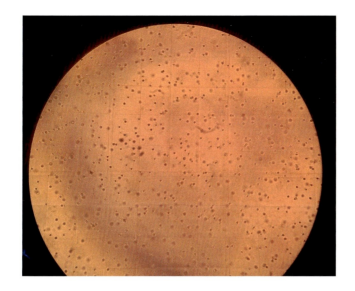

Fig. 20.4 Counting fungal spores under a microscope. Photo: Bakhit.

Fig. 20.5 Place the seeds in a suspension of 10^7 spores per milliliter. Photo: Bakhit.

Fig. 20.6 Cultivation of seeds in Petri dishes after inoculation. Photo: Bakhit.

20.12 Effect of *Trichoderma* species on increasing growth and antioxidant activity

Since biotic organisms are always exposed to biotic and abiotic stresses, they have to make an effective balanced state between growth, reproduction, and immune system under adverse environmental conditions. Plants contain both enzymatic (superoxide dismutase, peroxidase, and catalase) and nonenzymatic (glutathione, ascorbic acid, α-tocopherol, and phenolic compounds) mechanisms that inhibit free radicals before any possible attack to membranes or other seed components (Khajeh et al., 2021; Naddaf et al., 2021). Also, by controlling free radicals, they prevent oxidation. During the germination process, a significant increase is reported in respiratory activity and in the production of reactive oxygen species (ROS) (Tančić-Živanov et al., 2020; Naderi et al., 2021).

In seed physiology, reactive oxygen species are usually considered as toxic molecules, yet based on evidence, it can be concluded that ROS can also act as signaling molecules that are activated in a wide range of defense responses to various stimuli. Thus, ROS cellular levels have to be strictly controlled at both production and consumption levels. Cellular antioxidant mechanisms, including detoxifying enzymes and antioxidant compounds, potentially remove the resulted toxic ROS under stressful conditions, or control ROS concentration by regulating different signaling pathways. In order to be used as signaling molecules, the nontoxic level of ROS should be maintained. Evolutionarily, it is assumed that when cells learn to cope with ROS toxicity, they can use ROS as signaling molecules (Tančić-Živanov et al., 2020).

A great challenge in pepper production is boosting seed germination and germination vigor and promoting healthy plants with strong root systems. As they grow, *Trichoderma* species colonize roots to provide seasonal benefits to plants. That is why *Trichoderma* species are widely used as promoters of plant growth and its defense mechanisms (Ahmed et al., 2003; Tančić-Živanov et al., 2020).

Under conditions of combined toxicity of aluminum and copper, *Trichoderma* and *Azospirillum* can significantly increase the activity of root catalase enzyme better than other priming treatments (Afrouz, 2018). In line with previous findings on the effects of these experiments on the effects of leaf and root catalase, it was reported that growth-promoting fungi can reduce the effect of heavy metal stress through a defense mechanism, which in turn can prevent any increase in SOD enzyme, since the activity of antioxidant enzymes is not the only plant defense mechanism that possibly acts in reducing oxidative damage. Through proline synthesis, bacteria can also play a significant role in reducing damages caused by free radicals made by heavy metals and oxidative stress (Fazeli-Nasab and Sayyed, 2019; Ismail, 2021). Furthermore, a decrease in SOD enzyme activity was reported in barley seedlings that were inoculated with kdc bacteria (Omar et al., 2009).

In an experiment conducted on the effectiveness of seed biopriming with *Trichoderma* strains to promote the growth of pepper plants in the early stages and to evaluate their effects on germination, it was shown that *Trichoderma* strain had a positive effect on the root weight of pepper seedlings and even the germination increased significantly by 22%, yet it had no effect on the length of roots and stems. *Trichoderma* strains form an accumulation of reactive oxygen species (ROS) that acts as a signal molecule used to promote germination and emergence energy. Furthermore, a positive correlation was observed between pyrogallol peroxidase, superoxide dismutase, catalase activity, and germination in plants treated with these strains (Tančić-Živanov et al., 2020).

Analyzing the effect of *Trichoderma harzianum* on the activity of tomato antioxidant enzymes under drought stress indicated that *Trichoderma harzianum* can increase the activity of ascorbate peroxidase in tomatoes compared to control (Mastouri et al., 2010). Treating corn with *Trichoderma lixii* for an hour can lead to increased length, heavier and drier roots, shoots, relative water content, soluble protein, proline, chlorophyll, and carotenoid content and in turn reduce lipid peroxidation, hydrogen peroxide, and lipid peroxidation under salinity stress (Pehlivan et al., 2017). Treatment of wheat plant with *Trichoderma harzianum* for 24h reduced proline, malondialdehyde, and hydrogen peroxide. But it also increased total phenolic content and phenylalanine ammonia lyase activity in dry conditions (Shukla et al., 2015).

Treatment of wheat with *Trichoderma harzianum* under salinity stress increased germination, root and branch length, chlorophyll and membrane stability indices, and proline and phenolic content (Zvereva et al., 2012). Among improvements done in abiotic stress caused by biopriming, *Trichoderma* was used to control salinity and drought stress. In the case of using *Trichoderma lixii* in maize and *Trichoderma harzianum* in wheat, it was reported that better physiological and morphological parameters compared to the untreated control were observed. The application of *Trichoderma harzianum* which acts as a biopriming agent against drought stress in wheat was further studied. Finally, it was reported that not using water for a period of 4–13 days can lead to increased concentrations of stress enzymes and metabolites such as phenolics and, on the contrary, decrease hydrogen peroxide concentrations, malondialdehyde, and proline (Shukla et al., 2015).

20.13 Role of *Trichoderma* in dealing with biological stresses (biotic)

Extensive use of different *Trichoderma* species in plants has long been done to improve plant growth and control plant pathogens (Rezaloo et al., 2020; Altomare et al., 1999). Although *Trichoderma* is useful in promoting plant stress tolerance, especially in biological stress, its use as a growth stimulant has also been proven. Various *Trichoderma* species include fungi which can be found in all diverse habitats. The remarkable success of these fungi has received widespread attention in the biological control of many plant pathogenic fungi, including the pathogens of seed rot and seedling death of *Rhizoctonia solani*, various species of *Fusarium*, *Sclerotinia rolfsii*, and various species of *Pythium* and *Phytophthora*. The application of *Trichoderma* for chickpea seeds that act as a biopriming agent can significantly increase plant growth parameters (Altomare et al., 1999; Singh et al., 2016a). Other usages of *Trichoderma* biopriming include increasing wheat growth, promoting nitrogen uptake and recycling, and enhancing agronomic and physiological use efficiencies. It has also performed well even below 75% of the recommended fertilizer dose (Meena et al., 2016). In another experiment, it was found that *Trichoderma* could increase seedling growth and vigor better than *Trichoderma* (Ananthi et al., 2014).

Microbial inoculation can also provide several comprehensive and promising solutions to agricultural and environmental problems. This is done through inducing plant growth, protecting against disease, increasing availability of nutrients, and promoting their uptake into plants. *Trichoderma* species are the most common saprophytic fungi that are popular for suppressing plant diseases and are dominant in the global market as biopesticides (Singh et al., 2016a). However, many researches conducted over recent decades have emphasized various properties of *Trichoderma* which have made its use beyond its production. So far, different species of *Trichoderma* are identified which can induce plant growth, manage biological and nonbiological stresses, and improve efficiency of nutrient uptake and absorption (Mehetre and Mukherjee, 2015). Likewise, by improving the efficiency of nutrient consumption in plants, beneficial microbes have attracted a lot of international attention in terms of soil fertility (Meena et al., 2016; Deb et al., 2015). The species, *Trichoderma* spp., can increase crop production by improving the absorption of mineral fertilizers (Deb et al., 2015).

Besides its antimicrobial effect, *Trichoderma* can further stimulate the biological activity of a fixed antagonist microbial population, thereby enhancing plant growth. Likewise, the application of *Trichoderma* via seed biopriming increased enzyme activity by releasing specific metabolites in maize (López-Coria et al., 2016). In a research done on soybean seeds, *Trichoderma harzianum*, *Trichoderma virens*, and *Pseudomonas aeruginosa* were treated separately and in inoculated with *Colletotrichum truncatum*. The results showed that after 12 h, the bacterial population on the seeds increased by over four times along with a decrease in infection rate (about 2.97%), and at the same time, natural seedlings were formed (Begum et al., 2010).

Soaking corn seeds with *Trichoderma harzianum* for 24 h in pots could control *Fusarium verticillioides* disease and enhance germination and seedling parameters, 1000-seed weight germination, and yield (Chandra Nayaka et al., 2010). Similarly, in another experiment, *Trichoderma*, *Pseudomonas*, and *Glomus* were used. It was observed that tomato wilting was significantly reduced by up to 74% and seed germination was increased (Srivastava et al., 2010). Treatment of bean plant with *Trichoderma viride, Trichoderma harzianum, Bacillus subtilis, Trichoderma hamatum, Bacillus cereus*, and *Pseudomonas fluorescen* for 16 h under the control of *Rhizoctonia solani, Fusarium solani*, and *Sclerotium rolfsii* also increased the seed shelf life in seed storage (El-Mougy and Abdel-Kader, 2008).

In numerous papers, it is reported that disease control by *Trichoderma* was superior to chemical fungicides. Since there is a 33% reduction in disease infection, the use of *Trichoderma harzianum* with fungicide carboxin thiram at a rate of 2 g/kg of seeds can lead to a 12%–14% increase in germination, and in turn, the prevalence of the disease reduced by 44.1%–60.3% (Dubey et al., 2007). Inoculation of seeds with factors (*Trichoderma harzianum* and *Pseudomonas fluorescens*) may be useful in protection against seed-borne and soil-borne plant pathogens and increase the percentage and rate of seed germination and seedling growth (Sarkar and Rakshit, 2021). Seed inoculation with *Trichoderma* is used as an alternative method to control most seed-borne and soil-borne pathogens (Usha et al., 2012; Reddy et al., 2014).

20.14 Induction of plant resistance to nonbiological stresses (abiotic)

Plant response to stress depends on the plant's metabolic and morphological activity, growth stage, and potential yield (Jomeh Ghasem Abadi et al., 2019; Forouzandeh et al., 2019; Fazeli-Nasab et al., 2019b). Under field conditions, plants may face some heavy metal stress at several stages of growth. It can leave a direct effect on some important physiological parameters such as leaf area and chlorophyll content (Fazeli-Nasab et al., 2019a; Fazeli-Nasab and Fooladvand, 2019). Leaf onset and leaf surface development are often reduced by stress, and in some cases, they may even stop. In wheat genotypes, chlorophyll content and relative water content were reduced by an increase in stress intensity (Rascio et al., 1998). Besides, the same reduction of chlorophyll content was reported based on increased soil stress (Yadav and Jaiswal, 2001).

Nonbiological stresses are among the most important factors used in reducing crop yields. Since pants are unable to move, several responses are expressed at the molecular, biochemical, and physiological levels under stress to maintain their stability (Kerchev et al., 2020). Simultaneous occurrence of biological and abiotic stresses which is quite common in plant environments often results in reduced yields. Due to the limitation of cost-effective options, biopriming with different species of *Trichoderma* can be a suitable solution for most agricultural challenges (Hossain et al., 2018; Swain et al., 2021). In addition, the potential benefits of plant growth can be achieved through performing proper management of natural resources and strengthening environmental sustainability.

Such direct impact of various stresses often leads to a decrease in agricultural output, so it is required to identify these stresses. Since the changes mostly occur in plant system, a suitable approach should be taken to solve the issue easily. In general, abiotic stresses are extreme temperatures (heat and cold), drought (limited rainfall and dry winds), heavy metals, and salinity (Reddy et al., 2014; Krasensky and Jonak, 2012; Sarkar et al., 2018). However, significant biological stresses including viruses, fungi, bacteria, weeds, insects, and other pests and pathogens (Suzuki et al., 2014) have reduced the yield of important crops by an average of 50% (Wang et al., 2003; Isah, 2019).

An increase in abiotic stress like rising temperatures can lead to the spread of pathogens and weaken plant defense mechanisms (Suzuki et al., 2014). Genetic manipulation used to produce stress-resistant products is an extensive field that covers a vast range of researches on breeding programs, but due to the fact that they are time-consuming and involve large investments, and cheap and reasonable options are rare, it seems urgent to provide and adopt farmers with economically agricultural solutions; thus, biopriming can be a seed treatment process using microbial materials to accelerate germination and improve plant establishment in adverse conditions (Singh et al., 2016b).

Beneficial effects of microbe-plant interaction may also cover inducing plant growth and promoting disease resistance and stress tolerance to abiotic stresses. In fact, using metabolic regulation in their systems, the inoculated plants are enabled to adapt themselves to environmental stresses. In turn, it can lead to the accumulation of several compatible organic solutions such as sugars, polyamines, betaines, quaternary ammonium compounds, polyhydric alcohols, proline, and other amino acids (Ray et al., 2016). Proper development of resistance induction in treated plants with microbes can also help them to overcome pathogen contamination and control plant diseases (Jabnoun-Khiareddine et al., 2015). Several microorganisms such as *Trichoderma* promote deep root growth and provide suitable water and nutrition absorption in the plant under adverse conditions (Keswani et al., 2016; Bisen et al., 2016).

Another abiotic stress is heat and cold stress. Treatment of corn seeds with *T. harzianum* can increase growth against cold and heat stresses (Shoresh and Harman, 2008). *Trichoderma* can make the plant overcome abiotic stresses such as drought, salinity, cold, and heat. Treatment of tomato seeds with *T. harzianum* can also significantly increase germination in the osmotic potential (0.3 mP), mainly due to their high resistance in dry conditions (Harman and Shoresh, 2007). In another study evaluating the effect of T22 on maize using the proteomics method, 205 proteins were produced in the presence of *Trichoderma*, among which the most common proteins produced in carbohydrate cycle, especially proteins available in synthesis of glycolytic and tricarboxylic acids which are effective in drought stress (Harman et al., 2008).

Treatment of cocoa seedlings with *Trichoderma* results in building proteins called tonoplasts which act as membrane channels in plants to pump out water randomly (Bae et al., 2009). The expression of such proteins and the formation of membranes can also induce a reaction against drought. Furthermore, stomatal changes and reduced opening and closing of the stomata compared to uninoculated plants led to drought resistance in cocoa seedlings (Dar et al., 2017). In corn plants inoculated with *Trichoderma*, some genes such as ERD1 and RD29A are expressed that activate the enzyme histone acyltransferase and increase the plant's tolerance to drought stress (Verma et al., 2017; Varma et al., 2012).

Trichoderma can induce growth in soils exposed to both construction materials and sewage. Even after 12 weeks, these plants not only grow further against pathogens but also own a greater tolerance range (Adams et al., 2007).

20.15 Biological mechanisms of *Trichoderma*

Following the daily- growing demand for organic food production, the use of biofertilizers and biotoxins can be a suitable alternative for maintaining environmentally friendly high production. Since *Trichoderma* species are resistant to most chemical pesticides, they can be classified as good biological controllers and integrated controllers (Harman, 2011). In summary, reactions between *Trichoderma* and the pathogen can also cover three mechanisms including mycoparasitism, antibiosis, and competition which cause *Trichoderma* to invade a wide range of phytopathogenic fungi (Vinale et al., 2008; Alfiky and Weisskopf, 2021). One of the main biocontrol mechanisms of *Trichoderma* is mycoparasitism, which includes host diagnosis, growth toward its hyphae, contact, twisting, infiltration, and finally slipping and destruction of the fungus (Konappa et al., 2020c; Swain and Mukherjee, 2020).

In host detection which is the first step in mycoparasitism, *Trichoderma* fungi receive chemical stimuli released by the host and are absorbed into the host through chemotropics. *Trichoderma* species can release hydrolytic enzymes which in turn break down the host's cell wall and release gene inducers involved in mycoparasitism to detect the host (Contreras-Cornejo et al., 2020; Konappa et al., 2020c). After attaching *Trichoderma* to the host, these fungi produce hydrolytic enzymes that cause lysis of host hyphae and eventually decay (Konappa et al., 2020c; Swain and Mukherjee, 2020). These enzymes which are in fact cell wall degrading enzymes include chitinases, glucanases, and proteases. Here, the most important ones are chitinases. Some species of *Trichoderma* use chitinases to attack other fungi, insects, and nematodes (Poveda et al., 2020; Contreras-Cornejo et al., 2020).

Furthermore, *Trichoderma* chitinases in plants act as an inducer of the immune system. About 30 chitinases are introduced in *Trichoderma* that have different protein compositions and gene structures, mostly with antifungal activities (Mazrou et al., 2020). Another group of key enzymes in *Trichoderma* is glucanases. Beta-1 and 3-glucan (laminarin) and beta-1 and 4-glucan (cellulose) along with chitin are important constituents of the fungal cell wall. Glucanases also include beta-1 and 3-glucanases, beta-1 and 6-glucanases, and beta-1 and 4-glucanases (Alfiky and Weisskopf, 2021; Ghasemi et al., 2020).

A lot of *Trichoderma* strains which have caused a significant increase in cellulase in the presence of the pathogen can play an important role in pathogen control as well (Alfiky and Weisskopf, 2021; Konappa et al., 2020c). Overall, *Trichoderma* proteases play two main roles: (a) attacking lipids and cell wall proteins to lysis of host hyphae and (b) neutralizing enzymes produced by pathogens (Gautam and Naraian, 2020; Antoun et al., 1998). Antibiosis is a mechanism of *Trichoderma* used against pathogens. Similarly, *Trichoderma* inhibits the activity of other fungi by producing volatile and extracellular antimicrobial metabolites (Alfiky and Weisskopf, 2021; Xu et al., 2020).

The protective activity of *Trichoderma* against plant pathogens can build a set of secondary metabolites which includes various natural compounds (Sood et al., 2020). The biocontrol power of *T. harzianum* is caused by the production of antibiotic pyrone (Swehla et al., 2020; Sood et al., 2020). Synthesis of secondary metabolites by *Trichoderma* which also depends on the strains of this fungus has compounds with different antifungal properties that are chemically divided into different groups. Antibiotics produced by *Trichoderma* are further divided into two groups: light and nonpolar volatile antibiotics and nonvolatile and polar antibiotics. These are produced in high concentrations in soil environments and are able to affect a wide range of pathogens (Sood et al., 2020; Bedine Boat et al., 2020).

Trichoderma's competition mechanism is used to access carbon, nitrogen, and space resources and to limit and control plant pathogens. Similarly, by colonizing the flowers and removing pathogens, *T. harzianum* can control *B. cinerea* on grapes (Halifu et al., 2020). The predominant mechanism which is used by *T. harzianum* to control *F. oxysporum* was competition over food supply. Biotic components of soil can also play an important role in *Trichoderma* activity against plant pathogens (Mohamed and Haggag, 2006; El-Mohamedy et al., 2010) Application of GFP mutants showed that the amount of soil microbial biomass could increase hypha growth and production of *T. harzianum* spores (Tasik and Widyastuti, 2015; Bae and Knudsen, 2005). Some species of *Trichoderma*, such as *T. viride* and *T. harzianum* which have properties such as mycoparasitism and antibiotic and saprophytic ability to compete can significantly reduce the population of pathogenic fungi (Soltani et al., 2006). Yet, in natural conditions, they are not so effective in controlling fungal diseases mainly because of their low reproductive units in soil (Cavalcante et al., 2008).

Major biological control mechanisms of *Trichoderma* species are mycoparasitism, antibiotics, and competition for nutrients and soil space. They make terioderma invasive to various phytopathogenic fungi; however, despite biological control, *Trichoderma* species can promote plant growth and enhance plant defense mechanisms. *Trichoderma harzianum* fully colonizes, grows roots, and provides minimal seasonal benefits to plants. A set of morphological and bio chemical changes initiated by *Trichoderma* spp. Colonization and infiltration into plant root tissues are regarded as a part of the plant defense response. Specific strains of *Trichoderma* species which can also stimulate immune responses in its host plants are commonly classified among the best induced systemic resistance factors (ISRs). Besides, *Trichoderma* spp. affects root colonization and frequently increases root growth, crop productivity, resistance to abiotic stress, and nutrient uptake and utilization (Saba et al., 2012; Shoresh et al., 2010).

20.16 Colonization as interactions between *Trichoderma* and plant

Trichoderma are able to colonize the roots of both monocotyledonous and dicotyledonous plants. During colonization, *Trichoderma* hyphae can spirally penetrate the root cortex and wrap around the root. Some

Trichoderma strains can only colonize the root surface, whereas others can penetrate the surface of the root epidermis. However, this penetration which is just limited to the first and second layers induces and surrounds plant cells to store substances required for cell wall synthesis and building of phenolic compounds. Formation of these substances can in turn prevent the overgrowth of *Trichoderma* hyphae into the roots of plants (Brotman et al., 2008; Vinale et al., 2008; Carro-Huerga et al., 2020).

As resulted in some studies, plant colonization by *Trichoderma* strains leads to production and increases in amount of certain compounds and enzymes which are related to the plant's defense mechanisms. These compounds and enzymes cover different kinds of enzymes like peroxidases, chitinases, beta-1 and 3-glucanases, and lipoxygenase. Plant colonization induced by *Trichoderma* strains makes changes in plant metabolism and consequently provides the accumulation of antimicrobial compounds in the plant (Malinich et al., 2019; Anasontzis et al., 2019).

The production of phytoalexin is often stimulated by regional resistance in plant which is caused by *T. virens*, so Trichoderma strains cannot only stimulate building of antibiotic compounds, but also promote the plant to produce antibiotic compounds. Then, by colonizing plants, they can induce changes in the mechanism of plant metabolism. Such stimulation to produce and increase the compounds can in turn induce and promote plant resistance to defend systematically against pathogens in adverse natural conditions (Jaroszuk-Ściseł et al., 2019; Guzmán-Guzmán et al., 2019; Xu et al., 2020).

T. harzianum used in greenhouse products represented that this strain can have significant chemical and useful effects on controlling plant diseases and can be used as an alternative to chemical fungicides, since it has a longer shelf life than materials (Elad, 2000; Maketon et al., 2008; Akrami et al., 2009) *T. harzianum* also provided a significant increase in growth of lettuce, tomato, and pepper. Based on the results, the yield increased by 30% compared to plants which were not treated with *Trichoderma* (Vinale et al., 2004; Alavi et al., 2019). The root length of T22-treated corn was about 50–75 cm longer than that of untreated corn. This result was obtained through increased tolerance and resistance to pathogens and strains, as well as success in competition with soil microorganisms (Harman, 2006). A hydrophobin-like protein identified in T22-treated roots can increase root growth and root colonization and, finally, enhance soil utilization and promote plant growth (Seidl-Seiboth et al., 2011; El Enshasy et al., 2020).

In addition, *Trichoderma* promotes plant and root growth both by increasing absorption of nutrients and ions such as Cu, P, Fe, Mg, and Na, and by solubilizing a range of nutrients. T22 and RR17BC strains of *T. harzianum* and WW10TC4 strains of *T. avovidea* can also induce nitrate reductase expression in plants and convert nitrate to ammonium. This process is really essential for nitrogen metabolism (Harman et al., 2008; Seidl-Seiboth et al., 2011; Shoresh et al., 2010). In general, *Trichoderma* fungi are able to dissolve insoluble nutrients for plants in soil (leaving an indirect effect on plant). In this way, the plants are further induced to absorb more nutrients (direct effect) (Shoresh et al., 2010; Al-Ani, 2019; Bononi et al., 2020). Yet, the genetics and molecular basis of these effects are still unknown. The mechanisms mentioned earlier can all improve the growth characteristics of plants and provide conditions that lead to the formation of strong and vigorous plants.

Coexistence of fungi and growth-promoting bacteria with the host plant also has a fundamental function in fertility and sustainability of soil ecosystems (Gosling et al., 2006). Coexisting fungi can form coexisting communities with most plant species, and regarded as a type of biofertilizer, they are important for increasing agricultural production. Frequently, symbiosis in rhizosphere plays a mediating role between plant roots and soil mass to help the plant absorb water and nutrients (especially phosphorus) from the soil. Besides their effect on crop growth, the coexistence of *Trichoderma* can be significant in maintaining ecological balance in soil (Mishra and Abidi, 2010; Attarzadeh et al., 2019). Using plant growth stimulants can increase plant growth rate and affect the allocation and transfer of biomass between roots and stems, so by absorbing and transferring more nutrients, the dry weight of shoots increases, and naturally the percentage of establishment rises (Fazeli-Nasab and Sayyed, 2019).

Colonization of fungi can facilitate plant growth and development around the roots of host plants and in turn form a system that has positive effects on plant growth. Some of these improving effects are increased plant access to nutrients, especially phosphorus (Cardoso and Kuyper, 2006), photosynthesis of leaves (Harman et al., 2021), increased efficiency of water use in host plant (Oljira et al., 2020), changing concentration of plant hormones and amount of chlorophyll (Zehra et al., 2017).

As concluded by numerous researchers, coexistence of natural growth stimulants can improve the uptake of both motile and immobile elements by plant roots and increase the rate of plant establishment and yield (Liu et al., 2007; Bi et al., 2019; Abdollahi et al., 2019). Moreover, it was concluded (Shahhosini et al., 2013) that the coexistence of growth-promoting fungi improved effects on growth and yield of maize in dehydrated conditions. Accordingly, the mechanisms by which these fungi can improve water uptake in adverse water conditions can be divided into the following five groups (Song, 2005): (1) improving soil properties around the roots such as aggregation and soil improvement, (2) increasing root absorption levels and promoting water uptake efficiency, (3) increasing absorption of phosphorus and other nutrients, (4) activating host plant defense system and reducing oxidation risks caused by stressful conditions, and (5) inducing mobility of expression in host plant genes.

20.17 Conclusion

The use of *Trichoderma* strains in soil or its direct application to seedling roots is considered as an economical and effective method. Besides offering an antimicrobial effect, *Trichoderma* can stimulate biological activity of microbial population in a resident antagonist to enhance plant growth. Some possible explanations for these conditions cover control of certain pathogens followed by rapid growth and uptake of nutrients.

Trichoderma strains which are able to provide formation and accumulation of ROS that is regarded as an important factor in plant growth and development are mostly desirable for germination and improved growth of seeds. Also, there is a positive correlation between pyrogallol peroxidase, superoxide dismutase, catalase activity, and germination in plants treated with these strains.

Microbial inoculation of seeds with different species of *Trichoderma* is one of the most reasonable methods that require a little inoculation. Microbial inoculation of seeds leads to increased plant growth, better production of various phytohormones, and higher nutritional value of plant products and finally enhances the development of defense responses in plants against biological and nonbiological stresses. However, since there exist various seeds in terms of shape, size, and nature of seed coat, it is essential to pay much attention to standardization and homogeneity of microbial concentration and inoculation time for each product. Limited number of reports shows that some can increase plant growth even at very low microbial concentrations. Although it does not reduce the microbial concentration, it can save the cost of seed inoculation. Different aspects such as developing optimal formulations and understanding defense responses, as well as improving plant growth, are needed to be studied more thoroughly in the future. Currently, microbial inoculation of seeds is the best tool to overcome environmental stresses in a sustainable manner, since there is an increasing demand for chemicals. Fortunately, the effective use of *Trichoderma* can play an important role in controlling diseases and increasing the growth and growing indices of seeds and plants. However, better future researches are required on *Trichoderma* concentrations and the ways of its inoculation and application. Also, further experiments should be performed in pure and field conditions to evaluate the possible effect of *Trichoderma* strains as chemical substitutes.

References

Abdollahi, S., Ali Asgharzad, N., Zahtab Selmasi, S., Khoshru, B., 2019. Effects of endophytic fungus piriformospora indica on growth indices and nutrient uptake by anise plant (*Pimpinella anisum*) under water deficit stress conditions. J. Agric. Sci. Sustain. Prod. 29, 51–64.

Adams, P., De-Leij, F.A., Lynch, J., 2007. *Trichoderma harzianum* Rifai 1295-22 mediates growth promotion of crack willow (*Salix fragilis*) saplings in both clean and metal-contaminated soil. Microb. Ecol. 54, 306–313.

Afrouz, M., 2018. Evaluating the Effect of Seed Biopriming With *Trichoderma* an Plant Growth Promoting *Rhizobacteria* (Pgprs) on Physiological and Biochemical Aspects of Corn Seedling Growth under Copper and Aluminum Stress. Masters thesis, University of Mohaghegh Ardabili.

Afzal, I., Ahmad, N., Basra, S., Ahmad, R., Iqbal, A., 2002. Effect of different seed vigour enhancement techniques on hybrid maize (*Zea mays* L.). Pak. J. Agric. Sci. 39, 109–112.

Ahanger, M.A., Akram, N.A., Ashraf, M., Alyemeni, M.N., Wijaya, L., Ahmad, P., 2017. Plant responses to environmental stresses—from gene to biotechnology. AoB Plants 9, 1–17.

Ahmad, P., Prasad, M.N.V., 2011. Abiotic Stress Responses in Plants: Metabolism, Productivity and Sustainability. Springer Science & Business Media.

Ahmed, A.S., Ezziyyani, M., Sánchez, C.P., Candela, M.E., 2003. Effect of chitin on biological control activity of *Bacillus* spp. and *Trichoderma harzianum* against root rot disease in pepper (*Capsicum annuum*) plants. Eur. J. Plant Pathol. 109, 633–637.

Ahooi, S., Ajdanian, L., Nemati, H., Aroiee, H., 2021. Evaluation of developmental changes in celery plant under treatment of vermicompost and fungus *Trichoderma hirizianum* isolate Bi. J. Hortic. Sci. 34.

Akrami, M., Ibrahimov, A.S., Zafari, D., Valizadeh, E., 2009. Control *Fusarium* rot of bean by combination of by *Trichoderma harzianum* and *Trichoderma asperellum* in greenhouse condition. Agric. J. 4, 121–123.

Al-Ani, L.K.T., 2019. A patent survey of *Trichoderma* spp. (from 2007 to 2017). In: Singh, H.B., Keswani, C., Singh, S.P. (Eds.), Intellectual Property Issues in Microbiology. Springer Singapore, Singapore.

Alavi, S.M., Masoumiasl, A., Zare, N., Asghari Zakaria, R., Sheikhzade Mosaddegh, P., 2019. The role of ecotype, explant and plant growth regulators on cell suspension culture of *Ferulago angulata* (Schlecht.) Boiss. J. Hortic. Sci. 33, 525–536.

Alexopoulos, C.J., Mims, C.W., Blackwell, M., 1996. Introductory Mycology. John Wiley and Sons.

Alfiky, A., Weisskopf, L., 2021. Deciphering *Trichoderma*–plant–pathogen interactions for better development of biocontrol applications. J. Fungi 7, 61.

Altomare, C., Norvell, W., Björkman, T., Harman, G., 1999. Solubilization of phosphates and micronutrients by the plant-growth-promoting and biocontrol fungus *Trichoderma harzianum* Rifai 1295-22. Appl. Environ. Microbiol. 65, 2926–2933.

Amozadeh, S., Fazeli-Nasab, B., 2012. Improvements methods and mechanisms to salinity tolerance in agricultural crops. In: The First National Agricultural Conference in Difficult Environments. Islamic Azad University, Ramhormoz Branch.

Ananthi, M., Selvaraju, P., Sundaralingam, K., 2014. Effect of bio-priming using biocontrol agents on seed germination and seedling vigour in chilli (*Capsicum annuum* L.) 'Pkm 1'. J. Hortic. Sci. Biotechnol. 89, 564–568.

Anasontzis, G.E., Lebrun, M.H., Haon, M., Champion, C., Kohler, A., Lenfant, N., Martin, F., O'Connell, R.J., Riley, R., Grigoriev, I.V., 2019. Broad-specificity Gh131 B-glucanases are a hallmark of fungi and oomycetes that colonize plants. Environ. Microbiol. 21, 2724–2739.

Anhar, A., Sari, N.P., Advinda, L., Putri, D.H., Handayani, D., 2019. Effect of the indigenous *Trichoderma* application on germination of black glutinous rice seed. J. Phys. Conf. Ser., 012065.

Anjum, Z., Hayat, S., Ghazanfar, M.U., Ahmad, S., Adnan, M., Hussian, I., 2020. Does seed priming with *Trichoderma* isolates have any impact on germination and seedling vigor of wheat. Int. J. Bot. Stud. 5, 65–68.

Antoun, H., Beauchamp, C.J., Goussard, N., Chabot, R., Lalande, R., 1998. Potential of *Rhizobium* and *Bradyrhizobium* species as plant growth promoting *Rhizobacteria* on non-legumes: effect on radishes (*Raphanus sativus* L.). In: Molecular Microbial Ecology of the Soil. Springer.

Ashraf, M., Foolad, M.R., 2005. Pre-sowing seed treatment—a shotgun approach to improve germination, plant growth, and crop yield under saline and non-saline conditions. Adv. Agron. 88, 223–271.

Attarzadeh, M., Balouchi, H., Rajaie, M., Dehnavi, M.M., Salehi, A., 2019. Growth and nutrient content of *Echinacea purpurea* as affected by the combination of phosphorus with arbuscular *Mycorrhizal* fungus and *Pseudomonas florescent* bacterium under different irrigation regimes. J. Environ. Manag. 231, 182–188.

Azad, H., Fazeli-Nasab, B., Sobhanizade, A., 2017. A study into the effect of jasmonic and humic acids on some germination characteristics of rosselle (*Hibiscus sabdariffa*) seed under salinity stress. Iran. J. Seed Res. 4, 1–18.

Azarmi, R., Hajieghrari, B., Giglou, A., 2011. Effect of *Trichoderma* isolates on tomato seedling growth response and nutrient uptake. Afr. J. Biotechnol. 10, 5850–5855.

Bae, Y., Knudsen, G.R., 2005. Soil microbial biomass influence on growth and biocontrol efficacy of *Trichoderma harzianum*. Biol. Control 32, 236–242.

Bae, H., Sicher, R.C., Kim, M.S., Kim, S.-H., Strem, M.D., Melnick, R.L., Bailey, B.A., 2009. The beneficial endophyte *Trichoderma hamatum* isolate Dis 219b promotes growth and delays the onset of the drought response in *Theobroma cacao*. J. Exp. Bot. 60, 3279–3295.

Baker, R., Elad, Y., Chet, I., 1984. The controlled experiment in the scientific method with special emphasis on biological control. Phytopathology 74, 1019–1021.

Bedine Boat, M.A., Sameza, M.L., Iacomi, B., Tchameni, S.N., Boyom, F.F., 2020. Screening, identification and evaluation of *Trichoderma* spp. for biocontrol potential of common bean damping-off pathogens. Biocontrol Sci. Tech. 30, 228–242.

Begum, M., Sariah, M., Puteh, A., Abidin, M.Z., Rahman, M., Siddiqui, Y., 2010. Field performance of bio-primed seeds to suppress *Colletotrichum truncatum* causing damping-off and seedling stand of soybean. Biol. Control 53, 18–23.

Bennett, A.J., Whipps, J.M., 2008. Beneficial microorganism survival on seed, roots and in rhizosphere soil following application to seed during drum priming. Biol. Control 44, 349–361.

Bhargava, B., Gupta, Y.C., Dhiman, S.R., Sharma, P., 2015. Effect of seed priming on germination, growth and flowering of snapdragon (*Antirrhinum majus* L.). Natl. Acad. Sci. Lett. 38, 81–85.

Bi, Y., Zhang, J., Song, Z., Wang, Z., Qiu, L., Hu, J., Gong, Y., 2019. Arbuscular mycorrhizal fungi alleviate root damage stress induced by simulated coal mining subsidence ground fissures. Sci. Total Environ. 652, 398–405.

Bisen, K., Keswani, C., Patel, J., Sarma, B., Singh, H., 2016. *Trichoderma* spp.: efficient inducers of systemic resistance in plants. In: Microbial-Mediated Induced Systemic Resistance in Plants. Springer.

Bononi, L., Chiaramonte, J.B., Pansa, C.C., Moitinho, M.A., Melo, I.S., 2020. Phosphorus-solubilizing *Trichoderma* spp. from amazon soils improve soybean plant growth. Sci. Rep. 10, 2858.

Brotman, Y., Briff, E., Viterbo, A., Chet, I., 2008. Role of Swollenin, an expansin-like protein from *Trichoderma*, in plant root colonization. Plant Physiol. 147, 779–789.

Cao, L., Jiang, M., Zeng, Z., Du, A., Tan, H., Liu, Y., 2008. Trichoderma Atroviride F6 improves phytoextraction efficiency of mustard (*Brassica juncea* (L.) Coss. var. foliosa Bailey) in Cd, Ni contaminated soils. Chemosphere 71, 1769–1773.

Cardoso, I.M., Kuyper, T.W., 2006. Mycorrhizas and tropical soil fertility. Agric. Ecosyst. Environ. 116, 72-84.
Carro-Huerga, G., Compant, S., Gorfer, M., Cardoza, R.E., Schmoll, M., Gutiérrez, S., Casquero, P.A., 2020. Colonization of *Vitis vinifera* L. by the endophyte *Trichoderma* sp. strain T154: biocontrol activity against *Phaeoacremonium minimum*. Front. Plant Sci. 11, 1170.
Casimiro, I., Marchant, A., Bhalerao, R.P., Beeckman, T., Dhooge, S., Swarup, R., Graham, N., Inzé, D., Sandberg, G., Casero, P.J., 2001. Auxin transport promotes *Arabidopsis* lateral root initiation. Plant Cell 13, 843-852.
Cavalcante, R.S., Lima, H.L., Pinto, G.A., Gava, C.A., Rodrigues, S., 2008. Effect of moisture on *Trichoderma conidia* production on corn and wheat bran by solid state fermentation. Food Bioprocess Technol. 1, 100-104.
Chandra Nayaka, S., Niranjana, S., Uday Shankar, A., Niranjan Raj, S., Reddy, M., Prakash, H., Mortensen, C., 2010. Seed biopriming with novel strain of *Trichoderma harzianum* for the control of toxigenic *Fusarium verticillioides* and fumonisins in maize. Arch. Phytopathol. Plant Protect. 43, 264-282.
Chen, L.H., Huang, X.Q., Zhang, F.G., Zhao, D.K., Yang, X.M., Shen, Q.R., 2012. Application of *Trichoderma harzianum* Sqr-T037 bio-organic fertiliser significantly controls *fusarium* wilt and affects the microbial communities of continuously cropped soil of cucumber. J. Sci. Food Agric. 92, 2465-2470.
Chepsergon, J., Mwamburi, L., Kassim, M.K., 2014. Mechanism of drought tolerance in plants using *Trichoderma* spp. Int. J. Sci. Res. 3, 1592-1595.
Contreras-Cornejo, H.A., Macías-Rodríguez, L., Cortés-Penagos, C., López-Bucio, J., 2009. *Trichoderma virens*, a plant beneficial fungus, enhances biomass production and promotes lateral root growth through an auxin-dependent mechanism in *Arabidopsis*. Plant Physiol. 149, 1579-1592.
Contreras-Cornejo, H.A., Macías-Rodríguez, L., Del-Val, E., Larsen, J., 2016. Ecological functions of *Trichoderma* spp. and their secondary metabolites in the *Rhizosphere*: interactions with plants. FEMS Microbiol. Ecol. 92, fiw036.
Contreras-Cornejo, H.A., Macías-Rodríguez, L., Del-Val, E., Larsen, J., 2020. Interactions of *Trichoderma* with plants, insects, and plant pathogen microorganisms: chemical and molecular bases. In: Co-Evolution of Secondary Metabolites, pp. 263-290.
Dar, N.A., Amin, I., Wani, W., Wani, S.A., Shikari, A.B., Wani, S.H., Masoodi, K.Z., 2017. Abscisic acid: a key regulator of abiotic stress tolerance in plants. Plant Gene 11, 106-111.
Deb, S., Bhadoria, P.B.S., Mandal, B., Rakshit, A., Singh, H.B., 2015. Soil organic carbon: towards better soil health, productivity and climate change mitigation. Clim. Chang. Environ. Sustain. 3, 26-34.
Doni, F., Anizan, I., Radziah, C.C., Salman, A.H., Rodzihan, M.H., Yusoff, W.M.W., 2014. Enhancement of rice seed germination and vigour by "Trichoderma" spp. Res. J. Appl. Sci. Eng. Technol. 7, 4547-4552.
Dubey, S.C., Suresh, M., Singh, B., 2007. Evaluation of *Trichoderma* species against *Fusarium oxysporum* F. sp. ciceris for integrated management of chickpea wilt. Biol. Control 40, 118-127.
Duman, I., 2006. Effects of seed priming with peg or K3po4 on germination and seedling growth in lettuce. Pak. J. Biol. Sci. 9, 923-928.
El Enshasy, H.A., Ambehabati, K.K., Hanapi, S.Z., Dailin, D.J., Elsayed, E.A., Sukmawati, D., Malek, R.A., 2020. *Trichoderma* spp.: a unique fungal biofactory for healthy plant growth. In: Sharma, S.K., Singh, U.B., Sahu, P.K., Singh, H.V., Sharma, P.K. (Eds.), Rhizosphere Microbes: Soil and Plant Functions. Springer Singapore, Singapore.
Elad, Y., 2000. *Trichoderma harzianum* T39 preparation for biocontrol of plant diseases-control of botrytis cinerea, *Sclerotinia sclerotiorum* and *Cladosporium fulvum*. Biocontrol Sci. Tech. 10, 499-507.
El-Mohamedy, R.S., Ziedan, E., Abdalla, A., 2010. Biological soil treatment with *Trichoderma harzianum* to control root rot disease of grapevine (*Vitis vinifera* L.) in newly reclaimed lands in Nobaria Province. Arch. Phytopathol. Plant Protect. 43, 73-87.

El-Mougy, N., Abdel-Kader, M., 2008. Long-term activity of bio-priming seed treatment for biological control of faba bean root rot pathogens. Australas. Plant Pathol. 37, 464–471.

Fathollahy, S., Mozaffari, A., 2020. Investigation the effect of seed biopriming with plant growth promoting rhizobacteria (Pgpr) on antioxidant enzymes activity of seedling and germination indices of two wheat cultivar under salt stress conditions. Seed Sci. Technol. 9, 27–44.

Fazelienasab, B., Omidi, M., Amiritokaldani, M., 2004. Effects of abscisic acid on callus induction and regeneration of different wheat cultivars to mature embryo culture. In: News Directions for a Diverse Planet: Proceedings of the 4th International Brisbane, Australia, p. 26.

Fazeli-Nasab, B., 2021. Biological evaluation of coronaviruses and the study of molecular docking, linalool, and thymol as orf1ab protein inhibitors and the role of Sars-Cov-2 virus in bioterrorism. J. Ilam Univ. Med. Sci. 28, 77–96.

Fazeli-Nasab, B., Amozadeh, M., 2012. Molecular basis of plant tolerance to environmental stresses. Tech. J. Eng. Appl. Sci. 2, 128–130.

Fazeli-Nasab, B., Fooladvand, Z., 2019. The effects of plant growth regulators and explants on callus induction in Ajowan. Iran. J. Cell. Mol. Res. 32, 63–75.

Fazeli-Nasab, B., Sayyed, R., 2019. Plant growth-promoting rhizobacteria and salinity stress: a journey into the soil. In: Plant Growth Promoting Rhizobacteria for Sustainable Stress Management. Springer.

Fazeli-Nasab, B., Masour, O., Mehdi, A., 2012. Estimate of callus induction and volume immature and mature embryo culture and response to in-vitro salt resistance in presence of Nacl and Aba in salt tolerant wheat cultivars. Int. Agric. Crop Sci 4, 8–16.

Fazeli-Nasab, B., Davari, A., Nikoei, M., 2016. The effect of Kinetin on seed germination and seedling growth under salt stress in Sistan *C. copticum*. In: Second International & Fourteenth National Iranian Crop Science Congress. University of Guilan, Rasht, Iran.

Fazeli-Nasab, B., Moshtaghi, N., Forouzandeh, M., 2019a. Effect of solvent extraction on phenol, flavonoids and antioxidant activity of some Iranian Native Herbs. Sci. J. Ilam Univ. Med. Sci. 27, 14–26.

Fazeli-Nasab, B., Sayyed, R.Z., Farsi, M., Ansari, S., El-Enshasy, H.A., 2019b. Genetic assessment of the internal transcribed spacer region (Its1.2) in *Mangifera indica* L. Landraces. Physiol. Mol. Biol. Plants 26, 107–117.

Fazeli-Nasab, B., Khajeh, H., Rahmani, A.F., 2021. Effects of culture medium and plant hormones in organogenesis in olive (Cv. Kroneiki). J. Plant Bioinform. Biotechnol. 1, 1–13.

Finch-Savage, W., Dent, K., Clark, L., 2004. Soak conditions and temperature following sowing influence the response of maize (*Zea mays* L.) seeds to on-farm priming (pre-sowing seed soak). Field Crop Res. 90, 361–374.

Forouzandeh, M., Mohkami, Z., Fazeli-Nasab, B., 2019. Evaluation of biotic elicitors foliar application on functional changes, physiological and biochemical parameters of fennel (*Foeniculum vulgare*). Int. J. Plant Prod. 25, 49–65.

Gad, H.A., Al-Anany, M.S., Abdelgaleil, S.A., 2020. Enhancement the efficacy of spinosad for the control *Sitophilus oryzae* by combined application with diatomaceous earth and *Trichoderma harzianum*. J. Stored Prod. Res. 88, 101663.

Gautam, R.L., Naraian, R., 2020. *Trichoderma*, a factory of multipurpose enzymes: cloning of enzymatic genes. In: Fungal Biotechnology and Bioengineering. Springer.

Ghasemi, S., Safaie, N., Shahbazi, S., Shams-Bakhsh, M., Askari, H., 2020. The role of cell wall degrading enzymes in antagonistic traits of *Trichoderma* virens against *Rhizoctonia solani*. Iran. J. Biotechnol. 18, 18–28.

Gosling, P., Hodge, A., Goodlass, G., Bending, G., 2006. Arbuscular *Mycorrhizal* fungi and organic farming. Agric. Ecosyst. Environ. 113, 17–35.

Gravel, V., Antoun, H., Tweddell, R.J., 2007. Growth stimulation and fruit yield improvement of greenhouse tomato plants by inoculation with *Pseudomonas putida* or *Trichoderma atroviride*: possible role of indole acetic acid (Iaa). Soil Biol. Biochem. 39, 1968–1977.

Guler, N.S., Pehlivan, N., Karaoglu, S.A., Guzel, S., Bozdeveci, A., 2016. *Trichoderma atroviride* Id20g inoculation ameliorates drought stress-induced damages by improving antioxidant defence in maize seedlings. Acta Physiol. Plant. 38, 132.

Guzmán-Guzmán, P., Porras-Troncoso, M.D., Olmedo-Monfil, V., Herrera-Estrella, A., 2019. *Trichoderma* species: versatile plant symbionts. Phytopathology 109, 6–16.

Hajeeboland, R., Asgharzadeh, N., Mehrfar, Z., 2004. Ecological study of *Azotobacter* in two pasture lands of the North-West Iran and its inoculation effect on growth and mineral nutrition of wheat (*Triticum aestivum* L. Cv. Omid) plants. J. Water Soil Sci. 8, 75–90.

Hajieghrari, B., Mohammadi, M., 2016. Growth-promoting activity of indigenous *Trichoderma* isolates on wheat seed germination, seedling growth and yield. Aust. J. Crop. Sci. 10, 1339–1347.

Halifu, S., Deng, X., Song, X., Song, R., Liang, X., 2020. Inhibitory mechanism of *Trichoderma* virens Zt05 on *Rhizoctonia solani*. Plan. Theory 9, 912.

Harman, G.E., 2006. Overview of mechanisms and uses of *Trichoderma* spp. Phytopathology 96, 190–194.

Harman, G.E., 2011. *Trichoderma*—not just for biocontrol anymore. Phytoparasitica 39, 103–108.

Harman, G.E., Shoresh, M., 2007. The mechanisms and applications of symbiotic opportunistic plant symbionts. In: Novel Biotechnologies for Biocontrol Agent Enhancement and Management. Springer.

Harman, G.E., Howell, C.R., Viterbo, A., Chet, I., Lorito, M., 2004a. *Trichoderma* species—opportunistic, avirulent plant symbionts. Nat. Rev. Microbiol. 2, 43–56.

Harman, G.E., Petzoldt, R., Comis, A., Chen, J., 2004b. Interactions between *Trichoderma harzianum* strain T22 and maize inbred line Mo17 and effects of these interactions on diseases caused by *Pythium ultimum* and *Colletotrichum graminicola*. Phytopathology 94, 147–153.

Harman, G.E., Björkman, T., Ondik, K., Shoresh, M., 2008. Changing paradigms on the mode of action and uses of *Trichoderma* spp. Outlooks Pest Manag. 19, 24–29.

Harman, G., Doni, F., Khadka, R.B., Uphoff, N., 2021. Endophytic strains of *Trichoderma* increase plants' photosynthetic capability. J. Appl. Microbiol. 130, 529–546.

Hossain, M.A., Li, Z.-G., Hoque, T.S., Burritt, D.J., Fujita, M., Munné-Bosch, S., 2018. Heat or cold priming-induced cross-tolerance to abiotic stresses in plants: key regulators and possible mechanisms. Protoplasma 255, 399–412.

Hoyos-Carvajal, L., Orduz, S., Bissett, J., 2009. Growth stimulation in bean (*Phaseolus vulgaris* L.) by *Trichoderma*. Biol. Control 51, 409–416.

Isah, T., 2019. Stress and defense responses in plant secondary metabolites production. Biol. Res. 52, 39.

Islam, M.M., Hossain, D.M., Rahman, M.M.E., Suzuki, K., Narisawa, T., Hossain, I., Meah, M.B., Nonaka, M., Harada, N., 2016. Native *Trichoderma* strains isolated from Bangladesh with broad spectrum antifungal action against fungal phytopathogens. Arch. Phytopathol. Plant Protect. 49, 75–93.

Islam, M.M., Hossain, D.M., Nonaka, M., Harada, N., 2017. Biological control of tomato collar rot induced by *Sclerotium rolfsii* using *Trichoderma* species isolated in Bangladesh. Arch. Phytopathol. Plant Protect. 50, 109–116.

Islam, M.M., Shahid, S.B., Akter, A., Hossain, M.S., Bhuiyan, M.S.U., 2020. Effect of indigenous *Trichoderma strains* on growth of tomato seedlings. J. Exp. Agric. Int. 10, 1–6.

Ismail, S.M., 2021. Cholinesterase and aliesterase as a natural enzymatic defense against chlorpyrifos in field populations of *Spodoptera littoralis* (Boisdüval, 1833) (Lepidoptera, Noctüidae). J. Plant Bioinform. Biotechnol. 1, 41–50.

Jabnoun-Khiareddine, H., El-Mohamedy, R., Abdel-Kareem, F., Abdallah, R.A.B., Gueddes-Chahed, M., Daami-Remadi, M., 2015. Variation in chitosan and salicylic acid efficacy towards soil-borne and air-borne fungi and their suppressive effect of tomato wilt severity. J. Plant Pathol. Microbiol. 6, 325.

Jaroszuk-Ściseł, J., Tyśkiewicz, R., Nowak, A., Ozimek, E., Majewska, M., Hanaka, A., Tyśkiewicz, K., Pawlik, A., Janusz, G., 2019. Phytohormones (auxin, gibberellin) and acc deaminase in vitro synthesized by the Mycoparasitic *Trichoderma* Demtkz3a0 strain and changes in the level of auxin and plant resistance markers in wheat seedlings inoculated with this strain conidia. Int. J. Mol. Sci. 20, 4923.

Jomeh Ghasem Abadi, Z., Fakheri, B., Fazeli-Nasab, B., 2019. Study of the molecular diversity of internal transcribed spacer region (Its1.4) in some lettuce genotypes. J. Crop Breed. 11, 29–39.

Kamal, R.K., Athisayam, V., Gusain, Y.S., Kumar, V., 2018. *Trichoderma*: a most common biofertilizer with multiple roles in agriculture. Biomed. J. Sci. Tech. Res. 4, 1–3.

Kamle, M., Borah, R., Bora, H., Jaiswal, A.K., Singh, R.K., Kumar, P., 2020. Systemic acquired resistance (Sar) and induced systemic resistance (Isr): role and mechanism of action against phytopathogens. In: Fungal Biotechnology and Bioengineering. Springer.

Kavamura, V.N., Esposito, E., 2010. Biotechnological strategies applied to the decontamination of soils polluted with heavy metals. Biotechnol. Adv. 28, 61–69.

Kerchev, P., van der Meer, T., Sujeeth, N., Verlee, A., Stevens, C.V., Van Breusegem, F., Gechev, T., 2020. Molecular priming as an approach to induce tolerance against abiotic and oxidative stresses in crop plants. Biotechnol. Adv. 40, 107503.

Keswani, C., Bisen, K., Singh, S., Sarma, B., Singh, H., 2016. A proteomic approach to understand the tripartite interactions between plant-*Trichoderma*-pathogen: investigating the potential for efficient biological control. In: Plant, Soil and Microbes. Springer.

Khajeh, H., Fazeli, F., Mazarie, A., 2021. Effects of culture medium and concentration of different growth regulators on organogenesis damask rose (*Rosa damascena* Mill). J. Plant Bioinform. Biotechnol. 1, 14–27.

Khan, A.A., Sinha, A., Rathi, Y., 2005. Plant growth promoting activity of *Trichoderma harzianum* on rice seed germination and seedung vigour. Indian J. Agric. Res. 39, 256–262.

Khan, N., Bano, A., Ali, S., Babar, M.A., 2020. Crosstalk amongst phytohormones from planta and Pgpr under biotic and abiotic stresses. Plant Growth Regul. 90, 189–203.

Khoshmanzar, E., Aliasgharzad, N., Neyshabouri, M., Khoshru, B., Arzanlou, M., Lajayer, B.A., 2020. Effects of *Trichoderma* isolates on tomato growth and inducing its tolerance to water-deficit stress. Int. J. Environ. Sci. Technol. 17, 869–878.

Konappa, N., Krishnamurthy, S., Arakere, U.C., Chowdappa, S., Ramachandrappa, N.S., 2020a. Efficacy of indigenous plant growth-promoting rhizobacteria and *Trichoderma* strains in eliciting resistance against bacterial wilt in a tomato. Egypt. J. Biol. Pest Control 30, 106.

Konappa, N., Krishnamurthy, S., Dhamodaran, N., Arakere, U.C., Chowdappa, S., Ramachandrappa, N.S., 2020b. Opportunistic avirulent plant symbionts *Trichoderma*: exploring its potential against soilborne phytopathogens. In: Trichoderma: Agricultural Applications and Beyond. Springer.

Konappa, N., Krishnamurthy, S., Dhamodaran, N., Arakere, U.C., Ramachandrappa, N.S., Chowdappa, S., 2020c. Beneficial effects of *Trichoderma* on plant–pathogen interactions: understanding mechanisms underlying genes. In: Trichoderma: Agricultural Applications and Beyond. Springer.

Krasensky, J., Jonak, C., 2012. Drought, salt, and temperature stress-induced metabolic rearrangements and regulatory networks. J. Exp. Bot. 63, 1593–1608.

Küçük, Ç., Cevheri, C., Mutlu, A., 2019. Stimulation of barley (*Hordeum vulgare* L.) growth with local *Trichoderma* sp. isolates. Appl. Ecol. Environ. Res. 17, 4607–4614.

Li, Z., Xu, J., Gao, Y., Wang, C., Guo, G., Luo, Y., Huang, Y., Hu, W., Sheteiwy, M.S., Guan, Y., 2017. The synergistic priming effect of exogenous salicylic acid and H_2O_2 on chilling tolerance enhancement during maize (*Zea mays* L.) seed germination. Front. Plant Sci. 8, 1153.

Liu, A., Plenchette, C., Hamelin, C., 2007. Soil Nutrient and Water Providers: How Arbuscular Mycorrhizal Mycelia Support Plant Performance in a Resource-Limited World. Haworth Food & Agricultural Products Press.

Lo, C.-T., Lin, C.-Y., 2002. Screening strains of *Trichoderma* spp for plant growth enhancement in Taiwan. Plant Pathol. Bull. 11, 215–220.

Loc, N.H., Huy, N.D., Quang, H.T., Lan, T.T., Thu Ha, T.T., 2020. Characterisation and antifungal activity of extracellular chitinase from a biocontrol fungus, *Trichoderma asperellum* Pq34. Mycology 11, 38–48.

López-Coria, M., Hernández-Mendoza, J., Sánchez-Nieto, S., 2016. *Trichoderma asperellum* induces maize seedling growth by activating the plasma membrane H+-Atpase. Mol. Plant-Microbe Interact. 29, 797–806.

Maithani, D., Singh, H., Sharma, A., 2021. Stress alleviation in plants using Sar and Isr: current views on stress signaling network. In: Microbes and Signaling Biomolecules Against Plant Stress. Springer.

Maketon, M., Apisitsantikul, J., Siriraweekul, C., 2008. Greenhouse evaluation of *Bacillus subtilis* Ap-01 and *Trichoderma harzianum* Ap-001 in controlling tobacco diseases. Braz. J. Microbiol. 39, 296–300.

Malinich, E.A., Wang, K., Mukherjee, P.K., Kolomiets, M., Kenerley, C.M., 2019. Differential expression analysis of *Trichoderma virens* Rna reveals a dynamic transcriptome during colonization of *Zea mays* roots. BMC Genomics 20, 1–19.

Manske, G.B., Lüttger, A., Behl, R., Vlek, P.G., 1995. Nutrient efficiency based on Va Mycorrhizae (Vam) and total root length of wheat cultivars grown in India. Angew. Bot. 69, 108–110.

Mantja, K., Musa, Y., Ala, A., Rosmana, A., 2015. Indigenous *Trichoderma* isolated from maize *Rhizosphere* with potential for enhancing seedling growth. Int. J. Sci. Res. 4, 1814–1818.

Martínez-Medina, A., Alguacil, M.D.M., Pascual, J.A., Van Wees, S.C., 2014. Phytohormone profiles induced by *Trichoderma* isolates correspond with their biocontrol and plant growth-promoting activity on melon plants. J. Chem. Ecol. 40, 804–815.

Mastouri, F., Björkman, T., Harman, G.E., 2010. Seed treatment with *Trichoderma harzianum* alleviates biotic, abiotic, and physiological stresses in germinating seeds and seedlings. Phytopathology 100, 1213–1221.

Mayo-Prieto, S., Campelo, M., Lorenzana, A., Rodríguez-González, A., Reinoso, B., Gutiérrez, S., Casquero, P., 2020. Antifungal activity and bean growth promotion of *Trichoderma strains* isolated from seed vs soil. Eur. J. Plant Pathol. 158, 817–828.

Mazhabi, M., Nemati, H., Rouhani, H., Tehranifar, A., Moghadam, E., Kaveh, H., Rezaee, A., 2011. The effect of *Trichoderma* on polianthes qualitative and quantitative properties. J. Anim. Plant Sci. 21, 617–621.

Mazrou, Y., Makhlouf, A., Hassan, M., Baazeem, A., Hamad, A., Farid, M., 2020. Influence of chitinase production on the antagonistic activity of *Trichoderma* against plant-pathogenic fungi. J. Environ. Biol. 41, 1501–1510.

Meena, S.K., Rakshit, A., Meena, V.S., 2016. Effect of seed bio-priming and N doses under varied soil type on nitrogen use efficiency (Nue) of wheat (*Triticum aestivum* L.) under greenhouse conditions. Biocatal. Agric. Biotechnol. 6, 68–75.

Mehetre, S.T., Mukherjee, P.K., 2015. *Trichoderma* improves nutrient use efficiency in crop plants. In: Nutrient Use Efficiency: From Basics to Advances. Springer.

Mehrabi, A.A., Omidi, M., Fazelinasab, B., 2011. Effect of salt stress on seed germination, seedling growth and callus culture of rapeseed (*Brassica napus* L.). Iran. J. Field Crop Sci. 42, 81–90.

Mishra, L., Abidi, A., 2010. Phosphorus-zinc interaction: effects on yield components and biochemical composition and bread making qualities of wheat. World Appl. Sci. J. 10, 568–573.

Mohamed, H.A.-L.A., Haggag, W.M., 2006. Biocontrol potential of salinity tolerant mutants of *Trichoderma harzianum* against *Fusarium oxysporum*. Braz. J. Microbiol. 37, 181–191.

Monte, E., 2001. Understanding *Trichoderma*: between biotechnology and microbial ecology. Int. Microbiol. 4, 1–4.

Naddaf, M.E., Rabiei, G., Ganji Moghadam, E., Mohammadkhani, A., 2021. In vitro production of Ppv-free sweet cherry (*Prunus avium* Cv. Siahe-Mashhad) by meristem culture and micro-grafting. J. Plant Bioinform. Biotechnol. 1, 51–59.

Naderi, D., Jami, R., Rehman, F.U., 2021. A review of Rna motifs, identification algorithms and their function on plants. J. Plant Bioinform. Biotechnol. 1, 28–40.

Naeimi, S., Khosravi, V., Varga, A., Vágvölgyi, C., Kredics, L., 2020. Screening of organic substrates for solid-state fermentation, viability and bioefficacy of *Trichoderma harzianum* As12-2, a biocontrol strain against rice sheath blight disease. Agronomy 10, 1258.

Negi, S., Bharat, N.K., Kumar, M., 2019. Effect of seed biopriming with indigenous Pgpr, *Rhizobia* and *Trichoderma* sp. on growth, seed yield and incidence of diseases in French Bean (*Phaseolus vulgaris* L.). Legum. Res. LR-4135, 1–9.

Okoth, S.A., Otadoh, J.A., Ochanda, J.O., 2011. Improved seedling emergence and growth of maize and beans by *Trichoderma harziunum*. Trop. Subtrop. Agroecosyst. 13, 65–71.

Oljira, A.M., Hussain, T., Waghmode, T.R., Zhao, H., Sun, H., Liu, X., Wang, X., Liu, B., 2020. *Trichoderma* enhances net photosynthesis, water use efficiency, and growth of wheat (*Triticum aestivum* L.) under salt stress. Microorganisms 8, 1565.

Omar, M., Osman, M., Kasim, W., Abd El-Daim, I., 2009. Improvement of salt tolerance mechanisms of barley cultivated under salt stress using *Azospirillum brasilense*. In: Salinity and Water Stress. Springer.

Pehlivan, N., Yesilyurt, A.M., Durmus, N., Karaoglu, S.A., 2017. *Trichoderma lixii* Id11d seed biopriming mitigates dose dependent salt toxicity in maize. Acta Physiol. Plant. 39, 79.

Poveda, J., Eugui, D., Abril-Urias, P., 2020. Could *Trichoderma* be a plant pathogen? successful root colonization. In: Trichoderma. Springer.

Rahman, M.A., Sultana, R., Begum, M.F., Alam, M.F., 2012. Effect of culture filtrates of *Trichoderma* on seed germination and seedling growth in chili. Int. J. Biosci. 2, 46–55.

Rangel, D.E., Finlay, R.D., Hallsworth, J.E., Dadachova, E., Gadd, G.M., 2018. Fungal strategies for dealing with environment-and agriculture-induced stresses. Fungal Biol. 122, 602–612.

Rascio, A., Russo, M., Platani, C., Di Fonzo, N., 1998. Drought intensity effects on genotypic differences in tissue affinity for strongly bound water. Plant Sci. 132, 121–126.

Rasouli, H., Popović-Djordjević, J., Sayyed, R.Z., Zarayneh, S., Jafari, M., Fazeli-Nasab, B., 2020. Nanoparticles: a new threat to crop plants and soil rhizobia? In: Hayat, S., Pichtel, J., Faizan, M., Fariduddin, Q. (Eds.), Sustainable Agriculture Reviews 41: Nanotechnology for Plant Growth and Development. Springer International Publishing, Cham.

Ray, S., Singh, V., Singh, S., Sarma, B.K., Singh, H.B., 2016. Biochemical and histochemical analyses revealing endophytic *Alcaligenes faecalis* mediated suppression of oxidative stress in abelmoschus esculentus challenged with *Sclerotium rolfsii*. Plant Physiol. Biochem. 109, 430–441.

Raymaekers, K., Ponet, L., Holtappels, D., Berckmans, B., Cammue, B.P., 2020. Screening for novel biocontrol agents applicable in plant disease management—a review. Biol. Control 144, 104240.

Reddy, B.N., Saritha, K.V., Hindumathi, A., 2014. In vitro screening for antagonistic potential of seven species of *Trichoderma* against different plant pathogenic fungi. Res. J. Biol. 2, 29–36.

Reino, J.L., Guerrero, R.F., Hernández-Galán, R., Collado, I.G., 2008. Secondary metabolites from species of the biocontrol agent *Trichoderma*. Phytochem. Rev. 7, 89–123.

Rezaloo, Z., Shahbazi, S., Askari, H., 2020. Biopriming with *Trichoderma* on germination and vegetative characteristics of sweet corn, sugar beet and wheat. Iran. J. Seed Sci. Technol. 8, 199–210.

Rojo, F.G., Reynoso, M.M., Ferez, M., Chulze, S.N., Torres, A.M., 2007. Biological control by *Trichoderma* species of *Fusarium solani* causing peanut brown root rot under field conditions. Crop Prot. 26, 549–555.

Saba, H., Vibhash, D., Manisha, M., Prashant, K., Farhan, H., Tauseef, A., 2012. Trichoderma—a promising plant growth stimulator and biocontrol agent. Mycosphere 3, 524–531.

Saber, W.I., Ghoneem, K.M., Rashad, Y.M., Al-Askar, A.A., 2017. *Trichoderma harzianum* Wky1: an indole acetic acid producer for growth improvement and anthracnose disease control in sorghum. Biocontrol Sci. Tech. 27, 654–676.

Saddiq, M.S., Iqbal, S., Afzal, I., Ibrahim, A.M., Bakhtavar, M.A., Hafeez, M.B., Jahanzaib, Maqbool, M.M., 2019. Mitigation of salinity stress in wheat (*Triticum aestivum* L.) seedlings through physiological seed enhancements. J. Plant Nutr. 42, 1192–1204.

Sakakibara, H., 2006. Cytokinins: activity, biosynthesis, and translocation. Annu. Rev. Plant Biol. 57, 431–449.

Sarkar, D., Rakshit, A., 2021. Bio-priming in combination with mineral fertilizer improves nutritional quality and yield of red cabbage under Middle Gangetic Plains, India. Sci. Hortic. 283, 110075.

Sarkar, D., Pal, S., Mehjabeen, M., Singh, V., Singh, S., Pul, S., Garg, J., Rakshit, A., Singh, H., 2018. Addressing stresses in agriculture through bio-priming intervention. In: Advances in Seed Priming. Springer.

Sasidharan, S., Tuladhar, P., Raj, S., Saudagar, P., 2020. Understanding its role bioengineered *Trichoderma* in managing soil-borne plant diseases and its other benefits. In: Fungal Biotechnology and Bioengineering. Springer.

Seidl-Seiboth, V., Gruber, S., Sezerman, U., Schwecke, T., Albayrak, A., Neuhof, T., von Döhren, H., Baker, S.E., Kubicek, C.P., 2011. Novel hydrophobins from *Trichoderma* define a new hydrophobin subclass: protein properties, evolution, regulation and processing. J. Mol. Evol. 72, 339–351.

Seyed Sharifi, R., 2016. Application of biofertilizers and zinc increases yield, nodulation and unsaturated fatty acids of soybean. Zemdirbyste-Agriculture 103, 251–258.

Shabala, S., 2017. Plant Stress Physiology. Cabi.

Shaban, H., Fazeli-Nasab, B., Alahyari, H., Alizadeh, G., Shahpesandi, S., 2015. An overview of the benefits of compost tea on plant and soil structure. Adv. Biores. 6, 154–158.

Shahhosini, Z., Gholami, A., Asghari, H.R., 2013. The effects of *Mycorrhizal symbiosis* on yield and some growth characteristics of maize under water deficit condition. Iran. J. Field Crop Sci. 44, 249–260.

Shahi, H., Mahdinezhad, N., Haddadi, F., Fazeli Nasab, B., 2019. Assessment of relative gene expression of Wdhn5 in wild and agronomy wheat genotypes under drought stress conditions. Genet. Eng. Biosaf. J. 8, 189–199.

Shirazi, E., Fazeli-nasab, B., Ramshin, H.-A., Fazel-Najaf-Abadi, M., Izadi-darbandi, A., 2016. Evaluation of drought tolerance in wheat genotypes under drought stress at germination stage. J. Crop Breed. 8, 207–219.

Shoresh, M., Harman, G.E., 2008. The molecular basis of shoot responses of maize seedlings to *Trichoderma harzianum* T22 inoculation of the root: a proteomic approach. Plant Physiol. 147, 2147–2163.

Shoresh, M., Harman, G.E., Mastouri, F., 2010. Induced systemic resistance and plant responses to fungal biocontrol agents. Annu. Rev. Phytopathol. 48, 21–43.

Shukla, N., Awasthi, R., Rawat, L., Kumar, J., 2015. Seed biopriming with drought tolerant isolates of *Trichoderma harzianum* promote growth and drought tolerance in *Triticum aestivum*. Ann. Appl. Biol. 166, 171–182.

Singh, A., Srivastava, S., Singh, H., 2007. Effect of substrates on growth and shelf life of *Trichoderma harzianum* and its use in biocontrol of diseases. Bioresour. Technol. 98, 470–473.

Singh, V., Upadhyay, R., Sarma, B., Singh, H., 2016a. Seed bio-priming with *Trichoderma asperellum* effectively modulate plant growth promotion in pea. Int. J. Agric. Environ. Biotechnol. 9, 361–365.

Singh, V., Upadhyay, R.S., Sarma, B.K., Singh, H.B., 2016b. *Trichoderma asperellum* spore dose depended modulation of plant growth in vegetable crops. Microbiol. Res. 193, 74–86.

Singh, B.N., Dwivedi, P., Sarma, B.K., Singh, G.S., Singh, H.B., 2019a. A novel function of N-signaling in plants with special reference to *Trichoderma* interaction influencing plant growth, nitrogen use efficiency, and cross talk with plant hormones. 3 Biotech 9, 1–13.

Singh, R.K., Buckseth, T., Tiwari, J.K., Sharma, A.K., Chakrabarti, S., 2019b. Recent advances in production of healthy planting material for disease management in potato. Biotech Today 9, 7–15.

Smith, A.M., Zeeman, S.C., Smith, S.M., 2005. Starch degradation. Annu. Rev. Plant Biol. 56, 73–98.

Sofo, A., Tataranni, G., Xiloyannis, C., Dichio, B., Scopa, A., 2012. Direct effects of *Trichoderma harzianum* strain T-22 on micropropagated shoots of Gisela6® (*Prunus cerasus× Prunus canescens*) rootstock. Environ. Exp. Bot. 76, 33–38.

Soltani, H., Zafari, D., Rouhani, H., 2006. A study on biological control of the crown, root and tuber fungal diseases of potato by *Trichoderma harzianum* under in-vivo and field condition in Hamadan. Agric. Res. 5, 13–25.

Song, H., 2005. Effects of VAM on host plant in the condition of drought stress and its mechanisms. Electron. J. Biol. 1, 44–48.

Sood, M., Kapoor, D., Kumar, V., Sheteiwy, M.S., Ramakrishnan, M., Landi, M., Araniti, F., Sharma, A., 2020. Trichoderma: the "secrets" of a multitalented biocontrol agent. Plan. Theory 9, 762.

Speckbacher, V., Ruzsanyi, V., Martinez-Medina, A., Hinterdobler, W., Doppler, M., Schreiner, U., Böhmdorfer, S., Beccaccioli, M., Schuhmacher, R., Reverberi, M., 2020. The lipoxygenase Lox1 is involved in light-and injury-response, conidiation, and volatile organic compound biosynthesis in the Mycoparasitic fungus *Trichoderma atroviride*. Front. Microbiol. 11, 2004.

Srivastava, R., Khalid, A., Singh, U., Sharma, A., 2010. Evaluation of arbuscular mycorrhizal fungus, fluorescent *Pseudomonas* and *Trichoderma harzianum* formulation against *Fusarium oxysporum* f. sp. lycopersici for the management of tomato wilt. Biol. Control 53, 24–31.

Sriwati, R., Chamzurn, T., Soesanto, L., Munazhirah, M., 2019. Field application of *Trichoderma* suspension to control cacao pod rot (*Phytophthora palmivora*). Agrivita 41, 175–182.

Sun, Y.-M., Horng, C.-Y., Chang, F.-L., Cheng, L.-C., Tian, W.-X., 2010. Biosorption of lead, mercury and cadmium ions by *Aspergillus terreus* immobilized in a natural matrix. Pol. J. Microbiol. 59, 37–44.

Suzuki, N., Rivero, R.M., Shulaev, V., Blumwald, E., Mittler, R., 2014. Abiotic and biotic stress combinations. New Phytol. 203, 32–43.

Swain, H., Mukherjee, A.K., 2020. Host-pathogen–*Trichoderma* interaction. In: Trichoderma. Springer.

Swain, H., Adak, T., Mukherjee, A.K., Sarangi, S., Samal, P., Khandual, A., Jena, R., Bhattacharyya, P., Naik, S.K., Mehetre, S.T., 2021. Seed biopriming with *Trichoderma* strains isolated from tree bark improves plant growth, antioxidative defense system in rice and enhance straw degradation capacity. Front. Microbiol. 12, 240.

Swehla, A., Pandey, A.K., Nair, R.M., 2020. Bioactivity of *Trichoderma harzianum* isolates against the fungal root rot pathogens with special reference to *Macrophomina phaseolina* causing dry root rot of mungbean. Indian Phytopathol. 73, 787–792.

Tancic, S.L., 2013. Impact of *Trichoderma* spp. on soybean seed germination and potential antagonistic effect on *Sclerotinia sclerotiorum*. In: Pesticides and Phytomedicine/Pesticidi i fitomedicina, p. 28.

Tančić-Živanov, S., Medić-Pap, S., Danojević, D., Prvulović, D., 2020. Effect of *Trichoderma* spp. on growth promotion and antioxidative activity of pepper seedlings. Braz. Arch. Biol. Technol. 63, e20180659.

Tashakori Fard, E., Taghavi Ghasemkheyli, F., Pirdashti, H., Ghanbary, T., Bahmanyar, M., 2017. Symbiotic effect of Trichoderma atroviride on growth characteristics and yield of two cultivars of rapeseed (*Brassica napus* L.) in a contaminated soil treated with copper nitrate. Iran. J. Field Crops Res. 15, 74–86.

Tasik, S., Widyastuti, S.M., 2015. Mekanisme parasitisme *Trichoderma harzianum* terhadap *Fusarium oxysporum* pada semai acacia mangium. J. Hama Penyakit Tumbuhan Trop. 15, 72–80.

Taylor, A.G., 2020. Seed storage, germination, quality and enhancements. In: Wien, H.C., Stutzel, H. (Eds.), The Physiology of Vegetable Crops, second ed, pp. 1–30.

Usha, E., Reddy, S., Manuel, S.G., Kale, R.D., 2012. In-vitro control of fusarium oxysporum by *Aspergillus* sp. and *Trichoderma* sp. isolated from vermicompost. J. Bio Innov. 1, 142–147.

Varma, A., Bakshi, M., Lou, B., Hartmann, A., Oelmueller, R., 2012. *Piriformospora indica*: a novel plant growth-promoting *Mycorrhizal* fungus. Agric. Res. 1, 117–131.

Verma, I., Parihar, N.N., Sanjay, D.H., Rai, P.K., 2017. Effect of biologicals and chemicals seed treatments on growth, yield and yield attributing traits in maize (*Zea mays* L.). J. Pharmacogn. Phytochem. 6, 1955–1959.

Vinale, F., Ambrosio, G.D., Abadi, K., Scala, F., Marra, R., Turrà, D., Woo, S.L., Lorito, M., 2004. Application of *Trichoderma harzianum* (T22) and *Trichoderma atroviride* (P1) as plant growth promoters, and their compatibility with copper oxychloride. J. Zhejiang Univ. (Sci.) 30, 425.

Vinale, F., Sivasithamparam, K., Ghisalberti, E.L., Marra, R., Woo, S.L., Lorito, M., 2008. *Trichoderma*-plant-pathogen interactions. Soil Biol. Biochem. 40, 1–10.

Wang, W., Vinocur, B., Altman, A., 2003. Plant responses to drought, salinity and extreme temperatures: towards genetic engineering for stress tolerance. Planta 218, 1–14.

Weindling, R., 1932. *Trichoderma* lignorum as a parasite of other soil fungi. Phytopathology 22, 837–845.

Worrall, E.A., Hamid, A., Mody, K.T., Mitter, N., Pappu, H.R., 2018. Nanotechnology for plant disease management. Agronomy 8, 285.

Xu, Y., Zhang, J., Shao, J., Feng, H., Zhang, R., Shen, Q., 2020. Extracellular proteins of *Trichoderma guizhouense* elicit an immune response in maize (*Zea mays*) plants. Plant Soil 449, 1–17.

Yadav, R., Jaiswal, A., 2001. Morpho-physiological changes and variable yield of wheat genotypes under moisture stress conditions. Indian J. Plant Physiol. 6, 390–394.

Yan, L., Khan, R.A.A., 2021. Biological control of bacterial wilt in tomato through the metabolites produced by the biocontrol fungus, *Trichoderma harzianum*. Egypt. J. Biol. Pest Control 31, 1–9.

Yazdani, M., Pirdashti, H., Tajik, M.A., Bahmanyar, M.A., 2012. Effect of *Trichoderma* spp. and different organic manures on growth and development in soybean [*Glycine max* (L.) Merril.]. J. Crop. Prod. 1, 65–82.

Yu, Z., Wang, Z., Zhang, Y., Wang, Y., Liu, Z., 2021. Biocontrol and growth-promoting effect of *Trichoderma asperellum* Tasphu1 isolate from *Juglans mandshurica* rhizosphere soil. Microbiol. Res. 242, 126596.

Zaied, K., El-Hady, A., Afify, A.H., Nassef, M., 2003. Yield and nitrogen assimilation of winter wheat inoculated with new recombinant inoculants of *Rhizobacteria*. Pak. J. Biol. Sci. 6, 344–358.

Zehra, A., Meena, M., Dubey, M.K., Aamir, M., Upadhyay, R., 2017. Activation of defense response in tomato against fusarium wilt disease triggered by *Trichoderma harzianum* supplemented with exogenous chemical inducers (Sa and Meja). Braz. J. Bot. 40, 651–664.

Zhang, H., Godana, E.A., Sui, Y., Yang, Q., Zhang, X., Zhao, L., 2020. Biological control as an alternative to synthetic fungicides for the management of grey and blue mould diseases of table grapes: a review. Crit. Rev. Microbiol. 46, 450–462.

Zvereva, E., Vandyukova, I., Vandyukov, A., Katsyuba, S., Khamatgalimov, A., Kovalenko, V., 2012. Ir and Raman spectra, hydrogen bonds, and conformations of N-(2-hydroxyethyl)-4, 6-dimethyl-2-Oxo-1, 2-dihydropyrimidine (drug xymedone). Russ. Chem. Bull. 61, 1199–1206.

Bacterial biofertilizers for bioremediation: A priority for future research

Asfa Rizvi[a], Bilal Ahmed[b], Shahid Umar[a], and Mohammad Saghir Khan[c]

[a]Department of Botany, School of Chemical and Life Sciences, Jamia Hamdard, New Delhi, India, [b]School of Chemical Engineering, Yeungnam University, Gyeongsan, Republic of Korea, [c]Department of Agricultural Microbiology, Faculty of Agricultural Sciences, Aligarh Muslim University, Aligarh, Uttar Pradesh, India

21.1 Introduction

The human populations worldwide continue to grow worryingly causing various problems including food insecurity (Salcedo Gastelum et al., 2020; Igiri et al., 2018). So, to feed them healthy food, there is a need to optimize the crop production using fertile agrosystems. One of the essential factors of healthy soil is the prevalence of beneficial soil microbiota that essentially enhances the nutrient pool through biogeochemical cycling of the soil constituents. Therefore, the biofertilizers (microbial inoculants) must be friendly to the soil environment (Mercado-Blanco et al., 2018). However, majority of arable soils worldwide do not have optimal cultivation conditions due to abiotic stresses, for example, organic pollutants (Tomar et al., 2019), heavy metals (Rizvi et al., 2020a), drought (Wang et al., 2020a,b), salinity (Zahra et al., 2020), and extremes of temperature (Narayanan, 2018). Of these, discharge of metals from different anthropogenic activities (Bankole et al., 2019; Adamiec et al., 2016) poses a major threat to the sustainability of crop production systems. Heavy metals in general persist in the environment due to their biologically nondegradative feature and following uptake adversely affect the agronomically useful biofertilizers and their associated physiological functions (Sharma et al., 2020; Saif and Khan, 2017). Besides microbial compositions, the toxic metals beyond a certain threshold limit inhibit the growth and yields of food crops by altering their metabolism (Goyal et al., 2020; Rizvi and Khan, 2017b, 2018). Some of the most common adverse impacts of toxic metals include the genotoxicity, disruption of physiological process, such as lipid peroxidation, photosynthesis inactivation of the respiration process, disruption of protein synthesis, and carbohydrate metabolism (Rizvi and Khan, 2017a; Jaishankar et al., 2014; Van Assche and Clijsters, 1990), resulting into the death of plants (Waqas et al., 2019; Amari et al., 2017). Due to these, the abatement of polluted soils becomes imperative so that the stressed soils that are rendered unsuitable for cultivation be made cultivable again. In this regard, efforts have made to correct the polluted soils and hence to protect rhizospheres and to reestablish the fertility of soils. The major options adopted so far include the physical, chemical, and biological (bioremediation) process (Khan et al., 2009). The physical and chemical methods though, are, swift but are perplexing due to factors such as cost, operation facilities, and their disparaging effect on biological properties of soils resulting in secondary pollution (Ullah et al., 2015; Ali et al., 2013). Bioremediation that entails the use of biofertilizers notably the nitrogen fixers, phosphate solubilizers/mobilizers, potassic/zinc solubilizers, etc. (Esertas et al., 2020; Afzal et al., 2017; Bhojiya and Joshi, 2016) capable of detoxifying metals and endowed with unique plant growth modifying abilities has been found cheap and ecologically pleasant (Ouertani et al., 2020). Biofertilizers once established in soils and/or colonize the root surface alter the mobility and toxicity of pollutants and consequently avert further contamination. Accordingly, the metal-tolerant biofertilizers in recent times have attracted attention as a vital microbiological resource in soil remediation along with crop cultivation practices (Jain et al., 2020; Oves et al., 2019). When applied, such biofertilizers have revealed an encouraging impact on many food crops growing under stress. For instance, the metal-tolerant biofertilizers when applied to previously polluted soils or soils contaminated by design have significantly declined the metal toxicity and upgraded the yield of cereals (Rizvi et al., 2020a). Apart from their role in soil remediation and circumventing abiotic stress to food crops, biofertilizers also enhance the nutrient pool of soils (soil fertility) and hence the crop production in different agroecosystems by various direct and indirect mechanisms (Zaidi et al., 2017b). The biofertilizers comprising dual properties of metal remediation ability and plant growth-promoting activities are thus extremely treasured organisms in cultivation systems (Zerrouk et al., 2020). Therefore, the future application of different types of biofertilizers encompassing one or many useful activities such as remediation and capabilities to enhance the overall performance of crops in polluted soils place them above the most traditional microbiome in agronomic practices to fulfill "food/feed" demands across the globe.

21.2 Heavy metal contamination of agronomic soils: An overview

Soil is a nonrenewable resource that supports agroecosystems and serves as a reservoir for crop production (Hou et al., 2020). There is a need to double the global agricultural production by 2050 to feed the constantly increasing human populations. The anthropogenic activities and industrialization on the contrary cause soil contamination leading eventually to the losses in food production worldwide. Among different organic and inorganic pollutants, heavy metals are the most important soil contaminant due to toxicity, ubiquity, nonbiodegradability, and bioavailability to crops. Together, these properties make heavy metals a major threat to global food security. The major sources that add metals to agronomic soils include: (i) surface runoff from mine tailings (Rzymski et al., 2017), (ii) application of agrochemicals (Gonçalves Jr et al., 2014), (iii) industrial discharge (Al Moharbi et al., 2020), (iv) metal plating (Bai et al., 2016), (v) paint manufacturing (Woldeamanuale and Hassen, 2017), (vi) use of sewage sludge (Akcil et al., 2015) and poor-quality (polluted) water in crop cultivation practices, (vii) specks of dust and aerosols discharged from mining and smelting activities (Shu et al., 2021; Ray and Dey, 2020), (viii) vehicle use (Zhang et al., 2015), and (ix) cement manufacture (Dong et al., 2015) and electronic-waste processing unit (Debnath et al., 2018). Besides these, the natural sources such as volcanic activity, metal corrosion, metal evaporation from soil and water and sediment resuspension, soil erosion, and geological weathering also cause soil pollution. The metals so present in the soil alter the physicochemical properties of soil such as the pH, color, porosity, and natural chemistry and hence the quality of both soil and water. Also, the deposited metals in soils adversely affect the composition and functions of soil microbiome (Abdelbary et al., 2019), soil health, and crop productivity (Haider et al., 2021) and pose risks to human health through food chain (Kwunonwo et al., 2020). The mounting issue of soil pollution has therefore drawn the attention of workers across the globe to protect the declining cultivable lands and to restore the resources in its natural state (Glick, 2010). Conclusively, the remediation of metal-contaminated soils becomes necessary to (i) reduce the associated risks, (ii) make the land resource available for crop production, (iii) upgrade/maintain food security, and (iv) trim down the land problems arising from cultivation practices.

21.3 Bioremediation: Concepts and prospects

Scientists working in different areas across nations have realized the major agroecological threat and are trying to find ways to solve this global issue. In this regard, based on the contamination types, various physical (extraction, stabilization, immobilization, soil washing, landfills, and excavation) and chemical (coagulation, chemical precipitation, electrodialysis, evaporative recovery, floatation, flocculation, ion exchange, nanofiltration, and reverse osmosis) approaches have been attempted to restore the polluted environment. Sadly, majority of such methods are costly, nonspecific, exhibit relatively low competence, draining for the soil-plant system, and cannot be applied over a large polluted area (Gupta and Diwan, 2017). Additionally, the distortion of the physical, chemical, and biological properties of soils and emergence of secondary pollution are other major concerns. Due to these difficulties, physicochemical methods have been found unfit for agriculture. So, bioremediation has been accepted and advocated widely as a modern technique to counteract the problems associated with other remediation methods so as to decontaminate the polluted environment in an eco-friendly and sustainable manner (Abedinzadeh et al., 2019; Jin et al., 2018). Traditionally, bioremediation is an attractive process whereby hazardous wastes are biologically degraded under controlled conditions to an innocuous state, or to levels below concentration limits established by the regulatory authorities. This technique which involves the use of living organisms, especially microorganisms (microbial remediation) and plants (phytoremediation) to correct the polluted environment (soils, sediments, water, and air), can be categorized into (a) in situ bioremediation: the treatment of xenobiotics at the contaminated sites itself and (b) ex situ bioremediation: contaminated soil is treated in laboratories/at other sites. The ex situ method is, however, cost-effective, exhaustive, and dangerous for the environment and hence is generally not considered for remediation of polluted environment. Among naturally abundant resources, biosensor soil microbiomes, especially bacteria (Esertas et al., 2020; Afzal et al., 2017), fungi (Liaquat et al., 2020; Kumar et al., 2019), algae (Bordoloi et al., 2020; Salama et al., 2019), and yeasts (Sun et al., 2020), have been considered as the potential organisms for detoxification of pollutants and promoting the crop production under stressed soils (Khan et al., 2017; Pandey and Bhatt, 2016). Despite the fact that numerous bacterial communities with metal resistance abilities have been reported, yet there is a growing need to discover novel biofertilizer that can facilitate the growth and yield of plants in contaminated soils. Also, the increase in cultivation area seems challenging, so the approaches are needed to magnify the crop productivity per unit area. Since the use of agrochemicals and intensive cultivation practices are harming the land resources and environmental quality, microbiological resources could be helpful in resolving these issues.

Principally, the soil microbiome plays an important role in metal bioremediation due largely to their easy cultivation, produces no secondary pollution, and displays higher efficiency at low metal concentrations (Tarekegn et al., 2020; Bojórquez and Voltolina, 2016). Overall, biofertilizer-based bioremediation is a cheap, simple, and environmentally friendly remediation technology. However, the technique still has certain limitations like how indigenous soil microbiome would cope with the fluctuating environment after they are introduced into the polluted soils. Apart from directly influencing stressed environment, the biofertilizers also aid indirectly in the remediation process through phytoextraction by modifying the (i) solubility, availability, and transport of metals; (ii) nutrient pool of soil; (iii) hydrology of soil; and (iv) chelator molecules (Ma et al., 2015). For example, siderophores, a low molecular weight (400–1500 Da) iron-complexing compound, secreted by many biofertilizers like *Pseudomonas* (Kügler et al., 2020), *Azotobacter* (Ferreira et al., 2019), *Bacillus* (Esertas et al., 2020), rhizobia (Zhao et al., 2020), and *Azospirillum* (Vijayalakshmi et al., 2019) do play a significant role in mobilization and accumulation of metals. Despite constraints, bioremediation technologies currently used in many countries mostly adopt in situ biotreatment approach. Here, we discuss the role of different biofertilizers in the remediation of metal-contaminated soils.

21.4 Biofertilizer technology in bioremediation

21.4.1 Biofertilizers: Definition and categories

Broadly, biofertilizers are the formulated product prepared from one or many physiologically active microbes that stimulate the growth of plants by enhancing the nutrient availability in the rhizosphere (Singh et al., 2018; Kumar, 2016). Also, biofertilizers have been defined as the formulations consisting of beneficial microbial strains immobilized or trapped on inert carrier materials that can be employed to enhance plant growth and increase soil fertility (Aloo et al., 2020a,b). Biofertilizers, which include variously distributed soil microbiota with varying physiological activities (Fig. 21.1), are also known by other names such as bioformulations or microbial inoculants, microbial cultures, bioinoculants, bacterial inoculants, or bacterial fertilizers (El-Ghamry et al., 2018). Of the rhizosphere microbiota, the plant growth-promoting rhizobacteria (PGPR) are the most widely agronomically explored organisms (Glick, 2014) since long back. Field studies have revealed that the yield of food crops can be increased by approximately 25% applying biofertilizers while reducing the use of nitrogenous and phosphatic fertilizers by about 25%–50% and 25%,

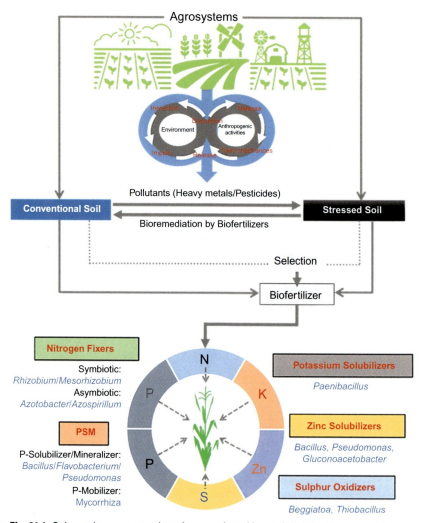

Fig. 21.1 Schematic representation of categories of bacterial biofertilizers used in the remediation of metal-contaminated environment.

respectively (Ghany et al., 2013). Biofertilizers are used to protect rhizosphere and to optimize agricultural production worldwide (Aasfar et al., 2021; Nosheen et al., 2021) but their role in reducing pollution and increasing plant nutrients in contaminated soils have recently been emphasized (Asoegwu et al., 2020). The success of bioremediation through bacterial biofertilizers, for example, nitrogen fixers, and nutrient mobilizers such as phosphate/potassium/zinc solubilizers, iron sequesters, and sulfur oxidizers (Naveed et al., 2020; Thomas and Singh, 2019), however, depends on biotic factors such as the competence and remediating capability of the introduced or indigenous biofertilizers and abiotic factors such as nutrients, soil moisture, pH, and temperature (Ivshina et al., 2001). The recent developments in the bacterial biofertilizer-mediated remediation of metal-contaminated soil are discussed in the following section, whereas the impact of biofertilizer on the crop production in conventional soils is discussed separately.

21.4.2 Remediation of metal-contaminated environment by biofertilizer: Recent advances

The application of plant-microbe interactions for the restoration of polluted soil is an important and well-adopted microbial technology (Hansda et al., 2014). Among bioremediation options, approximately 35% people prefer soil microbiome for the management of polluted soils while only 16% people choose plant-based (phytoremediation) technology for the removal of toxic metals from contaminated sites. Bioremediation employing microbe-based preparation is significant due largely to their exceptional qualities of accumulation, sorption, and enzymatic reduction (Banerjee et al., 2018; Long et al., 2018). Generally, biosorption, extracellular precipitation, biotransformation, sequestration, and flush out (efflux pumping) methods are some of the strategies adopted by soil microbiome to detoxify the contaminated environment (Mishra et al., 2017). The recent developments in the metal-removing strategies adopted by bacterial biofertilizers are described in the following section.

21.4.2.1 Biosorption

Biofertilizers facilitate both growth and metal uptake by plants and therefore act as promising biosorbents for metal remediation through a process called biosorption (Li et al., 2018a,b; Ayangbenro and Babalola, 2017). Biosorption is a passive process involving metabolism-independent extracellular adsorption, where metal ions are passively adsorbed onto the live or dead cell surface components of the biosorbing organisms (Inoue et al., 2017; Huang et al., 2014). In general, the biosorption of metals by biofertilizer involves three steps: physical entrapment (physical adsorption), (ii) ion exchange, and (iii) complexation with functional groups (Bai et al., 2014) that work independently or synergistically (Srinath et al., 2002). Mechanistically, the biosorption process includes a solid phase (sorbent or biosorbent; biological material) and a liquid phase (solvent, normally water) comprising a dissolved species to be sorbed (sorbate and metal ions). Since the biological materials (the sorbent) have a higher affinity for the metals (sorbate species), the metal ions are attracted toward the cell surface molecules (surface adsorption) and are removed by various mechanisms. The sorption properties of biofertilizers are influenced by the outer cell shield. Metals when picked up by organisms are linked by the active functional groups (carboxyl, phosphonate, amine, and hydroxyl) of the surface layers of the bacterial cells (Zang et al., 2018) which play a fundamental role in the sorption and complexation of pollutants (Vijayaraghavan and Yun, 2008). After adsorption on to cell surface, the metals are removed through desorption process. During the desorption process, metals linked by biofertilizer microbes are removed by weak mineral acid solutions (like HCl) or chelating compounds, like, EDTA. Metal ions like Cu^{2+}, Cr^{3+}, Ni^{2+}, Pb^{2+}, Zn^{2+}, Cd^{2+}, and Co^{2+} strongly bind to the microbial biomass at acidic pH ranging between 5 and 7 but further lowering the pH to two causes the liberation of metals from biosorbents. Apart from pH, the initial metal ion concentration, temperature, and microbial biomass also affect the biosorption process greatly (Rizvi et al., 2020b). Among organisms, bacteria are considered the most significant biosorbents due to their ubiquity, high surface-to-volume ratios (size), potential active chemisorption sites on the cell wall, ability to grow under controlled conditions, and resilience to environmental conditions (Srivastava et al., 2015). Among bacterial biofertilizers, the Gram-positive organisms are considered better biosorbing agents than Gram-negative bacteria due to thick peptidoglycan layer that contains numerous active sorption sites (Van Hullebusch et al., 2003). Among biosorbents, the biomass prepared from bacteria like *Azotobacter* (Rizvi and Khan, 2019a), rhizobia (Jobby et al., 2019), *Stenotrophomonas maltophilia*, and *Bacillus subtilis* (Wierzba, 2015) is widely used in metal cleanup programs due to the large and well-defined surface area of microbiological biomass, its high binding affinity, low cost, and environmentally friendly nature. However, among different biosorbents, the nonliving biomaterials (dead cells) such as metal-sequestering compounds are preferred because they remain insensitive to any toxic pollutants, do not require too much care and maintenance, and are economical besides they can easily be regenerated and reused. On the contrary, the living biosorbents are sensitive to toxic metals and frequently need nutrients and hence may enhance the BOD and COD of the effluent (Kurniawan et al., 2019). So, the living biofertilizers are generally not recommended in metal cleanup programs.

In a study by Jobby et al. (2019), *Sinorhizobium* sp. SAR1 (JX174035.1) recovered from *Sesbania sesban* root nodules tolerated the highest Cr concentration (1 mM) and showed excess secretion of extra polymeric substances (EPS), as revealed by scanning electron micrographs (SEMs) suggesting that the EPS might have helped the nodule bacterium to adapt to the Cr stress. Moreover, the energy-dispersive X-ray spectroscopy (EDX) data though did not show any peak of Cr, but the biosorption experiments revealed that the isolate SAR1 exhibited maximum adsorption of Cr (285.71 mg/g). The isotherm studies showed a better fit to Langmuir isotherm while the Weber and Morris plot concluded that the adsorption was regulated by film diffusion mechanism. The role of cell wall components and EPS in Cr adsorption to the biomass of *Sinorhizobium* was observed under FTIR images. These findings are likely to help in searching for more biofertilizers for effective remediation of contaminated soils. Likewise, the living and nonliving biosorbents of *Cupriavidus necator* GX5, *Sphingomonas* sp. GX15, and *Curtobacterium* sp. GX31 in an in vitro experiment differed in Cd(II) adsorption capacities as revealed by SEM-EDX, FTIR, and adsorption experiments (Li et al., 2018a,b). However, the dead cells demonstrated higher adsorption capacity than the live cells of GX31. Also, for GX5 and GX15, the load of the dead biomass was even greater than that of the living biomass at $20\,mg\,L^{-1}$ of Cd(II), yet the absorbing capacity of the two cultures was identical when treated with $100\,mg\,L^{-1}$ of Cd(II). Minor variation in spectra revealed under FTIR suggested that more functional groups of the dead biosorbents were involved in Cd(II) binding which further implied that the hydroxyl, amino, amide, and carboxyl groups played an important role in complexation with Cd(II). Conclusively, the dead cells especially those of GX31 proved more potent biosorbing material for Cd(II) remediation. Similarly, studies by Naz et al. (2016a,b) revealed that the biofertilizer species belonging to the genera *Pseudomonas*, *Arthrobacter*, *Exiguobacterium*, *Citrobacter*, and *Enterobacter* isolated from soils near a sugar industry (Peshawar, Pakistan) had considerable metal biosorption abilities. *Pseudomonas* sp. among genera reduced 37% Pb, 32% Ni, 29% Cu, and 32% Cr and was found as the most effective biosorbing organism, whereas *Enterobacter* sp. reduced 19% Pb, 7% Ni, 14% Cu, and 21% Cr and proved to be least effective. This finding suggests that *Pseudomonas* sp. may be used to remediate the contaminated environment. Despite all, the biosorption in general has certain advantages over conventional metal cleanup methods, for example, this is a low-cost and highly efficient technology which minimizes the chemical and biological sludge, regeneration of biosorbent, and possibility of metal recovery (Kratchovil and Volesky, 1998). However, the early saturation, the limited scope for biological process improvement, and the inability of biological materials to alter the valency of metals are some of the limitations of this process.

21.4.2.2 Bioaccumulation

Bioaccumulation is an active process that involves the uptake and transport across the cell membrane and metabolism-dependent intracellular accumulation of toxic pollutants (Pérez-Rama et al., 2010). It occurs when the absorption rate of the contaminant by microbes is higher. Bioaccumulation is a toxicokinetic process that affects the survivability of living organisms but depends on (i) structural and compositional properties of biofertilizers, (ii) genetic and physiological adaptations, (iii) species of metals, and (iv) the spectrum of metal toxicity (Blackwell et al., 1995). Microbial cell density also affects the metal accumulation process that, however, decreases considerably with a simultaneous increase in the biomass concentration due to the electrostatic interactions of the functional moieties of the cell wall such as polysaccharides, lipids, and proteins. Higher concentration of cells in suspension causes their linkage and, hence, reduces the number of metal-binding sites (Santana-Casiano et al., 1995). Apart from these, the high concentration of metals can also be destructive for living cells and hence affects adversely the accumulation process (Al-Asheh and Duvnjak, 1995). However, the biofertilizer populations can generally resist certain concentrations of metals (Khanna et al., 2019) beyond which such chemicals become toxic and endanger the organism (Tiwari and Lata, 2018). The sensitivity of biofertilizers to toxicants, however, varies with the species and surface characteristics of microbiomes and associated chemicals. Ideally, the organisms selected for bioaccumulation studies should exhibit (i) maximum tolerance ability to more than one contaminant and (ii) superior biotransformational capabilities so that the toxic chemicals are converted to nontoxic forms. This process is compared to biosorption for the reason that metal removal from cells and its recovery is connected with the cellular structure transformation. Practically, the bioaccumulation process includes the localization/accumulation of metals within specific tissues/cells, complexing with protein (e.g., metallothionein), and efflux pumping of toxicants (Srinath et al., 2002; Gadd, 1990). As an example, Shao et al. (2019) conducted an in vitro experiment to determine the impact of Pb(II) and Cr(VI) on growth, generation of reactive oxygen species (ROS), activities of superoxide dismutase (SOD), catalase (CAT), and distribution of bioaccumulated metals in Pb(II)- and Cr(VI)-tolerant P-solubilizing bacterium *Bacillus* sp. strain MRP-3. The ROS amplified from 1.4-fold to 1.8-fold under Pb(II) stress but decreased from 1.6-fold to 1.1-fold under Cr(VI) stress compared to control. The SOD activity was ROS-dependent, whereas the CAT activity increased from 11.4 to 21.8 U mg^{-1} exposed to Pb and 11.4 to 32.9 U mg^{-1} under Cr stress. The extracellular accumulation of Pb as revealed by SEM with energy-dispersive X-ray spectroscopy (STEM-EDS) differed between 61.7% and 95.9%, whereas extracellular accumulation of Cr accounted meagerly up to 3.6%. The functional groups associated with extracellular accumulation were, however, not located in the EPS, as confirmed by attenuated total reflection/Fourier-transform infrared (ATR-FTIR) spectroscopy analysis. Summarily, both biosorption and bioaccumulation processes adopted by biofertilizer can be used in metal remediation strategies, but they differ from each other in many respects (Table 21.1).

Table 21.1 The differences between biosorption and bioaccumulation processes.

Biosorption	Bioaccumulation
• Passive process	• Active toxicokinetic process
• Metabolism-independent extracellular adsorption; ions can be bound to the surface	• Metabolism-dependent intracellular accumulation; ions accumulate inside the cell
• Rapid process	• Slow process
• Energy-independent	• Energy-dependent
• Feasible for large-scale application	• Narrow range
• Low cost	• High cost
• Both live and dead biomass can be used	• Only live biomass is used
• Dry biomass remains insensitive to any toxic pollutant	• Live cells are sensitive to toxicants
• Does not require too much care, maintenance, and cultivation conditions	• Requires nutrients and proper growth conditions
• Reversible process	• Partially reversible process

Modified from Timková, I., Sedláková-Kaduková, J., Pristaš, P., 2018. Biosorption and bioaccumulation abilities of actinomycetes/streptomycetes isolated from metal contaminated sites. Separations 5, 54.

21.4.2.3 Biotransformation

Detoxification or transformation of the toxic metal state and making it unavailable by bacterial population is yet other mechanisms for metal removal from contaminated environment (Chaturvedi et al., 2021; Deepa and Mishra, 2020). Microbiological transformations of metals include oxidation, reduction, methylation, and demethylation and are mediated by microbial enzymes (Hossan et al., 2020; Jobby et al., 2018). Microorganisms reduce the metals' state and change their solubility. For instance, biofertilizers like *Azotobacter* (Belogolova et al., 2019), *Pseudomonas* (Oves et al., 2013), and rhizobia (Jobby et al., 2019) have been reported to reduce metal ions to nontoxic forms. This reaction may occur in vacuoles, on the cell surface and in the extracellular environment, which is important for metal recovery. As an example, Upadhyay et al. (2017) isolated a P-solubilizing bacterium MNU16 from coal mine soil and identified it as *Bacillus subtilis* MNU16 by 16S rRNA sequencing. The strain MNU16 tolerated chromium at 50 mg/L of Cr(VI) and reduced 75% of Cr(VI) to 13.23 mg/L within 72 h. The electron-dense precipitate images of *B. subtilis* MNU16 under TEM validated the reduction of Cr(VI) to Cr(III). The fluorescence microscopy and flow cytometry data presented an identical pattern and clearly revealed the lesser toxic effect of Cr (vi) up to 200 mg/L. Likewise, Jobby et al. (2019) recovered 22 *Rhizobium* and

Sinorhizobium strains from *Sesbania sesban* root nodules. Of these nodule biofertilizers, *Sinorhizobium* sp. SAR1 (JX174035.1) tolerated Cr maximally (1 mM) and therefore was chosen for further studies. The excess secretion of EPS, viewed under SEM was suggested as a probable reason for its adaptation to high Cr concentration. However, the EDX data did not show any Cr peak. Additionally, strain SAR1 (JX174035.1) maximally adsorbed Cr (285.71 mg/g) and the isotherm demonstrated a better fit to Langmuir isotherm. The FTIR data established the role of cell wall components and EPS in Cr adsorption onto the *Sinorhizobium* biomass. These and other related findings therefore suggest that the metal resistance/reducing efficacy of biofertilizers including both P-biofertilizers and N-biofertilizers provides an opportunity to develop an efficient biofertilizer-based bioremediation approach for abatement of contaminated soils.

21.4.2.4 Bioleaching of metals

Bioleaching, an essential field in the biohydrometallurgy, is a low-cost, environmentally friendly, and efficient process (Arshadi et al., 2019a). Broadly, bioleaching refers to the conversion of solid metals into water-soluble forms by the microorganisms. Bacteria-based bioleaching (Zhao and Wang, 2019) employing their metabolic products involves the mobilization of positive metal ions from insoluble ores often by biological dissolution or complexation processes (Jin et al., 2018; Valix, 2017). By oxidation and reduction reactions, insoluble metals are converted to soluble forms and transferred to the solution (Arshadi et al., 2019b). Bioleaching is generally applied for the wastes that comprise low concentrations of elements for which other extraction methods fail or cannot be effective. In general, the microbial secretions, such as low molecular weight organic acids (citric acid, gluconic acid, and oxalic acid), can dissolve metals (e.g., Cd) and soil particles containing metal minerals. The bioleaching is influenced greatly by pH, temperature, pulp density, chemical reactions (formation of precipitates), density and composition of biofertilizers, bacterial growth, nutrient concentration, and particle size, stirring frequency (rpm); oxygen content, and total bioleaching time (Rouchalova et al., 2020; Amiri et al., 2011). For instance, the leaching of metals increases with the supply of sufficient nutrients. As an example, the leaching rate of Cd in the absence of nutrients was only 9% which was increased to 36% when glucose and other nutrients were provided to microorganisms growing in soil columns after 38 days of growth (Chanmugathas and Bollag, 1998). Rouchalova et al. (2020) detected high concentrations of Cu, Pb, Zn, and Fe in the sludge sediment originating from the mining industry located in Zlaté Hory, Czech Republic, applying bioleaching process using acidophilic and mesophilic bacteria *Acidithiobacillus ferrooxidans*. The experiments were conducted under pilot conditions in a bioreactor at a pulp density of 2.5% and 4.2% (w/v). These metals were extracted from the leachates by flame atomic absorption spectrometry (F-AAS) while the residual metal concentrations were detected by X-ray fluorescence (XRF) spectrometry. The AAS results revealed the highest Fe (76.48%), Cu (82.01%), and Pb (88.90%) concentrations while Zn was dissolved for all fractions > 90%. The maximum dissolution of Zn (98.7%), Fe (85.4%), and Cu (96.44%) occurred at a pH of 1.8. Likewise, other soil bacteria, for example, *Citrobacter*, generate free inorganic P and form an insoluble metal P coat that entraps a large volume of toxic metals (Marchenko et al., 2015). Summarily, the microbe-based metal-extraction processes are typically eco-friendly. They do not use large amounts of energy as compared to roasting and smelting and so do not produce harmful gases, and hence, it is more compatible with antipollution regulatory bodies.

21.4.3 Factors affecting bacterial remediation of heavy metals

The tendency of heavy metals to be stimulatory or inhibitory to microorganisms depends on the species; extent and total metal concentrations; and the density, composition, and physiological functions of biofertilizers. On the contrary, the environmental variables such as temperature, pH, and organic acids can modify and change the transformation, transportation, valence states of metals that either alone or simultaneously affect the bioavailability of metals to biofertilizer populations. These factors together play a significant role in the biofertilizer-based detoxification/remediation of heavy metals from contaminated environment. For instance, an upsurge in temperature enhances the rate of adsorbate diffusion across the external boundary layer while the solubility of metals increases with an increase in temperature, which consequently improves the bioavailability of metals. From microbes' perspective, the activity of microbes like their metabolism and enzyme activity increases with rise in temperature but to a certain suitable range which eventually accelerates the bioremediation process. However, the stability of the microbe-metal complex depends on the sorption sites, microbial cell wall configuration, and ionization of functional moieties of the bacterial cell wall (Tarekegn et al., 2020; Ali Redha, 2020).

21.4.4 Bacterial biofertilizers for remediation of metal-contaminated environments

Management of metal-contaminated soils for safe cultivation has always been a grave task. The biofertilizers, especially nitrogenous and P-solubilizing groups, have made this task easier when applied to soils since they act as a potential agent to both bioremediate contaminated soil and optimize crop yields under normal soils by various mechanisms. However, such microbes may be influenced by several factors such as the competitive ability of the bioremedial agents and abiotic factors. Successful removal of metals by the addition of bacteria-based biofertilizers has been reported (Rizvi and Khan, 2019b) and are discussed.

21.4.4.1 Bioremediation by nitrogenous biofertilizers

The symbiotic interactions between legume plants and nodule bacteria (rhizobia) have attracted the attention for the restoration of metal-contaminated sites. On the contrary, the excess of metals adversely affects the symbiotic process, thereby decreasing the number of nodules, the frequency of nodulation, and concurrently the rate of BNF (Stambulska et al., 2018; Besharati and Memar Kouche-Bagh, 2017). So, to optimize the legumes production under stressed environment, there is an urgent need to discover metal-resistant rhizobia and/or to engineer the existing rhizobial biofertilizers to enhance their metal-tolerant capabilities (Fagorzi et al., 2018). And indeed, rhizobia have been reported to reduce the toxicity of metals to plants and allow them to grow normally under metal-contaminated soils (Edulamudi et al., 2021; Bellabarba et al., 2019). Fortunately, legumes growing in contaminated areas such as mine deposits (Sujkowska-Rybkowska et al., 2020), metal-polluted soils (Abdelkrim et al., 2019), and serpentine soils (Rubio-Sanz et al., 2018) have served as a biofactory for metal-resistant rhizobial strains. Some of the rhizobial genera, *Mesorhizobium* (Hao et al., 2015), *Rhizobium* (Abdel-Lateif, 2017), *Sinorhizobium* (Renitta et al., 2019; Jobby et al., 2015), and *Bradyrhizobium* (Seraj et al., 2020), etc., have been found as metal detoxifying N biofertilizers (Gopalakrishnan et al., 2015). Ferreira et al. (2018) isolated native nitrogen-fixing bacteria from coal mine wasteland (Brazil) using *Macroptilium atropurpureum* (DC) Urb and *Vicia sativa* L. as trap plants. The metal-tolerant N fixers were identified employing 16S rRNA sequence analysis as *Rhizobium, Bradyrhizobium,* and *Burkholderia*; all genera showed a variable tolerance to heavy metals: Cr > Cd > Zn > Ni > Cu. In greenhouse, the *M. atropurpureum* inoculated with *Bradyrhizobium* spp. strains UFSM-B53 and UFSM-B64 and *Rhizobium* sp. UFSM-B74 displayed better root growth, high dry matter accumulation, maximum number of nodules, nodule dry biomass, and exceptional symbiotic efficiency. These findings clearly highlighted the putative role of native N biofertilizers while growing under metal stress in coal-mining degraded areas. Similarly, *S. meliloti-* and *S. medicae*-inoculated *Medicago sativa* plants, grown under field conditions, had active nodulation, and metals were accumulated within the root nodules. The asymbiotic N biofertilizers like species of *Azotobacter* (Abo-Amer et al., 2014) and associative nitrogen fixers like *Azospirilla* (Ma et al., 2015) in a similar way have been reported as metal remediating agents. These findings suggest that the natural N biofertilizers could serve as a valuable tool for land restoration and phytostabilization of metals.

Rhizobia in general aid in the detoxification of metals by plants through phytostabilization, phytoimmobilization, phytoextraction, precipitation, chelation, biosorption, and bioaccumulation (Edulamudi et al., 2021; Belechheb et al., 2020; Lajayer et al., 2019). Among different organs, nodule of legumes acts as storage systems and provides enough place for plants to store metals and therefore lower the danger of direct exposure. Symbiotic/asymbiotic N biofertilizers can also contribute to phytoremediation through sequestration by excreting extracellular polymeric substances (Gupta and Diwan, 2017). For example, Rizvi and Khan (2019b) observed that the metal-tolerant biofertilizer agent *Pseudomonas aeruginosa* CPSB1 and *A. chroococcum* CAZ3 secreted 1306.7 and 1660 $\mu g\,mL^{-1}$ EPS, respectively, when grown in the presence of 200 and 100 $\mu g\,mL^{-1}$ Pb, respectively, with glucose as C source. Moreover, the SEM and EDX revealed the binding of metal ions to bacterial EPS while the functional group involved in metal chelation varied considerably as detected by FT-IR. The metal ions were adsorbed onto EPS, and hence, EPS played an important role in detoxification. Dhevagi et al. (2021) in other in vitro experiment isolated a Pb-tolerant asymbiotic N-fixing biofertilizer (*Azotobacter salinestris*) from sewage water irrigated soil and found that the maximum removal of Pb (61.5%) occurred after 72 h. The interaction effect between levels of Pb and contact time was significant. The Pb biosorption was confirmed by the changes in stretching intensities of functional groups and appearance of strong OH group. The biosorptive ability of *A. salinestris* was further confirmed by the micrograph images observed under SEM and EDX. Due to this novel trait, the synthesis of EPS by bacterial biofertilizers is likely to serve as an important emerging eco-friendly approach for heavy metal detoxification and hence could practically be applied as a bioremediation strategy in the metal-contaminated environments. Numerous studies have also examined the coculture effect of rhizobia with other PGPR on plant growth under stressed environment (Ju et al., 2020; Kong et al., 2017). For instance, the coinoculation of soybean with *B. japonicum* E109 and *A. brasilense* Az39 significantly influenced plant growth and arsenic phytostabilization in arsenic-contaminated conditions (Armendariz et al., 2019). Also, *B. juncea* inoculated with *Sinorhizobium* sp. Pb002 grew well in Pb-contaminated soil and enhanced the whole biomass, plant survival, and Pb uptake by *B. juncea* plants (Di Gregorio et al., 2006). In a similar experiment, green gram plants inoculated with *Bradyrhizobium* sp. (vigna) or *Rhizobium* sp.

had increased seed yield and grain protein even in the presence of excessive Ni or Zn (Wani et al., 2007, 2008).

21.4.4.2 Phosphate biofertilizers for heavy metal remediation

PSM: Definition and solubilization: An overview

Phosphorus is a major nutrient next only to N (Maharajan et al., 2018) whose deficiency restricts the overall growth of plants severely (Lambers and Plaxton, 2018). Phosphorus impacts different metabolic features such as cell division, tissue developments, nucleic acid structure, photosynthesis, energy, and biosynthesis of sugars of plants growing under conventional and stressed soils (Nesme et al., 2018; Khan et al., 2014). The low P availability therefore becomes a critical issue in plant metabolism worldwide (Kaur and Reddy, 2015). To fulfill P-demands, P-biofertilizers are considered a promising option over P-fertilizers due to reasons such as cost, environmental hazards (Abawari et al., 2020). Due to these, the soil microbiome generally belonging to P-solubilizing category often termed as PSM: a group of beneficial microorganisms capable of hydrolyzing organic and inorganic phosphorus compounds from insoluble P-sources (Mendoza-Arroyo et al., 2020; Kalayu, 2019) have been discovered to supplement chemical P-fertilizers. Briefly, the bacterial solubilization of soil P occurs through the secretion of organic acids (Table 21.2) which can either directly dissolve the mineral P as a result of anion exchange of PO_4^{2-} by acid anion or can chelate both iron and aluminum ions associated with phosphate. Changes in biotic factors such as microbial composition and functions and plants secretion, farming practices (Melo et al., 2018; Neal et al., 2017), long-term fertilizer application, and environmental variables, however, influence the solubilization and availability of P to plants (Macias-Benitez et al., 2020; Sumarsih et al., 2017).

Rationale for the use of metal-tolerant P-biofertilizers

The use of metal-tolerant bacterial P-biofertilizers in crop production is fortified for various reasons: (i) easy and low-cost inoculant production, (ii) exhibits tolerance to toxic metals, (iii) mobilizes metals to make plant uptake easier, (iv) causes no environmental hazards, (v) enhances soil P pool and protect rhizospheres, (vi) provides active biostimulants, (vii) guard plants from phytopathogens, and (viii) compatible to a range of food crops (Saranya et al., 2018; Rafique et al., 2017). Phosphate biofertilizers belonging to both Gram-negative such as *Pseudomonas, Stenotrophomonas, Massilia, Ochrobactrum, Acinetobacter, Cupriavidus* (Wan et al., 2020; Rizvi and Khan, 2017b), *Pantoea* and *Enterobacter* (Chen and Liu, 2019), *Azotobacter* (Rizvi and Khan, 2018) and Gram-positive like *Bacillus* (Wang et al., 2017) and *Arthrobacter* (Wan et al., 2020) sequester/immobilize metals and

Table 21.2 Organic acids secreted by bacterial phosphate biofertilizers.

P-biofertilizers	Organic acids	References
Bacillus, Burkholderia, Paenibacillus sp.	Gluconic oxalic, citric, tartaric, succinic, formic, and acetic acid	Chawngthu et al. (2020)
Pantoea, Pseudomonas, Serratia, and *Enterobacter*	Oxalic, citric, gluconic succinic, and fumaric acids	Rfaki et al. (2020)
Bacillus megaterium Y95, Y99, Y924, Y1412	Succinic acid, oxalic acid, and citric acid	Zheng et al. (2018)
Bacillus sp., *Pseudomonas* sp., *Proteus*	Citric acid, glutaric acid, glyoxylic acid, ketoglutaric acid, ketobutyric acid, malic acid, malonic acid, succinic acid, fumaric acid, tartaric acid, and gluconic acid	Selvi et al. (2017)
Pseudomonas veronii PSB12	Gluconic acid, propanedioic acid, propionate, butyrate, succinate, formate, citrate, and pentanoic acid	Chen et al. (2016)
Rhizobium tropici UFLA03-08, *Acinetobacter* sp. UFLA03-09, *Paenibacillus kribbensis* UFLA03-10, *P. kribbensis* UFLA03-106, *Paenibacillus* sp. UFLA03-116	Propionic acid, gluconic acid, tartaric acid, and malic acid	Marra et al. (2015)

negate the toxicity to plants (Chen et al., 2019). Due to these and their role in enhancing food production, P-biofertilizers are regarded as the most environmentally reliable microbial technology in metal-enriched cultivation practices (Zhu et al., 2019; Susilowati and Syekhfani, 2014).

Remedial impact of metal-tolerant bacterial phosphate biofertilizers on crops: Success story

The use of metal-resistant P-biofertilizer represents a valuable technology for augmenting biomass production besides assisting plants to cope with metal toxicity (Nazli et al., 2020). Following this concept, numerous bacterial P-solubilizers consisting of metal remediating potentials have been recovered from various rhizosphere/nonrhizosphere regions. Some notable metal-tolerant P-solubilizers are *Azotobacter* (Sumbul et al., 2020), *Achromobacter xylosoxidans* (Oves et al., 2019), *Bacillus* (Shao et al., 2019), *Pseudomonas* and *Pantoea* (Chen and Liu, 2019), and *Cronobacter* (Saranya et al., 2018). These P-biofertilizers in later experiments have been reported to successfully remediate the contaminated soils while reducing the metal mobility and hence the toxicity to plants. Due to their ability to (i) survival at high metal concentration, (ii) negate metal toxicity, and (iii) secrete different phytostimulants (Table 21.3), the metal-tolerant

Table 21.3 Metal tolerance and plant growth-stimulating features of phosphate biofertilizers.

Phosphate biofertilizers	Origin	Metal tolerance		Biostimulants	References
		Metals	Tolerance		
Serratia proteamaculans, Pseudomonas sp., P. veronii, P. psychrophile, Bacillus cereus	Roots of Acacia mangium, Eucalyptus camaldulensis, Pityrogramma calomelanos[a]	Pb	1875	Siderophores Nil	Yongpisanphop and Babel (2020)
Achromobacter xylosoxidans OS2	Metal-polluted soil[b]	Cd	300	Ammonia, IAA, HCN	Oves et al. (2019)
		Cr	1000		
		Ni	1200		
		Cu	1400		
		Zn	1500		
Cellulosimicrobium sp.	Rhizosphere[a]	Cr	800	IAA, siderophores	Tirry et al. (2018)
		Zn	1500		
		Cu	400		
		Ni	500		
		Pb	1500		
		Co	500		
Ensifer adhaerens	Chickpea root surface[b]	Cd	250	IAA, siderophore, HCN, NH_3	Oves et al. (2017)
		Cr	500		
		Cu	800		
		Zn	800		
		Ni	1000		
Pseudomonas aeruginosa (CPSB1)	Chili rhizosphere[a]	Cu	1400	IAA, siderophores, NH_3, ACC deaminase, HCN	Rizvi and Khan (2017a)
		Cd	1000		
		Cr	1000		
Pseudomonas aeruginosa	Metal-polluted chili rhizosphere[b]	Ni	400	IAA, siderophores, HCN, NH_3, ACC deaminase	Rizvi and Khan (2017b)
		Zn	800		
		Pb	1000		
		Cr	400		
Achromobacter piechaudii (E6S)	Stems of the Zn/Cd hyperaccumulator plant Sedum plumbizincicola[a]	Cd, Zn, Pb	100 of each metal	IAA, ACC deaminase,	Ma et al. (2016)

Tolerance level of heavy metals are expressed as [a]mg/L, [b]μg/L, and [c]mM, respectively; *IAA*, indole-3-acetic acid; *HCN*, hydrogen cyanide; *NH3*, ammonia; *ACC deaminase*, 1-aminocyclopropane-1-carboxylate deaminase.

bacterial P-biofertilizers have been used in remediation programs and cultivation practices to optimize crop production under derelict environment. For example, Tirry et al. (2018) discovered a multimetal (Cr, Zn, Cu, Ni, Pb, and Co)-resistant P-solubilizer *Cellulosimicrobium* sp. with Cr reducing, IAA and siderophore secretion capabilities. The greenhouse experiments demonstrated a significant improvement in the growth of alfalfa raised in Cr-, Zn-, and Cu-spiked soils and enhanced metal uptake by the *Cellulosimicrobium*-inoculated plants suggesting that the P-solubilizer *Cellulosimicrobium* sp. could play an important role in both bioremediation and plant growth promotion leading to the better management of environmental pollution. In a recent study, Jain et al. (2020) discovered potential P-solubilizing Zn-tolerant bacteria (ZTB), isolated from Zn-contaminated soils which had variable Zn tolerance efficiency and PGP activities. The ZTB also accumulated metals as revealed by AAS and SEM-EDS. The release of PGP biomolecules such as IAA, GA_3, NH_3, HCN, siderophores, ACC deaminase, phytase, K, and Si solubilization potential confirmed that the ZTB strains be applied as an efficient biofertilizer for augmenting crop production in stressed environment. Under greenhouse experiment, maize, grown in the presence of varying levels of Zn suffered from Zn toxicity and had poor growth. The inoculation of ZTB strains, however, relieved the Zn toxicity and concurrently boosted the measured plant growth parameters. Moreover, a noteworthy upsurge in superoxide dismutase, peroxidase, phenylalanine ammonia lyase, catalase, and polyphenol oxidase activity was recorded for ZTB-inoculated maize plants, growing under Zn stress. These findings together provide a new opportunity to maximize the use of P-biofertilizers in metal remediation vis-a-vis crop production in metal-contaminated soils. Arshad et al. (2019) in an in vitro experiment recovered two Cr-tolerant (up to $700\,mg\,L^{-1}$) P-solubilizing bacterial strains (A5 and A6) and identified them as *Pseudomonas plecoglossicida* (Gram-negative) and *Staphylococci saprophyticus* (Gram-positive), respectively, using 16S rRNA gene sequencing analysis. In greenhouse trials, the length of roots and shoots of spinach though did not differ significantly among bacterial strains but the IAA-positive *P. plecoglossicida* maximally increased the percentage of seed germination from 17% to 46% and whole dry biomass by 44% over control. Among the two Cr-tolerant PS strains, strain A5 (*P. plecoglossicida*) displayed better results and hence could be developed as a P-bioinoculant for remediation of Cr-contaminated soils and production of other crops. The studies conducted in vitro by Marzban et al. (2016) revealed that P-solubilizing metal-resistant (Cd, Cu, and Pb) bacterial strain Ws3 of *P. putida* secreted a significant amount of IAA (200 mg/L) while thriving well in sewage water irrigated agricultural fields. The P-solubilization and IAA excretion were, however, reduced when this bacterium was grown at 500 mg $CuCl_2$/l and 300 mg/L each of $PbCl_2$ or $CdCl_2$. When used as inoculant at weekly intervals for 40 days, the P-biofertilizer improved the growth condition and increased the plant biomass vis-a-vis declining the metal accumulation with greatest effect on both aerial (the shoot) and underground organ (the root) of growing maize plants. The results suggest that the P-biofertilizer *P. putida* could prevent the uptake of metals by plant tissues leading eventually to the reduction in toxicity of metals to plant organs. Similarly, the IAA- and siderophore-positive P-biofertilizer *Pseudomonas* showing resistance to Zn, Cs, Pb, As, and Hg, recovered from lake sediment (Li and Wusirika, 2011), caused a meaningful enhancement in Cu accumulation and significantly amplified the whole maize biomass. Arunakumara et al. (2015) isolated *Klebsiella oxytoca* JCM1665 from metal-contaminated soils which could solubilize insoluble P efficiently both in the presence and absence of metals (Co, Pb, and Zn). Inoculation of JCM1665 enhanced the growth of *Helianthus annuus* by 49%, 22%, and 39% in Co-, Pb-, and Zn-contaminated soils, respectively, relative to noninoculated plants. The P-solubilizer also increased the accumulation and translocation of Co, Pb, and Zn from roots to shoots. Water-soluble fraction of Co, Pb, and Zn in soil was improved by 51%, 24%, and 76%, respectively, in inoculated soils compared to nonbacterized soils.

21.5 Biofertilizers—A general perspective

The rapid advancement in agricultural activities has greatly optimized food production worldwide in order to counteract the massively increasing food/feed demand. To achieve this, the farm practitioners depend heavily on chemical fertilizers (Bhat, 2019; Kunawat et al., 2019). It is estimated that about 53 billion tons of NPK fertilizers are applied annually to supply nutrients to plants (FAO et al., 2017) but only a small fraction of such nutrients is used by plants and a greater percentage is lost through precipitation. However, such agrochemicals exhibit many adverse impacts (Aloo et al., 2019) on soil microbial compositions and their integrated functions causing soil health problems (Mahanty et al., 2017). So, an inexpensive and environmentally friendly approach like use of biofertilizers has emerged as an alternative to agrochemicals (Glick, 2020) which is safe for agrosystems also. Especially, the microbial formulations among many biofertilizers options provide effective solutions to the fertilizers problems for both restoring soil fertility and augmenting crop production (García-Fraile et al., 2015).

21.5.1 Bacteria-based biofertilizers in cultivation practices: An overview

21.5.1.1 Nitrogen fixers

Plants take up the N in the form of nitrates and ammonium ions (Gouda et al., 2018), both of which are often limiting in soil. So, to supplement the synthetic N fertilizers (Zhu et al., 2020) the emergence of N biofertilizers (Table 21.4) has indeed played an important role in reducing dependence on chemical N fertilizers (Lesueur et al., 2016). The symbiotic nitrogen-fixing (SNF) bacteria such as *Rhizobium*, *Azorhizobium*, *Bradyrhizobium*, *Mesorhizobium*, and *Sinorhizobium* transform atmospheric N into usable N form and supply it to plants which they use to synthesize vitamins, amino acids, nucleic acids, and other nitrogenous compounds. Also, some non-SNF organisms, such as *Acetobacter* and *Azotobacter*, colonize the roots and provide N to growing plants (Zaidi et al., 2015, 2017a). As an example, the coinoculation of wheat plants with *Azotobacter* and *Pseudomonas* enhanced the grain yield, protein content, and harvest index while

Table 21.4 Examples of nitrogen-fixing biofertilizers impacting different food crops.

Biofertilizers	Inoculated crops	References
Azotobacter, Azospirillum	Potato	Aloo et al. (2020a,b) and Singh et al. (2017)
R. japonicum, Bradyrhizobium, Mesorhizobium	Soybean, chickpea	Gunnabo et al. (2020), Yousaf et al. (2019), and Htwe et al. (2019)
Beijerinckia indica, Gluconacetobacter diazotrophicus, Azospirillum, Herbaspirillum seropedicae, H. rubrisubalbicans, B. tropica	Sugarcane	Nunes Oliveira et al. (2017) and Schultz et al. (2014)
A. chroococcum	Rice	Banik et al. (2019), Shabanamol et al. (2018), and Hossain et al. (2015)
Rhizobium sp., *Azospirillum*		
A. chroococcum, A. vinelandii, B. megaterium, B. licheniformis, A. brasilense	Maize	Mandić et al. (2019) and Zeffa et al. (2019)
Rhizobium sp., *Bradyrhizobium, Streptomyces griseoflavus*	Green gram	Choudhary et al. (2019)
A. brasilense, A. chroococcum	Wheat	Galindo et al. (2020) and Bageshwar et al. (2017)
Azospirillum, Azotobacter, Acinetobacter calcoaceticus	Tomato	Castillo et al. (2019) and Reddy et al. (2018)
Azotobacter, Azospirillum, Streptomyces and *Bacillus*	Banana	Widyantoro et al. (2019) and Proboningrum and Widono (2019)

reducing the input of chemical fertilizer by 25%–50% under field conditions (Herrmann and Lesueur, 2013). Similarly, several species of *Azospirillum* (*A. zeae, A. thiophilum, A. rugosum, A. picis, A. oryzae, A canadense, A. mazonense,* and *A. melinis*) have been found to supply N to plants (Ding et al., 2017). In general, the *Azospirillum* promotes plant growth and yield by modifying the cell wall elasticity or the morphology of the root, or both and through the production of phytohormones (Fukami et al., 2018; Lesueur et al., 2016). Likewise, endophytic and associative N_2 fixation in nonleguminous crops such as wheat (Kanitkar et al., 2020), rice (Fitriatin et al., 2019), maize (Zeffa et al., 2019), strawberry (Rueda et al., 2016), and sugarcane (Arthee and Marimuthu, 2017) using free-living diazotrophs like *Azotobacter, Azospirillum, Gluconacetobacter,* and *Burkholderia* has been reported.

21.5.1.2 Nutrient solubilizers

Phosphate solubilizers

Phosphorus (P) among nutrients is the second most essential macronutrient though 95% and 99% of soil P are insoluble, immobilized, or precipitated and hence not available to plants (Verma et al., 2019; Khan et al., 2007). As a result, only a small fraction of the total soil P is used by plants (Malhorta et al., 2018; Alori et al., 2017). However, the PSM (Emami et al., 2018) supply P due to their P-solubilization/mobilization abilities (Liang et al., 2020). Generally, the P-solubilizers make P available to plants through enzymes (organic P mineralization) or solubilize the insoluble P by acidification (Rafi et al., 2019). As an example, Zeng et al. (2017) reported that the P-solubilizing activities of *Pseudomonas frederiksbergensis* positively correlated with the production of organic acids. Accordingly, as microbial inoculant, many P-biofertilizers have been found to stimulate and facilitate plant growth under both traditional (Khan et al., 2007) and stressed soils (Zolfaghari et al., 2020) in different agroecosystems (Table 21.5). Indeed, the discovery of P-biofertilizers has opened up a new prospect for maintaining soil P pool and subsequently to optimize crop production (Ingle and Padole, 2017).

Potassium solubilizers

Potassium (K) is the third major plant nutrient (Proença et al., 2017) that regulates many enzyme activities and coordinates the root shoot ratio (Kour et al., 2019). More than 90% of soil K is, however, not available to plants due to its rapid fixation ability (Bahadur et al., 2019), and hence, this is considered a major bottleneck in crop production worldwide. The low K availability increases the susceptibility to pathogens and reduction in growth and yield of plants

Table 21.5 Phosphate biofertilizers influencing crop production in different agrosystems.

Crop	Scientific name	P-biofertilizers	Source
Eggplant	*Solanum melongena*	*Enterobacter* sp.	Li et al. (2019)
Cabbage	*Brassica rapa*	*Bacillus subtilis*	Kang et al. (2019)
Potato	*Solanum tuberosum*	*B. megaterium*, *Bacillus* spp., *Pseudomonas* spp., *Serratia* spp., *Serratia* sp., *Citrobacter* sp., *Klebsiella* sp., *Pseudomonas* sp.	Aloo et al. (2020a,b) and Abd El-Moaty et al. (2018)
Maize	*Zea mays*	*B. mojavensis*, *P. aeruginosa*, *Alcaligenes faecalis*, *P. syringae*, *B. cereus*, *Lysinibacillus fusiformis*, *P. fluorescens*, *Bacillus* spp., *Klebsiella* sp., *E. ludwigii*, *Pantoea* spp., *P. aeruginosa*, *E. asburiae*, *Acinetobacter brumalii*, *B. pumilus*, *Acinetobacter* sp.	Akintokun et al. (2019), Sandhya et al. (2017), Rafique et al. (2017), and de Abreu et al. (2017)
Wheat	*Triticum aestivum*	*P. putida*, *Azospirillum*, *Serratia marcescens*, *Pseudomonas* sp., *P. mosselii*, *Stenotrophomonas maltophilia*, *Chryseobacterium*, *Flavobacterium*, *P. Mexica*	Emami et al. (2019), Batool and Iqbal (2019), Emami et al. (2018), and Sood et al. (2018)
Rice	*Oryza sativa*	*S. marcescens*, *Pseudomonas* sp., *Rahnella aquatilis*, *Enterobacter* sp., *P. fluorescens*, *P. putida*, *P. agglomerans*, *P. orientalis*, *Paenibacillus kribbensis*, *B. aryabhattai*, *K. pneumoniae*	Shabanamol et al. (2018), Yaghoubi et al. (2018), and Bakhshandeh et al. (2015)
Sugarcane	*Saccharum officinarum*	*Herbaspirillum* spp., *Bacillus* spp., *Burkholderia mallei*, *B. cepacia*, *Proteus vulgaris*, *Pasteurella multocida*, *K. pneumoniae*, *K. oxytoca*, *E. cloacae*, *C. freundii*, *G. diazotrophicus*	Awais et al. (2019) and Saini et al. (2015)
Soybean	*Glycine max*	*R. japonicum*, *E. sakazakii*, *P. straminae*, *Acinetobacter calcoaceticus*, *B. acidiceler*, *B. megaterium*, *B. pumilus*, *B. safensis*, *B. simplex*, *Lysinibacillus fusiformis*, *Paenibacillus cineris*, *P. graminis*	Yousaf et al. (2019) and Kadmiri et al. (2018)
Chickpea	*Cicer arietinum*	Phosphate solubilizing bacteria	Muleta et al. (2021) and Zaidi et al. (2017a,b)
Pea	*Pisum sativum*	*Pseudomonas baetica*, *P. lutea*, *P. azotoformans*, *P. jessenii*, and *P. frederiksbergensis*	Bahena et al. (2015)
Tomato	*Solanum lycopersicum*	*Arthrobacter* and *Bacillus*, *P. gessardi*, *P. koreensis*, *P. brassicacearum*, *P. marginalis*, *Acinetobacter calcoaceticus*, *Rahnella aquatica*, *Pseudomonas* sp.	Tchakounté et al., 2020, Castillo et al. (2019), and Nassal et al., 2018

(Thornburg et al., 2020; Xu et al., 2020). However, artificial K fertilizers are often used to enrich K in agricultural soils, but due to high cost, it is not easily available to farmers (Ahmad et al., 2016). So, like other biofertilizers, the discovery of K-biofertilizer has provided a solution to the chemical potassic fertilizer (Sattar et al., 2019). As an example, the K-biofertilizers have been found to significantly improve the germination, nutrient uptake, growth, and yield of crops under both greenhouse and field conditions (Nosheen et al., 2021). Some biofertilizers such as *B. mucilaginous*, *A. chroococcum*, and *Rhizobium* spp. have been reported to solubilize K through organic acids and, when inoculated, have resulted in the enhanced growth of wheat (Wang et al., 2020a,b), chili (Zhao et al., 2019), maize (Meena et al., 2018), and sorghum (Mekdad and El-Sherif, 2016). Recently, the biopriming of wheat with a K-solubilizing strain of *B. edaphicus* showed a significant increase in plant organs (roots and shoots) relative to uninoculated plants (Etesami et al., 2017). Some examples highlighting the role of K-biofertilizers are given in Table 21.6. The recent progresses in the field of K-biofertilizers strongly suggest that the K-biofertilizers can significantly reduce the use of chemical fertilizers in an environmentally friendly manner (Rani and Sengar, 2020). The diversity, solubilizing abilities, and mechanisms of K-biofertilizers, though, have extensively been reviewed (Sindhu et al., 2016) but still there is a need to know about the efficacy and mode of action of K-biofertilizers (Teotia et al., 2016).

Table 21.6 Some examples of potassic biofertilizers augmenting crop production.

Bacteria	Crop	References
B. circulans, Klebsiella sp., *Citrobacter* sp., *Serratia* sp.	Potato	Aloo et al. (2020a,b)
Paenibacillus kribbensis, Pseudomonas, Bacillus, Stenotrophomonas, Methylobacterium, Arthrobacter, Pantoea, Achromobacter, Acinetobacter, Exiguobacterium, Staphylococcus	Wheat	Laxita and Shruti (2020) and Verma et al. (2015)
B. mojavensis, P. aeruginosa, Alcaligenes faecalis, P. syringae, B. cereus, B. licheniformis, B. subtilis, K. oxytoca	Maize	Imran et al. (2020) and Parmar et al. (2016)
Bacillus, Pseudomonas sp.	Sorghum and chili	Sangeeth et al. (2012)
P. gessardi, P. koreensis, P. brassicacearum, P. marginalis, Acinetobacter calcoaceticus, Rahnella aquatica	Tomato	Oo et al. (2020)
Paenibacillus glucanolyticus, B. megaterium, B. mucilaginosus	Black pepper	Sangeeth et al. (2012)
P. agglomerans, Rahnella aquatilis, P. orientalis, Pantoea ananatis, Enterobacter sp.	Rice	Oo et al. (2020) and Shabanamol et al. (2018)

Zinc solubilizers

Zinc is an essential plant nutrient that influences several physiological processes of plants (Samreen et al., 2017). Globally, Zn is deficient in most agricultural soils due to nutrient mining during crop harvesting and increased use of NPK fertilizers (Sindhu et al., 2019). While growing under Zn-deficient soils, the plants exhibit the Zn deficiency and the symptoms include (i) reduction in leaf size and chlorosis and (ii) increase in plant susceptibility to heat, light stress, and pathogenic attack (Dubey et al., 2020). To alleviate Zn deficiency, synthetic Zn fertilizers are frequently applied (Liu et al., 2020) but their cost and complex forming ability are major problems (Sindhu et al., 2019). In this regard, rhizobacterial strains, for example, the species of *Stenotrophomonas, Mycobacterium, Enterobacter, Pseudomonas, Xanthomonas, Azospirillum, Thiobacillus, Agrobacterium,* and *Rhizobium* sp., isolated from banana, chili, bean, groundnuts, maize, sorghum, tomato, and barley rhizospheres have shown Zn solubilizing activity (Ijaz et al., 2019; Zamana et al., 2018; Yaghoubi et al., 2017). As Zn biofertilizer (ZSB), species of *Azospirillum* and *Azotobacter* (El-Khateeb and Metwaly, 2019) has shown a significant improvement in Zn uptake in wheat plants relative to uninoculated plants (Naz et al., 2016a,b). Additionally, several ZSB, like *P. fragi, Pantoea dispersa, P. agglomerans, E. cloacae,* and *Rhizobium* sp., recovered from wheat and sugarcane rhizospheres demonstrated a considerable increase in Zn contents and growth of wheat grown in greenhouse conditions (Kamran et al., 2017). Recently, other ZSB have also found to enhance the growth and Zn content in *Capsicum annuum* fruit (Bhatt and Maheshwari, 2020) and maize and pulses grown under both greenhouse and field conditions (Hussain et al., 2020; Khande et al., 2017).

Iron sequesters

Iron (Fe) is the fourth most abundant nutrient element in soil which plays an important role in plant growth (Schmidt et al., 2020). Like many plant nutrient, majority of the agrarian lands are Fe-deficient and therefore its unavailability limits plant growth (Arora and Verma, 2017). Some plant growth-promoting rhizobacteria synthesize low molecular weight compounds often known as siderophores (Ghazy and El-Nahrawy, 2021; Srinivasan et al., 2019) which have a high affinity in Fe-starved environments (Mhlongo et al., 2018). While doing this, the siderophores act as strong Fe-chelators and bind with the rhizosphere Fe (Singh et al., 2018). In addition, the siderophore-producing bacteria act as a biocontrol agent and prevent the proliferation of plant pathogens by restricting Fe availability (Olanrewaju et al., 2017). Several rhizobacterial genera such as *Stenotrophomonas, Serratia, Pseudomonas, Bacillus,* and *Microbacterium* isolated from the wheat rhizosphere have shown siderophore production and following application significantly enhanced the Fe uptake in greenhouse-grown wheat plants (Emami et al., 2019). Also, *Bacillus* sp., *Pseudomonas* sp., *Rhizobium* sp., *Mesorhizobium* sp., and *Azotobacter* sp. isolated from various leguminous and nonleguminous plants have shown siderophore secreting potential (Verma and Pal, 2020).

Sulfur-oxidizing microbes

Sulfur (S) is an essential nutrient which regulates many plant enzyme activities such as superoxide dismutase, ascorbate peroxidase, monodehydroascorbate reductase, dehydroascorbate reductase, and glutathione reductase. In contrast, the sulfur deficiency adversely affects nitrogen metabolism leading to chlorosis, low lipid percentage, and poor plant growth and yield (Saha et al., 2018). Sulfur in soil exists in two forms: organic and inorganic. The organic S is converted into inorganic forms by sulfur-oxidizing microbes belonging to the genera *Bacillus* spp., *Pseudomonas* spp., and *Klebsiella* spp. (Yousef et al., 2019) and consequently enhance plant growth (Chaudhary et al., 2019). In an experiment, Pourbabaee et al. (2020) observed a considerable increase in height, yield, and nitrogen uptake of *Thiobacillus*-inoculated maize plants. Similarly, an increase in height, fresh and dry leaf mass, bulb weight, and diameter of garlic plant is reported (Hejazirad et al., 2017). The application of sulfur-oxidizing biofertilizers has been recommended for onion, oats, ginger, grape, garlic, and cauliflower under alkaline soil conditions (Macik et al., 2020; da Silva et al., 2012).

21.5.2 The current state of biofertilizers and crop production

Globally, around 170 organizations in 24 different countries are commercially producing microbe-based biofertilizers (Table 21.7) and many of them have well-established industries that produce, market, and distribute biofertilizers both at large and small scales (Bharti et al., 2017). Despite these, the commercialization of biofertilizers is low which, however, is expanding steadily worldwide (Verma et al., 2019) despite the fact that the cost of agrochemicals in some countries is low (Timmusk et al., 2017). In 2014, the biofertilizer market was only about 5% of the total chemical fertilizer market (BCC Research Global Market for Biopesticides, 2014) which according to the Market Data Forecast (2018–2023) was projected at USD 396.07 million in 2018 (an annual growth rate of 9.5%) and to approximately USD 623.51 million by 2023 (Timmusk et al., 2017). According to some predictions, the biofertilizer market share is likely to reach USD 1.66 billion by 2022 and will upsurge further at a compounding annual growth rate (CAGR) of 13.2% until 2022. The most advanced and prevalent

Table 21.7 Examples of some countries involved in commercial production of biofertilizers.

Country	Product	Organisms	Manufacturers	Crop	Source
Australia	Bio-N	*Azotobacter* sp.	Nutri-Tech solution	Not mentioned	Adeleke et al. (2019)
	Myco-Tea	*A. chroococcum, B. polymyxa*	Nutri-Tech solution	Tea	Adeleke et al. (2019)
	Twin N	*Azorhizobium* sp., *Azoarcus* sp., *Azospirillum* sp.	Mapleton Int. Ltd	Not mentioned	Adeleke et al. (2019)
Canada	Rhizocell GC Nodulator	*B. amyloliquefaciens* IT 45, *B. japonicum*	Lallen and plant care BASF Inc.	Beans, maize, carrot, rice, cotton	Odoh et al. (2019)
China	CBF	*Bacillus mucilaginosus, B. subtilis*	China BioFertilizer AG	Various cereals	Celador-Lera et al. (2018)
Germany	FZB 24 fl, BactofilA 10	*B. amylcliquefaciens, B. megaterium, P. fluorescens*	AbiTEP GmbH	Vegetables, cereals	Odoh et al. (2019)
India	Ajay Azospirillum	*Azospirillum*	Ajay Biotech	Cereals	Celador-Lera et al. (2018)
	Greenmax AgroTech Life Biomix, Biodinc, G max PGPR, Biomax	*Azotobacter, P. fluorescens*		Various crops	Odoh et al. (2019)
Russia	Azobacterium	*Azotobacter brasilense*	JSC Industrial Innovations	Wheat, barley, maize	Celador-Lera et al. (2018)
United Kingdom	Ammnite A 100	*Azotobacter, Bacillus, Rhizobium, Pseudomonas*	Cleveland biotech	Cucumber, tomato, pepper	Odoh et al. (2019)
	Legume Fix	*Rhizobium* sp., *B. japonicum*	Legume Technology	Common bean, Soybean	Adeleke et al. (2019)
	Twin N	*Azorhizobium* sp., *Azoarcus* sp., *Azospirillum* sp.	Mapleton Int. Ltd	Not mentioned	Adeleke et al. (2019)

Continued

Table 21.7 Examples of some countries involved in commercial production of biofertilizers—cont'd

Country	Product	Organisms	Manufacturers	Crop	Source
United States	Inogro	30 bacterial species	FLozyme Corporation	Rice	Celador-Lera et al. (2018)
	Vault NP	*B. japonicum*	Becker Underwood	Not mentioned	Adeleke et al. (2019)
	Chickpea Nodulator	*Mesorhizobium ciceri*	Becker Underwood	Chickpea	Adeleke et al. (2019)
	Cowpea Inoculant	Rhizobia	Becker Underwood	Cowpea	Adeleke et al. (2019)
	PHC Biopak	*B. azotofixans*, *B. licheniformis*, *B. megaterium*, *B. polymyxa*, *B. subtilis*, *B. thuringiensis*	Plant Health Care Inc.	Not mentioned	Adeleke et al. (2019)
	Complete Plus	*Bacillus* strains	Plant Health Care	Various crops	Mustafa et al. (2019)
	QuickRoots	*B. amylcliquefaciens*	Monsanto	Wheat and common bean	Celador-Lera et al. (2018)

market of biofertilizer is Europe where growth from $2566.4 million in 2012 reached to $4582.2 million until 2017: at an annual growth rate of 12.3% (PRWEB, 2014). Among biofertilizers, the nitrogenous biofertilizers dominate the biofertilizer market worldwide (Grand View Research, 2015) and cover about 78% of the international biofertilizer market, whereas the P-biofertilizers and other microbial formulations occupy ≈ 15% and 7%, respectively (Owen et al., 2015; Transparency Market Research, 2017). The rhizobial biofertilizers have been agronomically practiced for many decades now which has substantially reduced the need for mineral fertilizers in many countries while generating good revenue and adding to the economy of many countries as well (Rodríguez-Navarro et al., 2011). As an example, the United States and Canada alone generated largest revenue in 2011 accounting for about 73% of the total revenue collection from biofertilizers, with an expected CAGR of approximately 5.3% up to 2018. The economic gain from biofertilizer technology has further opened up new areas of revenue generation in many countries prompting scientists to discover novel biofertilizers endowed with the best plant growth-enhancing abilities.

21.5.3 Challenges and future prospects of biofertilizers

The application of bacterial biofertilizers in remediation of contaminated environment and in enhancing food production in a sustainable manner is gaining impetus and substantial success has also been recorded worldwide. However, considering its historical background, scientists have failed miserably to replace synthetic fertilizers by biofertilizers due to many reasons. There are major challenges with the adoption and application of microbe-based biofertilizers: (i) lack of awareness, (ii) insufficient availability of suitable carriers, (iii) lack of proper storage facilities, and (iv) fluctuating environmental conditions. Despite these, the application of bacterial biofertilizers has become one of the most vital components that have reduced the dependence on chemical fertilizer to a certain extent. Due to the high market demand for biofertilizers, studies on developing good-quality biofertilizer formulations have increased which is likely to change the overall shape of sustainable crop production systems. So, some suitable regulatory and legal guidelines are needed for adoption and application of biofertilizers which presently are strict and perhaps a deterrent in their utilization. Besides this, enhancing the shelf lives of microbial formulation is another challenge that needs timely attention. Apart from these, the scientific research should also focus on (i) how to optimize the inexpensive growth conditions for mass production and novel bioformulations, (ii) how biofertilizers be made to tolerate stressful environmental conditions, and (iii) how microbes can provide maximum benefit to plants and hence, to farmers.

21.6 Conclusion and future perspectives

The mono- or coculture of metal-tolerant bacterial biofertilizer remediates contaminated environment and can optimize crop production under stressed conditions. The metal mobilizing activity of biofertilizers further assists in the phytoextraction of metals from contaminated soils. The overall enhancement in growth and yield of food crops following metal-tolerant/metal-resistant biofertilizer applications under stressed soils could therefore be due to (i) metal tolerance and metal remediation abilities, (ii) assistance to phytoremediation technologies, (iii) availability of major and micronutrients to plants, (iv) supply of phytostimulants, and (v) protection of plants from phytopathogens. Integrating all, the soil-plant-microbe interface has obvious agronomic relevance in the future. The *in planta* application of bacterial biofertilizers having dual properties (multimetal resistance and PGP) offers an inexpensive option for prospecting the microbe-based remediation of metal-contaminated environment and advancing the "food and feed" concept with a lesser dependence on physicochemical remediation technologies and conventional synthetic fertilizers. However, further in-depth studies are needed to enhance the bioremediation efficacy and host range of biofertilizer organisms to quickly remove/eliminate the pollutants from contaminated environments. In days to come, there is every possibility that the practice of incorporating biofertilizer-based bioremediation technology in traditional cultivation practices will definitely grow and projections are that the demand for microbial formulations will increase enormously. More field studies are, however, required to better pinpoint the global impact of natural variables and of cotoxicants persisting in the concerned environment. Summarily, the global researchers, agricultural institutions, universities, and commercial manufacturers across the countries need to work hand in hand to fast-track the development of bacteria-based bioremediation technology and popularize their usage and adoption for remediation and cultivation in stressed lands.

Acknowledgment

Dr. Rizvi acknowledges the DST-SERB, Government of India, for the National Post Doctoral Fellowship that supported the collection and compilation of literature for this manuscript.

References

Aasfar, A., Bargaz, A., Yaakoubi, K., Hilali, A., Bennis, I., Zeroual, Y., Meftah Kadmiri, I., 2021. Nitrogen fixing *Azotobacter* species as potential soil biological enhancers for crop nutrition and yield stability. Front. Microbiol. 12, 354.

Abawari, R.A., Tuji, F.A., Yadete, D.M., 2020. Phosphate solubilizing bio-fertilizers and their role in bio-available P nutrient: an overview. Int. J. Appl. Agric. Sci. 6, 162.

Abd El-Moaty, N.M., Khalil, H.M.A., Gomaa, H.H., Ismail, M.A., El-Dougdoug, K.A., 2018. Isolation, characterization, and evaluation of multi-trait plant growth promoting rhizobacteria for their growth promoting. Middle East J. Appl. Sci. 8, 554–566.

Abdelbary, S., Elgamal, M.S., Farrag, A., 2019. Trends in heavy metals tolerance and uptake by *Pseudomonas aeruginosa*. In: *Pseudomonas aeruginosa*-An Armory Within. IntechOpen.

Abdelkrim, S., Jebara, S.H., Saadani, O., Chiboub, M., Abid, G., Mannai, K., Jebara, M., 2019. Heavy metal accumulation in *Lathyrus sativus* growing in contaminated soils and identification of symbiotic resistant bacteria. Arch. Microbiol. 201, 107–121.

Abdel-Lateif, K.S., 2017. Isolation and characterization of heavy metals resistant *Rhizobium* isolates from different governorates in Egypt. Afr. J. Biotechnol. 16, 643–647.

Abedinzadeh, M., Etesami, H., Alikhani, H.A., 2019. Characterization of rhizosphere and endophytic bacteria from roots of maize (*Zea mays* L.) plant irrigated with wastewater with biotechnological potential in agriculture. Biotechnol Rep 21, e00305.

Abo-Amer, A.E., Abu-Gharbia, M.A., Soltan, E.S.M., Abd El-Raheem, W.M., 2014. Isolation and molecular characterization of heavy metal-resistant *Azotobacter chroococcum* from agricultural soil and their potential application in bioremediation. Geomicrobiol J. 31, 551–561.

Adamiec, E., Jarosz-Krzemińska, E., Wieszała, R., 2016. Heavy metals from non-exhaust vehicle emissions in urban and motorway road dusts. Environ. Monit. Assess. 188, 369.

Adeleke, R.A., Raimi, A.R., Roopnarain, A., Mokubedi, S.M., 2019. Status and prospects of bacterial inoculants for sustainable management of agroecosystems. In: Giri, B., Prasad, R., Wu, Q.S., Varma, A. (Eds.), Biofertilizers for Sustainable Agriculture and Environment. Springer International Publishing, Cham, pp. 137–172.

Afzal, A.M., Rasool, M.H., Waseem, M., Aslam, B., 2017. Assessment of heavy metal tolerance and biosorptive potential of *Klebsiella variicola* isolated from industrial effluents. AMB Express 7, 184.

Ahmad, M., Nadeeem, S.M., Naveed, M., Zahid, Z.A., 2016. Potassium-solubilizing bacteria and their application in Agriculture. In: Meena, V., Maurya, B., Verma, J., Meena, R. (Eds.), Potassium Solubilizing Microorganisms for Sustainable Agriculture. Springer, New Delhi, pp. 293–313.

Akcil, A., Erust, C., Ozdemiroglu, S., Fonti, V., Beolchini, F., 2015. A review of approaches and techniques used in aquatic contaminated sediments: metal removal and stabilization by chemical and biotechnological processes. J. Clean. Prod. 86, 24–36.

Akintokun, A.K., Ezaka, E., Akintokun, P.O., Taiwo, L.B., 2019. Isolation, screening and response of maize to plant growth promoting rhizobacteria inoculants. Sci. Agric. Bohem. 50, 181–190.

Al Moharbi, S.S., Devi, M.G., Sangeetha, B.M., Jahan, S., 2020. Studies on the removal of copper ions from industrial effluent by *Azadirachta indica* powder. Appl Water Sci 10, 1–10.

Al-Asheh, S., Duvnjak, Z., 1995. Adsorption of copper and chromium by *Aspergillus carbonarius*. Biotechnol. Prog. 11, 638–642.

Ali Redha, A., 2020. Removal of heavy metals from aqueous media by biosorption. Arab. J. Basic Appl. Sci. 27, 183–193.

Ali, H., Khan, E., Sajad, M.A., 2013. Phytoremediation of heavy metals-concepts and applications. Chemosphere 91, 869–881.

Aloo, B.N., Makumba, B.A., Mbega, E.R., 2019. The potential of *Bacilli* rhizobacteria for sustainable crop production and environmental sustainability. Microbiol. Res. 219, 26–39.

Aloo, B.N., Mbega, E.R., Makumba, B.A., 2020a. Rhizobacteria-based technology for sustainable cropping of potato (*Solanum tuberosum* L.). Potato Res. 63, 157–177.

Aloo, B.N., Mbega, E.R., Makumba, B.A., Hertel, R., Danel, R., 2020b. Molecular identification and in vitro plant growth-promoting activities of culturable Potato (*Solanum tuberosum* L.) rhizobacteria in Tanzania. Potato Res. https://doi.org/10.1007/s11540-020-09465-x.

Alori, E.T., Glick, B.R., Babalola, O.O., 2017. Microbial phosphorus solubilization and its potential for use in sustainable agriculture. Front. Microbiol. 8, 971.

Amari, T., Ghnaya, T., Abdelly, C., 2017. Nickel, cadmium and lead phytotoxicity and potential of halophytic plants in heavy metal extraction. S. Afr. J. Bot. 111, 99–110.

Amiri, F., Mousavi, S.M., Yaghmaei, S., 2011. Enhancement of bioleaching of a spent Ni/Mo hydroprocessing catalyst by *Penicillium simplicissimum*. Sep. Purif. Technol. 80, 566–576.

Armendariz, A.L., Talano, M.A., Nicotra, M.F.O., Escudero, L., Breser, M.L., Porporatto, C., Agostini, E., 2019. Impact of double inoculation with *Bradyrhizobium japonicum* E109 and *Azospirillum brasilense* Az39 on soybean plants grown under arsenic stress. Plant Physiol. Biochem. 138, 26–35.

Arora, N.K., Verma, M., 2017. Modified microplate method for rapid and efficient estimation of siderophore produced by bacteria. 3 Biotech 7, 381.

Arshad, M., Javaid, A., Manzoor, M., Hina, K., Ali, M.A., Ahmed, I., 2019. Isolation and identification of chromium-tolerant bacterial strains and their potential to promote plant growth. In: E3S Web of Conferences. vol. 96. EDP Sciences, p. 01005.

Arshadi, M., Nili, S., Yaghmaei, S., 2019a. Ni and Cu recovery by bioleaching from the printed circuit boards of mobile phones in non-conventional medium. J. Environ. Manag. 250, 109502.

Arshadi, M., Yaghmaei, S., Mousavi, S.M., 2019b. Optimal electronic waste combination for maximal recovery of Cu-Ni-Fe by *Acidithiobacillus ferrooxidans*. J. Clean. Prod. 240, 118077.

Arthee, R., Marimuthu, P., 2017. Studies on endophytic *Burkholderia* sp. From sugarcane and its screening for plant growth promoting potential. J. Exp. Biol. 5, 242–257.

Arunakumara, K.K.I.U., Walpola, B.C., Yoon, M.H., 2015. Bioaugmentation-assisted phytoextraction of Co, Pb and Zn: an assessment with a phosphate-solubilizing bacterium isolated from metal-contaminated mines of Boryeong area in South Korea. Biotechnol. Agron. Soc. Environ. 19, 143–152.

Asoegwu, C.R., Awuchi, C.G., Nelson, K., Orji, C.G., Nwosu, O.U., Egbufor, U.C., Awuchi, C.G., 2020. A review on the role of biofertilizers in reducing soil pollution and increasing soil nutrients. Himalayan J. Agric. 1, 34–38.

Awais, M., Tariq, M., Ali, Q., Khan, A., Ali, A., Nasir, I.A., Husnain, T., 2019. Isolation, characterization and association among phosphate solubilizing bacteria from sugarcane rhizosphere. Cytol. Genet. 53, 86–95.

Ayangbenro, A.S., Babalola, O.O., 2017. A new strategy for heavy metal polluted environments: a review of microbial biosorbents. Int. J. Environ. Res. Public Health 14, 94.

Bageshwar, U., Srivastava, M., Pardha-Saradhi, P., Paul, S., Sellamuthu, G., Jaat, R., Das, H., 2017. An environment friendly engineered Azotobacter can replace substantial amount of urea fertilizer and yet sustain same wheat yield. Appl. Environ. Microbiol. 83, e00590-17.

Bahadur, I., Maurya, B., Roy, P., Kumar, A., 2019. Potassium-solubilizing bacteria (KSB): A microbial tool for K-solubility, cycling, and availability to plants. In: Kumar, A., Meena, V. (Eds.), Plant Growth Promoting Rhizobacteria for Agricultural Sustainability. Springer, Singapore, pp. 257–265.

Bahena, M.H.R., Salazar, S., Velázquez, E., Laguerre, G., Peix, A., 2015. Characterization of phosphate solubilizing rhizobacteria associated with pea (*Pisum sativum* L.) isolated from two agricultural soils. Symbiosis 67, 33–41.

Bai, J., Yang, X., Du, R., Chen, Y., Wang, S., Qiu, R., 2014. Biosorption mechanisms involved in immobilization of soil Pb by *Bacillus subtilis* DBM in a multi-metal-contaminated soil. J. Environ. Sci. 26, 2056–2064.

Bai, D., Ying, Q., Wang, N., Lin, J., 2016. Copper removal from electroplating wastewater by coprecipitation of copper-based supramolecular materials: preparation and application study. J. Chem., 5281561.

Bakhshandeh, E., Rahimian, H., Pirdashti, H., Nematzadeh, G., 2015. Evaluation of phosphate-solubilizing bacteria on the growth and grain yield of rice (*Oryza sativa* L.) cropped in northern Iran. J. Appl. Microbiol. 119, 1371–1382.

Banerjee, A., Jhariya, M.K., Yadav, D.K., Raj, A., 2018. Micro-remediation of metals: a new frontier in bioremediation. In: Handbook of Environmental Materials Management. Springer.

Banik, A., Dash, G.K., Swain, P., Kumar, U., Mukhopadhyay, S.K., Dangar, T.K., 2019. Application of rice (*Oryza sativa* L.) root endophytic diazotrophic *Azotobacter* sp. strain Avi2 (MCC 3432) can increase rice yield under green house and field condition. Microbiol. Res. 219, 56–65.

Bankole, M.T., Abdulkareem, A.S., Mohammed, I.A., Ochigbo, S.S., Tijani, J.O., Abubakre, O.K., Roos, W.D., 2019. Selected heavy metals removal from electroplating wastewater by purifed and polyhydroxylbutyrate functionalized carbon nanotubes adsorbents. Sci. Rep. 9, 4475.

Batool, S., Iqbal, A., 2019. Phosphate solubilizing rhizobacteria as alternative of chemical fertilizer for growth and yield of *Triticum aestivum* (Var. Galaxy 2013). Saudi J. Biol. Sci. 26, 1400–1410.

BCC, 2014. Research Global Market for Biopesticides; Market Research Reports; Wellesley, MA. pp. 1–137.

Belechheb, T., Bakkali, M., Laglaoui, A., Arakrak, A., 2020. Characterization and efficiency of rhizobial isolates nodulating *Cytisus monspessulanus* in the northwest of Morocco in relation to environmental stresses. In: Phyto-Microbiome in Stress Regulation. Springer, Singapore, pp. 63–72.

Bellabarba, A., Fagorzi, C., diCenzo, G.C., Pini, F., Viti, C., Checcucci, A., 2019. Deciphering the symbiotic plant microbiome: translating the most recent discoveries on rhizobia for the improvement of agricultural practices in metal-contaminated and high saline lands. Agronomy 9, 529.

Belogolova, G.A., Baenguev, B.A., Gordeeva, O.N., Sokolova, M.G., Pastukhov, M.V., Poletaeva, V.I., Vaishlya, O.B., 2019. Rhizobacteria effect on bioaccumulation and biotransformation of arsenic and heavy metal compounds in the technogenous soils. In: IOP Conf Ser: Earth and Env Sci. vol. 381. IOP Publishing, p. 012007.

Besharati, H., Memar Kouche-Bagh, S., 2017. Effect of lead pollution stress on biological nitrogen fixation of alfalfa plant. Env. Stresses Crop Sci. 10, 163–171.

Bharti, N., Sharma, S.K., Saini, S., Verma, A., Nimonkar, V., Prakash, O., 2017. Microbial plant probiotics: problems in application and formulation. In: Kumar, V., Kumar, M., Sharma, S., Prasad, R. (Eds.), Probiotics and Plant Health. Springer, Singapore, pp. 317–335.

Bhat, M.A., 2019. Plant growth promoting rhizobacteria (PGPR) for sustainable and eco-friendly agriculture. Acta Sci. Agric. 3, 23–25.

Bhatt, K., Maheshwari, D.K., 2020. Zinc solubilizing bacteria (*Bacillus megaterium*) with multifarious plant growth promoting activities alleviates growth in *Capsicum annuum* L. 3 Biotech 10, 36.

Bhojiya, A.A., Joshi, H., 2016. Study of potential plant growth-promoting activities and heavy metal tolerance of *Pseudomonas aeruginosa* HMR16 isolated from Zawar, Udaipur, India. Curr Trends Biotechnol Pharm 10, 161–168.

Blackwell, K.J., Singleton, I., Tobin, J.M., 1995. Metal cation uptake by yeast: a review. Appl. Microbiol. Biotechnol. 43, 579–584.

Bojórquez, C., Voltolina, D., 2016. Removal of cadmium and lead by adapted strains of Pseudomonas aeruginosa and Enterobacter cloacae. Rev. Int. Contam. Ambient. 32, 407–412.

Bordoloi, N., Tiwari, J., Kumar, S., Korstad, J., Bauddh, K., 2020. Efficiency of algae for heavy metal removal, bioenergy production, and carbon sequestration. In: Emerging Eco-friendly Green Technologies for Wastewater Treatment. Springer, Singapore, pp. 77–101.

Castillo, A.R., Gerding, M., Oyarzua, P., Zagal, E., Gerding, J., Fischer, S., 2019. Plant growth-promoting rhizobacteria able to improve NPK availability: selection, identification and effects on tomato growth. Chil. J. Agric. Res. 79, 473–485.

Celador-Lera, L., Jiménez-Gómez, A., Menéndez, E., Rivas, R., 2018. Biofertilizers based on bacterial endophytes isolated from cereals: potential solution to enhance these crops. In: Meena, V.S. (Ed.), Stress Management and Agricultural Sustainability: Role of Rhizospheric Microbes in Soil. Springer, Singapore, pp. 175–203.

Chanmugathas, P., Bollag, J.M., 1998. A column study of the biological mobilization and speciation of cadmium in soil. Arch. Environ. Contam. Toxicol. 17, 229–237.

Chaturvedi, S., Khare, A., Khurana, S.P., 2021. Toxicity of hexavalent chromium and its microbial detoxification through bioremediation. In: Removal of Emerging Contaminants Through Microbial Processes. Springer, Singapore, pp. 513–542.

Chaudhary, S., Thakur, N., Goyal, S.B.S., 2019. Role of sulphur oxidizing bacteria inoculated mustard in livestock production via effect on mustard growth. Pharm. Innov. J., 49–52.

Chawngthu, L., Hnamte, R., Lalfakzuala, R., 2020. Isolation and characterization of rhizospheric phosphate solubilizing bacteria from wetland paddy field of Mizoram, India. Geomicrobiol J. 37, 366–375.

Chen, Q., Liu, S., 2019. Identification and characterization of the phosphate-solubilizing bacterium *Pantoea* sp. *S32* in reclamation soil in Shanxi, China. Front. Microbiol. 10, 2171.

Chen, W., Yang, F., Zhan, L., Wang, J., 2016. Organic acid secretion and phosphate solubilizing efficiency of *Pseudomonas* sp. PSB12: effects of phosphorus forms and carbon sources. Geomicrobiol. J. 33, 870–877.

Chen, H., Zhang, J., Tang, L., Su, M., Tian, D., Zhang, L., 2019. Enhanced Pb immobilization via the combination of biochar and phosphate solubilizing bacteria. Environ. Int. 127, 395–401.

Choudhary, M., Patel, B.A., Meena, V.S., Yadav, R.P., Ghasal, P.C., 2019. Seed bio-priming of green gram with Rhizobium and levels of nitrogen and sulphur fertilization under sustainable agriculture. Legum. Res. 42, 205–210.

da Silva, M.F., de Souza, A.C., de Oliveira, P.J., Xavier, G.R., Rumjanek, N.G., de Barros Soares, L.H., Reis, V.M., 2012. Survival of endophytic bacteria in polymer-based inoculants and efficiency of their application to sugarcane. Plant Soil 356, 231–243.

de Abreu, C.S., Figueiredo, J.E.F., Oliveira, C.A., dos Santos, V.L., Gomes, E.A., Ribeiro, V.P., Lana, U.G.P., Marriel, I.E., 2017. Maize endophytic bacteria as mineral phosphate solubilizers. Genet. Mol. Res. 16, 1–13.

Debnath, B., Chowdhury, R., Ghosh, S.K., 2018. Sustainability of metal recovery from E-waste. Front. Environ. Sci. Eng. 12, 1–12.

Deepa, A., Mishra, B.K., 2020. Microbial biotransformation of hexavalent chromium [Cr (VI)] in tannery wastewater. In: Microbial Bioremediation & Biodegradation. Springer, Singapore, pp. 143–152.

Dhevagi, P., Priyatharshini, S., Ramya, A., Sudhakaran, M., 2021. Biosorption of lead ions by exopolysaccharide producing *Azotobacter* sp. J. Environ. Biol. 42, 40–50.

Di Gregorio, S., Barbafieri, M., Lampis, S., Sanangelantoni, A.M., Tassi, E., Vallini, G., 2006. Combined application of Triton X-100 and *Sinorhizobium* sp. Pb002 inoculum for the improvement of lead phytoextraction by *Brassica juncea* in EDTA amended soil. Chemosphere 63, 293–299.

Ding, J., Jiang, X., Guan, D., Zhao, B., Ma, M., Zhou, B., Cao, F., Yang, X., Li, L., Li, J., 2017. Influence of inorganic fertilizer and organic manure application on fungal communities in a long-term field experiment of Chinese Mollisols. Appl. Soil Ecol. 111, 114–122.

Dong, Z., Bank, M.S., Spengler, J.D., 2015. Assessing metal exposures in a community near a cement plant in the Northeast US. Int. J. Environ. Res. Public Health 12, 952–969.

Dubey, R., Gupta, D.K., Sharma, G.K., 2020. Chemical stress on plants. In: New Frontiers in Stress Management for Durable Agriculture. Springer, Berlin/Heidelberg, Germany, pp. 101–128.

Edulamudi, P., Antony Masilamani, A.J., Vanga, U.R., Divi, V.R.S.G., Konada, V.M., 2021. Nickel tolerance and biosorption potential of rhizobia associated with horse gram *Macrotyloma uniflorum* (Lam.) Verdc. Int. J. Phytoremediation, 1–7. https://doi.org/10.1080/15226514.2021.1884182.

El-Ghamry, A., Mosa, A.A., Alshaal, T., El-Ramady, H., 2018. Nanofertilizers vs. biofertilizers: new insights. Environ. Biodivers. Soil Sec. 2, 51–72.

El-Khateeb, N.M., Metwaly, M.M., 2019. Influence of some bio-fertilizers on wheat plants grown under graded levels of nitrogen fertilization. Int. J. Environ. 8, 43–56.

Emami, S., Alikhani, H.A., Pourbabaei, A.A., Etesami, H., Motashare, Z.B., Sarmadian, F., 2018. Improved growth and nutrient acquisition of wheat genotypes in phosphorus deficient soils by plant growth promoting rhizospheric and endophytic bacteria. Soil Sci. Plant Nutr. 64, 719–727.

Emami, S., Alikhani, H.A., Pourbabaei, A.A., Etesami, H., Motessharezadeh, B., 2019. Effect of rhizospheric and endophytic bacteria with multiple plant growth promoting traits on wheat growth. Environ. Sci. Pollut. Res. 26, 29804.

Esertas, Ü.Z., Uzunalioğlu, E., Güzel, Ş., 2020. Determination of bioremediation properties of soil-borne *Bacillus* sp. 5O5Y11 and its effect on the development of *Zea mays* in the presence of copper. Arch. Microbiol. 202, 1817–1829.

Etesami, H., Emami, S., Alikhani, H.A., 2017. Potassium solubilizing bacteria (KSB): mechanisms, promotion of plant growth, and future prospects a review. J. Soil Sci. Plant Nutr. 17, 897–911.

Fagorzi, C., Checcucci, A., DiCenzo, G.C., Debiec-Andrzejewska, K., Dziewit, L., Pini, F., Mengoni, A., 2018. Harnessing rhizobia to improve heavy-metal phytoremediation by legumes. Genes 9, 542.

FAO, UNICEF, WFP, WHO, 2017. The State of Food Security and Nutrition in the World. Food and Agricultural Organisation of the United States. https://www.fao.org/publications/sofi/2021/en/.

Ferreira, P.A.A., Dahmer, S.D.F.B., Backes, T., Silveira, A.D.O., Jacques, R.J.S., Zafar, M., Pauletto, E.A., Santos, M.A.O.D., Silva, K.D., Giachini, A.J., Antoniolli, Z.I., 2018. Isolation, characterization and symbiotic efficiency of nitrogen-fixing and heavy metal-tolerant bacteria from a coalmine wasteland. Rev. Bras. Ciênc. Solo 42, e0170171.

Ferreira, C.M., Vilas-Boas, Â., Sousa, C.A., Soares, H.M., Soares, E.V., 2019. Comparison of five bacterial strains producing siderophores with ability to chelate iron under alkaline conditions. AMB Express 9, 1–12.

Fitriatin, B.N., Nabila, M.E., Sofyan, E.T., Yuniarti, A., Turmuktini, T., 2019. Effect of beneficial soil microbes and inorganic fertilizers on soil nitrogen, chlorophyll and yield of upland rice on ultisols. In: IOP Conf Ser: Earth and Env Sci. vol. 393. IOP Publishing, p. 012013.

Fukami, J., Cerezini, P., Hungria, M., 2018. *Azospirillum*: benefits that go far beyond biological nitrogen fixation. AMB Express 8, 1–12.

Gadd, G.M., 1990. Heavy metal accumulation by bacteria and other microorganisms. Experientia 46, 834–840.

Galindo, F.S., Buzetti, S., Rodrigues, W.L., 2020. Inoculation of *Azospirillum brasilense* associated with silicon as a liming source to improve nitrogen fertilization in wheat crops. Sci. Rep. 10, 6160.

García-Fraile, P., Menéndez, E., Rivas, R., 2015. Role of bacterial biofertilizers in agriculture and forestry. AIMS Bioeng. 2, 183–205.

Ghany, T.A.M., Alawlaqi, M.M., Al Abboud, M.A., 2013. Role of biofertilizers in agriculture: a brief review. Mycopathologia 11, 95–101.

Ghazy, N., El-Nahrawy, S., 2021. Siderophore production by *Bacillus subtilis* MF497446 and *Pseudomonas koreensis* MG209738 and their efficacy in controlling *Cephalosporium maydis* in maize plant. Arch. Microbiol. 203, 1195–1209.

Glick, B.R., 2010. Using soil bacteria to facilitate phytoremediation. Biotechnol. Adv. 28, 367–374.

Glick, B.R., 2014. Bacteria with ACC deaminase can promote plant growth and help to feed the world. Microbiol. Res. 169, 30–39.

Glick, B.R., 2020. Introduction to plant growth-promoting bacteria. In: Beneficial Plant-Bacterial Interactions. Springer, Berlin/Heidelberg, Germany, pp. 1–37.

Gonçalves Jr., A.C., Nacke, H., Schwantes, D., Coelho, G.F., 2014. Heavy metal contamination in brazilian agricultural soils due to application of fertilizers. Env. Risk Assess. Soil Contam. 4, 105–135.

Gopalakrishnan, S., Sathya, A., Vijayabharathi, R., Varshney, R.K., Gowda, C.L., Krishnamurthy, L., 2015. Plant growth promoting rhizobia: challenges and opportunities. 3 Biotech 5, 355–377.

Gouda, S., Kerry, R.G., Das, G., Paramithiotis, S., Patra, J.K., 2018. Revitalization of plant growth promoting rhizobacteria for sustainable development in agriculture. Microbiol. Res. 206, 131–140.

Goyal, D., Yadav, A., Prasad, M., Singh, T.B., Shrivastav, P., Ali, A., Dantu, P.K., Mishra, S., 2020. Effect of heavy metals on plant growth: an overview. Contam. Agric. 79–101.

Grand View Research, 2015. Biofertilizers Market Analysis by Product (Nitrogen Fixing, Phosphate Solubilizing), by Application (Seed Treatment, Soil Treatment) and Segment Forecasts to 2022. Grand View Research. https://www.marketresearch.com/Grand-View-Research-v4060/Biofertilizers-Size-Share-Trends-Product-13365113/.

Gunnabo, A.H., van Heerwaarden, J., Geurts, R., 2020. Symbiotic interactions between chickpea (*Cicer arietinum* L.) genotypes and *Mesorhizobium* strains. Symbiosis 82, 235–248.

Gupta, P., Diwan, B., 2017. Bacterial exopolysaccharide mediated heavy metal removal: A review on biosynthesis, mechanism and remediation strategies. Biotechnol. Rep. 13, 58–71.

Haider, F.U., Liqun, C., Coulter, J.A., Cheema, S.A., Wu, J., Zhang, R., Wenjun, M., Farooq, M., 2021. Cadmium toxicity in plants: Impacts and remediation strategies. Ecotoxicol. Environ. Saf. 211, 111887.

Hansda, A., Kumar, V., Anshumali, A., Usmani, Z., 2014. Phytoremediation of heavy metals contaminated soil using plant growth promoting rhizobacteria (PGPR): a current perspective. Recent Res. Sci. Technol. 6, 131–134.

Hao, X., Xie, P., Zhu, Y.G., Taghavi, S., Wei, G., Rensing, C., 2015. Copper tolerance mechanisms of *Mesorhizobium amorphae* and its role in aiding phytostabilization by *Robinia pseudoacacia* in copper contaminated soil. Environ. Sci. Technol. 49, 2328–2340.

Hejazirad, P., Gholami, A., Pirdashty, H., Abbasiyan, A., 2017. Evaluation of *Thiobacillus* bacteria and mycorrhizal symbiosis on yield and yield components of garlic (*Allium sativum*) at different levels of sulphur. Agroecology 9, 76–87.

Herrmann, L., Lesueur, D., 2013. Challenges of formulation and quality of biofertilizers for successful inoculation. Appl. Microbiol. Biotechnol. 97, 8859–8873.

Hossain, M.M., Iffat, J., Salina, A., Rahman, M.N., Rahman, S.B., 2015. Effects of *Azospirillum* isolates isolated from paddy fields on the growth of rice plants. Res. Biotechnol. 6, 15–22.

Hossan, S., Hossain, S., Islam, M.R., Kabir, M.H., Ali, S., Islam, M.S., Imran, K.M., Moniruzzaman, M., Mou, T.J., Parvez, A.K., Mahmud, Z.H., 2020. Bioremediation of hexavalent chromium by chromium resistant bacteria reduces phytotoxicity. Int. J. Environ. Res. Public Health 17, 6013.

Hou, P., Liu, Y., Liu, W., Liu, G., Xie, R., Wang, K., Ming, B., Wang, Y., Zhao, R., Zhang, W., Wang, Y., 2020. How to increase maize production without extra nitrogen input. Resour. Conserv. Recycl. 160, 104913.

Htwe, A.Z., Moh, S.M., Soe, K.M., Moe, K., Yamakawa, T., 2019. Effects of biofertilizer produced from *Bradyrhizobium* and *Streptomyces griseoflavus* on plant growth, nodulation, nitrogen fixation, nutrient uptake, and seed yield of mungbean, cowpea, and soybean. Agronomy 9, 77.

Huang, F., Guo, C.L., Lu, G.N., Yi, X.Y., Zhu, L.D., Dang, Z., 2014. Bioaccumulation characterization of cadmium by growing *Bacillus cereus* RC-1 and its mechanism. Chemosphere 109, 134-142.

Hussain, A., Zahir, Z.A., Ditta, A., Tahir, M.U., Ahmad, M., Mumtaz, M.Z., Hayat, K., Hussain, S., 2020. Production and implication of bio-activated organic fertilizer enriched with zinc-solubilizing bacteria to boost up maize (*Zea mays* L.) production and biofortification under two cropping seasons. Agronomy 10, 39.

Igiri, B.E., Okoduwa, S.I., Idoko, G.O., Akabuogu, E.P., Adeyi, A.O., Ejiogu, I.K., 2018. Toxicity and bioremediation of heavy metals contaminated ecosystem from tannery wastewater: a review. J. Toxicol., 2568038.

Ijaz, M., Ali, Q., Ashraf, S., Kamran, M., Rehman, A., 2019. Development of future bioformulations for sustainable agriculture. In: Kumar, V., Prasad, R., Kumar, M., Choudhary, D.K. (Eds.), Microbiome in Plant Health and Disease: Challenges and Opportunities. Springer, Singapore, pp. 421-446.

Imran, M., Shahzad, S.M., Arif, M.S., Yasmeen, T., Ali, B., Tanveer, A., 2020. Inoculation of potassium solubilizing bacteria with different potassium fertilization sources mediates maize growth and productivity. Pak. J. Agric. Sci. 57, 1045-1055.

Ingle, K.P., Padole, D.A., 2017. Phosphate solubilizing microbes: an overview. Int. J. Curr. Microbiol. App. Sci. 6, 844-852.

Inoue, K., Parajuli, D., Ghimire, K.N., Biswas, B.K., Kawakita, H., Oshima, T., Ohto, K., 2017. Biosorbents for removing hazardous metals and metalloids. Materials 10, 857.

Ivshina, I.B., Kuyukina, M.S., Ritchkova, M.I., Philp, J.C., Cunningham, C.J., Christofi, N., 2001. Oleophilic biofertilizer based on a *Rhodococcus* surfactant complex for the bioremediation of crude oil-contaminated soil. In: AEHS Contaminated Soil Sediment and Water: International Issue, pp. 20-24.

Jain, D., Kour, R., Bhojiya, A.A., Meena, R.H., Singh, A., Mohanty, S.R., Rajpurohit, D., Ameta, K.D., 2020. Zinc tolerant plant growth promoting bacteria alleviates phytotoxic effects of zinc on maize through zinc immobilization. Sci. Rep. 10, 13865.

Jaishankar, M., Tseten, T., Anbalagan, N., Mathew, B.B., Beeregowda, K.N., 2014. Toxicity, mechanism and health effects of some heavy metals. Interdiscip. Toxicol. 7, 60-72.

Jin, Y., Luan, Y., Ning, Y., Wang, L., 2018. Effects and mechanisms of microbial remediation of heavy metals in soil: a critical review. Appl. Sci. 8, 1336.

Jobby, R., Jha, P., Desai, N., 2015. *Sinorhizobium*, a potential organism for bioremediation of nickel. Int. J. Adv. Res. 3, 706-717.

Jobby, R., Jha, P., Yadav, A.K., Desai, N., 2018. Biosorption and biotransformation of hexavalent chromium [Cr(VI)]: a comprehensive review. Chemosphere 207, 255-266.

Jobby, R., Jha, P., Gupta, A., Gupte, A., Desai, N., 2019. Biotransformation of chromium by root nodule bacteria *Sinorhizobium* sp. SAR1. PLoS One 14, e0219387.

Ju, W., Liu, L., Jin, X., Duan, C., Cui, Y., Wang, J., Ma, D., Zhao, W., Wang, Y., Fang, L., 2020. Co-inoculation effect of plant-growth-promoting rhizobacteria and rhizobium on EDDS assisted phytoremediation of Cu contaminated soils. Chemosphere 254, 126724.

Kadmiri, I.M., Chaouqui, L., Azaroual, S.E., Sijilmassi, B., Yaakoubi, K., Wahby, I., 2018. Phosphate-solubilizing and auxin-producing rhizobacteria promote plant growth under saline conditions. Arab. J. Sci. Eng. 43, 3403-3415.

Kalayu, G., 2019. Phosphate solubilizing microorganisms: Promising approach as biofertilizers. Int. J. Agron., 4917256.

Kamran, S., Shahid, I., Baig, D.N., Rizwan, M., Malik, K.A., Mehnaz, S., 2017. Contribution of zinc solubilizing bacteria in growth promotion and zinc content of wheat. Front. Microbiol. 8, 2593.

Kang, S.M., Hamayun, M., Khan, M.A., Iqbal, A., Lee, I.J., 2019. *Bacillus subtilis* JW1 enhances plant growth and nutrient uptake of Chinese cabbage through gibberellins secretion. J. Appl. Bot. Food Qual. 92, 172–178.

Kanitkar, S., Raut, V.M., Kulkarni, M., Vyas, A.K., Das, A., Kadam, M., 2020. The use of vitormone (*Azotobacter chroococcum*) a liquid bio-fertilizer along with chemical fertilizer on crop growth and yield of wheat (*Triticum aestivum* L). Int. J. Res. Appl. Sci. Biotechnol. 7, 80–88.

Kaur, G., Reddy, M.S., 2015. Effects of phosphate-solubilizing bacteria, rock phosphate and chemical fertilizers on maize-wheat cropping cycle and economics. Pedosphere 25, 428–437.

Khan, M.S., Zaidi, A., Wani, P.A., 2007. Role of phosphate solubilizing microorganisms in sustainable agriculture—a review. Agron. Sustain. Dev. 27, 29–43.

Khan, M.S., Zaidi, A., Wani, P.A., Oves, M., 2009. Role of plant growth promoting rhizobacteria in the remediation of metal contaminated soils. Environ. Chem. Lett. 7, 1–19.

Khan, M.S., Zaidi, A., Musarrat, J., 2014. Phosphate solubilizing microorganisms. In: Principles and Application of Microphos Technology. Springer-Verlag, Switzerland, p. 297.

Khan, W.U., Ahmad, S.R., Yasin, N.A., Ali, A., Ahmad, A., Akram, W., 2017. Application of *Bacillus megaterium* MCR-8 improved phytoextraction and stress alleviation of nickel in *Vinca rosea*. Int. J. Phytoremediation 19, 813–824.

Khande, R., Sushil, K.S., Ramesh, A., Mahaveer, P.S., 2017. Zinc solubilizing *Bacillus* strains that modulate growth, yield and zinc biofortification of soybean and wheat. Rhizosphere 4, 126–138.

Khanna, K., Jamwal, V.L., Gandhi, S.G., Ohri, P., Bhardwaj, R., 2019. Metal resistant PGPR lowered Cd uptake and expression of metal transporter genes with improved growth and photosynthetic pigments in *Lycopersicon esculentum* under metal toxicity. Sci. Rep. 9, 1–14.

Kong, Z., Deng, Z., Glick, B.R., Wei, G., Chou, M., 2017. A nodule endophytic plant growth-promoting Pseudomonas and its effects on growth, nodulation and metal uptake in *Medicago lupulina* under copper stress. Ann. Microbiol. 67, 49–58.

Kour, D., Rana, K.L., Yadav, A.N., Yadav, N., Kumar, M., Kumar, V., Vyas, P., Dhaliwal, H.S., Saxena, A.K., 2019. Microbial biofertilizers: bioresources and eco-friendly technologies for agricultural and environmental sustainability. Biocatal. Agric. Biotechnol. 23, 101487.

Kratchovil, D., Volesky, B., 1998. Advances in the biosorption of heavy metals. Trends Biotechnol. 16, 291–300.

Kügler, S., Cooper, R.E., Boessneck, J., Küsel, K., Wichard, T., 2020. Rhizobactin B is the preferred siderophore by a novel *Pseudomonas* isolate to obtain iron from dissolved organic matter in peatlands. Biometals 33, 415–443.

Kumar, A., 2016. Phosphate solubilizing bacteria in agriculture biotechnology: diversity, mechanism and their role in plant growth and crop yield. Int. J. Adv. Res. 4, 116–124.

Kumar, A., Chaturvedi, A.K., Yadav, K., Arunkumar, K.P., Malyan, S.K., Raja, P., Kumar, R., Khan, S.A., Yadav, K.K., Rana, K.L., Kour, D., 2019. Fungal phytoremediation of heavy metal-contaminated resources: current scenario and future prospects. In: Recent Advancement in White Biotechnology Through Fungi. Springer, Cham, pp. 437–461.

Kunawat, K., Sharma, P., Sirari, A., Singh, U., Saharan, K., 2019. Synergism of *Pseudomonas aeruginosa* (LSE-2) nodule endophyte with *Bradyrhizobium* sp. (LSBR-3) for improving plant growth, nutrient acquisition and soil health in soybean. World. J. Biotechnol. 35, 47.

Kurniawan, S.B., Imron, M.F., Purwanti, I.F., 2019. Biosorption of chromium by living cells of *Azotobacter* s8, *Bacillus subtilis* and *Pseudomonas aeruginosa* using batch system reactor. J. Ecol. Eng. 20, 184–189.

Kwunonwo, U.C., Odika, P.O., Onyia, N.I., 2020. A review of the health implications of heavy metals in food chain in Nigeria. Sci. World J., 6594109.

Lajayer, B.A., Moghadam, N.K., Maghsoodi, M.R., Ghorbanpour, M., Kariman, K., 2019. Phytoextraction of heavy metals from contaminated soil, water and atmosphere using ornamental plants: mechanisms and efficiency improvement strategies. Environ. Sci. Pollut. Res. 26, 8468–8484.

Lambers, H., Plaxton, W.C., 2018. P: back to the roots. Annu. Plant Rev. 48, 3–22.

Laxita, L., Shruti, S., 2020. Isolation and characterization of potassium solubilizing microorganisms from south Gujarat region and their effects on wheat plant. Mukt Shabd J. 9, 7483–7496.

Lesueur, D., Deaker, R., Herrmann, L., Bräu, L., Jansa, J., 2016. The production and potential of biofertilizers to improve crop yields. In: Arora, N.K., Menhaz, S., Balestrini, R. (Eds.), Bioformulations for Sustainable Agriculture. Springer, New Delhi, pp. 71–92.

Li, K., Wusirika, R., 2011. Effect of multiple metal resistant bacteria from contaminated lake sediments on metal accumulation and plant growth. J. Hazard. Mater. 189, 531–539.

Li, X., Li, D., Yan, Z., Ao, Y., 2018a. Adsorption of cadmium by live and dead biomass of plant growth-promoting rhizobacteria. RSC Adv. 8, 33523–33533.

Li, X., Li, D., Yan, Z., Ao, Y., 2018b. Biosorption and bioaccumulation characteristics of cadmium by plant growth-promoting rhizobacteria. RSC Adv. 8, 30902–30911.

Li, H., Ding, X., Chen, C., Zheng, X., Han, H., Li, C., Gong, J., Xu, T., Li, Q.X., Ding, G.C., Li, J., 2019. Enrichment of phosphate solubilizing bacteria during late developmental stages of eggplant (*Solanum melongena* L.). FEMS Microbiol. Ecol. 95, fiz023.

Liang, J.L., Liu, J., Jia, P., Yang, T.T., Zeng, Q.W., Zhang, S.C., Liao, B., Shu, W.S., Li, J.T., 2020. Novel phosphate-solubilizing bacteria enhance soil phosphorus cycling following ecological restoration of land degraded by mining. ISME J. 14, 1600–1613.

Liaquat, F., Munis, M.F.H., Haroon, U., Arif, S., Saqib, S., Zaman, W., Khan, A.R., Shi, J., Che, S., Liu, Q., 2020. Evaluation of metal tolerance of fungal strains isolated from contaminated mining soil of Nanjing, China. Biology 9, 469.

Liu, D.Y., Zhang, W., Liu, Y.M., Chen, X.P., Zou, C.Q., 2020. Soil application of zinc fertilizer increases maize yield by enhancing the kernel number and kernel weight of inferior grains. Front. Plant Sci. 11, 188.

Long, J., Gao, X., Su, M., Li, H., Chen, D., Zhou, S., 2018. Performance and mechanism of biosorption of Ni(II) from aqueous solution by non-living *Streptomyces roseorubens* SY. Colloids Surf. A Physicochem. Eng. Asp. 548, 125–133.

Ma, Y., Oliviera, R.S., Nai, F., Rajkumar, M., Luo, Y., Rocha, I., Freitas, H., 2015. The hyperaccumulator *Sedum plumbizincicola* harbors metal-resistant endophytic bacteria that improve its phytoextraction capacity in multi-metal contaminated soil. J. Environ. Manag. 156, 62–69.

Ma, Y., Zhang, C., Oliveira, R.S., Freitas, H., Luo, Y., 2016. Bioaugmentation with endophytic bacterium E6S homologous to *Achromobacter piechaudii* enhances metal rhizoaccumulation in host *Sedum plumbizincicola*. Front. Plant Sci. 7, 75.

Macias-Benitez, S., Garcia-Martinez, A.M., Caballero Jimenez, P., Gonzalez, J.M., Tejada Moral, M., Parrado Rubio, J., 2020. Rhizospheric organic acids as biostimulants: monitoring feedbacks on soil microorganisms and biochemical properties. Front. Plant Sci. 11, 633.

Macik, M., Gryta, A., Frac, M., 2020. Biofertilizers in agriculture: an overview on concepts, strategies and effects on soil microorganisms. Adv. Agron. 162, 31–87.

Mahanty, T., Bhattacharjee, S., Goswami, M., Bhattacharyya, P.N., Das, B., Gosh, A., Tribedi, P., 2017. Biofertilizers: a potential approach for sustainable agriculture development. Environ. Sci. Pollut. Res. 24, 3315–3333.

Maharajan, T., Ceasar, S.A., Krishna, A., Ramakrishnan, T.P., Duraipandiyan, M.V., Naif Abdulla, A.D., 2018. Utilization of molecular markers for improving the P efficiency in crop plants. Plant Breed. 137, 10-26.

Malhorta, H., Vandana, S.S., Pandey, R., 2018. Phosphorus nutrition: plant growth in response to deficiency and excess. In: Hasanuzzaman, M., Fujita, M., Oku, H., Nahar, K., Hawrylak-Nowak, B. (Eds.), Plant Nutrients and Abiotic Stress Tolerance. Springer, Singapore, pp. 170-190.

Mandić, V., Krnjaja, V., Djordjević, S., Djordjević, N., Bijelić, Z., Simić, A., Dragičević, V., 2019. Effects of bacterial seed inoculation on microbiological soil status and maize grain yield. Maydica 63, 8.

Marchenko, A.M., Pshinko, G.N., Demchenko, V.Y., Goncharuk, V.V., 2015. Leaching heavy metal from deposits of heavy metals with bacteria oxidizing elemental sulphur. J. Water Chem. Technol. 37, 311-316.

Market Data Forecast, 2018-2023. Microbial Soil Inoculants Market By Type (Plant Growth Promoting Microorganisms (PGPMs)), Bio-control Agents, and Plant-resistance Stimulants By Crop Type (Cereals & Grains, Oilseeds & Pulses, Fruits & Vegetables, and Other Crops), by Source (Bacterial, Fungal, and Others), by Region—Global Industry Analysis, Size, Share, Growth, Trends, and Forecasts.

Marra, L.M., de Oliveira-Longatti, S.M., Soares, R.F.S.C., de Lima, J.M., Olivares, F.L., Moreira, M.S.F., 2015. Initial pH of medium affects organic acids production but do not affect phosphate solubilization. Braz. J. Microbiol. 46, 367-375.

Marzban, A., Ebrahimipour, G., Karkhane, M., Teymouri, M., 2016. Metal resistant and phosphate solubilizing bacterium improves maize (*Zea mays*) growth and mitigates metal accumulation in plant. Biocatal. Agric. Biotechnol. 8, 13-17.

Meena, S.V., Maurya, B.R., Meena, S.K., Mishra, P.K., Bisht, J.K., Pattanayak, A., 2018. Potassium solubilization: strategies to mitigate potassium deficiency in agricultural soils. Glob. J. Biol. Agric. Health Sci. 7, 1-3.

Mekdad, A.A.A., El-Sherif, A.M.A., 2016. The effect of nitrogen and potassium fertilizers on yield and quality of sweet sorghum varieties under arid regions conditions. Int. J. Curr. Microbiol. App. Sci. 5, 011-823.

Melo, J., Carvalho, L., Correia, P., de Souza, S.B., Dias, T., Santana, M., 2018. Conventional farming disrupts cooperation among phosphate solubilising bacteria isolated from *Carica papaya's* rhizosphere. Appl. Soil Ecol. 124, 284-288.

Mendoza-Arroyo, G.E., Chan-Bacab, M.J., Aguila-Ramírez, R.N., Ortega-Morales, B.O., Canché Solís, R.E., Chab-Ruiz, A.O., Cob-Rivera, K.I., Dzib-Castillo, B., Tun-Che, R.E., Camacho-Chab, J.C., 2020. Inorganic phosphate solubilization by a novel isolated bacterial strain *Enterobacter* sp. ITCB-09 and its application potential as biofertilizer. Agriculture 10, 383.

Mercado-Blanco, J., Abrantes, I., Caracciolo, A.B., Bevivino, A., Ciancio, A., Grenni, P., 2018. Belowground microbiota and the health of tree crops. Front. Microbiol. 9, 1-27.

Mhlongo, M.I., Piater, L.A., Madala, N.E., Labuschagne, N., Dubery, I.A., 2018. The chemistry of plant-microbe interactions in the rhizosphere and the potential for metabolomics to reveal signaling related to defense priming and induced systemic resistance. Front. Plant Sci. 9, 112.

Mishra, J., Singh, R., Arora, N.K., 2017. Alleviation of heavy metal stress in plants and remediation of soil by rhizosphere microorganisms. Front. Microbiol. 8, 1706.

Muleta, A., Tesfaye, K., Selassie, T.H.H., Cook, D.R., Assefa, F., 2021. Phosphate solubilization and multiple plant growth promoting properties of *Mesorhizobium* species nodulating chickpea from acidic soils of Ethiopia. Arch. Microbiol. https://doi.org/10.1007/s00203-021-02189-7.

Mustafa, S., Kabir, S., Shabbir, U., Batool, R.S., 2019. Plant growth promoting rhizobacteria in sustainable agriculture: from theoretical to pragmatic approach. Symbiosis 78, 115-123.

Narayanan, S., 2018. Effects of high temperature stress and traits associated with tolerance in wheat. Open Access J. Sci. 2, 177–186.

Nassal, D., Spohn, M., Eltlbany, N., Jacquiod, S., Smalla, K., Marhan, S., Kandeler, E., 2018. Effects of phosphorus-mobilizing bacteria on tomato growth and soil microbial activity. Plant Soil 427, 17–37.

Naveed, M., Mustafa, A., Azhar, S.Q.T.A., Kamran, M., Ahmad, Z.Z., Núñez-Delgado, A., 2020. *Burkholderia phytofirmans* PsJN and tree twigs derived biochar together retrieved Pb-induced growth, physiological and biochemical disturbances by minimizing its uptake and translocation in mung bean (*Vigna radiata* L.). Environ. Sci. Pollut. Res. 257, 109974.

Naz, I., Ahmad, H., Khokhar, S.N., Khan, K., Shah, A.H., 2016a. Impact of zinc solubilizing bacteria on zinc contents of wheat. Am. Eurasian J. Agric. Environ. Sci. 16, 449–454.

Naz, T., Khan, M.D., Ahmed, I., Rehman, S.U., Rha, E.S., Malook, I., Jamil, M., 2016b. Biosorption of heavy metals by *Pseudomonas* species isolated from sugar industry. Toxicol. Ind. Health 32, 1619–1627.

Nazli, F., Mustafa, A., Ahmad, M., Hussain, A., Jamil, M., Wang, X., Shakeel, Q., Imtiaz, M., El-Esawi, M.A., 2020. A review on practical application and potentials of phytohormone-producing plant growth-promoting rhizobacteria for inducing heavy metal tolerance in crops. Sustainability 12, 9056.

Neal, A.L., Rossmann, M., Brearley, C., Akkari, E., Guyomar, C., Clark, I.M., 2017. Land-use influences phosphatase gene micro-diversity in soils. Environ. Microbiol. 19, 2740–2753.

Nesme, T., Metson, G.S., Bennett, E.M., 2018. Global P flows through agricultural trade. Glob. Environ. Chang. 50, 133–141.

Nosheen, S., Ajmal, I., Song, Y., 2021. Microbes as biofertilizers, a potential approach for sustainable crop production. Sustainability 13, 1868.

Nunes Oliveira, F.L., Silva Oliveira, W., Pereira Stamford, N., Nova Silva, E.V., Santiago Freitas, A.D., 2017. Effectiveness of biofertilizer enriched in N by *Beijerinckia indica* on sugarcane grown on an ultisol and the interactive effects between biofertilizer and sugarcane filter cake. J. Soil Sci. Plant Nutr. 17, 1040–1057.

Odoh, C.K., Eze, C.N., Akpi, U.K., Unah, V.U., 2019. Plant growth promoting rhizobacteria (PGPR): a novel agent for sustainable food production. Am. J. Agric. Biol. Sci. 14, 35–54.

Olanrewaju, O.S., Glick, B.R., Babalola, O.O., 2017. Mechanisms of action of plant growth promoting bacteria. World J. Microbiol. Biotechnol. 33, 197.

Oo, K., Win, T., Khai, A., Fu, P., 2020. Isolation, screening and molecular characterization of multifunctional plant growth promoting rhizobacteria for a sustainable agriculture. Am. J. Plant Sci. 11, 773–792.

Ouertani, R., Ouertani, A., Mahjoubi, M., Bousselmi, Y., Najjari, A., Cherif, H., Chamkhi, A., Mosbah, A., Khdhira, H., Sghaier, H., Chouchane, H., 2020. New plant growth-promoting, chromium-detoxifying *Microbacterium* species isolated from a tannery wastewater: performance and genomic insights. Front. Bioeng. Biotechnol. 8, 521.

Oves, M., Khan, M.S., Zaidi, A., 2013. Chromium reducing and plant growth promoting novel strain *Pseudomonas aeruginosa* OSG41 enhance chickpea growth in chromium amended soils. Eur. J. Soil Biol. 56, 72–83.

Oves, M., Khan, M.S., Qari, H.A., 2017. Ensifer adhaerens for heavy metal bioaccumulation, biosorption, and phosphate solubilization under metal stress condition. J. Taiwan Inst. Chem. Eng. 80, 540–552. https://doi.org/10.1016/j.jtice.2017.08.026.

Oves, M., Khan, M.S., Qari, H.A., 2019. Chromium reducing and phosphate solubilizing *Achromobacter xylosoxidans* from the heavy metal contaminated soil of the Brass city Moradabad, India. Int. J. Environ. Sci. Technol., 1–18.

Owen, D., Williams, A.P., Griffith, G.W., Withers, P.J.A., 2015. Use of commercial bioinoculants to increase agricultural production through improved phosphorus acquisition. Appl. Soil Ecol. 86, 41–54.

Pandey, N., Bhatt, R., 2016. Role of soil associated *Exiguobacterium* in reducing arsenic toxicity and promoting plant growth in *Vigna radiata*. Eur. J. Soil Biol. 75, 142–150.

Parmar, K.B., Mehta, B.P., Kunt, M.D., 2016. Isolation, characterization and identification of potassium solubilizing bacteria from rhizosphere soil of maize (*Zea mays*). Int. J. Sci. Environ. Technol. 5, 3030–3037.

Pérez-Rama, M., Torres, E., Suárez, C., Herrero, C., Abalde, J., 2010. Sorption isotherm studies of Cd (II) ions using living cells of the marine microalga *Tetraselmis suecica* (Kylin) Butch. J. Environ. Manag. 91, 2045–2050.

Pourbabaee, A.A., Koohbori, D.S., Seyed, H.H.M., Alikhani, H.A., Emami, S., 2020. Potential application of selected sulphur-oxidizing bacteria and different sources of sulphur in plant growth promotion under different moisture conditions. Commun. Soil Sci. Plant Anal. 51, 735–745.

Proboningrum, A., Widono, S., 2019. Effectivity and compatibility of Azotobacter and Bacillus for biological control agents of fusarium wilt on banana seedlings. In: IOP Conference Series: Earth Env Sci. vol. 250. OP Publishing, p. 012003.

Proença, D.N., Schwab, S., Baldani, J.I., Morais, P.V., 2017. Diversity and function of endophytic microbial community of plants with economical potential. In: De Azevedo, J.L., Quecine, M.C. (Eds.), Diversity and Benefits of Microorganisms From the Tropics. Springer, Cham, pp. 209–243.

PRWEB, 2014. Europe bio fertilizer market is expected to reach $4,582.2 million in 2017 new report by MicroMarket Monitor. PRWEB. Available online: http://www.micromarketmonitor.com/market/europe-bio-fertilizer-4637178345.html.

Rafi, M.M., Krishnaveni, M.S., Charyulu, P.B.B.N., 2019. Phosphate-solubilizing microorganisms and their emerging role in sustainable agriculture. In: Buddolla, V. (Ed.), Recent Developments in Applied Microbiology and Biochemistry. Academic Press, Dordrecht, pp. 223–233.

Rafique, M., Sultan, T., Ortas, I., Chaudhary, H., 2017. Enhancement of maize plant growth with inoculation of phosphate-solubilizing bacteria and biochar amendment in soil. Soil Sci. Plant Nutr. 63, 460–469.

Rani, V., Sengar, R.S., 2020. Use of potassium bio-fertilizers technology for sustainable agriculture in dry areas. J. Pharmacogn. Phytochem. 9, 1944–1946.

Ray, S., Dey, K., 2020. Coal mine water drainage: The current status and challenges. J. Inst. Eng. (India): Ser. D 101, 165–172.

Reddy, S., Singh, A.K., Masih, H., Benjamin, J.C., Ojha, S.K., Ramteke, P.W., Singla, A., 2018. Effect of *Azotobacter* sp. and *Azospirillum* sp. on vegetative growth of Tomato (*Lycopersicon esculentum*). J. Pharmacogn. Phytochem. 7, 2130–2137.

Renitta, J., Rujuta, V., Pamela, J., Neetin, D., 2019. Biosorption of copper and nickel by *Sinorhizobium* sp. SAR1: effect of chemical and physical pre-treatment and binary metal solutions. Res. J. Chem. Environ. 23, 1–6.

Rfaki, A., Zennouhi, O., Aliyat, F.Z., Nassiri, L., Ibijbijen, J., 2020. Isolation, selection and characterization of root-associated rock phosphate solubilizing bacteria in Moroccan wheat (*Triticum aestivum* L.). Geomicrobiol J. 37, 230–241.

Rizvi, A., Khan, M.S., 2017a. Biotoxic impact of heavy metals on growth, oxidative stress and morphological changes in root structure of wheat (*Triticum aestivum* L.) and stress alleviation by *Pseudomonas aeruginosa* strain CPSB1. Chemosphere 185, 942–952.

Rizvi, A., Khan, M.S., 2017b. Cellular damage, plant growth promoting activity and chromium reducing ability of metal tolerant *Pseudomonas aeruginosa* CPSB1 recovered from metal polluted chilli (*Capsicum annuum*) rhizosphere. Int. J. Agric. Res. Crop Sci. 1, 45–55.

Rizvi, A., Khan, M.S., 2018. Heavy metal induced oxidative damage and root morphology alterations of maize (*Zea mays* L.) plants and stress mitigation by metal tolerant nitrogen fixing *Azotobacter chroococcum*. Ecotoxicol. Environ. Saf. 157, 9–20.

Rizvi, A., Khan, M.S., 2019a. Putative role of bacterial biosorbent in metal sequestration revealed by SEM-EDX and FTIR. Indian J. Microbiol. 59, 246–249.

Rizvi, A., Khan, M.S., 2019b. Heavy metal-mediated toxicity to maize: oxidative damage, antioxidant defence response and metal distribution in plant organs. Int. J. Environ. Sci. Technol. 16, 4873–4886.

Rizvi, A., Zaidi, A., Ahmad, B., Ameen, F., Al Kahtani, M.D.F., Khan, M.S., 2020a. Heavy metal induced stress on wheat: phytotoxicity and microbiological management. RSC Adv. 10, 38379–38403.

Rizvi, A., Ahmed, B., Zaidi, A., Khan, M.S., 2020b. Biosorption of heavy metals by dry biomass of metal tolerant bacterial biosorbents: an efficient metal clean-up strategy. Environ. Monit. Assess. 192, 801.

Rodríguez-Navarro, D.N., Margaret Oliver, I., Albareda Contreras, M., Ruiz-Sainz, J.E., 2011. Soybean interactions with soil microbes, agronomical and molecular aspects. Agron. Sustain. Dev. 31, 173–190.

Rouchalova, D., Rouchalova, K., Janakova, I., Cablik, V., Janstova, S., 2020. Bioleaching of iron, copper, lead, and zinc from the sludge mining sediment at different particle sizes, pH, and pulp density using *Acidithiobacillus ferrooxidans*. Minerals 10, 1013.

Rubio-Sanz, L., Brito, B., Palacios, J., 2018. Analysis of metal tolerance in *Rhizobium leguminosarum* strains isolated from an ultramafic soil. FEMS Microbiol. Lett. 365, fny010.

Rueda, D., Valencia, G., Soria, N., Rueda, B.B., Manjunatha, B., Kundapur, R.R., Selvanayagam, M., 2016. Effect of *Azospirillum* spp. and *Azotobacter* spp. on the growth and yield of strawberry (*Fragaria vesca*) in hydroponic system under different nitrogen levels. J. Appl. Pharm. Sci. 6, 48–54.

Rzymski, P., Klimaszyk, P., Marszelewski, W., Borowiak, D., Mleczek, M., Nowiński, K., Pius, B., Niedzielski, P., Poniedziałek, B., 2017. The chemistry and toxicity of discharge waters from copper mine tailing impoundment in the valley of the Apuseni Mountains in Romania. Environ. Sci. Pollut. Res. 24, 21445–21458.

Saha, B., Saha, S., Roy, P.D., Padhan, D., Pati, S., Hazra, G.C., 2018. Microbial transformation of sulphur: an approach to combat the sulphur deficiencies in agricultural soils. In: Role of Rhizospheric Microbes in Soil. Springer, Berlin/Heidelberg, Germany, pp. 77–97.

Saif, S., Khan, M.S., 2017. Assessment of heavy metals toxicity on plant growth promoting rhizobacteria and seedling characteristics of *Pseudomonas putida* SFB3 inoculated greengram. Acta Sci. Agric. 1, 47–56.

Saini, R., Dudeja, S.S., Giri, R., Kumar, V., 2015. Isolation, characterization, and evaluation of bacterial root and nodule endophytes from chickpea cultivated in Northern India. J. Basic Microbiol. 55, 74–81.

Salama, E.S., Roh, H.S., Dev, S., Khan, M.A., Abou-Shanab, R.A., Chang, S.W., Jeon, B.H., 2019. Algae as a green technology for heavy metals removal from various wastewater. World J. Microbiol. Biotechnol. 35, 1–19.

Salcedo Gastelum, L.A., Díaz Rodríguez, A.M., Félix Pablos, C.M., Parra Cota, F.I., Santoyo, G., Puente, M.L., 2020. The current and future role of microbial culture collections in world food security. Front. Sustain. Food Syst. 4, 291.

Samreen, T., Shah, H.U., Ullah, S., Javid, M., 2017. Zinc effect on growth rate, chlorophyll, protein and mineral contents of hydroponically grown mungbean plant (*Vigna radiata*). Arab. J. Chem. 10, S1802–S1807.

Sandhya, V., Shrivastava, M., Ali, S.Z., Prasad, V.S.K., 2017. Endophytes from maize with plant growth promotion and biocontrol activity under drought stress. Russ. Agric. Sci. 43, 22–34.

Sangeeth, K., Bhai, R.S., Srinivasan, V., 2012. *Paenibacillus glucanolyticus*, a promising potassium solubilizing bacterium isolated from black pepper (*Piper nigrum* L.) rhizosphere. J. Spices Aromat. Crops 21, 118–124.

Santana-Casiano, J.M., Gonzalez-Davila, M., Perez-Peña, J., Millero, F.J., 1995. Pb^{2+} interactions with the marine phytoplankton *Dunaliella tertiolecta*. Mar. Chem. 48, 115–129.

Saranya, K., Sundaramanickam, A., Shekhar, S., Meena, M., Sathishkumar, R.S., Balasubramanian, T., 2018. Biosorption of multi-heavy metals by coral associated phosphate solubilising bacteria *Cronobacter muytjensii* KSCAS2. J. Environ. Manag. 222, 396–401.

Sattar, A., Naveed, M., Ali, M., Zahir, Z.A., Nadeem, S.M., Yaseen, M., Meena, V.S., Farooq, M., Singh, R., Rahman, M., 2019. Perspectives of potassium solubilizing microbes in sustainable food production system: a review. Appl. Soil Ecol. 133, 146–159.

Schmidt, W., Thomine, S., Buckhout, T.J., 2020. Iron nutrition and interactions in plants. Front. Plant Sci. 10, 1670.

Schultz, N., Silva, J.A.D., Sousa, J.S., Monteiro, R.C., Oliveira, R.P., Chaves, V.A., Pereira, W., Silva, M.F.D., Baldani, J.I., Boddey, R.M., Reis, V.M., 2014. Inoculation of sugarcane with diazotrophic bacteria. Rev. Bras. Ciênc. Solo 38, 407–414.

Selvi, K.B., Paul, J.J., Vijaya, V., Saraswathi, K., 2017. Analyzing the efficacy of phosphate solubilizing microorganisms by enrichment culture techniques. Biochem Mol Biol J 3, 1–7.

Seraj, M.F., Rahman, T., Lawrie, A.C., Reichman, S.M., 2020. Assessing the plant growth promoting and arsenic tolerance potential of *Bradyrhizobium japonicum* CB1809. Environ. Manag. 66, 930–939.

Shabanamol, S., Divya, K., George, T.K., Rishad, K.S., Sreekumar, T.S., Jisha, M.S., 2018. Characterization and in planta nitrogen fixation of plant growth promoting endophytic diazotrophic *Lysinibacillus sphaericus* isolated from rice (*Oryza sativa*). Physiol. Mol. Plant Pathol. 102, 46–54.

Shao, W., Li, M., Teng, Z., Qiu, B., Huo, Y., Zhang, K., 2019. Effects of Pb (II) and Cr (VI) stress on phosphate-solubilizing bacteria (*Bacillus* sp. strain MRP-3): Oxidative stress and bioaccumulation potential. Int. J. Environ. Res. Public Health 16, 2172.

Sharma, A., Kapoor, D., Wang, J., Shahzad, B., Kumar, V., Bali, A.S., Jasrotia, S., Zheng, B., Yuan, H., Yan, D., 2020. Chromium bioaccumulation and its impacts on plants: an overview. Plants 9, 100.

Shu, J., Lei, T., Deng, Y., Chen, M., Zeng, X., Liu, R., 2021. Metal mobility and toxicity of reclaimed copper smelting fly ash and smelting slag. RSC Adv. 11, 6877–6884.

Sindhu, S.S., Parmar, P., Phour, M., Sehrawat, A., 2016. Potassium-solubilizing microorganisms (ksms) and its effect on plant growth improvement. In: Meena, V., Maurya, B., Verma, J., Meena, R. (Eds.), Potassium Solubilizing Microorganisms for Sustainable Agriculture. Springer, New Delhi, pp. 175–185.

Sindhu, S.S., Sharma, R., Sindhu, S., Phour, M., 2019. Plant nutrient management through inoculation of zinc solubilizing bacteria for sustainable agriculture. In: Giri, B., Prasad, R., Wu, Q.S., Varma, A. (Eds.), Biofertilizers for Sustainable Agriculture and Environment; 55. Soil Biology, Springer, Cham, pp. 173–201.

Singh, M., Biswas, S.K., Nagar, D., Lal, K., Singh, J., 2017. Impact of bio-fertilizer on growth parameters and yield of potato. Int. J. Curr. Microbiol. App. Sci. 6, 1717–1724.

Singh, M., Singh, D., Gupta, A.D., Pandey, K.D., Singh, K.P., Kumar, A., 2018. Plant growth promoting rhizobacteria. In: PGPR Amelioration in Sustainable Agriculture. Elsevier, pp. 41–66.

Sood, G., Kaushal, R., Panwar, G., Dhiman, M., 2018. Effect of indigenous plant growth-promoting rhizobacteria on wheat (*Triticum aestivum* L.) productivity and soil nutrients. Commun. Soil Sci. Plant Anal. 50, 141–152.

Srinath, T., Verma, T., Ramteka, P.W., Garg, S.K., 2002. Chromium (VI) biosorption and bioaccumulation by chromate resistant bacteria. Chemosphere 48, 427–435.

Srinivasan, T., Kumar, A.G., Ganesh, P.S., 2019. PGPR siderophore and its role in antimicrobial activity in plants—a review. J. Agric. Forest Meteorol. Res. 2, 73–76.

Srivastava, S., Agrawal, S.B., Mondal, M.K., 2015. A review on progress of heavy metal removal using adsorbents of microbial and plant origin. Environ. Sci. Pollut. Res. 22, 15386–15415.

Stambulska, U.Y., Bayliak, M.M., Lushchak, V.I., 2018. Chromium (VI) toxicity in legume plants: modulation effects of rhizobial symbiosis. Biomed. Res. Int. 2018, 8031213.

Sujkowska-Rybkowska, M., Kasowska, D., Gediga, K., Banasiewicz, J., Stępkowski, T., 2020. *Lotus corniculatus*-rhizobia symbiosis under Ni, Co and Cr stress on ultramafic soil. Plant Soil 451, 459–484.

Sumarsih, E., Nugroho, B., Widyastuti, R., 2017. Study of root exudate organic acids and microbial population in the rhizosphere of oil palm seedling. J. Trop. Soils 22, 29–36.

Sumbul, A., Ansari, R.A., Rizvi, R., Mahmood, I., 2020. *Azotobacter*: a potential biofertilizer for soil and plant health management. Saudi J. Biol. Sci. 27, 3634.

Sun, G.L., Reynolds, E.E., Belcher, A.M., 2020. Using yeast to sustainably remediate and extract heavy metals from waste waters. Nat. Sustain. 3, 303–311.

Susilowati, L.E., Syekhfani, 2014. Characterization of phosphate solubilizing bacteria isolated from Pb contaminated soils and their potential for dissolving tricalcium phosphate. J. Degrad. Min. Lands Manag. 1, 57–62.

Tarekegn, M.M., Salilih, F.Z., Ishetu, A.I., 2020. Microbes used as a tool for bioremediation of heavy metal from the environment. Cogent Food Agric. 6, 1783174.

Tchakounté, G.V.T., Berger, B., Patz, S., Becker, M., Fankem, H., Taffouo, V.D., Ruppel, S., 2020. Selected rhizosphere bacteria help tomato plants cope with combined phosphorus and salt stresses. Microorganisms 8, 1844.

Teotia, P., Kumar, V., Kumar, M., Shrivastava, N., Varma, A., 2016. Rhizosphere microbes: potassium solubilization and crop productivity-present and future aspects. In: Meena, V., Maurya, B., Verma, J., Meena, R. (Eds.), Potassium Solubilizing Microorganisms for Sustainable Agriculture. Springer, New Delhi, pp. 315–325.

Thomas, L., Singh, I., 2019. Microbial biofertilizers: types and applications. In: Biofertilizers for Sustainable Agriculture and Environment. Springer, Berlin/Heidelberg, Germany, pp. 1–19.

Thornburg, T.E., Liu, J., Li, Q., Xue, H., Wang, G., Li, L., Fontana, J.E., Davis, K.E., Liu, W., Zhang, B., Zhang, Z., 2020. Potassium deficiency significantly affected plant growth and development as well as microrna-mediated mechanism in wheat (*Triticum aestivum* L.). Front. Plant Sci. 11, 1219.

Timmusk, S., Behers, L., Muthoni, J., Muraya, A., Aronsson, A.C., 2017. Perspectives and challenges of microbial application for crop improvement. Front. Plant Sci. 8, 49.

Tirry, N., Joutey, N.T., Sayel, H., Kouchou, A., Bahafid, W., Asri, M., El Ghachtouli, N., 2018. Screening of plant growth promoting traits in heavy metals resistant bacteria: prospects in phytoremediation. J. Genet. Eng. Biotechnol. 16, 613–619.

Tiwari, S., Lata, C., 2018. Heavy metal stress, signaling, and tolerance due to plant-associated microbes: an overview. Front. Plant Sci. 9, 452.

Tomar, R.S., Singh, B., Jajoo, A., 2019. Effects of organic pollutants on photosynthesis. In: Ahmad, P., Ahanger, M.A., Alyemeni, M.N., Alam, P. (Eds.), Photosynthesis, Productivity and Environmental Stress. Wiley Online, pp. 1–26.

Transparency Market Research, 2017. Biofertilizers Market (Nitrogen Fixing, Phosphate Solubilizing and Others) for Seed Treatment and Soil Treatment Applications—Global Industry Analysis, Size, Share, Growth, Trends and Forecast, 2013-2019.

Ullah, A., Heng, S., Munis, M.F.H., Fahad, S., Yang, X., 2015. Phytoremediation of heavy metals assisted by plant growth promoting (PGP) bacteria: a review. Environ. Exp. Bot. 117, 28–40.

Upadhyay, N., Vishwakarma, K., Singh, J., Mishra, M., Kumar, V., Rani, R., Mishra, R.K., Chauhan, D.K., Tripathi, D.K., Sharma, S., 2017. Tolerance and reduction of chromium (VI) by *Bacillus* sp. MNU16 isolated from contaminated coal mining soil. Front. Plant Sci. 8, 778.

Valix, M., 2017. Bioleaching of electronic waste: milestones and challenges. In: Current Developments in Biotechnology and Bioengineering. Elsevier, pp. 407–442.

Van Assche, F., Clijsters, H., 1990. Effects of metals on enzyme activity in plants. Plant Cell Environ. 13, 195–206.

Van Hullebusch, E.D., Zandvoort, M.H., Lens, P.N., 2003. Metal immobilisation by biofilms: mechanisms and analytical tools. Rev. Environ. Sci. Biotechnol. 2, 9–33.

Verma, T., Pal, P., 2020. Isolation and screening of rhizobacteria for various plant growth promoting attributes. J. Pharmacogn. Phytochem. 9, 1514–1517.

Verma, P., Yadav, A.N., Khannam, K.S., Panjiar, N., Kumar, S., Saxena, A.K., Suman, A., 2015. Assessment of genetic diversity and plant growth promoting attributes of psychrotolerant bacteria allied with wheat (*Triticum aestivum*) from the Northern hills zone of India. Ann. Microbiol. 65, 1885–1899.

Verma, M., Mishra, J., Arora, N.K., 2019. Plant growth-promoting rhizobacteria: diversity and applications. In: Sobti, R., Arora, N.K., Kothari, R. (Eds.), Environmental Biotechnology: For Sustainable Future. Springer, Singapore, pp. 129–173.

Vijayalakshmi, N.R., Swamy, M., Naik, N.M., 2019. *In vitro* screening and production of plant growth promoting substances by *azospirillum* isolates from rhizoplane of foxtail millet [*Setaria italica* (L.) Beauv]. Int. J. Pure Appl. Biosci. 7, 224–229.

Vijayaraghavan, K., Yun, Y.S., 2008. Bacterial biosorbents and biosorption. Biotechnol. Adv. 26, 266–291.

Wan, W., Qin, Y., Wu, H., Zuo, W., He, H., Tan, J., Wang, Y., He, D., 2020. Isolation and characterization of phosphorus solubilizing bacteria with multiple phosphorus sources utilizing capability and their potential for lead immobilization in soil. Front. Microbiol. 11, 752.

Wang, Z., Xu, G., Ma, P., Lin, Y., Yang, X., Cao, C., 2017. Isolation and characterization of a phosphorus-solubilizing bacterium from rhizosphere soils and its colonization of Chinese Cabbage (*Brassica campestris* ssp. chinensis). Front. Microbiol. 8, 1270.

Wang, C., Linderholm, H.W., Song, Y., Wang, F., Liu, Y., Tian, J., Xu, J., Song, Y., Ren, G., 2020a. Impacts of drought on maize and soybean production in northeast China during the past five decades. Int. J. Environ. Res. Public Health 17, 2459.

Wang, J., Li, R., Zhang, H., Wei, G., Li, Z., 2020b. Beneficial bacteria activate nutrients and promote wheat growth under conditions of reduced fertilizer application. BMC Microbiol. 20, 1–12.

Wani, P.A., Khan, M.S., Zaidi, A., 2007. Effect of metal tolerant plant growth promoting *Bradyrhizobium* sp. (vigna) on growth, symbiosis, seed yield and metal uptake by greengram plants. Chemosphere 70, 36–45.

Wani, P.A., Khan, M.S., Zaidi, A., 2008. Effect of metal-tolerant plant growth-promoting *Rhizobium* on the performance of pea grown in metal-amended soil. Arch. Environ. Contam. Toxicol. 55, 33–42.

Waqas, M.A., Kaya, C., Riaz, A., Li, Y.E., 2019. Potential mechanisms of abiotic stress tolerance in crop plants induced by *Thiourea*. Front. Plant Sci. 10, 1336.

Widyantoro, A., Hadiwiyono, Subagiya, 2019. Antagonism and compatibility of biofertilizer bacteria toward *Fusarium oxysporum* F.sp. Cubense. Asian J. Agric. Biol. 7, 263–268.

Wierzba, S., 2015. Biosorption of lead(II), zinc(II) and nickel(II) from industrial wastewater by *Stenotrophomonas maltophilia* and *Bacillus subtilis*. Pol. J. Chem. Technol. 17, 79–87.

Woldeamanuale, T.B., Hassen, A.S., 2017. Toxicity study of heavy metals pollutants and physico-chemical characterization of effluents collected from different paint industries in Addis Ababa, Ethiopia. J. Forensic Sci. Criminal Inves. 5, 555685.

Xu, X., Du, X., Wang, F., Sha, J., Chen, Q., Tian, G., Zhu, Z., Ge, S., Jiang, Y., 2020. Effects of potassium levels on plant growth, accumulation and distribution of carbon, and nitrate metabolism in apple dwarf rootstock seedlings. Front. Plant Sci. 11, 904.

Yaghoubi, K.M., Pirdashti, H., Rahimian, H., Nematzadeh, G., Ghajar, S.M., 2017. Potassium solubilising bacteria (KSB) isolated from rice paddy soil: from isolation, identification to K use efficiency. Symbiosis 76, 23.

Yaghoubi, K.M., Pirdashti, H., Rahimian, H., Nematzadeh, G., Ghajar, S.M., 2018. Nutrient use efficiency and nutrient uptake promoting of rice by potassium solubilizing bacteria (KSB). Cereal Res. Commun. 46, 1–12.

Yongpisanphop, J., Babel, S., 2020. Characterization of Pb-tolerant plant-growth-promoting endophytic bacteria for biosorption potential, isolated from roots of pb excluders grown in different habitats. Environ. Nat. Resour. J. 18, 268–274.

Yousaf, S., Zohaib, A., Anjum, S.A., Tabassum, T., Abbas, T., Irshad, S., Javed, U., Farooq, N., 2019. Effect of seed inoculation with plant growth promoting rhizobacteria on yield and quality of soybean. Pak. J. Agric. Sci. 32, 177–184.

Yousef, N., Mawad, A., Aldaby, E., Hassanein, M., 2019. Isolation of sulfur oxidizing bacteria from polluted water and screening for their efficiency of sulfide oxidase production. Global NEST J. 21, 259–264.

Zahra, N., Raza, Z.A., Mahmood, S., 2020. Effect of salinity stress on various growth and physiological attributes of two contrasting maize genotypes. Braz. Arch. Biol. Technol. 63, e20200072.

Zaidi, A., Ahmad, E., Khan, M.S., Saif, S., Rizvi, A., 2015. Role of plant growth promoting rhizobacteria in sustainable production of vegetables: current perspective. Sci. Hortic. 193, 231–239.

Zaidi, A., Khan, M.S., Ahmad, E., Saif, S., Rizvi, A., 2017a. Growth stimulation, nutrient quality and management of vegetable diseases using plant growth promoting rhizobacteria. In: Kumar, V., Kumar, M., Sharma, S., Prasad, R. (Eds.), Probiotics in Agroecosystem. Springer, Singapore, pp. 313–328.

Zaidi, A., Khan, M.S., Rizvi, A., Saif, S., Ahmad, B., Shahid, M., 2017b. Role of phosphate-solubilizing bacteria in legume improvement. In: Microbes for Legume Improvement. Springer, Cham, pp. 175–197.

Zamana, Q., Aslama, Z., Yaseenb, M., Ihsanc, M.Z., Khaliqa, A., Fahadd, S., Barshirb, S., Ramzanic, P.M.A., Naeeme, M., 2018. Zinc biofortification in rice: leveraging agriculture to moderate hidden hunger in developing countries. Arch. Agron. Soil Sci. 64, 147–161.

Zang, J., Yin, H., Chen, L., Liu, F., Chen, H., 2018. The role of different functional groups in a novel adsorption complexation–reduction multi step kinetic model for hexavalent chromium retention by undissolved humic acid. Environ. Pollut. 237, 740–746.

Zeffa, D.M., Perini, L.J., Silva, M.B., de Sousa, N.V., Scapim, C.A., Oliveira, A.L.M.D., Amaral Júnior, A.T.D., Azeredo Goncalves, L.S., 2019. *Azospirillum brasilense* promotes increases in growth and nitrogen use efficiency of maize genotypes. PLoS One 14, e0215332.

Zeng, Q., Wu, X., Wen, X., 2017. Identification and characterization of the rhizosphere phosphate solubilizing bacterium *Pseudomonas frederiksbergensis* JW-SD2 and its plant growth promoting effects on poplar seedlings. Ann. Microbiol. 67, 219–230.

Zerrouk, I.Z., Rahmoune, B., Auer, S., Rösler, S., Lin, T., Baluska, F., Dobrev, P.I., Motyka, V., Ludwig-Müller, J., 2020. Growth and aluminum tolerance of maize roots mediated by auxin-and cytokinin-producing *Bacillus toyonensis* requires polar auxin transport. Environ. Exp. Bot. 18, 104064.

Zhang, L., Shi, Z., Zhang, J., Jiang, Z., Wang, F., Huang, X., 2015. Spatial and seasonal characteristics of dissolved heavy metals in the east and west Guangdong coastal waters, South China. Mar. Pollut. Bull. 95, 419–426.

Zhao, F., Wang, S., 2019. Bioleaching of electronic waste using extreme acidophiles. In: Electronic Waste Management and Treatment Technology. Butterworth-Heinemann, pp. 153–174.

Zhao, Y., Zhang, M., Yang, W., Di, H.J., Ma, L., Liu, W., Li, B., 2019. Effects of microbial inoculants on phosphorus and potassium availability, bacterial community composition, and chili pepper growth in a calcareous soil: a greenhouse study. J. Soils Sediments 19, 3597–3607.

Zhao, J., Zhao, X., Wang, J., Gong, Q., Zhang, X., Zhang, G., 2020. Isolation, identification and characterization of endophytic bacterium *Rhizobium oryzihabitans* sp. nov. from rice root with biotechnological potential in agriculture. Microorganisms 8, 608.

Zheng, B.X., Ibrahim, M., Zhang, D.P., Bi, Q.F., Li, H.Z., Zhou, G.W., Ding, K., Peñuelas, J., Zhu, Y.G., Yang, X.R., 2018. Identification and characterization of inorganic-phosphate-solubilizing bacteria from agricultural fields with a rapid isolation method. AMB Express 8, 47.

Zhu, X., Lv, B., Shang, X., Wang, J., Li, M., Yu, X., 2019. The immobilization effects on Pb, Cd and Cu by the inoculation of organic phosphorus-degrading bacteria (OPDB) with rapeseed dregs in acidic soil. Geoderma 350, 1-10.

Zhu, S., Liu, L., Xu, Y., Yang, Y., Shi, R., 2020. Application of controlled release urea improved grain yield and nitrogen use efficiency: a meta-analysis. PLoS One 15, e0241481.

Zolfaghari, R., Rezaei, K., Fayyaz, P., Naghiha, R., Namvar, Z., 2020. The effect of indigenous phosphate-solubilizing bacteria on *Quercus brantii* seedlings under water stress. J. Sustain. For. https://doi.org/10.1080/10549811.2020.1817757.

Biopesticides: A key player in agro-environmental sustainability

H. R Archana[a,*], K Darshan[b,*], M Amrutha Lakshmi[c,*], Thungri Ghoshal[d,*], Bishnu Maya Bashayal[d,*], and Rashmi Aggarwal[d,*]

[a]Division of Seed Science and Technology, ICAR-Indian Agricultural Research Institute, New Delhi, India, [b]Forest Protection Division, ICFRE-Tropical Forest Research Institute (TFRI), Jabalpur, Madhya Pradesh, India, [c]ICAR-Indian Institute of Oilpalm Research, Pedavagi, Andhra Pradesh, India, [d]Division of Plant Pathology, ICAR-Indian Agricultural Research Institute, New Delhi, India

22.1   Introduction

Adequate nutrition is crucial to rehabilitate global starvation. How to meet the stipulated goal for global food demand has consistently, without fail, been ardently focused upon in the realms of food research and development. Both manmade and natural calamities together complicate matters toward attaining sustainable food development and eradicating malnutrition, a global target achievable by 2030 (FAO, 2018). Clearly, a growing population, assembling industrial infrastructure, along with the substantial climate drift will deliberately continue in reducing the net potential area of arable land mass until the end of the 21st century (Zhang and Cai, 2011). Again, provisioning to this sustainable need, recently estimated a call for a 25% to 70% increase in food production to essentially meet the crop mandate by 2050 (Hunter et al., 2017). Statistical studies on a global scale have accounted for a crop loss of 51% to 82% due to abiotic stresses (Oshunsanya et al., 2019), wherein the latest studies of FAO depicted 40% damage caused by pests and causal diseases. The biota of a plant system facilitates a plethora of phytopathogens that have coevolved by interfering with the crop's cultivation as the facades of manmade agricultural systems. So, their effects on crops are very hard to untangle from the multifaceted web of interactions of crop systems.

[*] All authors contributed equally

The World Health Organization (WHO) assesses that around that 2 lakhs people are killed worldwide, every year, because of pesticide poisoning. Moreover, the use of synthetic chemicals has also been blamed because of their residual toxicity and teratogenicity ability to create hormonal imbalance in humans (Feng and Zheng, 2007). Repetitive pesticide use has resulted in residue hazards, disturbing the natural ecosystem through disruption of natural enemies and pollinators, groundwater contamination, evolution of pest resistance, and a resurgence in pesticide-treated population and outbreaks of secondary pests (Dubey et al., 2011). Evidently, we are clearly acquainted with the hazardous environmental effects due to the application of chemical pesticides for crop-disease management, a practice for the last 25 years that is highly acerbic to both animals and human health (Pretty, 2012). Exploiting the biological measures and its formulations for crop protection with its environmental rewards is prominently notable. Therefore, biological pesticides or biopesticides are a compelling alternative to the perilous effects of traditional chemical pesticides. Formulated from living organisms or their derivatives such as phytochemicals, microbial products, or by-products (semiochemicals), biopesticides may be applied on crop plants to combat different harmful pathogens and control their detrimental effects on plants. Meeting the needs of agriculture with harmony and reliance on the available natural resources is basic to maintain ecological balance, a goal met by using biopesticides. The source of biopesticides, their formulation and molecular pattern of interaction, and their role in the successful establishment for sustainable eco-agriculture to fulfill the global demand for food are discussed in detail in this chapter.

22.2 What is sustainable environment and agriculture?

Modern agricultural practices were adapted to accomplish the demand for food for a growing human population. It necessitated a shift from traditional farming and small-scale domestication of plants and animals to mass-scale cultivation to generate high productivity, surplus storage, and sufficient distribution. Industrial agricultural procedures, also termed *intensive farming*, however, provided abundant crop produce at profitable rates, yet severely increased the environmental load, causing terminal damages. Monocropping, for example, a prominently practiced form of intensive farming, is a method of cultivating a single crop year after year on the same plot without crop rotation, requiring a heavy dose of chemical fertilizers and pesticides, hence, putting pressure on the soil and water resources. Again, evidentially, the immoderate input of synthetic pesticides was a prerequisite to meet the yield statistics of the Green Revolution, which enhanced agrarian productivity to facilitate the global food mandate (Rahman, 2015). This led to the causticity of the environment and its eco-receptors, including human health. The punitive effects on the environment due to these conventional agricultural practices prompted a preferable shift to sustainable agricultural practices during the 1960s, to equilibrate the nexus between adequately meeting the communal economic growth and preserving the environment. Consequently, focusing on the undisputed target, a blueprint of 17 Sustainable Development Goals (SDGs) and 169 targets, including ecological stewardship and food security, was drafted (Fenibo et al., 2020). It is a well-known fact that judicious use of environmental resources would help sustain the human race now and in the future. Accordingly, the United Nations aimed its goals to channelize resources in ways bearable to the environment, such as clean energy and optimum economic growth instead of high produce, via the adequate distribution of crops to decline hunger and poverty, thereby maintaining global health (Rathoure, 2019). Conserving the natural reservoirs supports the environment, resultantly providing clean health and human well-being. Not only safeguarding resources, but conservation is also aimed at waste management to prevent harmful discharge by recycling and reusing waste, ensuring nearly zero waste to landfills, henceforth reducing pollution. The intensive nature of industrial farming was not felt immediately, in contrast to the forward-looking nature being the key element of sustainable development practices. The practice of environmental renewability helps to ensure that the utilities today are met without jeopardizing tomorrow's requirements.

Sustainable agriculture is farming in renewable ways focusing on meeting present food and textile needs, with a minimal effect on the environment. Along with workable cultivation, the sustainable goals include minimal or zero use of chemical fertilizers, water conservation practices, and encouraging diversity in cultivating crops and the ecosystem (Velten et al., 2015). It further directs in helping farmers improve their techniques and living quality, by adopting profitable practices that are ecologically sound and provide decent revenue in its products for a demanding community (Brodt et al., 2011). Therefore, with the aforementioned methods, sustainable agriculture best suits and complements the schemes of modern agriculture. These viable applications prove worthwhile for both producers and their products, by constituting most techniques from organic farming. Amenably, the practices are applicable on multiple formats of farms and large and small ranches, extending optimal performances by harnessing both innovative practical technologies and restoring the best methods from early practices. Scientists and their studies have indicated that agrarian protocols should follow sustainable strategies against degrading

natural resources and polluting the available environment, which may lead to reducing the land's cultivable property. Dr. John E. Ikerd, Extension Professor at the University of Missouri, pointed that sustainable conventions must be in harmony with being ecologically sound, economically viable, as well as socially responsible. To avoid undesirable ecological pressure and to increase the Earth's natural resource base and improve soil fertility, adopting sustainable practices is a necessary alternative. Therefore, established on a multipronged goal, sustainable agriculture purports to:
- Capitalize on profitable farm income
- Inspire environmental management
- Augment the quality of life for farm families and communities
- Aim to increase the yield in food and fiber production

Outlining sustainable agriculture in the 17 SDGs, 8 of them are based on the conservation and careful use of critical resources and the principles of green chemistry (Fenibo et al., 2020). Discarding and replacing the abundant use of harmful chemical inputs, due to its cost-effectivity and rapid yield effect, with the steady results of sustainable practices would be a serious challenge. Among the green chemical agents applied, biopesticides have shown the potential to substitute for chemical pesticides with equal agricultural productivity. The seemingly positive usage of biobased pesticides via integrated pest management (IPM) has proven to be a very effective option for adopting most of the dimensions of sustainable agriculture (Fahad et al., 2020). Therefore, the substantial increase in biopesticide-driven IPM with the requisite education for its optimal usage will help develop skills and research to promote the conventions of sustainable agriculture.

22.3 Why must chemical pesticides not be used?

Agricultural crops are repetitively exposed and/or threatened by pests/pathogens that not only harm the growth but also the quality of the crop yield. To protect the host against such pest attacks and its causal diseases, farmers use cost-effective pesticides to cater to mass cultivation areas. Noticeably, the easy access of pesticides and its copious usage have markedly improved mass production of crop. Farmers have evenly exercised the use of synthetic pesticides due to its rapid and effective results (Pimentel, 2009). Nevertheless, synthetic pesticides pose grave disadvantages as future repercussions cause illness to human health in addition to the environment (Fig. 22.1). The extended use of these synthetic pesticides may also lead to the evolution of resistant pathogenic strains, a key serious reason why the use

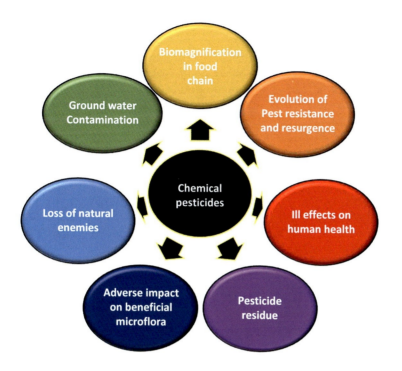

Fig. 22.1 Adverse effect caused by the use of synthetic pesticides.

of several synthetic fungicides was banned (Hollomon, 2015a, b). The conventional chemical insecticides and herbicides lack a targeted approach, with as much as 95% of its application going to waste, causing damage by dispersing in the surrounding air, water, and soil causing harm to beneficial nontargeted species (Yadav and Devi, 2017).

The unfavorable phenomenon is called *biomagnification*, a harmful condition of bioamplification or biological magnification of pesticides and metals in organisms, while progressively climbing higher in the food chain. Magnification of persistent chemicals have been detected in different animal and plant products; subsequent increased in toxin concentrations from one trophic level to the next is due to its accumulation from different food sources (Newman and Unger, 2003). The beneficial soil inhabitants are poorly affected as the chemicals contaminate the soil, altering both short- and long-term processes of the soil equilibrium. These compounds infringe soil ecology and productivity by eroding the soil microflora and adding salts to the soil constituents. Erosion reduces the soil microflora affecting the biochemical activity and diversity, altering the microbial organizational structure (Martinez-Toledo et al., 1998). Grave damage has been reported by the WHO estimating a global mortality of 200,000 annually due to pesticide poisoning, critically putting routinely exposed farmers at greater health risk, including cancer. The residual remains of chemicals in the food and surrounding milieu have been found to have a longer period of degradation (Dubey et al., 2011; Pretty, 2012; Feng and Zheng, 2007). Repeated exposure to pesticides such as Malathion and Trichlorfon proved to be a cause of reproductive abnormalities in humans (Ghorab and Khalil, 2015). Reports also read that Chloropyrifos caused genetic damage, gene mutations, and chronic illness such as asthma, hypertension, and cancer (Dey et al., 2015; Alavanja et al., 2013). Prolonged exposure to pesticides eventually caused resistant mutants of the target pests, ensuring immunity against routine pesticides, requiring a change in use. As a result, approximately 200 species of weeds and 500 species of arthropods developed resistance against herbicides (Heap, 2014) and insecticides (Roberts and Andre, 1994), respectively. The World Resources Institute (WRI) reported that more than 500 insect and mite species had developed immunity against one or more insecticides (World Resources Institute (WRI), 1994). According to Singh et al. (2018), there are serious concerns of residual remnants of synthetic pesticides accumulated in the sprayed food produce that later enter the food chain through consumers. The accumulated pesticide residues in food products and its associated broad spectrum of human health hazards causes short-term disorders to chronic illnesses. In addition, the routine use of chemical pesticides has effectively led to the extinction of potential pollinators, by causing fatal damage to these beneficial insects (Vanbergen and The Insect Pollinators Initiative, 2013).

These injuries caused on the ecosystem, due to prolific chemical usage, can be reduced by using varying alternative measures, such as adopting practices of organic farming, application of natural and biopesticides, and amending pesticide-related laws and their strict implementation during plantation. Consequently, it is imperative to conserve and protect pollinators against any manmade threats, as they are minor yet vital players of both farming and the natural biota. Therefore, acquiring alternative green practices prove beneficial and effective against chemical pesticides for long-term protection and sustenance of environmental balance. The appropriate selection of the required pesticides through IPM strategies provides distinctive protection to crops against attacking pests. This criticality led to a major shift to biopesticide formulation and its mass application practices, a major instrument of sustainable agricultural practice.

22.4 What are biopesticides?

The solution to a naturally occurring problem lies in its natural surroundings, which anticipates our unearthing. Therefore, biological pesticides, an alternative to chemical pesticides, are derived from nature and applied without harming the natural environment, for controlling any vermin attack. Biopesticides are living beings, or their natural derivatives are used to protect the cultivating crop plant against damaging pests. The living organism may include plants, animals, minerals, nematodes, and microorganisms such as fungi, bacteria, and viruses along with their by-products that neutralize attacking pest populations through environmentally nontoxic processes (Samada and Tambunan, 2020). The FAO defined biopesticides as biocontrol agents (BCAs) constituted of plant products, semiochemicals (pheromones), organisms that include predators, parasites, and some species of entomo-pathogenic nematodes, as well as their secondary metabolites. Initially, these BCAs were difficult to amplify to a larger amount and hence were used in smaller volumes.

22.5 Sources and types of biopesticides

The major sources of biopesticides mostly constitute plants and microorganisms, which are rich in bioactive and antimicrobial compounds (Nefzi et al., 2016) that include phenols, alcohols, alkaloids, steroids, quinones, saponins, and terpenes (Mizubuti et al., 2007). They can be classified naturally or based on their sources, as seen in Fig. 22.2.

22.5.1 Biochemical pesticides

The biochemical pesticides could comprise plant extracts that lure insects or insect-derived pheromones that hamper mating processes. Mostly found naturally, their environmentally nonharmful approach while targeting the pest organism (and usually being species-specific) is how they differ from the conventionally used chemical pesticides (Ritter Stephen, 2009; Chandler et al., 2011). The active component can contain a single or mixture of molecules of plant essential oils (EOs) or a mixture of isomers like that of insect pheromones. Noticeably, the

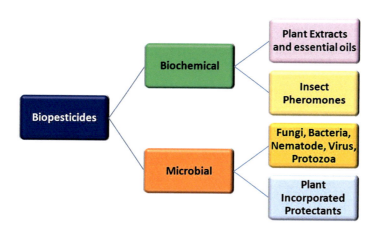

Fig. 22.2 Classification of biopesticides.

most active component of biochemical pesticides are found naturally; however, their synthetic analogues may be used in the development of marketable products, often to facilitate the product and/or process workability, such as with insect pheromones.

22.5.1.1 Plant extracts and oils

The botanical pesticides are plant-originating and mostly constituted from their EOs or other plant extracts. It's been an ancient practice by traditionally using naturally available plant extracts as botanical pesticides for varying agricultural pest control, such as safeguarding both the crop cultivation and its storage from insect pests. Neem, cotton, pyrethrum, and tobacco are some of the most common sources of botanical pesticides and have been found to be effective against several diverse sucking and chewing insect pests (Lengai and Muthomi, 2018) (Table 22.1). These plants can function as a part of the functioning ecosystem without adversity. Varied antifungal, antibacterial, and

Table 22.1 List of botanicals effective against insect pests/plant pathogens.

	Botanicals against insect pests		
Botanicals	Source extracted from	Target pests	References
Pyrethrum/Pyrethrins	Dried *Chrysanthemum cinerariifolium* and *Chrysanthemum coccineum* flower heads	Ants, aphids, ticks, and flies	Lengai and Muthomi (2018)
Rotenone	Roots of *Derris elliptica* or *Lonchocarpus* species	Leaf-feeding insects, including some beetles and caterpillars	Lengai and Muthomi (2018)
Ryania	Ground stems of *Rayania speciosa*	Caterpillar and thrips	Salma et al. (2011)
D-limonene and myrcene	*Citrus sinensis* derivative	Cereal leaf beetle	Zarubova et al. (2014)
Sabadilla lily seeds	*Schoenocaulon officinale*	Leaf hoppers, stink bugs, etc.	Salma et al. (2011)
Alkaloids	Woods of *Ryania speciosa* (Flacourtiaceae), *Ailanthus altissima*, *Quassia amara*, Seeds of *Schoenocaulon officinale*	Leafhoppers, aphids, and mites including many bacterial and fungal diseases	Marcic et al. (2012)

Table 22.1 List of botanicals effective against insect pests/plant pathogens—cont'd

Botanicals against plant pathogens

Botanicals	Target pathogens	References
Eucalyptus globulus, *Nerium oleander*, *Cinnamomum verum*, and *Allium sativum*	*Pseudominas syringae* pv. tomato, *Bacillus subtilis*, and juveniles of root knot nematodes	Salim et al. (2016)
Extract of *Datura metel*	*Phaeoisariopsis personata*, *Puccinia arachidis* and *Rhizoctonia solani*	Kishore and Pande (2005)
Turmeric (*Curcuma longa*), Lemon (*Citrus limon*), Garlic (*A. sativum*), Ginger (*Zingiber officinale*) and Pepper (*Capsicum frutescens*)	*Pythium ultimum*, *Alternaria solani* and *Fusarium oxysporum* f. sp. *lycopersici*	Muthomi et al. (2017)
Castor (*Ricinus communis*) seed extracts	*Aspergillus niger* and *Penicillium oxalicum* of yams (*Dioscoreaalata*)	Patrice et al. (2017)
Sundavathal (*Solanum indicum*) seed extracts and turmeric rhizome extracts (*C. longa*)	Rice brown spot (*Helminthosporium oryzae*)	Harish et al. (2004)
Cold-pressing the peel of orange and lemon	*Penicillium chrysogenum*, *A. flavus* and *A. niger*	Viuda-Martos et al. (2008)
Ethanolic extracts of Moss species and two leafy liverwort species	*Alternaria alternata*	Shirzadian et al. (2009)
Aqueous extracts of three Asteraceous allelopathic species	*Aspergillus niger*	Bajwa et al. (2001)
Extracts of *Rumex acetosella* roots	Barley powdery mildew	Choi et al. (2004)
Antifungal activity of *Aloe vera*	*Fusarium oxysporum*, *Rhophitulus solani* and *Colletotrichum coccodes*	De Rodriguez et al. (2005)
Thevetia peruviana extracts	*A. niger* and *Penicillium* spp.	Ravi-Kumar et al. (2007)
Xanthium strumarium, *Laurisnobilis*, *Salvia officinalis* and *Styrax officinalis*	*Phytopthora infestans*	Yanar et al. (2011)
Plant extracts of Ginger, Polyalthi and Clerodendrum	*Rhizoctonia solani* and *Colletotrichum capsici*	Choudhury et al. (2017)
Methanol extracts of *Terminalia chebula* and *Polyalthia longifolia* and chloroform extracts of *Z. officinale*	*Colletotrichum musae* and *Xanthomonas campestris* pv. Campestris	Bhutia et al. (2016)
Melia azedarach, *Phyllanthus niruri* and *Terminelia sepulae*	*Colletotrichum gloeosporioides*	Darshan et al. (2019a, b)

antiviral properties have been identified in several botanical agents that can be used as substitutes for chemicals for the regulation of crop diseases (Pretali et al., 2016). Those botanical pesticides that are chemically synthesized closely resemble the natural plant derivatives and hence are easily degradable by soil-borne microbes, maintaining the biodiversity of predators, animal, and human health by reducing pollution (Grainge and Ahmed, 1988). They contain several active ingredients in reduced concentrations for initiating biocidal properties against different pathogenic and/or pest populations (Kalaycioglu et al., 1997; Harish et al., 2008). Currently, botanical fungicides are most widely accepted for controlling plant pathogens. The reason to favor botanicals over chemicals includes reduced mammalian toxicity, increased pest target specificity, and biodegradability.

22.5.1.2 Essential oils

The naturally occurring EOs defend a plant by their varying antibacterial, antiviral, antifungal, and insecticidal properties including herbivores, reducing their appetency for the host plant. EOs are a composite combination of volatile carbon-based compounds synthesized as plant secondary metabolites. Many plant phylums, such as myrtaceae, lauraceae, rutaceae, lamiaceae, asteraceae, apiaceae, cupressaceae, poaceae, zingiberaceae, and piperaceae are exploited for insecticidal properties (Ebadollahi and Jalali Sendi, 2015) (Table 22.3). Plants naturally produce secondary metabolites such as alcohols, terpenes, and aromatic compounds for defending against herbivores and pathogens. These volatile compounds can prevent the feeding of herbivores and pathogens by directly causing toxicity in the targeted species or indirectly involve other predators and parasitoids in response to feeding damage (Wittstock and Gershenzon, 2002). These complexes play multiple roles such as attracting pollinators, a plant defense mechanism, or aid in communicating with other plants (Pichersky and Gershenzon, 2002). Recently, researchers have demonstrated such compounds showing larvicidal and antifeedant activity and warning effects on oviposition and acting as a repellent. Azadirachtin constituting 40% of neem seed oil indicates insecticidal and antifungal activity (Isman et al., 1991). The strong odor of the EOs in plants may also act as repellents for their neighboring crop plants and is also effective against many plant pathogens (e.g., French marigold and coriander) (Dubey et al., 2011) (Table 22.2).

22.5.1.3 Insect pheromones

Pheromones are attractant chemicals released from living organisms to send messages to individuals within a species, generally to its opposite sex, and are also found structurally similar to chemicals used in flavors and fragrances. Hundreds of pheromones have been

Table 22.2 List of various essential oils effective against agriculture pests/pathogens.

Essential oils effective against agriculture pests		
Essential oils	**Target pests**	**References**
Seeds of *Foeniculum vulgare*	Larvae of *Spodoptera littoralis* and *Culex quinquefasciatus*	Pavela et al. (2016)
Leaves of *Citrus limonum*	Coleopteran storage pests	Wang et al. (2015)
Nutmeg oil	*Sitophilus zeama* and *Triboliumcastaneum*	Huang and Ho (1998)
Annona squamosa	*Anopheles stephensi*	Saxena et al. (1993)
Seed extract of *Coffea arabica*	*Callosobruchus chinensis*	Rizvi et al. (1980)
Rhizomes of *Acorus calamus*	Broad spectrum	Varma and Dubey (1999)
Neem	Biting midge (*Culicoidesimicola*)	Braverman et al. (1999)
Souroranges (*Citrus aurantium*)	Bean weevils (*Callosobruchus phaseoli*)	Jacobson (1982)
Essential oils effective against plant pathogens		
Source of essential oils	**Target pathogens**	**References**
Bidens pilosa	*Fusarium* spp. and *Fusarium solani*	Deba et al. (2008)
Cuminum cyminum and *Zataria multiflora*	*Fusarium spp*	Naeini et al. (2010)
Neetle (*Urtica dioica*), eucalyptus (*Eucalyptus* sp.), thyme (*Thymus vulgaris*), and rute (*Ruta graveolens*)	*Alternaria alternata* of potato	Hadizadeh et al., 2009
Carum carvi, Pelargonium roseum, T. vulgaris and *Pimentadioica*	*Fusarium verticillioides, Penicillium expansum* and *Aspergillus flavus*	Zabka et al. (2009)
Eucalyptus globulus	*Aspergillus parasiticus* and *A. flavus*	Vilela et al. (2009)
Thyme, sage, eucalyptus, nutmeg and cassia	*A. alternata*	Feng and Zheng (2007)
Ocimum gratissimum	*Botryosphaeria rhodina* and *Alternaria* sp.	Faria et al. (2006)
Fennel, caraway, geranium and eucalyptus and lemongrass	*F. oxysporum, B. cinerea,* and *P. italicum*	Abo-El Seoud et al. (2005)
Cinnamon and Thyme Oil	*Colletotrichum gloeosporioides*	Darshan et al. (2019a, b)

chemically elucidated in insect species, including their sex pheromones, such as in the codling moth. Pheromones are classified under a broader category called semiochemicals. A semiochemical is defined as a message-bearing substance produced by a plant or animal, or a synthetic analogue of that substance, which evokes a behavioral response in individuals of the same or other species (USEPA (U.S. Environmental Protection Agency), 2008; Chandler et al., 2011). The semiochemicals are used for various functions including attracting others to a known food source or trail, locating a mate, or sending an alarm. The insect pheromones do not kill a target pest directly but indirectly destroy the pest population due to their action.

Following are the applications of semiochemicals in managing insect pests:
- To attract and trap an insect in a lethal pesticide-containing trap.
- To reduce pest population through interrupted mating, kill by attracting, mass trapping, and repellent techniques, for example, efforts to control the pink bollworm *Pectinophora gossypiella* (Saunders), by disruption of mating began with the sex attractant hexalure (McLaughlin et al., 1972).
- Monitoring pest populations as being part of larger IPM systems, mainly to find the optimal duration along with the amount of pesticide for its application (Thakore, 2006; USEPA (U.S. Environmental Protection Agency), 2008).
- To monitor the population of native species for coordinating the judgment of insecticide treatments.
- To evaluate the efficacy of pest management strategies by posttreatment valuation.
- To improve earlier methods of insect counts for deriving a conclusion at the time of screening.
- To increase the success of biological control by raising the predation/parasitism rates of predators and parasitoids (Lewis et al., 1975).
- When used in combination with insect traps, sex pheromones can be used to determine the different insect pests populating a present crop and what plant defense measures or further arrangements can be essentially taken to guarantee minimal crop damage (Adams et al., 2017).

22.5.2 Microbial biopesticides

Microorganism-based biopesticide forms the most substantial portion of biopesticide products (Koul, 2012). They include both naturally occurring or genetically altered bacteria, fungi, algae, viruses, or protozoan; a total of almost 125 species of natural adversaries have been globally marketed as biopesticides (Srivastava and Dhaliwal, 2010). They produce toxins causing disease in the pest or prevent other microbial pathogenic interaction with the protecting host plant through competition or other enacting modes (USEPA (U.S. Environmental Protection Agency), 2008). Most microbial biopesticides are found in the microbiome of the agricultural fields where they reside in combination with both pathogenic and beneficial organisms, especially the rhizosphere as a major source of various forms of significant microorganisms. Supplementary sources include manure, hay, straw, and cow shed as well (Beric et al., 2012).

22.5.2.1 Bacterial biopesticides

The different microbial biopesticides include common bacterial species such as *Bacillus, Burkholderia, Pseudomonas, Chromobacterium, Xanthomonas, Yersinia, Streptomyces,* and *Serratia.* They reflect obligate, facultative, or crystalliferous in properties and are the most used forms of microbial biopesticides. They are used against harmful insect pests and to control pathogenicity caused by harmful bacteria, fungi, or viruses. While acting as an insecticide, they target attacking species-specific moths and butterflies, beetles, flies, and mosquitoes. These bacteria are ingested by the targeted pest, i.e., apparently coming in contact with the pest body where they indict damage or lethality (USEPA (U.S. Environmental Protection Agency), 2008; Kawalekar, 2013).

22.5.2.2 Fungal biopesticides

The different fungal strains used as biopesticides include species of *Metarhizium anisopliae, Beauveria bassiana, Verticillium* sp., *Trichoderma* sp., *Chaetomium* sp., *Aspegillus* sp., and *Hirsutella*. These fungal biopesticides bioactively deter harmful insect pests and their causal diseases, including many fungi, bacteria, nematodes, and weeds. Their action is often parasitic or may secrete bioactive metabolites like enzymes, i.e., contingent to both the pesticidal fungus applied and the targeted pest. For example, *Beauveria bassiana* germinate, grow, and spread their spores in the targeted insect body, draining nutrients, releasing toxins, and causing its death (Omukoko, 2019). Showing antagonistic properties, *Trichoderma* grows inside the main tissues of harmful disease-causing fungus, secreting enzymes that degrade the fungal cell wall, feeding on the targeted host fungal cell constituents, and replicating its spores (USEPA (U.S. Environmental Protection Agency), 2008; Kawalekar, 2013). In vitro and in vivo studies have shown that cell-free culture filtrate of *Trichoderma reesei* and *C. globosum* inhibited the growth of spot blotch pathogen, *B. sorokiniana* (Mandal et al., 1999). *Chaetomium* spp. are reported as antagonists of soil and foliar pathogens such as *Pythium ultimum, Phytopthora citrophtora, Alternaria brassicola, Fusarium* spp., *Bipolaris sorokiniana,* and *Rhizoctonia solani* (Aggarwal et al., 2004; Jiang et al., 2017).

22.5.2.3 Viral biopesticides

Among the viruses studied to be used as biopesticides, baculoviruses (*Nuclear polyhedrosis virus*, NPV and *Granulosis virus*, GV) have been tested to be highly specific to the pest and hence the most efficient insect pest BCA against serious pests like *Spodoptera* spp. and *Heliothis* spp. that cause severe damage in cotton, fruit, and vegetable crops (Kour et al., 2019; Yadav et al., 2020). Baculoviruses (viral biopesticides) are pathogens that attacks insects and other arthropods. Contrasted with the other group members of this category, Baculoviruses are microscopic obligate parasitic elements that replicate their DNA completely depending on the host machinery (USEPA (U.S. Environmental Protection Agency), 2008). Composed of double-stranded DNA, the genetic material is damaged in the sun or the host's gut, hence it's protected by a polyhedron protein coat (USEPA (U.S. Environmental Protection Agency), 2008; Kawalekar, 2013). Lately, phages were added to this group of biopesticides or BCAs for biocontrol of diseases and seed treatment.

22.5.2.4 Other microbial biopesticides

Several other live organisms are used as active BCAs for IPM systems (USEPA (U.S. Environmental Protection Agency), 2008). The single-celled, animal-like protozoa have been rarely used as biopesticides until 2002. Microscopic worms classified as nematodes are typically parasitic and are commonly used as insecticides, with *Steinernema* and *Heterorhabditis* being the two main genera of entomopathogenic nematodes used (Kachhawa, 2017). The secreted compounds of multicellular predators such as insects can be used in effective BCAs; however, their application to an area lacking their natural predators may cause ecological instability. Therefore, macroscopic predators are beyond the scope of application as biopesticides.

22.5.2.5 Plant-incorporated protectants

As a natural pesticide choice, natural microbial and biochemical pesticides are the type commonly used by farmers and growers to control an existing pest problem, because they can be applied like synthetic pesticides but without toxic damage. But there exists another category called plant-incorporated protectants (PIPs) that are produced by modifying its genetic material by using genetic engineering methods. They are also known as genetically modified crops (Parker and Sander, 2017). Pesticidal activity here is due to natural material that targets particular organisms and interferes with the natural biochemistry of the pest.

22.6 Mode of action of biopesticides on different pests/pathogens

22.6.1 Mode of action of plant-based biopesticides against crop pests and pathogens

22.6.1.1 Mode of action of plant extracts and oils on insect pests

Plant-based biopesticides includes mainly neem-based pesticides, rotenone, pyrethrum, and plant EOs (eucalyptus) that are extracted from their sources using different methods and are highly efficacious against many severe pest attacks (Chengala and Singh, 2017) (Table 22.3). Botanical pesticides generally act as a neurotoxin, repellent oviposition, and feeding deterrent for most insect pests. Different

Table 22.3 Plant extracts source and its mechanism of action against insect pests.

Botanicals	Source	Active Ingredients	Mechanisms	Mode of Application	References
Azadirachtin (neem based botanical pesticide)	Neem tree (*Azadirachta indica*)	Azadirachtin, Salannin. Melandriol, and other limonoids	Mitotic inhibitor, imbalance in hormonal system, food poison, feeding & oviposition deterrent, and affects metamorphosis and reproduction, mortality	Neem Extract Cakes, Neem Oil, Kernel Extracts	El-Wakeil (2013); Campos et al. (2016); Qiao et al. (2014)
Rotenone	*Derris* spp., *Lonchocarpus* spp., and *Tephrosia* spp	Rotenone	Contact and food poison, cellular respiratory enzyme inhibitor, stomach poison	Dried root powder, spray	Ware and Whitacre (2004); El-Wakeil (2013)
Pyrethrum	*Chrysanthemum cinerariaefolium*	Pyrethrin I & II, Cinerin I & II, Jasmolin I & II	Disrupts the sodium and potassium ion exchange processes in the nerve fibers, contact poison	Flower extracts as a spray or dust	Rattan (2010); Sola et al. (2014); El-Wakeil (2013)
Eucalyptus essential oil	*Eucalyptus* spp.	8—cineole (eucalyptol), citronellol, citronellyl acetate, p—cymene, limonene, linalool, and α — pinene	Antifeedent, repellant, ovicidal, larvicidal, pupicidal and adulticidal	Spray or dust	Batish et al. (2008)

natural compounds such as allicin from garlic bulbs cause choking due to its adverse effects on the insects' neurotransmitter receptors (Chaubey, 2016). Phenolics and terpenoids from different plant sources form hydrophobic and ionic bonds disrupting multiple insect proteins and causing physiological disorder (Freeman and Beattie, 2008). Many compounds constituted from plant extracts also interfere with insect cell receptors resulting in nervous disruption and coordination failure, finally killing the insects (Rattan, 2010). Crude neem and its derivatives have various pesticidal properties including fungicidal, nematocidal, bactericidal, and antifeedant effect (Mehlhorn et al., 2011) (Table 22.3). Similarly, Rotenone, a natural isoflavone, can be used as a broad-spectrum botanical pesticide extracted from the stem and root tissues of the Fabaceae family such as tropical legumes Derris (*Derris elliptica, Derris involuta*), *Tephrosia virginiana*, and *Lonchocarpus* (*Lonchocarpus utilis, Lonchocarpus urucu*) (Weinzierl, 2000). Rotenone acts as a toxin inside the insect gut by inhibiting the cell respiratory enzymes (Guleria and Tiku, 2009). Pyrethrum is an important plant-based pesticide isolated from flowers of *Chrysanthemum cinerariaefolium* (El-Wakeil, 2013). The mode of action of pyrethrums is that it rapidly penetrates into the nervous system of insects, disrupting the sodium-potassium ion exchange process in the nerve fibers, disrupting the nerve impulse transmission. These insecticides are highly toxic and act instantly causing rapid knockdown paralysis in insects (El-Wakeil, 2013) (Table 22.3).

EOs interfere with the basic functionality of the pest insect including its metabolic, biochemical, physiological, and behavioral pattern. The pest insects, upon exposure, either inhale, ingest, or absorb these EOs. The speedy action against some pests indicates that it affects the nervous system and interferes with the neuromodulator octopamine (Enan, 2005) or GABA-gated chloride channels (Priestley et al., 2003; Khater et al., 2011). Some EOs have a larvicidal effect causing delayed larval development and restraining its growth to an adult insect, including those of medicinal and veterinary importance (Khater, 2003; Khater et al., 2009; Khater et al., 2011). Alkaloids isolated from *Annona squamosa* have shown both larvicidal (negating growth) and chemosterilant properties against the insect pest *Anopheles stephensi* at concentrations of 50–200 ppm. (Saxena et al., 1993). The seed extract from *Coffea arabica* contains a compound called 1,3,-Trimethylxanthine, which was found to be highly effective against *Callosobruchus chinensis*, causing sterility in almost all treated individuals at a concentration of 1.5%. Noteworthy, chemosterilants are highly important for IPM programs to reduce the incidence of pest resistance. The compound β-asarone extracted from rhizomes of *Acorus calamus* possesses antigonadial activity that actively hinders the ovarian development of different insects (Varma and Dubey, 1999). Darshan et al.

(2018, 2019a) tested the antifungal activity of cinnamon and thyme oil against *Colletotrichum gloeosporioides* causing anthracnose disease of papaya. The results revealed that in preharvest spray with *Bacillus* sp. (BSP1) (5%)+postharvest dipping with EC-formulated cinnamon oil (0.1%) recorded higher content of total sugar content (14.22%) and reducing sugar (11.23%) over control and also maintained quality of papaya fruit during 14 days of the experimental period.

22.6.1.2 Mode of action of botanicals against plant pathogens

Plant extracts in various solvents have been proven to be highly enriched with bioactive and antioxidant compounds. These botanical pesticides oppress the growth of phytopathogenic fungi by inhibiting their hyphal growth, inducing structural modifications on the mycelia by partitioning the fungal cell membranes, permeating it and causing leakage of the cell contents and damage (Paul and Sharma, 2002). They also inhibit aflatoxin and fumonisin production in some fungi (*Aspergillus* spp. and *Fusarium* spp.), resulting in reduced fungal mycotoxin-related pathogenicity (Martinez-Luis et al., 2012). For example, extracts from the Asteraceae plant family are constituted of compounds such as flavonoids, coumarins, alkaloids, and terpenoids causing high fungal toxicity. The terpenes, phenols, alcohols, alkaloids, tannins, and other secondary metabolites from efficient botanical pesticides damage cell walls, cell membranes, and cell organelles of fungi (Yoon et al., 2013). These also hamper fungal spore germination, germ tube development, delay spore formation, and disturb fungal cell metabolism by interrupting the synthesis of essential enzymes, DNA, and other cell protein (Martinez-Luis et al., 2012). When subjected to the plant extracts, *Fusarium verticillioides* failed to synthesize fumonisin and ergosterol in addition to a loss of its cellular components (da Silva Bomfim et al., 2015). For example, ginger (*Zingiber officiale*) extracts decreased the cytoplasmic content cells of *Fusarium*, besides altering the morphology of their microconidia (Yamamoto-Ribeiro et al., 2013).

Botanical pesticides may be chemically attributed with several antibacterial properties including inhibitory growth (Aljamali, 2013). Botanical pesticides inhibit important cellular processes such as synthesis of proteins, increasing membrane permeability, leading to leakage of cellular leakage, and finally death (Lengai et al., 2020). For example, methanolic extracts from *Aloe vera* hampered the growth of *Escherichia coli* and *Bacillus subtilis*, and acetone extracts inhibited *Pseudomonas aeruginosa*. Phytochemicals from *A. vera* reflect antimicrobial properties causing the denaturation of microbial proteins disrupting their functionality. *A. vera* contains cinnamic acid, which weakens the glucose uptake capacity and ATP synthesis (Djeussi et al., 2013). EOs from *Thymus vulgaris* possess antimicrobial activity against *Bacillus cereus, Klebsiella pneumonia, Staphylococcus aureus, Salmonella typhimurim,* and *Ecshrichia coli* (Kon and Rai, 2012). Also, constituents of thymol lead to an increase in bacterial cell membrane's permeability and depolarizing, causing interference in the basic cell mechanisms (Tian et al., 2021).

Botanical pesticides help plants to fight against viral pathogen's development by inducing antiviral protein production (Lengai et al., 2020). These restrict penetration of virus into the host cell, interrupting their multiplication, inactivate enzymes, and cause hemagglutination, which is essential for virus attachment (Rajasekharan et al., 2013). Viruses are unable to attach to the plant cell and replicate, which is important in establishing antiviral action (Bhanuprakash et al., 2008). Lab studies indicated that extracts from cotton seed oil sludge reduced the disease levels caused by Southern *Rice Black Streaked Dwarf Virus* and *Rice Stripe Virus* in rice (*Oryza sativa*) under field conditions. The antiviral activity was also attributed with the presence of gossypol and β-sitestrol, compounds present in the cotton seed oil sludge (Zhao et al., 2015). *Thuja orientalis* reduced the viral infection in watermelon (*Citrus lanatus*), which is attributed to the blocking of genetic material and inhibition of viral multiplication (Elbeshehy et al., 2015).

Some botanical pesticides have an impact on other rhizospheric microorganisms, which in turn affects nematode eggs and juveniles (Khan et al., 2008). In addition, some active compounds specifically kill egg masses, the second stage of juveniles, and decrease gall formation and ultimately disturb the nematode activity in infested soils (Kepenekci et al., 2016). Combined application of *Lantana camara* and *Trichoderma harzianum* reduced the reproduction rate in root knot nematodes and gall formation in tomato (Feyisa et al., 2015). Nevertheless, botanical pesticides are the richest source of phytochemicals such as glycosides, tannins, and alkaloids that significantly affects egg hatching, locomotion, and mortality of J2 stage juveniles.

22.6.2 Mode of action of microbial biopesticides used against insects/plant pathogens

Bacteria, fungi, virus, and nematodes use different modes of action to kill insect pests.

22.6.2.1 Bacterial biopesticides against insects

A bacteria bioagent must come into contact with the targeted pest, and they will need to be ingested. Bacteria hinder the digestion metabolism in insect pests by developing endotoxins unique to the insect pest in question. *Bacillus thuringiensis* subspecies and strains are the most commonly used microbial biopesticides (Bt). Bacterial biopesticides are extensively applied in the field of agriculture, forestry, and medicine to combat insect pests. Endotoxins or Cry proteins, having insecticidal properties against lepidopteran organisms, are synthesized into crystalline inclusions for use as biocontrol. When assimilated by the insect larvae, it induces gut paralysis causing it to reduce feeding and eventual death due to starvation and midgut epithelium damages (Betz et al., 2000). *B. thuringiensis* and Cry proteins are very competent and sustainable alternatives to chemical pesticides for controlling insect pests due to their targeted approach and green action mode.

22.6.2.2 Fungal biopesticides against insects

Entomopathogenic fungi are also considered to play a key role as biological control agent of insect pest populations. A very diverse array of fungal species is found from different classes that infect insects. The insect pathogenic studies first began in the 1980s, and their aim was to find the ways of controlling disease in the silkworm. Gilbert and Gill explained that this disease gave a hint of using insects infecting fungi for the management of insect pests. In recent times, around 90 genera and more than 700 species are considered as insect-infecting fungi that represent nearly all the major divisions of fungi such as ascomycota, zygomycota, deuteromycota, oomycote, and chytridiomycota (Kachhawa, 2017). The entomopathogens enter the insect primarily via the integument, but it may also infect the insect via ingestion, wounds, or the trachea. As spores are stored on the integument surface, the germinative tube begins to develop, and the fungi begin to excrete enzymes such as proteases, chitinases, quitobiases, lipases, and lipoxygenases. These enzymes aid in the process of mechanical penetration initiated by the appressorium, a specialized structure produced in the germinative tube, by degrading the insect's cuticle. The fungi grow as hyphal bodies within the insect, which disseminate through the haemocoel and invade various muscle tissues, fatty bodies, Malpighian tubes, mitochondria, and hemocytes, eventually killing the insect 3 to 14 days after infection. When the insect dies and all the nutrients are depleted, fungi begin to develop micelles and infiltrates all of the dead host's organs. Finally, hyphae from the interior of the insect penetrate the cuticle and emerge at the surface, where they initiate spore formation under the right conditions.

22.6.2.3 Virus biopesticides against insects

Baculovirus is a highly pathogenic virus that has been commonly used as a BCA against a variety of serious insect pests in its natural form (Knox et al., 2015). Baculoviruses must be consumed by the insect pest larvae to infect them. They enter the insect's body through the midgut and spread throughout the body, though the viral infection can be confined to the insect midgut or fat body in some insects.

Baculoviruses are either nucleopolyhedroviruses (NPVs) and granuloviruses (GVs). In NPVs, occlusion bodies contain several virus particles, while in GVs, occlusion bodies usually contain only a single virus particle. Baculoviruses are also occluded, which means that the virus particles are contained in a protein matrix. In baculovirus biology, the existence of occlusion bodies is critical because it enables the virus to live outside of the host. Insect larvae infected with Baculovirus initiates a cascade of molecular and cellular appendages that eventually leads to insect death due to the formation of massive amounts of polyhedral occlusion bodies containing rod-shaped virions (Chang et al., 2003).

22.6.2.4 Nematode biopesticides against insects

Various entomopathogenic nematodes that were discovered and developed as BCAs against insects belong to mainly two genera, *Steinernema* and *Heterorhabditis* (Copping and Menn, 2000). The parasitic cycle in nematodes is initiated in their third stage of infective juveniles (IJ). These nonfeeding juveniles of nematodes infect a suitable insect host by entering through the insect anus, mouth, and spiracles. Nematodes invade the hemocoel after entering the host and then release their symbiotic bacteria into the intestine. The bacteria then cause septicemia, which kills the host within 24 to 48 h. The bacteria rapidly manipulate the absorption of IJs nematodes, decaying the host tissues. Within the host cadaver, the nematodes complete nearly two to three generations. Entomopathogenic nematodes are widely used to protect plants from serious insect pests and diseases, and there have been several attempts to use IJs to biocontrol insect pest populations in the field by spraying. Despite this, little is known about indigenous nematode's ability to influence insect pest populations.

22.6.2.5 Protozoal biopesticides against insects

Protozoan diseases of insects are common in nature and notably play a role in regulating insect pests' populations (Maddox, 1987). They produce spores that infect several insect species. Germinating spores released from the protozoan sporoplasm intrude the target insect host cells, spreading the protozoan infection and causing tissue destruction. Nosemalocustae species have been most commercially used against harming grasshoppers (Patrick, 2004). The first reported microsporidium, *N. bombycis*, is a pathogen of silkworm pebrine that was found in Europe, North America, and Asia in the mid-19th century (Tounou, 2007). The spores formed by *Nosema* spp., when assimilated by the host pest, grows in the midgut. When the protozoan-infected host tissues excrete and are consumed by a susceptible host insect pest, the sporulation process regains in the insect body, causing in epizootic infection.

22.6.2.6 Mechanisms used by microbial biopesticides in controlling pathogens

Chemicals play a major role in controlling pests and disease, due to which production level has increased, but overuse has resulted in an ecological imbalance. Hence, researchers are deviating toward the potential use of beneficial microbes. Microbial pesticides act on pathogens by varied modes of action such as antagonism, hyperparasitism, antibiosis, and predation (Suprapta, 2012). However, these mechanisms of biological control are probably never mutually exclusive. Understanding the mechanism(s) of underlying mechanism of the biocontrolling process is a prerequisite to determine any BCA. Gray and Smith (2005) reported different PGPR association ranges based on the degree of bacterial proximity and intimacy of association with the host plant root, generally segregated into: (i) extracellular (ePGPR), i.e., existing in the rhizosphere of the root; (ii) on the rhizoplane (surface of the root); (iii) in the intercellular spaces of the root cortex; and (iv) intracellular (iPGPR), i.e., dwelling inside root cells, generally within specialized nodular structures. Examples of ePGPR include *Agrobacterium, Arthrobacter, Azotobacter, Azospirillum, Bacillus, Burkholderia, Caulobacter, Chromobacterium, Erwinia, Flavobacterium, Micrococcous, Pseudomonas*, and *Serratia*, etc., which are potential biological control agents that suppress the growth and colonizing of deleterious bacteria and fungi causing them harm, through competition (for substrate and nutrients), mycoparasitism, antibiosis, or a combination of these (Zhang and Yang, 2007). Examples of iPGPR are *Azorhizobium, Allorhizobium, Rhizobium, Mesorhizobium*, and *Bradyrhizobium* of the family Rhizobiaceae. Most rhizobacteria belonging to this group are Gram-negative rods; only few are Gram-positive rods, cocci, or pleomorphic. Also, numerous actinomycetes populating the rhizosphere microbial communities display highly beneficial traits, facilitating host plant growth. Among them, *Micromonospora* sp., *Streptomyces* spp., *Streptosporangium* sp., and *Thermobifida* sp. have shown extended potential as BCAs against different root fungal pathogens.

Antibiosis refers to pathogen death due to low molecular weight compounds such as specific toxins, antibiotics, or enzymes produced by the antagonist. Evidence for a role of antibiotics in biocontrol mechanism of plant disease control by both fungi and bacteria includes strong correlations between the antagonists' ability of antibiotic production and its biocontrolling efficiency (Park et al., 2005) (Table 22.4). Antibiosis is, therefore, the most prevalent antagonistic mechanism of *Chaetomium globosum* isolates against the pathogen *B. sorokiniana*, in agreement with studies by Aggarwal et al. (2004). A large number of antibiotics are formed by actinomycetes (about 8700 different antibiotics), bacteria (about 2900), and fungi (about 4900)

Table 22.4 The list of microbial biopesticides showing antibiosis mechanisms in controlling pathogens.

Biocontrol Agents	Antibiosis	Target Pathogen	References
Agrobacterium radiobacter	Agrocin 84	Agrobacterium tumefaciens	Sharma et al. (2017)
Bacillus subtilis	Iturin group	Most fungi	Jiang et al. (2020)
Pseudomonas fluorescence	2,4-diacetyl phloroglucinol (Phl)	Pseudomonas syringae pv. Tomato	Weller et al. (2004)
Streptomyces hygroscopicus var. geldanis	HCN	Thielaviopsisbasicola	Yi et al. (2010); Fouda et al. (2020)
	Oomycin A	Pythiumultimum	
	Pyoluteorin and Phl	P. ultimum, R. solani, Erwinia carotovora sub sp. atrocepia	
	Geldanamycin	R. solani	
Chaetomiumglobosum	Chaetoglobosin A, chaetomin, BHT, mollicelin G and cochliodinol	P. ultimum	Di Pietro et al. (1992); Soytong et al. (2001); Biswas et al. (2012)
Trichoderma virens	Gliovirin	P. ultimum	Burns and Benson (2000)

(Berdy, 2005). Aggarwal et al. (2013) tried to characterize the antifungal metabolites of *C. globosum* and assess their antagonistic behavior against fungal plant pathogens, cementing the prevalence of metabolites as one of the primary drivers of its biocontrol behavior. These antifungal metabolites are reported to suppress the growth of many seeds and soil-borne phytopathogens (Aggarwal et al., 2004, 2011, 2013). Competition between microorganisms colonizing the root or seed surfaces infers competing for nutrition, colonization space, or infection sites. Biocontrolling traits favoring competition include ecological plasticity, growth velocity, and development. External factors such as soil type, pH, temperature, and humidity, among others, also affect the action (Infante et al., 2009). Nutrient rivalry is thought to play an important role for disease suppression in general, though definitive evidence is difficult to come by. Biocontrol by nutrient competition occurs when a BCA competes for a specific substance reducing its bioavailability for the colonizing pathogen, hence restricting its development. Antibiotic bacteria, in particular, make better use of available food, leaving pathogens with inadequate nutrients for optimal growth and resulting in pathogenic starvation (Lin et al., 2014). Many BCAs adopt a mechanism of hyperparasitism against the targeted phytopathogen. Serially, sensing of the host/prey fungus, attraction, binding, coiling around, and host cell lysis by hydrolytic enzymes, often in combination with secondary metabolites, are all part of a typical mycoparasitic relationship mechanism. Hydrolytic enzymes such as -(1,6)-glucanases, chitinases, and proteases, as well as cellulases, xylanases, esterases, alkaline phosphatase, and lipase, lyse the host cell wall. The proposed mechanism of biocontrol by enhancing cell wall degradation was analyzed, and an investigation of ISR activity was done in wheat against *B. sorokiana* and *Puccinia graminis tritici* (Aggarwal, 2015). Hyperparasitism by mycoparasites during interactions with other fungi has frequently been proposed as a mode of action (Inayati et al., 2020). *C. globosum* mycoparasitizes with *B. sorokiniana* and produces antifungal metabolites that aid pathogen's cell wall lysis due to conidial deformation and hyphal coiling in the contact region (Moya et al., 2016). Hyperparasitism is a key mechanism used by *Trichoderma asperellum, T. atroviride, T. virens,* and *T. harzianum* for killing phytopathogenic fungi. The BCA *Pasteuria penetrans* feeds on *Meloidogyne* spp. root-knot nematodes. Darshan et al. (2020) initiated the first effort worldwide to unravel the biocontrol mechanisms of the fungus strain Cg2 against the phytopathogenic fungus *B. sorokiniana* isolate BS112 using the RNA-seq approach. They identified various genes of biological function involved in the biosynthesis of secondary metabolites, polyketide synthase, antibiotic, hydrolytic enzymes, and putative fungistatic metabolites. Complementing with the omics, Darshan et al. (2021) deciphered the network of interconnected pathways of *C. globosum* antagonistic-related genes against *Bipolaris sorokiniana*. These studies and the observed results pose a strong foundation for continued research in this field.

22.7 Role of biopesticides in sustainable environment and agricultural production

The need to shift away from readily available chemical pesticides was primarily due to their harmful and toxic effects on the environment, a serious drawback that could be overcome by naturally available biopesticides. Most bioproducts derived from nature decompose quickly, posing little environmental risk (Kawalekar, 2013). They also enrich agricultural soils by introducing beneficial microbial species (Javaid et al., 2016). These biopesticides are not only eco-friendly due to their biodegradable properties, but they also have equivalent efficacy in biocontrolling crop pests when compared to synthetic pesticides (Leng et al., 2011). Consumers' growing preference for organic and chemical-free food products has resulted in an increase in the production and acceptance of biopesticides, which have since proven to be a viable alternative to chemical pesticides while also promoting environmentally friendly practices (Akinrinnola and Okunlola, 2014). Again, biopesticides are generally characterized by short preharvest intervals and a targeted action, specifically against the attacking pest, allowing for safer usage on fresh fruits and vegetables while minimizing the risk to other species such as mammals and birds (Shiberu and Getu, 2016).

Hence, biopesticides are promisingly effective green methods for IPM that substantially contribute to the promotion of sustainable agricultural practices (Nawaz et al., 2016). Apparently, most biopesticides are easily derived from biological sources in the environment, which leads to many lasting rewards upon application, such as: (i) cost effectiveness and easy bioavailability; (ii) having a brief re-entry period ensures applicants' safety and also nonharmful to the applied food during consumption; and (iii) reducing the use of chemical pesticides in IPM. These biopesticides have also been used for other purposes such as food and feed along with scientific studies; due consideration of the farmer's field experiences for combating pest problems will help achieve sustainable agricultural practices. Furthermore, stakeholders are presently aiming biopesticides as a robust replacement for the chemically synthesized pesticides, which is an appreciative breakthrough in the field of agriculture research.

22.7.1 Biopesticides and environment

The benefits and drawbacks of microbial pesticides are primarily based on three modes of action against the harmful pests targeted: (a) a competition for nutrition, habitation, and existence; (b) formation of physical barriers; and (c) suppression through the release of metabolites and chemicals. The manufacturing process and final production cost are formed based on their mode of action, which are ecological, physical, and biochemical (Hubbard et al., 2014). The endospore-forming Gram-positive bacterium *B. subtilis,* whose spores are resistant to heat, chemicals, and radiation, gives its constituent pesticidal formulation a longer shelf life, increasing its demand over other pesticides on the market. These release toxic metabolites that have antimicrobial properties, constraining the growth of other microbial populations present in the nutrient-limited niche of soil. The alkaloids found in secondary metabolites of plants intoxicate a variety of attacking pests. Organophosphate and carbamate insecticides' modes of action are obscured by them (Regnault-Roger and Philogène, 2008). Initially used alkaloids such as "nicotine" were soon found to be a human nerve toxin. Pyrethrin, like DDT, is a widely used phytochemical that has a very high biocidal activity by inhibiting the sodium/potassium channels of nerve axons. Many flying insects become hyperactive and convulsive when exposed to pyrethrin I and II esters. Because pyrethrin activity causes problems with half-lives in ultraviolet radiation, synthetically produced pyrethroids solve the problem with comparable efficacy and negligible environmental side effects.

Pyrethroids, when working synergistically with piperonyl butoxide, paralyze insect pests for a shorter period (Rattan, 2010). Also, neem and its allele-chemical derivative show varied implicated actions against its pests, which include repellency, growth inhibition, negative impact on oviposition, chemosterilization, disruption during mating, etc. (Schmutterer and Ascher, 1995). This method is nontoxic and has gained popularity due to its environmentally friendly mode. EOs derived from neem, lavender (*Lavandula*), mint (*Mentha*), *Eucalyptus*, etc. inhibit the enzyme acetyl cholin esterase in insects (Keane and Ryan, 1999). Xu et al. (2014) described a new densovirus (HaDNV-1) that reduced the efficacy of the currently used biopesticides HaNPV and Bt toxin. The findings of this study demonstrated the need for more efficient biopesticides and, as a result, the need for more profound research into microbial interactions.

The biological pesticides provide valuable tools to cultivars by delivering highly effective pest and disease management solutions while having no negative environmental impact because their active and inert ingredients are generally recognized as safe (GRAS). Conclusively, biopesticides have very low toxicity to birds, fish, bees, and other wildlife, as well as beneficial soil microorganisms. They aid in the maintenance of beneficial insect populations, degrade easily in the environment, and have the potential to reduce the use of conventional pesticides through effective use in resistance management programs (Laengle and Cass, 2011). Several microbial antagonists have been used to combat plant pathogens. Among the microbial antagonists, the genus *Trichoderma* and *Pseudomonas* were extensively studied and tested (Santoso et al., 2007). Biopesticides typically have negligible maximum residue levels (MRLs), implying that no residue is left in the soil, resulting in sustainable agriculture (Biopesticide Industry Aliance (BPIA), 2009). Aside from the microbial content, the carrier media is made up of a variety of organic materials, such as animal broth or organic waste product (Soesanto et al., 2011). The media used is biodegradable (Gupta and Dikshit, 2010), so it does not leave a residue in the soil. It provides the necessary nutrients to promote soil microbial growth and can also serve as a food source for other soil microbes. According to Soesanto et al. (2010), using *Pseudomonas fluorescens* P60 against *Fusarium oxysporum* f.sp. *lycopersici* on tomato can increase the antagonist by up to tenfold. Complex biological interactions between biocontrol pseudomonads, present plant pathogens, their targeted hosts, and other members of the microbial biome are well known. Numerous studies have revealed that a variety of crop and soil factors can influence the population and efficacy of biocontrol pseudomonads enriched in agricultural soils (McSpadden Gardener, 2007).

22.7.2 Microbial biopesticides in integrated crop management (ICM)

Plant protectants based on biological agents are primarily used for biocontrolling plant pests (Copping and Menn, 2000). There has been an increase in the use of biopesticides, exemplified in the UK and the Netherlands, that managing invertebrate pests while protecting edible crops is currently done primarily through the use of natural antagonists supplied by specialized biopesticide industries (Van Lenteren, 2003). These microorganisms are extensively present in nature and play an important role in naturally regulating their target hosts. So they can be used as pest management tools and have a wide range of desirable properties for ICM (Hajek and Eilenberg, 2018). They are typically characterized as: (1) leaving little or no toxic residue, (2) having significantly lower development and registration costs than synthetic pesticides (Hajek and Eilenberg, 2018), and (3) not naturally infecting vertebrates, making them safe for humans, livestock, and vertebrate wildlife. Their application to crops can be done using the same equipment as used for applying the conventional chemical pesticides. They are also synthesized in a way similar to conventional pesticides for enhanced efficacy.

Two main approaches are used in therapeutic microbial pest control. The first approach is the "classical" method of introducing a nonendemic, host-specific natural enemy to suppress an alien (i.e., invasive) pest (Kuris, 2003). Classically introducing natural BCAs to these new niches will help them with their permanent adaptation and spread within their new surroundings, assisting in biocontrolling the invasive pest species population. Such a method does not necessitate the manufacture of BCAs in large quantities as they are generally executed through nationalized or state programs and work positively in many cases. However, the lack of extensive monitoring (quantitative and objective, as required by ecological theory) of such programs makes it difficult for cost and profit evaluation (Thomas and Reid, 2007). The second approach, which involves increasing the amount of natural BCA in the new niche, does not anticipate a long-lasting or permanent establishment. Customarily, BCAs supplied as commercial products are used in two forms: inoculative and inundative. The inoculative application is a basis of pest control through the acting BCAs introduced along with their successive progenies (Hajek and Eilenberg, 2018). As a result, the agents are expected to be preserved within the newly introduced pest environment niche.

Apparently, microbial BCAs constitute the most abundant and diverse assembly of living beings on Earth, offering a huge scope of beneficial exploitation. Many microbes still await its discovery, and those that have been discovered must now determine their untapped and unknown potential. Therefore, several technical and ecological challenges remain unexplored for their application as microbial control agents and form valuable constituents of ICM (Chandler et al., 2011). Distinctively, inundative applications accomplish a rapid method of pest management by mass application of individuals of the released agent only, with no anticipation of control by their progeny. Therefore, the efficacy of the inundative method is dose-dependent. Its application method is similar to that of chemical pesticides, and as a result, it is possibly the most broadly used method of microbial control. In reality, both inoculation and inundation form a continuum, with the control agent surviving for varying periods depending on biological factors such as host availability, environmental ecological stability, and cropping system (Whipps, 2001). It is known that increasing biological control of microbial products can make significant contributions to pest management as part of ICM (Dent and Binks, 2020). Presently, the U.S. Department of Agriculture initiated four agrarian motivated movements, aimed at the promotion of biocontrol for reducing problem in plant diseases. The mandate, which principally aims to convert 75% of U.S. agriculture to IPM, necessitates the use of biocontrol technologies to reduce the use of chemical pesticides. Having foreknown the antiquity of fungicides, biological control can thus stand as an alternative strategy for the control of plant diseases. Other IPM methods for crop plant ailment control, however, are still needed in many environmental settings, because the agro-ecosystem is a variable dynamic function involving several factors influencing disease and crop development (Maloy, 1993).

22.7.3 Constraints in the synthesis of microorganism-based biopesticides

The availability of microorganisms as a product or formulation, which facilitates the technology's transfer from lab to land, is critical to the success of biopesticides in suppressing pests and diseases. Biopesticide development and use are constrained by some of the same issues that stifle global progress (Keswani et al., 2016).

The major constraints include:
(a) A lack of appropriate screening protocols for identifying promising biopesticide candidates
(b) Lacks adequate knowledge on the BCA's and pest's ecology
(c) No proper optimization of fermentation technology and mass production protocols for BCA's production
(d) Inconsistently filed performance
(e) Poor shelf life of bioformulations
(f) Lacks patent protection
(g) Prohibitive and complex registration procedures
(h) Gaps in awareness and training
(i) Limits in progress of advanced technology
(j) Possibility of contamination with human pathogens

The development of a biopesticide is an extensive process that includes selecting the effectual strains from the microbial population, screening, mass production, and assigning a suitable carrier and other inert components for a stable bioformulation following assessing the shelf life and efficacy. The biopesticides should focus on a high-quality production of a stable formulation to ensure an increase in the interest of farmers, as low grades of product may jeopardize the whole reputation of the biopesticide market. The potentiality of the biopesticides in the control of pests should be conducted both at lab and field conditions in different soil types with diversified microbial communities and climatic conditions (Roberts and Lohrke, 2003). Supposedly, there are many reporting contamination and a lower CFU count for maintaining its shelf life of the BCA in the bioformulation, resulting in poor quality and inconsistent performances on its targeted pest (Arora et al., 2010). Evidentially, several dependent factors determine the shelf life that may include the method of production, the carrier substances used in the formulation, its packaging, transport, and storage. It would lead to the development of viable biopesticide formulations. Regrettably, not many efficient production methods are available based on solid substrates. Therefore, developing cost-effective technology for mass production of viable propagules of the bioagents is acutely concerning. The foremost limitations that hinder the mass implementation of biopesticides include lack of suitable guidance with simple illustrations for its appropriate application, briefer shelf life than desirable of the BCA, and inapt technologies and equipment for application.

Motivating growers through publicity, field demonstrations, farmers days, bio-village adoption, and conducting periodic trainings for commercial producers and farmers to increase/improve supply can all help to popularize BCAs. For industrial linkages, technical assistance in quality control and registration should be made available to entrepreneurs. Regular monitoring is necessary to maintain quality, and continuous research support should be extended to standardize dosage, storage, and delivery systems. The government has provided positive policy support for the use of more BCAs in crop protection.

22.8 Market of biopesticides

The biopesticides act as plant protectants globally (bioinsecticides, bionematicides, biofungicides) becoming the key elements of insect control and field crop improvement. The global biopesticides market size is projected to grow at a CAGR of 14.7% from an estimated value of USD $4.3 billion in 2020 to reach USD $8.5 billion by 2025. The increase in the infestation of pests, pest resistance and resurgence, and the ban on key active ingredients are driving the growth of the biopesticides market. With increased demand for organic food crops and the harmful effects of chemical-based farming, farmers have started to adopt biopesticides. Europe is estimated to be the fastest-growing market in the forecast period (Dublin, 2020). The market has been gaining wide importance among farmers to produce residue-free food products with the usage of microbial-based bioinsecticides. Biological solutions have proved to be an effective alternative to conventional chemicals and even work optimally when applied in combination. Increasing biotic and abiotic stress has resulted in the emergence of invasive pests, which has resulted in reduced crop yields. Most of the bioinsecticides have been commercialized and produced at a large scale, such as *B. thuringiensis*, which has proved to be effective against controlling insect pests. Key players in the bioinsecticides market include BASF SE (Germany), Bayer AG (Germany), Novozymes A/S (Denmark), Certis USA L.L.C (US), Marrone Bio Innovations (US), Syngenta AG (Switzerland), Nufarm (Australia), and Andermatt Biocontrol AG (Switzerland) (Anonymous, 2020a, b).

In India, the usage of biopesticides is growing at a faster pace than that of the chemical variety. According to the Indian Ministry of Agriculture, in the last 10 years, consumption of biopesticides increased by 23%, while that of chemical pesticides grew only by 2%. Biopesticides currently account for 5% of the Indian pesticide market, with at least 15 microbial species and 970 microbial formulations registered through the Central Insecticides Board and Registration Committee (CIBRC). According to Kumar et al. (2019), as of 2017, more than 200 products based on entomopathogenic fungi like *Beauveria brongniartii, B. bassiana, Hirsutella thompsonii, Metarhizium anisopilae, Lecanicillium lecanii,* and nematocidal fungi such as *Pochonia chlamydosporia* and *Purpureocillium lilacinum* are registered for use against various arthropods and plant parasitic nematodes. Concerning bacteria, products based on *B. thuringiensis* (Bt) subsp. *Kurstaki,* which are registered against bollworms, loopers, and other lepidopterans, are around 30. Two viruses are registered, namely *Spodoptera litura* NPV (5 products) and *Helicoverpa armigera* NPV (22 products), for use against armyworms and bollworms. Among entomopathogenic nematode species, there are four products that are sold in the Indian market. The main reason for such a massive boost in biopesticides is government support for the biofertilizer industries and rising awareness among the farmers. Government also supports usage of eco-friendly agri-inputs by introducing certain schemes and subsidiaries. Organic farming in India has become very popular, acting as the main driver for the biopesticide market. About 2.78 million hectares of farmland was under organic cultivation as of March 2020, according to the Union Ministry of Agriculture and Farmers' Welfare.

India ranks eighth in terms of World's Organic Agricultural land and first in terms of total number of organic producers as per 2020 data (Anonymous (2020b)). It produced around 2.75 million MT (2019-20) of certified organic products. Such vast organic farming is fueling the biopesticide market in India, and it is anticipated to continue in the future. Following are a few success stories relating biopesticide and its agent application in the Indian agricultural setting:

(1) Biocontrolling diamondback moths, and *Helicoverpa* on cotton, pigeon-pea, and tomato using *B. thuringiensis*.
(2) Biocontrolling mango hoppers mealy bugs and coffee pod borer by *Beauveria*; the control of white fly on cotton with neem products.
(3) The biocontrol of *Helicoverpa* on gram by N.P.V. and of sugarcane borers by *Trichogramma*.
(4) Biocontrol of rots and wilts in different host crop by *Trichoderma*-based bioproducts (Teng et al., 2006).

22.9 Future prospects and conclusion

The application of biological pesticides is a major contributor in promoting and ensuring sustainable agricultural practices and, in a way, substantially protecting both human health and environment. The biopesticide flourishing market is on par matching the increasing requirements of farmers for inculcating healthier practices during crop cultivation. The market has provided novel bioproducts of improved efficacy that may be used solely, in rotation to, or in combination with the traditionally used chemicals. With the compelling green advantages and safe and easy application methods, the pest biocontrolling sector has instigated higher academic focus and industrial investment to augment stable bioformulations further determining their optimized production. This counts for developing novel biobased compositions against unfamiliar targets or arming the already existing biocontrolling products with additional technologies for improving their bioactivity. Progressive molecular studies help provide newer insights in understanding the available surfeit and diverse population of insect microbial interaction in the plant biome (Abdelfattah et al., 2018; Bennett et al., 2018). Scientific studies in the field of invertebrate pathology are further unfolding novel microbially derived bioactive components. Also, advancement in the legislative frameworks for promoting greener agro-ecosystems has fueled the incorporation of biocontrolling strategies in IPM, resulting its rapid market growth.

It is evidentially cogent that biological pesticides are essential, yet contribute only to a subsequent part of the larger solution when compared to the vast field of sustainable agriculture. Shifting the chemical dependency of farmers to newer green practices further necessitates the discovery of additional biological tools for encouraging organic agricultural practices. Therefore, in spite being just a part of the greener solution, biopesticides play a pivotal role by offering competent tools for generating sustainable agriculture products. They have been a convincing alternative source of the acrid chemical pesticides in common use, but under strict scrutiny regarding their aftereffects on plant and human health and on the environment (Villaverde et al., 2016). Hence, application of several biopesticides have proved its profound effectivity in biocontrolling the harmful pests of agrarian crops and improving their yield. Enhanced production, newer formulation, and advancement in greener technologies are resulting in highly efficacious modes for crop pest control. Such has resulted due to the increased number of dedicated laboratories and firms that focused their study work on discovering and developing methods of biological liquid cultures and in vivo systems with increased production efficiency at cheaper expenses (Shapiro-Ilan et al., 2014). Discovering natural toxins against pests with incorporating recombinant DNA technologies and proteomics have carved newer strategies for replacing the traditional methods of presenting the toxin to the targeted insect pests with those of higher efficiency. We are already familiar with the numerous applications of baculoviruses as pesticides for several commercial cropping systems worldwide. For the betterment of the future ecosystem, modern society focuses on the "green consumerism" leading to reduced use of synthetic substances in food and significant promotion of plant-based products, recommended as safe for plant pest management in both industrialized and developing nations (Dubey et al., 2010; Dimetry, 2012).

A need of the hour has led to refocus research on developing recognized botanicals rather than newer plants (Cavoski et al., 2012). Redirecting research on traditional bioproducts like neem and its derivate products for its commercial use in targeting both known and unfamiliar pests is also gaining importance. Biological control may not be solely effective compared to integrating other pest-controlling techniques during its application for IPM. Globally, the adoption of sustainable management practices calls for internationally collaborating in promoting biological strategies in agricultural systems (Greathead, 1991). A gradual increase is envisaging a larger futuristic use of phytochemicals as a coercive alternative to chemical pesticides due to their cost effectivity, eco-safety, and bioavailability.

References

Abdelfattah, A., Malacrino, A., Wisniewski, M., Cacciola, S.O., Schena, L., 2018. Metabarcoding: a powerful tool to investigate microbial communities and shape future plant protection strategies. Biol. Control 120, 1–10.

Abo-El Seoud, M.A., Sarhan, M.M., Omar, A.E., Helal, M.M., 2005. Biocides formulation of essential oils having antimicrobial activity. Arch. Phytopathol. Plant Protect. 38 (3), 175–184.

Adams, C.G., Schenker, J.H., McGhee, P.S., Gut, L.J., Brunner, J.F., Miller, J.R., 2017. Maximizing information yield from pheromone-baited monitoring traps: estimating plume reach, trapping radius, and absolute density of *Cydia pomonella* (Lepidoptera: Tortricidae) in Michigan apple. J. Econ. Entomol. 110 (2), 305–318.

Aggarwal, R., 2015. *Chaetomium globosum*: A potential biocontrol agent and its mechanism of action. Indian Phytopathol. 68 (1), 8–24.

Aggarwal, R., Gupta, S., Singh, V.B., Sharma, S., 2011. Microbial detoxification of pathotoxin produced by spot blotch pathogen *Bipolaris sorokiniana* infecting wheat. J. Plant Biochem. Biotechnol. 20 (1), 66–73.

Aggarwal, R., Kharbikar, L.L., Sharma, S., Gupta, S., Yadav, A., 2013. Phylogenetic relationships of *Chaetomium* isolates based on the internal transcribed spacer region of the rRNA gene cluster. Afr. J. Biotechnol. 12 (9), 914–920.

Aggarwal, R., Tewari, A.K., Srivastava, K.D., Singh, D.V., 2004. Role of antibiosis in the biological control of spot blotch (*Cochliobolus sativus*) of wheat by *Chaetomium globosum*. Mycopathologia 157, 369–377.

Akinrinnola, O., Okunlola, A.I., 2014. Effectiveness of botanical formulations in vegetable production and bio-diversity preservation in Ondo State, Nigeria. J. Hortic. For. 6 (1), 6–13.

Alavanja, M.C.R., Ross, M.K., Bonner, M.R., 2013. Increased cancer burden among pesticide applicators and others due to pesticide exposure. Cancer J. Clin. 63, 120–142.

Aljamali, N.M., 2013. Study effect of medical plant extracts in comparison with antibiotic against bacteria. J. Sci. Innov. Res. 2, 843–845.

Anonymous, 2020a. Bioinsecticides Market by Organism Type (*Bacteria thuringiensis*, *Beauveria bassiana*, and *Metarhizium anisopliae*), Type (Microbials and Macrobials), Mode of Application, Formulation, Crop Type, and Region—Global Trends and Forecast to 2025. marketsandmarkets.com.

Anonymous, 2020b. Organic Food for Health and Nutrition. PIB, Delhi.

Arora, N.K., Khare, E., Maheshwari, D.K., 2010. Plant growth promoting rhizobacteria: Constraints in bioformulation, commercialization, and future strategies. In: Plant Growth and Health Promoting bacteria. Springer, Berlin, Heidelberg, pp. 97–116.

Bajwa, R., Akhtar, N., Javaid, A., 2001. Antifungal activity of allelopathic plant extracts. I. Effect of aqueous extracts of three allelopathic Asteraceous species on growth of *aspergilli*. Pak. J. Biol. Sci. 4, 503–507.

Batish, D.R., Singh, H.P., Kohli, R.K., Kaur, S., 2008. Eucalyptus essential oil as a natural pesticide. For. Ecol. Manag. 256 (12), 2166–2174.

Bennett, A.E., Orrell, P., Malacrino, A., Pozo, M.J., 2018. Fungal-mediated above-belowground interactions: the community approach, stability, evolution, mechanisms, and applications. In: Aboveground-Belowground Community Ecology. Springer, Cham, pp. 85–116.

Berdy, J., 2005. Bioactive microbial metabolites. J. Antibioti. 58 (1), 1–26.

Beric, T., Kojic, M., Stankovic, S., Topisirovic, L., Degrassi, G., Myers, M., Venturi, V., Fira, D., 2012. Antimicrobial activity of *Bacillus* sp. natural isolates and their potential use in the biocontrol of phytopathogenic bacteria. Food Technol. Biotechnol. 50 (1), 25–31.

Betz, F.S., Hammond, B.G., Fuchs, R.L., 2000. Safety and advantages of *Bacillus thuringiensis*-protected plants to control insect pests. Regul. Toxicol. Pharmacol. 32 (2), 156–173.

Bhanuprakash, V., Hosamani, M., Balamurugan, V., Gandhale, P., Naresh, R., Swarup, D., Singh, R.K., 2008. *In vitro* antiviral activity of plant extracts on goat pox virus replication. Indian J. Exp. Biol. 46, 120–127.

Bhutia, D.D., Zhimo, Y., Kole, R., Saha, J., 2016. Antifungal activity of plant extracts against *Colletotrichum musae*, the post-harvest anthracnose pathogen of banana cv. Martaman. Nutr. Food Sci. 46.

Biopesticide Industry Aliance (BPIA), 2009. Biopesticides in a Program with Traditional Chemicals Offer Growers Sustainable Solutions. http://www.bioworksinc.com/industry-news/biopesticide_grower_whitepaper.pdf. (19 May 2021).

Biswas, S.K., Aggarwal, R., Srivastava, K.D., Gupta, S., Dureja, P., 2012. Characterization of antifungal metabolites of Chaetomium globosum Kunze and their antagonism against fungal plant pathogens. J. Biol. Control 26 (1), 70-74.

Braverman, Y., Chizov-Ginzburg, A., Mullens, B.A., 1999. Mosquito repellent attracts *Culicoides imicola* (Diptera: Ceratopogonidae). J. Med. Entomol. 36 (1), 113-115.

Brodt, S., Six, J., Feenstra, G., Ingels, C., Campbell, D., 2011. Sustainable agriculture. Natl. Educ. Knowl 3 (1).

Burns, J.R., Benson, D.M., 2000. Biocontrol of damping-off of *Catharanthus roseus* caused by *Pythium ultimum* with *Trichoderma virens* and binucleate *Rhizoctonia* fungi. Plant Dis. 84 (6), 644-648.

Campos, E.V., de Oliveira, J.L., Pascoli, M., de Lima, R., Fraceto, L.F., 2016. Neem oil and crop protection: from now to the future. Front. Plant Sci. 7, 1494.

Cavoski, I., Chami, Z.A., Bouzebboudja, F., Sasanelli, N., Simeone, V., Mondelli, D., Miano, T., Sarais, G., Ntalli, N.G., Caboni, P., 2012. *Melia azedarach* controls *Meloidogyne incognita* and triggers plant defense mechanisms on cucumber. Crop Prot. 35, 85-90.

Chandler, D., Bailey, A.S., Tatchell, G.M., Davidson, G., Greaves, J., Grant, W.P., 2011. The development, regulation and use of biopesticides for integrated pest management. Philos. Trans. R. Soc., B 366 (1573), 1987-1998.

Chang, J.H., Choi, J.Y., Jin, B.R., Roh, J.Y., Olszewski, J.A., Seo, S.J., O'Reilly, D.R., Je, Y.H., 2003. An improved baculovirus insecticide producing occlusion bodies that contain *Bacillus thuringiensis* insect toxin. J. Invertebr. Pathol. 84 (1), 30-37.

Chaubey, M.K., 2016. Fumigant and contact toxicity of *Allium sativum* (Alliaceae) essential oil against *Sitophilus oryzae* L.(Coleoptera: Dryophthoridae). Entomol. Appl. Sci. Lett. 3, 43-48.

Chengala, L., Singh, N., 2017. Botanical pesticides—a major alternative to chemical pesticides: a review. Int. J. Life Sci. 5 (4), 722-729.

Choi, G.J., Lee, S.W., Jang, K.S., Kim, J.S., Cho, K.Y., Kim, J.C., 2004. Effects of chrysophanol, parietin, and nepodin of Rumex crispus on barley and cucumber powdery mildews. Crop Prot. 23 (12), 1215-1221.

Choudhury, D., Saha, S., Nath, R., Kole, R.K., Saha, J., 2017. Management of chilli anthracnose by botanicals fungicides caused by *Colletotrichum capsici*. J. Pharmacogn. Phytochem. 6 (4), 997-1002.

Copping, L.G., Menn, J.J., 2000. Bio-pesticides: a review of their action, applications and efficacy. Pest Manag. Sci. 56, 651-676.

Darshan, K., Aggarwal, R., Bashyal, B.M., Mohan, M.H., 2021. Deciphering the network of interconnected pathways of *Chaetomium globosum* antagonistic related genes against *Bipolaris sorokiniana* using RNA seq approach. J. Biol. Control. 34 (4), 258-269.

Darshan, K., Aggarwal, R., Bashyal, B.M., Singh, J., Shanmugam, V., Gurjar, M.S., Solanke, A.U., 2020. Transcriptome profiling provides insights into potential antagonistic mechanisms involved in *Chaetomium globosum* against *Bipolaris sorokiniana*. Front. Microbiol. 11.

Darshan, K., Vanitha, S., Kamalakannan, A., 2019a. Antifungal activity of cinnamon and thyme oil against *Colletotrichum gloeosporioides* causing anthracnose disease of papaya. J. Mycol. Plant Pathol. 49 (3), 273.

Darshan, K., Vanitha, S., Kamalakannan, A., Kavanashree, K., Manasa, R., 2018. *In vitro* and *in vivo* evaluation of biocontrol agents against post-harvest anthracnose of papaya caused by *Colletotrichum gleosporioides* (Penz.). Green Farming 9 (5), 811-818.

Darshan, K., Vanitha, S., Venugopala, K.M., Parthasarathy, S., 2019b. Strategic eco-friendly management of post-harvest fruit rot in papaya caused by *Colletotrichum gloeosporioides*. J. Biol. Control. 33 (3), 225–235.

De Rodrıguez, D.J., Hernández-Castillo, D., Rodrıguez-Garcıa, R., Angulo-Sánchez, J.L., 2005. Antifungal activity in vitro of *Aloe vera* pulp and liquid fraction against plant pathogenic fungi. Ind. Crop. Prod. 21 (1), 81–87.

Deba, F., Xuan, T.D., Yasuda, M., Tawata, S., 2008. Chemical composition and antioxidant, antibacterial and antifungal activities of the essential oils from *Bidens pilosa* Linn. var. *Radiata*. Food Control 19 (4), 346–352.

Dent, D., Binks, R.H., 2020. Insect Pest Management. Cabi.

Dey, K.R., Choudhury, P., Dutta, B.K., 2015. Impact of pesticide use on the health of farmers: a study in Barak Valley, Assam (India). J. Environ. Chem. Ecotoxicol. 10, 269–277.

Di Pietro, A., Gut-Rella, M., Pachlatko, J.P., Schwinn, F.J., 1992. Role of antibiotics produced by *Chaetomium globosum* in biocontrol of *Pythium ultimum*, a causal agent of damping-off. Phytopathology 82 (2), 131–135.

Dimetry, N.Z., 2012. Prospects of botanical pesticides for the future in integrated pest management programme (IPM) with special reference to neem uses in Egypt. Arch. Phytopathol. Plant Protect. 45 (10), 1138–1161.

Djeussi, D.E., Noumedem, J.A., Seukep, J.A., Fankam, A.G., Voukeng, I.K., Tankeo, S.B., Nkuete, A.H., Kuete, V., 2013. Antibacterial activities of selected edible plants extracts against multidrug-resistant gram-negative bacteria. BMC Complement. Altern. Med. 13 (1), 1–8.

Dubey, N.K., Kumar, A., Singh, P., Shukla, R., 2010. Exploitation of natural compounds in eco-friendly management of plant pests. In: Recent Developments in Management of Plant Diseases. Springer, Dordrecht, pp. 181–198.

Dubey, N.K., Shukla, R., Kumar, A., Singh, P., Prakash, B., 2011. Global scenario on the application of natural products in integrated pest management programmes. In: Natural Products in Plant Pest Management. vol. 1, pp. 1–20.

Dublin, 2020. Global Biopesticides Market (2020 to 2025) - Increase in Acceptance of Organic Food Presents Opportunities. Research and Market.

Ebadollahi, A., Jalali Sendi, J., 2015. A review on recent research results on bio-effects of plant essential oils against major coleopteran insect pests. Toxin Rev. 34 (2), 76–91.

Elbeshehy, E.K., Metwali, E.M., Almaghrabi, O.A., 2015. Antiviral activity of *Thuja orientalis* extracts against watermelon mosaic virus (WMV) on *Citrullus lanatus*. Saudi J. Biol. Sci. 22 (2), 211–219.

El-Wakeil, N.E., 2013. Retracted article: botanical pesticides and their mode of action. Gesunde Pflanzen 65 (4), 125–149.

Enan, E.E., 2005. Molecular and pharmacological analysis of an octopamine receptor from American cockroach and fruit fly in response to plant essential oils. Arch. Insect Biochem. Physiol. 59 (3), 161–171. Published in Collaboration with the Entomological Society of America.

Fahad, S., Saud, S., Akhter, A., Bajwa, A.A., Hassan, S., Battaglia, M., Adnan, M., Wahid, F., Datta, R., Babur, E., Danish, S., 2020. Bio-based integrated pest management in rice: an agro-ecosystems friendly approach for agricultural sustainability. J. Saudi Soc. Agric. Sci. 20.

FAO, et al., 2018. The State of Food Security and Nutrition in the World 2018. Building Climate Resilience for Food Security and Nutrition. FAO, Rome, https://doi.org/10.1109/JSTARS.2014.2300145.

Faria, T.D.J., Ferreira, R.S., Yassumoto, L., Souza, J.R.P.D., Ishikawa, N., KBarbosa, A.D.M., 2006. Antifungal activity of essential oil isolated from *Ocimum gratissimum* L. (eugenol chemotype) against phytopathogenic fungi. Braz. Arch. Biol. Technol. 49 (6), 867–871.

Feng, W., Zheng, X., 2007. Essential oils to control *Alternaria alternata* in vitro and in vivo. Food Control 18 (9), 1126–1130.

Fenibo, E.O., Ijoma, G.N., Matambo, T., 2020. Biopesticides in Sustainable Agriculture: Current Status and Future Prospects.

Feyisa, B., Lencho, A., Selvaraj, T., Getaneh, G., 2015. Evaluation of some botanicals and *Trichoderma harzianum* for the management of tomato root-knot nematode (*Meloidogyne incognita* (Kofoid and white) chit wood). Adv. Crop Sci. Technol., 2–10.

Fouda, A., Hassan, S.E.D., Abdo, A., MEl-Gamal, M.S., 2020. Antimicrobial, antioxidant and Larvicidal activities of spherical silver nanoparticles synthesized by endophytic *Streptomyces* spp. Biol. Trace Elem. Res. 195 (2), 707–724.

Freeman, B.C., Beattie, G.A., 2008. An Overview of Plant Defences against Pathogens and Herbivores. The Plant Health Instructor.

Ghorab, M., Khalil, M.S., 2015. Toxicological effects of organophosphates pesticides. Int. J. Environ. Monit. Anal. 4, 218–220.

Grainge, M., Ahmed, S., 1988. Hand Book of Plants With Pest Control Properties. John Wiley, New York, p. 41.

Gray, E.J., Smith, D.L., 2005. Intracellular and extracellular PGPR: commonalities and distinctions in the plant-bacterium signalling processes. Soil Biol. Biochem. 37 (3), 395–412.

Greathead, D.J., 1991. Biological control in the tropics: present opportunities and future prospects. Int. J. Trop. Insect Sci. 12 (1), 3–8.

Guleria, S., Tiku, A.K., 2009. Botanicals in pest management: current status and future perspectives. In: Integrated Pest Management: Innovation-Development Process. Springer, Dordrecht, pp. 317–329.

Gupta, S., Dikshit, A.K., 2010. Biopesticides: an ecofriendly approach for pest control. J. Biopestic. 3 (Special Issue), 186.

Hadizadeh, I., Peivastegan, B., Hamzehzarghani, H., 2009. Antifungal activity of essential oils from some medicinal plants of Iran against *Alternaria alternate*. Am. J. Appl. Sci. 6 (5), 857–861.

Hajek, A.E., Eilenberg, J., 2018. Natural Enemies: An Introduction to Biological Control. Cambridge University Press.

Harish, S., Saravanakumar, D., Radjacommare, R., Ebenezar, E.G., Seetharaman, K., 2008. Use of plant extracts and biocontrol agents for the management of brown spot disease in rice. BioControl 53, 555–567.

Harish, S., Saravanan, T., Radjacommare, R., Ebenezar, E.G., Seetharaman, K., 2004. Mycotoxic effect of seed extracts against *Helminthosporium oryzae* Breda de hann, the incitant of rice brown spot. J. Biol. Sci. 4, 366–369.

Heap, I., 2014. Herbicide resistant weeds. In: Integrated Pest Management. Springer, Dordrecht, pp. 281–301.

Hollomon, D.W., 2015a. Fungicide resistance: facing the challenge—a review. Plant Prot. Sci. 51 (4), 170–176.

Hollomon, D.W., 2015b. Fungicide resistance: 40 years on and still a major problem. In: Fungicide Resistance in Plant Pathogens. Springer, Tokyo, pp. 3–11.

Huang, Y., Ho, S.H., 1998. Toxicity and antifeedant activities of cinnamaldehyde against the grain storage insects, *Tribolium castaneum* (Herbst) and *Sitophilus zeamais* Motsch. J. Stored Prod. Res. 34 (1), 11–17.

Hubbard, M., Hynes, R.K., Erlandson, M., Bailey, K.L., 2014. The biochemistry behind biopesticide efficacy. Sustain. Chem. Proc. 2 (1), 1–8.

Hunter, M.C., Smith, R.G., Schipanski, M.E., Atwood, L.W., Mortensen, D.A., 2017. Agriculture in 2050: recalibrating targets for sustainable intensification. Bioscience 67 (4), 386–391.

Inayati, A., Sulistyowati, L., Aini, L.Q., Yusnawan, E., 2020. Mycoparasitic activity of indigenous *Trichoderma virens* strains against Mungbean soil borne pathogen *Rhizoctonia solani*: Hyperparasite and hydrolytic enzyme production. AGRIVITA, J. Agric. Sci. 42 (2), 229–242.

Infante, D.M., Allan, J.D., Linke, S., Norris, R.H., 2009. Relationship of fish and macroinvertebrate assemblages to environmental factors: implications for community concordance. Hydrobiologia 623 (1), 87–103.

Isman, M.B., Koul, O., Arnason, J.T., Stewart, J., Salloum, G.S., 1991. Developing a neem-based insecticide for Canada. Mem. Entomol. Soc. Can. 123 (S159), 39–46.

Jacobson, M., 1982. Plants, insects, and man—their interrelationships. Econ. Bot. 36 (3), 346–354.

Javaid, M.K., Ashiq, M., Tahir, M., 2016. Potential of biological agents in decontamination of agricultural soil. Scientifica 2016.

Jiang, C., Li, Z., Shi, Y., Guo, D., Pang, B., Chen, X., Shao, D., Liu, Y., Shi, J., 2020. *Bacillus subtilis* inhibits *aspergillus carbonarius* by producing iturin A, which disturbs the transport, energy metabolism, and osmotic pressure of fungal cells as revealed by transcriptomics analysis. Int. J. Food Microbiol. 330, 108783.

Jiang, C., Song, J., Zhang, J., Yang, Q., 2017. Identification and characterization of the major antifungal substance against *fusarium sporotrichioides* from *Chaetomium globosum*. World J. Microbiol. Biotechnol. 33 (6), 108. https://doi.org/10.1007/s11274-017-2274-x.

Kachhawa, D., 2017. Microorganisms as a biopesticides. J. Entomol. Zool. Stud. 5 (3), 468–473.

Kalaycioglu, A., Oner, C., Erden, G., 1997. Observation of the antimutagenic potencies of plant extracts and pesticides in the *Salmonella typhimurium* strains TA 98 and TA 100. Turk. J. Bot. 21, 127–130.

Kawalekar, J.S., 2013. Role of biofertilizers and biopesticides for sustainable agriculture. J. Bio Innov. 2 (3), 73–78.

Keane, S., Ryan, M.F., 1999. Purification, characterisation, and inhibition by monoterpenes of acetylcholinesterase from the waxmoth, *galleria mellonella* (L.). Insect Biochem. Mol. Biol. 29 (12), 1097–1104.

Kepenekci, I., Erdogus, Derdogan, P., 2016. Effects of some plant extracts on root-knot nematodes *in vitro* and *in vivo* conditions. Turkiye Entomoloji Dergisi 40 (1).

Keswani, C., Sarma, B.K., Singh, H.B., 2016. Synthesis of policy support, quality control, and regulatory management of biopesticides in sustainable agriculture. In: Agriculturally Important Microorganisms. Springer, Singapore, pp. 3–12.

Khan, Z., Kim, S.G., Jeon, Y.H., Khan, H.U., Son, S.H., Kim, Y.H., 2008. A plant growth promoting *Rhizobacterium*, *Paenibacillus polymyxa* strain GBR-1, suppresses root-knot nematode. Bioresour. Technol. 99 (8), 3016–3023.

Khater, H.F., 2003. Biocontrol of Some Insects. Zagazig University, Benha, Egypt.

Khater, H.F., Hanafy, A., Abdel-Mageed, A.D., Ramadan, M.Y., El-Madawy, R.S., 2011. Control of the myiasis-producing fly, *Lucilia sericata*, with Egyptian essential oils. Int. J. Dermatol. 50 (2), 187–194.

Khater, H.F., Ramadan, M.Y., El-Madawy, R.S., 2009. Lousicidal, ovicidal and repellent efficacy of some essential oils against lice and flies infesting water buffaloes in Egypt. Vet. Parasitol. 164 (2–4), 257–266.

Kishore, G.K., Pande, S., 2005. Integrated applications of aqueous leaf extract of *Datura metel* and chlorothalonil improved control of late leaf spot and rust of groundnut. Australas. Plant Pathol. 34 (2), 261–264.

Knox, C., Moore, S.D., Luke, G.A., Hill, M.P., 2015. Baculovirus-based strategies for the management of insect pests: a focus on development and application in South Africa. Biocontrol Sci. Tech. 25 (1), 1–20.

Kon, K., Rai, M., 2012. Antibacterial Activity of *Thymus vulgaris* Essential Oil Alone and in Combination with Other Essential Oils.

Koul, O., 2012. Microbial biopesticides: opportunities and challenges. Biocontrol News Inform. 33 (2), 1R.

Kour, D., Rana, K.L., Yadav, N., Yadav, A.N., Singh, J., Rastegari, A.A., Saxena, A.K., 2019. Agriculturally and industrially important fungi: current developments and potential biotechnological applications. In: Recent Advancement in White Biotechnology Through Fungi. Springer, Cham, pp. 1–64.

Kumar, K.K., Sridhar, J., Murali-Baskaran, R.K., Senthil-Nathan, S., Kaushal, P., Dara, S.K., Arthurs, S., 2019. Microbial biopesticides for insect pest management in India: current status and future prospects. J. Invertebr. Pathol. 165, 74–81.

Kuris, A.M., 2003. Did biological control cause extinction of the coconut moth, Levuana iridescens, in Fiji? In: Marine Bioinvasions: Patterns, Processes and Perspectives. Springer, Dordrecht, pp. 133–141.

Laengle, T., Cass, L., 2011. Categories of Biopesticides and Related Products.

Leng, P., Zhang, Z., Pan, G., Zhao, M., 2011. Applications and development trends in biopesticides. Afr. J. Biotechnol. 10 (86), 19864–19873.

Lengai, G.M., Muthomi, J.W., 2018. Biopesticides and their Role in Sustainable Agricultural Production.

Lengai, G.M., Muthomi, J.W., Mbega, E.R., 2020. Phytochemical activity and role of botanical pesticides in pest management for sustainable agricultural crop production. Sci. African 7, e00239.

Lewis, W.J., Jones, R.L., Nordlund, D.A., Sparks, A.N., 1975. Kairomones and their use for management of entomophagous insects: I. evaluation for increasing rates of parasitization by *Trichogramma* spp. in the field. J. Chem. Ecol. 1 (3), 343–347.

Lin, L., Pantapalangkoor, P., Tan, B., Bruhn, K.W., Ho, T., Nielsen, T., Skaar, E.P., Zhang, Y., Bai, R., Wang, A., Doherty, T.M., 2014. Transferrin iron starvation therapy for lethal bacterial and fungal infections. J. Infect. Dis. 210 (2), 254–264.

Maddox, J.V., 1987. Protozoan diseases. In: Epizootiology of Insect Diseases. vol. 1, pp. 417–452.

Maloy, O.C., 1993. Plant Disease Control: Principles and Practice. John Wiley and Sons, Inc.

Mandal, S., Srivastava, K.D., Aggarwal, R., Singh, D.V., 1999. Mycoparasitic action of some fungi on spot blotch pathogen (Drechslera sorokiniana) of wheat. Indian Phytopathol. 52, 39–43.

Marcic, D., Prijovic, M., Drobnjakovic, T., Medo, I., Peric, P., Milenkovic, S., 2012. Greenhouse and field evaluation of two biopesticides against *Tetranychus urticae* and *Panonychus ulmi* (Acari: Tetranychidae). Pestic. Phytomed. 27 (2), 313–320.

Martinez-Luis, S., Cherigo, L., Arnold, E., Spadafora, C., Gerwick, W.H., Cubilla-Rios, L., 2012. Antiparasitic and anticancer constituents of the endophytic fungus *aspergillus* sp. strain F1544. Nat. Prod. Commun. 7 (2), 1934578X1200700207.

Martinez-Toledo, M.V., Salmeron, V., Rodelas, B., Pozo, C., Gonzalez-Lopez, J., 1998. Effects of the fungicide Captan on some functional groups of soil microflora. Appl. Soil Ecol. 7 (3), 245–255.

McLaughlin, J.R., Shorey, H.H., Gaston, L.K., Kaae, R.S., Stewart, F.D., 1972. Sex pheromones of Lepidoptera. XXXI. Disruption of sex pheromone communication in *Pectinophora gossypiella* with hexalure. Environ. Entomol. 1 (5), 645–650.

McSpadden Gardener, B.B., 2007. Diversity and ecology of biocontrol *Pseudomonas* spp. in agricultural systems. Phytopathology 97 (2), 221–226.

Mehlhorn, H., Abdel-Ghaffar, F., Al-Rasheid, K.A., Schmidt, J., Semmler, M., 2011. Ovicidal effects of a neem seed extract preparation on eggs of body and head lice. Parasitol. Res. 109 (5), 1299–1302.

Mizubuti, G.S.E., Junior, V.L., Forbes, G.A., 2007. Management of late blight with alternative products. Pest Technol. 2, 106–116.

Moya, P., Pedemonte Roman, D., Susana, A., Franco, M.E.E., Sisterna, M.N., 2016. Antagonism and Modes of Action of *Chaetomium Globosum* Species Group, Potential Biocontrol Agent of Barley Foliar Diseases.

Muthomi, J.W., Lengai, G.M.W., Wagacha, J.M., Narla, R.D., 2017. In vitro activity of plant extracts against some important plant pathogenic Fungi of tomato. Aust. J. Crop. Sci. 6, 683–689.

Naeini, A., Ziglari, T., Shokri, H., Khosravi, A.R., 2010. Assessment of growth-inhibiting effect of some plant essential oils on different *fusarium* isolates. J. Mycol. Méd. 20 (3), 174–178.

Nawaz, M., Mabubu, J.I., Hua, H., 2016. Current status and advancement of biopesticides: microbial and botanical pesticides. J. Entomol. Zool. Stud. 4 (2), 241–246.

Nefzi, A., Abdallah, B.A.R., Jabnoun-Khiareddine, H., Saidiana-Medimagh, S., Haouala, R., Remadi, M., 2016. Antifungal activity of aqueous and organic extracts from *Withania somnifera* l. against *fusarium oxysporum* f. sp. *radicis-lycopersici*. J. Microbial. Biochem. Technol. 8, 144–150.

Newman, M.C., Unger, M.A., 2003. Fundamentals of Ecotoxicology. CRC Press, Boca Raton, FL.

Omukoko, C.A., 2019. Review on an agricultural insect pests control using entomopathogenic fungus *Beauveria bassiana*. Afr. J. Horti. Sci. 16, 13–20.

Oshunsanya, S.O., Nwosu, N.J., Li, Y., 2019. Abiotic stress in agricultural crops under climatic conditions. In: Sustainable agriculture, forest and environmental management. Springer, Singapore, pp. 71–100.

Park, J.H., Choi, G.J., Jang, K.S., Lim, H.K., Kim, H.T., Cho, K.Y., Kim, J.C., 2005. Antifungal activity against plant pathogenic fungi of *Chaetoviridins* isolated from *Chaetomium globosum*. FEMS Microbiol. Lett. 252 (2), 309–313.

Parker, K.M., Sander, M., 2017. Environmental fate of insecticidal plant-incorporated protectants from genetically modified crops: knowledge gaps and research opportunities. Environ. Sci. Technol. 51.

Patrice, A.K., Seka, K., Francis, Y.K., Theophile, A.S., Fatoumata, F., Diallo, H.A., 2017. Effects of three aqueous plant extracts in the control of Fungi associated with post-harvest of yam (*Dioscorea alata*). Int. J. Agron. Agric. Res. 3, 77–87.

Patrick, C.D., 2004. Grasshoppers and Their Control. Texas Farmer Collection.

Paul, P.K., Sharma, P.D., 2002. *Azadirachta indica* leaf extract induces resistance in barley against leaf stripe disease. Physiol. Mol. Plant Pathol. 61 (1), 3–13.

Pavela, R., Zabka, M., Bednar, J., Triska, J., Vrchotova, N., 2016. New knowledge for yield, composition and insecticidal activity of essential oils obtained from the aerial parts or seeds of fennel (*Foeniculum vulgare* mill.). Ind. Crop. Prod. 83, 275–282.

Pichersky, E., Gershenzon, J., 2002. The formation and function of plant volatiles: perfumes for pollinator attraction and defense. Curr. Opin. Plant Biol. 5 (3), 237–243.

Pimentel, D., 2009. Pesticides and pest control. In: Integrated Pest Management: Innovation-Development Process.

Pretali, L., Bernardo, L., Butterfield, T.S., Trevisan, M., Lucini, L., 2016. Botanical and biological pesticides elicit a similar induced systemic response in tomato (*Solanum lycopersicum*) secondary metabolism. Phytochemistry 130, 56–63.

Pretty, J.N., 2012. The Pesticide Detox: Towards a more Sustainable Agriculture. Earthscan.

Priestley, C.M., Williamson, E.M., Wafford, K.A., Sattelle, D.B., 2003. Thymol, a constituent of thyme essential oil, is a positive allosteric modulator of human GABAA receptors and a homo-oligomeric GABA receptor from *Drosophila melanogaster*. Br. J. Pharmacol. 140 (8), 1363–1372.

Qiao, J., Zou, X., Lai, D., Yan, Y., Wang, Q., Li, W., Deng, S., Xu, H., Gu, H., 2014. Azadirachtin blocks the calcium channel and modulates the cholinergic miniature synaptic current in the central nervous system of Drosophila. Pest Manag. Sci. 70 (7), 1041–1047.

Rahman, S., 2015. Green revolution in India: environmental degradation and impact on livestock. Asian J. Water, Environ. Pollut. 12 (1), 75–80.

Rajasekharan, S., Rana, J., Gulati, S., Sharma, S.K., Gupta, V., Gupta, S., 2013. Predicting the host protein interactors of Chandipura virus using a structural similarity-based approach. Pathog. Dis. 69 (1), 29–35.

Rathoure, A.K. (Ed.), 2019. Zero Waste: Management Practices for Environmental Sustainability: Management Practices for Environmental Sustainability. CRC Press.

Rattan, R.S., 2010. Mechanism of action of insecticidal secondary metabolites of plant origin. Crop Prot. 29 (9), 913–920.

Ravi-Kumar, K., Venkatesh, K.S., Umesh-Kumar, S., 2007. The 53-kDa proteolytic product of precursor starch-hydrolyzing enzyme of *aspergillus Niger* has taka-amylase-like activity. Appl. Microbiol. Biotechnol. 74 (5), 1011–1015.

Regnault-Roger, C., Philogène, B.J., 2008. Past and current prospects for the use of botanicals and plant allelochemicals in integrated pest management. Pharm. Biol. 46 (1-2), 41–52.

Ritter Stephen, K., 2009. Pinpointing trends in pesticide use. Chem. Eng. News 87 (7), 35.

Rizvi, S.J.H., Pandey, S.K., Mukerji, D., Mathur, S.N., 1980. 1, 3, 7-Trimethylxanthine, a new chemosterilant for stored grain pest, *Callosobruchus chinensis* (L.). Z. Angew. Entomol. 90 (1-5), 378–381.

Roberts, D.R., Andre, R.G., 1994. Insecticide resistance issues in vector-borne disease control. Am. J. Trop. Med. Hyg. 50 (6_Suppl), 21–34.

Roberts, D.P., Lohrke, S.M., 2003. United States Department of Agriculture-Agricultural Research Service research programs in biological control of plant diseases. Pest Manag. Sci.: Formerly Pesticide Science 59 (6-7), 654–664.

Salim, H.A., Salman, I.S., Ishtar, I.M., Hatam, H.H., 2016. Evaluation of some plant extracts for their nematicidal properties against root-knot nematode, *Meloidogyne* sp. J. Genet. Environ. Resour. Conserv. 3, 241–244.

Salma, M., Ratul, C.R., Jogen, C.K., 2011. A review on the use of biopesticides in insect pest management. Int. J Sci Adv Tech. 1, 169–178.

Samada, L.H., Tambunan, U.S.F., 2020. Biopesticides as promising alternatives to chemical pesticides: a review of their current and future status. OnLine J. Biol. Sci. 20 (2), 66–76.

Santoso, S.E., Soesanto, L., Haryanto, T.A.D., 2007. Biological suppression of fusarium wilt on shallot by *Trichoderma harzianum*, *Trichoderma koningii*, and *Pseudomonas fluorescens* P 60. J. Hama dan Penyakit Tumbuhan Tropika 7 (1), 53–61.

Saxena, R.C., Harshan, V., Saxena, A., Sukumaran, P., Sharma, M.C., Kumar, M.L., 1993. Larvicidal and chemosterilant activity of *Annona squamosa* alkaloids against Anopheles stephensi. J. Am. Mosq. Control Assoc. 9, 84.

Schmutterer, H., Ascher, K.R.S., 1995. Neem tree (*Azadirachta indica* A. Juss.) and other meliaceous plants. VCH.

Shapiro-Ilan, D.I., Han, R., Qiu, X., 2014. Production of entomopathogenic nematodes. In: Mass Production of Beneficial Organisms. Academic Press, pp. 321–355.

Sharma, A., Gupta, A., Khosla, K., Mahajan, R., Mahajan, P., 2017. Antagonistic potential of native agrocin-producing non-pathogenic *Agrobacterium tumefaciens* strain UHFBA-218 to control crown gall in peach. Phytoprotection 97 (1), 1–11.

Shiberu, T., Getu, E., 2016. Assessment of selected botanical extracts against *Liriomyza* species (Diptera: Agromyzidae) on tomato under glasshouse condition. Int. J. Fauna Biol. Stud. 1, 87–90.

Shirzadian, S., Azad, H.A., Khalghani, J., 2009. Introductional study of antifungal activities of bryophyte extracts. Appl. Entomol. Phytopathol. 77 (1).

da Silva Bomfim, N., Nakassugi, L.P., Oliveira, J.F.P., Kohiyama, C.Y., Mossini, S.A.G., Grespan, R., Nerilo, S.B., Mallmann, C.A., Abreu Filho, B.A., Machinski Jr., M., 2015. Antifungal activity and inhibition of fumonisin production by *Rosmarinus officinalis* L. essential oil in *fusarium verticillioides* (Sacc.) Nirenberg. Food Chem. 166, 330–336.

Singh, N.S., Sharma, R., Parween, T., Patanjali, P.K., 2018. Pesticide contamination and human health risk factor. In: Modern age Environmental Problems and Their Remediation. Springer, Cham, pp. 49–68.

Soesanto, L., Mugiastuti, E., Rahayuniati, R.F., 2010. Kajian mekanisme antagonis *Pseudomonas fluorescens* P60 terhadap *fusarium oxysporum* f. sp. *lycopersici* pada tanaman tomat in vivo. J. Hama dan Penyakit Tumbuhan Tropika 10 (2), 108–115.

Soesanto, L., Mugiastuti, E., Rahayuniati, R.F., 2011 June. Morphological and physiological features of *Pseudomonas fluorescens* P60. In: 4th International Seminar of Indonesian Society for Microbiology, pp. 22-24.

Sola, P., Mvumi, B.M., Ogendo, J.O., Mponda, O., Kamanula, J.F., Nyirenda, S.P., Belmain, S.R., Stevenson, P.C., 2014. Botanical pesticide production, trade and regulatory mechanisms in sub-Saharan Africa: making a case for plant-based pesticidal products. Food security, 6(3), pp. 369-3. Mechanism of action of insecticidal secondary metabolites of plant origin. Crop Prot. 29 (9), 913-920.

Soytong, K., Kanokmedhakul, S., Kukongviriyapa, V., Isobe, M., 2001. Application of Chaetomium species (Ketomium) as a new broad spectrum biological fungicide for plant disease control. Fungal Divers. 7, 1-15.

Srivastava, K.P., Dhaliwal, G.S., 2010. A Textbook of Applied Entomology. Kalyani Publishers, New Delhi, p. 113.

Suprapta, D.N., 2012. Potential of microbial antagonists as biocontrol agents against plant fungal pathogens. J. ISSAAS 18 (2), 1-8.

Teng, Y.P., Liang, Z.S., Chen, R., 2006. Preliminary study of *Trichoderma* against the root rot disease of *Astragallus*. Acta. Agric. Boreali-Occident. Sin. 15 (2), 69-71.

Thakore, Y., 2006. The biopesticide market for global agricultural use. Ind. Biotechnol. 2 (3), 194-208.

Thomas, M.B., Reid, A.M., 2007. Are exotic natural enemies an effective way of controlling invasive plants? Trends Ecol. Evol. 22 (9), 447-453.

Tian, L., Wang, X., Liu, R., Zhang, D., Wang, X., Sun, R., Guo, W., Yang, S., Li, H., Gong, G., 2021. Antibacterial mechanism of thymol against *Enterobacter sakazakii*. Food Control 123, 107716.

Tounou, A.K., 2007. The Potential of Paranosema (Nosema) locustae (Microsporidia: Nosematidae) and Its Combination With *Metarhizium anisopliae* var. *acridum* (Deuteromycotina: Hyphomycetes) for the Control of Locusts and Grasshoppers in West Africa. Cuvillier Verlag.

USEPA (U.S. Environmental Protection Agency), 2008. Draft Nanomaterial Research Strategy. EPA/600/S-08/002, U.S. Environmental Protection Agency, Office of Research and Development, Washington, DC.

Van Lenteren, J.C., 2003. Greenhouses without pesticides: a vision for the future. IOBC WPRS Bull. 26 (9), 3-4.

Vanbergen, A.J., The Insect Pollinators Initiative, 2013. Threats to an ecosystem service: pressures on pollinators. Front. Ecol. Environ. 11 (5), 251-259.

Varma, J., Dubey, N.K., 1999. Prospectives of botanical and microbial products as pesticides of tomorrow. Curr. Sci., 172-179.

Velten, S., Leventon, J., Jager, N., Newig, J., 2015. What is sustainable agriculture? A systematic review. Sustainability 7 (6), 7833-7865.

Vilela, G.R., de Almeida, G.S., D'Arce, M.A.B.R., Moraes, M.H.D., Brito, J.O., da Silva, M.F.D.G., Silva, S.C., de Stefano Piedade, S.M., Calori-Domingues, M.A., da Gloria, E.M., 2009. Activity of essential oil and its major compound, 1, 8-cineole, from *Eucalyptus globulusLabill.*, against the storage fungi *aspergillus flavus* link and *aspergillus parasiticus* Speare. J. Stored Prod. Res. 45 (2), 108-111.

Villaverde, J.J., Sandín-España, P., Sevilla-Morán, B., López-Goti, C., Alonso-Prados, J.L., 2016. Biopesticides from natural products: current development, legislative framework, and future trends. Bioresources 11 (2), 5618-5640.

Viuda-Martos, M., Ruiz-Navajas, Y., Fernández-López, J., Pérez-Álvarez, J.A., 2008. Functional properties of honey, propolis, and royal jelly. J. Food Sci. 73 (9), R117-R124.

Wang, X.Q., Hao, Y., Chen, S., Jiang, Q., Yang, Q., Li, Q., 2015. The effect of chemical composition and bioactivity of several essential oils on *Tenebrio molitor* (Coleoptera: Tenebrionidae). J. Insect Sci. 15, 116-122.

Ware, G.W., Whitacre, D.M., 2004. An introduction to insecticides. In: The Pesticide Book. 6.

Weinzierl, R.A., 2000. Botanical insecticides, soaps, and oils. In: Biological and Biotechnological Control of Insect Pests, pp. 101–121.

Weller, D.M., Van Pelt, J.A., Mavrodi, D.V., Pieterse, C.M.J., Bakker, P.A.H.M., Van Loon, L.C., 2004. Induced systemic resistance (ISR) in Arabidopsis against *Pseudomonas syringae* pv. Tomato by 2, 4- diacetylphloroglucinol (DAPG)-producing *Pseudomonas fluorescens*. Phytopathology 94, S108.

Whipps, J.M., 2001. Microbial interactions and biocontrol in the rhizosphere. J. Exp. Bot. 52 (suppl_1), 487–511.

Wittstock, U., Gershenzon, J., 2002. Constitutive plant toxins and their role in defense against herbivores and pathogens. Curr. Opin. Plant Biol. 5 (4), 300–307.

World Resources Institute (WRI), 1994. In Collaboration With the United Nations Environment Programme and the United Nations Development Programme, World Resources. World Resources Institute, Washington, DC.

Xu, P., Liu, Y., Graham, R.I., Wilson, K., Wu, K., 2014. Densovirus is a mutualistic symbiont of a global crop pest (*Helicoverpa armigera*) and protects against a baculovirus and Bt biopesticide. PLoS Pathog. 10 (10), e1004490.

Yadav, I.C., Devi, N.L., 2017. Pesticides classification and its impact on human and environment. Environ. Sci. Eng. 6, 140–158.

Yadav, S.K., Kumawat, K.C., Deshwal, H.L., Kumar, S., 2020. Evaluation of Sequences of Insecticides, Biopesticides and Bioagents against Major Insect Pests of Okra.

Yamamoto-Ribeiro, M.M.G., Grespan, R., Kohiyama, C.Y., Ferreira, F.D., Mossini, S.A.G., Silva, E.L., de Abreu Filho, B.A., Mikcha, J.M.G., Junior, M.M., 2013. Effect of *Zingiber officinale* essential oil on *fusarium verticillioides* and fumonisin production. Food Chem. 141 (3), 3147–3152.

Yanar, Y., Kadioglu, I., Gökçe, A., Demirtas, I., Gören, N., Çam, H., Whalon, M., 2011. In vitro antifungal activities of 26 plant extracts on mycelial growth of *Phytophthora infestans* (Mont.) de Bary. Afr. J. Biotechnol. 10 (14), 2625–2629.

Yi, L., Xiao, C., Ma, G., Wu, Q., Dou, Y., 2010. Biocontrol effect and inhibition activity of antagonistic actinomycetes strain TA21 against *Thielaviopsis basicola*. Chin. J. Biol. Control 26 (2), 186–192.

Yoon, M.Y., Cha, B., Kim, J.C., 2013. Recent trends in studies on botanical fungicides in agriculture. Plant Pathol. J. 29 (1), 1.

Zabka, M., Pavela, R., Slezakova, L., 2009. Antifungal effect of *Pimenta dioica* essential oil against dangerous pathogenic and toxinogenic fungi. Ind. Crop. Prod. 30 (2), 250–253.

Zarubova, L., Lenka, K., Pavel, N., Miloslav, Z., Ondrej, D., Skuhrovec, J., 2014. Botanical pesticides and their human health safety on the example of *Citrus sinensis* essential oil and *Oulema melanopus* under laboratory conditions. Mendel Net, 330–336.

Zhang, X., Cai, X., 2011. Climate change impacts on global agricultural land availability. Environ. Res. Lett. 6 (1), 014014.

Zhang, H., Yang, Q., 2007. Expressed sequence tags-based identification of genes in the biocontrol agent *Chaetomium cupreum*. Appl. Microbiol. Biotechnol. 74 (3), 650–658.

Zhao, G.X., Xu, L.H., Pan, H., Lin, Q.R., Huang, M.Y., Cai, J.Y., Ouyang, D.Y., He, X.H., 2015. The BH3-mimetic gossypol and noncytotoxic doses of valproic acid induce apoptosis by suppressing cyclin-A2/Akt/FOXO3a signaling. Oncotarget 6 (36), 38952.

23

Plant-pathogen interaction: Mechanisms and evolution

U.M. Aruna Kumara[a], P.L.V.N. Cooray[b], N. Ambanpola[b], and N. Thiruchchelvan[c]

[a]Department of Agricultural Technology, Faculty of Technology, University of Colombo, Homagama, Sri Lanka, [b]Board of Study in Plant Protection, Postgraduate Institute of Agriculture, Faculty of Agriculture, University of Peradeniya, Preadeniya, Sri Lanka, [c]Department of Agricultural Biology, Faculty of Agriculture, University of Jaffna, Kilinochchi, Sri Lanka

23.1 Introduction

The biosphere is the thin life-supporting substrate of the Earth's surface that extends from the atmosphere to the deep-sea vents of the ocean a few kilometers in depth. All of the ecosystems on Earth together are called the biosphere where living organisms interact with their nonliving abiotic factors (Thompson et al., 2020). Living organisms are represented by flora and fauna, but there are organisms that cannot be visualized by the naked eye called microorganisms (Thompson et al., 2020). Flora or plants are primary producers that can produce their own food by a physiological mechanism called photosynthesis. This special ability is governed by the chloroplast and chlorophylls, which can absorb radiant energy from the sun. Microorganisms are minute breathing things found all over the place and are too small to be seen by the unaided eye. They live in various environments such as water, soil, and the air (Orr and Ralebitso-Senior, 2016).

Organisms interact with each other in an ecosystem for their survival (Thompson et al., 2020). These interactions can be beneficial or harmful (Upson et al., 2018). These relationships are more likely for their energy requirements in the process of food production in terms of cellular respiration to build up energy for cellular functions (Frantzeskakis et al., 2018). Plants have coevolved symbiotically with broad microbial consortia through evolutionary adaptation. This mechanism helped plants dominate on the Earth's crust as land plants (Thompson et al., 2020). Fungal and bacterial associates like lichens provide the main support for development of soil on bedrock. Without soil as a substrate, it's impossible for plants to initiate their lives (Balogh-Brunstad et al., 2017), which means that without support of microbial life, there would be no plant life on Earth. Therefore, plant microbial interactions go back billions of years (Upson et al., 2018).

Legume N_2-fixing nodules and arbuscular mycorrhizae are the examples of mutualism, and biotrophic and necrotrophic microorganisms are worthy examples of antagonists, which represent the symbiotic associations or interactions between plants and associated microbial communities (Hawkes and Connor, 2017). The endophytic and epiphytic microbiomes act as external elicitors to express host genes to mitigate biotic and abiotic stress. In addition to that, the exchange of genetic materials between each other are found in symbiosis association (Matveeva et al., 2018). Plant-pathogen interactions involve the secretion of microbial toxic proteins to digest host tissues for their survival. Suppression of plant defense mechanisms, interrupting vascular movement of fluid, and regulation of plant gene expressions are some of the interferences done by pathogenic microorganisms in plant tissues (Borrelli et al., 2018). *Agrobacterium tumefaciens* is a good example of this situation. Wounding in the host plant by bacteria induces the production of olein by the host plant as a food supplement for bacteria (Babich et al., 2020). In addition to that, the pathogen has ability to change the plant or host defense mechanism to develop an immune compromised situation in the plant to better escape from the plant's defense mechanism (Borrelli et al., 2018). This chapter mainly focuses on mechanisms and the evolution of plant-pathogen interactions.

23.2 Plants

The beginning of plant life on Earth is one of the most significant pieces of evolutionary evidence in Earth's history. The body of plants was greatly changed due to the colonization of plants on the terrestrial segment of the biosphere. This introduction of plants to the Earth's crust initiated biogeochemical cycles such as carbon and nitrogen cycles (Morris et al., 2015). As a result, this reduced the atmospheric CO_2 levels and increased the oxygen level in the atmosphere, which created new habitats for fauna and fungal communities (Labandeira, 2013). In addition, significant changes resulted in soil's colloidal and stabilized sediment systems, which in turn improved the river system and landscape of the Earth (Gibling and Davies, 2012).

Therefore, calculation of the timeline of phytoterrestrialization of the planet is difficult due to the unavailability of methodologies other than the molecular clock method (Morris et al., 2018). By using the identified fossil record, molecular evolution of plants can be estimated. Unfortunately, the correlation between the four principal lineages of land plants, namely, hornworts, liverworts, mosses, and tracheophytes, have yet been resolved by using available scientific methodologies (Morris et al., 2018). Various plant scientists all over the world have tried to identify the robust timeline of land plant origin with the help of phylogenetic analysis (Wickett et al., 2014).

Plants are photoautotrophs, in that they generate their own food through photosynthesis. Heterotrophs are the other organisms; they do not make their own food and depend on other organisms for their survival (Thompson et al., 2020). Most of the heterotrophs utilize plants as their food supplement. Plants are primary producers in many ecosystems. In addition to that, living organisms utilize oxygen for their cellular respiration to generate energy for cellular functions. The major contribution to the atmospheric organisms is photosynthesis. Because of this, plants act as the host for many organisms (Thompson et al., 2020). Plants are composed of macromolecules such as carbohydrates, lipids, and proteins; these macromolecules are a good source of nutrients for many heterotrophic organisms. Plant lacks mobile defender cells or a somatic adaptive immune system. Plants have evolved structural, chemical, and protein-based defense mechanisms to perceive microbials in the innate immunity of each cell and on systemic signals emanating from infection sites, and to translate that perception into an adaptive resistance gene response (Ausubel, 2005; Chisholm et al., 2006).

23.3 Plant pathogens

Plant pathogens are microorganisms such as bacteria, fungi, viruses, viroids, phytoplasma, nematodes, and parasitic higher plants. These organisms inhabit and survive on plants as a host for their various requirements including nutritional requirements and sites for their reproduction and replication. These organisms can alter the regular function of plants such as growth, development, and reproduction (Chisholm et al., 2006). Eventually these organisms became a pathogen or parasite to the plants. Some pathogens have a broad host range or host specificity (Mikkelsen et al., 2005). Potential ability of a microorganism to cause disease in a plant as a pathogen is referred to as virulence. Avirulance is the condition where there is no harm to the plant's health by a microbe; in fact, there could be a symbiosis (Surico, 2013).

The plant-pathogen interaction can be represented as a triangle, where the susceptible host, virulent pathogen, and favorable environment are the three legs. Breaking one of these legs is the way to control plant diseases caused by biotic agents (Ravichandra, 2013). Most of these organisms act as saprophytes; they utilize the decaying part of plants or convert those materials into energy or food by various mechanisms. Most of the plant-pathogen interactions are due to nutritional requirements of the plant-pathogenic microorganisms. The pathogen invades a host plant through a direct method by enzymatic cleavage of cell wall and plasma membrane or invades through a natural opening such as stomata or through a wound; it then starts to weaken the plant's defense mechanism and induces the release of nutrients for their survival (Jibril et al., 2016). Biotrophs are the group of plant-pathogenic microorganisms that feed on living plant tissues and do not cause death to their host. *Blumeria graminis,* which causes a powdery mildew in grape vine and *Xanthomonas oryzae* in rice, is a good example of a biotrophic foliar pathogen. Necrotrophs are pathogens that produce toxins or release tissue-degrading enzymes, which devastate the plant's defenses and stimulates immediate release of nutrients. *Botrytis cinerea* and *Erwinia carotovora* cause gray mold and bacterial soft-rot, respectively, and are plant-pathogenic necrotrophs (Jibril et al., 2016). Some of the plant pathogens such as *Magnaporthe grisea* are good examples of hemibiotrophs, in that they invade a host plant as a biotroph and eventually become a nectrotroph during the last phase of disease development (Freeman and Beattie, 2008).

Filamentous and branched true fungi are parasites that cause many diseases in plants. Fungal pathogens are the most predominating phytopathogenic group (Ryder and Talbot, 2015). Phytopathogenic fungi are the plant pathogens that colonize on the surface or inside the plant as exopytes or endopytes, respectively. General necrosis or localized cell death is the most common symptom of plant-pathogenic fungi. In addition to that, stunting, distortions, and abnormal changes in plant tissues are some of other characteristic symptoms (Gagne-Bourgue et al., 2013). Availability of live or dead body parts of the fungi such as hyphae, mycelia, fruiting bodies, and reproductive structure such as spores on plant structures are recognized as signs. These signs are a disease diagnosis tool in fungal infestations (Jibril et al., 2016).

Viruses are submicroscopic infectious pathogens that are too small to be seen with a light microscope. Viruses are composed of nucleic acid and protein capsule. This nucleic acid or genetic material provides chemical information for a few proteins that are essential for completion of their life cycle. A virus needs a biosystem or biological system for reproduction or replication. Therefore, viruses are called obligate parasites. Viruses are host-specific, but they can use any type of living organisms as their host including animals, plants, fungi, and bacteria (Gergerich and Dolja, 2006; Jibril et al., 2016).

Plant-parasitic nematodes (PPNs) cause major yield losses in agricultural crops (Ali et al., 2017; Strange and Scott, 2005). PPNs are known to infect almost all economically important crops like rice, wheat, maize, soybean, potato, tomato, and sugar beets, and are responsible for more than 20% of the global yearly yield losses (Jung and

Wyss, 1999). Economically, these losses reflect to be more than $150 billion (Abad et al., 2008).

PPNs are small multicellular organisms, normally 300 to 1000 μm in length, but some nematodes are up to 4 mm long; the greatest diameter of their body ranges from 15 to 35 μm (Agrios, 2005a, b). Nematodes are microscopic in nature, invisible to the naked eye, but they can be observed easily even using the low power of a microscope (Agrios, 2005a, b); (Decraemer and Hunt, 1979). Nematodes are, in general, worm-like, unsegmented smooth bodies, without appendages. However, in some species, mature females are swollen and become pear-shaped or spheroid bodies (Agrios, 2005a, b). In the known 197 PPNs genera, more than 4300 species have been reported, which indicate 7% of the phylum Nematoda (Decraemer and Hunt, 1979).

PPNs use different genetic and molecular tools to infect plants, and their infection results in symptoms such as the formation of root galls, leaf necrosis, leaf chlorosis, stunted and patchy growth, possible wilting, and predisposition (Ali et al., 2019, 2015; Webster, 1969). Symptoms on the aboveground parts is difficult to identify, even that realistic nematode infestation occurred (McDonald and Nicol, 2005). Cyst nematodes (*Globodera* spp. and *Heterodera* spp.) and root knot nematodes (RKN) (*Meloidogyne* spp.) are economically important (Saeed et al., 2019). Sedentary endoparasitic nematodes are mostly responsible for causing the greatest damage to many crops (Tamilarasan and Rajm, 2013). While cyst nematodes are specific in host plants selection, RKN can infect a very wide range of economically important crops (Sikora et al., 2005). RKN and cyst nematodes induce a feeding site, which acts as a continuous source of nutrition to the sedentary nematode stages but in a different way such as "giant cells" developed by RKN and "syncytia" by cyst in the plant roots (Jones, 1981; Saeed et al., 2019).

23.4 Plant-pathogen interaction

Plant-microbial interaction is a potential area to study and realize the relationships between plant and microbes, whether beneficial or pathogenic. The relationships could be used in crop improvement programs to overcome the biotic effect on plant's growth and development (Imam et al., 2016). Association between plants and microbes are high drivers that flourish underneath the soil in the terrestrial environment as well as within plants as inter- or intracellular (Bulgarelli et al., 2013; Vorholt, 2012). Host specificity or single plant species provide or act as a host for a number of microbes or pathogens. As well as a single plant, species act as a host for a large number of microbes or pathogens. The specific nature of plant-microbial or pathogenic interactions created this diversity through evolutionary adaptation over millions of years. However, most of the occurrences of plants and microbial interactions could be a pathogenic relation (Imam et al., 2016). The degree of virulence of plant pathogens can deviate from normal to epidemic conditions resulting in severe yield loss in a crop production system and adversely affect global food security (Imam et al., 2014). These plant-microbial interactions have an effect on plant health. Therefore, it is important to understand the many hindrances in plant-microbial interactions to study specific signal transduction pathways in beneficial or pathogenic interactions (Imam et al., 2015).

The advancement in "omics" technologies such as sequencing technologies has improved research and development activities on plant-pathogen interactions. These researches revealed interesting molecular mechanisms in the interactions among plant and microbes or pathogens (Imam et al., 2016). Interaction among plants and disease-causing agents or pathogens has a genetic correlation through their evolution. Recent findings suggested that this phenomenon might be due to horizontal gene transfer through millions of years. To cause a disease or microorganism to be pathogen for a given plant, there should be three requirements: the plant should be susceptible; the microorganism should be virulent; and there must be favorable environmental conditions. Despite matching all these conditions, there would not be disease development. This representation generally is called a disease triangle. Microorganisms can be virulent or avirulant, or the plant can be susceptible or resistant under plant microbial interactions. Therefore, this situation divides microorganisms into three clusters: pathogenic or parasitic, mutualistic and commensalitic microbes. These interactions are genetically driven, and virulence and susceptibility in the host-pathogen interactions can be altered by even a single amino acid alteration in the regulatory proteins (Carroll et al., 2011).

Global climatic changes adversely affect crop production systems worldwide. Therefore, there is a necessity to optimize crop production systems for high yield in available limited fertile land. Disease dissemination among crop production systems has created a direct loss in the ultimate yield. Therefore, exploitation of interactions between plant and microbes is the most promising solution, and it would be a significant part in crop disease management (Reid, 2011). Recent advancements in researches on plant and microbe interactions in the field of the oldest symbiosis between plants and mycorrhizae, nitrogen fixation in plants, and pathogenesis have given fruitful information on common and diverged signaling network mechanisms in plant-pathogen interactions (Wirthmueller et al., 2013). These findings elucidate the potential resistant genes or QTLs against pathogens in one of the better crop cultivars or wild crop relatives. These findings will be utilized in the development of resistance cultivars through molecular breeding and genetic engineering approaches. Development of resistance against plant-pathogenic microorganisms limit the use of agrochemicals on crop protection and eventually support for the good health of environment (Gust et al., 2010). Microbiota interact with the plant in various ways, and these interactions need to be studied comprehensively to identify the physiological and metabolic changes created during these interactions. When plants reduce their immunity due to an imbalance of nutrient requirements, environmental conditions and genetic status may lead to the conversion of microbiota into pathobiota. Moreover, research and development activities relating to energy and resources use efficiency under pathogenic invasion to develop resistance, which needs to be studied comprehensively (Haggag et al., 2015).

23.5 Mechanisms involved in plant-pathogen interactions

Plant pathogens use diverse life strategies for the invasion process. Resistance and tolerance are the two main mechanisms described in plant-pathogen interactions. Resistance is the ability to bind pathogen multiplication by the host, and tolerance is the ability to reduce the effect of infection. The first part of this section discussed the terms related to plant-pathogen interactions such as biotrophs, necrotrophs, hemibiotrophs, virulence, avirulence, resistance, and susceptibility. The second part of the section discussed different plant pathogens such as fungi, fungal like organisms, bacteria, mollicutes, viruses, viroids, nematodes, and parasitic higher plants. Under the plant, the immune system mainly discussed the zigzag model of the plant's immune system and components of pattern-triggered immunity and effector-triggered immunity. At last, we shall discuss system-acquired resistance and RNA-based antiviral immunity, respectively.

Plant pathogens can classified as biotrophs, necrotrophs, and hemibiotrophs. Biotrophic pathogens feed on live tissues and do not threaten the life of the host plant (Ali et al., 2017; De Silva et al., 2016). The saprophytic bacterial species such as *Pseudomonas putida* that do not cause disease on any plant species are referred to as nonpathogens (Lee and Rose, 2010). Virulence is the ability of a microorganism to cause diseases in the plant, eventually called a plant pathogen, and its opposite nature is called avirulence (Ellis and Jones, 1998). Host resistance or susceptibility and virulence or avirulence nature of a pathogen is a predetermining characteristic in crop production systems. Single resistant gene differences could be the determining factor of the host plant, whether they have a resistant or susceptible genotype against the particular pathogen. Sometimes they may contain multiple resistance genes in a single species. Resistance genes show differences among each other according to the resistance gene specificity (Surico, 2013).

Parasitic plants can be classified into two classes: hemiparasites and holoparasites. Hemiparasites have a facultative parasitic nature. It receives water, minerals, and nutrients from their hosts but retain some of their photosynthetic ability. Holoparasites have little or no photosynthetic capability and deprive nutrients and water from the host plants (Poulin, 2011). The nematodes are one of most common soil-borne root pathogens, but a few species feed primarily upon shoot tissues. All plant-pathogenic nematodes live part of their life cycle in the soil and occur in greatest abundance in the top 15 to 30 cm of soil (Abtahi and Bakooie, 2017). Nematodes obtain food from plants with a specialized spear called stylet, and its size and shape is used to infer their mode of feeding and the classification.

Fungi and fungal-like organisms (FLOs) are responsible for a large range of serious plant diseases, more than any other group of plant pest. There are four main types of fungi appearing on plants: obligate pathogenic, obligate saprophytic, facultative pathogenic, and facultative saprophytic. Fungi can be beneficial as well as pathogenic to the plant (Agrios, 2005a, b). Mycorrhiza is a typical example of a symbiotic association between angiosperm and fungus. In addition to that, some other fungi are involved in biogeocycles and convert decaying materials into useful soil nutrients. Fungi and FLOs are contains lack chlorophyll and are able to overwinter in soil or on plant debris. They enter plants through natural openings such as stomata and through wounds. They produce specialized hyphal structures called appressoria, and through mechanical and enzymatic activity will directly lead to penetrate the plant tissues (Ryder and Talbot, 2015). They damage plants by killing cells and/or causing biochemical stress. *Magnaporthe oryzae, Botrytis cinerea, Puccinia* spp., *Fusarium graminearum, Fusarium oxysporum, Blumeria graminis, Mycosphaerella graminicola, Colletotrichum* spp., *Ustilago maydis,* and *Melampsora lini* are identified as most pathogenic fungal infectious organisms based on scientific/economic importance (Dean et al., 2012).

Bacteria are generally identified as highly pathogenic microorganisms to the plants. Rod-shaped bacteria (Bacilli) mostly infect plants (Aguilar-Marcelino et al., 2020). Plant-pathogenic bacteria (PPB) are classified into three families: Xantomonadaceae, Pseudomonaceae, and Enterobacteriaceae (Agrios, 2005a, b). Tropical and subtropical climatic regions in the world have the most detrimental effect from bacterial pathogens on its crop production systems (Ashbolt, 2004). *Pseudomonas syringae, Ralstonia solanacearum, A. tumefaciens, X. oryzae pv. Oryzae, Xanthomonas campestris, Xanthomonas axonopodis, Erwinia amylovora, Xylella fastidiosa, Dickeya (dadantii and solani)*, and *Pectobacterium carotovorum* are identified as the most pathogenic bacterial infectious organisms based on scientific/economic importance (Mansfield et al., 2012). Some specific pathogenicity factors, namely cell wall-degrading enzymes, exopolysaccharides, effector proteins, phytohormones, and toxins, have the ability to spread diseases among plants (Gagne-Bourgue et al., 2013). Gas or stomata openings, water or hydathodes pores, and wounds are the basic openings of the plant that facilitate bacterial invasion, and multiplication takes place in the intercellular apoplast spaces (Mansfield et al., 2012).

The mollicutes are very small, Gram-positive, wall-less bacteria that must be examined by an electron microscope. *Spiroplasma, Mycoplasma,* and *Acholeplasma* classify mollicutes. The mollicutes are phloem-restricted pathogens (spiroplasmas, mycoplasma-like organisms) or surface contaminants (*Spiroplasma* spp., *Mycoplasma* spp., *Acholeplasma* spp., *Phytoplasma* spp.) (Bové, 1988). The plant-pathogenic mollicutes are transmitted by insect vectors (Perilla-Henao and Casteel, 2016).

All viruses are obligate parasites that totally depend on the host's cellular machinery, and a host's range of viruses vary from very narrow to very broad. Plant susceptibility or resistance to viruses is determined primarily by the plant genotype. Plants can prevent virus infection both actively and passively. The failure of the production of one or more host factors required for reproduction of virus leads to developing passive defense against plant-pathogenic viruses. In contrast, activation of a defense responsive gene in a plant may result in active defense against pathogenic viruses (Ding, 2010).

There are two branches of the plant immune system (Jones and Dangl, 2006). Pattern-triggered immunity (PTI) is one branch, and it uses transmembrane pattern recognition receptors (PRRs) that respond to slowly evolving microbial or pathogen-associated molecular patterns (MAMPS or PAMPs), such as flagellin (Boller and Felix, 2009). Disease resistance (*R*) protein-mediated effector-triggered immunity (ETI) is the second one (Jones and Dangl, 2006; Spoel and Dong, 2012). There are similarities among animals and plants that correspond to the identification of pathogens. In animals, the first line of defense against infectious disease is innate immunity mediated by the Toll-like receptor (TLR) family on membranes of leukocytes including macrophages and dendritic cells, which are based on the recognition of structurally conserved molecules derived from microbes. The innate immunity of plants is mediated by the surface receptors containing transmembrane proteins in the plasma membrane.

Jones and Dangl (2006) suggested the four-phased "zigzag" model (Fig. 23.1) for the plant immune system. In phase 1, plants recognize PAMPs, DAMPs, and viral dsRNA, resulting in PAMP-triggered

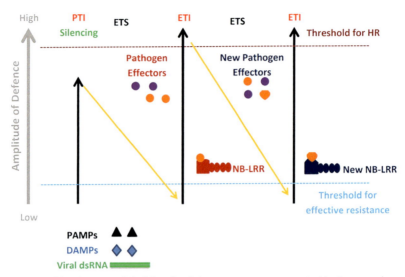

Fig. 23.1 The zigzag model of the plant's immune system presented in Jones and Dangl (2006).

immunity (PTI) that can prevent further colonization. In phase 2, successful pathogen effectors that interfere with the PTI enable pathogen nutrition and dispersal, or otherwise develop pathogen virulence, resulting in effector-triggered susceptibility (ETS). In phase 3, NB-LRR proteins specifically recognize a particular effector directly or indirectly and activate the effector-triggered immunity (ETI). ETI is an accelerated and amplified PTI response, resulting in disease resistance and, usually, a threshold for hypersensitive cell death (HR). In phase 4, natural selection drives the pathogen's capability to avoid ETI by acquiring additional effectors. Coevolution results in new plant NB-LRR alleles that can recognize one of the newly acquired effectors, resulting again in ETI (Jones and Dangl, 2006).

The PTI is typically activated by cell surface-localized transmembrane pattern recognition receptors (PRRs) of conserved molecules characteristic to different pathogen classes called either microbe-associated or pathogen-associated molecular patterns (MAMPs/PAMPs). Some PRRs recognize host-derived "danger" signals (damage-associated molecular patterns; DAMPs) released during infection. The lipopolysaccharide (LPS) envelope, peptidoglycans (PGN), flagellin, bacterial elongation factor (EF), methylated bacterial DNA fragments, fungal cell wall-derived glucans, chitins, and proteins, and quorum-sensing factors are examples of PAMPs (Boller and Felix, 2009; Nürnberger and Kemmerling, 2009). Plant peptides or cell wall fragments released during infection or wounding are examples of DAMPs. PAMPs or DAMPs act as ligands that bind to PRRs and activate PRRs to trigger physiological changes resulting in PAMP-triggered immunity (PTI). Examples of PTI initiate physiological changes are bursts of calcium and reactive oxygen species (ROS) and activation of mitogen-associated and calcium-dependent protein kinases (MAPKs and CDPKs), leading to massive transcriptional reprogramming. Mutualistic symbiotic organisms and beneficial microbes also need to overcome PTI for colonization in their hosts.

Receptor-like kinases (RLKs) and receptor-like proteins (RLPs) are two main classes of PRRs. Other than functional classification, PRRs can be classified (Fig. 23.2) based on their variability of extracellular domains, such as leucine-rich repeat (LRR), lysine motif (LysM), lectin, and epidermal growth factor (EGF)-like domains, and thus can recognize a wide range of ligands.

Plant RLK is analogous to animal receptor tyrosine kinases. RLKs contain three main domains: an ectodomain (ECD), a transmembrane domain, and a cytoplasmic kinase domain. But RLPs lack a cytoplasmic kinase domain with respect to RLKs. The highly variable ECDs are important for the recognition of wide range of ligands. Most defense-related RLKs and RLPs are triggered by non-self-molecules.

FLS2 can recognize bacterial-different PAMP motif within flagellin (flg22, flgII-28) (Gómez-Gómez and Boller, 2002). Flagellin is the main component of the bacterial flagellum, and flagellum-based motility is important for bacterial pathogenicity in plants (Zipfel and Felix, 2005). FLS2 ligand recognition specificities vary in different plant species and are not restricted to defense responses triggering. For example, *Arabidopsis* FLS2 (*At*FLS2) recognizes flg22 of plant

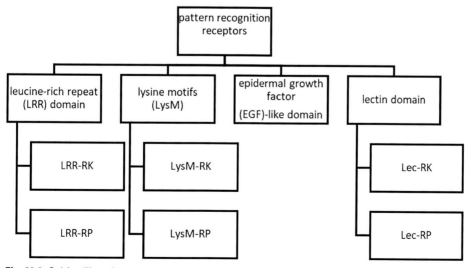

Fig. 23.2 Subfamilies of pattern recognition receptors (PRR).

growth-promoting *Burkholderia phytofirmans* (Chinchilla et al., 2006). *At*FLS2 rapidly forms heterodimers upon ligand binding with another LRR-RK coreceptor such as BAK1 or SERK and Brassinolide (BL) receptor BRI1 (Heese et al., 2007). Ligand-induced receptor complex formation depends on the presence or absence of ligand. For example, BIR2 acts as a negative regulator of PAMP-triggered immunity by limiting BAK1-receptor complex formation. In the mutants and absence of ligands, BIR2 poses a differential impact on BAK1-regulated processes, such as hyperresponsiveness (HR) that enhances cell death and resistance to pathogens (Halter et al., 2014).

Highly conserved bacterial elongation factor Tu (EF-Tu) acts as a PAMP in *Arabidopsis thaliana* and other Brassicaceae members. As a response to pathogenic bacterial infection in *Arabidopsis* plants, *At*EFR binds elf18, thereby initiating AtEFR-AtBAK1 heterodimerization and inducing an oxidative burst and biosynthesis of ethylene (Kunze et al., 2004). *At*FLS2 and *At*EFR are considered very similarly functioning LRR-RKs. AtPEPR1 and AtPEPR2 both recognize endogenous AtPep1 peptide that is usually categorized as DAMP, which involves microbial infection, wounding, and herbivores (Greeff et al., 2012). PEPR1 possesses cGMP production and catalytic function in vitro. This cGMP triggers cyclic nucleotide-activated Ca2+ channels (Ali et al., 2007).

Tomato plant (*Solanum lycopersicum*) *Sl*Eix1 and *Sl*Eix2 bind *Trichoderma* fungal cell wall-derived ethylene-inducing xylanase (Eix) and form heterodimers in a ligand-dependent manner. Other than expression of pathogen-related (PR) proteins as defense responses, Eix induces ethylene biosynthesis, ROS, medium alkalinization, and hypersensitive response (Sharfman et al., 2011). *Sl*BAK1 interacts constitutively with *Sl*Eix1, but not with *Sl*Eix2, and attenuates *Sl*Eix2 signaling while only *Sl*Eix2 mediates defense responses (Bar et al., 2010). The *Sl*Ve1 immune receptor governs resistance to vascular wilt fungi strains *Verticillium dahliae* and *Verticillium albo-atrum* in a *Sl*BAK1-dependent manner (Castroverde et al., 2016; De Jonge et al., 2012). Pathogenic microbes release a number of effector proteins into the plant cytosol or apoplast and use a variety of mechanisms for infection and colonization.

Plant membrane RLKs and RLPs containing extracellular lysin motif (LysM) domain proteins serve as pattern recognition receptors of different *N*-acetylglucosamine (GlcNAc)-containing ligands. Mainly, there are three types of glycans that contain GlcNAc-ligands molecules produced by microorganisms important for the perception: bacterial peptidoglycans, rhizobacterial Nod factors (NFs), and fungal chitin (Buendia et al., 2018; Gust et al., 2012). These receptors are characterized functionally associated with the hydrolytic breakdown of GlcNAc-containing microbial glycans and activated different plant responses after microorganism perception leading either to the establishment of beneficial symbiosis or defense responses against pathogens. LysM domain is found in most living organisms, except the *Archaea* (Zhang et al., 2009).

A plasma membrane glycoprotein chitin elicitor receptor kinase 1 (CERK1) is the best studied *Arabidopsis* LysM-RLK (Greeff et al., 2012). This gene is important for the induction of all chitooligosaccharide-responsive genes and leads to resistance against fungal pathogens. LysM CERK1 may not function alone in chitin perception, and like many other receptors it also requires a partner protein, but unlike FLS2 and EFR, CERK1 is BAK1-independent. CERK1 is a unique mediated chitin signaling pathway, but it may share a downstream pathway with the FLS2/flagellin- and EFR/EF-Tu-mediated signaling pathways (Wan et al., 2008). Receptor protein chitin elicitor binding protein (CEBiP) is a critically important component for chitin signaling in rice. CEBiP contains extracellular LysM domains but lacks an intracellular signaling domain. Rice cells require LysM-RLK and OsCERK1, in addition to CEBiP, for chitin signaling in a ligand-dependent manner (Shimizu et al., 2010). LysM domain receptor proteins *Arabidopsis* LYM1 and LYM3 are structurally similar to rice CEBiP and physically interact with different PGNs bacteria but not chitin recognition. However, LYM1 and LYM3 proteins respond to fungus chitin (Willmann et al., 2011).

Pseudomonas syringae pathogen can destabilize *Arabidopsis* PRRs AtCERK1 by utilization of the effector protein that leads to dampening basal immune responses triggered on chitin. The fungal pathogen *Cladosporium fulvum* on tomato secretes the LysM-domain-containing effector protein extracellular protein 6 (Ecp6), which competes for ligand binding with plant chitin receptors and blocks chitin-inducible plant defenses (Sánchez-Vallet et al., 2013). The fungi *Mycosphaerella graminicola* and *Magnaporthe oryzae* also have Ecp6 homolog proteins. Ecp6-like LysM effector proteins are widely distributed throughout the fungal kingdom (Gust et al., 2012).

Lectins are proteins of nonimmune origin that contain at least one noncatalytic domain that binds to specific free sugars or glycans present either in a free form or as part of glycolipids and glycoproteins without altering the structure of the carbohydrate. These carbohydrate-binding lectin proteins have been identified in viruses, fungi, and bacteria to animals and plants (Lannoo and Van Damme, 2010). Plant lectins originate from the organisms or from damaged plant cell wall structures, which are very diverse with respect to carbohydrate recognition domains, protein structures, and glycan binding specificities (Van Dammes et al., 2011). Some RLPs and RLKs contain an extracellular lectin domain (Vaid et al., 2012).

Plasma membrane lectin receptor-like kinases (LecRLKs) found in higher plants are involved in stress tolerance and other diverse functions as plant growth and development. LecRLK proteins contain three main domains: a lectin domain, transmembrane region, and a cytoplasmic Serine/Threonine kinase domain. There are three types of lectin receptor kinases G, C, and L based on conserved motifs within the lectin domain. Plasminogen apple nematode motifs (PAN) and EGF motifs are two examples of conserved motifs (Lv et al., 2020). G-type lectin receptor kinases are involved in self-incompatibility in flowering plants (Sherman-Broyles et al., 2007), while C-type lectin motifs can mediate innate immune responses but are rare in plants (Cambi et al., 2005). L-type lectin receptor kinases are widespread in vascular plants, and *Arabidopsis* contains 36 out of the 45 currently identified L-type lectin receptor kinases (Wang and Bouwmeester, 2017). *Arabidopsis* L-type lectin receptor kinases incorporated within the cell wall-associated defense showed increased susceptibility to the bacterium *P. syringae* and the oomycete pathogens *Phytophthora brassicae* and *P. capsici* (Balagué et al., 2017). Some L-type lectin receptor kinases were found to associate with FLS2, emphasizing its bacterial immunity and some closed stomata that leads to elevation of oxidative burst and resistance to bacterial pathogens (Desclos-Theveniau et al., 2012).

Pattern recognition of invading organisms leads to a series of cellular and physiological responses, which collectively contributes to plant resistance. Cellular and physiological responses triggered by patterns are calcium influx, plasma membrane depolarization and extracellular alkalinization, apoplastic ROS bursts, nitride oxide production, phosphatidic acid production, actin filament remodeling, and stomatal closure (Yu et al., 2017).

Calcium ions are a major secondary messenger molecule in plants under different stress conditions and a component of various structures in cell wall and cellular membranes (Kader and Lindberg, 2010). The increase of Cytosolic Ca^{2+} concentration ($[Ca^+]_{cyt}$) is a common cellular response generated due to the recognition of different MAMPs or DAMPs. The $[Ca^+]_{cyt}$ increases due to Ca^+ mobilization from organelles and/or the influx of Ca^{2+} from the extracellular space. The calcium-dependent NADPH oxidase activation results in hydrogen peroxide (H_2O_2), which may activate calcium channels from the plasma membrane and increase $[Ca^{2+}]_{cyt}$. The $[Ca^{2+}]_{cyt}$ increase leads to activation of mitogen-activated protein kinase, microtubule depolymerization, defense gene activation, and cell death (Lecourieux et al., 2002). flg22, fungal cryptogein, and β-glucan trigger $[Ca^{2+}]_{cyt}$ increase, but chitin induce $[Ca^{2+}]_{cyt}$. PGN weakly triggers a $[Ca^{2+}]_{cyt}$ increase than flg2. DAMPs also induce a typical $[Ca^{2+}]_{cyt}$ (Yu et al., 2017). Activation of FLS2 and EFR leads to opening of the BAK1-dependent calcium-associated plasma membrane anion channels. It is the initial step of the pathogen defense pathways (Jeworutzki et al., 2010).

PAMP perception induces rapid Ca^{2+} and H^+ influxes and Cl^-, NO^{3-}, and K^+ effluxes across the plasma membrane, which leads to membrane depolarization and extracellular alkalinization (Bolwell et al., 2002). Depolarization and extracellular alkalinization in *Arabidopsis* were recorded after flg22 or elf18 treatment. Extracellular alkalinization induced by PGN is slower than flg22 but is more persistent (Yu et al., 2017).

The flg22 treatment triggers a higher peak value of Cl^- efflux than of H^+ in plasma membrane depolarization due to the involvement of anion channels. Plasma membrane H^+-ATPases are important for the establishment of cellular membrane potential in plants (Elmore and Coaker, 2011). Plasma membrane H^+-ATPases cooperate with the signaling protein RPM1-Interacting4 (RIN4) to regulate stomatal structures during bacterial invasion such as *P. syringae* on leaf tissue, which directly regulates guard cell turgor pressure (Zhou et al., 2015). Plant integrin-linked kinase 1 (ILK1) positively regulates flg22-induced plasma membrane depolarization and activation of cation channels required for K^+ efflux. These ions channels are activated by Ca^{2+}-dependent ROS and hydroxyl radicals in a pathogen attack, resulting in dramatic K^+ efflux. The K^+ loss simulates cytosolic proteases and endonucleases, leading to programmed cell death (Demidchik, 2014).

ROS play an important role in plant signaling as a response to invading microbial pathogens. ROS include singlet oxygen ($^1O_2$), superoxide anion ($O_2\cdot^-$), H_2O_2, and hydroxyl radicals ($\cdot OH$). ROS production in the apoplast starts as a response to the activation of plasma membrane immune receptors. The apoplastic ROS are produced by cell wall peroxidases, plasma membrane-localized NADPH oxidases (respiratory burst oxidase homologs; RBOHs), and amine oxidases in plants. Activation of these receptors increased cellular responses including extracellular alkalinization, ion channel activities, a transient increase of cytosolic Ca^{2+}, activation of calcium-dependent protein kinases (CPKs), a transient ROS burst, and activation of MAP kinase (MAPK) cascades (Kärkönen and Kuchitsu, 2015). According to research-based results, the ROS burst is initiated within about 4 to 6 min after PAMP treatments in various plant species, reaches its peak in approximately 10 to 15 min, then gradually deceases to a resting state after 30 min (Yu et al., 2017).

In the preactivation state, FLS2 is associated with two closely related cytoplasmic protein kinases: BIK1 and PBL1. The perception of ligand flg22 with FLS2 recruits coreceptor BAK1 and leads to transphosphorylation of FLS2/BAK1 and BIK1/PBL1. Activated complexes lead to transient bursts of cytosolic calcium, apoplastic ROS, and disease resistance to both bacterial and fungal pathogens. RBOHDs activation depend on intracellular $[Ca^{2+}]$ (Podgórska et al., 2017). ROS and Ca^+ are important for the cell-to-cell, long-distance propagation calcium waves and ROS waves that mediate systemic signaling during pathogenic attack as well as other biotic and abiotic stresses. Antioxidants such as glutathione, ascorbate, and tocopherol form redox buffers important for the determination of the lifetime and specificity of the ROS signal and nitric oxide (NO) (Qi et al., 2017). To conclude, ROS participate by strengthening plant cell walls by oxidative cross-linking of polymers, acting as versatile signaling molecules, making a toxin barrier against subsequent pathogen infections, and mediating multiple responses.

Plant PRR and R proteins mediate signaling pathways that rely on rapid production of reactive nitrogen species (RNS). NO mainly mediates biological function through accumulation of RNS. NO is produced from nitrite through enzymatic reduction. Nitrate reductase (NR) and nitrite-NO reductase (Ni-NOR) enzymes in root cells are the two enzymes that participate in this catalytic function. Other than reductive NO biosynthetic pathways, there are oxidative NO biosynthetic pathways such as the nonidentified nitric oxide synthase (NOS)-like enzyme and NO production from polyamines (PAs) mediated by enzymes. Copper amine oxidase (CuAO) is a very good example of the PA (Bellin et al., 2013). PAMP-induced NO production is connected to Ca^{2+} signaling, because Ca^{2+} channel blockers inhibit PAMP-induced NO production (Ma et al., 2008). NO, a free radical gas, can cause stomatal closure, rapid burst, and act as secondary messenger and antimicrobial agent. Under different stress conditions, NO can perform both cytotoxic and protective functions by signaling mechanisms and interaction with ROS. The cytotoxic and protective properties of NO could depend on the concentration of NO, the type of cells, and time of exposure. NO burst is detected within a few minutes upon treatment with PAMPs such as flg22, PGN, cryptogein, xylanase, and LPS, or with DAMPs such as eATP and AtPEPs (Scheler et al., 2013).

S-nitrosylation is the covalent attachment of a nitric oxide group (−NO) to cysteine thiol within a protein to form an S-nitrosothiol (SNO). S-nitrosylation has diverse regulatory roles in bacteria, yeast, plants, and all mammalian cells. NO-mediated physiological functions are executed via protein S-nitrosylation reactions, which function as a fundamental mechanism for cellular signaling (Feng et al., 2019). *Arabidopsis thaliana* salicylic acid binding protein 3 (AtSABP3) has SA binding properties, and carbonic anhydrase (CA) activity required to develop resistance against pathogens undergoes S-nitrosylation. Therefore, AtSABP3 protein is one of the first targets for S-nitrosylation in plants by most pathogens (Wang et al., 2009).

Phosphatidic acid (PA) is a type of phospholipid incorporated in lipid biosynthesis and is important as a key cellular signaling lipid molecule. In plants, PA formation is triggered by various biotic and abiotic stress factors, such as pathogen infection, drought, salinity, wounding, and cold, and PA signal production is fast (Yunus et al., 2015). PA is involved in MAP kinase (MAPK) activation, defense gene induction, regulation of ROS production, and actin remodeling (Testerink and Munnik, 2011).

PA is mainly present at the membrane. Most *Arabidopsis* phospholipase D (PLD) is localized to the plasma membrane, whereas some are present in vesicles, intracellular membranes, or the tonoplast (Yao and Xue, 2018). Two distinct PA biosynthetic pathways exist in plants. The first one is direct hydrolysis of structural phospholipids by PLD to PA. The second pathway is a combination of phospholipase C (PLC) and diacylglycerol kinase (DGK) to phosphorylate diacylglycerol (DAG) to PA (Testerink and Munnik, 2005). In general, pathogenic elicitors such as the flg22 and chitosan activate the PLC-DGK pathway and induce PA production in tomato plants. In rice cells, both PLC and PLD activities are elevated upon chitosan ligand treatment. PA undergoes cross-function with other signaling molecules. For example, NO burst is required for PA production, and inhibition of either PLC or DGK activity decreases ROS production. PA or DAG treatment directly induces ROS production in rice cells, but PA stimulates RBOHD activity to increase ROS production (Yu et al., 2017).

Stomata are microscopic pores on the leaf epidermis formed by two guard cells, which regulate water transpiration and CO_2 uptake by leaves. Stomatal guard cells contain various receptors to sense different abiotic and biotic stresses. For example, stomatal openings are a major route for pathogen entry. In infection, plants recognize PAMPs, and this leads to stomatal closure (Agurla et al., 2018). Abscisic acid (ABA) is a stress hormone in plants that accumulates under different abiotic and biotic stresses. A typical example of the effect of ABA on leaves is to defend against pathogens by restricting their entry by closing stomata pores (Bharath et al., 2021).

The bacterial flg22 triggers rapid stomatal closure after about 15 min. in guard cells of *Arabidopsis* leaves. The Open Stomata 1 (OST1) mutants lack a key ABA-signaling protein kinase but remained flg22-responsive when linked to the activation of the two S-type anion channels, SLAC1 and SLAH3. Those anion channels function as the main switches for PAMP-induced stomatal closure. They release anions into the stomatal guard cell wall to depolarize the plasma membrane. This depolarization forces out K^+ through K^+ efflux channels and as a result shrinking, and so the stomatal pores close. The initial steps in flg22 and ABA signaling pathways are different, but they merge at the level of OST1 (Melotto et al., 2006; Roelfsema et al., 2012).

The eukaryotic actin cytoskeleton is required for numerous cellular processes, including cell rigidity, cytoplasmic streaming among different compartments, development and movement, gene expression, and signal transduction as a response to biotic and abiotic stresses. It is the key signaling network in plants that mediates cellular homeostasis and the activation of specific signaling as a response to pathogen perception (Porter and Day, 2016). However, with the improvement of plant resistant-associated PTI, most adapted pathogens produce various effector proteins, and they increase susceptibility of the plant by inhibiting the activity of protein-mediated PTI (Jones and Dangl, 2006). Therefore, plants have developed a second layer of innate immunity called effector-triggered immunity (ETI). ETI is developed by polymorphic intracellular receptors containing nucleotide-binding (NB) and leucine-rich repeat (LRR) domains. Effector recognition occurs at both the plasma membrane and cytoplasm. R proteins directly or indirectly recognize pathogen effectors from diverse kingdoms and activate defense programs often in localized programmed cell death (PCD) at the infection site, also called hypersensitive response (HR), which locally limits pathogen spread (Khan et al., 2016). R proteins mediate immunity against obligate biotrophic or hemibiotrophic pathogens but not against necrotrophic pathogens.

There are different resistance genes in viral, bacterial, fungal, nematode, and insect pathogens. Resistance gene-encoded proteins contain at least three core domains: a C-terminal leucine-rich repeat domain (LRR), a central nucleotide binding site (NBS) domain, and an N-terminal domain that either homologous to the animal interleukin-1 receptors (TIR) or a potential coiled-coil (CC) domain. Therefore, there are two types of resistant proteins (Fig. 23.3) based on the variability of N-terminal domain: coiled-coil (CC)-NB-LRR proteins or toll/interleukin 1 receptor-like (TIR)-NB-LRR proteins. Some R proteins have a C-terminal extension as a WRKY transcription factor domain following the LRR at the C-terminal (Wu et al., 2014).

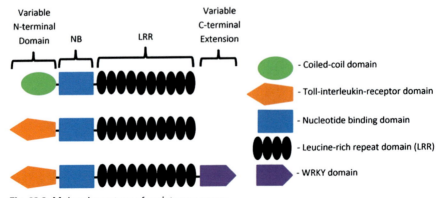

Fig. 23.3 Molecular nature of resistance genes.

PTI initiates basal disease resistance for the fundamental infection but adapted pathogens that secrete effectors to destabilize PTI and manipulate host metabolism (Effector-Triggered Susceptibility; ETS). During coevolution of plant and pathogens, plants evolved resistance proteins to detect pathogen effectors or their activity (Effector-Triggered Immunity, ETI) (Takken et al., 2006). According to this situation, basal disease resistance can be defined by the following equation:

$$\text{Basal disease resistance (innate immunity)} = (\text{PTI} + \text{ETI}) - \text{ETS}.$$

ETI can induce the production of mobile immune signals in plants, such as salicylic acid (SA), methyl salicylic acid (MeSA), azelaic acid (AzA), glycerol-3-phosphate (G3P), and abietane diterpenoid dehydroabietinal (DA). These mobile immune signals are transported from the infection site to uninfected tissues, and this induced immune mechanism is called systemic acquired resistance (SAR). SAR is the broad-spectrum plant disease resistance against pathogenic bacteria, fungi, oomycetes, and viruses with no specificity to the initial infection induced after a local avirulent pathogen infection (Gao et al., 2015; Park et al., 2007). These chemical signals then lead to expression of the antimicrobial pathogenesis-related (PR) genes in the uninfected tissue, which protects plants from subsequent pathogen attack (Durrant and Dong, 2004). SAR is also induced by the defense hormone SA or SA synthetic analogs benzothiadiazole S-methyl ester (BTH) and 2,6-dichloroisonicotinic acid (INA). SAR can be introduced as immune "memory" in plants, which can last for weeks to months, and possibly even through the whole growing season (Kothari and Patel, 2004).

The ETI can trigger SAR through synthesis of signal molecules as SA. SAR signaling is associated with the ETI-induced changes in amino acid homeostasis. Pathogenic microbial infections lead to the biosynthesis of a large number of phenylalanine derivatives as phenolic compounds. Transcription factor TBF1 (TL1-binding factor 1) is rapidly translated from its mRNA upon pathogen challenge as the earliest triggering responses for SAR (Saijo et al., 2009).

The pyridine nucleotides NAD (NAD^+ and NADH) and NADP ($NADP^+$ and NADPH) mediate various fundamental biological processes such as calcium homeostasis, energy metabolism, generation of oxidative stress, immunological functions, gene expressions, aging, and cell death. Plant invaders induce the hypersensitive response of plants and, as a result, NAD^+ and $NADP^+$ leaked into the intercellular fluid at concentrations sufficient to induce PR gene expression. This process depends on Ca^{2+} signaling and SA production (Lee, 2001). Mobile immune signals such as abietane diterpenoid dehydroabietinal, methyl salicylic acid, azelaic acid, and glycerol-3-phosphate (G3P) are incorporated with the regulation of the expression of different genes, which encodes different proteins important for major plant physiological activities (Fu and Dong, 2013).

In plant RNA-based antiviral immunity, plants detect viral double-stranded RNA (dsRNA) to trigger RNA silencing. Viral dsRNA is recognized as a PAMP by plant PRRs and is processed into small interfering RNAs (siRNAs) by the plant cell ribonuclease Dicer. Then viral siRNAs are loaded in plant Argonaute (AGO) proteins to lead viral mRNA cleavage (Couto and Zipfel, 2016; Han, 2019). Many viral suppressors of plant viral dsRNA silencing pathways target the RNA components or the protein components. The activity of viral suppressors of RNA silencing contributes to ensure productive virus replication in plant hosts. Plant geminiviral V2 protein is an example of a viral suppressor. It may compete for binding to 5′ overhang-containing dsRNA with an *A. thaliana* protein SGS3, which is essential for secondary siRNA synthesis. RNases also act as viral suppressors of viral dsRNA silencing. For example, plant Crinivirus-encoded RNase3 blocks RNA silencing by degrading siRNAs. Plant Cucumoviral 2b protein binds to both Argonaute 1 (AGO1) and siRNAs that might explain the inhibition of synthesis of viral secondary siRNAs (Ding, 2010).

The molecular mechanisms in plant-pathogen interactions can be successfully identified by genetic engineering technologies, for example, Calvitti et al. (2012) established an *Aedes albopictus*-Wolbachia symbiotic association by artificial transfer of the wPip strain from *Culex pipiens* for effective pest control. Although some of the pathogen genomes are relatively small and can be sequenced with much ease, mapping entire genomes of the host species for every occurrence of modification in host-pathogen interaction can be a challenging task. However, sequencing the exomes can serve as a cost-effective unbiased alternative and can provide valuable information on drafting strategies to combat disease pathogenesis. Use of transcriptomic technologies with advanced tools can be used to identify molecular mechanisms involved in host pathogen interactions (O'Brien et al., 2011; Venu et al., 2011).

Twenty-five new strains of plant-pathogenic *Pseudomonas syringae* have been identified by next-generation sequencing technology (Lindeberg et al., 2008). Recent studies revealed that *P. syringae* genomes are highly dynamic, and extensive polymorphism is found in the distribution of type III secreted effectors (T3SEs) and other virulence-associated genes, even among strains within the same pathovar (Hu et al., 2010). Identifying the mechanism involved in plant-pathogen interactions to prolong development in sequencing technologies is very important. Therefore, addressing the disadvantages in existing sequencing technologies is important (Hu et al., 2010).

Exome sequencing technology will overcome the problems associated in assembling the whole genome to identify the virulence mechanisms in plant-pathogen interactions (Calvitti et al., 2012).

The rhizosphere is the most populated area of microbes that interact with plants. Plant-arbuscular mycorrhizal (AM) symbiosis is of prime importance. These beneficial interactions can improve plant growth and development even under biotic and abiotic stress conditions while releasing plant exudates and microbial secretions into the root zone. Yet the molecular mechanisms involved in these complex plant root-soil microbiota interactions remain unfamiliar (Balasubramanian et al., 2021).

23.6 Evolution of plant-pathogen interactions

Plant-pathogen interaction is a complex process involving pathogen- and plant-derived molecules. With an advancement by one partner, for example, due to increased pathogen virulence, strong selective pressure is placed on the plant host to increase or change its specific defense response. The pathogen will face increased selective pressures to make compensatory improvements if the host resistance is successful after it changes. Therefore, pathogen characteristics such as pathogenicity, virulence, and plant defense evolved with time (Anderson et al., 2010). Some pathogens have evolved a necrotrophic lifestyle, which feeds on dead or dying cells, while others have evolved a biotrophic lifestyle, which involves feeding on live plant tissue, whereas others became hemibiotrophs, with a biotrophic phase and a necrotrophic phase (Glazebrook, 2005).

A defense mechanism in plants is PAMP-triggered immunity (PTI) where nonspecific defenses are induced due to the recognition of microbial-derived PAMPs by PRRs of the plants. Perception of bacterial flagellin by plant PPR, called FLS2, is one example of a well-known PAMP-PRR system. All higher plants that are sequenced have FLS2 homologs conserved in their genome (Boller and He, 2009). Hence it is evident that PTI seems to have evolved from early times. In *A. thaliana*, EFR is a pattern-recognition receptor (PRR) that binds to the bacterial protein EF-Tu. This EFR and EF-Tu involved PTI system is only found in Brassicaceae, which shows the continuation of evolution of the PTI system in plants (Zipfel et al., 2006).

Several factors are involved in the evolution of plant pathogens, which are mutations, sexual recombination, lateral gene transfer, whole genome exchange, and chromosomal instability. Mutation is the main factor responsible in originating a majority of new pathotypes of the fungus *Puccinia graminis* (Burdon and Roelfs, 1985; Watson, 1981). These new pathotypes differ from previous types through point or small deletions, or the insertion of transposable elements occurred in avirulence genes. From existing pathogenic variations, sexual recombination generates a wide variety of new virulence combinations. Pathogens with sexual recombination ability provide the most successful evolutionary responses to evolving tolerance patterns in plant populations due to the new allelic combinations acquired (McDonald and Linde, 2002). Bacteria are highly adaptable in exchanging genetic material among organisms with broad genetic distances. Fungi have a similar potential for gene transfer to bacteria, but to the some degree (Ochman and Moran, 2001). Some fungi plant pathogens contain virus-like particles or double-stranded ribonucleic acids that reduce pathogenicity. These are the results of lateral gene transfer in fungi. These dsRNA viruses are the cause that change hyper- to hypovirulence in a number of fungal pathogens (Boland, 1992). Lateral gene transfer is identified in *Pyrenophora tritici-repentis*, a fungi causing yellow spot in wheat where its toxin-producing sequence (ToxA) is transferred from a separate pathogen, *Septoria nodorum* (Oliver and Solomon, 2008). Combining two, although very different, strains of the same pathogen result in completely new lineages with markedly different virulence spectra on established hosts occurs due to whole genome transfer. This new pathogen evolved to have a broad range of hosts within the parent host range, for example, a combination of *Melampsora medusae* and *Melampsora larici populina*, producing a novel *Melampsora* species (Spiers and Hopcroft, 1994).

Pathogens have also evolutionary changes to manufacture plant hormones or control host hormone biosynthesis to inhibit plant defense responses and spread disease. Phytopathogens use a variety of methods to change endogenous hormones of hosts. Pathogen-derived biosynthetic genes for cytokinin and auxin are found in galls induced by *A. tumefaciens* and are regulated by plant-derived promoters incorporated into the plant genome (Zambryski, 1992). Effectors produced by pathogens have also evolved with time to change the host plant hormone levels. For example, the type III effector AvrRpt2 of *P. syringae* has been shown to modify auxin physiology in the host, and elevated auxin levels suppress plant defenses and encourage disease (Chen et al., 2007).

Each gene for resistance in the host has a corresponding gene for pathogenicity in the pathogen (Flor, 1971). Pathogenicity and virulence genes in pathogens that have high mutation rates could suppress the host plant resistant genes. If there are no resistance genes in the host population, there is no selection pressure on the pathogen population (Crill, 2001). Virulency is described as the degree of damage caused to a host by pathogen infection, and pathogenicity is a parasite's ability to infect a host and cause disease. Both virulence and pathogenicity are subject to selection, and changes in pathogen populations can influence pathogen evolution and host-pathogen coevolution (Sacristan and Garcia-Arenal, 2008).

The identification of conserved pathogen-associated molecular patterns, which activates basal defenses, is one defense approach of the plant. Another line of defense is the detection of pathogenicity effectors that may have evolved to inhibit PTI (Chisholm et al., 2006; Jones and Dangl, 2006). Two major models of host-parasite interaction are the gene-for-gene (GFG) and the matching-allele (MA) models. According to the gene-for-model, R proteins in plants identify corresponding avirulence (Avr) factors that are proteins produced in pathogens (Jones and Dangl, 2006). The host identification of the Avr factor activates defense responses, which restrict the pathogen's spread from the infection site, frequently linked to localized host cell death or hypersensitivity (HR). If the pathogen lacks the Avr allele, or if the host lacks the resistance R allele, the parasite is not recognized by the host, resistance is not caused, and the host is infected. Pathogens changing their Avr factors to prevent R-dependent identification, and hosts developing new R protein specificities to recognize the corresponding Avr factors, is the coevolution of plant and pathogen according to the gene-for-gene module (Parker and Gilbert, 2004; Thrall et al., 2001). In the MA model, to initiate an infection, a precise genetic match between the parasite and the host is necessary (Agrawal and Lively, 2002). Plant-pathogen interaction that aligns with this model is the plant fungal interaction where host specific toxins (HST) act as avirulent factors to the host. These HSTs need specific target sites (Wolpert et al., 2002). The wheat pathogens *Pythium tritici repentis* and *Phaeosphaeria nodorum* produce the HST ToxA, which interacts with Tsn1 and leads to necrosis. The ToxA protein-encoding gene in emergent virulent strains of *P. tritici-repentis* wheat pathogen may have been horizontally transferred from the ancient wheat pathogen *P. nodorum*, according to one theory (Friesen et al., 2006). These are the evolutionary aspects in plant-pathogen interactions.

Based on the scientific evidence, plant-microbial or plant-pathogen interactions evolutionarily range from beneficial symbioses to harmful pathogens. Therefore, growing research in the field of ecology and evolutionary genetics of plant-pathogen interaction is of timely importance (Matveeva et al., 2018). Throughout history, plants coevolve symbiotically with various types of microbial consortiums, which has given tremendous support to plants for their terrestrialization. The well-studied mutualists of legume N_2-fixing nodules, arbuscular mycorrhizae, and antagonists biotrophic and necrotrophic provide sufficient evidence to prove endless support given by fungal and bacterial agents to sustain the plant in harsh environmental conditions since the beginning of life on Earth (Matveeva et al., 2018).

As mentioned by Wang and Bouwmeester (2017), most of the genetic correlation in plant-pathogen interactions can be elucidated through novel research approaches such as next generation sequencing (NGS). Based on NGS profiling, it's possible to differentiate gene and transcriptomic behavior during symbiosis interaction among plant and microbes. For example, *Phomopsis liquidambari* studied by Zhou et al. (2015) is established in endophytic and saprophytic systems with rice (*Oryza sativa* L.). The results revealed that most of the genes responsible for amino acids and carbohydrate metabolism, fatty acid biosynthesis, and secondary metabolism were upregulated during the endophytic fungal association with rice. In contrast, most of the genes involved in pathways of xenobiotic biodegradation and metabolism are upregulated in saprophytic association with rice. This information provides a general idea about genetic regulation of organisms to adapt various environmental conditions for their survival or long-lasting potential. NGS helps to find these footprints of such exchange of gene regulations under different associations between plant and microbes (Matveeva et al., 2018).

23.7 Conclusions

Based on our understanding, plant-pathogen relationships and interactions are more vital to study as an important area of concern. The effect created by plant pathogens on crop production and ultimate food security is becoming a great threat to mankind. Therefore, development of host resistance or breeding of crop cultivars with induced immunity against plant-pathogenic microorganisms is a big challenge faced by the agricultural scientific community worldwide. However, we still have several aspects and problems to overcome in this decade to get closer to answering the various questions related to these interactions for the development of pathogen-resistant crops for human sustainability. Recent knowledge about plant-pathogen interactions and induced stress on plant growth and development is a limiting factor. Various technological aspects such as NGS, phenomics, metabolomics, and bioinformatics are trying to fill this knowledge gap to get a clear picture about these interactions. Developing understanding of the sensors and signaling pathways involved among plants and pathogens is important to reduce the time required for development of new strategies to improve plant health. Therefore, we have to concentrate the research in the field of molecular plant pathology to study the plant immune response toward virulence created by plant pathogens and development of models to study real-life scenarios.

References

Abad, P., Gouzy, J., Aury, J.-M., Castagnone-Sereno, P., Danchin, E.G.J., Deleury, E., Perfus-Barbeoch, L., Anthouard, V., Artiguenave, F., Blok, V.C., Caillaud, M.-C., Coutinho, P.M., Dasilva, C., De Luca, F., Deau, F., Esquibet, M., Flutre, T., Goldstone, J.V., Hamamouch, N., Hewezi, T., Jaillon, O., Jubin, C., Leonetti, P., Magliano, M., Maier, T.R., Markov, G.V., McVeigh, P., Pesole, G., Poulain, J., Robinson-Rechavi, M., Sallet, E., Ségurens, B., Steinbach, D., Tytgat, T., Ugarte, E., van Ghelder, C., Veronico, P., Baum, T.J., Blaxter, M., Bleve-Zacheo, T., Davis, E.L., Ewbank, J.J., Favery, B., Grenier, E., Henrissat, B., Jones, J.T., Laudet, V., Maule, A.G., Quesneville, H., Rosso, M.-N., Schiex, T., Smant, G., Weissenbach, J., Wincker, P., 2008. Genome sequence of the metazoan plant-parasitic nematode Meloidogyne incognita. Nat. Biotechnol. 26, 909–915. https://doi.org/10.1038/nbt.1482.

Abtahi, F., Bakooie, M., 2017. Medicinal plant diseases caused by nematodes. Med. Plants Environ. Chall., 329–344. https://doi.org/10.1007/978-3-319-68717-9_18.

Agrawal, A., Lively, C.M., 2002. Infection genetics: gene-for-gene versus matching-alleles models and all points in between. Evol. Ecol. Res. 4, 79–90.

Agrios, G.N., 2005a. Plant Pathology, fifth ed. Elsevier, https://doi.org/10.1016/C2009-0-02037-6.

Agrios, G.N., 2005b. Plant diseases caused by prokaryotes: bacteria and mollicutes. In: Plant Pathology. Elsevier, pp. 615–703, https://doi.org/10.1016/B978-0-08-047378-9.50018-X.

Aguilar-Marcelino, L., Mendoza-de-Gives, P., Al-Ani, L.K.T., López-Arellano, M.E., Gómez-Rodríguez, O., Villar-Luna, E., Reyes-Guerrero, D.E., 2020. Using molecular techniques applied to beneficial microorganisms as biotechnological tools for controlling agricultural plant pathogens and pest. Mol. Aspects Plant Benefic. Microbes Agric., 333–349. https://doi.org/10.1016/b978-0-12-818469-1.00027-4.

Agurla, S., Gahir, S., Munemasa, S., Murata, Y., Raghavendra, A.S., 2018. Mechanism of stomatal closure in plants exposed to drought and cold stress. Adv. Exp. Med. Biol. 1081, 215–232. https://doi.org/10.1007/978-981-13-1244-1_12.

Ali, M.A., Abbas, A., Azeem, F., Javed, N., Bohlmann, H., 2015. Plant-nematode interactions: from genomics to metabolomics. Int. J. Agric. Biol. 17, 1071–1082. https://doi.org/10.17957/ijab/15.0037.

Ali, M.A., Azeem, F., Li, H., Bohlmann, H., 2017. Smart parasitic nematodes use multifaceted strategies to parasitize plants. Front. Plant Sci. 8, 1–21. https://doi.org/10.3389/fpls.2017.01699.

Ali, R., Ma, W., Lemtiri-Chlieh, F., Tsaltas, D., Leng, Q., Von Bodman, S., Berkowitz, G.A., 2007. Death don't have no mercy and neither does calcium: Arabidopsis CYCLIC NUCLEOTIDE GATED CHANNEL2 and innate immunity. Plant Cell 19, 1081–1095. https://doi.org/10.1105/tpc.106.045096.

Ali, M.A., Shahzadi, M., Zahoor, A., Dababat, A.A., Toktay, H., Bakhsh, A., Nawaz, M.A., Li, H., 2019. Resistance to cereal cyst nematodes in wheat and barley: an emphasis on classical and modern approaches. Int. J. Mol. Sci. 20, 1–18. https://doi.org/10.3390/ijms20020432.

Anderson, J.P., Gleason, C.A., Foley, R.C., Thrall, P.H., Burdon, J.B., Singh, K.B., 2010. Plants versus pathogens: an evolutionary arms race. Funct. Plant Biol. 37, 499–512. https://doi.org/10.1071/FP09304.

Ashbolt, N.J., 2004. Microbial contamination of drinking water and disease outcomes in developing regions. Toxicology 198, 229–238. https://doi.org/10.1016/j.tox.2004.01.030.

Ausubel, F.M., 2005. Are innate immune signaling pathways in plants and animals conserved? Nat. Immunol. 6, 973–979. https://doi.org/10.1038/ni1253.

Babich, O., Sukhikh, S., Pungin, A., Ivanova, S., Asyakina, L., Prosekov, A., 2020. Modern trends in the in vitro production and use of callus, suspension cells and root cultures of medicinal plants. Molecules 25, 1–18. https://doi.org/10.3390/molecules25245805.

Balagué, C., Gouget, A., Bouchez, O., Souriac, C., Haget, N., Boutet-Mercey, S., Govers, F., Roby, D., Canut, H., 2017. The Arabidopsis thaliana lectin receptor kinase LecRK-I.9 is required for full resistance to pseudomonas syringae and affects jasmonate signalling. Mol. Plant Pathol. 18, 937–948. https://doi.org/10.1111/mpp.12457.

Balasubramanian, V.K., Jansson, C., Baker, S.E., Ahkami, A.H., 2021. Molecular mechanisms of plant-microbe interactions in the rhizosphere as targets for improving plant productivity. In: Rhizosphere Biology: Intraction between Microbes and Plants, pp. 295–338, https://doi.org/10.1007/978-981-15-6125-2_14.

Balogh-Brunstad, Z., Keller, C., Shi, Z., Wallander, H., Stipp, S., 2017. Ectomycorrhizal Fungi and mineral interactions in the rhizosphere of scots and red pine seedlings. Soils 1, 5. https://doi.org/10.3390/soils1010005.

Bar, M., Sharfman, M., Ron, M., Avni, A., 2010. BAK1 is required for the attenuation of ethylene-inducing xylanase (Eix)-induced defense responses by the decoy receptor LeEix1. Plant J. 63, 791–800. https://doi.org/10.1111/j.1365-313X.2010.04282.x.

Bellin, D., Asai, S., Delledonne, M., Yoshioka, H., 2013. Nitric oxide as a mediator for defense responses. Mol. Plant-Microbe Interact. 26, 271–277. https://doi.org/10.1094/MPMI-09-12-0214-CR.

Bharath, P., Gahir, S., Raghavendra, A.S., 2021. Abscisic acid-induced stomatal closure: an important component of plant defense against abiotic and biotic stress. Front. Plant Sci. 12, 1–18. https://doi.org/10.3389/fpls.2021.615114.

Boland, G.J., 1992. Hypovirulence and double-stranded RNA in Sclerotinia sclerotiorum. Can. J. Plant Pathol. 14, 10–17. https://doi.org/10.1080/07060669209500900.

Boller, T., Felix, G., 2009. A renaissance of elicitors: perception of microbe-associated molecular patterns and danger signals by pattern-recognition receptors. Annu. Rev. Plant Biol. 60, 379–407. https://doi.org/10.1146/annurev.arplant.57.032905.105346.

Boller, T., He, S.Y., 2009. Innate immunity in plants: an arms race between pattern recognition receptors in plants and effectors in microbial pathogens. Science 324, 742–744. https://doi.org/10.1126/science.1171647.

Bolwell, G.P., Bindschedler, L.V., Blee, K.A., Butt, V.S., Davies, D.R., Gardner, S.L., Gerrish, C., Minibayeva, F., 2002. The apoplastic oxidative burst in response to biotic stress in plants: a three-component system. J. Exp. Bot. 53, 1367–1376. https://doi.org/10.1093/jxb/53.372.1367.

Borrelli, V.M.G., Brambilla, V., Rogowsky, P., Marocco, A., Lanubile, A., 2018. The enhancement of plant disease resistance using CRISPR/Cas9 technology. Front. Plant Sci. 9. https://doi.org/10.3389/fpls.2018.01245.

Bové, J.M., 1988. Plant mollicutes: phloem-restricted agents and surface contaminants. Acta Hortic. https://doi.org/10.17660/actahortic.1988.225.25.

Buendia, L., Girardin, A., Wang, T., Cottret, L., Lefebvre, B., 2018. LysM receptor-like kinase and LysM receptor-like protein families: an update on phylogeny and functional characterization. Front. Plant Sci. 9, 1–25. https://doi.org/10.3389/fpls.2018.01531.

Bulgarelli, D., Schlaeppi, K., Spaepen, S., van Themaat, E.V.L., Schulze-Lefert, P., 2013. Structure and functions of the bacterial microbiota of plants. Annu. Rev. Plant Biol. 64, 807–838. https://doi.org/10.1146/annurev-arplant-050312-120106.

Burdon, J.J., Roelfs, A., 1985. Isozyme and virulence variation in asexually reproducing populations of Puccinia graminis and P. recondita on wheat. Phytopathology 75, 907–913.

Calvitti, M., Moretti, R., Skidmore, A.R., Dobson, S.L., 2012. Wolbachia strain wPip yields a pattern of cytoplasmic incompatibility enhancing a Wolbachia-based suppression strategy against the disease vector Aedes albopictus. Parasit. Vectors 5, 1–9. https://doi.org/10.1186/1756-3305-5-254.

Cambi, A., Koopman, M., Figdor, C.G., 2005. How C-type lectins detect pathogens. Cell. Microbiol. 7, 481–488. https://doi.org/10.1111/j.1462-5822.2005.00506.x.

Carroll, R.K., Shelburne, S.A., Olsen, R.J., Suber, B., Sahasrabhojane, P., Kumaraswami, M., Beres, S.B., Shea, P.R., Flores, A.R., Musser, J.M., 2011. Naturally occurring single amino acid replacements in a regulatory protein alter streptococcal gene expression and virulence in mice. J. Clin. Invest. 121, 1956-1968. https://doi.org/10.1172/JCI45169.

Castroverde, C.D.M., Nazar, R.N., Robb, J., 2016. Verticillium Ave1 effector induces tomato defense gene expression independent of Ve1 protein. Plant Signal. Behav. 11, 1-3. https://doi.org/10.1080/15592324.2016.1245254.

Chen, Z., Agnew, J.L., Cohen, J.D., He, P., Shan, L., Sheen, J., Kunkel, B.N., 2007. Pseudomonas syringae type III effector AvrRpt2 alters Arabidopsis thaliana auxin physiology. Proc. Natl. Acad. Sci. U. S. A. 104, 20131-20136. https://doi.org/10.1073/pnas.0704901104.

Chinchilla, D., Bauer, Z., Regenass, M., Boller, T., Felix, G., 2006. The Arabidopsis receptor kinase FLS2 binds flg22 and determines the specificity of flagellin perception. Plant Cell 18, 465-476. https://doi.org/10.1105/tpc.105.036574.

Chisholm, S.T., Coaker, G., Day, B., Staskawicz, B.J., 2006. Host-microbe interactions: shaping the evolution of the plant immune response. Cell 124, 803-814. https://doi.org/10.1016/j.cell.2006.02.008.

Couto, D., Zipfel, C., 2016. receptor signalling in plants. Nat. Rev. Immunol. https://doi.org/10.1038/nri.2016.77.

Crill, J.P., 2001. The role of host resistance in the evolution of plant pathogens. African Plant Prot. 7, 1-19.

De Jonge, R., Van Esse, H.P., Maruthachalam, K., Bolton, M.D., Santhanam, P., Saber, M.K., Zhang, Z., Usami, T., Lievens, B., Subbarao, K.V., Thomma, B.P.H.J., 2012. Tomato immune receptor Ve1 recognizes effector of multiple fungal pathogens uncovered by genome and RNA sequencing. Proc. Natl. Acad. Sci. U. S. A. 109, 5110-5115. https://doi.org/10.1073/pnas.1119623109.

De Silva, N.I., Lumyong, S., Hyde, K.D., Bulgakov, T., Phillips, A.J.L., Yan, J.Y., 2016. Mycosphere essays 9: defining biotrophs and hemibiotrophs. Mycosphere 7, 545-559. https://doi.org/10.5943/mycosphere/7/5/2.

Dean, R., Van Kan, J.A.L., Pretorius, Z.A., Hammond-Kosack, K.E., Di Pietro, A., Spanu, P.D., Rudd, J.J., Dickman, M., Kahmann, R., Ellis, J., Foster, G.D., 2012. The top 10 fungal pathogens in molecular plant pathology. Mol. Plant Pathol. 13, 414-430. https://doi.org/10.1111/j.1364-3703.2011.00783.x.

Decraemer, W., Hunt, D.J., 1979. Structure and classification. In: Perry, R.N., Moens, M. (Eds.), California Insects. University of California Press, pp. 16-33, https://doi.org/10.1525/9780520906136-005.

Demidchik, V., 2014. Mechanisms and physiological roles of K+ efflux from root cells. J. Plant Physiol. 171, 696-707. https://doi.org/10.1016/j.jplph.2014.01.015.

Desclos-Theveniau, M., Arnaud, D., Huang, T.Y., Lin, G.J.C., Chen, W.Y., Lin, Y.C., Zimmerli, L., 2012. The Arabidopsis lectin receptor kinase LecRK-V.5 represses stomatal immunity induced by pseudomonas syringae pv. Tomato DC3000. PLoS Pathog. 8. https://doi.org/10.1371/journal.ppat.1002513.

Ding, S., 2010. RNA-based antiviral immunity. Nat. Rev. Immunol. 10, 632-644. https://doi.org/10.1038/nri2824.

Durrant, W.E., Dong, X., 2004. Systemic acquired resistance. Annu. Rev. Phytopathol. 42, 185-209. https://doi.org/10.1146/annurev.phyto.42.040803.140421.

Ellis, J., Jones, D., 1998. Structure and function of proteins controlling strain-specific pathogen resistance in plants. Curr. Opin. Plant Biol. 1, 288-293. https://doi.org/10.1016/1369-5266(88)80048-7.

Elmore, J.M., Coaker, G., 2011. The role of the plasma membrane H+-ATPase in plant-microbe interactions. Mol. Plant 4, 416-427. https://doi.org/10.1093/mp/ssq083.

Feng, J., Chen, L., Zuo, J., 2019. Protein S-Nitrosylation in plants: current progresses and challenges. J. Integr. Plant Biol. 61, 1206-1223. https://doi.org/10.1111/jipb.12780.

Flor, H.H., 1971. Current status of the gene-for-gene concept. Annu. Rev. Phytopathol. 9, 275–296. https://doi.org/10.1146/annurev.py.09.090171.001423.

Frantzeskakis, L., Kracher, B., Kusch, S., Yoshikawa-Maekawa, M., Bauer, S., Pedersen, C., Spanu, P.D., Maekawa, T., Schulze-Lefert, P., Panstruga, R., 2018. Signatures of host specialization and a recent transposable element burst in the dynamic one-speed genome of the fungal barley powdery mildew pathogen. BMC Genomics 19, 381. https://doi.org/10.1186/s12864-018-4750-6.

Freeman, B., Beattie, G., 2008. An overview of plant defenses against pathogens and herbivores. In: The Plant Health Instructor, pp. 1–12, https://doi.org/10.1094/PHI-I-2008-0226-01.

Friesen, T.L., Stukenbrock, E.H., Liu, Z., Meinhardt, S., Ling, H., Faris, J.D., Rasmussen, J.B., Solomon, P.S., McDonald, B.A., Oliver, R.P., 2006. Emergence of a new disease as a result of interspecific virulence gene transfer. Nat. Genet. 38, 953–956. https://doi.org/10.1038/ng1839.

Fu, Z.Q., Dong, X., 2013. Systemic acquired resistance: turning local infection into global defense. Annu. Rev. Plant Biol. 64, 839–863. https://doi.org/10.1146/annurev-arplant-042811-105606.

Gagne-Bourgue, F., Aliferis, K.A., Seguin, P., Rani, M., Samson, R., Jabaji, S., 2013. Isolation and characterization of indigenous endophytic bacteria associated with leaves of switchgrass (Panicum virgatum L.) cultivars. J. Appl. Microbiol. 114, 836–853. https://doi.org/10.1111/jam.12088.

Gao, Q.M., Zhu, S., Kachroo, P., Kachroo, A., 2015. Signal regulators of systemic acquired resistance. Front. Plant Sci. 6, 1–12. https://doi.org/10.3389/fpls.2015.00228.

Gergerich, R.C., Dolja, V.V., 2006. Introduction to plant viruses, the invisible foe. In: The Plant Health Instructor, pp. 1–16, https://doi.org/10.1094/PHI-I-2006-0414-01.

Gibling, M.R., Davies, N.S., 2012. Palaeozoic landscapes shaped by plant evolution. Nat. Geosci. 5, 99–105. https://doi.org/10.1038/ngeo1376.

Glazebrook, J., 2005. Contrasting mechanisms of defense against biotrophic and necrotrophic pathogens. Annu. Rev. Phytopathol. 43, 205–227. https://doi.org/10.1146/annurev.phyto.43.040204.135923.

Gómez-Gómez, L., Boller, T., 2002. Flagellin perception: a paradigm for innate immunity. Trends Plant Sci. 7, 251–256. https://doi.org/10.1016/S1360-1385(02)02261-6.

Greeff, C., Roux, M., Mundy, J., Petersen, M., 2012. Receptor-like kinase complexes in plant innate immunity. Front. Plant Sci. 3, 1–7. https://doi.org/10.3389/fpls.2012.00209.

Gust, A.A., Brunner, F., Nürnberger, T., 2010. Biotechnological concepts for improving plant innate immunity. Curr. Opin. Biotechnol. 21, 204–210. https://doi.org/10.1016/j.copbio.2010.02.004.

Gust, A.A., Willmann, R., Desaki, Y., Grabherr, H.M., Nürnberger, T., 2012. Plant LysM proteins: modules mediating symbiosis and immunity. Trends Plant Sci. 17, 495–502. https://doi.org/10.1016/j.tplants.2012.04.003.

Haggag, W.M., Abouziena, H.F., Abd-El-Kreem, F., El Habbasha, S., 2015. Agriculture biotechnology for management of multiple biotic and abiotic environmental stress in crops. J. Chem. Pharm. Res. 7, 882–889.

Halter, T., Imkampe, J., Mazzotta, S., Wierzba, M., Postel, S., Bücherl, C., Kiefer, C., Stahl, M., Chinchilla, D., Wang, X., Nürnberger, T., Zipfel, C., Clouse, S., Borst, J.W., Boeren, S., De Vries, S.C., Tax, F., Kemmerling, B., 2014. The leucine-rich repeat receptor kinase BIR2 is a negative regulator of BAK1 in plant immunity. Curr. Biol. 24, 134–143. https://doi.org/10.1016/j.cub.2013.11.047.

Han, G., 2019. Tansley Review Origin and Evolution of the Plant Immune System., https://doi.org/10.1111/nph.15596.

Hawkes, C.V., Connor, E.W., 2017. Translating Phytobiomes from theory to practice: ecological and evolutionary considerations. Phytobiomes J. 1, 57–69. https://doi.org/10.1094/PBIOMES-05-17-0019-RVW.

Heese, A., Hann, D.R., Gimenez-Ibanez, S., Jones, A.M.E., He, K., Li, J., Schroeder, J.I., Peck, S.C., Rathjen, J.P., 2007. The receptor-like kinase SERK3/BAK1 is a central regulator of innate immunity in plants. Proc. Natl. Acad. Sci. U. S. A. 104, 12217-12222. https://doi.org/10.1073/pnas.0705306104.

Hu, B., Du, J., Zou, R., Yuan, Y., 2010. An environment-sensitive synthetic microbial ecosystem. PLoS One 5. https://doi.org/10.1371/journal.pone.0010619, e10619.

Imam, J., Alam, S., Mandal, N.P., Shukla, P., Sharma, T.R., Variar, M., 2015. Molecular identification and virulence analysis of AVR genes in rice blast pathogen, Magnaporthe oryzae from eastern India. Euphytica 206, 21-31. https://doi.org/10.1007/s10681-015-1465-5.

Imam, J., Alam, S., Mandal, N.P., Variar, M., Shukla, P., 2014. Molecular screening for identification of blast resistance genes in north east and eastern Indian rice germplasm (Oryza sativa L.) with PCR based makers. Euphytica 196, 199-211. https://doi.org/10.1007/s10681-013-1024-x.

Imam, J., Singh, P.K., Shukla, P., 2016. Plant microbe interactions in post genomic era: perspectives and applications. Front. Microbiol. 7, 1-15. https://doi.org/10.3389/fmicb.2016.01488.

Jeworutzki, E., Roelfsema, M.R.G., Anschütz, U., Krol, E., Elzenga, J.T.M., Felix, G., Boller, T., Hedrich, R., Becker, D., 2010. Early signaling through the arabidopsis pattern recognition receptors FLS2 and EFR involves Ca2+-associated opening of plasma membrane anion channels. Plant J. 62, 367-378. https://doi.org/10.1111/j.1365-313X.2010.04155.x.

Jibril, S.M., Jakada, B.H., Kutama, A.S., Umar, H.Y., 2016. Plant and pathogens: pathogen recognision, invasion and plant defense mechanism. Int. J. Curr. Microbiol. Appl. Sci. 5, 247257. https://doi.org/10.20546/ijcmas.2016.506.028.

Jones, M.G.K., 1981. Host cell responses to endoparasitic nematode attack: structure and function of giant cells and syncytia. Ann. Appl. Biol. 97, 353-372. https://doi.org/10.1111/j.1744-7348.1981.tb05122.x.

Jones, J.D.G., Dangl, J.L., 2006. The plant immune system. Nature 444, 323-329. https://doi.org/10.1038/nature05286.

Jung, C., Wyss, U., 1999. New approaches to control plant parasitic nematodes. Appl. Microbiol. Biotechnol. 51, 439-446. https://doi.org/10.1007/s002530051414.

Kader, M.A., Lindberg, S., 2010. Cytosolic calcium and pH signaling in plants under salinity stress. Plant Signal. Behav. 5, 233-238. https://doi.org/10.4161/psb.5.3.10740.

Kärkönen, A., Kuchitsu, K., 2015. Reactive oxygen species in cell wall metabolism and development in plants. Phytochemistry 112, 22-32. https://doi.org/10.1016/j.phytochem.2014.09.016.

Khan, M., Subramaniam, R., Desveaux, D., 2016. Of guards, decoys, baits and traps: pathogen perception in plants by type III effector sensors. Curr. Opin. Microbiol. 29, 49-55. https://doi.org/10.1016/j.mib.2015.10.006.

Kothari, I.L., Patel, M., 2004. Plant immunization. Indian J. Exp. Biol. 42, 244-252.

Kunze, G., Zipfel, C., Robatzek, S., Niehaus, K., Boller, T., Felix, G., 2004. The N terminus of bacterial elongation factor Tu elicits innate immunity in Arabidopsis plants. Plant Cell 16, 3496-3507. https://doi.org/10.1105/tpc.104.026765.

Labandeira, C.C., 2013. A paleobiologic perspective on plant-insect interactions. Curr. Opin. Plant Biol. 16, 414-421. https://doi.org/10.1016/j.pbi.2013.06.003.

Lannoo, N., Van Damme, E.J.M., 2010. Biochimica et Biophysica Acta nucleocytoplasmic plant lectins. BBA-Gen. Subjects 1800, 190-201. https://doi.org/10.1016/j.bbagen.2009.07.021.

Lecourieux, D., Mazars, C., Pauly, N., Ranjeva, R., Pugin, A., 2002. Analysis and effects of cytosolic free calcium increases in response to elicitors in *Nicotiana plumbaginifolia* cells. Plant Cell 14, 2627-2641. https://doi.org/10.1105/tpc.005579.

Lee, H.C., 2001. Physiological functions of cyclic ADP-ribose and NAADP as calcium messengers. Annu. Rev. Pharmacol. Toxicol. 41, 317-345.

Lee, S.-J., Rose, J.K.C., 2010. Mediation of the transition from biotrophy to necrotrophy in hemibiotrophic plant pathogens by secreted effector proteins. Plant Signal. Behav. 5, 769–772. https://doi.org/10.4161/psb.5.6.11778.

Lindeberg, M., Myers, C.R., Collmer, A., Schneider, D.J., 2008. Roadmap to new virulence determinants in *Pseudomonas syringae*: insights from comparative genomics and genome organization. Mol. Plant-Microbe Interact. 21, 685–700. https://doi.org/10.1094/MPMI-21-6-0685.

Lv, D., Wang, G., Xiong, L.-R., Sun, J.-X., Chen, Y., Guo, C.-L., Yu, Y., He, H.-L., Cai, R., Pan, J.-S., 2020. Genome-wide identification and characterization of lectin receptor-like kinase gene family in cucumber and expression profiling analysis under different treatments. Genes (Basel). 11, 1032. https://doi.org/10.3390/genes11091032.

Ma, W., Smigel, A., Tsai, Y.C., Braam, J., Berkowitz, G.A., 2008. Innate immunity signaling: cytosolic Ca2+ elevation is linked to downstream nitric oxide generation through the action of calmodulin or a calmodulin-like protein. Plant Physiol. 148, 818–828. https://doi.org/10.1104/pp.108.125104.

Mansfield, J., Genin, S., Magori, S., Citovsky, V., Sriariyanum, M., Ronald, P., Dow, M., Verdier, V., Beer, S.V., Machado, M.A., Toth, I., Salmond, G., Foster, G.D., 2012. Top 10 plant pathogenic bacteria in molecular plant pathology. Mol. Plant Pathol. 13, 614–629. https://doi.org/10.1111/j.1364-3703.2012.00804.x.

Matveeva, T., Provorov, N., Valkonen, J.P.T., 2018. Editorial: cooperative adaptation and evolution in plant-microbe systems. Front. Plant Sci. 9, 1–2. https://doi.org/10.3389/fpls.2018.01090.

McDonald, B.A., Linde, C., 2002. Pathogen population genetics, evolutionary potential, and durable resistance. Annu. Rev. Phytopathol. 40, 349–379. https://doi.org/10.1146/annurev.phyto.40.120501.101443.

McDonald, A.H., Nicol, J.M., 2005. Nematode parasites of cereals. In: Luc, M., Sikora, R.A., Bridge, J. (Eds.), Plant Parasitic Nematodes in Subtropical and Tropical Agriculture, second ed. CABI Internationals, Wallingford, pp. 131–191, https://doi.org/10.1079/9780851997278.0131.

Melotto, M., Underwood, W., Koczan, J., Nomura, K., He, S.Y., 2006. Plant stomata function in innate immunity against bacterial invasion. Cell 126, 969–980. https://doi.org/10.1016/j.cell.2006.06.054.

Mikkelsen, L., Elphinstone, J., Jensen, D.F., 2005. Literature review on detection and eradication of plant pathogens in sludge, soils and treated biowaste. In: Desk Study on Bulk Density. The Royal Veterinary and Agricultural University.

Morris, J.L., Leake, J.R., Stein, W.E., Berry, C.M., Marshall, J.E.A., Wellman, C.H., Milton, J.A., Hillier, S., Mannolini, F., Quirk, J., Beerling, D.J., 2015. Investigating Devonian trees as geo-engineers of past climates: linking palaeosols to palaeobotany and experimental geobiology. Palaeontology 58, 787–801. https://doi.org/10.1111/pala.12185.

Morris, J.L., Puttick, M.N., Clark, J.W., Edwards, D., Kenrick, P., Pressel, S., Wellman, C.H., Yang, Z., Schneider, H., Donoghue, P.C.J., 2018. The timescale of early land plant evolution. Proc. Natl. Acad. Sci. 115, E2274–E2283. https://doi.org/10.1073/pnas.1719588115.

Nürnberger, T., Kemmerling, B., 2009. Chapter 1 PAMP-triggered basal immunity in plants. Adv. Bot. Res. 51, 1–38. https://doi.org/10.1016/S0065-2296(09)51001-4.

O'Brien, H.E., Thakur, S., Guttman, D.S., 2011. Evolution of plant pathogenesis in *pseudomonas syringae*: a genomics perspective. Annu. Rev. Phytopathol. 49, 269–289. https://doi.org/10.1146/annurev-phyto-072910-095242.

Ochman, H., Moran, N.A., 2001. Genes lost and genes found: evolution of bacterial pathogenesis and symbiosis. Science 292, 1096–1099. https://doi.org/10.1126/science.1058543.

Oliver, R.P., Solomon, P.S., 2008. Recent fungal diseases of crop plants: is lateral gene transfer a common theme? Mol. Plant-Microbe Interact. 21, 287–293. https://doi.org/10.1094/MPMI-21-3-0287.

Orr, C.H., Ralebitso-Senior, T.K., 2016. Summation of the microbial ecology of biochar application. In: Biochar Application. Elsevier, pp. 293–311, https://doi.org/10.1016/B978-0-12-803433-0.00012-6.

Park, S.W., Kaimoyo, E., Kumar, D., Mosher, S., Klessig, D.F., 2007. Methyl salicylate is a critical mobile signal for plant systemic acquired resistance. Science 318, 113–116. https://doi.org/10.1126/science.1147113.

Parker, I.M., Gilbert, G.S., 2004. The evolutionary ecology of novel plant-pathogen interactions. Annu. Rev. Ecol. Evol. Syst. 35, 675–700. https://doi.org/10.1146/annurev.ecolsys.34.011802.132339.

Perilla-Henao, L.M., Casteel, C.L., 2016. Vector-borne bacterial plant pathogens: interactions with hemipteran insects and plants. Front. Plant Sci. 7, 1–15. https://doi.org/10.3389/fpls.2016.01163.

Podgórska, A., Burian, M., Szal, B., 2017. Extra-cellular but extra-ordinarily important for cells: Apoplastic reactive oxygen species metabolism. Front. Plant Sci. 8, 1–20. https://doi.org/10.3389/fpls.2017.01353.

Porter, K., Day, B., 2016. From filaments to function: the role of the plant actin cytoskeleton in pathogen perception, signaling and immunity. J. Integr. Plant Biol. 58, 299–311. https://doi.org/10.1111/jipb.12445.

Poulin, R., 2011. The many roads to parasitism. A tale of convergence. In: Advances in Parasitology, first ed. lsevier Ltd, https://doi.org/10.1016/B978-0-12-385897-9.00001-X.

Qi, J., Wang, J., Gong, Z., Zhou, J.M., 2017. Apoplastic ROS signaling in plant immunity. Curr. Opin. Plant Biol. 38, 92–100. https://doi.org/10.1016/j.pbi.2017.04.022.

Ravichandra, N., 2013. Fundamentals of Plant Pathology. PHI Learning, New Delhi.

Reid, A., 2011. Microbes helping to improve crop productivity. Microbe Magazine 6, 435–439. https://doi.org/10.1128/microbe.6.435.1.

Roelfsema, M.R.G., Hedrich, R., Geiger, D., 2012. Anion channels: master switches of stress responses. Trends Plant Sci. 17, 221–229. https://doi.org/10.1016/j.tplants.2012.01.009.

Ryder, L.S., Talbot, N.J., 2015. Regulation of appressorium development in pathogenic fungi. Curr. Opin. Plant Biol. 26, 8–13. https://doi.org/10.1016/j.pbi.2015.05.013.

Sacristan, S., Garcia-Arenal, F., 2008. The evolution of virulence and pathogenicity in plant pathogen populations. Mol. Plant Pathol. 9, 369–384 https://doi.org/10.1111/j.1364-3703.2007.00460.x.

Saeed, M., Mukhtar, T., Rehman, M.A., 2019. Temporal fluctuations in the population of Citrus nematode (Tylenchulus semipenetrans) in the Pothowar region of Pakistan. Pak. J. Zool. 51, 2257–2263. https://doi.org/10.17582/journal.pjz/2019.51.6.2257.2263.

Saijo, Y., Tintor, N., Lu, X., Rauf, P., Pajerowska-Mukhtar, K., Häweker, H., Dong, X., Robatzek, S., Schulze-Lefert, P., 2009. Receptor quality control in the endoplasmic reticulum for plant innate immunity. EMBO J. 28, 3439–3449. https://doi.org/10.1038/emboj.2009.263.

Sánchez-Vallet, A., Saleem-Batcha, R., Kombrink, A., Hansen, G., Valkenburg, D.J., Thomma, B.P.H.J., Mesters, J.R., 2013. Fungal effector Ecp6 outcompetes host immune receptor for chitin binding through intrachain LysM dimerization. elife 2013, 1–16. https://doi.org/10.7554/eLife.00790.

Scheler, C., Durner, J., Astier, J., 2013. Nitric oxide and reactive oxygen species in plant biotic interactions. Curr. Opin. Plant Biol. 16, 534–539. https://doi.org/10.1016/j.pbi.2013.06.020.

Sharfman, M., Bar, M., Ehrlich, M., Schuster, S., Melech-Bonfil, S., Ezer, R., Sessa, G., Avni, A., 2011. Endosomal signaling of the tomato leucine-rich repeat receptor-like protein LeEix2. Plant J. 68, 413–423. https://doi.org/10.1111/j.1365-313X.2011.04696.x.

Sherman-Broyles, S., Boggs, N., Farkas, A., Liu, P., Vrebalov, J., Nasrallah, M.E., Nasrallah, J.B., 2007. S locus genes and the evolution of self-fertility in Arabidopsis thaliana. Plant Cell 19, 94–106. https://doi.org/10.1105/tpc.106.048199.

Shimizu, T., Nakano, T., Takamizawa, D., Desaki, Y., Ishii-Minami, N., Nishizawa, Y., Minami, E., Okada, K., Yamane, H., Kaku, H., Shibuya, N., 2010. Two LysM receptor molecules, CEBiP and OsCERK1, cooperatively regulate chitin elicitor signaling in rice. Plant J. 64, 204–214. https://doi.org/10.1111/j.1365-313X.2010.04324.x.

Sikora, R.A., Bridge, J., Starr, J.L., 2005. Management practices: an overview of integrated nematode management technologies. In: Lic, M., Sikora, R., Bridge, J. (Eds.), Plant Parasitic Nematodes in Subtropical and Tropical Agriculture. CABI ebooks, pp. 793–825, https://doi.org/10.1079/9780851997278.0793.

Spiers, A.G., Hopcroft, D.H., 1994. Comparative studies of the poplar rusts Melampsora medusae, M. larici-Populina and their interspecific hybrid M.medusae-Populina. Mycol. Res. 98, 889–903. https://doi.org/10.1016/S0953-7562(09)80260-8.

Spoel, S.H., Dong, X., 2012. How do plants achieve immunity? Defence without specialized immune cells. Nat. Rev. Immunol. 12, 89–100. https://doi.org/10.1038/nri3141.

Strange, R.N., Scott, P.R., 2005. Plant disease: a threat to global food security. Annu. Rev. Phytopathol. 43, 83–116. https://doi.org/10.1146/annurev.phyto.43.113004.133839.

Surico, G., 2013. The concepts of plant pathogenicity, virulence/avirulence and effector proteins by a teacher of plant pathology. Phytopathol. Mediterr. 52, 399–417.

Takken, F.L., Albrecht, M., Tameling, W.I.L., 2006. Resistance proteins: molecular switches of plant defence. Curr. Opin. Plant Biol. 9, 383–390. https://doi.org/10.1016/j.pbi.2006.05.009.

Tamilarasan, S., Rajm, M., 2013. Engineering crop plants for nematode resistance through host-derived RNA interference. Cell Dev. Biol., 2. https://doi.org/10.4172/2168-9296.1000114.

Testerink, C., Munnik, T., 2005. Phosphatidic acid: a multifunctional stress signaling lipid in plants. Trends Plant Sci. 10, 368–375. https://doi.org/10.1016/j.tplants.2005.06.002.

Testerink, C., Munnik, T., 2011. Molecular, cellular, and physiological responses to phosphatidic acid formation in plants. J. Exp. Bot. 62, 2349–2361. https://doi.org/10.1093/jxb/err079.

Thompson, J.N., Thompson, M.B., Gates, D.M., 2020. Biosphere. Encyclopaedia Britannica Inc.

Thrall, P.H., Burdon, J.J., Young, A., Thrall, P.H., Burdon, J.J., Young, A., 2001. Variation in resistance and virulence among demes of a plant host-pathogen Metapopulation. J. Ecol. 89, 736–748.

Upson, J.L., Zess, E.K., Białas, A., Wu, C., Kamoun, S., 2018. The coming of age of EvoMPMI: evolutionary molecular plant-microbe interactions across multiple timescales. Curr. Opin. Plant Biol. 44, 108–116. https://doi.org/10.1016/j.pbi.2018.03.003.

Vaid, N., Pandey, P.K., Tuteja, N., 2012. Genome-wide analysis of lectin receptor-like kinase family from Arabidopsis and rice. Plant Mol. Biol. 80, 365–388. https://doi.org/10.1007/s11103-012-9952-8.

Van Dammes, E.J.M., Fouquaert, E., Lannoo, N., Vandenborre, G., Schouppe, D., Peumans, W.J., 2011. Novel concepts about the role of lectins in the plant cell. Adv. Exp. Med. Biol. 705, 271–294. https://doi.org/10.1007/978-1-4419-7877-6_13.

Venu, R.C., Zhang, Y., Weaver, B., Carswell, P., Mitchell, T.K., Meyers, B.C., Boehm, M.J., Wang, G.-L., 2011. Large scale identification of genes involved in plant-fungal interactions using Illumina's sequencing-by-synthesis technology. Methods Mol. Biol. Methods Mol. Biol. 722, 167–178. https://doi.org/10.1007/978-1-61779-040-9_12.

Vorholt, J.A., 2012. Microbial life in the phyllosphere. Nat. Rev. Microbiol. 10, 828–840. https://doi.org/10.1038/nrmicro2910.

Wan, J., Zhang, X.C., Neece, D., Ramonell, K.M., Clough, S., Kim, S.Y., Stacey, M.G., Stacey, G., 2008. A LysM receptor-like kinase plays a critical role in chitin signaling and fungal resistance in Arabidopsis. Plant Cell 20, 471–481. https://doi.org/10.1105/tpc.107.056754.

Wang, Y., Bouwmeester, K., 2017. L-type lectin receptor kinases: new forces in plant immunity. PLoS Pathog. 13, 1–7. https://doi.org/10.1371/journal.ppat.1006433.

Wang, Y.Q., Feechan, A., Yun, B.W., Shafiei, R., Hofmann, A., Taylor, P., Xue, P., Yang, F.Q., Xie, Z.S., Pallas, J.A., Chu, C.C., Loake, G.J., 2009. S-nitrosylation of AtSABP3 antagonizes the expression of plant immunity. J. Biol. Chem. 284, 2131–2137. https://doi.org/10.1074/jbc.M806782200.

Watson, I.A., 1981. Wheat and its rust parasites in Australia. In: Wheat Science—Today and Tomorrow, pp. 129–147.

Webster, J.M., 1969. The host-parasite relationships of plant-parasitic nematodes. Adv. Parasitol. 7, 1–40. https://doi.org/10.1016/S0065-308X(08)60433-9.

Wickett, N.J., Mirarab, S., Nguyen, N., Warnow, T., Carpenter, E., Matasci, N., Ayyampalayam, S., Barker, M.S., Burleigh, J.G., Gitzendanner, M.A., Ruhfel, B.R., Wafula, E., Der, J.P., Graham, S.W., Mathews, S., Melkonian, M., Soltis, D.E., Soltis, P.S., Miles, N.W., Rothfels, C.J., Pokorny, L., Shaw, A.J., DeGironimo, L., Stevenson, D.W., Surek, B., Villarreal, J.C., Roure, B., Philippe, H., DePamphilis, C.W., Chen, T., Deyholos, M.K., Baucom, R.S., Kutchan, T.M., Augustin, M.M., Wang, J., Zhang, Y., Tian, Z., Yan, Z., Wu, X., Sun, X., Wong, G.K.-S., Leebens-Mack, J., 2014. Phylotranscriptomic analysis of the origin and early diversification of land plants. Proc. Natl. Acad. Sci. 111, E4859–E4868. https://doi.org/10.1073/pnas.1323926111.

Willmann, R., Lajunen, H.M., Erbs, G., Newman, M.A., Kolb, D., Tsuda, K., Katagiri, F., Fliegmann, J., Bono, J.J., Cullimore, J.V., Jehle, A.K., Götz, F., Kulik, A., Molinaro, A., Lipka, V., Gust, A.A., Nürnberger, T., 2011. Arabidopsis lysin-motif proteins LYM1 LYM3 CERK1 mediate bacterial peptidoglycan sensing and immunity to bacterial infection. Proc. Natl. Acad. Sci. U. S. A. 108, 19824–19829. https://doi.org/10.1073/pnas.1112862108.

Wirthmueller, L., Maqbool, A., Banfield, M.J., 2013. On the front line: structural insights into plant-pathogen interactions. Nat. Rev. Microbiol. 11, 761–776. https://doi.org/10.1038/nrmicro3118.

Wolpert, T.J., Dunkle, L.D., Ciuffetti, L.M., 2002. Host-selective toxins and avirulence determinants: what's in a name? Annu. Rev. Phytopathol. 40, 251–285. https://doi.org/10.1146/annurev.phyto.40.011402.114210.

Wu, L., Chen, H., Curtis, C., Fu, Z.Q., 2014. Go in for the kill: how plants deploy effector-triggered immunity to combat pathogens. Virulence 5, 710–721. https://doi.org/10.4161/viru.29755.

Yao, H.Y., Xue, H.W., 2018. Phosphatidic acid plays key roles regulating plant development and stress responses. J. Integr. Plant Biol. 60, 851–863. https://doi.org/10.1111/jipb.12655.

Yu, X., Feng, B., He, P., Shan, L., 2017. From chaos to harmony: responses and signaling upon microbial pattern recognition. Ann. Rev. Phytopathol. 55, 109–137. https://doi.org/10.1146/annurev-phyto-080516-035649.

Yunus, I.S., Cazenave-Gassiot, A., Liu, Y.C., Lin, Y.C., Wenk, M.R., Nakamura, Y., 2015. Phosphatidic acid is a major phospholipid class in reproductive organs of arabidopsis thaliana. Plant Signal. Behav. 10. https://doi.org/10.1080/15592324.2015.1049790.

Zambryski, P.C., 1992. Chronicles from the agrobacterium-plant cell DNA transfer story. Annu. Rev. Plant Physiol. Plant Mol. Biol. 43, 465–490. https://doi.org/10.1146/annurev.pp.43.060192.002341.

Zhang, X.C., Cannon, S.B., Stacey, G., 2009. Evolutionary genomics of LysM genes in land plants. BMC Evol. Biol. 9, 1–13. https://doi.org/10.1186/1471-2148-9-183.

Zhou, Z., Wu, Y., Yang, Y., Du, M., Zhang, X., Guo, Y., Li, C., Zhou, J.M., 2015. An arabidopsis plasma membrane proton ATPase modulates JA signaling and is exploited by the pseudomonas syringae effector protein AvrB for stomatal invasion. Plant Cell 27, 2032–2041. https://doi.org/10.1105/tpc.15.00466.

Zipfel, C., Felix, G., 2005. Plants and animals: a different taste for microbes? Curr. Opin. Plant Biol. 8, 353–360. https://doi.org/10.1016/j.pbi.2005.05.004.

Zipfel, C., Kunze, G., Chinchilla, D., Caniard, A., Jones, J.D.G., Boller, T., Felix, G., 2006. Perception of the bacterial PAMP EF-Tu by the receptor EFR restricts agrobacterium-mediated transformation. Cell 125, 749–760. https://doi.org/10.1016/j.cell.2006.03.037.

24

Global biofertilizer market: Emerging trends and opportunities

Sanjay Kumar Joshi and Ajay Kumar Gauraha
Agri-Business and Rural Management, Indira Gandhi Agricultural University, Raipur, India

24.1 Introduction: The market and opportunities

Globally, agribusinesses are shifting to natural and sustainable alternatives from chemical sources of nutrients with growing concerns over the role of agricultural practices in climate change. That is where the use of microbes to play the role of chemical fertilizers can be a pivotal area of focus. All kinds of nitrogenous, phosphatic, and potassic fertilizers when manufactured emit all kinds of greenhouse gases. When they are applied to crops as nutrients, a significant portion goes to air and water as runoff. For transportation needs, chemical fertilizers need bulk transportation mechanisms like trucks, tankers, ships, and railways. On the contrary, microbial fertilizers not only establish a symbiotic relationship, which in turn result in nutrient balance by establishing nutrient cycles but can also be even mailed to farmers since they are needed in small quantities (Mulvany, 2020).

> If we want to make sure our farmers are resilient, that we can scale our food system to power the planet for the next couple of generations and protect the environment, we have to reinvent fertilizer—Karsten Temme, CEO, Pivot Bio.

The global biofertilizer market was reported at $1.06 billion in the year 2016 and is expected to touch the $2.1 billion mark at a cumulative average growth rate of 12.04% by the year 2022 (prnewswire, 2017). Biofertilizers are formulated by utilizing live microorganisms, when applied to soil, seed, or plant surfaces, that improve plant growth by increasing the supply of several essential nutrients to plants. They increase soil fertility through various processes, such as P solubilizing, N fixation, and synthesizing growth-promoting substances. These completely natural processes do not cause any harmful impact on the environment. The area under agriculture, as well as arable land in many parts of the world, has been declining over time due to heavy industrialization, mining, and urbanization. The rampant use of chemical fertilizers has also contributed to the reduced fertility levels rendering lesser arable land fit for agriculture. Farmers are thus opting to use organic manure and biofertilizers to maintain soil fertility and crop yield; change in consumer preferences toward healthy food and the recent shift toward the widespread adoption of organic farming are propelling the growth of the biofertilizer market (MicroMarketMonitor.com, 2015).

The ever-increasing world population has put significant pressure on land and other natural resources. The exponential rate of crop production to feed the burgeoning world population leads to rapid loss of plant nutrients, thereby impacting nutrient content in crops. Therefore, to help replenish this loss of nutrients, farmers are engaged in the application of chemical fertilizers. However, biological activities in the soil are decelerated due to excessive use of chemical fertilizers, which leads to impaired soil quality. As a result, farmers are steadily moving toward a more sustainable farming solution, which, in most cases, comes out as practicing organic farming. As the awareness of health hazards associated with the use of chemicals in food products is increasing, the organic farming industry has also started growing at a promising rate. Government initiatives also encourage farmers to adopt practices involved in organic farming such as using biofertilizers. It is predicted that the biofertilizer market will register a CAGR of almost 14% by 2023 (MicroMarketMonitor.com, 2015). In the year 2016, the market size of the global biofertilizer market reached USD 1106.4 million, and it is projected to grow at the rate of 14.2% to reach USD 3124.5 million by the end of 2024. The global nitrogen-fixing market is also anticipated to reach USD 2242.4 million by 2024 at a CAGR of 14%. However, the market for the phosphate-solubilizing segment is expected to grow by a CAGR of 14.4% for the same period (MarketWatch, 2021).

24.2 The market dynamics and growth drivers

Rising food safety issues, residual levels in food items, and other environmental issues have put the consumers' awareness level at an all-time high. This increased awareness fueled by the pandemic has induced the ever-increasing preferences for chemical-free food items. Thus, organic food products are becoming no-nonsense choices. Owing to this, most countries in the western part of the globe have witnessed a rise in organic food retail sales along with market growth of biological inputs such as biofertilizers.

In the eastern part of the globe where the majority of human population resides, it has always been a challenge to meet the increasing demand for food, which has resulted in more consumption of fertilizers. But the major concerns of the region are air and soil contamination and their ill effects on human beings. Due to this, the major emphasis of the governments of this region has been on the use of environmental-friendly fertilizers, especially biofertilizers and manures.

24.3 Growth-share matrix

Fig. 24.1 below represents the most attractive to most risky markets from the point of view of investments and may help businesses for strategic decision-making in terms of market entry, growth projects, and exit plans. This matrix has been prepared based on the BCG matrix. It, however, positions geographies into quadrants instead of brands and still poses meaningful insights. The growth-share matrix below presents the zones that define the geographical position based on market size, relative growth rate, and other anticipated factors of growth. As can be understood from the figure below, the markets are classified into four different quadrants based on growth rate and market size as explained below:

24.3.1 Dogs or extremely risky quadrant

This quadrant indicates the market with a relatively low market size and growth rate. A part of North American market players and geography falls into it. These are the markets that neither are growing nor can survive. The best strategy for such cases is to divest and get out of business.

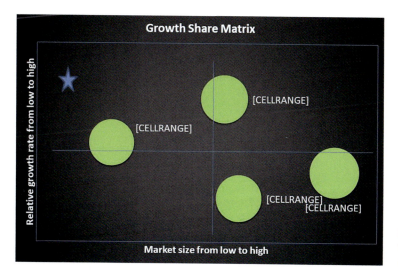

Fig. 24.1 The most attractive to most risky markets from the point of view of investments.

24.3.2 A question mark or play safe quadrant

The second quadrant on the right falls under this category. Question marks are the businesses that operate with a relatively low growth rate in a high-market-size environment. EU and the western region fall under this category. The strategy for such markets is to identify the promising businesses or territories and invest in them as they are going to be the leaders of tomorrow for the business.

24.3.3 Cash cows or consolidation quadrant

The top right quadrant is classified as a cash cow. These are the businesses that operate under a high growth market with a high market share. The business here should focus on retaining their position as long as they can and think of mergers and acquisitions if necessary. Asia Pacific region is one such market for biofertilizer firms.

24.3.4 Stars or growth quadrant

This quadrant is full of firms with a high growth rate and relatively low market size. A part of North America and Asia Pacific falls into this category. As evident from the data that North American and Asian markets are the promising growth markets of tomorrow for biofertilizer business, firms should invest heavily into this region to earn profits in the future.

24.4 Biofertilizer market Segmentation

The biofertilizer market can be segmented based on microorganism type, crop type, type of biofertilizer products, mode of application, and the most prevalent region where they are used. The microorganism segment comprises microorganisms like *Azotobacter, Azospirillum,* phosphate-solubilizing bacteria, cyanobacteria, and *Rhizobium*. Based on the crop ecosystem, the global biofertilizer market can be categorized into cereals and grains, fruits and vegetables, pulses and oilseeds, and others. Similarly, the biofertilizer product market can also be segmented as nitrogen fixing, phosphate solubilizing, and others. Based on how they are applied, the market can be segmented based on seed treatment, soil treatment, and others (Fig. 24.2).

24.5 The challenges

The lack of awareness of biofertilizers and their direct comparison with their chemical counterparts in terms of performance has been the biggest challenge for the biofertilizer market. The established

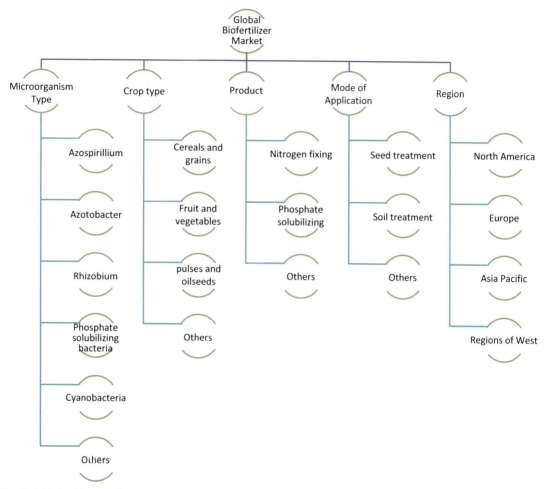

Fig. 24.2 Biofertilizer market segments based on seed treatment, soil treatment, and others.

distribution network of conventional fertilizer companies also poses a challenge for the biofertilizer products to get into the supply chain.

24.6 Market ecosystem

The increasing consumer preferences for organic fruits and vegetables have become the prime factor for market growth and dynamics of the biofertilizer market. Based on how they are applied, the soil treatment segment is the one that is projected to witness higher growth as compared to others in the biofertilizer market. Based on type, the nitrogen-fixing segment is estimated to grow at a higher rate in the biofertilizer market (Fig. 24.3).

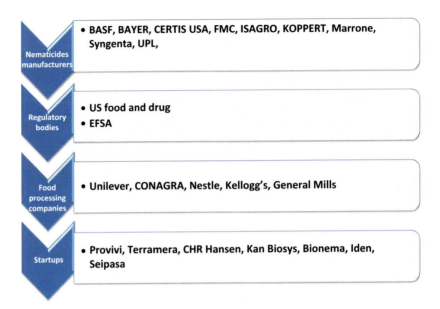

Fig. 24.3 Profile of key players in the industry.

24.7 Biofertilizers and Indian agriculture

The availability and affordability of fossil fuel-based chemical fertilizers at the farm level in India have been ensured only through imports and subsidies. In such a scenario, biofertilizers have emerged as a highly potent alternative to chemical fertilizers because of their eco-friendly, easy-to-apply, non-toxic, and cost-effective nature. They also make available the nutrients to plants that are naturally abundant in soil or atmosphere and act in supplementing the deficit of agrochemicals. In addition, they are a type of product that is likely to be commercially promising in the long run with the onset of adequate information available to producers and farmers through experience and communication. In India, the government has been trying to increase the application of biofertilizers alongside modern agrochemicals, thus reducing the use of agrochemicals and also minimizing the dependency on importing the raw materials to produce them.

The distribution of biofertilizers and its adoption rate had not steadily grown over time; rather, it slowed down in the late 1990s. Firstly, in industrial growth parameters like this, one can imagine that a faster and possibly accelerating growth performance can be expected from industries starting from a small base as the input finds greater acceptance in the market. Secondly, as more and more new entries emerge in the market, the average capacity comes down, resulting in the kind of industry formed and led by a large number of smaller units. While size adjustment in infant industries is a normal phenomenon, it must be kept in mind that success in the value chain of an agro-input also calls for substantial sales networking and a detailed understanding of the field reality in agriculture (Ghosh, 2004).

24.8 Market potential and constraints

Risky investment followed by high competition and inadequately trained personnel for marketing is the topmost constraint in the marketing of biofertilizers and biopesticides (Tidke, 2011). One of the main barriers that producers and investors face is inadequate demand and seasonal inconsistency of the existing demand. It has to be noted, however, that technology in itself is nascent and evolving. The rice-dominated eastern region remains conservative to adopt, and the wheat-rice growing northern region has not shown much interest either (Luft and Korin, 2009). The central and state governments in India have been promoting the use of biofertilizers by extending support through various project fundings, extensions, and subsidies on sales with varying degrees of emphasis (Alam, 2000). Over time, as the industry develops from infancy to its growth with public guidance, the only reasons responsible are going to be (a) increasing sales volumes and diffusion across the country and (b) better role of profit-motivated private enterprises (Mazid and Khan, 2014).

Economically, biofertilizers are profitable to farmers as they contribute to higher nutrient use efficiency, better break-even margins, and comparatively reduced quantities of chemical fertilizers (Mishra and Dash, 2014). Table 24.1 below depicts the shreds of evidence of the profitability of biofertilizers in selected countries where the successful application of biofertilizers has resulted in a significant reduction in the use of chemical fertilizers in the selected crops. This can be quite significant to policymakers to understand farmers' decisions related to the incorporation of biofertilizers into their fields (Masso et al., 2015). As can be seen, the PSB, *Azotobacter*, and K-solubilizing bacteria have significantly contributed to a reduction in the use of chemical fertilizers to the tune of 25% to 50%. Biofertilizers based on P and K solubilization are the most promising products for South Asian markets as the soils in the region are deficit of the two nutrients and thus can help evade the geography from nutrient deficit menace that may hit the area in not so distant future which will render very little patches of the land fit for agriculture. However, as mentioned earlier in the chapter, the current leading biofertilizer products are based on nitrogen fixation and the same is going to be the trend for a couple of years.

Looking at the table, it is evident that the application of biofertilizers poses immense opportunities in terms of better economic return because they are cheaper than chemical fertilizers with a benefit-cost ratio of 10:1 (Tiwari et al., 2004). Despite their ability to contribute to sustainable yield enhancement and an economic benefit that cannot

Table 24.1 Extent of fertilizer use reduction after various biofertilizer applications on various crops.

Country	Biofertilizer	Reduction of chemical fertilizer rate (%)	Name of crop
India	PSB	25	Sugarcane
India, West Bengal	*Azotobacter* and PSB	25	Mustard
Thailand	*Bacillus cereus*	50	Rice
Iran	*Azotobacter*	25	Black cumin
Pakistan	K-solubilizing bacteria, *Azotobacter*, *Azospirillum*, *Azoarcus*, and *Zoogloea*	50	Maize
Colombia	*Azospirillum brasilense*, *A. Amazonense* and *Azotobacter*	75	Cotton and rice
	Rhizobium (Rh) and *Bacillus megaterium*		Snap bean
Egypt	*Azospirillum* and *Azotobacter*	100	Okra
	Azotobacter sp., *Azospirillum* sp. and	50	Flax
	PSB containing *Bacillus* sp. *Azotobacter* and *Azospirillum*	100	Maize

be ignored, biofertilizers are not widely adopted and are available to small and marginal farmers worldwide. In geographies like Asia especially India, growers are recognizing the benefits of biofertilizers gradually. Apart from the obvious use in fruits and vegetables, they are now being used in traditional crops like soybean and sugarcane too. This indicates a huge opportunity lying in the future of a growing market like India where the overall agrochemical market is estimated to be more than 16,000 crores. Sustained monsoons and growing demand for natural products from consumers are all the reasons due to which the demand and use of biostimulants are rising at the rate of 15% annually in India.

24.9 Impact of COVID-19 on biofertilizer market and the future

COVID-19 has impacted businesses all over, and biofertilizer businesses are no exception. However, they have been able to manage their global operations and supply chains. The companies with multiple manufacturing facilities had little impact on the production cycles.

As with every sector, the pandemic has also negatively impacted the global biofertilizer market due to emergent situations and changes in socioeconomic conditions. This has resulted in a dwindling supply chain of raw materials. Scattered supplies of raw materials have also impacted the manufacturers of biofertilizers across economies just like chemical fertilizers. Uneven availability of logistics and market uncertainties have emerged as a significant problem at present, which indicates that the same is going to be the trend in the future as well leading to a demand fall for biofertilizers. The same, however, cannot be said about the future as the situations are bound to improve after the much-needed interruptions by various stakeholders and governments to improve the situation in general and supply chain in particular leading to easier predictions of biofertilizer markets.

References

Alam, G., 2000. A Study of Biopesticides and Biofertilisers in Haryana, India. International Institute for Environment and Development.

Ghosh, N., 2004. Promoting biofertilisers in Indian agriculture. Econ. Polit. Wkly. 39 (52), 5617–5625. Available at: https://www.epw.in/journal/2004/52/review-agriculture-review-issues-specials/promoting-biofertilisers-indian. (Accessed 30 June 2021).

Luft, G., Korin, A., 2009. Turning Oil into Salt: Energy Independence through Fuel Choice. Booksurge Llc.

MarketWatch, 2021. BioFertilizers Market Size 2021 Share, Growth By Top Company, Business Opportunity, Regional Analysis, Application, Driver, Trends & Forecasts By 2030. Available at: https://www.marketwatch.com/press-release/biofertilizers-market-size-2021-share-growth-by-top-company-business-opportunity-regional-analysis-application-driver-trends-forecasts by-2030-2021-06-11?tesla=y. (Accessed 16 May 2021).

Masso, C., et al., 2015. Worldwide contrast in application of bio-fertilizers for sustainable agriculture: lessons for sub-Saharan Africa. J. Biol. Agric. Healthc. 5 (12), 34–50.

Mazid, M., Khan, T.A., 2014. Future of bio-fertilizers in Indian agriculture: an overview. Int. J. Agric Food Res. 3 (3), 10–23. https://doi.org/10.24102/ijafr.v3i3.132.

MicroMarketMonitor.com, 2015. North America Bio Fertilizer Market by Application, Type, Source, Geography—2019. Available at: http://www.micromarketmonitor.com/market/north-america-bio-fertilizer-5250154124.html. (Accessed 16 July 2021).

Mishra, P., Dash, D., 2014. Rejuvenation of bio-fertilizers for sustainable agriculture and economic development. Consilience: J. Sustain. Dev. 11 (1), 41–61.

Mulvany, L., 2020. Bill Gates Invests in Clean Air-to-Nitrogen Fertilizer Startup—Bloomberg.

prnewswire, 2017. Global Bio-Fertilizers Market, 2022—$2.1 Billion Market By Product Type, Crop Type, Microorganism, Regions and Vendors—Research and Markets. Available at: https://www.prnewswire.com/news-releases/global-bio-fertilizers-market-2022---21-billion-market-by-product-type-crop-type-microorganism-regions-and-vendors---research-and-markets-300453398.html. (Accessed 14 May 2021).

Tidke, V., 2011. Market Potential and Marketing Strategies for Biofertilizers and Biopesticides in Nasik District (MS). JAU, Junagadh. Available at: https://krishikosh.egranth.ac.in/handle/1/5810005165. (Accessed 30 June 2021).

Tiwari, P., Adholeya, A., Prakash, A., Arora, D.K., 2004. Commercialization of arbuscular mycorrhizal biofertilizer. In: Arora, D.K. (Ed.), Fungal Biotechnology in Agricultural, Food, and Environmental Applications. vol. 21, pp. 195–203.

25

Organic agriculture for agro-environmental sustainability

Neelam Thakur[a], Simranjeet Kaur[a], Tanvir Kaur[b], Preety Tomar[a], Rubee Devi[b], Seema Thakur[c], Nidhi Tyagi[c], Rajesh Thakur[c], Devinder Kumar Mehta[d], and Ajar Nath Yadav[b]

[a]Department of Zoology, Akal College of Basic Sciences, Eternal University, Baru Sahib, Himachal Pradesh, India, [b]Department of Biotechnology, Dr. Khem Singh Gill Akal College of Agriculture, Eternal University, Baru Sahib, Sirmour, Himachal Pradesh, India, [c]Krishi Vigyan Kendra Kandaghat, Solan, India, [d]Dr. Y.S. Parmar University of Horticulture and Forestry, Nauni-Solan, India

25.1 Introduction

Man, *Homo sapiens* L., existed as a species on this earth approximately about 1–2 million years ago. During that period, he collected wild plants as food in the form of juicy fruits, seeds, stems, and roots along with wild animals that he may possibly grasp by fishing and hunting. He was so known as a hunter-gatherer in our cultural prehistory which still continues today in some aboriginals. In the late early Neolithic or Mesolithic age (10,000 years ago), humans started to domesticate or cultivate plants. His activities and life were modernized by agriculture, which facilitated him to dump his nomadic lifestyle and settle down in enduring habitations. The practice of farming started with the slow alteration of wild varieties into cultivated plants (Panda and Khush, 1995). Rural societies at that time continually laid emphasis on the acquirement of textiles, food, and other resources of plants and animals in origin. Agricultural production increased with the use of high-yielding varieties and the extensive use of NPK fertilizers which also inflated the harmful effects on the primary source. Currently, health problems related to intensive modern agriculture, like pesticide residues in food products and groundwater contamination, are a significant matter of concern (Kour et al., 2021a; Kumar et al., 2021). The negative impacts caused by modern agriculture led to the origin of new ideas in farming that are sustainable and imply the judicious use of resources, and this new method can be broadly termed as organic farming.

In April 1991, the Netherland conference on agriculture and environment/FAO adopted the idea of Sustainable Agriculture and Rural Development (SARD) as an innovative strategy. The main objectives are to provide employment to rural people, income generation, food security, and natural resource conservation as well as environmental protection. According to United Nations Development Programme (UNDP) in agriculture, sustainable development represents conservation of national food security and work to improve the living standard of rural people who mainly depend on agricultural farming for a living and at the same time assured that the natural resource on which they and their children depend is no more depleted but in turn these resources are regenerated.

Organic farming is a type of production approach which avoids the use of artificial synthetic chemical fertilizers, growth regulators, pesticides, and livestock feed additives. In developed countries, the majority of consumers believed organic foods to be healthier and safer as compared to traditionally produced foods (Funk and Kennedy, 2016). Consumers all over the world recommended organic farming as a better option for climate protection and environment and animal welfare (Seufert et al., 2017). Awareness among people toward organic farming is still low in developing countries, but at the same time in European countries, preferences for organic products are gaining a strong position (Probst et al., 2012). Today, more than 100 countries widely support organic farming (Seufert et al., 2017), and private and governmental standards are characteristically based on the principles developed by IFOAM.

Organic systems to a maximum extent rely on crop rotations, animal manures, crop residues, green manures, and off-farm organic wastes and at the same time deal with biological pest management to improve soil productivity and tillage, supply nutrients to plants, and manage the population of insects, pests, and weeds. Traditional agricultural production results in environmental harms such as biodiversity loss, climate change, water pollution, and soil degradation (Foley et al., 2011). Organic farming is based on the idea of giving "Back to Nature" in which the land is cultivated and crops are raised in a way that keeps the physiological state of the soil alive, healthy, and free from pollutants. Organic agriculture is a complete production and management system that emphasizes the health of agroecosystem, diversity of flora and fauna, biological processes of organisms associated, and physiological activities of soil. The term "organic" was employed the first time in reference to farming by Northbourne (2003) in the book "Look to the Land, 1940." The International Food Standards (IFS), Codex Alimentarius, describes organic agriculture as a holistic production management system that promotes the enhancement of agroecosystem health, along with diversity and soil biological activities. It laid emphasis on the utilization of management practices through the use of off-farm inputs and also took into consideration the need for domestically adapted systems.

The International Federation of Organic Agriculture Movement (FiBL & IFOAM), the worldwide umbrella organization for organic agriculture, was founded in 1972 in Versailles, near Paris, by efforts of five organizations from three continents. Organic agriculture defined by FiBL & IFOAM states that "organic agriculture embraces all agricultural systems that promote the economically, environmentally, and socially complete production of food and fibers." Organic agriculture dramatically reduces harmful external inputs into the environment by restricting the utilization of chemosynthetic fertilizers and pesticides. Agrosynthetic insecticides/pesticides are not utilized in organic farming. Planned renovation of the farm with an organic organization is essential for organic production. This organic management includes a balanced addition of nitrogen-fixing leguminous crops, multiannual crop rotation, green manuring crops, and deep rooting to increase soil fertility. Artificial nitrogen fertilizers are not recommended in organic farming as they have a high potential to upset natural plant growth prototypes and results in weakened soil fertility. By-products of organic farm animals/livestock farming are only imported from an established organic unit into a new organic unit. The harvest or by-product of organic farm animals are manure, urine, compost, and slurry. In organic farming, all the inputs come from own farm, and whenever possible, both animal and plant production are integrated. Several studies suggested that organic foods have a low level of pesticide residues (Barański et al., 2014; Dangour et al., 2010; Huber et al., 2011). Organic agriculture adheres to globally accepted principles, which are enforced among local socioeconomic, climatic, and cultural settings. As a logical consequence, IFOAM stresses and supports the development of independent systems on local and regional levels (Weidmann et al., 2007) (Fig. 25.1).

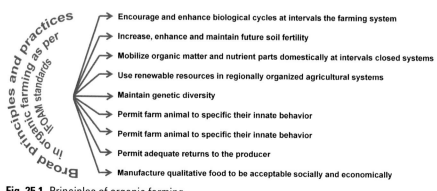

Fig. 25.1 Principles of organic farming.

25.2 Standards of organic agriculture

In organic agriculture, there are a set of standards that define a practice as organic. These standards address the varied aspects of organic production (Fig. 25.2).

25.3 Status of organic agriculture

Organic agriculture practices mainly focus on increasing the farmland as well as crop productivity and maintaining soil health by applying organic waste materials, such as cow dung and other farm wastes along with other biological materials including microbial fertilizers to increase the productivity of the soil and have no harmful effect on the environment (Guruswamy and Gurunathan, 2010; Makadia and Patel, 2015; Narayanan and Narayanan, 2005). The importance of organic agriculture has been considered all over the world to meet the need for sustainable farming.

25.3.1 Worldwide

Over the last three decades, since 1985, the land area for organic agriculture is increasing gradually. According to the survey conducted by the Research Institute of Organic Agriculture in 2015, the land for organic agricultural practices constitutes about 50.9 million hectares. The highest land area for organic agriculture is about 22.7 m hectares in Australia. The second-largest area is in Argentina making up 3.1 m hectares of organic agricultural land followed by 2 m hectares of area in the United States. An extensive increase in organically cultured land has been observed in several African countries including Zimbabwe, Kenya, Madagascar, and Ivory Coast. But the growth rate of 69.8 m hectares in the organically cultivated land has been observed by 2017 with the highest land area in Australia (35.65 million hectares). By

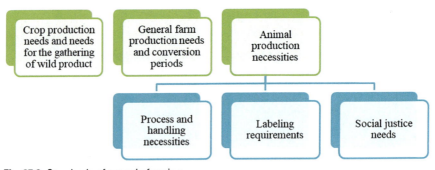

Fig. 25.2 Standards of organic farming.

2018, about 71.5 m hectares of land areas being practiced for organic agriculture were observed with the largest organically cultivated area in the Oceania region (36 m hectares) (Fig. 25.3; Table 25.1).

Throughout the world, organically cultivated farmland constitutes about 1.5% of the world's total agricultural land. The largest organic

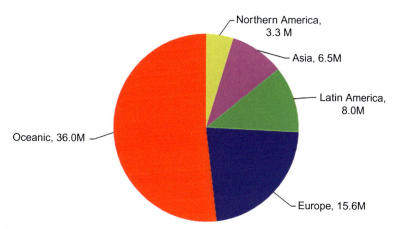

Fig. 25.3 Continental organic areas. Source: FiBL & IFOAM Survey, 2018, Accessed in July 2020.

Table 25.1 Global scenario of organic agriculture.

Country	Organic area (farmland) (ha)	Organic area share of total farmland (%)	Organic producers	Number of affiliations of IFOAM
Afghanistan	786	00	10	00
Albania	746.54	0.06	82	00
Algeria	772	00	64	00
Andorra	2.0	0.01	1.0	00
Argentina	3,629,968.00	2.44	1366.00	00
Armenia	693.7	0.04	35	00
Australia	35,687,799.00	8.78	1829.00	00
Austria	637,805.00	24.66	25'795.00	00
Azerbaijan	37,630.00	0.79	305	00
Bahamas	48.56	0.35	1.0	00
Bangladesh	503.9	0.01	9335.00	00
Belarus	1655.63	0.02	24	00
Belgium	89,025.00	6.81	2264.00	00

Continued

Table 25.1 Global scenario of organic agriculture—cont'd

Country	Organic area (farmland) (ha)	Organic area share of total farmland (%)	Organic producers	Number of affiliations of IFOAM
Belize	219.65	0.14	150	00
Benin	16,453.83	0.44	4030.00	00
Bhutan	6632.00	1.26	4354.00	00
Bolivia (Plurinational State)	114,305.63	0.3	12,114.00	00
Bosnia and Herzegovina	896.4	0.04	251.00	00
Botswana	00	00	2.0	00
Brazil	1,188,254.81	0.42	17,507.97	00
Bulgaria	162,332.37	3.49	6471.00	00
Burkina Faso	56,663.02	0.47	26,627.00	00
Burundi	163.82	0.01	16	00
Cambodia	27,550.30	0.51	5788.00	00
Cameroon	1088.53	0.01	499	00
Canada	1,311,571.81	2.01	5791.00	00
Cape Verde	495.00	0.59	1.0	00
Chad	00	00	1.0	00
Channel Islands	180.00	1.89	00	00
Chile	16,305.00	0.10	1609.00	00
China	3,135,000.00	0.61	6308.00*	00
Colombia	22,314.08	0.05	3496.00	00
Comoros	2142.00	1.61	680.00	00
Cook Islands	24.00	1.60	58.00	00
Costa Rica	8964.34	0.49	50.00	00
Croatia	103,166.00	6.57	4374.00	00
Cuba	6181.20	0.10	510.00	00
Cyprus	6022.40	5.36	1249.00	00
Czech Republic	538,893.66	12.82	4601.00	00
Côte d'Ivoire	50,573.53	0.06	2776.00	00
Democratic Republic of the Congo	60,624.30	0.23	30,170.00	00
Denmark	256,711.00	9.80	3637.00	00
Dominica	239.77	0.96	00	00
Dominican Republic	169,026.00	7.19	16,119.00	00
Ecuador	41,792.90	0.75	12,912.00	00
Egypt	116,000.00	3.10	970.00	00
El Salvador	1678.80	0.10	380.00	00
Estonia	206,590.00	21.58	1948.00	00
Eswatini	185.70	0.02	2.00	00

Table 25.1 Global scenario of organic agriculture—cont'd

Country	Organic area (farmland) (ha)	Organic area share of total farmland (%)	Organic producers	Number of affiliations of IFOAM
Ethiopia	186,155.10	0.51	203,602.00	00
Falkland Islands (Malvinas)	31,937.00	2.88	4.00	00
Faroe Islands	251.15	8.37	1.00	00
Fiji	41,154.07	9.68	67.00	00
Finland	297,442.00	13.03	5129.00	00
France	2,035,024.00	7.34	41,632.00	00
French Guiana (France)	3103.00	10.14	75.00	00
French Polynesia	1512.19	3.32	12.00	00
Gambia	19.8	0.00	1.00	00
Georgia	1451.83	0.06	1075.00	00
Germany	1,521,314.00	9.09	31,713.00	79
Ghana	29,662.99	0.19	3228.00	00
Greece	492,627.00	6.03	29,594.00	00
Grenada	84.00	00	23.00	00
Guadeloupe (France)	272.00	0.51	63.00	00
Guatemala	14,000.00	0.37	6346.00	00
Guinea	10.00	0.00	1.00	00
Guinea-Bissau	834.65	0.05	1.00	00
Haiti	4403.40	0.24	4661.00	00
Honduras	29,273.53	0.90	6023.00	00
Hungary	209,382.00	4.50	3929.00	00
Iceland	24,855.20	1.33	29.00	00
India	1,938,220.79	1.08	1,149,371.00	55
Indonesia	251,630.98	0.44	18,162.00	00
Iran (Islamic Republic of)	11,915.88	0.03	20.00	00
Iraq	62.50	00	00	00
Ireland	118,699.00	2.39	1725.00	00
Israel	6665.70	1.24	349.00	00
Italy	1,958,045.00	15.79	69,317.00	00
Jamaica	374.00	0.08	127.00	00
Japan	10,792.00	0.24	3678.00	00
Jordan	1446.00	0.14	23.00	00
Kazakhstan	192,133.60	0.09	63.00	00
Kenya	154,488.00	0.56	37,295.00	00
Kiribati	1600.00	4.71	900.00	00
Kosovo	160.00	0.04	150.00	00
Kuwait	21.70	0.01	1.00	00

Continued

Table 25.1 Global scenario of organic agriculture—cont'd

Country	Organic area (farmland) (ha)	Organic area share of total farmland (%)	Organic producers	Number of affiliations of IFOAM
Kyrgyzstan	22,117.60	0.21	1107.00	00
Lao People's Democratic Republic	7668.25	0.32	1342.00	00
Latvia	280,383.00	15.44	4178.00	00
Lebanon	1241.35	0.19	111.00	00
Lesotho	0.83	00	3.00	00
Liberia	2.00	00	00	00
Liechtenstein	1413.00	38.52	46.00	00
Lithuania	239,691.00	8.27	2476.00	00
Luxembourg	5782.00	4.41	103.00	00
Madagascar	48,757.00	0.12	32,367.00	00
Malawi	12,398.90	0.21	295.00	00
Malaysia	9575.70	0.12	29.00	00
Mali	12,654.55	0.03	12,272.00	00
Malta	47.20	0.46	19.00	00
Martinique (France)	398.00	1.26	64.00	00
Mauritius	2.53	00	22.00	00
Mayotte	35.00	0.29	3.00	00
Mexico	183,225.00	0.17	27,000.00	00
Moldova	17,151.42	0.70	135.00	00
Mongolia	635.77	00	13.00	00
Montenegro	4454.68	1.93	328.00	00
Morocco	9917.00	0.03	277.00	00
Mozambique	14,933.46	0.03	269.00	00
Myanmar	12,305.20	0.10	48.00	00
Namibia	66.00	00	8.00	00
Nepal	11,851.20	00	1622.00	00
Netherlands	57,904.00	3.13	1696.00	00
New Zealand	88,871.00	0.80	876.00	00
Nicaragua	34,786.96	0.69	8193.00	00
Niger	253.9	00	2.00	00
Nigeria	57,116.95	0.08	1091.00	00
Niue	43.19	0.86	1.00	00
North Macedonia	4409.00	0.35	775.00	00
Norway	46,377.00	4.66	2057.00	00
Oman	42.64	0.00	5.00	00
Pakistan	64,885.30	0.18	415.00	00

Table 25.1 Global scenario of organic agriculture—cont'd

Country	Organic area (farmland) (ha)	Organic area share of total farmland (%)	Organic producers	Number of affiliations of IFOAM
Palestine	4870.15	1.63	1440.00	00
Panama	5929.00	0.26	18.00	00
Papua New Guinea	49,573.45	4.17	12,742.00	00
Paraguay	42,818.00	0.2	5187.00	00
Peru	311,460.99	1.28	103,554.00	00
Philippines	218,569.80	1.76	12,366.00	00
Poland	484,676.17	3.36	19,224.00	00
Portugal	213,118.00	5.85	5213.00	00
Puerto Rico	14.40	0.01	5.00	00
Republic of Korea	24,700.00	1.41	15,500.00	00
Romania	326,260.00	2.50	7908.00	00
Russian Federation	606,974.98	0.28	40.00	00
Rwanda	2130.00	0.12	3870.00	00
Reunion (France)	1272.00	2.58	306.00	00
Samoa	97,655.56	34.51	2038.00	00
Sao Tome and Principe	10,934.20	22.45	3564.00	00
Saudi Arabia	18,631.00	0.01	6.00	00
Senegal	7989.41	0.09	10,369.00	00
Serbia	19,254.58	0.55	373.00	00
Seychelles	00	00	1.00	00
Sierra Leone	99,230.30	2.51	304.00	00
Singapore	2.60	0.39	00	00
Slovakia	188,986.00	9.97	439.00	00
Slovenia	47,848.28	9.85	3738.00	00
Solomon Islands	4714.33	4.37	1098.00	00
South Africa	82,818.38	0.01	237.00	00
Spain	2,246,475.00	9.64	39,505.00	00
Sri Lanka	77,169.04	2.82	1416.00	00
Sudan	76,941.40	0.11	3.00	00
Suriname	93.55	0.11	39.00	00
Sweden	608,758.00	19.85	5801.00	00
Switzerland	160,991.94	15.39	7032.00	00
Syrian Arab Republic	19,987.00	0.14	2458.00	00
Taiwan	8759.03	1.10	3556.00	00
Tajikistan	8806.41	0.19	953.00	00
Thailand	95,065.91	0.43	58,490.00	00
Timor-Leste	63,882.00	16.81	4.00	00

Continued

Table 25.1 Global scenario of organic agriculture—cont'd

Country	Organic area (farmland) (ha)	Organic area share of total farmland (%)	Organic producers	Number of affiliations of IFOAM
Togo	41,323.04	1.08	38,414.00	00
Tonga	684.50	2.07	1060.00	00
Tunisia	286,623.00	2.85	7456.00	00
Turkey	646,247.00	1.68	79,563.00	00
Uganda	262,282.00	1.82	00	00
Ukraine	309,100.00	0.72	501.00	00
United Arab Emirates	4687.00	1.21	95.00	00
United Kingdom	457,377.00	2.66	3544.00	00
United States Virgin Islands	26.11	0.65	00	00
United Republic of Tanzania	278,467.12	0.70	148,610.00	00
United States of America	2,023,430.00	0.59	18,166.00	48
Uruguay	2,147,083.00	14.86	12.00	00
Uzbekistan	943	00	1.00	00
Vanuatu	25,648.20	13.72	47.00	00
Viet Nam	237,693.00	2.19	17,169.00	00
Zambia	1228.00	0.01	286.00	00
Zimbabwe	415.00	00	511.00	00

Source: FiBL & IFOAM survey, 2018 accessed in July 2020.

agricultural land is in Australia with a 35.6 m hectare area followed by 3.6 m hectares in Argentina and 3.1 m hectares in China. Till now, organic agriculture has been employed by more than 180 countries in the world. The highest organic share of 38.5% and 34.4% was reported in Liechtenstein and Samoa (Fig. 25.4; Table 25.2).

Among the continents, by 2018, the total certified agricultural land area in Africa is about 2 m hectares with 806,000 producers. In the Asian continent, about 1.3 m producers manage almost 6.5 m hectare area for organic farming. About 15.6 m hectares of organic agriculture land area was present in the European continent that is being handled by more than 418,000 producers. In Northern America, about 3.3 m hectare land is used for organic agriculture among which 2 m hectares is contributed by the United States and 1.3 m hectare share is managed by Canadian producers. Even in Latin America, 8 m hectare land is used to grow organic products by almost 228,000 producers of the continent. Almost 36.0 m hectare of the land area is managed organically by 21,000 producers in the Oceania continent (Australia, Pacific Islands, and New Zealand). Among these, only Australia constitutes about 99% of the organically farmed land (Willer et al., 2020) (Fig. 25.5, Table 25.3).

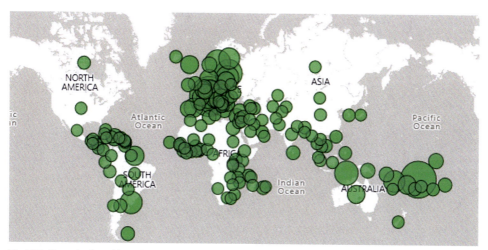

Fig. 25.4 Organically cultivated areas globally. Source: FiBL & IFOAM Survey, 2018, Accessed in July 2020.

Table 25.2 Continental data on organic agriculture.

Region	Number of countries with organic agricultural data	Countries share on providing data (%)
Asia	41	84
Africa	44	79
Europe	47	96
North America	03	60
Latin America	33	72
Oceania	13	52
World	181	79

Source: FiBL & IFOAM survey, 2018 accessed in July 2020.

Fig. 25.5 Growth rate of the organically cultivated area from 1999 to 2018. Source: FiBL & IFOAM Survey, 2018, Accessed in July 2020.

Table 25.3 Glimpse of organic agriculture statistics between 2018 and 2019.

Area	
Cultivated area (organic + in-conversion)	1938,220.79 ha
Wild harvest collection area	1,490,417.98 ha
Total area (cultivated + wild harvest)	34,28,638.00 ha or 3.56 million ha
Production	
Farm production	2,607,385.00 tons
Wild harvest production	37,930.67 tons
Total production	2,645,315.67 tons

25.3.2 National status of organic agriculture

In India, the thought of organic agriculture is much familiar from the ancient period during the time of Vedas or even before Vedas. Several short descriptions about organic cultivation were found in the ancient books related to the time of Ramayana, Rigveda, Arthashastra, and Mahabharata. Years back, numerous programs were initiated by Mahatma Gandhi on organic farming at various places that leads to the green revolution. At that time, about 279 programs were run based on organic farming. A large number of people go behind organic agriculture after the "Sevagram Declaration in 1994." In the last 20–25 years, organic farming is highly flourished in India. In the year 2001, National Programme on Organic Production (NPOP) started by the government and National Standards for Organic Production (NSOP) were also on track; after that, the organic farming sector started growing day by day. In 2001–02, the Government of India inaugurated Agricultural & Processed Food Products Export Development Authority (APEDA) under the Agricultural and Processed Food Products Export Development Authority Act. First Indian Organic Certification Agency (INDOCERT) was established in 2002. In 2003, the total authorized area for organic agriculture was about 76,236 ha. ICCOA arranged the first "Indian Organic Trade Fair" in the year 2005. After that, there is a progressive increase in the organic farm area and it reached 1,030,311 ha in 2008 (Menon, 2009) (Table 25.4).

According to APEDA, India is in the 9th position with regard to the world's organic agriculture and is in the top position in the matter of producers. About 3.56 m hectare area was registered in 2018 under National Programme for Organic Production, among which 50% is wild and 50% is organically cultivated. The largest certified area for organic cultivation is in Madhya Pradesh. The second largest is Rajasthan followed by Maharashtra and Uttar Pradesh. In 2016, a notable achievement was observed in Sikkim as more than 76,000 ha area (almost entire cultivated land area) was converted into the organically cultivated area with documentation (APEDA, accessed in July 2020) (Table 25.5).

Table 25.4 Organically certified area (cultivated + wild) from 2014 to 2019.

S. no.	State name	2014–15	2015–16	2016–17	2017–18	2018–19
1	Andaman & Nicobar Islands	321.28	00	00	00	7484.00
2	Andhra Pradesh	100,623.81	93,350.72	172,783.03	184,748.65	37,409.71
3	Arunachal Pradesh	3688.61	72,485.26	72,311.27	6179.68	9246.93
4	Assam	16,256.02	28,493.24	23,930.39	28,071.81	28,234.66
5	Bihar	247.10	91.70	679.20	695.8	3519.50
6	Chhattisgarh	32,405.10	180,924.93	179,752.13	191,464.66	206,180.71
7	Goa	15,621.24	16,957.59	15,762.43	15,698.97	20,964.80
8	Gujarat	49,862.00	80,421.40	70,495.05	85,400.71	94,708.68
9	Haryana	6783.21	4889.20	5031.75	6912.39	5998.58
10	Himachal Pradesh	1,370,744.0	1,358,449.24	14,376.72	170,153.46	203,847.50
11	Jammu & Kashmir	50,111.22	54,515.010	181,608.31	180,870.34	187,002.88
12	Jharkhand	71,383.60	77,048.73	36,813.94	51,187.934	58,116.87
13	Karnataka	92,157.00	133,647.27	81,948.80	105,515.02	104,962.37
14	Kerala	23,123.00	44,788.49	43,701.87	34,160.14	40,911.23
15	Lakshadweep	895.52	895.52	895.52	895.51	895.51
16	Madhya Pradesh	1,926,369.01	2,275,567.10	2,292,697.39	1,156,881.4	918,303.08
17	Maharashtra	217,649.19	266,299.23	292,391.78	304,074.81	261,571.74
18	Manipur	168.20	251.40	241.40	5397.9	7460.82
19	Meghalaya	4489.29	4609.42	9629.59	40,335.66	48,409.74
20	Mizoram	764.24	213.80	210.0	998.95	7039.89
21	Nagaland	8362.43	6186.93	4699.93	8839.86	8268.56
22	New Delhi	69.13	23.03	9.23	9.23	0.71
23	Odisha	91,056.40	109,224.04	99,736.16	117,910.30	127,851.77
24	Pondicherry	2.84	2.83	2.83	2.83	2.83
25	Punjab	19,293.58	17,577.20	17,648.53	18,000.76	25,524.58

Continued

Table 25.4 Organically certified area (cultivated + wild) from 2014 to 2019—cont'd

S. no.	State name	2014–15	2015–16	2016–17	2017–18	2018–19
26	Rajasthan	483,090.67	553,447.70	539,522.12	442,133.72	632,701.22
27	Sikkim	76,392.38	75,851.21	75,218.27	76,076.17	75,798.91
28	Tamil Nadu	12,536.97	19,529.79	10,775.68	20,070.50	26,546.82
29	Telangana	2902.83	10,355.58	9687.84	8919.82	8759.52
30	Tripura	203.56	203.56	203.56	2251.19	2534.51
31	Uttar Pradesh	107,529.11	106,292.39	101,459.94	192,734.40	205,980.81
32	Uttarakhand	92,480.23	99,900.38	93,586.41	104,134.66	41,409.55
33	West Bengal	16,266.61	17,890.41	5176.02	5811.48	20,989.65
Total		4,893,851	5,710,384	4,452,987	3,566,538	3,428,638.77

Table 25.5 State-wise data of organically cultivated farm area in India (2018–19).

S. no.	State name	Organic area (Ha.)	In conversion area (in Ha.)	Total area under certification process (ha)
1	Madhya Pradesh	379,996.68	294,055.16	674,051.85
2	Maharashtra	158,097.13	92,837.19	250,934.33
3	Rajasthan	110,240.21	113,751.25	223,991.46
4	Odisha	73,124.0	22,615.60	95,739.70
5	Gujarat	60,185.39	33,655.88	93,841.27
6	Karnataka	57,018.08	26,080.69	83,098.78
7	Sikkim	73,654.88	2144.03	75,798.91
8	Uttar Pradesh	44,802.36	18,035.77	62,838.13
9	Meghalaya	1612.69	46,797.04	48,409.74
10	Kerala	19,232.89	19,171.34	38,404.23
11	Uttarakhand	20,052.26	16,606.2	36,658.55
12	Andhra Pradesh	13,763.38	18,747.33	32,510.71
13	Assam	15,223.47	12,951.19	28,174.66
14	Jammu & Kashmir	17,558.75	7444.13	25,002.88
15	Jharkhand	2977.16	21,339.70	24,316.87
16	Tamil Nadu	4314.61	18,144.93	22,459.54
17	Chhattisgarh	7356.53	13,869.17	21,225.71
18	Goa	10,696.36	2612.45	13,308.82
19	Himachal Pradesh	8527.13	4537.90	13,065.03
20	Arunachal Pradesh	627.15	8619.78	9246.93
21	Punjab	317.75	8590.83	8908.58
22	Telangana	6322.92	2436.60	8759.52
23	Nagaland	2751.17	5517.38	8268.56
24	Andaman & Nicobar Islands	0.000	7484.00	7484.0
25	Mizoram	0.000	7039.89	7039.89
26	West Bengal	4984.20	1305.45	6289.65
27	Haryana	2291.85	3686.63	5978.48
28	Manipur	241.40	5219.42	5460.82
29	Bihar	1.20	3518.30	3519.50
30	Tripura	203.56	2330.95	2534.51
31	Lakshadweep	895.51	0.00	895.51
32	Pondicherry	2.83	0.00	2.83
33	New Delhi	0.71	0.00	0.71
Total		1,097,074.39	841,146.39	1,938,220.79

By analyzing the reports provided by the Ministry of Agriculture & Farmers Welfare, Govt. of India (GOI), a major increase in the certified organic area has been observed. According to GOI, in 2009–10, the area registered for organic cultivation was about 4.55 lakh hectares which was raised up to 7.23 lakh hectares in 2013–14 with more than 6-lakh producers. To boost the organic farming system in India, National Mission for Sustainable Agriculture (NMSA) along with "Paramparagat Krishi Vikas Yojana" has been implemented to raise the organic registered area.

According to APEDA, about 1.70 MT organic products of all varieties were produced in 2017–18. Among them, sugarcane, pulses, tea, spices, coffee, vegetables, oil seeds, cereals, fruits, dry fruits, and medicinal plants are the major ones even though organic cotton fibers and other products were also processed. Organic products are not only produced but also exported to other countries. In 2017–18, the quantity of export was 4.58 lakh MT and the exported food product accounts for 515.44 million USD (Table 25.6).

25.3.3 Special reference to northeast states in India

In the northeast hill region of India, organic farming is an economic, eco-friendly, and suitable method for food production. The states of northeast like Arunachal Pradesh, Sikkim, Assam, Mizoram,

Table 25.6 Major highlights of organic food market (FiBL and IFOAM, 2020).

Year	Highlights
1999	In International Trade Centre (ITC), Geneva, for the first time organic market data was published
2001	Ecovia Intelligence formed for analyzing the organic produces
	Organic food sales reached $20 billion
	In Germany, Bio-Siegel was launched as a national organic logo
2002	USDA National Organic Programme (NOP) implemented
2005	National Organic Standards were introduced by China
2008	Organic food sale crossed $50 billion Globally
2009	EU implemented revised organic regulation
	Brazil introduced national organic standards
2010	Introduction of current EU organic logo
2012	US-EU organic equivalency agreement
2014	Surge in demand for organic baby food gives China the largest organic infant formula market
2016	US passes GM labeling bill, whilst non-GMO project product sale reached $20 billion
2017	Amazon buys the world's most leading natural & organic food retailer whole foods market

Manipur, Nagaland, Tripura, and Meghalaya are ritually organic. In India, about 8% of the total geographical area was covered by the northeast region which is about 2.6 lakh sq. km. About 4% of the population in India resides in this geographical area (Babu et al., 2017) as this area is separated from the main land so the culturing practices during the green revolution do not reach them or are accomplished very slowly. So, people in this region still use the ancient and traditional method of farming "Jhum" Jhuming. They use nutrient-rich soil formed by putrefaction of organic matters. Although this technique offers less production with low income, by combining this with scientific farming practices, the productivity increases and the farmers get much more income. After longer experimentation from 2004 to 2017 by the ICAR research complex for NEH region, Umiam, the farming practice resulted in increased productivity (Das et al., 2019).

In Sikkim, all agricultural land is registered for organic farming and it becomes the first organic state in India. The approach implemented by the government to gradually reduce chemical fertilizers and pesticides and finally ban them on sale becomes a boon to the producers of Sikkim. Due to this change, approximately 66,000 farmers' families get benefitted. This approach boosted the tourism sector in the state even the "Future Policy Gold Award" (gold prize) by the UN Food and Agriculture Organization (FAO) was also awarded to the state.

In Meghalaya, under the rice and maize field, the effect of organic nutrient resources was tested on soil health and its productivity by ICAR Research Complex for NEH Region, Umiam, Meghalaya, since 2005. Farmyard manure (FYM), local compost (LC), vermicompost, integrated nutrient sources (one-third each of FYM + VC + LC), and control were the five nutrient sources used. Application of integrated nutrient sources as organic manure is found to be the best management practice followed by FYM that increased productivity as well as maintained soil health (Munda et al., 2014).

Many women farmers of Mizoram indulged themselves in activities, such as organic farming, horticulture, and floriculture. By doing this, they became independent, helping the family financially and also supporting the economy of the state. The state government helps them by running suitable training programs of composting and utilization of animal manure and crop residue, green manure, and vermiculture for the promotion of organic farming. By applying these techniques, Ms. Laltanpuii succeeded in achieving higher productivity as well as higher earnings. She grew mustard and collected about 500 kg of mustard leaves from a 10-acre land and makes about rupees 1.5 lakh annually. She further said that the Government of Mizoram under the mission "Organic Value Chain Development for the Northeast Region" is more concerned about farming of organic ginger, tea, Mizo chili, and turmeric (AIR, 2020). Producers of organic cultivation in Nagaland were known as "Naga farmers" and the Nagaland Government has registered the organic produces under the brand name "Naga Organics" to help the producers in marketing. In Naga farmers, a young farmer "Lanaukun Imchen" in Nagaland (Dimapur) grows organic mint, ginger, tea, and other spices in about 24.7 acres of land area. He works on the integrated farming system with a mixed cropping system. For the last 5 years, he has been promoting the organic farming system and is awarded as the best Indian organic farmer by the Union Minister of the State for Agriculture and Farmer's Welfare, Krishna Raj, on December 28, 2019, in New Delhi. He is selling his produces online as well as exporting them to other countries also (Ambrocia, 2019).

In the Manipur state also, farmers are doing organic agriculture traditionally that becomes the major source of their livelihood. A farmer "Mayengbam Shyamchandra Meitei" has been awarded several awards because of his major contribution into integrated organic-based farming in promoting integrated organic agriculture. Meitei prepared organic fertilizers from his vermicompost plant. He said that "a country cannot dream about an economically self-reliant state without bringing green revolution." This integrated agricultural system involves the use of natural resource management, and the earnings he gets are helpful for his livelihood. He further said that the government also provides financial help to promote organic farming (Ani, 2019).

25.4 Pros and cons of organic agriculture

Pros: Health is the most pronounced word of today's time, whether of humans or the environment it has become a major concern (Kirchmann et al., 2008). Environmental health has been ruined as the industrial revolution resulted in heavy pollution of Mother Earth. Conventional farming, in which a huge amount of various chemicals are being utilized and industrial wastes are released into the land, water, and air, has severely affected the environment (Kaur et al., 2020; Yadav et al., 2021). Diversity decline of flora and fauna, diseases widespread among human race, and loss of fertility and nutrient content of soil are some affects that are caused by human activities. It is known that these entire problems are caused by the conventional methods of agriculture. Conventional agriculture not only affects the environment but also affects the health of human. Organic farming in comparison to conventional farming is a more sustainable way of farming (Fig. 25.6). This farming method uses the natural resources of the earth to get several benefits. Food grown by this farming practice produces organic foods (foods free from chemicals), which are more in demand mainly in industrialized countries as people are more conscious about their health (Willer et al., 2009). Organic farming also has a beneficial impact on soil health by reducing the harmful pollutants

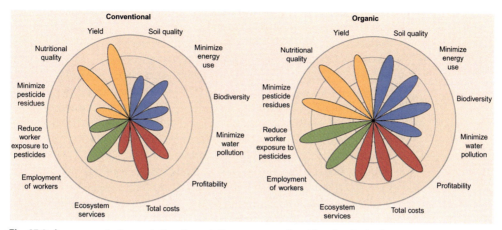

Fig. 25.6 Assessment of organic farming relative to conventional farming in the four major areas of sustainability. Adapted with permission from Reganold, J.P., Wachter, J.M., 2016. Organic agriculture in the twenty-first century. Nat. Plants 2, 15221. https://doi.org/10.1038/nplants.2015.221.

by decomposing them and also increasing the flora, fauna, and microbe biodiversity (Nandwani and Nwosisi, 2016). Some health benefits of organic produce are:

- It contains fewer amounts of nitrites, nitrates, and pesticidal residues than usual crops.
- Organic products have a good amount of phenolic compounds, essential amino acids, vitamin C, and sugar present in them.
- A lot of magnesium, iron, phosphorus, chromium, molybdenum, iodine, selenium, boron, copper, manganese, calcium, potassium, vanadium, sodium, and zinc are present.
- Vegetables and fruits obtained with the help of organic farming have a relatively higher quality and storage. They show less signs of decay, mass losses, and decomposition progressions.

Cons: Organic farming, the alternative to conventional agricultural practices (practices done with the use of synthetic chemical-based products), is one of the beneficial techniques that produces food grain by using natural resources of the planet. This sustainable method has several benefits, but besides that, it also has several cons, and one of the major challenges is yield (Kirchmann et al., 2008). Nowadays, yield is the major concern of an agriculturist as the population of the planet is drastically increasing and the need for agricultural products cannot be fulfilled by organic farming methods (Kaur et al., 2020). Organic farming, without a doubt, increases the soil organic matter and fertility for better farming, but this technique is not able to provide the basic nutrients that are required by the plants for their appropriate growth. This farming system does not have the ability to avail nutrients as it just can help in repairing damaged soil. The other major con of organic farming is weed population and pest attacks. Crop production is severely affected by pathogen attacks, i.e., insects, microbes, and weeds (Pattanapant and Shivakoti, 2013). Unlike conventional agriculture methods, in organic farming, there are no such weeds and pathogen inhibitors that control their growth and increase the productivity of cultivated crops. Organic farming progress is also limited by cons like insufficient labor as organic farming requires more manpower as compared to conventional agriculture, and at the same time, we have less agriculture fields available with farmers as compared to the earlier one because of industrialization (Pornpratansombat, 2006). This method also has some other limitations such as insufficient information and research techniques, limited amount of available organic inputs and organic market, weak infrastructure, and complications of organic standards (Nandwani and Nwosisi, 2016). The issue of affordable and flexible access to certification is also one of the constraints.

Some challenges that need to be addressed are:
- Maintaining flexibility in organic standards and processes of certification
- Development of agronomy for weeds, soil fertility, and animal health
- Expansion of research in this field
- Preservation and enhancement of storage of produce
- High prices of produces and problems of inconsistency in availability and quality
- Capacity building programs and training of relevant personnel
- Regulation and marketing strategies

25.5 Some of the products used in organic farming

In agriculture field, in order to practice organic farming, various types of products could be applied instead of chemically synthesized products. Cow dung, waste fodder, milk, plant growth-promoting microbes, and green manure are some different types of products used in agriculture.

25.5.1 Cow dung for soil fertilization and conditioning

Cow's dung is the major source of biofertilizers and is used for energy generation in numerous developing countries. It is a very effective alternative to chemical fertilizers via enhancing the long-term productivity while preserving soil quality and enhancing the microbial population. Cow dung compost and vermicompost increase the amount of soil organic matter and increase water absorption and investment capacity with increased cation exchange capacity (Raj et al., 2014). This is one of the renewable and sustainable sources of energy as dung cakes or biogas replaces the reliance on firewood, fossil fuel, charcoal, and fuel wood. In the mixed farming system, most livestock products are used and animals are fed for living from local resources, such as grass, crop residues, fodder trees, and shrubs. Farm animals (bullocks, buffaloes, and cows) also give us dung and urine to enrich the soil, while crop residues and fodder make up a bulk amount of feed for those animals (Kesavan and Swaminathan, 2008). Besides this, proper and safe application of cow dung will not only improve yield production but also prevent humans from diseases and reduce the chances of pathogenic diseases. Consequently, the improper use of cow dung should be avoided and used as organic compost to establish a healthy and sustainable agricultural system.

25.5.2 Waste fodder in composting

Disposal of organic waste material from various sources including household, agriculture, and industries has a serious problem causing environmental hazards. In India, around 320 million tons of agricultural wastes are produced annually (Suthar et al., 2005). At the research farm located in Maharashtra, Shibala, a field experiment was conducted to evaluate proper waste management methods. Various composting methods are used such as vermicomposting methods to increase productivity, quality, and yield. The vegetable and agricultural waste materials are used to prepare several composts and vermicomposts. In this process, for plants of maize, an analysis of growth as well as dry matter chemicals (N, P, K, Ca, water-soluble, and sugar reduction) was carried out. The total yield after harvesting was also estimated. In this experiment, Naikwade et al. (2012) recorded that maize treated with organic manure has maximum yield production with high nutrient quality and also maintains soil fertility. The use of vegetable waste as a starting material for composting and vermicomposting has better potential as compared to agricultural waste. Composting and vermicomposting are the most suitable methods used in organic forming or organic agriculture with which we convert the waste to wealth.

25.5.3 Milk for disease and weed control

In organic agriculture field, milk has been used against powdery mildew (*Sphaerotheca fuliginea*) disease in crop plants. After the application of milk as a spray in diseased plants or field, there is a reduction in diseases resulting in good health of plants. Earlier, Bettiol (1999) has used cow's fresh milk against zucchini squash (*Cucurbita pepo*) under greenhouse conditions. The efficiency of fresh cow milk was tested against powdery mildew on zucchini squash in five greenhouse conditions. Plants were sprayed with milk at different concentrations, i.e., 5, 10, 20, 30, 40, and 50%. Apart from this, fungicides (fenarimol 0.1 mL/L or benomyl 0.1 g/L) were applied once a week with control treatment. After that, the powdery mildew frequency was visually assessed weekly on individual leaves. High milk concentration was more effective as compared to traditional fungicides. This study showed that milk is a very effective tool against powdery mildew diseases in organic agriculture.

In another investigation, Crisp et al. (2006) reported that grapevine powdery mildew caused by the necator fungus *Erysiphe* (Uncinula) necator is a major disease affecting grape's yield and quality. This disease is controlled mainly by the daily application of sulfur plus synthetic fungicides in traditional vineyards and sulfur plus botanical and mineral oil in organic agriculture. In this research, milk is identified as a potential substitute for synthetic fungicides and sulfur against powdery mildew. *Lactoferrin* of milk initiated the development of free radicals that is correlated with the regulation of powdery mildew. In this study, Ferrandino and Smith (2007) also discussed the effects of milk spray on foliar symptoms and yield components of pumpkins (*C. pepo* cv. Howden) under a field experiment under a wide range of environmental conditions and inoculum pressure. In another report, to suppress powdery mildew (*Sphaerotheca pannosa* var. *rosae*) on rose (*Rosa* sp. L.), anhydrous milk fat and soybean oil emulsions (Chee et al., 2011) and milk and compost tea were used. The plants of pumpkins under field conditions are used to make recyclable fiber glasses (DeBacco, 2011).

In another experiment, Sudisha et al. (2011) reported that raw cow's milk is useful against downy mildew disease (*Sclerospora graminicola*) of pearl millet crop and Wurms and Chee (2011) also used anhydrous milk fat and soybean oil emulsions against powdery mildew (*Podosphaera leucotricha*) in apple seedling. After that, Savocchia et al. (2011) also demonstrated that powdery mildew of grapevine was controlled by the application of botanical oil. In the polyhouse experiment during the seasons of 2016 and 2017, the effectiveness of raw cow's milk and whey against cucumber powdery mildew [*Sphaerotheca fuliginea* (Schlecht.) *Pollacci*] was noticed. Different concentrations of milk were sprayed on cucumber plants: 5%, 10%, 20%, 30%, 40%, and 50% in water and this treatment was used four times within 1 week of both seasons. This experiment showed that, at varying doses, raw milk and whey successfully controlled the diseases (Kamel et al., 2017).

25.5.4 Microbes (PGPRs) for all-round development

Organic farming is considered the most victorious technique for achieving a sustainable agrosystem, and because of their importance and demands, organic farming has attained a lot of popularity in the current scenario. They help in maintaining a sustainable ecosystem, increased fertility of soil and crop production, usage of renewable resources, harmonious balance among animal husbandry, boosting biological cycles within the farming system, genetic diversity of plants or production system as well as the protection of flora and fauna of the particular area. An increased demand among people for chemical-free and nutritious food provides a better and healthier future (Rajib et al., 2013). In organic agriculture, microorganisms play an important role in plant growth promotion and maintaining soil fertility. Microbes are fungi, bacteria (nitrogen fixing, P solubilization, K solubilization, and Zn solubilization, and photosynthetic bacteria), yeast, and lactic acid bacteria (Kour et al., 2020; Rana et al., 2020; Yadav et al., 2017a). In organic agriculture, microbes are used as bioinoculants/consortium and in nutrient cycling and plant production and protection and to maintain soil health and fertility (Devi et al., 2020; Kour et al., 2021b).

Nowadays, we use synthetic fertilizers, chemical pesticides, and other technological inputs for crop protection with the minimum usage of soil natural resources and their microbial population. These diverse microbes are beneficial and show biocontrol activity against weeds, crop diseases, and pests and at the same time play an important role in soil fertility. Some biofertilizers are *Azotobacter, Beijerinckia, Clostridium, Klebsiella, Anabaena, Nostoc, Rhizobium, Frankia, Anabaena azollae,* and *Azospirillum* (nitrogen-fixing microbes) (Bhattacherjee and Dey, 2014) and *Bacillus megaterium* var. *phosphaticum, Bacillus subtilis, Bacillus circulans, Pseudomonas striata, Penicillium* sp., and *Aspergillus awamori* (P-solubilizing microbes) used in organic farming (Yadav et al., 2018). Microbial biopesticide products are obtained from microorganisms which are advantageous and may be applied against harmful insect pests and plant diseases year after year. This biopesticide in the market also covers a broad spectrum and approximately 90% of whole biopesticides have the capacity to improve agricultural sustainability and public health (Thakur et al., 2020).

25.5.5 Green manure for soil health

Degradation of soil health is one of the most important problems faced by farmers. Instead of that, most of the field is barren. Indefinite uses of chemical fertilizers deteriorate the physical, chemical, and biological properties of soil. Solutions to overcome this problem or issue are the idea of organic farming. Organic farming relies on organic compost, such as compost from farm yards, green compost, and other composts. Therefore, green compost is one of the most important composts used in organic farming. Green compost is the use of crops that increases the fertility of the soil and is a crop that is grown for soil benefits. Green compost crop boosts the microbial activity, nitrogen, soil, and organic fuel and green compost also tenders some advantages (Dubey et al., 2015). Green compost crops result in an increase of bioactivity in the field and such crops strengthen the quality of the soil. Cultivation of green compost helps reduce soil erosion and the losses caused by leaching and secondarily provide high nutrients to plants. Green compost crops are also reported to suppress weeds, minimize pests and diseases, and provide supplementary animal forage. The most important green manure crops are sunn hemp, daincha, Pillipesara, cluster beans, *Sesbania rostrata*, cowpea, mung bean, green leaf compost Gliricidia (*Gliricidia sepium*), Pongamia (*Pongamia pinnata*), neem (*Azadirachta indica*), Gulmohar (*Delonix regia*), Peltophorum (*Peltophorum ferrugineum*), Mahua (*Madhuca indica*), Tarwar (*Cassia auriculata*), Ipomoea (*Ipomoea* sp.), and Water hyacinth (*Eicchornea* spp.).

25.6 Impact of organic agriculture

Agriculture sustainability, the heart of the current debate, concludes the use of planet's natural resources, as the earth has been ruined by synthetic resources (chemicals) by humans. Organic farming is one such method in which natural resources are being utilized for agricultural practices and eliminates chemical inputs (Fig. 25.7). This technique is often perceived to have a beneficial impact on agriculture as well as on the environment which has been detailed in the following section.

25.6.1 Agriculture

In agriculture, the input of organic matter is a well-known practice that has a beneficial impact on plant production by providing necessary nutrients (macronutrients and micronutrients) to the soil.

25.6.1.1 Crop productivity

Plants require various kinds of nutrient supplements like nitrogen, phosphorus, and potassium. Over the past five decades, nutrient requirement for crop production was fulfilled by using chemical fertilizers. These chemicals which were prepared synthetically increase the productivity of crop but also cause a negative impact on the environment like environmental pollution, degradation of soil fertility, depletion of soil nutrients, and soil biodiversity (Yadav, 2021;

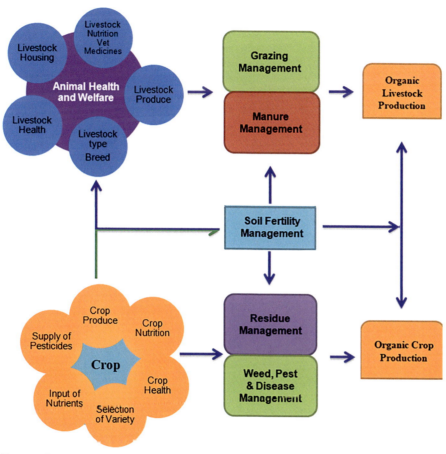

Fig. 25.7 Organic agriculture for soil fertility management.

Yadav et al., 2017b). Organic farming practices that use organic matter have been known to provide various required nutrients that enhance crop productivity without affecting the environment (Yadav et al., 2013). Numerous reports have been published that have reported an increase in crop productivity by following organic farming methods. A study by Sharma and Mittra (1990) and Tamaki et al. (1999) has reported that organic farming has increased the yield of rice. In another study, Ranganathan and Selvaseelan (1997) have reported an increase in mushroom and rice productivity. Singh et al. (2001a) and Singh et al. (2001b) have reported that farming with daincha has increased the productivity of the chickpea plant. In another report, earthworm use in the field has also been responsible for increasing plant productivity (Edwards and Lofty, 1974).

25.6.1.2 Crop protection

Crop productivity is severely affected by pathogens, which is a major hurdle. Pesticides that have been prepared by using organic resources can lower the pathogens in a sustainable way without affecting the environment and soil biodiversity (Nandwani and Nwosisi, 2016). The main approach used in organic agriculture is to deal with the exact cause of a problem rather than indulging the symptoms. Therefore, management is at a much higher priority as compared to control. A healthy crop plant is less susceptible to disease and pest infestation. Therefore, the principal aim for organic farmers is to generate conditions that keep plants healthy. Communication among living organisms and the environment is very crucial for plant health. Under favorable conditions, the own protection mechanism of plants to fight against infection is sufficient. This is the reason why a well-controlled ecosystem may be successful in reducing the intensity of diseases or pest population. Many crop varieties showed efficient defense mechanisms against pests and diseases and have a lesser infection risk than others.

The health of a plant mainly depends on the fertility of a particular soil. When well-balanced nutrition is available, the plant develops into a healthy plant and as a result is less susceptible to infections. Climatic conditions like sufficient water supply and suitable temperatures are additional factors crucial for a plant to grow healthily. If the conditions are not favorable or suitable, then the plant becomes stressed. Stress conditions weaken the defense mechanisms of a particular plant and in turn make effortless targets for diseases and pests. The most important point in organic farming is to grow healthy plants that are free of many diseases and pest problems. The population density of insects, mites, bacteria, fungi, nematodes, and others increases according to favorable environmental conditions. Under a favorable condition, the population density of insects and diseases will increase, and under an unfavorable condition, the density will decrease.

This interaction is very significant for the population dynamics of various pests against their predators. Whenever the pests find appropriate conditions to nurture, it boosts their population. The predator that depends for feeds on the pests finds extra food and results in a greater number than before. The amplified predator population results in the reduction of pest inhabitants, as pests serve food for the predator. Natural enemies are also known as "friends of farmers" as they help the farmers to manage diseases and pest population in the field crops. Natural enemies of diseases and pests do not cause any harm to people or plants. We can be dividing them into four groups: parasitoids (parasitizing on pest), predators (eating pest), pathogens (causing disease to pest), and nematodes (entomopathogenic nematodes) that attack and kill insects.

25.6.1.3 Environment

The inputs are not harmful nor do they have a negative impact on the environment. Organic farming does not give rise to unacceptable pollution of ground or surface water, soil, and air. By-products of organic farming are not harmful to human beings as they are without any harmful agents which are prepared by humans (man-made chemicals).

Biodiversity

Conventional agricultural practices have been known as a major driver of diversity decline because in this kind of practices a large amount of hazardous chemical pesticides are being used (Fuller et al., 2005). These pesticides not only decline pest population but also reduce soil biodiversity of flora, fauna, and microbes. Organic farming has also been known to enhance the biodiversity of the soil by increasing the organic matter to increase the fertility of the soil (Gabriel et al., 2010). Multiple studies have compared the diversity of flora, fauna, and microbes in convention and organic farming and all the reports have reported that by following the organic farming practices diversity of flora, fauna, and microbes has been increased. Studies also reported that weed species belonging to the Fabaceae, Brassicaceae, and Polygonaceae family have been used in organic farming which also results in high flora and fauna (Hald, 1999). Some other reports by Schmid et al. (2011), Tu et al. (2006a) and Tu et al. (2006b) showed an increase in microbial diversity in soil after opting for organic farming.

25.6.1.4 Climate and global warming

In subsequent decades, billions of people especially in developing countries will face changes in climate patterns that may result in problems like severe water shortages, flooding, rising temperatures, shifting of crop growing seasons, increased food shortages, and distribution of illness vectors. The anticipated temperature rise of 1 to 2.5°C by 2030 will have deleterious effects, particularly in terms of reduced crop yield. The elevated drought periods in several parts of the globe and erratic rainfalls can endanger yield stability and therefore put world food production in jeopardy.

Organic production strategies are emphasized on soil carbon retention capacity. This presumably helps resist the climatic challenges, especially in countries which are most susceptible to global climatic changes. Soil erosion is effectively reduced by organic agriculture (Fig. 25.8).

25.6.1.5 Soil and soil fertility

Soil acts as a living organism because it is a habitat for plants, animals, and microbes, which are all interlinked with each other.

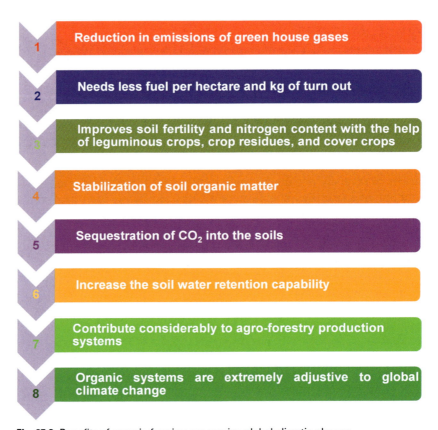

Fig. 25.8 Benefits of organic farming concerning global climatic change.

Mineral particles hold nutrients, which are little by little released by weathering. Roots of plants and a number of microorganisms actively liquefy nutrients from the particles of minerals and utilize them for growth. Plants require minerals to construct organic matter to carry out their physiological processes. Besides mineral particles, soil also encloses a lesser or bigger amount of humus or organic matter, resulting from putrefaction of biomass. Organic matters are present in the topmost layer of the soil, which is an area under the process of continuous transformation. The lively component of soil organic matter may be more decomposed by soil organisms. Resultant structures recombine themselves and form stable humus compositions, which remain in the soil for several years. This long-term top soil organic matter or humus adds a lot to improve soil fertility. This also leads to an increase in soil nutrients. These farming methods also help in declining soil pollution by enhancing the diversity of the soil that is involved in various chemical breakdowns (Leifeld and Fuhrer, 2010). The organic farming system is dependent on proper utilization of organic materials present in the soil. This reliance is in order to enhance the biological, physical, and chemical properties of soil for the elevation of crop production (Watson et al., 2002). Soil organic matter aids to fabricate a soft soil structure by means of a lot of pores. This leads to better aeration, better infiltration of rain or irrigation water, and easier penetration of roots. The visible and invisible parts of organic matter act as a tiny sponge that holds water and as a glue to form stable crumbs, respectively as organic matters have decomposing biomass that provides a sound-balanced blend of many nutrients which plants need for their proper growth. The decomposing activity also acts as a slow supplier of nutrients to the field crops.

25.7 Regulatory policies in organic agriculture

There are international bodies that regulate and formulate the standards, requirements, and all other aspects of organic agriculture. Each country also has its own national laws and policies to regulate the organic farming systems. The international bodies are discussed below:

Food Agriculture Organization (FAO): Organic agriculture gained the attention of FAO in the late 1990s on the recommendation of the committee on agriculture. Organic farming officially became a part of FAO in March 1999. In 2007, during the International Conference on Organic Agriculture and Food Security organized in Rome, the member countries of FAO recognized the potential of organics to create stable and sustainable agriculture. The primary aim of FAO was to disperse the information regarding organic farming systems. The idea of the FAO website was propelled in the year 2000 in regard to regularly updated information on organic culture. FAO is also dealing with synchronization of requirements and standards of organic culture with the help of the United Nations Conference on Trade and Development (UNCTAD) and IFAOM. For this, the International Taskforce on Harmonization and Equivalence in Organic Agriculture was established in 2002. This task force has representation in all aspects from ministries and actors to other prominent personalities in the field of trade and agriculture.

FAO also helps its member countries with assistance in the development of the regulatory framework, legislation, market potential, and other expertises' advice in the field of organic agriculture. Development Law Service of FAO has been of great help in the review, drafting, legislating, and establishing certification in various member countries.

Codex Alimentarius Commission (Codex): It was established as a resolution by FAO in 1961 and 1963 by World Health Organization (WHO) after increasing concerns pertaining to food quality, safety, and trade barriers. Its prime motive was the protection of human health and impartiality in the food trade by promoting harmony and publicizing the food standard practices. The Codex guidelines for processing, production, labeling, and marketing of organic produce were adopted in 1999, and for livestock production, the adoption of rules took place in 2001. This was followed by amendments and revisions in 2003, 2004, 2007, 2008, 2009, and 2010. This establishment was undertaken to prevent false claims and to elevate the trade among nations. The guidelines stress on labeling, inspection, quality insurance, and certification as an essential part of organic systems.

International Federation of Organic Agriculture Movements (FiBL & IFOAM): This is an international nongovernmental organization that was established in 1972 with its headquarters located in Bonn, Germany. The organization provides supervision and assistance to all organic groups worldwide.

IFOAM Organic Guarantee System is designed to assist in the production of eminent organic standards as well as international guarantee of certification and standards. IFOAM Organic Guarantee System has the following components:
- IFOAM Family of Standards
- IFOAM Standard for Organic Production and Processing
- IFOAM Community of Best Practice Standards
- Global Organic Mark
- IFOAM Accreditation and Global Organic System Accreditation (GOSA)

International Organization for Standardization (ISO):- It was established in 1946 by representatives from 25 countries to enhance international coordination and amalgamation of industrial standards. It is a nongovernmental association with currently 162 member countries and is supervised by a secretariat, located in Geneva.

International Task Force on Harmonization and Equivalence in Organic Agriculture (ITF): This is a joint venture of FAO, UNCTAD, and IFOAM from 2003 and 2008. It comprises personalities working for intergovernmental agencies, government agencies, private sectors, as well as public organizations which are associated with organic agriculture and its standards, regulation, certification, and trade.

GOMA Project: This is a follow-up project after the conclusion of ITF in 2008. This project is led in cooperation by FAO, UNCTAD, and IFOAM. It is managed by a committee consisting of representatives from organic agriculture and law firms of all partners and is funded by the Norwegian Agency for Development Cooperation. The purpose of this project is to provide assistance for trading of organic products in different organic systems by the application of ITF commendations. It also promotes the use of two tools developed by ITF Guide for Assessing Equivalence of Standards and Technical Regulations and the International Requirements for Organic Certification Bodies (IROCB). These can be used by private and non-private sectors for the identification of proper organic standards and certification.

International environmental treaties

Convention to Combat Desertification and Convention on Biological Diversity also has some regulatory legislation on organic agriculture systems. They provide certain guiding principles which should be taken into consideration while developing the national laws and policy on organic farming. International Trade Laws can also play a role in the development of national policies regarding organic systems. Mainly, the two main agreements of WTO that are most relevant in this context are GATT and TBT Agreement (Weidmann et al., 2007).

25.7.1 Regulatory regarding organic farming

COUNTRY: Argentina
- *Regulatory Body/Program: National Law 25.127*
  - Applicability-Decree 260/2001: Resolutions
  - Organic Production, Processing, and Handling Rules-Decree 260/2001: Resolutions
  - Organic Labeling and Requirements-National Law 25.127: Resolution
  - Accreditation and Certification System-National Law 25.127, Decree 260/2001: Resolution
  - Administrative Provisions: Resolutions

COUNTRY: European Union
- *Regulatory Body/Program: EU ORGANIC REGULATION (2007)*
  - Scope and Definition Title I
  - Objectives and Principles for Organic Production Title II
  - Organic Production and Processing Title III
  - Organic Labeling and Requirements Title IV
  - Control System Title V
  - Trade with Third Countries Title VI

COUNTRY: India
- *Regulatory Body/Program: NATIONAL PROGRAMME ON ORGANIC PRODUCTION*
  - Applicability—Sections 1 and 2
  - Organic Production and Handling Standards—Section 3
  - Organic Labeling and Requirements—Sections 4 and 5
  - Accreditation and Certification System—Section 6
  - Administrative Provisions—Section 7

COUNTRY: USA
- *Regulatory Body/Program: ORGANIC FOODS PRODUCTION ACT (1990-2005)*
- *NATIONAL ORGANIC PROGRAM (2002-2006)*

- Applicability-Subpart B
- Organic Production and Handling Standards-Subpart C
- Organic Labeling and Requirements-Subpart D
- Accreditation and Certification System-Subparts E & F
- Administrative Provisions-Subpart G

COUNTRY: JAPAN
- *Regulatory Body/Programs:*

LAW N°175 MINISTERIAL ORDINANCE N° 62
- General Accreditation and Certification System
- General Labeling and Requirements

MAFF NOTIFICATION
- JAS Organic Standards-List of Permitted Substances
- Organic Labeling and Requirements

COUNTRY: Canada
- *Regulatory Body/Programs:*
- *AGRICULTURAL PRODUCTION ACT*
- *ORGANIC PRODUCTS REGULATION*
  - Accreditation and Certification System
  - Labeling Requirements
- *NATIONAL ORGANIC STANDARD (2006 version)*
  - Canadian Organic Standard
    - Scope of Application
    - Section 1
    - General Principles-Introduction
    - Organic Production, Processing, and Handling Rules— Sections 5–8
  - Canadian List of Permitted Substances

25.8 Market potential of organic produce

In the agricultural industry, the movement of organic products in the main market was started in the 1990s. The marketing of organic products was approximately $18 billion in 2000 that raised to $33 billion in 2005 and reached $50 billion in 2008 (Van Elzakker and Eyhorn, 2004, 2010; Yussefi and Willer, 2007). The value of these organic products elevated more and more and reached $81.6 billion in 2015 throughout the world. In the European continents, the highest grains in produce were recorded in rice, wheat, corn, sorghum, buckwheat, barley, and sorghum. The highest production of grains is confined to the countries China and United States (Popović et al., 2017) while Lithuania and Romania had the biggest market share of organic wheat in 2014. In Europe, among the 11.6 m hectare total organic crop, grains account for about 1.9 m hectares.

The demand of organically prepared fruits was also growing globally. Italy was the largest producer of organic fruits followed by Turkey, the United States, France, Spain, Poland, and Germany. A recorded increase in the organic product market has been observed in 2015. About a 25% increase in the organic food market of Spain has been observed followed by Iceland and Sweden about 23 and 20%, respectively. In 2018, the organic food market reached $100 billion (approximately 97 billion euros). The foremost leading organic market is in the United States accounting for 40.6 billion euros followed by Germany and France accounting for 10.9 and 9.1 billion euros, respectively. In 2018, the highest organic market share was in Denmark that is about 11.5% of its total food market. By 2022, the organic food market is supposed to reach about $327,600. Moreover, the awareness among people and standard of living increase the demand for organic foods (FiBL and IFOAM, 2020).

25.9 Conclusion and future prospects

Food and nutritional security are, therefore, a significant international concern. It is calculable that by the year 2020, the worldwide population can reach the 8-billion mark. The galloping explosion of the population created during the last 5–6 decades needs not solely food security but also nutritional security. The role of cereals, pulses, fruits, vegetables, and other food commodities in nutritional security is huge; therefore, the assembly of crops needs to be enhanced. Under the current situation of world warming and global climatic change, organic farming has the dual objective of system property and environmental protection.

Organic farming, particularly of vegetables, is gaining momentum worldwide; thanks for increasing awareness and concern on adverse effects of indiscriminate use of chemical fertilizers and pesticides and machinery on food quality, soil health, human health, and environment. The organic agriculture system has the robust potential for building a resilient food system within the face of uncertainties through farm diversification and building soil fertility with organic residues. Certified organic crop merchandise supply high-income choices for farmers and thus will function as promoters for eco-friendly farming applied worldwide. The long-run success of organic crop production would mostly depend on the scale of the farm and provide nonchemical inputs that are to be totally saved by a well-proven package of practices addressing the objectives of manufacturing crops organically. Consequently, these organic farming practices got to communicate modification within the ancient conception of farming.

References

AIR, N, 2020. Mizoram Women & Organic Farming—A Report.
Ambrocia, M., 2019. How Nagaland is Working Towards Becoming an Organic State. https://www.eastmojo.com/news/2019/08/22/how-nagaland-is-working-towards-becoming-an-organic-state/.

Ani, 2019. Manipur Farmers Adopt Integrated Farming for Better Sustainability. https://newslivetv.com/manipur-farmers-all-praise-for-new-integrated-farming-technique/.

Babu, S., Singh, R., Yadav, G.S., 2017. Organic Farming: Problems and Prospects in North East India., https://doi.org/10.13140/RG.2.2.24681.49769.

Barański, M., Średnicka-Tober, D., Volakakis, N., Seal, C., Sanderson, R., Stewart, G.B., et al., 2014. Higher antioxidant and lower cadmium concentrations and lower incidence of pesticide residues in organically grown crops: a systematic literature review and meta-analyses. Br. J. Nutr. 112, 794–811.

Bettiol, W., 1999. Effectiveness of cow's milk against zucchini squash powdery mildew (*Sphaerotheca fuliginea*) in greenhouse conditions. Crop Prot. 18, 489–492.

Bhattacherjee, R., Dey, U., 2014. A way towards organic farming; a review. Afr. J. Microbiol. Res. 8, 2332–2342.

Chee, A.A., Wurms, K., George, M., 2011. Control of powdery mildew (*Sphaerotheca pannosa* var *rosae*) on rose (*Rosa* L sp) using anhydrous milk fat and soybean oil emulsions. N. Z. Plant Prot. 64, 195–200.

Crisp, P., Wicks, T., Troup, G., Scott, E., 2006. Mode of action of milk and whey in the control of grapevine powdery mildew. Australas. Plant Pathol. 35, 487–493.

Dangour, A.D., Lock, K., Hayter, A., Aikenhead, A., Allen, E., Uauy, R., 2010. Nutrition-related health effects of organic foods: a systematic review. Am. J. Clin. Nutr. 92, 203–210.

Das, A., Layek, J., Ramkrushna, G., Babu, S., Devi, M.T., Dey, U., et al., 2019. Integrated organic farming system: an innovative approach for enhancing productivity and income of farmers in north eastern hill region of India. Indian J. Agric. Sci. 89, 1267–1272.

DeBacco, M., 2011. Compost Tea and Milk to Suppress Powdery Mildew (*Podosphaera xanthii*) on Pumpkins and Evaluation of Horticultural Pots Made from Recyclable Fibers Under Field Conditions. University of Connecticut, United States, pp. 1–64 (Masters Thesis).

Devi, R., Kaur, T., Kour, D., Rana, K.L., Yadav, A., Yadav, A.N., 2020. Beneficial fungal communities from different habitats and their roles in plant growth promotion and soil health. Microb. Biosyst. 5, 21–47. https://doi.org/10.21608/mb.2020.32802.1016.

Dubey, L., Dubey, M., Jain, P., 2015. Role of green manuring in organic farming. Plant Arch. 15, 23–26.

Edwards, C., Lofty, J., 1974. Ttte invertebrate fauna of the Park Grass Plots. I. Soil fauna. Roth. Fjcp. Stat. Rep. 1974 (Part 2), 133–154.

Ferrandino, F.J., Smith, V.L., 2007. The effect of milk-based foliar sprays on yield components of field pumpkins with powdery mildew. Crop Prot. 26, 657–663.

FiBL & IFOAM, 2020. https://wwwfiblorg/enhtml.

Foley, J.A., Ramankutty, N., Brauman, K.A., Cassidy, E.S., Gerber, J.S., Johnston, M., et al., 2011. Solutions for a cultivated planet. Nature 478, 337–342.

Fuller, R., Norton, L., Feber, R., Johnson, P., Chamberlain, D.E., Joys, A.C., et al., 2005. Benefits of organic farming to biodiversity vary among taxa. Biol. Lett. 1, 431–434.

Funk, C., Kennedy, B., 2016. The New Food Fights: US Public Divides Over Food Science. Pew Research Center. http://www.pewinternet.org/2016/12/01/the-new-food-fights/.

Gabriel, D., Sait, S.M., Hodgson, J.A., Schmutz, U., Kunin, W.E., Benton, T.G., 2010. Scale matters: the impact of organic farming on biodiversity at different spatial scales. Ecol. Lett. 13, 858–869.

Guruswamy, K., Gurunathan, K.B., 2010. Technology management for Indian farmers to enhance their profitability. Int. J. Comm. Bus. Manag. 3, 199–205.

Hald, A., 1999. Weed vegetation (wild flora) of long established organic versus conventional cereal fields in Denmark. Ann. Appl. Biol. 134, 307–314.

Huber, M., Knottnerus, J.A., Green, L., van der Horst, H., Jadad, A.R., Kromhout, D., et al., 2011. How should we define health? Br. Med. J. 343, d4163.

Kamel, S., Ketta, H., Emeran, A., 2017. Efficacy of raw cow milk and whey against cucumber powdery mildew disease caused by *Sphaerotheca fuliginea* (Schlecht.) Pollacci under plastic house conditions. Egypt J. Biol. Pest Control 27, 135–142.

Kaur, T., Rana, K.L., Kour, D., Sheikh, I., Yadav, N., Kumar, V., et al., 2020. Microbe-mediated biofortification for micronutrients: present status and future challenges. In: Rastegari, A.A., Yadav, A.N., Yadav, N. (Eds.), Trends of Microbial Biotechnology for Sustainable Agriculture and Biomedicine Systems: Perspectives for Human Health. Elsevier, Amsterdam, pp. 1–17, https://doi.org/10.1016/B978-0-12-820528-0.00002-8.

Kesavan, P., Swaminathan, M., 2008. Strategies and models for agricultural sustainability in developing Asian countries. Philos. Trans. R. Soc. B: Biol. Sci. 363, 877–891.

Kirchmann, H., Thorvaldsson, G., Bergström, L., Gerzabek, M., Andrén, O., Eriksson, L.-O., et al., 2008. Fundamentals of organic agriculture—past and present. In: Kirchmann, H., Bergström, L. (Eds.), Organic Crop Production—Ambitions and Limitations. Springer, Netherlands, Dordrecht, pp. 13–37, https://doi.org/10.1007/978-1-4020-9316-6_2.

Kour, D., Rana, K.L., Yadav, A.N., Yadav, N., Kumar, M., Kumar, V., et al., 2020. Microbial biofertilizers: bioresources and eco-friendly technologies for agricultural and environmental sustainability. Biocatal. Agric. Biotechnol. 23. https://doi.org/10.1016/j.bcab.2019.101487, 101487.

Kour, D., Kaur, T., Devi, R., Yadav, A., Singh, M., Joshi, D., et al., 2021a. Beneficial microbiomes for bioremediation of diverse contaminated environments for environmental sustainability: present status and future challenges. Environ. Sci. Pollut. Res. https://doi.org/10.1007/s11356-021-13252-7.

Kour, D., Rana, K.L., Kaur, T., Yadav, N., Yadav, A.N., Kumar, M., et al., 2021b. Biodiversity, current developments and potential biotechnological applications of phosphorus-solubilizing and -mobilizing microbes: a review. Pedosphere 31, 43–75. https://doi.org/10.1016/S1002-0160(20)60057-1.

Kumar, M., Yadav, A.N., Saxena, R., Paul, D., Tomar, R.S., 2021. Biodiversity of pesticides degrading microbial communities and their environmental impact. Biocatal. Agric. Biotechnol. 31. https://doi.org/10.1016/j.bcab.2020.101883, 101883.

Leifeld, J., Fuhrer, J., 2010. Organic farming and soil carbon sequestration: what do we really know about the benefits? Ambio 39, 585–599.

Makadia, J., Patel, K., 2015. Prospects, status and marketing of organic products in India—a review. Agric. Rev. 36, 73–76.

Menon, M., 2009. Organic Agriculture and Market Potential in India. The world of Organic Agriculture Statistics and Emerging Trends Willer, Report-IFOAM, Bonn, and FiBL, Frick. pp. 142–307.

Munda, G., Das, A., Patel, D., 2014. Organic Farming in Hill Ecosystems—Prospects and Practices. ICAR Research Complex for NEH Region, pp. 1–15.

Naikwade, P., Sankpal, S., Jadhav, B., 2012. Management of waste by composting, vermicomposting and it's use for improvement of growth, yield and quality of fodder maize. ARPN J. Sci. Technol. 2, 184–194.

Nandwani, D., Nwosisi, S., 2016. Global trends in organic agriculture. In: Nandwani, D. (Ed.), Organic Farming for Sustainable Agriculture. Springer International Publishing, Cham, pp. 1–35, https://doi.org/10.1007/978-3-319-26803-3_1.

Narayanan, S., Narayanan, S., 2005. Organic Farming in India: Relevance, Problems and Constraints. National Bank for Agriculture and Rural Development Mumbai, pp. 1–93.

Northbourne, L., 2003. Look to the Land. 1940. Dent, London.

Panda, N., Khush, G., 1995. Host Plant Resistance to Insects. CAB international, Wallingford, UK.

Pattanapant, A., Shivakoti, G.P., 2013. Opportunities and constraints of organic agriculture in Chiang Mai province, Thailand. Asia-Pac. Dev. J. 16, 115–147.

Popović, A., Golijan, J., Sečanski, M., Čamdžija, Z., 2017. Current status and future prospects of organic cereal production in the world. Agron. J. 18, 199–207.

Pornpratansombat, P., 2006. Thai organic farming: explicit knowledge and empirical study. Kasetsart J. Soc. Sci. 27 (2), 264–277.

Probst, L., Houedjofonon, E., Ayerakwa, H.M., Haas, R., 2012. Will they buy it? The potential for marketing organic vegetables in the food vending sector to strengthen vegetable safety: a choice experiment study in three West African cities. Food Policy 37, 296–308.

Raj, A., Jhariya, M.K., Toppo, P., 2014. Cow dung for eco-friendly and sustainable productive farming. Environ. Sci. 3, 201–202.

Rajib, R., Upasana, B., Svetla, S., Jagatpati, T., 2013. Organic farming for crop improvement and sustainable agriculture in the era of climate change. OnLine J. Biol. Sci. 13, 50–65.

Rana, K.L., Kour, D., Kaur, T., Devi, R., Yadav, A.N., Yadav, N., et al., 2020. Endophytic microbes: biodiversity, plant growth-promoting mechanisms and potential applications for agricultural sustainability. Antonie Van Leeuwenhoek 113, 1075–1107. https://doi.org/10.1007/s10482-020-01429-y.

Ranganathan, D.S., Selvaseelan, D.A., 1997. Mushroom spent rice straw compost and composted coir pith as organic manures for rice. J. Indian Soc. Soil Sci. 45, 510–514.

Savocchia, S., Mandel, R., Crisp, P., Scott, E., 2011. Evaluation of 'alternative' materials to sulfur and synthetic fungicides for control of grapevine powdery mildew in a warm climate region of Australia. Australas. Plant Pathol. 40, 20–27.

Schmid, F., Moser, G., Müller, H., Berg, G., 2011. Functional and structural microbial diversity in organic and conventional viticulture: organic farming benefits natural biocontrol agents. Appl. Environ. Microbiol. 77, 2188–2191.

Seufert, V., Ramankutty, N., Mayerhofer, T., 2017. What is this thing called organic?—how organic farming is codified in regulations. Food Policy 68, 10–20.

Sharma, A., Mittra, B., 1990. Complementary effect of organic, bio-and mineral fertilisers in rice based cropping system. Fert. News 35, 43–51.

Singh, K., Prasad, B., Sinha, S., 2001a. Effect of integrated nutrient management on a Typic Haplaquant on yield and nutrient availability in a rice-wheat cropping system. Aust. J. Agric. Res. 52, 855–858.

Singh, K., Sharma, I., Srivastava, V., 2001b. Effect of FYM, fertilizer and plant density on productivity of Rice-Wheat sequence. J. Res. Birsa Agric. Univ. 13, 159–162.

Sudisha, J., Kumar, A., Amruthesh, K.N., Niranjana, S.R., Shetty, H.S., 2011. Elicitation of resistance and defense related enzymes by raw cow milk and amino acids in pearl millet against downy mildew disease caused by *Sclerospora graminicola*. Crop Prot. 30, 794–801.

Suthar, S., Watts, J., Sandhu, M., Rana, S., Kanwal, A., Gupta, D., et al., 2005. Vermicomposting of kitchen waste by using *Eisenia fetida* (Savigny). Asian J. Microbiol. Biotechnol. Environ. Sci. 7, 541.

Tamaki, M., Itani, T., Yamamoto, Y., 1999. Effects of organic and inorganic fertilizers on the growth of rice plants under different light intensities. Jpn. J. Crop. Sci. 68, 16–20.

Thakur, N., Kaur, S., Tomar, P., Thakur, S., Yadav, A.N., 2020. Microbial biopesticides: current status and advancement for sustainable agriculture and environment. In: Rastegari, A.A., Yadav, A.N., Yadav, N. (Eds.), New and Future Developments in Microbial Biotechnology and Bioengineering. Elsevier, pp. 243–282.

Tu, C., Louws, F.J., Creamer, N.G., Mueller, J.P., Brownie, C., Fager, K., et al., 2006a. Responses of soil microbial biomass and N availability to transition strategies from conventional to organic farming systems. Agric. Ecosyst. Environ. 113, 206–215.

Tu, C., Ristaino, J.B., Hu, S., 2006b. Soil microbial biomass and activity in organic tomato farming systems: effects of organic inputs and straw mulching. Soil Biol. Biochem. 38, 247–255.

Van Elzakker, B., Eyhorn, F., 2004. Developing sustainable value chains with smallholders. In: The Organic Business Guide, Online Report.

van Elzakker, B., Eyhorn, F., 2010. The Organic Business Guide. Developing Sustainable Value Chains with Smallholders. vol. 160 IFOAM and collaborating organisations (Helvetas, Agro Eco Louis Bolk Institute, ICCO, UNEP), pp. 1–48.

Watson, C., Atkinson, D., Gosling, P., Jackson, L., Rayns, F., 2002. Managing soil fertility in organic farming systems. Soil Use Manag. 18, 239–247.

Weidmann, G., Kilcher, L., Garibay, S.V., 2007. IFOAM Training Manual for Organic Agriculture in the Arid and Semiarid Tropics. https://orgprints.org/id/eprint/25232/.

Willer, H., Rohwedder, M., Wynen, E., 2009. Organic agriculture worldwide: current statistics. In: Willer, H., Kilcher, L. (Eds.), The World of Organic Agriculture. Statistics and Emerging Trends 2009. FIBL-IFOAM Report. IFOAM, Bonn; FiBL, Frick; ITC, Geneva, pp. 25–58.

Willer, H., Schlatter, B., Trávníček, J., Kemper, L., Lernoud, J., 2020. The World of Organic Agriculture. Statistics and Emerging Trends 2020. pp. 1–337. https://orgprints.org/id/eprint/37222/.

Wurms, K., Chee, A.A., 2011. Control of powdery mildew (*Podosphaera leucotricha*) on apple seedlings using anhydrous milk fat and soybean oil emulsions. N. Z. Plant Prot. 64, 201–208.

Yadav, A.N., 2021. Beneficial plant-microbe interactions for agricultural sustainability. J. Appl. Biol. Biotechnol. 9, 1–4. https://doi.org/10.7324/JABB.2021.91ed.

Yadav, S., Babu, S., Yadav, M., Singh, K., Yadav, G., Pal, S., 2013. A review of organic farming for sustainable agriculture in Northern India. Int. J. Agron. 2013, 1–8.

Yadav, A.N., Kumar, R., Kumar, S., Kumar, V., Sugitha, T., Singh, B., et al., 2017a. Beneficial microbiomes: biodiversity and potential biotechnological applications for sustainable agriculture and human health. J. Appl. Biol. Biotechnol. 5, 45–57. https://doi.org/10.7324/JABB.2017.50607.

Yadav, A.N., Verma, P., Singh, B., Chauhan, V.S., Suman, A., Saxena, A.K., 2017b. Plant growth promoting bacteria: biodiversity and multifunctional attributes for sustainable agriculture. Adv. Biotechnol. Microb. 5, 1–16.

Yadav, S.K., Soni, R., Rajput, A.S., 2018. Role of microbes in organic farming for sustainable agro-ecosystem. In: Panpatte, D., Jhala, Y., Shelat, H., Vyas, R. (Eds.), Microorganisms for Green Revolution. Springer, pp. 241–252.

Yadav, A.N., Kour, D., Kaur, T., Devi, R., Yadav, A., Dikilitas, M., et al., 2021. Biodiversity, and biotechnological contribution of beneficial soil microbiomes for nutrient cycling, plant growth improvement and nutrient uptake. Biocatal. Agric. Biotechnol. 33. https://doi.org/10.1016/j.bcab.2021.102009, 102009.

Yussefi, M., Willer, H., 2007. Organic farming worldwide 2007: overview & main statistics. In: Willer, H., Yussefi, M. (Eds.), The World of Organic Agriculture-Statistics and Emerging Trends 2007. International Federation of Organic Agriculture Movements IFOAM, Bonn, pp. 9–16.

26

Contributing effects of vermicompost on soil health and farmers' socioeconomic sustainability

Pallabi Mishra[a] and Debiprasad Dash[b]
[a]Department of Business Administration, Utkal University, Bhubaneswar, India, [b]KVK Bhadrak, OUAT, Bhubaneswar, India

26.1 Introduction

India is a land of farmers. Agriculture is the most predominant profession in our country. In spite of the increase in literacy level and more comfortable jobs and businesses, the young India is gradually bending down towards agricultural profession. The entrepreneurial mind-set is again traveling back to our roots with a purpose of sustainability. With this aim of sustainable development, the approach is more towards organic agriculture rather than conventional agriculture. Patil et al. (2014) have differentiated between conventional agriculture and organic agriculture. The study reveals that organic agriculture is more sustainable than conventional agriculture. The costs of inputs are less, leading to lower financial risks in the case of organic approach. The experiments were performed in different regional conditions and there was a variation in the results. Depending on regional environment and crops cultivated, organic farming can be a viable agricultural practice in the state. Organic agriculture has the ability to boost net returns, lower crop failure risks and reduce the harsh impact on the environment. In dry regions, the balance of nitrogen in most of the crop rotations with organic cultivation is negative, suggesting a need to increase nitrogen supply. In both dry and wetter areas, net losses in the event of crop failure are significantly lower in organic agriculture than in traditional agriculture. With better nutrient management, potential sustainability in both wet and dry fields can be increased.

With the advent of Green Revolution, it is seen that there is a degradation of the soil matter (Barmon and Tarafder, 2019). This is due to the repeated agriculture of the same crop in the same field over and over again (Ranamukhaarachchi and Begum, 2005; Rahman and Ranamukhaarachchi, 2003). "The lingering and extreme use of inorganic pesticides, fertilizers, and groundwater in crop production exerts severe human and soil health hazards with environmental pollution" (Kumar et al., 2015; White et al., 2012). Further Rahman and Barmon (2019) have concluded that the repeated use of the same crop and similar agricultural practices have reduced the potency of soil. Yu et al. (2014) have studied that the rotation of legumes and jute cropping system has depleted the organic matter of the soil in China. Depletion of soil organic matter degrades the ecosystem and loses the resilience of the ecosystem as well (Feller et al., 2012; Bronick and Lal, 2005). There is a serious socioeconomic problem created by Green Revolution due to a drastic declination of organic content in soil and its multifunctionality (Ashraf et al., 2016). This has resulted in a decline of the fertility of soil and the socioeconomic condition of mainly marginal and small farmers' categories and their families (Srivastava et al., 2016). Considering these hazards, farmers are encouraged to take up organic techniques of agriculture (Chao et al., 2003; Yadav et al., 2013). Organic agriculture is sustainable economically and a practicable method of crop production has less environmental pollution paving the way to better sustainability of agriculture, soil, environment, and mankind.

Declining soil health and reducing crop productivity are due to the lowering of organic matter among several other causes. Over the years, the traditional organic amendments, viz. cattle manure, are also decreasing due to a decrease in cattle population and further domestic uses such as fuel and plaster coating of the thatched mud houses (Indoria et al., 2018). "As per a conventional estimation, around 600 to 700 million tons of agricultural wastes including 272 million tons of crop residues are generated each year in India" (Suthar, 2009). But most of it is not utilized and can be recycled to vermicompost (biofertilizer rich in nutrients) for the practices of restoration of land in a sustainable way. A lower carbon-nitrogen ratio of vermicompost indicates the suitability for use as a soil amendment (Nurhidayati et al., 2018). In other way, the productivity of rice as a staple food in many countries largely depends on soil and the nutrients/foods supplied to the soil ecosystem. Hence, it is necessary to sustain and enhance the soil and rice productivity restoring soil health with increased rice yield adopting a sustainable approach to meet the upward demand of surging population. The chemical fertilizers are mostly used for the cultivation of rice which gradually reduces the fertility of the soil (Biswas et al., 2017). Repeated use of inorganic NPK fertilizers results in the mining of micronutrients causing deficiency, deterioration of physicochemical and biological properties of soil and, thus, crop production under an unsustainable approach. The supply mechanism of nutrients in soil for rice production should be viable economically, ecofriendly, and accepted socially without affecting the gross crop production with the restoration of soil health. The incorporation of organic manures for crop production is an important management practice to restore soil and rice productivity. Vermicomposting is an alternative way to recycle these huge unutilized organic wastes into good-quality vermicompost for healthy plants and soil, and it saves from the negative environmental impact of these wastes in the society.

Vermicompost is the product of a process called vermicomposting where earthworms help in converting organic biodegradable waste materials into humus-like material. Globally, numerous researchers have found out that the nutrient content of vermicompost is higher than the conventional compost. In fact, vermicompost can augment the physicochemical and biological fertility of soil (Lim et al., 2015). Vermicompost is the technique of utilizing earthworms to convert organic waste to create compost rich in nutrients (Kumar et al., 2015). Vermicompost has the capacity to retain the moisture and organic matter in the topsoil layer (Barmon and Tarafder, 2019). Soil treated with vermicompost is physically better aerated and has more bulk density, porosity, and water retention. There is a better crop yield due to improvement in chemical properties such as electrical conductivity, pH, and organic matter content (Yadav et al., 2013). There are other materials in vermicompost that induces the growth of plants. Although vermicomposts have shown significant growth in plants, yet their application at high concentrations could impede the plant growth due to the excess concentration of soluble salts available in vermicompost. Therefore, vermicomposts should be used moderately to obtain maximum plant yield (Lim et al., 2015). Vermicompost has also been used to treat various wastes. In the case of treatment of solid waste by various composting methods, the method of vermicomposting is safer for an effective way of sanitization of solid wastes. It enhances the soil health by increasing the microbial activity that further improves nutrient availability in soil (Ashraf et al., 2016; Bhattacharya et al., 2016). The humic compounds in vermicompost increase the process of humification that speeds up the organic matter conversion (Singh and Singh, 2017; Saha et al., 2012). Sharma and Banik (2014) have stated that the use of vermicompost to crops minimizes the requisite of chemical fertilizers. It further adds nutrients essential to enhance crop yield and improves the soil health. The cultivation cost was high in the case of vermicompost as farmers had to buy them but economic returns were higher due to better yield.

A lot of studies have been conducted on vermicompost and its uses. Experiments have been performed in various fields to study the effect of vermicompost physically, chemically, and biologically on soil and crops. But the experiences of the farmers have not been taken care of by researchers. There is a dearth of research on the opinion and experience of farmers practicing organic agriculture using vermicompost as manure for their crops. Another aspect where there is a gap in research is the social and economic sustainability of farmers who are using vermicompost. The benefits of using vermicompost over other chemical manures not only in context to soil health and plant growth but also the socioeconomic sustainability of farmers need to be researched upon. Keeping in mind the aforementioned gaps found the following objectives have been framed. The first objective is to study the effect of vermicompost on paddy-grown soil health followed by the role of vermicompost usage on farmers' socioeconomic sustainability.

26.2 Literature review

The objectives mentioned already have been illustrated as follows.

26.3 Effect of vermicompost on paddy-grown soil health

Rice is the staple food for most Indians and people globally. It is perhaps the third most consumed grain in the world. About 90% of the market supply and demand of rice comes from Asia. This demand needs to be fulfilled with an annual increase of 1% with the ever-increasing global population. Cultivation of rice contributes to greenhouse gas emission. Hence, the sustainability in rice production is as much necessary as economic contribution of the crop (Bhaduri et al., 2020). One of the most dominant cropping systems in tropical areas with proper irrigation facilities is rice-rice cropping system (Mohanty et al., 2013). Both wet rainfed areas and dry irrigated seasons, where other crops are not possible to be cultivated, are taken under rice crop only. Due to repeated practice, the removal of a high amount of essential nutrients and organic matter from the soil in long run leads to a sharp decline in yield of crop, productivity, and subsequently retarding soil quality (Mohanty et al., 2013). Massive depletion in the soil organic composition is also due to low input use from the outside farm and low clay contents in the areas under rice crop. The extensive use of chemical fertilizers in paddy has been recognized as a major reason leading to an increase in hazardous emissions from rice fields (Nayak et al., 2012). Moreover, chemical fertilizers used on a long-term basis result in low productivity of rice and soil health deterioration (Jha et al., 2020). On the contrary, the application of carbon-rich inputs like manure of farmyard, cow dung, composts, etc., benefits highly for improvement in physicochemical and biological healthiness of arable soil (Das et al., 2018). As such, due to replacements of inorganic chemical fertilizers by organic fertilizers under rice cultivation, remarkable use efficiencies of soil quality sustenance and productivity of crop have been observed (Goswami et al., 2017). Restoration of soil quality ensuring productivity and also greatly stabilizing fragile agroecosystems is only possible through organic-based nutrient management (Das et al., 2018). Several organic acids, hormones, and useful microorganisms available in vermicompost facilitate sustainable nutrient mobilization, hence boosting crop growth (Lazcano and Domínguez, 2011).

Sahariah et al. (2020) have researched the effect of vermicompost made from municipal solid waste (MSW) on soil crop interface. An assessment under an intensive rice cropping system was taken on the impact of vermicompost on soil health and soil organic composition where 20% to 40% of the nitrogen fertilization as per recommendation was substituted by vermicompost for the paddy crop. "It was found that degree of humidification, humic acid, soil organic carbon (SOC) storage, and fulvic acid C content in soil increased gradually by 55%–60% under NPK60 + VC and NPK80 + VC treatments in a span of 2 years. There was a spectacular improvement in NPK60 + VC (2.79 folds)- and NPK80 + VC (2.25 folds)-treated soil. The carbon pool management index in soil was greatest under NPK60 + VC (2.1) treatment followed by NPK100 + VC (1.96) and NPK80 + VC (1.87) treatments." An increased grain yield and crop biomass were observed under vermicompost treatment. The results revealed that the production of rice was better because of the improvement of soil organic composition and fractions of humified carbon in soil. The research indicated that microbial health of soil and soil organic composition balance, which significantly correlated with rice production, was sustained due to the application of vermicompost. The recommended amount of vermicompost for paddy is 1 ton per acre after transplanting (Chanu et al., 2018). They further found that vermicompost helps in control of pests and pathogens in addition to minimizing the risk of diseases.

Jha et al. (2020) found that the use of vermicompost and other organic manures increased the yield of rice and grain attributing characters of paddy. There was also an increase in a number of tillers per square meter and a number of grain spikes. Mondal et al. (2020) reported that the application of recommended doses of NPK fertilizer along with 25% vermicompost showed great potential and increased the growth, morphophysiology, and yield of paddy along with improved soil health under the old alluvial soil agroclimatic zone, and the overall study revealed that incorporation of vermicompost benefitted in upward increase in biodiversity of soil, organic carbon, available nitrogen, phosphorus, potassium, and micronutrient status in soil.

Green Revolution, though achieved a quantum jump in cereal production, greatly relied on chemical fertilizers to enhance the yield potentiality, but it took a heavy toll on soil health in the subsequent decades causing yield stagnation (Sheehy et al., 2007). Organic manures are often preferred as a low-cost and easily accessible nutrient source for improving the soil's physical-chemical-biological parameters and better access to soil nutrients for long run (Kassam et al., 2011). So far, more emphasis was given on the integrated approach; however, the present study highlights the significance of sole vermicompost for achieving sustainable soil health and crop yield as well as grain quality of rice.

The treatment on vermicompost was reported to have great potentiality in increasing the performance of paddy based on growth, morphological physiology, yield, and soil health under the agroclimatic old alluvial soil zone. It is also obvious from the study that increasing soil biodiversity, organic carbon, available form of nitrogen, phosphorus, potassium, and micronutrient status in soil was due to the application of vermicompost. On the contrary, it helps in enhancing the economic conditions of farmers and spending at the same time less money to import chemical fertilizers from other countries by government. The research shows absolutely the importance of organic agriculture; therefore, vermicompost as a natural fertilizer may be put to good use under cultivation for increased production and sustainable agricultural systems (Mondal et al., 2020).

Results reveal that productivity due to the application of vermicompost is significantly higher than the use of chemicals alone among the users of vermicompost as expected. But the gain in profitability is not different significantly due to the higher cost of vermicompost. Productivity was significantly increased when the use of vermicompost along with other conventional inputs and its users are relatively more efficient technically (Rahman and Barmon, 2019). The growth of rice was better under the continuous practice of organic agriculture than with conventional agriculture. The application of vermicompost increases the grain and straw yield of rice significantly and saves up to 50% of recommended doses of NPK fertilizer in upland rice condition (Kumar et al., 2015). The chlorophyll content in paddy was remarkably high in the case of vermicompost-treated fields in both wet and dry seasons. An increasing trend was found in soil respiration, soil biomass carbon, and enzymes like urease and phosphatase by using vermicompost in paddy fields. The microbial quotient and the microbial metabolic quotient in soil were significantly high leading to better biological soil health. There was also a significant rise in tiller production of rice. There was a significantly greater biomass yield during the wet season. A remarkable increase in crop growth rate and relative growth rate was also noted by using vermicompost. The overall grain yield during wet season was more than dry season (Sahariah et al., 2020).

"Vermicompost enriched with *A. chroococcum* resulted in the highest improvement in plant growth, leaf chlorophyll content, grain yield, and nitrate reductase activity of rice, followed by enrichment with *A. brasilense*." Application of enriched vermicomposts improved significantly nutrient contents in plant and available nitrogen, phosphorus, potassium, and organic carbon in the soil after harvest where the best result was obtained in the treatment treated with vermicompost enriched with *Azotobacter*. As a whole, the inferences suggest that a weed available abundantly in the northeast part of India as locally available plant biomasses, particularly *Ipomoea carnea*, can be easily converted into vermicompost as a source of nutrient which is environment-friendly by adopting the process of vermicomposting, further enriching the vermicompost with microorganisms such as *A. chroococcum* and *A. brasilense* which promotes the growth of the plant (Mahanta et al., 2012).

Soil fertility is improved and restored by the earthworms, and crop productivity is boosted due to the use of excretory products of earthworms called "vermicast." Beneficial soil microbes are excreted, and proteins, polysaccharides, and other nitrogenous compounds are secreted into the soil by the earthworms. Aeration and soil fragmentation are promoted, and "soil turning" and dispersion in farmlands are brought about due to these worms. Moreover, volume of air-soil can be increased from 8% to 30% by the worm activity. "One acre of land can contain up to 3 million earthworms, the activities of which can bring up to 8–10 tons of 'topsoil' to the surface (in the form of vermicast) every year. The presence of worms regenerate compacted soils and improves water penetration in such soils by over 50%" (Sinha et al., 2010). Repeated application of vermicompost restores fertility of soil by enhancing physicochemical and biological properties of soils with low fertility. Earthworms are usually bred in a mixture of cow dung, soil, agricultural and farm residues during the process of vermicomposting (Negi and Maikhuri, 2013).

From the above-said literature review, the following parameters in Table 26.1 are chosen for the study based on expert views.

26.4 Role of vermicompost usage on farmers' socioeconomic sustainability

Sustainability is the development of present and future growth and maintenance of mankind and its environment. "The word sustainability is derived from the Latin word *sustinere* (*tenure,* to hold; *sus,* up)." "Sustainability means to hold up, to maintain for present and into the future" as defined by Wikipedia. A sustainable business needs to attain social, economic, and, environmental sustainability (Mishra and Chhatoi, 2017; Svensson and Wagner, 2015).

Table 26.1 Parameters to measure soil health.

Sl. No.	Conditions/parameters	Source
1	Organic carbon	Kumar et al., 2015; Indoria et al., 2018; Sahariah et al., 2015
2	Nutrient status	Singh and Singh, 2017; Banik et al., 2004; Jeyabal and Kuppuswamy, 2001
3	Color of the soil	Sahariah et al., 2020; Goswami et al., 2017
4	Weed problem	Bharamappanavara et al., 2010
5	Beneficial insects	Sinha et al., 2010
6	Population of earthworm	Sinha et al., 2010
7	Soil structure	Kumar et al., 2015; Mishra and Dash, 2014; Nayak et al., 2012
8	Soil aggregation	Kumar et al., 2015; Sinha et al., 2010
9	Water holding capacity	Sahariah et al., 2020; Barmon and Tarafder, 2019; Kumar et al., 2015; Goswami et al., 2017; Sinha et al., 2010
10	Yield	Sreedevi and Veerabhadra Rao, 2020; Jha et al., 2020; Mishra and Dash, 2014
11	Pathogen control	Singh and Singh, 2017; Rorat et al., 2017; Park et al., 2011; Kavamura and Esposito, 2010
12	Plant growth	Manivannan et al., 2009
13	Leaf chlorophyll content	Manivannan et al., 2009
14	Nutrient status in postharvest soil	Jha et al., 2020; Sahariah et al., 2020; Goswami et al., 2017; Mohanty et al., 2013; Lazcano and Domínguez, 2011
15	Soil porosity	Sahariah et al., 2020; Kumar et al., 2015; Singh and Singh, 2017; Goswami et al., 2017
16	Soil aeration	Kumar et al., 2015; Singh and Singh, 2017; Tripathy et al., 2014
17	Water drainage	Singh and Singh, 2017
18	Soil hardness	Sinha et al., 2010

In the study conducted by Sahariah et al. (2020), it was seen that the highest benefit/cost ratio was observed in the case of NPK60 + vermicompost (benefit/cost ratio = 5.55) followed by NPK80 + vermicompost (benefit/cost ratio = 5.44). The use of vermicompost improved the benefit/cost ratio that indicated a growth of socioeconomic sustainability of farmers. The improved yield and productivity contributed to the sustainability of farmers (Bejbaruah et al., 2013).

A spectacular crop growth and yield of paddy crop during both the wet and dry seasons were due to the reduction in chemical fertilizers and increasing use of vermicompost and FYM application and the crop productivity of paddy was higher in the rainy season than during the rest part of the year. Plant growth is affected through the supply of growth-promoting hormones like gibberellin, auxin, and cytokinin due to vermicompost (Lazcano and Domínguez, 2011). As such, vermicompost application facilitates microbial proliferation in soil promoting better roots and acquisition of nutrients that lead to greater crop yields (Goswami et al., 2017; Kizilkaya et al., 2012). The improvement of crop yield leads to increased sales of the paddy grains, thereby fetching better prices for the farmers. This contributes to their economic conditions, making them more sustainable.

Annolfo et al. (2017) selected the following indicators for their study. The yield is the measure of the output amount per hectare produced in the farm followed by profitability of farm which is the difference between gross income and gross expenses of farm. Labor demand is the level of demand for labor at farm fields. Labor productivity is the ratio between yield output and inputs used. It is also the total number of working hours or total employment. Income stability is the labor return of a farmer's labor over time. A 100% increase in productivity of labor was followed by labor demand, farm profitability, and yield by using vermicompost.

On the contrary, Barmon and Tarafder (2019) have studied the influence of using vermicompost as manure on individual and household income. Household income is the sum of the total income of all the members of a household or family. The individual and household income of farmers using vermicompost with irrigation facilities increased 1.19 times their previous income. This study was performed in Bangladesh on modern variety of paddy. They further found that although the yield/hectare was significantly high, the labor demand was the same as before using vermicompost. The study further revealed that the revenue generation was higher in the case of the paddy grown with vermicompost leading to increased farm profitability. The benefit/cost ratio was significantly higher in the case of paddy grown in vermicompost-treated fields over paddy grown in chemically treated fields. The farmers who did not use vermicompost were relatively poor than their counterparts as they sold their labors to other crop-producing farmers. On the contrary, Rahman and Barmon (2019) argued that although there was an increase in the productivity of the farm, the profitability was not increased due to higher cost of vermicompost. The net return from the production of rice was significantly higher with better yield (Chanu et al., 2018). There was a significant upward surge in family and individual income. The labor and input cost decreased as the farmers got a better price for their produce. They further found that the health, education, and living standards of the farmers' families improved over time by using vermicompost in agriculture. The socioeconomic condition also improved. Purkayastha (2012) stated that vermicomposting is a method of generating additional income sources and economic empowerment and assuring a sustainable livelihood approach with environmental benefits. Further vermicomposting is a successful model for rural and less socioeconomic resourceful communities.

Uddin and Dhar (2017) have found significant upgradation in the socioeconomic sustainability of farmers using vermicompost. Due to better yield and productivity, the profit was more. The study further stated that due to better income generation, the women of the household could be involved in other income-generating and socioeconomic activities and spare more time for their family. In spite of all positive aspects, the farmers still need more skill training programs for motivation. The aforementioned literature led to the following conditions/parameters to measure the usage role of vermicompost on the socioeconomic conditions of the farmers as shown in Table 26.2.

Table 26.2 Parameters to measure the socioeconomic sustainability of farmers.

Sl. No.	Conditions/parameters	Sources
1	Vermicompost quantity used	Barmon and Tarafder, 2019
2	Labor cost per day	Sreedevi and Veerabhadra Rao, 2020
3	Number of tillers	Jha et al., 2020; Mishra and Dash, 2014
4	Rice yield/hectare	Sahariah et al., 2020; Kumar et al., 2015; Goswami et al., 2017; Annolfo et al., 2017; Kizilkaya et al., 2012
5	Labor demand	Sreedevi and Veerabhadra Rao, 2020; Barmon and Tarafder, 2019; Annolfo et al., 2017
6	Labor productivity	Sreedevi and Veerabhadra Rao, 2020; Singh and Singh, 2017; Annolfo et al., 2017
7	Farm profitability	Barmon and Tarafder, 2019; Annolfo et al., 2017
8	Income stability	Annolfo et al., 2017; Uddin and Dhar, 2017
9	Skills and training	Uddin and Dhar, 2017; Bharamappanavara et al., 2010
10	Family income level	Barmon and Tarafder, 2019; Uddin and Dhar, 2017
11	Individual income level	Barmon and Tarafder, 2019; Uddin and Dhar, 2017
12	Fixed assets (land, house, electronic items, mobiles, machines, etc.)	Below et al., 2012
13	Education for the kids	Below et al., 2012
14	Position in the society	Carlsson et al., 2007
15	Family expenditure	Below et al., 2012
16	Benefit/cost ratio	Uddin and Dhar, 2017; Mishra and Dash, 2014
17	Living standard	Uddin and Dhar, 2017
18	Health	Uddin and Dhar, 2017
19	Status	Uddin and Dhar, 2017
20	Children are out of home for higher studies	Below et al., 2012

26.5 Methodology

This research is based upon primary data and analytical in nature. Data collected were disproportionate over the following sample domains through purposive sampling. Generalization of findings may not be inferred from the limited sample size. About 119 respondents chosen for the survey belonged to various demographic profiles. The respondents were farmers with an experience of 2–5 years of using vermicompost as a fertilizer. The research is based on their experience of using vermicompost in paddy-grown fields. A 3-point Likert scale structured close-ended questionnaire was designed with 38 variables. These 38 variables are distributed over 2 factors of the study. The factors are soil health with 18 variables and socioeconomic sustainability with 20 variables. Paddy was taken for the study as it is the mostly grown crop in Odisha. The paddy crop under study was taken during the Kharif season grown in rainfed condition with the availability of lift irrigation system. The analysis of the collected data was performed by descriptive and inferential statistics.

26.6 Brief profile of the villages taken for the study

For the purpose discussed earlier, data were collected over four sample domains—Padi, Kuanrda, Gopalpur, Nuapokhari, Badamahinsigotha, Bodak Mahajib, Gandakura villages of Bonth block; Solagaon, Bandhatia, Radhaballavpur, Salampur of Dhamnagar block; Jhinkiria, and Bodak of Tihidi block; and Naichhanpur of Basudevpur block of Bhadrak district, Odisha. The agroecological situation of these villages comes under the northeastern coastal plain zone. The soil type is mostly alluvial. The irrigation is by canals and lift irrigation (LI) points. The most grown crops are rice followed by black gram, green gram, sunflower, and vegetables. Apart from agriculture, pisciculture, dairy agriculture, poultry, mushroom, and goatery are other enterprises. The annual rainfall is around 1400–1500 mm, where the mean max temperature is 32–33 degree centigrade and the mean min temperature is 21–22 degree centigrade.

26.7 Results and discussion

The data collected were put to analysis using descriptive statistical tools and analysis of variance (ANOVA) which lead to the following results. The computation is made in SPSS 23 and MS-Excel 2010.

The detailed demographic profile of the respondents with regard to age, gender, education, occupation, marital status, caste, source of income, individual and family income level, family or household expenditure, category of farmers, family size, earning members in a family, and dependants in the family is shown in Table 26.3. All the

Table 26.3 Demographic profile of respondents.

Demographic profile		No. of respondents	% of Respondents
Gender	Female	0	0
	Male	119	100
	Transgender	0	0
Age(in years)	15–19	0	0
	20–24	0	0
	25–44	31	26.1
	45–64	81	68.1
	65–79	7	5.9
	80 >	0	0
Educational background	No formal education/have not been to school ever	0	0
	under matric	49	41.2
	10+2	45	37.8
	Undergraduate	0	0
	Graduate	15	12.6
	Postgraduate	4	3.4
	Technical/professional course	6	5
Marital status	Never married	4	3.4
	Married	115	96.6
	Widowed	0	0
	Divorced/separated	0	0
Caste	General	27	22.7
	SEBC	42	35.3
	OBC	34	28.6
	SC	16	13.4
	ST	0	0
	Others (Please mention)	0	0
	Agriculture	100	84
	Job	4	3.4
	Agriculture + business	15	12.6
	Others (Please mention)	0	0
Source of income	< 2.5 lks	71	59.7
	2.5–5 lks	48	40.3

Category	Range		
Individual Income level	5–7.5 lks		0
	7.5–10 lks		0
	10–12.5 lks		0
	12.5 lks >		0
Family Income level	< 2.5 lks	71	59.7
	2.5–5 lks	48	40.3
	5–7.5 lks	0	0
	7.5–10 lks	0	0
	10–12.5 lks	0	0
	12.5 lks >	0	0
Family expenditure Level	< 2.5 lks	75	63
	2.5–5 lks	44	37
	5–7.5 lks	0	0
	7.5–10 lks	0	0
	10–12.5 lks	0	0
	12.5 lks >	0	0
Family size	< 2	0	
	2–4	52	43.7
	4–6	46	38.7
	6–8	13	10.9
	8–10	0	0
	10 >	8	6.7
Farmer category (available agricultural land in hectares)	Marginal (below 1.00 ha)	40	33.6
	Small (1.00–2.00 ha)	39	32.8
	Semi-medium (2.00–4.00 ha)	32	26.9
	Medium (4.00–10.00 ha)	8	6.7
	Large (10.00 ha and above)	0	0

respondents taken for the study were males. 68.1% of the respondents were aged between 45 and 64, 26.1% were between 25 and 44, and 5.9% were aged between 65 and 79. There were no respondents in other age groups. Coming to the educational background, 41.2% had attended formal school but not completed their 10th. About 37.8% of the respondents had completed +2, 12.6% were graduates, and 3.4% had completed their postgraduation. In marital status, 96.6% were married and only 3.4% were still single. The caste included 35.3% socially economic backward class (SEBC) followed by 28.6% other backward class (OBC), 22.7% general, and 13.4% scheduled caste (SC). The source of income was mostly from agriculture, i.e., 84%. 12.6% of the respondents were involved in some other business along with agriculture and 3.4% had a major source of income from their jobs. The individual income level was less than 2.5 lakhs for 59.7% and between 2.5 and 5 lakhs per annum for 40.3% of the respondents. The family income was exactly the same as individual income. Coming to the family expenditure, 63% spent within 2.5 lakhs, whereas 37% between 2.5 and 5 lakhs per year. The family size was 2 to 4 members for 43.7%, 4 to 6 for 38.7%, 6 to 8 for 10.9%, and more than 10 for 6.7%. 52.9% of the respondents had 2 to 4 dependants in the family, 30.3% had 4 to 6, 10.1% had less than 2, 4.2% had 6 to 8 and 2.5% had 8 to 10 dependants in the family. The categories of farmers were 33.6% marginal, 32.8% small, 26.9% semimedium, and 6.7% medium.

Table 26.4 shows the descriptive statistics of soil health variables/parameters taken for the study. 18 variables were chosen for the study based on extant literature and expert views. The variables have been arranged in descending order on the basis of their mean values as given in the table. All the variables had equal ranks except soil structure, color of soil, weed problems, pathogen control, and soil hardness.

Table 26.5 represents the descriptive statistics of socioeconomic sustainability variables/parameters taken for the study. 20 variables were chosen for the study based on literature review and expert opinion. The variables have been arranged in descending order on the basis of their mean values as given in the table. All the variables had equal ranks except living standard, labor productivity, fixed assets, kids' education, family expenditure, individual expenditure, children out for higher studies, labor demand, and labor cost per day.

The following hypotheses have been proposed to be tested. One-way ANOVA has been used to test the hypotheses.

$H_0 1$: There is no significant difference between different variables of vermicompost effect on soil health of paddy.

The result of ANOVA of variables of vermicompost effect on soil health is displayed in Table 26.6. The mean square between the groups is 2505.33. The obtained value of "F" for the difference of vermicompost effect on soil health on its variables is $2505.333 < 0.05$, indicating

Table 26.4 Descriptive statistics of vermicompost application on soil health parameters.

Parameters/variables	N Statistic	Minimum Statistic	Maximum Statistic	Mean Statistic	Std. deviation Statistic
Water holding capacity	119	3	3	3.00	0.000
Water drainage	119	3	3	3.00	0.000
Soil aeration	119	3	3	3.00	0.000
Soil porosity	119	3	3	3.00	0.000
Nutrient content in postharvest soil	119	3	3	3.00	0.000
Leaf chlorophyll content	119	3	3	3.00	0.000
Plant growth	119	3	3	3.00	0.000
Better yield	119	3	3	3.00	0.000
Soil aggregation	119	3	3	3.00	0.000
Beneficial insects	119	3	3	3.00	0.000
Population of earthworm	119	3	3	3.00	0.000
Nutrient status	119	3	3	3.00	0.000
Organic carbon	119	3	3	3.00	0.000
Soil structure	119	1	3	2.93	0.362
Color of soil	119	1	3	2.92	0.403
Weed problems	119	1	3	1.11	0.386
Pathogen control	119	1	2	1.03	0.181
Soil hardness	119	1	1	1.00	0.000

that numerator of F ratio is greater than the denominator. The test was carried out at a total df of 2141 where 17 (n1) and 2124 (n2) degrees of freedom at 95% confidence level for the mean score of all the 18 variables. The significance value is 0.001 which is less than 0.05. Since the p-value $<$ significance level (0.05), the difference in the mean value of different variables of soil health is statistically significantly different. So the null hypothesis—there is no significant difference between different variables of vermicompost effect on soil health of paddy—is rejected with 95% confidence.

$H_0 2$: There is no significant difference between the usage of vermicompost on different variables of socioeconomic sustainability of farmers.

The result of ANOVA summary of variables of socioeconomic sustainability is presented in Table 26.7. The between-group mean square is 26.05742 and the within-group mean square is 0.189886. The value of "F" obtained for the difference of socioeconomic sustainability on variables is 137.226 $<$ 0.05 indicating that numerator of F ratio is greater

Table 26.5 Descriptive statistics of vermicompost effect on variables of socioeconomic sustainability.

Parameters/variables	N Statistic	Minimum Statistic	Maximum Statistic	Mean Statistic	Std. deviation Statistic
Status	119	3	3	3.00	0.000
Health	119	3	3	3.00	0.000
Benefit/cost	119	3	3	3.00	0.000
Position in society	119	3	3	3.00	0.000
Family income	119	3	3	3.00	0.000
Individual income	119	3	3	3.00	0.000
Skills and training	119	3	3	3.00	0.000
Income stability	119	3	3	3.00	0.000
Farm profitability	119	3	3	3.00	0.000
Rice yield	119	3	3	3.00	0.000
Tillers	119	3	3	3.00	0.000
Vermicompost quantity used	119	3	3	3.00	0.000
Living standard	119	2	3	2.97	0.181
Labor productivity	119	2	3	2.95	0.220
Fixed assets	119	2	3	2.92	0.266
Kids' education	119	2	3	2.72	0.450
Family expenditure	119	1	3	2.12	0.984
Children out for higher studies	119	2	2	2.00	0.000
Labor demand	119	1	3	1.85	0.980
Labor cost per day	119	1	3	1.59	0.858

Table 26.6 ANOVA summary of vermicompost effect on soil health.

		Sum of squares	Df	Mean square	F	Sig.
Soil health	Between groups	1124.266	17	66.13327	2505.333	0.000
	Within groups	56.06723	2124	0.026397		
	Total	1180.333	2141			

than the denominator. The test was carried out at a total df of 2498 where 19 (n1) and 2478 (n2) degrees of freedom at 95% confidence level for the mean score of all the 20 variables. The significant value is 0.000 being less than 0.05. Since the p-value < significance level (0.05), the difference in the mean value of variables of socioeconomic

Table 26.7 ANOVA of vermicompost usage role on socioeconomic sustainability of farmers.

		Sum of squares	Df	Mean square	F	Sig.
Socioeconomic sustainability	Between groups	521.1485	19	26.05742	137.2266	0.000
	Within groups	470.5378	2478	0.189886		
	Total	991.6863	2498			

sustainability is statistically significantly different. So the null hypothesis—there is no significant difference between the usage of vermicompost on different variables of socioeconomic sustainability of farmers—is rejected with 95% confidence.

The effect of vermicompost was positive on rice-cultivated soil. It led to improvement of soil nutrient status, nutrient content rise in postharvest soil, and a better soil structure. This is similar to the study by Bhaduri et al. (2020) who proposed nutrient availability, soil microbial activity, and grain quality as a long-term strategy for soil sustainability. There is an increasing trend in soil aeration, soil porosity, and soil aggregation after the application of vermicompost to the soil. Further, it was found from the experience of the farmers that the yield was better with the increase in the population of earthworms. It was observed that the chlorophyll content of the leaf improved similar to the findings of Manivannan et al. (2009). There was a decrease in weed problems, soil hardness, and pathogens as given by Sinha et al. (2010) which depicts the usefulness of vermicompost for soil. The use of vermicompost had a positive effect on soil health in the case of paddy-grown areas taken for the study.

Analyzing the socioeconomic sustainability variables, the result was mixed. There was a decrease in labor demand as well as labor cost per day adding to better economic conditions of farmers which matched with the findings of Annolfo et al. (2017). An increasing trend was found in the benefit/cost ratio, rice yield, and tillers. The labor productivity also improved. More training was imparted by the Agriculture Department and the Krishi Vigyan Kendras under the Indian Council of Agricultural Research (ICAR) to these farmers to impart knowledge on the know-how of vermicomposting and agriculture. The health of the farmers and their families improved as they no longer faced the side effects of chemical fertilizers. There was an increase in individual and family income leading to better living standards and position in the society. This was similar to the findings of Barmon and Tarafder (2019). Since most of the farmers were marginal and small, the asset enhancement was not much as they had less avenues. This result was contrary to the study by Kumar et al. (2019) who found that the marginal and small category farmers had a miserable socioeconomic condition. The kids went to school with an increased income and the farmers also spent more on food and education. Some farmers even sent their children out of the village for higher studies showing their interest in literacy.

26.8 Conclusion

Vermicompost is the transformation of biodegradable waste to biofertilizer which is used in sustainable organic agriculture. Vermicomposting is an integrated approach that improves soil health and productivity of crop without harming the environment. The farmers' friend earthworm and certain microbes disintegrate all the waste converting them to biofertilizers used for agriculture. The zero-waste production further improves physical, chemical, and biological soil health. Along with improved soil health, the use of vermicompost as a fertilizer even had a positive effect on the socioeconomic conditions of the farmers. Organic agriculture can produce better food for us without harming the environment and soil, and the use of vermicompost is an inclusive way of building sustainability.

References

Annolfo, R., Gemmill-Herren, B., Graeub, B., Garibaldi, L.A., 2017. A review of social and economic performance of agroecology. Int. J. Agric. Sustain. *156*, 632–644.

Ashraf, I., Ahmad, I., Nafees, M., Yousaf, M.M., Ahmad, B., 2016. A review on organic agriculture for sustainable agricultural production. Pure Appl. Biol. *52*, 277.

Banik, P., Bejbaruah, R., Farm, A.E., 2004. Effect of vermicompost on rice Oryza sativa yield and soil-fertility status of rainfed humid sub-tropics. Indian J. Agric. Sci. 74, 488–491.

Barmon, B.K., Tarafder, S.K., 2019. Impacts of vermicompost manure on MV paddy production in Bangladesh: a case study of Jessore district. Asia Pac. J. Rural. Dev. *291*, 52–76.

Bejbaruah, R., Sharma, R.C., Banik, P., 2013. Split application of vermicompost to rice Oryza sativa L.: its effect on productivity, yield components, and N dynamics. Org. Agric. *32*, 123–128.

Below, T.B., Mutabazi, K.D., Kirschke, D., Franke, C., Sieber, S., Siebert, R., Tscherning, K., 2012. Can farmers' adaptation to climate change be explained by socio-economic household-level variables? Glob. Environ. Chang. *221*, 223–235.

Bhaduri, D., Shahid, M., Chatterjee, D., Nayak, A.K., 2020. Organic Nutrient Management in Rice-Based System for Sustenance of Soil Health, Grain Quality and Productivity. Extended Summaries, p. 514.

Bharamappanavara, S.C., Mundinamani, S.M., Nithya, V.G., Naik, B.K., 2010. Sustainable approach of on-farm demonstrations (OFD) and its socio-economic impact on farmers in the tank commands of north eastern Karnataka. Organised by Department of Economic Sciences, p. 91.

Bhattacharya, S.S., Kim, K.H., Ullah, M.A., Goswami, L., Sahariah, B., Bhattacharyya, P., Cho, S.B., Hwang, O.H., 2016. The effects of composting approaches on the emissions of anthropogenic volatile organic compounds: a comparison between vermicomposting and general aerobic composting. Environ. Pollut. 208, 600-607.

Biswas, S., Hazra, G.C., Purakayastha, T.J., Saha, N., Mitran, T., Roy, S.S., Basak, N., Mandal, B., 2017. Establishment of critical limits of indicators and indices of soil quality in rice–rice cropping systems under different soil orders. Geoderma 292, 34-48.

Bronick, C.J., Lal, R., 2005. Soil structure and management: a review. Geoderma 124 (1-2), 3-22.

Carlsson, F., Nam, P.K., Linde-Rahr, M., Martinsson, P., 2007. Are Vietnamese farmers concerned with their relative position in society? J. Dev. Stud. 43 (7), 1177-1188.

Chanu, L.J., Hazarika, S., Choudhury, B.U., Ramesh, T., Balusamy, A., Moirangthem, P., Yumnam, A., Sinha, P.K., 2018. A Guide to Vermicomposting-Production process and Socio Economic Aspects. Extension Bulletin No: 81.

Chao, H.I., Zibilske, L.M., Ohno, T., 2003. Effects of earthworm casts and composts on coil microbial activity and plant nutrient availability. Soil Biol. Biochem. 35 (2), 295-302.

Das, S., Teja, K.C., Mukherjee, S., Seal, S., Sah, R.K., Duary, B., Kim, K.H., Bhattacharya, S.S., 2018. Impact of edaphic factors and nutrient management on the hepatoprotective efficiency of Carlinoside purified from pigeon pea leaves: an evaluation of UGT1A1 activity in hepatitis induced organelles. Environ. Res. 161, 512-523.

Feller, C., Blanchart, E., Bernoux, M., Lal, R., Manlay, R., 2012. Soil fertility concepts over the past two centuries: the importance attributed to soil organic matter in developed and developing countries. Arch. Agron. Soil Sci. 58 (sup 1), S3-S21.

Goswami, L., Nath, A., Sutradhar, S., Bhattacharya, S.S., Kalamdhad, A., Vellingiri, K., Kim, K.H., 2017. Application of drum compost and vermicompost to improve soil health, growth, and yield parameters for tomato and cabbage plants. J. Environ. Manag. 200, 243-252.

Indoria, A.K., Sharma, K.L., Reddy, K., Srinivasarao, C., Srinivas, K., Balloli, S.S., Osman, M., Pratibha, G., Raju, N.S., 2018. Alternative sources of soil organic amendments for sustaining soil health and crop productivity in India—impacts, potential availability, constraints and future strategies. Curr. Sci. 115, 2052-2062.

Jeyabal, A., Kuppuswamy, G., 2001. Recycling of organic wastes for the production of vermicompost and its response in rice-legume cropping system and soil fertility. Eur. J. Agron. 15 (3), 153-170.

Jha, A.K., Mehta, B.K., Chatterjee, K., 2020. Effect of Integration of Different Sources of Plant Nutrients on Yield of Transplanted Rice in Sahibganj District of Jharkhand. Extended Summaries, pp. 598-600.

Kassam, A., Stoop, W., Uphoff, N., 2011. Review of SRI modifications in rice crop and water management and research issues for making further improvements in agricultural and water productivity. Paddy Water Environ. 9 (1), 163-180.

Kavamura, V.N., Esposito, E., 2010. Biotechnological strategies applied to the decontamination of soils polluted with heavy metals. Biotechnol. Adv. 28 (1), 61-69.

Kizilkaya, R., Izzet, A.K.Ç.A., Aşkin, T., Yilmaz, R., Olekhov, V., Samofalova, İ., Mudrykh, N., 2012. Effect of soil contamination with azadirachtin on dehydrogenase and catalase activity of soil. Eurasian J. Soil Sci. 1 (2), 98-103.

Kumar, S., Chintala, R., Rohila, J.S., Schumacher, T., Goyal, A., Mbonimpa, E., 2015. Soil and crop management for sustainable agriculture. Sustain. Agric. Rev. 16, 63-84.

Kumar, H.M., Chauhan, N.B., Patel, D.D., Patel, J.B., 2019. Predictive factors to avoid farming as a livelihood. J. Econ. Struct. 8 (1), 1-18.

Lazcano, C., Domínguez, J., 2011. The use of vermicompost in sustainable agriculture: impact on plant growth and soil fertility. Soil Nutr. 10 (1–23), 187.

Lim, S.L., Wu, T.Y., Lim, P.N., Shak, K.P.Y., 2015. The use of vermicompost in organic agriculture: overview, effects on soil and economics. J. Sci. Food Agric. 95 (6), 1143-1156.

Mahanta, K., Jha, D.K., Rajkhowa, D.J., Manoj-Kumar, 2012. Microbial enrichment of vermicompost prepared from different plant biomasses and their effect on rice (*Oryza sativa* L.) growth and soil fertility. Biol. Agric. Hortic. 28 (4), 241–250.

Manivannan, S., Balamurugan, M., Parthasarathi, K., Gunasekaran, G., Ranganathan, L.S., 2009. Effect of vermicompost on soil fertility and crop productivity-beans (Phaseolus vulgaris). J. Environ. Biol. 30 (2), 275–281.

Mishra, P., Chhatoi, B.P., 2017. Sustainability of khadi: the SEEC approach. Pratibimba 17 (2), 67–75.

Mishra, P., Dash, D., 2014. Rejuvenation of biofertilizer for sustainable agriculture and economic development. Consilience 11, 41–61.

Mohanty, S., Nayak, A.K., Kumar, A., Tripathi, R., Shahid, M., Bhattacharyya, P., Raja, R., Panda, B.B., 2013. Carbon and nitrogen mineralization kinetics in soil of rice-rice system under long term application of chemical fertilizers and farmyard manure. Eur. J. Soil Biol. 58, 113–121.

Mondal, T., Datta, J.K., Mondal, N.K., 2020. Recycling of municipal solid waste into valuable organic fertilizer towards rejuvenation of crop physiology, yield and soil health. Arch. Agron. Soil Sci., 1–12.

Nayak, A.K., Gangwar, B., Shukla, A.K., Mazumdar, S.P., Kumar, A., Raja, R., Kumar, A., Kumar, V., Rai, P.K., Mohan, U., 2012. Long-term effect of different integrated nutrient management on soil organic carbon and its fractions and sustainability of rice-wheat system in Indo Gangetic Plains of India. Field Crop Res. 127, 129–139.

Negi, V.S., Maikhuri, R.K., 2013. Socio-ecological and religious perspective of agrobiodiversity conservation: issues, concern and priority for sustainable agriculture, central Himalaya. J. Agric. Environ. Ethics 26 (2), 491–512.

Nurhidayati, N., Machfudz, M., Murwani, I., 2018. Direct and residual effect of various vermicompost on soil nutrient and nutrient uptake dynamics and productivity of four mustard Pak-Coi (Brassica rapa L.) sequences in organic agriculture system. Int. J. Recycl. Org. Waste Agric. 7 (2), 173–181.

Park, J.H., Lamb, D., Paneerselvam, P., Choppala, G., Bolan, N., Chung, J., 2011. Role of organic amendments on enhanced bioremediation of heavy metalloid contaminated soils. J. Hazard. Mater. 185 (2–3), 549–574.

Patil, S., Reidsma, P., Shah, P., Purushothaman, S., Wolf, J., 2014. Comparing conventional and organic agriculture in Karnataka, India: where and when can organic agriculture be sustainable? Land Use Policy 37, 40–51.

Purkayastha, R.D., 2012. Forming community enterprises using vermicomposting as a tool for socio-economic betterment. In: Proceedings of the 2012 International Conference on Economics, Business and Marketing Management. vol. 29.

Rahman, S., Barmon, B.K., 2019. Greening modern rice agriculture using vermicompost and its impact on productivity and efficiency: an empirical analysis from Bangladesh. Agriculture 9 (11), 239.

Rahman, M.M., Ranamukhaarachchi, S.L., 2003. Fertility status and possible environmental consequences of Tista floodplain soils in Bangladesh. Sci. Technol. Asia, 11–19.

Ranamukhaarachchi, S.L., Begum, M.M.R.S.N., 2005. Soil fertility and land productivity under different cropping systems in highlands and medium highlands of Chandina sub-district, Bangladesh. Asia-Pac. J. Rural. Dev. 15 (1), 63–76.

Rorat, A., Wloka, D., Grobelak, A., Grosser, A., Sosnecka, A., Milczarek, M., Jelonek, P., Vandenbulcke, F., Kacprzak, M., 2017. Vermiremediation of polycyclic aromatic hydrocarbons and heavy metals in sewage sludge composting process. J. Environ. Manag. 187, 347–353.

Saha, S., Dutta, D., Ray, D.P., Karmakar, R., 2012. Vermicompost and soil quality. In: Farming for Food and Water Security. Springer, Dordrecht, pp. 243–264.

Sahariah, B., Das, S., Goswami, L., Paul, S., Bhattacharyya, P., Bhattacharya, S.S., 2020. An avenue for replacement of chemical fertilization under rice-rice cropping pattern: sustaining soil health and organic C pool via MSW-based vermicomposts. Arch. Agron. Soil Sci. 66 (10), 1449–1465.

Sahariah, B., Goswami, L., Kim, K.H., Bhattacharyya, P., Bhattacharya, S.S., 2015. Metal remediation and biodegradation potential of earthworm species on municipal solid waste: a parallel analysis between Metaphire posthuma and Eisenia fetida. Bioresour. Technol. 180, 230–236.

Sharma, R.C., Banik, P., 2014. Vermicompost and fertilizer application: effect on productivity and profitability of baby corn (Zea Mays L.) and soil health. Compost Sci. Util. 22 (2), 83–92.

Sheehy, J.E., Mitchell, P.L., Hardy, B., 2007. Charting New Pathways to C4 Rice. International Rice Research Institute, Los Banos: Philippines.

Singh, A., Singh, G.S., 2017. Vermicomposting: a sustainable tool for environmental equilibria. Environ. Qual. Manag. 27 (1), 23–40.

Sinha, R.K., Agarwal, S., Chauhan, K., Valani, D., 2010. The wonders of earthworms & its vermicompost in farm production: Charles Darwin's 'friends of farmers', with potential to replace destructive chemical fertilizers. Agric. Sci. 1 (02), 76.

Sreedevi, P., Veerabhadra Rao, K., 2020. Performance evaluation of paddy transplanter in north coastal districts of Andhra Pradesh. In: Combining High-Throughput Phenotyping and Genome-Wide Association Studies for Genetic Dissection of Nitrogen Use Efficiency in Rice, pp. 572–573.

Srivastava, P., Singh, R., Tripathi, S., Raghubanshi, A.S., 2016. An urgent need for sustainable thinking in agriculture—an Indian scenario. Ecol. Indic. 67, 611–622.

Suthar, S., 2009. Impact of vermicompost and composted farmyard manure on growth and yield of garlic (Allium stivum L.) field crop. Int. J. Plant Prod. 31, 27–38.

Svensson, G., Wagner, B., 2015. Implementing and managing economic, social and environmental efforts of business sustainability. Manag. Environ. Qual., 195.

Tripathy, S., Bhattacharyya, P., Mohapatra, R., Som, A., Chowdhury, D., 2014. Influence of different fractions of heavy metals on microbial ecophysiological indicators and enzyme activities in century old municipal solid waste amended soil. Ecol. Eng. 70, 25–34.

Uddin, M.T., Dhar, A.R., 2017. Conservation agriculture practice in Bangladesh: farmers' socioeconomic status and soil environment perspective. Int. J. Econ. Manag. Eng. 11 (5), 1251–1259.

White, P.J., Crawford, J.W., Álvarez, M.C.D., Moreno, R.G., 2012. Soil management for sustainable agriculture. In: European Geosciences Union General Assembly 2011, Vienna, Austria, 3-8 April 2011. Applied and Environmental Soil Science. 2012.

Yadav, S.K., Babu, S., Yadav, M.K., Singh, K., Yadav, G.S., Pal, S., 2013. A review of organic agriculture for sustainable agriculture in northern India. Int. J. Agron. 13, 1–8.

Yu, Y.L., Xue, L.H., Yang, L.Z., 2014. Winter legumes in rice crop rotations reduces nitrogen loss and improves rice yield and soil nitrogen supply. Agron. Sustain. Dev. 34 (3), 633–640.

Index

Note: Page numbers followed by *f* indicate figures and *t* indicate tables.

A

Abiotic stress, 116, 319, 364, 436–437, 463–464, 547
  crop loss, 395–396
  drought stress, 397–398
  economic loss, 395–396
  induced systemic tolerance (IST), 396–397
  management
    drought stress, 18–20
    high temperature stress, 20
    low-temperature stress, 20–21
    salinity stress, 16–17
  metal toxicity, 399–400
  microbe-mediated stress, in crop plants (*see* Plant growth-promoting microorganism-mediated abiotic stress)
  microbial bioinoculants for, 331–332
  plant-microbe interactions and mitigation of, 253–261
  resource management and adaptation strategies, in plants, 395–396
  rhizobacteria-induced stress resilience, 396–397
  salinity, 398
  temperature, 399
  tolerance, microbial inoculants for, 296–301
  yield loss, 395–396
Abscisic acid (ABA), 14, 139, 331–332, 404, 671
Acacia mearnsii, 164
Acalypha indica, 271–272
Acculospora kentinensis, 243–244
Acetebacter diazotrophicus, 14–15
Achromobacter piechaudii ARV8, 134
Acidithiobacillus ferrooxidans, 576–577
Acids
  abscisic acid (ABA), 14, 139, 331–332, 404, 671
  fatty acids, 187–188
  gibberellic acid, 137–138
  gluconic acid, 515–516
  indole-3-acetic acid (IAA), 14–15, 259–260, 287, 297–298, 333, 529, 534
  monounsaturated fatty acids, 187–188
  organic acid (OA)
    by bacterial phosphate biofertilizers, 581*t*
    phosphate-solubilizing bacterial biofertilizers, 432–434, 432*t*
    phosphate-solubilizing fungi and actinomycetes, 432–434, 433*t*
    production, 11
    of P solubilization, 506
  phosphatase, 513–515
  phosphatidic acid (PA), 670–671
  polyunsaturated fatty acids, 195
  production, 10
  salicylic acid, 139–140
Acorus calamus, 628–629
Actinobacteria, 70
Actinomycetes, 25
  phosphate solubilization, 422, 424–425
  organic acids secretion, 432–434, 433*t*
  sources, 427, 430*t*
Actinorhizal associations, 128
Acyl homoserine lactone (AHL), 115
Aedes albopictus-Wolbachia symbiotic association, 674
Aerobic diazotrophs, 208–209
Aeroponic system, 243–244
AFPs. *See* Antifreeze proteins (AFPs)
Aggressive marketing strategies, 302
Agrarian techniques, 109–110
Agricultural & Processed Food Products Export Development Authority (APEDA), 710, 714
Agriculture, 54–55, 210, 215–218, 535, 722–727
  climate and global warming, 725
  crops, 253
    productivity, 15–16, 722–723
    protection, 724
  environment, 725
  microbial consortium in, 116–117
  organic agriculture (*see* Organic agriculture)
  soil and soil fertility, 725–727
  for sustainability, 287–301
    biocontrol agents, 291–292
    biofertilizers, 288–291
    biotic and abiotic stress tolerance, 296–301
    different microbial groups in soil ecosystem, 292–296
    microorganisms in, 295*t*
  sustainable environment and, 614–616
  U.S. Department of Agriculture, 638–639
Agrobacterium spp.
  radiobacter, 132–133
  tumefaciens, 10–11, 467, 656, 676
Agrobacterium transformation, 276–277

759

Agroecosystem, 700–701
  ecological niches, microbial diversity in, 97–100
  Jhum agroecosystem, microbial diversity in, 95–97
  microbial diversity in, 97–100
  micro-soil-plant agro-ecosystem, 246
  phosphate (P) biofertilizers in, 423–424
  sustainable agroecosystems, 246
Agro-environmental sustainability, organic agriculture for, 699–701
  certified area, 711–712*t*
  continental data on, 709*t*
  glimpse of, 710*t*
  global scenario of, 703–708*t*
  highlights of food market, 714*t*
  impact of, 722–727
  market potential of, 730–731
  national status of, 710–714
  principles of, 701*f*
  products used in, 718–722
    cow dung for soil fertilization, 718–719
    green manure for soil health, 721–722
    microbes (PGPRs) for all-round development, 721
    milk for disease and weed control, 719–720
    waste fodder in composting, 719
  pros and cons of, 716–718
  regulatory policies in, 727–730
  standards of, 702, 702*f*
  state-wise data of, 713*t*
  status of, 702–716
Agronomic soils, heavy metal contamination of, 567
Agrosynthetic insecticides/pesticides, 701
AHL. *See* Acyl homoserine lactone (AHL)
Alcaligenes fecalis, 101
Alexopoulos, 526
Aliphatic hydrocarbon-contaminated soil, 141–142

Alnus nepalensis, 97
Aloe vera, 629–630
Alpha-amylase, 534
Alternanthera philoxeroides, 141–142
AM. *See* Arbuscular mycorrhizae
Amaranthus spp.
  hypochondriacus, 135
  mangostanus, 135
Amensalism, 48
AMF. *See* Arbuscular mycorrhizal fungi (AMF)
1-Aminocyclopropane-1-carboxylic acid (ACC), 296–297, 333, 404–405, 434–436
Ampelomyces quisqualis, 295–296
Analysis of variance (ANOVA), 750–753
  effect on soil health, 752*t*
  socioeconomic sustainability of farmers, 753*t*
Analytical techniques
  in metabolomics, 368, 369*t*
    limitation, 372
    mass spectrometry (MS), 371–372
    nuclear magnetic resonance (NMR), 370
    popularity of, 368–369, 370*f*
  in proteomics
    gel free-MS-based proteomics, 358, 360–361
    2DE-gel-based proteomics, 358–360, 360*f*
Anecdotal records, 159
Anopheles stephensi, 628–629
ANOVA. *See* Analysis of variance (ANOVA)
Antagonism, 161–163, 165, 169, 173
Antarctic bacterium *Pseudoalteromonas*, 195
Antibiosis, 164, 325–326, 526–527, 548, 633–635
Antibiotic metabolites, 325–326
Antibiotics, 22
Antifreeze proteins (AFPs), 189

Antifungal activity of halophytic rhizospheric bacteria, 262*f*
Antifungal volatile organic compounds, 165
Antimicrobial metabolites, 49
Antimicrobial pathogenesis-related (PR) genes, 673
Antimicrobial peptides (AMPs), 325–326, 465–466
Antioxidant enzymes, 406–408
Antirrhinum majus, 531
APEDA. *See* Agricultural & Processed Food Products Export Development Authority (APEDA)
Apoplastic ROS, 669
Aquimonas voraii, 66–68
Arabidopsis FLS2 (*At*FLS2), 665–666
Arabidopsis *NPR1* gene, 468–470
Arabidopsis thaliana, 49–50, 53, 666, 674
Arbuscular mycorrhizae (AM), 18–19, 129, 199–200, 493–494, 656
Arbuscular mycorrhizal fungi (AMF), 225, 229, 288–290, 334
  classification of, 226–227*t*
  integration of, 240–241
  interaction of, 233–234
  production and commercialization of, 243–245
  suppression of, 230–233
Arbutoid mycorrhizae, 129–130
Array-based sequencing, 347
Artificial nitrogen fertilizers, 701
Ascomycetes, 130
Ascomycota, 485–487
Aspergillus spp.
  brasilense, 139
  flavus, 17, 58
Asteraceae plant family, 629
Atmospheric nitrogen fixation, 51*f*, 208
ATP-dependent process of biological nitrogen fixation, 4

ATP-dependent transport system, 188–189
Augmentative biocontrol (ABC), 156
Augmented bioremediation bacteria, 301
Auxin, 14, 138–139
Auxin-like hormones, 534
Avirulence, 661–662
Avirulent genes (Avr), 468, 677
Axenic cultivation of AM fungi, 244–245
Azolla filiculoides, 52
Azospirilla group, 97–98
Azospirillum brasilense, 516–517
Azotobacter spp., 322–323
  brasilense, 743
  chroococcum, 743

B
Bacillus-based biocontrol agents, 58
Bacillus-based formulations, 58
Bacillus spp.
  amyloliquefaciens, 13, 374
  aryabhattai, 11–12
  magaterium, 8–9
  subtilis, 26, 531
    Ap113, 327–330
    FB17, 355
    GB03, 133–134
    IB22, 139
  thuringiensis, 466–467, 630–631
  velezensis, 26, 355
  xiamenensis, 22–23
Bacteria, 4
  actinobacteria, 70
  augmented bioremediation bacteria, 301
  bacteria-based bioleaching, 576–577
  biopesticides, 624–625
    against insects, 630–631
  consortia, 49, 116–117
  culture, 539f
  cyanobacteria, 128, 210
  diversity in hot water springs, 66–68
  dizaotrophic bacteria, 7–8
  endophytic bacteria, 99
  filamentous photosynthetic cyanobacteria, 128
  genetically improved entomopathogenic bacteria, 327
  gram-positive soil actinobacteria, 70
  halophilic bacteria, 255
  mutualistic bacteria, 54
  nitrogen-fixing rhizobacteria, 288–290
  phosphate biofertilizers organic acids by, 581t
  plant growth-promoting bacteria (PGPB), 114, 240–241, 288–290, 537
  plant-pathogenic bacteria (PPB), 662–663
  plant stress homeostasis-regulating bacteria (PSHB), 288–290
  potassium-solubilizing bacteria (KSB), 484–485, 486t
  potential phosphate-solubilizing bacteria, 9
  QS molecules, 49
  rhizobacteria, 464–465
  salt-tolerant halophytic rhizospheric bacteria, 256–259t
  species, 4
  strains, 10–11, 68
    GPTSA100-9T, 66–68
  sulfur-oxidizing bacteria, 52
  symbiotic bacteria, 4, 50
  symbiotic nitrogen fixation (SNF) bacteria, 464–465, 585–586
  Zn-tolerant bacteria (ZTB), 581–584
Bacteria-based biofertilizers in cultivation practices
  nitrogen fixers, 585–586, 585t
  nutrient solubilizers, 586–590
    iron sequesters, 589–590
    phosphate solubilizers, 586, 587t
  potassium solubilizers, 586–588, 588t
  sulfur-oxidizing microbes, 590
  zinc solubilizers, 589
Bacteria-based bioleaching, 576–577
Bacterial biofertilizers, 593–594
  for bioremediation, 565–566
  biosorption vs. bioaccumulation processes, 575t
  concepts and prospects of, 567–569
  definition and categories of, 569–571, 570f
  factors of, 577
  heavy metal contamination of agronomic soils, 567
  metal-contaminated environment, remediation of, 571–584
  perspective of, 584–594
  phosphate biofertilizers, organic acids by, 581t
Bacteriocins, 22
Bacterium BPM3, 68
Baculoviruses. *See* Viral biopesticides
Barley powdery mildew fungus, 364
Basidiomycota, 485–487
BCAs. *See* Biological control agents (BCAs)
BcGV1. *See* Botrytis cinerea genomovirus 1 (BcGV1)
Bean yellow dwarf virus (BeYDV), 471–472
Beauveria bassiana, 625
Beneficial biotic organisms, 525
Beneficial fungi, 175–176
Beneficial microbial population, 23
Beneficial microbiomes
  implication of, 27
  in organic agriculture, 27
  in plant growth promotion, 2–15, 5–7t
  nitrogen fixation, 3–8

Beneficial microbiomes (*Continued*)
  phosphorus solubilization, 8–9
  phytohormone production, 14–15
  potassium solubilization, 10–11
  siderophore production, 12–13
  zinc solubilization, 11–12
  for sustainable crop production and protection, 23–26
    biofertilizers, 23–25, 24f
    biopesticides, 25–26
Beneficial microorganisms, 51–52, 301–302
Beneficial rhizobacteria, 139–140
Beneficial soil inhabitants, 617–618
Beneficial soil microbes, 743
Bergey's Manual of Systematic Bacteriology, 70–71
Beta vulgaris L, 351
Bioaccumulation process, 573–574, 575t
Biochemical composition of root exudates, 136t
Biochemical pesticides, 619–624
  essential oils (EOs), 622, 623t
  insect pheromones, 622–624
  plant extracts and oils, 620–622, 620–621t
Biochemical substances, 528
Biocontrol agents, 21–22, 58, 132–133, 165, 169, 229–231, 233–234, 240–241, 638
  activity of halophytes, 259–260
  microbial inoculants as, 291–292
  Nagoya Protocol, 174–175
Biodiversity, 725
Bioenergy crops, 297–298
Biofertilizer-based bioremediation, 569
Biofertilizers, 23–25, 24f, 27, 142, 143–144t, 253–254, 306, 502–503, 721. *See also* Fertilizers

bacterial biofertilizers (*see* Bacterial biofertilizers)
bioinoculants of, 481–482
bioremediation by nitrogenous biofertilizers, 578–580
categories of, 424–425, 481–482
challenges and prospects of, 593–594
commercial production of, 591–592t
constituents, 481–482
and crop production, 590–593
definition of, 421–422, 481–482, 569–571
diazotroph-based biofertilizers, 215
and Indian agriculture, 694–695
market segmentation, 692, 693f
market value, 424–425
metal-tolerant P-biofertilizers, 565–566, 580–581, 582t
microbe-based biofertilizers, 593–594
microbial bioinoculants, 320, 321t, 322f, 330–331
microbial inoculants as, 288–291
nature and functions, 481–482
nitrogen-fixing biofertilizers, 424–425, 585–586, 585t
perspective of, 584–594
phosphate (P) biofertilizers, 508–509, 509t, 580–584
  in agrosystems, 423–424
  application methods and limitations, 430–431
  challenges and future prospects, 447
  constraints in, 517–518
  eco-physiological traits, 436–437, 437f
  field-level constraints in, 517
  financial constraints in, 517
  food production system, application in, 437–438
  for heavy metal remediation, 580–584

Indian Agricultural Research Institute (IARI) microphos, 424–425
  manufacturers of, 424–425, 426t
  marketing of, 509–512, 510t
  market value, 424–425
  metal tolerance and plant growth-stimulating features of, 582t
  microphos production, phases and carriers of, 427
  mineral phosphate solubilization (mps), 424–425
  molecular engineering of, 445–447
  on native soil microorganisms, 516–517
  nutrient management, role in, 424–425
  phosphobacterin, 424–425
  phyto-beneficial features, 434–436, 435t
  P-solubilizing microbes as (*see* Phosphate-solubilizing microorganisms (PSMs))
  raw material constraints in, 517
  seedling dipping of, 430–431
  seed surface inoculation of, 430–431
  technical constraints in, 517
potash biofertilizers (*see* Potash biofertilizers)
potassium-solubilizing biofertilizers (KSB), 442–443
psychrophilic PGP microbes, 199–200
remediation of metal-contaminated environment, 570f, 571–577
  bioaccumulation process, 573–574, 575t
  bioleaching of metals, 576–577
  biosorption process, 571–573, 575t

Index **763**

biotransformation process, 573–574, 575t
symbiotic/asymbiotic N biofertilizers, 579–580
Biofilm, 110–111
Bioinformatics tools, in proteomics, 361
Bioinoculant, 287
Bioleaching of metals, 576–577
Biolistic techniques, 276–277
Biological control agents (BCAs), 160, 163, 165
  of forest pathogens, 155–156
    adoption of, 161–162
    challenges of, 169–175
    Heterobasidion annosum, 166–169
    importance of, 156–160
    interactions contributing to, 162–163
    mechanisms of, 164–166
  in forestry species, 170–172t
  laborious laboratory screening and field screening of, 173
  sensitivity of, 169–173
Biological mechanisms of *Trichoderma*, 547–549
Biological nitrogen fixation, 23, 50, 296–297
Biological nitrogen-fixing diazotrophs, 215–217
Biological pesticides, 637
Biological stresses, 536
Biomagnification, 617–618
Biopesticides, 25–26, 306, 613–614, 618–619
  bacterial biopesticides, 624–625
  classification of, 619f
  and environment, 636–637
  fungal biopesticides, 625
    against insects, 631
  market of, 640–642
  microbial biopesticides, 624–626 (*see also* Microbial biopesticides)
  microorganism-based biopesticides forms, 624
  synthesis of, 639–640

mode of action, 626–635
role of, 635–640
sustainable environment and agriculture, 614–616
viral biopesticides, 625–626, 631–632
  against insects, 631–632
Bioprospecting PGPR, 300
Bioremediation
  augmented bioremediation bacteria, 301
  biofertilizer-based bioremediation, 569
  biofertilizer technology in, 569–584
  concepts and prospects, 567–569
  ex situ bioremediation, 567–568
  of heavy metals, 577
  of metal-contaminated environments, 577–584
  by nitrogenous biofertilizers, 578–580
  in situ bioremediation, 567–568
Biosecurity risk, 159
Biosorption process, 571–573, 575t
Biosphere, 655
Biostimulants, 117, 110t
Biosurfactants, 326
Biotechnology, 302
  molecular methods, 286
Biotic organisms, 542
Biotic stress, 253, 319, 364, 463–464
  causative agents of, 253
  management, 21–23
  microbial bioinoculants for, 331–332
  for plants, 157
  tolerance, microbial inoculants for, 296–301
Biotransformation process, 573–574, 575t
Biotrophic pathogens, 48, 661–662
Blumeria graminis, 657–658
Botanical pesticides, 620–622, 620–621t, 626–630
Botrytis cinerea, 164, 657–658

Botrytis cinerea genomovirus 1 (BcGV1), 164
Bradyrhizobium spp.
  elkanii, 134
  japonicum, 210
Brassicaceae, 675
Breeding techniques, 20
Brevibacillus laterosporus, 68
Bromophenol blue dye method, 427
Burkholderia spp.
  multivorans, 515–516
  phytofirmans, 133–134, 300
  pyrrocinia, 26
  vietnamiensis, 327–330

C

CAGR. *See* Compound annual growth rate (CAGR)
Calcium ions, 668
Callosobruchus chinensis, 628–629
Carbamate insecticides, 636–637
Carbon dioxide (CO_2), 300–301
Carboxylate siderophores, 12–13
Cash cows, 692
Cauliflower mosaic viral (CMV), 276–277
Causative agents of biotic stress, 253
CDMs. *See* Cellulose-degrading microorganisms (CDMs)
cDNA-AFLP technique, 353
CEBiP. *See* Chitin elicitor binding protein (CEBiP)
Cell membrane response, 187–188
Cellular antioxidant mechanisms, 542
Cellulose-degrading microorganisms (CDMs), 96
Cellulosimicrobium cellulans, 135
Cell wall-degrading enzymes (CWDEs), 326
Central Insecticides Board and Registration Committee (CIBRC), 641–642
CFU. *See* Colony-forming units (CFU)
Chaetomium cupreum, 295–296

Chelation, 25
Chemical-based fertilizers, 4–7
Chemical fertilizers, 304–305, 421–425
Chemical pesticides, 156, 616–618
Chemical signaling, 49
Chemosynthetic fertilizers, 701
Chick-pea-*Rhizobium* system, 213
China-based study, 159–160
Chitinase gene, 327–330
Chitin elicitor binding protein (CEBiP), 667
Chloris gayana, 243
Chlorophyll depletion, 531
Chromosomal symbiosis islands, 209–210
Chrysanthemum spp.
  cinerariaefolium, 626–628
  morifolium, 16–17
CIBRC. *See* Central Insecticides Board and Registration Committee (CIBRC)
Cinnamomum zeylanicum, 273
Citrus lanatus, 630
Cladosporium fulvum, 667
"Classical" method, 638
Climate change, 157–158, 300–301
Cloned genes, 516*t*
Clonostachys, 326
Clustered regulatory interspaced palindrome repeats-associated protein 9 (CRISPR/cas9) technology, 306, 471–472, 472–473*t*
Clustering methods, 377
CMV. *See* Cauliflower mosaic viral (CMV)
Codex Alimentarius Commission (Codex), 700–701, 727–728
Coexisting fungi, 551
Coffea arabica, 628–629
Cold acclimation proteins (caps), 189–191
Cold-active enzymes, 191, 193–195
Cold-active xylanases, 193–195
Cold shock domain (CSD), 189–191
Cold shock proteins (CSPs), 189–191

Cold shock response, 189–191
Cold stress, 187, 190*t*
Cold tolerance in microorganisms, 186–191
  antifreeze proteins, 189
  cell membrane response, 187–188
  cold acclimation proteins and cold shock response, 189–191
  cryoprotectants, 188–189
  RNA degradosome, 191
Cold-tolerant enzymes, 191–200
  agricultural aspects, 197–200
    plant growth-promoting (PGP) microbes, 198
    psychrophilic PGP microbes, 198–200
  environmental aspects, 195–197
  industrial and medical aspects, 193–195
  mutagenized microorganisms, 194*t*
Cold-tolerant mutants in cold ecosystems, diversity of, 186
Colletotrichum gloeosporioides, 628–629
Colonization, 527, 549
  Trichoderma vs. plant, 549–552
Colony-forming units (CFU), 70–71
Commercial biofertilizer, 27
Comparative genomics, 344
Competition, 526–528, 532–533, 536, 547–550
Complex anatomy, 174
Complex microbial enzyme systems, 100–101
Compound annual growth rate (CAGR), 175
Consolidation quadrant, 692
Consortium, 320–321
Constraints
  in fungal biopesticides, 625
  in microorganism-based biopesticides, 639–640
Contaminated habitats, 100–102

Continental data on organic agriculture, 709*t*
Continental organic areas, 703*f*
Continuous cultures, 244
Controlling pathogens, 633–635, 634*t*
Conventional agriculture methods, 716–718, 717*f*, 725
Conventional gene delivery systems, 277*f*
Conventional methods of AMF, 243–244
Cost-per-base sequencing, 347–348
Counting fungal spores, 541*f*
COVID-19 on biofertilizer market, 696–697
Cow dung for soil fertilization, 718–719
Cronartium ribicola, 157–158
Crop production, 23–26, 54–55, 722–723
Crop protection, 724
Crotolaria pallida, 98–99
Cryoprotectants, 188–189
Cryphonectria parasitica, 157–158
CRY proteins, 466–467
CSD. *See* Cold shock domain (CSD)
CSPs. *See* Cold shock proteins (CSPs)
Cucurbita pepo, 719–720
Culex pipiens, 674
Cultured host plants, 243–244
Curtobacterium albidum, 17
Cyanobacteria, 128, 210
Cyst nematodes, 659
Cytokinins, 14, 139

D
Damage-associated molecular patterns (DAMPs), 664–665
DDRT-PCR technique, 353
DED. *See* Dutch elm disease (DED)
Degradation process, 195–196
Demographic profile of respondents, 747–750, 748–749*t*

Denaturing gradient gel
    electrophoresis (DGGE), 71
Dephosphorylation mechanism,
    187
Desiccation tolerance testing,
    491–492
Desorption electrospray
    ionization-mass
    spectrometry, 372
Desorption sonic spray ionization,
    372
Detoxification, 575–576
Deuteromycetes, 130
DGGE. See Denaturing gradient
    gel electrophoresis (DGGE)
Diazotroph-based biofertilizers, 215
Diazotrophic biofertilizers, 216t
Diazotrophs, 50, 210–211
    genomic and transcriptomics of,
        212–214
    to nonlegumes, 217–218
Dibru-Saikhowa Biosphere
    Reserve in Assam, 69–70
Dideoxy method, 346–347
Dietzia natronolimnaea strain
    STR1, 298–299
Differential method, 378
Diospyros kaki, 273
Direct biocontrol screening, 165
Disastrous environmental issues,
    110
Disease biocontrol, 164–165
Disease triangle, 660
Distilled water, fungal colonies in,
    540f
Diverse farming systems, 97–98
Diversity, of phosphate-
    solubilizing
    microorganisms, 505, 506t
Dizaotrophic bacteria, 7–8
DNA-based methods, 55
DNA-DNA based hybridization,
    349–350
DNA double-strand breaks
    (DSBs), 471–472
Dominant bacterial phyla, 46
Dos Santos model, 209
Dothideomycetes isolate YCB36a,
    166

Dothistroma septosporum, 159
Drought stress, 18–20, 297–298,
    397–398, 547
Drought tolerance, 296–297
Dutch elm disease (DED), 165–166

E

Echinochloa crusgalli, 135
Ecological niches, microbial
    diversity in
    agroecosystem, 97–100
    bacterial diversity in hot water
        springs, 66–68
    contaminated habitats, 100–102
    forest ecosystem, 69–95
    jhum agroecosystem, 95–97
    metagenome analysis of
        Phumdi, 68–69
Ecosystems, 126, 129–130, 135–137
    agroecosystem, 700–701
    microbial diversity in, 97–100
    forest ecosystems, 157, 159
    microbial diversity in, 69–95
    micro-soil-plant agro-
        ecosystem, 246
    sustainable agro-ecosystems,
        246
    terrestrial ecosystem, 52
Ectomycorrhizae (ECM), 129–130,
    493–494
Effector-triggered immunity (ETI),
    468, 663–664, 671–673
Effector-triggered susceptibility
    (ETS), 663–664
Electrospray ionization (ESI), 371
Eminent Dutch research groups,
    163
Endogenous metabolites, 366
Endomycorrhizae, 128–129
Endophytes, 56
Endophytic actinomycetes, 56
Endophytic bacteria, 99
Endophytic diazotrophs, 7–8, 210
Endophytic microbes, 320,
    332–333
Endophytic microbiomes, 656
Energy transfer mechanisms, 8
Enhanced plant defense
    mechanisms, 55–56

Enhanced plant growth, 53
Enrichment culture technique,
    427
Enterobacter lignolyticus, 98
Entomopathogenic fungi, 631
Entomopathogenic nematodes,
    632
Environmental DNA (eDNA),
    159–160
Environmental risk assessments
    for biocontrol agents,
    174–175
Environmental stress, 253,
    536–539
Enzyme engineering, 193
Enzyme production, 259–260
EOs. See Essential oils (EOs)
Epiphytes, 46–47
Epiphytic microbiomes, 656
Epiphytotic, 164, 175–176
EPSs. See Exopolysaccharides
    (EPSs)
Ericoid mycorrhizae, 130
Erlenmeyer flask, 539
Erwinia carotovora, 657–658
Escherichia coli, 304
Essential oils (EOs), 622, 623t,
    629–630
Ethylene, 14, 53, 333, 434–436
Ethylene-responsive factors
    (ERFs), 471–472
ETI. See Effector-triggered
    immunity (ETI)
ETS. See Effector-triggered
    susceptibility (ETS)
Eukaryotic actin cytoskeleton,
    671–672
Euphorbia nivulia stem latex, 272
European study group, 161
Evolve, 675–676
Exogenous metabolites, 366
Exome sequencing technology,
    674–675
Exopolysaccharides (EPSs), 255
    of P solubilization, 507–508
Expressed sequence tags (ESTs),
    344–345, 353
Expression profiling, 344–345
Expression proteomics, 345, 357

Ex situ bioremediation, 567–568
Extenuating abiotic stresses, 133–135
Extracellular exopolysaccharides (EPSs), 409
Extracellular PGPR (ePGPR), 464–465
Extremely risky quadrant, 691

F

F-AAS. *See* Flame atomic absorption spectrometry (F-AAS)
Fatty acids, 187–188
Fertility, 218
Fertilizers. *See also* Biofertilizers
   artificial nitrogen fertilizers, 701
   chemical-based fertilizers, 4–7
   chemical fertilizers, 304–305, 421–425
   chemosynthetic fertilizers, 701
   fossil fuel-based chemical fertilizers, 694
   NPK fertilizers, 699
   synthetic fertilizers, 306
   traditional fertilizers, 51–52
Festuca arundinacea, 141–142
Field-level constraints, in phosphate biofertilizers, 517
Filamentous photosynthetic cyanobacteria, 128
Financial constraints, in phosphate biofertilizers, 517
First Indian Organic Certification Agency (INDOCERT), 710
Fisher's method, 377
Flame atomic absorption spectrometry (F-AAS), 576–577
Flavobacterium indicum, 66–68
Foliar pathogens, 625
Food, 285–286, 288–290, 297–298, 731
Forest-based business enterprises, 156
Forest ecosystems, 157, 159
   microbial diversity in, 69–95
Forest pathogens, 157

biological control, 155–156
   adoption of, 161–162
   challenges of, 169–175
   Heterobasidion annosum, 166–169
   importance of, 156–160
   interactions contributing to, 162–163
   mechanisms of, 164–166
   fungal pathogens, 157–158
   human pathogens, 174
   necrotrophic pathogens, 48
   soil-borne pathogens, 132–133
   submicroscopic infectious pathogens, 658
Forestry crops, 156, 165
Forestry microbial biocontrol, global trade in, 175
Fossil fuel-based chemical fertilizers, 694
Four-phased "zigzag" model, 663–664, 664*f*
Frateuria aurantia, 10–11
Functional characterization, 159–160
Functional proteomics, 345, 358
Fungal biocontrol, 163
Fungal bioinoculants, 326
Fungal biopesticides, 625
   against insects, 631
Fungal colonies
   in distilled water, 540*f*
   growth of, 540*f*
Fungal genera, 69–70
Fungal pathogens, 157–158
Fungal siderophores, 133
Fungal species, 9
Fungal strains, 17
Fungi
   arbuscular mycorrhizal fungi (AMF), 225, 229, 288–290, 334
   classification of, 226–227*t*
   integration of, 240–241
   interaction of, 233–234
   production and commercialization of, 243–245
   suppression of, 230–233

axenic cultivation of AM fungi, 244–245
beneficial fungi, 175–176
coexisting fungi, 551
entomopathogenic fungi, 631
mycorrhizal fungi colonization, 20, 133, 235–236, 238–240, 243–244, 503–505
phosphorus-solubilizing fungi, 9
phytopathogenic fungi, 658
plant growth-promoting fungi (PGPF), 529
potassium-solubilizing fungi (KSF), 484–487, 486*t*
Trichoderma, 323–330, 525–529
   asperelloides, 165
   asperellum, 99–100, 530–531
   atroviride, 19
   avovidea, 550–551
   biological mechanisms of, 547–549
   in dealing with biological stresses (biotic), 544–545
   harzianum, 16, 132–134, 351, 351–352*t*, 355, 527–528, 531–532, 540*f*, 543, 545, 549–551, 630
   in increasing germination indices, 532–534
   on increasing growth and antioxidant activity, 542–543
   lixii, 543
   longibrachiatum, 16–17
   and plant, colonization as, 549–552
   in stimulating plant growth, 529–532
   virens, 550
   viride, 531
   in vitro/axenic cultivation of AM fungi, 244–245
Funneliformis mosseae, 235–236
Fusarium spp.
   circinatum, 159
   graminearum, 56
   oxysporum, 133
   pallidoroseum, 19

proliferatum, 295–296
sambucinum, 238–239
verticillioides, 545, 629

G

GABA-gated chloride channels, 628–629
Gaeumannomyces graminis var. *tritici* (Ggt), 327, 355
Gardenia jasminoides, 273
Gas chromatography, 212
Gas chromatography-mass spectrometry (GC-MS), 369t, 371
Gel-based inoculants, 58
GenBank, 344
Gene expression, 344–345
Gene-for-gene (GFG), 468, 677
Genera, 4
Generally recognized as safe (GRAS), 637
Genes
  antimicrobial pathogenesis-related (PR) genes, 673
  arabidopsis *NPR1* gene, 468–470
  avirulent genes (Avr), 468
  chitinase gene, 327–330
  cloned genes, 516t
  Hup genes, 213
  Nif genes, 213
  nonexpressor of pathogenesis-related genes 1 (*NPR1*), 468–470
  OsERF922 gene, 471–472
  phosphatidylinositol-phospholipase C (PI-PLC) gene, 470
  resistance genes, 661–662, 672f
  SlDMR1 gene, 470
  SlPMR4 gene, 470
  susceptibility (S) genes, 470
  ToxA protein-encoding gene, 677
  trypsin inhibitor gene, 466–467
Genetically improved entomopathogenic bacteria, 327
Genetically improved microbial bioinoculants, 327, 328–329t

Genetic engineering (GE), 286, 303–304
  in nanoparticle, 276–277
  in strain improvement, 304–306
  techniques, 323
Genetic/genome-editing methods, 55
  of phosphate solubilization by PSMs, 513–516
  techniques, 320–321
  technology, 471–472
Genetic manipulation, 546
Genetic resource management, 286
Genome mining, 214
Genome-scale analysis, 378
Genomics, 344
  agricultural microbiological research, application in, 351, 351–352t
  diversity analysis, 349–351
  genome sequencing, 346
  next-generation sequencing (NGS), 344–345, 350t
    advantages, 347
    array-based sequencing, 347
    glass solid phase surface, synthesis on, 348
    high-throughput pyrosequencing on beads, 347
    sequencing by ligation on beads, 347–348
    third-generation sequencing, 348–349
  of organisms, manipulation of, 303–304
  Sanger sequencing, 346–347
  sequencing techniques, 346
  studies, 192
Gibberella fujikuroi, 137–138
Gibberellic acid, 137–138
Gibberellins, 14
Global agriculture, application in, 215–217
Global biofertilizer market, 424–425, 689–690
  challenges of, 692–693
  COVID-19 on, 696–697

ecosystem, 693
growth-share matrix, 691–692
and Indian agriculture, 694–695
market dynamics and growth drivers, 690–691
potential and constraints, 695–696, 696t
segmentation, 692, 693f
Global biological control market, 175
Global climatic change, 725, 726f
Global trade in forestry microbial biocontrol, 175
Glomus spp.
  etunicatum, 18–20
  fasciculatum, 230–231
  intraradices, 18–19, 238–239
  mosseae, 18–20, 133, 243
  tortuosum, 17
  versiform, 18–19
Glucanases, 548
Gluconacetobacter diazotrophicus, 210, 217–218
Gluconic acid, 515–516
Glycine betaine, 188–189
Gram-positive bacterium, 9
Gram-positive soil actinobacteria, 70
GRAS. *See* Generally recognized as safe (GRAS)
Green consumerism, 642–643
Green manure for soil health, 721–722
Green nanoparticle, 276
Green Revolution, 614–615, 737–738, 741–742
Growth drivers, 690–691
Growth quadrant, 692
Growth-share matrix, 691–692
Guide RNA (gRNA), 471–472

H

Halophilic bacteria, 255
Halophytes, biocontrol activity of, 259–260
Halophytic rhizospheric bacteria, antifungal activity of, 262f
Halotolerant bacteria, mechanisms of, 255

Halotolerant plant growth-promoting rhizospheric bacteria (HT-PGPR), 255
Heat-liable activity, 192
Heat stress, 300, 395–396, 400–401
Heavy metal
   contamination of agronomic soils, 567
   remediation, phosphate biofertilizers for, 580–584
   stress, mitigation of, 299–300
   toxicity, 401
HeliScope, 349
Hemibiotrophs, 48
Hemiparasites, 662
Hessian fly, 470
Heterobasidion annosum, 166–169
Heterotrophic diazotrophs, 211
Heterotrophs, 657
High-resolution mass spectrometry (HR-MS), 371
High temperature stress, 20
High-throughput methods, 54
High-throughput pyrosequencing on beads, 347
High-throughput sequencing (HTS), 159–160
High-throughput techniques, 58–59
Holobiont, 343–344
Holoparasites, 662
Homology-directed repair (HDR), 471–472
Host plant, 656–659, 661–662, 676–677
Host specific toxins (HST), 677
Host-symbiont-soil nutrient status, 241
Hot water springs
   of Assam, 66–68
   bacterial diversity in, 66–68
HST. *See* Host specific toxins (HST)
HT-PGPR. *See* Halotolerant plant growth-promoting rhizospheric bacteria (HT-PGPR)
HTS. *See* High-throughput sequencing (HTS)
Human pathogens, 174

Humus, 137
Hup genes, 213
Hydrogenase, 208
Hydrolytic enzyme production, 261*t*
Hydroxamates, 12–13
Hydroxamate siderophores, 12–13
Hyperparasitism, 325–326, 633–635

I

IAA. *See* Indole-3-acetic acid (IAA)
ICAR. *See* Indian Council of Agricultural Research (ICAR)
ICM. *See* Integrated crop management (ICM)
IFOAM. *See* International Federation of Organic Agriculture Movement (IFOAM)
IFOAM Organic Guarantee System, 728
IFS. *See* International Food Standards (IFS)
Illumina genome analyzer, 348
Illumina MiSeq platform, 71
Improved physiochemical nature of soil, 135–137
Indian Agricultural Research Institute (IARI) microphos, 424–425
Indian Council of Agricultural Research (ICAR), 753–754
Indian Ministry of Agriculture, 641–642
Indian Organic Trade Fair, 710
Indirect mechanisms of biocontrol, 165–166
Indo-Burma biodiversity hotspot, 65–66
Indole-3-acetic acid (IAA), 14–15, 259–260, 287, 297–298, 333, 529, 534
Induced systemic resistance (ISR), 49–50, 165–166, 325, 355
Induced systemic tolerance (IST), 331–332, 396–397, 409–410
Industrial agricultural procedures, 614–615

Industrial microbiology, 302
Infective propagules (IP), 243
Infusion scaffolds of carbon nanotubes, 276–277
Initial inoculum density, 242
INMS. *See* Integrated nutrient management systems (INMS)
Inoculants
   beneficial microbial inoculants, 54–55
   microbial inoculants, 58
   natural rhizobial inoculants, 58
Inoculation, 116–117
   microbials, 58, 304–305, 544
      in agriculture for sustainability, 287–301
   of seeds, 552
   Petri dishes after, 541*f*
   seed surface, of P biofertilizers, 430–431
   sequence of symbionts, 242
   strain inoculation, 15
Inoculative form, 638
Inorganic phosphorus, 503
Inositol phosphate (soil phytate), 503
Insecticidal proteins, 466–467
Insects
   bacterial biopesticides against, 630–631
   fungal biopesticides against, 631
   nematode biopesticides against, 632
   pheromones, 622–624
   protozoal biopesticides against, 632
   viral biopesticides against, 631–632
In situ bioremediation, 567–568
Integrated crop management (ICM), 638–639
Integrated forest disease management, 176–177
Integrated nutrient management systems (INMS), 142
Integrated omics. *See* Multi-omics approach

Integrated pest management (IPM), 616
Integrated plant nutrient system (IPNS), 518–519
Integrin-linked kinase 1 (ILK1), 669
Intensive farming, 614–615
Interaction, plant-pathogen, 655–656, 659–661
 evolution of, 675–678
 mechanisms in, 661–675
International Conference on Organic Agriculture and Food Security, 727
International environmental treaties, 729
International Federation of Organic Agriculture Movement (IFOAM), 701, 727–728
International Food Standards (IFS), 700–701
International Organization for Biological Control (IOBC), 160–161, 175–176
International Organization for Standardization (ISO), 728
International Requirements for Organic Certification Bodies (IROCB), 728–729
International Taskforce on Harmonization and Equivalence in Organic Agriculture, 727–728
International Trade Laws, 729
Intracellular PGPR (iPGPR), 464–465
Inundative form, 638
In vitro/axenic cultivation of AM fungi, 244–245
In-vitro/chemical labeling techniques, 360–361
In vitro plant growth-promoting traits, 261t
In-vivo labeling techniques, 360–361
IOBC. See International Organization for Biological Control (IOBC)

Ion mobility spectroscopy (IMS), 371–372
IP. See Infective propagules (IP)
IPM. See Integrated pest management (IPM)
IPNS. See Integrated plant nutrient system (IPNS)
Ipomoea spp.
 batatas, 243–244
 carnea, 743
IROCB. See International Requirements for Organic Certification Bodies (IROCB)
Iron (Fe), 12
Iron sequesters, 589–590
Isobaric mass tags (TMT), 360–361
Isoelectric focusing (IEF), 358–360
ISR. See Induced systemic resistance (ISR)
ITF Guide for Assessing Equivalence of Standards and Technical Regulations, 728–729
iTRAQ, 360–361

J

Jatropha curcas, 271–272
Jhum agroecosystem, microbial diversity in, 95–97
Jhum land soil, 96

K

Kobresia myosuroides, 129–130
Kodiak, 58
Kosakonia sacchari, 97

L

Labeling techniques, 360–361
Labor productivity, 745
Lantana camara, 630
Large-scale gene expression analysis, 344–345
Large-scale production of spores, 244–245
Lectins, 667
Leghemoglobin, 127–128
Legume N_2-fixing nodules, 656
Lichens, 110–111

Liquid chromatography-tandem mass spectrometry (LC-MS/MS), 362–363, 371
Listeria monocytogenes, 188–189
Living organisms, 655
Loop (laboratory tube), 540
Low-resolution mass spectrometry (LR-MS), 371
Low-temperature stress, 20–21, 300
Lyophilized (freeze-dried) inoculants, 58
Lysiloma acapulcensis, 276
Lysinibacillus endophyticus, 15

M

Magnaporthe spp.
 grisea, 657–658
 oryzae, 133, 471–472
MAMPs. See Microbe-associated molecular patterns (MAMPs)
Managed microbial consortia, 111–113
 synthetic consortia, 112–113
Mannheim process, 483
Market
 of biopesticides, 640–642
 constraints, 695–696
 in phosphate biofertilizers, 518
 dynamics, 690–691
 ecosystem, 693
 potential, 695–696
 of organic products, 730–731
Massively parallel signature sequencing (MPSS), 354
Mass spectrometry (MS)
 metabolomics, 369t, 371–372
 proteomics
  gel free-MS-based proteomics, 358, 360–361
  2DE-MS-based proteomics, 358–360, 360f, 362–363
Matching-allele (MA) models, 677
Matrix-assisted laser desorption/ionization time-of-flight mass spectrometry (MALDI-TOF/MS), 362–363

Maximum residue levels (MRLs), 637
"Mayengbam Shyamchandra Meitei", 716
Medicago spp.
  sativa, 269
  truncatula, 113–114, 214, 238–239
Meganucleases, 471–472
Meloidogyne sp., 225–227, 233–234
  incognita, 132–133
Metabolic products, 304
Metabolite quenching, 368
Metabolites, 332–334, 345
Metabolomics, 345
  agricultural microbiological research, application in, 373–374, 374–375t
  analytical techniques, 368, 369t
    limitation, 372
    mass spectrometry (MS), 371–372
    nuclear magnetic resonance (NMR), 370
    popularity of, 368–369, 370f
  data processing and interpretation, 372–373
  definition, 366
  endogenous metabolites, 366
  exogenous metabolites, 366
  microbial metabolomics analysis, workflow for, 366–367, 367f
  primary metabolites, 366
  sample preparation and extraction, 368
  secondary metabolites, 366
  targeted/nontargeted metabolites, 366
  types of, 366
Metagenome analysis of Phumdi, 68–69
Metagenomics approach, 323
Metal-plant interaction, 333
Metal-resistant P-biofertilizer, 581–584
Metal-tolerant P-biofertilizers, 565–566, 580–581, 582t
Metal toxicity, 399–401

Meta-omics technologies, 344
Metaproteomics, 345, 493
Metataxonomic studies, 45–46
Metatranscriptomics, 344–345
Methanosarcina barkeri, 211
Methylobacterium sp., 14–15
Microarray technology, 344–345
Microarray well-refined technology, 354
Microbe-associated molecular patterns (MAMPs), 664–665
Microbe-based biofertilizers, 593–594
Microbe-based metal-extraction processes, 576–577
Microbes, 1–2, 28. *See also* Microorganism
  for all-round development, 721
  characterization of, 3f
  list of, 143–144t
  for sustainable crop production, 1
  synthesize, 137–138
  vital role in multiple ecosystem services, 65–66
Microbes-mediated phytohormone biosynthesis, 139–140
Microbial-based agricultural enhancements, 288
Microbial bioinoculants, 334
  as biocontrol agents, 320, 321t, 322f, 323–330
  as biofertilizers, 320, 321t, 322f, 330–331
  for biotic and abiotic stresses, 331–332
  definition of, 320
  functions of, 320
  manipulation
    genetic traits, 322–323
    importance of, 320–321
  pathogens, plants, and plant microbiome composition and diversity, 320–321, 324f
  phytohormones and metabolites production, 332–334

Microbial biopesticides, 624–626, 721
  bacterial biopesticides, 624–625
    against insects, 630–631
  in controlling pathogens, 633–635, 634t
  fungal biopesticides, 625
    against insects, 631
  in integrated crop management (ICM), 638–639
  microorganism-based biopesticides
    forms, 624
    synthesis of, 639–640
  nematode biopesticides against insects, 632
  plant-incorporated protectants, 626
  protozoal biopesticides against insects, 632
  viral biopesticides, 625–626
    against insects, 631–632
Microbial biotechnology, 286
  for strain improvement, 285–287, 302–303
  for sustainable agriculture, 301–306
    genetic engineering in strain improvement, 304–306
    genome of organisms, 303–304
    modification of microbial genomes, 306
    naturally occurring variants, 303
Microbial-derived PAMPs, 675
Microbial diversity, 65–66
  in ecological niches
    agroecosystem, 97–100
    bacterial diversity in hot water springs, 66–68
    contaminated habitats, 100–102
    forest ecosystem, 69–95
    jhum agroecosystem, 95–97
    metagenome analysis of Phumdi, 68–69
  exploitation of, 65–66
  literature synthesis on, 65–66

Microbial/pathogen-associated
molecular patterns
(MAMPs/PAMPs), 468
Microbial/pathogen triggered
immunity (MTI/PTI), 468
Microbials, 175. *See also* Microbes;
Microorganism
antagonists, 637
BCAs, 638–639
cells, 116–117
communities, 109–111
consortia, 109–111, 112f, 114
as biostimulants, 117, 118t
genera, 46–47, 50–51
genomes, modification of, 306
inoculation, 58, 304–305, 544
in agriculture for
sustainability, 287–301
of seeds, 552
interactions within consortium,
114–116
products, 2, 508–509
species, 13, 46
Microbiological tools, 287
Microbiomes, 45–46, 48, 56,
565–567, 569, 571, 580
beneficial microbiomes
implication of, 27
in organic agriculture, 27
in plant growth promotion
(*see* Beneficial
microbiomes, in plant
growth promotion)
for sustainable crop
production and protection,
23–26, 24f
for mitigation of biotic and
abiotic stresses, 15–23
abiotic stress management
(*see* Abiotic stress,
management)
biotic stress management,
21–23
plant-associated microbiomes,
46
plant microbiome, in translation
and commercialization, 58
Micrococcus roseus (MTCC 678),
187–188

Microflora of mycorrhizal
rhizosphere, 239–240
Microorganism, 48–49, 100–101,
185, 207–208, 210–211, 214,
287–290, 296–297, 301,
306–307, 502, 655, 657, 660,
666–667. *See also* Microbes;
Microbials
in agriculture for sustainability,
295t
beneficial microorganisms,
51–52, 301–302
cellulose-degrading
microorganisms (CDMs), 96
cold tolerance in, 186–191
antifreeze proteins, 189
cell membrane response,
187–188
cold acclimation proteins
and cold shock response,
189–191
cryoprotectants, 188–189
RNA degradosome, 191
mechanisms of cold tolerance
in (*see* Cold tolerance in
microorganisms)
microorganism-based
biopesticides
forms, 624
synthesis of, 639–640
mutagenized microorganisms,
194t
native soil microorganisms,
phosphate biofertilizers on,
516–517
phosphate-solubilizing
microorganisms (PSMs)
(*see* Phosphate-solubilizing
microorganisms (PSMs))
phosphorus-mobilizing
microorganisms, 503–505
plant growth-promoting
microorganisms (PGPMs),
46, 113–114
polyphosphate-accumulating
microorganisms, 507
potassium-solubilizing
microorganisms (KSMs),
486t, 493–494

actinomycetes, 484
potassium-solubilizing
bacteria (KSB), 484–485
potassium-solubilizing fungi
(KSF), 484–487
psychrophilic microorganisms,
186
Microphos technology, 424–425
Microscopic worms, 626
Micro-soil-plant agro-ecosystem,
246
Milk for disease, 719–720
Mineral acquisition, 130–132
Mineral-dissolving compounds, 8
Mineralization process, 503
Mineral phosphate solubilization
(mps), 424–425, 446–447,
516t
MinION technology, 349
Mitigation
of drought stress, 297–298
of heat stress, 300
of heavy metals stress, 299–300
MLSA. *See* Multilocus sequence
analysis (MLSA)
MLST. *See* Multilocus sequence
typing (MLST)
Mobile immune signals, 673–674
Modern biotechnology, 301–302
Modified Strullu-Romand (MSR)
medium, 244
Molecular engineering, of P
biofertilizers, 445–447
Molecular mechanisms, 660,
674–675
Mollicutes, 663
Monilinia vaccinii-corymbosi,
290–291
Monocropping, 614–615
Monotropoid mycorrhizae, 130
Monounsaturated fatty acids,
187–188
Monoxenic cultures, 244
Mucoromycota, 485–487
Multilocus enzyme
electrophoresis (MLEE),
350–351
Multilocus sequence analysis
(MLSA), 173

Multilocus sequence typing (MLST), 173, 350–351
Multi-omics approach, 345
  in agricultural microbiology, 379
  element based data integration, 377
  integrated analysis, 377
  mathematical analysis, 378
  pathway based data integration, 378
  postanalysis integration, 377
  research publication trends, 376, 376f
  system modeling, 377
  transcriptomics, integration of and metabolomics, 376–377
    and proteomics, 376–377
Multivariate analysis, 377
Muriate of potash (MOP), 483
Mutagenized microorganisms, 194t
Mutation, 675–676
Mutualisms, 127–128
  actinorhizal associations, 128
  plant-cyanobacterial mutualisms, 128
  rhizobia-legume mutualism, 127–128
Mutualistic bacteria, 54
Mutualistic behavior, 50
Mycoparasitism, 326, 526–527, 547–549
Mycorrhizae, 128–130, 662
  arbuscular mycorrhizae (AM), 18–19, 129, 199–200, 493–494, 656
  arbutoid mycorrhizae, 129–130
  colonization, 241
  ectomycorrhizae (ECM), 129–130, 493–494
  endomycorrhizae, 128–129
  ericoid mycorrhizae, 130
  monotropoid mycorrhizae, 130
  orchid and monotropoid mycorrhizae, 130
Mycorrhizae in plant-parasitic nematodes management, 225–229
  arbuscular mycorrhizal fungi (AMF)
    classification of, 226–227t
    integration of, 240–241
    interaction of, 233–234
    production and commercialization of, 243–245
    suppression of, 230–233
    host-symbiont-soil nutrient status, 241
    initial inoculum density, 242
    mycorrhizal mode of action, 234–240
      biochemical and physiological disturbances, 236–239
      microflora of mycorrhizal rhizosphere, 239–240
      physical interferences, 235–236
Mycorrhiza-induced chitinases isoforms, 239
Mycorrhizal fungi colonization, 20, 133, 235–236, 238–240, 243–244, 503–505
Mycorrhizal plants, 18–19
Mycorrhizal rhizosphere, microflora of, 239–240
Mycorrhizosphere, 229, 234–235, 239–240

N

Naga farmers, 715–716
Nagoya Protocol, 174–176
Nanoparticles (NPs), 269–271
  effect of, 274t, 275f
  genetic engineering in, 276–277
  green nanoparticle, 276
  in plant growth and stress tolerance, 273–275
  in plants, 271–273
Nanotechnology for plant growth promotion, 269–271
  nanoparticles (NPs), 269–271
    effect of, 274t, 275f
    genetic engineering in, 276–277
    green nanoparticle, 276
    in plant growth and stress tolerance, 273–275
    in plants, 271–273
National Botanical Research Institute P (NBRIP) medium, 427
National Mission for Sustainable Agriculture (NMSA), 714
National Programme for Organic Production (NPOP), 710
National Standards for Organic Production (NSOP), 710
National status of organic agriculture, 710–714
Native soil microorganisms, phosphate biofertilizers on, 516–517
Naturally occurring variants, 303
Natural polysaccharides, 255
Natural rhizobial inoculants, 58
NCDs. See Noncyanobacterial diazotrophs (NCDs)
Necrosis, 658
Necrotrophic pathogens, 48
Nematodes, 225–227, 228f, 659, 662
  biopesticides against insects, 632
  genera, 46
NEMiD. See North-east microbial database (NEMiD)
Neopestalotiopsis piceana, 99–100
Neuromodulator octopamine, 628–629
Next-generation sequencing (NGS), 176–177, 350t, 351, 678
  advantages, 347
  array-based sequencing, 347
  glass solid phase surface, synthesis on, 348
  high-throughput pyrosequencing on beads, 347
  methods, 58–59
  sequencing by ligation on beads, 347–348
  third-generation sequencing, 348–349

NFT. *See* Nutrient film technique (NFT)
NGS. *See* Next-generation sequencing (NGS)
Niche exclusion, 325
Nicotine, 636–637
Nif genes, 213
Nitrogenase enzyme, 208–209
Nitrogen fixation, 3–8
  in diazotrophs, 208–209
  in ocean, 210–212
Nitrogen-fixing biofertilizers, 424–425, 585–586, 585t
Nitrogen-fixing rhizobacteria, 288–290
Nitrogenous biofertilizers, bioremediation by, 578–580
NMSA. *See* National Mission for Sustainable Agriculture (NMSA)
Nod factors, 127–128
Nonbiological stresses (abiotic)
  plant resistance to, 545–547
Noncyanobacterial diazotrophs (NCDs), 211, 218
Nonexpressor of pathogenesis-related genes 1 (*NPR1*), 468–470
Nonhomologous end joining (NHEJ), 471–472
Nonnutritional mechanisms, 234–235
Nonsymbiotic diazotrophs, 213
Nontargeted metabolomics, 366
North-East (NE) India, 65
  ecological niches and associated microbial diversity in, 67f
  ethnic diversity in, 97–98
  microbial diversity in, 72–94t
North-east microbial database (NEMiD), 102
Norwegian Agency for Development Cooperation, 728–729
Nosemalocustae species, 632
Nostoc azollae, 128
NPK fertilizers, 699

NPOP. *See* National Programme for Organic Production (NPOP)
NPs. *See* Nanoparticles (NPs)
NSOP. *See* National Standards for Organic Production (NSOP)
Nuclear magnetic resonance (NMR), 369t, 370
Nucleic acid staining, 212
Nucleopolyhedroviruses (NPVs), 631–632
Nutrient Agar (NA) culture, 539f
Nutrient film technique (NFT), 243–244
Nutrient mobilization, 50–52, 116–117
Nutrients, availability of, 236–237
Nutrient solubilizers, 586–590
  iron sequesters, 589–590
  phosphate solubilizers, 586, 587t
  potassium solubilizers, 586–588, 588t
  sulfur-oxidizing microbes, 590
  zinc solubilizers, 589
Nutritional security, 731

O

Ocimum sanctum, 21
Omics technologies, 193, 197–198, 343, 346, 660
  genomics, 344
    agricultural microbiological research, application in, 351, 351–352t
    diversity analysis, 349–351
    genome sequencing, 346
    next-generation sequencing (NGS), 344–345, 347–349, 350t
    Sanger sequencing, 346–347
  integrated/multi-omics, 345
    in agricultural microbiology, 379
    element based data integration, 377
    integrated analysis, 377
    mathematical analysis, 378

  metabolomics and transcriptomics, integration of, 376–377
  pathway based data integration, 378
  postanalysis integration, 377
  research publication trends, 376, 376f
  system modeling, 377
  transcriptomics and proteomics, integration of, 376–377
metabolomics, 345
  agricultural microbiological research, application in, 373–374, 374–375t
  analytical techniques, 368–372, 369t
  data processing and interpretation, 372–373
  definition, 366
  exogenous/endogenous metabolites, 366
  microbial metabolomics analysis, workflow for, 366–367, 367f
  primary/secondary metabolites, 366
  sample preparation and extraction, 368
  targeted/nontargeted metabolites, 366
  types of, 366
meta-omics technologies, 344
proteomics
  agricultural microbiological research, application in, 364, 365t
  analysis, steps in, 358, 359f, 361–363
  analytical techniques, 358–361
  bioinformatics tools, 361
  definition, 357
  metaproteomics, 345
  types of, 345, 357–358
transcriptomics, 344–345
  agricultural microbiological research, application in, 355, 356t

Omics technologies (Continued)
  cDNA-AFLP technique, 353
  DDRT-PCR technique, 353
  definition, 353
  expressed sequence tags (ESTs), 353
  massively parallel signature sequencing (MPSS), 354
  microarray well-refined technology, 354
  RNA sequencing, 354–355
  Sanger sequencing, 353
  serial analysis of gene expression (SAGE), 354
Open Stomata 1 (OST1) mutants, 671
Ophiostoma ulmi, 165–166
Opportunistic endophytes, 46–47
Orchid mycorrhizae, 130
Organic acid (OA)
  by bacterial phosphate biofertilizers, 581*t*
  phosphate-solubilizing bacterial biofertilizers, 432–434, 432*t*
  phosphate-solubilizing fungi and actinomycetes, 432–434, 433*t*
  production, 11
  of P solubilization, 506
Organic agriculture, 97–98, 699–701, 737, 754
  beneficial microbiomes in, 27
  certified area, 711–712*t*
  continental data on, 709*t*
  glimpse of, 710*t*
  global scenario of, 703–708*t*
  highlights of food market, 714*t*
  impact of, 722–727
  market potential of, 730–731
  national status of, 710–714
  principles of, 701*f*
  products used in, 718–722
    cow dung for soil fertilization, 718–719
    green manure for soil health, 721–722
    microbes (PGPRs) for all-round development, 721

milk for disease and weed control, 719–720
waste fodder in composting, 719
pros and cons of, 716–718
regulatory policies in, 727–730
standards of, 702, 702*f*
state-wise data of, 713*t*
status of, 702–716
sustainable agriculture (*see* Sustainable agriculture)
Organically certified area, 711–712*t*
Organic farming. *See* Organic agriculture
Organic food market, 714*t*
Organic manures, 741–742
Organic phosphorus, 503
"Organic Value Chain Development for the Northeast Region", 715–716
Organophosphate, 636–637
Oryza sativa, 273–274
OsERF922 gene, 471–472
Osmolytes, 405–406, 406*t*
Oxidative stress, 400–401

P

Pac Bio sequencing, 348
Paddy-grown soil health, vermicompost on, 737–740
Paenibacillus glucanolyticus, 10
PAHs. *See* Polycyclic aromatic hydrocarbons (PAHs)
PAMP-induced stomatal closure, 671
PAMPs. *See* Pathogen-associated molecular patterns (PAMPs)
PAMP-triggered immunity (PTI), 675
Paracoccus sphaerophysae, 13
"Paramparagat Krishi Vikas Yojana", 714
Parasitic plants, 662
Parasitism, 48, 164, 326
Paspalum notatum, 243–244
Pasteuria Bioscience Company, 227–228

Pasteuria penetrans, 633–635
Pathogen-associated molecular patterns (PAMPs), 664–665
Pathogenesis-related proteins (PR proteins), 238–239
Pathogenic microbial infections, 673
Pathogens, 109–110, 242, 525, 528–530, 532, 538, 544–550, 552
  biotrophic pathogens, 48, 661–662
  controlling pathogens, 633–635, 634*t*
  foliar pathogens, 625
  forest pathogens (*see* Forest pathogens)
  potent human pathogens, 70
Pattern recognition receptors (PRR), 665*f*
PCR-based quantitative methods, 305
Pearson's correlation coefficient, 377
Penicillium antarcticum, 195
Peroxidase (POX) activity, 239, 406–408
Pesticidal activity, 626
Pesticides
  agrosynthetic insecticides/ pesticides, 701
  bacterial biopesticides, 624–625
  against insects, 630–631
  biochemical pesticides, 619–624
  essential oils (EOs), 622, 623*t*
  insect pheromones, 622–624
  plant extracts and oils, 620–622, 620–621*t*
  biological pesticides, 637
  biopesticides (*see* Biopesticides)
  botanical pesticides, 620–622, 620–621*t*, 626–630
  chemical pesticides, 156, 616–618
  microorganism-based biopesticides, constraints in, 639–640
  synthetic pesticides, 617*f*
Petri dish culture, 244

PGP. *See* Plant growth-promoting (PGP)
PGPB. *See* Plant growth-promoting bacteria (PGPB)
PGPMs. *See* Plant growth-promoting microorganisms (PGPMs)
PGPR. *See* Plant growth-promoting rhizobacteria (PGPR)
Phaeocrytopus gauemanii, 157–158
Phaeosphaeria nodorum, 677
Phaseolus vulgaris, 8–9
Phenazine-1-carboxylic acid (PCA), 327
Phenylalanine ammonia lyase (PAL), 468–470
Pheromones, 622–623
Phlebiopsis gigantea, 165, 168
Phomopsis liquidambari, 678
Phosphate acquisition efficiency (PAE), 441–442
Phosphate (P) biofertilizers, 508–509, 509t, 580–584
  in agrosystems, 423–424
  application methods and limitations, 430–431
  challenges and future prospects, 447
  constraints in, 517–518
  eco-physiological traits, 436–437, 437f
  food production system, application in, 437–438
  for heavy metal remediation, 580–584
  Indian Agricultural Research Institute (IARI) microphos, 424–425
  manufacturers of, 424–425, 426t
  marketing of, 509–512, 510t
  market value, 424–425
  metal tolerance and plant growth-stimulating features of, 582t
  microphos production, phases and carriers of, 427

  mineral phosphate solubilization (mps), 424–425
  molecular engineering of, 445–447
  on native soil microorganisms, 516–517
  nutrient management, role in, 424–425
  phosphobacterin, 424–425
  phyto-beneficial features, 434–436, 435t
  of P solubilization, 507
  P-solubilizing microbes as (*see* Phosphate-solubilizing microorganisms (PSMs))
  for sustainable agriculture
    cereal crops, PSB inoculation effect on, 437–443, 439t
    crop yield and quality, under stressful conditions, 443–445, 444–445t
    maize crop, plant growth regulators, experimental conditions, and agronomic response of, 439–441, 440t
    vegetables and pulse crops, PSB inoculation effect on, 437–438, 438t
Phosphate-mobilizing biofertilizer, 424–425
Phosphate solubilization, 214, 259–260, 503–505, 508, 586, 587t
Phosphate-solubilizing actinomycetes, 422, 424–425
  organic acids secretion, 432–434, 433t
  sources, 427, 430t
Phosphate-solubilizing bacteria (PSB), 130–131, 424–425
  conventional phosphate solubilizers, 427, 428–429t
  global fertilizer market, 424–425
  inoculation effect on
    cereal crops, 437–443, 439t
    vegetables and pulse crops, 437–438, 438t

  insoluble phosphate (TCP) solubilization, zone of, 427, 431f
  metal-tolerant, 436–437
  nitrogen-fixing bacteria, 427, 430t
  organic acids secretion, 432–434, 432t
  phosphobacterin, 424–425
  stress-tolerant, 427, 428–429t, 436–437
Phosphate-solubilizing fungi (PSF), 422, 424–425
  organic acids secretion, 432–434, 433t
  sources, 427, 430t
Phosphate-solubilizing microorganisms (PSMs), 502
  actinomycetes, 422, 424–425
  organic acids secretion, 432–434, 433t
  sources, 427, 430t
  in agriculture, 503–505
  bacteria (*see* Phosphate-solubilizing bacteria (PSB))
  crop yield and quality, under stressful conditions, 443–445, 444–445t
  definition, 421–424
  detection methods, 427
  diversity of, 505, 506t
  fungi (*see* Phosphate-solubilizing fungi (PSF))
  identification and functional analysis, 427
  mechanisms of P solubilization, 505–508
    exopolysaccharides (EPSs), 507–508
    organic acids, 506
    phosphatases and phytases, 507
    siderophores, 508
  metal-tolerant P solubilizers, 422
  microphos production, phases and carriers of, 427
  phosphate solubilization by, 513–516

Phosphate-solubilizing
    microorganisms (PSMs)
    (Continued)
  on plant growth and yield, 513
  plant growth-promoting
      activities of, 513, 514t
  soil application, 430–431
  in soils, 506t
  solubilization mechanisms,
      432–434
  stress-tolerant, 422, 443–445,
      444–445t
  survival, biotic factors
      influence on, 431–432
Phosphate use efficiency (PUE),
    434–436
Phosphatidic acid (PA), 670–671
*Phosphatidylinositol-
    phospholipase C (PI-PLC)*
    gene, 470
Phosphobacterin, 424–425
Phosphorus, 501–502
  solubilization, 8–9
Phosphorus-mobilizing
    microorganisms, 503–505
Phosphorus-solubilizing fungi, 9
Phosphorus-solubilizing microbes
    (PSB), 8
Phosphorylation mechanism, 187
Photoautotrophs, 657
Photosynthesis, 655
Phumdi, metagenome analysis of,
    68–69
Phyllanthus amarus, 271–272
Phyllobacterium brassicacearum
    strain STM196, 297–298
Phyllosphere region, 45–46
Phylospheric microbes, 319
Physicochemical stresses, 536
Phytases, of P solubilization, 507
Phytoalexins, 465–466
Phytobiome research, 45–46
  functional roles of sustainable
      agriculture
    crop productivity, 54–55
    enhanced plant defense
        mechanisms, 55–56
    nutrient mobilization, 50–52,
        51f

plant growth and health
    effect, 53–54
  plant-microbe interactions
      and, 57t
  plant-associated microbiomes,
      46–48
  plant-microbial interactions,
      48–50, 57t
  plant microbiome, in
      translation and
      commercialization, 58
Phytohormones, 14, 53, 137–138,
    332–334, 404
  gibberellins, 137–138
  production, 14–15
Phytopathogenic fungi, 658
Phytopathogens, 327, 328–329t,
    676
Phytophthora-specific primers,
    159–160
Phytophthora spp.
  cinnamomi, 240
  infestans, 156–157
Phytostimulation, 137–140
Pikovskaya agar plate method,
    259–260
Pine wilt disease (PWD), 158
Pinus spp.
  pinea, 168
  radiata, 159
Piriformospora indica, 19
Pisum sativum, 139
Planomicrobium chinense, 15
Plant-arbuscular mycorrhizal
    (AM) symbiosis, 675
Plant-associated microbiomes,
    46–48
Plant-cyanobacterial mutualisms,
    128
Plant growth-promoting (PGP),
    2–15, 5–7t, 214, 269–271
  activities of PSMs, 513, 514t
  biofertilizers, 424–425
  diazotrophs, 207–208
    beneficial mechanisms, 214
    diazotrophic biofertilizers,
        216t
    genomic and transcriptomics
        of diazotrophs, 212–214

global agriculture,
    application in, 215–217
  nitrogen fixation, 208–212
  to nonlegumes, 217–218
  properties, 215t
  terrestrial nitrogen-fixing
      diazotrophs, 209–210
microbes, 16–17, 20–23, 198,
    332–333
nanoparticles (NPs), 269–271
  effect of, 274t, 275f
  genetic engineering in,
      276–277
  green nanoparticle, 276
  in plant growth and stress
      tolerance, 273–275
  in plants, 271–273
nitrogen fixation, 3–8
phosphorus solubilization, 8–9
phytohormone production,
    14–15
potassium solubilization, 10–11
siderophore production, 12–13
traits, analysis of, 259–260
zinc solubilization, 11–12
Plant growth-promoting bacteria
    (PGPB), 114, 240–241,
    288–290, 537
Plant growth-promoting fungi
    (PGPF), 529
Plant growth-promoting
    microorganism-mediated
    abiotic stress, 409–410
  alleviation of, 402–403t
  drought stress, 400
  heat stress, 400–401
  metal toxicity, 401
  salt stress, 400
  stress tolerance, 396–397, 401
    1-aminocyclopropane-
        1-carboxylate (ACC)
        deaminase activity,
        404–405
    antioxidant enzymes,
        biosynthesis of, 406–408
    biofilm formation, 409
    extracellular
        exopolysaccharides (EPSs)
        production, 409

nutrient uptake, 408–409
osmolytes, production of, 405–406, 406t
phytohormone production, 404
Plant growth-promoting microorganisms (PGPMs), 46, 113–114
Plant growth-promoting rhizobacteria (PGPR), 54, 197–198, 207–208, 288–291, 320, 322–323, 331–332, 364, 569–571
  abiotic stress tolerance, in plants, 396–397, 410
  1-aminocyclopropane-1-carboxylate (ACC) deaminase activity, 404–405
  antioxidant enzymes, production and modulation of, 406–408
  biofilm formation, 409
  extracellular exopolysaccharides (EPSs) production, 409
  induced systemic tolerance (IST), 396–397, 409–410
  nutrient uptake, 408–409
  osmolytes, production of, 405–406, 406t
  phytohormone production, 404
  and activities, 254
  definition, 464–465
  in plant growth and development, 291
  rhizobia, 464–465
  siderophores, 466–467
  strain RB1, 141
  types of, 464–465
Plant-incorporated protectants (PIPs), 626
Plant-microbe interactions, 463–464
  CRISPR/Cas9 technology, 471–472, 472–473t
  defense genes
    Agrobacterium tumefaciens, 467

avirulent genes (Avr), 468
effector-triggered immunity (ETI), 468
gene-for-gene concept, 468
host defense genes, overexpression of, 468–470
host-resistant genes (R), 468
ecological considerations, 465
mutualisms, 127–128
  actinorhizal associations, 128
  plant-cyanobacterial mutualisms, 128
  rhizobia-legume mutualism, 127–128
rhizobacteria, 464–465
RNA interference (RNAi) mediated gene silencing, 470
secondary metabolites
  antimicrobial compounds, 465–466
  toxin production, in host, 466–467
Plant-microbial associations, 53
Plant-microbial interactions, 48–50, 57t
Plant parasitic nematodes (PPNs), 658–659
  mycorrhizal mode of, 234–240
  suppression of, 230–233
Plant-pathogenic bacteria (PPB), 662–663
Plant-probiotic bacterium, 99–100
Plants, 656–657
  biotechnology, 54–55
  diseases, 21–22
    resistance, CRISPR/Cas9 applications in, 471–472, 472–473t
  endophytes, 56
  extracts and oils, 620–622, 620–621t
  on insect pests, 626–629, 627t
  genetic engineering, 277f
  growth
    hormones, 254
    nanoparticles (NPs) in, 273–275
  salt stress on, 254–255

growth-stimulating features of phosphate biofertilizers, 582t
growth under stressed condition, 116
immune system, 663–664
invaders, 673–674
microbiome, 319
  engineering results, 110
  in translation and commercialization, 58
microbiota, 111
pathogens, 156–157, 657–659
  botanicals against, 629–630
  interaction of, 468, 655–656, 659–678
  management of, 240–242
photosynthates, 237–238
protectants, 638
REs, 126
resistance to nonbiological stresses (abiotic), 545–547
rhizospheric microorganism interactions, 253–254
Plant stress homeostasis-regulating bacteria (PSHR), 288–290
Plasma membrane lectin receptor-like kinases (LecRLKs), 667–668
Platycladus orientalis (oriental thuja), 14–15
Play safe quadrant, 692
Pollutants, 565–567, 571–574, 594
Polyacrylamide gel electrophoresis (PAGE), 353
Polychlorinated biphenyl (PCB) compounds, 141
Polycyclic aromatic hydrocarbons (PAHs), 100–101
Polymerase chain reaction, 55
Polyphenol oxidase (PPO), 406–408
Polyphosphate-accumulating microorganisms, 507
Polyphosphate kinases (PPKs), 515–516
Polyunsaturated fatty acids, 195

Poncirus trifoliata, 20
Population genetic studies, 159–160
Porostereum spadiceum, 15
Potash biofertilizers, 482–483, 494
  commercial potash-based biofertilizers, 487–490, 488–489t
  development of, 487–490
  field applications and crop improvement, 493–494
  formulation of, 490
    bioformulation processes, 491
    consortium, selection and development of, 490–491
    desiccation tolerance testing, 491–492
    field efficacy, 492–493
    storage stability testing, 492
  muriate of potash (MOP), 483
  potassium chloride, 483
  potassium-solubilizing microorganisms (KSMs), 486t, 493–494
    actinomycetes, 484
    potassium-solubilizing bacteria (KSB), 484–485
    potassium-solubilizing fungi (KSF), 484–487
  sulfate of potash (SOP), 483
  sustainable agriculture, 484
Potassium, 51
  cycle, 482–483, 482f
  solubilizers, 10–11, 586–588, 588t
Potassium chloride (KCl), 482–483
Potassium-solubilizing bacteria (KSB), 484–485, 486t
Potassium-solubilizing biofertilizers (KSB), 442–443
Potassium-solubilizing fungi (KSF), 484–487, 486t
Potassium-solubilizing microorganisms (KSMs), 486t, 493–494
  actinomycetes, 484
  potassium-solubilizing bacteria (KSB), 484–485

potassium-solubilizing fungi (KSF), 484–487
Potato Dextrose Agar (PDA) for fungal culture, 539f
Pot culture techniques, 243
Potent human pathogens, 70
Potential biological control agents, 633
Potential phosphate-solubilizing bacteria, 9
PPKs. *See* Polyphosphate kinases (PPKs)
PPNs. *See* Plant-parasitic nematodes (PPNs)
PQQ. *See* Pyrrolo quinoline quinone (PQQ)
Principle mechanism, 8
Promicromonospora sp., 14–15
Propionibacterium acidipropionici, 364
Proteases, 164
  inhibitors, 466–467
Protein Prophet, 363
Protein-protein interactions (PPIs) analysis, 363
Proteins, 345
Proteomics
  agricultural microbiological research, application in, 364, 365t
  analysis, steps in, 359f
    database searching, 358, 363
    identified protein, verification of, 358, 363
    protein extraction and separation, 358, 362
    protein identification, 358, 362–363
    protein-protein interactions (PPIs) analysis, 358, 363
    sample preparation, 362
    statistical analysis, 358, 363
  analytical techniques
    gel free-MS-based proteomics, 358, 360–361
    2DE-gel-based proteomics, 358–360, 360f
    bioinformatics tools, 361
    definition, 357

expression proteomics, 345, 357
functional proteomics, 345, 358
metaproteomics, 345
structural proteomics, 345, 358
Protists, 47
Protozoal biopesticides against insects, 632
PSB. *See* Phosphorus-solubilizing microbes (PSB)
Pseudomonas cepacia, 22–23
Pseudomonas spp., 325
  aeruginosa, 9, 133
  azotoformans, 10–11
  extremaustralis, 138–139
  fluorescens, 99, 132–133, 287, 351, 351–352t
  fragi, 101
  frederiksbergensis, 296–297, 586
  koreensis, 15
  protegens, 327
  putida, 240, 661–662
    MTCC5279, 296–297
    NBR10987, 133–134
  savastanoi, 49
  solanacearum, 240
  strains, 22
  stutzeri, 7–8, 195
  synxantha, 327
  syringae, 187–188, 306, 667, 674–675
  vancouverensis, 296–297
Pseudopestalotiopsis theae, 99–100
PSHB. *See* Plant stress homeostasis-regulating bacteria (PSHB)
PSMs. *See* Phosphate-solubilizing microorganisms (PSMs)
P solubilization by PSMs, mechanisms of, 505–508
Psychrophiles, 185–187, 191–192
Psychrophilic microbes, 191
Psychrophilic microorganisms, 186
Psychrophilic PGP microbes
  characterization of, 199
  formulation of, 199–200
  isolation of, 198
  selection of, 199
Psychrotrophs, 186–187

Puccinia graminis, 675–676
Putative diazotrophs, 211
PWD. *See* Pine wilt disease (PWD)
Pyoluteorin (Plt), 327
Pyrenophora tritici-repentis, 675–676
Pyrethrin, 636–637
Pyrethroids, 636–637
Pyridine nucleotides NAD, 673–674
Pyrrolo quinoline quinone (PQQ), 515–516
Pythium tritici-repentis, 677

Q

QIIME data analysis package, 71
qRT-PCR analysis of samples, 54
Quorum sensing (QS), 49, 112–113

R

Ralstonia solanacearum, 56, 161–162, 164
Rapid seed germination, 532
Raw material constraints, in phosphate biofertilizers, 517
Reactive nitrogen species (RNS), 670
Reactive oxygen species (ROS), 18, 254–255, 400, 406–408, 468–470, 542–543
Receptor-like kinases (RLKs), 665
Regulatory policies in organic agriculture, 727–730
REs. *See* Root exudates (REs)
Research Institute of Organic Agriculture, 702–703
Resistance, 661–662, 672f
Reverse transcriptase (RT) enzyme, 353
Rhizobacteria, 464–465
Rhizobacterial genera, 589–590
Rhizobia, 4, 464–465, 579–580
Rhizobia-legume mutualism, 127–128
Rhizobia-mediated biological nitrogen fixation, 130–131
Rhizobiome, 464–465
Rhizoctonia solani, 238–241, 544

Rhizoglomus irregulare, 355
Rhizophagus irregularis, 236, 240–241
Rhizoremediation, 140–142
Rhizosheath, 255
Rhizosheath soil, 135–137
Rhizospheres, 301, 675
 effect, 343–344
 microbes, 116–117, 130–132, 319–320, 332–333
 microbiome, 135–137
 microorganisms, 130–131, 630
 region, 45–46
Rhizospheric bacteria, 140–142, 253–255, 259–260
Rhododendron arboreum, 71
Rice blast, 471–472
Rice genera, 7–8
Rio Summit, 174–175
RLKs. *See* Receptor-like kinases (RLKs)
RNA
 degradosome, 191
 sequencing, 354–355
RNA interference (RNAi) mediated gene silencing, 470
RNS. *See* Reactive nitrogen species (RNS)
Roche (454 pyrosequencing) technique, 347
Rock phosphate (RP), 423–424
Root-associated tissues morphology, 235–236
Root colonizers, 47–48
Root endophytes, 47–48
Root exudates (REs), 126
 biochemical composition of, 136t
Root-feeding nematodes, 225–227
Root organ culture, 244–245
ROS. *See* Reactive oxygen species (ROS)
Rotenone, 626–628
Rotylenchulus reniformis, 237

S

Saccharomyces cerevisiae, 217
Salicylic acid, 139–140

Salinity stress, 16–17, 253–255, 263, 298–299, 398
Salt overly sensitive (SOS) pathway, 406–408
Salt stress, 333
 on plant growth, 254–255
Salt-tolerant halophytic rhizospheric bacteria, 256–259t
Sanger sequencing, 346–347, 353
Saprophytic bacterial species, 661–662
SAR. *See* Stramenopiles-Alveolata-Rhizaria (SAR); Systemic acquired resistance (SAR) pathway
SARD. *See* Sustainable Agriculture and Rural Development (SARD)
Sclerospora graminicola, 720
Sedentary endoparasitic nematodes, 659
Seed biological treatments (biopriming)
 application of, 537–539
 based on fungi microstructure, 539–540
Seed germination
 factors of, 535
 significance of, 532–534
Seedling dipping, of P biofertilizers, 430–431
Seedling establishment, 535
 environmental stresses on, 535–536
 in seed production, 534
Seed physiology, 542
Seed production, seedling establishment in, 534
Seed surface inoculation, of P biofertilizers, 430–431
Semiochemicals, 622–623
Septoria nodorum, 675–676
Serial analysis of gene expression (SAGE), 354
Serial dilution plating method, 70–71
Serial plate dilution method, 427
Serratia marcescens, 101, 133, 174

Sesbania rostrata, 127
"Sevagram Declaration in 1994", 710
Shikimic acid pathway, 468–470
Shingobacterium antarcticum (MTCC 675), 187–188
Short-interfering RNA (siRNA), 276–277
SI. *See* Solubilization index (SI)
Siderophores, 22–23, 164, 466–467
 production, 12–13
 of P solubilization, 508
Signal molecule acyl homoserine lactone, 115
Silver nanoparticles, 270–271, 276
Single microbial antagonist, 169
Single-molecule real-time (SMRT) technology, 348
Single-nucleotide polymorphism (SNP), 350–351
Sinorhizobium meliloti, 113–114, 306
siRNA. *See* Short-interfering RNA (siRNA)
16S rRNA sequencing methods, 51–52
SlDMR1 gene, 470
SlPMR4 gene, 470
Small interfering RNAs (siRNAs), 674
Smart agricultural practices, 54
SNF bacteria. *See* Symbiotic nitrogen fixation (SNF) bacteria
S-nitrosylation, 670
Socioeconomic sustainability
 of farmers, 743–746, 746*t*
 variables of, 752*t*, 753–754
SOD. *See* Superoxide dismutase (SOD)
Soil-borne nematodes, 233–234
Soil-borne pathogens, 132–133
Soil-bound organic molecules, 52
Soil health and productivity, 125–127
 biochemical composition of root exudates, 136*t*
 biocontrol agent, 132–133
 biofertilizers, 142

extenuating abiotic stresses, 133–135
improved physiochemical nature of soil, 135–137
mineral acquisition, 130–132
mutualisms, 127–128
 actinorhizal associations, 128
 plant-cyanobacterial mutualisms, 128
 rhizobia-legume mutualism, 127–128
mycorrhizae, 128–130
 arbuscular mycorrhizae (AM), 129
 ectomycorrhizae, 129–130
 ericoid mycorrhizae, 130
 orchid and monotropoid mycorrhizae, 130
phytostimulation, 137–140
rhizoremediation, 140–142
Soil-plant-microbes interactions, 125–127
 biochemical composition of root exudates, 136*t*
 biocontrol agent, 132–133
 biofertilizers, 142
 extenuating abiotic stresses, 133–135
 improved physiochemical nature of soil, 135–137
 mineral acquisition, 130–132
 mutualisms, 127–128
 actinorhizal associations, 128
 plant-cyanobacterial mutualisms, 128
 rhizobia-legume mutualism, 127–128
 mycorrhizae, 128–130
 arbuscular mycorrhizae (AM), 129
 ectomycorrhizae, 129–130
 ericoid mycorrhizae, 130
 orchid and monotropoid mycorrhizae, 130
 phytostimulation, 137–140
 rhizoremediation, 140–142
Soils, 207–208, 210–211, 218
 application, of P biofertilizers, 430–431

bacteria identification, 65–66
ecosystem, microbial groups in
 endophytic fungi, 293–294
 entomopathogenic fungi, 295–296
 mycoparasitic fungi, 295–296
 mycorrhizal fungi, 292–293
 rhizobia, 292
 rhizospheric fungi, 294–295
fertility, 288–290, 725–727, 743
 cow dung for, 718–719
 management, 723*f*
health, 501–505, 513
 parameters of, 744*t*, 751*t*
inoculants, 320
microbes, 126
microbial biomass, 95–96
microbial community, 304–305
microbiome, 47, 569
microbiota, 95–96, 116–117
microorganisms, 503–505
pathogens, 625
phosphorus, status and availability of, 503
phytate, 503
productivity, 125–126
salinity, 16, 254, 398
virome, 48
Solanum spp.
 lycopersicum, 666
 tuberosum, 156–157
Solid-phase microextraction (SPME), 165
SOLID technology, 347–348
Solubilization Index (SI), 259–260
Spearman's correlation coefficient, 377
Spectrum of specificity, 169
Sphaerophysa salsalu, 13
Sphaerotheca fuliginea, 719–720
SPME. *See* Solid-phase microextraction (SPME)
16S ribosomal RNA (16S rRNA), 349–351
Standards of organic agriculture, 702, 702*f*
Staphylococcus aureus NiOs, 271–272

Statistical Analysis System (SAS) software, 363
Statistical package for social sciences (SPSS), 363
Status of organic agriculture, 702–716
Stenotrophomonas maltophilia
   MHF ENV 22, 141
   RSD6, 99–100
Stomatal guard cells, 671
ST-PGPR, 263
Strain improvement
   genetic engineering in, 304–306
   microbial biotechnology for, 285–287
Strain inoculation, 15
Stramenopiles-Alveolata-Rhizaria (SAR), 47
Streptomyces fradiae, 15
Stresses, 536
   abiotic stress (*see* Abiotic stress)
   biological stresses, 536
   biotic stress, 253, 319, 364, 463–464
      management, 21–23
      microbial bioinoculants for, 331–332
      for plants, 157
      tolerance, microbial inoculants for, 296–301
   cold stress, 187, 190*t*
   drought stress, 18–20, 297–298, 397–398, 547
   environmental stress, 253, 536–539
   extenuating abiotic stresses, 133–135
   heat stress, 300, 395–396, 400–401
   high temperature stress, 20
   low-temperature stress, 20–21, 300
   oxidative stress, 400–401
   physicochemical stresses, 536
   salinity stress, 16–17, 298–299
   tolerance, NPs in, 273–275
   water stress, 397–398, 400
Structural proteomics, 345, 358

Subfamilies of pattern recognition receptors (PRR), 665*f*
Submicroscopic infectious pathogens, 658
Sulfate of potash (SOP), 483
Sulfur-oxidizing bacteria, 52
Sulfur-oxidizing microbes, 590
Superoxide dismutase (SOD), 20, 406–408
Surface-enhanced laser desorption/ionization time-of-flight mass spectrometry (SELDI-TOF/MS), 362–363
Susceptibility (S) genes, 470
Sustainability, 2, 24*f*
   assessment, 717*f*
Sustainable agriculture, 45–46, 47*f*, 48, 54–55, 109–110, 113–114, 253–254, 502, 615–616
   benefits of, 437–438
   definition of, 437–438
   environment and, 614–616
   functional roles of phytobiome, 50–56
      crop productivity, 54–55
      enhanced plant defense mechanisms, 55–56
      nutrient mobilization, 50–52, 51*f*
      plant growth and health effect, 53–54
      plant-microbe interactions and, 57*t*
   phosphate (P) biofertilizers
      cereal crops, PSB inoculation effect on, 437–443, 439*t*
      crop yield and quality, under stressful conditions, 443–445, 444–445*t*
      maize crop, plant growth regulators, experimental conditions, and agronomic response of, 439–441, 440*t*
      vegetables and pulse crops, PSB inoculation effect on, 437–438, 438*t*
   potash biofertilizers, 484
Sustainable Agriculture and Rural Development (SARD), 700

Sustainable agroecosystems, 246
Sustainable crop production and protection, 23–26, 24*f*
   biofertilizers, 23–25, 24*f*
   biopesticides, 25–26
Sustainable Development Goals (SDGs), 614–616
Symbionts, inoculations sequence of, 242
Symbiotic/asymbiotic N biofertilizers, 579–580
Symbiotic bacteria, 4, 50
Symbiotic diazotrophs, 213
Symbiotic nitrogen fixation (SNF) bacteria, 464–465, 585–586
Sym plasmid, 127
Synthetic consortia, 112–113
Synthetic fertilizers, 306
Synthetic pesticides, 617*f*
Systemic acquired resistance (SAR) pathway, 165–166, 468–470, 673

T

Talaromyces flavus, 26
Taq polymerase, 66–68
Targeted metabolomics, 366
TBF1-cassette, 468–470
Tea rhizosphere, 98
Technical constraints, in phosphate biofertilizers, 517
Technology
   CRISPR/Cas9 technology, 471–472, 472–473*t*
   exome sequencing technology, 674–675
   genetic/genome-editing, 471–472
   microarray, 344–345
   microarray well-refined technology, 354
   microbial biotechnology (*see* Microbial biotechnology)
   microphos technology, 424–425
   modern biotechnology, 301–302
   omics technologies (*see* Omics technologies)

Technology (Continued)
  single-molecule real-time (SMRT) technology, 348
  SOLID technology, 347–348
Temperature, 399
Terminalia chebula, 271–272
Terrestrial ecosystem, 52
Terrestrial nitrogen-fixing diazotrophs, 209–210
TH. *See* Thermal hysteresis (TH)
Thaumatin-like protein (TLP), 468–470
Therapeutic microbial pest control, 638
Thermal hysteresis (TH), 189
Thermostable PGPR, 300
Thermo-tolerant bacterium, 9
Thermus aquaticus, 66–68
Thuja orientalis, 630
Thymus vulgaris, 629–630
Thysanolaena maxima (TM) rhizosphere, 96
Tobacco plant rhizosphere, 10
Toll-like receptor (TLR), 663
Tonoplasts, 547
ToxA protein-encoding gene, 677
Toxic compounds, 466–467
Traditional fertilizers, 51–52
Transcription activator-like effector nucleases (TALENs), 471–472
Transcription factor TBF1 (TL1-binding factor 1), 673
Transcriptomics, 344–345
  agricultural microbiological research, application in, 355, 356t
  cDNA-AFLP technique, 353
  DDRT-PCR technique, 353
  definition, 353
  expressed sequence tags (ESTs), 353
  massively parallel signature sequencing (MPSS), 354
  microarray well-refined technology, 354
  RNA sequencing, 354–355
  Sanger sequencing, 353
  serial analysis of gene expression (SAGE), 354
Translation research, 271
Transmembrane sensory domain, 187
Tree-associated microbial species, 160
Trichoderma fungi, 323–330, 525–529
  asperelloides, 165
  asperellum, 99–100, 530–531
  atroviride, 19
  avovidea, 550–551
  biological mechanisms of, 547–549
  in dealing with biological stresses (biotic), 544–545
  harzianum, 16, 132–134, 351, 351–352t, 355, 527–528, 531–532, 540f, 543, 545, 549–551, 630
  in increasing germination indices, 532–534
  on increasing growth and antioxidant activity, 542–543
  lixii, 543
  longibrachiatum, 16–17
  and plant, colonization as, 549–552
  in stimulating plant growth, 529–532
  virens, 550
  viride, 531
Triple superphosphate (TSP), 441–442
Triticum spp.
  aestivum, 19
  harzianum, 19
Trypsin inhibitor gene, 466–467
Tsukamurella tyrosinosolvens, 9
2DE-gel-based proteomics, 358–360, 360f
Two-dimensional-gas chromatography-mass spectrometry (2D-GC-MS), 371
Two-dimensional gel electrophoresis-mass spectrometry (2-DE/MS), 358–360, 360f, 362–363
Tylenchulus semipenetrans (Citrus nematode), 230–231
Type III secreted effectors (T3SEs), 674–675

U

Ulmus minor, 166
UNCCD, 125–126
UNCTAD. *See* United Nations Conference on Trade and Development (UNCTAD)
UNDP. *See* United Nations Development Programme (UNDP)
UN Food and Agriculture Organization (FAO), 16, 140, 618–619, 715, 727
Uninterrupted cell proliferation, 127–128
Union Ministry of Agriculture and Farmers' Welfare, 641–642
United Nations Conference on Trade and Development (UNCTAD), 727
United Nations Development Programme (UNDP), 700
United Nations General Assembly, 301–302
United Nations Strategic Plan for Forests, 155
UN Sustainable development goals, 176–177
Upstream open-reading frames (uORFs$_{TBF1}$), 468–470
Uromycladium acaciae, 164
U.S. Department of Agriculture, 638–639

V

VAM fungi. *See* Vesicular arbuscular mycorrhizal (VAM) fungi
Variovorax paradoxus, 139
Vascular embolism, 165–166

Vector-foreign DNA complex, 304
Vegetative foliar tissues, 111
Vermicompost
　application on soil health
　　parameters, 751*t*
　effect on soil health, 752*t*
　on farmers' socioeconomic
　　sustainability, 743–746
　lower carbon-nitrogen ratio of,
　　738–739
　on paddy-grown soil health,
　　737–743
Vesicular arbuscular mycorrhizal
　(VAM) fungi, 230–231
Vigna radiata, 15, 139–140
VIP3A protein, 466–467
Viral biopesticides, 625–626,
　631–632
　against insects, 631–632
Viral dsRNA, 674
Viral genera, 48
Virulence, 661–662, 676–677

Volatile organic compounds
　(VOCs), 53, 115–116

W
Waste fodder in composting, 719
Waste management systems,
　196–197
Water samples, 95
Water stress, 397–398, 400
Water use efficiency (WUE),
　297–298
Weed control, 719–720
Weighted Gene Coexpression
　Network Analysis
　(WGCNA), 378
Western blot technique, 363
World Health Organization
　(WHO), 614, 727–728
World Resources Institute (WRI),
　617–618
WUE. *See* Water use efficiency
　(WUE)

X
Xanthomonas oryzae, 657–658
X-ray (XAS) synchrotron
　spectroscopy, 274–275

Z
Zea mays, 274–275
Zero-waste production, 754
"Zigzag" model, 663–664, 664*f*
Zinc finger nucleases (ZENs),
　471–472
Zinc solubilization, 11–12,
　259–260, 589
zinc-solubilizing microbes (ZSB),
　11
Zizania latifolia, 99–100
Zn-tolerant bacteria (ZTB),
　581–584
ZSB. *See* zinc-solubilizing
　microbes (ZSB)
ZTB. *See* Zn-tolerant bacteria
　(ZTB)

Printed in the United States
by Baker & Taylor Publisher Services